# CELESTIAL NAVIGATION

## step by step

Second Edition

WARREN
NORVILLE

INTERNATIONAL MARINE PUBLISHING COMPANY
CAMDEN, MAINE

Typeset by The Ellsworth American, Ellsworth, Maine
Printed and bound by Fairfield Graphics, Fairfield, Pennsylvania



Published by International Marine Publishing Company
21 Elm Street, Camden, Maine 04843
(207) 236-4342

**Library of Congress Cataloging in Publication Data**

Norville, Warren.
Celestial navigation step by step.

Includes index.
1. Nautical astronomy. 2. Navigation. I. Title.
VK555.N86 1984 623.89 83-47888
ISBN 0-87742-177-3

*To Jim Allen*
*Who taught me, when 16 years old, to*
*do a day's work at sea.*

# CONTENTS

# PREFACE

I began the Preface to the first (1973) edition of this book by saying, "You can learn celestial navigation. If you can count your change after buying this book you know all the mathematics you will need. Yet, this is a comprehensive work that covers thoroughly the practical use of celestial navigation at sea."

The statement holds true for this revised edition as well. True, I have added a chapter covering the working of celestial observations by use of the tables in Ageton's Method, which you probably know as H.O. 211. I have also added chapters on great circle sailing and on working sights by use of basic and scientific calculators—the kind you buy over the counter in almost any shopping mall. Although these may seem like profound additions, the mathematics required is still only the ability to add and subtract, multiply and divide. Yet, if you understand any trigonometry at all, there is sufficient reference to the direct trigonometric solutions to satisfy your curiosity.

This work is a compilation of ideas gained in teaching celestial navigation to classes of adults at the University of South Alabama and at Spring Hill College. These classes were sponsored by the Department of Continuing Education. The students came from varied educational and experiential backgrounds. Most finished the class with confidence that they could work a sight and plot a position. Many were so enthusiastic that they bought sextants. Even if they did not plan to go offshore, they had learned two things: Celestial navigation can be a lot of fun, and a good sextant is a handy tool for a sailor even if he never goes far to sea. Many of my students had never seen a set of parallel rulers before their first evening in class.

The first five chapters explain the meaning of the terms used in navigation, and explain the concepts behind the practical side of modern marine navigation. Chapters 6 through 19 are "how to" lessons. The student should study these chapters with continuous reference to the Appendixes and the excerpts from the sight reduction tables and almanacs printed there. He should have his instruments for position plotting at his side, also. He should follow each step of each chapter as if he were following an instructor as he outlined the procedure in class. Then each step should be repeated until he feels he has mastered it. In each chapter, the steps of the initial example are presented with a detailed explanation of what is involved. The same or a similar problem is then worked using the format that would be used in actual practice.

All chapters in the first edition have been retained, and

expanded where necessary to cover areas of need disclosed in teaching and in making deliveries of vessels to destinations in east longitude and south latitude. Thus, the second edition benefits from nine years of additional experience. Exercise problems and examples have been reworked to conform to the 1983 almanacs.

I mentioned that three chapters have been added to the text. Chapter 17 explains how to work sights by Ageton's Method, because "H.O. 211," as many still call this system, continues to be very popular, universal in application, and especially suited for use when solving problems in great circle sailing. Chapter 18 has been added because great circle sailing is an extremely important part of offshore voyage planning, and because you already know how to work a great circle sailing problem if you know how to work a sight. Chapter 19, on the use of the handheld calculator to reduce a celestial observation, has been added because almost everyone who aspires to be a navigator has a calculator suitable for this purpose. With nothing but a calculator, an almanac, and accurate time, you can work any sight.

The book is arranged to allow you to put your proficiency to use as it is developed. Master the material in Chapter 6, "How to Work a Sun Sight," and you can cross any ocean, although it would be much safer, and more fun, too, to know how to shoot the stars and planets.

If your plans lead around the world, by all means complete Chapters 13 and 14. Then, if you are seriously interested in becoming a well-rounded practical navigator—one who can step aboard a vessel and make a safe passage with nothing more than a Bowditch and an almanac under his arm—complete the book.

The first edition of this work was based on the tables in H.O. 214, *Tables of Computed Altitude and Azimuth,* despite the fact that these tables were already scheduled for replacement by the new tables, *Sight Reduction Tables for Marine Navigation,* known then by the short title of H.O. 229 and now as Pub. No. 229. Nevertheless, this book survived and prospered through six printings and 10 years to arrive at the time for this revision. It prospered not because there are still many navigators who have, love, and use H.O. 214, although that is certainly the case, but because the use of the tables is only one part of the sight reduction process. There are only three ways to get a set of the H.O. 214 tables today: Talk someone like me out of their set—a moral impossibility. Steal a set—mine are under lock and key. Hope someone wills you theirs—can you wait? For this reason, the most important revision made here was to substitute the tables in *Sight Reduction Tables for Air Navigation,* Pub. No. 249, for the tables in H.O. 214. You can get the three volumes of Pub. No. 249 off the shelf at any chart agent's store today. The spirit of H.O. 214 lives with us in Pub. No. 249.

The format of Volumes 2 and 3 of Pub. No. 249 is almost the same as that of H.O. 214. Anyone familiar with the tables in H.O. 214 can use the tables in Pub. No. 249 after only a moment's review. The differences are minor, except for one thing. Volumes 2 and 3 of Pub. No. 249 cannot be used to work sights of bodies whose declinations exceed 30° north or south. Although many navigational stars have declinations greater than 30°, the sun, moon, and planets' declinations never exceed 30°. Consequently, you can go anywhere in the world with reasonable assurance of getting sights you can work with Volume 2 or 3 of Pub. No. 249. For positions between latitudes 39° north and south, you will need Volume 2. For positions between latitudes 40° and 90° north or south, you will need Volume 3. Volumes 2 and 3 are identical except for the difference in tabulated latitudes.

Volume 1 of Pub. No. 249 is a specialized set of tables of precomputed altitudes of selected stars that reduces the work of the navigator immensely. The use of Volume 1 is explained in the text.

The arguments used to enter the tables in Pub. No. 229 are identical to those used in Pub. No. 249. The actual consultation of the tables, whether Pub. No. 249 or Pub. No. 229, comes very near the end of the problem. Most of the work in reducing a sight lies in using the almanac and correcting time. This is what is taught in the "how to" lessons. However, to keep this work as current as possible, Chapter 15 gives an explanation of the tables in Pub. No. 229 with examples.

Page excerpts from Pub. No. 249 and Pub. No. 229 are included in the Appendixes. These excerpts are included through the courtesy of the U.S. Navy, Office of Oceanography, now the Defense Mapping Agency Hydrographic/Topographic Center.

For the reader's convenience, page excerpts from the *Nautical Almanac* and the *Air Almanac* for the year 1983 are also included in the Appendixes. These excerpts are included by the courtesy of Her Majesty's Almanac Office and The Almanac Office, U.S. Naval Observatory.

None of the definitions in the first edition was incorrect insofar as usage in navigational parlance is concerned, but a case could have been made by a purist that one or two definitions were not strictly accurate as given. In the second edition, emphasis has been placed on giving stricter definitions of such terms as *Greenwich hour angle, local hour angle,* and *meridian angle,* despite the fact that looser definitions persist in current government navigation publications.

In making these revisions, *the needs of the user of this book were kept paramount.* The direct simplicity and comprehensiveness of the first edition have not been lost.

One other point should be stressed. You can no more learn to navigate from reading a book than you can learn to fly by reading a book. To learn to use the instruments required takes practice—practice using a sextant, taking time, and working a sight at every opportunity. You will learn more that way, and enjoy it, too.

Warren Norville
Mobile, Alabama

# ACKNOWLEDGMENTS

It is fitting to begin by acknowledging again those who helped in the first edition of this book. Here is what I said in 1973:

It is with hesitation that I attempt to recognize the contributions so many have made to make this work possible. I hesitate because I know I cannot name them all, and I want no one to feel his or her contribution is not appreciated. It is with real gratitude I say "thank you" to my many friends whose indications of interest and whose remarks of encouragement meant so much while this work was in progress. I cannot mention all of you, but I am grateful to each of you.

I do especially want to recognize some whose assistance was essential to the success of this work: Dr. Kenneth W. Coons, owner and skipper of the yacht *Asperida II*, for reviewing the manuscript and giving me the benefit of his comments; Mr. Roby Bevan and Mr. Lawrence Gamotis, who so meticulously checked my calculations and my definitions; Dr. F. H. Mitchell, University of South Alabama, for reviewing the manuscript in its early stages and giving me the benefit of his remarks; and most particularly, Dr. William A. Hoppe, Associate Dean, College of Arts and Sciences, University of South Alabama, for his assistance and encouragement.

The appearance of this revised edition does nothing to diminish my appreciation for the help I received from those mentioned above. Rather, this appreciation is felt more intensely, because it is on the work and help of those friends that this revised edition is built. Nevertheless, a revision of this order is not something one undertakes entirely on his own. As is often the case, many others have made significant contributions to this revised work. I realize there are many who will go unmentioned; they will not go unappreciated. They are among my friends and shipmates and students, whose help ranged from one or more cogent comments to patient listening as, in a classroom, I exercised my techniques of explanation on their captive souls. Then there are some whom I must mention by name: Captain J. James Lester, Extra Master, British Board of Trade, Master Mariner, United States Coast Guard, and best of friends and shipmates, who not only checked and initialed each and every page of this work, but carried the bulky manuscript with him by air to the People's Republic of China, and found time to do so much while attending to his regular duties as master of an ocean drillship operating off the coast there; Doug Goudie, friend, student, and fine sailor, for proofing many of the examples in this new work; my son, Warren Norville, Jr., for his invaluable advice in the development of the final chapters of this edition—especially those explaining the basic formulas used in sight reduction—and for checking my explanations and computations in the use of the scientific calculator; and Captain A. J. "Al" Tatman, USCG (Retired) and now an ardent small-boat sailor, for reviewing the manuscript and giving me the benefit of his suggestions.

I am especially grateful to Texas Instruments Incorporated, and in particular to their good people, Bill Kenton and Jon Campbell. Mr. Kenton did not give up until he had tracked me down by telephone in response to a letter I had written, and then spent the better part of an hour answering my questions. When I had finished interrogating him, he put me in touch with Mr. Campbell, who read my remarks and checked my calculations in Chapter 19, "Sight Reduction by Calculator." To both these gentlemen, I offer a heartfelt thank-you.

And most important, to my friends at International Marine Publishing Company, whose interest over the years and whose encouragement at all times made undertaking an enterprise such as this a pleasure instead of a task, I say thank you. Then there is that unsung and often unrecognized essential person, the editor, Jon Eaton, whose close attention to detail often brought this broad-reaching author back to a straight course. To Jon I am especially grateful.

The wise navigator uses all reliable aids available to him, and seeks to understand their uses and limitations. He learns to evaluate his various aids when he has means for checking their accuracy and reliability, so that he can adequately interpret their indications when his resources are limited. He stores in his mind the fundamental knowledge that may be needed in an emergency. Machines may reflect much of the *science* of navigation, but only a competent human can practice the *art* of navigation.

*American Practical Navigator,* 1962 ed.

# CELESTIAL NAVIGATION

## step by step

# 1 WHAT IT'S ALL ABOUT

Navigation is an art and a science. Equal importance must be given to both facets. This book can get you well along in the practical application of the science. It can even give you a few pointers and various tricks that may help you master the art. Practice and experience will blend this science and this art into a technique that will enable you to find your way on any ocean or anywhere your experience and your craft's capabilities prudently permit you to go.

This is navigation for people. There are enough texts today on navigation for professionals to keep us all confused forever. These books are pretty heavy going for the beginner.

This book is for the yachtsman, fisherman, shrimpboat skipper, or any other person who wants to get more out of his vessel by knowing how to better find his way around. The techniques used and the routine they are combined with in the course of the day are called by navigators "A Day's Work in Navigation at Sea," or, for short, "A Day's Work at Sea."

The navigator's routine begins before dawn. He opens his almanac and, based on his dead reckoning position:

1. He computes the time of local morning twilight, and computes a dead reckoning (DR) position at that time.
2. He determines what stars, planets, or other heavenly bodies may be visible. He selects six or seven and computes what their altitudes and azimuths will be at the time he intends to "shoot" them.
3. At the right time he shoots each body and notes the exact time of the sight and the sextant altitude.
4. When he has observed all the bodies he intends to, he works his sights to obtain a line of position from each.
5. He advances or retards each line of position to a common time and plots them. This gives him a fix. Technically speaking, this is a running fix, but, since the time intervals are so short between the first and last sight, most navigators consider them simultaneous unless they are in a fast-moving vessel.
6. He notes the latitude and longitude of the fix, enters this in his navigator's notebook, and plots the fix on the chart.
7. He compares this fix and the course and

distance made good with the vessel's intended track, and determines any changes of course necessary to continue to his destination. He then reports this position and any recommendations to the skipper. If the skipper agrees, a course or speed change may be authorized.

8. The vessel's DR (dead reckoning) track is now advanced from this fix.

The navigator now has a few hours' respite. He has time for a cup of coffee and a gam with the watch. If twilight was early, he may even catch a few winks before breakfast.

9. About two hours after sunrise, when the sun is near 30° altitude, he shoots the sun. He computes this sight and plots the resultant line of position.
10. Around 0800 ZT (zone time) he will get a "time tick" and wind the chronometer, unless it is battery powered. In either case, he will note the chronometer error, see if it is gaining or losing, and by how much, and log this in the chronometer notebook.
11. About two hours later he may observe the sun again. After computing and plotting this sight, he will advance his earlier sun line to this time and plot his running fix. Sometimes the moon and even Venus may be available for observation at this time. The navigator should try a sight on these bodies also. These nearly simultaneous observations can give an excellent fix. The DR plot is advanced from this fix.
12. During midmorning the navigator will take the azimuth of the sun. To save extra computation, he may have an assistant make this observation while he himself takes his midmorning sun sight. He may observe the azimuth himself immediately after taking the sight, knowing that the azimuth of a celestial body changes, on average, one degree every four minutes. He compares the computed azimuth with the observed azimuth. The difference gives him the compass error for the vessel's heading at that time.
13. After plotting his forenoon fix, the navigator will compute the noon DR and the zone time of local apparent noon (LAN).
14. About 10 minutes before LAN, he goes on deck. At LAN, when the sun is on the meridian, he takes his noon observation, reduces it for latitude, and plots the latitude. He advances his best morning sun line to determine his noon position by a running fix. This is traditionally the high point of the day's navigation. From it, by comparison with the previous day's LAN position, the navigator gets his distance run from noon to noon, and course and speed made good. This is the time interval used for record purposes in the highly competitive days of the clippers. It is the traditional "day's run," noon to noon. At this time he will also compute the course to steer and the distance remaining. If the latter is considerable, and circumstances allow, he will determine both course and distance by great circle sailing.
15. The navigator takes his afternoon sun sights after the sun has changed in azimuth enough to give a good cross with the noon latitude line. The procedures here are very similar to those used for the morning observations.
16. After his last afternoon fix the navigator computes the zone time of local evening twilight, selects his stars for observation, plots his DR for that time, and precomputes the altitudes and azimuths of the stars he selects so as to find them readily in the few minutes of twilight available.
17. A few minutes before twilight, "star time," the navigator goes on deck with his stopwatch and sextant, and as soon as his selected stars appear, he begins shooting stars.
18. After taking these star sights, he works them for lines of position. He plots these and notes his fix just as he did for the morning stars.
19. During the entire day he maintains a dead reckoning plot of his vessel's progress.

If you can perform competently all the duties listed above, you may call yourself a navigator. Many navigators on sailing yachts will content themselves with just the morning, noon, and afternoon sun sights. I personally prefer to do the entire routine, except that I may pass up the afternoon sun line. You never know what sight may be your last for several days. It is important to know how to do them all. Besides, it is fun, and on a vessel at sea, it is a useful way to make time pass.

# 2 NAUTICAL ASTRONOMY

Becoming a good navigator requires considerable study and practice, not because navigation is so hard, but because it is a field so unfamiliar to most people. Learning the terms and procedures is a must, but you may be surprised how much you know already.

Take longitude, for example. Longitude measures the angular distance in degrees or parts of a degree east or west of a starting point called the *prime meridian.* The earth is a sphere for all practical purposes, so there is no naturally defined starting point from which to measure longitude. The English chose the meridian that passes through the position of their observatory at Greenwich, England. We use this as our point of beginning, also. Longitude is measured 180° west of the prime meridian for the Western Hemisphere, and this is called *west longitude.* Longitude is also measured 180° east of the prime meridian for the Eastern Hemisphere, and this is called *east longitude.*

Meridians of longitude are great circles that converge at the poles. A *great circle* is that circle enscribed on the surface of a sphere by a plane that passes through the center of the sphere. If a *meridian* passes through Greenwich, England, we call it the *prime meridian, Greenwich meridian*, or just *Greenwich.* If it passes through your position and you are the observer, it is called the *observer's meridian.* The part of the meridian (half of the great circle) of longitude that passes through your *zenith* is termed the *upper branch* of the meridian. The part that passes through the *nadir* of your zenith is termed the *lower branch.* (Zenith and nadir are defined later in this chapter.) If you are at longitude 88° W, 88° W is the upper branch of your meridian and 92° E is the lower branch. A point on a meridian of longitude may be anywhere on that arc between the earth's poles. This means we need another kind of coordinate to measure angular distance between the poles. We call this coordinate a *parallel of latitude.*

We measure latitude as the angular distance north or south of the equator. The *equator* is a great circle on the earth's surface described by a plane perpendicular to the earth's axis. The earth's axis passes through the poles. We designate the equator as 0° latitude. The North Pole is 90° north latitude. The South Pole is 90° south latitude. The equator lies equidistant between the earth's poles. The points at which the axis of the earth's rotation passes through the earth's surface are called the *geographic poles.* The projections of the earth's axis through the poles to the celestial sphere are the *celestial poles.* The pole with the same designation (North or South) as the latitude of the observer is the *elevated pole.*

Since a plane passing through the center of a sphere describes a great circle, a plane passing through a sphere at any place other than the center describes a *small circle.* If these other planes are also perpendicular to the axis and consequently parallel to the plane that describes the equator, the circles they describe are called *parallels of latitude.* Thus latitude is the angular measurement on the earth's surface from the center of the earth north or south of the equator. If the parallel is south of the equator, it measures *south latitude.* If the parallel is north of the equator, it measures *north latitude.*

With these coordinates you can now fix a point exactly on the earth's surface. Mobile, Alabama, is 88° west of Greenwich, England. In other words, Mobile is on the 88th meridian west of the prime meridian, or in 88° west longitude. Mobile is also 30° north of the equator. It is in 30° north latitude. So we say Mobile, Alabama, is at latitude 30° north and longitude 88° west. There is only one place where these two measurements coincide. This place is Mobile, Alabama. Latitude 30° N and longitude 88° W constitute the *terrestrial coordinates* of Mobile, Alabama, or the *geographic position* of Mobile, Alabama. We abbreviate this *GP.*

We can define the geographic position of Mobile, Alabama, much more precisely than this. Mobile is really not at exactly latitude 30° north and longitude 88° west. These are the nearest even-degree coordinates and actually pinpoint a GP several miles east and about 40 miles south of Mobile. Whole degrees are used to make the illustration easier. The precise GP of Mobile, Alabama, measured from Barton Academy, an historic school building in Mobile, listed in Appendix S, "Maritime Positions," in Bowditch, is latitude 30° 41.0′ N and longitude 88° 07.0′ W.

It is necessary to be this precise in the practice of navigation. A degree of latitude is equal to a distance of 60 miles on the earth's surface. A degree of longitude will vary in distance from 60 miles on the earth's surface at the equator to zero miles at the poles. There are 360 degrees in a circle, 60 minutes of arc in a degree, and 60 seconds of arc in one minute of arc. A minute of arc of latitude equals one mile. A minute of arc of any great circle on the surface of the earth equals one mile. We are talking about *nautical miles*, of course. A nautical mile is 6,080 feet. A second of arc is approximately 101 feet. Since 101 feet is more precise than practical navigation demands, positions today are noted in degrees, minutes, and tenths of a minute.

This is how we locate a place or geographic position on the earth's surface. How do we measure the position of bodies in the heavens? This requires a little imagination. Suppose you are at the center of the earth and every body in the heavens and every point on the earth is projected to a spot on the surface of a sphere whose center is also the earth's center. We must imagine that the surface of this sphere is an infinite distance from the center of the earth. We call this sphere the *celestial sphere.* Even though this sphere is imaginary, we can develop a concept of celestial coordinates that is valid enough for our needs. Therefore, every body in the heavens and every point on the earth has a position on the celestial sphere that corresponds to its relative position in space or on the earth's surface.

## CELESTIAL EQUATOR

The circle on the celestial sphere equidistant from the celestial poles at all points is the *celestial equator.* The celestial equator is a projection of the earth's equator onto the celestial sphere. It is in the same plane as the earth's equator.

## CELESTIAL POLES

Each of the earth's poles can be projected onto the celestial sphere. These are the *celestial poles.* They take the names "north" and "south" in the same manner as the earth's poles. The celestial pole that has the same name as the latitude of an observer is the *elevated celestial pole* or *elevated pole.*

## GREENWICH HOUR ANGLE

The meridian of Greenwich, England, has a corresponding meridian on the celestial sphere. This is the *celestial zero meridian.* The angular distance of any body west of this meridian is the body's *Greenwich hour angle*, abbreviated *GHA.* Greenwich hour angle is always measured west through 360°; with this important difference, Greenwich hour angle corresponds to longitude on the earth's surface.

## LOCAL HOUR ANGLE

*Local hour angle* (abbreviated *LHA*) measures the angular distance of a body *west* of the observer's meridian. This is the classic and strictly correct definition of local hour angle. All you need to do to find the LHA of a body is to find the difference between the Greenwich hour angle of the body and your longitude. This difference, when reckoned in a westerly direction, is the LHA of the body. At one time, LHA was computed easterly also. Today, this value, which is really the reciprocal of the LHA, is called "t," the *meridian angle*. Meridian angle can be west, too; when "t" is west, "t" equals LHA. Unlike LHA, "t" is never greater than 180°. So much for the technically correct. Your longitude as an observer may be a dead reckoning longitude, an assumed longitude, or your actual longitude, depending on how you approach the problem. We will see more of this later. It is important to note that LHA or "t" is one of the arguments needed to enter the tables of Pub. No. 249.

## DECLINATION

To fix a body's position in the heavens, you need two coordinates. The GHA of a body tells how far west the body is from the zero meridian. You must also know the body's position relative to the celestial equator. The angular distance north or south of the celestial equator is the *declination* of a body. Declination corresponds to latitude on the earth's surface. Just as in latitude, if a body is south of the celestial equator, its declination is south. If a body is north of the celestial equator, its declination is north.

## ZENITH

Your zenith is the point on the celestial sphere directly over your head. The declination of your zenith has the same name and value as your latitude. If your latitude is 30° north, your zenith will have a declination of 30° north. Also, the Greenwich hour angle of your zenith will have the same name and value as your longitude if you are in west longitude. If you are in east longitude, subtract your longitude from 360° to get the Greenwich hour angle of your zenith. Still another relationship to know is that the geographic position of your zenith is your position.

## HORIZON

The part of the celestial sphere visible to an observer has been likened to a giant, inverted bowl. The rim of this bowl is the *horizon*. Believe it or not we have three different horizons. In navigation we really reduce all our observations to coincide with the earth's center. Our calculations are worked as if we took our observations from the center of the earth, which is also the center of the celestial sphere. A plane that passes through the earth's center and is perpendicular to the radius from the earth's center to the observer's zenith intersects the celestial sphere at the *celestial horizon*. Of course, we take our sights from the surface of the earth. A plane that touches a sphere's surface at one point is tangent to that sphere. A plane that touches the earth's surface at one point is tangent to the surface of the earth. From geometry we know that a tangent is perpendicular to a radius at the point of tangency. If this plane's point of tangency (the point where it touches the earth's surface) is at your geographic position, this plane is also perpendicular to the radius of the zenith. This plane, which intersects the surface of the earth at your geographic position, GP, is the plane that intersects the celestial sphere at the *terrestrial horizon*. This plane is parallel to the plane of the celestial horizon and separated from it by a distance equal to the radius of the earth.

Placing the observer at an imaginary point at the center of the earth requires a correction to sextant altitude to accommodate the difference in altitude of a body observed from two points—the surface of the earth and the center of the earth. This correction is called, somewhat loosely, *horizontal parallax*. Horizontal parallax (more technically, geocentric parallax) is a factor that is important only when taking sights of bodies in the solar system. Parallax is greatest when a body is on the horizon, and is then correctly termed horizontal parallax. It may vary from less than a minute for the minor planets and the sun to almost a degree for the moon.

The third horizon we are concerned with in navigation is the *observer's horizon* or *visible horizon*. The observer's eye is not exactly on the surface of the earth but several feet above it. For this reason the *visible horizon* will be somewhere below the celestial horizon and the terrestrial horizon. The correction for this is called *dip*. Figure 1 shows the celestial horizon and the terrestrial horizon. The visible horizon will be explained more completely in Chapter 4.

## AZIMUTH

When the direction toward a celestial body from a point on the earth is reckoned from true north (000°/360°T) in a clockwise, right-hand, or easterly manner, that direction is the *true azimuth*. An azimuth is the celestial bear-

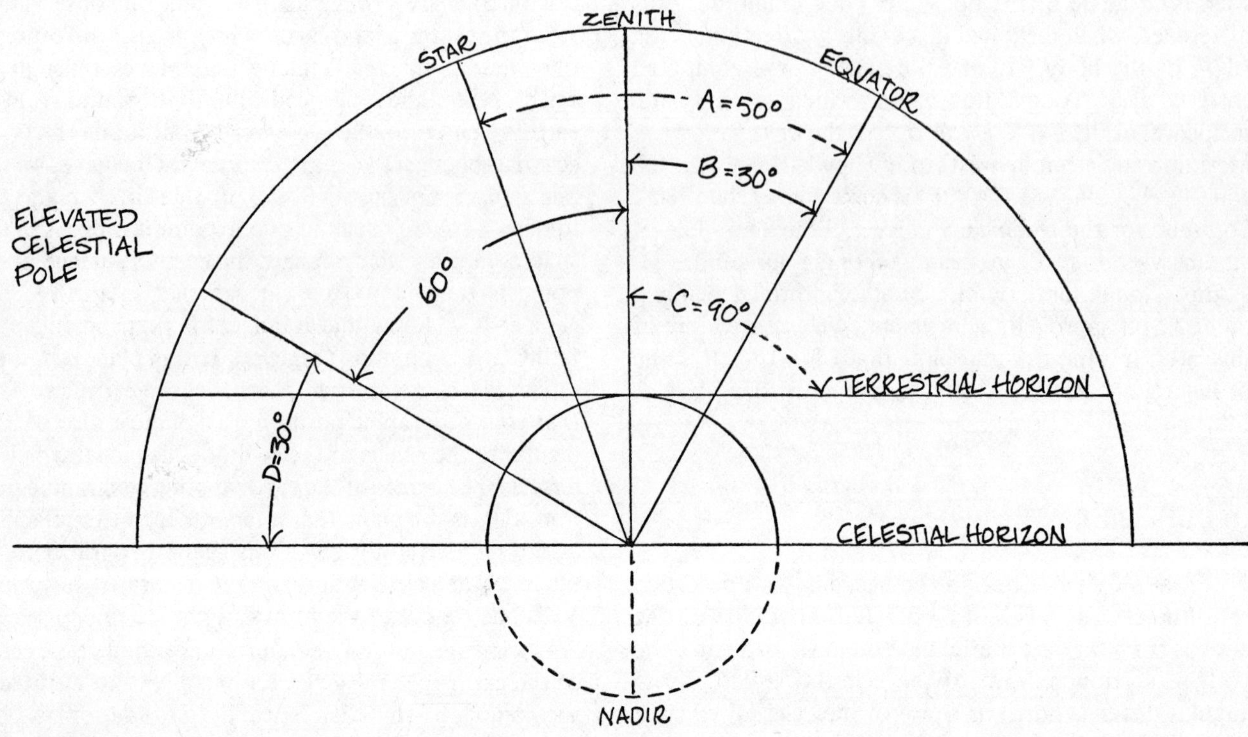


**Figure 1.** *The large semicircle represents the visible half of the celestial sphere. The small circle is the earth. The dashed lines represent that portion below the horizon. The circles are concentric. The celestial pole, zenith, and the equator are labeled. Zenith is the point directly over your head. The terrestrial horizon is tangent to the earth's surface at that point where a line from the zenith to the center of the earth passes through the earth's surface. Thus the zenith is perpendicular to the terrestrial horizon. By extension, the zenith is perpendicular to the celestial horizon, since the two horizons are parallel. Angle A of 50° is the declination of the star in the celestial sphere. Angle B is the latitude, 30°. Angle C of 90° is the angle made by the terrestrial horizon and the line from the zenith.*

*Observe the diagram carefully. If the angle between the equator and your zenith is 30° (your latitude), then the angle between the zenith and the elevated celestial pole must be 60°, because the elevated pole is perpendicular to the equator. Now, if this is so, the angle between the elevated pole and the celestial horizon must be 30° (D). This is an important relationship, because it is the basis for obtaining latitude by observations of Polaris.*

ing of a body. If you see the sun over your compass on a bearing corrected to 081°T, 081° is the true azimuth of the sun at that time.

## NADIR

Nadir is a word you will hear frequently. A line dropped from your zenith, through your geographic position, and consequently through the earth's center, intercepts the celestial sphere on the opposite side of your zenith. This point of intersection on the other side of the celestial sphere is the *nadir* of your zenith. Nadir represents the lowest point relative to your position, while zenith represents the highest. You have probably heard people speak of their spirits being at zenith when they are very happy and at nadir when they are very blue.

The mathematics of navigation involves a solution of one side of a spherical triangle. Fortunately for the modern practical navigator, this part of the problem is already worked and tabulated in various publications such as the *Sight Reduction Tables for Air Navigation*, Pub. No. 249, which is published by the Defense Mapping Agency Hydrographic/Topographic Center. It tabulates various solutions of spherical triangles having local hour angles of 0° to 360°, declinations from 0° to 29°, and latitudes from 0° to 89°. The tables in Volumes 2 and 3 of Pub. No. 249 can be used to reduce sights of the sun, the planets, and the moon, because the declinations of these bodies do not exceed 29°. In addition, many stars have declinations of 29° or less. For those stars whose declinations exceed 29°, Volume 1 of Pub. No. 249, *Sight Reduction Tables for Air Navigation (Selected Stars)*, may be used, but only for the seven selected stars in the tables. To reduce sights of other stars whose declinations exceed 29°, it is necessary to use the tables in Pub. No. 229, *Sight Reduction Tables for Marine Navigation*, or Ageton's Method (formerly H.O. 211 and still available as Table 35 in Bowditch), or some other method.With precomputed tables, all the navigator needs to do is add and subtract tabulated values where appropriate. The people who computed these tables had another problem. They were faced with actually working out the trigonometric solutions. To do this, they had to relate each factor to the surface of the same sphere. This can be either the celestial sphere or the earth's surface. Different navigators have alternate preferences.

Since we are looking for a position on the earth, it seems to me that the earth's surface is a good place to lay out our spherical triangle. To do this we use the elevated pole and the geographic position of the body observed and solve for our position. This is an oversimplification, but it brings us to the concept I want you to learn, which is: For every point on the earth's surface we can project a corresponding point on the celestial sphere, and for every point in space we can project a corresponding point on the earth's surface. If the point in space is a celestial body, the point on the earth's surface that corresponds to its location in space is the body's geographic position—the point at which the body will be at zenith. If the body is a star, this point is called the star's *substellar point*. If the body is the sun, this point is called the sun's *subsolar point*. If the body is a planet, this point is called the *subplanetary point*. If the body is the moon, this point is called the *sublunar point*.

These are important concepts to know, for they will help you visualize positions of bodies in space relative to your position on earth. As you become more experienced in navigation, you will find that these concepts will help you approximate the altitude and azimuth of a body without precise calculations. For example, if you know the moon is available for a daytime observation, and the moon's GHA is east of your longitude and its declination is south of your latitude, you know its GP is south and east of you too. With experience, you will learn to take the GHA and declination of a body and estimate its altitude and azimuth with surprising accuracy.

These are most of the major terms you will need to know. The definitions I have given you in some cases may be more analogous than strictly scientific. For further study, you should get *The American Practical Navigator* by Bowditch, *Dutton's Navigation and Piloting* by Maloney, or *A Primer of Navigation* by Mixter. These are all excellent books, and anyone who seriously wants to be a navigator should own a copy of each of them.*

---

**Dutton's* and Mixter are available from the publisher of this book: International Marine Publishing Company, 21 Elm Street, Camden, Maine.

# 3 TIME

The most fundamental idea a navigator must understand is the relationship of time to longitude. Longitude and time have been the Waterloo of many would-be navigators. This need not be if you will just consider hours, minutes, and seconds of time as another system of measuring circles and sectors of circles, a system that can be substituted for degrees, minutes, and seconds of arc. Just as you can measure distances on the earth's surface in miles, yards, and feet, you can also measure them in kilometers, meters, and centimeters.

There are 360 degrees in a circle. These degrees can be divided further into minutes and seconds of arc. Thus, to go around the world from a point to the same point requires a traverse of 360°. In the measurement of longitude, these 360° by custom are divided equally into 180° west longitude and 180° east longitude. Just as easily, the earth's circumference, from a point around the earth to the same point, can be measured in hours that can be divided into minutes and seconds of time. The sun is overhead at Greenwich, England, at noon. The next day, 24 hours later, the sun is overhead at Greenwich again. You understand right away what is meant when someone says this is so, because the earth has undergone 24 hours of rotation. The earth has also undergone a 360° rotation. So, with equal facility we can say there are 24 hours to a circle and 360° to a circle. It is consequently possible to carry the relationship further and say that 15° equals one hour; one minute of arc equals four seconds of time and one minute of time equals 15 minutes of arc. You can convert time into arc just as you can convert feet into meters. The easiest way to do this is by use of the *Nautical Almanac*. The *Nautical Almanac* is an annual publication that tabulates astronomical data for the year of issue along with other information needed by a navigator to work problems pertaining to celestial navigation. The complete relationship for conversion purposes is tabulated on page *i* of the yellow pages in the back of your *Nautical Almanac* in the table entitled "Conversion of Arc into Time."

The concept of this relationship between time and arc is reserved for circles on the earth's surface and on the celestial sphere. The relationship is further limited to the position and change in position primarily of the celestial bodies. Because of this, you may find the Greenwich hour angle and sidereal hour angle of various bodies tabulated in hours, minutes, and seconds. (Do not worry if you do not understand the meaning of sidereal hour angle now. This term is defined and explained in the chapter on how to work a star sight.) It wasn't too many years ago that the *Nautical Almanac* tabulated these values in hours, minutes, and seconds. It is important to note that degrees of latitude or declination are never converted into time.

What is "time"? It is really a very tangible thing, but something very difficult to tie down with a good definition. The prime unit of time is the day. This is the period (I could say the time) it takes for the sun to make a complete circle of the earth. We know the sun does not go around the earth; we really mean that a day is the time it takes the sun to *appear* to cross the meridian of our nadir, 180° from our zenith, twice in succession. When the sun crosses the meridian of our nadir, it is our local apparent midnight. We measure days now from midnight to midnight. The sun, though, does not take exactly the same time to return to the meridian of our nadir each day. This is the result of several factors, among which is the effect of the position of the earth in its orbit. To compensate for this, we must compute an average position of the sun. This is called the *mean sun*. Mean time is reckoned according to the mean sun.

*Local mean time* is the mean time of the observer's meridian. In the case of Greenwich, England, it is known as *Greenwich mean time,* abbreviated *GMT*. To tabulate the positions of the heavenly bodies properly and in a useful manner, we must use the same time reference for all. Obviously, if we are going to reckon our angular distances from Greenwich, England, it is logical to use *Greenwich mean time*. The current year's *Nautical Almanac* will supply you with the celestial coordinates, at any instant of GMT (to the nearest second), of all the celestial bodies used in navigation.

Due to the rotation of the earth on its axis, the GMTs of such phenomena as sunrise, noon, and sunset vary with the observer's position. Since we regulate our lives by such things as the time of daylight, it is necessary to determine a local time that approximates the times of these occurrences at Greenwich. This is called *zone time* or *ZT*. If the night is half over at midnight or the sun almost on the meridian at noon at Greenwich, we want the sun to be almost on the nadir of our local meridian when our ship's clock says it is midnight and almost on our local meridian when our ship's clock says it is noon. We want the sun's position in the heavens to conform as nearly as possible to local time at our position.

If we are on a ship at sea and our almanac tells us that sunrise, for example, is at 0620 on the Greenwich meridian for our latitude on this date, we adjust our ship's clocks so that sunrise is near 0620 zone time on board our vessel. If the GMT of sunrise on our ship is three hours later than it is on the same latitude at the Greenwich meridian, we say our vessel is keeping time three hours later than Greenwich. We say we are in a *plus-3 time zone*. This means that to get GMT we must add three hours to our zone time. Of course we won't change our clocks every time we cross a meridian, because the changes would be too frequent. Every time we see that the zone time of a phenomenon begins to vary considerably from that tabulated, we will add an hour to our zone time if going east and subtract an hour if going west. To disturb a ship's routine least, time changes are usually made during the wee hours of the midwatch.

Ashore we keep *standard time*, which coincides in many ways with zone time. It differs from zone time because people living in the same economic, social, and political neighborhood need to keep the same time.

The mean sun crosses each successive whole-degree meridian four minutes after the one previous. In other words, if the mean sun crosses the Greenwich meridian at noon, it will cross the meridian at 1° west longitude at 1204 GMT. The mean sun will cross the meridian of 7° 30.0′ west longitude at 1230 GMT. It will cross the meridian of 15° west longitude at 1300 GMT, and it will cross the meridian of 30° west longitude at 1400 GMT. Fifteen degrees west longitude is one hour away from Greenwich. When it is 1300 GMT, it is 1200 noon ZT at the 15th meridian west of Greenwich. The meridian of 30° west longitude is two hours away from Greenwich.

It is easy to see that, for every 15 degrees west we go, the Greenwich time of local noon becomes one hour later—so also does the time of any other celestial phenomenon. These meridians of 15-degree intervals are called *zone meridians* and the local mean time that occurs at these meridians is called *zone time*. The time zone extends 7° 30′ on each side of the zone meridian. At sea, time aboard ship is kept to coincide with zone time. Ashore, the practice is not as consistent. It would not be very wise to make one area of a state keep different time from another area of the same state simply because it happened to lap slightly over into another time zone (although it does occur in some states). Also, it is more logical to change a time at the point of some natural barrier of division that comes nearest to 7° 30′ on either side of the zone meridian. Time that comes as near as possible to the zone time, but is adjusted to meet economic, geographic, or political situations, is called *standard time*. In the continental United States, we keep four different standard times—Eastern, Central, Mountain, and Pacific. Examination of a map showing standard time zones in the United States will show how these divisions approximate the zone meridians but vary as necessary to suit local requirements.

The mean sun is an imaginary concept conceived to assist us in the reckoning of mean time. True sun, or *ap-*

*parent sun*, varies in position as much as 15 minutes of time ahead of or behind the mean sun. This is caused mainly by the effect of the earth's position in its orbit. The difference in the position of the real or true sun and the mean sun is the *equation of time*. If the actual position of the sun at noon Greenwich mean time is 1° 30.0′ east of Greenwich, the equation of time is six minutes. The time position of the true sun is 1154. It is 1154 *apparent time*.

Apparent time is a measure of the sun's true position. To get the Greenwich apparent time at any instant, convert the Greenwich hour angle (GHA) of the sun in degrees, minutes, and seconds at that instant to hours, minutes, and seconds of time. In practice today, the equation of time is seldom used. Apparent time, however, is important in determining the time the sun will be on an observer's meridian. This is the time of *local apparent noon*, when observations for latitude are made. Apparent time is the true measure of the sun's position in time. It varies for each meridian, and no two places can have the same apparent time unless they are on the same meridian. At a given instant of Greenwich mean time, the zone time is the same at 85° west longitude and 95° west longitude. The apparent time at these two meridians, however, differs by approximately 40 minutes.

The daily routine of a ship, whether it is a man-of-war or a yacht, is carried on according to zone time. The data required for navigation are tabulated in the *Nautical Almanac* according to Greenwich mean time. For this reason, the navigator constantly needs to find Greenwich time when zone time is known. When the time at Greenwich passes midnight, the Greenwich date will change too. The full problem then is to find the time at Greenwich and the date at Greenwich.

For example, suppose you are underway in the Gulf of Mexico at longitude 90° west. You have calculated the zone time of evening twilight on 25 August 1983 to be 1830. You want this information to plan your evening star sights. You also need the GHA of Aries. To find the GHA of Aries, you must enter the *Nautical Almanac* with the Greenwich mean time and date as entering arguments. Don't worry about how you get the GHA of Aries. We will cover this in another chapter. Right now you must learn how to find Greenwich time and date.

Remember that an hour of time equals 15° of longitude. Divide 90° (your longitude) by 15°, which gives you a quotient of six. This means that 90° is six hours away from 0°, which is the longitude at Greenwich. You are in west longitude. Time in west longitude is earlier than Greenwich time. The Greenwich day starts before the local day. The sun rises in the east. It is that simple. So if Greenwich time is later than local time and the difference is six hours, to get Greenwich time you must add six hours to the local time. In this situation, Greenwich time is said to be *best*. This six hours is the *zone description*. Since it must be added to local time to get Greenwich time, it is said to be *zone plus six*. The date may change too. Set up the problem like this:

| | | |
|---|---|---|
| Local Date | Zone Time | $18^h$-$30^m$ |
| 25 Aug. | Zone Description + | $6^h$-$00^m$ |
| Greenwich Date | | |
| 26 Aug. | Greenwich Time | $24^h$-$30^m$ |
| | or | $00^h$-$30^m$ |

There are 24 hours in a day. Greenwich time is actually $00^h$-$30^m$. This is 30 minutes past Greenwich midnight, so the *Greenwich date changes*. Another way to do it is as follows:

| | | |
|---|---|---|
| Local Date | Zone Time | $18^h$-$30^m$ |
| 25 Aug. | Zone Description + | $6^h$-$00^m$ |
| | Greenwich Time | $24^h$-$30^m$ |
| | | $-$ $24^h$-$00^m$ |
| Greenwich Date | | |
| 26 Aug. | GMT | $00^h$-$30^m$ |

Greenwich date changes. It is a day later.

Had you been in 90° east longitude, the problem would have been solved similarly. Remember that, since you are east of Greenwich, the sun rises at your position before it rises at Greenwich. Your day starts sooner. Zone time will be later than Greenwich time. To find Greenwich time you must subtract six hours. There is still six hours of difference because you are 90° from Greenwich. Greenwich time is earlier. We say Greenwich time is *least*. The zone description here is *minus six*. Using the same zone time of 1830 and the same date of 25 August, let's see how the problem works out:

| | | |
|---|---|---|
| Local Date | Zone Time | $18^h$-$30^m$ |
| 25 Aug. | Zone Description $-$ | $6^h$-$00^m$ |
| Greenwich Date | | |
| 25 Aug. | GMT | $12^h$-$30^m$ |

Greenwich date did not change.

Suppose you had been trying to find the Greenwich mean time of morning twilight at your position on 25 August. Zone time of morning twilight is 0517. You are in 90° west longitude. What is the Greenwich time and date of local morning twilight? The solution:

| | | |
|---|---|---|
| Local Date | ZT | $05^h$-$17^m$ |
| 25 Aug. | Zone + | $6^h$-$00^m$ |
| Greenwich Date | | |
| 25 Aug. | GMT | $11^h$-$17^m$ |

This time Greenwich date did not change.

Had you been at 90° east longitude, you would have worked the problem as follows:

| | | |
|---|---|---|
| Local Date | ZT | $05^h$-$17^m$ |
| 25 Aug. | Zone − | $6^h$-$00^m$ |
| Greenwich Date | | |
| 24 Aug. | GMT | $23^h$-$17^m$ |

Greenwich date changes. It is a day earlier in Greenwich.

Had your longitude been 85° W or 95° W, the solution would have been exactly the same. The time zone extends 7° 30.0′ either side of the zone meridian, which in this case happens to be 90° W. At either 85° W or 95° W, your zone description will still be +6. Therefore, you must still add 6 hours to your zone time to get Greenwich mean time. You will be in time zone +6 from 82° 30.0′ W to 97° 30.0′ W.

Finding Greenwich time and date can be a real brain twister if you let it. Yet the problem is quite simple if you keep in mind that in east longitude you subtract the zone description from the zone time, and in west longitude you add the zone description to the zone time to get Greenwich time. If the time change at Greenwich passes through midnight, the Greenwich date will change too. We can sum up the problem with a memory aid that will appear frequently in different ways as long as you study celestial navigation:

*When longitude is east*
*Greenwich time is least*
*When longitude is west*
*Greenwich time is best*

# 4 INSTRUMENTS OF CELESTIAL NAVIGATION

Several precision instruments are required to collect the data needed for navigation from an observation of a celestial body. The fundamental basis of any navigation problem is the keeping of as accurate a dead reckoning track as possible. To do this requires an instrument to accurately determine direction, which we call our course, and some means of estimating speed or distance run. The compass is the instrument used for direction. The means of determining speed, and hence distance, may vary from an elaborate pitometer or taffrail log on the finer yachts to a seaman's eye estimate on smaller craft. The information from these instruments furnishes the navigator with what he needs to solve his speed, time, and distance problems. The solution to one of these problems, when applied to a known point of beginning, which is known as the *departure*, and true course steered, will give the vessel's *dead reckoning* or DR position.

Dead reckoning is basic to celestial navigation. The remainder of the process involves checking the accuracy of the DR plot by the use of azimuths and bearings to determine compass error, and obtaining lines of position to get a fix to compare with the DR position. The latest fix is then used as a new departure.

Lines of position, LOPs, may be obtained by bearings on fixed charted landmarks, as in piloting. Lines of position may also be obtained by use of radio bearings, as in radio navigation. When out of sight of land and accurate radio range, the third method of obtaining LOPs is by celestial observation. This often is the only means available. Now this is by no means intended to put down celestial navigation as a poor third choice. Quite the contrary. Next to accurate bearings on well-charted landmarks, celestial LOPs are probably the most accurate means of getting a fix available to the operator of a small vessel.

There is no reason why LOPs from each of these sources—fixed landmarks, radio bearings, and celestial observations—cannot be used in combination. On our Atlantic crossing in *Asperida II*, we used two sun lines, an RDF bearing from a Moroccan radio station, and a visual bearing on a mountain peak called Bu Namane on the coast of Morocco. These four bearings from three different sources gave us a fix that warned us of our proximity to the low-lying lee shore on the North African coast in time to avoid disaster.

Our interest in this book is in learning how to obtain celestial lines of position. I mention the other navigation methods because they are equally important. We will leave the details of the compass and taffrail log to a study of piloting and dead reckoning. Use of radio direction finders and Loran are properly studied in radio navigation. Our interest is in the instruments of celestial navigation.

The unknown in navigation is always your own vessel's position. Finding your position in celestial navigation requires solving a spherical triangle. Remember from Chapter 2 on nautical astronomy that the spherical triangle is composed of arcs of a great circle connecting the elevated pole, the GP of the body observed, and the observer's position. (Solutions to the spherical triangle have already been done for you in the tables of Pub. No. 249.) The location of the elevated pole is fixed. The position of the body observed is tabulated for the instant of observation in the *Nautical Almanac*. To know this instant of time requires an accurate timekeeping device. The timepiece used is called a *chronometer*. Since the time used to tabulate values in the *Nautical Almanac* is Greenwich mean time, the chronometer is always set to Greenwich mean time, GMT.

The chronometer is really a fine clock. It should keep accurate time, but more important, any gain or loss in time must be consistent. If it loses or gains three seconds a day, the chronometer must lose or gain exactly three seconds every day to be any good. This loss or gain is called the *chronometer rate*, or just *rate* for short. The accumulated error due to this loss or gain over a given number of days is called the *chronometer error*. If the navigator knows the error of his chronometer on a given past date, and he also knows its rate, he can calculate his chronometer error for any date he wants if the rate is constant. If the rate is not constant, it is said to be erratic, and the chronometer must be sent to the shop.

So important is precise time to the nearest second that large vessels carry three chronometers. This way they can check them against each other and see when one chronometer becomes erratic. Time signals also are received periodically by radio to check the chronometer error. In the old days, these signals were visual, and a vessel had to wait until she arrived at a port giving time signals to check her chronometer.

Today, time signals are broadcast by radio. The best station, if you can get it, is WWV, broadcasting from Fort Collins, Colorado. WWV is sponsored by the National Bureau of Standards of the United States Government. Frequencies on which WWV broadcasts are 2.5 MHz, 5 MHz, 10 MHz, and 15 MHz. WWV-H is a subsidiary of WWV and broadcasts from Hawaii. Many other broadcast stations also transmit a time signal every hour. This is particularly true of foreign stations broadcasting on the short-wave band. You can also get GMT by telephoning area code 312, telephone number 976-1616.

Chronometers are expensive and quite bulky items. These factors can be insurmountable obstacles if the yacht is small and the owner's budget limited. With radio time signals available, a fine chronometer is not absolutely necessary. A good watch—or better, two good watches—and a radio capable of receiving WWV will fill your needs quite well. I used to use a surplus Waltham hack mounted in gimbals like a small chronometer. A hack is a comparison watch used on larger vessels to regulate a ship's clock. It is set by the ship's chronometer, corrected for chronometer error, and, by applying zone description, is adjusted to read zone time. Because of the need for precise times in the radio shack, engine room, and other vital areas of the ship, plus the need for these clocks to be synchronized closely, a hack must be a precision instrument. This is an excellent timepiece, but not nearly as large or expensive as a chronometer. For a backup timepiece, I also used an Elgin pocketwatch with a plain dial, well marked in large, black numbers. Of course it had a second hand. The radio is a multiband portable of the best quality. The total investment in all of this equipment does not exceed one-third the cost of a real chronometer, yet fills the need quite well. Watches used in this manner should be checked against a time signal every day, and the rate should be determined. Then, if for some reason you cannot get a radio time signal for a few days, you can still determine your chronometer error. Chronometers and watches used as chronometers should be wound every day at the same time even if they are the so-called eight-day type. This keeps a more constant spring pressure, and thus aids the timepiece in keeping a constant rate.

Today I use a marine quartz chronometer in a case. My second chronometer is a certified wrist chronometer with a "tuning fork" movement. Today, quartz-movement timepieces are the only rational choices. Their rates are normally very small, only seconds a year.

A battery-operated quartz chronometer or watch needs only to be checked every day, and the batteries renewed as required.

Another instrument is needed to determine your position relative to the body you are observing. You get your measure of this position by observing the vertical angle of this body above the celestial horizon. As you will see in later chapters, this really gives you a locus of positions you may be on. These positions will plot as a circle; the angular distance of the body above the celestial horizon will be the same when observed from anywhere on the circumference of this circle. This angular distance above the celestial horizon is called the body's *altitude*. The instrument for measuring the altitude is called a *sextant*. When the navigator first observes the altitude of a body, he observes its altitude above the visible horizon. This is the *sextant altitude* of the body. This altitude must be corrected for various errors to get the true altitude above the celestial horizon. The corrected altitude is called the *observed altitude*. Sextant altitude is abbreviated $H_s$. Observed altitude is abbreviated $H_o$.

An explanation of the principle of the sextant is not necessary here. However, I do believe you would benefit from a thorough reading of sextant theory in Bowditch, *Dutton's*, or Mixter. Your job is to learn to use the sextant and use it skillfully. This will take a lot of practice, especially on a small vessel where the vessel's motion and the observer's low height of eye complicate the problem. Skill is the answer to this and the only way to develop skill is to practice. If you are serious about wanting to be a navigator, you will need a good sextant. Then, every time you are afloat and a good horizon is available, you should practice taking sights.

Taking a sight, except for meridian altitude and Polaris observations,* requires an exact notation of the time of the sight. This time must be to the nearest second. Some navigators will have an assistant who will note the exact time the navigator sings out "mark!" to indicate the body is on the horizon. On small yachts and commercial vessels, an extra hand may not be available. The navigator then will have to take his own time. He holds in his hand a stopwatch. When he brings the body to the horizon, he starts the stopwatch. He then goes to the chronometer and reads it first in seconds, then in minutes, and then in hours. When he notes the seconds, he stops his stopwatch at the same instant. If he takes the chronometer time read and subtracts from it the time the stopwatch ran, he will have the chronometer time of the sight. If he applies the chronometer error, he will have the Greenwich mean time of his sight.

Let's take a typical problem. Say you are the navigator of a yacht at sea. You observe the sun's lower limb. You start your stopwatch at the instant the lower rim of the sun's disk is tangent to the horizon. You go below, note your chronometer time to be $35^s$-$11^m$-$14^h$, and stop the stopwatch when you note the $35^s$. Your stopwatch showed an elapsed time of $01^m$-$13^s$. Your chronometer is $2^m$-$14^s$ slow. What is the Greenwich mean time of your sight? You should set up the problem this way:

| | |
|---|---|
| C | $14^h$-$11^m$-$35^s$ |
| W | $01^m$-$13^s$ |
| C−W | $14^h$-$10^m$-$22^s$ |
| CE | $02^m$-$14^s$ |
| GMT | $14^h$-$12^m$-$36^s$ |

C = chronometer
W = watch, stopwatch
C − W = chronometer minus stopwatch
CE = chronometer error
GMT = Greenwich mean time

What do we mean by bringing the body down to the horizon? To answer this will require an explanation of how the mariner's sextant is made. At the center of the arc of the sextant is a pivoted mirror. This "glass" is the *index mirror*. On the edge of the sextant away from the observer is another glass that is half silvered and half clear. This is the *horizon mirror*. To take a sight, the navigator looks through the telescope of the sextant at the horizon. The telescope is so aligned that the navigator sees the horizon through the left side of the horizon glass, while the silvered part of the horizon glass is on the right. To take a sight, the navigator moves the index mirror until the body he wishes to observe is reflected in the horizon mirror. He then adjusts this reflection until the body appears to just touch the horizon. The beginning navigator will be surprised to see that he will have to keep adjusting the index mirror. He can actually see the body move in relation to the horizon. When the navigator has the body on the horizon, he swings his sextant around the axis of his line of sight several degrees in each direction. This is called *swinging the arc*. The body should swing above the horizon and touch the horizon at the point where the sextant is in a vertical position. At this instant the navigator marks the time. The sextant must be vertical to get a true altitude observation. The navigator reads his altitude from a pointer on an arm mounted under the index mirror. This is the *index arm* of the sextant. This pointer indicates the degrees of altitude on the arc of the sextant. Closer reading is provided by a micrometer drum or vernier. The sextant should be read to the nearest tenth of a minute of arc.

The altitude you read off your sextant must be corrected for several errors. The sextant itself may be off in its reading. This error is generally due to maladjustment of the mirrors. To determine this error, move the index arm so the pointer reads zero. Then look through the telescope at the horizon. The true horizon and its reflected image should coincide and stay in coincidence

*Meridian altitude observations are explained in Chapter 11, Polaris observations, in Chapter 12.

when the arc is swung. If they are not in coincidence, the index arm should be moved until the true horizon and its reflected image do coincide. Now read the sextant. The difference between this reading and zero is the *index error* (see Figure 2).

If the index pointer points to the right of the zero mark, it is pointing below the zero mark. It is, in the mariner's vernacular, "off the arc." The sextant measures from 0° to at least 120°, and a measurement below zero is outside this range. In this case the sextant will measure too low. Most sextants, however, do have arcs graduated to several degrees below the zero mark, and can thus measure angles less than zero. This reading below the zero mark must be added to any sextant to get the correct observed altitude. So if the index pointer points to a value on the arc to the right of the zero, it is less than zero and said to be "off the arc." It must be added to any reading of the sextant, that is, put "on" the sextant altitude. So we say "if it is off, it is on."

When the pointer points to the left of the zero, it means the sextant will measure angles to be higher than they really are. Since the pointer is inside the 0° to 120° range, the error is said to be "on the arc." Because this error makes the sextant read too high, it must be subtracted from or taken "off" the sextant altitude to get a correct observed altitude. Thus, if the index error is "on" the arc, it is "off" the sum of sextant altitude corrections, and vice versa. "If it is on, it is off, and if it is off, it is on."

One important consideration must be borne in mind when reading the index error "off" the arc or less than zero. In this case, the reading of the micrometer drum or vernier must be subtracted from 60′ to get the correct minutes and tenths of the index error.

Most experienced navigators will not tolerate an index error in a sextant. Unless the instrument is damaged, it is not at all difficult to reduce the index error to zero. To learn to do this, the beginner should consult Bowditch or *Dutton's* and then ask for help from someone who knows how to adjust a sextant for index error.

Observations of celestial bodies are computed as if they were made at the center of the body. When taking an observation of the sun or the moon with a mariner's sextant, it is not possible to measure the exact center of either of these bodies. The navigator will make his observation by bringing the reflection of the sun or moon down until either the upper or lower edge of the body is just tangent to the horizon. If the lower rim of the sun or moon is brought into tangency with the horizon, the navigator is said to be observing the *lower limb* of the sun or moon. If the upper rim of the sun or moon is brought into tangency with the horizon, the sight is said to be an observation of the sun or moon's *upper limb*. Except when the sun's disk is partially obscured, observations of the sun are made of the lower limb. Observations of the moon may be either of the upper limb or lower limb, depending on the phase of the moon at the time of observation. Regardless of whether the upper limb or lower limb is observed, a correction to the sextant altitude must be applied to adjust the altitude to the body's center. This correction is called *semidiameter*. The semidiameter corrections are found in the *Nautical Almanac*. For an observation of the lower limb, the correction is always positive. For an observation of the upper limb, the correction is always negative. These corrections are applied *only* to observations of the sun and the moon. Stars and planets appear as true mathematical points in the horizon mirror. Consequently, no semidiameter correction is required for the stars and planets.

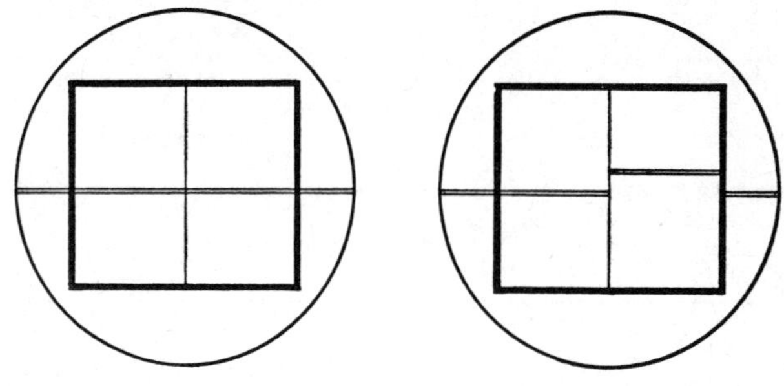

**Figure 2.** *Index error. In both cases, the sextant is set at zero. The reflected and direct images of the horizon in the left illustration align; no index error is present. The reflected and direct images in the right illustration do not align; index error is present.*

The earth's atmosphere is a denser medium than the near vacuum of space. If you have ever observed that a stick with one end under water seems to bend, you will understand that light rays entering a denser medium are always bent, unless the angle at which they strike the surface of the denser medium is 90 degrees. Light entering the earth's atmosphere is bent in the same way. This causes another error in a sextant observation. This error is *refraction*. It is always negative. Refraction is tabulated as a combined correction in the *Nautical Almanac*.

In addition to index, semidiameter, and refraction errors, the navigator must also compensate for *parallax* and *dip*. The error of parallax is caused by the fact that observations of celestial bodies are made from the surface, not the center, of the earth. The correction for parallax is very large, nearly a degree in some cases, for observations of the moon. For observations of the sun or planets

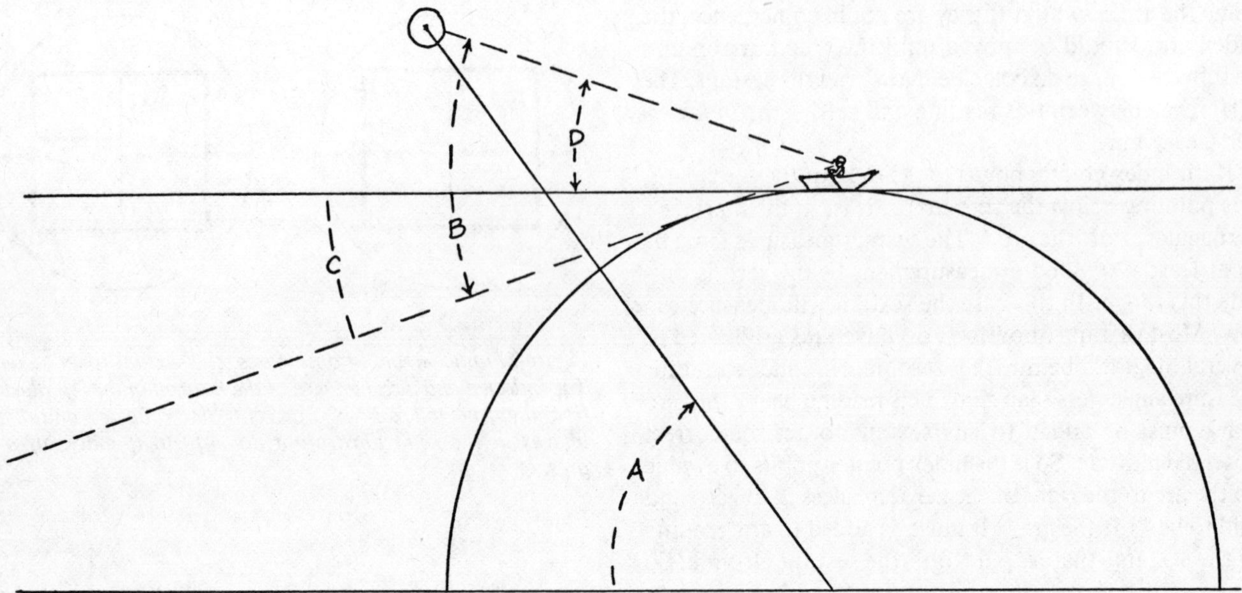


**Figure 3.** *Sextant altitude corrections. Angle A is the corrected or observed altitude, $H_o$. Angle B is the sextant altitude, $H_s$. Angle C is the error caused by height of eye above the earth's surface, the dip correction. The difference between angles A and D is the parallax. Parallax and dip are greatly exaggerated for the sake of illustration. Except for horizontal parallax of the moon, the value of these sextant altitude corrections would seldom exceed a few minutes.*

the error is much smaller but still significant. Parallax is of no practical significance in observations of the stars. Consequently, parallax is not a correction to be applied to sextant altitudes of stars. Horizontal parallax of the moon is tabulated in the daily pages of the *Nautical Almanac.* Parallax corrections to sextant altitudes of the sun and planets are tabulated in the *Nautical Almanac* as combined corrections including refraction. The correction is always positive.

Dip is the correction for the error caused by the navigator's eye level being above the earth's surface by an amount equal to the height of his vessel's deck plus his own height. This correction is sometimes called *height of eye.* It is tabulated in the *Nautical Almanac* as dip. Dip is always negative.

There are other sources of minor error in sextant observations that are not usually taken into consideration by the navigator. Again, it would behoove you to read the sections on sextant altitude corrections in Bowditch or *Dutton's.*

Below is a summary of the corrections to sextant altitude.

### Index Error

Index error is an error in the instrument itself. The navigator must determine the index error by observing the horizon or a star. Index error may be positive or negative, depending on the individual sextant. Good practice requires this error to be reduced to zero by adjustment whenever possible.

### Semidiameter

Semidiameter applies only to observations of the sun or moon. It is applied to adjust the altitude of the rim of the body to the altitude of the center of the body. The value of this correction is found in the *Nautical Almanac.* Semidiameter is always positive when applied to lower limb sights. It is always negative when the upper limb is observed.

### Refraction

Correction for refraction is necessitated by the bending of light rays from the observed body when they enter the earth's atmosphere. The correction is always negative. Refraction is normally a small correction, but under extremes of temperature, humidity, and air pollution, the error due to refraction may be gross and indeterminate. Such conditions occur along arid coasts when the air is full of desert loess, as in the Red Sea, and in areas where temperature inversions exist. Corrections for these nonstandard conditions are found on page A4 of the *Nautical Almanac*. Refraction may also be substantial when bodies are observed at low altitude. Corrections for low-altitude observations may be found in Pub. No. 249 and the *Nautical Almanac*.

**Parallax**

Parallax is error caused by taking sights from the surface, rather than the center, of the earth. The correction for parallax is always positive and is applied to the sun, moon, and the minor planets Venus and Mars, but not stars.

**Dip**

Dip is caused by the height of eye of the observer above the earth's surface. The correction for dip is always negative.

Sextant altitude corrections are very important. The essence of the whole problem of sight reduction (working your sight to get a line of position) is to compare an observed altitude of a celestial body with a computed altitude. The observed altitude, $H_o$, is a product of your true position on the earth's surface. You use your dead reckoning position, or an assumed position that evolves from your DR, to compute an altitude for the time of the sight. Comparing this computed altitude, $H_c$, with your observed altitude, $H_o$, and plotting the results, furnishes you with a line of position for that sight. The accuracy of this line of position depends on the accuracy of your calculations and the accuracy of your observed altitude. Hence the emphasis on applying sextant altitude corrections properly.

# 5 LINES OF POSITION

The paradox of navigation is that so many people think it is such a mysterious thing, and yet they do not realize how much navigation they really do on an almost professional level every time they get afloat. They may not use the most sophisticated instruments, but this really shows how good they are. For example, let us look at a rather ordinary situation that could happen to any two people we might know.

Two fellows were in a small fishing skiff. They had been having fabulous luck, but it was time to return to shore. Since they were in a new spot, they wanted to be able to find this spot again. They were wise enough to know that the technique of Boudreaux and Batiste, who marked the spot by making an X on the bottom of the boat, would not do.

One of them had a pocket compass. He sighted over it to a tall tree on a point on the riverbank. He told his companion, "We can come back to where that tree bears 125° on my pocket compass."

His companion thought a moment, and he said, "That won't tell us if we are over our fishing hole. Why don't you sight on that other tree over there too?" A line from this tree made a large angle when it intersected the line of bearing from the first one.

The first man replied, "Good! It is 020° on my pocket compass."

What had they done? They had obtained a pair of lines of position. Anybody who has ever fished seriously has done this. It is just common sense. Maybe they didn't use a compass, but they carefully eyeballed the landmarks. This is good navigation.

These two fellows knew they were on both these lines they had observed. Maybe they had a map or a chart. They could draw the bearings on the chart if they could pinpoint the trees they took the bearings on. Until they got the line off the second tree, they were really in a fix to "fix" the place again. They could not tell where they were on the first line, but only that they were on it. But, when they got the second line of position, the only place they could be on both lines was at the point where they crossed. Now they had a *fix*, a navigational term for the intersection of two or more lines of position. Had they been able to get a bearing on a third object that intersected those from the other two, they would have had an even better fix. Two lines of position will give you a fix. Three, four, or more will give you a more reliable fix.

In coastal and inshore navigation, we use bearings on charted objects to get a fix. We prefer lighthouses,

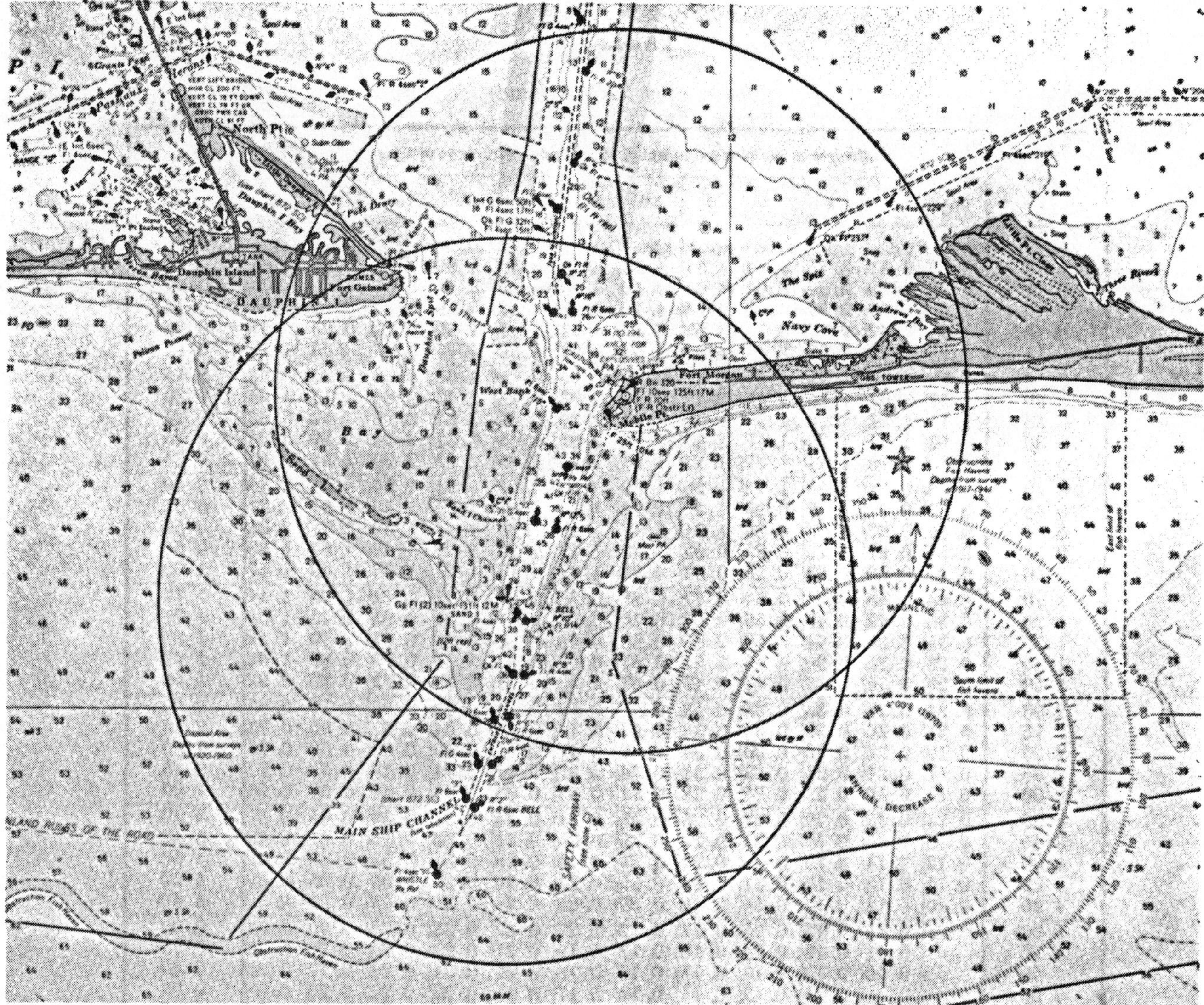


**Figure 4.** *Using vertical angles to determine your distances off two lighthouses. The circles cross at your position. The same principle is used in celestial navigation with circles of equal altitude.*

beacons, or other charted aids to navigation. We can use prominent church steeples, buildings, silos, water tanks, or other more mundane things if they are accurately charted. We call this phase of navigation *piloting*. It is a highly important part of navigation that any navigator must master.

In celestial navigation we also use lines of position. Technically, they are not straight lines, but arcs of circles of equal altitude. Circles of equal altitude? How does this work? Let us take an example from piloting. Look at Figure 4.

Suppose one foggy morning you find yourself off the mouth of Mobile Bay. You hope that maybe through a rift in the fog you can get a glimpse of Sand Island Light long enough to get an observation. You have your sextant in your hand because you want to get a vertical angle on the light. The height of Sand Island Light is both charted and given in the *Light List*. With this information and the observed vertical angle, you can get your distance off using Table 9 in Bowditch (see Figure 5). If you have someone else take a bearing on the light at the same time you observe the vertical angle, you can get a fix. Otherwise, all you will know is that you are on a circle whose radius is equal to your distance from Sand

## TABLE 9

Distance by Vertical Angle

| Angle | Difference in feet between height of object and height of eye of observer | | | | | | | | | | | Angle |
|---|---|---|---|---|---|---|---|---|---|---|---|---|
| | 100 | 120 | 140 | 160 | 180 | 200 | 250 | 300 | 350 | 400 | 450 | |
| ° ′ | *Miles* | *Miles* | *Miles* | *Miles* | *Miles* | *Miles* | *Miles* | *Miles* | *Miles* | *Miles* | *Miles* | ° ′ |
| 0 00 | 11.6 | 12.7 | 13.7 | 14.7 | 15.6 | 16.4 | 18.4 | 20.1 | 21.7 | 23.2 | 24.6 | 0 00 |
| 0 01 | 10.5 | 11.6 | 12.6 | 13.6 | 14.4 | 15.3 | 17.2 | 19.0 | 20.6 | 22.1 | 23.5 | 0 01 |
| 0 02 | 9.5 | 10.6 | 11.6 | 12.5 | 13.4 | 14.2 | 16.1 | 17.9 | 19.5 | 21.0 | 22.4 | 0 02 |
| 0 03 | 8.6 | 9.6 | 10.6 | 11.6 | 12.4 | 13.2 | 15.1 | 16.9 | 18.5 | 19.9 | 21.3 | 0 03 |
| 0 04 | 7.8 | 8.8 | 9.8 | 10.7 | 11.5 | 12.3 | 14.2 | 15.9 | 17.5 | 19.0 | 20.3 | 0 04 |
| 0 05 | 7.1 | 8.1 | 9.0 | 9.9 | 10.7 | 11.5 | 13.4 | 15.0 | 16.6 | 18.0 | 19.4 | 0 05 |
| 0 06 | 6.5 | 7.4 | 8.3 | 9.2 | 10.0 | 10.8 | 12.5 | 14.2 | 15.7 | 17.1 | 18.5 | 0 06 |
| 0 07 | 6.0 | 6.9 | 7.7 | 8.5 | 9.3 | 10.1 | 11.8 | 13.4 | 14.9 | 16.3 | 17.7 | 0 07 |
| 0 08 | 5.5 | 6.4 | 7.2 | 8.0 | 8.7 | 9.5 | 11.2 | 12.7 | 14.2 | 15.6 | 16.9 | 0 08 |
| 0 09 | 5.1 | 5.9 | 6.7 | 7.5 | 8.2 | 8.9 | 10.5 | 12.1 | 13.5 | 14.9 | 16.1 | 0 09 |
| 0 10 | 4.7 | 5.5 | 6.3 | 7.0 | 7.7 | 8.4 | 10.0 | 11.5 | 12.9 | 14.2 | 15.4 | 0 10 |
| 0 11 | 4.4 | 5.2 | 5.9 | 6.6 | 7.3 | 7.9 | 9.5 | 10.9 | 12.3 | 13.6 | 14.8 | 0 11 |
| 0 12 | 4.1 | 4.8 | 5.5 | 6.2 | 6.8 | 7.5 | 9.0 | 10.4 | 11.7 | 13.0 | 14.2 | 0 12 |
| 0 13 | 3.9 | 4.6 | 5.2 | 5.9 | 6.5 | 7.1 | 8.5 | 9.9 | 11.2 | 12.4 | 13.6 | 0 13 |
| 0 14 | 3.6 | 4.3 | 4.9 | 5.5 | 6.1 | 6.7 | 8.1 | 9.5 | 10.7 | 11.9 | 13.1 | 0 14 |
| 0 15 | 3.4 | 4.1 | 4.7 | 5.3 | 5.8 | 6.4 | 7.8 | 9.0 | 10.3 | 11.4 | 12.6 | 0 15 |
| 0 20 | 2.7 | 3.2 | 3.7 | 4.2 | 4.6 | 5.1 | 6.3 | 7.4 | 8.4 | 9.4 | 10.4 | 0 20 |
| 0 25 | 2.2 | 2.6 | 3.0 | 3.4 | 3.8 | 4.2 | 5.2 | 6.2 | 7.1 | 8.0 | 8.9 | 0 25 |
| 0 30 | 1.8 | 2.2 | 2.5 | 2.9 | 3.2 | 3.6 | 4.4 | 5.3 | 6.1 | 6.9 | 7.7 | 0 30 |
| 0 35 | 1.6 | 1.9 | 2.2 | 2.5 | 2.8 | 3.1 | 3.9 | 4.6 | 5.3 | 6.0 | 6.7 | 0 35 |
| 0 40 | 1.4 | 1.7 | 1.9 | 2.2 | 2.5 | 2.7 | 3.4 | 4.1 | 4.7 | 5.4 | 6.0 | 0 40 |
| 0 45 | 1.2 | 1.5 | 1.7 | 2.0 | 2.2 | 2.5 | 3.1 | 3.6 | 4.2 | 4.8 | 5.4 | 0 45 |
| 0 50 | 1.1 | 1.3 | 1.6 | 1.8 | 2.0 | 2.2 | 2.8 | 3.3 | 3.8 | 4.4 | 4.9 | 0 50 |
| 0 55 | 1.0 | 1.2 | 1.4 | 1.6 | 1.8 | 2.0 | 2.5 | 3.0 | 3.5 | 4.0 | 4.5 | 0 55 |
| 1 00 | .9 | 1.1 | 1.3 | 1.5 | 1.7 | 1.9 | 2.3 | 2.8 | 3.2 | 3.7 | 4.1 | 1 00 |
| 1 10 | .8 | 1.0 | 1.1 | 1.3 | 1.4 | 1.6 | 2.0 | 2.4 | 2.8 | 3.2 | 3.6 | 1 10 |
| 1 20 | .7 | .8 | 1.0 | 1.1 | 1.3 | 1.4 | 1.8 | 2.1 | 2.4 | 2.8 | 3.1 | 1 20 |
| 1 30 | .6 | .8 | .9 | 1.0 | 1.1 | 1.2 | 1.6 | 1.9 | 2.2 | 2.5 | 2.8 | 1 30 |
| 1 40 | .6 | .7 | .8 | .9 | 1.0 | 1.1 | 1.4 | 1.7 | 2.0 | 2.2 | 2.5 | 1 40 |
| 1 50 | .5 | .6 | .7 | .8 | .9 | 1.0 | 1.3 | 1.5 | 1.8 | 2.0 | 2.3 | 1 50 |
| 2 00 | .5 | .6 | .7 | .8 | .8 | .9 | 1.2 | 1.4 | 1.6 | 1.9 | 2.1 | 2 00 |
| 2 30 | | .5 | .5 | .6 | .7 | .8 | .9 | 1.1 | 1.3 | 1.5 | 1.7 | 2 30 |
| 3 00 | | | | .5 | .6 | .6 | .8 | .9 | 1.1 | 1.3 | 1.4 | 3 00 |
| 3 30 | | | | | .5 | .5 | .7 | .8 | .9 | 1.1 | 1.2 | 3 30 |
| 4 00 | | | | | | .5 | .6 | .7 | .8 | .9 | 1.1 | 4 00 |
| 4 30 | | | | | | | .5 | .6 | .7 | .8 | .9 | 4 30 |
| 5 00 | | | | | | | .5 | .6 | .7 | .8 | .8 | 5 00 |
| 6 00 | | | | | | | | .5 | .5 | .6 | .7 | 6 00 |
| 7 00 | | | | | | | | | .5 | .5 | .6 | 7 00 |
| 8 00 | | | | | | | | | | .5 | .5 | 8 00 |
| 10 00 | | | | | | | | | | | | 10 00 |

**Figure 5.** *An excerpt from Table 9 of Bowditch.*

Island Light. Sand Island Light is the center of this circle. Now suppose your assistant was not fast enough to get the bearing on Sand Island Light. The rift widened, however, so you were able to get a vertical angle on Mobile Point Light. You go to Table 9 in Bowditch with the vertical angle and the height of the light, and find your distance off Mobile Point Light. Now you know you are on a circle whose radius is your distance off Mobile Point Light. You know you are on two circles at the same time. Let's call these circles "altitude circles." Again, the only place you can be on both of these circles is at a point where they intersect. But they intersect in two places! How can you tell which intersection you are on? Usually common sense and good judgment will come in here. In the situation illustrated, you know you are not just off the beach at Dauphin Island. If you are turned around and confused, an approximate bearing will help, as will soundings and other tricks that are not part of celestial navigation but are more properly part of piloting. Usually, one intersection will be so far from your DR that it is obvious you must be on the other.

You use precisely the same technique in celestial navigation. You don't use Table 9, of course. You do measure the altitude of a body with your sextant. You know you are somewhere on a circle from which this body appears at the observed altitude. Since it will be at this altitude from any point on the circle at the instant of time that you make your observation, the circle is designated a *circle of equal altitude*. This circle may have a diameter of many thousands of miles! By itself, this doesn't tell us much at all. You can limit your probable arc by the use of an azimuth. An azimuth is the horizontal direction of a celestial point from a terrestrial point. This begins to really tell us something. Because the circle has such a large diameter, you can also plot the arc, which will be your line of position, as a straight line without introducing any measurable inaccuracy.

How does an azimuth help you find out which arc of this circle to plot? This is certainly a fair question. The answer is not to be given in a few words, but the solution of the problem posed is not a hard one. It just takes a little doing. Remember that the navigator always keeps his DR plot up to date. To work his sight, he computes what the altitude of the body *would be* if he were at his DR position at the instant of the sight, or, as we are going to do first, computes what the altitude would be from the nearest assumed position (AP) that will give him a whole-number degree of latitude and local hour angle to work with. By nearest, I mean nearest to his DR. We use the assumed position to make the arithmetic easier. This eliminates the use of decimals and fractions.

Pub. No. 249, which is the primary method taught in this book, uses only the assumed position. Ageton's Method (H.O. 211), which can be used to reduce any observation of any celestial body regardless of declination, uses the DR position. Pub. No. 229, which, like Ageton's Method, is not limited by declination, uses the assumed position. To enter Pub. No. 249, we need three arguments. They are *latitude, local hour angle* (or *t),* and the body's *declination*. From the tables in Pub. No. 249, we get computed altitude and azimuth. The azimuth tells us which direction the body is from our assumed position. For all practical purposes, it tells us which way the body is from our true position too. The navigator takes his computed altitude and compares it with his observed altitude. The difference between his computed altitude and his observed altitude is the difference in radius of the altitude circle of his assumed position and the altitude circle of his true position. This difference is called the *altitude intercept*. It is the difference in nautical miles. If the computed altitude is greater than the observed altitude, then the circle of true position, the one you are actually on, is farther away from the body than the assumed position. If the computed altitude is less than the observed altitude, the circle of equal altitude of your true position is of lesser radius or closer to the body.

The center of both these circles is the *GP* (geodetic or geographic position) of the observed body. When dealing with a specific body, this GP is more likely called the substellar, subsolar, sublunar, or subplanetary point, as appropriate. To help you remember the relation between these two circles of equal altitude, we have a rule. It is *Computed Greater Away*. By implication, we get *Computed Lesser Toward*. We don't memorize this, though, because we have a good memory aid for Computed Greater Away. It is *Coast Guard Academy* or *C.G.A.* Another memory aid is the Japanese-sounding phrase Ho Mo To, *Ho Mo*re *To*ward. Ho in this case stands for observed altitude.

Why is this relationship between these circles of equal altitude true? Let us go back to Sand Island Light. Take several vertical angles that might be observed angles on the light. Then go into Table 9 of Bowditch. Notice that as the observed angle gets larger, the distance from the light gets smaller.

You can prove this to yourself another way. Take your sextant and go outside. If you do not have a sextant, make yourself an astrolabe out of a dime-store protractor. Pick out a tall, slender object like a tree or a flagpole. Observe the angle formed by its top and base from your position. Move closer to the tree or flagpole, or whatever it is you chose. You will see that the angle is larger when observed from this closer point. Now go back beyond the point where you made your first observation. You will notice that the angle is smaller. As you move away the angle decreases. If you move closer the angle increases. Thus, if you observe an angle greater than the one you compute, you are nearer the GP, whether it is the base of a tree or the substellar point of a star. If the observed angle is less than the computed angle, you are farther away.

The azimuth is a radius running through your assumed position, and therefore perpendicular to a tangent of this altitude circle. Your LOP is *perpendicular* to your azimuth.

To plot your azimuth, simply consider it as the true course you would steer if you wanted to go from your assumed position to the GP of the body you observed. This is exactly what it is. It is even a great circle course. So if the azimuth of the body observed was 218°, lay out a line from your AP 218°T, and measure along it a distance equal in nautical miles to the altitude intercept. At this point draw your LOP perpendicular to the azimuth. This is the case when your observed altitude, $H_o$, is more, thus toward, Ho Mo To. Had your computed altitude, $H_c$, been greater than your observed altitude, your LOP would be farther from your body's GP than your assumed position. In this case, you must steer away from the GP of the observed body to get to your LOP. To plot the LOP in this case, lay a course from your AP that is the reciprocal, or opposite, of your azimuth, 038° in this case, and again measure along it a distance in nautical miles equal to your altitude intercept and draw your LOP perpendicular to the azimuth through this point. When you plot an azimuth or its reciprocal, it is a good idea to draw a dashed line to represent the azimuth. This dashed line is less likely to be mistaken for an LOP later. LOPs are always plotted as solid lines.

Now, that is how you plot the LOP resulting from a celestial sight. Plotting can seem to be a problem. Perhaps a simple explanation can help. Suppose you have a sight that reduces to an altitude intercept of 15.0 miles "toward" and an azimuth of 138°. How do you plot your LOP?

In reducing this sight, you selected the latitude of, say, 26° N as being nearest your DR latitude. To make the LHA come out even, you used an assumed longitude of 91° 23.6′ W. The first step in plotting a line of position is to plot this assumed position on your chart or plotting sheet. Then from this assumed position, lay off your azimuth of 138° and measure off 15 miles. Draw a line perpendicular to this azimuth of 138° through the 15-mile point. This perpendicular is your LOP. Had the intercept been "away," you would extend your azimuth in the direction of 318° (138° + 180°), and lay off your 15-mile intercept. Remember that your assumed position, azimuth, and altitude intercept will be different for each sight. You must use the appropriate AP, altitude intercept, and azimuth for each sight. An example of how to plot an LOP is shown in Figure 6.

Sometimes it is not possible to get simultaneous observations. To get a fix from the sun only, you may have to wait several hours for the azimuth to change enough to give you a good wide angle of intersection or cross. Some navigators use only the sun, but they are far from being good navigators, and in my opinion there is some question of their competence. I certainly do not advocate that you navigate by use of sun sights only. During the day, though, the sun is usually the only body available. Sometimes the moon may be usable and even Venus can be observed under good conditions when it is in the right place in the heavens. But usually you will find it necessary to get a fix from two sun lines reduced from observations several hours apart. Obviously, you must make an adjustment in time before you can get a fix. We get a fix by knowing we are on two or more lines of position at the same time. The point where they cross is our fix. The problem is to equate the time for two lines from observations made hours apart.

How do you do this? You can advance your first line of position along your course to the time of your second sight, or the latest sight if more than two are involved. You may retard the later sights to the time of your first sight, or you may adjust all of them to the time of a required position report. The usual custom is to advance all sights to the latest one. This is not hard. Just remember that these LOPs will advance at the same rate of speed your vessel advances and will be affected by all the factors affecting your vessel's DR. The second thing to remember is that they always keep their same orientation to the elevated pole. They always point in the same direction. They remain 90° to the observed azimuth of the body. It is as if you had this line fixed at a certain distance and direction from your vessel on some kind of a physical extension.

In actual practice, we begin with a point where the LOP to be advanced crosses our course line. If the LOP does not cross the course line because it is nearly parallel, you lay down another line, parallel to the course, that will cross the LOP. This crossing is your point to advance. If the LOP is exactly parallel to your course, it will not need to be advanced, as it is really your true course line. Advance this point exactly as you would your DR. If your ship changes course, your point will change course. If this point was three miles ahead of your original DR at the time of your first sight, it will be three miles ahead of your DR at the time of your second sight, whether you change course or not. If only two sights are involved,

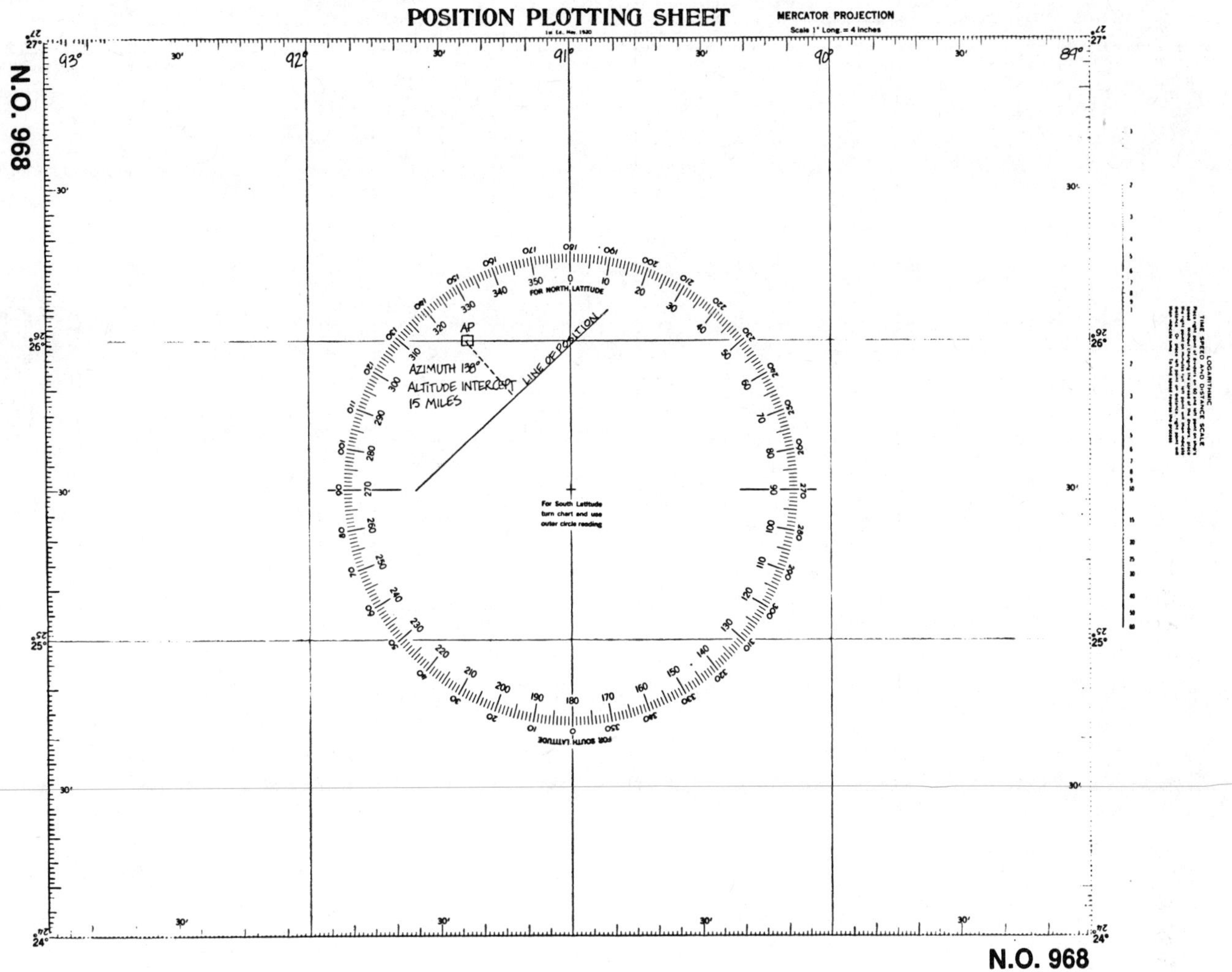


**Figure 6.** *Plotting a line of position (LOP).*

draw a line parallel to your first LOP through this point three miles ahead of the DR at the time of your second sight. This is your advanced LOP. Label it with the time of the first sight, the time it is advanced to, and the name of the body observed: 0800/1034 Sun.

For example, you make an observation of the sun at 0822 ship's time in the morning. You plot the resulting LOP. It crosses your course two miles behind your DR. If your vessel is making 10 knots, in two hours you will advance your DR 20 miles. To advance your LOP, simply advance the point of crossing 20 miles to keep it two miles astern of your DR. Draw a line through it parallel to your 0822 sun line. Label this line "0822/1022 Sun." If you change course during the two hours, your point just changes course too. Keep in mind that the LOP keeps the same direction when advanced. It remains parallel to the original LOP. The fix that results from this technique is called a *running fix* to distinguish it from a fix from simultaneous observations, which is called simply a *fix.*

Now for a more graphic example. See Figure 7. For the purpose of our illustration, let us assume that the 0802 sun line in Figure 7 is the same as the one resulting from the sight you plotted in Figure 6. You should actually try to plot, for yourself, the lines of position illustrated in these two figures. At the time of the 0802 sight, assume your DR was Lat. 25° 47.8′ N and Long. 91° 19.3′ W. Your course is 090° true. Your speed is 10 knots. You plotted your 0802 sun line when you reduced it. It is shown here as it would appear on your plotting sheet. It falls about 5.8 miles in front of your DR. You make a second observation of the sun at 1000. In reducing the 1000 sight, you use an assumed position of Lat. 26° 00.0′

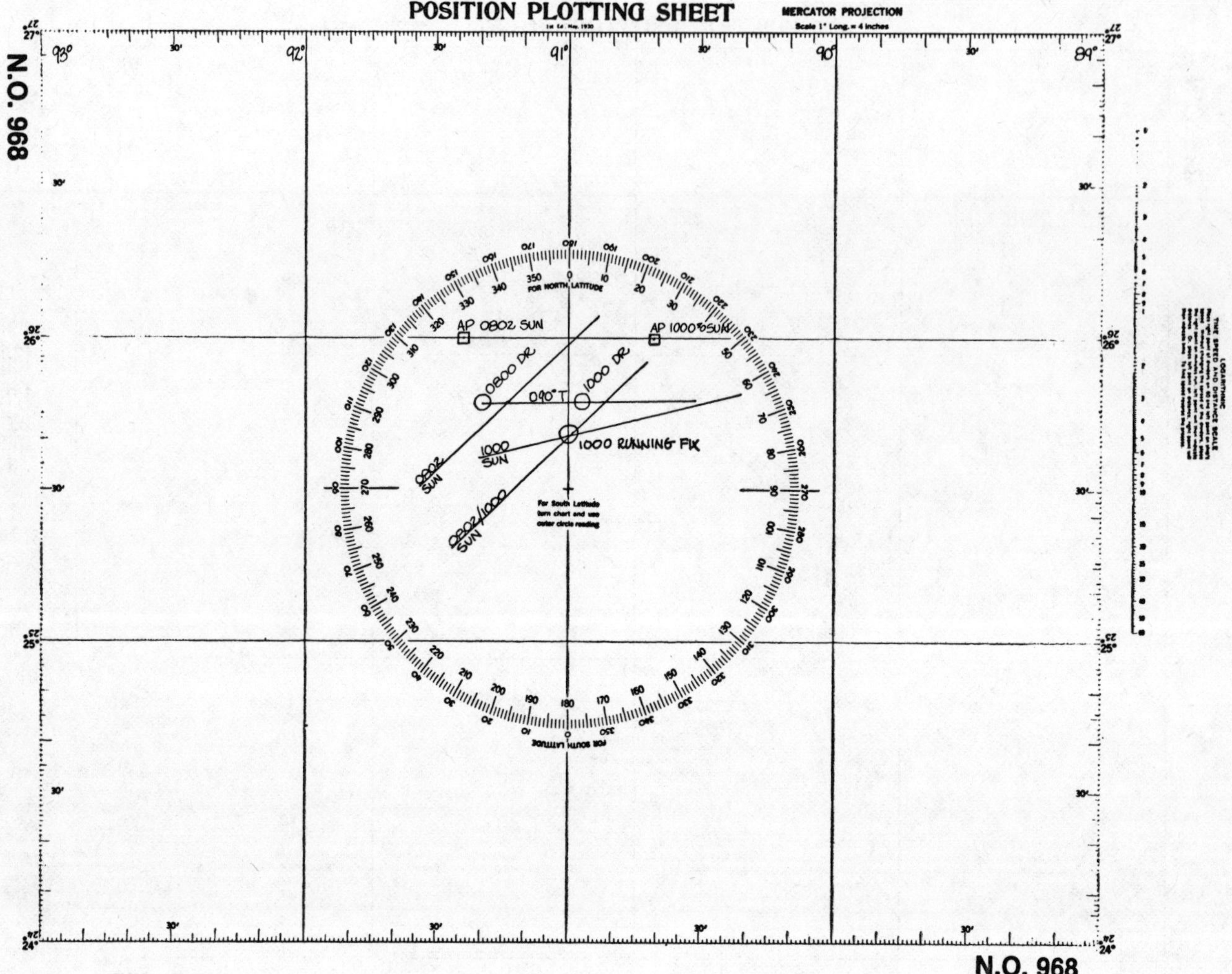


**Figure 7.** *Using two sun lines to obtain a running fix.*

N, Long. 90° 40.8′ W. Your altitude intercept is 14.8 miles toward on an azimuth of 167°.

The first thing you should do is plot your 1000 DR. The two minutes of difference between two hours and one hour and 58 minutes is not significant, so advance your 0802 DR 20 miles (two hours at 10 knots) along your course. To advance your 0802 line, simply move it 20 miles along your course also, measuring from the point at which the 0802 sun line originally intersected your course. Notice that this second point is 5.8 miles ahead of your 1000 DR. Now plot your 1000 sun line. The point at which it crosses your advanced 0802 sun line, labeled 0802/1000 Sun, is your running fix. Label it as such. Notice that all lines appearing on the plotting sheet are labeled. This must be done so that you can identify any line on the plotting sheet at any time.

The running fix is a good fix, but it is susceptible to error. It is no better than the information you use to advance your DR. Thus it is not as reliable as a fix from simultaneous observations. The longer the time before you get the second sight, the more time any error of course or speed has to affect the accuracy of your advanced LOP. If your course and speed are very accurate, your advanced LOP will be very accurate. If not, the error will be in proportion to the inaccuracies. For this reason, it is not possible to say how far in time it is feasible to advance an LOP. Several hours is the rule, but if circumstances require and conditions are good, a much longer time may give usable results that are certainly better than no results at all. This is a situation in which the navigator's experience and judgment will be put to the test.

# 6 HOW TO WORK A SUN SIGHT

Assume your ship left port on Monday, 17 October 1983. The navigator took his departure from a prominent landmark at the last known point of reference with the land. The problem is stated thus:

At 1645 ZT on 17 October the yacht *War Hat* took her departure from Sand Island Light at the mouth of Mobile Bay. Sand Island Light bore 310° true. Distance off by vertical sextant angle was three miles. Your departure was your last known fix with reference to the land. By plotting we see that this position was Lat. 30° 10.8′ N, Long. 88° 00.4′ W.

Wind was south by west at 26 knots, sea moderate, sky overcast. Course was set at 145° true. Speed by taffrail log was five knots. At 0645 Tuesday, 18 October, the wind backed to the southeast and the yacht was put about and faired away on a course of 209° true. Wind continued fresh and speed by taffrail log was six knots. The sky remained overcast all day and through the night.

At approximately 0800 on the morning of the 19th, the sun appeared through a break in the clouds. The navigator was able to observe the sun's lower limb. Sextant altitude was 25° 50.6′. The chronometer read $14^h$-$07^m$-$29^s$. The watch read $1^m$-$37^s$. The chronometer was fast $3^m$-$12^s$. Height of eye was 12 feet. Compute and plot the line of position from this sun sight.

The idea of stating the problem this way is to show how it is really done. Vessels go to sea. They may go days without a sight and must navigate by dead reckoning. Then one day a chance appears and the navigator grabs an observation. It may be a crucial one if he has gone a long time between sights. This is especially true if his dead reckoning shows him nearing land or some other hazard. This can happen often. For this reason, it is important that the navigator always maintain an up-to-the-moment dead reckoning plot and that he take as many sights as necessary to fix his position several times a day, as outlined in Chapter 1.

You should keep your plot on a plotting sheet instead of a chart. Charts are more expensive, and extensive plotting, with its ensuing erasures, could obscure the very vital information you may need later. Deep-sea navigators keep their plots on a plotting sheet when off soundings, and transfer their positions to the chart several times a day to check their progress. Plotting sheets are Mercator projections. (The plotting illustrated in Figures 6 and 7, in Chapter 5, was done on a plotting sheet.) They are essentially blank pieces of paper with a true compass rose and the parallels of latitude and meridians of longitude printed on them. No other navigational data are provided. A convenient logarithmic scale is provided to help solve the speed-time-distance problems of dead reckoning. Latitudes, north and south, are printed on the plotting sheet and labeled. Meridians of longitude are also printed on the plotting sheet, but these meridians are not labeled. You must select a plotting sheet that covers the latitude you are working in. If you are in 31° north latitude, you must select a plotting sheet that has that latitude printed on it. You would use this same sheet if you were in 31° south latitude. You write in the value of the meridians to suit the longitudes you are in. If you are in 88° W longitude, you select a meridian to be 88° W, and number those to the left of it 89° and 90° and those to the right 87° and 86° and so on. If you are in 88° E longitude, you select a meridian to be 88° E, and number the meridians to the right 89° and 90°, those to the left 87° and 86°.

Plotting sheets come in several series. The 920 series, numbered 920 through 936, covers 78° of north or south latitude. Sheet 920 covers 4° south latitude to 4° north latitude; sheet 936 covers 74° to 78° north or south latitude. (Sheet 936 is limited to a four-degree spread in latitude because Mercator distortion requires a greater spread between parallels of latitude on the printed sheet in higher latitudes.) The 920 series sheets are large sheets designed for use on large vessels. If they are all you can get, it is usually more convenient to cut them in two along a meridian of longitude. There are two compass roses on each 920 series plotting sheet near the left center and the right center. When you cut, be careful to leave a latitude scale and a compass rose on each piece to measure distances and directions.

For small vessels, the 960 series plotting sheets are better. Although they use the same numerical scale as the larger 920 series sheets, they are on a much smaller piece of paper. The 960 series sheets begin with sheet 960, covering 2° south latitude to 2° north latitude. The highest sheet in the series is sheet 975, which covers latitudes 45° to 49° north or south. The 960 series plotting sheets fit the chart tables of most vessels nicely.

If you want a still smaller plotting sheet, universal plotting sheets are available. Universal plotting sheets are published by the Defense Mapping Agency Aerospace Center. These sheets are about the size of two pieces of legal stationery. They are truly excellent for use on vessels with limited space for plotting. The scale of the universal plotting sheet is about one-half that of the others, however. This may cause some loss of resolution, what we commonly call "accuracy," but not enough to matter if you are careful. There are enough other factors to cause loss of close resolution in small-craft navigation to make universal plotting sheets very compatible with the needs of the small-craft navigator making an offshore passage. Because of their smaller scale, universal plotting sheets represent a larger area of the earth's surface. This gives you the advantage of being able to keep more of the plot on one sheet. Parallels of latitude are printed on the universal plotting sheet. Meridians of longitude must be drawn in, but the procedure is simple, and it is explained adequately on page 88 of the 1977 edition of Bowditch, Volume I.

One word of caution. To use the 920 and 960 series of plotting sheets in south latitudes you have to turn them *upside down*. Note the admonition in the center of the compass roses of these 920 and 960 series plotting sheets. You buy plotting sheets from the same place you buy your charts. You can make your own plotting sheets, too. Your chart agent may have pads of graph paper with a compass rose printed in the center. If not, any kind of pad will do. The 1977 edition of Bowditch, Volume I, tells you on pages 86 and 87 how to construct a simple plotting sheet in four easy steps.

How does the navigator work his sun line? The procedure is divided into 11 steps. To follow the steps, see the sight reduction form in this chapter. This form is basically the same one that will be used to reduce any sight for a line of position.

**Step 1**

The navigator keeps a continuous dead reckoning plot. His point of departure is Sand Island Light. He plots his 0800 DR position for the morning of October 19. To do this you will have to take your plotting sheet and set it up. (Remember, you are now the navigator, so when I say "you" or "the navigator," I mean you.)

On your plotting sheet, the longitude is not numbered. Since you are going to be sailing mostly south, number the middle longitude line 88° and number the others in order to the left and right. Since you are in west longitude, label the lines to the left of 88° in increasing order, and those to the right in decreasing order. If you knew you were to be heading east, you would put 88° on the left-hand side and number east in decreasing order. You would do just the opposite if bound west. The 0800 DR plots out to be approximately Lat. 27° 00.0′ N, Long. 88° 37.2′ W.

It will be good practice for you to plot this entire problem beginning with your departure on Monday the 17th. This will help you fix in your mind the complete picture of a day's work for a navigator. Please do not copy the illustration, but to provide you with a check on your own work, the entire problem covered in this chapter is illustrated in Figures 8 and 9. Two small plotting sheets are used. You can get a plotting sheet large enough to contain the entire problem, but small vessels use the smaller plotting sheets because they are more manageable. This work is intended for the skippers of small vessels. Hence we will illustrate our work accordingly.

One more thing: Provision should be made for keeping your work forms. In the event of accident or violation of marine laws and regulations, they may have legal weight. Professional navigators keep a *navigator's notebook*, which is their personal property and not part of the vessel's records.

**Step 2**

Record your sight on the form. You should record the date and zone time; then the chronometer time, C, to the nearest second; your stopwatch reading, W; and your sextant altitude, $H_s$.

Note that the date is written in the order of day/month/year. This is the way sailors, aviators, and military people write the date. It is also the way most people in other countries write the date.

It is important to be neat. It will help you to avoid errors. You are now ready for Step 3.

**Step 3**

Correct your time for GMT (Greenwich mean time) by subtracting stopwatch time, W, from the chronometer reading, C, and by applying CE (chronometer error).

**Step 4**

Open your 1983 *Nautical Almanac* to the daily pages and find October 19. Excerpts of the daily pages of the 1983 *Nautical Almanac* are in appendix B of this book. Refer to them as you go along just as if you were working this sight. This is important! You will note that each pair of facing pages covers three days. For instance, October 19, 20, and 21 are on one pair of pages. The first day across the top is the 19th. The sun column is on the left side of the right-hand page. To the left of it is the time column, which is divided into 24 hours for each day. Go down this column to 14 hours in the October 19 date, and extract the hour angle for 14 hours of GMT on the 19th. This will be in a subcolumn under "Sun" headed "GHA." Extract 33° 44.2′ and put it in the form. Under the subheading "Dec" for declination, extract S 9° 54.1′ and enter it in the form.

SUN (☉), Lower/~~Upper~~ Limb **Step 1**
Date 19/10/83, ZT 0800
DR Lat. 27°-00.0′N Long. 88°-37.2′ W

| | | |
|---|---|---|
| **Step 2** | C | 14h-07m-29s |
| | W | 01m-37s |
| **Step 3** | C-W | 14h-05m-52s |
| | CE | 03m-12s f |
| | GMT | 14h-02m-40s |
| **Step 4** | GHA | 33°-44.2′ |
| **Step 5** | Corr. (min. & sec.) | 0°-40.0′ |
| | GHA | 34°-24.2′ |
| **Step 6** | **Assumed Long. | 88°-24.2′ |
| | *LHA (t) | 54°-00.0′E |
| | ***Assumed Lat. | 27°-00.0′N |
| **Step 4** | Declination | S 9°-54.1′ |
| **Step 5** | d difference +0.9 | 0.0′ |
| | *Declination | S 9°-54.1′ |
| **Step 7** | $H_t$ | 26°-30.0′ |
| **Step 8** | Correction d=-35 | −32.0′ |
| | $H_c$ | 25°-58.0′ |
| | $H_o$ | 26°-01.5′ |
| **Step 10** | **Altitude Intercept | 3.5′t |
| **Step 7** | Z | N 117° E |
| | From 360° (if req.) | |
| | **Zn | 117° |
| **Step 2** | $H_s$ | 25°-50.6′ |
| **Step 9** | Main Corr. +14.3 | |
| | Addn'l 0.0 | |
| | Dip −3.4 | |
| | Total Corr. +10.9 | +10.9 |
| | $H_o$ | 26°-01.5′ |

*Required to enter Pub. No. 249
**Required to plot line of position
***Required to enter Pub. No. 249 and to plot line of position

No interpolation for declination is necessary because the time is so near the even hour. Had an interpolation for declination been required, it could have been done two ways. Inspection will show that the declination is increasing southward. Thus any correction for minutes, or fractions of the hour, is added. A mental interpolation would be one way and will usually suffice, but it is good practice to go to the bottom of the "Dec" column and

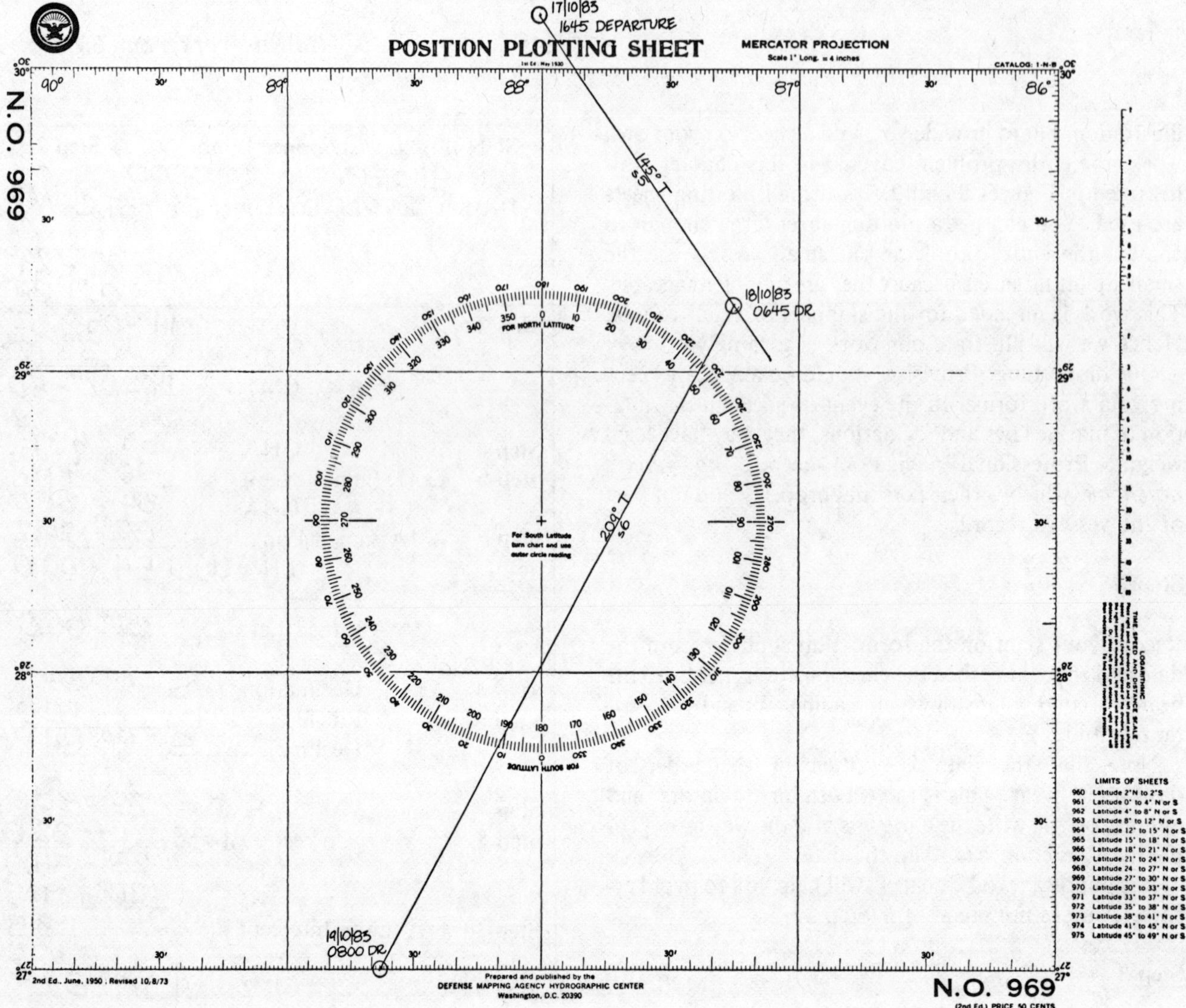


**Figure 8.**

**Figures 8 and 9.** *Look at both plots very carefully. The navigator set up the plotting sheets with 86° W as the longitude of the first meridian on the right-hand side. He then numbered the meridians to the left 87°, 88°, 89°, and 90°. The latitudes are already printed on the plotting sheets. He transferred the 0800 DR of the 19th from the old plotting sheet, Figure 8, and continued the day's work on a new sheet, Figure 9. Just after 0730 on the morning of the 20th, the plot ran off the left-hand side of the sheet. What did the navigator do? He simply transferred the 0730 DR of 20 October to the right-hand side of the sheet, and renumbered the first meridian 91°. He could do this because he knew his intended track was southwest. This flexibility of the plotting sheet reduces the need for carrying too many sheets, and if you want them for record or souvenir purposes, it reduces the number of sheets you must store. As was done here, it is quite permissible to start your plot outside the grid on the margin. The 0915 and 1125 DRs for 20 October have been left out of Figure 9 to reduce the clutter.*

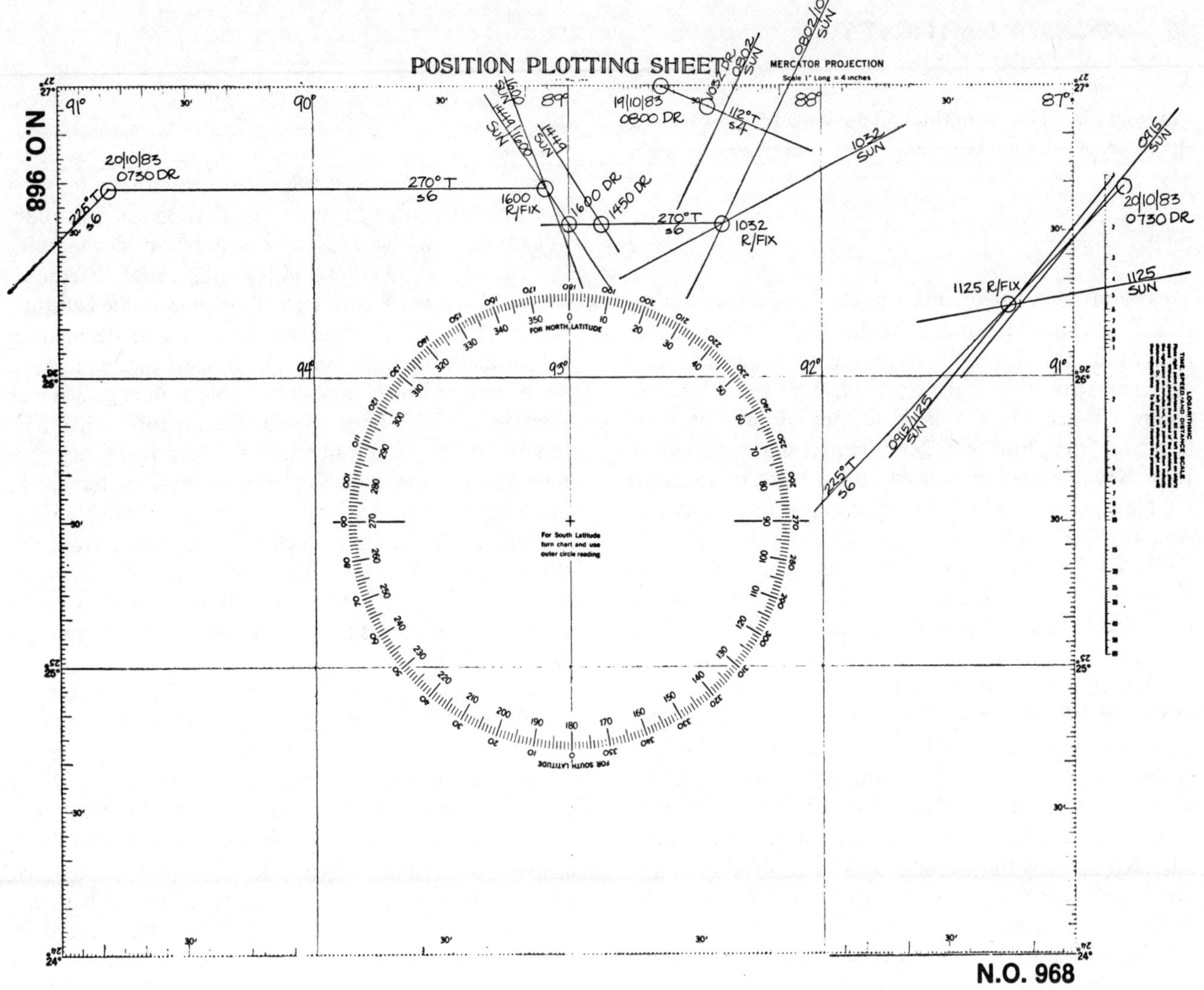


**Figure 9.**

opposite the letter "d" extract the value 0.9. Place this just to the right of the "d difference" in the form.

**Step 5**

It is necessary to make a correction to the sun's GHA for $2^m$-$40^s$. To do this, go to the yellow pages in the back of your *Almanac* entitled "Increments and Corrections," reproduced in black and white in appendix B in this book. Find the page with $2^m$ in large type at the top (page iii, left column, *Nautical Almanac*). In the time column headed $2^m$, go down to $40^s$ and opposite this, under "Sun and Planets," extract the correction 0° 40.0′. Add this to the GHA from the sun's daily page column and get the sun's GHA at the time of the sight. Had interpolation for declination been necessary, the correction would have been found by taking the value of "d" at the bottom of the "Dec" column of the daily pages and applying it in the "v or d" column of the $2^m$ box in the yellow pages. In this case "d," which is the hourly change in declination, is 0.9. The correction is 0.0.

**Step 6**

From your DR plot, you find your 0800 DR on the 19th to be Lat. 27° 00.0′ N and Long. 88° 37.2′ W. To use Pub. No. 249, you enter with three arguments. They are local hour angle or meridian angle (t) to a whole degree, latitude to a whole degree, and the tabulated declination for the time of your sight.

It is obvious that a GHA of 34° 24.2′ subtracted from a longitude of 88° 37.2′ will not produce an LHA or t of a whole degree. To produce an LHA or t of a whole degree, you must use an assumed longitude. To make your plot easy and accurate, select the nearest longitude to your DR that will give a whole-number LHA or t when the GHA of the body (in this case, the sun) is subtracted from it. Thus your assumed longitude is 88° 24.2′ W.

Subtract the GHA from this and get your LHA or t. Since the GHA is less than your longitude, your meridian angle, t, is east.

**Step 7**

To round out your assumed postion, you need an assumed latitude. This should be the nearest whole degree of latitude to your DR. You now have the necessary arguments to enter Pub. No. 249. You should use an assumed latitude of 27°, because your DR latitude is 27° 00.0′ N. Open Pub. No. 249 (the necessary pages from Pub. No. 249 are reproduced in this book in appendix C). Now go back to Step 4 and pick up your declination, which is S 9° 54.1′.

The 27° latitude section of Pub. No. 249 covers several pages. Turn toward the back until you come to a column under declination headed 9° (pages 162 to 167 in Pub. No. 249). You will notice that some pages are titled "Declination Same Name as Latitude," and some pages are titled "Declination Contrary Name to Latitude." Your assumed latitude is north and the sun's declination is south. South and north are contrary. You will use the 9° column on page 164 of Pub. No. 249, the "Contrary" page. On either side of the page is a column headed "LHA." This is the column for t or local hour angle. Go down this column to "54," then go across to the 9° declination column. The number 54 is your t and 9° is your declination. There are three subheadings in this column. You will use all three. Under "$H_c$," which is really $H_t$ or tabulated altitude, you extract $H_t$ 26° 30.0′. Values in the sight reduction tables of Pub. No. 249 are given to the nearest whole minute of arc for tabulated altitudes. Write the $H_t$ of 26° 30′ as 26° 30.0′ in order to use the decimal value of your observed altitude, $H_o$. Under the column headed by "d," extract the value, –35. Look at $H_t$ under the declination 10° column adjacent to the right. There, $H_t$ is 25° 55.0′. $H_t$ is decreasing as the declination increases, so the sign of d is minus. Had $H_t$ been increasing as declination increased, the sign of d would have been plus. Under the column headed "Z" (the symbol for the azimuth angle), extract 117°. Azimuth angles are reckoned from the elevated pole, which is the same as your latitude and thus north, and in the direction of the meridian angle, t, which is east in this case. Azimuths are reckoned clockwise from north and are rounded off to the nearest whole degree, so here Z (azimuth angle) will equal Zn (azimuth), or 117°, because the elevated pole is north and t is east. Had t been west, the azimuth angle would have been reckoned west from north, and you would have had to subtract Z from 360° to get Zn.

**Step 8**

There is no need to interpolate for azimuth. It does not vary much for a degree or so of declination. Altitude, though, will vary here 55′ for one degree of declination. This will plot as 55 miles and is a substantial difference. You have to make a correction to your tabulated altitude ($H_t$) for the difference between tabulated declination and the actual declination at the time of the sight. Turn to the back of Pub. No. 249 to Table 5 on page 248 (see appendix C of this book). Go across the top to the "35" column and down the side to "54" and read your correction at the junction. The value 54 is the difference in declination and 35 is your d. The correction is 32′. Since the altitude decreases as declination opposite from the latitude increases, you must subtract this from $H_t$ (tabulated altitude). Inspection will tell you this, but the value of d is given as plus or minus, as appropriate, in Pub. No. 249.

Where do you make this inspection? Go back to page 164 of Pub. No. 249 and go down the 9° declination column to the LHA 54 line. This is how you got your $H_t$ of 26° 30.0′. Remember what I said a little earlier about looking at the $H_t$ for 10° of declination? Your declination is really S 9° 54.1′, so the $H_c$ you are computing must lie somewhere between the tabulated value of $H_t$ in the 9° column and the value of $H_t$ in the 10° column. $H_t$ for a declination of 10° is 25° 55.0′. This is less than the 26° 30.0′ you extracted from the 9° column. This is what is meant by saying the tabulated altitude, $H_t$, is decreasing as the declination increases. And since $H_t$ is decreasing, the correction must be subtracted. Had $H_t$ been increasing, the correction would have had to be added. The time to make this inspection, had you needed to, is when you first extract $H_t$. That is the time to inspect the adjacent column bracketing your actual declination. If $H_t$ is increasing, label d plus, and if $H_t$ is decreasing, label d minus. I am telling you this because Pub. No. 249 is almost identical to the now-obsolete H.O. 214. In H.O. 214 the sign of the declination difference is not given. You may use H.O. 214 some day. There are many sets of these tables still in use. Remember to make this inspection, and you should have no trouble. Now you are ready for Step 9.

**Step 9**

Correct your sextant altitude, $H_s$. This will give you your observed altitude, which we label $H_o$. Remember that there are five sextant altitude corrections. You can combine three of these corrections by referring to the inside of the hard cover, front, of the *Nautical Almanac* (the table is also reprinted in appendix B). In the left-hand column in the "Oct.-Mar." inner box under the column "App. Alt.," read down to the 25° 26′/ 26° 36′ bracket and extract under "Lower Limb" the value +14.3′. In the right-hand column labeled "Dip," on the same page, go down the column to 11.9′/ 12.6′ under "Ht. of eye" and extract −3.4′. Apply these corrections to your $H_s$ of 25° 50.6′. Your $H_o$ is 26° 01.5′.

**Step 10**

Find the difference between $H_c$ and $H_o$. This is your altitude intercept. It is 3.5′.

Your computed altitude is less than the observed altitude. This means you are nearer the sun's subsolar point than is your assumed position. This is why you put the "t" after your intercept to indicate it is "toward." To plot the LOP, run a line at 117° true through your assumed position and step off 3.5 miles toward the sun (actually the sun's GP or subsolar point) and draw a perpendicular to the 117° line. This is your 0802 sun line, which is your LOP from this sight. (Note that the time of the DR position was 0800. The navigator was striving for an 0800 sight, but things seldom work out that closely.)

**Step 11**

Plot your line of position and label it.

With only minor differences and exceptions to suit special requirements—such as sights of Polaris and observations of bodies on your meridian—this is exactly the same format we use to work any other sight. We call this process *sight reduction*.

This single line of position or LOP did not give you a fix. It did give you some very useful information, though. It told you that you were many miles east of your course (see Figures 8 and 9). Let's work another sight to get a fix.

At 0800 you noticed your speed to be four knots and holding, and you changed course to 112° true. At about 1035 on 19/10/83, the navigator observed another sight of the sun's lower limb. This is the sight:

| | |
|---|---|
| C | $16^h$-$38^m$-$15^s$ |
| W | $02^m$-$49^s$ |
| $H_s$ | 49° 56.7′ |

Reduce the sight and plot the resulting LOP. Then advance the 0802 LOP to the time of this sight. Your fix should be:

| | |
|---|---|
| Lat. | 26° 31.0′ N |
| Long. | 88° 24.2′ W |

Practice makes perfect in navigation just as it does in anything else. We will continue our voyage with the following practice problems.

At 1032 course was changed to 270°T. No observation of the sun was possible at local apparent noon (this comes in another lesson). Speed was increased to six knots. At about 1450 ZT, the navigator observed the sun's lower limb as follows:

| | |
|---|---|
| C | $20^h$-$54^m$-$15^s$ |
| W | $02^m$-$19^s$ |
| $H_s$ | 31° 19.6′ |

At 1600 ZT another observation of the sun's lower limb was made as follows:

| | |
|---|---|
| C | $22^h$-$06^m$-$25^s$ |
| W | $01^m$-$53^s$ |
| $H_s$ | 16° 55.7′ |

Work the two sights above and plot your running fix. You will note that the two lines of position do not cross at a very wide angle. They are oriented more north and south than east and west. This makes your fix somewhat more subject to error than it would be had you been able to get a greater change in azimuth. In this case, your latitude is not as certain because of the orientation of the LOPs.

The yacht continued through the night on course 270°T. At 0730 the next day, course was changed to 225°T. At about 0920 ZT, the navigator shot the sun's lower limb for a morning sun line. This is the sight. Work it:

| | |
|---|---|
| C | $15^h$-$20^m$-$10^s$ |
| W | $01^m$-$39^s$ |
| $H_s$ | 37° 27.4′ |

After reducing this sight, plot the resulting LOP.

At about 1125, another sight was taken of the sun's lower limb:

| | |
|---|---|
| C | $17^h$-$29^m$-$15^s$ |
| W | $01^m$-$23^s$ |
| $H_s$ | 52° 43.9′ |

The complete work involved in reducing these practice sights beginning with the 1032 sight the morning of the 19th can be found in appendix A. You should work and plot each sight before consulting the answers in the Appendix.

# 7 HOW TO WORK A PLANET SIGHT

Your yacht continued on course 225°T. No LAN or afternoon sun was observed because of overcast skies. Late in the afternoon, the clouds began to break up. The navigator was able to observe a "star" through a break in the overcast. He identified it as the planet Jupiter. His observation was as follows. Jupiter 20/10/83, 1825 ZT.

| | |
|---|---|
| C | $00^h$-$29^m$-$25^s$ |
| W | $01^m$-$12^s$ |
| $H_s$ | 19° 04.3′ |

The chronometer has a rate of $5^s$ per day losing.

With only a few modifications, you work this sight in 11 steps exactly as you did your sun sight. Let's go through these steps for your planet sight and see how much they are alike and where they differ. The form on page 33 is almost exactly the same, but not quite.

### Step 1

Plot your DR position to the time of the sight. The DR position obtained will be approximately Lat. 25° 45.0′ N, Long. 91° 47.0′ W.

### Step 2

Record your sight on the form.

### Step 3

Correct your time to get the GMT at the instant your sight was taken.

Note: It is after midnight. The Greenwich date has changed. It is 21 October 1983 in Greenwich.

So far the procedure is the same as the one for working a sun sight. Now you are ready to continue on to Step 4.

### Step 4

Open your *Nautical Almanac* to the daily pages. Be very careful because the Greenwich date has changed. It is now 21 October in Greenwich, England, but still 20 October at your position. This can occur with any celestial body, including the sun, at certain longitudes.

Solving for a planet is no different in method from solving for the sun. What is different is where you find your data. Of course you cannot use the sun columns when solving for Jupiter. You must use the Jupiter col-

PLANET JUPITER **Step 1**
Date 20|10|83 ZT 1825
DR Lat. 25°-45′ N Long. 91°-47.0′ W

|  |  |  |
|---|---|---|
| **Step 2** | C | 00h-29m-25s |
| | W | 1m-12s |
| **Step 3** | C-W | 00h-28m-13s |
| | CE | 3m-07s f |
| | GMT | 00h-25m-06s |
| **Step 4** | GHA | 140°-12.4′ |
| **Step 5** | Corr. (min. & sec.) | 6°-16.5′ |
| | v correction +2.0 | 0.9′ |
| | GHA | 146°-29.8′ |
| **Step 6** | **Assumed Long. | 91°-29.8′ |
| | *LHA (t-W) | 55°-00.0′ |
| | ***Assumed Lat. | 26°-00.0′ N |
| **Step 4** | Declination | S 21°-32.5′ |
| **Step 5** | d difference +0.1 | 0.0′ |
| | *Declination | S 21°-32.5′ |
| **Step 7** | $H_t$ | 18°-55.0′ |
| **Step 8** | Correction d-38 | −21.0 |
| | $H_c$ | 18°-34.0′ |
| | $H_o$ | 18°-58.1′ |
| **Step 10** | **Altitude Intercept | 24.1′ t |
| **Step 7** | Z | N 126° W |
| | From 360° (if req.) | 360° |
| | **Zn | 234° |
| **Step 2** | $H_s$ | 19°-04.3′ |
| **Step 9** | Main Corr. −2.8 | |
| | Addn'l | |
| | Dip −3.4 | |
| | Total Corr. | −06.2 |
| | $H_o$ | 18°-58.1′ |

*Required to enter Pub. No. 249
**Required to plot line of position
***Required to enter Pub. No. 249 and to plot line of position

umn, which is near the center on the left-hand pages of the *Almanac*. Other than this, Step 4 for Jupiter is exactly the same as for the sun.

Under the date "21 Friday," at 00 hours, extract the GHA of 140° 12.4′ and the declination of S 21° 32.5′ for the even hour. The value of "d" is +0.1. There is an additional correction to GHA for the reduction of planet sights that is not required for the sun. At the very bottom of the GHA column, just to the left of the "d" value used to correct declination, is the letter "v," and opposite this is the value 2.0. Extract this value and put it into your form at the place indicated. This "v" correction to the GHA is required for all the planets. It is always positive except for Venus, when it may be either positive or negative. For Venus the sign is given in the *Nautical Almanac* if the value is negative. You are working with Jupiter, so the sign will be positive.

**Step 5**

This step is exactly the same as that for reducing a sun sight except for the small "v" correction. Go to the yellow pages, "Increments and Corrections," and get the correction for 25 minutes and 6 seconds. Use the "Sun and Planets" column to extract the main GHA correction in minutes and seconds, and the "v" correction for GHA and "d" correction for declination.

**Step 6**

To get your whole-number degree of LHA, select an assumed longitude just as you do for reducing a sun sight. Your DR is Long. 91° 47.0′ W. Your assumed longitude should be 91° 29.8′ W. Local hour angle is 55°; t is W.

**Step 7**

You are now ready to go into Pub. No. 249. Remember that you need three arguments to enter these tables. The arguments are assumed latitude to a whole degree, LHA or t to a whole degree, and declination as tabulated in the *Nautical Almanac* for the time of the sight in degrees, minutes, and tenths of a minute. The nearest even latitude to your DR latitude is 26°, so open Pub. No. 249 at the 26° latitude index. Since Jupiter has a south declination, turn to the 21° declination column on the "Contrary" page. Extract $H_t$ 18° 55.0′, d −38, and Z 126° on the 55° LHA line.

**Step 8**

Apply your d correction as taken from Pub. No. 249, Table 5. This is −21.

**Step 9**

Correct your sextant altitude, $H_s$, to get your observed altitude, $H_o$. You will find these corrections on the inside hard cover of the *Nautical Almanac* in the center box titled "Stars and Planets" (reproduced in appendix B of this book). Find your altitude bracket and extract −2.8′. In the dip correction table extract the correction for a height of eye of 12 feet. Dip is −3.4′.

**Step 10**

Find the difference between $H_o$ and $H_c$, which is your altitude intercept.

**Step 11**

Plot your 1825 Jupiter LOP and label it. But before you do, look at Step 7 in the work form. The navigator subtracted the tabulated azimuth angle, Z, of 126 degrees from 360 degrees. Why did he do this? Go back to the definition of azimuth in Chapter 2: *Azimuths* are reckoned from the North Pole in an easterly or clockwise direction. Now thumb back to the end of Step 7 in Chapter 6. Here *azimuth angles* are defined as being reckoned from the elevated pole in the direction of the meridian angle, t. At this point you are in north latitude and west longitude (a position that makes the relationship between azimuth angle and azimuth uncomplicated). In the 0802 sun sight in Chapter 6, t is east. The tabulated azimuth angle from Pub. No. 249 is 117 degrees, which is written Z = N 117° E. But azimuth is also reckoned from north through east. Therefore Z = Zn, and Zn = 117°. The 1449 and 1600 sights in Chapter 6 were worked with a t of west. Remember when t is west, t equals LHA; when t is east, t equals 360° minus LHA. We will see more of this relationship when we discuss azimuths in east longitude and south latitude. For now all you need to know is, if the body is east of you (its GHA smaller when compared to your longitude west), t is east, and Z = Zn. If the body is west of you (GHA larger when compared to your longitude west), Z must be subtracted from 360° to get Zn. In Pub. No. 249, Pub. No. 229, and some of the other precomputed or inspection-type tables, a notation on each page tells when in north latitude to make the tabulated Z equal to Zn and when Z must be subtracted from 360°. (Please do not think of south latitude for now.) Look at the upper left-hand corner of the tables. There is a notation that says:

N. Lat. LHA greater than 180° .......Zn = Z
LHA less than 180° .......Zn = 360° − Z

The notation always refers to LHA, which is always west. As I said above, to find the LHA when the body is east of you, subtract t from 360°. The remainder will be the LHA and will be greater than 180°. When t is west, t is the same as LHA and is less than 180°. Both t and LHA are tabulated in the "LHA" column of Pub. No. 249. Assuming that you have a t of 45° east, what is the corresponding LHA? To get this from the tables, look up 45° in the LHA column on either side of the tables. On the same line across the page in the opposite LHA column is the value 315°; LHA is 315°, which is greater than 180°, so Z = Zn. Had t been 22° east, you would have found on the same line across the page the value 338°, which would be the LHA. This is all well and good, but you will be better able to understand other methods, and your knowledge will be broader, if you learn to convert Z to Zn by the rules: *Azimuth angle is reckoned from the elevated pole in the direction of the meridian angle and azimuth is reckoned from the North Pole in an easterly direction.* For now, while you are in the Northern Hemisphere, the application of these rules only requires you to remember that if the GHA of the body observed is less than your longitude, Z = Zn; if the GHA is greater than your longitude, Zn = 360° − Z.

Jupiter was the only planet visible as an "evening star" at this time. This is because of the positions of the other planets in the heavens. The Jupiter sight plots very close to the 1825 DR. This made the navigator happy. He did take another sight of Jupiter before working the first one. This is good practice when only one body is available for observation through a lucky break in the clouds. The second sight gives you something to fall back on in case you make an error in taking or recording the first one. Your second Jupiter sight is as follows:

20/10/83, 1830 ZT, Jupiter

| | |
|---|---|
| C | $00^h$-$34^m$-$30^s$ |
| W | $01^m$-$08^s$ |
| $H_s$ | 18° 09.7′ |

You should work this sight.

The yacht continued on the same course at the same speed through the night. The weather cleared during the midwatch. At morning twilight you obtained a sight on the planet Venus. The DR at 0500 was Lat. 24° 59.0′ N, Long. 92° 38.0′ W:

21/10/83, 0505 ZT, Venus

| | |
|---|---|
| C | $11^h$-$12^m$-$10^s$ |
| W | $01^m$-$18^s$ |
| Lat. | 25° 00.0′ N |
| $H_s$ | 30° 57.0′ |

Work this sight. (The solutions to this sight and the second Jupiter sight above are in appendix A.)

An inspection of the daily pages of the *Nautical*

*Almanac* shows Mars would also be in a good position for observation had not Mars been almost in conjunction with Venus. Their GHAs and declinations are almost identical. LOPs obtained from observing both of these planets would have plotted as parallel lines. The navigator chose Venus because it is a much brighter body. Since Venus will remain in position for observation the entire morning, the navigator will also plan to observe Venus along with his sun sights, if atmospheric conditions permit.

During this same period of morning twilight you made several star sights. The navigator picks his stars considering the probability of their being available, not obscured by clouds or above a poor horizon, and different enough in their azimuths to form a good "cross" when the lines of position are plotted.

# 8 HOW TO WORK A MOON SIGHT

"Don't fool with the moon! It is a very difficult body to observe accurately. Besides, with all the other bodies in the heavens, you don't need to use the moon."

When you begin to navigate seriously, you will hear the above statement and similar ones over and over again. They will be made by people who are usually uninformed and always lacking in experience. *The moon is a very important body in celestial navigation.* Admittedly, getting good results from a moon sight does require more attention to detail. However, this additional effort is not at all difficult, and mastering the reduction of a moon sight should be a welcome challenge to any serious navigator.

The moon is important because it is frequently available for observation in the daytime. Modern celestial navigation depends on lines of position to give the navigator a fix. Only one line of position will result from any one observation. Unless another body is available, the navigator, during the day, must wait long enough for the sun (since the sun is usually the only other body available to observe during the daytime) to change its azimuth enough to give a good "cross" with an earlier sun line. This results in a running fix. Although a running fix is a perfectly good fix, it still suffers from any error in dead reckoning that might have crept in during the time interval between sights. The requirement for precise navigation increases as conditions deteriorate. Since poor conditions usually mean a loss of accuracy in dead reckoning navigation, it is important to be able to get a true fix by simultaneous observation of two or more celestial bodies. With the occasional exception of the planet Venus, the moon will be the only body you can observe simultaneously with the sun.

Let's define "simultaneous observation." Strictly speaking, things that are done simultaneously are done at the same instant. For our purpose here, "simultaneous" simply means that not enough time elapses between sights for an error in dead reckoning to have any significant effect on the solution of the sights. On the average yacht or small commercial vessel, speeds seldom exceed 8 or 10 knots. Thus, the interval between "simultaneous" sights can be as long as several minutes, as it may well be in the case of a round of star observations.

It is important that you know the days during which the moon will be in a favorable position for daytime observations. This will depend on the *phase* of the moon. Generally speaking, the moon will be available for daytime observations several days before and after its first quarter and several days before and after its last quarter. So the moon may not be around when you need it most, but if it is, it will certainly pay you to know how to use it.

I am going to illustrate how crucial a good moon sight can be. This is almost a true story, but I must change the date and time so that you can work the problem with the page excerpts from the *Nautical Almanac* in appendix B.

I was navigator on the yacht *Asperida II* out of Gibraltar in November 1968, bound for Barbados, West Indies, via Las Palmas, Gran Canaria. We were hit by a storm that, coupled with the crucial effect of wind and current near the Canary Islands, made us uncertain of our position.

It happened like this. The yacht had been buffeted for three days by a severe storm, with sustained winds of whole gale force and squalls in excess of force 12. In the previous 24 hours, some way had been made on a course intended to carry the yacht to Las Palmas, Gran Canaria. Three days had elapsed since the last good position. The navigator had been able to get sun lines on two occasions the afternoon before. Nevertheless, the position of the vessel was not accurately known. The proximity of the Canary Islands and their rockbound coasts, plus the strength of the wind and current, made beating up to Las Palmas almost impossible. If the yacht had been blown too far south, the situation would have been critical. An accurate position was needed and needed immediately. I could not wait for sufficient time to elapse to get a running fix from the sun, because in that interval the yacht might well have been swept past the islands.

For the purpose of our illustration, we will say the date was 10 November 1983. At about 1445 zone time, the navigator obtained an observation of the sun. As soon as he had recorded this sun sight, he observed the moon's upper limb. The chronometer read $15^h$-$48^m$-$43^s$. Elapsed time by stopwatch was $00^m$-$27^s$. Sextant altitude was 31° 00.2′. The chronometer was $1^m$-$13^s$ fast. The navigator worked the moon sight. See the accompanying sight reduction form.

**Step 1**

Plot the best DR position you can under these circumstances. It is approximately Lat. 28° 32.0′ N, Long. 14° 59.0′ W.

**Step 2**

Record your sight.

MOON, ~~Lower~~/Upper Limb **Step 1**
Date 10/11/83 ZT 1445
DR Lat. 28°-32′ N Long. 14°-59.0′ W

|  | | |
|---|---|---|
| **Step 2** | C | $15^h$-$48^m$-$43^s$ |
| | W | $00^m$-$27^s$ |
| **Step 3** | C-W | $15^h$-$48^m$-$16^s$ |
| | CE | $01^m$-$13^s$ f |
| | GMT | $15^h$-$47^m$-$03^s$ |
| **Step 4** | GHA | 336°-10.6′ |
| **Step 5** | Corr. (min. & sec.) | 11°-13.6′ |
| | v corr. +10.6 | 8.4 |
| | GHA | 347°-32.6′ |
| | **Assumed Long. | 14°-32.6′ W |
| | *LHA | 333°-00.0′ |
| | ***Assumed Lat. | 29°-00.0′ N |
| **Step 4** | Declination | S 24°-18.9′ |
| **Step 5** | d difference −3.6 | −2.9 |
| | *Declination | S 24°-16.0′ |
| **Step 7** | $H_t$ | 30°-59.0′ |
| **Step 8** | Correction d −53 | −14.0 |
| | $H_c$ | 30°-45.0′ |
| | $H_o$ | 31°-27.3′ |
| **Step 10** | **Altitude Intercept | 42.3 t |
| **Step 7** | Z | 151° |
| | From 360° (if req.) | |
| | **Zn | 151° |
| **Step 2** | $H_s$ | 31°-00.2′ |
| **Step 9** | Main Corr. | +58.5 |
| | Addn'l, HP 54.8′ | |
| | ****L corr. | |
| | ****U corr. | +2.0 |
| | Upper Limb | (−30.0′) |
| | Dip | −3.4 |
| | Total Corr. +27.1 | +27.1 |
| | $H_o$ | 31°-27.3′ |

*Required to enter Pub No. 249
**Required to plot line of position
***Required to enter Pub. No. 249 and to plot line of position
****Use correction U or L depending on whether upper or lower limb was observed.

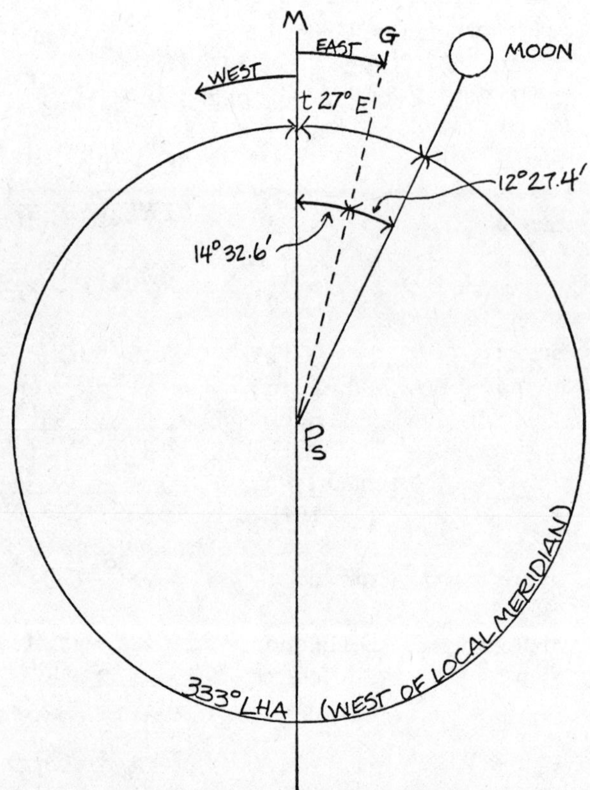


**Figure 10.** *Determining the LHA of the moon. M is the observer's meridian, G is the Greenwich meridian, and $P_s$ is the South Pole.*

**Step 3**

Correct your time for GMT.

**Step 4**

Open your *Nautical Almanac* to 10 November 1983. The moon data are in a box at about the center of the right-hand page. There are five columns of data tabulated for the moon. In the first moon column, GHA, extract the GHA of the moon for 15 hours, just as you would for any other sight. It is GHA 336° 10.6′. In the column next to the GHA column is the value "v," and the column is so headed. From this column opposite $15^h$ take +10.6′. This "v" correction is similar to the "v" correction you used in working the planet sights. Because the moon is so near the earth, it changes its angular position in relation to the earth much more rapidly than the sun or planets. This requires a substantial "v" correction for the moon's GHA. Since the moon's "v" changes so rapidly, its value is tabulated for each hour instead of the planets' three-day intervals.

Also, because of the moon's proximity to the earth, its declination changes quite rapidly, and a much larger correction is needed for horizontal parallax. Basically, these two corrections are the only additional requirements to reduce a moon sight. Strictly speaking, they are not unique to the moon, but only more critical. With other bodies, the corrections are sometimes small enough to be ignored or combined with other corrections. With the moon they each represent a substantial correction and must be applied individually.

Let's continue reducing the sight. Keep on the page for November 10, read across the 15-hour line, and under "Dec" extract S 24° 18.9′. Under the "d" column extract d 3.6′. By inspection, we determine that "d" is negative, since the moon's declination is decreasing. In the last column under "HP," we see that the moon's horizontal parallax is HP 54.8′, and we enter that number in the form.

**Step 5**

Correct your GHA of $15^h$ from the daily pages for the additional $47^m$-$03^s$ of time. Here you go to the yellow pages of the *Nautical Almanac* as you do for any other sight. Turn to the yellow page with $47^m$ in the upper right-hand corner. You should notice that a separate column exists for the moon. Opposite $03^s$ in the moon column of the $47^m$ box extract 11° 13.6′. You must also add a "v" correction to the moon's GHA, so go back to Step 4 and pick up the "v" of 10.6′. Opposite this figure in the "v or d" correction boxes extract v correction 8.4′. Add these to the GHA from the daily pages and get GHA 347° 32.6′. In this step you should also correct for your increment of declination. From Step 4 pick up the value of "d" of 3.6. Remember this correction is negative. The correction is 2.9. Subtract this from your tabulated declination and get S 24° 16.0′.

**Step 6**

Now it is time to find your LHA of the moon. Again, this is like any other sight. Select an assumed longitude to give you a whole degree of local hour angle. This assumed longitude should be as near as possible to your DR longitude. The procedure here will vary a little because you are near the prime meridian.

The moon is east of Greenwich. To get the LHA, you subtract your assumed longitude from the moon's GHA. This will give you the moon's LHA as west. (See Chapter 13.) Figure 10 will help you visualize the relationship between GHA and LHA in this problem. You should select an assumed longitude of 14° 32.6′ W. Because of the manner in which Pub. No. 249 is tabulated, you do not have to find the meridian angle (t), which in this case is 27° E. You can enter Pub. No. 249 with the LHA of 333°. The LHA of 333° and the meridian angle east of 27° are on the same line across the page in Pub. No. 249. The meridian angle east is always the difference between the LHA and 360°. Since the sum of the LHA and meridian angle east across any line in Pub. No. 249 always equals 360°, you can enter Pub. No. 249 with either the LHA or the meridian angle east (t east), depending on which is first derived when applying the assumed longitude.

**Step 7**

The nearest whole latitude to your DR is 29°. With this, you now have an even assumed latitude, an even LHA, and a declination. You are ready to enter Pub. No. 249. This is done exactly as before. Using the 29° index, open your tables. Your declination and latitude have contrary names. Use the right-hand page. Extract your $H_t$ of 30° 59.0′. Your azimuth of the moon is Z 151°. Your d is −53. The altitude is decreasing as the declination of the moon increases south, so d is negative.

**Step 8**

Correct your tabulated altitude for the difference in declination. The correction for 16′ of declination and a d of −53 is −14.0′. Your computed altitude $H_c$ is 30° 45.0′.

**Step 9**

Correct your sextant altitude. For moon sights, this is another step that can produce errors. At this time, it would be wise to pause and read sections 13 through 17 of your *Nautical Almanac*, which you will find in appendix B. Then turn to the inside of the back cover of the *Nautical Almanac*. Both pages, xxxiv and xxxv, give altitude corrections for the moon. You will notice a dip correction table on page xxxiv exactly like the one on the inside front cover of the *Nautical Almanac*. Apply your dip correction first. It is 3.4′.

From the upper table on page xxxiv, enter the column headed "30° - 34°." Go down this column until you come to "31," which is your sextant altitude to the nearest even degree. On either side of the table is a column headed "App. Alt." This column is divided in six increments of 10′ between each degree range. Under 31° and opposite 00 minutes extract your main correction, 58.5. Continue down the 30° - 34° column. About halfway down the page the column heading changes to "L U." These two letters stand for lower and upper limb. You are working an observation of the moon's upper limb. You will find your correction under "U." Notice also that the columns at either side are now headed "HP." This is the HP or horizontal parallax column. Go down it until you find the value of HP you extracted from the daily pages. It is 54.8. From the "U" column, you extract +2.0 opposite the HP of 54.9, which is the tabulated value nearest HP 54.8. Read carefully the small explanation of these tables in the lower right-hand corner of page xxxiv. This tells you that all the above corrections are to be added to your sextant altitude. It also tells you that 30.0′ is to be *subtracted* from observations of the moon's upper limb. Your observed altitude is 31° 27.3′.

**Step 10**

Compute your altitude intercept by finding the difference between your $H_o$ and your $H_c$.

**Step 11**

Plot your resulting LOP.

**SUMMARY**

Several comments are in order at this point. It is obvious that most of the procedure for working a moon sight is just the same as that for working any other sight. A little extra attention must be given to finding GHA and declination. The method is no different, but corrections we sometimes generalize or ignore with other bodies must be applied carefully to a moon sight. Sextant altitude corrections are more involved, but again they are not difficult if you will follow carefully the instructions on page xxxiv of the *Nautical Almanac*.

The 1545 sun sight was also reduced. The resulting line of position from this sight crossed our moon line at Lat. 28° 03.5′ N, Long. 14° 46.1′ W. This is a little over 31 miles south-southeast of the DR. Had we steered for La Luz by setting a course SW as indicated by our DR, we would have been swept past the islands. Under the weather conditions at the time, it would have been best to continue on to the Cape Verde Islands hundreds of miles away had this happened. As it was, our simultaneous sights of the sun and moon showed our position to be about 35 miles east-southeast of La Luz. We could set a course NW, adding a little northing to offset the effect of the current, and still gain La Luz. Had we waited an hour or so longer, it may well have been too late.

I chose this example to show you why simultaneous observations can be so important. A few other lessons can be gained by close analysis. It should be obvious that you should stay with your navigation. Had these sights been delayed at all, it might have been too late. The 42-miles intercept is a bit large, but, under adverse conditions such as those encountered here, it is not at all impossible to have such a large intercept. It would have been good practice to work these sights again using 28° N as our assumed latitude, but *only after we had changed our course to NW*. Our intercept was large because our DR was off 31 miles. This is quite possible, too, under the conditions experienced in this storm. This is not a good example of dead reckoning, but good dead reckoning can be almost impossible in heavy weather in a small vessel.

This example was a sight of the moon's upper limb. Had it been of the moon's lower limb, the procedure would have been the same, but you would look in the appropriate "L U" subcolumn, of course. The only real difference is that you do not have to subtract 30′ from the sextant altitude of a lower limb sight of the moon.

In my eagerness to show how important a moon sight can be, I may have led you to believe it is only used under adverse conditions. Nothing could be further from the truth. When the moon is in a good position I always observe it at the same time I observe the sun. Then too, on hazy mornings or hazy evenings, the moon and one or two bright planets such as Venus and Jupiter may be all that show through the haze. With a little practice, you will use the moon with as much ease and skill as you use any other body.

# 9 HOW TO WORK A STAR SIGHT

Several factors help a navigator decide what celestial body he wants to observe for a line of position. The most fundamental requirement is a body's availability. If you can't see it, you can't shoot it. On the other hand, if a particular celestial body is the only body in the sky that you can see, you will shoot it no matter how poorly positioned it might be.

Another consideration is the difference in azimuth of the bodies available. If you take sights on several bodies whose azimuths are nearly the same, or almost reciprocals of each other, your LOPs will be so nearly parallel that a slight error in your observation or calculations may make them cross miles away from your true position. There are some special benefits to be gained by observing bodies with reciprocal azimuths, but we will cover these benefits in a later chapter. For most ordinary purposes, you want bodies whose azimuths are different enough to cause the LOPs you get from observing them to cross at a wide angle. If only one body is available, you must wait for its azimuth to change enough to give you a good cross. This means that the best you can get is a running fix.

A running fix is a good fix if your dead reckoning is good, but the fix all navigators dream of is a "cartwheel" from simultaneous observations of several bodies. The lines of position cross at the hub. With the exception of those rare days when the sun, moon, and Venus are in good position at the same time, the only times you find celestial bodies for such observations are at morning and evening twilight when the sky is clear. Then you can shoot the planets and the stars. Since the planets are wanderers in the heavens (and from this get their name), it follows that the most dependable bodies are the stars. These are the bodies the navigator can depend on to give him that "pinpoint" that tells him exactly where he was at the time he took his sights. The warm glow of satisfaction you get from this must be experienced to be understood.

For this reason, "star sights" and "star time" have almost usurped the traditional time of local apparent noon and the noon sight as the time a navigator looks forward to for putting his house in order, when rough seas and stormy skies have given him a bad day.

Thus stars have a very important place in navigation, a place perhaps no more or no less important than any other, but still a place that is unique. You must know how to navigate by the stars to be a good navigator.

How do you work a star sight? Almost like any other. Again, the best way to show you is to take an example and go through the procedure step by step. Really there is only one fundamental step that is different. It may be important to explain this in part before we begin, and, if necessary, we can digress to explain further as we proceed.

Suppose you were called on to publish a nautical almanac. The powers that handed you this problem told you that you must publish data that would give any navigator the celestial coordinates of the sun, moon, Venus, Mars, Jupiter, and Saturn for any time (to the nearest second) on any day of the year. In addition, you were told that you must also provide this information for the 57 most commonly used stars. And by the way, you also must include about 170 other stars that the navigator might need, knowing that he would want to be able to extract the same data for these to the same degree of accuracy, of course.

It goes without saying that the size of this almanac must be small enough to be handy. Yet you would have to have about 230 columns instead of six! How can it be done?

The astronomers who formulate the almanacs are faced with the same problem. They solve it by taking advantage of one fact. The stars really change very little in relation to one another or in the overall celestial sphere. The declination may move hardly 1′ of arc in a month. The same can be said for the differences in hour angle between stars. This relationship remains so constant over these long periods that the old navigators referred to the stars as "fixed stars." The only motion of these "fixed stars" we have to consider for short periods of time is the apparent motion caused by the diurnal rotation of the earth on its axis.

If this is the case, why don't we select some point in the heavens and tabulate its motion by the day, hour, minute, and second in reference to Greenwich? Then if we can tabulate the declination of all of the stars we are likely to use, plus their angular distance west of this reference point, we can determine the exact celestial position of any star at any instant by adding this angular distance west of our reference point to the Greenwich hour angle of our point. This is exactly what the astronomers have done. The point of reference they selected is the point on the celestial sphere where the sun crosses the celestial equator on its trip north each spring. More specifically, this point is the intersection of the plane of the ecliptic and the vernal equinox. This point is called *the first point of Aries.* Its symbol is the ram's head. It is tabulated in the *Almanac* as "Aries." Since it is on the celestial equator, Aries has no declination, so only Greenwich hour angle is tabulated.

Now all we need is the angular distance of the various stars west of this point, and their angular distance north or south of the celestial equator, to find their celestial position at any time. This angular distance of a star west of Aries is called its *sidereal hour angle* or *SHA*. From this it should be apparent that if you add the SHA of a star to the Greenwich hour angle of Aries, the sum will be the Greenwich hour angle of the star. The angular distance of a star north or south of the celestial equator is its declination. The adding of the SHA of a particular star (the SHA is different for each star) to the GHA of Aries, which is used for all stars, to find a star's GHA is the main difference in the method of working star sights as opposed to planet or sun sights.

By now you are probably completely confused and even becoming alarmed. You shouldn't be.

Let's work some sights. You must shoot fast and accurately to get several in the short period of time available. You get the following:

| 21/10/83, Regulus | | |
|---|---|---|
| | C | $11^h$-$14^m$-$15^s$ |
| | W | $00^m$-$21^s$ |
| | $H_s$ | 43° 53.6′ |
| | Sirius | |
| | C | $11^h$-$16^m$-$05^s$ |
| | W | $00^m$-$18^s$ |
| | $H_s$ | 48° 09.4′ |
| | Rigel | |
| | C | $11^h$-$18^m$-$10^s$ |
| | W | $00^m$-$23^s$ |
| | $H_s$ | 47° 27.2′ |
| | Aldebaran | |
| | C | $11^h$-$22^m$-$40^s$ |
| | W | $00^m$-$24^s$ |
| | $H_s$ | 53° 29.9′ |
| | Alphard | |
| | C | $11^h$-$24^m$-$30^s$ |
| | W | $00^m$-$17^s$ |
| | $H_s$ | 42° 44.5′ |

We will work the Regulus observation step by step just as we did the sun and planets. See the accompanying sight reduction form.

**Step 1**

Plot your DR. You have already done this in doing your work to reduce the Venus sight in Chapter 7. Call your DR at 0505 Lat. 24° 59.0′ N, Long. 92° 38.0′ W.

STAR REGULUS **Step 1**
Date 21|10|83 ZT 0510
DR Lat. 24°-59′ N Long. 92°-38′ W

|  | | |
|---|---|---|
| **Step 2** | C | 11h-14m-15s |
| | W | 0m-21s |
| **Step 3** | C-W | 11h-13m-54s |
| | CE | 3m-02s f |
| | GMT | 11h-10m-52s |
| **Step 4** | GHA Aries | 194°-19.5′ |
| **Step 5** | Corr. (min. & sec.) | 2°-43.4′ |
| | GHA Aries | 197°-02.9′ |
| **Step 5A** | SHA star | 208°-07.8′ |
| | Sum | 405°-10.7′ |
| | Less 360° (if req.) | 360°-00.0′ |
| | GHA star | 45°-10.7′ |
| **Step 6** | **Assumed Long. | 92°-10.7′ W |
| | *LHA (t) | 47°-00.0′ E |
| | ***Assumed Lat. | 25°-00.0′ N |
| **Step 5A** | *Declination | N 12°-03.0′ |
| **Step 7** | $H_t$ | 43°-50.0′ |
| **Step 8** | Correction d+23 | + 1.0 |
| | $H_c$ | 43°-51.0′ |
| | $H_o$ | 43°-49.2′ |
| **Step 10** | **Altitude Intercept | 1.8a |
| | Z | N 097° E |
| | From 360° (if req.) | |
| | **Zn | 097° |
| **Step 2** | $H_s$ | 43°-53.6′ |
| **Step 9** | Main Corr. −1.0 | |
| | Addn'l | |
| | Dip −3.4 | |
| | Total Corr. | −04.4′ |
| | $H_o$ | 43°-49.2′ |

*Required to enter Pub. No. 249
**Required to plot line of position
***Required to enter Pub. No. 249 and to plot line of position

**Step 2**

Record your sight.

**Step 3**

Correct your time to get the GMT at the instant your sight was taken. Up to now nothing is different from reducing a sun sight.

**Step 4**

Open your *Nautical Almanac* to the daily pages. It is Friday, 21 October 1983. On the left-hand page, the first column on the left is headed "Aries." Go down this column into the "Friday, 21" section, and opposite $11^h$ extract GHA Aries 194° 19.5′. We always use Aries for any star. Not much different from a sun sight is it?

**Step 5**

Turn to the yellow pages, "Increments and Corrections." Find the $10^m$ box and run down the left-hand side to $52^s$. Up to now, Step 5 has been the same as for a sun sight. Now we change. Note that a separate column exists for "Aries." It is to the right of the "Sun Planets" column. From this "Aries" column, opposite $52^s$, extract the correction, 2° 43.4′. Add this to the GHA of Aries to get the corrected GHA of Aries, 197° 02.9′.

Now the big change. To keep our numerical sequence we will number this as Step 5A.

**Step 5A**

On the right-hand side of the left-hand daily pages of your *Nautical Almanac* is a box entitled "Stars." In this box are three columns headed "Name," "SHA," and "Dec." The names are in alphabetical order. Run down the list of names until you come to Regulus. Extract SHA 208° 07.8′ and Dec. N 12° 03.0′. Add your GHA of Aries and SHA of Regulus. You get 405° 10.7′. Whoa! you say. It can't be 405°! In a way you are right, but all this means is that we have gone completely around the world. To get things into their right perspective just subtract 360°. Any time the GHA of a star exceeds 360°, just subtract 360° from the GHA and everything will be back on track. The GHA is 45° 10.7′. Let's go on to Step 6.

**Step 6**

We are back in familiar waters. This is just the same as a sun sight. To get an assumed longitude, select the longitude nearest your DR that will give you a whole-number degree of meridian angle when the GHA of the body is subtracted from it. Your assumed longitude is 92° 10.7′. Your meridian angle of Regulus is 47° 00.0′

E. Your nearest even latitude is 25° 00.0′ N, which is your assumed latitude.

**Step 7**

You now have an even meridian angle of 47°, an even assumed latitude of 25° N, and a tabulated declination for Regulus of N 12° 03.0′. These are the three arguments needed to enter Pub. No. 249. Open Pub. No. 249 at the 25° latitude pages. Turn to the 12° declination column. Regulus has a north declination and our latitude is north, so use the "Same Name" page. Go down the LHA column to "47," your meridian angle (t) east, and opposite in the 12° declination column extract $H_t$ 43° 50.0′, d +23, and Z 97°.

**Step 8**

Extract the value of your "d" correction from Table 5 of your Pub. No. 249 tables and apply it to your $H_t$. The correction is +1.0. Your $H_c$ is 43° 51.0′.

**Step 9**

Correct your sextant altitude, $H_s$. Go to the inside of the front hard cover of your *Nautical Almanac*. In the center box headed "Stars and Planets," extract the correction for an $H_s$ of 43° 53.6′. It is −1.0′. Your dip correction is −3.4′. Total correction is −4.4′. Your $H_o$ is 43° 49.2′.

**Step 10**

Subtract your $H_o$ from your $H_c$ to get your altitude intercept of 1.8a.

**Step 11**

Plot your 0510 Regulus LOP and label it.

**SUMMARY**

Remember that Step 1 is to bring your DR up to the time of the sight. Now let's work the Sirius sight according to the format most navigators use in practice. See the Sirius sight reduction form.

The reductions of the star sights of Rigel, Aldebaran, and Alphard are worked completely in appendix A. You should work these sights before consulting the solutions in the Appendix, and then compare your computations.

STAR SIRIUS
Date 21/10/83, ZT 0515
DR Lat. 24°-59.0′ N Long. 92°-38′ W

| | |
|---|---|
| C | 11h-16m-05s |
| W | 0m-18s |
| C-W | 11h-15m-47s |
| CE | 03m-02s f |
| GMT | 11h-12m-45s |
| | |
| GHA Aries | 194°-19.5′ |
| Corr. (min. & sec.) | 3°-11.8′ |
| GHA Aries | 197°-31.3′ |
| SHA star | 258°-53.7′ |
| Sum | 456°-25.0′ |
| Less 360° (if req.) | 360°-00.0′ |
| GHA star | 96°-25.0′ |
| **Assumed Long. | 92°-25.0′ W |
| *LHA (t-W) | 4°-00.0′ |
| | |
| ***Assumed Lat. | 25°-00.0′ N |
| | |
| *Declination | S 16°-41.4′ |
| | |
| $H_t$ | 48°-49.0′ |
| Correction d−60 | −41.0 |
| $H_c$ | 48°-08.0′ |
| $H_o$ | 48°-05.1 |
| **Altitude Intercept | 2.9 a |
| | |
| Z | N 174° W |
| From 360° (if req.) | 360° |
| **Zn | 186° |
| | |
| $H_s$ | 48°-09.4′ |
| Main Corr. −0.9 | |
| Addn'l | |
| Dip −3.4 | |
| Total Corr. | −4.3 |
| $H_o$ | 48°-05.1′ |

*Required to enter Pub. No. 249
**Required to plot line of position
***Required to enter Pub. No. 249 and to plot line of position

Altogether, you have worked six sights to get a fix from these observations, including the Venus sight from Chapter 7. This is a big order. To the beginner, it will seem that it may take all day to work six or seven sights. This is not so. Of course, many navigators reduce the problem by taking only four or five sights. I like to get at least three well-positioned bodies. If possible, I try for at least five stars, and I like six or seven. If a planet is available, I usually can't resist a shot at it. Yet I can usually work these all out in an hour or so, including the time required to plot the resulting LOPs.

To do this you must become systematic. I do it in this way. First I will correct all my sextant altitudes, $H_s$, to obtain observed altitudes, $H_o$. Next I correct my times, and get the GMT of each observation. Then I go to the daily pages and extract the GHA of Aries for the even hour of GMT, and enter this into the calculation for each star. If I also shot a planet, I will get its GHA at this time. Then I go to the yellow pages and extract the corrections for minutes and seconds of time for each sight. After adding these and getting the GHA of Aries for each sight, I go back to the daily pages and extract the SHA and declination for each star. I add these to the GHA of Aries in each calculation. This gives me the GHA of each star. I then select an appropriate assumed longitude for each star to get a local hour angle or meridian angle for each. Now I am ready to enter Pub. No. 249. I go into the proper latitude section and extract the appropriate $H_t$, d, and azimuth for each calculation. Then I go to Table 5 and extract each d correction. I apply each of these to the appropriate $H_t$, get my altitude intercepts, and I am ready to plot. Thus I handle the almanac only four times, or more exactly, once in four different places. I pick up Pub. No. 249 only once and use two different pages. Had I worked each sight through before starting the next, I would have picked up the almanac 24 times. It pays to organize your work.

We have been taking these sights over a period of time between 15 and 20 minutes. Is our fix a running fix? It certainly is, and, had we been making 30 knots, it would be necessary to advance or retard each LOP to a common time. At 30 knots, our vessel would have gone 10 miles in 20 minutes! This is enough to keep us from getting a "pinpoint" fix. Actually we are only going six knots in our yacht *War Hat*. The distance run in 20 minutes at six knots is only two miles. At this speed I do not try to advance or retard LOPs resulting from a series of sights taken over a 15- or 20-minute period. I call the time of my fix the midpoint in the time interval it took me to shoot the stars, rounded off to the nearest five minutes — in this case 0515. I admit this is not the purest practice, but it is sufficiently accurate for most cases.

# 10 STAR TIME AND PRECOMPUTED ALTITUDE

I could have headed this chapter "Computed Zone Time of Various Celestial Phenomena." You probably would have groaned in despair if I had done so. I guess this is part of the problem in teaching navigation. The big words make it seem hard when it really isn't.

Let us take a useful example of a time of a "celestial phenomenon." Suppose you and a friend are anchored in Dauphin Island Bay. It is night, supper is over, and you are planning tomorrow's fishing in the Gulf of Mexico. It is the evening of July 25, 1983. You and your friend decide you must be on the fishing grounds a half hour before sunrise. It will take an hour and 15 minutes to run out there. What time should you get underway?

You are going to compute your data from your position in Dauphin Island Bay. This position is Lat. 30° 15.3′ N, Long. 88° 05.3′ W. Go about this step by step as in other lessons.

Open your *Nautical Almanac* to the daily pages for 25 July. Just to the right of the center of the page you will see columns headed "Twilight," "Sunrise," and "Moonrise." The center column of this page is headed "Lat" for latitude. To find the time of sunrise for 25 July , go down the latitude column to "N 30" (your latitude) and extract $05^{h}$-$14^{m}$.

There is no need to interpolate for latitude, since your latitude (30° 15.3′ N) is so near 30°. You do need to interpolate for longitude. Time of the phenomenon is tabulated to the zone meridian, which in this case is 90° W. Your longitude is 88° 05.3′. Don't worry about the .3′. Decimals are rounded off when calculating things such as star time, sunrise, and sunset. To interpolate you need the difference in longitude.

| | |
|---|---|
| Zone Meridian | 90° 00′ |
| Your Long. | 88° 05′ |
| Diff. Long. | 1° 55′ |

Convert this difference in longitude of 1° 55′ to time. To do this, turn to the first page of the yellow pages in the back of the *Nautical Almanac*. The table is "Conversion of Arc to Time."

Go down the degree columns until you come to "1°" and extract $04^{m}$.

Then look at the right-hand side of the page, where there is a box for converting minutes of arc to time. Run down these columns until you reach "55′" and extract $3^{m}$-$40^{s}$, which we will round off to $04^{m}$.

Our total correction for difference of longitude is:

$04^m$
$04^m$
———
$08^m$

Since our longitude, 88°, is east of the zone meridian, 90°, we must subtract the correction because sunrise will be earlier:

| | |
|---|---|
| | $05^h$-$14^m$ |
| | $08^m$ |
| Time of sunrise | $05^h$-$06^m$ |

The time to run to your fishing grounds is $1^h$-$15^m$. The time to get up and be underway:

$05^h$-$06^m$
$01^h$-$15^m$
———
$03^h$-$51^m$

or 0351 less $30^m$ equals 0321.

Suppose you, as the navigator on the yacht *War Hat*, had wanted to know what time morning nautical twilight would be at your DR position on the morning of 21 October. Again, we do this step by step:

**Step 1**

Plot your DR. A little judgment will come into play here first to approximate the time of nautical twilight. This can generally be done by inspection. The few minutes of error will not be serious. You determine you need a DR at about 0505, which plots out to be Lat. 24° 59.0′ N and Long. 92° 38.0′ W.

**Step 2**

Enter your *Nautical Almanac* for 21 October 1983. On the right-hand daily page opposite "N 20°" in the "Twilight" box, extract under the "Naut." column the time, $05^h$-$08^m$.

**Step 3**

This time your latitude is halfway between the two tabulated latitudes of N 20° and N 30°. You must interpolate as follows:

| | |
|---|---|
| DR Lat. | 24° 59.0′ |
| Tab. Lat. | 20° 00′ |
| Diff. Lat. | 4° 59.0′ |

Since 4° 59.0′ is so near 5°, call your difference of latitude 5°.

**Step 4**

It is also necessary to extract the time of nautical twilight at 30° N. From the same table you see it is $05^h$-$13^m$. Subtracting the time at 20° N ($05^h$-$08^m$) gives a difference in time for a 10° difference in latitude of $05^m$:

$05^h$-$13^m$
$05^h$-$08^m$
———
$00^h$-$05^m$

Since the difference between your DR latitude and either tabulated latitude is 5°, the time difference at your DR is $02^m$-$30^s$. Call it $2^m$. Since the time of nautical twilight is getting later as the latitude increases, you must add this:

| | |
|---|---|
| At Lat. 20°N | $05^h$-$08^m$ |
| Lat. Diff. | $00^h$-$02^m$ |
| Zone Time of Naut. Twilight at Lat. 25° N | $05^h$-$10^m$ |

**Step 5**

The time $05^h$-$10^m$ is for the zone meridian. You must interpolate for longitude again. Your DR longitude is 92° 38.0′ W. The zone meridian is 90°.

| | |
|---|---|
| | 92° 38.0′ |
| | 90° 00.0′ |
| Diff. Long. | 2° 38.0′ |

The difference of longitude is converted to time:

$00^h$-$08^m$
$00^h$-$02^m$-$32^s$
———
$00^h$-$10^m$-$32^s$
or
$00^h$-$11^m$

Since you are west of the zone meridian, you must add this to the zone time of the phenomenon:

| | |
|---|---|
| | $00^h$-$11^m$ |
| | $05^h$-$10^m$ |
| ZT of Naut. Twilight | $05^h$-$21^m$ |

Suppose our navigator wanted to know the ZT of civil twilight. He would have done it in this way:

| | | | |
|---|---|---|---|
| DR Lat. | 24° 59.0′ | | |
| Tab. Lat. | 20° 00.0′ | | |
| Diff. Lat. | 4° 59.0′ | | |
| Call it | 5° 00.0′ | | |
| Lat. 30 | $05^h$-$40^m$ | | |
| Lat. 20 | $05^h$-$34^m$ | | |
| Diff. Time | $00^h$-$06^m$ | diff. at tab. lat. | |
| Diff. Time | $00^h$-$03^m$ | for DR lat. | |
| Zone Time | $05^h$-$37^m$ | | ($05^h$-$34^m + 03^m$) |
| DR Long. | 92° 38.0′ | | |
| Zone Mer. | 90° 00.0′ | | |
| Diff. Long. | 2° 38.0′ | | |
| Diff. Long. Time | $00^h$-$11^m$ | | |
| ZT at DR | $05^h$-$48^m$ | | ($05^h$-$37^m + 11^m$) |

To solve this problem the navigator first found the ZT of civil twilight for Lat. 30° N and Lat. 20° N in the *Nautical Almanac*. He subtracted the lesser from the greater and obtained a difference in time of $00^h$-$06^m$. He then found his difference in latitude between 20° N and 25° N, his DR latitude. Since this was 5°, and the tables are tabulated for latitude increments of 10°, he divided $06^m$ by 2 and obtained an answer of $03^m$. Since he was working up from 20° N, he added this $03^m$ to the time of twilight at 20° N, $05^h$-$34^m$, and got $05^h$-$37^m$ as the time of twilight at the zone meridian and the DR latitude. It was then necessary to interpolate for difference in longitude to obtain the time of twilight at his DR meridian. This amounted to another $11^m$. The DR longitude was west of the zone meridian, so he added this to the ZT of civil twilight at the zone meridian. This gave him the ZT of civil twilight at his DR, $05^h$-$48^m$.

Time of twilight concerns the navigator because, in order to take star sights, he must have enough darkness to see the stars, but he must also have enough daylight to see the horizon clearly. Twilight in the morning and in the evening are the only times during the day when this occurs.

The navigator usually computes the time of nautical twilight for the morning star sights and the time of civil twilight for the evening star sights. A definition of each type of twilight, and an explanation of the sequence in which they occur, will show you the reason for this. There are three defined types of twilight: astronomical twilight, nautical twilight, and civil twilight. Astronomical twilight is defined as twilight that has its darker limit when the center of the sun is 18 degrees below the celestial horizon. The time of astronomical twilight is not of frequent interest to the navigator and is not listed in the *Nautical Almanac*. At astronomical twilight it is too dark to see the horizon distinctly enough to get accurate star (or planet) sights. The *Nautical Almanac* does list the times of the darker limits of nautical and civil twilight. Nautical twilight has its darker limit when the sun's center is 12 degrees below the celestial horizon. At this time, in the morning, the horizon becomes well-enough defined to take sights. You can see the dimmer stars well enough to shoot. Civil twilight is defined as that twilight that has its darker limit when the sun's center is 6 degrees below the celestial horizon. At this time the horizon is very distinct, and the brighter stars (and planets) are still quite visible. Of course, all my comments as to the quality of the horizon presume good visibility, with no haze, clouds, or other interference. As daylight comes, the first twilight to occur is astronomical twilight, which lasts until the time of the darker limit of nautical twilight, when it is time to begin shooting the dimmer stars. Morning nautical twilight ends with the beginning of civil twilight, which ends with sunrise. In the evening, the order is reversed. Civil twilight begins with sunset and ends when the sun's center falls 6 degrees below the celestial horizon. Nautical twilight follows, and ends when the sun's center reaches a point 12 degrees below the celestial horizon, when it is likely too dark to see the visible horizon for taking sights. Astronomical twilight follows, and ends at the onset of night, when the sun's center is 18 degrees below the celestial horizon. Because you want to start taking sights as soon as conditions allow, the time of nautical twilight is of prime interest in the morning, and the time of civil twilight is of prime interest in the evening. In considering all this, other factors may have an effect. Some navigators claim it is possible on very bright moonlit nights to get good sights through the entire time the moonlight is available, long before or after twilight. I do not make a practice of this, but I do take sights at such times as a check on their accuracy, and I would sure not hesitate to use them, with caution, if I badly needed a sight. There are even times when Venus, Mars, Jupiter, or even Saturn is in the right position to be observed after the onset of night. And by no means overlook the moon itself under such conditions.

The zone time of twilight, whether nautical twilight in the morning or civil twilight in the evening, is of most concern to the navigator. For evening phenomena, exactly the same procedure is used as for morning phenomena, except that values will be extracted from the lower right-hand side of the right-hand daily page involved. It is important to be accurate to within a minute, but no smart navigator waits until the last minute to appear on deck at twilight. You should always give yourself 10 minutes' leeway.

Another use of the advance knowledge of the time of twilight is to give the navigator a time for which to precompute the altitudes and azimuths of the stars he intends to observe next star time. This time must be accurate, since the navigator will use this information to find the stars he selected.

Even after you have solved the problem of identifying a star and locating it in the heavens, it can be a real test of your skill with a sextant to get it down to the horizon. This problem is magnified many times by any motion of the vessel or erratic steering on the part of the helmsman. If partial cloud cover obscures your star from time to time, you really have a problem. However, if you know the approximate altitude of a star and its azimuth, think how easy it can be to preset your sextant to this altitude. Go on deck and look out along the line of the azimuth. Unless your calculations are way off, you will see the star in your horizon glass. All you need do is bring it up or down to exact coincidence with the horizon, and observe the exact time. You have your sight. This is the trick navigators use to take a number of star sights in the limited time of twilight.

How do you precompute these altitudes and azimuths? Exactly the same way you compute the actual sight when you observe it. For example, let us consider the star Sirius as one you selected to observe. You would set up the problem as follows:

Date 21/10/83, Star Sirius
Estimated time of sight $11^h$-$15^m$-$00^s$ GMT

| | | |
|---|---|---|
| GHA Aries | 194° 19.5′ | |
| Corr. | 3° 45.6′ | |
| GHA Aries | 198° 05.1′* | |
| SHA star | 258° 53.7′ | |
| Sum | 456° 58.8′ | |
| Less 360° | −360° 00.0′ | |
| GHA star | 96° 58.8′ | |
| Assumed Long. | 92° 58.8′ | W |
| LHA (t west) | 4° 00.0′ | |
| Assumed Lat. | 25° 00.0′ | N |
| Dec. | S 16° 41.4′ | |
| $H_t$ | 48° 49.0′ | |
| d −60 | −41.0′ | |
| $H_c$ | 48° 08.0′ | |
| Z | N 174° | W |
| from | 360° | |
| Zn | 186° | |

You have computed the altitude and azimuth for Sirius. You can and should do this for each star you intend to shoot. Remember to organize your steps in your *Nautical Almanac* and Pub. No. 249 as described in Chapter 9.

By now you are no doubt saying, "Isn't this twice the work?" Well, it isn't. You use the same solution to get the altitude intercept for your actual sight. All you usually have to do is adjust your assumed longitude for the always slight difference in time. When you set the problem up, leave space for the adjustments. Space at the bottom of the precomputed solution is excellent. We will continue with the same example.

Assume your actual time of observation was C $11^h$-$16^m$-$05^s$, W $0^m$-$18^s$. The problem works out as follows:

| | | |
|---|---|---|
| C | $11^h$-$16^m$-$05^s$ | |
| W | $00^m$-$18^s$ | |
| C − W | $11^h$-$15^m$-$47^s$ | |
| CE | $03^m$-$02^s$ | f |

*Note: At this point, if you were using a graphic star finder, you would apply your assumed longitude and get the LHA of Aries. You are using Pub. No. 249, so continue.

| | | |
|---|---|---|
| GMT | $11^h$-$12^m$-$45^s$ | |
| GMT | $11^h$-$15^m$-$00^s$ | (estimated for precomputation) |
| Diff. | $2^m$-$15^s$ | |
| Diff. | 0° 33.8′ | (in longitude from yellow pages, "Increments and Corrections," "Aries" column.) |
| Adjustment to Assumed Long. | 0° 33.8′ | |
| Assumed Long. | 92° 58.8′ | (precomputation) |
| Assumed Long. | 92° 25.0′ | (time of sight**) |
| $H_s$ | 48° 09.4′ | |
| Main corr. −0.9 | | |
| Dip −3.4 | −4.3′ | |

**Because the time of the actual sight was $2^m$-$15^s$ earlier than the time used for precomputation, the adjustment to assumed longitude must be subtracted. The assumed longitude becomes less than it was.

| | |
|---|---|
| $H_o$ | 48° 05.1′ |
| $H_c$ | 48° 08.0′ |
| | 2.9 away |

In other words, convert your difference in time and apply it to your assumed longitude.

Since the actual time of the sight was earlier, the adjustment must be subtracted from the assumed longitude used in the precomputations. Occasionally, the time between the time used in your precomputation and the time of your actual sight will vary enough to change your LHA a degree or so. In this event, an additional adjustment to your $H_c$ may be required. You can minimize such adjustments by precomputing your sights with an increment of time between them equal to the approximate amount of time it takes you to take two consecutive sights. Experience will tell you what time interval to use. Then start your observations as near as possible to the time used in your precomputations, and make your observations in the same sequence as your precomputations. You will be surprised how little additional adjustment will be required to make your calculations conform to the time of the sights.

# 11 THE NOON SIGHT AND LATITUDE BY MERIDIAN ALTITUDE

The noon sight to determine latitude is probably the oldest celestial observation used in navigation. Scholars tell us that the noon sight was developed by the Arabs and used extensively by them. The Arabs contributed much to navigation. Their environment required the science of navigation on the trackless desert as well as on the trackless sea.

Simplicity of the required calculations, the lack of a need for an exact time, and great reliability make the noon sight quite important to the mariner—so much so that the noon sight is a tradition of the sea. On board ship, everyone who can use a sextant is expected to participate in taking the sun at local apparent noon, and when I was in the Navy assigned to a merchant vessel, no less than five of us marched solemnly out on the wing of the bridge for this purpose each day.

To take the sun at noon you must know at what time the sun will be on your meridian. This is local apparent noon (LAN). Computing this time on a moving vessel is very similar to figuring the time of sunrise or sunset or other celestial phenomena of this type. The problem is really easier, though, because you need not interpolate for difference of latitude. A body will be on an observer's meridian at the same time regardless of the observer's latitude.

There are several ways to compute the time of LAN. When the GHA of the sun equals the longitude of the observer, it should be obvious that the sun is on the observer's meridian. By definition, this is the time of local apparent noon. This is the highest point the sun will reach in the heavens that day. It is also the time at which the sun will have an azimuth of 000°T or 180°T, depending on your latitude. To find this time, consult your almanac for the date involved. Another method is to apply the equation of time from your almanac. This will give you your zone time of LAN on the zone meridian. Correct this for difference of longitude and you will have the time of LAN at your position.

Let's work this out both ways and see how they compare.

Your last fix was at 0515, when you got your round of stars. You continue on course 225°T at a speed of six knots. No sun was available during the morning, so you reckon your noon DR from your 0515 fix.

**Step 1**

Plot your 1200 DR at six knots for $6^h$-$45^m$. You will go 40.5 miles. Your 1200 DR is Lat. 24° 33.1′ N and Long. 92° 43.4′ W. The date is 21 October 1983.

**Step 2**

You decide you will compute the time of local apparent noon by the GHA method. You open your *Nautical Almanac* to the daily pages and to the date of 21 October 1983. Extract the nearest GHA of the sun that is *less* than your longitude. At the same time, extract the even hour of this GHA. Note your values as follows:

GMT 17$^h$-00$^m$-00$^s$ GHA 78° 49.8′

**Step 3**

Subtract this GHA 78° 49.8′ from your DR longitude. Note the difference in longitude in degrees.

| | |
|---|---|
| DR Long. | 92° 43.4′ |
| GHA | 78° 49.8′ |
| Diff. Long. | 13° 53.6′ |

**Step 4**

Turn to page i of the yellow pages in your *Nautical Almanac* (reproduced in appendix B of this book). With the aid of this table, convert 13° 53.6′ to time:

00$^h$-52$^m$-00$^s$
03$^m$-34$^s$

00$^h$-55$^m$-34$^s$

**Step 5**

Add this to your even hours of GMT to get the GMT of LAN:

17$^h$-00$^m$-00$^s$
00$^h$-55$^m$-34$^s$

17$^h$-55$^m$-34$^s$

Subtract plus 6, which is your zone description, to get ZT of LAN:

17$^h$-55$^m$-34$^s$
6$^h$-00$^m$-00$^s$

ZT of LAN 11$^h$-55$^m$-34$^s$

At this zone time, the sun will be on your meridian. The smart navigator will not wait until the last minute to appear on deck. You should go topside with your sextant 10 minutes or so ahead of this time and observe the sun every minute or two until it stops rising. Note the time and altitude at the sun's highest point. This is the time and altitude at LAN.

If you want to compute the time of LAN by the equation of time method, here is how you do it:

**Step 1**

Plot your 1200 DR.

**Step 2**

Open the *Nautical Almanac* to the daily pages for 21 October 1983. In the lower right-hand corner of the right-hand page is a small box. A small column on the left end of this box headed "Day" gives the dates "19," "20," and "21." Opposite "21" under "Equation of Time" in the "12$^h$" column extract 15$^m$-17$^s$.

**Step 3**

Note the local mean time of noon, which is 12$^h$-00$^m$-00$^s$. Since the time of meridian passage given in the *Nautical Almanac* is before noon, the sign of the equation of time is plus (+). Since you are working from mean time to apparent time, the sign is reversed. So subtract the equation of time and get:

12$^h$-00$^m$-00$^s$
15$^m$-17$^s$

11$^h$-44$^m$-43$^s$

**Step 4**

The time 11$^h$-44$^m$-43$^s$ is the time of LAN on the zone meridian. You must correct for difference in longitude:

| | |
|---|---|
| Zone Long. | 90° 00.0′ |
| DR Long. | 92° 43.4′ |
| Diff. Long. | 2° 43.4′ |

**Step 5**

Convert 2° 43.4′ to time. Since 92° is west of the zone meridian, you must add. From page i:

$08^m$-$00^s$
$02^m$-$52^s$
$11^h$-$44^m$-$43^s$

ZT of LAN $11^h$-$55^m$-$35^s$

Of course, you can also extract from the box on the lower right-hand corner of the right-hand daily page the time of meridian passage for October 21. It is $11^h$-$45^m$. Then convert 2° 43.4′ to time and get the ZT of LAN.

$11^h$-$45^m$
$10^m$-$52^s$
$11^h$-$55^m$-$52^s$

There are thus really three ways to compute ZT of LAN. Admittedly, the last is the simplest, but it doesn't require the student navigator to exercise an understanding of the principle of LAN. Besides, the other methods have their place. All navigators of the old school use the equation of time. Navigators who are addicted to the *Air Almanac* and related publications use the GHA method. The way things are tabulated in the *Air Almanac*, the GHA method is too easy to make using another worthwhile. Then, too, the GHA method is the only method of computing the meridian passage of the moon and the stars and planets.

You know how to compute the time of local apparent noon, your LAN. What do you do next? You take your noon sight and you work it. How? Like this.

Go on deck 10 minutes or so before your computed zone time of local apparent noon. Start shooting the sun. At the moment the sun is no longer gaining in sextant altitude, note your time to the nearest minute, and record your sextant altitude. With this information, work your problem. The basic formula is L equals Z plus D. I hate to use formulas, but this is so simple it is self-defeating not to. The nomenclature is:

L = latitude
Z = zenith distance
D = declination

Stated in words, latitude equals the algebraic sum of the zenith distance and the declination. Z is 90° 00.0′ minus the altitude, $H_o$, of the sun. D is the declination as tabulated in the daily pages of your *Nautical Almanac*.

Let's work the problem. At 1155 ZT you observe the sun's lower limb to be $H_s$ 54° 46.2′. The sun is on the meridian. What is your latitude?

**Step 1**

Record your sight. You already have your DR.

ZT $11^h$-$56^m$-$00^s$ $H_s$ 54° 46.2′

**Step 2**

Correct sextant altitude, $H_s$, to get observed altitude, $H_o$.

| | | |
|---|---|---|
| Main Corr. | +15.5′ | 54° 46.2′ |
| Dip | −3.4′ = | +12.1′ |
| | $H_o$ | 54° 58.3′ |

**Step 3**

Find your zenith distance by subtracting $H_o$ from 90° (89° 60.0′).

| | |
|---|---|
| | 89° 60.0′ |
| | −54° 58.3′ |
| Z | 35° 01.7′ |

**Step 4**

Add your declination algebraically. Since your latitude is N and declination is S, the sign is minus.

| | |
|---|---|
| Z | 35° 01.7′ |
| Dec. | −10° 40.7′ |
| Lat. | 24° 21.0′ |

This is really all there is to it. Remember your formula, L = Z + D. This is an algebraic formula, but it is so simple, it makes those of us who flunked algebra feel good, because even we can work it.

Since this is an algebraic formula, some of the values in it will be positive and some negative. To determine this, the following rules apply in the Northern Hemisphere:

1. Latitude N, declination N, but declination less than latitude, then all values are positive and should be added.
2. Latitude N, declination N, but declination greater than latitude, then the zenith distance is negative and must be subtracted.
3. Latitude N, declination S, then declination is negative and must be subtracted.

Now, of course, this is conversely true of situations in the Southern Hemisphere. You may not go around the world or to the South Seas, but in case you do, the same three cases apply. To state the same situations more generally:

1. Latitude same name as declination, but declination less than latitude, then all values are positive and should be added.
2. Latitude same name as declination, but declination greater than latitude, then the zenith distance is negative and must be subtracted.
3. Latitude contrary name to declination, then declination is negative and must be subtracted.

To understand why these rules are true, let's look at three illustrations. Diagrams such as those accompanying these explanations can be difficult. They can be difficult *unless* you sit down with a piece of paper and a protractor, measure the angles, and draw out the diagrams. Try it. You may find understanding the diagram to be a brain teaser, but not too hard. Drawing the diagrams out in this manner will surely help you understand the three cases of L = Z + D used to solve meridian altitude observations.

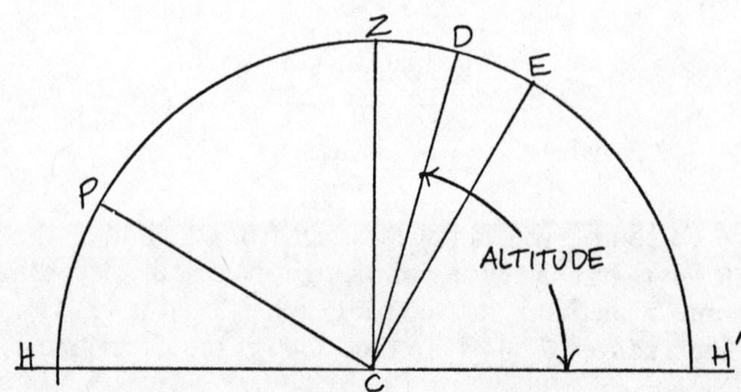


**Figure 11.** *Meridian altitude diagram, Case I.*

## CASE I: LATITUDE AND DECLINATION SAME NAME, BUT DECLINATION LESS THAN LATITUDE

For the purpose of following the diagram in Figure 11, assume your DR latitude is 30° north, and the declination of the body you are observing is 15° north. Latitude is the unknown in meridian altitude observations, and finding latitude is the purpose of meridian altitude observations. The principles illustrated will apply in any case in which the latitude is the same name as, and greater than, the declination.

The line HH′ is the celestial horizon. C is the center of the earth. C is also your position, because, for the purpose of celestial observations, the center of the earth is considered the observer's position. P is the elevated pole. In this example, P is the North Pole. Z is your zenith. (In all navigation literature, by convention, Z may denote azimuth angle, zenith distance, or zenith.) Remember that the declination of the zenith equals your latitude? If you can find the declination of your zenith you have found your latitude. D is the declination of the body you observed. E is the celestial equator.

By definition of zenith, angle HCZ is 90°. Also by definition, angle PCE is 90°. Declination measures the distance of a body north or south of the celestial equator. Angle DCE, the body's declination, is 15 °. You get the value of DCE from the appropriate daily pages of the *Nautical Almanac*. By inspection, it is apparent angle DCH′ is the altitude, $H_o$, of the body you are observing. Angle DCH′ subtracted from angle ZCH′ will give you the zenith distance. Angle ZCH′ is 90° for the same reason angle ZCH is 90°. If you subtract angle DCH′ from angle ZCH′, the remainder is angle ZCD. Angle ZCD is your zenith distance. By looking at the diagram, you can see angle ZCD plus angle DCE equals angle ZCE. Angle ZCE is your latitude because ZCE is the declination of your zenith, and the declination of your zenith equals your latitude. Now, if ZCE equals L, ZCD equals Z, and DCE equals D, you get the formula L = Z + D. All values in Case I are positive.

Let's see how this works out. You are looking for angle ZCE, the latitude. You get DCH′, the $H_o$, by corrected sextant observation. Angle DCE, the declination, is taken from the *Nautical Almanac*. Assume the $H_o$ you obtained is 75°. Z equals 90° minus $H_o$. Z then must equal 90° minus 75° or 15°. By substituting these values of 15° for Z and north 15° from the *Nautical Almanac* for D, you get L = 15° + 15°. This adds up to L = 30°. L is north because it has the same name as D, and D is north.

## CASE II: LATITUDE AND DECLINATION SAME NAME, BUT DECLINATION GREATER THAN LATITUDE

This time, let's assume your DR latitude is 15° north. Again latitude is the unknown. You open your *Nautical Almanac* and find the declination of the body you observed to be north 30°. The principles illustrated here for Case II will apply for any situation in which latitude is the same name as, but less than, the declination. Remember, latitude sights, or more properly meridian altitude sights, work for any body you observe on the meridian, so you could observe a body whose declination is 30° even though the sun's declination never exceeds 23° 27′ north or south.

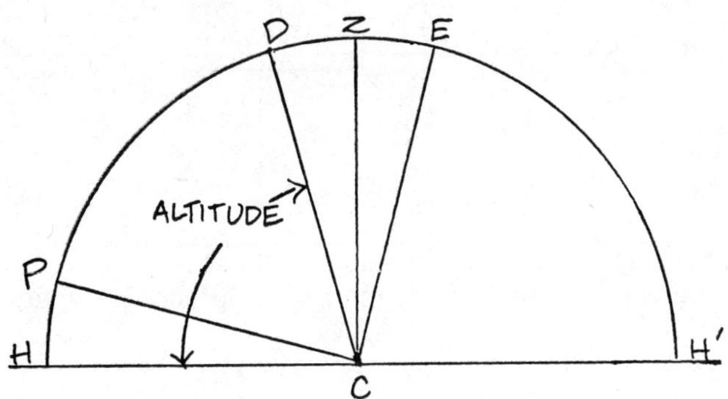


**Figure 12.** *Meridian altitude diagram, Case II.*

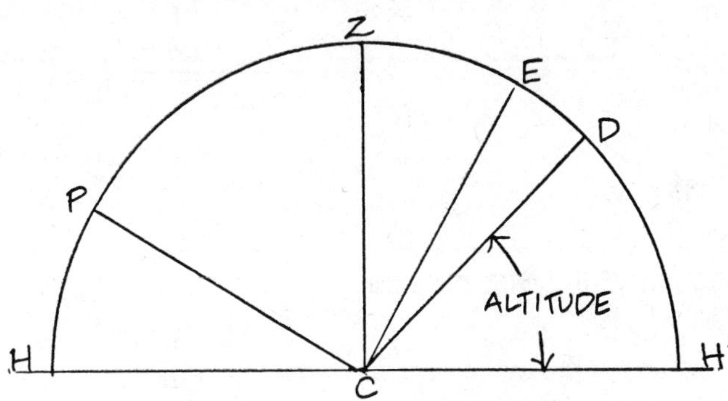


**Figure 13.** *Meridian altitude diagram, Case III.*

In Case II, the letters represent the same parts of the celestial sphere as in Case I. (See Figure 12.) Again HCZ equals 90° by definition. Angle PCE equals 90° by definition. Angle ZCE is your latitude. Angle DCE is the body's declination, north 30°. Angle HCD is the observed altitude, $H_0$, of the body. By looking at the diagram, you can see angle HCD plus angle DCZ is 90°, and the difference between angles HCZ and HCD is angle DCZ. Angle DCZ is the zenith distance. Zenith distance is 90° minus $H_0$. You get the declination of the body you observed from the *Nautical Almanac*. Angle ZCE equals angle DCE minus angle DCZ. But ZCE is L, DCZ is Z, and DCE is D; from this comes the formula $L = Z + D$. Remember, this is an algebraic expression. Declination, D, is positive in sign because D is north, and latitude, L, is north, and they are thus the same name. But from the diagram you see Z must be subtracted to get latitude. Z must be subtracted because the azimuth of the body you observed was toward the elevated pole. So in Case II, $L = -Z + D$.

Assume $H_0$ was observed to be 75°. Then $Z = 90° - 75°$. $Z = 15°$, and is negative. The declination, D, was found to be north 30° by consulting the *Nautical Almanac*. Latitude will be found by the expression $L = -15° + 30°$, or $L = +15°$. The plus sign indicates the latitude and declination are the same name, and north. In Case II meridian observations, you will always face the elevated pole. In Case I and Case III meridian observations you will always face the equator.

CASE III: LATITUDE AND DECLINATION CONTRARY NAMES

See Figure 13. Again angle HCZ is 90°. Angle PCE is 90°. Angle ZCE is your latitude, and the angle you want to determine. Angle ECD is the declination of the body you observed; since D is on the opposite side of E from Z, it is of contrary name. Let's assume the declination, which you extracted from the *Nautical Almanac*, is south 15°. Your DR latitude is 30° north. As in Case I, angle DCH′ is the observed altitude, $H_0$, of the body observed. Angle DCH′ subtracted from angle ZCH′ leaves angle ZCD. Angle ZCD is the zenith distance. Angle ZCE is your latitude and angle ZCE plus angle DCE equals angle ZCD. So if you observed $H_0$ to be 45°, by subtracting your $H_0$ of 45° from 90° to get ZCD, the zenith distance, you find ZCD equals 45°, too. A ZCD of 45° less angle ECD of 15° equals an angle ZCE of 30°. Thus, latitude is 30°. In the formula, $L = Z + D$, D is negative because D is contrary name to latitude. The formula in this case is expressed $L = Z + (-D)$. Thus $L = 45° - 15°$. $L = 30°$ N. L is north because L is opposite in sign to the declination and greater, and declination is south.

Let's look at some problems and see how our formula works out:

**Situation No. 1.** DR Lat. 39° 48.2′ N, DR Long. 60° 54.9′ W; ZT 1159; $H_s$ 73° 20.7′; sun's lower limb; 26 June 1983.

**Step 1**

Record your sight.

**Step 2**

Correct your $H_s$ for $H_0$.

| | |
|---|---|
| Main corr. | +15.9′ |
| Dip | −3.4′ |
| | 12.5′ |

| | |
|---|---|
| $H_s$ | 73° 20.7′ |
| | +12.5′ |
| $H_o$ | 73° 33.2′ |

**Step 3**

Find your zenith distance.

| | |
|---|---|
| $H_o$ | 73° 33.2′ |
| | 89° 60.0′ |
| Z | 16° 26.8′ |

**Step 4**

The declination of the sun at 1159 ZT (1559 GMT) was N 23° 21.9′ and has the same name as the latitude, so the sign is plus.

| | |
|---|---|
| Z | 16° 26.8′ |
| Dec. N | 23° 21.9′ |
| Lat. | 39° 48.7′ N |

**Situation No. 2.** DR Lat. 11° 41.2′ N, DR Long. 60° 54.9′ W; ZT 1159; $H_s$ 78° 08.3′; 26 June 1983.

**Step 1**

Record your sight.

**Step 2**

Correct your $H_s$ for $H_o$.

| | |
|---|---|
| Main corr. | +16.0′ |
| Dip | − 3.4′ |
| | +12.6′ |

| | |
|---|---|
| $H_s$ | 78° 08.3′ |
| | +12.6′ |
| $H_o$ | 78° 20.9′ |

**Step 3**

Find your zenith distance.

| | |
|---|---|
| $H_o$ | 78° 20.9′ |
| | 89° 60.0′ |
| Z | 11° 39.1′ |

**Step 4**

Your declination has the same name as the latitude, but is greater, so the sign of the zenith distance is minus.

| | |
|---|---|
| Z | 11° 39.1′ |
| Dec. N | 23° 21.9′ |
| Lat. | 11° 42.8′ N |

The difference from your DR latitude is worked in purposely for the sake of realism.

Most experienced navigators use what is called a *latitude constant*. All they have to do is apply the difference between their observed altitude, $H_o$, and a precomputed altitude they worked up when they computed the time of LAN to get their DR latitude at LAN. Even the old rule of *Computed Greater Away* will work. Just be sure which way the sun bears, 000° or 180°, so you can visualize the celestial relationship involved.

Suppose at 1030 ZT on the morning of 21 October you wanted to precompute your noon sight. Here is how you would do it.

**Step 1**

Plot your noon DR, which is Lat. 24° 33.1′ N, Long. 92° 43.4′ W.

**Step 2**

Add your DR latitude and your declination.

| | |
|---|---|
| DR Lat. | 24° 33.1′ |
| Dec. S | 10° 40.8′ |
| | 35° 13.9′ |

**Step 3**

Subtract the sum from 90°. The result is your $H_c$.

| | |
|---|---|
| | 89° 60.0′ |
| | 35° 13.9′ |
| $H_c$ | 54° 46.1′ |

**Step 4**

Observe the sun at LAN. Record your $H_s$ and correct it for $H_o$.

| | | |
|---|---|---|
| | $H_s$ | 54° 46.2′ |
| +15.5′ | | |
| − 3.4′ = | | +12.1′ |
| | $H_o$ | 54° 58.3′ |
| Difference 12.2′t | | |

Here is where common sense and the ability to visualize things are necessary. Shut your eyes and think. If you are in north latitude and the sun's declination is south, the sun must bear 180° when it is on your meridian. So, if your observed altitude shows you to be *closer to* the sun's GP by 12.2 miles than what your computed altitude showed, the difference must be *subtracted from* your DR latitude. To go 12.2 miles toward a body that is south of you when you are in north latitude decreases your latitude. Hence:

| | |
|---|---|
| DR Lat. | 24° 24.0′ N |
| | −12.2′ |
| Lat. | 24° 11.8′ N |

Not only does this precomputation give the navigator a rapid solution to his problem at noon, it also gives him a close approximation to his $H_s$. This permits him to preset his sextant to this altitude. On overcast days when the sun is just peeping through the clouds, the time saved may permit you to get an observation when you otherwise would not have had the time to do so.

Sometimes the sun will refuse to show its face precisely at LAN. Then, a few minutes later, it will beam forth thinking it has foiled you completely in taking your noon sight. This need not be so. Tables 29 and 30 of Bowditch will provide you with a solution to this problem. These tables work as well for any other celestial body. It is not within the scope of this book to cover this procedure. If, however, you are confronted with the problem, careful reading of the explanations to these tables will prepare you to use them. The technique will also work for sights taken a few moments before noon. We call these *ex-meridian altitude sights.*

Another name for the noon sight is *meridian altitude* of the sun. The problem is often called "latitude by meridian altitude of the sun." The same technique will work for any body in the heavens. You first compute the time at which the body will be on your meridian. You do this in exactly the same way you compute meridian passage (LAN) of the sun by using the sun's GHA. You must use the GHA of the body involved, of course. When you derive this information, observe the body for $H_s$. Correct your altitude for $H_o$. Find your zenith distance. Apply your declination using exactly the same rules you used for the sun. The result will be latitude by meridian altitude of the body.

## LATITUDE BY OBSERVATION AT LOWER TRANSIT

If at twilight a body has a GHA approximately equal to your longitude, it may be on the meridian at *star time.* The Sirius sight in Chapter 9 is an example of this. When I figure the time of twilight and precompute the altitudes and azimuths of the stars I plan to shoot, I always check to see if any planet or star may be in such a position. I can tell by inspection of the *Nautical Almanac* if a planet will be on the meridian at this time. The SHA and time of meridian passage of the four navigational planets—Venus, Mars, Jupiter, and Saturn—are given in the lower left-hand corner of the appropriate left-hand page in the daily pages of the *Nautical Almanac.* A bit more effort is necessary to determine when the stars cross your meridian, but some stars offer two possibilities of a meridian altitude shot even in lower latitudes. How so?

When a body's declination has the same name as your latitude, and the declination is greater than your colatitude, the body will never set. That is, it will not dip below the visible horizon at any time during the 24-hour day. An immediate example of this is Polaris. Polaris has a declination north of approximately 89°. Because of this, Polaris will always remain above the horizon when viewed from any point on the earth's surface above 1° north latitude. At the other extreme is the summer sun (and planets) in polar regions—the *midnight sun.* At all latitudes between, you find numerous bodies that will not set and will always be in position to be observed. Although there is no convenient southern pole star, this situation applies in both north and south latitudes. As you move toward the pole from the equator the number of bodies that are always in position for observation increases, until, at the pole, they all ride the celestial merry-go-round and remain above the horizon the entire 24-hour day.

What is colatitude? you ask. It is not another unintelligible term I am throwing at you. Colatitude is simply the difference that results when you subtract your latitude from 90°. While I am at it, I should also define codeclination, although you likely surmised what it is as

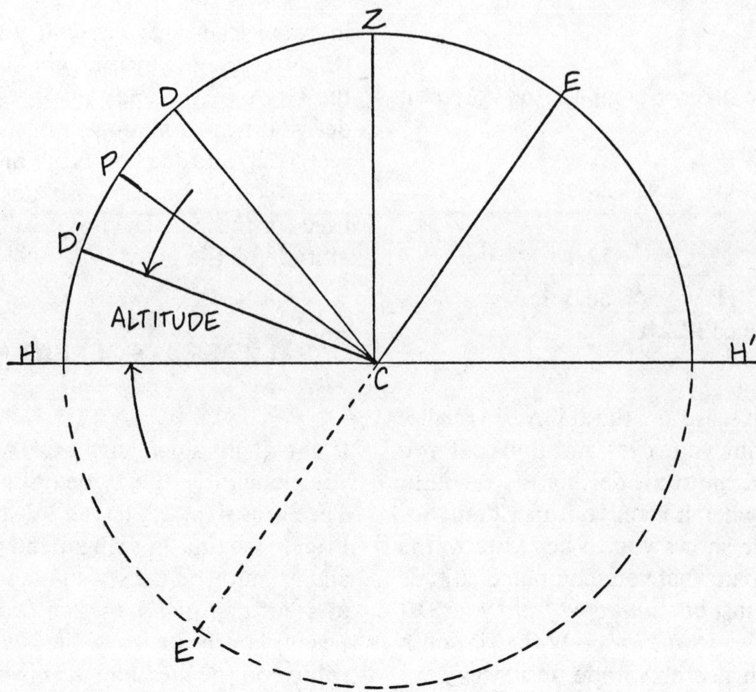


**Figure 14-A.** *Latitude by meridian altitude observation at lower transit. Declination same name as and greater than latitude.*

you were reading this sentence: It is the difference between the declination of a body and 90°. What does all this add up to? The fact that some bodies—always stars in nonpolar regions—will make two meridian passages in a day. The second meridian transit is called meridian passage at *lower transit*. When a body crosses the meridian that has the same GHA as your zenith, the body is in *upper transit*. In Cases I, II, and III, the body—the sun—was observed in upper transit. When a body is at lower transit, its GHA will equal the GHA of your zenith plus 180°. In north latitude the azimuth at lower transit will always be 000°/360°; in south latitude the azimuth at lower transit will always be 180°. If you could focus a camera on the elevated pole and make a time exposure over a 24-hour period of all the stars near the pole, the stars that pass above the horizon at lower transit would make a circle around the pole on the film. For this reason, astronomers call these stars *circumpolar stars.* Figures 14-A and 14-B diagram the geometry involved in reducing a sight of a body at lower transit to find your latitude.

Assume your latitude is 35° north. The declination of the star you observed is north 75°. The angle DCE in Figure 14-A is the declination as diagramed at upper transit. I have drawn the complete cross section of the celestial sphere at your meridian, showing that part of the celestial meridian below the horizon as a dashed semicircle. The extension of the celestial equator below the horizon is CE′. Then the angle D′CE′ is the declination of the star at lower transit. From Figure 1 (Chapter 2) and Figures 11,12, and 13 in this chapter, you know that the altitude of the elevated pole is equal to your latitude, in this case 35°. This is the angle PCH. The angle D′CH is the altitude of the body when observed at lower transit. But angle PCH is the sum of the angles D′CH and D′CP. The codeclination, angle D′CP, equals 90° minus 75°, or 15°. If your latitude is 35°, the angle D′CH, the altitude of the body, is 20°. So if you were to observe a star at D′ at an altitude of 20°, and add 15°, the codeclination of the star, the total would be 35°, the latitude. Thus we can say that for bodies at lower transit, the observer's latitude equals the sum of the codeclination and the observed altitude. The formula looks like this: $L = (90° - \text{Dec.}) + H_o$. Let's work an example.

At zone time 1823 on 6 February 1983, from a DR latitude of 33° 17.6′ N, longitude 37° 41.3′ W, you observed the star Kochab at lower transit. The sextant altitude was 17° 29.3′. What was your latitude? The declination of Kochab from the *Nautical Almanac* was N 74° 13.5′.

**Step 1**

Record your sight.

**Step 2**

Correct your $H_s$ to $H_o$ as you have always done.

| | |
|---|---|
| $H_o$ | 17° 22.9′ |

**Step 3**

Find the codeclination.

| | |
|---|---|
| | 89° 60.0′ |
| Dec. | 74° 13.5′ |
| Codec. | 15° 46.5′ |

**Step 4**

Add the codeclination and the observed altitude.

| | |
|---|---|
| $H_o$ | 17° 22.9′ |
| Codec. | 15° 46.5′ |
| Lat. | 33° 09.4′ |

A lot of otherwise clever navigators think lower transit observations are techniques used only in very high latitudes or in polar navigation. I used the example of a lower transit sight of Kochab at a DR latitude of 33° 17.6′ N to show you that this is by no means so. In addition, reducing a lower transit sight for latitude is easier than the three cases of upper transit sights covered earlier in this chapter. Admittedly, more and more stars will be available for observation at lower transit as you get nearer the pole, but you could have observed Kochab from the Gulf of Mexico, had you needed to! The altitude would have been 13° or less, but in a pinch, even this is better than nothing. Had you been at 45° N, which is right on the caravan route to Merry England, Kochab's $H_o$ would have been near 28°, which is a high enough altitude for any sight. And latitude 45° is only halfway to the pole. My point is, lower transit observations in the right places are as useful as any other observation in everyday navigation at sea. This is just another technique the navigator should have in his bag of tricks. Someday you may need them all.

How did I know Kochab would be on the meridian at lower transit at star time on 6 February 1983? Although I told you earlier, maybe I should show you the arithmetic involved. The zone time of star time was 1825. Twilight usually lasts 20 to 30 minutes at 33° N latitude in February. The zone description at longitude 37° 41.3′ W is +2. Here is what I did:

| | |
|---|---|
| ZT | $18^h$-$25^m$-$00^s$ |
| ZD | +2 |
| GMT | $20^h$-$25^m$-$00^s$ |

With the GMT I found the GHA of Aries.

| | |
|---|---|
| GHA | 82° 38.0′ |

Then I added 180° to the DR longitude, and subtracted the GHA of Aries from the total. (I added 180° to the DR longitude because the lower branch of the DR meridian is 180° from the upper branch. I am looking for a body, in this case a star, that may be on the lower branch of my meridian at star time.)

| | |
|---|---|
| DR Long. | 37° 41.3′ |
| | 180° 00.0′ |
| Total | 217° 41.3′ |
| GHA Aries | 82° 38.0′ |
| SHA required | 135° |

With the required SHA I ran down the star column in the daily pages for 6 February and found any star that had an SHA near 135°. Bear in mind, the GHA of Aries will change a degree every four minutes. With 20 to 30 minutes of star time, you have better than five degrees of leeway—maybe nearer 10 degrees—in SHA. Kochab's SHA on 6 February was 137° 18.8′. Close enough. The next question: Does the declination have the same name as my latitude, and is it greater than my colatitude? It should be sufficiently greater to cause the body to be at least at a usable altitude, if not an optimum altitude. The figures with Kochab looked like this:

| | | |
|---|---|---|
| GHA Aries | 82° 38.0′ | at GMT $20^h$-$25^m$ |
| SHA Kochab | 137° 18.8′ | |
| | 219° 56.8′ | |
| Less | 180° 00.0′ | |
| | 39° 56.8′ | |
| DR Long. | 37° 41.3′ | |
| Difference | 2° 15.5′ | |

Convert the difference of longitude to time, add or sub-

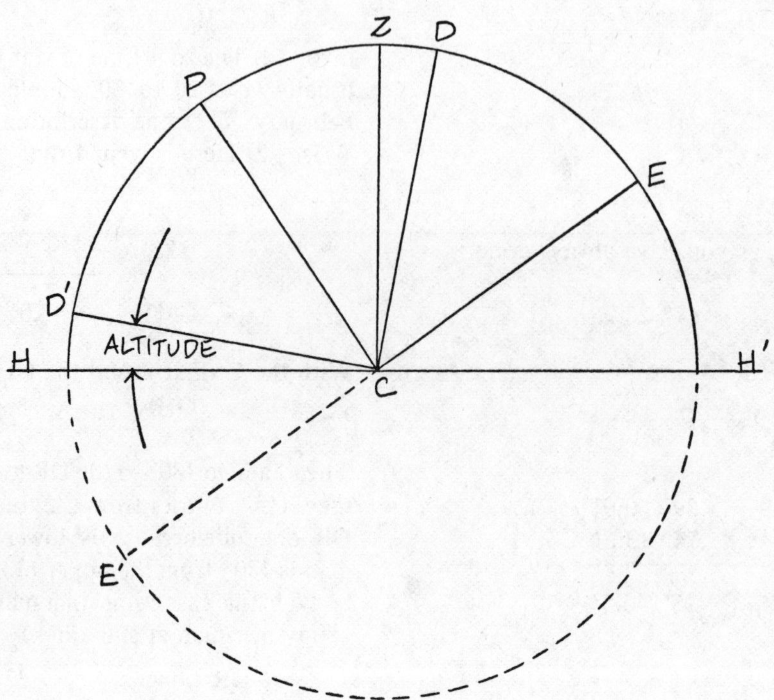


**Figure 14-B.** *Latitude by meridian altitude observation at lower transit. Latitude same name as and greater than declination.*

tract it from the GMT at which you found the GHA of Aries in your precomputations, and apply the zone description:

| | | |
|---|---|---|
| Difference | $00^h$-$09^m$ | (early) |
| GMT | $20^h$-$25^m$ | |
| Lower Transit | $20^h$-$16^m$ | |
| ZD | $+2^h$-$00^m$ | (sign reversed) |
| ZT transit | $18^h$-$16^m$ | |

You subtracted the difference of longitude (converted to difference in time) from the 1825 ZT because Kochab was on the lower branch of the meridian at 39° 56.6′ W at 1825. The +2 zone description was subtracted because you were converting Greenwich time to zone time.

When the latitude exceeds 45°, celestial bodies having declinations that are the same name as but less than the latitude may be above the horizon at lower transit. For this to happen, the body's codeclination must also be less than the latitude, the difference, as in other lower-transit observations, being the altitude. See Figure 14-B. P is the elevated pole, and, as described before, the angle PCH has the same value as the latitude. C is the center of the celestial sphere (and of the earth), and H is the horizon. D is the declination of the body at upper transit. The body's azimuth at upper transit will be 180° if P is north, 360° if P is south. Z is the zenith of your position. D′ is the position of the body at lower transit. The angles PCD and PCD′ have the same value and are the body's codeclination. You can get the declination from the almanacs and subtract this from 90° to get the codeclination, which will tell you if the body should be sufficiently above the horizon at lower transit to be useful for observation. Angle PCD′ is the angle of interest here, because it is apparent from inspection of Figure 14-B that the sum of angle PCD′ and angle D′CH, which is the altitude, equals angle PCH, which is the latitude.

Today some armchair navigators do not give much attention to meridian altitude observations of bodies other than the sun. They contend such observations are not necessary with lines of position derived from sight reduction tables or formulas. I emphatically disagree with them! Meridian altitude observations are simple and accurate. Should you be in the position of *running under the sun*, as vessels going from north to south or south to north latitude find themselves doing when crossing the equator, the sun may be too high for a good observation at local apparent noon. During the morning it will be so close to the prime vertical (the vertical circle running due east and due west across the celestial sphere and passing through your zenith) that the easterly azimuth will change but a few degrees between sunrise and noon. Minutes after meridian passage, at near 90° in altitude, it will flop over to a westerly azimuth, and not change much until sunset. You cannot get a running fix from sun sights in a situation like this. A meridian passage observation of another body—the moon during the daytime or the stars or planets at twilight—can be a godsend that enables you to determine your latitude with confidence. And because meridian passage observations do not require chronometer time, you can find your way by parallel sailing if your chronometers are out of order and time ticks are few and far between. That is how our ancestors did it.

# 12 LATITUDE BY POLARIS

Remember Chapter 2 on nautical astronomy? We went back to our plane geometry and by diagram showed that the elevation of the observer's pole equaled the observer's latitude. If the North Star, Polaris, is right on the pole, we can get our latitude simply by measuring the altitude of Polaris. Unfortunately, Polaris is not right on the North Pole, but may be a degree or so away. However, this is close enough to make reducing Polaris sights quite simple. Here is how you do it:

At about 0500 ZT, 21 October 1983, you took a sight on the star Polaris while taking your other stars as described in Chapter 9. What is your latitude by Polaris?

**Step 1**

Record your sight.

| | |
|---|---|
| $H_s$ | 25° 14.7′ |
| C | $11^h$-$07^m$-$15^s$ |
| W | $00^m$-$29^s$ |

**Step 2**

Correct your time for GMT.

| | |
|---|---|
| C | $11^h$-$07^m$-$15^s$ |
| W | $00^m$-$29^s$ |
| C – W | $11^h$-$06^m$-$46^s$ |
| CE | $03^m$-$02^s$ fast |
| GMT | $11^h$-$03^m$-$44^s$ |

**Step 3**

Open your *Nautical Almanac* to the daily pages for 21 October 1983. In the far left column of the left-hand page, extract the GHA of Aries for $11^h$-$00^m$-$00^s$. It is 194° 19.5′.

**Step 4**

To correct for $03^m$-$44^s$, open your *Nautical Almanac* to the yellow pages, "Increments and Corrections." Turn to page iii, which contains the corrections for $03^m$-$00^s$ to $03^m$-$60^s$. In the $3^m$ box extract the correction for $03^m$-$44^s$ in the Aries column.

| | |
|---|---|
| | 194° 19.5′ |
| | 0° 56.2′ |
| GHA Aries | 195° 15.7′ |

**Step 5**

Determine the LHA of Aries. Extreme accuracy is not too important. The adjustments are tabulated only to the nearest degree. Use your DR longitude.

| | | |
|---|---|---|
| | 195° 15.7′ | |
| DR Long. | 92° 38.0′ | W |
| LHA Aries | 102° 37.7′ | W |

You should use an LHA of 103° for adjustments to the altitude of Polaris.

**Step 6**

At this point, it is a good idea to correct your sextant altitude. Do this as you would for any other star.

| | | |
|---|---|---|
| | $H_s$ | 25° 14.7′ |
| Star −2.0′ | | |
| Dip −3.4′ | | |
| −5.4′ | | −05.4′ |
| | $H_o$ | 25° 09.3′ |

**Step 7**

Open your *Nautical Almanac* to the tables entitled "Polaris (Pole Star) Tables, 1983." These tables begin on page 274 of the 1983 *Nautical Almanac*. VERY IMPORTANT: Note at the beginning of the footnote at the bottom of page 274 that 1° must be subtracted from the corrected sextant altitude, $H_o$.

| |
|---|
| 25° 09.3′ |
| − 1° 00.0′ |
| 24° 09.3′ |

The other adjustments to $H_o$ are always added. There are three. The first adjusts for the LHA of Aries, the second for your DR latitude, and the third for the month of the year. The LHA of Aries is an argument to obtain all of these adjustments.

**Step 8**

Adjust your $H_o$ for the LHA of Aries. This adjustment is in the first table at the top of the Polaris pages. Across the top of each page are 12 columns headed by the LHA of Aries in 10-degree increments. The first column of the first page covers 0° to 9° 60′, the second column, 10° to 19° 60′, and on through 350° to 359° 60′. Since your LHA of 103° is between 100° and 109°, you should select this column. Go down this column until you reach the line that corresponds to "3°" in the vertical column on the left under "LHA Aries." Extract +0° 42.0′.

**Step 9**

Continue down this 100°-109° LHA column to the next table. Keep going until you come to the line that corresponds to your DR latitude as tabulated in the far left column, headed "Lat." Your DR latitude lies halfway between 20° and 30° N. You could interpolate visually, but just call the correction +00.4′.

**Step 10**

Keep on going in the same 100°-109° LHA column to the third table. When you get to the line that corresponds to your month, October, extract +00.3′.

**Step 11**

Add the sum of these corrections to $H_o$ (from which 1.0° has already been subtracted) to obtain your latitude:

| | |
|---|---|
| $H_o$ | 24° 09.3′ |
| | 42.0′ |
| | 00.4′ |
| | 00.3′ |
| Lat. | 24° 52.0′ N |

To summarize, your step-by-step procedure should look like this in your navigator's notebook:

| | | |
|---|---|---|
| Step 1: | $H_s$ | 25° 14.7′ |
| Record your sight. | C | $11^h$-$07^m$-$15^s$ |
| | W | $00^m$-$29^s$ |

| | | | |
|---|---|---|---|
| Step 2: Correct your time. | C−W | $11^h$-$06^m$-$46^s$ | |
| | CE | $03^m$-$02^s$ | fast |
| | GMT | $11^h$-$03^m$-$44^s$ | |
| Step 3: Extract GHA of Aries. | | 194° 19.5′ | |
| Step 4: Correct GHA. | Corr. | 0° 56.2′ | |
| | GHA Aries | 195° 15.7′ | |
| Step 5: Apply your DR Long. to get LHA. | | 92° 38.0′ | |
| | LHA Aries | 102° 37.7′ | |
| Step 6: Correct your sextant altitude. | Main | −2.0′ | |
| | Dip | 3.4′ | |
| | | −5.4′ | |
| | $H_s$ | 25° 14.7′ | |
| | $H_o$ | 25° 09.3′ | |

| | |
|---|---|
| Step 7: From Polaris tables of the *Nautical Almanac.* | −1° 00.0′ |
| | 24° 09.3′ |
| Step 8: First table. | + 0° 42.0′ |
| Step 9: Second table. | + 0° 00.4′ |
| Step 10: Third table. | + 0° 00.3′ |
| Step 11: Add for latitude. | 24° 52.0′ N |

The latitude has to be north. Polaris is not visible south of the equator. For that matter, you must be well north of the equator to see Polaris at all, since the altitude of Polaris at the equator approximates zero. Unfortunately, you do not have an equivalent star at the South Pole to make things as easy for you.

# 13 LHA IN EAST LONGITUDE OR NEAR THE PRIME MERIDIAN

Until now we have been dealing with problems of sight reduction in which your DR position was west of Greenwich, so far into west longitude that the GHA of the body used, and thus the longitude of its GP, was well into the Western Hemisphere also. We did get a glimpse at a problem of how to find LHA when you are near the prime meridian. We saw this in the chapter on reducing a moon sight. It is important that we take a closer look at this procedure and the method of finding LHA when you are in east longitude. I have purposely waited until now to present this to you.

You will probably do all of your navigating in or near your home waters. Nevertheless, anyone who is a true sailor has deep in his heart a desire to sail around the world, or at least make a long, exotic cruise. If you are lucky enough to fulfill that desire, you will want to know how to work a sight when your DR puts you in east longitude or near the Greenwich meridian. It is really not at all hard. You should be so familiar by now with manipulating the degrees and minutes of GHA and longitude that you will wonder why I even brought it up. However, it will not hurt to go through a few sample problems.

On 25 October 1983 at 1445 ZT, you find your vessel at DR latitude 14° 18.0′ N, longitude 62° 41.6′ E. You want to know the Greenwich hour angle of the sun. Had you taken a sight, you would have recorded this sight and corrected your time for the exact Greenwich mean time of your sight. Sixty degrees east longitude would place your vessel in time zone −4. Therefore the GMT of your sight would be $10^h$-$45^m$-$00^s$. Open your *Nautical Almanac* to the daily page for 25 October. Extract the GHA of the sun for $10^h$ GMT, and add the correction for $45^m$.

| | |
|---|---|
| GHA | 333° 57.8′ |
| Corr. | 11° 15.0′ |
| GHA | 345° 12.8′ |

This is precisely how you have found GHA in all the other chapters.

Greenwich hour angle is tabulated through 360° from the prime or Greenwich meridian west. Up to now you have been working in west longitude. If the GHA of the body was east of your position, you simply subtracted the GHA from the longitude of your position. The result was the meridian angle east or t east. If the GHA of the

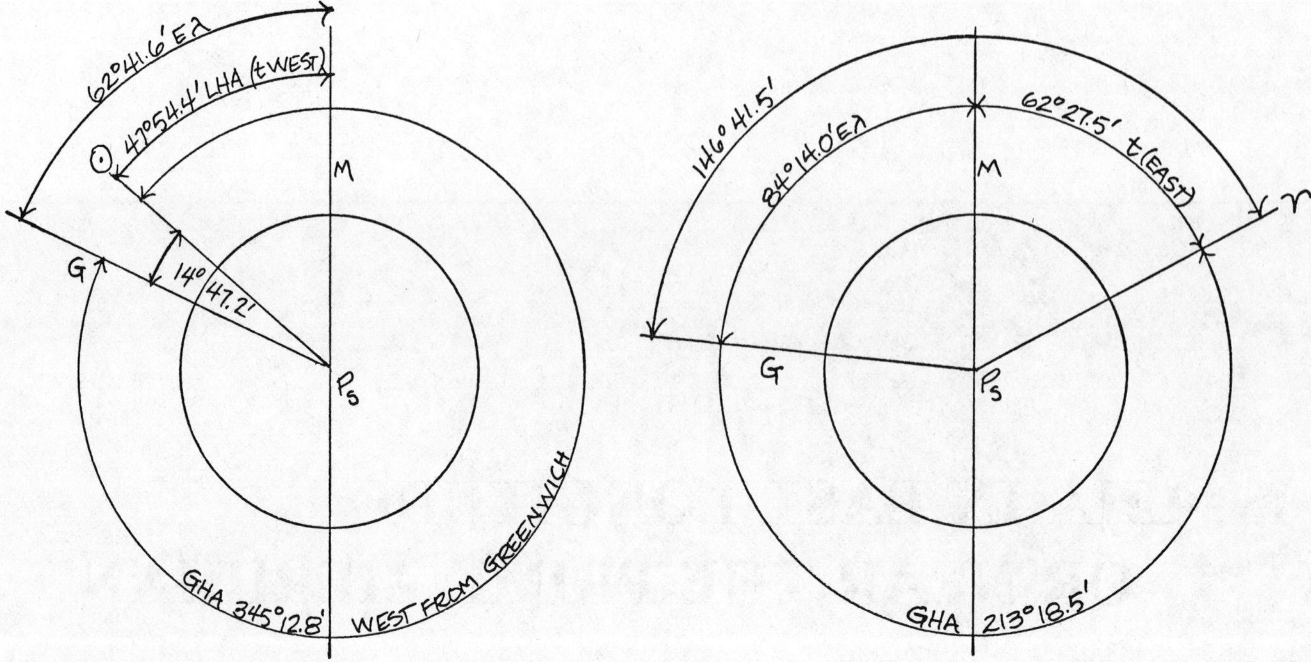


**Figure 15.** *Visualizing LHA when the difference between the observed body's GHA and 360° is less than the observer's longitude east. The Greek letter lambda, λ, is an abbreviation for longitude. The small circle with a dot in the center is the astronomer's symbol for the sun. M is your meridian, and G is the Greenwich meridian. $P_s$ is the South Pole.*

**Figure 16.** *Visualizing LHA when the difference between the observed body's GHA and 360° is greater than the observer's longitude east. The symbol for Aries, ♈, is the astronomer's symbol for the Ram's Head. More precisely, it is the symbol for the first point of Aries, which navigators refer to as Aries.*

body was west of your position, you needed only to subtract your longitude from the GHA of the body to get the LHA of the body, because LHA is always west. (Remember, LHA and t west are the same.)

In east longitude, one more step is required. With a DR longitude of 62° 41.6′ E and a GHA of 345° 12.8′, you should add your GHA to your longitude and subtract 360°.

| | | |
|---|---|---|
| GHA | 345° 12.8′ | |
| Long. | 62° 41.6′ | E |
| Sum | 407° 54.4′ | |
| | 360° 00.0′ | |
| LHA | 47° 54.4′ | W |

In Figure 15 the center of the circle is the South Pole. The meridian angle, t, is west, and is thus the LHA, because the difference between the sun's GHA and 360° is less than the longitude east. The sun is only 14° 47.2′ east of Greenwich. You are 62° 41.6′ east of Greenwich. So, the sun must be west of you. This means the meridian angle, t, is west. When t is west, t equals LHA.

When the difference between the GHA of a body and 360° is greater than your longitude east, the meridian angle of the body will be east. This can be illustrated by an example of finding the LHA of Aries. Suppose on 25 October 1983 at 1800 ZT you want to find the LHA of Aries. You are in 84° 14.0′ east longitude, which puts you in time zone −6. By subtracting $6^h$-$00^m$ from 1800, your time at Greenwich is found to be 1200. The GHA of Aries at 1200 GMT this date is:

| | | |
|---|---|---|
| GHA | 213° 18.5′ | |
| from | 360° 00.0′ | |
| | 146° 41.5′ | E |
| Long. | 84° 14.0′ | E |
| t | 62° 27.5′ | E |

See Figure 16 for a graphic solution to this problem.

Since I have confronted you with the problem of LHA

in east longitude, I might as well spring another surprise on you. In tables such as Pub. No. 249 and Pub. No. 229, local hour angles are tabulated through 360° west of the observer's meridian. You do not need to use easterly meridian angles in these tables for an entering argument, although we have been doing so because it is simpler when reducing antemeridian sights (sights of bodies observed before their meridian passage) in west longitude. Here we are talking about east longitude. In east longitude, finding LHA is much easier than finding meridian angle. When you are in east longitude, you need only add your GHA to your longitude. If the result is greater than 360°, subtract 360° from it. For illustration:

The first problem (see Figure 15), Long. 62° 41.6′ E, GHA of the sun 345° 12.8′.

| | | |
|---|---|---|
| GHA | 345° 12.8′ | |
| Long. | 62° 41.6′ | E |
| Sum | 407° 54.4′ | |
| | − 360° 00.0′ | |
| LHA | 47° 54.4′ | |

The second problem (see Figure 16), Long. 84° 14.0′ E, GHA of Aries 213° 18.5′.

| | | |
|---|---|---|
| GHA | 213° 18.5′ | |
| Long. | 84° 14.0′ | E |
| LHA | 297° 32.5′ | |

This is all there is to finding LHA in east longitude. It is really so simple I hesitated to bring it up. Yet, this is an area that can confuse navigators. You must understand this to be a well-rounded navigator. I hope you will be lucky enough to use the information someday.

The problem of finding LHA when your DR is close enough to the Greenwich meridian for you to be in a different hemisphere than the body's GP may call for a little more ability to visualize the time relationship of celestial bodies to longitude and GHA. The moon sight in Chapter 8 is an example of this. Some more examples follow:

You are in 21° 14.3′ W longitude. The GHA of the sun is 342° 19.8′. What is the LHA of the sun?

| | | |
|---|---|---|
| GHA | 342° 19.8′ | |
| Long. | 21° 14.3′ | W |
| LHA | 321° 05.5′ | |

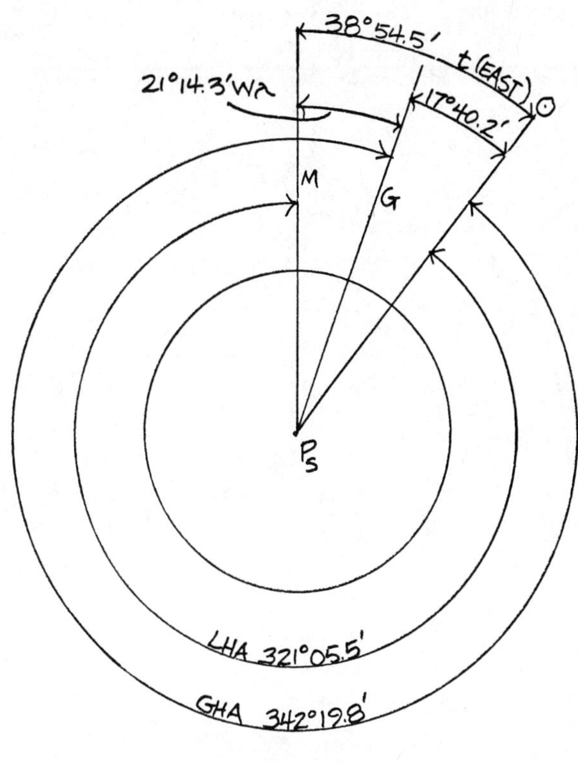


**Figure 17.**

The LHA is west. You are west of Greenwich. The GHA is tabulated west. You are nearer the GP longitude of the body, in this case the sun, by an amount equal to your longitude west. Had you wanted to find t east for entering Pub. No. 249, you would subtract the LHA from 360°, but why go through this extra step? Especially when t east is not needed to enter Pub. No. 249.* See Figure 17.

| | | |
|---|---|---|
| LHA | 321° 05.5′ | |
| from | 360° 00.0′ | |
| t | 38° 54.5′ | E |

Had your DR longitude been 21° 14.3′ E, to find the LHA of the sun you would simply add your GHA and longitude as above. If the sum exceeded 360°, you would subtract 360°. For example (see Figure 18):

*You will see in Chapter 17 that finding the meridian angle, t, is always necessary in east or west longitude when working a sight using Ageton's Method.

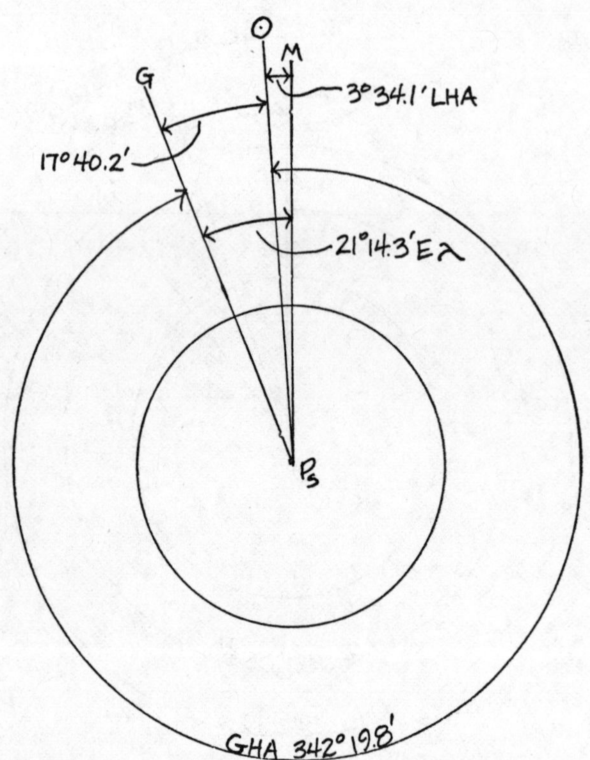


Figure 18.

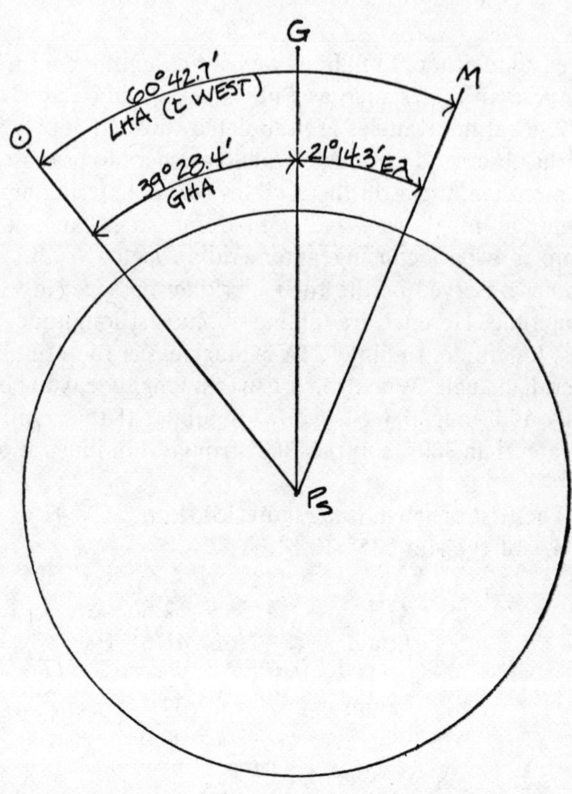


Figure 19.

| | |
|---|---|
| GHA | 342° 19.8′ |
| Long. | 21° 14.3′ E |
| | 363° 34.1′ |
| | 360° 00.0′ |
| LHA | 3° 34.1′ (t W) |

Let's take an example in which the GHA of the sun is 39° 28.4′. Your DR longitude is 21° 14.3′ E. What is your LHA? (See Figure 19.)

| | |
|---|---|
| GHA | 39° 28.4′ |
| Long. | 21° 14.3′ E |
| LHA | 60° 42.7′ (t W) |

Under the old system of employing local hour angles east, the sequence used by some navigators to get the LHA when near the prime meridian could cause confusion. The preceding examples will eliminate this problem.

Let us reduce several sights assuming your vessel's DR position to be in east longitude. At sea on the morning of 21 October 1983, in DR latitude 17° 16.0′ N, longitude 86° 29.0′ E, you observe the star Regulus as follows:

| | | |
|---|---|---|
| C | $23^h$-$08^m$-$15^s$ | |
| W | $00^m$-$27^s$ | |
| CE | $00^m$-$04^s$ | fast |
| $H_s$ | 41° 14.6′ | |
| Height of eye | 18 feet | |

Reduce this sight for a line of position. See the accompanying sight reduction form.

This is a good time to compare several aspects of working sights when the DR is in west longitude to working sights when the DR is in east longitude. The first and most obvious difference is that in east longitude you choose an assumed longitude that will give an LHA in even degrees when *added* to the GHA. In west longitude, which is where we were in all the examples before this chapter, you choose an assumed longitude that will give an even degree of LHA as its *difference* from the GHA. In an antemeridian sight of a body in west longitude, the GHA will always be less than the longitude. For this

| | |
|---|---|
| STAR | REGULUS |
| Date | 21/10/83 ZT 0505 |
| DR Lat. | 17°-16.0' N Long. 86°-29.0' E |
| C | 23h-08m-15s |
| W | 0m-27s |
| C-W | 23h-07m-48s |
| CE | 00m-04s f |
| **** GMT | 23h-07m-44s |
| (Note: Greenwich date is 20/10/83) | |
| GHA Aries | 13°-49.9' |
| Corr. (min. & sec.) | 1°-56.3' |
| GHA Aries | 15°-46.2' |
| SHA star | 208°-07.8' |
| Sum | 223°-54.0' |
| Less 360° (if req.) | -0°-0' |
| GHA star | 223°-54.0' |
| **Assumed Long. | 86°-06.0' E |
| *LHA | 310°-00.0' |
| ***Assumed Lat. | 17°-00.0' N |
| *Declination | N 12°-03.0' |
| $H_t$ | 41°-27.0' |
| Correction +13 | + 1.0 |
| $H_c$ | 41°-28.0' |
| $H_o$ | 41°-09.4' |
| **Altitude Intercept | 18.6 a |
| Z | N 089° E |
| From 360° (if req.) | |
| **Zn | 089° |
| $H_s$ | 41°-14.6' |
| Main Corr. -1.1 | |
| Addn'l | |
| Dip -4.1 | |
| Total Corr. | -05.2' |
| $H_o$ | 41°-09.4' |

*Required to enter Pub. No. 249
**Required to plot line of position
***Required to enter Pub. No. 249 and to plot line of position
**** Notice the change in date. It is not yet the 21st in Greenwich. So you must look up GHA of Aries for the 20th at 2300 hours.

| | |
|---|---|
| SUN (☉), Lower/~~Upper~~ Limb | |
| Date | 20/10/83 ZT 0936 |
| DR Lat. | 25°-43.0' N Long. 157°-42.7' E |
| C | 23h-38m-25s |
| W | 1m-31s |
| C-W | 23h-36m-54s |
| CE | 0m-04s f |
| GMT | 23h-36m-50s |
| (Note: Greenwich date is 19/10/83) | |
| GHA | 168°-45.3' |
| Corr. (min. & sec.) | 9°-12.5' |
| GHA | 177°-57.8' |
| **Assumed Long. | 158°-02.2' E |
| *LHA | 336°-00.0' |
| ***Assumed Lat. | 26°-00.0' N |
| Declination | S 10°-02.3' |
| d difference +0.9 | + 0.5' |
| *Declination | S 10°-02.8' |
| $H_t$ | 47°-06.0' |
| Correction d -51 | -02.0' |
| $H_c$ | 47°-04.0' |
| $H_o$ | 47°-03.5' |
| **Altitude Intercept | 0.5' a |
| Z | N 144° E |
| From 360° (if req.) | |
| **Zn | 144° |
| $H_s$ | 46°-51.1' |
| Main Corr. +15.3 | |
| Addn'l | |
| Dip -2.9 | |
| Total Corr. | +12.4' |
| $H_o$ | 47°-03.5' |

*Required to enter Pub. No. 249
**Required to plot line of position
***Required to enter Pub. No. 249 and to plot line of position

SUN (⊙), ~~Lower~~/Upper Limb
Date 21|10|83 ZT 1503
DR Lat. 27°-19.8' N Long. 152°-28.0' E

| | |
|---|---|
| C | $05^h-06^m-05^s$ |
| W | $1^m-38^s$ |
| C-W | $05^h-04^m-27^s$ |
| CE | $0^m-17^s$ f |
| GMT | $05^h-04^m-10^s$ |
| GHA | 258°-48.6' |
| Corr. (min. & sec.) | 1°-02.5' |
| GHA | 259°-51.1' |
| **Assumed Long. | 152°-08.9' E |
| *LHA | 412°-00.0' |
| ****Less 360° | 360°-00.0' |
| LHA (t-W) | 52°-00.0' |
| ***Assumed Lat. | 27°-00.0' N |
| Declination | S 10°-29.2' |
| d difference +0.9 | 0.1 |
| *Declination | S 10°-29.3' |
| $H_t$ | 27°-29.0' |
| Correction d −37 | −18.0' |
| $H_c$ | 27°-11.0' |
| $H_o$ | 26°-31.6' |
| **Altitude Intercept | 39.4 a |
| Z | N 119° W |
| From 360° (if req.) | 360° |
| **Zn | 241° |
| $H_s$ | 26°-53.1' |
| Main Corr. −17.9 | |
| Addn'l | |
| Dip − 3.6 | |
| Total Corr. | −21.5' |
| $H_o$ | 26°-31.6' |

*Required to enter Pub. No. 249
**Required to plot line of position
***Required to enter Pub. No. 249 and to plot line of position
**** Since the sum of the body's (the sun's) GHA and the assumed longitude east is greater than 360°, you must subtract 360° from the sum. The remainder is the LHA or t west.

STAR RASALHAGUE
Date 21|10|83 ZT 1755
DR Lat. 26°-51' N Long. 168°-19.6' E

| | |
|---|---|
| C | $06^h-57^m-20^s$ |
| W | $01^m-12^s$ |
| C-W | $06^h-56^m-08^s$ |
| CE | $00^m-08^s$ S |
| GMT | $06^h-56^m-16^s$ |
| GHA Aries | 119°-07.2' |
| Corr. (min. & sec.) | 14°-06.3' |
| GHA Aries | 133°-13.5' |
| SHA star | 96°-27.7' |
| Sum | 229°-41.2' |
| Less 360° (if req.) | 0°-00.0' |
| GHA star | 229°-41.2' |
| **Assumed Long. | 168°-18.8' E |
| *LHA | 398°-00.0' |
| Less 360° | 360°-00.0' |
| LHA (t-W) | 38°-00.0' |
| ***Assumed Lat. | 27°-00.0' N |
| *Declination | N 12°-34.4' |
| $H_t$ | 51°-22.0' |
| Correction d+28 | +16.0' |
| $H_c$ | 51°-38.0' |
| $H_o$ | 51°-44.5' |
| **Altitude Intercept | 6.5 t |
| Z | N 105° W |
| From 360° (if req.) | 360° |
| **Zn | 255° |
| $H_s$ | 51°-48.5' |
| Main Corr. −0.8 | |
| Addn'l | |
| Dip −3.2 | |
| Total Corr. | −04.0' |
| $H_o$ | 51°-44.5' |

*Required to enter Pub. No. 249
**Required to plot line of position
***Required to enter Pub. No. 249 and to plot line of position

reason it is one easy step to find the meridian angle east (t east). Just subtract the body's GHA from the assumed longitude. It would take an extra step to find LHA—you must subtract t from 360°. But, again, there is no reason to find LHA because you can enter the tables with t east. The tables provide values of t on one side of the columns and values of LHA on the other side. When using t, it is necessary to know that t is east in order to find the true azimuth. This is why I have insisted in previous chapters that we find t east for antemeridian sights (we were in west longitude). The primary reason for knowing t east is to be able to convert azimuth angle, Z, to true azimuth, Zn. A notation in the upper left-hand corner of the upper page in each pair of facing pages in Pub. No. 249, Volumes 2 and 3, tells you that when the LHA is greater than 180°, Z = Zn. (When LHA is greater than 180°, t will equal 360° − LHA, and t will be east.) But we did not bother to determine LHA. Hence the need to name t east or west. Besides, when you rely on a notation such as this, you are working from rote, and not from understanding. Isn't it better to understand what you are doing, rather than to depend blindly on one or more rules? As you get deeper into the study and practice of navigation, you will see the importance of what I am saying, but for now, let us get back to our comparison of the differences in reducing sights in east and west longitude.

With sights taken before meridian passage—antemeridian sights—in east longitude, it takes an extra step to find t east. This is just the opposite of sights in west longitude. We find LHA first, and because we are using LHA (and LHA is always west) we can read the notation in the upper left-hand corner of Pub. No. 249 to find true azimuth. This notation says, in effect, that in north latitude, when LHA is greater than 180°, Z equals Zn; when LHA is less than 180°, you must subtract Z from 360° to get Zn. We will discuss finding true azimuth in general and introduce finding true azimuth in south latitude in the next chapter. For now, in order to get the concepts necessary to reduce a sight in east longitude (and in north latitude) fixed in our minds, let us work a few more sights.

At 0936 ZT on 20 October 1983, you find your vessel at DR latitude 25° 43′ N, longitude 157° 42.7′ E. You observe the sun's lower limb as follows:

| | | |
|---|---|---|
| C | $23^h$-$38^m$-$25^s$ | |
| W | $01^m$-$31^s$ | |
| CE | $00^m$-$04^s$ | fast |
| $H_s$ | 46° 51.1′ | |
| Height of eye | 9 feet | |

The complete reduction of this sight is given in the second form on page 69.

Let us see how things work out with a postmeridian sight. At 1503 ZT on 21 October 1983, you find your boat at DR latitude 27° 19.8′ N, longitude 152° 28.0′ E. You observe the sun's upper limb over the silver lining of a dense cloud as follows:

| | | |
|---|---|---|
| C | $05^h$-$06^m$-$05^s$ | |
| W | $01^m$-$38^s$ | |
| CE | $00^m$-$17^s$ | fast |
| $H_s$ | 26° 53.1′ | |
| Height of eye | 14 feet | |

(If you have not caught on before now, it should be apparent that I am also giving you some practice with the *Nautical Almanac*. This should cause you no problems, and it will be good additional practice.) The reduction of this afternoon sun sight is given in the first form on page 70.

If you have followed me this far, and there is no reason why you should not have, an evening star sight well into east longitude will illustrate still another slight modification of the steps used to determine LHA to reduce a sight. (A similar situation can accompany morning sights in west longitude.) Reduce the following sight of the star Rasalhague, and you will see what I mean. The date is 21 October 1983, and your DR position is latitude 26° 51.0′ N, longitude 168° 19.6′ E. The sight is as follows:

| | | |
|---|---|---|
| ZT | 1755 | |
| C | $06^h$-$57^m$-$20^s$ | |
| W | $01^m$-$12^s$ | |
| CE | $00^m$-$08^s$ | slow |
| $H_s$ | 51° 48.5′ | |
| Height of eye | 11 feet | |

What was the difference in the steps used to find LHA? There was only one. We subtracted 360° only after finding the sum of the GHA of Aries, the SHA of the star, and our assumed longitude. You only need to subtract 360° from the sum if the sum exceeds 360°, but it always will when postmeridian sights are taken in east longitude. The very fact that the sum of the GHA, SHA, and assumed longitude exceeds 360° tells us that the difference is LHA, and the meridian angle, t, is west.

# 14 HOW TO FIND TRUE AZIMUTH

The true celestial azimuth of a body is necessary to plot the line of position resulting from a sight reduction. If your azimuth is incorrect, your line of position is incorrect. Unless you can determine the true azimuth correctly, you cannot plot your LOP. It is equally important to know how to determine true azimuth in order to find compass error.

Finding true azimuth is a problem because it is reckoned from true north, easterly in a clockwise direction. The azimuth angles tabulated in Pub. No. 249 are reckoned from the *elevated pole* in the direction of the *meridian angle*. Thus, if you are south of the equator and observe a body after it has crossed your meridian, the azimuth angle in Pub. No. 249 would be tabulated as *south* so many degrees *west*.

This problem has been simplified in tables such as Pub. No. 229 and Pub. No. 249, Volumes 2 and 3. In these tables, a simple set of rules given at the top and bottom of each page tells you what the true azimuth is. Nevertheless, the ability to visualize the methods used to find true azimuth will be a big help in case you ever find yourself confused.

The ability to convert azimuth angle to true azimuth by reference to the name (north or south) of the elevated pole and the name (east or west) of the meridian angle, t, is essential for working sights by the use of Ageton's tables, which appear in the 1975 edition of Bowditch, Volume II, as Table 35. (Table 35 is reproduced in its entirety in appendix G of this book.) These tables were originally published as H.O. 211. Although now out of print as a separate publication of the Defense Mapping Agency, Ageton's Method refuses to die, and even seems to be enjoying considerable vigor as a viable method of reducing sights. You will see why when we look at the steps required to work a sight by H.O. 211, as this method of sight reduction is still called. What is important for now is that, to determine true azimuth when working from H.O. 211, you have to know the rule: *Azimuth angle is always measured from the elevated pole in the direction of the meridian angle.*

This book is principally a treatise on the use of Pub. No. 249 as a means of computing problems in celestial navigation. Up to now, we have been dealing with the reduction of sights taken in the Northern Hemisphere. Let's take a closer look at what we do with an azimuth angle read from Pub. No. 249 when we are in the Northern Hemisphere. It doesn't matter if you are in west or east longitude—the rule is the same. If the meridian angle, t, is east, the tabulated azimuth angle is the true azimuth. If the meridian angle is west, the tabulated azimuth angle must be subtracted from 360° to get true azimuth. This follows the basics you learned in earlier chapters. The rule is that the tabulated azimuth angle in Pub. No. 249 is always reckoned from the elevated pole in the direction of the meridian angle.

This same rule works in the Southern Hemisphere. To apply the rule when t is east, you subtract the tabulated azimuth angle from 180° to get the true azimuth.

Hold on you say! That isn't the same way we have been doing it.

Well it is. Just analyze the situation. If the elevated pole is north and t is east, and the numerical value of north on the compass rose is either 000° or 360°, you just add the tabulated azimuth angle to the azimuth of the elevated pole, because the azimuth, true azimuth I mean, increases in a clockwise direction. You add the tabulated azimuth angle to 000°, which comes out to the same value as the tabulated azimuth angle. If you like to do useless arithmetic you can add it to 360° instead of 000°, and subtract 360° from the sum.

The other situation in the Northern Hemisphere is the postmeridian sight, the sight of a body with meridian angle, t, west. Again, in the Northern Hemisphere, the numerical value of the elevated pole on the compass rose is 000° or 360°. This time, though, the tabulated azimuth angle is reckoned from the pole west. This means the tabulated azimuth angle must be subtracted from 360° to get the true azimuth. This time you can't do any useless arithmetic, since you can't subtract from 000°.

This puts the rule in a little different light. It doesn't really change anything, but, instead of thinking of the North Pole as just north, you are thinking of it as either 000° or 360°. You can do the same thing with the South Pole. Think of it as 180°. This makes the South Pole easier to work with, because the only number to remember is 180°, the true azimuth of the South Pole.

The solution is simple. To find true azimuth when the meridian angle, t, of a body is east and the elevated pole is south, subtract the tabulated azimuth angle in Pub. No. 249 from 180°, the numerical value of the true azimuth of the South Pole on the compass rose.

To find true azimuth when t is west and the elevated pole is south, add the tabulated azimuth angle in Pub. No. 249 to 180°.

Here are some examples:

The navigator takes a morning observation of the sun at sea. The DR of his vessel is Lat. 35° 17.8′ S, Long. 43° 37.6′ W. From Pub. No. 249, while reducing the sight, he extracts a tabulated azimuth angle of 110°. What is the true azimuth, Zn, of the sun? The elevated pole is south, 180°. The meridian angle, t, of a morning sun sight is always east. Therefore, to find true azimuth, subtract the tabulated azimuth angle, 110°, from 180°. Zn equals 070°.

That afternoon, the navigator took a midafternoon observation of the sun. Obviously, the ship is still in south latitude and west longitude. While reducing the sight this time, he extracted a tabulated azimuth angle of 097° from Pub. No. 249. What is the sun's true azimuth? The elevated pole is south, 180°. The meridian angle is the LHA of the afternoon sun, which is always west. Therefore, to find true azimuth, add the tabulated azimuth angle, 097°, to 180°. Zn equals 277°.

This is all there is to it. Except for working with the elevated pole at 180° instead of 360°, finding the true azimuth of a body when your position is south of the equator is the same as when you are north of the equator.

Another use of the celestial azimuth is to check your compass error. When you take a sight, if you can, have someone on board observe the compass azimuth of the body you observe. You can compare this with the computed azimuth, Zn, to get your compass error.

If it is not possible for an assistant to observe the azimuth, you can do it yourself. Don't tarry; as soon as you have recorded your sight, hurry back up on deck and observe the compass azimuth of the body you just shot. The azimuths of celestial bodies change about one degree every four minutes of time. Unless several minutes elapse between your sight and your observed azimuth, the computed true azimuth, Zn, resulting from your sight will be accurate enough for comparison with your observed compass azimuth to determine your compass error.

Suppose you took a sight of the sun and computed the Zn to be 103°. Your compass azimuth was 089°. What is your compass error? Obviously, it is 14°.

I am going to refer you back to those lessons in compass correction I hope you had before you tackled celestial navigation. When a correction has to be added to a compass course, bearing, or azimuth, that error is an easterly error. So, your compass error is 14° *east.*

Since I have brought up the subject of compass error, you should remember that the error results from two sources. One is the error of *variation*, and is the result of the earth's magnetic pole not being at the same place as the earth's geographic pole. Variation changes as your ship's position changes on the surface of the earth. The variation for a particular point on the earth's surface can be found on the nautical charts.

The other source of compass error is *deviation*. This is error due to magnetic forces within the vessel. Your ship's deviation will change with each change of your ship's heading. Deviation may also change for a number

| True Azimuth (Computed) | Course (Steered) | Variation (From Chart) | Magnetic Azimuth | Deviation | Compass Azimuth (Observed) |
|---|---|---|---|---|---|
| 103° | 000° | 5° E | 098° | 9° E | 089° |
| 103° | 015° | 5° E | 098° | 6° E | 092° |
| 103° | 030° | 5° E | 098° | 5° E | 093° |
| 103° | 045° | 5° E | 098° | 4° E | 094° |
| 103° | 060° | 5° E | 098° | 1° E | 097° |
| 103° | 075° | 5° E | 098° | 0° E | 098° |
| 103° | 090° | 5° E | 098° | 1° W | 099° |
| 103° | 105° | 5° E | 098° | 3° W | 101° |
| 103° | 120° | 5° E | 098° | 5° W | 103° |

of other reasons. Many things affect your vessel's magnetic properties. The change in deviation may be sudden, gross, and spectacular, as in a near miss from lightning. It may be slow and subtle, the result of remaining on one heading a long time. This, along with the vibrations of a moving vessel, may cause the molecules in any ferrous metal on board to realign themselves in reference to other magnetic forces. Using up stores can change your vessel's magnetism. All these factors affect your vessel's deviation.

Deviation is the error you must compute by comparing your compass azimuth with the magnetic azimuth. (You find magnetic azimuth by applying the variation shown on the chart for your position to the celestial azimuth you computed.)

Always bear in mind that deviation is not constant except for a limited time on one heading. Deviation will change every time your vessel's heading changes. Celestial observations can be used to make up a deviation card to tell you what your deviation is on various headings. This is not the usual use of celestial azimuths at sea, however. On a passage, you are most interested in knowing your compass error for the course you are steering. This is the day-by-day use of the celestial azimuth.

I am going to review the methods of correcting the compass. In the first example, you observed the compass azimuth of the sun to be 089°. You computed the celestial azimuth to be 103°. Your compass error was therefore 14° E. If your chart showed the variation at your ship's position to be 5° E, what was your deviation for the course you were on when you took the azimuth?

The solution is essentially what mariners call an "uncorrecting" problem. The rule for correcting the compass from a compass course to a true course is to add easterly error and subtract westerly error. The steps are first to correct your compass course by applying deviation to get your magnetic course, and then, apply variation to get your true course. In short, the sequence to correct a compass course to a true course is compass, deviation, magnetic, variation, and true. Remember the letters C D M V T. We have a memory aid for this: "Can Dead Men Vote Twice."

Let's go back to the problem. You are going from a true azimuth of 103° to a compass azimuth of 089°. To do this requires a reversal of the correcting sequence. This is why sailors call this sequence "uncorrecting" the compass. We reverse everything. The sign of the error is reversed. Easterly error is subtracted when "uncorrecting" and westerly error is added. The sequence is reversed also. You must work from true through variation to magnetic through deviation to compass. The letters to remember are T V M D C. We have a memory aid for this too: "Timid Virgins Make Dull Companions."

We come to the point where other distinctions are necessary. Variation is constant at a given point on the earth's surface at a given time. *Deviation changes for every change in the heading of your vessel.*

The table above will give the solution to the problem stated above and illustrate what I am trying to say. Suppose you want to make a deviation table for your vessel. Since this is a hypothetical situation, we will assume you could run through nine course changes in four minutes. This is really not possible, but for the sake of illustration, if you could, your deviation table would look like the table here. The data would continue on around the compass rose to 360° in increments of 15° for course steered.

You would observe the compass azimuth and the compass course. You would compute your true azimuth from your tables—Pub. No. 249 or your azimuth tables—and apply your variation from the chart. This would give you your magnetic azimuth. Note that true azimuth and magnetic azimuth are constant throughout the table. The difference between your magnetic azimuth and your observed compass azimuth is your deviation.

As I said earlier, the more usual use of the celestial azimuth is for the navigator to check the deviation for the particular heading the vessel is on during a passage. If the course is changed substantially, over 15°, the deviation should be rechecked. For instance, study the following:

Your yacht is on a passage to windward, steering a course of 225°C. You observe an azimuth of the sun to be 133°, and compute it to be 140°. What is the deviation of the yacht's compass on course 225°C? The chart shows the variation at your position to be 8° E. What is your yacht's true course?

**Solution:**

| | | |
|---|---|---|
| Observed azimuth | 133° | |
| Computed azimuth | 140° | (true azimuth) |
| Compass error | 7° E | |
| Variation | 8° E | |
| Deviation | 1° W | |

To find the true course:

| | |
|---|---|
| Compass | 225° |
| Deviation | 1° W |
| Magnetic | 224° |
| Variation | 8° E |
| True | 232° |

At 1100 the wind backs, and you are able to come to a course of 200°C. You observe the azimuth of the sun to be 141°, and compute it to be 154°. What is the deviation on course 200°C, and what is the yacht's true course?

**Solution:**

| | |
|---|---|
| Observed azimuth | 141° |
| Computed azimuth | 154° |
| Compass error | 13° E |
| Variation | 8° E |
| Deviation | 5° E |

To find the true course:

| | |
|---|---|
| Compass | 200° |
| Deviation | 5° E |
| Magnetic | 205° |
| Variation | 8° E |
| True | 213° |

The deviation changed 6° in this course change. This may not seem like a big change, but it could seriously affect your dead reckoning. This example should show you why it is important to check your deviation each time you change course.

Even if you have a deviation card for your ship, you should check the deviation daily. On a long passage, deviation will change even if you remain on the same course. Changes in the magnetism of your vessel will cause changes in the deviation. This will put your deviation card out of date. It doesn't take a lot of vision to understand that if you want to get where you are going, it helps a lot to know what course you are on. You can't know this if you do not know your deviation.

Use your celestial azimuth to check your compass error. Even in pilot waters, unless you can get lined up with a range, it is often easier to check your compass error for deviation on a particular course by using a celestial azimuth of a body.

As a matter of convenience, I have assumed you use the Zn from a sight reduction to check your compass error. Of course, in pilot waters, you will not be taking sights in most cases. (There are cases where you might.) You can compute an azimuth from Pub. No. 249 or Pub. No. 229 anytime you need a true azimuth. For ease of use, though, there are special tables of precomputed azimuths to facilitate compass correcting. I will not recommend any specific set of tables, but I do suggest you look into selecting a good book of azimuth tables.

# 15 HOW TO WORK A STAR SIGHT BY PUB. NO. 229

In the introduction to this book, I told you that most of what you learned in this study would apply equally to the use of the tables in Pub. No. 249 and the tables in Pub. No. 229. As you become familiar with the steps involved in reducing a sight, you will become aware of several sources of the information required to complete the various steps. You can get GHA and declination from the *Air Almanac* as well as the *Nautical Almanac*. These data may be tabulated differently, but since these celestial coordinates are a function of time and date, the method of tabulation cannot be extremely different. In the same way, Pub. No. 229 does not differ so terribly much from Pub. No. 249, nor do the other precomputed sight reduction tables. They all depend on local hour angle, declination, and latitude for entering arguments.

A few words of comparison between Pub. No. 249 and Pub. No. 229 will illustrate what I mean. I do not intend to give an item-by-item comparison, but I will show you how similar the sight reduction process is by working again one of the star sights we worked in Chapter 9. Pub. No. 249 is indexed by latitude. The tables on a given page are entered with the arguments of declination and local hour angle. Pub. No. 229 is indexed with local hour angle as the first argument. The tables do have page numbers, but since a different LHA is provided on each page, LHA is the most convenient guide to open the book by. Latitude is found across the top and bottom of the page. Declination is tabulated in the vertical column at either edge of the page. Your $H_t$ is extracted at the intersection of the latitude column and the declination line. An interpolation table inside the front and back covers is provided for the correction required by the difference in declination. Declination is tabulated only to the even degree in the main tables. The value "d" in the main tables is called the *altitude difference*. This altitude difference is tabulated across the top of the interpolation table. The difference in declination is tabulated in the vertical column on the left-hand side of the table. The correction is applied in accordance with the sign of the altitude difference, "d," which should be extracted at the time the main tables are consulted.

As we have done before, let's work a problem. To have an exact comparison with the methods used in Pub. No. 249, we will rework the Regulus sight we used as an example in Chapter 9.

The star Regulus was observed on the morning of 21 October 1983 at a dead reckoning position of Lat. 24° 59.0′ N, Long. 92° 38.0′ W. Sextant altitude was 43°

53.6′. The chronometer read $11^h$-$14^m$-$15^s$, and elapsed time by stopwatch was $0^m$-$21^s$. The chronometer was fast $3^m$-$02^s$. Reduce this sight for a line of position using Pub. No. 229. See the accompanying sight reduction form.

**Step 1**

Plot your DR.

**Step 2**

Record your sight on the work form. The footnotes on the form referring to Pub. No. 249 apply equally to Pub. No. 229.

**Step 3**

Correct the chronometer time to get the GMT at the instant the sight was taken.

**Step 4**

Open your *Nautical Almanac* to the daily pages for Friday, 21 October 1983. On the left-hand page go down the Aries column to "Friday, 21," and opposite $11^h$ extract GHA Aries 194° 19.5′.

**Step 5**

Turn to the yellow pages in the back of your *Nautical Almanac*. Find the $10^m$ box and run down the left-hand side to $52^s$. Read across the box to the Aries column and extract the correction of 2° 43.4′. Add this to the GHA of Aries you extracted from the daily pages.

**Step 5A**

It should be obvious by now that the procedure we have used is exactly the same as for sight reduction by Pub. No. 249. It is exactly the same as in Chapter 9. You use the same arguments to enter both tables, Pub. No. 249 and Pub. No. 229. Consequently, you get this information exactly the same way. Go to the daily pages of your *Nautical Almanac*, and from the right-hand side of the left-hand page for 21 October, extract the SHA of Regulus, 208° 07.8′, and the declination, N 12° 03.0′. Add the SHA of Regulus to the GHA of Aries. This totals 405° 10.7′. Subtract 360° from this total to get 45° 10.7′, the GHA of Regulus.

STAR REGULUS
Date 21|10|83 ZT 0510
DR Lat. 24°-59.0′ N Long. 92°-38.0′ W

| | |
|---|---|
| C | $11^h$-$14^m$-$15^s$ |
| W | $0^m$-$21^s$ |
| C-W | $11^h$-$13^m$-$54^s$ |
| CE | $3^m$-$02^s$ f |
| GMT | $11^h$-$10^m$-$52^s$ |
| GHA Aries | 194°-19.5′ |
| Corr. (min. & sec.) | 2°-43.4′ |
| GHA Aries | 197°-02.9′ |
| SHA star | 208°-07.8′ |
| Sum | 405°-10.7′ |
| Less 360° (if req.) | -360°-00.0′ |
| GHA star | 45°-10.7′ |
| **Assumed Long. | 92°-10.7′ |
| *LHA (t-E) | 47°-00.0′ E |
| ***Assumed Lat. | 25°-00.0′ N |
| *Declination | N 12°-03.0′ |
| $H_t$ | 43°-49.5′ |
| Correction d +23 +1.0 +0.2 | +1.2 |
| $H_c$ | 43°-50.7′ |
| $H_o$ | 43°-49.2′ |
| **Altitude Intercept | 1.5 a |
| Z | N 97.4° E |
| From 360° (if req.) | |
| **Zn | 097° |
| $H_s$ | 43°-53.6′ |
| Main Corr. -1.0 | |
| Addn'l | |
| Dip -3.4 | |
| Total Corr. | -04.4′ |
| $H_o$ | 43°-49.2′ |

*Required to enter Pub. No. 249
**Required to plot line of position
***Required to enter Pub. No. 249 and to plot line of position

**Step 6**

Select an assumed longitude, just as you would for a solution by Pub. No. 249. To produce an even degree of LHA or meridian angle, the assumed longitude should be 92° 10.7′ W. Subtract from this the GHA of Regulus and obtain the meridian angle "t" of Regulus. Meridian angle in west longitude is east when GHA is less than the assumed longitude. To get the true LHA of Regulus, we should have to subtract 47° from 360°. This would give us our LHA of 313°. When you open Pub. 229, you will see the page is indexed "47°—313°."

You will need to know the LHA, 313°, to find your true azimuth if you use the legend at the top of the page in Pub. No. 229. If you use a t of 47° east, you must apply the rule for finding true azimuth. Look at the upper right-hand corner of the "47°—313°" page of Pub. No. 229. You will see, in small print, a legend telling you that if your latitude is north and your LHA is greater than 180°, then Zn = Z. Z is the tabulated azimuth angle. And this legend also tells you that if LHA is less than 180°, then Zn = 360° − Z. From now on, it will be best to find the LHA even though you could have used a t east of 47° to get your $H_c$. You will need the LHA of 313° to find your true azimuth, Zn, if you use the legends on the corners of the pages.

**Step 7**

Now you have an even degree of LHA, and you have your declination of Regulus, N 12° 03.0′. You should select 25° 00.0′ as your assumed latitude. At this point, you could open either Pub. No. 249 or Pub. No. 229. The entering arguments are the same. Since this is an example of how to use Pub. No. 229, open Pub. No. 229 to the page headed "LHA 47°—313°." (For convenience, a copy of this page is reproduced in appendix F of this book.) This page has a number, too, but it is really easier to find a page in Pub. No. 229 by LHA than by page number.

On the left-hand page is the legend "Latitude Same Name as Declination." You are using Volume 2 of Pub. No. 229. Volume 2 covers latitudes from 15° to 30°. In Volume 2, or any other volume of Pub. No. 229, there are two places where you will find facing pages headed for any given LHA. The latitude columns will differ. The latitude columns for the first eight degrees of latitude covered in any volume of Pub. No. 229 will be in the first half of the volume, and the latitude columns for the second eight degrees of latitude covered will be in the second half of the volume.

For example, had your assumed latitude been 21° in the problem you just worked to reduce the Regulus sight, you would have found your tabulated altitude and azimuth angle in the first part of Volume 2 of Pub. No. 229. Since the assumed latitude you used was 25°, you were required to look for your tabulated altitude and azimuth angle in the second, or back, half of Volume 2. In Volume 2 of Pub. No. 229, latitudes 15° through 22° are tabulated in the front part of the book. Latitudes 23° through 30° are tabulated in the back part of the book.

The volume is not actually divided into two equal sections. The back half simply begins somewhere near the middle of the book. There are two pages of diagrams for graphic interpolation of declination separating the front and back portions of the book. Until we go into the more refined uses of Pub. No. 229, you will not use the diagram pages for anything but separators.

Since your assumed latitude is between 23° and 30°, the page headed "47°—313° LHA," having your assumed latitude, 25°, will be in the back half of Volume 2 of Pub. No. 229. Open to the page headed "47°—313°, Latitude Same Name as Declination," and find the column headed 25°. The "Latitude Same Name as Declination" page is the left-hand page. The right-hand page is the "Latitude Contrary Name to Declination" page. Go down the 25° latitude column on the left-hand page. Your assumed latitude is the same name as the declination of Regulus. When you reach the 12° line of the "Dec" column on the edge of the table, follow this line across the table until you come to the 25° latitude column, and extract your $H_t$ (called *tabular altitude* in the introduction to Pub. No. 229, and labeled $H_c$ in the tables) of 43° 49.5′, a "d" of +23.3 (be sure to note the sign of "d"), and a Z of 97.4°.

**Step 8**

It is necessary to interpolate for the difference in declination between N 12° 03.0′, the declination of Regulus, and the tabulated declination of 12° 00.0′. This difference is, of course, 03.0′. To do this, turn to the interpolation tables in Pub. No. 229 on the inside front or back cover as appropriate. The inside front cover tabulates the corrections for 0.0′ to 31.9′ of declination. The inside back cover tabulates the corrections for declination between 28.0′ and 59.9′. In these tables, the declination increment is the argument in the vertical column on the left. The altitude difference, "d," from the main tables, is tabulated across the top of the tables in units of ten in the left-hand box. To the right is a tabulation for units of 0′ to 9′ across the top of the table, with the decimal tenths in a vertical column to the left of this subtabulation to provide the horizontal coordinate for using these tables.

The difference in declination we are concerned with is 03.0′. Open the inside front cover, since the difference is less than 31.9′. Find the declination increment column, headed "Dec. Inc.," containing 3.0′. This happens to be the column at the far left. Across the top of the table select the 20′ column in the box labeled "Tens," since your "d" is 23.3. Go down this column until you reach the 03.0′ line of declination increment. Extract 1.0 as your "first contribution" to your tabulated altitude correction. Remember the sign is plus. Now go to the box just to the right, in the same tables, labeled "Units," and the column headed 3′. Go down this column until you come to the line for .3′. Be *sure* you go down the column until you come to the line for .3′ *in that group of decimals*, .1′ to .9′, opposite the 3.0′ to 3.9′ of declination increments. An inspection of this part of the table will show why this is so important. At the intersection of the 3′ column and the .3′ line, in the 3.0′ to 3.9′ group of declination increments, extract 0.2, your "second contribution." Again, the sign is plus. Add these two contributions to get your correction of +1.2. Add this to the tabulated altitude, $H_t$, and get your computed altitude, $H_c$, 43° 50.7′.

In most cases, the above procedure is all there is to it. Had there been a dot after the altitude difference of 23.3, a correction for "double second difference" would have had to be applied to the tabulated altitude also. This correction is usually very small. To find the double second correction, you must go to the main table and extract the altitude difference for the declination you are concerned with. Then extract the altitude difference for the higher declination and subtract one from the other. Now go back to the interpolation table. The remainder is your entering argument. You will find the double second difference tabulated on the right of each interpolation table. Go down the column on the left of the "Double Second Diff. and Corr." table to where you see the remainder of the two altitude differences bracketed by a slightly smaller and a slightly greater value. The double second difference correction will be opposite these figures. Apply this correction to the tabulated altitude when required.

**Step 9**

Correct your sextant altitude, $H_s$, for observed altitude, $H_o$. You do this the same way for any method of sight reduction.

**Step 10**

Subtract $H_o$ from $H_c$ to get your altitude intercept. Remember *computed greater away*, C.G.A.

**Step 11**

Plot your line of position.

Compare this solution with the method given in Chapter 9 for reducing a star sight by use of Pub. No. 249. There is little difference either in the altitude intercept or in the method of deriving it. Just a small variation in Step 7 and Step 8. This difference is primarily due to the change in the method of tabulation.

You may wonder why they went to the trouble to compute the Pub. No. 229 tables to replace H.O. 214. One reason is that by doing the calculations by computer, a source of error found in H.O. 214 was eliminated. Also, the new format is considered an improvement that will reduce the time needed to work a sight reduction problem, and reduce the probability of error.

Pub. No. 229 is now on the shelves of your chart agent. When you get a copy, study it carefully. What I have said should help you follow the method used in this publication. I hope it will. I suggest you carefully read the instructions in the front of these tables also. Then practice using the tables at every opportunity.

A parting word for H.O. 214 and the other tables: H.O. 214 will be with us a long time, because many navigators already have these tables, love them, and will not want to spend the money to replace them until they are so dog-eared as to be no longer usable. Also, if you are to call yourself a competent navigator, you should be able to use whatever set of tables you find aboard your ship, provided they are not too out of date.

# 16 HOW TO USE THE AIR ALMANAC

By now it should be apparent to you that the problem of obtaining a line of position from the reduction of an observation can be broken down into four major components:

1. Make your observation and note the exact time and altitude.
2. Determine the LHA or t east and the declination of the body at the time of observation at your assumed position.
3. With your LHA or t east, declination, and assumed latitude, enter the sight reduction tables and extract the computed altitude and azimuth.
4. Find the difference between the computed altitude and the observed altitude and plot the resulting LOP using the computed azimuth.

Working a sight requires information that must be obtained to a certain degree of exactitude. Up to now in this book, we have been getting the tabulated portion of this information from the *Nautical Almanac*, the *Sight Reduction Tables for Air Navigation* (Pub. No. 249) series, and the *Sight Reduction Tables for Marine Navigation* (Pub. No. 229) series. I have said repeatedly that the use of the sight reduction tables depends on three arguments: LHA or t, declination, and assumed latitude. These arguments are obtained by extracting data from your *Nautical Almanac* and applying these data to your assumed position. You are then ready to enter the tables.

If we can get the tabulated data for our three arguments from another accurate source, they should be just as good as the data from the *Nautical Almanac*. The *Air Almanac* is just such another publication.

All right. I know just what you are thinking. You are saying: Why? Why have two almanacs for the same purpose? The reason is this: To navigate competently requires a judicious balance between speed and "accuracy." I really should say speed and resolution, for this is what I mean. If you are going fast, but are high enough to see for miles around, as you would be in an airplane, it is much more important to be able to get an accurate fix within a resolution of 10 miles before the data are too old to be valuable, than it is to get a more precise fix long after you have missed your checkpoint.

Air navigation publications are tabulated in such a manner and to such a degree of resolution that they lend themselves to speedy solution of the sight reduction problem at the expense of "accuracy," or resolution. Tabulations in the marine navigation publications are carried to the nearest tenth of a minute of arc. Tabulations in the air navigation publications, with the exception of data for the sun and Aries, are carried to the nearest whole minute of arc. This will give you enough resolution, when coupled with the usually better sights the marine navigator can get, to put your fix much closer to the true position than the 10 miles referred to above for air navigators. Theoretically, your resolution should be within one mile. One mile is just a little more than the width of a dull pencil on a plotting sheet and is well within the requirements of the small vessel offshore. For that matter, it is within the requirements of most vessels offshore, and is better than most navigators expect from any method. Air navigation publications are used to navigate merchant ships and even naval vessels.

You say this still hasn't answered the question of why do we use these publications for surface navigation? We do this because the same features of these publications that provide speedy solutions to navigation problems also provide easy solutions. The *Nautical Almanac* is still the best almanac when precision is required. It is also best for the beginner because it gives him more exercise in the sight reduction procedures. It is appropriate to mention that you may use the *Air Almanac* with Pub. No. 229, or the *Sight Reduction Tables for Air Navigation* with the *Nautical Almanac*, which is what we have been doing through most of this book.

Let's work some sights with the *Air Almanac* and the Pub. No. 249 tables. We will work a sun sight, and then a star sight.

You are at sea on the afternoon of 14 June 1983. Your DR position is Lat. 24° 39.7′ N, Long. 89° 21.4′ W. You observe the sun's lower limb at about 1520 ZT. The sextant altitude is 44° 38.8′. The chronometer read $21^h$-$16^m$-$45^s$, stopwatch, $0^m$-$28^s$. The chronometer is slow $2^m$-$19^s$. Reduce this sight for a line of position using the *Air Almanac* and Pub. No. 249. Page excerpts from the *Air Almanac* will be found in appendix D. We will use the same step-by-step procedure on the same form we used in Chapter 6.

SUN (⊙), Lower/~~Upper~~ Limb
Date 14/6/83, ZT 1520
DR Lat. 24°-39.7′ N Long. 89°-21.4′ W

| | |
|---|---|
| C | $21^h$-$16^m$-$45^s$ |
| W | $00^m$-$28^s$ |
| C-W | $21^h$-$16^m$-$17^s$ |
| CE | $2^m$-$19^s$ S |
| GMT | $21^h$-$18^m$-$36^s$ |
| GHA | 137°-27.4′ |
| Corr. (min. & sec.) | 2°-09.0′ |
| GHA | 139°-36.4′ |
| **Assumed Long. | 89°-36.4′ |
| *LHA (t-W) | 50°-00.0′ |
| ***Assumed Lat. | 25°-00.0′ N |
| Declination | N 23°-16.1′ |
| d difference | |
| *Declination | |
| $H_t$ | 44°-32.0′ |
| Correction d+13 | + 3.0′ |
| $H_c$ | 44°-35.0′ |
| $H_o$ | 44°-49.6 |
| **Altitude Intercept | 14.6′ t |
| Z | N 082° W |
| From 360° (if req.) | 360° |
| **Zn | 278° |
| $H_s$ | 44°-38.8′ |
| Main Corr. Semi ⌀ +15.8 | |
| Addn'l refraction −1.0 | |
| Dip −4.0 | |
| Total Corr. | +10.8 |
| $H_o$ | 44°-49.6′ |

*Required to enter Pub. No. 249
**Required to plot line of position
***Required to enter Pub. No. 249 and to plot line of position

**Step 1**

Plot your DR position up to the time of your sight.

**Step 2**

Record your sight on the work form. Note the local date and zone time, your chronometer time, C, your stopwatch reading, W, and your sextant altitude.

**Step 3**

Apply the stopwatch reading and the chronometer error, CE, to obtain Greenwich mean time.

Pick up the *Air Almanac.* Before going any further with the reduction of this sun sight, turn to the section following the daily pages and the Navigational Star Chart. Read carefully the preface and the explanation of the tables. Really, you should do this with any book or publication you use if you want to get the most out of it. After you complete this reading, return to reducing your sun sight.

**Step 4**

Open the *Air Almanac* to the daily page for 14 June. At this time, several differences in the way this publication is organized should be obvious. When you first picked the almanac up, the front cover showed that it covered only part of the year 1983 instead of the entire year, as the *Nautical Almanac* does. The *Air Almanac* is published in two volumes each year, covering the months of January through June in the first volume, July through December in the second. The daily pages are organized differently also. When you open to the pages containing the date of 14 June, you will see the right-hand page (page 329) tabulates the astronomical data for $00^h$-$00^m$ GMT to $12^h$-$00^m$ GMT. To find the data for the afternoon hours of GMT on 14 June 1983, it is necessary to turn to the next page (page 330). This is a left-hand page. This puts all the data for one day on both sides of a single sheet. When that day, Greenwich date, is over, the navigator just tears the page out and throws it away. Thus, the first page is always the current page.

You are now on the correct page to extract the GHA and declination of the sun at $21^h$-$10^m$ of GMT for this date. In the *Air Almanac*, data are tabulated for each hour and in 10-minute increments between the hours on the daily pages. So go down the GMT column on the far left to $21^h$-$10^m$, and extract the GHA of the sun, 137° 27.4′, and the sun's declination, N 23° 16.1′. Because the data are tabulated in 10-minute intervals of time, interpolation for declination is never necessary. The sun column is next to the GMT column. The tenth-of-a-minute values are tabulated in smaller type because the air navigator usually disregards them. Since they are available, I suggest you use these tenths in surface navigation. It doesn't add much extra work, and it closes the resolution gap.

**Step 5**

It is necessary to make a correction for the additional $8^m$-$36^s$ of time, and add it to the sun's GHA. To do this, open the front cover of the *Air Almanac* and on the inside of the hard cover find a box on the right, headed "Interpolation of GHA." In this box are two time columns. One is for the sun, Aries, and the planets, labeled "Sun, etc.," and the other column is for the moon. Go down the "Sun, etc." column to $08^m$-$33^s$, and just below to the right extract 2° 09′ and put it in the form as a correction to GHA. Add these up and get the GHA of the sun. There is no correction for declination, so this is all there is to Step 5.

Had you wanted a more precise correction for $8^m$-$36^s$, you could have turned to pages A100 and A101 of the white pages in the back. The tables are headed "Interpolation of GHA Sun," and they are tabulated to a precision of $1^s$. Most navigators, however, find the table inside the front hard cover adequate.

**Step 6**

Your 1520 DR is Lat. 24° 39.7′ N, Long. 89° 21.4′ W. You now must select the longitude nearest your DR longitude that will give you an even degree of local hour angle. The assumed longitude to select is obviously 89° 36.4′ W. Put this in the form and subtract it from the sun's GHA. The LHA is 50° 00.0′.

Before going to Step 7, I am going to digress from the sight reduction problem and say again what I said in the Preface, but in a bit more detail. Pub. No. 249 is published in three volumes. Volume 1 is entitled *Sight Reduction Tables for Air Navigation (Selected Stars),* and the current edition is further labeled *Epoch 1985*. The tables in Volume 1 must be revised for each epoch, which is a period of five years. I will have more to say about Pub. No. 249, Volume 1, later in this chapter. You will be very pleasantly surprised at the ease of working selected star sights by this method.

Pub. No. 249, Volume 2, is entitled *Sight Reduction Tables for Air Navigation (Latitudes 0°-39°) (Declinations 0°-29°)*. Volume 3 carries the same title as Volume 2, except that the latitudes covered range from 40° to 89°. Declinations covered range from the zenith to the horizon and below. Volumes 2 and 3 are very similar in the arrangement of their tabulations to H.O. 214, the basic publication used for sight reduction in the first edition of this book. There are minor differences that will appear as you continue to work this sight. The data in Volumes 2 and 3 are permanent in nature.

Now let's go back to the sun sight and Step 7.

**Step 7**

You have the sun's LHA and declination, just as if you had gotten them from the *Nautical Almanac*. To enter the tables, you still need an assumed latitude. The nearest even latitude to your DR is 25° 00.0′ N. Select this as your assumed latitude and enter it in the form. Step 7 is precisely the same as it has always been.

**Step 8**

Step 8 is exactly the same as it has always been.

**Step 9**

Correct your sextant altitude, $H_s$, to get your observed altitude, $H_o$. Remember the five corrections to the sextant altitude of the sun we listed in Chapter 4? The most important one, because of its size, is semidiameter. In the *Air Almanac*, semidiameter, SD, is found in the lower right-hand corner of the daily pages. For 14 June 1983, the sun's SD is 15.8′. Because this was an observation of the sun's lower limb, the correction must be added. Put it in the form opposite "Main Corr." The next correction to be applied is refraction. The corrections for refraction are tabulated in the *Air Almanac* on the last page, page A104, in the table on the top of the page. Read the explanation to this table on the inside back cover. Your sextant altitude is between 33° and 63°. Your height above sea level is under 1,000 feet, so use the "0" column of the table. Opposite, on the left, extract the refraction correction of 1′. Put this in the form opposite "Addn'l." Note that the correction is negative. The marine navigator disregards corrections for temperature and Coriolis force under most conditions.

No correction is tabulated for parallax except for the moon.

The correction for dip is found inside the back hard cover. For a height of eye of 12 feet, the value of the dip correction is 4.0′, and it is negative.

Add these corrections algebraically and apply the sum to your $H_s$. This will give you your observed altitude, $H_o$, of 44° 49.6′.

**Step 10**

Find the difference between your $H_c$ and your $H_o$ by subtracting the lesser from the greater. This is your altitude intercept, 14.6′. The observed altitude, $H_o$, is the greater, so the line of position will plot toward the sun's GP. Put a small "t" after the 14.6′, exactly as before.

**Step 11**

Plot your 1519 sun line.

To further illustrate the use of the *Air Almanac*, let us work a star sight. As an added wrinkle, why don't we try a star sight using Volume 1 rather than Volume 2 or 3 of Pub. No. 249? I did not introduce Volume 1 earlier, because I wanted you to master the use of Volumes 2 and 3 first. You may use Volume 2 or Volume 3 of Pub. No. 249 to reduce star sights with the *Air Almanac* as well as the *Nautical Almanac*, but when the selected stars in Volume 1 of Pub. No. 249 are available for observation, no navigator would think of using Volume 2 or Volume 3.

Volume 1 of Pub. No. 249 lists various selected stars in groups of seven. These stars are selected from among those available to give the navigator the best possible cross. Those three stars with a diamond symbol by their names are selected to give the best possible cross of the seven. They are suitably situated for a three-star fix. To reduce a sight of a selected star in Volume 1 of Pub. No. 249, the navigator need only compute the local hour angle of Aries for the time of his sight, and then turn to the page in Volume 1, Pub. No. 249, having his assumed latitude. The $H_c$ and Zn can be read directly from the tables without additional computation.

You may notice that some of the azimuths, Zn, of the selected stars in Volume 1 of Pub. No. 249 are reciprocals. Despite what I have said all along about wanting to get observations of bodies whose azimuths will cross at wide angles, there are advantages to getting simultaneous sights of two bodies with reciprocal azimuths. If conditions are good, and the observations are taken and reduced with precise accuracy, the resulting LOPs will plot superimposed on each other. In effect you will get one LOP from two observations. I can hear you say, Why do this? What is the idea?

The idea is this: At sea, observations are almost never taken under perfect conditions. Frequently the most serious fault affecting sights on a small boat is the lack of a good horizon. Sometimes, however, the horizon to leeward may be better than the horizon to windward. Or, at star time, the horizon may be obscured in one or more arcs due to cloud shadow or other causes. In cases like these, the very fact that one or more sights are poor will be disclosed to the navigator when he plots two LOPs resulting from sights on bodies with reciprocal azimuths, because any error will cause the LOPs to plot as separate LOPs. The navigator must then recall the appearance of the horizon at the time of his sights, and apply his judgment to determine which sights are most reliable. Plotting LOPs from observations of two bodies at reciprocal azimuths gives you the means to do this.

The navigator will first use Volume 1 of Pub. No. 249 as a star finder, and to precompute the altitude and azimuth of his stars to shoot at twilight. He then observes these stars and uses Volume 1 to reduce the sights.

Here is how he does it:

Assume you are navigator on a vessel at sea on 14 June 1983. You estimate that your DR will be Lat. 25° 03.6′ N, Long. 90° 11.6′ W at 1920 zone time. You compute the time of evening twilight to be 1919 ZT by referring to page A87 of the *Air Almanac*. You are in zone plus 6, so the Greenwich mean time of twilight at your position will be $01^h$-$19^m$, 15 June. You can call this $01^h$-$20^m$ and be near enough for the purpose of precomputing your star sights from Pub. No. 249, Volume 1.

Open your *Air Almanac* to the daily page for 15 June 1983, A.M. Opposite the GMT of $01^h$-$20^m$, extract the GHA of Aries, 282° 45.9′, from the Aries column.

Use an assumed longitude of 89° 54.9′ W. The LHA of Aries is 194°. Open Volume 1 of Pub. No. 249 to the Lat. 25° N pages. Opposite the LHA of Aries of 194°, you will see listed the computed altitude, $H_c$, and true azimuth, Zn, of seven stars. These are the stars you should observe at evening twilight.

This is all there is to it. The procedure is so easy you can precompute the altitudes and azimuths of seven stars by inspection with the aid of a piece of scratch paper!

One word of caution. As you proceed into the 1985 epoch, corrections for the years 1986 through 1989 must be applied to your line of position. For information on this correction after 1985, turn to Table 5, "Precession and Nutation Correction," beginning on page 322 of Volume 1 of Pub. No. 249. These corrections are small, and in daily practice are usually ignored. Nevertheless, you should be aware of them.

By now it may seem redundant, but let's work a star sight using Volume 1 of Pub. No. 249. You are at your DR position of Lat. 25° 03.6′ N, Long. 90° 11.6′ W. At about 1925 ZT you observed the star Spica. Sextant altitude was 53° 29.7′. The chronometer read $01^h$-$22^m$-$45^s$. The stopwatch showed an elapsed time of $00^m$-$28^s$. The chronometer was slow $2^m$-$19^s$. Break out a sight reduction form for star sights, and reduce this sight.

**Steps 1-6**

Because by now you should know the method, I have combined Steps 1 through 3, recording the sight and correcting the time for GMT, and Steps 4,5, and 6, finding GHA and LHA.

You must completely disregard the SHA and declination of Spica, because these data will not be needed to reduce the sight of a star by Volume 1 of Pub. No. 249. This would be Step 5A in Chapter 9.

**Step 7**

Select an assumed latitude of 25° 00′ N. To enter Volume 1 of Pub. No. 249, you need only the LHA of Aries and an assumed latitude. You read the computed altitude, $H_c$, of Spica, 53° 16.0′, and the true azimuth, Zn, of 168°, directly from the tables.

STAR SPICA
Date 14/6/83 ZT 1925
DR Lat. 25°-03.6′ N Long. 90°-11.6′ W

| | |
|---|---|
| C | $01^h$-$22^m$-$45^s$ |
| W | $00^m$-$28^s$ |
| C–W | $01^h$-$22^m$-$17^s$ |
| CE | $2^m$-$19^s$ S |
| GMT | $01^h$-$24^m$-$36^s$ |
| (Note: Greenwich date is 15/6/83) | |
| GHA Aries | 282°-45.9′ |
| Corr. (min. & sec.) | 1°-09.0′ |
| GHA Aries | 283°-54.9′ |
| SHA star | |
| Sum | |
| Less 360° (if req.) | |
| GHA star | |
| **Assumed Long. | 89°-54.9′ |
| *LHA | 194°-00.0′ |
| ***Assumed Lat. | 25°-00.0′ N |
| *Declination | |
| $H_t$ | |
| Correction | |
| $H_c$ | 53°-16.0′ |
| $H_o$ | 53°-24.7′ |
| **Altitude Intercept | 8.7′ t |
| Z | |
| From 360° (if req.) | |
| **Zn | 168° |
| $H_s$ | 53°-29.7′ |
| Main Corr. | −0.0′ |
| Addn'l | −1.0′ |
| Dip | −4.0′ |
| Total Corr. | −05.0′ |
| $H_o$ | 53°-24.7′ |

*Required to enter Pub. No. 249
**Required to plot line of position
***Required to enter Pub. No. 249 and to plot line of position

*This method works only if you use the selected stars listed in Volume 1.*

**Step 8**

This step is omitted.

**Step 9**

Correct your sextant altitude, $H_s$, for observed altitude, $H_o$. Since this is a star sight, only the corrections for refraction and dip need be applied when using the *Air Almanac*. The correction for refraction is found on the last page of the *Air Almanac*. In the "0" column, find the values that bracket your $H_s$ of 53° and extract the correction of −1′. The dip correction is found on the back of the hard cover. For your height of eye of 12 feet, the correction is −4.0′.

**Step 10**

Find the altitude intercept by subtracting $H_c$ from $H_o$. It is 8.7 t.

**Step 11**

Plot and label your 1925 Spica LOP.

This is a complete solution of the sight reduction problem by use of Volume 1, Pub. No. 249. Look at the form, and see how many spaces remain blank. This is an indication of the time that may be saved by using this method.

Like all short cuts, Volume 1 of Pub. No. 249 has its limitations. The stars selected are only seven out of the hundreds that are listed in the almanacs. They are the best to use when they are available, but there is no guarantee that all seven won't maliciously seek seclusion behind some cloud at star time while their brothers beam forth with a twinkling grin. So don't lose your knack for working star sights by the regular methods employed in Pub. No. 229 and in Volumes 2 and 3 of Pub. No. 249.

Stars that are workable with Volumes 2 and 3 of Pub. No. 249 are marked with the symbol "†" in the *Air Almanac*. Stars used in Volume 1 of Pub. No. 249 are marked with an asterisk, "*." The asterisk does not mean you may use these stars at any time to reduce sights by use of Volume 1. The asterisk simply indicates the stars from among which the seven stars tabulated for certain combinations of assumed latitude and LHA may be selected.

My students often ask me which almanac—the *Nautical Almanac* or the *Air Almanac*—I prefer. This is a difficult question to answer objectively. There are advantages to both for the marine navigator. The *Nautical Almanac* is published in one volume for the full year. You need buy and update only one book when you use the *Nautical Almanac*. With the *Air Almanac*, you must buy two volumes, one for the first six months of the year and one for the last six months. On the other hand, the *Air Almanac* presents some information in an easier-to-use manner than does the *Nautical Almanac*. I do like the Navigational Star Chart appearing right behind page 364, the last page of the daily pages, in the *Air Almanac*. (This chart is reproduced in appendix D.) I would rather work a moon sight by use of the *Air Almanac*. Latitude by Polaris is much simpler when worked with the *Air Almanac*. But these differences are minor. I am equally at home with either almanac. The best solution, if you cannot decide on one or the other, is to have both aboard, as they do on merchant and naval vessels. That way you can become proficient with both. And you will have the Navigational Star Chart in the *Air Almanac*.

Learn to use your *Nautical Almanac* and your *Air Almanac* with equal facility. Swap back and forth from time to time. It will keep you sharp. Practice constantly. You will be surprised how much skill you will develop. New worlds will be opened to you.

The good navigator is capable of working sights by the use of several sight reduction methods. Even training yourself to at least understand the cosine-haversine formula is not a waste of time, although admittedly not necessary.

# 17 HOW TO WORK A SIGHT BY AGETON'S METHOD (H.O. 211)

Merchantmen have chartrooms big enough to hold dances. Often the only people there are the mate on watch and maybe the captain. Naval vessels have chartrooms that are even bigger on capital ships. On naval vessels, however, there is usually scant elbowroom in the chart-house. Nevertheless, there is plenty of room to store the seven volumes of H.O. 214, the six volumes of Pub. No. 229, the three volumes of Pub. No. 249, and any other publication you may care to carry. Aboard a small oceangoing vessel, such as a fisherman or a yacht, it is a different matter. Put all these tomes aboard and your ship develops a list to whichever side the bookcase may be on. If you make one-way deliveries across oceans, as some of us do, lugging these heavy books through airports can be exhausting. The trend among small-boat navigators has been toward Pub. No. 249. In one volume of Pub. No. 249, Volume 2 or 3, you get the same latitude coverage that you get in three volumes of H.O. 214 or two of Pub. No. 229. The saving in space—I restrained myself from saying volume—is severalfold, and so too for the weight. Of course, the tables in Pub. No. 249 are limited to bodies having declinations of not more than 30°. Volume 1 of Pub. No. 249 is supposed to ameliorate this to some extent by tabulating selected stars with greater declinations. The idea is great, but too often it is these seven selected stars that hide behind seven little cottonball clouds in the heavens for the duration of star time. I am not contradicting my remarks in the Preface. Not having anything aboard but the needed volumes of Pub. No. 249 would not, except in very special circumstances, be a reason for not sailing. On the other hand, Murphy's Law is always present, waiting for that night you have been looking forward to—the one that promises a landfall in the wee hours of the morning. A good star fix is crucial to a safe landfall. But no stars! The cottonballs are there to hide the selected stars. The stars in the direction of the equator are hiding behind a belt of haze, while in the direction of the elevated pole, those hardy constellations that rule the higher latitudes beam forth in all their glory.

You could cuss. It helps. But better, you could break out your sextant, shoot those polar friends, and make them tell all by use of Ageton's tables. If you are lucky enough to have one of the little booklets published by the Navy Hydrographic Office years ago as H.O. 211, your only problem may be locating this gem. In its original form, H.O. 211 is a remarkably small booklet, six by 9¼ inches by 3/16 inch thick. You can slip it into a jacket pocket, but once you have used it, locating it will not be a problem. You will always know where it is, because with this tiny booklet you can go anywhere in the world, reduce any sight you may ever take, and work any problem in great circle sailing you ever need work. For this excellent set of tables we are indebted to Lieutenant (later, Admiral) Arthur A. Ageton, United States Navy. Let's not call these tables "H.O. 211" anymore; Admiral Ageton deserves recognition.

I would be remiss if I did not mention another set of perhaps equally excellent sight reduction tables, published in a booklet almost exactly the same size as H.O. 211. This is Dreisonstok's Method, designed by Lieutenant Commander J.Y. Dreisonstok, U.S.N., and originally published by the Navy Hydrographic Office as H.O. 208. Some argue that Dreisonstok's tables are superior to Ageton's. In my opinion, this is one of those arguments navigators engage in over a glass of sherry. I am happy with both, although I tend to use Ageton's tables out of habit. During World War II a navigator could tell whether he likely had a later date of rank than his colleagues in other ships by noticing whether they preferred Dreisonstok to Ageton or vice versa. The old hats used Dreisonstok because it came out earlier.

Neither of these tables is offered any longer as a separate publication by the Defense Mapping Agency Hydrographic/Topographic Center, successor to the Hydrographic Office, but both of them can be bought from commercial publishers.

Ageton's tables are included, as Table 35, in Volume II of Pub. No. 9, *American Practical Navigator*, which we all know as Bowditch. (Appendix G in this book is the entire Table 35, taken from Bowditch.) Since no competent navigator would go to sea without his Bowditch, you will always have Ageton's tables with you. It is still nice to have one of those handy little copies of H.O. 211, because they are so easy to manage. I always toss in my copy of Ageton's tables (H.O. 211 in its original form) when I pack my publications for an offshore voyage.

Ageton's tables must have declined in popularity for a while after the development of the several sets of precomputed tables such as H.O. 214 and similar, successor publications. The precomputed tables seemed easier, and the users felt that they had a better grasp of what they really were doing. It was all an illusion. The only evident advantage of the precomputed tables is that you get an early indication of gross error when you first enter tables such as Pub. No. 249. If the tabulated altitude, $H_t$, is not reasonably close to the corresponding value of $H_o$, you know there is something wrong somewhere. With Ageton's tables, this comparison comes a bit later in the sight reduction process. This very fact could be a hidden advantage of Ageton's Method: Be more careful with your fingers, and the problem is solved. As for getting a better grasp of what they are doing, I wonder how many navigators who work sights from precomputed tables every day have any idea why they do it as they do. Anyone who uses Ageton's tables will sooner or later investigate just what he is doing.

If I told you I was going to show you how to reduce a sight by use of basic trigonometric formulas, you would quickly put this book back on the shelf. But with Ageton, that is exactly what you are doing, with one simple modification to make it easy. If you had plane trigonometry in high school (middle school, if you are a preppie), you may recall that in solving basic trigonometric problems, you multiply the sine or another function of an angle by a side of the triangle involved. The angles are measured in degrees and the sides in a linear measure such as feet, meters, miles, or the like. In spherical trigonometry, the sides are the degrees of the angle. The sines, cosines, tangents, secants, and other trigonometric functions may be taken from tables, and they can come out of the tables with one whole lot of numbers to the right of the decimal point. Multiplying these long numbers is tedious, and it is easy to make mistakes. Adding them would be much easier. This brings us to the use of logarithms. If you have a table to give you the logarithms of the trigonometric functions involved, you can eliminate multiplication and division entirely. Remember that to multiply numbers, you add their logarithms. To divide numbers, you subtract their logarithms. The tables in Ageton's Method are tables of logarithms modified to make it easy for you to work either sights of celestial bodies or other solutions of the celestial triangle, such as great circle problems, by a series of formulas. More specifically, Ageton's tables are simply tables of the logsecants and logcosecants of angles between 0° and 180°, with one modification: One of the biggest problems faced by those of us who are not mathematicians when working with logfunctions of angles is the characteristic. This is the part of the logarithm to the left of the decimal point. Ageton solved this problem by taking the logsecant and logcosecant of every angle between 0° and 180° and multiplying each of them by 100,000. This eliminates the decimal and the characteristic. The result is much easier arithmetic and no confusion. Ageton's Method requires you to add or subtract four two-line columns of four- or five-digit numbers, on top of the addition and subtraction required for sight reduction by use of precomputed tables. This is certainly not an excessive demand, but admittedly a bit more work. Why go to this trouble?

There are several reasons, most of which I have already

mentioned. Why carry around 10 pounds of books in your seabag when you can tote a book weighing a few ounces that will do the same thing and much more? Another big reason for using Ageton's tables is that *you plot your lines of position from your DR position*. Adding four sets of two five-digit figures takes much less time than plotting an assumed position. Let's work a sight by Ageton's Method.

The same morning that you shot the round of stars we worked in Chapter 9, you also got an observation of Capella. But Capella's declination is near 46° north, and you cannot reduce a sight on Capella by use of Pub. No. 249. You do have your Bowditch with you. The observation of Capella was as follows:

| | |
|---|---|
| C | $11^h$-$20^m$-$35^s$ |
| W | $00^m$-$21^s$ |
| CE | $03^m$-$02^s$ f |
| $H_s$ | 59° 41.8′ |

Work this sight by Ageton's Method, Table 35 of Bowditch. The procedure is exactly the same if you use a copy of H.O. 211. Table 35 is reproduced in its entirety in appendix G. We will reduce this sight step by step, just as we have been doing.

**Step 1**

Plot your DR. It is the same, with minor adjustments, as the DR you used to work the morning planet sight and star sights earlier by Pub. No. 249. Your DR for the purpose of reducing this sight is Lat. 24° 58.7′ N, Long. 92° 37.7′ W.

**Step 2**

Record your sight. No change here.

**Step 3**

Correct your time to get the exact GMT of the sight. Still no difference from what you have been doing.

**Step 4**

Step 4 is exactly the same as always. Open whatever almanac you happen to be using to the daily pages for 21 October 1983. (I am using the *Nautical Almanac* in this example.) From the Aries column extract the GHA of Aries, which is 194° 19.5′ for the even-hour time of $11^h$ GMT. No change.

**Step 5**

From the yellow pages find the correction for $17^m$-$12^s$ difference in time. The correction is 4° 18.7′. Add this to the GHA of Aries for the even hour, and get the GHA of Aries at time of sight, 198° 38.2′. By now you may be wondering when the changes come. Please stay with me.

**Step 5A**

Step 5A is also old hat. Precisely as you have done it before. These six steps are required for any star sight reduction with any set of tables, except that Step 5A is not needed for use with Pub. No. 249, Volume 1. From the list of stars on the right-hand side of the left-hand daily page for 21 October in the *Nautical Almanac*, you extract the SHA of Capella, which is 281° 07.8′ (and, while you are at it, the declination, which is N 45° 58.9′). Add the SHA of Capella to the GHA of Aries, and get 479° 46.0′. Since there are only 360° in a circle, subtract 360° from this sum, and get a GHA of Capella of 119° 46.0′. As I said, no change whatsoever.

**Step 6**

Here come the differences. They are not enough different to cause you any trouble. To reduce a sight by Ageton's Method you do not need an assumed position, because you do not need to round off decimals to make the arithmetic easier. It is as easy to work with decimals in these tables as it is to work with whole numbers. For this reason, and others, *you work the problem from your DR, and plot the resulting LOP from your DR*. So subtract your DR longitude from the GHA of Capella: Long. 92° 37.7′ W subtracted from 119° 46.0′ leaves an LHA of 27° 08.3′. This is the LHA because Capella is west of your DR longitude (119° is west of 92°), and thus the meridian angle, t, is west. The differences to this point are so small that you can work the sight this far with the same form you have been using for the Pub. No. 249 sights. Just remember to use your DR and not an assumed position.

**Step 7**

At this point you must change the form you have been using to reduce your sight. Look over the form in the example accompanying this problem, and imagine there are no numbers filled in below the DR longitude. Drop down two spaces, as was done in the form, and write in the meridian angle, t, of 27° 08.3′, and label it W. With

BODY CAPELLA (Upper/Lower limb, if Sun or Moon)
Date 21|10|83 ZT 0505
DR Lat. 24°-58.7' N Long. 92°-37.7' W

| | | | |
|---|---|---|---|
| C | 11h-20m-35s | Hs | 59°-41.8' |
| W | 00m-21s | Main Corr. | −0.6 |
| C-W | 11h-20m-14s | Addn'l | |
| CE | 3m-02s f | Dip | −3.4 |
| GMT | 11h-17m-12s | Total Corr. | −04.0' |
| | | Ho | 59°-37.8' |

| | |
|---|---|
| GHA (~~Body~~/Aries) | 194°-19.5' |
| Corr. (min. & sec.) | 4°-18.7' |
| GHA | 198°-38.2' |
| SHA (if Star) | 281°-07.8' |
| Sum | 479°-46.0' |
| Less 360° if req. | 360°-00.0' |
| GHA | 119°-46.0' |
| DR Long. | 92°-37.7' W |

| | | Add | Subtract | Add | Azimuth Subtract |
|---|---|---|---|---|---|
| LHA (t west) | 27°-08.3' W | A 34097 | | | |
| Declination | N 45°-58.9' | | | | |
| d difference | -0- | | | | |
| Declination | N 45°-58.9' | B 15810 | A 14319 | | |
| R | | A 49907 | B 2298 | B 2298 | A 49907 |
| K | 49°-18.5' N | | A 12021 | | |
| DR Lat. | 24°-58.7' N | | | | |
| (K ~ L) | 24°-19.8' | | | B 4040 | |
| Hc | 59°-47.5' | | | A 6338 | B 29831 |
| | | | | | A 20076 |
| Ho | 59°-37.8' | | | | |
| Alt. Intercept | 9.7 a | | | | |

Z N 039° W
From 360° (N. Lat.) if req.; 360°
subtracted from or added to 180° (S. Lat.): ______
Zn 321°

Ageton's Method you always use the meridian angle, east or west as the case may be. (This may explain why the term "LHA east" came into common usage even though it is not strictly correct.) Forms are nice, and I do recommend them if you do not allow yourself to get married to them. It is important that you know how to work any sight without a printed form, by laying out the form in your navigator's notebook as you work. To lay out this form in your notebook, move to the right on the line just above t, and in the same relative spaces shown in the form used in the example, write *Add, Subtract, Add, Azimuth Subtract*. You do not need to draw the vertical lines, but if they help, there is no reason not to. You are ready to go into the tables in appendix G.

**Steps 7A and 8**

Open the tables and look at them carefully. They consist of five columns to a page. The columns are headed by values of angular measurements, each successive heading increasing by an increment of 30 minutes over the one preceeding. Under each column head are two subcolumns headed, of all things, A and B. There are 60 lines in each column, with every other line representing a whole minute of arc and being so marked. The first line is 0′, followed by an unmarked line, then 1′, and so on through 30 minutes of arc. The unmarked lines represent half-minute increments. This is all there is to the tables. There is no equivalent to Table 5 of Pub. No. 249 in Ageton's tables, because interpolation is never necessary for practical navigation. Notice, at the foot of the tables, another series of headings, increasing in 30-minute increments from right to left. If you subtract the angular measurement at the head of a column plus the 30 minutes of arc tabulated within the column from 180°, you will get a number equal to the heading at the foot of the column. The angular measurement at the foot of a column is the reciprocal of the measurement—plus 30′—at the head of that column. Let's work the Capella sight.

The meridian angle, t, is 27° 08.3′. Turn the pages of the tables until you come to a column headed 27°. The tables in Bowditch and H.O. 211 are identical in arrangement. The figures in the first line of the table give the values of A and B for 27° 00′. To get the value for 27° 08.3′, run down the left-hand side of the column to the 8′ line. Run across this line to the A column under 27°, and get the value of A, which is 34097. Label this value "A" and write it next to t (and on the same line) under the first *Add* column in the form.

When you extracted the declination of 45° 58.9′ from the almanac, you should have written it in the form at the place the corrected declination would have gone for any other body. There is no correction needed for the declination of a star. So turn the pages to a column headed, this time, 45° 30′. You do this because 45° 58.9′ lies between 45° 30′ and 46° 00′. Make a mental subtraction of 30′ from 58.9′; you get 28.9′. Run down the left-hand minute column to the 29′ line. The figure 28.9′ is nearer 29′ than 28.5′, which is the unmarked line between the 28′ line and the 29′ line. Now run across the page on the 29′ line to the B subcolumn below 45° 30′, and extract a B value of 15810. Notice the B columns are printed in heavier type than the A columns. This makes it easier to keep track of what you are doing in the tables. Write the B value of 15810 in the first *Add* column on the same line as the declination and under the A value of 34907 that you extracted for t. While you have the tables open and are on the 29′ line under 45° 30′, extract the value of A, which is 14319, and write this value under the first *Subtract* column, labeling it "A." You have opened the tables twice.

Add the values of A34097 and B15810 in the first *Add* column. The sum is 49907. The number 49907 is the A value of an angle in the divided navigational triangle, the angle Ageton has labeled "R." This time you must go hunting. Thumb through the tables until you find the number nearest 49907 somewhere in the A subcolumns. If you promise to hunt for it, I will tell you where it is; the nearest number tabulated is 49909, in the 18° 00′ column. But the value of R in degrees is not what we need. All we want is the value of B that corresponds to the A of 49907, now 49909. This B value is 2298.

At this point you have more values to enter in the *Add, Subtract* columns in your work form. In the first *Subtract* column, on the same line as A49907, enter B2298; under the third column, which is the second *Add* column, enter B2298 again. In the last column, which is the second *Subtract* column, enter A49907 again. You do this to save coming back to the tables later.

Subtract the B value of 2298 from A, 14319, in the first *Subtract* column. The remainder is 12021, which is the A value of another angle, K, in the divided navigational triangle. Thumb through the tables to a column that contains A12021. This is the column headed 49°, and the nearest tabulated value to 12021 under A is 12020 on the 18.5′ line. K is 49° 18.5′.

Subtract your DR latitude, 24° 58.7′ N, from K, 49° 18.5′. There is a rule that tells you whether to add or subtract DR latitude from K. The rule states that you always give K the same name as the declination; when K and the latitude are the same name you subtract, and when K and the latitude are different names, you add them to get the value written in the form as "(K∼ L)." If you read the last part of this chapter to discover why you do this, after you follow through the step-by-step solutions I am giving you here, you will understand the relationship of K and latitude. In the next example, which will be a sun sight, K and the latitude will have different names, and will therefore be added. But let's get back on course. You subtracted your DR latitude from K and got 24° 19.8′. Now look up the B value of 24° 19.8′ in the tables. B is 4040, because 19.8′ is nearest to 20′ in the left-hand column. Enter B4040 under B2298 in the second *Add* column in the form and on the same line as (K∼ L). Add B2298 to B4040 and get A6338. Now search the tables for an A of 6338. You will find it under the 59° 30′ column, on the 17.5′ line, as precisely that. Your computed altitude, $H_c$, is 59° 47.5′, because 59° 30′ + 17.5′ = 59° 47.5′.

The second subtraction column in the form, the one you labeled *Azimuth Subtract* in your notebook, is, of course, the solution for azimuth angle. When you found the value A6338, you should have taken the B value of 29831 opposite A in the 59° 30′ column on the 17.5′ line, and entered it under the A value of 49907 you entered earlier. By subtraction, you find the A value for the azimuth angle is 20076. Go into the tables and look for the value in the A subcolumns nearest 20076. It is 20074, and you find it under the 39° 00′ column, on the 02.5′ line. This is the same column entered from the foot as 140° 30′, a very important observation. Why?

Read the instruction at the head of the tables on each page that says: "Always Take 'Z' From Bottom of Table, Except When 'K' is Same Name and Greater Than Latitude, in Which Case Take 'Z' From Top of Table." A glance back up your work confirms that K is greater than the DR latitude and has the same name. Z must be taken from the top of the page. Extract a Z of 39° 02.5′, and round it off to Z = 039°. The azimuth angle, Z, must be converted to true azimuth, Zn. This is done exactly as we have been doing it, in the manner I explained in Chapter 14. The meridian angle, t, is west, and the elevated pole is north. Label Z N 039° W, and subtract 039° from 360°. The true azimuth is 321°.

**Step 9**

Correct your sextant altitude, $H_s$, and get an observed altitude, $H_o$, of 59° 37.8′. This is no different than the way you have always done it.

**Step 10**

Find the difference between $H_c$ and $H_o$. Your altitude intercept is 9.7 away. No change.

**Step 11**

Plot your 0505 Capella LOP *from your DR*, and label it. You measure the altitude intercept toward or away from the DR, along the direction or the reciprocal direction of the azimuth, just as you did when plotting from an assumed position.

If you have followed me this far, you may be saying to yourself, this procedure seems more than just a little different to me. Except for the use of your DR instead of an assumed position to work the sight and plot your LOP from, the only big changes are in Steps 7 and 8. After you have worked a few sights with Ageton's tables, you will see that it has taken me much longer to tell you how to do it than it actually takes to do it. To illustrate, let's work the sun sight we took on the morning of 20 October 1983 at 0915, and compare this with our original solution by Pub. No. 249. The sight was taken at DR Lat. 26° 31.0′ N, Long. 90° 55.2′ W. The sight was as follows:

| | |
|---|---|
| C | $15^h$-$20^m$-$10^s$ |
| W | $01^m$-$39^s$ |
| CE | $03^m$-$07^s$ f |
| $H_s$ | 37° 27.4′ |

Your DR was Lat. 26° 31.0′ N, Long. 90° 55.2′ W. This time you are not going to use a preprinted form; you are going to work the sight in your navigator's notebook, just as you would do at sea. Please excuse me if I am repeating myself. I have no quarrel with the use of printed forms to reduce sights. I can see that they could be a boon to the person who reduces sights only at infrequent intervals, and to the beginning navigator. *But do not get married to any form.* I have seen people who claimed to be proficient navigators come unglued when an unwitting shipmate opened a hatch and allowed their precious forms to blow overboard. LAS—Lost At Sea! Forms do give you a sequence to follow when learning. I suggest you take the form and copy it in your notebook each time you reduce a sight. Soon you will see that you no longer need the form to do the work, and then you will have one less stack of papers to lug around with you. So let's line up that notebook and get with it. Please follow me as we go along.

I am going to digress for just one minute: See the symbols I have used in my longhand example of a sight as you would work it in a notebook? Instead of writing out a lot of frequently used terms, navigators use symbols. The symbol ☉ means the sun, the symbol $\underline{\odot}$ means the sun's lower limb, and the symbol $\overline{\odot}$ means the sun's upper limb. You are working a sight of the sun's lower limb. The symbol λ is the Greek letter lambda, lower case, and stands for longitude to distinguish it from latitude when abbreviated. There are other symbols we may use, but as you see them, their meaning will be obvious.

This is the point at which you can get a fair assessment of the difference in time and effort involved in working a sight by Ageton's Method rather than precomputed tables. I believe you will be pleasantly surprised.

**Step 1**

Same as before.

☉ ZT 0915, 20/10/83, DR L 26°-31.0′ N, λ 90°-55.2′ W

| | | | | | |
|---|---|---|---|---|---|
| C | 15ʰ-20ᵐ-10ˢ | | +15.0 | $H_s$ 37°-27.4′ | |
| W | 1ᵐ-39ˢ | | −3.4 | +11.6 | |
| C-W | 15ʰ-18ᵐ-31ˢ | | | $H_o$ 37°-39.0′ | |
| CE | 3ᵐ-07ˢ f | | | | |
| GMT | 15ʰ-15ᵐ-24ˢ | | | | |

| | | ADD | SUBT. | ADD | Z/Zn SUBT. |
|---|---|---|---|---|---|
| GHA | 48°-47.1′ | | | | |
| Corr. | 3°-51.0′ | | | | |
| GHA ☉ | 52°-38.1′ | | | | |
| DR λ | 90°-55.2′ W | | | | |
| t | 38°-17.1′ E | A 20792 | | | |
| DEC. | S 10°-16.7′ | | | | |
| d diff. | +0.2′ | | | | |
| DEC. | S 10°-16.9′ | B 703 | A 74832 | | |
| R | | A 21495 | B 10087 | B 10087 | A 21495 |
| K | 13°-01.0′ S | | A 64745 | | |
| DR L | 26°-31.0′ N | | | | |
| (K~L) | 39°-32.0′ | | | B 11280 | |
| $H_c$ | 37°-41.5′ | | | A 21367 | B 10165 |
| $H_o$ | 37°-39.0′ | | | | A 11330 |
| | 2.5a | | | | |

Z = N 129°-37′ E

Zn = 130°

### Steps 2, 3, 4, and 5

Same as before. The GHA of the sun is 52° 38.1′.

### Step 6

Find t in degrees, minutes, and tenths of a minute of arc by subtracting the GHA of the sun from your *DR longitude.* The meridian angle, t, is 38° 17.1′ E. No more work.

### Steps 7 and 8

Here you find the major differences in reducing a sight by Ageton's Method. The results of this step give you what you obtained through Steps 7 and 8 in the reduction of a sight by use of the precomputed tables.

To the right of t 38° 17.1′ E, on the line above, set up four columns, and head them *Add, Subtract, Add,* and *Z/Zn Subtract.* Since I write large, I find it helps to draw vertical lines.

Now open your tables and extract the A value for t under the 38° 00′ column on the 17′ line. Label it "A 20792."

Turn to the page with the 10° 00′ column and get the B value for the sun's declination of S 10° 16.9′ on the 17′ line. The B value is 703.

While in the 10° 00′ column on the 17′ line also extract the A value for 10° 17′ and enter it in the next column, the *Subtract* column, in your notebook. A is 74832.

Add the A20792 for t and the B703 for declination in the first *Add* column. The sum is 21495. Label this A. This is the A value for angle R. You do not need R itself, but only this A of 21495 so you can find where 21495, or the number nearest to it, appears in the A subcolumns. This search takes only a few seconds. When you find the place, extract the B value of R on the same line in the same column. B is 10087. Write it in the first *Subtract* column, the second column you set up in your notebook. While you are at it, write B10087 again on the same line but under the third column in your notebook, which is the second *Add* column.

Then write the A value of R again in the *Z/Zn Subtract* column. You now have, under the four columns and on the same line as R, A21495, B10087, B10087, A21495.

Subtract B from A in the first *Subtract* column, and get the A value of angle K, 64745. Find A64745 in the tables; it is under the 13° A subcolumn, on the 01.0′ line. K is 13° 01.0′ S.

Find (K∼ L). Remember the rule. K contrary name to latitude, add K and the latitude. K takes the same name as the declination. Add your DR latitude of 26° 31.0′ N to K, 13° 01.0′ S, and get a (K∼ L) of 39° 32.0′. Look up 39° 32.0′ in the tables and extract the B value of 11280. This goes in the second *Add* column in your notebook, the third column you set up.

Add the B10087 of R you entered in this column earlier to the B11280 of (K∼ L), and get A21367. A21367 is the A value of $H_c$. Thumb through the tables until you find this number, or the one nearest to it, under an A subcolumn. It is under the 37° 30′ column on the 11.5′ line. $H_c$ is 37° 41.5′. Enter it in your worksheet.

While you have the A value of 21367 located, extract the corresponding B value of 10165 and enter it in the *Z/Zn Subtract* column in your notebook. Subtract this from the A value of R, 21495, which you entered earlier, and get A11330, the A value of the azimuth angle, Z. Again look at the rule given at the top of the tables. You see Z must be taken from the *bottom* of the page on which you find A11330. This is in the column *above* 129° 30′. On the *right-hand* side of the tables go up the minute column to the 07.0′ line. Z is 129° 37.0′. Since your DR latitude is north, and t is east, Z is N 130° E when rounded off. So Z equals Zn, 130°.

### Step 9

Correct your $H_s$, 37° 27.4′, to an $H_o$ of 37° 39.0′. No change.

### Step 10

Subtract $H_o$ from $H_c$, and get the altitude intercept, 2.5′ away.

### Step 11

Plot your 0915 sun LOP from your DR position.

Ageton's Method provides for working sights of any declination from any DR. This allows the navigator to shoot bodies whose LHA and meridian angle both may be greater than 90°. We have already seen one situation in which a star was in position to observe when its LHA and t were both greater than 90°, when we reduced the lower transit observation of Kochab in Chapter 11. In middle and upper latitudes you routinely get observations of bodies whose LHA and t may both exceed 90°. Above the Arctic Circle or below the Antarctic Circle you may shoot the sun, moon, or planets in this situation. Reducing such a sight by Ageton's Method is no different from

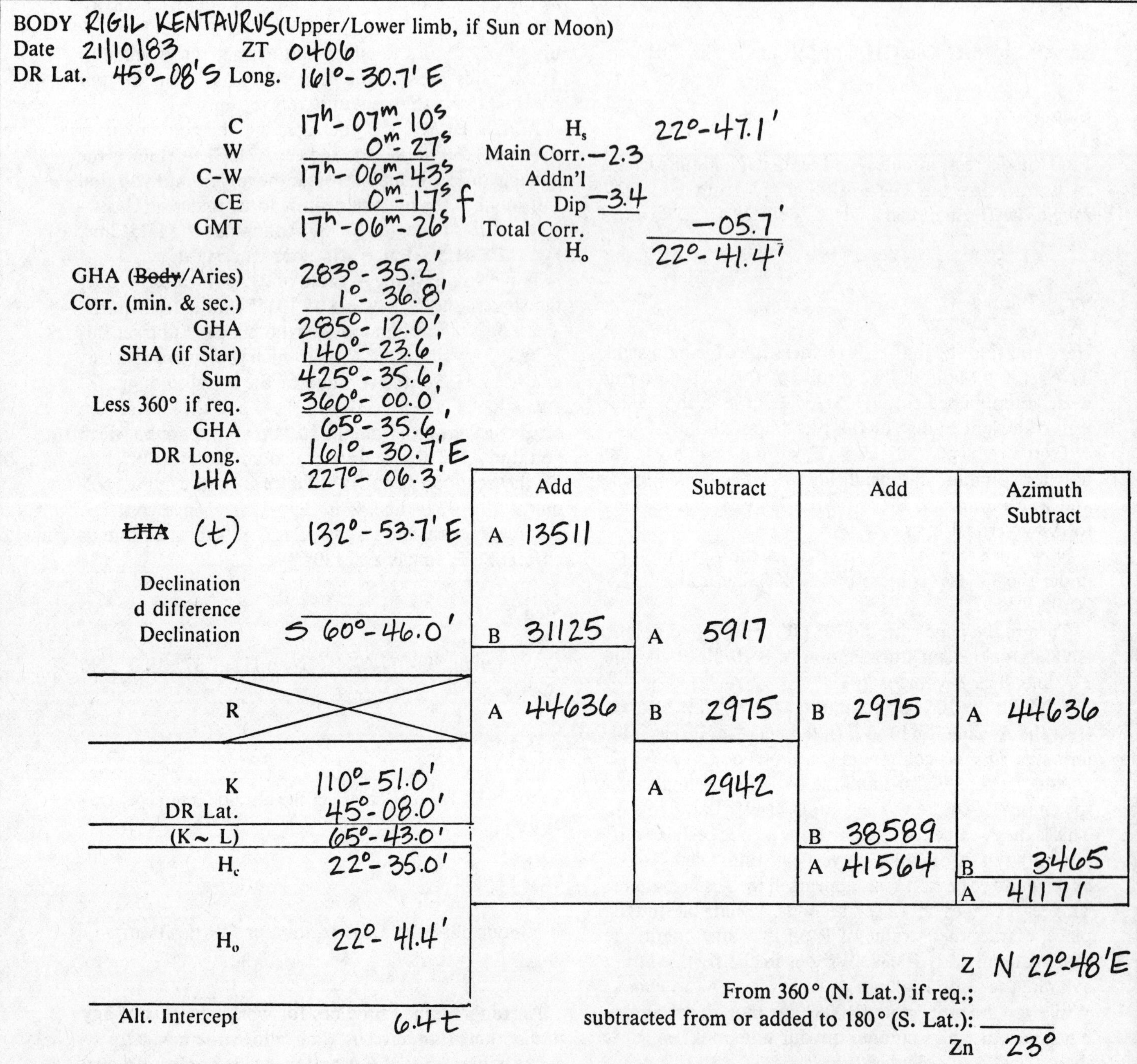

BODY RIGIL KENTAURUS (Upper/Lower limb, if Sun or Moon)
Date 21/10/83 ZT 0406
DR Lat. 45°-08'S Long. 161°-30.7' E

| | | | |
|---|---|---|---|
| C | 17ʰ-07ᵐ-10ˢ | $H_s$ | 22°-47.1' |
| W | 0ᵐ-27ˢ | Main Corr. −2.3 | |
| C−W | 17ʰ-06ᵐ-43ˢ | Addn'l | |
| CE | 0ᵐ-17ˢ f | Dip −3.4 | |
| GMT | 17ʰ-06ᵐ-26ˢ | Total Corr. | −05.7' |
| | | $H_o$ | 22°-41.4' |

| | |
|---|---|
| GHA (~~Body~~/Aries) | 283°-35.2' |
| Corr. (min. & sec.) | 1°-36.8' |
| GHA | 285°-12.0' |
| SHA (if Star) | 140°-23.6' |
| Sum | 425°-35.6' |
| Less 360° if req. | 360°-00.0' |
| GHA | 65°-35.6' |
| DR Long. | 161°-30.7' E |
| LHA | 227°-06.3' |

| | | Add | Subtract | Add | Azimuth Subtract |
|---|---|---|---|---|---|
| ~~LHA~~ (t) | 132°-53.7' E | A 13511 | | | |
| Declination<br>d difference<br>Declination | S 60°-46.0' | B 31125 | A 5917 | | |
| R | | A 44636 | B 2975 | B 2975 | A 44636 |
| K | 110°-51.0' | | A 2942 | | |
| DR Lat. | 45°-08.0' | | | | |
| (K ~ L) | 65°-43.0' | | | B 38589 | |
| $H_c$ | 22°-35.0' | | | A 41564 | B 3465 |
| | | | | | A 41171 |
| $H_o$ | 22°-41.4' | | | | Z N 22°-48'E |
| Alt. Intercept | 6.4 t | | | | |

From 360° (N. Lat.) if req.; subtracted from or added to 180° (S. Lat.): ______
Zn 23°

working any other sight by Ageton's Method, except for one item. Look at the legend above the tables on any of the left-hand pages of Table 35 (or H.O. 211). "When Meridian Angle is *Greater* Than 90°, Take 'K' From *Bottom* of Table." The emphasis is mine. The observation of Rigil Kentaurus that follows is such a sight. Work it out, keeping in mind the instruction just quoted.

On 21 October 1983, as navigator of the schooner *Marapesa*, you observed the star Rigil Kentaurus at an $H_s$ of 22° 47.1'. Chronometer read 17ʰ-07ᵐ-10ˢ, elapsed watch time was 0ᵐ-27ˢ, and chronometer error was 0ᵐ-17ˢ fast. Height of eye was 12 feet. Reduce this sight for a line of position by Ageton's Method. The schooner's DR position was Lat. 45° 08' S, Long. 161° 30.7' E.

The illustration above is the complete sight reduction. See if you can work this sight by yourself.

Even this short explanation of how to reduce a sight by the use of Ageton's Method likely took more time than it would take to actually work the sight. For a fair comparison, compare the time differences between Steps 7 and 8 of Ageton's Method and whatever precomputed sight reduction method you might use, such as Pub. No. 249. Ageton's Method could actually take less time, but if not, the difference in time is not significant. The versatility of Ageton compared with 249 is the significant thing.

Want to know just what you have been doing while reducing the sights you have worked by Ageton's Method? If not, skip this section and take up the text again at the beginning of the next chapter. But before going on to the next chapter, may I suggest that, for comparison and practice, you rework using Ageton's Method at least one each of the sights of the sun, stars, and planets that we worked in earlier chapters by Pub. No. 249? Begin with Chapter 6. Do everything, including the plotting. A word of caution: The intercepts will not be the same, because you are not plotting from an assumed position. The azimuths should be nearly equal.

I emphatically urge you to hang in here for this discussion. The going will not be that tough. *And if you intend to use that little pocket calculator you carry around with you, what I am going to tell you will help a lot.* Although you have been doing it by rote, you have been working trigonometric formulas in this chapter to find a computed altitude and an azimuth. If you understand what you have been doing so far, you are well on your way to being able to work any sight reduction from the most basic trigonometric tables, such as Table 31 or Table 33 in Bowditch. Then, if your forms blow overboard and your Pub. No. 249 tables get soaked, you will not be in serious trouble, as long as you have an almanac and Bowditch. In fact, you will even then be able to choose between Ageton's tables and Table 31 or Table 33 for a method of sight reduction.

If you learn to navigate with only these publications, you can be much more flexible in the decisions you make at sea. When I was researching my book *Coastal Navigation Step By Step*, I took my daughter and one of her girl friends along as crew in *Li'l Tiger*. I wanted to make a coastal passage, but the weather forecast indicated little chance of this. I left my Pub. No. 249 and all the related paraphernalia at home. We took the bus down the coast to where *Li'l Tiger* was lying. Of course, as soon as we got there the forecast changed—we could run the beach outside, rather than following the protected, inside route. But suppose another change caught us in the open Gulf of Mexico? Winter storms there conform to a pattern, with a howling sou'easter followed by a sudden wind shift to the sou'west that veers into an even meaner nor'wester. If you do not get in before the sou'easter hits, it is suicide to try to run most of the entrances in a small vessel. And who wants to be on a lee shore in a sou'east gale? The thing to do is run offshore, usually tearing off to the southwest under reefed main and storm jib at best. Then, before you can get back, the nor'wester slams into you, usually forcing you to lie-to and drift off to the southeast. When it is all over, guessing where you are with an estimated position is just that—a guess. The name of the game is to find out where you are and get into safe waters, in case another front is heading your way. Good, accurate navigation is essential. Without my customary tables, I hesitated to make the outside passage until I reviewed my Bowditch. I learned to navigate with only a Bowditch and an almanac, but it had been years since I had used such basic techniques. I realized I could still do it with sufficient ease. We were not concerned about the ability of *Li'l Tiger* and her crew. There is a long-term almanac in Bowditch.

Before we go into the explanation of what Admiral Ageton set up for us in his tables, let's take a quick lesson in trigonometry. For some of you, it will be a review. For all of you it is an essential prerequisite to getting the best everyday use of that little calculator you have become addicted to. Furthermore, trigonometry can help you every day in almost any situation, ashore or in your boat. After Hurricane Frederick I had a dead tree in my yard that had to come down. It was a tall pine. Like everyone else, I had spent all I could afford on professional tree removers, and I had neither the climbing tools nor the inclination to shinny up this thing and cut it up piece by piece where it stood. On the other hand, I didn't want to drop it on my next-door neighbor's roof. The best of neighbors have their limits. I had to find out how tall the tree was. So I broke out my sextant, went out in the yard, positioned myself 50 feet from the base of the tree, and took a shot at the top. When I got the top on line with the base of the trunk in the horizon glass, I read the sextant. With this information, I was able to figure the height of the tree. A good thing, too, because normally we cut trees flush with the ground. The top four feet of this one would have landed in my neighbor's yard, without leaving any margin for error. I cut it off six feet above the ground and kept it all on my side of the fence. How did I figure all this out? With trigonometry.

Suppose it is 50 feet from the cabintop to the mast truck on a sloop you are rigging. It is 16 feet from the leading edge of the mast to the stemhead fitting. How long is the headstay? Simple trigonometry will give you the answer faster than I can type this.

Figure 20 is a plane right triangle. This means that angle

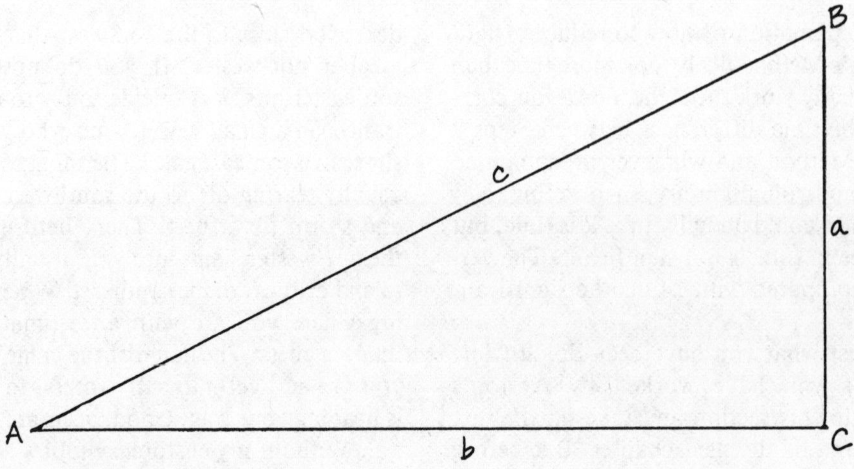


**Figure 20.** *A plane right triangle.*

C is 90°. Let's assume angle A is 34°. Because the sum of the angles in a plane triangle must equal 180°, angle B must be 56°.

In trigonometry we learn that there is a relationship between the sizes of the three angles in a plane right triangle and the three sides of the triangle. The proportional lengths of the sides (we sometimes call the sides "legs") varies with the sizes of the two acute (less-than-90°) angles. We call the ratios of the lengths of the sides functions and give them names like sine, cosine, tangent, cotangent, secant, and cosecant.

In Figure 20, side a divided by side c is the sine of angle A. We write this, sin $A = \frac{a}{c}$ (notice I dropped the "e" at the end of sine). This means sin A equals side a divided by side c. Of course I know I may be getting too simple, but some of you may need more memory refreshing than others. We can write all the functions of any plane right triangle this way:

$$\sin A = \frac{a}{c} \qquad \cot A = \frac{b}{a}$$

$$\cos A = \frac{b}{c} \qquad \sec A = \frac{c}{b}$$

$$\tan A = \frac{a}{b} \qquad \csc A = \frac{c}{a}$$

The terms sin, cos, tan, cot, sec, and csc are abbreviations for sine, cosine, tangent, cotangent, secant, and cosecant. This is all I needed to know to find out how tall that tree was. It is all you need to know to find out how long that headstay is. How do you do it?

I measured off 50 feet from the base of the tree, leg b. I measured an angle with my sextant, angle A, and found it to be 47° 16.6′. From what I just told you, tan $A = \frac{a}{b}$ · Side a is the height of the tree. I looked in Table 31 of Bowditch and found the 47° column, the same way you found the A and B values in Table 35. I found 47° at the foot of the column on page 230 (Volume II, 1975 edition) of Bowditch. Since I was entering the table at the bottom, I found the column I wanted at the bottom; all the functions are tabulated. I found the tangent, "tan," column and ran up to the 16′ line, where I read 1.08243. On the 17′ line above I read 1.08306. There followed a quick eyeball interpolation. The difference between 243 and 306 is 63. Six tenths of 63 is 37.8. I added 0.000378 to 1.08243 and got the tangent of 47° 16.6′, 1.08281. Multiplying 50 feet by 1.08281 gave me 54.14. The tree was 54 feet 2 inches tall.

To find the length of the headstay, you again use two known measurements. You know that side a, which is 16 feet, divided by side b, which is 50 feet, equals the tangent of angle A. Angle A is the angle the headstay makes with the mast. Divide 50 into 16 and get 0.32000. Look up 0.32000 in Table 31 of Bowditch. It is the tangent of 17° 44.4′. Actually 44′ would be close enough. The sine of angle A, which you now know to be 17° 44′, is 0.30459. Glance across the table on the 44′ line to get the sine of angle A. Since sin $A = \frac{a}{c}$, and a = 16 feet, sin $A = \frac{16}{c}$. There you have it: side c (the length of the headstay) is equal to 16 divided by 0.30459. That blooming headstay is 52.53 feet long, assuming the angle at the base of the mast is 90°.

The next time you want to rig a boat and do not have

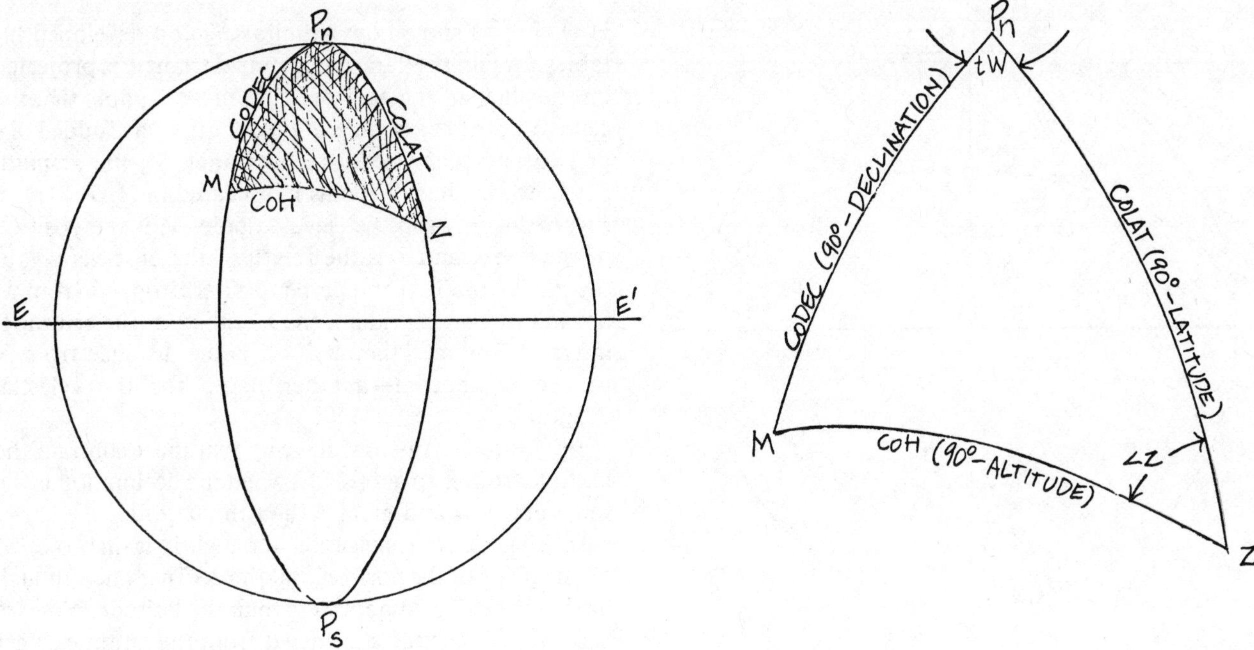


**Figure 21.** *The undivided spherical triangle as it appears in astronomy and navigation on the celestial sphere.*

**Figure 22.** *The undivided navigational triangle, removed from the celestial sphere.*

any drawings, don't spend half a day climbing around the old rigging; use your sextant or a tape measure and Table 31 of Bowditch.

The spherical triangle is not quite as simple, in a way, but it may actually be easier to visualize. The legs of the spherical triangle, like the angles, are measured in degrees. This is because the legs are really arcs of a great circle. They can also, however, be measured in linear units, as we are going to do when we talk about great circle sailing in the next chapter.

The shaded portion of Figure 21 is the undivided spherical triangle, the astronomical triangle, as it would appear on the celestial sphere. The same triangle projected back to earth is the navigational triangle. In the figure, $P_n$ is the pole (north), M is the position of the body observed, and Z is the zenith of the observer. The triangle is called the "undivided navigational triangle" because it has not been divided into two right spherical triangles. As in plane trigonometry, right triangles are easier to cope with than oblique triangles. In this undivided triangle, the side MZ is the coaltitude (CoH), side $P_nZ$ is the colatitude, and side $P_nM$ is the codeclination. Do you recall that in Chapter 11 we defined codeclination as 90° minus the declination? Then colatitude is 90° minus the latitude, and coaltitude is 90° minus the altitude.

Figure 22 is the same undivided navigational triangle, enlarged to show the meridian angle, t, which is west because M is west of the observer's meridian. Angle Z (the symbol ∠ means angle) is the azimuth angle. The figure shows why we have been naming t east or west as the case may be, and why we name azimuth angle by the name of the elevated pole using the direction of t. We are gathering together a lot of reasons why we have been doing things the way we have in this and other chapters.

Figures 21 and 22 are included because they illustrate the concepts you likely have struggled over if you have ever tried to dig out the whys and wherefores of the solution to the navigational triangle in one of the standard texts. This in no way is intended to disparage these works; I think, in the last analysis, they are nothing less than marvelous. They do assume the reader has a working knowledge of the mathematics at hand. Look at Figure 23. The navigational triangle has been divided into two right triangles by dropping a perpendicular from M to the zenith meridian. The perspective is also changed in Figure 23 to show the now-divided triangle as it would project on the celestial sphere if the center point of the projection were at the point where the observer's meridian, the zenith, crossed the celestial equator.

The perspective is more easily visualized in the next illustration, Figure 24. This is the same drawing used in

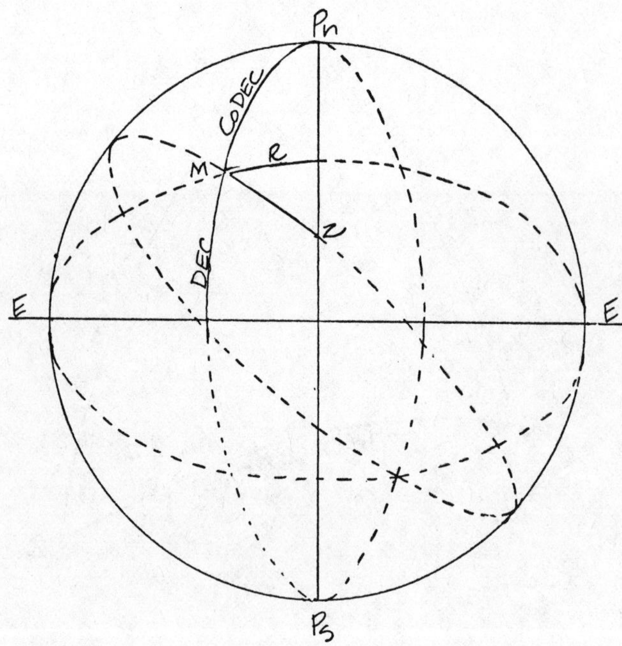


**Figure 23.** *The spherical triangle, as seen from the point where the equator crosses the meridian of the observer. The triangle has been divided by dropping the perpendicular R from M to the meridian of the observer. Each leg of the triangle is an arc of a great circle, including R. The zenith meridian* appears *straight only because it is the zenith meridian!*

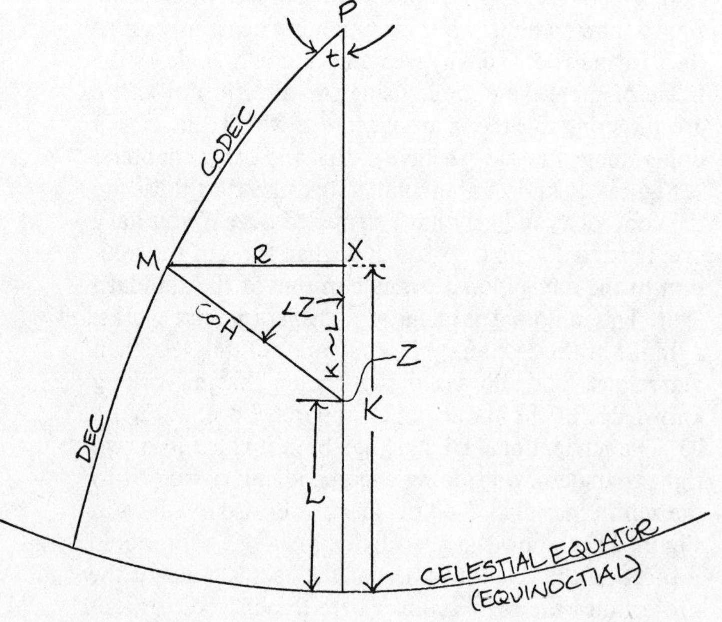


**Figure 24.** *Declination same name as and greater than latitude.*

H.O. 211 to show how Admiral Ageton developed his tables. In Figure 24 the navigational triangle is projected on the plane of the horizon. This means simply that the center of the projection is at the zenith. I have added one or two embellishments, such as angle Z, the azimuth angle, to the illustration as it appeared in H.O. 211. In Figure 24, P is still the elevated pole, M is the position of the body, and Z is the zenith of the observer—you, the navigator. R is the perpendicular dropped from M normal to your meridian, the meridian of the zenith. R intersects the meridian at X. K is the distance from X to the equinoctial—another name for the celestial equator.

In Figure 24 it is obvious why you must subtract the latitude from K to get (K~ L) when the declination is the same name as and greater than the latitude.

In Figure 25, the navigational triangle is again projected on the plane of the horizon, but the declination, although having the same name, is less than the latitude. You see here why K must be subtracted from the latitude to get (K~ L).

In Figure 26, the navigational triangle is projected on the plane of the horizon as in Figures 24 and 25, but the declination is contrary name to the latitude. Latitude must be added to K to get (K~ L).

In these last three illustrations it is also obvious why K always takes the same name as the declination of the body.

The basic formulas for reducing sights (precomputed for you in Pub. Nos. 249 and 229, but awaiting your attention in Chapter 19!) are predicated on the undivided navigational triangle, and must be solved according to four cases. The navigator must keep in mind the various relationships of t and the declination to know when to add t, when to subtract t, and when to add or subtract the declination. One of the greatest advantages of Ageton's Method, which is based on the divided navigational triangle, is that it is uniform under all conditions. You do not need to worry about adding something when you should have subtracted it.

Admiral Ageton gave us the four separate formulas that he used in developing his tables. The formula for computed altitude is:

$$\csc H_c = \sec R \sec (K \sim L)$$

The formula for the azimuth angle is:

$$\csc Z = \frac{\csc R}{\sec H_c}$$

These formulas are simple enough, and, as said above, uniform under any combination of declination, latitude, and t, but before they can be worked, you have to get

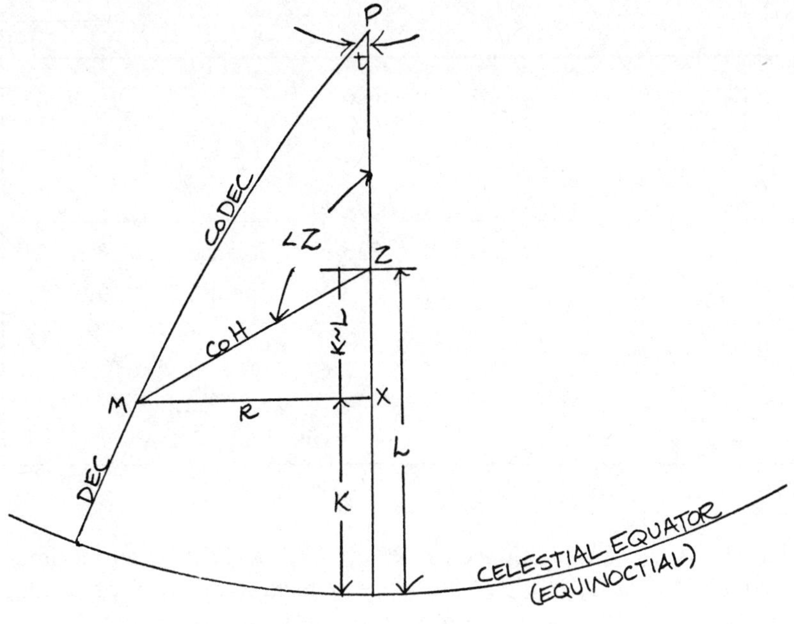


**Figure 25.** *Latitude same name as and greater than declination.*

R and (K~ L). Ageton gave us two additional formulas to find these values. Because these values, R and (K~ L), are terms of the first two formulas above, you have to find R and (K~ L) before you can find $H_c$ and Z, the azimuth angle. The formulas are:

$$\csc R = \csc t \sec d$$

$$\csc K = \frac{\csc d}{\sec R}$$

Since you need R to find K, the formulas are worked in the order given.

$$\text{Csc } R = \csc t \sec d$$

can be expressed, the cosecant of R is equal to the cosecant of the meridian angle multiplied by the secant of the declination. You know how to find t and the declination from consultation of your almanac and the application of the procedures in Steps 1 through 5 (and 5A, for stars). Using Ageton's formulas, we will work a sight on the star Kaus Australis. The sight was taken at DR Lat. 32° 16.3′ S, Long. 31° 41.3′ W. Chronometer read $20^h$-$32^m$-$15^s$, watch $0^m$-$48^s$, and the chronometer error was $0^m$-$04^s$ slow. Sextant altitude was 64° 51.2′. With this information, you find t to be 30° 08.5′ W, declination to be S 34° 23.7′, and $H_o$, 64° 47.3′. Here is how you would work this sight with Ageton's formulas if you did not have Ageton's tables.

Open Volume II of Bowditch to Table 31, "Natural Trigonometric Functions," and extract the values of the

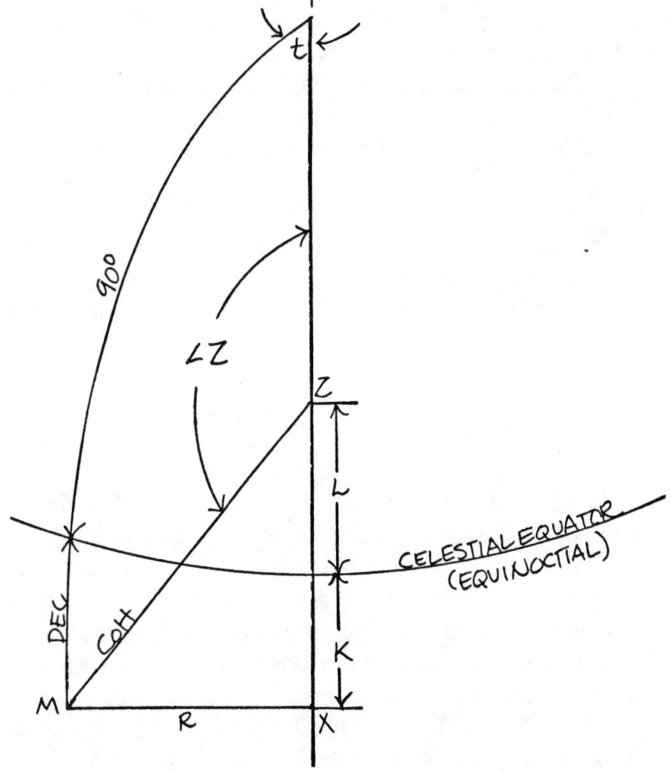


**Figure 26.** *Latitude and declination contrary name.*

| Add | | Subtract | | Add | | Subtract | |
|---|---|---|---|---|---|---|---|
| log csc t | 10.29918 | | | | | | |
| log sec d | 10.08347 | log csc d | 10.24803 | | | | |
| log csc R | 20.38265 | log sec R | 10.04090 | log sec R | 10.04090 | log csc R | 10.38265 |
| K | 38° 21.9′ S | log csc K | 0.20713 | | | | |
| DR Lat. | 32° 16.3′ S | | | log sec | | | |
| (K∼ L) | 6° 05.6′ | | | (K∼ L) | 10.00246 | | |
| $H_c$ | 64° 49.0′ | | | log csc $H_c$ | 20.04336 | log sec $H_c$ | 10.37108 |
| $H_o$ | 64° 47.3′ | | | | | log csc Z | 0.01157 |
| | 1.7 a | | | | | | |
| | | | | | | Z | S 76° 50′ W |
| | | | | | | | 180° |
| | | | | | | Zn | 257° |

functions you need to solve for R in the formula. (Table 31 is excerpted in appendix H.) By now you should be able to look up things in tables, so we can press on. By substituting specific values in the formula csc R = csc t sec d, you get csc R = 1.99148 (you did a bit of eyeball interpolation) × 1.21182. If you love multiplication, you should have one whale of a good time for the next several minutes (unless you have one of those $9 calculators). If you don't, there is a better way.

Use logarithms. Remember that to multiply numbers using their logarithms you add the logarithms; to divide, you subtract. Now our formula looks like this:

$$\log \csc R = \log \csc t + \log \sec d$$

This time go to Table 33 of Bowditch, "Logarithms of Trigonometric Functions" (see appendix I), and get the values to fill out the formula. (Call it an equation, if you really want to impress someone.) From Table 33 you get this:

$$\log \csc R = 10.29918 + 10.08347$$
$$\log \csc R = 20.38265$$

Go into Table 33, and search for 20.38265. You find it under the cosecant column under the angle value 24° between the 28′ and 29′ lines. But you do not want the value of R because you are going to work the next formula—I mean equation—for K. You want the logsecant of R. But hold on just one minute, I can hear you say! The number we obtained for the logcosecant of R was *20*.38265. The number I found in the tables was *10*.38265. The numbers 10 and 20 to the left of the decimal in these logarithms are the characteristics of the logarithms. Remember I said they could be troublesome? This is why Ageton multiplied the logsecants and logcosecants by 100,000 to devise his tables. That zapped those troublesome characteristics. He termed the logcosecant multiplied by 100,000 "A" and the logsecant times 100,000 "B." But since we have come this far reducing our sight of Kaus Australis the hard way, let's go through with it. Not only will it take you a long way toward an appreciation of Ageton, but it is not so difficult that you will not enjoy working a sight this way every once in a while, in case you ever find yourself with only the most basic tables.

Going back to the first formula, csc R = csc t sec d, set it up like this to work it with logarithms:

| | | |
|---|---|---|
| + | log csc t | 10.29918 |
| | log sec d | 10.08347 |
| | log csc R | 20.38265 |

The next formula is $\csc K = \dfrac{\csc d}{\sec R}$, which, to work logarithmically, sets up this way:

| | | |
|---|---|---|
| − | log csc d | 10.24803 |
| | log sec R | 10.04090 |
| | log csc K | 0.20713 |

Thus, K = 38° 22′, if you don't let that characteristic, 10, to the left of the decimal worry you. (If this does worry you, as it always does me, study Article 110 in Volume II of Bowditch.) We have two more formulas to work out, but by now you should be wondering if there is not some way you can eliminate some of this thumbing back and forth between pages in Table 33. You looked up the log sec of the declination to get the log csc of R by adding it to the log csc of t. Then, doggone it, you had to go back and look up the log csc of declination so you could subtract the log sec of R from it. You still have to work out log csc $H_c$ = log sec R + log sec (K∼ L). How can you set this thing up to save all this extra effort? It should not take you too long to come up with a notebook routine that looks like the one above.

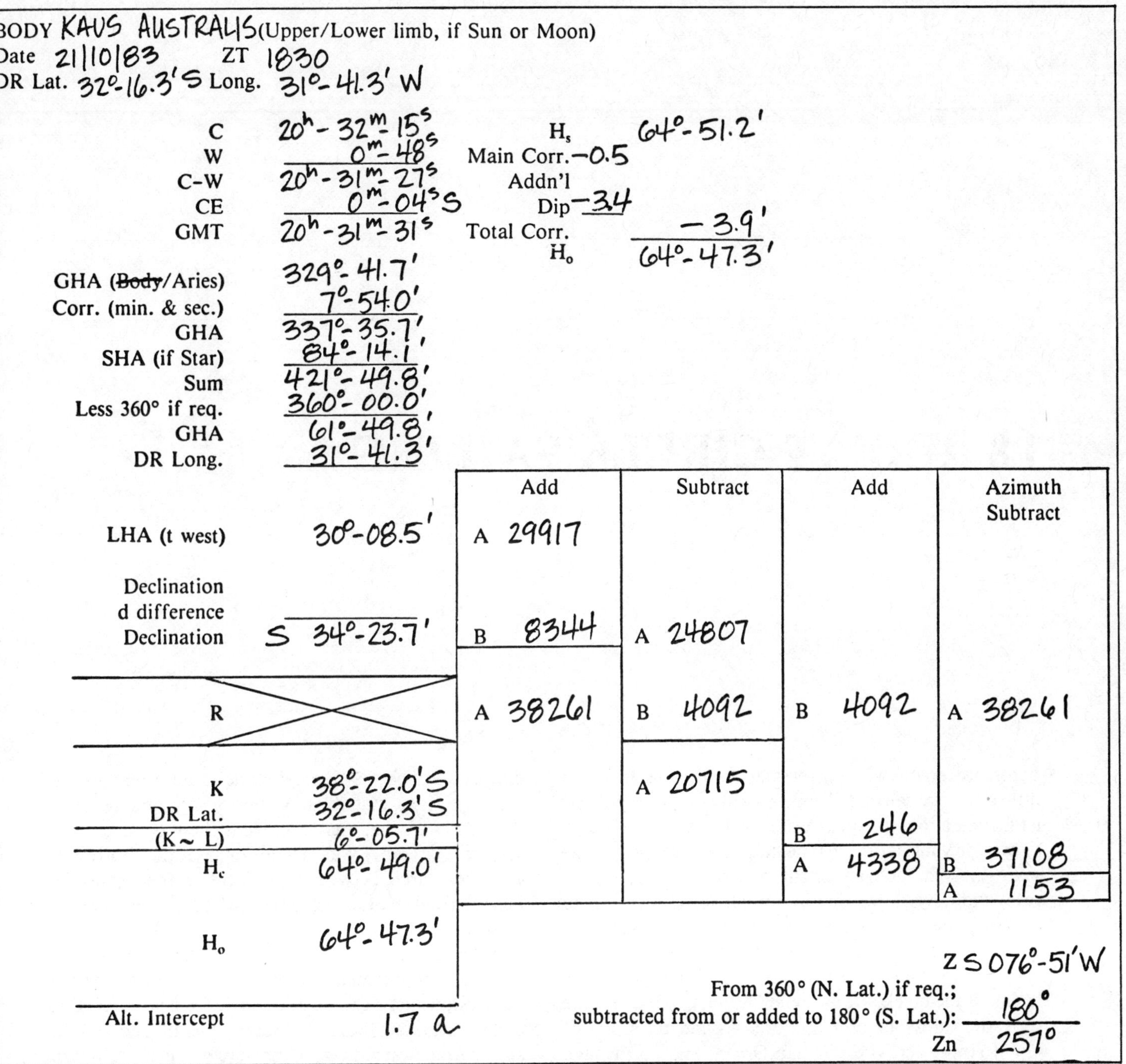

BODY KAUS AUSTRALIS (Upper/Lower limb, if Sun or Moon)
Date 21/10/83 ZT 1830
DR Lat. 32°-16.3′ S Long. 31°-41.3′ W

| | | | |
|---|---|---|---|
| C | 20h-32m-15s | $H_s$ | 64°-51.2′ |
| W | 0m-48s | Main Corr. | −0.5 |
| C-W | 20h-31m-27s | Addn'l | |
| CE | 0m-04s S | Dip | −3.4 |
| GMT | 20h-31m-31s | Total Corr. | −3.9′ |
| | | $H_o$ | 64°-47.3′ |

| | |
|---|---|
| GHA (~~Body~~/Aries) | 329°-41.7′ |
| Corr. (min. & sec.) | 7°-54.0′ |
| GHA | 337°-35.7′ |
| SHA (if Star) | 84°-14.1′ |
| Sum | 421°-49.8′ |
| Less 360° if req. | 360°-00.0′ |
| GHA | 61°-49.8′ |
| DR Long. | 31°-41.3′ |

| | | Add | Subtract | Add | Azimuth Subtract |
|---|---|---|---|---|---|
| LHA (t west) | 30°-08.5′ | A 29917 | | | |
| Declination / d difference / Declination | S 34°-23.7′ | B 8344 | A 24807 | | |
| R | | A 38261 | B 4092 | B 4092 | A 38261 |
| K | 38°-22.0′ S | | A 20715 | | |
| DR Lat. | 32°-16.3′ S | | | | |
| (K ~ L) | 6°-05.7′ | | | B 246 | |
| $H_c$ | 64°-49.0′ | | | A 4338 | B 37108 |
| | | | | | A 1153 |
| $H_o$ | 64°-47.3′ | | | | |
| Alt. Intercept | 1.7 a | | | | |

Z S 076°-51′ W
From 360° (N. Lat.) if req.;
subtracted from or added to 180° (S. Lat.): 180°
Zn 257°

(Notice the arrangement also takes cares of the fourth formula: log csc Z = log csc R − log sec $H_c$.)

That was not as bad as I led you to believe, was it? And look what you got. See the accompanying example of the same sight on Kaus Australis worked by Ageton's tables. Remember, Ageton took the logsecants and logcosecants of angles between 0° 00.5′ and 180°, in increments of a half minute, and multiplied them by 100,000. This got rid of that nuisance of a characteristic, which you ended up ignoring anyhow. Then Admiral Ageton labeled the products of the logcosecants "A" and the products of the logsecants "B" and compiled them into a handy little set of tables that is duck soup to work with.

You now have a basic understanding of how and why Ageton's Method works, and, after working a sight using the admiral's formulas but not his tables, you have an appreciation of the work that Ageton's tables can save you. But if you have come this far, think about where you are. You can now work sights with nothing more than an almanac and Bowditch, if you ever have to.

You are also ready to use that scientific calculator.

# 18 GREAT CIRCLE SAILING

Before we look at working celestial observations with that little scientific calculator you carry in your shirt pocket, let us talk about great circle sailing. True, you could go to sea for your entire lifetime and never use great circle sailing. It is far more important to avoid head winds, head seas, and adverse weather in a small vessel than it is to cut off a few miles of distance. In a sailing vessel a fair wind may be much more important than distance. That is where the old clipper ship saying, "The long way 'round is the quickest way home," came from. The route took the ships out of Baltimore, around the Cape of Good Hope, across the Indian Ocean, and up through the Straits of Sunda to China for cargo; then across the Pacific, around the Horn, and home. Each voyage was a trip around the world. The route may have left something to be desired as far as weather is concerned in the Great Southern Ocean, but you could not find fault with the winds. The ship captains saved some distance on the long passages, however, by sailing a great circle.

The idea that great circle sailing takes a ship into high latitudes, where weather can be severe and winds and currents adverse, has led many small-boat sailors to think that a great circle is some kind of anachronism to talk about when a bunch of hayseeds are tuned in. When conditions are such that a great circle route offers none of the objections stated above, however, and no other hazards intervene, a great circle is the way to go. Why sail great circle? A great circle course is the shortest arc of a great circle between two places on the earth's surface. This arc is the shortest distance between the two places.

We sailed to Brazil from Mobile in the spring of 1982. At a point approximately at latitude 18° north, longitude 62° west, we set a great circle course for a point off Ponta do Calcanhar, Brazil. We saved several hundred miles and five days on passage. Great circle sailing was not the only factor involved, but it was a major one.

But you do not need to learn great circle sailing. *You already know how to work great circle sailing problems.* If you can reduce a sight by Pub. No. 249, you can sail a great circle with an adequate degree of usefulness. If you can work a sight by Pub. No. 229, you can do it well enough. And if you can work a sight by Ageton's Method, you can do great circle sailing problems as well as you are ever likely to have to.

By now you should be able to work a sight by any of these methods.

To work a great circle problem under the circumstances

you are most likely to encounter in ordinary voyage planning, simply use the point of departure as your DR position. The latitude and longitude of your destination take the place of the declination and GHA of a body observed. The azimuth is the initial true course. Computed altitude subtracted from 90° is the distance when the degrees of coaltitude are converted to minutes of arc. That is all there is to it, unless you are going halfway around the world or a distance at least great enough to involve a meridian angle greater than 90°. From a practical point of view this is such a rare situation that it can be left to be pondered by those who delve more deeply into the subject.

Let us see how a problem in great circle sailing can be worked with Pub. No. 249. The people who worked up Pub. No. 249 never intended it to be used for great circle sailing, but it works well enough provided the latitude of your destination does not exceed 29° north or south. When this condition is met, Pub. No. 249 will give you a sufficiently accurate initial course. Pub. No. 249 does not provide for interpolation for differences in latitude and longitude between your DR position and the nearest whole-degree position. You cannot use an assumed position for great circle sailing because this could cause an error of perhaps as much as 60 miles in the distance calculations. Nevertheless, you can use standard linear interpolation and reduce the error in distance considerably. Let's work the great circle sailing problem I worked in 1982 on the way to Brazil. Do it in your notebook, and use the table excerpts in appendix C.

**Step 1**

Set up the problem: Required is the great circle course and the distance from a fix at sea at latitude 18° 20.0′ N, longitude 61° 44.0′ W to a point off Ponta do Calcanhar, at latitude 5° 25.0′ S, longitude 34° 25.0′ W. Please follow the work in the example.

**Step 2**

Interpolate for the difference in meridian angle, t, of 19′ when the tabulated latitude is 18°, contrary name. You find that the $H_t$ at 18° N, contrary name, for a meridian angle of 27° 19′ is 54° 34.4′.

**Step 3**

Interpolate for the difference in t of 19′ at the tabulated latitude of 19° N, contrary name. $H_t$ at 19° N when t is 27° 19′ is 53° 57.0′.

**Step 4**

Interpolate for the difference in latitude when $H_t$ at Lat. 18° N is 54° 34.4′ and $H_t$ at Lat. 19° N is 53° 57.0′. You have completed three interpolations. The last, for latitude, gives an $H_t$ of 54° 21.9′.

**Step 5**

Interpolate for the difference in declination using the latitude of the destination, 5° 25.0′ S, as the declination. By eyeball interpolation you take a d factor of minus 40 to enter Table 5 (d lies between -39 and -41 in the tables). The d correction is − 17. Apply this to the last $H_t$, and call the result $H_c$. $H_c$ is 54° 04.9′.

**Step 6**

Subtract $H_c$ from 90° to get a coaltitude of 34° 55.1′. Multiply 34° by 60 to convert degrees to minutes of arc. (One minute of arc on a great circle of the earth is a nautical mile.) The product is 2,040. Add the 55.1′ of $H_c$ to 2,040; 2,095.1 miles is the distance required.

**Step 7**

You can eyeball the azimuth angle, Z. In the tables, when you were making the interpolations, the value of Z was nearest to 128°. Remember the rule for converting azimuth angle to azimuth. Z N 128° E becomes Zn 128°T. This is the initial course.

The great circle course will appear as a curved course on a Mercator projection. You cannot steer this curved course, so lay down a rhumb line of 128°T on your sailing chart, a Mercator projection, and run this rhumb line for your initial course. Periodically, at least daily, compute another great circle course from your last fix, and plot it as a rhumb line for your next course to steer. Despite the four-step interpolation, the work is easy enough even with Pub. No. 249.

Should the latitude and longitude of your departure and your destination happen to be even degrees, interpolation is not necessary. Should only the latitude of the destination fail to be in even degrees, the interpolation done for the declination in Table 5 is all you need. Interpolation is reduced by the number of factors in the problem that can be taken as whole-number degrees, without minutes of arc involved. For this reason, when plotting a new departure course, scan the current rhumb line to a point where at least one of the parameters in the problem, such as latitude, longitude, or, if you are lucky, both,

## Great circle course and distance by Pub. No. 249

**STEP 1** Departure L 18°-20.0′ N (DR L), λ 61°-44.0′ W (DR λ)
Destination L 5°-25.0′ S (Dec.), λ 34°-25.0′ W (GHA)
t 27°-19.0′ E

**STEP 2** Interpolation for difference of t of 19′ @ L 18°:

t 27°, $H_t$ 54°-49.0′ — 54°-49.0′
t 28°, $H_t$ 54°-03.0′
−46.0′

Diff. t = 19′, so 19/60 × −46 = −14.6′

$H_t$ 54°-34.4′

**STEP 3** Interpolation for difference of t of 19′ @ L 19°:

t 27°, $H_t$ 54°-11.0′ — 54°-11.0′
t 28°, $H_t$ 53°-27.0′
−44.0

Diff. t = 19′, so 19/60 × −44 = −14.0′

$H_t$ 53°-57.0′

**STEP 4** Interpolation for difference in latitude:

@18°N, $H_t$ 54°-34.4′ — 54°-34.4′
@19°N, $H_t$ 53°-57.0′
Diff. $H_t$ −37.4′

Diff. L 20′: 20/60 × −37.4′ — −12.5′

$H_t$ 54°-21.9′

**STEP 5** Interpolation for diff. in declination d −40, d correction — −17.0

$H_c$ 54°-04.9

**STEP 6** 90°-00.0

co$H_c$ 34°-55.1′

Distance = (34° × 60) + 55.1′, Dist. = 2095 miles

**STEP 7** Z = N 128° E, Zn = 128°T, Initial course = 128°T.

are even. If the distance is only a few miles from the departure point, run a Mercator DR plot to this point and compute the great circle course and distance from there. This approach works as well for Pub. No. 229 as it does for Pub. No. 249.

Pub. No. 229 does provide for interpolation of latitude and longitude of both the destination and the departure when these positions are not in even degrees. The interpolation is graphic by use of diagrams included in the tables. (See Figures 27 through 29. Interpolation could also be done in a manner similar to what you did in Pub. No. 249.) There are three diagrams, A, B, and C. Diagram C is a transparent overlay. There are four cases involved in solving the great circle problem by use of Pub. No. 229. In Case I, the latitudes of departure and destination are of the same name and the distance is less than 90° (5,400 miles). In Case II, the latitudes of departure and destination are of contrary name and the distance is less than 90°. Cases III and IV are for greater distances. Since the interpolation is graphic, it is only necessary to take the tabulated $H_t$, as if entering the tables for a sight reduction with even degrees of meridian angle and latitude. Take the azimuth angle as tabulated, and convert it to azimuth. This is the great circle initial course.

Extract $H_t$ as for a sight, and get $H_c$ by correcting for declination (using the difference between the latitude of your destination and the nearest tabulated latitude as though it were the difference in declination). Refine the distance calculations from the latitude of your departure to minutes and tenths of a minute of meridian angle by use of the diagrams. Subtract $H_c$ from 90°, and multiply by 60 to get the distance. Suppose you are at Honolulu, Hawaii, Lat. 21° 18.5′ N, Long. 157° 52.3′ W, and want to go to Pago Pago, American Samoa. What is the great circle course and distance? Work the problem by use of Pub. No. 229. (See appendix F.)

**Step 1**

| | | |
|---|---|---|
| Departure | Lat. 21° 18.5′ N | Long. 157° 18.5′ W |
| Destination | Lat. 14° 16.5′ S | Long. 170° 41.0′ W |
| | | t 13° 22.5′ W |

**Step 2**

From the page "LHA 13°" in Pub. No. 229, Volume 2, Lat. 21°, Dec. 14° S, contrary:

| | | |
|---|---|---|
| | | $H_t$ 52° 44.6′ |
| Dec. Inc. 16.5′, d −56.3 | (1st part −13.8) | |
| | (2nd part −1.8) | −15.6′ |
| | | $H_c$ 52° 29.0′ |

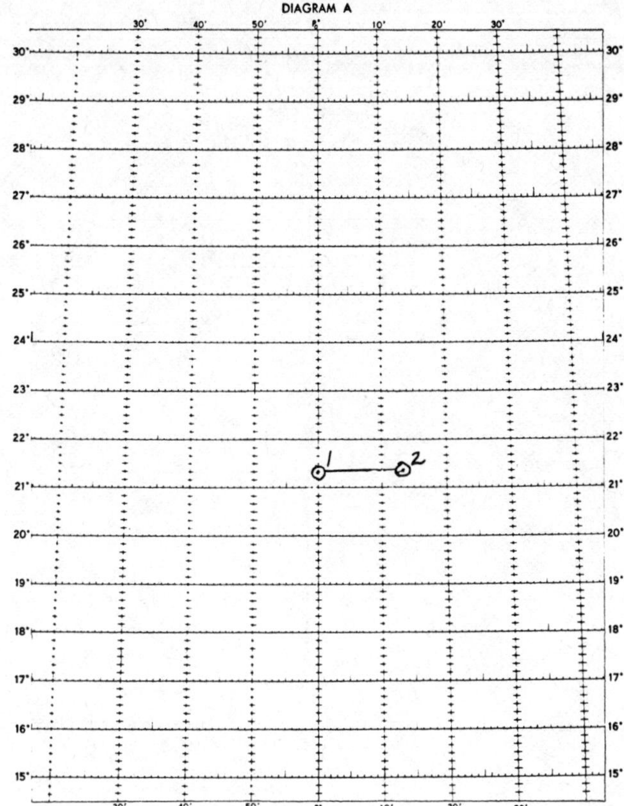


**Figure 27.**

**Step 3**

By use of the diagrams, find the correction to altitude by interpolation for the latitude of the departure and the difference of t.

The diagrams are in the center of the tables. On diagram A, mark lightly with a pencil the latitude of departure, and label it "1." Then mark the value of the meridian angle, t (see Figure 27), label this "2," and draw a line from 1 to 2.

Now take diagram C (Figure 28) and lay it over A with their centerlines coincident. Lay the 18.5-minute mark on C directly over the latitude of departure on A. Number this point "3" on C.

Draw a line on C, while holding it in place, to mark 2 on A. Label this point "4" on C.

In the tables you found the tabulated Z to be 158.9°. Since t is west, mark the 159° point on C at 159° counterclockwise from 0°.

Take C and put it on diagram B (Figure 29), with mark 4 on the line of B that includes $H_c$, the 26° -52° line. Be sure to align the centerline of C with the centerline of

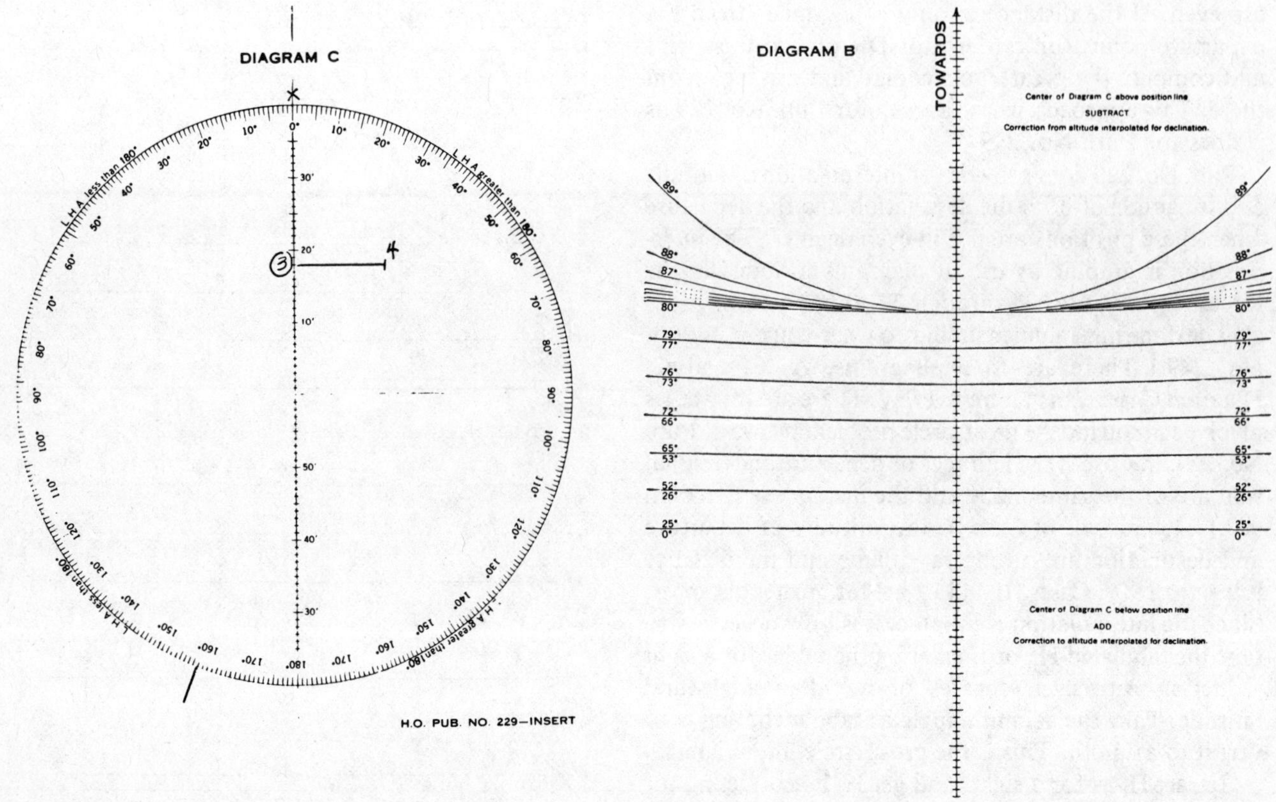


**Figure 28.**

**Figure 29.**

B. Read down the graduations of the centerline of B, and get an altitude correction to $H_c$ of 22.2′. Add this to $H_c$ 52° 29.0′ to get an interpolated $H_c$ of 52° 51.2′.

**Step 4**

Subtract $H_c$ from 90°, multiply 37 by 60, and add the 7.8′ to get the distance.

| | |
|---|---|
| $H_c$ | 52° 51.2′ |
| | 90° 00.0′ |
| co$H_c$ | 37° 08.8′ |

Distance = 2,228.8 miles. Initial course = 159°T.

To me, the use of the diagrams in Pub. No. 229 is a worse hassel than the linear interpolation in Pub. No. 249. But take heart. The most accurate and simplest way to solve a great circle sailing problem is by use of Ageton's Method. It is just as simple to go from Juneau, Alaska to Sydney, Australia by Ageton's Method with great circle sailing as it is to do a problem involving lesser distance. Nonstop voyages of this distance are, to say the least, rare, so let's continue to confine ourselves to passages of less than 5,400 miles.

You wake up one morning in Papeete, Tahiti, and realize it is time to leave. The best things never last forever. You decide a very long time at sea is in order. You clear for—you guessed it—Sydney. Before you go, you want to store the boat. If you can average 125 miles a day, how many days of stores will you require, not counting reserves? What course should you steer on departure?

Reach for Bowditch or, if you have one, that handy little volume, H.O. 211. The solution is in the accompanying worksheet. Follow it as we go through the steps.

Papeete is at latitude 17° 32.0′ S, longitude 149° 34.0′ W. Sydney is at 33° 51.0′ S latitude, 151° 13.0′ E longitude. On the work form for Ageton's Method, if you have one (lined out in your notebook if you don't), put the latitude of Papeete in the spot indicated for DR latitude. Put the longitude of Papeete in the spot indicated for DR longitude. Enter the latitude of Sydney on the line indicated for declination, and the longitude of Sydney on the line indicated for GHA. This is Step 1.

**Step 2**

Since you are going from west longitude to east longitude, add the longitude of departure, 149° 34.0′ W, and the longitude of your destination, 151° 13.0′ E. The sum is 300° 47.0′. Subtract this from 360° 00.0′. The remainder of 59° 13.0′ is t, the meridian angle. Look up 59° 13.0′ in the A subcolumn of Table 35, and extract 6595. (The entire table is reproduced in appendix G.) Put this in the form as shown, in the first *Add* column.

**Step 3**

Enter the latitude of Sydney on the declination line of the form, then extract the B value for this declination from Table 35. It is 8066. Enter it as shown in the first *Add* column. This is exactly the way you proceeded when you worked those sights by Ageton's Method. I am only going into this much detail here to show you that it is an identical process—Ageton's Method needs no interpolation for practical navigation. After the workout with Pub. No. 229 and Pub. No. 249, you ought to dance with joy. But back on course: Add these two figures, and get 14661, the A value of R.

**Step 4**

Find 14661 under the A subcolumn in Table 5, and extract the B value right next to it in the B subcolumn of the same main column. Enter this in the first *Subtract* column of your worksheet, as shown in the form, enter it again in the next *Add* column, and enter A again in the *Azimuth Subtract* column. All three entries are on a line across.

**Step 5**

Subtract the two numbers in the first *Subtract* column, and get the difference of 9966. Look this up in Table 35 to get the value of K from the A subcolumn. K is 52° 39.0′. Name it south, because K always takes the same name as the declination. Subtract the latitude of Papeete, your DR latitude for the sake of solving the problem, from K, and get 35° 07.0′ for your value of (K∼ L).

**Step 6**

Find the B value of 35° 07.0′ in Table 35. It is 8726. Enter this in the second *Add* column, as shown on a line with (K∼ L). Add this value to the B value of R, which you got earlier, to get an A value of 24173. Look this up in Table 35, and extract an $H_c$ of 34° 58′.

**Step 7**

Subtract the $H_c$ of 34° 58.0′ from 90° 00.0′ to get the coaltitude, 55° 02.0′. Multiply 55° by 60 and add the 2.0 minutes of arc to get the distance. Distance from Papeete to Sydney is 3,302 miles.

**Step 8**

While you were getting the value of $H_c$ that corresponds to an A value of 24173, you should also have taken the B value next to it in the tables. This B value is 8646. Put 8646 in the *Azimuth Subtract* column and subtract this from the A value of R you got earlier to get the A value of Z, the azimuth angle. A is 6015. You find—after remembering the rule to follow when K is the same name and greater than the latitude—that Z is S 60° 32′ W. Round this off to 61° and add it to 180° to get Zn, the azimuth.

You find the azimuth, Zn, to be 241°, which is the initial course, and the distance from Papeete to Sydney to be 3,302 miles. When you worked the sight by Ageton's Method the biggest slowdown was the time it took to write the numbers in the form. There was no interpolation.

By the way, your ETE (estimated time en route) will be $26^d$-$09^h$-$59^m$. Better put plenty of stores aboard, with an adequate margin of safety. Crews on some vessels making passages like this have been reduced to drinking the water on board.

Some remarks about great circle sailing are in order.

The rule of thumb is, nothing is gained by sailing the great circle route over distances less than 500 miles.

Should the great circle route take you into higher latitudes than you want to sail in, compromise by using *composite* sailing. To do this calculate the great circle course from your point of departure to your destination. When the great circle course first crosses the highest latitude you care to reach, alter course to sail along that parallel of latitude until you come to the point at which the great circle course again crosses the parallel. There, resume the great circle route.

Composite sailing works because the great circle course always curves toward the elevated pole from the point of departure. An example of this is seen in the last prob-

BODY GREAT CIRCLE FROM PAPEETE, TAHITI TO SYDNEY, AUSTRALIA ~~(Upper/Lower limb, if Sun or Moon)~~

Date 17/5/83 ZT

Dep. DR Lat. 17°-32′ S Long. 149°-34′ W (Papeete)

Dest. " 33°-51′ S " 151°-13′ E (Sydney)

| | | | |
|---|---|---|---|
| C | | $H_s$ | |
| W | | Main Corr. | |
| C-W | | Addn'l | |
| CE | | Dip | |
| GMT | | Total Corr. | |
| | | $H_o$ | |

| | |
|---|---|
| GHA (Body/Aries) | |
| Corr. (min. & sec.) | |
| GHA (λ Sydney) | 151°-13.0′ E |
| SHA (if Star) | |
| Sum | |
| Less 360° if req. | |
| GHA | |
| DR Long. (λ Papeete) | 149°-34.0′ W |
| | 300°-47.0′ |
| | 360°-00.0′ |
| LHA (t west) | 59°-13.0′ |
| Declination | |
| d difference | |
| Declination | S 33°-51.0′ |
| R | |
| K | 52°-39.0′ S |
| DR Lat. | 17°-32.0′ S |
| (K ~ L) | 35°-07.0′ |
| $H_c$ | 34°-58.0′ |
| | 90°-00.0′ |
| | co $H_c$ 55°-02.0′ |
| $H_o$ | (55° × 60) + 2.0′ = 3,302.0 |
| Alt. Intercept | |

| Add | Subtract | Add | Azimuth Subtract |
|---|---|---|---|
| A 6595 | | | |
| B 8066 | A 25413 | | |
| A 14661 | B 15447 | B 15447 | A 14661 |
| | A 9966 | | |
| | | B 8726 | B 8646 |
| | | A 24173 | A 6015 |

Z S 60°-32′ W

From 360° (N. Lat.) if req.;
subtracted from or added to 180° (S. Lat.): 180°

Zn 241°

Distance is 3,302.0 miles

Course - 241° T

lem. The Mercator course from Papeete to Sydney is 253°T; the great circle course is 242°T. The initial course to steer is 242°T.

Always plot great circle courses on your charts. First plot them on great circle charts. These are gnomonic projections, designed solely for great circle sailing. The great circle course plots as a straight line on a gnomonic chart. The course can be gotten directly from one of these charts, normally just by drawing a straight line between the departure and the destination on the chart. Then note the positions of points at regular intervals along the great circle route, and plot these on your Mercator chart. The great circle chart will not have the detailed data needed for safe navigation,so use the Mercator chart to make sure the great circle course does not take you through dangerous waters. Great circle distances can also be gotten from great circle charts, but I find it preferable to use Ageton's Method.

On a Mercator chart a great circle track will appear as a curve, crossing each parallel of latitude higher than the rhumb line course on the chart. The Mercator course will *appear* shorter, and will even measure shorter on the Mercator chart due to Mercator distortion. The great circle track is the shortest distance.

Should you be forced any distance off the great circle course, it is better to compute another great circle course than it is to attempt to return to the original great circle track.

Since there is no way anyone can steer a continuously curving course, the way to steer a great circle route is to steer a series of short rhumb line courses, changing course at intervals to approximate the great circle course. The general rule of thumb is to change course at intervals of five degrees of longitude. This must be varied depending on the amount of "northing" or "southing" the course delivers. In small vessels proceeding at slow speeds, changing course once a day usually serves the purpose well enough to approximate the great circle route unless you are in very high latitudes.

Other uses of great circle solutions include correcting radio direction bearings when considerable distances are involved between the position of the sending station and your boat. In middle and lower latitudes, most navigators do not bother with this conversion of radio bearings to Mercator bearings (in order to plot them correctly on a Mercator chart) unless the distance is over 200 miles. In higher latitudes, the correction is required for lesser distances. A quick method for the conversion of great circle radio bearings to Mercator bearings is given in Table 1 of Bowditch, Volume II, 1975 edition. (These tables are *not* reproduced in the appendixes of this book. You ought to invest in a Bowditch.) In higher latitudes, Table 1 should also be used for visual bearings over great distances, and let me assure you, it is not unusual to take a visual bearing of 100 miles or more in high latitudes. The correction is required due to the convergence of the meridians at the poles. The table is self-explanatory. If you do not have a copy of Table 1 handy, use one of the methods we have already discussed for working a great circle sailing problem.

It is more convenient and more accurate to solve course and distance problems over shorter distances by use of other sailings, such as *traverse sailing, middle-latitude sailing, parallel sailing,* and *Mercator sailing*, than by plotting. With the combined use of these techniques you can make an ocean passage and never use a chart for position finding or continuing course determination. Before World War II we were taught to take a departure from the land, maintain a DR by traverse sailing until we got a fix by celestial observation, and then find the new course and distance by the appropriate sailing problem solution—including great circle sailing if the distance to go warranted it. Only after all this was done did anyone plot anything on the chart, and then only to see if we were in or getting into dangerous waters.

I strongly urge that you learn the *sailings*. On a small vessel beating to windward, it is much quicker and much easier to work your day's work in a notebook, with your head up, than it is to plot courses over a jumping chart table with your head down. In these conditions the head over the chart table has been known to lead to the head over the bucket. To avoid this, there is one more thing you have to learn.

With all the above, it is still necessary to plot those Sumner Lines, which is the correct name for the celestial lines of position you have been getting in this book, by use of the equal altitude method proposed by Commander Marcq St.-Hilaire, a Frenchman. To find your longitude at sea without plotting you have to know how to work a time sight! We are not going to discuss sight reduction by time sight in this book, because we have simpler methods of getting LOPs, and finding positions at sea by use of celestial LOPs is the modern (post-World War II) convention. Sight reduction by time sight is explained in the 1977 edition of Bowditch, Volume I, and you might find it interesting. It could save charts and buckets, and who knows? I suspect that with calculator navigation, it is no big thing at all.

# 19 SIGHT REDUCTION BY CALCULATOR

The hand-held calculator is an amazing little gizmo. It has intimidated people who would not hesitate to tackle an involved equation and work it out with pencil and paper. Yet it has been a boon to those who are so inept with figures they cannot add two and two. I am supposing you do have the basic ability to add, subtract, multiply, and divide. Beyond that, it does not matter what your mathematical proficiency is. Calculators have something in them for everybody. If you ever get used to using one at sea, you will never leave it home. They are a great help in reducing the drudgery of the arithmetic involved in navigation. But they subtly do something else that is important. They make the navigator understand why he does things the way he does, what the reason is for the process used in arriving at the solutions of computed altitude, azimuth, or whatever it may be. The immediate advantage of knowing these basics is that although you may cry bitter tears if anything happens to your calculator, you will be able to break out that pencil and paper and the tables in Bowditch and do it the "hard way," which is not all that hard. What I am saying will be adequately clear to you as we get into this chapter.

Before we go any further, it is necessary to set out some premises:

We are only going to talk about the calculator for sight reduction. Please do not let this limit your thinking about the things you can do with your calculator. The hand-held calculator is probably more of a boon to the coastal navigator than it is to the deep-sea navigator. You can use a calculator for almost any type of navigation problem imaginable. The first thing any navigator will use his or her calculator for will likely be the speed/time/distance solutions needed to plot a DR or advance a line of position. With your calculator, interpolation is a cinch. You can use it to solve set and drift problems, to figure the range of visibility of lights, to calculate dip, and for a multitude of other things. So many things, in fact, that a book could be written on the subject, as many have been. These are the reasons why we are going to limit ourselves to using the calculator for sight reduction. By extension, this includes great circle sailing.

We are also going to limit our discussion to the types of calculators you can buy at the shopping mall down the street. There are two general types: The basic type, which multiplies, divides, adds, subtracts, and maybe finds square roots; and the "scientific" or "intermediate" type, which does everything the basic does but also has, along with a lot of mysterious things, trigonometric functions. Figure 30 illustrates a basic calculator of the type covered by this discussion. It is the Texas Instruments model TI-1100, made available in June 1983 at a purchase price of less than $10. An example of the type of scientific calculator I will refer to in this discussion is shown in Figure 31. This is the TI-30 SLR, produced by Texas Instruments. Note that it is light powered, which means you do not need to worry about the battery running down halfway between Australia and Cape Horn or between Mobile and Cozumel. This little jewel retails for about $20. It will do everything covered in this chapter and a lot more.

We are *not* going to talk about programmable calculators. They are great. They are also expensive, and their manufacturers furnish customers with, I would guess, adequate instruction books. I sure hope so. The book that came with my little scientific calculator has a lot in it that is over my head.

This brings up another point: I am *not* a mathematician! Please keep this in mind. I am not telling you this to cover myself in case I make any blunders. Quite the contrary. I make this admission to emphasize that you do not need to be quick with numbers or a whiz at trig to do the kind of navigation we are going to do here.

**Figure 30.** *A basic calculator. This little, inexpensive instrument adds, subtracts, multiplies, divides, does percentages, and extracts square roots. It is a handy tool for the navigator. The calculator shown is the TI-1100 produced by Texas Instruments. (Courtesy Texas Instruments, Incorporated)*

## USE OF THE BASIC CALCULATOR IN SIGHT REDUCTION

The best way to learn to use a calculator for sight reduction is to work sights in the way you are used to working them, and then work them again with the calculator. As you go along compare your results, and at the end, compare your answers. Use the calculator to whatever extent suits you. At first you may merely add, subtract, multiply, and divide the numbers you get out of tables. This is great for Ageton's Method. Later, if you turn into one of those calculator freaks, you may find yourself entering two figures—the meridian angle and the declination—and coming up with the altitude intercept after maybe a hundred or more operations. All perfect. You can be the judge of where you want to level off, but to begin with, let's use the basic calculator to do simple addition, subtraction, multiplication, and division. *WORK ALONG WITH ME!* I cannot overemphasize the importance of this admonition. You *cannot* learn to use any calculator by reading about it. *You must study and practice by doing the operations on the calculator as you go along.* Work along with me as we do the star sight the navigator took on the morning of 21 October 1983. This is the first example in Chapter 9.

To save time hunting back through the book, I am go-

**Figure 31.** *An intermediate or scientific calculator. In addition to matching the capabilities of the basic calculator, this instrument can work with trigonometric functions and perform numerous other operations. It can easily work through any task required in the conduct of a day's work at sea. The instrument shown is the TI-30 SLR, produced by Texas Instruments. (Courtesy Texas Instruments, Incorporated)*

ing to restate the problem: At 0510 ZT on 21 October 1983, the schooner *War Hat* was at DR Lat. 24° 59′ N, Long. 92° 38.0′ W. The navigator observed the star Regulus as follows:

| | |
|---|---|
| C | $11^h$-$14^m$-$15^s$ |
| W | $00^m$-$21^s$ |
| CE | $03^m$-$02^s$ f |
| $H_s$ | 43° 53.6′ |

With this information you found the following arguments with which to enter the tables of Pub. No. 249 and reduce the sight for an LOP.

| | |
|---|---|
| t | 47° E (t 47° E equals an LHA of 313°) |
| d | N 12° 03.0′, where d is declination |
| L | 25° 00.0′ N |
| $H_o$ | 43° 49.2′ |

You got this information the way you have been getting it through the entire book. I find it simpler to do the required arithmetic for the derivation of these arguments in my notebook, with the calculator set aside. In ordinary circumstances there is no reason to use an assumed position to reduce a sight with a calculator. We are doing it here to get results that are comparable with those we got when we reduced this sight by Pub. No. 249.

Before you work this sight with any calculator, you have to know the formula. It is, basically:

$$\sin h = \sin L \sin d + \cos L \cos d \cos LHA$$

You can substitute t for LHA. We reviewed the meaning of the trigonometric functions, such as sines, cosines, tangents, cotangents, secants, and cosecants, in the part of Chapter 17 I said you could skip. I urged you not to, though, especially if you wished to learn sight reduction with a calculator. Anyhow, the equation above simply means that if you multiply the sine of the latitude by the sine of the declination and add to this product the product of the cosine of the latitude, the cosine of the declination, and the cosine of the LHA (or meridian angle), you will get the sine of the altitude. Formulas—equations—will not be hard to handle if you think of them as a shorthand.

There are four basic cases involved with the use of this formula.* In *Case I,* t is less than 90°, and latitude and declination have the same name. This means that the LHA may be less than 90°, as in a postmeridian sight such as a sun sight in the afternoon, or the LHA may be greater than 270° (meaning that t would be east and less than 90°). This is why I put the LHA, 313°, in parentheses after t 47° E when I gave you the data to reduce the Regulus sight. Can you see now why I have been emphasizing the relationship between t and LHA throughout this book? The Regulus sight is a Case I sight. In Case I sights, the body observed will always be above the celestial horizon, and both the sine product and the cosine product in the formula will be positive.

In *Case II,* t is less than 90°, and latitude and declination have contrary names. Again this means the LHA may be less than 90° when you use LHA in the equation, or it may be between 270° and 360°, in which case t will be east and will equal the difference between the LHA and 360°. Remember? Now, the rule in Case II sight reductions is that declination is treated as a negative angle, because declination and latitude have contrary names. The declination in Case II sights will therefore have a value somewhere between 0° and −90°. The cosine of such an angle is positive, but the sine will be negative. (If you remember the unit circle from your trigonometry days, you may recall that an angle between 0° and −90° falls in the fourth quadrant of the unit circle.) This means that the product of the sine functions in the formula will be negative, and must therefore be subtracted from the product of the cosine functions, which will be positive. Actually this is algebraic addition, in case you are asking why two formulas are not given, one with a plus sign and one with a minus sign. The product of the sine functions is negative, and adding a negative value is equivalent to subtracting a positive value. If you get a negative value of sin h when you work out the equation, the body is below the celestial horizon. In Case II sights, the body may be above or below the celestial horizon. A "body" below the celestial horizon would be encountered in the solution of a great circle problem in marine navigation.

In *Case III,* t is greater than 90° and the latitude and declination have the same name. In Case III situations, the LHA lies between 90° and 270°. When the LHA is between 90° and 180°, the LHA and t will be equal, merely representing different names for the same thing. If the LHA is 107°, then t is 107° W. When the LHA lies between 180° and 270°, t will be east. Example: if the LHA is 231°, t is 129° E. The body can be above or below the celestial horizon. The sine product in the equation will be positive, because latitude and declination have the same name. The cosine product will be negative because t is greater than 90° and less than 180°, and the cosine of such an angle is negative. The two products on the right side of the equation must be subtracted, the lesser from the greater, and the result named plus or minus according to the sign of the greater.

* For more on these four cases, see Bowditch, Vol. II, 1975 edition, Article 706, "The Undivided Astronomical Triangle."

In *Case IV,* t is greater than 90° and the latitude and declination have contrary names. The LHA lies between 90° and 270°. Here again, as in Case III, the LHA and t are the same when the LHA is less than 180°, and t is east when the LHA lies between 180° and 270°. In Case IV sight reductions, the product of the sine functions is always negative, and the product of the cosine functions is always negative. It follows that sin h, and therefore $H_c$ itself, will be negative. The body will always be below the celestial horizon.

Let's state the formula again and work the Regulus sight, a Case I sight.

$$\sin h = \sin L \sin d + \cos L \cos d \cos t$$

This becomes:

$$\sin h = \sin 25° \sin 12° 03.0' + \cos 25° \cos 12° 03.0' \cos 47°$$

You could use the LHA of 313° as easily as t with an intermediate calculator, because the intermediate calculator gives trigonometric functions for any angle from 0° to 360°.

Since the basic calculator does not give trigonometric functions, it will be best to use t when the LHA exceeds 180°. You are going to have to look up values in a set of trigonometric tables. These tables, even in Bowditch, do not tabulate angles greater than 180°. This does not mean you cannot find functions of angles greater than 180°, but it does mean you have to find the equivalent angle in one of the two quadrants of the unit circle containing angles less than 180°, and know whether to give the function a plus or minus value. This requires a little more knowledge of trigonometry than this book should cover.* And why go to the extra work? If LHA is greater than 180°, simply use the value of the meridian angle, t, east.

Finally we are ready to work that Regulus sight. We are going to use the natural trigonometric function tables in Bowditch because, as a navigator, you should have a Bowditch handy. If you do not, you can use any trigonometry textbook. Break out your basic calculator and those tables. You find that sin h = sin L sin d + cos L cos d cos t becomes:

$$\sin h = 0.42262 \times 0.20877 \text{ added to } 0.90631 \times 0.97815 \times 0.68200$$

The way I stated this is ambiguous. Do you multiply the first two values, then add the product to the next value, and multiply this by the fourth value and then by the fifth, the cos t? Or what do you do? It is necessary to state an equation like this in correct mathematical terms. There is nothing ambiguous about setting up the same equation as follows:

$$\sin h = [(+0.42262)\times(+0.20877)] + [(+0.90631)\times(+0.97797)\times(+0.68200)]$$

This tells you beyond a doubt that you must multiply the two figures to the left of the central plus sign, and then multiply the three figures to the right of the plus sign, and at the end add the two products of all this multiplication together. And what do you get? You get the sine of h, and h is $H_c$ in the sight reduction.

If you knew you would have to sit in a corner and multiply all these five-place decimal figures together on paper, you would turn to Table 33 of Bowditch instead of Table 31, because logarithms are added when multiplication is needed, and adding figures such as these takes less effort than multiplying them. (Table 33 is excerpted in appendix I.) With a calculator, however, it takes no more effort to press the multiplication key, "×," than it does to press the plus key, "+." The plus signs in front of each number in this equation are not absolutely necessary, because when no sign is given in front of a number, the number is considered to be positive. But we are going to be dealing with both plus and minus quantities as we go through problems in each of the four cases, and consequently, we should be getting used to designating numbers plus or minus as necessary.

To work the equation with your basic calculator, perform the following operations:

Enter 0.42262
Press the multiplication key ×
Enter 0.20877
Press the = key
Jot down the result, 0.08823
Clear the calculator
Enter 0.90631
Press the × key
Enter 0.97797
Press ×
Enter 0.68200
Press =
Press +
Enter 0.08823
Press =
Get 0.69271, the sine of your $H_c$.

Go back to Table 31 of Bowditch or to that trig book you purloined, find 0.69271 under the sine column, and extract 43° 51'. Your $H_c$ is 43° 51'. Find the difference

*For more on the unit circle and trigonometric functions in various quadrants, see Bowditch, Vol. II, 1975 edition, Article 140.

between this $H_c$ and the $H_o$ of 43° 49.2′. The altitude intercept is 1.8′ away.

You have the altitude intercept, but you do not have an azimuth. You cannot plot an LOP without an azimuth. The *time-and-altitude azimuth formula* is the easiest to work on the calculator, but it has one disadvantage. When the body observed is near the prime vertical it is difficult to tell whether the azimuth is north or south of due east or due west. By definition, a body is on the prime vertical when its azimuth is either due east, 090°, or due west, 270°. The prime vertical is defined as the vertical circle that passes through the east and west points at your position. All vertical sextant angles are observed along a vertical circle. That is why you swing the arc when taking a sight. Thus, when taking a sight on a body on the prime vertical, the azimuth is either east (in an antemeridian sight) or west (in a postmeridian sight). The Regulus sight is an example of an azimuth too near due east to allow the use of a time-and-altitude azimuth formula without some care. From the Pub. No. 249 solution you know that Regulus has a Zn of 097°. That is within 7° of the prime vertical. The time-and-altitude azimuth formula can be used without worry, however, when the body observed is not too close to the prime vertical. The formula states:

$$\sin Z = \frac{\sin t \cos d}{\cos h}$$

Because of its simplicity, I do use the time-and-altitude azimuth formula when it is safe to do so. And how do I tell? Not by some deep delving into mysterious formulae, but by stepping out on deck and glancing over the compass at the body I want to shoot. If the corrected azimuth is at least 10 degrees more or less from 090° or 270°, I use the time-and-altitude azimuth formula.

Let us see how this works with our Regulus sight. When actual values are substituted, the formula reads:

$$\sin Z = \frac{\sin 47° \cos 12° 03.0'}{\cos 43° 50'}$$

Which becomes (using Table 31):

$$\sin Z = \frac{0.73135 \times 0.97797}{0.72136}$$

To work this on your basic calculator:

Enter 0.73135
Press ×
Enter 0.97797
Press ÷
Enter 0.72136
Press =
Get 0.99151, the sine of Z.

The sine of Z must be converted to an angle by going back to Table 31 as you have done before. Z is either 82° 32′ or 97° 28′, according to Table 31. Now you must convert Z to Zn, but how can you tell which value of Z to choose?

You already know the azimuth, Zn, is 097°, because you are working the Regulus sight worked earlier by Pub. No. 249 to guide you while learning to reduce sights by calculator. When you work sight reductions by calculator for keeps, you will not be backing up your solutions with a completed solution for comparison, although, if you get bogged down, a glance into one of the precomputed tables may set you straight. Maybe. But that takes enough extra effort to question the efficiency of using the calculator in the first place. But otherwise how are you going to tell, when an antemeridian sight gives you a possible Z of 82° or any other measure close to the prime vertical, whether it is the actual Zn or whether it must be subtracted from 180°? Or, if it is a postmeridian sight, whether Z must be added to 180° or subtracted from 360°? If there is any doubt, a glance over your compass, corrected for variation and deviation, may be enough. If the body is fairly close to the prime vertical, the corrected compass azimuth must be observed with an alidade-equipped bearing circle on the compass. There is no ambiguity unless the body observed is near the prime vertical. If the body is right on the prime vertical, the azimuth will be either 090° or 270°, and thus there is no ambiguity. I dare say the time-and-altitude azimuth formula is the one most used by navigators, except in the few ambiguous cases.

If you do not think the formula I just gave you is all that easy on a basic calculator, it can be stated another way:

$$\sin Z = \sin t \cos d \sec h$$

I set it up as I did before because I am leading you toward performing the same operations on a scientific calculator. It does not matter which formula you use, because the secant of h is equal to one divided by the cosine of h. Maybe you can bang out this second formula faster on your basic calculator. Table 31 does have secants tabulated. (Most scientific calculators do not.) Anyhow, the equation now becomes:

$$\sin Z = \sin 47° \cos 12° 03' \sec 43° 50'$$

And this becomes:

$$\sin Z = 0.73135 \times 0.97797 \times 1.38628$$

Run this through your calculator as follows:

Enter 0.73135
Press ×

Enter 0.97797
Press ×
Enter 1.38628
Press =
Get 0.991521, the sin of Z.

Look this up in the tables, and find that Z is either 97° 28′ or 82° 32′, as before.

Should you get an observation near the prime vertical and you are concerned about a possible ambiguity in the azimuth, you can solve the problem for azimuth by use of either of two formulas. They are the *time azimuth formula* and the *altitude azimuth formula.* Either of these formulas requires more manipulation of the calculator keys than the time-and-altitude azimuth formula. They do eliminate the problem of quadrant ambiguity. Let us take a look at the time azimuth formula.

There are several ways to express any trigonometric formula. You merely substitute equivalent terms, just as we substituted secant h for cosine h in the time-and-altitude azimuth formula. You may come across the time azimuth formula written as:

$$\tan Z = \frac{\cos d \sin LHA}{(\cos L \sin d) - (\sin L \cos d \cos LHA)}$$

Or as:

$$\tan Z = \frac{\sin LHA}{(\cos L \tan d) - (\sin L \cos LHA)}$$

Take your pick. They come out to the same thing. We could work both of them for comparison, but look at the two formulas. There are seven terms in the right-hand side of the first expression, and only five in the right-hand side of the second. Either one will give us Z. The fewer terms we have to enter in a calculator, the less work and the less chance of error. I often ask the Great Captain in the Sky, Lord, why did they make those keys so small, and You my fingers so large? This is such a problem with calculators, you may feel like throwing yours overboard before you get the hang of it. Please don't! Take your time, be careful, and jot down intermediate results as you go. But back on course. Let us set up the equation:

$$\tan Z = \frac{\sin 313^\circ}{(\cos 25^\circ \tan 12^\circ 03') - (\sin 25^\circ \cos 313^\circ)}$$

You are going to get a surprise when you go into the tables to look up the sine and cosine of 313°. Most tables do not list angles this large. Table 31 of Bowditch only lists angles up to 180°. You can get around this by substituting the meridian angle for the LHA in the equation. The equation now looks like this:

$$\tan Z = \frac{\sin 47^\circ}{(\cos 25^\circ \tan 12^\circ 03') - (\sin 25^\circ \cos 47^\circ)}$$

You are now nearly ready to extract the values of the functions from Table 31 of Bowditch, but you must be careful to give the correct sign, positive or negative, to each value. These are the rules to follow:

1) The latitude is always treated as a positive angle, and its sine and cosine functions are positive.
2) If the declination has the same name as the latitude, it is positive; if it has the opposite name, it is negative. Since the declination is always less than 90° (for those of you who recall the unit circle), the tangent of a positive declination is positive, and the tangent of a negative declination is always negative.
3) If the LHA is less than 90°, both the sine and the cosine of the meridian angle are positive. If the LHA is greater than 90° but less than 180°, the sine of the meridian angle is positive, but the cosine is negative. If the LHA is greater than 180° but less than 270°, both the sine and the cosine of the meridian angle are negative. If the LHA is greater than 270° but less than 360°, the sine of the meridian angle is negative, but the cosine is positive.

Now we are ready to use Table 31 of Bowditch. For our Regulus sight, the equation will appear as follows when the values are substituted:

$$\tan Z = \frac{-0.73135}{[(+0.90631) \times (0.21347)] - [(+0.42262) \times (+0.68200)]}$$

The quickest way to work this out on a basic calculator is to work out the values on the bottom first, so:

Enter 0.90631
Press ×
Enter 0.21347
Press =
Record 0.19347
Clear your calculator
Enter 0.42262
Press ×
Enter 0.68200
Press =
Get 0.28822
Now press −
Enter 0.19347, the number you recorded earlier
Press =
Get −0.09476 *(Be sure to note the minus sign, which, because of the order of the terms in the subtraction, does not appear in the display.)*
Record −0.09476
Enter the −0.73135 from the numerator of the equation

Press ÷
Enter −0.09476, the number you just recorded
Press =
Get +7.71792

Go into Table 31 or whatever table you are using and find an angle whose tangent is 7.71792. You will see that the angle in question is either 97° 23′ or 82° 37′, but the tables do not enable you to choose between the two. Another set of rules is necessary to locate the correct azimuth angle in the tables:

1) If LHA is less than 180° and the computed value of tan Z is negative, extract from Table 31 the obtuse (greater-than-90°) angle that has a tangent equal to tan Z. (Note that the signs of functions are not given in Table 31.)
2) If LHA is less than 180° and the computed value of tan Z is positive, extract from the tables the acute angle that has a tangent equal to tan Z.
3) If LHA is greater than 180° and the computed value of tan Z is negative, extract the acute angle that has a tangent equal to tan Z.
4) If LHA is greater than 180° and the computed value of tan Z is positive, extract the obtuse angle that has a tangent equal to tan Z.

Following these rules will lead you to the correct azimuth angle in Table 31. Then you must convert azimuth angle to true azimuth just as we have done throughout the book.

Let's finish the derivation of the azimuth for our Regulus sight. Our LHA is 313°, which is greater than 180°, and our value of tan Z is positive. Therefore, we choose 97° 23′ rather than 82° 37′ as the value of the azimuth angle. Since the sight was taken in north latitude, and the meridian angle is east, the azimuth angle is N 97° 23′ E, and the true azimuth is 097°.

Now you have an altitude intercept of 1.8′ away and a Zn of 097°. You can plot your Regulus LOP.

Compared to the time-and-altitude azimuth formula,

$$\sin Z = \frac{\sin t \cos d}{\cos h},$$

the time azimuth formula is quite a bit more effort. Practice with using the tables and manipulating the keys of your calculator will make this somewhat easier, but not enough to warrant using the time azimuth formula unless you simply have to. So, in addition to practice, work smart; if a body is close to the prime vertical, and you can afford to, wait a couple of hours for the azimuth of the body to change enough to enable you to use the time-and-altitude azimuth formula. Or choose another body, if available, when you have to have that compass check, and observe it for azimuth.

## USE OF THE SCIENTIFIC CALCULATOR IN SIGHT REDUCTION

A serious navigator should invest that little added money and get an intermediate, scientific calculator. It will save you a lot of time working the formulas you have just enjoyed working with your basic calculator, *if you do not make any mistakes.* There are a lot of key operations with even the basic calculator, and one mistake in sequence or entry could send you back to home plate. With the scientific calculator, there are many more chances to press the keys in any given solution. And this means there are many more times you can goof. A beginner should not try to get the final answer by going through one series of key punches. You can begin by doing little more than using the calculator to find functions of the angles you need to fill out the formula, and then following much the same sequence of key pressing that you followed with the basic. As you get more proficient, you can streamline the operation. In time, with one series of operations you will get both an $H_c$ and a Z. For your own sanity, and for record purposes, too, it is better to keep a log of what operations you do in your navigator's notebook as you go along. Make note of the product of sin L × sin d before going on, in the beginning.

Let us work that same Regulus sight. We will do it step by step, beginning with Step 6. Steps 1 through 5 never change.

### Step 6

Select your formula for $H_c$. It is:

$$\sin h = \sin L \sin d + \cos L \cos d \cos t$$

For use with a calculator you can write this as:

$$H_c = \sin^{-1} [(\sin L \sin d) + (\cos L \cos d \cos t)]$$

Translated into plain English, this second expression means that $H_c$ is the angle whose sine is equal to the product of the sine of the latitude and the sine of the declination added to the product of the cosine of the latitude, the cosine of the declination, and the cosine of the meridian angle. It seems I said that same thing when we talked about basic calculators, but not quite. If you want to sound like a budding Einstein, you can say $H_c$ is the arc sine (or inverse sine) of ... and add all the rest.

### Step 7

Substitute for the terms of the formula the actual values you need for the solution. In the way you have always proceeded in Steps 1 to 5, you found t = 47°, L = 25°, d = N 12° 03′. Write this in your notebook:

$$H_c = \sin^{-1} [(\sin 25° \sin 12° 03') + (\cos 25° \cos 12° 03' \cos 47°)]$$

A digression is in order at this point. You cannot work with degrees and minutes of arc. You must convert 03′ into a decimal fraction of a degree. To do this, divide 03 by 60 on your calculator; enter 3, press ÷, enter 60, press +, enter 12, press =, and read 12.05. Now d is 12.05°. You are ready to go.

As I have urged you before, follow me as we work this sight with the intermediate calculator. It is only proper, however, that I warn you not to become disturbed if your calculator does not give you *exactly* the same answers I show here. The calculator converts the numbers you enter into natural logarithms. The mantissas (the portions to the right of the decimal point) of these logarithms are carried out six, seven, or more places (depending on how many digits are to the left of the decimal), and then rounded off. (It is important you realize you are not getting six, seven, or more digit resolution because of this. When you convert the minutes and tenths of a minute of an angle to a decimal fraction of a degree, you actually begin with a value accurate only to the nearest tenth of a minute. So the slight variations you may get with your calculator four, five, or more places to the right of the decimal point will not cause significant errors.) The small variations are what is known as the *algorithm oddity/rounding-off error*. Depending on the sequence in which you ask for different information from your calculator, you may get slight variations in the answer given. For example, run through the sequence outlined in Step 8 that follows.

When you convert 0.6927091, the sine of $H_c$, to angle $H_c$, you will get an angle of 43.844939°. Now just for fun, ask the calculator for the sine of 43.844939°. You will get 0.6927091 again, but now ask the calculator for the angle that this is the sine of, and you may get 43.844943 instead of 43.844939. The difference is insignificant, but if you are trying to get *exactly* the same answer over and over again, it can be quite frustrating. So, when the last two digits, six or seven places to the right of the decimal, do not agree with those in this book, it will likely mean that your calculator is producing a rounding-off error. You are not the one who is making the mistake.

You can demonstrate this further if your curiosity is not satisfied. Enter 6, then press $y^x$, enter 2, and press =. The display equals 36. Now do this again, but by use of natural logarithms: Enter 6, press ln x, press +, enter 6 again, press ln x, and press =. The display reads 3.58319, the natural logarithm of 36. To convert this number, press INV/2nd, and then press ln x. But you did not get 36. You got 36.000002! That 2 should not be there, but it is. But look where it is! Six places to the right of the decimal. This "error" is insignificant in any navigation problem you will ever encounter. And, this algorithm oddity/rounding-off error is found in all brands of calculators of this mathematical level I know of.

**Step 8**

Enter 25, press the sine key (and wait a second or two for the calculator to come up with the sine of 25°), press ×, enter 12.05, press sine, pause, press +, enter 25, press cos, pause, press ×, enter 12.05, press cos, pause until display shows 0.9779658, press ×, enter 47, press cos, pause, press =, and get 0.6927091. This is the sine of $H_c$. How do you get $H_c$ out of your calculator? Several paragraphs above I said that "$H_c = \sin^{-1}$ ..." means that $H_c$ is the angle whose sine is, or $H_c$ is the *inverse* sine of, the expression on the right-hand side of our equation. Look at the keyboard of your calculator. Somewhere, if it is a scientific calculator, is a key marked "INV/2nd." To get the angle whose sine is in the display, as 0.6927091 is in your display at this moment, press INV/2nd, then press sin. The display will read 43.844939°. This is $H_c$ in degrees and a decimal fraction of a degree.

**Step 9**

Convert .844939 to minutes: Press −, enter 43, press =, press ×, enter 60, press =, and get 50.69634. Call this 51′ and add 43° to it. $H_c$ now reads 43° 51′, just as it did when you worked this same problem with a basic calculator. Compare this value to $H_o$ to get your altitude intercept, as always. But leave this step until later. For now, 43.844939° stays in the calculator.

An improvement over the basic calculator, isn't it? That was quick and simple. I assumed you had studied the keyboard of your calculator and were familiar with the keys on it. Another assumption was that you had read the instruction book that came with your calculator. If not, I suggest you give it a thorough going-over. You will need it a little later, but before we go into sights involving a declination and latitude of contrary name, you have to get an azimuth to plot this Regulus LOP. This time we will use the time-and-altitude azimuth formula,

$$\sin Z = \frac{\cos d \sin t}{\cos h},$$

which can be stated as easily, sin Z = sin t cos d sec h.

I explained that the secant of an angle equals its cosine divided into 1. When you studied your keyboard, you may have noticed your scientific calculator has keys for sine, marked "sin," cosine, marked "cos," and "tan" for tangent, but no cotangent, secant, or cosecant keys. How do you enter a secant or a cosecant? The secant is the cosine divided into 1, as I said a moment ago. The cosecant is the sine divided into 1. You can state this as $\sec A = \frac{1}{\cos A}$ and $\csc A = \frac{1}{\sin A}$. But you do not have to enter 1, then press divide, and enter whatever the calculator told you the sine or cosine was. Your calculator should have a key marked "1/x." If you want the cosecant of 47°, enter 47°, press sin, press 1/x, and the display will show the cosecant after a second or so of computation. It is very important not to rush the calculator, but to give it time to find the functions you ask for. Are you ready to work out that azimuth?

**Step 10**

This is Step 10 because finding Zn is as much a part of modern sight reduction as finding $H_c$. Write down the formula in your navigator's notebook. Remember, you only use the time azimuth formula or altitude azimuth formula if you must. You decide the azimuth of Regulus is far enough away from the prime vertical to cause you no uncertainty. Write down :

$$\sin Z = \sin t \cos d \sec h$$

**Step 11**

Fill in the formula for the specific values you will be working with, and note them in your notebook:

$$\sin Z = \sin 47° \cos 12.05° \sec 43.844939°$$

It does not matter in straight multiplication what number you multiply first. You could as easily set this equation up as:

$$\sin Z = \sec h \cos d \sin t$$

I am telling you this because the $H_c$, 43.844939°, is still in your calculator awaiting your pleasure.

**Step 12**

Work out the azimuth on your calculator: Press cos, press 1/x, pause, press ×, enter 12.05, press cos, press ×, enter 47, press sin, press =, and get 0.9916537, the sine of Z. Now press INV/2nd, press sin, and get 82.540554°. It takes a moment for the calculator to come up with this. Z is 82.540554°. To find Zn you must subtract Z from 180°: You know this because, when you sighted Regulus over your compass, you saw that its bearing was a trifle south of east. Now press −, enter 180, press =, and get -97.459456. Ignore the minus sign, and round this off to 97°, your Zn. You are ready to plot your Regulus LOP.

After comparing the reduction of a sight by use of a scientific calculator with the same reduction using a basic calculator, maybe I ought to wait for you to run down to the corner and buy a scientific calculator, if you do not already own one. Why? Because in reducing a sight with the scientific calculator, you never have to touch Table 31 of Bowditch ("Natural Trigonometric Functions") or its equivalent. When you get back we will work another sight, this time a Case II sight reduction in which the declination and latitude are of contrary name and t is less than 90°.

Are you ready?

This time why don't you work that sun sight you took at 0800 on 19 October 1983? That was the first sight you got after departing Mobile in *War Hat*. To save you from having to thumb back to Chapter 6, I will state again the information you developed from the *Nautical Almanac* to reduce this sight. From the first five steps in the sight reduction (the information obtained in these steps can come in part either from the *Air Almanac* or the *Nautical Almanac*) you find:

$$\begin{aligned} t &= 54° \text{ E} \\ d &= \text{S } 9° \ 54.1' \\ L &= 27° \ 00.0' \text{ N} \\ H_o &= 26° \ 01.5' \end{aligned}$$

**Step 6**

Write the formula in your navigator's notebook:

$$H_c = \sin^{-1} [(\sin L \sin d) + (\cos L \cos d \cos t)]$$

**Step 7**

Substitute the specific values for the terms in the equation. But first convert any minutes of arc to decimal degrees: d = S 9° 54.1′. Enter 54.1, press ÷, enter 60, press =, and get 0.9016667. d = 9.9016667. Since declination is contrary name to latitude, d becomes −9.9016667°. There are no other values that are not whole degrees. Set up your equation:

$$H_c = \sin^{-1} [(\sin 27 \times \sin -9.9016667) + (\cos 27 \times \cos -9.9016667 \times \cos 54)]$$

**Step 8**

This time I am not going to tell you each time to pause after you ask the calculator for functions, but remember to give it the second or so it needs to compute the answer. See how long it takes you to do this:

Enter 27
Press sin
Press ×
Enter 9.9016667
Press +/− (This key is on your keyboard, but you may have to look for it)
Press sin
Press +
Enter 27
Press cos
Press ×
Enter 9.9016667 (No need to press +/− this time since cos −9.9016667 is equal to cos +9.9016667)
Press cos
Press ×
Enter 54
Press cos
Press =
You get 0.4378521
Press INV/2nd
Press sin
Get 25.966913, the $H_c$ you are looking for.

**Step 9**

Logically, Step 9 belongs here, but do it at the very last, and for now leave $H_c$ in the calculator. In Step 9, you will convert the decimal fraction .966913 to minutes of arc by multiplying by 60 on your calculator, and get 58.01478. Round this off to 58′. $H_c$ is 25° 58′, $H_o$ is 26° 01.3′. The altitude intercept is 3.3′ toward.

You now need an azimuth.

**Step 10**

There is no chance of quadrant ambiguity here. You can see by looking in the direction of the sun that it is nowhere near the prime vertical. Use the time-and-altitude azimuth formula:

$$Z = \sin^{-1} (\sec h \cos d \sin t)$$

Remember, this is just an alternate way of writing the familiar formula. Enter this in your notebook.

**Step 11**

Enter the specific values in your equation, and put this in your notebook:

$$\sin Z = \sec 25.966913 \cos 9.9016667 \sin 54$$

**Step 12**

Run this through your calculator (25.966913 must be entered into your calculator, if you cleared it to perform Step 9. Assume you left 25.966913 in the calculator):

Press cos
Press 1/x
Press ×
Enter 9.9016667
Press cos
Press ×
Enter 54
Press sin
Press =
You get 0.88864566, the sine of Z
Press INV/2nd
Press sin
You get 62.431301

Z = 62.431301°. Press −, enter 180, and get −117.5687. Round this off to the nearest whole degree, and Zn = 118°.

Compare the results of this sun sight and the Regulus sight you worked above with the earlier reductions you did with Pub. No. 249. Close, aren't they? Actually, the calculator results are much more precise because Pub. No. 249 lets you work solutions only to the nearest minute of arc. In fact, the calculator gives you far better resolution than you will ever need in practical navigation.

Since you did not perform Step 9, but left $H_c$ in the calculator, do it now, and plot your LOP. For purposes of comparison, I am trying to keep my numeration of the steps as close as possible to that used in the beginning of this book.

Let us do a Case III sight. Case III sights are sights of a body whose meridian angle and LHA are greater than 90°, and whose latitude and declination have the same name. We have worked a Case III sight already in this book. It is the Rigil Kentaurus sight you worked in Chapter 17.

On the morning of 21 October 1983, ZT 0406, DR Lat. 45° 08.0′ S, Long. 161° 30.7′ E, you observed the star Rigil Kentaurus as follows:

| | |
|---|---|
| C | $17^h$-$07^m$-$10^s$ |
| W | $00^m$-$27^s$ |

| | | |
|---|---|---|
| CE | $00^m$-$17^s$ fast | |
| $H_s$ | 22° 47.1′ | |

Reduce the sight for a line of position. This time go through the whole process in your notebook:

| | | |
|---|---|---|
| C | $17^h$-$07^m$-$10^s$ | |
| W | $00^m$-$27^s$ | |
| C−W | $17^h$-$06^m$-$43^s$ | |
| CE | $00^m$-$17^s$ f | |
| GMT | $17^h$-$06^m$-$26^s$ | |
| GHA Aries | 283° 35.2′ | |
| Corr. | 1° 36.8′ | |
| GHA Aries | 285° 12.0′ | |
| SHA Star | 140° 23.6′ | Dec. S 60° 46.0′ |
| Sum | 425° 35.6′ | |
| Less | 360° 00.0′ | |
| GHA Star | 65° 35.6′ | |
| DR Long. | 161° 30.7′ | |
| LHA | 227° 06.3′ | |

You can use your calculator to do the arithmetic in Steps 1 through 5. To correct time when C is $17^h$-$07^m$-$10^s$, W is $0^m$-$27^s$, and CE is $0^m$-$17^s$, do this: Mentally convert C to $17^h$-$06^m$-$70^s$. Now enter 70, press −, enter 27, press = (so you get the $43^s$ to record in your notebook), press −, enter 17, press =. The display reads 26; GMT is $17^h$-$06^m$-$26^s$.

Correcting the tabulated GHA of Aries or any other GHA is even simpler, because everything is added. Add only the minutes of arc and mentally convert any sum over 60 minutes to additional degrees and minutes, as follows: The GHA of Aries on the daily pages is 283° 35.2′, and the correction is 1° 36.8′. Enter 35.2, press +, enter 36.8, press =. Read 72, which converts to 1° 12′. Add this and the 1° of the correction to 283°, and get a GHA of Aries of 285° 12′. Continue in this fashion until you find LHA or t.

Doing the first five steps of the sight reduction process this way is not much easier than doing it with a pencil in the notebook. On the other hand, it is so easy when doing so much addition and subtraction to make one little error that will throw the whole sight off, that it may pay to do even these simple operations with your calculator.

This time you are going to use your DR position to reduce the sight of Rigil Kentaurus. You can use an assumed position if you prefer, and avoid some conversions of minutes of arcs to decimal fractions of degrees. Unless by sheer coincidence the declination is a whole number of degrees, there will always be at least one conversion. Use of your DR position or an assumed position will work equally well. I think, as you get used to using your calculator, you will tend to work from your DR.

You got the LHA, 227° 06.3′, from Steps 1 through 5, as always. Convert LHA to t: t = 360° − LHA; t = 360° − 227° 06.3′; t = 132° 53.7′.

### Step 6

Write down in your notebook the formula you are going to use:

$$H_c = \sin^{-1} [(\sin L \sin D) + (\cos L \cos d \cos t)]$$

### Step 7

Complete the equation with the values specific to the sight you are reducing:

$$H_c = \sin^{-1} [(\sin 45° \, 08' \sin 60° \, 46') + (\cos 45° \, 08' \cos 60° \, 46' \cos 132° \, 53.7')]$$

Convert the minutes of arc into decimal fractions of a degree for each value in the equation. L, d, and t now read:

t = 132.895°
L = 45.133333°
d = 60.766667°

Your equation is now stated as:

$$H_c = \sin^{-1} [(\sin 45.133333° \sin 60.766667°) + (\cos 45.133333° \cos 60.766667° \cos 132.895°)]$$

### Step 8

Run this equation through your calculator:

Enter 45.133333
Press sin
Press ×
Enter 60.766667
Press sin
Press +
Enter 45.133333
Press cos
Press ×
Enter 60.766667
Press cos
Press ×
Enter 132.895
Press cos
Press =

You get 0.3839804 in your display. This is the sine of $H_c$. To convert sin $H_c$ to $H_c$, press INV/2nd, then press sin. The display shows 22.580455. $H_c$ is 22.580455°.

**Step 9**

In the tabular solution you did earlier in the book, you found the altitude intercept at this time. So as not to interrupt the flow of information (data) in your calculator, jot down 22.580455 in your notebook, and press STO/ x̄ (the memory). Then go on to Step 10.

**Step 10**

$H_c$ is in your calculator. Write the formula for azimuth angle in your notebook:

$$Z = \sin^{-1}(\sec h \cos d \sin t)$$

**Step 11**

Enter the required values in your equation:

$$Z = \sin^{-1}(\sec 22.580455 \cos 60.766667 \sin 132.895)$$

**Step 12**

Your calculator is on, so begin your solution:

Press cos
Press 1/x
Press ×
Enter 60.766667
Press cos
Press ×
Enter 132.895
Press sin
Press =

You get 0.3874831, the sine of the azimuth angle. Now press INV/2nd, press sin, and get 22.797978. Z = 22.797978°. Round this off to 23°, and you have Z = N 023° E. Z = Zn, so Zn = 023°.

At this time, go back to Step 9 and convert 22.580455° to degrees and minutes of arc. You find that $H_c$ is 22° 34.8′. You corrected the $H_s$ of 22° 47.1′ to an $H_o$ of 22° 41.4′. Subtract $H_c$ from $H_o$ and get an altitude intercept of 6.6′ toward.

After some practice, you will be able to reduce a sight faster and with greater precision with your scientific calculator than you can with any set of tables. One day you may be able to do it as follows, using the Rigil Kentaurus sight as an example:

Enter 45.133333
Press sin
Press ×
Enter 60.766667
Press sin
Press +
Enter 45.133333
Press cos
Press ×
Enter 60.766667
Press cos
Press ×
Enter 132.895
Press cos
Press =

Stop only long enough to jot down 0.3839804. (You could use the memory instead, but then you would have no record.)

Press INV/2nd
Press sin
Jot down 22.580455
Press cos
Press 1/x
Press ×
Enter 60.766667
Press cos
Press ×
Enter 132.895
Press sin
Press =
Press INV/2nd
Press sin
Get 22.797978

Round this off to 23°, and convert Z to Zn as necessary. Go back to the number, 22.580455, and convert this to degrees and minutes of arc for the $H_c$ to subtract from your $H_o$. You have gotten your altitude intercept and your azimuth. Plot the Rigil Kentaurus LOP from your DR.

The reason for jotting down 0.3839804 and 22.580455 was to save you the trouble of going all the way back to the beginning if for some reason the results of your fingers flitting over the keys of your calculator were not plausible numbers. If the calculated azimuth, Zn, were in the southwest quadrant when you had to look east northeast to see it, almost certainly your fingers hit the wrong key. Clear your calculator and go back to 22.580455. But before you run the equation for Z through your calculator, look at your sextant.

The $H_s$ should be reasonably close to 22.580455. If it is not, go back to home plate and do the whole shebang over. If $H_c$ is in the ballpark, do the azimuth angle equation over. Work slowly, carefully, and deliberately. You will be surprised at how proficient you will become.

Do not worry about undetected mistakes. The odds are that any mistakes will be so preposterous as to be self-evident. If for some odd reason the incorrect solutions seem all right on paper, they likely will not plot anywhere within reason. The plot is your final check on the accuracy of solutions for sights reduced by calculator or by any other method. And bear in mind, the blunder may not be in what you did with the calculator. It can come anywhere—reading the sextant, taking time, or extracting the wrong data from the almanac. If all tries with the calculator still do not provide any reasonable results, go back to the very beginning and do it all over. The same things happen with any other kind of sight reduction.

## USE OF THE SCIENTIFIC CALCULATOR WITH AGETON'S FORMULAS

You should now be ready to navigate anywhere in the world with only an almanac and a scientific calculator. But do not leave your tables home. At least bring your Bowditch. The formulas you use in your calculator can be worked on paper by the use of logarithms; it is not all that difficult a thing. So Table 33 of Bowditch is nice to have, in case your calculator malfunctions. But Ageton's Method is in Bowditch, too. Why not use it?

Can one reduce sights by Ageton's formulas with a calculator? Sure, and it may be a good alternate way of doing it. Look back to Chapter 17 and see the sight of Kaus Australis you reduced by Ageton's formulas. You did this one by the use of logarithmic functions, adding and subtracting exactly as you did in reducing sights by Ageton's tables. To reduce this or any sight with your calculator, set up the form exactly as you do when you use Ageton's tables, but where it says *Add,* substitute *Multiply,* and where it says *Subtract,* substitute *Divide.* Remember, the A values in Ageton's tables are logcosecants and the B values are logsecants (multiplied by 100,000). With your calculator, you do not need to resort to logfunctions because you do not need to avoid multiplication. Also, do not forget that you must convert minutes and tenths of a minute of arc to decimal fractions of degrees for calculator operations. Thus, the t of 30° 08.5′ becomes 30.1427°. Instead of looking up the values in tables, ask your calculator: What is the cosecant of t, when t is 30.1427°? To do this, enter 30.1427 in the calculator, press sin, then press 1/x. (The cosecant of an angle equals one divided by the sine of that angle. You could enter 1, press ÷, enter 30.1427, press sin, press =, and get the same thing. The calculator is set up to save you this extra effort.) The display will be 1.9914156, the cosecant of 30.1427°. You can check yourself by looking up the cosecant of 30.1427° in Table 31 of Bowditch. To do this, you must convert the decimal part of the angle back to minutes by multiplying 0.1427 by 60; you get 08.562′ instead of 08.5′, because we rounded off to get 0.1427. From Table 31 you find that the inverse cosecant of 1.9914156 lies between 30° 08.0′ and 30° 09.0′. No need to interpolate; this is close enough to show that you are operating your calculator correctly. Write the value 1.9914156 in the form on the line for t.

The declination of Kaus Australis is S 34° 23.7′. This converts to 34.395°. To get the secant of 34.395°, enter 34.395, press cos, press 1/x (the secant of an angle is one divided by the cosine); the display will read 1.211821, the secant of 34.395°. Do each step this way. Use your calculator instead of thumbing back and forth through a set of tables. When reducing a sight by Ageton's formulas with a calculator, solve each step in progression: First solve for R, then for K, and then for $H_c$ and Z. Jot it all down in the form as you go along. Then, if you make a blunder, you will not have to start all over, and you will have a record in your navigator's notebook.

The next thing you know, you will be running the formula, R = csc t sec d, through your calculator, your fingers lightly tripping over the keys like those of a piano virtuoso, while you jot down R. (You will need to note R to find the secant of R, in order to keep going.) Then you can solve for K, and so on. Let's work a sight all the way through with the calculator. Glance back at the Capella sight in Chapter 17, and note the values you will need: t is 27° 08.3′ west, d is N 45° 58.9′, and the DR latitude is 24° 58.7′ N. We'll work this sight on a scientific calculator. Do it in the way I'm outlining at first, then someday you may be able to start with a t of 27° 08.3′ and run through the three formulas required for $H_c$ without making one notation. This level of skill is difficult to achieve, however, and unless you keep in continuous practice, you will soon lose the ability to work a sight this way without an error. If you make one error, no matter how slight (there are no *little* errors in calculator work), you must start all over. So for now, and likely from now on, follow the procedure I will outline here:

### Step 1

Values of t, d, and $H_o$ are found exactly as before. Even here you can use the calculator to check your addition and subtraction, but you may find this part easy enough to do by hand. So, beginning with t, enter 8.3, press ÷, enter 60, press +, enter 27, press =, press sin, press 1/x. The display should read 2.1923075. Put this in the first *Multiply* column in your form. Press STO/ x̄ to put this in the calculator's memory. Now enter 58.9, press ÷, enter 60, press +, enter 45, press =, press cos, press 1/x. The display will read 1.4390798. Put this

Capella:
21/10/83, ZT 0505, DR L 24°-58.7′ N, λ 92°-37.7′ W

| | | |
|---|---|---|
| C | $11^h$-$20^m$-$35^s$ | |
| W | $00^m$-$21^s$ | |
| C–W | $11^h$-$20^m$-$14^s$ | |
| CE | $3^m$-$02^s$ | f |
| GMT | $11^h$-$17^m$-$12^s$ | |

| | | |
|---|---|---|
| Main Corr. | −0.6 | $H_s$ 59°-41.8′ |
| Dip | −3.4 | − 4.0 |
| | | $H_o$ 59°-37.8′ |

csc R = csc t sec d

csc K = $\frac{\csc d}{\sec R}$

csc $H_c$ = sec R sec (K~L)

csc Z = $\frac{\csc R}{\sec H_c}$

A is cosecant
B is secant

| | | | MULT. | DIV'D | MULT. | DIV'D |
|---|---|---|---|---|---|---|
| GHA ♈ | 194°-19.5′ | | | | | |
| Corr. | 4°-18.7′ | | | | | |
| GHA ♈ | 198°-38.2′ | | | | | |
| SHA | 281°-07.8′ | | | | | |
| | 479°-46.0′ | | | | | |
| | − 360°-00.0′ | | | | | |
| GHA | 119°-46.0′ | | | | | |
| DR λ | 92°-37.7′ | W | | | | |
| LHA | 27°-08.3′ | (t W) | A 2.1923075 | | | |
| Dec. | N 45°-58.9′ | | B 1.4390798 | A 1.3905933 | | |
| R | 18.479583° | | A 3.1549054 | B 1.0543667 | B 1.0543667 | A 3.1549054 |
| K | 49.306957° | | | A 1.3188897 | | |
| L | 24.978333° | | | | | |
| (K~L) | 24.328624° | | | | B 1.0974568 | |
| $H_c$ | 59.792983° | | | | A 1.1571219 | B 1.9875745 |
| | | | | | | A 1.5873143 |
| $H_o$ | 59.630000 | | | | | |
| | 0.162983° | | | | | |
| | 9.78′ a | | | | | |

| | |
|---|---|
| Z | N 39.04975° W |
| | 360.0000 |
| Zn | 320.95° |
| Zn | 321° |

opposite declination in the form, as sec d. Now press ×, press RCL to recall the cosecant of t from memory, and multiply csc t by the secant of d. Press =, and the display should read 3.1549054. This is the cosecant of R. Write it down. To ask the calculator for the angular value of R, press 1/x, press INV/2nd, press sin. The display will read 18.479583, which is R. Note the angle R of 18.479583° in the form. You will need it later.

### Step 2

The next step is to solve for K to find (K~L). The formula for K is $\csc K = \frac{\csc d}{\sec R}$. R is in the calculator display. To find sec R, press cos, press 1/x, and the display reads 1.0543667, the secant of R. Put this in your form. Press STO/ x̄. If you glance ahead to the more compact worksheet for this sight reduction, you will see that d is noted both as 45° 58.9′ and as 45.981667°. I often skip the second notation; it is too easy to enter 58.9, press ÷, enter 60, press +, enter 45, and press = to get 45.981667. After you get a little practice, you will see that your fingers can move accurately over the keys of the calculator many times faster than you can read these directions. But back on track: 45.981667 is in the calculator. Press sin, press 1/x, and get the csc d of 1.3905933 in the display. Put this in the form. Now press ÷, press RCL, and press =. The display reads 1.3188897, the cosecant of K. To get the angle K, press 1/x, press INV/2nd, press sin, and read 49.306957 in the display. K is 49.306957°. Note this in the form, and press STO/ x̄. (Each time after the first in a series of operations that STO/ x̄ is pressed, a new value supplants the previously stored value. Thus, your calculator need have no more than one memory.)

### Step 3

Find (K~L). K is in the calculator's memory. Enter the DR latitude of 24° 58.7′ in degrees and decimal fractions of a degree by entering 58.7, then pressing ÷ and entering 60. Press +, enter 24, and press =. L now reads as 24.978333°. Now press −, press RCL to recall K, and press =. The display reads −24.328624. Ignoring the minus sign (which is there only because, for convenience, you subtracted the larger value, K, from the smaller value, L, rather than vice versa), you have 24.328624, the value of (K~L). Note this in the form.

### Step 4

You are ready to find $H_c$ by Ageton's formula, $\csc H_c = \sec R \sec (K{\sim}L)$. Since (K~L) is in your calculator, press cos, press 1/x, and make a note of the display, 1.0974568, which is the sec (K~L). Press STO/ x̄, and enter the value of R, 18.479583°, after locating it above. Press cos, press 1/x, and note the display of 1.0543667. Now press ×, press RCL, press =, and note the display of 1.1571219, the csc $H_c$. To extract $H_c$ from the calculator, press 1/x, press INV/2nd, and press sin. The display reads 59.792983. $H_c$ is 59.792983°.

### Step 5

The altitude intercept is the difference between $H_c$ and $H_o$. $H_c$ is already in the calculator, so you need to enter $H_o$, which is 59° 37.8′. To do this, first put $H_c$ in the memory by pressing STO/ x̄. Then enter 37.8, press ÷, enter 60, press +, enter 59, and press =. $H_o$ is 59.63°. Now press −, press RCL, and press =. The display will read −0.162983. This is the altitude intercept in *decimal fractions of a degree.* You must have the altitude intercept in minutes of arc (nautical miles) to plot. So press ×, enter 60, press =. The display reads −9.77898. $H_c$ is greater. Remember the mnemonic—Coast Guard Academy, CGA — Computed Greater Away. The altitude intercept is 9.8 miles away. This is different from the solution by Ageton's tables by one tenth of a minute. The calculator is more precise, Ageton's tables working only to the nearest 0.5′.

### Step 6

By this point the third column of the form drawn in your notebook should be complete. The final step in the sight reduction process by Ageton's Method is finding the azimuth, Zn. You must first solve for azimuth angle, Z. Ageton's formula for Z is $\csc Z = \frac{\csc R}{\sec H_c}$. Your $H_c$ of 59.792983 should still be in the memory. Press RCL. To get the secant of $H_c$, press cos, then press 1/x. The display will read 1.9875745. Press STO/ x̄ to put this in memory, enter the R of 18.479583, press sin, and press 1/x. The display reads 3.1549054. Now press ÷, press RCL, and press =. The display reads 1.5873143. This is the csc Z. Ask your calculator for the azimuth angle, Z, by pressing 1/x, then INV/2nd, then sin. The display will read 39.04975. This is Z in degrees and a decimal fraction of a degree. Now press −, enter 360 (t is west, latitude is north, so Z is N 39.04975° W), and press =. The display is −320.95025. Ignore the minus sign, which is there only because you found it convenient to subtract 360° from the azimuth angle instead of subtracting the azimuth angle from 360°. Zn is 320.95. Round this off to 321° and plot the Capella LOP.

At this point, go back to the Capella sight in Chapter 17 and compare the two sight reductions—one by Ageton's tables, the other by calculator—and you will see that the results are almost identical. The calculator sight reduction is more precise. Ageton's tables are not interpolated, and values are tabulated to the nearest 0.5′. The calculator carries solutions to several places to the right of the decimal point as a matter of course.

I suggested that you use the same form when reducing a sight by using Ageton's Method with a calculator that we used in Chapter 17 with Table 35 from Bowditch, so you could see that it is almost precisely the same procedure. The only exception is that you can use natural trigonometric functions instead of logarithmic functions (or logfunctions multiplied by 100,000), as we discussed in depth in Chapter 17. You can use natural functions with a calculator because it is as easy to multiply many-digit numbers on a calculator as it is to add them. Simply press the × key instead of the + key.

There is one problem with following the format used with tabular solutions by Ageton's Method when using a calculator: Those six- and seven-place decimals (digits to the right of the decimal point in a number) take up a lot of room on a notebook page. You must write small. After you understand the process, there is another system of notation you may like better. (See the accompanying form.) The steps with the calculator are exactly the same. By now you should be aware that the first column in the standard form for sight reduction by Ageton's Method solves for R, the second column, for K. Then you find (K∼L) and solve for $H_c$ in the third column. The fourth column solves for Z to find Zn. This makes a solution by calculator, when using Ageton's Method, easy to break down into a step-by-step process. Let us run through that Capella sight again, as if we had not already done it before: Repetition is the mother of learning.

**Step 1**

Solve for R. The formula for R is: csc R = csc t sec d. The equation becomes csc R = csc 27° 08.3′ × sec 45° 58.9′. Put the formula and the equation in your notebook. With the calculator, convert t to degrees and a decimal fraction of a degree, by entering 8.3, pressing ÷, entering 60, pressing +, entering 27, and pressing =. The display reads 27.138333. This is t in degrees and decimal parts of a degree. Now find the cosecant of t by pressing sin, then 1/x. The display reads 2.1923075, the csc t. Start to fill in your equation by writing:

$$\csc R = 2.1923075 \times$$

Times what? Times the sec d. Convert d to degrees and decimal parts of a degree, but before you do, put 2.1923075 in the calculator's memory by pressing STO/ x̄. The declination of Capella is 45° 58.9′. Enter 58.9′, press ÷, enter 60, press +, enter 45, and press =. The display tells you d is 45.981667°. To find the sec d, press cos, then press 1/x. The display reads 1.4390798. Add this to the equation you just started above. You now have:

$$\csc R = 2.1923075 \times 1.4390798$$

Find R by multiplying these two functions, csc t and sec d. To do this, press ×, press RCL, and press =. (The csc t of 2.1923075 was in the memory.) The display reads 3.1549054. Write this in your notebook this way:

$$\csc R = 3.1549054$$

Ask your calculator for the angular value of R by pressing 1/x, then INV/2nd, then sin. The display reads 18.479583. This is R. Jot it down as I have done in the accompanying worksheet. You will need R to find the secant of R in solving for K. The angle R, 18.479583, remains in the calculator.

**Step 2**

Solve for K. The formula is $\csc K = \dfrac{\csc d}{\sec R}$. Put this in your notebook. If you write down the formula for each step each time you use it, you will soon have them committed irrevocably to memory. R is in your calculator. To find sec R, press cos, then press 1/x. The display reads 1.0543667, the sec R. Begin to fill in the equation in your notebook by writing:

$$\csc K = \frac{\quad\quad}{1.0543667}$$

Put the sec R of 1.0543667 in the calculator's memory by pressing STO/ x̄.

Now find the csc d. Either go back and pick up the value of d in degrees and decimals or work it out again by entering 58.9, pressing ÷, entering 60, pressing +, entering 45, and pressing =. It is so easy to run your fingers through the steps to convert degrees and minutes to degrees and decimals that I do it over each time. I find I make fewer mistakes; d is 45.981667. To get csc d, press sin, then press 1/x. The display reads 1.3905933. Complete the equation by putting this value above the sec R to indicate that it is the dividend of the fraction. Now the equation looks like this:

$$\csc K = \frac{1.3905933}{1.0543667}$$

Capella: 21|10|83, ZT 0505, DR L 24°-58.7′N, λ 92°-37.7′ W

| | | | | | |
|---|---|---|---|---|---|
| C | 11ʰ-20ᵐ-35ˢ | | −0.6 | $H_s$ | 59°-41.8′ |
| W | 00ᵐ-21ˢ | | −3.4 | | −4.0 |
| C-W | 11ʰ-20ᵐ-14ˢ | | | $H_o$ | 59°-37.8′ |
| CE | 3ᵐ-02ˢ f | | | | |
| GMT | 11ʰ-17ᵐ-12ˢ | | | | |

| | | |
|---|---|---|
| GHA ♈ | 194°-19.5′ | |
| Corr. | 4°-18.7′ | |
| GHA ♈ | 198°-38.2′ | |
| SHA | 281°-07.8′ | Dec. N 45°-58.9′/45.981667° |
| | 479°-46.0′ | |
| | −360°-00.0′ | |
| GHA | 119°-46.0′ | |
| DR λ | 92°-37.7′ | |
| LHA | 27°-08.3′ (t W) 27.138333° | |

$\csc R = \csc t \sec d$
$\csc K = \frac{\csc d}{\sec R}$
$\csc H_c = \sec R \sec(K \sim L)$
$\csc Z = \frac{\csc R}{\sec H_c}$

STEP 1
$\csc R = \csc t \sec d$
$\csc R = \csc 27°\text{-}08.3' \times \sec 45°\text{-}58.9'$
$\csc R = 2.1923075 \times 1.4390798$
$\csc R = 3.1549054$
$R = 18.479583°$

STEP 2
$\csc K = \frac{\csc d}{\sec R}$
$\csc K = \frac{1.3905933}{1.0543667}$
$\csc K = 1.3188897$
$K = 49.306957°$

STEP 3
$L = 24.978333°$
$(K \sim L) = 24.328624°$

STEP 4
$\csc H_c = \sec R \sec(K \sim L)$
$\csc H_c = 1.0543667 \times 1.0974568$
$\csc H_c = 1.1571219$

STEP 5
$H_c = 59.792983°$
$H_o = 59.63$
$-0.162983 = 9.78a$

STEP 6
$\csc Z = \frac{\csc R}{\sec H_c}$
$\csc Z = \frac{3.1549054}{1.9875745}$
$\csc Z = 1.5873143$
$Z = N\ 39.05°\ W$
$Zn = 321°$

With the csc d of 1.3905933 in the display, tell the calculator to complete the arithmetic by pressing ÷, then RCL, then =. The display reads 1.3188897, the csc K. Put this in your notebook as follows:

$$\csc K = 1.3188897$$

Find K by pressing 1/x, then INV/2nd, then sin. The K of 49.306957 is in the display.

**Step 3**

Find (K∼ L). K is 49.306957°, ready for manipulation in the calculator and in the display. Since K is north, the same name as the latitude, subtract the lesser from the greater to find (K∼ L). First put the value of K, 49.306957, in memory by pressing STO/ x̄. The latitude is 24° 58.7′ N. Convert this to degrees and decimal parts of a degree by entering 58.7, then pressing ÷, entering 60, pressing +, entering 24, and pressing =. The display reads 24.978333. Now press −, press RCL, and press =. The display reads −24.328624. Ignore the minus sign, which is there only because you found it convenient to subtract K from the latitude instead of subtracting the latitude from K. The value 24.328624 is (K∼ L).

**Step 4**

Solve for $H_c$. Ageton's formula is:

$$\csc H_c = \sec R \sec (K \sim L)$$

Since you have already used the sec R once in finding K, glance back up the page and pick up the sec R of 1.0543667, and begin to write the equation in your notebook:

$$\csc H_c = 1.0543667 \times$$

The equation is incomplete. Find the sec (K∼ L). Enter the (K∼ L) of 24.328624°, press cos, then press 1/x. The sec (K∼ L) is 1.0974568, and is in the display. Complete the equation in your notebook:

$$\csc H_c = 1.0543667 \times 1.0974568$$

Then press ×, enter 1.0543667, and press =. The display reads 1.1571219. Write in your notebook:

$$\csc H_c = 1.1571219$$

Now ask the calculator for $H_c$: Press 1/x, press INV/2nd, and press sin. The display reads 59.792983. This is $H_c$.

**Step 5**

Find the altitude intercept. The $H_c$ of 59.792983 is in the calculator. Put it in memory by pressing STO/ x̄. Then enter $H_o$, 59° 37.8′, in degrees and decimals by entering the minutes, 37.8, pressing ÷, entering 60, pressing +, entering 59, and pressing =. The display reads 59.63. To make the subtraction, press −, then RCL, then =. The display reads −0.162983. This is the altitude intercept in degrees. Convert this to minutes of arc (nautical miles) by pressing ×, entering 60, and pressing =. The altitude intercept is in the display as −9.77898. Round this off to 9.78 miles. $H_c$ is greater than $H_o$, so the altitude intercept is 9.78 miles away.

**Step 6**

Solve for Z. Ageton's formula is :

$$\csc Z = \frac{\csc R}{\sec H_c}$$

First recall $H_c$ from memory by pressing RCL. Then press cos, then 1/x. The display reads 1.9875745. Press STO/ x̄ to put sec $H_c$ in the calculator's memory. Now go all the way back to Step 1, pick up the csc R of 3.1549054, and enter it. Then press ÷, then RCL, then =. The display reads 1.5873143. You can write these two steps in your notebook as:

$$\csc Z = \frac{3.1549054}{1.9875745}$$

$$\csc Z = 1.5873143$$

Ask the calculator for Z by pressing 1/x, then INV/2nd, then sin. The display reads 39.04975. Z is N 39.049° W. Press −, enter 360, and press =. Zn is 320.95025; Zn is 321°.

**Step 7**

You are ready to plot the Capella LOP.

Why would anyone want to use a calculator to work a sight by Ageton's formulas? I am sure this must have seemed a tedious and involved routine after reducing sights earlier in this chapter by using basic formulas. Remember, however, that in sight reduction by the basic formulas we studied earlier we had to be aware of quadrantal ambiguity when determining the azimuth angle. There is no such ambiguity with Ageton's formulas. This is a tremendous advantage. As for the tedium involved, when you get used to reducing observations by Ageton's formulas with your calculator, you will be able to work a sight for $H_c$ by starting with t, manipulating the keys, and using the memory to avoid having to stop at all until the display reads $H_c$. *But I do not recommend this!*

Thc chance of error is too great to ever allow any such efficiency to be attained by the average navigator. I know because at times, just for the fun of it, I will run a sight through the calculator without stopping to take notes between steps. My average results are quite poor. As I said above, one error, when you are not making notes in some kind of format as you go along, will mean going all the way back to the beginning and starting over. Why take this chance, when it is easy to make notations as you go along? You should have a record of your navigational calculations anyhow. After you have reduced a few sights by Ageton's formulas and calculator you will find it is not only easy but quite fast. It takes anyone who uses a calculator regularly only minutes to run through the sight reduction process as I have outlined it in this chapter, even though it may take hours to follow the same process through these pages. My point is that the time saved by being overefficient is not commensurate with the time lost when you make an error. To repeat, one error takes you back to home plate.

You are going to make errors, if for no other reason, simply because you are going to get a bit rusty between voyages. Check yourself by comparing the displays you get when asking your calculator for some function of an angle with the values tabulated in Table 31 of Bowditch. If you are certain you are doing operations correctly, check your calculator; the instruments go haywire at times. The odds are, however, that you are the one who is making some maddening slip that is hard for even you to detect. Calculator navigation can be efficient, accurate, and fun.

We are not going into Case IV sight reduction by calculator. These sights are always solutions for negative altitudes, and therefore of interest to marine navigators primarily for great circle solutions. Coaltitudes in these cases are always greater than 90° (5,400 nautical miles). Only large vessels making ocean passages over such distances are concerned with such problems. Should you ever need to work a great circle sailing problem, you will not find it too hard. For now, it is enough to be able to work the problems you will use frequently. If you can do the work set forth in the various chapters in this book, you can be proud of your navigating skills.

But one word of caution: You will never know it all. The fascination of navigation is the fact that there are so many useful things to know. But you need not keep a line on the wharf while learning. With good sense, good seamanship, and only the ability to reduce a sun sight by Pub. No. 249, you can go a long way. But the more you know, the farther you can go, and with greater safety to your ship and her people.

Fair winds and following seas!

# APPENDIXES

# A SOLUTIONS TO EXERCISE PROBLEMS

SUN (☉), Lower/~~Upper~~ Limb
Date 19/10/83 ZT 1035
DR Lat. 26°-56.3′ N Long. 88°-27.2′ W

| | | |
|---|---|---|
| C | | 16h-38m-15s |
| W | | 02m-49s |
| C-W | | 16h-35m-26s |
| CE | | 03m-12s f |
| GMT | | 16h-32m-14s |
| GHA | | 63°-44.5′ |
| Corr. (min. & sec.) | | 8°-03.5′ |
| GHA | | 71°-48.0′ |
| **Assumed Long. | | 88°-48.0′ |
| *LHA | (t) | 17°-00.0′ E |
| ***Assumed Lat. | | 27°-00.0′ N |
| Declination | S | 9°-55.9′ |
| d difference | +0.9 | +00.5 |
| *Declination | | 9°-56.4′ |
| $H_t$ | | 50°-24.0′ |
| Correction | d-55 | -51.0′ |
| $H_c$ | | 49°-33.0′ |
| $H_o$ | | 50°-08.7′ |
| **Altitude Intercept | | 35.7 t |
| Z | | N 153° E |
| From 360° (if req.) | | |
| **Zn | | 153° |
| $H_s$ | | 49°-56.7′ |
| Main Corr. | +15.4 | |
| Addn'l | | |
| Dip | -3.4 | |
| Total Corr. | | +12.0 |
| $H_o$ | | 50°-08.7′ |

*Required to enter Pub. No. 249
**Required to plot line of position
***Required to enter Pub. No. 249 and to plot line of position

SUN (☉), Lower/~~Upper~~ Limb
Date 19/10/83 ZT 1450
DR Lat. 26°-31′ N Long. 88°-52′ W

| | | |
|---|---|---|
| C | | 20h-54m-15s |
| W | | 2m-19s |
| C-W | | 20h-51m-56s |
| CE | | 3m-12s f |
| GMT | | 20h-48m-44s |
| GHA | | 123°-44.9′ |
| Corr. (min. & sec.) | | 12°-11.0′ |
| GHA | | 135°-55.9′ |
| **Assumed Long. | | 88°-55.9′ |
| *LHA | (t-w) | 47°-00.0′ |
| ***Assumed Lat. | | 27°-00.0′ N |
| Declination | S | 9°-59.6′ |
| d difference | +0.9 | +0.7 |
| *Declination | S | 10°-00.3′ |
| $H_t$ | | 31°-18.0′ |
| Correction | d-39 | -0.0 |
| $H_c$ | | 31°-18.0′ |
| $H_o$ | | 31°-30.9′ |
| **Altitude Intercept | | 12.9 t |
| Z | | N 123° W |
| From 360° (if req.) | | 360° |
| **Zn | | 237° |
| $H_s$ | | 31°-19.6′ |
| Main Corr. | +14.7 | |
| Addn'l | | |
| Dip | -3.4 | |
| Total Corr. | | +11.3 |
| $H_o$ | | 31°-30.9′ |

*Required to enter Pub. No. 249
**Required to plot line of position
***Required to enter Pub. No. 249 and to plot line of position

SUN (⊙), Lower/~~Upper~~ Limb
Date 19|10|83 ZT 1600
DR Lat. 26°-31' N Long. 89°-00.2' W

| | |
|---|---|
| C | 22h-06m-25s |
| W | 1m-53s |
| C-W | 22h-04m-32s |
| CE | 3m-12s f |
| GMT | 22h-01m-20s |
| GHA | 153°-45.2' |
| Corr. (min. & sec.) | 0°-20.0' |
| GHA | 154°-05.2' |
| **Assumed Long. | 89°-05.2' |
| *LHA (t-W) | 65°-00.0' |
| ***Assumed Lat. | 27°-00.0' N |
| Declination | S 10°-01.4' |
| d difference +0.9 | − 0.0 |
| *Declination | S 10°-01.4' |
| $H_t$ | 16°-59.0' |
| Correction d-33 | − 1.0 |
| $H_c$ | 16°-58.0' |
| $H_o$ | 17°-05.4' |
| **Altitude Intercept | 7.4 t |
| Z | N 111° W |
| From 360° (if req.) | 360° |
| **Zn | 249° |
| $H_s$ | 16°-55.7' |
| Main Corr. +13.1 | |
| Addn'l | |
| Dip −3.4 | |
| Total Corr. | +09.7' |
| $H_o$ | 17°-05.4' |

*Required to enter Pub. No. 249
**Required to plot line of position
***Required to enter Pub. No. 249 and to plot line of position

SUN (⊙), Lower/~~Upper~~ Limb
Date 20|10|83 ZT 0915
DR Lat. 26°-31.0' N Long. 90°-55.2' W

| | |
|---|---|
| C | 15h-20m-10s |
| W | 1m-39s |
| C-W | 15h-18m-31s |
| CE | 3m-07s f |
| GMT | 15h-15m-24s |
| GHA | 48°-47.1' |
| Corr. (min. & sec.) | 3°-51.0' |
| GHA | 52°-38.1' |
| **Assumed Long. | 90°-38.1' |
| *LHA (t) | 38°-00.0' E |
| ***Assumed Lat. | 27°-00.0' N |
| Declination | S 10°-16.7' |
| d difference +0.9 | + 0.2 |
| *Declination | S 10°-16.9' |
| $H_t$ | 37°-47.0' |
| Correction d-44 | − 12.0 |
| $H_c$ | 37°-35.0' |
| $H_o$ | 37°-39.0' |
| **Altitude Intercept | 4.0 t |
| Z | N 130° E |
| From 360° (if req.) | |
| **Zn | 130° |
| $H_s$ | 37°-27.4' |
| Main Corr. +15.0 | |
| Addn'l | |
| Dip −3.4 | |
| Total Corr. | + 11.6 |
| $H_o$ | 37°-39.0' |

*Required to enter Pub. No. 249
**Required to plot line of position
***Required to enter Pub. No. 249 and to plot line of position

The chronometer error is now $3^{m}$-$07^{s}$. Recall that earlier I told you the chronometer was losing $5^{s}$ per day. You must apply the rate daily to get your CE when a time signal is not obtained.

SUN (⊙), Lower/~~Upper~~ Limb
Date 20/10/83 ZT 1125
DR Lat. 26°-23' N Long. 91°-09' W

| | |
|---|---|
| C | 17h-29m-15s |
| W | 1m-23s |
| C-W | 17h-27m-52s |
| CE | 3m-07s f |
| GMT | 17h-24m-45s |
| GHA | 78°-47.3' |
| Corr. (min. & sec.) | 6°-11.3 |
| GHA | 84°-58.6' |
| **Assumed Long. | 90°-58.6' |
| *LHA (t) | 6°-00.0' E |
| ***Assumed Lat. | 26°-00.0' N |
| Declination | S 10°-18.5' |
| d difference +0.9 | +0.4 |
| *Declination | S 10°-18.9' |
| $H_t$ | 53°-32.0' |
| Correction -59 | -19.0 |
| $H_c$ | 53°-13.0' |
| $H_o$ | 52°-56.0' |
| **Altitude Intercept | 17.0 a |
| Z | N 170° E |
| From 360° (if req.) | |
| **Zn | 170° |
| $H_s$ | 52°-43.9' |
| Main Corr. +15.5 | |
| Addn'l | |
| Dip -3.4 | |
| Total Corr. | +12.1 |
| $H_o$ | 52°-56.0' |

*Required to enter Pub. No. 249
**Required to plot line of position
***Required to enter Pub. No. 249 and to plot line of position

PLANET JUPITER
Date 20/10/83 ZT 1830
DR Lat. 25°-45.0' N Long. 91°-47.0' W

| | |
|---|---|
| C | 00h-34m-30s |
| W | 1m-08s |
| C-W | 00h-33m-22s |
| CE | 3m-07s f |
| GMT | 00h-30m-15s |
| GHA | 140°-12.4' |
| Corr. (min. & sec.) | 7°-33.8' |
| v correction +2.0 | 1.0 |
| GHA | 147°-47.2' |
| **Assumed Long. | 91°-47.2' |
| *LHA (t-W) | 56°-00.0' |
| ***Assumed Lat. | 26°-00.0' |
| *Declination | S 21°-32.5' |
| d difference +0.1 | +0.1 |
| *Declination | S 21°-32.6' |
| $H_t$ | 18°-11.0' |
| Correction d -37 | -20.0 |
| $H_c$ | 17°-51.0' |
| $H_o$ | 18°-03.4' |
| **Altitude Intercept | 12.4 t |
| Z | N 125° W |
| From 360° (if req.) | 360° |
| **Zn | 235° |
| $H_s$ | 18°-09.7' |
| Main Corr. -2.9 | |
| Addn'l | |
| Dip -3.4 | |
| Total Corr. | -6.3 |
| $H_o$ | 18°-03.4' |

*Required to enter Pub. No. 249
**Required to plot line of position
***Required to enter Pub. No. 249 and to plot line of position

| PLANET | VENUS |
|---|---|
| Date | 21/10/83 ZT 0505 |
| DR Lat. | 24°-59.0' N Long. 92°-38.0' W |
| C | 11h-12m-10s |
| W | 1m-18s |
| C-W | 11h-10m-52s |
| CE | 3m-02s f |
| GMT | 11h-07m-50s |
| GHA | 31°-31.7' |
| Corr. (min. & sec.) | 1°-57.5' |
| v correction +0.4 | +0.1 |
| GHA | 33°-29.3' |
| **Assumed Long. | 92°-29.3' |
| *LHA (t) | 59°-00.0' E |
| ***Assumed Lat. | 25°-00.0' N |
| *Declination | N 6°-13.5' |
| d difference −0.5 | −0.1 |
| *Declination | N 6°-13.4' |
| $H_t$ | 30°-33.0' |
| Correction d+26 | + 6.0 |
| $H_c$ | 30°-39.0' |
| $H_o$ | 30°-52.3' |
| **Altitude Intercept | 13.3' t |
| Z | N 098° E |
| From 360° (if req.) | |
| **Zn | 098° |
| $H_s$ | 30°-57.0' |
| Main Corr. | −1.6 |
| Addn'l | +0.3 |
| Dip | −3.4 |
| Total Corr. | − 4.7 |
| $H_o$ | 30°-52.3' |

*Required to enter Pub. No. 249
**Required to plot line of position
***Required to enter Pub. No. 249 and to plot line of position

| STAR | RIGEL |
|---|---|
| Date | 21/10/83 ZT 0515 |
| DR Lat. | 24°-59' N Long. 92°-38' W |
| C | 11h-18m-10s |
| W | 00m-23s |
| C-W | 11h-17m-47s |
| CE | 3m-02s f |
| GMT | 11h-14m-45s |
| GHA Aries | 194°-19.5' |
| Corr. (min. & sec.) | 3°-41.9' |
| GHA Aries | 198°-01.4' |
| SHA star | 281°-33.7' |
| Sum | 479°-35.1' |
| Less 360° (if req.) | 360°-00.0' |
| GHA star | 119°-35.1' |
| **Assumed Long. | 92°-35.1' |
| *LHA (t-W) | 27°-00.0' |
| ***Assumed Lat. | 25°-00.0' N |
| *Declination | S 8°-13.0' |
| $H_t$ | 47°-48.0' |
| Correction d−47 | −10.0 |
| $H_c$ | 47°-38.0' |
| $H_o$ | 47°-22.9' |
| **Altitude Intercept | 15.1 a |
| Z | N 138° W |
| From 360° (if req.) | 360° |
| **Zn | 222° |
| $H_s$ | 47°-27.2' |
| Main Corr. | −0.9 |
| Addn'l | |
| Dip | −3.4 |
| Total Corr. | −04.3' |
| $H_o$ | 47°-22.9' |

*Required to enter Pub. No. 249
**Required to plot line of position
***Required to enter Pub. No. 249 and to plot line of position

Note: The chronometer error has changed again. Another day has passed. The $5^s$ per day rate must be applied.

STAR ALDEBARAN
Date 21/10/83 ZT 0515
DR Lat. 24°-59' N Long. 92°-38' W

| | |
|---|---|
| C | 11h- 22m- 40s |
| W | 00m- 24s |
| C-W | 11h- 22m- 16s |
| CE | 3m- 02s f |
| GMT | 11h- 19m- 14s |
| GHA Aries | 194°-19.5' |
| Corr. (min. & sec.) | 4°-49.3' |
| GHA Aries | 199°-08.8' |
| SHA star | 291°-15.2' |
| Sum | 490°-24.0' |
| Less 360° (if req.) | 360°-00.0' |
| GHA star | 130°-24.0' |
| **Assumed Long. | 92°-24.0' |
| *LHA (t-W) | 38°-00.0' |
| ***Assumed Lat. | 25°-00.0' N |
| *Declination | N 16°-28.7' |
| $H_t$ | 53°-25.0' |
| Correction d+21 | +10.0 |
| $H_c$ | 53°-35.0' |
| $H_o$ | 53°-25.8' |
| **Altitude Intercept | 9.2 a |
| Z | N 097° W |
| From 360° (if req.) | 360° |
| **Zn | 263° |
| $H_s$ | 53°-29.9' |
| Main Corr. | -0.7 |
| Addn'l | |
| Dip | -3.4 |
| Total Corr. | -04.1' |
| $H_o$ | 53°-25.8' |

*Required to enter Pub. No. 249
**Required to plot line of position
***Required to enter Pub. No. 249 and to plot line of position

STAR ALPHARD
Date 21/10/83 ZT 0520
DR Lat. 24°-59' N Long. 92°-38' W

| | |
|---|---|
| C | 11h- 24m- 30s |
| W | 00m- 17s |
| C-W | 11h- 24m- 13s |
| CE | 03m- 02s f |
| GMT | 11h- 21m- 11s |
| GHA Aries | 194°-19.5' |
| Corr. (min. & sec.) | 5°-18.6' |
| GHA Aries | 199°-38.1' |
| SHA star | 218°-18.5' |
| Sum | 417°-56.6' |
| Less 360° (if req.) | 360°-00.0' |
| GHA star | 57°-56.6' |
| **Assumed Long. | 92°-56.6' |
| *LHA (t) | 35°-00.0' E |
| ***Assumed Lat. | 25°-00' N |
| *Declination | S 8°-35' |
| $H_t$ | 42°-34.0' |
| Correction d-43 | -25.0 |
| $H_c$ | 42°-09.0' |
| $H_o$ | 42°-40.1' |
| **Altitude Intercept | 31.1 t |
| Z | N 130° E |
| From 360° (if req.) | |
| **Zn | 130° |
| $H_s$ | 42°-44.5' |
| Main Corr. | -1.0 |
| Addn'l | |
| Dip | -3.4 |
| Total Corr. | -04.4' |
| $H_o$ | 42°-40.1' |

*Required to enter Pub. No. 249
**Required to plot line of position
***Required to enter Pub. No. 249 and to plot line of position

# B PAGE EXCERPTS FROM THE *NAUTICAL ALMANAC*

Material in this appendix is taken from the *Nautical Almanac* by permission of the Controller of Her Majesty's Stationery Office, Royal Greenwich Observatory, Sussex, England, and by permission of the Almanac Office, U.S. Naval Observatory, Washington, D.C.

# ALTITUDE CORRECTION TABLES 10°-90°—SUN, STARS, PLANETS

## SUN

| OCT.—MAR. App. Alt. | Lower Limb | Upper Limb | APR.—SEPT. App. Alt. | Lower Limb | Upper Limb |
|---|---|---|---|---|---|
| ° ′ | ′ | ′ | ° ′ | ′ | ′ |
| 9 34 | +10·8 | −21·5 | 9 39 | +10·6 | −21·2 |
| 9 45 | +10·9 | −21·4 | 9 51 | +10·7 | −21·1 |
| 9 56 | +11·0 | −21·3 | 10 03 | +10·8 | −21·0 |
| 10 08 | +11·1 | −21·2 | 10 15 | +10·9 | −20·9 |
| 10 21 | +11·2 | −21·1 | 10 27 | +11·0 | −20·8 |
| 10 34 | +11·3 | −21·0 | 10 40 | +11·1 | −20·7 |
| 10 47 | +11·4 | −20·9 | 10 54 | +11·2 | −20·6 |
| 11 01 | +11·5 | −20·8 | 11 08 | +11·3 | −20·5 |
| 11 15 | +11·6 | −20·7 | 11 23 | +11·4 | −20·4 |
| 11 30 | +11·7 | −20·6 | 11 38 | +11·5 | −20·3 |
| 11 46 | +11·8 | −20·5 | 11 54 | +11·6 | −20·2 |
| 12 02 | +11·9 | −20·4 | 12 10 | +11·7 | −20·1 |
| 12 19 | +12·0 | −20·3 | 12 28 | +11·8 | −20·0 |
| 12 37 | +12·1 | −20·2 | 12 46 | +11·9 | −19·9 |
| 12 55 | +12·2 | −20·1 | 13 05 | +12·0 | −19·8 |
| 13 14 | +12·3 | −20·0 | 13 24 | +12·1 | −19·7 |
| 13 35 | +12·4 | −19·9 | 13 45 | +12·2 | −19·6 |
| 13 56 | +12·5 | −19·8 | 14 07 | +12·3 | −19·5 |
| 14 18 | +12·6 | −19·7 | 14 30 | +12·4 | −19·4 |
| 14 42 | +12·7 | −19·6 | 14 54 | +12·5 | −19·3 |
| 15 06 | +12·8 | −19·5 | 15 19 | +12·6 | −19·2 |
| 15 32 | +12·9 | −19·4 | 15 46 | +12·7 | −19·1 |
| 15 59 | +13·0 | −19·3 | 16 14 | +12·8 | −19·0 |
| 16 28 | +13·1 | −19·2 | 16 44 | +12·9 | −18·9 |
| 16 59 | +13·2 | −19·1 | 17 15 | +13·0 | −18·8 |
| 17 32 | +13·3 | −19·0 | 17 48 | +13·1 | −18·7 |
| 18 06 | +13·4 | −18·9 | 18 24 | +13·2 | −18·6 |
| 18 42 | +13·5 | −18·8 | 19 01 | +13·3 | −18·5 |
| 19 21 | +13·6 | −18·7 | 19 42 | +13·4 | −18·4 |
| 20 03 | +13·7 | −18·6 | 20 25 | +13·5 | −18·3 |
| 20 48 | +13·8 | −18·5 | 21 11 | +13·6 | −18·2 |
| 21 35 | +13·9 | −18·4 | 22 00 | +13·7 | −18·1 |
| 22 26 | +14·0 | −18·3 | 22 54 | +13·8 | −18·0 |
| 23 22 | +14·1 | −18·2 | 23 51 | +13·9 | −17·9 |
| 24 21 | +14·2 | −18·1 | 24 53 | +14·0 | −17·8 |
| 25 26 | +14·3 | −18·0 | 26 00 | +14·1 | −17·7 |
| 26 36 | +14·4 | −17·9 | 27 13 | +14·2 | −17·6 |
| 27 52 | +14·5 | −17·8 | 28 33 | +14·3 | −17·5 |
| 29 15 | +14·6 | −17·7 | 30 00 | +14·4 | −17·4 |
| 30 46 | +14·7 | −17·6 | 31 35 | +14·5 | −17·3 |
| 32 26 | +14·8 | −17·5 | 33 20 | +14·6 | −17·2 |
| 34 17 | +14·9 | −17·4 | 35 17 | +14·7 | −17·1 |
| 36 20 | +15·0 | −17·3 | 37 26 | +14·8 | −17·0 |
| 38 36 | +15·1 | −17·2 | 39 50 | +14·9 | −16·9 |
| 41 08 | +15·2 | −17·1 | 42 31 | +15·0 | −16·8 |
| 43 59 | +15·3 | −17·0 | 45 31 | +15·1 | −16·7 |
| 47 10 | +15·4 | −16·9 | 48 55 | +15·2 | −16·6 |
| 50 46 | +15·5 | −16·8 | 52 44 | +15·3 | −16·5 |
| 54 49 | +15·6 | −16·7 | 57 02 | +15·4 | −16·4 |
| 59 23 | +15·7 | −16·6 | 61 51 | +15·5 | −16·3 |
| 64 30 | +15·8 | −16·5 | 67 17 | +15·6 | −16·2 |
| 70 12 | +15·9 | −16·4 | 73 16 | +15·7 | −16·1 |
| 76 26 | +16·0 | −16·3 | 79 43 | +15·8 | −16·0 |
| 83 05 | +16·1 | −16·2 | 86 32 | +15·9 | −15·9 |
| 90 00 | | | 90 00 | | |

## STARS AND PLANETS

| App. Alt. | Corrn |
|---|---|
| ° ′ | ′ |
| 9 56 | −5·3 |
| 10 08 | −5·2 |
| 10 20 | −5·1 |
| 10 33 | −5·0 |
| 10 46 | −4·9 |
| 11 00 | −4·8 |
| 11 14 | −4·7 |
| 11 29 | −4·6 |
| 11 45 | −4·5 |
| 12 01 | −4·4 |
| 12 18 | −4·3 |
| 12 35 | −4·2 |
| 12 54 | −4·1 |
| 13 13 | −4·0 |
| 13 33 | −3·9 |
| 13 54 | −3·8 |
| 14 16 | −3·7 |
| 14 40 | −3·6 |
| 15 04 | −3·5 |
| 15 30 | −3·4 |
| 15 57 | −3·3 |
| 16 26 | −3·2 |
| 16 56 | −3·1 |
| 17 28 | −3·0 |
| 18 02 | −2·9 |
| 18 38 | −2·8 |
| 19 17 | −2·7 |
| 19 58 | −2·6 |
| 20 42 | −2·5 |
| 21 28 | −2·4 |
| 22 19 | −2·3 |
| 23 13 | −2·2 |
| 24 11 | −2·1 |
| 25 14 | −2·0 |
| 26 22 | −1·9 |
| 27 36 | −1·8 |
| 28 56 | −1·7 |
| 30 24 | −1·6 |
| 32 00 | −1·5 |
| 33 45 | −1·4 |
| 35 40 | −1·3 |
| 37 48 | −1·2 |
| 40 08 | −1·1 |
| 42 44 | −1·0 |
| 45 36 | −0·9 |
| 48 47 | −0·8 |
| 52 18 | −0·7 |
| 56 11 | −0·6 |
| 60 28 | −0·5 |
| 65 08 | −0·4 |
| 70 11 | −0·3 |
| 75 34 | −0·2 |
| 81 13 | −0·1 |
| 87 03 | 0·0 |
| 90 00 | |

**App. Alt. — Additional Corrn**

**1983**

**VENUS**

|  | App. Alt. (°) | Additional Corrn (′) |
|---|---|---|
| Jan. 1-May 10 | 0 – 42 | +0·1 |
| May 11-June 23 | 0 – 47 | +0·2 |
| June 24-July 19 | 0 – 46 | +0·3 |
| July 20-Aug. 3 | 0 – 11 | +0·4 |
| | 11 – 41 | +0·5 |
| Aug. 4-Aug. 12 | 0 – 6 | +0·5 |
| | 6 – 20 | +0·6 |
| | 20 – 31 | +0·7 |
| Aug. 13-Sept. 7 | 0 – 4 | +0·6 |
| | 4 – 12 | +0·7 |
| | 12 – 22 | +0·8 |
| Sept. 8-Sept. 16 | 0 – 6 | +0·5 |
| | 6 – 20 | +0·6 |
| | 20 – 31 | +0·7 |
| Sept. 17-Oct. 2 | 0 – 11 | +0·4 |
| | 11 – 41 | +0·5 |
| Oct. 3-Oct. 30 | 0 – 46 | +0·3 |
| Oct. 31-Dec. 17 | 0 – 47 | +0·2 |
| Dec. 18-Dec. 31 | 0 – 42 | +0·1 |

**MARS**

|  | App. Alt. (°) | Additional Corrn (′) |
|---|---|---|
| Jan. 1-Dec. 31 | 0 – 60 | +0·1 |

## DIP

| Ht. of Eye (m) | Corrn (′) | Ht. of Eye (ft.) |
|---|---|---|
| 2·4 | −2·8 | 8·0 |
| 2·6 | −2·9 | 8·6 |
| 2·8 | −3·0 | 9·2 |
| 3·0 | −3·1 | 9·8 |
| 3·2 | −3·2 | 10·5 |
| 3·4 | −3·3 | 11·2 |
| 3·6 | −3·4 | 11·9 |
| 3·8 | −3·5 | 12·6 |
| 4·0 | −3·6 | 13·3 |
| 4·3 | −3·7 | 14·1 |
| 4·5 | −3·8 | 14·9 |
| 4·7 | −3·9 | 15·7 |
| 5·0 | −4·0 | 16·5 |
| 5·2 | −4·1 | 17·4 |
| 5·5 | −4·2 | 18·3 |
| 5·8 | −4·3 | 19·1 |
| 6·1 | −4·4 | 20·1 |
| 6·3 | −4·5 | 21·0 |
| 6·6 | −4·6 | 22·0 |
| 6·9 | −4·7 | 22·9 |
| 7·2 | −4·8 | 23·9 |
| 7·5 | −4·9 | 24·9 |
| 7·9 | −5·0 | 26·0 |
| 8·2 | −5·1 | 27·1 |
| 8·5 | −5·2 | 28·1 |
| 8·8 | −5·3 | 29·2 |
| 9·2 | −5·4 | 30·4 |
| 9·5 | −5·5 | 31·5 |
| 9·9 | −5·6 | 32·7 |
| 10·3 | −5·7 | 33·9 |
| 10·6 | −5·8 | 35·1 |
| 11·0 | −5·9 | 36·3 |
| 11·4 | −6·0 | 37·6 |
| 11·8 | −6·1 | 38·9 |
| 12·2 | −6·2 | 40·1 |
| 12·6 | −6·3 | 41·5 |
| 13·0 | −6·4 | 42·8 |
| 13·4 | −6·5 | 44·2 |
| 13·8 | −6·6 | 45·5 |
| 14·2 | −6·7 | 46·9 |
| 14·7 | −6·8 | 48·4 |
| 15·1 | −6·9 | 49·8 |
| 15·5 | −7·0 | 51·3 |
| 16·0 | −7·1 | 52·8 |
| 16·5 | −7·2 | 54·3 |
| 16·9 | −7·3 | 55·8 |
| 17·4 | −7·4 | 57·4 |
| 17·9 | −7·5 | 58·9 |
| 18·4 | −7·6 | 60·5 |
| 18·8 | −7·7 | 62·1 |
| 19·3 | −7·8 | 63·8 |
| 19·8 | −7·9 | 65·4 |
| 20·4 | −8·0 | 67·1 |
| 20·9 | −8·1 | 68·8 |
| 21·4 | | 70·5 |

| Ht. of Eye | Corrn |
|---|---|
| m | ′ |
| 1·0 | −1·8 |
| 1·5 | −2·2 |
| 2·0 | −2·5 |
| 2·5 | −2·8 |
| 3·0 | −3·0 |
| See table ← | |
| m | ′ |
| 20 | −7·9 |
| 22 | −8·3 |
| 24 | −8·6 |
| 26 | −9·0 |
| 28 | −9·3 |
| 30 | −9·6 |
| 32 | −10·0 |
| 34 | −10·3 |
| 36 | −10·6 |
| 38 | −10·8 |
| 40 | −11·1 |
| 42 | −11·4 |
| 44 | −11·7 |
| 46 | −11·9 |
| 48 | −12·2 |
| ft. | ′ |
| 2 | −1·4 |
| 4 | −1·9 |
| 6 | −2·4 |
| 8 | −2·7 |
| 10 | −3·1 |
| See table ← | |
| ft. | ′ |
| 70 | −8·1 |
| 75 | −8·4 |
| 80 | −8·7 |
| 85 | −8·9 |
| 90 | −9·2 |
| 95 | −9·5 |
| 100 | −9·7 |
| 105 | −9·9 |
| 110 | −10·2 |
| 115 | −10·4 |
| 120 | −10·6 |
| 125 | −10·8 |
| 130 | −11·1 |
| 135 | −11·3 |
| 140 | −11·5 |
| 145 | −11·7 |
| 150 | −11·9 |
| 155 | −12·1 |

App. Alt. = Apparent altitude = Sextant altitude corrected for index error and dip.

For daylight observations of Venus, see pages 259 and 260.

| | G.M.T. d h | ARIES G.H.A. ° ′ | VENUS −3.3 G.H.A. ° ′ | VENUS Dec. ° ′ | MARS +1.4 G.H.A. ° ′ | MARS Dec. ° ′ | JUPITER −1.5 G.H.A. ° ′ | JUPITER Dec. ° ′ | SATURN +0.7 G.H.A. ° ′ | SATURN Dec. ° ′ |
|---|---|---|---|---|---|---|---|---|---|---|
| | 6 00 | 135 33.7 | 154 21.2 | S 9 32.4 | 148 49.9 | S 6 35.0 | 249 57.1 | S20 41.5 | 282 33.3 | S10 35.6 |
| | 01 | 150 36.2 | 169 20.7 | 31.1 | 163 50.5 | 34.2 | 264 59.2 | 41.5 | 297 35.7 | 35.6 |
| | 02 | 165 38.7 | 184 20.3 | 29.9 | 178 51.2 | 33.4 | 280 01.3 | 41.6 | 312 38.1 | 35.6 |
| | 03 | 180 41.1 | 199 19.8 | ·· 28.7 | 193 51.8 | ·· 32.6 | 295 03.4 | ·· 41.6 | 327 40.6 | ·· 35.6 |
| | 04 | 195 43.6 | 214 19.4 | 27.5 | 208 52.5 | 31.9 | 310 05.5 | 41.7 | 342 43.0 | 35.6 |
| | 05 | 210 46.0 | 229 18.9 | 26.3 | 223 53.2 | 31.1 | 325 07.6 | 41.7 | 357 45.4 | 35.6 |
| | 06 | 225 48.5 | 244 18.5 | S 9 25.1 | 238 53.8 | S 6 30.3 | 340 09.7 | S20 41.8 | 12 47.9 | S10 35.6 |
| | 07 | 240 51.0 | 259 18.1 | 23.9 | 253 54.5 | 29.5 | 355 11.8 | 41.8 | 27 50.3 | 35.6 |
| | 08 | 255 53.4 | 274 17.6 | 22.7 | 268 55.2 | 28.7 | 10 13.9 | 41.9 | 42 52.7 | 35.6 |
| S | 09 | 270 55.9 | 289 17.2 | ·· 21.5 | 283 55.8 | ·· 28.0 | 25 16.0 | ·· 41.9 | 57 55.2 | ·· 35.6 |
| U | 10 | 285 58.4 | 304 16.8 | 20.3 | 298 56.5 | 27.2 | 40 18.1 | 42.0 | 72 57.6 | 35.6 |
| N | 11 | 301 00.8 | 319 16.3 | 19.1 | 313 57.2 | 26.4 | 55 20.2 | 42.0 | 88 00.0 | 35.6 |
| D | 12 | 316 03.3 | 334 15.9 | S 9 17.9 | 328 57.8 | S 6 25.6 | 70 22.3 | S20 42.1 | 103 02.5 | S10 35.6 |
| A | 13 | 331 05.8 | 349 15.4 | 16.6 | 343 58.5 | 24.8 | 85 24.4 | 42.1 | 118 04.9 | 35.6 |
| Y | 14 | 346 08.2 | 4 15.0 | 15.4 | 358 59.2 | 24.1 | 100 26.5 | 42.2 | 133 07.3 | 35.6 |
| | 15 | 1 10.7 | 19 14.6 | ·· 14.2 | 13 59.8 | ·· 23.3 | 115 28.7 | ·· 42.3 | 148 09.8 | ·· 35.6 |
| | 16 | 16 13.2 | 34 14.1 | 13.0 | 29 00.5 | 22.5 | 130 30.8 | 42.3 | 163 12.2 | 35.6 |
| | 17 | 31 15.6 | 49 13.7 | 11.8 | 44 01.1 | 21.7 | 145 32.9 | 42.4 | 178 14.7 | 35.6 |
| | 18 | 46 18.1 | 64 13.3 | S 9 10.6 | 59 01.8 | S 6 20.9 | 160 35.0 | S20 42.4 | 193 17.1 | S10 35.6 |
| | 19 | 61 20.5 | 79 12.8 | 09.4 | 74 02.5 | 20.1 | 175 37.1 | 42.5 | 208 19.5 | 35.6 |
| | 20 | 76 23.0 | 94 12.4 | 08.2 | 89 03.1 | 19.4 | 190 39.2 | 42.5 | 223 22.0 | 35.6 |
| | 21 | 91 25.5 | 109 12.0 | ·· 07.0 | 104 03.8 | ·· 18.6 | 205 41.3 | ·· 42.6 | 238 24.4 | ·· 35.6 |
| | 22 | 106 27.9 | 124 11.5 | 05.7 | 119 04.5 | 17.8 | 220 43.4 | 42.6 | 253 26.8 | 35.5 |
| | 23 | 121 30.4 | 139 11.1 | 04.5 | 134 05.1 | 17.0 | 235 45.5 | 42.7 | 268 29.3 | 35.5 |
| | 7 00 | 136 32.9 | 154 10.7 | S 9 03.3 | 149 05.8 | S 6 16.2 | 250 47.6 | S20 42.7 | 283 31.7 | S10 35.5 |
| | 01 | 151 35.3 | 169 10.2 | 02.1 | 164 06.5 | 15.5 | 265 49.7 | 42.8 | 298 34.2 | 35.5 |
| | 02 | 166 37.8 | 184 09.8 | 9 00.9 | 179 07.1 | 14.7 | 280 51.8 | 42.8 | 313 36.6 | 35.5 |
| | 03 | 181 40.3 | 199 09.4 | 8 59.7 | 194 07.8 | ·· 13.9 | 295 53.9 | ·· 42.9 | 328 39.0 | ·· 35.5 |
| | 04 | 196 42.7 | 214 08.9 | 58.4 | 209 08.5 | 13.1 | 310 56.0 | 42.9 | 343 41.5 | 35.5 |
| | 05 | 211 45.2 | 229 08.5 | 57.2 | 224 09.1 | 12.3 | 325 58.2 | 43.0 | 358 43.9 | 35.5 |
| | 06 | 226 47.6 | 244 08.1 | S 8 56.0 | 239 09.8 | S 6 11.6 | 341 00.3 | S20 43.0 | 13 46.3 | S10 35.5 |
| | 07 | 241 50.1 | 259 07.6 | 54.8 | 254 10.5 | 10.8 | 356 02.4 | 43.1 | 28 48.8 | 35.5 |
| | 08 | 256 52.6 | 274 07.2 | 53.6 | 269 11.1 | 10.0 | 11 04.5 | 43.2 | 43 51.2 | 35.5 |
| M | 09 | 271 55.0 | 289 06.8 | ·· 52.4 | 284 11.8 | ·· 09.2 | 26 06.6 | ·· 43.2 | 58 53.7 | ·· 35.5 |
| O | 10 | 286 57.5 | 304 06.4 | 51.1 | 299 12.5 | 08.4 | 41 08.7 | 43.3 | 73 56.1 | 35.5 |
| N | 11 | 302 00.0 | 319 05.9 | 49.9 | 314 13.1 | 07.6 | 56 10.8 | 43.3 | 88 58.5 | 35.5 |
| D | 12 | 317 02.4 | 334 05.5 | S 8 48.7 | 329 13.8 | S 6 06.9 | 71 12.9 | S20 43.4 | 104 01.0 | S10 35.5 |
| A | 13 | 332 04.9 | 349 05.1 | 47.5 | 344 14.5 | 06.1 | 86 15.0 | 43.4 | 119 03.4 | 35.5 |
| Y | 14 | 347 07.4 | 4 04.6 | 46.3 | 359 15.1 | 05.3 | 101 17.1 | 43.5 | 134 05.9 | 35.5 |
| | 15 | 2 09.8 | 19 04.2 | ·· 45.0 | 14 15.8 | ·· 04.5 | 116 19.3 | ·· 43.5 | 149 08.3 | ·· 35.5 |
| | 16 | 17 12.3 | 34 03.8 | 43.8 | 29 16.5 | 03.7 | 131 21.4 | 43.6 | 164 10.7 | 35.5 |
| | 17 | 32 14.8 | 49 03.4 | 42.6 | 44 17.1 | 02.9 | 146 23.5 | 43.6 | 179 13.2 | 35.5 |
| | 18 | 47 17.2 | 64 02.9 | S 8 41.4 | 59 17.8 | S 6 02.2 | 161 25.6 | S20 43.7 | 194 15.6 | S10 35.5 |
| | 19 | 62 19.7 | 79 02.5 | 40.1 | 74 18.5 | 01.4 | 176 27.7 | 43.7 | 209 18.1 | 35.5 |
| | 20 | 77 22.1 | 94 02.1 | 38.9 | 89 19.2 | 6 00.6 | 191 29.8 | 43.8 | 224 20.5 | 35.5 |
| | 21 | 92 24.6 | 109 01.7 | ·· 37.7 | 104 19.8 | 5 59.8 | 206 31.9 | ·· 43.8 | 239 22.9 | ·· 35.5 |
| | 22 | 107 27.1 | 124 01.2 | 36.5 | 119 20.5 | 59.0 | 221 34.0 | 43.9 | 254 25.4 | 35.5 |
| | 23 | 122 29.5 | 139 00.8 | 35.2 | 134 21.2 | 58.3 | 236 36.1 | 43.9 | 269 27.8 | 35.5 |
| | 8 00 | 137 32.0 | 154 00.4 | S 8 34.0 | 149 21.8 | S 5 57.5 | 251 38.3 | S20 44.0 | 284 30.3 | S10 35.5 |
| | 01 | 152 34.5 | 169 00.0 | 32.8 | 164 22.5 | 56.7 | 266 40.4 | 44.0 | 299 32.7 | 35.5 |
| | 02 | 167 36.9 | 183 59.5 | 31.6 | 179 23.2 | 55.9 | 281 42.5 | 44.1 | 314 35.1 | 35.5 |
| | 03 | 182 39.4 | 198 59.1 | ·· 30.3 | 194 23.8 | ·· 55.1 | 296 44.6 | ·· 44.1 | 329 37.6 | ·· 35.5 |
| | 04 | 197 41.9 | 213 58.7 | 29.1 | 209 24.5 | 54.3 | 311 46.7 | 44.2 | 344 40.0 | 35.5 |
| | 05 | 212 44.3 | 228 58.3 | 27.9 | 224 25.2 | 53.6 | 326 48.8 | 44.2 | 359 42.5 | 35.5 |
| | 06 | 227 46.8 | 243 57.9 | S 8 26.7 | 239 25.8 | S 5 52.8 | 341 50.9 | S20 44.3 | 14 44.9 | S10 35.5 |
| | 07 | 242 49.3 | 258 57.4 | 25.4 | 254 26.5 | 52.0 | 356 53.1 | 44.3 | 29 47.4 | 35.4 |
| T | 08 | 257 51.7 | 273 57.0 | 24.2 | 269 27.2 | 51.2 | 11 55.2 | 44.4 | 44 49.8 | 35.4 |
| U | 09 | 272 54.2 | 288 56.6 | ·· 23.0 | 284 27.9 | ·· 50.4 | 26 57.3 | ·· 44.4 | 59 52.2 | ·· 35.4 |
| E | 10 | 287 56.6 | 303 56.2 | 21.8 | 299 28.5 | 49.6 | 41 59.4 | 44.5 | 74 54.7 | 35.4 |
| S | 11 | 302 59.1 | 318 55.8 | 20.5 | 314 29.2 | 48.9 | 57 01.5 | 44.5 | 89 57.1 | 35.4 |
| D | 12 | 318 01.6 | 333 55.3 | S 8 19.3 | 329 29.9 | S 5 48.1 | 72 03.6 | S20 44.6 | 104 59.6 | S10 35.4 |
| A | 13 | 333 04.0 | 348 54.9 | 18.1 | 344 30.5 | 47.3 | 87 05.8 | 44.6 | 120 02.0 | 35.4 |
| Y | 14 | 348 06.5 | 3 54.5 | 16.8 | 359 31.2 | 46.5 | 102 07.9 | 44.7 | 135 04.5 | 35.4 |
| | 15 | 3 09.0 | 18 54.1 | ·· 15.6 | 14 31.9 | ·· 45.7 | 117 10.0 | ·· 44.8 | 150 06.9 | ·· 35.4 |
| | 16 | 18 11.4 | 33 53.7 | 14.4 | 29 32.5 | 44.9 | 132 12.1 | 44.8 | 165 09.4 | 35.4 |
| | 17 | 33 13.9 | 48 53.2 | 13.1 | 44 33.2 | 44.1 | 147 14.2 | 44.9 | 180 11.8 | 35.4 |
| | 18 | 48 16.4 | 63 52.8 | S 8 11.9 | 59 33.9 | S 5 43.4 | 162 16.3 | S20 44.9 | 195 14.2 | S10 35.4 |
| | 19 | 63 18.8 | 78 52.4 | 10.7 | 74 34.6 | 42.6 | 177 18.5 | 45.0 | 210 16.7 | 35.4 |
| | 20 | 78 21.3 | 93 52.0 | 09.5 | 89 35.2 | 41.8 | 192 20.6 | 45.0 | 225 19.1 | 35.4 |
| | 21 | 93 23.8 | 108 51.6 | ·· 08.2 | 104 35.9 | ·· 41.0 | 207 22.7 | ·· 45.1 | 240 21.6 | ·· 35.4 |
| | 22 | 108 26.2 | 123 51.2 | 07.0 | 119 36.6 | 40.2 | 222 24.8 | 45.1 | 255 24.0 | 35.4 |
| | 23 | 123 28.7 | 138 50.7 | 05.8 | 134 37.2 | 39.4 | 237 26.9 | 45.2 | 270 26.5 | 35.4 |
| | Mer. Pass. | h m 14 51.4 | $v$ −0.4 | $d$ 1.2 | $v$ 0.7 | $d$ 0.8 | $v$ 2.1 | $d$ 0.1 | $v$ 2.4 | $d$ 0.0 |

**STARS**

| Name | S.H.A. ° ′ | Dec. ° ′ |
|---|---|---|
| Acamar | 315 35.9 | S40 22.7 |
| Achernar | 335 44.1 | S57 19.7 |
| Acrux | 173 35.0 | S63 00.1 |
| Adhara | 255 30.5 | S28 57.1 |
| Aldebaran | 291 15.9 | N16 28.5 |
| Alioth | 166 40.6 | N56 02.9 |
| Alkaid | 153 16.9 | N49 23.6 |
| Al Na'ir | 28 13.2 | S47 02.8 |
| Alnilam | 276 09.7 | S 1 12.8 |
| Alphard | 218 18.6 | S 8 35.1 |
| Alphecca | 126 30.7 | N26 46.1 |
| Alpheratz | 358 07.8 | N28 59.8 |
| Altair | 62 31.2 | N 8 49.2 |
| Ankaa | 353 38.8 | S42 24.2 |
| Antares | 112 54.9 | S26 23.6 |
| Arcturus | 146 16.8 | N19 16.1 |
| Atria | 108 17.9 | S68 59.6 |
| Avior | 234 26.9 | S59 27.3 |
| Bellatrix | 278 56.7 | N 6 20.0 |
| Betelgeuse | 271 26.2 | N 7 24.2 |
| Canopus | 264 06.0 | S52 41.4 |
| Capella | 281 08.5 | N45 59.0 |
| Deneb | 49 47.8 | N45 13.0 |
| Denebola | 182 57.0 | N14 39.9 |
| Diphda | 349 19.3 | S18 05.0 |
| Dubhe | 194 19.3 | N61 50.5 |
| Elnath | 278 41.8 | N28 35.7 |
| Eltanin | 90 57.3 | N51 29.2 |
| Enif | 34 10.2 | N 9 47.6 |
| Fomalhaut | 15 49.8 | S29 43.0 |
| Gacrux | 172 26.6 | S57 00.9 |
| Gienah | 176 16.0 | S17 26.8 |
| Hadar | 149 20.9 | S60 17.2 |
| Hamal | 328 27.1 | N23 22.9 |
| Kaus Aust. | 84 14.9 | S34 23.6 |
| Kochab | 137 18.8 | N74 13.2 |
| Markab | 14 01.8 | N15 06.7 |
| Menkar | 314 39.3 | N 4 01.3 |
| Menkent | 148 34.9 | S36 17.0 |
| Miaplacidus | 221 43.8 | S69 38.8 |
| Mirfak | 309 13.7 | N49 48.2 |
| Nunki | 76 27.3 | S26 19.2 |
| Peacock | 53 56.2 | S56 47.5 |
| Pollux | 243 55.7 | N28 04.1 |
| Procyon | 245 23.7 | N 5 16.1 |
| Rasalhague | 96 28.2 | N12 34.2 |
| Regulus | 208 07.8 | N12 03.0 |
| Rigel | 281 34.2 | S 8 13.4 |
| Rigil Kent. | 140 23.5 | S60 45.6 |
| Sabik | 102 39.4 | S15 42.3 |
| Schedar | 350 07.5 | N56 26.8 |
| Shaula | 96 53.7 | S37 05.5 |
| Sirius | 258 53.9 | S16 41.7 |
| Spica | 158 55.6 | S11 04.3 |
| Suhail | 223 09.1 | S43 21.8 |
| Vega | 80 55.0 | N38 45.8 |
| Zuben'ubi | 137 31.1 | S15 58.3 |

| | S.H.A. ° ′ | Mer. Pass. h m |
|---|---|---|
| Venus | 17 37.8 | 13 44 |
| Mars | 12 32.9 | 14 03 |
| Jupiter | 114 14.7 | 7 16 |
| Saturn | 146 58.8 | 5 05 |

| G.M.T. | | ARIES | VENUS −4.0 | | MARS +1.8 | | JUPITER −2.0 | | SATURN +0.8 | | STARS | | |
|---|---|---|---|---|---|---|---|---|---|---|---|---|---|
| | | G.H.A. | G.H.A. | Dec. | G.H.A. | Dec. | G.H.A. | Dec. | G.H.A. | Dec. | Name | S.H.A. | Dec. |
| d | h | ° ′ | ° ′ | ° ′ | ° ′ | ° ′ | ° ′ | ° ′ | ° ′ | ° ′ | | ° ′ | ° ′ |
| 24 | 00 | 271 34.9 | 131 34.0 | N17 10.8 | 185 30.0 | N24 00.6 | 30 34.5 | S19 55.0 | 64 51.8 | S 8 16.3 | Acamar | 315 36.0 | S40 22.1 |
| | 01 | 286 37.3 | 146 34.3 | 09.9 | 200 30.6 | 00.7 | 45 37.3 | 54.9 | 79 54.3 | 16.3 | Achernar | 335 43.9 | S57 19.1 |
| | 02 | 301 39.8 | 161 34.6 | 09.0 | 215 31.2 | 00.7 | 60 40.0 | 54.9 | 94 56.8 | 16.3 | Acrux | 173 35.1 | S63 00.7 |
| | 03 | 316 42.3 | 176 34.9 | ·· 08.2 | 230 31.8 | ·· 00.8 | 75 42.7 | ·· 54.9 | 109 59.3 | ·· 16.3 | Adhara | 255 30.9 | S28 56.9 |
| | 04 | 331 44.7 | 191 35.3 | 07.3 | 245 32.4 | 00.9 | 90 45.4 | 54.8 | 125 01.8 | 16.3 | Aldebaran | 291 16.1 | N16 28.6 |
| | 05 | 346 47.2 | 206 35.6 | 06.4 | 260 33.1 | 00.9 | 105 48.1 | 54.8 | 140 04.3 | 16.3 | | | |
| | 06 | 1 49.7 | 221 35.9 | N17 05.6 | 275 33.7 | N24 01.0 | 120 50.8 | S19 54.7 | 155 06.8 | S 8 16.3 | Alioth | 166 40.5 | N56 03.3 |
| | 07 | 16 52.1 | 236 36.2 | 04.7 | 290 34.3 | 01.1 | 135 53.6 | 54.7 | 170 09.3 | 16.3 | Alkaid | 153 16.6 | N49 24.1 |
| | 08 | 31 54.6 | 251 36.5 | 03.8 | 305 34.9 | 01.1 | 150 56.3 | 54.7 | 185 11.8 | 16.3 | Al Na'ir | 28 12.1 | S47 02.4 |
| F | 09 | 46 57.1 | 266 36.8 | ·· 03.0 | 320 35.5 | ·· 01.2 | 165 59.0 | ·· 54.6 | 200 14.3 | ·· 16.3 | Alnilam | 276 10.0 | S 1 12.7 |
| R | 10 | 61 59.5 | 281 37.1 | 02.1 | 335 36.1 | 01.3 | 181 01.7 | 54.6 | 215 16.8 | 16.3 | Alphard | 218 18.9 | S 8 35.1 |
| I | 11 | 77 02.0 | 296 37.5 | 01.2 | 350 36.7 | 01.3 | 196 04.4 | 54.5 | 230 19.3 | 16.3 | | | |
| D | 12 | 92 04.4 | 311 37.8 | N17 00.3 | 5 37.3 | N24 01.4 | 211 07.2 | S19 54.5 | 245 21.8 | S 8 16.3 | Alphecca | 126 30.1 | N26 46.3 |
| A | 13 | 107 06.9 | 326 38.1 | 16 59.5 | 20 37.9 | 01.5 | 226 09.9 | 54.5 | 260 24.3 | 16.3 | Alpheratz | 358 07.3 | N28 59.7 |
| Y | 14 | 122 09.4 | 341 38.4 | 58.6 | 35 38.5 | 01.5 | 241 12.6 | 54.4 | 275 26.8 | 16.3 | Altair | 62 30.3 | N 8 49.4 |
| | 15 | 137 11.8 | 356 38.8 | ·· 57.7 | 50 39.1 | ·· 01.6 | 256 15.3 | ·· 54.4 | 290 29.3 | ·· 16.3 | Ankaa | 353 38.2 | S42 23.6 |
| | 16 | 152 14.3 | 11 39.1 | 56.9 | 65 39.7 | 01.6 | 271 18.0 | 54.4 | 305 31.8 | 16.3 | Antares | 112 54.1 | S26 23.8 |
| | 17 | 167 16.8 | 26 39.4 | 56.0 | 80 40.4 | 01.7 | 286 20.7 | 54.3 | 320 34.3 | 16.3 | | | |
| | 18 | 182 19.2 | 41 39.7 | N16 55.1 | 95 41.0 | N24 01.8 | 301 23.5 | S19 54.3 | 335 36.8 | S 8 16.3 | Arcturus | 146 16.5 | N19 16.3 |
| | 19 | 197 21.7 | 56 40.1 | 54.3 | 110 41.6 | 01.8 | 316 26.2 | 54.2 | 350 39.2 | 16.3 | Atria | 108 16.1 | S69 00.0 |
| | 20 | 212 24.2 | 71 40.4 | 53.4 | 125 42.2 | 01.9 | 331 28.9 | 54.2 | 5 41.7 | 16.3 | Avior | 234 28.0 | S59 27.4 |
| | 21 | 227 26.6 | 86 40.7 | ·· 52.5 | 140 42.8 | ·· 01.9 | 346 31.6 | ·· 54.2 | 20 44.2 | ·· 16.3 | Bellatrix | 278 57.0 | N 6 20.1 |
| | 22 | 242 29.1 | 101 41.1 | 51.6 | 155 43.4 | 02.0 | 1 34.3 | 54.1 | 35 46.7 | 16.3 | Betelgeuse | 271 26.5 | N 7 24.3 |
| | 23 | 257 31.6 | 116 41.4 | 50.8 | 170 44.0 | 02.1 | 16 37.0 | 54.1 | 50 49.2 | 16.3 | | | |
| 25 | 00 | 272 34.0 | 131 41.8 | N16 49.9 | 185 44.6 | N24 02.1 | 31 39.7 | S19 54.1 | 65 51.7 | S 8 16.3 | Canopus | 264 06.9 | S52 41.2 |
| | 01 | 287 36.5 | 146 42.1 | 49.0 | 200 45.2 | 02.2 | 46 42.5 | 54.0 | 80 54.2 | 16.3 | Capella | 281 08.9 | N45 58.9 |
| | 02 | 302 38.9 | 161 42.4 | 48.1 | 215 45.8 | 02.2 | 61 45.2 | 54.0 | 95 56.7 | 16.3 | Deneb | 49 46.8 | N45 13.0 |
| | 03 | 317 41.4 | 176 42.8 | ·· 47.3 | 230 46.4 | ·· 02.3 | 76 47.9 | ·· 53.9 | 110 59.2 | ·· 16.3 | Denebola | 182 57.0 | N14 40.1 |
| | 04 | 332 43.9 | 191 43.1 | 46.4 | 245 47.0 | 02.4 | 91 50.6 | 53.9 | 126 01.7 | 16.3 | Diphda | 349 18.9 | S18 04.6 |
| | 05 | 347 46.3 | 206 43.5 | 45.5 | 260 47.7 | 02.4 | 106 53.3 | 53.9 | 141 04.2 | 16.3 | | | |
| | 06 | 2 48.8 | 221 43.8 | N16 44.6 | 275 48.3 | N24 02.5 | 121 56.0 | S19 53.8 | 156 06.7 | S 8 16.3 | Dubhe | 194 19.8 | N61 50.8 |
| | 07 | 17 51.3 | 236 44.2 | 43.8 | 290 48.9 | 02.5 | 136 58.7 | 53.8 | 171 09.2 | 16.3 | Elnath | 278 42.0 | N28 35.6 |
| S | 08 | 32 53.7 | 251 44.5 | 42.9 | 305 49.5 | 02.6 | 152 01.4 | 53.7 | 186 11.7 | 16.3 | Eltanin | 90 56.3 | N51 29.4 |
| A | 09 | 47 56.2 | 266 44.9 | ·· 42.0 | 320 50.1 | ·· 02.6 | 167 04.2 | ·· 53.7 | 201 14.1 | ·· 16.3 | Enif | 34 09.4 | N 9 47.8 |
| T | 10 | 62 58.7 | 281 45.2 | 41.1 | 335 50.7 | 02.7 | 182 06.9 | 53.7 | 216 16.6 | 16.3 | Fomalhaut | 15 49.1 | S29 42.5 |
| U | 11 | 78 01.1 | 296 45.6 | 40.3 | 350 51.3 | 02.7 | 197 09.6 | 53.6 | 231 19.1 | 16.3 | | | |
| R | 12 | 93 03.6 | 311 45.9 | N16 39.4 | 5 51.9 | N24 02.8 | 212 12.3 | S19 53.6 | 246 21.6 | S 8 16.3 | Gacrux | 172 26.6 | S57 01.4 |
| D | 13 | 108 06.1 | 326 46.3 | 38.5 | 20 52.5 | 02.9 | 227 15.0 | 53.6 | 261 24.1 | 16.3 | Gienah | 176 16.0 | S17 27.0 |
| A | 14 | 123 08.5 | 341 46.7 | 37.6 | 35 53.1 | 02.9 | 242 17.7 | 53.5 | 276 26.6 | 16.3 | Hadar | 149 20.3 | S60 17.8 |
| Y | 15 | 138 11.0 | 356 47.0 | ·· 36.8 | 50 53.8 | ·· 03.0 | 257 20.4 | ·· 53.5 | 291 29.1 | ·· 16.3 | Hamal | 328 26.9 | N23 22.9 |
| | 16 | 153 13.4 | 11 47.4 | 35.9 | 65 54.4 | 03.0 | 272 23.1 | 53.4 | 306 31.6 | 16.3 | Kaus Aust. | 84 13.8 | S34 23.6 |
| | 17 | 168 15.9 | 26 47.8 | 35.0 | 80 55.0 | 03.1 | 287 25.8 | 53.4 | 321 34.1 | 16.3 | | | |
| | 18 | 183 18.4 | 41 48.1 | N16 34.1 | 95 55.6 | N24 03.1 | 302 28.6 | S19 53.4 | 336 36.6 | S 8 16.3 | Kochab | 137 18.1 | N74 13.7 |
| | 19 | 198 20.8 | 56 48.5 | 33.2 | 110 56.2 | 03.2 | 317 31.3 | 53.3 | 351 39.1 | 16.3 | Markab | 14 01.1 | N15 06.8 |
| | 20 | 213 23.3 | 71 48.9 | 32.4 | 125 56.8 | 03.2 | 332 34.0 | 53.3 | 6 41.5 | 16.3 | Menkar | 314 39.3 | N 4 01.4 |
| | 21 | 228 25.8 | 86 49.2 | ·· 31.5 | 140 57.4 | ·· 03.3 | 347 36.7 | ·· 53.3 | 21 44.0 | ·· 16.3 | Menkent | 148 34.5 | S36 17.4 |
| | 22 | 243 28.2 | 101 49.6 | 30.6 | 155 58.0 | 03.3 | 2 39.4 | 53.2 | 36 46.5 | 16.3 | Miaplacidus | 221 45.4 | S69 39.1 |
| | 23 | 258 30.7 | 116 50.0 | 29.7 | 170 58.6 | 03.4 | 17 42.1 | 53.2 | 51 49.0 | 16.3 | | | |
| 26 | 00 | 273 33.2 | 131 50.3 | N16 28.8 | 185 59.2 | N24 03.4 | 32 44.8 | S19 53.2 | 66 51.5 | S 8 16.3 | Mirfak | 309 13.7 | N49 48.0 |
| | 01 | 288 35.6 | 146 50.7 | 28.0 | 200 59.8 | 03.5 | 47 47.5 | 53.1 | 81 54.0 | 16.3 | Nunki | 76 26.3 | S26 19.1 |
| | 02 | 303 38.1 | 161 51.1 | 27.1 | 216 00.5 | 03.5 | 62 50.2 | 53.1 | 96 56.5 | 16.4 | Peacock | 53 54.7 | S56 47.3 |
| | 03 | 318 40.5 | 176 51.5 | ·· 26.2 | 231 01.1 | ·· 03.6 | 77 52.9 | ·· 53.0 | 111 59.0 | ·· 16.4 | Pollux | 243 56.1 | N28 04.1 |
| | 04 | 333 43.0 | 191 51.9 | 25.3 | 246 01.7 | 03.6 | 92 55.6 | 53.0 | 127 01.5 | 16.4 | Procyon | 245 24.0 | N 5 16.2 |
| | 05 | 348 45.5 | 206 52.2 | 24.4 | 261 02.3 | 03.7 | 107 58.4 | 53.0 | 142 03.9 | 16.4 | | | |
| | 06 | 3 47.9 | 221 52.6 | N16 23.5 | 276 02.9 | N24 03.7 | 123 01.1 | S19 52.9 | 157 06.4 | S 8 16.4 | Rasalhague | 96 27.4 | N12 34.3 |
| | 07 | 18 50.4 | 236 53.0 | 22.7 | 291 03.5 | 03.8 | 138 03.8 | 52.9 | 172 08.9 | 16.4 | Regulus | 208 08.1 | N12 03.1 |
| | 08 | 33 52.9 | 251 53.4 | 21.8 | 306 04.1 | 03.8 | 153 06.5 | 52.9 | 187 11.4 | 16.4 | Rigel | 281 34.4 | S 8 13.2 |
| S | 09 | 48 55.3 | 266 53.8 | ·· 20.9 | 321 04.7 | ·· 03.9 | 168 09.2 | ·· 52.8 | 202 13.9 | ·· 16.4 | Rigil Kent. | 140 22.8 | S60 46.2 |
| U | 10 | 63 57.8 | 281 54.2 | 20.0 | 336 05.3 | 03.9 | 183 11.9 | 52.8 | 217 16.4 | 16.4 | Sabik | 102 38.5 | S15 42.3 |
| N | 11 | 79 00.3 | 296 54.6 | 19.1 | 351 06.0 | 03.9 | 198 14.6 | 52.7 | 232 18.9 | 16.4 | | | |
| D | 12 | 94 02.7 | 311 55.0 | N16 18.2 | 6 06.6 | N24 04.0 | 213 17.3 | S19 52.7 | 247 21.4 | S 8 16.4 | Schedar | 350 06.9 | N56 26.4 |
| A | 13 | 109 05.2 | 326 55.3 | 17.4 | 21 07.2 | 04.0 | 228 20.0 | 52.7 | 262 23.9 | 16.4 | Shaula | 96 52.7 | S37 05.6 |
| Y | 14 | 124 07.7 | 341 55.7 | 16.5 | 36 07.8 | 04.1 | 243 22.7 | 52.6 | 277 26.3 | 16.4 | Sirius | 258 54.3 | S16 41.6 |
| | 15 | 139 10.1 | 356 56.1 | ·· 15.6 | 51 08.4 | ·· 04.1 | 258 25.4 | ·· 52.6 | 292 28.8 | ·· 16.4 | Spica | 158 55.4 | S11 04.5 |
| | 16 | 154 12.6 | 11 56.5 | 14.7 | 66 09.0 | 04.2 | 273 28.1 | 52.6 | 307 31.3 | 16.4 | Suhail | 223 09.7 | S43 22.0 |
| | 17 | 169 15.0 | 26 56.9 | 13.8 | 81 09.6 | 04.2 | 288 30.8 | 52.5 | 322 33.8 | 16.4 | | | |
| | 18 | 184 17.5 | 41 57.3 | N16 12.9 | 96 10.2 | N24 04.3 | 303 33.5 | S19 52.5 | 337 36.3 | S 8 16.4 | Vega | 80 54.1 | N38 46.0 |
| | 19 | 199 20.0 | 56 57.7 | 12.0 | 111 10.8 | 04.3 | 318 36.2 | 52.5 | 352 38.8 | 16.4 | Zuben'ubi | 137 30.6 | S15 58.4 |
| | 20 | 214 22.4 | 71 58.1 | 11.2 | 126 11.5 | 04.3 | 333 38.9 | 52.4 | 7 41.3 | 16.4 | | S.H.A. | Mer. Pass. |
| | 21 | 229 24.9 | 86 58.6 | ·· 10.3 | 141 12.1 | ·· 04.4 | 348 41.6 | ·· 52.4 | 22 43.7 | ·· 16.4 | | ° ′ | h m |
| | 22 | 244 27.4 | 101 59.0 | 09.4 | 156 12.7 | 04.4 | 3 44.3 | 52.3 | 37 46.2 | 16.4 | Venus | 219 07.7 | 15 13 |
| | 23 | 259 29.8 | 116 59.4 | 08.5 | 171 13.3 | 04.5 | 18 47.0 | 52.3 | 52 48.7 | 16.4 | Mars | 273 10.6 | 11 37 |
| Mer. Pass. | | h m 5 48.8 | *v* 0.4 | *d* 0.9 | *v* 0.6 | *d* 0.1 | *v* 2.7 | *d* 0.0 | *v* 2.5 | *d* 0.0 | Jupiter | 119 05.7 | 21 49 |
| | | | | | | | | | | | Saturn | 153 17.7 | 19 33 |

| G.M.T. | | SUN G.H.A. | SUN Dec. | MOON G.H.A. | v | MOON Dec. | d | H.P. |
|---|---|---|---|---|---|---|---|---|
| d | h | ° ′ | ° ′ | ° ′ | ′ | ° ′ | ′ | ′ |
| 24 | 00 | 179 27.8 | N23 25.7 | 15 59.1 | 10.7 | S21 59.9 | 5.8 | 55.2 |
| | 01 | 194 27.7 | 25.6 | 30 28.8 | 10.6 | 22 05.7 | 5.7 | 55.2 |
| | 02 | 209 27.6 | 25.6 | 44 58.4 | 10.6 | 22 11.4 | 5.5 | 55.2 |
| | 03 | 224 27.4 | ·· 25.5 | 59 28.0 | 10.6 | 22 16.9 | 5.5 | 55.2 |
| | 04 | 239 27.3 | 25.5 | 73 57.6 | 10.5 | 22 22.4 | 5.4 | 55.2 |
| | 05 | 254 27.2 | 25.5 | 88 27.1 | 10.5 | 22 27.8 | 5.2 | 55.1 |
| | 06 | 269 27.0 | N23 25.4 | 102 56.6 | 10.6 | S22 33.0 | 5.1 | 55.1 |
| | 07 | 284 26.9 | 25.4 | 117 26.2 | 10.4 | 22 38.1 | 5.0 | 55.1 |
| | 08 | 299 26.8 | 25.4 | 131 55.6 | 10.5 | 22 43.1 | 4.9 | 55.1 |
| F | 09 | 314 26.6 | ·· 25.3 | 146 25.1 | 10.4 | 22 48.0 | 4.8 | 55.1 |
| R | 10 | 329 26.5 | 25.3 | 160 54.5 | 10.5 | 22 52.8 | 4.6 | 55.1 |
| I | 11 | 344 26.4 | 25.2 | 175 24.0 | 10.4 | 22 57.4 | 4.6 | 55.0 |
| D | 12 | 359 26.2 | N23 25.2 | 189 53.4 | 10.4 | S23 02.0 | 4.4 | 55.0 |
| A | 13 | 14 26.1 | 25.1 | 204 22.8 | 10.3 | 23 06.4 | 4.3 | 55.0 |
| Y | 14 | 29 26.0 | 25.1 | 218 52.1 | 10.4 | 23 10.7 | 4.2 | 55.0 |
| | 15 | 44 25.8 | ·· 25.1 | 233 21.5 | 10.3 | 23 14.9 | 4.1 | 55.0 |
| | 16 | 59 25.7 | 25.0 | 247 50.8 | 10.3 | 23 19.0 | 3.9 | 55.0 |
| | 17 | 74 25.6 | 25.0 | 262 20.1 | 10.3 | 23 22.9 | 3.9 | 54.9 |
| | 18 | 89 25.4 | N23 24.9 | 276 49.4 | 10.3 | S23 26.8 | 3.7 | 54.9 |
| | 19 | 104 25.3 | 24.9 | 291 18.7 | 10.3 | 23 30.5 | 3.6 | 54.9 |
| | 20 | 119 25.2 | 24.8 | 305 48.0 | 10.3 | 23 34.1 | 3.4 | 54.9 |
| | 21 | 134 25.1 | ·· 24.8 | 320 17.3 | 10.2 | 23 37.5 | 3.4 | 54.9 |
| | 22 | 149 24.9 | 24.7 | 334 46.5 | 10.3 | 23 40.9 | 3.2 | 54.9 |
| | 23 | 164 24.8 | 24.7 | 349 15.8 | 10.2 | 23 44.1 | 3.2 | 54.9 |
| 25 | 00 | 179 24.7 | N23 24.6 | 3 45.0 | 10.3 | S23 47.3 | 2.9 | 54.8 |
| | 01 | 194 24.5 | 24.6 | 18 14.3 | 10.2 | 23 50.2 | 2.9 | 54.8 |
| | 02 | 209 24.4 | 24.5 | 32 43.5 | 10.2 | 23 53.1 | 2.8 | 54.8 |
| | 03 | 224 24.3 | ·· 24.4 | 47 12.7 | 10.2 | 23 55.9 | 2.6 | 54.8 |
| | 04 | 239 24.1 | 24.4 | 61 41.9 | 10.2 | 23 58.5 | 2.5 | 54.8 |
| | 05 | 254 24.0 | 24.3 | 76 11.1 | 10.2 | 24 01.0 | 2.4 | 54.8 |
| | 06 | 269 23.9 | N23 24.3 | 90 40.3 | 10.2 | S24 03.4 | 2.3 | 54.8 |
| | 07 | 284 23.7 | 24.2 | 105 09.5 | 10.2 | 24 05.7 | 2.1 | 54.7 |
| S | 08 | 299 23.6 | 24.2 | 119 38.7 | 10.2 | 24 07.8 | 2.1 | 54.7 |
| A | 09 | 314 23.5 | ·· 24.1 | 134 07.9 | 10.2 | 24 09.9 | 1.9 | 54.7 |
| T | 10 | 329 23.3 | 24.1 | 148 37.1 | 10.2 | 24 11.8 | 1.8 | 54.7 |
| U | 11 | 344 23.2 | 24.0 | 163 06.3 | 10.2 | 24 13.6 | 1.6 | 54.7 |
| R | 12 | 359 23.1 | N23 23.9 | 177 35.5 | 10.2 | S24 15.2 | 1.6 | 54.7 |
| D | 13 | 14 22.9 | 23.9 | 192 04.7 | 10.2 | 24 16.8 | 1.4 | 54.7 |
| A | 14 | 29 22.8 | 23.8 | 206 33.9 | 10.2 | 24 18.2 | 1.3 | 54.6 |
| Y | 15 | 44 22.7 | ·· 23.7 | 221 03.1 | 10.2 | 24 19.5 | 1.2 | 54.6 |
| | 16 | 59 22.5 | 23.7 | 235 32.3 | 10.2 | 24 20.7 | 1.0 | 54.6 |
| | 17 | 74 22.4 | 23.6 | 250 01.5 | 10.2 | 24 21.7 | 1.0 | 54.6 |
| | 18 | 89 22.3 | N23 23.6 | 264 30.7 | 10.3 | S24 22.7 | 0.8 | 54.6 |
| | 19 | 104 22.1 | 23.5 | 279 00.0 | 10.2 | 24 23.5 | 0.7 | 54.6 |
| | 20 | 119 22.0 | 23.4 | 293 29.2 | 10.3 | 24 24.2 | 0.6 | 54.6 |
| | 21 | 134 21.9 | ·· 23.4 | 307 58.5 | 10.2 | 24 24.8 | 0.4 | 54.6 |
| | 22 | 149 21.8 | 23.3 | 322 27.7 | 10.3 | 24 25.2 | 0.3 | 54.5 |
| | 23 | 164 21.6 | 23.2 | 336 57.0 | 10.3 | 24 25.5 | 0.3 | 54.5 |
| 26 | 00 | 179 21.5 | N23 23.1 | 351 26.3 | 10.3 | S24 25.8 | 0.0 | 54.5 |
| | 01 | 194 21.4 | 23.1 | 5 55.6 | 10.3 | 24 25.8 | 0.0 | 54.5 |
| | 02 | 209 21.2 | 23.0 | 20 24.9 | 10.3 | 24 25.8 | 0.1 | 54.5 |
| | 03 | 224 21.1 | ·· 22.9 | 34 54.2 | 10.4 | 24 25.7 | 0.3 | 54.5 |
| | 04 | 239 21.0 | 22.9 | 49 23.6 | 10.4 | 24 25.4 | 0.4 | 54.5 |
| | 05 | 254 20.8 | 22.8 | 63 53.0 | 10.3 | 24 25.0 | 0.5 | 54.5 |
| | 06 | 269 20.7 | N23 22.7 | 78 22.3 | 10.5 | S24 24.5 | 0.6 | 54.4 |
| | 07 | 284 20.6 | 22.6 | 92 51.8 | 10.4 | 24 23.9 | 0.8 | 54.4 |
| | 08 | 299 20.4 | 22.6 | 107 21.2 | 10.4 | 24 23.1 | 0.9 | 54.4 |
| S | 09 | 314 20.3 | ·· 22.5 | 121 50.6 | 10.5 | 24 22.2 | 1.0 | 54.4 |
| U | 10 | 329 20.2 | 22.4 | 136 20.1 | 10.5 | 24 21.2 | 1.1 | 54.4 |
| N | 11 | 344 20.0 | 22.3 | 150 49.6 | 10.5 | 24 20.1 | 1.2 | 54.4 |
| D | 12 | 359 19.9 | N23 22.3 | 165 19.1 | 10.5 | S24 18.9 | 1.3 | 54.4 |
| A | 13 | 14 19.8 | 22.2 | 179 48.6 | 10.6 | 24 17.6 | 1.5 | 54.4 |
| Y | 14 | 29 19.7 | 22.1 | 194 18.2 | 10.6 | 24 16.1 | 1.6 | 54.4 |
| | 15 | 44 19.5 | ·· 22.0 | 208 47.8 | 10.6 | 24 14.5 | 1.7 | 54.4 |
| | 16 | 59 19.4 | 21.9 | 223 17.4 | 10.7 | 24 12.8 | 1.8 | 54.3 |
| | 17 | 74 19.3 | 21.9 | 237 47.1 | 10.7 | 24 11.0 | 1.9 | 54.3 |
| | 18 | 89 19.1 | N23 21.8 | 252 16.8 | 10.7 | S24 09.1 | 2.0 | 54.3 |
| | 19 | 104 19.0 | 21.7 | 266 46.5 | 10.7 | 24 07.1 | 2.2 | 54.3 |
| | 20 | 119 18.9 | 21.6 | 281 16.2 | 10.8 | 24 04.9 | 2.3 | 54.3 |
| | 21 | 134 18.7 | ·· 21.5 | 295 46.0 | 10.8 | 24 02.6 | 2.4 | 54.3 |
| | 22 | 149 18.6 | 21.4 | 310 15.8 | 10.8 | 24 00.2 | 2.5 | 54.3 |
| | 23 | 164 18.5 | 21.4 | 324 45.6 | 10.9 | 23 57.7 | 2.6 | 54.3 |
| | | S.D. 15.8 | *d* 0.1 | S.D. 15.0 | | 14.9 | | 14.8 |

| Lat. | Twilight Naut. | Twilight Civil | Sunrise | Moonrise 24 | Moonrise 25 | Moonrise 26 | Moonrise 27 |
|---|---|---|---|---|---|---|---|
| ° | h m | h m | h m | h m | h m | h m | h m |
| N 72 | □ | □ | □ | ■ | ■ | ■ | ■ |
| N 70 | □ | □ | □ | ■ | ■ | ■ | ■ |
| 68 | □ | □ | □ | ■ | ■ | ■ | ■ |
| 66 | □ | □ | □ | 23 16 | ■ | ■ | 01 11 |
| 64 | //// | //// | 01 32 | 21 59 | 23 10 | 23 47 | 24 01 |
| 62 | //// | //// | 02 10 | 21 23 | 22 27 | 23 10 | 23 34 |
| 60 | //// | 00 51 | 02 37 | 20 57 | 21 59 | 22 43 | 23 12 |
| N 58 | //// | 01 42 | 02 57 | 20 37 | 21 37 | 22 22 | 22 54 |
| 56 | //// | 02 11 | 03 14 | 20 20 | 21 19 | 22 05 | 22 40 |
| 54 | 00 47 | 02 34 | 03 28 | 20 06 | 21 04 | 21 51 | 22 27 |
| 52 | 01 33 | 02 52 | 03 41 | 19 53 | 20 51 | 21 38 | 22 16 |
| 50 | 02 01 | 03 07 | 03 52 | 19 42 | 20 39 | 21 27 | 22 06 |
| 45 | 02 47 | 03 37 | 04 14 | 19 19 | 20 15 | 21 04 | 21 45 |
| N 40 | 03 17 | 03 59 | 04 32 | 19 01 | 19 56 | 20 45 | 21 28 |
| 35 | 03 41 | 04 17 | 04 47 | 18 46 | 19 40 | 20 29 | 21 13 |
| 30 | 03 59 | 04 33 | 05 00 | 18 32 | 19 26 | 20 16 | 21 01 |
| 20 | 04 28 | 04 58 | 05 22 | 18 10 | 19 03 | 19 53 | 20 39 |
| N 10 | 04 51 | 05 18 | 05 41 | 17 50 | 18 42 | 19 33 | 20 21 |
| 0 | 05 10 | 05 36 | 05 59 | 17 32 | 18 23 | 19 14 | 20 03 |
| S 10 | 05 27 | 05 53 | 06 16 | 17 14 | 18 04 | 18 55 | 19 46 |
| 20 | 05 43 | 06 11 | 06 35 | 16 55 | 17 44 | 18 35 | 19 27 |
| 30 | 06 00 | 06 30 | 06 56 | 16 33 | 17 20 | 18 12 | 19 05 |
| 35 | 06 08 | 06 40 | 07 08 | 16 20 | 17 07 | 17 58 | 18 53 |
| 40 | 06 18 | 06 52 | 07 23 | 16 05 | 16 51 | 17 42 | 18 38 |
| 45 | 06 28 | 07 05 | 07 39 | 15 47 | 16 32 | 17 24 | 18 21 |
| S 50 | 06 40 | 07 22 | 08 00 | 15 25 | 16 08 | 17 00 | 17 59 |
| 52 | 06 45 | 07 29 | 08 10 | 15 14 | 15 57 | 16 49 | 17 49 |
| 54 | 06 51 | 07 37 | 08 21 | 15 03 | 15 44 | 16 36 | 17 37 |
| 56 | 06 57 | 07 46 | 08 34 | 14 49 | 15 29 | 16 21 | 17 24 |
| 58 | 07 04 | 07 57 | 08 48 | 14 33 | 15 12 | 16 04 | 17 08 |
| S 60 | 07 11 | 08 08 | 09 06 | 14 15 | 14 51 | 15 43 | 16 49 |

| Lat. | Sunset | Twilight Civil | Twilight Naut. | Moonset 24 | Moonset 25 | Moonset 26 | Moonset 27 |
|---|---|---|---|---|---|---|---|
| ° | h m | h m | h m | h m | h m | h m | h m |
| N 72 | □ | □ | □ | ■ | ■ | ■ | ■ |
| N 70 | □ | □ | □ | ■ | ■ | ■ | ■ |
| 68 | □ | □ | □ | ■ | ■ | ■ | ■ |
| 66 | □ | □ | □ | 00 33 | 00 08 | ■ | 01 45 |
| 64 | 22 32 | //// | //// | 01 11 | 01 25 | 02 01 | 03 09 |
| 62 | 21 54 | //// | //// | 01 38 | 02 02 | 02 43 | 03 46 |
| 60 | 21 28 | 23 13 | //// | 01 59 | 02 28 | 03 12 | 04 12 |
| N 58 | 21 07 | 22 23 | //// | 02 16 | 02 49 | 03 34 | 04 33 |
| 56 | 20 51 | 21 53 | //// | 02 31 | 03 06 | 03 52 | 04 49 |
| 54 | 20 36 | 21 31 | 23 17 | 02 44 | 03 20 | 04 07 | 05 04 |
| 52 | 20 24 | 21 13 | 22 31 | 02 55 | 03 33 | 04 20 | 05 16 |
| 50 | 20 13 | 20 58 | 22 03 | 03 05 | 03 44 | 04 32 | 05 27 |
| 45 | 19 51 | 20 28 | 21 18 | 03 25 | 04 07 | 04 56 | 05 50 |
| N 40 | 19 33 | 20 06 | 20 47 | 03 42 | 04 26 | 05 15 | 06 09 |
| 35 | 19 18 | 19 47 | 20 24 | 03 56 | 04 41 | 05 31 | 06 24 |
| 30 | 19 05 | 19 32 | 20 06 | 04 09 | 04 55 | 05 45 | 06 37 |
| 20 | 18 43 | 19 07 | 19 36 | 04 30 | 05 18 | 06 08 | 07 00 |
| N 10 | 18 24 | 18 47 | 19 14 | 04 49 | 05 38 | 06 29 | 07 20 |
| 0 | 18 06 | 18 29 | 18 55 | 05 06 | 05 57 | 06 48 | 07 38 |
| S 10 | 17 49 | 18 12 | 18 38 | 05 23 | 06 15 | 07 07 | 07 56 |
| 20 | 17 30 | 17 54 | 18 22 | 05 42 | 06 36 | 07 27 | 08 15 |
| 30 | 17 09 | 17 35 | 18 05 | 06 03 | 06 59 | 07 50 | 08 38 |
| 35 | 16 57 | 17 25 | 17 56 | 06 16 | 07 12 | 08 04 | 08 51 |
| 40 | 16 42 | 17 13 | 17 47 | 06 30 | 07 28 | 08 20 | 09 06 |
| 45 | 16 26 | 16 59 | 17 37 | 06 48 | 07 47 | 08 39 | 09 24 |
| S 50 | 16 05 | 16 43 | 17 25 | 07 09 | 08 10 | 09 03 | 09 46 |
| 52 | 15 55 | 16 36 | 17 20 | 07 19 | 08 21 | 09 14 | 09 56 |
| 54 | 15 44 | 16 28 | 17 14 | 07 31 | 08 34 | 09 27 | 10 08 |
| 56 | 15 31 | 16 19 | 17 08 | 07 44 | 08 49 | 09 42 | 10 22 |
| 58 | 15 17 | 16 08 | 17 01 | 07 59 | 09 06 | 09 59 | 10 38 |
| S 60 | 14 59 | 15 57 | 16 54 | 08 18 | 09 27 | 10 20 | 10 57 |

| Day | SUN Eqn. of Time 00ʰ | SUN Eqn. of Time 12ʰ | SUN Mer. Pass. | MOON Mer. Pass. Upper | MOON Mer. Pass. Lower | Age | Phase |
|---|---|---|---|---|---|---|---|
| | m s | m s | h m | h m | h m | d | |
| 24 | 02 08 | 02 15 | 12 02 | 23 44 | 11 19 | 13 | ○ |
| 25 | 02 21 | 02 27 | 12 02 | 24 35 | 12 10 | 14 | |
| 26 | 02 34 | 02 40 | 12 03 | 00 35 | 13 01 | 15 | |

| | G.M.T. d h | ARIES G.H.A. ° ′ | VENUS −4.2 G.H.A. ° ′ | VENUS Dec. ° ′ | MARS +1.9 G.H.A. ° ′ | MARS Dec. ° ′ | JUPITER −1.9 G.H.A. ° ′ | JUPITER Dec. ° ′ | SATURN +0.9 G.H.A. ° ′ | SATURN Dec. ° ′ |
|---|---|---|---|---|---|---|---|---|---|---|
| | 24 00 | 301 09.0 | 142 56.5 | N 6 32.7 | 193 09.7 | N23 16.5 | 62 00.9 | S19 40.2 | 94 08.0 | S 8 31.6 |
| | 01 | 316 11.5 | 157 58.3 | 31.9 | 208 10.4 | 16.3 | 77 03.4 | 40.2 | 109 10.4 | 31.6 |
| | 02 | 331 14.0 | 173 00.1 | 31.1 | 223 11.1 | 16.1 | 92 05.9 | 40.2 | 124 12.8 | 31.7 |
| | 03 | 346 16.4 | 188 01.9 | ·· 30.4 | 238 11.8 | ·· 15.9 | 107 08.4 | ·· 40.2 | 139 15.1 | ·· 31.7 |
| | 04 | 1 18.9 | 203 03.7 | 29.6 | 253 12.5 | 15.7 | 122 10.9 | 40.2 | 154 17.5 | 31.8 |
| | 05 | 16 21.4 | 218 05.5 | 28.8 | 268 13.1 | 15.5 | 137 13.4 | 40.2 | 169 19.9 | 31.8 |
| | 06 | 31 23.8 | 233 07.3 | N 6 28.0 | 283 13.8 | N23 15.4 | 152 16.0 | S19 40.2 | 184 22.3 | S 8 31.9 |
| | 07 | 46 26.3 | 248 09.1 | 27.3 | 298 14.5 | 15.2 | 167 18.5 | 40.2 | 199 24.7 | 31.9 |
| | 08 | 61 28.8 | 263 10.9 | 26.5 | 313 15.2 | 15.0 | 182 21.0 | 40.2 | 214 27.0 | 31.9 |
| S | 09 | 76 31.2 | 278 12.7 | ·· 25.7 | 328 15.9 | ·· 14.8 | 197 23.5 | ·· 40.3 | 229 29.4 | ·· 32.0 |
| U | 10 | 91 33.7 | 293 14.5 | 24.9 | 343 16.6 | 14.6 | 212 26.0 | 40.3 | 244 31.8 | 32.0 |
| N | 11 | 106 36.1 | 308 16.3 | 24.2 | 358 17.2 | 14.4 | 227 28.5 | 40.3 | 259 34.2 | 32.1 |
| D | 12 | 121 38.6 | 323 18.1 | N 6 23.4 | 13 17.9 | N23 14.2 | 242 31.0 | S19 40.3 | 274 36.6 | S 8 32.1 |
| A | 13 | 136 41.1 | 338 19.9 | 22.6 | 28 18.6 | 14.0 | 257 33.5 | 40.3 | 289 38.9 | 32.2 |
| Y | 14 | 151 43.5 | 353 21.7 | 21.9 | 43 19.3 | 13.9 | 272 36.0 | 40.3 | 304 41.3 | 32.2 |
| | 15 | 166 46.0 | 8 23.6 | ·· 21.1 | 58 20.0 | ·· 13.7 | 287 38.5 | ·· 40.3 | 319 43.7 | ·· 32.2 |
| | 16 | 181 48.5 | 23 25.4 | 20.3 | 73 20.7 | 13.5 | 302 41.0 | 40.3 | 334 46.1 | 32.3 |
| | 17 | 196 50.9 | 38 27.2 | 19.6 | 88 21.4 | 13.3 | 317 43.5 | 40.3 | 349 48.5 | 32.3 |
| | 18 | 211 53.4 | 53 29.1 | N 6 18.8 | 103 22.0 | N23 13.1 | 332 46.0 | S19 40.3 | 4 50.8 | S 8 32.4 |
| | 19 | 226 55.9 | 68 30.9 | 18.0 | 118 22.7 | 12.9 | 347 48.5 | 40.3 | 19 53.2 | 32.4 |
| | 20 | 241 58.3 | 83 32.7 | 17.3 | 133 23.4 | 12.7 | 2 51.0 | 40.3 | 34 55.6 | 32.5 |
| | 21 | 257 00.8 | 98 34.6 | ·· 16.5 | 148 24.1 | ·· 12.5 | 17 53.5 | ·· 40.3 | 49 58.0 | ·· 32.5 |
| | 22 | 272 03.3 | 113 36.4 | 15.7 | 163 24.8 | 12.3 | 32 56.0 | 40.3 | 65 00.4 | 32.5 |
| | 23 | 287 05.7 | 128 38.3 | 15.0 | 178 25.5 | 12.1 | 47 58.5 | 40.3 | 80 02.7 | 32.6 |
| | 25 00 | 302 08.2 | 143 40.1 | N 6 14.2 | 193 26.2 | N23 12.0 | 63 01.0 | S19 40.3 | 95 05.1 | S 8 32.6 |
| | 01 | 317 10.6 | 158 42.0 | 13.4 | 208 26.8 | 11.8 | 78 03.5 | 40.3 | 110 07.5 | 32.7 |
| | 02 | 332 13.1 | 173 43.9 | 12.7 | 223 27.5 | 11.6 | 93 06.0 | 40.3 | 125 09.9 | 32.7 |
| | 03 | 347 15.6 | 188 45.7 | ·· 11.9 | 238 28.2 | ·· 11.4 | 108 08.5 | ·· 40.3 | 140 12.2 | ·· 32.8 |
| | 04 | 2 18.0 | 203 47.6 | 11.2 | 253 28.9 | 11.2 | 123 11.0 | 40.3 | 155 14.6 | 32.8 |
| | 05 | 17 20.5 | 218 49.4 | 10.4 | 268 29.6 | 11.0 | 138 13.5 | 40.3 | 170 17.0 | 32.9 |
| | 06 | 32 23.0 | 233 51.3 | N 6 09.7 | 283 30.3 | N23 10.8 | 153 16.0 | S19 40.3 | 185 19.4 | S 8 32.9 |
| | 07 | 47 25.4 | 248 53.2 | 08.9 | 298 31.0 | 10.6 | 168 18.5 | 40.3 | 200 21.8 | 32.9 |
| | 08 | 62 27.9 | 263 55.1 | 08.1 | 313 31.6 | 10.4 | 183 21.0 | 40.3 | 215 24.1 | 33.0 |
| M | 09 | 77 30.4 | 278 57.0 | ·· 07.4 | 328 32.3 | ·· 10.2 | 198 23.5 | ·· 40.3 | 230 26.5 | ·· 33.0 |
| O | 10 | 92 32.8 | 293 58.8 | 06.6 | 343 33.0 | 10.0 | 213 26.0 | 40.3 | 245 28.9 | 33.1 |
| N | 11 | 107 35.3 | 309 00.7 | 05.9 | 358 33.7 | 09.8 | 228 28.5 | 40.3 | 260 31.3 | 33.1 |
| D | 12 | 122 37.8 | 324 02.6 | N 6 05.1 | 13 34.4 | N23 09.6 | 243 31.0 | S19 40.3 | 275 33.6 | S 8 33.2 |
| A | 13 | 137 40.2 | 339 04.5 | 04.4 | 28 35.1 | 09.4 | 258 33.5 | 40.3 | 290 36.0 | 33.2 |
| Y | 14 | 152 42.7 | 354 06.4 | 03.6 | 43 35.8 | 09.2 | 273 36.0 | 40.3 | 305 38.4 | 33.3 |
| | 15 | 167 45.1 | 9 08.3 | ·· 02.9 | 58 36.5 | ·· 09.0 | 288 38.5 | ·· 40.3 | 320 40.8 | ·· 33.3 |
| | 16 | 182 47.6 | 24 10.2 | 02.1 | 73 37.1 | 08.8 | 303 41.0 | 40.3 | 335 43.1 | 33.3 |
| | 17 | 197 50.1 | 39 12.1 | 01.4 | 88 37.8 | 08.7 | 318 43.5 | 40.3 | 350 45.5 | 33.4 |
| | 18 | 212 52.5 | 54 14.0 | N 6 00.6 | 103 38.5 | N23 08.5 | 333 46.0 | S19 40.3 | 5 47.9 | S 8 33.4 |
| | 19 | 227 55.0 | 69 15.9 | 5 59.9 | 118 39.2 | 08.3 | 348 48.5 | 40.3 | 20 50.3 | 33.5 |
| | 20 | 242 57.5 | 84 17.8 | 59.1 | 133 39.9 | 08.1 | 3 50.9 | 40.4 | 35 52.6 | 33.5 |
| | 21 | 257 59.9 | 99 19.8 | ·· 58.4 | 148 40.6 | ·· 07.9 | 18 53.4 | ·· 40.4 | 50 55.0 | ·· 33.6 |
| | 22 | 273 02.4 | 114 21.7 | 57.7 | 163 41.3 | 07.7 | 33 55.9 | 40.4 | 65 57.4 | 33.6 |
| | 23 | 288 04.9 | 129 23.6 | 56.9 | 178 42.0 | 07.5 | 48 58.4 | 40.4 | 80 59.8 | 33.7 |
| | 26 00 | 303 07.3 | 144 25.5 | N 5 56.2 | 193 42.7 | N23 07.3 | 64 00.9 | S19 40.4 | 96 02.1 | S 8 33.7 |
| | 01 | 318 09.8 | 159 27.5 | 55.4 | 208 43.3 | 07.1 | 79 03.4 | 40.4 | 111 04.5 | 33.7 |
| | 02 | 333 12.2 | 174 29.4 | 54.7 | 223 44.0 | 06.9 | 94 05.9 | 40.4 | 126 06.9 | 33.8 |
| | 03 | 348 14.7 | 189 31.3 | ·· 54.0 | 238 44.7 | ·· 06.7 | 109 08.4 | ·· 40.4 | 141 09.3 | ·· 33.8 |
| | 04 | 3 17.2 | 204 33.3 | 53.2 | 253 45.4 | 06.5 | 124 10.9 | 40.4 | 156 11.6 | 33.9 |
| | 05 | 18 19.6 | 219 35.2 | 52.5 | 268 46.1 | 06.3 | 139 13.4 | 40.4 | 171 14.0 | 33.9 |
| | 06 | 33 22.1 | 234 37.2 | N 5 51.7 | 283 46.8 | N23 06.1 | 154 15.9 | S19 40.4 | 186 16.4 | S 8 34.0 |
| | 07 | 48 24.6 | 249 39.1 | 51.0 | 298 47.5 | 05.9 | 169 18.4 | 40.4 | 201 18.8 | 34.0 |
| | 08 | 63 27.0 | 264 41.1 | 50.3 | 313 48.2 | 05.7 | 184 20.8 | 40.4 | 216 21.1 | 34.1 |
| T | 09 | 78 29.5 | 279 43.0 | ·· 49.5 | 328 48.9 | ·· 05.5 | 199 23.3 | ·· 40.4 | 231 23.5 | ·· 34.1 |
| U | 10 | 93 32.0 | 294 45.0 | 48.8 | 343 49.6 | 05.3 | 214 25.8 | 40.4 | 246 25.9 | 34.2 |
| E | 11 | 108 34.4 | 309 47.0 | 48.1 | 358 50.3 | 05.1 | 229 28.3 | 40.4 | 261 28.3 | 34.2 |
| S D | 12 | 123 36.9 | 324 48.9 | N 5 47.3 | 13 50.9 | N23 04.8 | 244 30.8 | S19 40.4 | 276 30.6 | S 8 34.2 |
| A | 13 | 138 39.4 | 339 50.9 | 46.6 | 28 51.6 | 04.6 | 259 33.3 | 40.4 | 291 33.0 | 34.3 |
| Y | 14 | 153 41.8 | 354 52.9 | 45.9 | 43 52.3 | 04.4 | 274 35.8 | 40.4 | 306 35.4 | 34.3 |
| | 15 | 168 44.3 | 9 54.8 | ·· 45.2 | 58 53.0 | ·· 04.2 | 289 38.3 | ·· 40.4 | 321 37.7 | ·· 34.4 |
| | 16 | 183 46.7 | 24 56.8 | 44.4 | 73 53.7 | 04.0 | 304 40.7 | 40.4 | 336 40.1 | 34.4 |
| | 17 | 198 49.2 | 39 58.8 | 43.7 | 88 54.4 | 03.8 | 319 43.2 | 40.4 | 351 42.5 | 34.5 |
| | 18 | 213 51.7 | 55 00.8 | N 5 43.0 | 103 55.1 | N23 03.6 | 334 45.7 | S19 40.5 | 6 44.9 | S 8 34.5 |
| | 19 | 228 54.1 | 70 02.8 | 42.3 | 118 55.8 | 03.4 | 349 48.2 | 40.5 | 21 47.2 | 34.6 |
| | 20 | 243 56.6 | 85 04.8 | 41.5 | 133 56.5 | 03.2 | 4 50.7 | 40.5 | 36 49.6 | 34.6 |
| | 21 | 258 59.1 | 100 06.8 | ·· 40.8 | 148 57.2 | ·· 03.0 | 19 53.2 | ·· 40.5 | 51 52.0 | ·· 34.7 |
| | 22 | 274 01.5 | 115 08.8 | 40.1 | 163 57.9 | 02.8 | 34 55.7 | 40.5 | 66 54.3 | 34.7 |
| | 23 | 289 04.0 | 130 10.8 | 39.4 | 178 58.6 | 02.6 | 49 58.1 | 40.5 | 81 56.7 | 34.8 |
| | Mer. Pass. | h m 3 50.8 | *v* 1.9 | *d* 0.8 | *v* 0.7 | *d* 0.2 | *v* 2.5 | *d* 0.0 | *v* 2.4 | *d* 0.0 |

**STARS**

| Name | S.H.A. ° ′ | Dec. ° ′ |
|---|---|---|
| Acamar | 315 35.7 | S40 22.0 |
| Achernar | 335 43.5 | S57 19.0 |
| Acrux | 173 35.4 | S63 00.6 |
| Adhara | 255 30.9 | S28 56.8 |
| Aldebaran | 291 15.9 | N16 28.6 |
| Alioth | 166 40.7 | N56 03.3 |
| Alkaid | 153 16.8 | N49 24.1 |
| Al Na'ir | 28 11.8 | S47 02.4 |
| Alnilam | 276 09.8 | S 1 12.6 |
| Alphard | 218 18.9 | S 8 35.1 |
| Alphecca | 126 30.2 | N26 46.4 |
| Alpheratz | 358 07.1 | N28 59.8 |
| Altair | 62 30.2 | N 8 49.5 |
| Ankaa | 353 37.9 | S42 23.6 |
| Antares | 112 54.1 | S26 23.8 |
| Arcturus | 146 16.5 | N19 16.3 |
| Atria | 108 16.2 | S69 00.1 |
| Avior | 234 28.1 | S59 27.3 |
| Bellatrix | 278 56.8 | N 6 20.2 |
| Betelgeuse | 271 26.3 | N 7 24.4 |
| Canopus | 264 06.8 | S52 41.0 |
| Capella | 281 08.6 | N45 58.8 |
| Deneb | 49 46.7 | N45 13.2 |
| Denebola | 182 57.1 | N14 40.1 |
| Diphda | 349 18.7 | S18 04.5 |
| Dubhe | 194 19.9 | N61 50.7 |
| Elnath | 278 41.8 | N28 35.6 |
| Eltanin | 90 56.3 | N51 29.6 |
| Enif | 34 09.3 | N 9 47.9 |
| Fomalhaut | 15 48.8 | S29 42.5 |
| Gacrux | 172 26.8 | S57 01.4 |
| Gienah | 176 16.0 | S17 27.0 |
| Hadar | 149 20.6 | S60 17.8 |
| Hamal | 328 26.6 | N23 23.0 |
| Kaus Aust. | 84 13.7 | S34 23.7 |
| Kochab | 137 18.6 | N74 13.8 |
| Markab | 14 00.9 | N15 06.9 |
| Menkar | 314 39.1 | N 4 01.5 |
| Menkent | 148 34.6 | S36 17.4 |
| Miaplacidus | 221 45.6 | S69 38.9 |
| Mirfak | 309 13.4 | N49 48.0 |
| Nunki | 76 26.3 | S26 19.1 |
| Peacock | 53 54.4 | S56 47.3 |
| Pollux | 243 56.0 | N28 04.1 |
| Procyon | 245 24.0 | N 5 16.2 |
| Rasalhague | 96 27.4 | N12 34.4 |
| Regulus | 208 08.1 | N12 03.1 |
| Rigel | 281 34.3 | S 8 13.1 |
| Rigil Kent. | 140 23.0 | S60 46.2 |
| Sabik | 102 38.5 | S15 42.3 |
| Schedar | 350 06.6 | N56 26.5 |
| Shaula | 96 52.6 | S37 05.6 |
| Sirius | 258 54.2 | S16 41.5 |
| Spica | 158 55.4 | S11 04.5 |
| Suhail | 223 09.8 | S43 21.9 |
| Vega | 80 54.1 | N38 46.2 |
| Zuben'ubi | 137 30.7 | S15 58.4 |

| | S.H.A. ° ′ | Mer. Pass. h m |
|---|---|---|
| Venus | 201 32.0 | 14 24 |
| Mars | 251 18.0 | 11 06 |
| Jupiter | 120 52.8 | 19 45 |
| Saturn | 152 56.9 | 17 37 |

| G.M.T. | SUN G.H.A. | SUN Dec. | MOON G.H.A. | v | MOON Dec. | d | H.P. |
|---|---|---|---|---|---|---|---|
| d h | ° ′ | ° ′ | ° ′ | ′ | ° ′ | ′ | ′ |
| 24 00 | 178 23.6 | N20 02.1 | 9 17.5 | 10.8 | S24 08.9 | 2.0 | 54.2 |
| 01 | 193 23.5 | 01.5 | 23 47.3 | 10.9 | 24 06.9 | 2.2 | 54.2 |
| 02 | 208 23.5 | 01.0 | 38 17.2 | 10.9 | 24 04.7 | 2.3 | 54.2 |
| 03 | 223 23.5 | ·· 00.5 | 52 47.1 | 10.9 | 24 02.4 | 2.4 | 54.2 |
| 04 | 238 23.5 | 20 00.0 | 67 17.0 | 10.9 | 24 00.0 | 2.5 | 54.2 |
| 05 | 253 23.5 | 19 59.5 | 81 46.9 | 11.0 | 23 57.5 | 2.6 | 54.2 |
| 06 | 268 23.5 | N19 59.0 | 96 16.9 | 11.0 | S23 54.9 | 2.7 | 54.2 |
| 07 | 283 23.5 | 58.4 | 110 46.9 | 11.1 | 23 52.2 | 2.9 | 54.2 |
| 08 | 298 23.4 | 57.9 | 125 17.0 | 11.0 | 23 49.3 | 2.9 | 54.1 |
| S 09 | 313 23.4 | ·· 57.4 | 139 47.0 | 11.1 | 23 46.4 | 3.1 | 54.1 |
| U 10 | 328 23.4 | 56.9 | 154 17.1 | 11.2 | 23 43.3 | 3.2 | 54.1 |
| N 11 | 343 23.4 | 56.4 | 168 47.3 | 11.2 | 23 40.1 | 3.3 | 54.1 |
| D 12 | 358 23.4 | N19 55.9 | 183 17.5 | 11.2 | S23 36.8 | 3.4 | 54.1 |
| A 13 | 13 23.4 | 55.3 | 197 47.7 | 11.2 | 23 33.4 | 3.5 | 54.1 |
| Y 14 | 28 23.4 | 54.8 | 212 17.9 | 11.3 | 23 29.9 | 3.6 | 54.1 |
| 15 | 43 23.3 | ·· 54.3 | 226 48.2 | 11.3 | 23 26.3 | 3.8 | 54.1 |
| 16 | 58 23.3 | 53.8 | 241 18.5 | 11.4 | 23 22.5 | 3.8 | 54.1 |
| 17 | 73 23.3 | 53.2 | 255 48.9 | 11.4 | 23 18.7 | 4.0 | 54.1 |
| 18 | 88 23.3 | N19 52.7 | 270 19.3 | 11.4 | S23 14.7 | 4.0 | 54.1 |
| 19 | 103 23.3 | 52.2 | 284 49.7 | 11.5 | 23 10.7 | 4.2 | 54.1 |
| 20 | 118 23.3 | 51.7 | 299 20.2 | 11.5 | 23 06.5 | 4.3 | 54.1 |
| 21 | 133 23.3 | ·· 51.1 | 313 50.7 | 11.5 | 23 02.2 | 4.4 | 54.1 |
| 22 | 148 23.3 | 50.6 | 328 21.2 | 11.6 | 22 57.8 | 4.5 | 54.1 |
| 23 | 163 23.3 | 50.1 | 342 51.8 | 11.6 | 22 53.3 | 4.5 | 54.1 |
| 25 00 | 178 23.2 | N19 49.6 | 357 22.4 | 11.7 | S22 48.8 | 4.7 | 54.0 |
| 01 | 193 23.2 | 49.0 | 11 53.1 | 11.7 | 22 44.1 | 4.8 | 54.0 |
| 02 | 208 23.2 | 48.5 | 26 23.8 | 11.8 | 22 39.3 | 4.9 | 54.0 |
| 03 | 223 23.2 | ·· 48.0 | 40 54.6 | 11.8 | 22 34.4 | 5.1 | 54.0 |
| 04 | 238 23.2 | 47.4 | 55 25.4 | 11.9 | 22 29.3 | 5.1 | 54.0 |
| 05 | 253 23.2 | 46.9 | 69 56.3 | 11.9 | 22 24.2 | 5.2 | 54.0 |
| 06 | 268 23.2 | N19 46.4 | 84 27.2 | 11.9 | S22 19.0 | 5.3 | 54.0 |
| 07 | 283 23.2 | 45.9 | 98 58.1 | 12.0 | 22 13.7 | 5.4 | 54.0 |
| 08 | 298 23.2 | 45.3 | 113 29.1 | 12.0 | 22 08.3 | 5.5 | 54.0 |
| M 09 | 313 23.2 | ·· 44.8 | 128 00.1 | 12.1 | 22 02.8 | 5.6 | 54.0 |
| O 10 | 328 23.2 | 44.3 | 142 31.2 | 12.1 | 21 57.2 | 5.7 | 54.0 |
| N 11 | 343 23.2 | 43.7 | 157 02.3 | 12.2 | 21 51.5 | 5.8 | 54.0 |
| D 12 | 358 23.1 | N19 43.2 | 171 33.5 | 12.2 | S21 45.7 | 6.0 | 54.0 |
| A 13 | 13 23.1 | 42.7 | 186 04.7 | 12.2 | 21 39.7 | 6.0 | 54.0 |
| Y 14 | 28 23.1 | 42.1 | 200 35.9 | 12.3 | 21 33.7 | 6.1 | 54.0 |
| 15 | 43 23.1 | ·· 41.6 | 215 07.2 | 12.4 | 21 27.6 | 6.2 | 54.0 |
| 16 | 58 23.1 | 41.0 | 229 38.6 | 12.4 | 21 21.4 | 6.2 | 54.0 |
| 17 | 73 23.1 | 40.5 | 244 10.0 | 12.4 | 21 15.2 | 6.4 | 54.0 |
| 18 | 88 23.1 | N19 40.0 | 258 41.4 | 12.5 | S21 08.8 | 6.5 | 54.0 |
| 19 | 103 23.1 | 39.4 | 273 12.9 | 12.6 | 21 02.3 | 6.6 | 54.0 |
| 20 | 118 23.1 | 38.9 | 287 44.5 | 12.6 | 20 55.7 | 6.6 | 54.0 |
| 21 | 133 23.1 | ·· 38.4 | 302 16.1 | 12.6 | 20 49.1 | 6.8 | 54.0 |
| 22 | 148 23.1 | 37.8 | 316 47.7 | 12.7 | 20 42.3 | 6.8 | 54.0 |
| 23 | 163 23.1 | 37.3 | 331 19.4 | 12.8 | 20 35.5 | 7.0 | 54.0 |
| 26 00 | 178 23.1 | N19 36.7 | 345 51.2 | 12.8 | S20 28.5 | 7.0 | 54.0 |
| 01 | 193 23.1 | 36.2 | 0 23.0 | 12.8 | 20 21.5 | 7.1 | 54.0 |
| 02 | 208 23.1 | 35.7 | 14 54.8 | 12.9 | 20 14.4 | 7.2 | 54.0 |
| 03 | 223 23.1 | ·· 35.1 | 29 26.7 | 12.9 | 20 07.2 | 7.3 | 54.0 |
| 04 | 238 23.1 | 34.6 | 43 58.6 | 13.0 | 19 59.9 | 7.4 | 54.0 |
| 05 | 253 23.1 | 34.0 | 58 30.6 | 13.0 | 19 52.5 | 7.4 | 54.0 |
| 06 | 268 23.1 | N19 33.5 | 73 02.6 | 13.1 | S19 45.1 | 7.6 | 54.0 |
| 07 | 283 23.1 | 32.9 | 87 34.7 | 13.2 | 19 37.5 | 7.6 | 54.0 |
| T 08 | 298 23.1 | 32.4 | 102 06.9 | 13.1 | 19 29.9 | 7.7 | 54.0 |
| U 09 | 313 23.1 | ·· 31.8 | 116 39.0 | 13.3 | 19 22.2 | 7.8 | 54.0 |
| E 10 | 328 23.1 | 31.3 | 131 11.3 | 13.3 | 19 14.4 | 7.9 | 54.0 |
| S 11 | 343 23.0 | 30.7 | 145 43.6 | 13.3 | 19 06.5 | 8.0 | 54.0 |
| D 12 | 358 23.0 | N19 30.2 | 160 15.9 | 13.4 | S18 58.5 | 8.0 | 54.0 |
| A 13 | 13 23.0 | 29.7 | 174 48.3 | 13.4 | 18 50.5 | 8.2 | 54.0 |
| Y 14 | 28 23.0 | 29.1 | 189 20.7 | 13.5 | 18 42.3 | 8.2 | 54.0 |
| 15 | 43 23.0 | ·· 28.6 | 203 53.2 | 13.5 | 18 34.1 | 8.3 | 54.0 |
| 16 | 58 23.0 | 28.0 | 218 25.7 | 13.6 | 18 25.8 | 8.3 | 54.0 |
| 17 | 73 23.0 | 27.5 | 232 58.3 | 13.6 | 18 17.5 | 8.5 | 54.0 |
| 18 | 88 23.0 | N19 26.9 | 247 30.9 | 13.7 | S18 09.0 | 8.5 | 54.0 |
| 19 | 103 23.0 | 26.4 | 262 03.6 | 13.7 | 18 00.5 | 8.6 | 54.0 |
| 20 | 118 23.0 | 25.8 | 276 36.3 | 13.7 | 17 51.9 | 8.7 | 54.0 |
| 21 | 133 23.0 | ·· 25.2 | 291 09.0 | 13.9 | 17 43.2 | 8.7 | 54.0 |
| 22 | 148 23.1 | 24.7 | 305 41.9 | 13.8 | 17 34.5 | 8.8 | 54.0 |
| 23 | 163 23.1 | 24.1 | 320 14.7 | 13.9 | 17 25.7 | 8.9 | 54.0 |
| | S.D. 15.8 | d 0.5 | S.D. 14.7 | | 14.7 | | 14.7 |

| Lat. | Twilight Naut. | Twilight Civil | Sunrise | Moonrise 24 | 25 | 26 | 27 |
|---|---|---|---|---|---|---|---|
| ° | h m | h m | h m | h m | h m | h m | h m |
| N 72 | ▭ | ▭ | ▭ | ■ | ■ | ■ | {00 35 / 23 34} |
| N 70 | ▭ | ▭ | ▭ | ■ | ■ | 23 38 | 23 09 |
| 68 | //// | //// | 01 31 | ■ | 23 30 | 23 04 | 22 50 |
| 66 | //// | //// | 02 15 | 22 58 | 22 47 | 22 40 | 22 34 |
| 64 | //// | 00 35 | 02 44 | 22 10 | 22 17 | 22 20 | 22 21 |
| 62 | //// | 01 42 | 03 06 | 21 39 | 21 55 | 22 04 | 22 10 |
| 60 | //// | 02 15 | 03 23 | 21 16 | 21 37 | 21 51 | 22 01 |
| N 58 | 00 32 | 02 39 | 03 37 | 20 57 | 21 22 | 21 40 | 21 53 |
| 56 | 01 33 | 02 58 | 03 50 | 20 41 | 21 09 | 21 30 | 21 45 |
| 54 | 02 04 | 03 14 | 04 01 | 20 28 | 20 58 | 21 21 | 21 39 |
| 52 | 02 26 | 03 27 | 04 10 | 20 16 | 20 48 | 21 13 | 21 33 |
| 50 | 02 44 | 03 38 | 04 19 | 20 06 | 20 39 | 21 06 | 21 28 |
| 45 | 03 17 | 04 02 | 04 37 | 19 44 | 20 20 | 20 50 | 21 16 |
| N 40 | 03 42 | 04 20 | 04 51 | 19 26 | 20 05 | 20 38 | 21 07 |
| 35 | 04 01 | 04 35 | 05 04 | 19 11 | 19 51 | 20 27 | 20 58 |
| 30 | 04 16 | 04 48 | 05 14 | 18 58 | 19 40 | 20 17 | 20 51 |
| 20 | 04 41 | 05 09 | 05 33 | 18 36 | 19 20 | 20 01 | 20 39 |
| N 10 | 05 00 | 05 26 | 05 48 | 18 17 | 19 03 | 19 47 | 20 28 |
| 0 | 05 15 | 05 41 | 06 03 | 17 59 | 18 47 | 19 33 | 20 17 |
| S 10 | 05 29 | 05 55 | 06 17 | 17 41 | 18 31 | 19 20 | 20 07 |
| 20 | 05 43 | 06 09 | 06 33 | 17 22 | 18 14 | 19 05 | 19 56 |
| 30 | 05 56 | 06 25 | 06 50 | 17 00 | 17 54 | 18 49 | 19 43 |
| 35 | 06 02 | 06 33 | 07 00 | 16 47 | 17 42 | 18 39 | 19 35 |
| 40 | 06 09 | 06 42 | 07 12 | 16 31 | 17 29 | 18 28 | 19 27 |
| 45 | 06 17 | 06 53 | 07 25 | 16 13 | 17 13 | 18 15 | 19 17 |
| S 50 | 06 26 | 07 06 | 07 42 | 15 51 | 16 53 | 17 59 | 19 05 |
| 52 | 06 30 | 07 11 | 07 49 | 15 40 | 16 44 | 17 51 | 18 59 |
| 54 | 06 34 | 07 18 | 07 58 | 15 28 | 16 33 | 17 43 | 18 53 |
| 56 | 06 38 | 07 24 | 08 08 | 15 14 | 16 22 | 17 33 | 18 46 |
| 58 | 06 43 | 07 32 | 08 19 | 14 58 | 16 08 | 17 22 | 18 38 |
| S 60 | 06 48 | 07 41 | 08 31 | 14 38 | 15 51 | 17 10 | 18 29 |

| Lat. | Sunset | Twilight Civil | Twilight Naut. | Moonset 24 | 25 | 26 | 27 |
|---|---|---|---|---|---|---|---|
| ° | h m | h m | h m | h m | h m | h m | h m |
| N 72 | ▭ | ▭ | ▭ | ■ | ■ | ■ | 03 10 |
| N 70 | ▭ | ▭ | ▭ | ■ | ■ | ■ | 04 05 |
| 68 | 22 36 | //// | //// | ■ | ■ | 02 38 | 04 38 |
| 66 | 21 55 | //// | //// | ■ | 01 31 | 03 21 | 05 02 |
| 64 | 21 27 | 23 25 | //// | 00 56 | 02 18 | 03 49 | 05 20 |
| 62 | 21 05 | 22 27 | //// | 01 35 | 02 49 | 04 11 | 05 35 |
| 60 | 20 48 | 21 55 | //// | 02 03 | 03 12 | 04 28 | 05 48 |
| N 58 | 20 34 | 21 32 | 23 29 | 02 24 | 03 30 | 04 43 | 05 59 |
| 56 | 20 22 | 21 13 | 22 36 | 02 42 | 03 45 | 04 55 | 06 08 |
| 54 | 20 11 | 20 58 | 22 07 | 02 56 | 03 58 | 05 06 | 06 16 |
| 52 | 20 02 | 20 45 | 21 45 | 03 09 | 04 10 | 05 16 | 06 24 |
| 50 | 19 53 | 20 33 | 21 27 | 03 20 | 04 20 | 05 24 | 06 30 |
| 45 | 19 36 | 20 10 | 20 54 | 03 44 | 04 42 | 05 42 | 06 45 |
| N 40 | 19 21 | 19 52 | 20 30 | 04 03 | 04 59 | 05 57 | 06 56 |
| 35 | 19 09 | 19 37 | 20 12 | 04 18 | 05 13 | 06 09 | 07 06 |
| 30 | 18 58 | 19 25 | 19 56 | 04 32 | 05 26 | 06 20 | 07 15 |
| 20 | 18 40 | 19 04 | 19 32 | 04 55 | 05 47 | 06 39 | 07 30 |
| N 10 | 18 25 | 18 47 | 19 13 | 05 15 | 06 05 | 06 55 | 07 43 |
| 0 | 18 10 | 18 32 | 18 57 | 05 34 | 06 23 | 07 10 | 07 55 |
| S 10 | 17 56 | 18 18 | 18 44 | 05 52 | 06 40 | 07 24 | 08 07 |
| 20 | 17 40 | 18 04 | 18 31 | 06 12 | 06 58 | 07 40 | 08 19 |
| 30 | 17 23 | 17 49 | 18 18 | 06 35 | 07 19 | 07 58 | 08 34 |
| 35 | 17 13 | 17 40 | 18 11 | 06 48 | 07 31 | 08 09 | 08 42 |
| 40 | 17 01 | 17 31 | 18 04 | 07 04 | 07 45 | 08 21 | 08 52 |
| 45 | 16 48 | 17 20 | 17 56 | 07 22 | 08 02 | 08 35 | 09 03 |
| S 50 | 16 32 | 17 08 | 17 48 | 07 45 | 08 22 | 08 52 | 09 16 |
| 52 | 16 24 | 17 02 | 17 44 | 07 56 | 08 32 | 09 00 | 09 22 |
| 54 | 16 15 | 16 56 | 17 40 | 08 08 | 08 42 | 09 09 | 09 29 |
| 56 | 16 06 | 16 49 | 17 35 | 08 22 | 08 55 | 09 19 | 09 36 |
| 58 | 15 55 | 16 41 | 17 31 | 08 39 | 09 09 | 09 30 | 09 45 |
| S 60 | 15 42 | 16 33 | 17 26 | 08 59 | 09 25 | 09 43 | 09 55 |

| Day | SUN Eqn. of Time 00ʰ | SUN Eqn. of Time 12ʰ | SUN Mer. Pass. | MOON Mer. Pass. Upper | MOON Mer. Pass. Lower | Age | Phase |
|---|---|---|---|---|---|---|---|
| | m s | m s | h m | h m | h m | d | |
| 24 | 06 26 | 06 26 | 12 06 | 24 11 | 11 46 | 14 | |
| 25 | 06 27 | 06 27 | 12 06 | 00 11 | 12 35 | 15 | ○ |
| 26 | 06 28 | 06 28 | 12 06 | 00 58 | 13 21 | 16 | |

| G.M.T. d | h | ARIES G.H.A. | VENUS −4.2 G.H.A. | VENUS Dec. | MARS +1.9 G.H.A. | MARS Dec. | JUPITER −1.4 G.H.A. | JUPITER Dec. | SATURN +0.8 G.H.A. | SATURN Dec. |
|---|---|---|---|---|---|---|---|---|---|---|
| | | ° ′ | ° ′ | ° ′ | ° ′ | ° ′ | ° ′ | ° ′ | ° ′ | ° ′ |
| 19 | 00 | 26 54.1 | 226 10.6 | N 6 43.1 | 223 14.7 | N 8 26.4 | 138 38.4 | S21 28.9 | 172 42.0 | S11 22.1 |
| | 01 | 41 56.6 | 241 11.0 | 42.6 | 238 15.7 | 25.8 | 153 40.3 | 29.0 | 187 44.2 | 22.2 |
| | 02 | 56 59.1 | 256 11.4 | 42.2 | 253 16.7 | 25.2 | 168 42.3 | 29.1 | 202 46.4 | 22.3 |
| | 03 | 72 01.5 | 271 11.8 | ·· 41.7 | 268 17.8 | ·· 24.7 | 183 44.3 | ·· 29.1 | 217 48.5 | ·· 22.4 |
| | 04 | 87 04.0 | 286 12.1 | 41.2 | 283 18.8 | 24.1 | 198 46.2 | 29.2 | 232 50.7 | 22.5 |
| | 05 | 102 06.4 | 301 12.5 | 40.7 | 298 19.8 | 23.5 | 213 48.2 | 29.3 | 247 52.9 | 22.6 |
| | 06 | 117 08.9 | 316 12.9 | N 6 40.3 | 313 20.9 | N 8 22.9 | 228 50.1 | S21 29.4 | 262 55.1 | S11 22.7 |
| W | 07 | 132 11.4 | 331 13.3 | 39.8 | 328 21.9 | 22.4 | 243 52.1 | 29.4 | 277 57.2 | 22.8 |
| E | 08 | 147 13.8 | 346 13.7 | 39.3 | 343 22.9 | 21.8 | 258 54.1 | 29.5 | 292 59.4 | 22.9 |
| D | 09 | 162 16.3 | 1 14.0 | ·· 38.8 | 358 24.0 | ·· 21.2 | 273 56.0 | ·· 29.6 | 308 01.6 | ·· 23.0 |
| N | 10 | 177 18.8 | 16 14.4 | 38.3 | 13 25.0 | 20.6 | 288 58.0 | 29.7 | 323 03.8 | 23.1 |
| E | 11 | 192 21.2 | 31 14.8 | 37.9 | 28 26.0 | 20.1 | 304 00.0 | 29.7 | 338 06.0 | 23.2 |
| S | 12 | 207 23.7 | 46 15.2 | N 6 37.4 | 43 27.1 | N 8 19.5 | 319 01.9 | S21 29.8 | 353 08.1 | S11 23.3 |
| D | 13 | 222 26.2 | 61 15.6 | 36.9 | 58 28.1 | 18.9 | 334 03.9 | 29.9 | 8 10.3 | 23.4 |
| A | 14 | 237 28.6 | 76 15.9 | 36.4 | 73 29.1 | 18.4 | 349 05.8 | 30.0 | 23 12.5 | 23.5 |
| Y | 15 | 252 31.1 | 91 16.3 | ·· 35.9 | 88 30.2 | ·· 17.8 | 4 07.8 | ·· 30.0 | 38 14.7 | ·· 23.6 |
| | 16 | 267 33.5 | 106 16.7 | 35.4 | 103 31.2 | 17.2 | 19 09.8 | 30.1 | 53 16.8 | 23.7 |
| | 17 | 282 36.0 | 121 17.0 | 34.9 | 118 32.2 | 16.6 | 34 11.7 | 30.2 | 68 19.0 | 23.8 |
| | 18 | 297 38.5 | 136 17.4 | N 6 34.4 | 133 33.3 | N 8 16.1 | 49 13.7 | S21 30.3 | 83 21.2 | S11 23.9 |
| | 19 | 312 40.9 | 151 17.8 | 33.9 | 148 34.3 | 15.5 | 64 15.6 | 30.3 | 98 23.4 | 24.0 |
| | 20 | 327 43.4 | 166 18.2 | 33.5 | 163 35.4 | 14.9 | 79 17.6 | 30.4 | 113 25.6 | 24.1 |
| | 21 | 342 45.9 | 181 18.5 | ·· 33.0 | 178 36.4 | ·· 14.3 | 94 19.5 | ·· 30.5 | 128 27.7 | ·· 24.2 |
| | 22 | 357 48.3 | 196 18.9 | 32.5 | 193 37.4 | 13.8 | 109 21.5 | 30.6 | 143 29.9 | 24.3 |
| | 23 | 12 50.8 | 211 19.3 | 32.0 | 208 38.5 | 13.2 | 124 23.5 | 30.6 | 158 32.1 | 24.4 |
| 20 | 00 | 27 53.3 | 226 19.6 | N 6 31.5 | 223 39.5 | N 8 12.6 | 139 25.4 | S21 30.7 | 173 34.3 | S11 24.5 |
| | 01 | 42 55.7 | 241 20.0 | 31.0 | 238 40.5 | 12.0 | 154 27.4 | 30.8 | 188 36.4 | 24.6 |
| | 02 | 57 58.2 | 256 20.3 | 30.5 | 253 41.6 | 11.5 | 169 29.3 | 30.9 | 203 38.6 | 24.7 |
| | 03 | 73 00.7 | 271 20.7 | ·· 30.0 | 268 42.6 | ·· 10.9 | 184 31.3 | ·· 30.9 | 218 40.8 | ·· 24.8 |
| | 04 | 88 03.1 | 286 21.1 | 29.5 | 283 43.6 | 10.3 | 199 33.3 | 31.0 | 233 43.0 | 24.9 |
| | 05 | 103 05.6 | 301 21.4 | 29.0 | 298 44.7 | 09.8 | 214 35.2 | 31.1 | 248 45.2 | 25.0 |
| | 06 | 118 08.0 | 316 21.8 | N 6 28.5 | 313 45.7 | N 8 09.2 | 229 37.2 | S21 31.2 | 263 47.3 | S11 25.1 |
| | 07 | 133 10.5 | 331 22.1 | 28.0 | 328 46.8 | 08.6 | 244 39.1 | 31.2 | 278 49.5 | 25.2 |
| T | 08 | 148 13.0 | 346 22.5 | 27.5 | 343 47.8 | 08.0 | 259 41.1 | 31.3 | 293 51.7 | 25.3 |
| H | 09 | 163 15.4 | 1 22.8 | ·· 27.0 | 358 48.8 | ·· 07.5 | 274 43.0 | ·· 31.4 | 308 53.9 | ·· 25.4 |
| U | 10 | 178 17.9 | 16 23.2 | 26.5 | 13 49.9 | 06.9 | 289 45.0 | 31.5 | 323 56.0 | 25.5 |
| R | 11 | 193 20.4 | 31 23.5 | 26.0 | 28 50.9 | 06.3 | 304 47.0 | 31.5 | 338 58.2 | 25.6 |
| S | 12 | 208 22.8 | 46 23.9 | N 6 25.5 | 43 51.9 | N 8 05.7 | 319 48.9 | S21 31.6 | 354 00.4 | S11 25.7 |
| D | 13 | 223 25.3 | 61 24.3 | 24.9 | 58 53.0 | 05.2 | 334 50.9 | 31.7 | 9 02.6 | 25.8 |
| A | 14 | 238 27.8 | 76 24.6 | 24.4 | 73 54.0 | 04.6 | 349 52.8 | 31.8 | 24 04.7 | 25.9 |
| Y | 15 | 253 30.2 | 91 24.9 | ·· 23.9 | 88 55.1 | ·· 04.0 | 4 54.8 | ·· 31.8 | 39 06.9 | ·· 26.0 |
| | 16 | 268 32.7 | 106 25.3 | 23.4 | 103 56.1 | 03.4 | 19 56.7 | 31.9 | 54 09.1 | 26.1 |
| | 17 | 283 35.2 | 121 25.6 | 22.9 | 118 57.1 | 02.9 | 34 58.7 | 32.0 | 69 11.3 | 26.2 |
| | 18 | 298 37.6 | 136 26.0 | N 6 22.4 | 133 58.2 | N 8 02.3 | 50 00.7 | S21 32.1 | 84 13.5 | S11 26.3 |
| | 19 | 313 40.1 | 151 26.3 | 21.9 | 148 59.2 | 01.7 | 65 02.6 | 32.1 | 99 15.6 | 26.4 |
| | 20 | 328 42.5 | 166 26.7 | 21.4 | 164 00.2 | 01.1 | 80 04.6 | 32.2 | 114 17.8 | 26.5 |
| | 21 | 343 45.0 | 181 27.0 | ·· 20.8 | 179 01.3 | ·· 00.6 | 95 06.5 | ·· 32.3 | 129 20.0 | ·· 26.6 |
| | 22 | 358 47.5 | 196 27.4 | 20.3 | 194 02.3 | 8 00.0 | 110 08.5 | 32.4 | 144 22.2 | 26.7 |
| | 23 | 13 49.9 | 211 27.7 | 19.8 | 209 03.4 | 7 59.4 | 125 10.4 | 32.4 | 159 24.3 | 26.8 |
| 21 | 00 | 28 52.4 | 226 28.0 | N 6 19.3 | 224 04.4 | N 7 58.8 | 140 12.4 | S21 32.5 | 174 26.5 | S11 26.9 |
| | 01 | 43 54.9 | 241 28.4 | 18.8 | 239 05.4 | 58.3 | 155 14.3 | 32.6 | 189 28.7 | 27.0 |
| | 02 | 58 57.3 | 256 28.7 | 18.3 | 254 06.5 | 57.7 | 170 16.3 | 32.7 | 204 30.9 | 27.1 |
| | 03 | 73 59.8 | 271 29.0 | ·· 17.7 | 269 07.5 | ·· 57.1 | 185 18.2 | ·· 32.7 | 219 33.0 | ·· 27.2 |
| | 04 | 89 02.3 | 286 29.4 | 17.2 | 284 08.6 | 56.5 | 200 20.2 | 32.8 | 234 35.2 | 27.3 |
| | 05 | 104 04.7 | 301 29.7 | 16.7 | 299 09.6 | 56.0 | 215 22.2 | 32.9 | 249 37.4 | 27.4 |
| | 06 | 119 07.2 | 316 30.0 | N 6 16.2 | 314 10.6 | N 7 55.4 | 230 24.1 | S21 33.0 | 264 39.6 | S11 27.5 |
| | 07 | 134 09.6 | 331 30.4 | 15.7 | 329 11.7 | 54.8 | 245 26.1 | 33.0 | 279 41.8 | 27.6 |
| | 08 | 149 12.1 | 346 30.7 | 15.1 | 344 12.7 | 54.2 | 260 28.0 | 33.1 | 294 43.9 | 27.7 |
| F | 09 | 164 14.6 | 1 31.0 | ·· 14.6 | 359 13.8 | ·· 53.7 | 275 30.0 | ·· 33.2 | 309 46.1 | ·· 27.8 |
| R | 10 | 179 17.0 | 16 31.4 | 14.1 | 14 14.8 | 53.1 | 290 31.9 | 33.3 | 324 48.3 | 27.9 |
| I | 11 | 194 19.5 | 31 31.7 | 13.5 | 29 15.8 | 52.5 | 305 33.9 | 33.3 | 339 50.5 | 28.0 |
| D | 12 | 209 22.0 | 46 32.0 | N 6 13.0 | 44 16.9 | N 7 51.9 | 320 35.8 | S21 33.4 | 354 52.6 | S11 28.1 |
| A | 13 | 224 24.4 | 61 32.3 | 12.5 | 59 17.9 | 51.4 | 335 37.8 | 33.5 | 9 54.8 | 28.2 |
| Y | 14 | 239 26.9 | 76 32.7 | 12.0 | 74 19.0 | 50.8 | 350 39.7 | 33.6 | 24 57.0 | 28.3 |
| | 15 | 254 29.4 | 91 33.0 | ·· 11.4 | 89 20.0 | ·· 50.2 | 5 41.7 | ·· 33.6 | 39 59.2 | ·· 28.4 |
| | 16 | 269 31.8 | 106 33.3 | 10.9 | 104 21.0 | 49.6 | 20 43.6 | 33.7 | 55 01.3 | 28.5 |
| | 17 | 284 34.3 | 121 33.6 | 10.4 | 119 22.1 | 49.1 | 35 45.6 | 33.8 | 70 03.5 | 28.6 |
| | 18 | 299 36.8 | 136 34.0 | N 6 09.8 | 134 23.1 | N 7 48.5 | 50 47.5 | S21 33.9 | 85 05.7 | S11 28.7 |
| | 19 | 314 39.2 | 151 34.3 | 09.3 | 149 24.2 | 47.9 | 65 49.5 | 33.9 | 100 07.9 | 28.8 |
| | 20 | 329 41.7 | 166 34.6 | 08.8 | 164 25.2 | 47.3 | 80 51.5 | 34.0 | 115 10.0 | 28.9 |
| | 21 | 344 44.1 | 181 34.9 | ·· 08.2 | 179 26.2 | ·· 46.8 | 95 53.4 | ·· 34.1 | 130 12.2 | ·· 29.0 |
| | 22 | 359 46.6 | 196 35.2 | 07.7 | 194 27.3 | 46.2 | 110 55.4 | 34.2 | 145 14.4 | 29.1 |
| | 23 | 14 49.1 | 211 35.5 | 07.1 | 209 28.3 | 45.6 | 125 57.3 | 34.2 | 160 16.6 | 29.2 |
| Mer. Pass. | | h m 22 04.8 | *v* 0.4 | *d* 0.5 | *v* 1.0 | *d* 0.6 | *v* 2.0 | *d* 0.1 | *v* 2.2 | *d* 0.1 |

**STARS**

| Name | S.H.A. ° ′ | Dec. ° ′ |
|---|---|---|
| Acamar | 315 35.1 | S40 22.1 |
| Achernar | 335 42.9 | S57 19.2 |
| Acrux | 173 35.6 | S63 00.3 |
| Adhara | 255 30.3 | S28 56.6 |
| Aldebaran | 291 15.2 | N16 28.7 |
| Alioth | 166 41.0 | N56 03.0 |
| Alkaid | 153 17.1 | N49 23.8 |
| Al Na'ir | 28 11.8 | S47 02.6 |
| Alnilam | 276 09.2 | S 1 12.5 |
| Alphard | 218 18.5 | S 8 35.0 |
| Alphecca | 126 30.5 | N26 46.3 |
| Alpheratz | 358 06.8 | N29 00.1 |
| Altair | 62 30.4 | N 8 49.6 |
| Ankaa | 353 37.6 | S42 23.7 |
| Antares | 112 54.5 | S26 23.8 |
| Arcturus | 146 16.8 | N19 16.2 |
| Atria | 108 17.2 | S69 00.1 |
| Avior | 234 27.5 | S59 27.0 |
| Bellatrix | 278 56.2 | N 6 20.3 |
| Betelgeuse | 271 25.7 | N 7 24.4 |
| Canopus | 264 06.0 | S52 40.8 |
| Capella | 281 07.8 | N45 58.9 |
| Deneb | 49 46.9 | N45 13.5 |
| Denebola | 182 57.1 | N14 40.0 |
| Diphda | 349 18.3 | S18 04.5 |
| Dubhe | 194 19.7 | N61 50.3 |
| Elnath | 278 41.1 | N28 35.7 |
| Eltanin | 90 57.0 | N51 29.7 |
| Enif | 34 09.3 | N 9 48.1 |
| Fomalhaut | 15 48.7 | S29 42.6 |
| Gacrux | 172 27.0 | S57 01.1 |
| Gienah | 176 16.0 | S17 26.8 |
| Hadar | 149 21.0 | S60 17.6 |
| Hamal | 328 26.1 | N23 23.2 |
| Kaus Aust. | 84 14.1 | S34 23.7 |
| Kochab | 137 19.9 | N74 13.5 |
| Markab | 14 00.8 | N15 07.1 |
| Menkar | 314 38.5 | N 4 01.7 |
| Menkent | 148 34.9 | S36 17.3 |
| Miaplacidus | 221 45.0 | S69 38.6 |
| Mirfak | 309 12.6 | N49 48.2 |
| Nunki | 76 26.5 | S26 19.2 |
| Peacock | 53 54.7 | S56 47.6 |
| Pollux | 243 55.4 | N28 04.0 |
| Procyon | 245 23.5 | N 5 16.2 |
| Rasalhague | 96 27.7 | N12 34.4 |
| Regulus | 208 07.8 | N12 03.0 |
| Rigel | 281 33.7 | S 8 13.0 |
| Rigil Kent. | 140 23.6 | S60 46.0 |
| Sabik | 102 38.8 | S15 42.3 |
| Schedar | 350 06.1 | N56 27.0 |
| Shaula | 96 53.0 | S37 05.7 |
| Sirius | 258 53.7 | S16 41.4 |
| Spica | 158 55.6 | S11 04.4 |
| Suhail | 223 09.4 | S43 21.6 |
| Vega | 80 54.5 | N38 46.3 |
| Zuben'ubi | 137 30.9 | S15 58.3 |

| | S.H.A. ° ′ | Mer. Pass. h m |
|---|---|---|
| Venus | 198 26.4 | 8 54 |
| Mars | 195 46.2 | 9 05 |
| Jupiter | 111 32.2 | 14 40 |
| Saturn | 145 41.0 | 12 24 |

| | G.M.T. d h | SUN G.H.A. ° ′ | SUN Dec. ° ′ | MOON G.H.A. ° ′ | v ′ | MOON Dec. ° ′ | d ′ | H.P. ′ |
|---|---|---|---|---|---|---|---|---|
| | 19 00 | 183 42.6 | S 9 41.5 | 32 30.4 | 15.5 | S 7 57.5 | 12.2 | 54.7 |
| | 01 | 198 42.7 | 42.4 | 47 04.9 | 15.5 | 7 45.3 | 12.3 | 54.7 |
| | 02 | 213 42.8 | 43.3 | 61 39.4 | 15.6 | 7 33.0 | 12.3 | 54.7 |
| | 03 | 228 42.9 | ·· 44.2 | 76 14.0 | 15.5 | 7 20.7 | 12.4 | 54.7 |
| | 04 | 243 43.1 | 45.1 | 90 48.5 | 15.5 | 7 08.3 | 12.4 | 54.7 |
| | 05 | 258 43.2 | 46.0 | 105 23.0 | 15.6 | 6 55.9 | 12.4 | 54.7 |
| | 06 | 273 43.3 | S 9 46.9 | 119 57.6 | 15.5 | S 6 43.5 | 12.5 | 54.8 |
| W | 07 | 288 43.4 | 47.8 | 134 32.1 | 15.6 | 6 31.0 | 12.5 | 54.8 |
| E | 08 | 303 43.5 | 48.7 | 149 06.7 | 15.5 | 6 18.5 | 12.5 | 54.8 |
| D | 09 | 318 43.6 | ·· 49.6 | 163 41.2 | 15.6 | 6 06.0 | 12.5 | 54.8 |
| N | 10 | 333 43.8 | 50.5 | 178 15.8 | 15.5 | 5 53.5 | 12.6 | 54.8 |
| E | 11 | 348 43.9 | 51.4 | 192 50.3 | 15.6 | 5 40.9 | 12.7 | 54.8 |
| S | 12 | 3 44.0 | S 9 52.3 | 207 24.9 | 15.6 | S 5 28.2 | 12.6 | 54.8 |
| D | 13 | 18 44.1 | 53.2 | 221 59.5 | 15.5 | 5 15.6 | 12.7 | 54.9 |
| A | 14 | 33 44.2 | 54.1 | 236 34.0 | 15.6 | 5 02.9 | 12.7 | 54.9 |
| Y | 15 | 48 44.3 | ·· 55.0 | 251 08.6 | 15.6 | 4 50.2 | 12.7 | 54.9 |
| | 16 | 63 44.5 | 55.9 | 265 43.2 | 15.5 | 4 37.5 | 12.8 | 54.9 |
| | 17 | 78 44.6 | 56.8 | 280 17.7 | 15.6 | 4 24.7 | 12.8 | 54.9 |
| | 18 | 93 44.7 | S 9 57.7 | 294 52.3 | 15.5 | S 4 11.9 | 12.8 | 54.9 |
| | 19 | 108 44.8 | 58.7 | 309 26.8 | 15.6 | 3 59.1 | 12.9 | 55.0 |
| | 20 | 123 44.9 | 9 59.6 | 324 01.4 | 15.5 | 3 46.2 | 12.8 | 55.0 |
| | 21 | 138 45.0 | 10 00.5 | 338 35.9 | 15.6 | 3 33.4 | 12.9 | 55.0 |
| | 22 | 153 45.2 | 01.4 | 353 10.5 | 15.5 | 3 20.5 | 12.9 | 55.0 |
| | 23 | 168 45.3 | 02.3 | 7 45.0 | 15.5 | 3 07.6 | 13.0 | 55.0 |
| | 20 00 | 183 45.4 | S10 03.2 | 22 19.5 | 15.6 | S 2 54.6 | 12.9 | 55.1 |
| | 01 | 198 45.5 | 04.1 | 36 54.1 | 15.5 | 2 41.7 | 13.0 | 55.1 |
| | 02 | 213 45.6 | 05.0 | 51 28.6 | 15.5 | 2 28.7 | 13.0 | 55.1 |
| | 03 | 228 45.7 | ·· 05.9 | 66 03.1 | 15.4 | 2 15.7 | 13.0 | 55.1 |
| | 04 | 243 45.8 | 06.8 | 80 37.5 | 15.5 | 2 02.7 | 13.0 | 55.1 |
| | 05 | 258 46.0 | 07.7 | 95 12.0 | 15.5 | 1 49.7 | 13.1 | 55.1 |
| | 06 | 273 46.1 | S10 08.6 | 109 46.5 | 15.4 | S 1 36.6 | 13.1 | 55.2 |
| | 07 | 288 46.2 | 09.5 | 124 20.9 | 15.4 | 1 23.5 | 13.0 | 55.2 |
| T | 08 | 303 46.3 | 10.4 | 138 55.3 | 15.5 | 1 10.5 | 13.1 | 55.2 |
| H | 09 | 318 46.4 | ·· 11.3 | 153 29.8 | 15.4 | 0 57.4 | 13.1 | 55.2 |
| U | 10 | 333 46.5 | 12.2 | 168 04.2 | 15.3 | 0 44.3 | 13.2 | 55.2 |
| R | 11 | 348 46.6 | 13.1 | 182 38.5 | 15.4 | 0 31.1 | 13.1 | 55.3 |
| S | 12 | 3 46.7 | S10 14.0 | 197 12.9 | 15.3 | S 0 18.0 | 13.1 | 55.3 |
| D | 13 | 18 46.8 | 14.9 | 211 47.2 | 15.4 | S 0 04.9 | 13.2 | 55.3 |
| A | 14 | 33 47.0 | 15.8 | 226 21.6 | 15.3 | N 0 08.3 | 13.1 | 55.3 |
| Y | 15 | 48 47.1 | ·· 16.7 | 240 55.9 | 15.2 | 0 21.4 | 13.2 | 55.3 |
| | 16 | 63 47.2 | 17.6 | 255 30.1 | 15.3 | 0 34.6 | 13.2 | 55.4 |
| | 17 | 78 47.3 | 18.5 | 270 04.4 | 15.2 | 0 47.8 | 13.2 | 55.4 |
| | 18 | 93 47.4 | S10 19.4 | 284 38.6 | 15.2 | N 1 01.0 | 13.2 | 55.4 |
| | 19 | 108 47.5 | 20.2 | 299 12.8 | 15.2 | 1 14.2 | 13.2 | 55.4 |
| | 20 | 123 47.6 | 21.1 | 313 47.0 | 15.2 | 1 27.4 | 13.2 | 55.4 |
| | 21 | 138 47.7 | ·· 22.0 | 328 21.2 | 15.1 | 1 40.6 | 13.2 | 55.4 |
| | 22 | 153 47.8 | 22.9 | 342 55.3 | 15.1 | 1 53.8 | 13.2 | 55.5 |
| | 23 | 168 47.9 | 23.8 | 357 29.4 | 15.1 | 2 07.0 | 13.2 | 55.5 |
| | 21 00 | 183 48.0 | S10 24.7 | 12 03.5 | 15.0 | N 2 20.2 | 13.2 | 55.5 |
| | 01 | 198 48.2 | 25.6 | 26 37.5 | 15.0 | 2 33.4 | 13.2 | 55.5 |
| | 02 | 213 48.3 | 26.5 | 41 11.5 | 15.0 | 2 46.6 | 13.2 | 55.5 |
| | 03 | 228 48.4 | ·· 27.4 | 55 45.5 | 14.9 | 2 59.8 | 13.2 | 55.6 |
| | 04 | 243 48.5 | 28.3 | 70 19.4 | 14.9 | 3 13.0 | 13.2 | 55.6 |
| | 05 | 258 48.6 | 29.2 | 84 53.3 | 14.9 | 3 26.2 | 13.2 | 55.6 |
| | 06 | 273 48.7 | S10 30.1 | 99 27.2 | 14.9 | N 3 39.4 | 13.2 | 55.6 |
| | 07 | 288 48.8 | 31.0 | 114 01.1 | 14.8 | 3 52.6 | 13.2 | 55.7 |
| | 08 | 303 48.9 | 31.9 | 128 34.9 | 14.7 | 4 05.8 | 13.2 | 55.7 |
| F | 09 | 318 49.0 | ·· 32.8 | 143 08.6 | 14.8 | 4 19.0 | 13.2 | 55.7 |
| R | 10 | 333 49.1 | 33.7 | 157 42.4 | 14.6 | 4 32.2 | 13.2 | 55.7 |
| I | 11 | 348 49.2 | 34.6 | 172 16.0 | 14.7 | 4 45.4 | 13.1 | 55.7 |
| D | 12 | 3 49.3 | S10 35.5 | 186 49.7 | 14.6 | N 4 58.5 | 13.2 | 55.8 |
| A | 13 | 18 49.4 | 36.3 | 201 23.3 | 14.6 | 5 11.7 | 13.1 | 55.8 |
| Y | 14 | 33 49.5 | 37.2 | 215 56.9 | 14.5 | 5 24.8 | 13.1 | 55.8 |
| | 15 | 48 49.6 | ·· 38.1 | 230 30.4 | 14.5 | 5 37.9 | 13.1 | 55.8 |
| | 16 | 63 49.7 | 39.0 | 245 03.9 | 14.4 | 5 51.0 | 13.1 | 55.8 |
| | 17 | 78 49.8 | 39.9 | 259 37.3 | 14.4 | 6 04.1 | 13.1 | 55.9 |
| | 18 | 93 49.9 | S10 40.8 | 274 10.7 | 14.4 | N 6 17.2 | 13.1 | 55.9 |
| | 19 | 108 50.0 | 41.7 | 288 44.1 | 14.3 | 6 30.3 | 13.0 | 55.9 |
| | 20 | 123 50.1 | 42.6 | 303 17.4 | 14.2 | 6 43.3 | 13.0 | 55.9 |
| | 21 | 138 50.2 | ·· 43.5 | 317 50.6 | 14.2 | 6 56.3 | 13.0 | 55.9 |
| | 22 | 153 50.3 | 44.4 | 332 23.8 | 14.2 | 7 09.3 | 13.0 | 56.0 |
| | 23 | 168 50.4 | 45.2 | 346 57.0 | 14.1 | 7 22.3 | 13.0 | 56.0 |
| | | S.D. 16.1 | d 0.9 | S.D. 14.9 | | 15.1 | | 15.2 |

| Lat. ° | Twilight Naut. h m | Twilight Civil h m | Sunrise h m | Moonrise 19 h m | Moonrise 20 h m | Moonrise 21 h m | Moonrise 22 h m |
|---|---|---|---|---|---|---|---|
| N 72 | 05 14 | 06 33 | 07 46 | 17 15 | 16 51 | 16 27 | 15 58 |
| N 70 | 05 16 | 06 27 | 07 32 | 17 09 | 16 52 | 16 35 | 16 16 |
| 68 | 05 17 | 06 22 | 07 20 | 17 04 | 16 53 | 16 42 | 16 30 |
| 66 | 05 18 | 06 17 | 07 11 | 17 00 | 16 54 | 16 47 | 16 41 |
| 64 | 05 18 | 06 14 | 07 03 | 16 56 | 16 54 | 16 52 | 16 51 |
| 62 | 05 19 | 06 10 | 06 56 | 16 53 | 16 55 | 16 56 | 16 59 |
| 60 | 05 19 | 06 07 | 06 50 | 16 50 | 16 55 | 17 00 | 17 06 |
| N 58 | 05 19 | 06 05 | 06 45 | 16 48 | 16 55 | 17 03 | 17 12 |
| 56 | 05 19 | 06 02 | 06 40 | 16 46 | 16 56 | 17 06 | 17 18 |
| 54 | 05 19 | 06 00 | 06 36 | 16 44 | 16 56 | 17 09 | 17 23 |
| 52 | 05 19 | 05 58 | 06 32 | 16 42 | 16 56 | 17 11 | 17 28 |
| 50 | 05 18 | 05 56 | 06 29 | 16 40 | 16 57 | 17 13 | 17 32 |
| 45 | 05 17 | 05 52 | 06 21 | 16 37 | 16 57 | 17 18 | 17 41 |
| N 40 | 05 16 | 05 48 | 06 15 | 16 34 | 16 58 | 17 22 | 17 48 |
| 35 | 05 14 | 05 44 | 06 10 | 16 31 | 16 58 | 17 25 | 17 55 |
| 30 | 05 13 | 05 40 | 06 05 | 16 29 | 16 58 | 17 29 | 18 01 |
| 20 | 05 08 | 05 34 | 05 56 | 16 25 | 16 59 | 17 34 | 18 11 |
| N 10 | 05 03 | 05 27 | 05 49 | 16 21 | 17 00 | 17 39 | 18 20 |
| 0 | 04 56 | 05 21 | 05 42 | 16 18 | 17 00 | 17 43 | 18 28 |
| S 10 | 04 48 | 05 13 | 05 34 | 16 15 | 17 01 | 17 48 | 18 37 |
| 20 | 04 38 | 05 04 | 05 26 | 16 11 | 17 01 | 17 53 | 18 46 |
| 30 | 04 24 | 04 53 | 05 17 | 16 07 | 17 02 | 17 58 | 18 56 |
| 35 | 04 15 | 04 46 | 05 12 | 16 05 | 17 03 | 18 02 | 19 03 |
| 40 | 04 04 | 04 38 | 05 06 | 16 02 | 17 03 | 18 05 | 19 09 |
| 45 | 03 51 | 04 28 | 04 59 | 15 59 | 17 04 | 18 10 | 19 18 |
| S 50 | 03 34 | 04 16 | 04 50 | 15 55 | 17 04 | 18 15 | 19 28 |
| 52 | 03 26 | 04 10 | 04 46 | 15 54 | 17 05 | 18 17 | 19 32 |
| 54 | 03 16 | 04 04 | 04 42 | 15 52 | 17 05 | 18 20 | 19 37 |
| 56 | 03 05 | 03 57 | 04 37 | 15 50 | 17 05 | 18 23 | 19 43 |
| 58 | 02 52 | 03 49 | 04 32 | 15 47 | 17 06 | 18 26 | 19 49 |
| S 60 | 02 37 | 03 39 | 04 26 | 15 45 | 17 06 | 18 30 | 19 56 |

| Lat. ° | Sunset h m | Twilight Civil h m | Twilight Naut. h m | Moonset 19 h m | Moonset 20 h m | Moonset 21 h m | Moonset 22 h m |
|---|---|---|---|---|---|---|---|
| N 72 | 15 42 | 16 55 | 18 13 | 02 17 | 04 12 | 06 07 | 08 09 |
| N 70 | 15 56 | 17 01 | 18 12 | 02 29 | 04 15 | 06 01 | 07 53 |
| 68 | 16 08 | 17 06 | 18 11 | 02 38 | 04 17 | 05 57 | 07 41 |
| 66 | 16 18 | 17 11 | 18 10 | 02 46 | 04 19 | 05 54 | 07 32 |
| 64 | 16 26 | 17 15 | 18 10 | 02 52 | 04 21 | 05 51 | 07 24 |
| 62 | 16 33 | 17 18 | 18 10 | 02 58 | 04 22 | 05 48 | 07 17 |
| 60 | 16 39 | 17 21 | 18 09 | 03 03 | 04 24 | 05 46 | 07 11 |
| N 58 | 16 44 | 17 24 | 18 09 | 03 07 | 04 25 | 05 44 | 07 05 |
| 56 | 16 49 | 17 27 | 18 10 | 03 11 | 04 26 | 05 42 | 07 01 |
| 54 | 16 53 | 17 29 | 18 10 | 03 14 | 04 27 | 05 41 | 06 57 |
| 52 | 16 57 | 17 31 | 18 10 | 03 17 | 04 28 | 05 39 | 06 53 |
| 50 | 17 00 | 17 33 | 18 10 | 03 20 | 04 28 | 05 38 | 06 49 |
| 45 | 17 08 | 17 38 | 18 12 | 03 26 | 04 30 | 05 35 | 06 42 |
| N 40 | 17 14 | 17 42 | 18 13 | 03 31 | 04 31 | 05 33 | 06 36 |
| 35 | 17 20 | 17 45 | 18 15 | 03 35 | 04 33 | 05 31 | 06 31 |
| 30 | 17 25 | 17 49 | 18 17 | 03 39 | 04 34 | 05 29 | 06 26 |
| 20 | 17 33 | 17 56 | 18 21 | 03 45 | 04 35 | 05 26 | 06 18 |
| N 10 | 17 41 | 18 02 | 18 27 | 03 51 | 04 37 | 05 23 | 06 11 |
| 0 | 17 48 | 18 09 | 18 34 | 03 56 | 04 38 | 05 21 | 06 05 |
| S 10 | 17 56 | 18 17 | 18 42 | 04 01 | 04 40 | 05 18 | 05 58 |
| 20 | 18 04 | 18 26 | 18 53 | 04 07 | 04 41 | 05 15 | 05 51 |
| 30 | 18 13 | 18 38 | 19 07 | 04 13 | 04 43 | 05 12 | 05 43 |
| 35 | 18 18 | 18 44 | 19 16 | 04 17 | 04 44 | 05 10 | 05 39 |
| 40 | 18 25 | 18 53 | 19 26 | 04 21 | 04 45 | 05 08 | 05 34 |
| 45 | 18 32 | 19 03 | 19 40 | 04 25 | 04 46 | 05 06 | 05 28 |
| S 50 | 18 40 | 19 15 | 19 57 | 04 31 | 04 47 | 05 03 | 05 21 |
| 52 | 18 45 | 19 21 | 20 06 | 04 33 | 04 48 | 05 02 | 05 17 |
| 54 | 18 49 | 19 27 | 20 16 | 04 36 | 04 48 | 05 01 | 05 14 |
| 56 | 18 54 | 19 35 | 20 27 | 04 39 | 04 49 | 04 59 | 05 10 |
| 58 | 18 59 | 19 43 | 20 40 | 04 42 | 04 50 | 04 58 | 05 06 |
| S 60 | 19 06 | 19 53 | 20 56 | 04 46 | 04 51 | 04 56 | 05 01 |

| Day | SUN Eqn. of Time 00ʰ m s | SUN Eqn. of Time 12ʰ m s | SUN Mer. Pass. h m | MOON Mer. Pass. Upper h m | MOON Mer. Pass. Lower h m | Age d | Phase |
|---|---|---|---|---|---|---|---|
| 19 | 14 50 | 14 56 | 11 45 | 22 28 | 10 07 | 13 | ○ |
| 20 | 15 01 | 15 07 | 11 45 | 23 10 | 10 49 | 14 | |
| 21 | 15 12 | 15 17 | 11 45 | 23 54 | 11 32 | 15 | |

| G.M.T. | | ARIES | VENUS −4.1 | | MARS +1.9 | | JUPITER −1.4 | | SATURN +0.8 | |
|---|---|---|---|---|---|---|---|---|---|---|
| d | h | G.H.A. ° ′ | G.H.A. ° ′ | Dec. ° ′ | G.H.A. ° ′ | Dec. ° ′ | G.H.A. ° ′ | Dec. ° ′ | G.H.A. ° ′ | Dec. ° ′ |
| 25 | 00 | 32 49.0 | 226 56.0 | N 5 25.5 | 225 44.6 | N 7 03.5 | 143 19.4 | S21 39.7 | 177 55.4 | S11 36.5 |
| | 01 | 47 51.4 | 241 56.3 | 24.9 | 240 45.7 | 02.9 | 158 21.3 | 39.7 | 192 57.6 | 36.6 |
| | 02 | 62 53.9 | 256 56.5 | 24.3 | 255 46.7 | 02.3 | 173 23.2 | 39.8 | 207 59.7 | 36.7 |
| | 03 | 77 56.3 | 271 56.8 | ·· 23.7 | 270 47.8 | ·· 01.7 | 188 25.2 | ·· 39.9 | 223 01.9 | ·· 36.8 |
| | 04 | 92 58.8 | 286 57.0 | 23.1 | 285 48.8 | 01.2 | 203 27.1 | 40.0 | 238 04.1 | 36.9 |
| | 05 | 108 01.3 | 301 57.2 | 22.5 | 300 49.9 | 00.6 | 218 29.1 | 40.0 | 253 06.3 | 37.0 |
| | 06 | 123 03.7 | 316 57.5 | N 5 21.9 | 315 50.9 | N 7 00.0 | 233 31.0 | S21 40.1 | 268 08.4 | S11 37.1 |
| | 07 | 138 06.2 | 331 57.7 | 21.3 | 330 52.0 | 6 59.4 | 248 33.0 | 40.2 | 283 10.6 | 37.2 |
| T | 08 | 153 08.7 | 346 58.0 | 20.7 | 345 53.0 | 58.9 | 263 34.9 | 40.3 | 298 12.8 | 37.3 |
| U | 09 | 168 11.1 | 1 58.2 | ·· 20.1 | 0 54.1 | ·· 58.3 | 278 36.8 | ·· 40.3 | 313 15.0 | ·· 37.4 |
| E | 10 | 183 13.6 | 16 58.5 | 19.5 | 15 55.1 | 57.7 | 293 38.8 | 40.4 | 328 17.1 | 37.5 |
| S | 11 | 198 16.1 | 31 58.7 | 18.9 | 30 56.2 | 57.1 | 308 40.7 | 40.5 | 343 19.3 | 37.6 |
| D | 12 | 213 18.5 | 46 58.9 | N 5 18.3 | 45 57.2 | N 6 56.5 | 323 42.6 | S21 40.6 | 358 21.5 | S11 37.7 |
| A | 13 | 228 21.0 | 61 59.2 | 17.7 | 60 58.3 | 56.0 | 338 44.6 | 40.6 | 13 23.6 | 37.8 |
| Y | 14 | 243 23.5 | 76 59.4 | 17.1 | 75 59.3 | 55.4 | 353 46.5 | 40.7 | 28 25.8 | 37.9 |
| | 15 | 258 25.9 | 91 59.6 | ·· 16.4 | 91 00.4 | ·· 54.8 | 8 48.5 | ·· 40.8 | 43 28.0 | ·· 38.0 |
| | 16 | 273 28.4 | 106 59.9 | 15.8 | 106 01.5 | 54.2 | 23 50.4 | 40.9 | 58 30.2 | 38.1 |
| | 17 | 288 30.8 | 122 00.1 | 15.2 | 121 02.5 | 53.6 | 38 52.3 | 40.9 | 73 32.3 | 38.2 |
| | 18 | 303 33.3 | 137 00.3 | N 5 14.6 | 136 03.6 | N 6 53.1 | 53 54.3 | S21 41.0 | 88 34.5 | S11 38.3 |
| | 19 | 318 35.8 | 152 00.6 | 14.0 | 151 04.6 | 52.5 | 68 56.2 | 41.1 | 103 36.7 | 38.4 |
| | 20 | 333 38.2 | 167 00.8 | 13.4 | 166 05.7 | 51.9 | 83 58.2 | 41.2 | 118 38.9 | 38.5 |
| | 21 | 348 40.7 | 182 01.0 | ·· 12.8 | 181 06.7 | ·· 51.3 | 99 00.1 | ·· 41.2 | 133 41.0 | ·· 38.6 |
| | 22 | 3 43.2 | 197 01.3 | 12.1 | 196 07.8 | 50.8 | 114 02.0 | 41.3 | 148 43.2 | 38.7 |
| | 23 | 18 45.6 | 212 01.5 | 11.5 | 211 08.8 | 50.2 | 129 04.0 | 41.4 | 163 45.4 | 38.8 |
| 26 | 00 | 33 48.1 | 227 01.7 | N 5 10.9 | 226 09.9 | N 6 49.6 | 144 05.9 | S21 41.4 | 178 47.6 | S11 38.9 |
| | 01 | 48 50.6 | 242 01.9 | 10.3 | 241 10.9 | 49.0 | 159 07.8 | 41.5 | 193 49.7 | 39.0 |
| | 02 | 63 53.0 | 257 02.2 | 09.7 | 256 12.0 | 48.4 | 174 09.8 | 41.6 | 208 51.9 | 39.0 |
| | 03 | 78 55.5 | 272 02.4 | ·· 09.0 | 271 13.0 | ·· 47.9 | 189 11.7 | ·· 41.7 | 223 54.1 | ·· 39.1 |
| | 04 | 93 58.0 | 287 02.6 | 08.4 | 286 14.1 | 47.3 | 204 13.7 | 41.7 | 238 56.3 | 39.2 |
| | 05 | 109 00.4 | 302 02.8 | 07.8 | 301 15.1 | 46.7 | 219 15.6 | 41.8 | 253 58.4 | 39.3 |
| | 06 | 124 02.9 | 317 03.1 | N 5 07.2 | 316 16.2 | N 6 46.1 | 234 17.5 | S21 41.9 | 269 00.6 | S11 39.4 |
| W | 07 | 139 05.3 | 332 03.3 | 06.5 | 331 17.2 | 45.5 | 249 19.5 | 42.0 | 284 02.8 | 39.5 |
| E | 08 | 154 07.8 | 347 03.5 | 05.9 | 346 18.3 | 45.0 | 264 21.4 | 42.0 | 299 05.0 | 39.6 |
| D | 09 | 169 10.3 | 2 03.7 | ·· 05.3 | 1 19.3 | ·· 44.4 | 279 23.3 | ·· 42.1 | 314 07.1 | ·· 39.7 |
| N | 10 | 184 12.7 | 17 03.9 | 04.7 | 16 20.4 | 43.8 | 294 25.3 | 42.2 | 329 09.3 | 39.8 |
| E | 11 | 199 15.2 | 32 04.2 | 04.0 | 31 21.4 | 43.2 | 309 27.2 | 42.3 | 344 11.5 | 39.9 |
| S | 12 | 214 17.7 | 47 04.4 | N 5 03.4 | 46 22.5 | N 6 42.6 | 324 29.1 | S21 42.3 | 359 13.7 | S11 40.0 |
| D | 13 | 229 20.1 | 62 04.6 | 02.8 | 61 23.6 | 42.1 | 339 31.1 | 42.4 | 14 15.8 | 40.1 |
| A | 14 | 244 22.6 | 77 04.8 | 02.1 | 76 24.6 | 41.5 | 354 33.0 | 42.5 | 29 18.0 | 40.2 |
| Y | 15 | 259 25.1 | 92 05.0 | ·· 01.5 | 91 25.7 | ·· 40.9 | 9 35.0 | ·· 42.6 | 44 20.2 | ·· 40.3 |
| | 16 | 274 27.5 | 107 05.2 | 00.9 | 106 26.7 | 40.3 | 24 36.9 | 42.6 | 59 22.4 | 40.4 |
| | 17 | 289 30.0 | 122 05.5 | 5 00.2 | 121 27.8 | 39.7 | 39 38.8 | 42.7 | 74 24.5 | 40.5 |
| | 18 | 304 32.4 | 137 05.7 | N 4 59.6 | 136 28.8 | N 6 39.2 | 54 40.8 | S21 42.8 | 89 26.7 | S11 40.6 |
| | 19 | 319 34.9 | 152 05.9 | 59.0 | 151 29.9 | 38.6 | 69 42.7 | 42.8 | 104 28.9 | 40.7 |
| | 20 | 334 37.4 | 167 06.1 | 58.3 | 166 30.9 | 38.0 | 84 44.6 | 42.9 | 119 31.0 | 40.8 |
| | 21 | 349 39.8 | 182 06.3 | ·· 57.7 | 181 32.0 | ·· 37.4 | 99 46.6 | ·· 43.0 | 134 33.2 | ·· 40.9 |
| | 22 | 4 42.3 | 197 06.5 | 57.1 | 196 33.0 | 36.8 | 114 48.5 | 43.1 | 149 35.4 | 41.0 |
| | 23 | 19 44.8 | 212 06.7 | 56.4 | 211 34.1 | 36.3 | 129 50.4 | 43.1 | 164 37.6 | 41.1 |
| 27 | 00 | 34 47.2 | 227 06.9 | N 4 55.8 | 226 35.2 | N 6 35.7 | 144 52.4 | S21 43.2 | 179 39.7 | S11 41.2 |
| | 01 | 49 49.7 | 242 07.1 | 55.2 | 241 36.2 | 35.1 | 159 54.3 | 43.3 | 194 41.9 | 41.3 |
| | 02 | 64 52.2 | 257 07.3 | 54.5 | 256 37.3 | 34.5 | 174 56.2 | 43.4 | 209 44.1 | 41.4 |
| | 03 | 79 54.6 | 272 07.6 | ·· 53.9 | 271 38.3 | ·· 33.9 | 189 58.2 | ·· 43.4 | 224 46.3 | ·· 41.5 |
| | 04 | 94 57.1 | 287 07.8 | 53.2 | 286 39.4 | 33.4 | 205 00.1 | 43.5 | 239 48.4 | 41.6 |
| | 05 | 109 59.6 | 302 08.0 | 52.6 | 301 40.4 | 32.8 | 220 02.0 | 43.6 | 254 50.6 | 41.7 |
| | 06 | 125 02.0 | 317 08.2 | N 4 51.9 | 316 41.5 | N 6 32.2 | 235 04.0 | S21 43.7 | 269 52.8 | S11 41.8 |
| | 07 | 140 04.5 | 332 08.4 | 51.3 | 331 42.5 | 31.6 | 250 05.9 | 43.7 | 284 55.0 | 41.9 |
| T | 08 | 155 06.9 | 347 08.6 | 50.7 | 346 43.6 | 31.0 | 265 07.8 | 43.8 | 299 57.1 | 42.0 |
| H | 09 | 170 09.4 | 2 08.8 | ·· 50.0 | 1 44.6 | ·· 30.5 | 280 09.8 | ·· 43.9 | 314 59.3 | ·· 42.1 |
| U | 10 | 185 11.9 | 17 09.0 | 49.4 | 16 45.7 | 29.9 | 295 11.7 | 43.9 | 330 01.5 | 42.2 |
| R | 11 | 200 14.3 | 32 09.2 | 48.7 | 31 46.8 | 29.3 | 310 13.6 | 44.0 | 345 03.7 | 42.3 |
| S | 12 | 215 16.8 | 47 09.4 | N 4 48.1 | 46 47.8 | N 6 28.7 | 325 15.6 | S21 44.1 | 0 05.8 | S11 42.4 |
| D | 13 | 230 19.3 | 62 09.6 | 47.4 | 61 48.9 | 28.1 | 340 17.5 | 44.2 | 15 08.0 | 42.5 |
| A | 14 | 245 21.7 | 77 09.8 | 46.8 | 76 49.9 | 27.6 | 355 19.4 | 44.2 | 30 10.2 | 42.6 |
| Y | 15 | 260 24.2 | 92 10.0 | ·· 46.1 | 91 51.0 | ·· 27.0 | 10 21.4 | ·· 44.3 | 45 12.4 | ·· 42.7 |
| | 16 | 275 26.7 | 107 10.2 | 45.5 | 106 52.0 | 26.4 | 25 23.3 | 44.4 | 60 14.5 | 42.8 |
| | 17 | 290 29.1 | 122 10.4 | 44.8 | 121 53.1 | 25.8 | 40 25.2 | 44.5 | 75 16.7 | 42.9 |
| | 18 | 305 31.6 | 137 10.5 | N 4 44.2 | 136 54.2 | N 6 25.2 | 55 27.2 | S21 44.5 | 90 18.9 | S11 43.0 |
| | 19 | 320 34.1 | 152 10.7 | 43.5 | 151 55.2 | 24.7 | 70 29.1 | 44.6 | 105 21.0 | 43.1 |
| | 20 | 335 36.5 | 167 10.9 | 42.9 | 166 56.3 | 24.1 | 85 31.0 | 44.7 | 120 23.2 | 43.2 |
| | 21 | 350 39.0 | 182 11.1 | ·· 42.2 | 181 57.3 | ·· 23.5 | 100 33.0 | ·· 44.8 | 135 25.4 | ·· 43.3 |
| | 22 | 5 41.4 | 197 11.3 | 41.6 | 196 58.4 | 22.9 | 115 34.9 | 44.8 | 150 27.6 | 43.4 |
| | 23 | 20 43.9 | 212 11.5 | 40.9 | 211 59.4 | 22.3 | 130 36.8 | 44.9 | 165 29.7 | 43.5 |
| Mer. Pass. | | h m 21 41.2 | v 0.2 | d 0.6 | v 1.1 | d 0.6 | v 1.9 | d 0.1 | v 2.2 | d 0.1 |

**STARS**

| Name | S.H.A. ° ′ | Dec. ° ′ |
|---|---|---|
| Acamar | 315 35.1 | S40 22.1 |
| Achernar | 335 42.9 | S57 19.2 |
| Acrux | 173 35.6 | S63 00.3 |
| Adhara | 255 30.3 | S28 56.7 |
| Aldebaran | 291 15.2 | N16 28.7 |
| Alioth | 166 40.9 | N56 02.9 |
| Alkaid | 153 17.1 | N49 23.8 |
| Al Na'ir | 28 11.8 | S47 02.7 |
| Alnilam | 276 09.2 | S 1 12.5 |
| Alphard | 218 18.5 | S 8 35.0 |
| Alphecca | 126 30.5 | N26 46.3 |
| Alpheratz | 358 06.8 | N29 00.1 |
| Altair | 62 30.4 | N 8 49.6 |
| Ankaa | 353 37.6 | S42 23.8 |
| Antares | 112 54.5 | S26 23.8 |
| Arcturus | 146 16.8 | N19 16.2 |
| Atria | 108 17.2 | S69 00.1 |
| Avior | 234 27.4 | S59 27.0 |
| Bellatrix | 278 56.1 | N 6 20.3 |
| Betelgeuse | 271 25.7 | N 7 24.4 |
| Canopus | 264 06.0 | S52 40.8 |
| Capella | 281 07.7 | N45 58.9 |
| Deneb | 49 47.0 | N45 13.5 |
| Denebola | 182 57.0 | N14 39.9 |
| Diphda | 349 18.3 | S18 04.6 |
| Dubhe | 194 19.6 | N61 50.3 |
| Elnath | 278 41.1 | N28 35.7 |
| Eltanin | 90 57.0 | N51 29.7 |
| Enif | 34 09.3 | N 9 48.1 |
| Fomalhaut | 15 48.7 | S29 42.6 |
| Gacrux | 172 27.0 | S57 01.1 |
| Gienah | 176 16.0 | S17 26.9 |
| Hadar | 149 21.0 | S60 17.6 |
| Hamal | 328 26.1 | N23 23.2 |
| Kaus Aust. | 84 14.1 | S34 23.7 |
| Kochab | 137 20.0 | N74 13.4 |
| Markab | 14 00.8 | N15 07.1 |
| Menkar | 314 38.5 | N 4 01.7 |
| Menkent | 148 34.9 | S36 17.3 |
| Miaplacidus | 221 44.9 | S69 38.6 |
| Mirfak | 309 12.5 | N49 48.2 |
| Nunki | 76 26.5 | S26 19.2 |
| Peacock | 53 54.8 | S56 47.6 |
| Pollux | 243 55.4 | N28 04.0 |
| Procyon | 245 23.4 | N 5 16.2 |
| Rasalhague | 96 27.7 | N12 34.4 |
| Regulus | 208 07.8 | N12 03.0 |
| Rigel | 281 33.6 | S 8 13.0 |
| Rigil Kent. | 140 23.6 | S60 46.0 |
| Sabik | 102 38.8 | S15 42.3 |
| Schedar | 350 06.1 | N56 27.0 |
| Shaula | 96 53.0 | S37 05.7 |
| Sirius | 258 53.6 | S16 41.4 |
| Spica | 158 55.5 | S11 04.4 |
| Suhail | 223 09.3 | S43 21.6 |
| Vega | 80 54.5 | N38 46.3 |
| Zuben'ubi | 137 30.9 | S15 58.3 |

| | S.H.A. ° ′ | Mer. Pass. h m |
|---|---|---|
| Venus | 193 13.6 | 8 52 |
| Mars | 192 21.8 | 8 55 |
| Jupiter | 110 17.8 | 14 22 |
| Saturn | 144 59.5 | 12 03 |

| G.M.T. d h | SUN G.H.A. ° ′ | SUN Dec. ° ′ | MOON G.H.A. ° ′ | v ′ | MOON Dec. ° ′ | d ′ | H.P. ′ |
|---|---|---|---|---|---|---|---|
| 25 00 | 183 57.0 | S11 49.4 | 326 13.9 | 8.7 | N20 55.8 | 8.2 | 57.5 |
| 01 | 198 57.1 | 50.3 | 340 41.6 | 8.6 | 21 04.0 | 8.0 | 57.5 |
| 02 | 213 57.2 | 51.1 | 355 09.2 | 8.5 | 21 12.0 | 7.9 | 57.5 |
| 03 | 228 57.3 | ·· 52.0 | 9 36.7 | 8.4 | 21 19.9 | 7.8 | 57.5 |
| 04 | 243 57.3 | 52.9 | 24 04.1 | 8.4 | 21 27.7 | 7.7 | 57.6 |
| 05 | 258 57.4 | 53.7 | 38 31.5 | 8.2 | 21 35.4 | 7.6 | 57.6 |
| 06 | 273 57.5 | S11 54.6 | 52 58.7 | 8.2 | N21 43.0 | 7.4 | 57.6 |
| 07 | 288 57.6 | 55.5 | 67 25.9 | 8.1 | 21 50.4 | 7.4 | 57.6 |
| T 08 | 303 57.7 | 56.3 | 81 53.0 | 8.0 | 21 57.8 | 7.2 | 57.6 |
| U 09 | 318 57.7 | ·· 57.2 | 96 20.0 | 8.0 | 22 05.0 | 7.1 | 57.7 |
| E 10 | 333 57.8 | 58.1 | 110 47.0 | 7.8 | 22 12.1 | 6.9 | 57.7 |
| S 11 | 348 57.9 | 58.9 | 125 13.8 | 7.8 | 22 19.0 | 6.9 | 57.7 |
| D 12 | 3 58.0 | S11 59.8 | 139 40.6 | 7.7 | N22 25.9 | 6.7 | 57.7 |
| A 13 | 18 58.0 | 12 00.6 | 154 07.3 | 7.6 | 22 32.6 | 6.5 | 57.7 |
| Y 14 | 33 58.1 | 01.5 | 168 33.9 | 7.6 | 22 39.1 | 6.5 | 57.7 |
| 15 | 48 58.2 | ·· 02.4 | 183 00.5 | 7.4 | 22 45.6 | 6.3 | 57.8 |
| 16 | 63 58.3 | 03.2 | 197 26.9 | 7.4 | 22 51.9 | 6.2 | 57.8 |
| 17 | 78 58.3 | 04.1 | 211 53.3 | 7.3 | 22 58.1 | 6.1 | 57.8 |
| 18 | 93 58.4 | S12 05.0 | 226 19.6 | 7.3 | N23 04.2 | 5.9 | 57.8 |
| 19 | 108 58.5 | 05.8 | 240 45.9 | 7.2 | 23 10.1 | 5.8 | 57.8 |
| 20 | 123 58.6 | 06.7 | 255 12.1 | 7.1 | 23 15.9 | 5.6 | 57.9 |
| 21 | 138 58.6 | ·· 07.5 | 269 38.2 | 7.0 | 23 21.5 | 5.5 | 57.9 |
| 22 | 153 58.7 | 08.4 | 284 04.2 | 6.9 | 23 27.0 | 5.4 | 57.9 |
| 23 | 168 58.8 | 09.3 | 298 30.1 | 6.9 | 23 32.4 | 5.3 | 57.9 |
| 26 00 | 183 58.8 | S12 10.1 | 312 56.0 | 6.8 | N23 37.7 | 5.0 | 57.9 |
| 01 | 198 58.9 | 11.0 | 327 21.8 | 6.8 | 23 42.7 | 5.0 | 58.0 |
| 02 | 213 59.0 | 11.8 | 341 47.6 | 6.7 | 23 47.7 | 4.8 | 58.0 |
| 03 | 228 59.1 | ·· 12.7 | 356 13.3 | 6.6 | 23 52.5 | 4.7 | 58.0 |
| 04 | 243 59.1 | 13.6 | 10 38.9 | 6.6 | 23 57.2 | 4.5 | 58.0 |
| 05 | 258 59.2 | 14.4 | 25 04.5 | 6.5 | 24 01.7 | 4.4 | 58.0 |
| 06 | 273 59.3 | S12 15.3 | 39 30.0 | 6.4 | N24 06.1 | 4.2 | 58.0 |
| W 07 | 288 59.3 | 16.1 | 53 55.4 | 6.4 | 24 10.3 | 4.1 | 58.1 |
| E 08 | 303 59.4 | 17.0 | 68 20.8 | 6.3 | 24 14.4 | 4.0 | 58.1 |
| D 09 | 318 59.5 | ·· 17.9 | 82 46.1 | 6.2 | 24 18.4 | 3.8 | 58.1 |
| N 10 | 333 59.5 | 18.7 | 97 11.3 | 6.2 | 24 22.2 | 3.6 | 58.1 |
| E 11 | 348 59.6 | 19.6 | 111 36.5 | 6.2 | 24 25.8 | 3.5 | 58.1 |
| S 12 | 3 59.7 | S12 20.4 | 126 01.7 | 6.1 | N24 29.3 | 3.3 | 58.1 |
| D 13 | 18 59.7 | 21.3 | 140 26.8 | 6.0 | 24 32.6 | 3.2 | 58.2 |
| A 14 | 33 59.8 | 22.1 | 154 51.8 | 6.0 | 24 35.8 | 3.1 | 58.2 |
| Y 15 | 48 59.9 | ·· 23.0 | 169 16.8 | 6.0 | 24 38.9 | 2.8 | 58.2 |
| 16 | 63 59.9 | 23.8 | 183 41.8 | 5.9 | 24 41.7 | 2.8 | 58.2 |
| 17 | 79 00.0 | 24.7 | 198 06.7 | 5.8 | 24 44.5 | 2.6 | 58.2 |
| 18 | 94 00.1 | S12 25.6 | 212 31.5 | 5.8 | N24 47.1 | 2.4 | 58.2 |
| 19 | 109 00.1 | 26.4 | 226 56.3 | 5.8 | 24 49.5 | 2.2 | 58.3 |
| 20 | 124 00.2 | 27.3 | 241 21.1 | 5.7 | 24 51.7 | 2.1 | 58.3 |
| 21 | 139 00.3 | ·· 28.1 | 255 45.8 | 5.7 | 24 53.8 | 2.0 | 58.3 |
| 22 | 154 00.3 | 29.0 | 270 10.5 | 5.6 | 24 55.8 | 1.8 | 58.3 |
| 23 | 169 00.4 | 29.8 | 284 35.1 | 5.6 | 24 57.6 | 1.6 | 58.3 |
| 27 00 | 184 00.5 | S12 30.7 | 298 59.7 | 5.6 | N24 59.2 | 1.5 | 58.3 |
| 01 | 199 00.5 | 31.5 | 313 24.3 | 5.5 | 25 00.7 | 1.3 | 58.4 |
| 02 | 214 00.6 | 32.4 | 327 48.8 | 5.5 | 25 02.0 | 1.2 | 58.4 |
| 03 | 229 00.6 | ·· 33.2 | 342 13.3 | 5.5 | 25 03.2 | 1.0 | 58.4 |
| 04 | 244 00.7 | 34.1 | 356 37.8 | 5.4 | 25 04.2 | 0.8 | 58.4 |
| 05 | 259 00.8 | 34.9 | 11 02.2 | 5.5 | 25 05.0 | 0.7 | 58.4 |
| 06 | 274 00.8 | S12 35.8 | 25 26.7 | 5.4 | N25 05.7 | 0.5 | 58.4 |
| 07 | 289 00.9 | 36.6 | 39 51.1 | 5.3 | 25 06.2 | 0.3 | 58.5 |
| T 08 | 304 01.0 | 37.5 | 54 15.4 | 5.4 | 25 06.5 | 0.2 | 58.5 |
| H 09 | 319 01.0 | ·· 38.3 | 68 39.8 | 5.3 | 25 06.7 | 0.1 | 58.5 |
| U 10 | 334 01.1 | 39.2 | 83 04.1 | 5.3 | 25 06.8 | 0.2 | 58.5 |
| R 11 | 349 01.1 | 40.0 | 97 28.4 | 5.3 | 25 06.6 | 0.3 | 58.5 |
| S 12 | 4 01.2 | S12 40.9 | 111 52.7 | 5.3 | N25 06.3 | 0.4 | 58.5 |
| D 13 | 19 01.3 | 41.7 | 126 17.0 | 5.2 | 25 05.9 | 0.7 | 58.6 |
| A 14 | 34 01.3 | 42.6 | 140 41.2 | 5.3 | 25 05.2 | 0.7 | 58.6 |
| Y 15 | 49 01.4 | ·· 43.4 | 155 05.5 | 5.2 | 25 04.5 | 1.0 | 58.6 |
| 16 | 64 01.4 | 44.3 | 169 29.7 | 5.3 | 25 03.5 | 1.1 | 58.6 |
| 17 | 79 01.5 | 45.1 | 183 54.0 | 5.2 | 25 02.4 | 1.3 | 58.6 |
| 18 | 94 01.6 | S12 46.0 | 198 18.2 | 5.2 | N25 01.1 | 1.4 | 58.6 |
| 19 | 109 01.6 | 46.8 | 212 42.4 | 5.3 | 24 59.7 | 1.6 | 58.7 |
| 20 | 124 01.7 | 47.7 | 227 06.7 | 5.2 | 24 58.1 | 1.8 | 58.7 |
| 21 | 139 01.7 | ·· 48.5 | 241 30.9 | 5.2 | 24 56.3 | 1.9 | 58.7 |
| 22 | 154 01.8 | 49.3 | 255 55.1 | 5.2 | 24 54.4 | 2.1 | 58.7 |
| 23 | 169 01.8 | 50.2 | 270 19.3 | 5.3 | 24 52.3 | 2.3 | 58.7 |
| | S.D. 16.1 | d 0.9 | S.D. 15.7 | | 15.8 | | 16.0 |

| Lat. ° | Twilight Naut. h m | Twilight Civil h m | Sunrise h m | Moonrise 25 h m | Moonrise 26 h m | Moonrise 27 h m | Moonrise 28 h m |
|---|---|---|---|---|---|---|---|
| N 72 | 05 39 | 06 59 | 08 18 | ▭ | ▭ | ▭ | ▭ |
| N 70 | 05 38 | 06 50 | 07 59 | ▭ | ▭ | ▭ | ▭ |
| 68 | 05 37 | 06 43 | 07 44 | ▭ | ▭ | ▭ | ▭ |
| 66 | 05 36 | 06 36 | 07 32 | 16 09 | ▭ | ▭ | 18 35 |
| 64 | 05 35 | 06 31 | 07 21 | 16 54 | 17 08 | 17 59 | 19 38 |
| 62 | 05 34 | 06 26 | 07 13 | 17 23 | 17 51 | 18 46 | 20 12 |
| 60 | 05 33 | 06 21 | 07 05 | 17 46 | 18 20 | 19 17 | 20 37 |
| N 58 | 05 32 | 06 18 | 06 59 | 18 04 | 18 42 | 19 40 | 20 57 |
| 56 | 05 31 | 06 14 | 06 53 | 18 20 | 19 01 | 19 58 | 21 13 |
| 54 | 05 30 | 06 11 | 06 48 | 18 33 | 19 16 | 20 14 | 21 27 |
| 52 | 05 29 | 06 08 | 06 43 | 18 45 | 19 29 | 20 28 | 21 39 |
| 50 | 05 27 | 06 05 | 06 39 | 18 55 | 19 41 | 20 40 | 21 50 |
| 45 | 05 25 | 05 59 | 06 29 | 19 17 | 20 05 | 21 04 | 22 13 |
| N 40 | 05 22 | 05 54 | 06 22 | 19 35 | 20 25 | 21 24 | 22 31 |
| 35 | 05 19 | 05 49 | 06 15 | 19 50 | 20 41 | 21 40 | 22 46 |
| 30 | 05 16 | 05 44 | 06 09 | 20 02 | 20 55 | 21 55 | 22 59 |
| 20 | 05 10 | 05 36 | 05 59 | 20 25 | 21 19 | 22 19 | 23 21 |
| N 10 | 05 03 | 05 28 | 05 49 | 20 44 | 21 40 | 22 40 | 23 41 |
| 0 | 04 55 | 05 19 | 05 41 | 21 02 | 22 00 | 22 59 | 23 59 |
| S 10 | 04 45 | 05 10 | 05 32 | 21 20 | 22 20 | 23 19 | 24 17 |
| 20 | 04 33 | 05 00 | 05 22 | 21 40 | 22 41 | 23 40 | 24 36 |
| 30 | 04 17 | 04 46 | 05 11 | 22 03 | 23 05 | 24 04 | 00 04 |
| 35 | 04 07 | 04 38 | 05 05 | 22 16 | 23 19 | 24 18 | 00 18 |
| 40 | 03 55 | 04 29 | 04 58 | 22 31 | 23 36 | 24 35 | 00 35 |
| 45 | 03 39 | 04 18 | 04 49 | 22 50 | 23 56 | 24 54 | 00 54 |
| S 50 | 03 20 | 04 04 | 04 39 | 23 13 | 24 21 | 00 21 | 01 19 |
| 52 | 03 10 | 03 57 | 04 34 | 23 24 | 24 33 | 00 33 | 01 31 |
| 54 | 02 59 | 03 49 | 04 29 | 23 36 | 24 47 | 00 47 | 01 44 |
| 56 | 02 46 | 03 41 | 04 23 | 23 51 | 25 03 | 01 03 | 02 00 |
| 58 | 02 31 | 03 31 | 04 16 | 24 08 | 00 08 | 01 22 | 02 18 |
| S 60 | 02 12 | 03 20 | 04 08 | 24 29 | 00 29 | 01 46 | 02 41 |

| Lat. ° | Sunset h m | Twilight Civil h m | Twilight Naut. h m | Moonset 25 h m | Moonset 26 h m | Moonset 27 h m | Moonset 28 h m |
|---|---|---|---|---|---|---|---|
| N 72 | 15 09 | 16 27 | 17 47 | ▭ | ▭ | ▭ | ▭ |
| N 70 | 15 28 | 16 36 | 17 48 | ▭ | ▭ | ▭ | ▭ |
| 68 | 15 43 | 16 44 | 17 49 | ▭ | ▭ | ▭ | ▭ |
| 66 | 15 55 | 16 51 | 17 51 | 13 16 | ▭ | ▭ | 16 59 |
| 64 | 16 05 | 16 56 | 17 52 | 12 32 | 14 17 | 15 30 | 15 56 |
| 62 | 16 14 | 17 01 | 17 53 | 12 03 | 13 34 | 14 43 | 15 21 |
| 60 | 16 22 | 17 06 | 17 54 | 11 41 | 13 06 | 14 12 | 14 56 |
| N 58 | 16 28 | 17 09 | 17 55 | 11 23 | 12 44 | 13 49 | 14 36 |
| 56 | 16 34 | 17 13 | 17 56 | 11 08 | 12 26 | 13 30 | 14 19 |
| 54 | 16 40 | 17 16 | 17 58 | 10 55 | 12 10 | 13 15 | 14 04 |
| 52 | 16 44 | 17 19 | 17 59 | 10 44 | 11 57 | 13 01 | 13 52 |
| 50 | 16 49 | 17 22 | 18 00 | 10 34 | 11 46 | 12 49 | 13 41 |
| 45 | 16 58 | 17 28 | 18 03 | 10 13 | 11 22 | 12 24 | 13 18 |
| N 40 | 17 06 | 17 34 | 18 05 | 09 56 | 11 02 | 12 04 | 12 59 |
| 35 | 17 13 | 17 39 | 18 08 | 09 42 | 10 46 | 11 48 | 12 44 |
| 30 | 17 19 | 17 43 | 18 11 | 09 29 | 10 33 | 11 33 | 12 30 |
| 20 | 17 29 | 17 52 | 18 18 | 09 08 | 10 09 | 11 09 | 12 07 |
| N 10 | 17 39 | 18 00 | 18 25 | 08 50 | 09 48 | 10 48 | 11 47 |
| 0 | 17 47 | 18 09 | 18 33 | 08 33 | 09 29 | 10 28 | 11 28 |
| S 10 | 17 56 | 18 18 | 18 43 | 08 16 | 09 10 | 10 08 | 11 09 |
| 20 | 18 06 | 18 29 | 18 55 | 07 58 | 08 50 | 09 47 | 10 48 |
| 30 | 18 17 | 18 42 | 19 12 | 07 37 | 08 27 | 09 23 | 10 25 |
| 35 | 18 24 | 18 50 | 19 22 | 07 25 | 08 13 | 09 08 | 10 11 |
| 40 | 18 31 | 19 00 | 19 34 | 07 11 | 07 57 | 08 51 | 09 55 |
| 45 | 18 40 | 19 11 | 19 50 | 06 55 | 07 38 | 08 31 | 09 35 |
| S 50 | 18 50 | 19 26 | 20 10 | 06 35 | 07 14 | 08 06 | 09 11 |
| 52 | 18 55 | 19 33 | 20 20 | 06 25 | 07 03 | 07 54 | 08 59 |
| 54 | 19 01 | 19 40 | 20 31 | 06 14 | 06 50 | 07 40 | 08 46 |
| 56 | 19 07 | 19 49 | 20 44 | 06 02 | 06 35 | 07 24 | 08 31 |
| 58 | 19 14 | 19 59 | 21 00 | 05 48 | 06 18 | 07 05 | 08 12 |
| S 60 | 19 21 | 20 11 | 21 20 | 05 32 | 05 57 | 06 41 | 07 50 |

| Day | SUN Eqn. of Time 00ʰ m s | SUN Eqn. of Time 12ʰ m s | SUN Mer. Pass. h m | MOON Mer. Pass. Upper h m | MOON Mer. Pass. Lower h m | Age d | Phase |
|---|---|---|---|---|---|---|---|
| 25 | 15 48 | 15 52 | 11 44 | 02 20 | 14 48 | 19 | |
| 26 | 15 55 | 15 59 | 11 44 | 03 16 | 15 45 | 20 | ◑ |
| 27 | 16 02 | 16 05 | 11 44 | 04 14 | 16 44 | 21 | |

| | G.M.T. | SUN G.H.A. | SUN Dec. | MOON G.H.A. | *v* | MOON Dec. | *d* | H.P. |
|---|---|---|---|---|---|---|---|---|
| | d h | ° ′ | ° ′ | ° ′ | ′ | ° ′ | ′ | ′ |
| | 9 00 | 184 03.4 | S16 37.9 | 131 33.4 | 8.9 | S25 03.5 | 1.4 | 55.7 |
| | 01 | 199 03.4 | 38.6 | 146 01.3 | 9.0 | 25 04.9 | 1.1 | 55.6 |
| | 02 | 214 03.4 | 39.3 | 160 29.3 | 9.1 | 25 06.0 | 1.1 | 55.6 |
| | 03 | 229 03.3 | ·· 40.0 | 174 57.4 | 9.0 | 25 07.1 | 0.9 | 55.6 |
| | 04 | 244 03.3 | 40.8 | 189 25.4 | 9.1 | 25 08.0 | 0.8 | 55.6 |
| | 05 | 259 03.2 | 41.5 | 203 53.5 | 9.1 | 25 08.8 | 0.6 | 55.5 |
| | 06 | 274 03.2 | S16 42.2 | 218 21.6 | 9.1 | S25 09.4 | 0.5 | 55.5 |
| W | 07 | 289 03.1 | 42.9 | 232 49.7 | 9.2 | 25 09.9 | 0.4 | 55.5 |
| E | 08 | 304 03.1 | 43.6 | 247 17.9 | 9.2 | 25 10.3 | 0.3 | 55.4 |
| D | 09 | 319 03.0 | ·· 44.4 | 261 46.1 | 9.2 | 25 10.6 | 0.1 | 55.4 |
| N | 10 | 334 03.0 | 45.1 | 276 14.3 | 9.3 | 25 10.7 | 0.0 | 55.4 |
| E | 11 | 349 02.9 | 45.8 | 290 42.6 | 9.3 | 25 10.7 | 0.2 | 55.4 |
| S | 12 | 4 02.9 | S16 46.5 | 305 10.9 | 9.3 | S25 10.5 | 0.3 | 55.4 |
| D | 13 | 19 02.8 | 47.2 | 319 39.2 | 9.4 | 25 10.2 | 0.4 | 55.3 |
| A | 14 | 34 02.8 | 48.0 | 334 07.6 | 9.4 | 25 09.8 | 0.5 | 55.3 |
| Y | 15 | 49 02.7 | ·· 48.7 | 348 36.0 | 9.4 | 25 09.3 | 0.7 | 55.3 |
| | 16 | 64 02.7 | 49.4 | 3 04.4 | 9.5 | 25 08.6 | 0.7 | 55.3 |
| | 17 | 79 02.6 | 50.1 | 17 32.9 | 9.5 | 25 07.9 | 1.0 | 55.2 |
| | 18 | 94 02.6 | S16 50.8 | 32 01.4 | 9.6 | S25 06.9 | 1.0 | 55.2 |
| | 19 | 109 02.5 | 51.5 | 46 30.0 | 9.6 | 25 05.9 | 1.2 | 55.2 |
| | 20 | 124 02.5 | 52.2 | 60 58.6 | 9.6 | 25 04.7 | 1.3 | 55.2 |
| | 21 | 139 02.4 | ·· 53.0 | 75 27.2 | 9.7 | 25 03.4 | 1.4 | 55.1 |
| | 22 | 154 02.4 | 53.7 | 89 55.9 | 9.7 | 25 02.0 | 1.6 | 55.1 |
| | 23 | 169 02.3 | 54.4 | 104 24.6 | 9.8 | 25 00.4 | 1.6 | 55.1 |
| | 10 00 | 184 02.2 | S16 55.1 | 118 53.4 | 9.8 | S24 58.8 | 1.8 | 55.1 |
| | 01 | 199 02.2 | 55.8 | 133 22.2 | 9.8 | 24 57.0 | 2.0 | 55.1 |
| | 02 | 214 02.1 | 56.5 | 147 51.0 | 9.9 | 24 55.0 | 2.0 | 55.0 |
| | 03 | 229 02.1 | ·· 57.2 | 162 19.9 | 10.0 | 24 53.0 | 2.2 | 55.0 |
| | 04 | 244 02.0 | 57.9 | 176 48.9 | 10.0 | 24 50.8 | 2.3 | 55.0 |
| | 05 | 259 02.0 | 58.7 | 191 17.9 | 10.0 | 24 48.5 | 2.4 | 55.0 |
| | 06 | 274 01.9 | S16 59.4 | 205 46.9 | 10.1 | S24 46.1 | 2.5 | 55.0 |
| | 07 | 289 01.9 | 17 00.1 | 220 16.0 | 10.2 | 24 43.6 | 2.7 | 54.9 |
| T | 08 | 304 01.8 | 00.8 | 234 45.2 | 10.2 | 24 40.9 | 2.8 | 54.9 |
| H | 09 | 319 01.7 | ·· 01.5 | 249 14.4 | 10.2 | 24 38.1 | 2.9 | 54.9 |
| U | 10 | 334 01.7 | 02.2 | 263 43.6 | 10.3 | 24 35.2 | 3.0 | 54.9 |
| R | 11 | 349 01.6 | 02.9 | 278 12.9 | 10.4 | 24 32.2 | 3.1 | 54.9 |
| S | 12 | 4 01.6 | S17 03.6 | 292 42.3 | 10.4 | S24 29.1 | 3.3 | 54.8 |
| D | 13 | 19 01.5 | 04.3 | 307 11.7 | 10.4 | 24 25.8 | 3.4 | 54.8 |
| A | 14 | 34 01.4 | 05.0 | 321 41.1 | 10.5 | 24 22.4 | 3.5 | 54.8 |
| Y | 15 | 49 01.4 | ·· 05.7 | 336 10.6 | 10.6 | 24 18.9 | 3.6 | 54.8 |
| | 16 | 64 01.3 | 06.4 | 350 40.2 | 10.6 | 24 15.3 | 3.7 | 54.8 |
| | 17 | 79 01.3 | 07.1 | 5 09.8 | 10.7 | 24 11.6 | 3.8 | 54.7 |
| | 18 | 94 01.2 | S17 07.8 | 19 39.5 | 10.7 | S24 07.8 | 4.0 | 54.7 |
| | 19 | 109 01.1 | 08.5 | 34 09.2 | 10.8 | 24 03.8 | 4.0 | 54.7 |
| | 20 | 124 01.1 | 09.2 | 48 39.0 | 10.8 | 23 59.8 | 4.2 | 54.7 |
| | 21 | 139 01.0 | ·· 10.0 | 63 08.8 | 10.9 | 23 55.6 | 4.3 | 54.7 |
| | 22 | 154 01.0 | 10.7 | 77 38.7 | 11.0 | 23 51.3 | 4.4 | 54.7 |
| | 23 | 169 00.9 | 11.4 | 92 08.7 | 11.0 | 23 46.9 | 4.5 | 54.6 |
| | 11 00 | 184 00.8 | S17 12.1 | 106 38.7 | 11.1 | S23 42.4 | 4.6 | 54.6 |
| | 01 | 199 00.8 | 12.8 | 121 08.8 | 11.1 | 23 37.8 | 4.7 | 54.6 |
| | 02 | 214 00.7 | 13.5 | 135 38.9 | 11.2 | 23 33.1 | 4.8 | 54.6 |
| | 03 | 229 00.6 | ·· 14.2 | 150 09.1 | 11.2 | 23 28.3 | 5.0 | 54.6 |
| | 04 | 244 00.6 | 14.9 | 164 39.3 | 11.3 | 23 23.3 | 5.0 | 54.6 |
| | 05 | 259 00.5 | 15.5 | 179 09.6 | 11.4 | 23 18.3 | 5.2 | 54.6 |
| | 06 | 274 00.5 | S17 16.2 | 193 40.0 | 11.4 | S23 13.1 | 5.2 | 54.5 |
| | 07 | 289 00.4 | 16.9 | 208 10.4 | 11.5 | 23 07.9 | 5.4 | 54.5 |
| | 08 | 304 00.3 | 17.6 | 222 40.9 | 11.5 | 23 02.5 | 5.5 | 54.5 |
| F | 09 | 319 00.3 | ·· 18.3 | 237 11.4 | 11.6 | 22 57.0 | 5.5 | 54.5 |
| R | 10 | 334 00.2 | 19.0 | 251 42.0 | 11.7 | 22 51.5 | 5.7 | 54.5 |
| I | 11 | 349 00.1 | 19.7 | 266 12.7 | 11.7 | 22 45.8 | 5.8 | 54.5 |
| D | 12 | 4 00.1 | S17 20.4 | 280 43.4 | 11.8 | S22 40.0 | 5.9 | 54.5 |
| A | 13 | 19 00.0 | 21.1 | 295 14.2 | 11.8 | 22 34.1 | 5.9 | 54.5 |
| Y | 14 | 33 59.9 | 21.8 | 309 45.0 | 11.9 | 22 28.2 | 6.1 | 54.4 |
| | 15 | 48 59.9 | ·· 22.5 | 324 15.9 | 12.0 | 22 22.1 | 6.2 | 54.4 |
| | 16 | 63 59.8 | 23.2 | 338 46.9 | 12.0 | 22 15.9 | 6.3 | 54.4 |
| | 17 | 78 59.7 | 23.9 | 353 17.9 | 12.1 | 22 09.6 | 6.3 | 54.4 |
| | 18 | 93 59.6 | S17 24.6 | 7 49.0 | 12.1 | S22 03.3 | 6.5 | 54.4 |
| | 19 | 108 59.6 | 25.3 | 22 20.1 | 12.2 | 21 56.8 | 6.6 | 54.4 |
| | 20 | 123 59.5 | 26.0 | 36 51.3 | 12.3 | 21 50.2 | 6.6 | 54.4 |
| | 21 | 138 59.4 | ·· 26.6 | 51 22.6 | 12.3 | 21 43.6 | 6.8 | 54.4 |
| | 22 | 153 59.4 | 27.3 | 65 53.9 | 12.4 | 21 36.8 | 6.8 | 54.4 |
| | 23 | 168 59.3 | 28.0 | 80 25.3 | 12.5 | 21 30.0 | 7.0 | 54.4 |
| | | S.D. 16.2 | *d* 0.7 | S.D. 15.1 | | 14.9 | | 14.8 |

| Lat. | Twilight Naut. | Twilight Civil | Sunrise | Moonrise 9 | Moonrise 10 | Moonrise 11 | Moonrise 12 |
|---|---|---|---|---|---|---|---|
| ° | h m | h m | h m | h m | h m | h m | h m |
| N 72 | 06 39 | 08 08 | 09 57 | ▬ | ▬ | ▬ | ▬ |
| N 70 | 06 31 | 07 49 | 09 15 | ▬ | ▬ | ▬ | 17 07 |
| 68 | 06 24 | 07 35 | 08 47 | ▬ | ▬ | ▬ | 16 13 |
| 66 | 06 19 | 07 23 | 08 26 | ▬ | ▬ | 15 55 | 15 40 |
| 64 | 06 14 | 07 13 | 08 09 | 14 47 | 15 10 | 15 15 | 15 15 |
| 62 | 06 10 | 07 04 | 07 55 | 13 54 | 14 29 | 14 47 | 14 56 |
| 60 | 06 06 | 06 56 | 07 44 | 13 22 | 14 01 | 14 26 | 14 40 |
| N 58 | 06 02 | 06 50 | 07 34 | 12 58 | 13 40 | 14 08 | 14 27 |
| 56 | 05 59 | 06 44 | 07 25 | 12 38 | 13 22 | 13 53 | 14 15 |
| 54 | 05 55 | 06 38 | 07 17 | 12 22 | 13 07 | 13 40 | 14 05 |
| 52 | 05 53 | 06 33 | 07 10 | 12 08 | 12 54 | 13 29 | 13 56 |
| 50 | 05 50 | 06 29 | 07 04 | 11 56 | 12 43 | 13 19 | 13 47 |
| 45 | 05 43 | 06 19 | 06 50 | 11 31 | 12 19 | 12 58 | 13 30 |
| N 40 | 05 38 | 06 10 | 06 39 | 11 11 | 12 00 | 12 41 | 13 16 |
| 35 | 05 32 | 06 02 | 06 29 | 10 54 | 11 44 | 12 26 | 13 03 |
| 30 | 05 27 | 05 55 | 06 20 | 10 40 | 11 30 | 12 14 | 12 53 |
| 20 | 05 16 | 05 43 | 06 06 | 10 15 | 11 06 | 11 52 | 12 34 |
| N 10 | 05 05 | 05 31 | 05 53 | 09 54 | 10 46 | 11 34 | 12 18 |
| 0 | 04 54 | 05 19 | 05 40 | 09 35 | 10 26 | 11 16 | 12 03 |
| S 10 | 04 40 | 05 06 | 05 28 | 09 15 | 10 07 | 10 59 | 11 48 |
| 20 | 04 24 | 04 51 | 05 15 | 08 54 | 09 47 | 10 40 | 11 32 |
| 30 | 04 03 | 04 33 | 04 59 | 08 29 | 09 23 | 10 18 | 11 13 |
| 35 | 03 49 | 04 23 | 04 50 | 08 15 | 09 09 | 10 05 | 11 02 |
| 40 | 03 33 | 04 10 | 04 40 | 07 59 | 08 53 | 09 51 | 10 50 |
| 45 | 03 13 | 03 54 | 04 28 | 07 39 | 08 34 | 09 33 | 10 35 |
| S 50 | 02 46 | 03 35 | 04 13 | 07 14 | 08 09 | 09 11 | 10 17 |
| 52 | 02 32 | 03 25 | 04 06 | 07 02 | 07 58 | 09 01 | 10 08 |
| 54 | 02 15 | 03 15 | 03 58 | 06 48 | 07 44 | 08 49 | 09 58 |
| 56 | 01 55 | 03 02 | 03 49 | 06 32 | 07 29 | 08 35 | 09 47 |
| 58 | 01 29 | 02 48 | 03 39 | 06 13 | 07 11 | 08 20 | 09 35 |
| S 60 | 00 47 | 02 31 | 03 28 | 05 50 | 06 48 | 08 01 | 09 20 |

| Lat. | Sunset | Twilight Civil | Twilight Naut. | Moonset 9 | Moonset 10 | Moonset 11 | Moonset 12 |
|---|---|---|---|---|---|---|---|
| ° | h m | h m | h m | h m | h m | h m | h m |
| N 72 | 13 29 | 15 18 | 16 47 | ▬ | ▬ | ▬ | ▬ |
| N 70 | 14 11 | 15 37 | 16 55 | ▬ | ▬ | ▬ | 19 36 |
| 68 | 14 40 | 15 52 | 17 02 | ▬ | ▬ | ▬ | 20 29 |
| 66 | 15 01 | 16 04 | 17 08 | ▬ | ▬ | 19 10 | 21 01 |
| 64 | 15 18 | 16 14 | 17 13 | 16 49 | 18 13 | 19 49 | 21 25 |
| 62 | 15 31 | 16 23 | 17 17 | 17 42 | 18 53 | 20 17 | 21 43 |
| 60 | 15 43 | 16 31 | 17 21 | 18 14 | 19 20 | 20 38 | 21 59 |
| N 58 | 15 53 | 16 37 | 17 25 | 18 38 | 19 42 | 20 55 | 22 12 |
| 56 | 16 02 | 16 43 | 17 28 | 18 57 | 19 59 | 21 09 | 22 23 |
| 54 | 16 10 | 16 49 | 17 32 | 19 13 | 20 14 | 21 22 | 22 32 |
| 52 | 16 17 | 16 54 | 17 35 | 19 27 | 20 27 | 21 33 | 22 41 |
| 50 | 16 24 | 16 58 | 17 37 | 19 39 | 20 38 | 21 42 | 22 49 |
| 45 | 16 37 | 17 09 | 17 44 | 20 04 | 21 01 | 22 03 | 23 05 |
| N 40 | 16 49 | 17 17 | 17 50 | 20 24 | 21 20 | 22 19 | 23 19 |
| 35 | 16 58 | 17 25 | 17 55 | 20 40 | 21 36 | 22 33 | 23 30 |
| 30 | 17 07 | 17 32 | 18 01 | 20 55 | 21 49 | 22 45 | 23 40 |
| 20 | 17 22 | 17 45 | 18 11 | 21 19 | 22 12 | 23 05 | 23 57 |
| N 10 | 17 35 | 17 57 | 18 22 | 21 40 | 22 32 | 23 23 | 24 11 |
| 0 | 17 47 | 18 09 | 18 34 | 22 00 | 22 51 | 23 39 | 24 25 |
| S 10 | 18 00 | 18 22 | 18 48 | 22 19 | 23 09 | 23 56 | 24 39 |
| 20 | 18 13 | 18 37 | 19 04 | 22 40 | 23 29 | 24 13 | 00 13 |
| 30 | 18 29 | 18 55 | 19 26 | 23 04 | 23 51 | 24 33 | 00 33 |
| 35 | 18 38 | 19 06 | 19 39 | 23 19 | 24 05 | 00 05 | 00 44 |
| 40 | 18 49 | 19 19 | 19 55 | 23 35 | 24 20 | 00 20 | 00 58 |
| 45 | 19 01 | 19 34 | 20 16 | 23 55 | 24 38 | 00 38 | 01 13 |
| S 50 | 19 16 | 19 54 | 20 44 | 24 19 | 00 19 | 01 00 | 01 33 |
| 52 | 19 23 | 20 04 | 20 58 | 24 31 | 00 31 | 01 11 | 01 42 |
| 54 | 19 31 | 20 15 | 21 15 | 24 44 | 00 44 | 01 23 | 01 52 |
| 56 | 19 40 | 20 27 | 21 36 | 00 08 | 01 00 | 01 37 | 02 03 |
| 58 | 19 50 | 20 42 | 22 03 | 00 27 | 01 18 | 01 53 | 02 16 |
| S 60 | 20 02 | 21 00 | 22 49 | 00 51 | 01 41 | 02 12 | 02 32 |

| Day | SUN Eqn. of Time 00ʰ | SUN Eqn. of Time 12ʰ | SUN Mer. Pass. | MOON Mer. Pass. Upper | MOON Mer. Pass. Lower | Age | Phase |
|---|---|---|---|---|---|---|---|
| | m s | m s | h m | h m | h m | d | |
| 9 | 16 14 | 16 12 | 11 44 | 15 47 | 03 21 | 05 | |
| 10 | 16 09 | 16 06 | 11 44 | 16 39 | 04 13 | 06 | ◑ |
| 11 | 16 03 | 16 00 | 11 44 | 17 28 | 05 03 | 07 | |

# POLARIS (POLE STAR) TABLES, 1983

## FOR DETERMINING LATITUDE FROM SEXTANT ALTITUDE AND FOR AZIMUTH

| L.H.A. ARIES | 0°– 9° | 10°– 19° | 20°– 29° | 30°– 39° | 40°– 49° | 50°– 59° | 60°– 69° | 70°– 79° | 80°– 89° | 90°– 99° | 100°– 109° | 110°– 119° |
|---|---|---|---|---|---|---|---|---|---|---|---|---|
| | $a_0$ | $a_0$ | $a_0$ | $a_0$ | $a_0$ | $a_0$ | $a_0$ | $a_0$ | $a_0$ | $a_0$ | $a_0$ | $a_0$ |
| ° | ° ′ | ° ′ | ° ′ | ° ′ | ° ′ | ° ′ | ° ′ | ° ′ | ° ′ | ° ′ | ° ′ | ° ′ |
| 0 | 0 18.5 | 0 14.4 | 0 11.6 | 0 10.3 | 0 10.5 | 0 12.2 | 0 15.3 | 0 19.8 | 0 25.5 | 0 32.1 | 0 39.6 | 0 47.7 |
| 1 | 18.0 | 14.0 | 11.4 | 10.3 | 10.6 | 12.4 | 15.7 | 20.3 | 26.1 | 32.9 | 40.4 | 48.5 |
| 2 | 17.5 | 13.7 | 11.2 | 10.2 | 10.7 | 12.7 | 16.1 | 20.8 | 26.7 | 33.6 | 41.2 | 49.4 |
| 3 | 17.1 | 13.4 | 11.1 | 10.2 | 10.9 | 13.0 | 16.5 | 21.4 | 27.4 | 34.3 | 42.0 | 50.2 |
| 4 | 16.7 | 13.1 | 10.9 | 10.2 | 11.0 | 13.3 | 17.0 | 21.9 | 28.0 | 35.1 | 42.8 | 51.0 |
| 5 | 0 16.3 | 0 12.8 | 0 10.8 | 0 10.2 | 0 11.2 | 0 13.6 | 0 17.4 | 0 22.5 | 0 28.7 | 0 35.8 | 0 43.6 | 0 51.9 |
| 6 | 15.8 | 12.5 | 10.6 | 10.2 | 11.3 | 13.9 | 17.9 | 23.1 | 29.4 | 36.6 | 44.4 | 52.7 |
| 7 | 15.5 | 12.3 | 10.5 | 10.3 | 11.5 | 14.2 | 18.3 | 23.6 | 30.0 | 37.3 | 45.2 | 53.6 |
| 8 | 15.1 | 12.0 | 10.4 | 10.3 | 11.7 | 14.6 | 18.8 | 24.2 | 30.7 | 38.1 | 46.1 | 54.4 |
| 9 | 14.7 | 11.8 | 10.4 | 10.4 | 12.0 | 14.9 | 19.3 | 24.8 | 31.4 | 38.9 | 46.9 | 55.3 |
| 10 | 0 14.4 | 0 11.6 | 0 10.3 | 0 10.5 | 0 12.2 | 0 15.3 | 0 19.8 | 0 25.5 | 0 32.1 | 0 39.6 | 0 47.7 | 0 56.1 |
| Lat. | $a_1$ | $a_1$ | $a_1$ | $a_1$ | $a_1$ | $a_1$ | $a_1$ | $a_1$ | $a_1$ | $a_1$ | $a_1$ | $a_1$ |
| ° | ′ | ′ | ′ | ′ | ′ | ′ | ′ | ′ | ′ | ′ | ′ | ′ |
| 0 | 0.5 | 0.6 | 0.6 | 0.6 | 0.6 | 0.5 | 0.5 | 0.4 | 0.4 | 0.3 | 0.2 | 0.2 |
| 10 | .5 | .6 | .6 | .6 | .6 | .6 | .5 | .4 | .4 | .3 | .3 | .3 |
| 20 | .5 | .6 | .6 | .6 | .6 | .6 | .5 | .5 | .4 | .4 | .3 | .3 |
| 30 | .6 | .6 | .6 | .6 | .6 | .6 | .5 | .5 | .5 | .4 | .4 | .4 |
| 40 | 0.6 | 0.6 | 0.6 | 0.6 | 0.6 | 0.6 | 0.6 | 0.5 | 0.5 | 0.5 | 0.5 | 0.5 |
| 45 | .6 | .6 | .6 | .6 | .6 | .6 | .6 | .6 | .6 | .5 | .5 | .5 |
| 50 | .6 | .6 | .6 | .6 | .6 | .6 | .6 | .6 | .6 | .6 | .6 | .6 |
| 55 | .6 | .6 | .6 | .6 | .6 | .6 | .6 | .6 | .6 | .7 | .7 | .7 |
| 60 | .6 | .6 | .6 | .6 | .6 | .6 | .7 | .7 | .7 | .7 | .8 | .8 |
| 62 | 0.7 | 0.6 | 0.6 | 0.6 | 0.6 | 0.6 | 0.7 | 0.7 | 0.7 | 0.8 | 0.8 | 0.8 |
| 64 | .7 | .6 | .6 | .6 | .6 | .6 | .7 | .7 | .8 | .8 | .9 | 0.9 |
| 66 | .7 | .6 | .6 | .6 | .6 | .6 | .7 | .8 | .8 | .9 | 0.9 | 1.0 |
| 68 | 0.7 | 0.6 | 0.6 | 0.6 | 0.6 | 0.7 | 0.7 | 0.8 | 0.9 | 0.9 | 1.0 | 1.0 |
| Month | $a_2$ | $a_2$ | $a_2$ | $a_2$ | $a_2$ | $a_2$ | $a_2$ | $a_2$ | $a_2$ | $a_2$ | $a_2$ | $a_2$ |
| | ′ | ′ | ′ | ′ | ′ | ′ | ′ | ′ | ′ | ′ | ′ | ′ |
| **Jan.** | 0.7 | 0.7 | 0.7 | 0.7 | 0.7 | 0.7 | 0.7 | 0.7 | 0.7 | 0.7 | 0.6 | 0.6 |
| **Feb.** | .6 | .6 | .7 | .7 | .7 | .7 | .8 | .8 | .8 | .8 | .8 | .8 |
| **Mar.** | .5 | .5 | .6 | .6 | .7 | .7 | .8 | .8 | .8 | .9 | .9 | .9 |
| **Apr.** | 0.3 | .4 | .04 | 0.5 | 0.5 | 0.6 | 0.7 | 0.7 | 0.8 | 0.8 | 0.9 | 0.9 |
| **May** | .2 | .2 | .3 | .3 | .4 | .5 | .5 | .6 | .7 | .8 | .8 | .9 |
| **June** | .2 | .2 | .2 | .2 | .3 | .3 | .4 | .5 | .5 | .6 | .7 | .8 |
| **July** | 0.2 | 0.2 | 0.2 | 0.2 | 0.2 | 0.2 | 0.3 | 0.3 | 0.4 | 0.5 | 0.5 | 0.6 |
| **Aug.** | .3 | .3 | .3 | .2 | .2 | .2 | .2 | .3 | .3 | .3 | .4 | .4 |
| **Sept.** | .5 | .5 | .4 | .4 | .3 | .3 | .3 | .3 | .3 | .3 | .3 | .3 |
| **Oct.** | 0.7 | 0.6 | 0.6 | 0.5 | 0.5 | 0.4 | 0.4 | 0.3 | 0.3 | 0.3 | 0.3 | 0.3 |
| **Nov.** | 0.9 | 0.8 | .8 | .7 | .7 | .6 | .5 | .5 | .4 | .4 | .3 | .3 |
| **Dec.** | 1.0 | 1.0 | 0.9 | 0.9 | 0.8 | 0.8 | 0.7 | 0.6 | 0.6 | 0.5 | 0.4 | 0.4 |
| Lat. | AZIMUTH | | | | | | | | | | | |
| ° | ° | ° | ° | ° | ° | ° | ° | ° | ° | ° | ° | ° |
| 0 | 0.4 | 0.3 | 0.1 | 0.0 | 359.8 | 359.7 | 359.6 | 359.5 | 359.4 | 359.3 | 359.2 | 359.2 |
| 20 | 0.4 | 0.3 | 0.1 | 0.0 | 359.8 | 359.7 | 359.5 | 359.4 | 359.3 | 359.2 | 359.2 | 359.1 |
| 40 | 0.5 | 0.3 | 0.2 | 0.0 | 359.8 | 359.6 | 359.4 | 359.3 | 359.2 | 359.1 | 359.0 | 359.0 |
| 50 | 0.6 | 0.4 | 0.2 | 0.0 | 359.7 | 359.5 | 359.3 | 359.2 | 359.0 | 358.9 | 358.8 | 358.8 |
| 55 | 0.7 | 0.5 | 0.2 | 0.0 | 359.7 | 359.5 | 359.3 | 359.1 | 358.9 | 358.7 | 358.7 | 358.6 |
| 60 | 0.8 | 0.5 | 0.3 | 0.0 | 359.7 | 359.4 | 359.1 | 358.9 | 358.7 | 358.6 | 358.5 | 358.4 |
| 65 | 0.9 | 0.6 | 0.3 | 0.0 | 359.6 | 359.3 | 359.0 | 358.7 | 358.5 | 358.3 | 358.2 | 358.1 |

Latitude = Apparent altitude (corrected for refraction) $- 1° + a_0 + a_1 + a_2$

The table is entered with L.H.A. Aries to determine the column to be used; each column refers to a range of 10°. $a_0$ is taken, with mental interpolation, from the upper table with the units of L.H.A. Aries in degrees as argument; $a_1$, $a_2$ are taken, without interpolation, from the second and third tables with arguments latitude and month respectively. $a_0$, $a_1$, $a_2$ are always positive. The final table gives the azimuth of *Polaris.*

# CONVERSION OF ARC TO TIME

| 0°–59° | | 60°–119° | | 120°–179° | | 180°–239° | | 240°–299° | | 300°–359° | | | 0′·00 | 0′·25 | 0′·50 | 0′·75 |
|---|---|---|---|---|---|---|---|---|---|---|---|---|---|---|---|---|
| ° | h m | ° | h m | ° | h m | ° | h m | ° | h m | ° | h m | ′ | m s | m s | m s | m s |
| 0 | 0 00 | 60 | 4 00 | 120 | 8 00 | 180 | 12 00 | 240 | 16 00 | 300 | 20 00 | 0 | 0 00 | 0 01 | 0 02 | 0 03 |
| 1 | 0 04 | 61 | 4 04 | 121 | 8 04 | 181 | 12 04 | 241 | 16 04 | 301 | 20 04 | 1 | 0 04 | 0 05 | 0 06 | 0 07 |
| 2 | 0 08 | 62 | 4 08 | 122 | 8 08 | 182 | 12 08 | 242 | 16 08 | 302 | 20 08 | 2 | 0 08 | 0 09 | 0 10 | 0 11 |
| 3 | 0 12 | 63 | 4 12 | 123 | 8 12 | 183 | 12 12 | 243 | 16 12 | 303 | 20 12 | 3 | 0 12 | 0 13 | 0 14 | 0 15 |
| 4 | 0 16 | 64 | 4 16 | 124 | 8 16 | 184 | 12 16 | 244 | 16 16 | 304 | 20 16 | 4 | 0 16 | 0 17 | 0 18 | 0 19 |
| 5 | 0 20 | 65 | 4 20 | 125 | 8 20 | 185 | 12 20 | 245 | 16 20 | 305 | 20 20 | 5 | 0 20 | 0 21 | 0 22 | 0 23 |
| 6 | 0 24 | 66 | 4 24 | 126 | 8 24 | 186 | 12 24 | 246 | 16 24 | 306 | 20 24 | 6 | 0 24 | 0 25 | 0 26 | 0 27 |
| 7 | 0 28 | 67 | 4 28 | 127 | 8 28 | 187 | 12 28 | 247 | 16 28 | 307 | 20 28 | 7 | 0 28 | 0 29 | 0 30 | 0 31 |
| 8 | 0 32 | 68 | 4 32 | 128 | 8 32 | 188 | 12 32 | 248 | 16 32 | 308 | 20 32 | 8 | 0 32 | 0 33 | 0 34 | 0 35 |
| 9 | 0 36 | 69 | 4 36 | 129 | 8 36 | 189 | 12 36 | 249 | 16 36 | 309 | 20 36 | 9 | 0 36 | 0 37 | 0 38 | 0 39 |
| 10 | 0 40 | 70 | 4 40 | 130 | 8 40 | 190 | 12 40 | 250 | 16 40 | 310 | 20 40 | 10 | 0 40 | 0 41 | 0 42 | 0 43 |
| 11 | 0 44 | 71 | 4 44 | 131 | 8 44 | 191 | 12 44 | 251 | 16 44 | 311 | 20 44 | 11 | 0 44 | 0 45 | 0 46 | 0 47 |
| 12 | 0 48 | 72 | 4 48 | 132 | 8 48 | 192 | 12 48 | 252 | 16 48 | 312 | 20 48 | 12 | 0 48 | 0 49 | 0 50 | 0 51 |
| 13 | 0 52 | 73 | 4 52 | 133 | 8 52 | 193 | 12 52 | 253 | 16 52 | 313 | 20 52 | 13 | 0 52 | 0 53 | 0 54 | 0 55 |
| 14 | 0 56 | 74 | 4 56 | 134 | 8 56 | 194 | 12 56 | 254 | 16 56 | 314 | 20 56 | 14 | 0 56 | 0 57 | 0 58 | 0 59 |
| 15 | 1 00 | 75 | 5 00 | 135 | 9 00 | 195 | 13 00 | 255 | 17 00 | 315 | 21 00 | 15 | 1 00 | 1 01 | 1 02 | 1 03 |
| 16 | 1 04 | 76 | 5 04 | 136 | 9 04 | 196 | 13 04 | 256 | 17 04 | 316 | 21 04 | 16 | 1 04 | 1 05 | 1 06 | 1 07 |
| 17 | 1 08 | 77 | 5 08 | 137 | 9 08 | 197 | 13 08 | 257 | 17 08 | 317 | 21 08 | 17 | 1 08 | 1 09 | 1 10 | 1 11 |
| 18 | 1 12 | 78 | 5 12 | 138 | 9 12 | 198 | 13 12 | 258 | 17 12 | 318 | 21 12 | 18 | 1 12 | 1 13 | 1 14 | 1 15 |
| 19 | 1 16 | 79 | 5 16 | 139 | 9 16 | 199 | 13 16 | 259 | 17 16 | 319 | 21 16 | 19 | 1 16 | 1 17 | 1 18 | 1 19 |
| 20 | 1 20 | 80 | 5 20 | 140 | 9 20 | 200 | 13 20 | 260 | 17 20 | 320 | 21 20 | 20 | 1 20 | 1 21 | 1 22 | 1 23 |
| 21 | 1 24 | 81 | 5 24 | 141 | 9 24 | 201 | 13 24 | 261 | 17 24 | 321 | 21 24 | 21 | 1 24 | 1 25 | 1 26 | 1 27 |
| 22 | 1 28 | 82 | 5 28 | 142 | 9 28 | 202 | 13 28 | 262 | 17 28 | 322 | 21 28 | 22 | 1 28 | 1 29 | 1 30 | 1 31 |
| 23 | 1 32 | 83 | 5 32 | 143 | 9 32 | 203 | 13 32 | 263 | 17 32 | 323 | 21 32 | 23 | 1 32 | 1 33 | 1 34 | 1 35 |
| 24 | 1 36 | 84 | 5 36 | 144 | 9 36 | 204 | 13 36 | 264 | 17 36 | 324 | 21 36 | 24 | 1 36 | 1 37 | 1 38 | 1 39 |
| 25 | 1 40 | 85 | 5 40 | 145 | 9 40 | 205 | 13 40 | 265 | 17 40 | 325 | 21 40 | 25 | 1 40 | 1 41 | 1 42 | 1 43 |
| 26 | 1 44 | 86 | 5 44 | 146 | 9 44 | 206 | 13 44 | 266 | 17 44 | 326 | 21 44 | 26 | 1 44 | 1 45 | 1 46 | 1 47 |
| 27 | 1 48 | 87 | 5 48 | 147 | 9 48 | 207 | 13 48 | 267 | 17 48 | 327 | 21 48 | 27 | 1 48 | 1 49 | 1 50 | 1 51 |
| 28 | 1 52 | 88 | 5 52 | 148 | 9 52 | 208 | 13 52 | 268 | 17 52 | 328 | 21 52 | 28 | 1 52 | 1 53 | 1 54 | 1 55 |
| 29 | 1 56 | 89 | 5 56 | 149 | 9 56 | 209 | 13 56 | 269 | 17 56 | 329 | 21 56 | 29 | 1 56 | 1 57 | 1 58 | 1 59 |
| 30 | 2 00 | 90 | 6 00 | 150 | 10 00 | 210 | 14 00 | 270 | 18 00 | 330 | 22 00 | 30 | 2 00 | 2 01 | 2 02 | 2 03 |
| 31 | 2 04 | 91 | 6 04 | 151 | 10 04 | 211 | 14 04 | 271 | 18 04 | 331 | 22 04 | 31 | 2 04 | 2 05 | 2 06 | 2 07 |
| 32 | 2 08 | 92 | 6 08 | 152 | 10 08 | 212 | 14 08 | 272 | 18 08 | 332 | 22 08 | 32 | 2 08 | 2 09 | 2 10 | 2 11 |
| 33 | 2 12 | 93 | 6 12 | 153 | 10 12 | 213 | 14 12 | 273 | 18 12 | 333 | 22 12 | 33 | 2 12 | 2 13 | 2 14 | 2 15 |
| 34 | 2 16 | 94 | 6 16 | 154 | 10 16 | 214 | 14 16 | 274 | 18 16 | 334 | 22 16 | 34 | 2 16 | 2 17 | 2 18 | 2 19 |
| 35 | 2 20 | 95 | 6 20 | 155 | 10 20 | 215 | 14 20 | 275 | 18 20 | 335 | 22 20 | 35 | 2 20 | 2 21 | 2 22 | 2 23 |
| 36 | 2 24 | 96 | 6 24 | 156 | 10 24 | 216 | 14 24 | 276 | 18 24 | 336 | 22 24 | 36 | 2 24 | 2 25 | 2 26 | 2 27 |
| 37 | 2 28 | 97 | 6 28 | 157 | 10 28 | 217 | 14 28 | 277 | 18 28 | 337 | 22 28 | 37 | 2 28 | 2 29 | 2 30 | 2 31 |
| 38 | 2 32 | 98 | 6 32 | 158 | 10 32 | 218 | 14 32 | 278 | 18 32 | 338 | 22 32 | 38 | 2 32 | 2 33 | 2 34 | 2 35 |
| 39 | 2 36 | 99 | 6 36 | 159 | 10 36 | 219 | 14 36 | 279 | 18 36 | 339 | 22 36 | 39 | 2 36 | 2 37 | 2 38 | 2 39 |
| 40 | 2 40 | 100 | 6 40 | 160 | 10 40 | 220 | 14 40 | 280 | 18 40 | 340 | 22 40 | 40 | 2 40 | 2 41 | 2 42 | 2 43 |
| 41 | 2 44 | 101 | 6 44 | 161 | 10 44 | 221 | 14 44 | 281 | 18 44 | 341 | 22 44 | 41 | 2 44 | 2 45 | 2 46 | 2 47 |
| 42 | 2 48 | 102 | 6 48 | 162 | 10 48 | 222 | 14 48 | 282 | 18 48 | 342 | 22 48 | 42 | 2 48 | 2 49 | 2 50 | 2 51 |
| 43 | 2 52 | 103 | 6 52 | 163 | 10 52 | 223 | 14 52 | 283 | 18 52 | 343 | 22 52 | 43 | 2 52 | 2 53 | 2 54 | 2 55 |
| 44 | 2 56 | 104 | 6 56 | 164 | 10 56 | 224 | 14 56 | 284 | 18 56 | 344 | 22 56 | 44 | 2 56 | 2 57 | 2 58 | 2 59 |
| 45 | 3 00 | 105 | 7 00 | 165 | 11 00 | 225 | 15 00 | 285 | 19 00 | 345 | 23 00 | 45 | 3 00 | 3 01 | 3 02 | 3 03 |
| 46 | 3 04 | 106 | 7 04 | 166 | 11 04 | 226 | 15 04 | 286 | 19 04 | 346 | 23 04 | 46 | 3 04 | 3 05 | 3 06 | 3 07 |
| 47 | 3 08 | 107 | 7 08 | 167 | 11 08 | 227 | 15 08 | 287 | 19 08 | 347 | 23 08 | 47 | 3 08 | 3 09 | 3 10 | 3 11 |
| 48 | 3 12 | 108 | 7 12 | 168 | 11 12 | 228 | 15 12 | 288 | 19 12 | 348 | 23 12 | 48 | 3 12 | 3 13 | 3 14 | 3 15 |
| 49 | 3 16 | 109 | 7 16 | 169 | 11 16 | 229 | 15 16 | 289 | 19 16 | 349 | 23 16 | 49 | 3 16 | 3 17 | 3 18 | 3 19 |
| 50 | 3 20 | 110 | 7 20 | 170 | 11 20 | 230 | 15 20 | 290 | 19 20 | 350 | 23 20 | 50 | 3 20 | 3 21 | 3 22 | 3 23 |
| 51 | 3 24 | 111 | 7 24 | 171 | 11 24 | 231 | 15 24 | 291 | 19 24 | 351 | 23 24 | 51 | 3 24 | 3 25 | 3 26 | 3 27 |
| 52 | 3 28 | 112 | 7 28 | 172 | 11 28 | 232 | 15 28 | 292 | 19 28 | 352 | 23 28 | 52 | 3 28 | 3 29 | 3 30 | 3 31 |
| 53 | 3 32 | 113 | 7 32 | 173 | 11 32 | 233 | 15 32 | 293 | 19 32 | 353 | 23 32 | 53 | 3 32 | 3 33 | 3 34 | 3 35 |
| 54 | 3 36 | 114 | 7 36 | 174 | 11 36 | 234 | 15 36 | 294 | 19 36 | 354 | 23 36 | 54 | 3 36 | 3 37 | 3 38 | 3 39 |
| 55 | 3 40 | 115 | 7 40 | 175 | 11 40 | 235 | 15 40 | 295 | 19 40 | 355 | 23 40 | 55 | 3 40 | 3 41 | 3 42 | 3 43 |
| 56 | 3 44 | 116 | 7 44 | 176 | 11 44 | 236 | 15 44 | 296 | 19 44 | 356 | 23 44 | 56 | 3 44 | 3 45 | 3 46 | 3 47 |
| 57 | 3 48 | 117 | 7 48 | 177 | 11 48 | 237 | 15 48 | 297 | 19 48 | 357 | 23 48 | 57 | 3 48 | 3 49 | 3 50 | 3 51 |
| 58 | 3 52 | 118 | 7 52 | 178 | 11 52 | 238 | 15 52 | 298 | 19 52 | 358 | 23 52 | 58 | 3 52 | 3 53 | 3 54 | 3 55 |
| 59 | 3 56 | 119 | 7 56 | 179 | 11 56 | 239 | 15 56 | 299 | 19 56 | 359 | 23 56 | 59 | 3 56 | 3 57 | 3 58 | 3 59 |

The above table is for converting expressions in arc to their equivalent in time; its main use in this Almanac is for the conversion of longitude for application to L.M.T. (*added* if *west*, *subtracted* if *east*) to give G.M.T. or vice versa, particularly in the case of sunrise, sunset, etc.

| $\overset{m}{0}$ | SUN PLANETS | ARIES | MOON | $v$ or $d$ | Corr$^n$ | $v$ or $d$ | Corr$^n$ | $v$ or $d$ | Corr$^n$ |
|---|---|---|---|---|---|---|---|---|---|
| s | ° ′ | ° ′ | ° ′ | ′ | ′ | ′ | ′ | ′ | ′ |
| 00 | 0 00·0 | 0 00·0 | 0 00·0 | 0·0 | 0·0 | 6·0 | 0·1 | 12·0 | 0·1 |
| 01 | 0 00·3 | 0 00·3 | 0 00·2 | 0·1 | 0·0 | 6·1 | 0·1 | 12·1 | 0·1 |
| 02 | 0 00·5 | 0 00·5 | 0 00·5 | 0·2 | 0·0 | 6·2 | 0·1 | 12·2 | 0·1 |
| 03 | 0 00·8 | 0 00·8 | 0 00·7 | 0·3 | 0·0 | 6·3 | 0·1 | 12·3 | 0·1 |
| 04 | 0 01·0 | 0 01·0 | 0 01·0 | 0·4 | 0·0 | 6·4 | 0·1 | 12·4 | 0·1 |
| 05 | 0 01·3 | 0 01·3 | 0 01·2 | 0·5 | 0·0 | 6·5 | 0·1 | 12·5 | 0·1 |
| 06 | 0 01·5 | 0 01·5 | 0 01·4 | 0·6 | 0·0 | 6·6 | 0·1 | 12·6 | 0·1 |
| 07 | 0 01·8 | 0 01·8 | 0 01·7 | 0·7 | 0·0 | 6·7 | 0·1 | 12·7 | 0·1 |
| 08 | 0 02·0 | 0 02·0 | 0 01·9 | 0·8 | 0·0 | 6·8 | 0·1 | 12·8 | 0·1 |
| 09 | 0 02·3 | 0 02·3 | 0 02·1 | 0·9 | 0·0 | 6·9 | 0·1 | 12·9 | 0·1 |
| 10 | 0 02·5 | 0 02·5 | 0 02·4 | 1·0 | 0·0 | 7·0 | 0·1 | 13·0 | 0·1 |
| 11 | 0 02·8 | 0 02·8 | 0 02·6 | 1·1 | 0·0 | 7·1 | 0·1 | 13·1 | 0·1 |
| 12 | 0 03·0 | 0 03·0 | 0 02·9 | 1·2 | 0·0 | 7·2 | 0·1 | 13·2 | 0·1 |
| 13 | 0 03·3 | 0 03·3 | 0 03·1 | 1·3 | 0·0 | 7·3 | 0·1 | 13·3 | 0·1 |
| 14 | 0 03·5 | 0 03·5 | 0 03·3 | 1·4 | 0·0 | 7·4 | 0·1 | 13·4 | 0·1 |
| 15 | 0 03·8 | 0 03·8 | 0 03·6 | 1·5 | 0·0 | 7·5 | 0·1 | 13·5 | 0·1 |
| 16 | 0 04·0 | 0 04·0 | 0 03·8 | 1·6 | 0·0 | 7·6 | 0·1 | 13·6 | 0·1 |
| 17 | 0 04·3 | 0 04·3 | 0 04·1 | 1·7 | 0·0 | 7·7 | 0·1 | 13·7 | 0·1 |
| 18 | 0 04·5 | 0 04·5 | 0 04·3 | 1·8 | 0·0 | 7·8 | 0·1 | 13·8 | 0·1 |
| 19 | 0 04·8 | 0 04·8 | 0 04·5 | 1·9 | 0·0 | 7·9 | 0·1 | 13·9 | 0·1 |
| 20 | 0 05·0 | 0 05·0 | 0 04·8 | 2·0 | 0·0 | 8·0 | 0·1 | 14·0 | 0·1 |
| 21 | 0 05·3 | 0 05·3 | 0 05·0 | 2·1 | 0·0 | 8·1 | 0·1 | 14·1 | 0·1 |
| 22 | 0 05·5 | 0 05·5 | 0 05·2 | 2·2 | 0·0 | 8·2 | 0·1 | 14·2 | 0·1 |
| 23 | 0 05·8 | 0 05·8 | 0 05·5 | 2·3 | 0·0 | 8·3 | 0·1 | 14·3 | 0·1 |
| 24 | 0 06·0 | 0 06·0 | 0 05·7 | 2·4 | 0·0 | 8·4 | 0·1 | 14·4 | 0·1 |
| 25 | 0 06·3 | 0 06·3 | 0 06·0 | 2·5 | 0·0 | 8·5 | 0·1 | 14·5 | 0·1 |
| 26 | 0 06·5 | 0 06·5 | 0 06·2 | 2·6 | 0·0 | 8·6 | 0·1 | 14·6 | 0·1 |
| 27 | 0 06·8 | 0 06·8 | 0 06·4 | 2·7 | 0·0 | 8·7 | 0·1 | 14·7 | 0·1 |
| 28 | 0 07·0 | 0 07·0 | 0 06·7 | 2·8 | 0·0 | 8·8 | 0·1 | 14·8 | 0·1 |
| 29 | 0 07·3 | 0 07·3 | 0 06·9 | 2·9 | 0·0 | 8·9 | 0·1 | 14·9 | 0·1 |
| 30 | 0 07·5 | 0 07·5 | 0 07·2 | 3·0 | 0·0 | 9·0 | 0·1 | 15·0 | 0·1 |
| 31 | 0 07·8 | 0 07·8 | 0 07·4 | 3·1 | 0·0 | 9·1 | 0·1 | 15·1 | 0·1 |
| 32 | 0 08·0 | 0 08·0 | 0 07·6 | 3·2 | 0·0 | 9·2 | 0·1 | 15·2 | 0·1 |
| 33 | 0 08·3 | 0 08·3 | 0 07·9 | 3·3 | 0·0 | 9·3 | 0·1 | 15·3 | 0·1 |
| 34 | 0 08·5 | 0 08·5 | 0 08·1 | 3·4 | 0·0 | 9·4 | 0·1 | 15·4 | 0·1 |
| 35 | 0 08·8 | 0 08·8 | 0 08·4 | 3·5 | 0·0 | 9·5 | 0·1 | 15·5 | 0·1 |
| 36 | 0 09·0 | 0 09·0 | 0 08·6 | 3·6 | 0·0 | 9·6 | 0·1 | 15·6 | 0·1 |
| 37 | 0 09·3 | 0 09·3 | 0 08·8 | 3·7 | 0·0 | 9·7 | 0·1 | 15·7 | 0·1 |
| 38 | 0 09·5 | 0 09·5 | 0 09·1 | 3·8 | 0·0 | 9·8 | 0·1 | 15·8 | 0·1 |
| 39 | 0 09·8 | 0 09·8 | 0 09·3 | 3·9 | 0·0 | 9·9 | 0·1 | 15·9 | 0·1 |
| 40 | 0 10·0 | 0 10·0 | 0 09·5 | 4·0 | 0·0 | 10·0 | 0·1 | 16·0 | 0·1 |
| 41 | 0 10·3 | 0 10·3 | 0 09·8 | 4·1 | 0·0 | 10·1 | 0·1 | 16·1 | 0·1 |
| 42 | 0 10·5 | 0 10·5 | 0 10·0 | 4·2 | 0·0 | 10·2 | 0·1 | 16·2 | 0·1 |
| 43 | 0 10·8 | 0 10·8 | 0 10·3 | 4·3 | 0·0 | 10·3 | 0·1 | 16·3 | 0·1 |
| 44 | 0 11·0 | 0 11·0 | 0 10·5 | 4·4 | 0·0 | 10·4 | 0·1 | 16·4 | 0·1 |
| 45 | 0 11·3 | 0 11·3 | 0 10·7 | 4·5 | 0·0 | 10·5 | 0·1 | 16·5 | 0·1 |
| 46 | 0 11·5 | 0 11·5 | 0 11·0 | 4·6 | 0·0 | 10·6 | 0·1 | 16·6 | 0·1 |
| 47 | 0 11·8 | 0 11·8 | 0 11·2 | 4·7 | 0·0 | 10·7 | 0·1 | 16·7 | 0·1 |
| 48 | 0 12·0 | 0 12·0 | 0 11·5 | 4·8 | 0·0 | 10·8 | 0·1 | 16·8 | 0·1 |
| 49 | 0 12·3 | 0 12·3 | 0 11·7 | 4·9 | 0·0 | 10·9 | 0·1 | 16·9 | 0·1 |
| 50 | 0 12·5 | 0 12·5 | 0 11·9 | 5·0 | 0·0 | 11·0 | 0·1 | 17·0 | 0·1 |
| 51 | 0 12·8 | 0 12·8 | 0 12·2 | 5·1 | 0·0 | 11·1 | 0·1 | 17·1 | 0·1 |
| 52 | 0 13·0 | 0 13·0 | 0 12·4 | 5·2 | 0·0 | 11·2 | 0·1 | 17·2 | 0·1 |
| 53 | 0 13·3 | 0 13·3 | 0 12·6 | 5·3 | 0·0 | 11·3 | 0·1 | 17·3 | 0·1 |
| 54 | 0 13·5 | 0 13·5 | 0 12·9 | 5·4 | 0·0 | 11·4 | 0·1 | 17·4 | 0·1 |
| 55 | 0 13·8 | 0 13·8 | 0 13·1 | 5·5 | 0·0 | 11·5 | 0·1 | 17·5 | 0·1 |
| 56 | 0 14·0 | 0 14·0 | 0 13·4 | 5·6 | 0·0 | 11·6 | 0·1 | 17·6 | 0·1 |
| 57 | 0 14·3 | 0 14·3 | 0 13·6 | 5·7 | 0·0 | 11·7 | 0·1 | 17·7 | 0·1 |
| 58 | 0 14·5 | 0 14·5 | 0 13·8 | 5·8 | 0·0 | 11·8 | 0·1 | 17·8 | 0·1 |
| 59 | 0 14·8 | 0 14·8 | 0 14·1 | 5·9 | 0·0 | 11·9 | 0·1 | 17·9 | 0·1 |
| 60 | 0 15·0 | 0 15·0 | 0 14·3 | 6·0 | 0·1 | 12·0 | 0·1 | 18·0 | 0·2 |

| $\overset{m}{1}$ | SUN PLANETS | ARIES | MOON | $v$ or $d$ | Corr$^n$ | $v$ or $d$ | Corr$^n$ | $v$ or $d$ | Corr$^n$ |
|---|---|---|---|---|---|---|---|---|---|
| s | ° ′ | ° ′ | ° ′ | ′ | ′ | ′ | ′ | ′ | ′ |
| 00 | 0 15·0 | 0 15·0 | 0 14·3 | 0·0 | 0·0 | 6·0 | 0·2 | 12·0 | 0·3 |
| 01 | 0 15·3 | 0 15·3 | 0 14·6 | 0·1 | 0·0 | 6·1 | 0·2 | 12·1 | 0·3 |
| 02 | 0 15·5 | 0 15·5 | 0 14·8 | 0·2 | 0·0 | 6·2 | 0·2 | 12·2 | 0·3 |
| 03 | 0 15·8 | 0 15·8 | 0 15·0 | 0·3 | 0·0 | 6·3 | 0·2 | 12·3 | 0·3 |
| 04 | 0 16·0 | 0 16·0 | 0 15·3 | 0·4 | 0·0 | 6·4 | 0·2 | 12·4 | 0·3 |
| 05 | 0 16·3 | 0 16·3 | 0 15·5 | 0·5 | 0·0 | 6·5 | 0·2 | 12·5 | 0·3 |
| 06 | 0 16·5 | 0 16·5 | 0 15·7 | 0·6 | 0·0 | 6·6 | 0·2 | 12·6 | 0·3 |
| 07 | 0 16·8 | 0 16·8 | 0 16·0 | 0·7 | 0·0 | 6·7 | 0·2 | 12·7 | 0·3 |
| 08 | 0 17·0 | 0 17·0 | 0 16·2 | 0·8 | 0·0 | 6·8 | 0·2 | 12·8 | 0·3 |
| 09 | 0 17·3 | 0 17·3 | 0 16·5 | 0·9 | 0·0 | 6·9 | 0·2 | 12·9 | 0·3 |
| 10 | 0 17·5 | 0 17·5 | 0 16·7 | 1·0 | 0·0 | 7·0 | 0·2 | 13·0 | 0·3 |
| 11 | 0 17·8 | 0 17·8 | 0 16·9 | 1·1 | 0·0 | 7·1 | 0·2 | 13·1 | 0·3 |
| 12 | 0 18·0 | 0 18·0 | 0 17·2 | 1·2 | 0·0 | 7·2 | 0·2 | 13·2 | 0·3 |
| 13 | 0 18·3 | 0 18·3 | 0 17·4 | 1·3 | 0·0 | 7·3 | 0·2 | 13·3 | 0·3 |
| 14 | 0 18·5 | 0 18·6 | 0 17·7 | 1·4 | 0·0 | 7·4 | 0·2 | 13·4 | 0·3 |
| 15 | 0 18·8 | 0 18·8 | 0 17·9 | 1·5 | 0·0 | 7·5 | 0·2 | 13·5 | 0·3 |
| 16 | 0 19·0 | 0 19·1 | 0 18·1 | 1·6 | 0·0 | 7·6 | 0·2 | 13·6 | 0·3 |
| 17 | 0 19·3 | 0 19·3 | 0 18·4 | 1·7 | 0·0 | 7·7 | 0·2 | 13·7 | 0·3 |
| 18 | 0 19·5 | 0 19·6 | 0 18·6 | 1·8 | 0·0 | 7·8 | 0·2 | 13·8 | 0·3 |
| 19 | 0 19·8 | 0 19·8 | 0 18·9 | 1·9 | 0·0 | 7·9 | 0·2 | 13·9 | 0·3 |
| 20 | 0 20·0 | 0 20·1 | 0 19·1 | 2·0 | 0·1 | 8·0 | 0·2 | 14·0 | 0·4 |
| 21 | 0 20·3 | 0 20·3 | 0 19·3 | 2·1 | 0·1 | 8·1 | 0·2 | 14·1 | 0·4 |
| 22 | 0 20·5 | 0 20·6 | 0 19·6 | 2·2 | 0·1 | 8·2 | 0·2 | 14·2 | 0·4 |
| 23 | 0 20·8 | 0 20·8 | 0 19·8 | 2·3 | 0·1 | 8·3 | 0·2 | 14·3 | 0·4 |
| 24 | 0 21·0 | 0 21·1 | 0 20·0 | 2·4 | 0·1 | 8·4 | 0·2 | 14·4 | 0·4 |
| 25 | 0 21·3 | 0 21·3 | 0 20·3 | 2·5 | 0·1 | 8·5 | 0·2 | 14·5 | 0·4 |
| 26 | 0 21·5 | 0 21·6 | 0 20·5 | 2·6 | 0·1 | 8·6 | 0·2 | 14·6 | 0·4 |
| 27 | 0 21·8 | 0 21·8 | 0 20·8 | 2·7 | 0·1 | 8·7 | 0·2 | 14·7 | 0·4 |
| 28 | 0 22·0 | 0 22·1 | 0 21·0 | 2·8 | 0·1 | 8·8 | 0·2 | 14·8 | 0·4 |
| 29 | 0 22·3 | 0 22·3 | 0 21·2 | 2·9 | 0·1 | 8·9 | 0·2 | 14·9 | 0·4 |
| 30 | 0 22·5 | 0 22·6 | 0 21·5 | 3·0 | 0·1 | 9·0 | 0·2 | 15·0 | 0·4 |
| 31 | 0 22·8 | 0 22·8 | 0 21·7 | 3·1 | 0·1 | 9·1 | 0·2 | 15·1 | 0·4 |
| 32 | 0 23·0 | 0 23·1 | 0 22·0 | 3·2 | 0·1 | 9·2 | 0·2 | 15·2 | 0·4 |
| 33 | 0 23·3 | 0 23·3 | 0 22·2 | 3·3 | 0·1 | 9·3 | 0·2 | 15·3 | 0·4 |
| 34 | 0 23·5 | 0 23·6 | 0 22·4 | 3·4 | 0·1 | 9·4 | 0·2 | 15·4 | 0·4 |
| 35 | 0 23·8 | 0 23·8 | 0 22·7 | 3·5 | 0·1 | 9·5 | 0·2 | 15·5 | 0·4 |
| 36 | 0 24·0 | 0 24·1 | 0 22·9 | 3·6 | 0·1 | 9·6 | 0·2 | 15·6 | 0·4 |
| 37 | 0 24·3 | 0 24·3 | 0 23·1 | 3·7 | 0·1 | 9·7 | 0·2 | 15·7 | 0·4 |
| 38 | 0 24·5 | 0 24·6 | 0 23·4 | 3·8 | 0·1 | 9·8 | 0·2 | 15·8 | 0·4 |
| 39 | 0 24·8 | 0 24·8 | 0 23·6 | 3·9 | 0·1 | 9·9 | 0·2 | 15·9 | 0·4 |
| 40 | 0 25·0 | 0 25·1 | 0 23·9 | 4·0 | 0·1 | 10·0 | 0·3 | 16·0 | 0·4 |
| 41 | 0 25·3 | 0 25·3 | 0 24·1 | 4·1 | 0·1 | 10·1 | 0·3 | 16·1 | 0·4 |
| 42 | 0 25·5 | 0 25·6 | 0 24·3 | 4·2 | 0·1 | 10·2 | 0·3 | 16·2 | 0·4 |
| 43 | 0 25·8 | 0 25·8 | 0 24·6 | 4·3 | 0·1 | 10·3 | 0·3 | 16·3 | 0·4 |
| 44 | 0 26·0 | 0 26·1 | 0 24·8 | 4·4 | 0·1 | 10·4 | 0·3 | 16·4 | 0·4 |
| 45 | 0 26·3 | 0 26·3 | 0 25·1 | 4·5 | 0·1 | 10·5 | 0·3 | 16·5 | 0·4 |
| 46 | 0 26·5 | 0 26·6 | 0 25·3 | 4·6 | 0·1 | 10·6 | 0·3 | 16·6 | 0·4 |
| 47 | 0 26·8 | 0 26·8 | 0 25·5 | 4·7 | 0·1 | 10·7 | 0·3 | 16·7 | 0·4 |
| 48 | 0 27·0 | 0 27·1 | 0 25·8 | 4·8 | 0·1 | 10·8 | 0·3 | 16·8 | 0·4 |
| 49 | 0 27·3 | 0 27·3 | 0 26·0 | 4·9 | 0·1 | 10·9 | 0·3 | 16·9 | 0·4 |
| 50 | 0 27·5 | 0 27·6 | 0 26·2 | 5·0 | 0·1 | 11·0 | 0·3 | 17·0 | 0·4 |
| 51 | 0 27·8 | 0 27·8 | 0 26·5 | 5·1 | 0·1 | 11·1 | 0·3 | 17·1 | 0·4 |
| 52 | 0 28·0 | 0 28·1 | 0 26·7 | 5·2 | 0·1 | 11·2 | 0·3 | 17·2 | 0·4 |
| 53 | 0 28·3 | 0 28·3 | 0 27·0 | 5·3 | 0·1 | 11·3 | 0·3 | 17·3 | 0·4 |
| 54 | 0 28·5 | 0 28·6 | 0 27·2 | 5·4 | 0·1 | 11·4 | 0·3 | 17·4 | 0·4 |
| 55 | 0 28·8 | 0 28·8 | 0 27·4 | 5·5 | 0·1 | 11·5 | 0·3 | 17·5 | 0·4 |
| 56 | 0 29·0 | 0 29·1 | 0 27·7 | 5·6 | 0·1 | 11·6 | 0·3 | 17·6 | 0·4 |
| 57 | 0 29·3 | 0 29·3 | 0 27·9 | 5·7 | 0·1 | 11·7 | 0·3 | 17·7 | 0·4 |
| 58 | 0 29·5 | 0 29·6 | 0 28·2 | 5·8 | 0·1 | 11·8 | 0·3 | 17·8 | 0·4 |
| 59 | 0 29·8 | 0 29·8 | 0 28·4 | 5·9 | 0·1 | 11·9 | 0·3 | 17·9 | 0·4 |
| 60 | 0 30·0 | 0 30·1 | 0 28·6 | 6·0 | 0·2 | 12·0 | 0·3 | 18·0 | 0·5 |

| 2ᵐ s | SUN PLANETS ° ′ | ARIES ° ′ | MOON ° ′ | v or d ′ | Corrⁿ ′ | v or d ′ | Corrⁿ ′ | v or d ′ | Corrⁿ ′ |
|---|---|---|---|---|---|---|---|---|---|
| 00 | 0 30·0 | 0 30·1 | 0 28·6 | 0·0 | 0·0 | 6·0 | 0·3 | 12·0 | 0·5 |
| 01 | 0 30·3 | 0 30·3 | 0 28·9 | 0·1 | 0·0 | 6·1 | 0·3 | 12·1 | 0·5 |
| 02 | 0 30·5 | 0 30·6 | 0 29·1 | 0·2 | 0·0 | 6·2 | 0·3 | 12·2 | 0·5 |
| 03 | 0 30·8 | 0 30·8 | 0 29·3 | 0·3 | 0·0 | 6·3 | 0·3 | 12·3 | 0·5 |
| 04 | 0 31·0 | 0 31·1 | 0 29·6 | 0·4 | 0·0 | 6·4 | 0·3 | 12·4 | 0·5 |
| 05 | 0 31·3 | 0 31·3 | 0 29·8 | 0·5 | 0·0 | 6·5 | 0·3 | 12·5 | 0·5 |
| 06 | 0 31·5 | 0 31·6 | 0 30·1 | 0·6 | 0·0 | 6·6 | 0·3 | 12·6 | 0·5 |
| 07 | 0 31·8 | 0 31·8 | 0 30·3 | 0·7 | 0·0 | 6·7 | 0·3 | 12·7 | 0·5 |
| 08 | 0 32·0 | 0 32·1 | 0 30·5 | 0·8 | 0·0 | 6·8 | 0·3 | 12·8 | 0·5 |
| 09 | 0 32·3 | 0 32·3 | 0 30·8 | 0·9 | 0·0 | 6·9 | 0·3 | 12·9 | 0·5 |
| 10 | 0 32·5 | 0 32·6 | 0 31·0 | 1·0 | 0·0 | 7·0 | 0·3 | 13·0 | 0·5 |
| 11 | 0 32·8 | 0 32·8 | 0 31·3 | 1·1 | 0·0 | 7·1 | 0·3 | 13·1 | 0·5 |
| 12 | 0 33·0 | 0 33·1 | 0 31·5 | 1·2 | 0·1 | 7·2 | 0·3 | 13·2 | 0·6 |
| 13 | 0 33·3 | 0 33·3 | 0 31·7 | 1·3 | 0·1 | 7·3 | 0·3 | 13·3 | 0·6 |
| 14 | 0 33·5 | 0 33·6 | 0 32·0 | 1·4 | 0·1 | 7·4 | 0·3 | 13·4 | 0·6 |
| 15 | 0 33·8 | 0 33·8 | 0 32·2 | 1·5 | 0·1 | 7·5 | 0·3 | 13·5 | 0·6 |
| 16 | 0 34·0 | 0 34·1 | 0 32·5 | 1·6 | 0·1 | 7·6 | 0·3 | 13·6 | 0·6 |
| 17 | 0 34·3 | 0 34·3 | 0 32·7 | 1·7 | 0·1 | 7·7 | 0·3 | 13·7 | 0·6 |
| 18 | 0 34·5 | 0 34·6 | 0 32·9 | 1·8 | 0·1 | 7·8 | 0·3 | 13·8 | 0·6 |
| 19 | 0 34·8 | 0 34·8 | 0 33·2 | 1·9 | 0·1 | 7·9 | 0·3 | 13·9 | 0·6 |
| 20 | 0 35·0 | 0 35·1 | 0 33·4 | 2·0 | 0·1 | 8·0 | 0·3 | 14·0 | 0·6 |
| 21 | 0 35·3 | 0 35·3 | 0 33·6 | 2·1 | 0·1 | 8·1 | 0·3 | 14·1 | 0·6 |
| 22 | 0 35·5 | 0 35·6 | 0 33·9 | 2·2 | 0·1 | 8·2 | 0·3 | 14·2 | 0·6 |
| 23 | 0 35·8 | 0 35·8 | 0 34·1 | 2·3 | 0·1 | 8·3 | 0·3 | 14·3 | 0·6 |
| 24 | 0 36·0 | 0 36·1 | 0 34·4 | 2·4 | 0·1 | 8·4 | 0·4 | 14·4 | 0·6 |
| 25 | 0 36·3 | 0 36·3 | 0 34·6 | 2·5 | 0·1 | 8·5 | 0·4 | 14·5 | 0·6 |
| 26 | 0 36·5 | 0 36·6 | 0 34·8 | 2·6 | 0·1 | 8·6 | 0·4 | 14·6 | 0·6 |
| 27 | 0 36·8 | 0 36·9 | 0 35·1 | 2·7 | 0·1 | 8·7 | 0·4 | 14·7 | 0·6 |
| 28 | 0 37·0 | 0 37·1 | 0 35·3 | 2·8 | 0·1 | 8·8 | 0·4 | 14·8 | 0·6 |
| 29 | 0 37·3 | 0 37·4 | 0 35·6 | 2·9 | 0·1 | 8·9 | 0·4 | 14·9 | 0·6 |
| 30 | 0 37·5 | 0 37·6 | 0 35·8 | 3·0 | 0·1 | 9·0 | 0·4 | 15·0 | 0·6 |
| 31 | 0 37·8 | 0 37·9 | 0 36·0 | 3·1 | 0·1 | 9·1 | 0·4 | 15·1 | 0·6 |
| 32 | 0 38·0 | 0 38·1 | 0 36·3 | 3·2 | 0·1 | 9·2 | 0·4 | 15·2 | 0·6 |
| 33 | 0 38·3 | 0 38·4 | 0 36·5 | 3·3 | 0·1 | 9·3 | 0·4 | 15·3 | 0·6 |
| 34 | 0 38·5 | 0 38·6 | 0 36·7 | 3·4 | 0·1 | 9·4 | 0·4 | 15·4 | 0·6 |
| 35 | 0 38·8 | 0 38·9 | 0 37·0 | 3·5 | 0·1 | 9·5 | 0·4 | 15·5 | 0·6 |
| 36 | 0 39·0 | 0 39·1 | 0 37·2 | 3·6 | 0·2 | 9·6 | 0·4 | 15·6 | 0·7 |
| 37 | 0 39·3 | 0 39·4 | 0 37·5 | 3·7 | 0·2 | 9·7 | 0·4 | 15·7 | 0·7 |
| 38 | 0 39·5 | 0 39·6 | 0 37·7 | 3·8 | 0·2 | 9·8 | 0·4 | 15·8 | 0·7 |
| 39 | 0 39·8 | 0 39·9 | 0 37·9 | 3·9 | 0·2 | 9·9 | 0·4 | 15·9 | 0·7 |
| 40 | 0 40·0 | 0 40·1 | 0 38·2 | 4·0 | 0·2 | 10·0 | 0·4 | 16·0 | 0·7 |
| 41 | 0 40·3 | 0 40·4 | 0 38·4 | 4·1 | 0·2 | 10·1 | 0·4 | 16·1 | 0·7 |
| 42 | 0 40·5 | 0 40·6 | 0 38·7 | 4·2 | 0·2 | 10·2 | 0·4 | 16·2 | 0·7 |
| 43 | 0 40·8 | 0 40·9 | 0 38·9 | 4·3 | 0·2 | 10·3 | 0·4 | 16·3 | 0·7 |
| 44 | 0 41·0 | 0 41·1 | 0 39·1 | 4·4 | 0·2 | 10·4 | 0·4 | 16·4 | 0·7 |
| 45 | 0 41·3 | 0 41·4 | 0 39·4 | 4·5 | 0·2 | 10·5 | 0·4 | 16·5 | 0·7 |
| 46 | 0 41·5 | 0 41·6 | 0 39·6 | 4·6 | 0·2 | 10·6 | 0·4 | 16·6 | 0·7 |
| 47 | 0 41·8 | 0 41·9 | 0 39·8 | 4·7 | 0·2 | 10·7 | 0·4 | 16·7 | 0·7 |
| 48 | 0 42·0 | 0 42·1 | 0 40·1 | 4·8 | 0·2 | 10·8 | 0·5 | 16·8 | 0·7 |
| 49 | 0 42·3 | 0 42·4 | 0 40·3 | 4·9 | 0·2 | 10·9 | 0·5 | 16·9 | 0·7 |
| 50 | 0 42·5 | 0 42·6 | 0 40·6 | 5·0 | 0·2 | 11·0 | 0·5 | 17·0 | 0·7 |
| 51 | 0 42·8 | 0 42·9 | 0 40·8 | 5·1 | 0·2 | 11·1 | 0·5 | 17·1 | 0·7 |
| 52 | 0 43·0 | 0 43·1 | 0 41·0 | 5·2 | 0·2 | 11·2 | 0·5 | 17·2 | 0·7 |
| 53 | 0 43·3 | 0 43·4 | 0 41·3 | 5·3 | 0·2 | 11·3 | 0·5 | 17·3 | 0·7 |
| 54 | 0 43·5 | 0 43·6 | 0 41·5 | 5·4 | 0·2 | 11·4 | 0·5 | 17·4 | 0·7 |
| 55 | 0 43·8 | 0 43·9 | 0 41·8 | 5·5 | 0·2 | 11·5 | 0·5 | 17·5 | 0·7 |
| 56 | 0 44·0 | 0 44·1 | 0 42·0 | 5·6 | 0·2 | 11·6 | 0·5 | 17·6 | 0·7 |
| 57 | 0 44·3 | 0 44·4 | 0 42·2 | 5·7 | 0·2 | 11·7 | 0·5 | 17·7 | 0·7 |
| 58 | 0 44·5 | 0 44·6 | 0 42·5 | 5·8 | 0·2 | 11·8 | 0·5 | 17·8 | 0·7 |
| 59 | 0 44·8 | 0 44·9 | 0 42·7 | 5·9 | 0·2 | 11·9 | 0·5 | 17·9 | 0·7 |
| 60 | 0 45·0 | 0 45·1 | 0 43·0 | 6·0 | 0·3 | 12·0 | 0·5 | 18·0 | 0·8 |

| 3ᵐ s | SUN PLANETS ° ′ | ARIES ° ′ | MOON ° ′ | v or d ′ | Corrⁿ ′ | v or d ′ | Corrⁿ ′ | v or d ′ | Corrⁿ ′ |
|---|---|---|---|---|---|---|---|---|---|
| 00 | 0 45·0 | 0 45·1 | 0 43·0 | 0·0 | 0·0 | 6·0 | 0·4 | 12·0 | 0·7 |
| 01 | 0 45·3 | 0 45·4 | 0 43·2 | 0·1 | 0·0 | 6·1 | 0·4 | 12·1 | 0·7 |
| 02 | 0 45·5 | 0 45·6 | 0 43·4 | 0·2 | 0·0 | 6·2 | 0·4 | 12·2 | 0·7 |
| 03 | 0 45·8 | 0 45·9 | 0 43·7 | 0·3 | 0·0 | 6·3 | 0·4 | 12·3 | 0·7 |
| 04 | 0 46·0 | 0 46·1 | 0 43·9 | 0·4 | 0·0 | 6·4 | 0·4 | 12·4 | 0·7 |
| 05 | 0 46·3 | 0 46·4 | 0 44·1 | 0·5 | 0·0 | 6·5 | 0·4 | 12·5 | 0·7 |
| 06 | 0 46·5 | 0 46·6 | 0 44·4 | 0·6 | 0·0 | 6·6 | 0·4 | 12·6 | 0·7 |
| 07 | 0 46·8 | 0 46·9 | 0 44·6 | 0·7 | 0·0 | 6·7 | 0·4 | 12·7 | 0·7 |
| 08 | 0 47·0 | 0 47·1 | 0 44·9 | 0·8 | 0·0 | 6·8 | 0·4 | 12·8 | 0·7 |
| 09 | 0 47·3 | 0 47·4 | 0 45·1 | 0·9 | 0·1 | 6·9 | 0·4 | 12·9 | 0·8 |
| 10 | 0 47·5 | 0 47·6 | 0 45·3 | 1·0 | 0·1 | 7·0 | 0·4 | 13·0 | 0·8 |
| 11 | 0 47·8 | 0 47·9 | 0 45·6 | 1·1 | 0·1 | 7·1 | 0·4 | 13·1 | 0·8 |
| 12 | 0 48·0 | 0 48·1 | 0 45·8 | 1·2 | 0·1 | 7·2 | 0·4 | 13·2 | 0·8 |
| 13 | 0 48·3 | 0 48·4 | 0 46·1 | 1·3 | 0·1 | 7·3 | 0·4 | 13·3 | 0·8 |
| 14 | 0 48·5 | 0 48·6 | 0 46·3 | 1·4 | 0·1 | 7·4 | 0·4 | 13·4 | 0·8 |
| 15 | 0 48·8 | 0 48·9 | 0 46·5 | 1·5 | 0·1 | 7·5 | 0·4 | 13·5 | 0·8 |
| 16 | 0 49·0 | 0 49·1 | 0 46·8 | 1·6 | 0·1 | 7·6 | 0·4 | 13·6 | 0·8 |
| 17 | 0 49·3 | 0 49·4 | 0 47·0 | 1·7 | 0·1 | 7·7 | 0·4 | 13·7 | 0·8 |
| 18 | 0 49·5 | 0 49·6 | 0 47·2 | 1·8 | 0·1 | 7·8 | 0·5 | 13·8 | 0·8 |
| 19 | 0 49·8 | 0 49·9 | 0 47·5 | 1·9 | 0·1 | 7·9 | 0·5 | 13·9 | 0·8 |
| 20 | 0 50·0 | 0 50·1 | 0 47·7 | 2·0 | 0·1 | 8·0 | 0·5 | 14·0 | 0·8 |
| 21 | 0 50·3 | 0 50·4 | 0 48·0 | 2·1 | 0·1 | 8·1 | 0·5 | 14·1 | 0·8 |
| 22 | 0 50·5 | 0 50·6 | 0 48·2 | 2·2 | 0·1 | 8·2 | 0·5 | 14·2 | 0·8 |
| 23 | 0 50·8 | 0 50·9 | 0 48·4 | 2·3 | 0·1 | 8·3 | 0·5 | 14·3 | 0·8 |
| 24 | 0 51·0 | 0 51·1 | 0 48·7 | 2·4 | 0·1 | 8·4 | 0·5 | 14·4 | 0·8 |
| 25 | 0 51·3 | 0 51·4 | 0 48·9 | 2·5 | 0·1 | 8·5 | 0·5 | 14·5 | 0·8 |
| 26 | 0 51·5 | 0 51·6 | 0 49·2 | 2·6 | 0·2 | 8·6 | 0·5 | 14·6 | 0·9 |
| 27 | 0 51·8 | 0 51·9 | 0 49·4 | 2·7 | 0·2 | 8·7 | 0·5 | 14·7 | 0·9 |
| 28 | 0 52·0 | 0 52·1 | 0 49·6 | 2·8 | 0·2 | 8·8 | 0·5 | 14·8 | 0·9 |
| 29 | 0 52·3 | 0 52·4 | 0 49·9 | 2·9 | 0·2 | 8·9 | 0·5 | 14·9 | 0·9 |
| 30 | 0 52·5 | 0 52·6 | 0 50·1 | 3·0 | 0·2 | 9·0 | 0·5 | 15·0 | 0·9 |
| 31 | 0 52·8 | 0 52·9 | 0 50·3 | 3·1 | 0·2 | 9·1 | 0·5 | 15·1 | 0·9 |
| 32 | 0 53·0 | 0 53·1 | 0 50·6 | 3·2 | 0·2 | 9·2 | 0·5 | 15·2 | 0·9 |
| 33 | 0 53·3 | 0 53·4 | 0 50·8 | 3·3 | 0·2 | 9·3 | 0·5 | 15·3 | 0·9 |
| 34 | 0 53·5 | 0 53·6 | 0 51·1 | 3·4 | 0·2 | 9·4 | 0·5 | 15·4 | 0·9 |
| 35 | 0 53·8 | 0 53·9 | 0 51·3 | 3·5 | 0·2 | 9·5 | 0·6 | 15·5 | 0·9 |
| 36 | 0 54·0 | 0 54·1 | 0 51·5 | 3·6 | 0·2 | 9·6 | 0·6 | 15·6 | 0·9 |
| 37 | 0 54·3 | 0 54·4 | 0 51·8 | 3·7 | 0·2 | 9·7 | 0·6 | 15·7 | 0·9 |
| 38 | 0 54·5 | 0 54·6 | 0 52·0 | 3·8 | 0·2 | 9·8 | 0·6 | 15·8 | 0·9 |
| 39 | 0 54·8 | 0 54·9 | 0 52·3 | 3·9 | 0·2 | 9·9 | 0·6 | 15·9 | 0·9 |
| 40 | 0 55·0 | 0 55·2 | 0 52·5 | 4·0 | 0·2 | 10·0 | 0·6 | 16·0 | 0·9 |
| 41 | 0 55·3 | 0 55·4 | 0 52·7 | 4·1 | 0·2 | 10·1 | 0·6 | 16·1 | 0·9 |
| 42 | 0 55·5 | 0 55·7 | 0 53·0 | 4·2 | 0·2 | 10·2 | 0·6 | 16·2 | 0·9 |
| 43 | 0 55·8 | 0 55·9 | 0 53·2 | 4·3 | 0·3 | 10·3 | 0·6 | 16·3 | 1·0 |
| 44 | 0 56·0 | 0 56·2 | 0 53·4 | 4·4 | 0·3 | 10·4 | 0·6 | 16·4 | 1·0 |
| 45 | 0 56·3 | 0 56·4 | 0 53·7 | 4·5 | 0·3 | 10·5 | 0·6 | 16·5 | 1·0 |
| 46 | 0 56·5 | 0 56·7 | 0 53·9 | 4·6 | 0·3 | 10·6 | 0·6 | 16·6 | 1·0 |
| 47 | 0 56·8 | 0 56·9 | 0 54·2 | 4·7 | 0·3 | 10·7 | 0·6 | 16·7 | 1·0 |
| 48 | 0 57·0 | 0 57·2 | 0 54·4 | 4·8 | 0·3 | 10·8 | 0·6 | 16·8 | 1·0 |
| 49 | 0 57·3 | 0 57·4 | 0 54·6 | 4·9 | 0·3 | 10·9 | 0·6 | 16·9 | 1·0 |
| 50 | 0 57·5 | 0 57·7 | 0 54·9 | 5·0 | 0·3 | 11·0 | 0·6 | 17·0 | 1·0 |
| 51 | 0 57·8 | 0 57·9 | 0 55·1 | 5·1 | 0·3 | 11·1 | 0·6 | 17·1 | 1·0 |
| 52 | 0 58·0 | 0 58·2 | 0 55·4 | 5·2 | 0·3 | 11·2 | 0·7 | 17·2 | 1·0 |
| 53 | 0 58·3 | 0 58·4 | 0 55·6 | 5·3 | 0·3 | 11·3 | 0·7 | 17·3 | 1·0 |
| 54 | 0 58·5 | 0 58·7 | 0 55·8 | 5·4 | 0·3 | 11·4 | 0·7 | 17·4 | 1·0 |
| 55 | 0 58·8 | 0 58·9 | 0 56·1 | 5·5 | 0·3 | 11·5 | 0·7 | 17·5 | 1·0 |
| 56 | 0 59·0 | 0 59·2 | 0 56·3 | 5·6 | 0·3 | 11·6 | 0·7 | 17·6 | 1·0 |
| 57 | 0 59·3 | 0 59·4 | 0 56·6 | 5·7 | 0·3 | 11·7 | 0·7 | 17·7 | 1·0 |
| 58 | 0 59·5 | 0 59·7 | 0 56·8 | 5·8 | 0·3 | 11·8 | 0·7 | 17·8 | 1·0 |
| 59 | 0 59·8 | 0 59·9 | 0 57·0 | 5·9 | 0·3 | 11·9 | 0·7 | 17·9 | 1·0 |
| 60 | 1 00·0 | 1 00·2 | 0 57·3 | 6·0 | 0·4 | 12·0 | 0·7 | 18·0 | 1·1 |

| $4^{m}$ s | SUN PLANETS ° ′ | ARIES ° ′ | MOON ° ′ | v or d ′ | Corrⁿ ′ | v or d ′ | Corrⁿ ′ | v or d ′ | Corrⁿ ′ |
|---|---|---|---|---|---|---|---|---|---|
| 00 | 1 00·0 | 1 00·2 | 0 57·3 | 0·0 | 0·0 | 6·0 | 0·5 | 12·0 | 0·9 |
| 01 | 1 00·3 | 1 00·4 | 0 57·5 | 0·1 | 0·0 | 6·1 | 0·5 | 12·1 | 0·9 |
| 02 | 1 00·5 | 1 00·7 | 0 57·7 | 0·2 | 0·0 | 6·2 | 0·5 | 12·2 | 0·9 |
| 03 | 1 00·8 | 1 00·9 | 0 58·0 | 0·3 | 0·0 | 6·3 | 0·5 | 12·3 | 0·9 |
| 04 | 1 01·0 | 1 01·2 | 0 58·2 | 0·4 | 0·0 | 6·4 | 0·5 | 12·4 | 0·9 |
| 05 | 1 01·3 | 1 01·4 | 0 58·5 | 0·5 | 0·0 | 6·5 | 0·5 | 12·5 | 0·9 |
| 06 | 1 01·5 | 1 01·7 | 0 58·7 | 0·6 | 0·0 | 6·6 | 0·5 | 12·6 | 0·9 |
| 07 | 1 01·8 | 1 01·9 | 0 58·9 | 0·7 | 0·1 | 6·7 | 0·5 | 12·7 | 1·0 |
| 08 | 1 02·0 | 1 02·2 | 0 59·2 | 0·8 | 0·1 | 6·8 | 0·5 | 12·8 | 1·0 |
| 09 | 1 02·3 | 1 02·4 | 0 59·4 | 0·9 | 0·1 | 6·9 | 0·5 | 12·9 | 1·0 |
| 10 | 1 02·5 | 1 02·7 | 0 59·7 | 1·0 | 0·1 | 7·0 | 0·5 | 13·0 | 1·0 |
| 11 | 1 02·8 | 1 02·9 | 0 59·9 | 1·1 | 0·1 | 7·1 | 0·5 | 13·1 | 1·0 |
| 12 | 1 03·0 | 1 03·2 | 1 00·1 | 1·2 | 0·1 | 7·2 | 0·5 | 13·2 | 1·0 |
| 13 | 1 03·3 | 1 03·4 | 1 00·4 | 1·3 | 0·1 | 7·3 | 0·5 | 13·3 | 1·0 |
| 14 | 1 03·5 | 1 03·7 | 1 00·6 | 1·4 | 0·1 | 7·4 | 0·6 | 13·4 | 1·0 |
| 15 | 1 03·8 | 1 03·9 | 1 00·8 | 1·5 | 0·1 | 7·5 | 0·6 | 13·5 | 1·0 |
| 16 | 1 04·0 | 1 04·2 | 1 01·1 | 1·6 | 0·1 | 7·6 | 0·6 | 13·6 | 1·0 |
| 17 | 1 04·3 | 1 04·4 | 1 01·3 | 1·7 | 0·1 | 7·7 | 0·6 | 13·7 | 1·0 |
| 18 | 1 04·5 | 1 04·7 | 1 01·6 | 1·8 | 0·1 | 7·8 | 0·6 | 13·8 | 1·0 |
| 19 | 1 04·8 | 1 04·9 | 1 01·8 | 1·9 | 0·1 | 7·9 | 0·6 | 13·9 | 1·0 |
| 20 | 1 05·0 | 1 05·2 | 1 02·0 | 2·0 | 0·2 | 8·0 | 0·6 | 14·0 | 1·1 |
| 21 | 1 05·3 | 1 05·4 | 1 02·3 | 2·1 | 0·2 | 8·1 | 0·6 | 14·1 | 1·1 |
| 22 | 1 05·5 | 1 05·7 | 1 02·5 | 2·2 | 0·2 | 8·2 | 0·6 | 14·2 | 1·1 |
| 23 | 1 05·8 | 1 05·9 | 1 02·8 | 2·3 | 0·2 | 8·3 | 0·6 | 14·3 | 1·1 |
| 24 | 1 06·0 | 1 06·2 | 1 03·0 | 2·4 | 0·2 | 8·4 | 0·6 | 14·4 | 1·1 |
| 25 | 1 06·3 | 1 06·4 | 1 03·2 | 2·5 | 0·2 | 8·5 | 0·6 | 14·5 | 1·1 |
| 26 | 1 06·5 | 1 06·7 | 1 03·5 | 2·6 | 0·2 | 8·6 | 0·6 | 14·6 | 1·1 |
| 27 | 1 06·8 | 1 06·9 | 1 03·7 | 2·7 | 0·2 | 8·7 | 0·7 | 14·7 | 1·1 |
| 28 | 1 07·0 | 1 07·2 | 1 03·9 | 2·8 | 0·2 | 8·8 | 0·7 | 14·8 | 1·1 |
| 29 | 1 07·3 | 1 07·4 | 1 04·2 | 2·9 | 0·2 | 8·9 | 0·7 | 14·9 | 1·1 |
| 30 | 1 07·5 | 1 07·7 | 1 04·4 | 3·0 | 0·2 | 9·0 | 0·7 | 15·0 | 1·1 |
| 31 | 1 07·8 | 1 07·9 | 1 04·7 | 3·1 | 0·2 | 9·1 | 0·7 | 15·1 | 1·1 |
| 32 | 1 08·0 | 1 08·2 | 1 04·9 | 3·2 | 0·2 | 9·2 | 0·7 | 15·2 | 1·1 |
| 33 | 1 08·3 | 1 08·4 | 1 05·1 | 3·3 | 0·2 | 9·3 | 0·7 | 15·3 | 1·1 |
| 34 | 1 08·5 | 1 08·7 | 1 05·4 | 3·4 | 0·3 | 9·4 | 0·7 | 15·4 | 1·2 |
| 35 | 1 08·8 | 1 08·9 | 1 05·6 | 3·5 | 0·3 | 9·5 | 0·7 | 15·5 | 1·2 |
| 36 | 1 09·0 | 1 09·2 | 1 05·9 | 3·6 | 0·3 | 9·6 | 0·7 | 15·6 | 1·2 |
| 37 | 1 09·3 | 1 09·4 | 1 06·1 | 3·7 | 0·3 | 9·7 | 0·7 | 15·7 | 1·2 |
| 38 | 1 09·5 | 1 09·7 | 1 06·3 | 3·8 | 0·3 | 9·8 | 0·7 | 15·8 | 1·2 |
| 39 | 1 09·8 | 1 09·9 | 1 06·6 | 3·9 | 0·3 | 9·9 | 0·7 | 15·9 | 1·2 |
| 40 | 1 10·0 | 1 10·2 | 1 06·8 | 4·0 | 0·3 | 10·0 | 0·8 | 16·0 | 1·2 |
| 41 | 1 10·3 | 1 10·4 | 1 07·0 | 4·1 | 0·3 | 10·1 | 0·8 | 16·1 | 1·2 |
| 42 | 1 10·5 | 1 10·7 | 1 07·3 | 4·2 | 0·3 | 10·2 | 0·8 | 16·2 | 1·2 |
| 43 | 1 10·8 | 1 10·9 | 1 07·5 | 4·3 | 0·3 | 10·3 | 0·8 | 16·3 | 1·2 |
| 44 | 1 11·0 | 1 11·2 | 1 07·8 | 4·4 | 0·3 | 10·4 | 0·8 | 16·4 | 1·2 |
| 45 | 1 11·3 | 1 11·4 | 1 08·0 | 4·5 | 0·3 | 10·5 | 0·8 | 16·5 | 1·2 |
| 46 | 1 11·5 | 1 11·7 | 1 08·2 | 4·6 | 0·3 | 10·6 | 0·8 | 16·6 | 1·2 |
| 47 | 1 11·8 | 1 11·9 | 1 08·5 | 4·7 | 0·4 | 10·7 | 0·8 | 16·7 | 1·3 |
| 48 | 1 12·0 | 1 12·2 | 1 08·7 | 4·8 | 0·4 | 10·8 | 0·8 | 16·8 | 1·3 |
| 49 | 1 12·3 | 1 12·4 | 1 09·0 | 4·9 | 0·4 | 10·9 | 0·8 | 16·9 | 1·3 |
| 50 | 1 12·5 | 1 12·7 | 1 09·2 | 5·0 | 0·4 | 11·0 | 0·8 | 17·0 | 1·3 |
| 51 | 1 12·8 | 1 12·9 | 1 09·4 | 5·1 | 0·4 | 11·1 | 0·8 | 17·1 | 1·3 |
| 52 | 1 13·0 | 1 13·2 | 1 09·7 | 5·2 | 0·4 | 11·2 | 0·8 | 17·2 | 1·3 |
| 53 | 1 13·3 | 1 13·5 | 1 09·9 | 5·3 | 0·4 | 11·3 | 0·8 | 17·3 | 1·3 |
| 54 | 1 13·5 | 1 13·7 | 1 10·2 | 5·4 | 0·4 | 11·4 | 0·9 | 17·4 | 1·3 |
| 55 | 1 13·8 | 1 14·0 | 1 10·4 | 5·5 | 0·4 | 11·5 | 0·9 | 17·5 | 1·3 |
| 56 | 1 14·0 | 1 14·2 | 1 10·6 | 5·6 | 0·4 | 11·6 | 0·9 | 17·6 | 1·3 |
| 57 | 1 14·3 | 1 14·5 | 1 10·9 | 5·7 | 0·4 | 11·7 | 0·9 | 17·7 | 1·3 |
| 58 | 1 14·5 | 1 14·7 | 1 11·1 | 5·8 | 0·4 | 11·8 | 0·9 | 17·8 | 1·3 |
| 59 | 1 14·8 | 1 15·0 | 1 11·3 | 5·9 | 0·4 | 11·9 | 0·9 | 17·9 | 1·3 |
| 60 | 1 15·0 | 1 15·2 | 1 11·6 | 6·0 | 0·5 | 12·0 | 0·9 | 18·0 | 1·4 |

| $5^{m}$ s | SUN PLANETS ° ′ | ARIES ° ′ | MOON ° ′ | v or d ′ | Corrⁿ ′ | v or d ′ | Corrⁿ ′ | v or d ′ | Corrⁿ ′ |
|---|---|---|---|---|---|---|---|---|---|
| 00 | 1 15·0 | 1 15·2 | 1 11·6 | 0·0 | 0·0 | 6·0 | 0·6 | 12·0 | 1·1 |
| 01 | 1 15·3 | 1 15·5 | 1 11·8 | 0·1 | 0·0 | 6·1 | 0·6 | 12·1 | 1·1 |
| 02 | 1 15·5 | 1 15·7 | 1 12·1 | 0·2 | 0·0 | 6·2 | 0·6 | 12·2 | 1·1 |
| 03 | 1 15·8 | 1 16·0 | 1 12·3 | 0·3 | 0·0 | 6·3 | 0·6 | 12·3 | 1·1 |
| 04 | 1 16·0 | 1 16·2 | 1 12·5 | 0·4 | 0·0 | 6·4 | 0·6 | 12·4 | 1·1 |
| 05 | 1 16·3 | 1 16·5 | 1 12·8 | 0·5 | 0·0 | 6·5 | 0·6 | 12·5 | 1·1 |
| 06 | 1 16·5 | 1 16·7 | 1 13·0 | 0·6 | 0·1 | 6·6 | 0·6 | 12·6 | 1·2 |
| 07 | 1 16·8 | 1 17·0 | 1 13·3 | 0·7 | 0·1 | 6·7 | 0·6 | 12·7 | 1·2 |
| 08 | 1 17·0 | 1 17·2 | 1 13·5 | 0·8 | 0·1 | 6·8 | 0·6 | 12·8 | 1·2 |
| 09 | 1 17·3 | 1 17·5 | 1 13·7 | 0·9 | 0·1 | 6·9 | 0·6 | 12·9 | 1·2 |
| 10 | 1 17·5 | 1 17·7 | 1 14·0 | 1·0 | 0·1 | 7·0 | 0·6 | 13·0 | 1·2 |
| 11 | 1 17·8 | 1 18·0 | 1 14·2 | 1·1 | 0·1 | 7·1 | 0·7 | 13·1 | 1·2 |
| 12 | 1 18·0 | 1 18·2 | 1 14·4 | 1·2 | 0·1 | 7·2 | 0·7 | 13·2 | 1·2 |
| 13 | 1 18·3 | 1 18·5 | 1 14·7 | 1·3 | 0·1 | 7·3 | 0·7 | 13·3 | 1·2 |
| 14 | 1 18·5 | 1 18·7 | 1 14·9 | 1·4 | 0·1 | 7·4 | 0·7 | 13·4 | 1·2 |
| 15 | 1 18·8 | 1 19·0 | 1 15·2 | 1·5 | 0·1 | 7·5 | 0·7 | 13·5 | 1·2 |
| 16 | 1 19·0 | 1 19·2 | 1 15·4 | 1·6 | 0·1 | 7·6 | 0·7 | 13·6 | 1·2 |
| 17 | 1 19·3 | 1 19·5 | 1 15·6 | 1·7 | 0·2 | 7·7 | 0·7 | 13·7 | 1·3 |
| 18 | 1 19·5 | 1 19·7 | 1 15·9 | 1·8 | 0·2 | 7·8 | 0·7 | 13·8 | 1·3 |
| 19 | 1 19·8 | 1 20·0 | 1 16·1 | 1·9 | 0·2 | 7·9 | 0·7 | 13·9 | 1·3 |
| 20 | 1 20·0 | 1 20·2 | 1 16·4 | 2·0 | 0·2 | 8·0 | 0·7 | 14·0 | 1·3 |
| 21 | 1 20·3 | 1 20·5 | 1 16·6 | 2·1 | 0·2 | 8·1 | 0·7 | 14·1 | 1·3 |
| 22 | 1 20·5 | 1 20·7 | 1 16·8 | 2·2 | 0·2 | 8·2 | 0·8 | 14·2 | 1·3 |
| 23 | 1 20·8 | 1 21·0 | 1 17·1 | 2·3 | 0·2 | 8·3 | 0·8 | 14·3 | 1·3 |
| 24 | 1 21·0 | 1 21·2 | 1 17·3 | 2·4 | 0·2 | 8·4 | 0·8 | 14·4 | 1·3 |
| 25 | 1 21·3 | 1 21·5 | 1 17·5 | 2·5 | 0·2 | 8·5 | 0·8 | 14·5 | 1·3 |
| 26 | 1 21·5 | 1 21·7 | 1 17·8 | 2·6 | 0·2 | 8·6 | 0·8 | 14·6 | 1·3 |
| 27 | 1 21·8 | 1 22·0 | 1 18·0 | 2·7 | 0·2 | 8·7 | 0·8 | 14·7 | 1·3 |
| 28 | 1 22·0 | 1 22·2 | 1 18·3 | 2·8 | 0·3 | 8·8 | 0·8 | 14·8 | 1·4 |
| 29 | 1 22·3 | 1 22·5 | 1 18·5 | 2·9 | 0·3 | 8·9 | 0·8 | 14·9 | 1·4 |
| 30 | 1 22·5 | 1 22·7 | 1 18·7 | 3·0 | 0·3 | 9·0 | 0·8 | 15·0 | 1·4 |
| 31 | 1 22·8 | 1 23·0 | 1 19·0 | 3·1 | 0·3 | 9·1 | 0·8 | 15·1 | 1·4 |
| 32 | 1 23·0 | 1 23·2 | 1 19·2 | 3·2 | 0·3 | 9·2 | 0·8 | 15·2 | 1·4 |
| 33 | 1 23·3 | 1 23·5 | 1 19·5 | 3·3 | 0·3 | 9·3 | 0·9 | 15·3 | 1·4 |
| 34 | 1 23·5 | 1 23·7 | 1 19·7 | 3·4 | 0·3 | 9·4 | 0·9 | 15·4 | 1·4 |
| 35 | 1 23·8 | 1 24·0 | 1 19·9 | 3·5 | 0·3 | 9·5 | 0·9 | 15·5 | 1·4 |
| 36 | 1 24·0 | 1 24·2 | 1 20·2 | 3·6 | 0·3 | 9·6 | 0·9 | 15·6 | 1·4 |
| 37 | 1 24·3 | 1 24·5 | 1 20·4 | 3·7 | 0·3 | 9·7 | 0·9 | 15·7 | 1·4 |
| 38 | 1 24·5 | 1 24·7 | 1 20·7 | 3·8 | 0·3 | 9·8 | 0·9 | 15·8 | 1·4 |
| 39 | 1 24·8 | 1 25·0 | 1 20·9 | 3·9 | 0·4 | 9·9 | 0·9 | 15·9 | 1·5 |
| 40 | 1 25·0 | 1 25·2 | 1 21·1 | 4·0 | 0·4 | 10·0 | 0·9 | 16·0 | 1·5 |
| 41 | 1 25·3 | 1 25·5 | 1 21·4 | 4·1 | 0·4 | 10·1 | 0·9 | 16·1 | 1·5 |
| 42 | 1 25·5 | 1 25·7 | 1 21·6 | 4·2 | 0·4 | 10·2 | 0·9 | 16·2 | 1·5 |
| 43 | 1 25·8 | 1 26·0 | 1 21·8 | 4·3 | 0·4 | 10·3 | 0·9 | 16·3 | 1·5 |
| 44 | 1 26·0 | 1 26·2 | 1 22·1 | 4·4 | 0·4 | 10·4 | 1·0 | 16·4 | 1·5 |
| 45 | 1 26·3 | 1 26·5 | 1 22·3 | 4·5 | 0·4 | 10·5 | 1·0 | 16·5 | 1·5 |
| 46 | 1 26·5 | 1 26·7 | 1 22·6 | 4·6 | 0·4 | 10·6 | 1·0 | 16·6 | 1·5 |
| 47 | 1 26·8 | 1 27·0 | 1 22·8 | 4·7 | 0·4 | 10·7 | 1·0 | 16·7 | 1·5 |
| 48 | 1 27·0 | 1 27·2 | 1 23·0 | 4·8 | 0·4 | 10·8 | 1·0 | 16·8 | 1·5 |
| 49 | 1 27·3 | 1 27·5 | 1 23·3 | 4·9 | 0·4 | 10·9 | 1·0 | 16·9 | 1·5 |
| 50 | 1 27·5 | 1 27·7 | 1 23·5 | 5·0 | 0·5 | 11·0 | 1·0 | 17·0 | 1·6 |
| 51 | 1 27·8 | 1 28·0 | 1 23·8 | 5·1 | 0·5 | 11·1 | 1·0 | 17·1 | 1·6 |
| 52 | 1 28·0 | 1 28·2 | 1 24·0 | 5·2 | 0·5 | 11·2 | 1·0 | 17·2 | 1·6 |
| 53 | 1 28·3 | 1 28·5 | 1 24·2 | 5·3 | 0·5 | 11·3 | 1·0 | 17·3 | 1·6 |
| 54 | 1 28·5 | 1 28·7 | 1 24·5 | 5·4 | 0·5 | 11·4 | 1·0 | 17·4 | 1·6 |
| 55 | 1 28·8 | 1 29·0 | 1 24·7 | 5·5 | 0·5 | 11·5 | 1·1 | 17·5 | 1·6 |
| 56 | 1 29·0 | 1 29·2 | 1 24·9 | 5·6 | 0·5 | 11·6 | 1·1 | 17·6 | 1·6 |
| 57 | 1 29·3 | 1 29·5 | 1 25·2 | 5·7 | 0·5 | 11·7 | 1·1 | 17·7 | 1·6 |
| 58 | 1 29·5 | 1 29·7 | 1 25·4 | 5·8 | 0·5 | 11·8 | 1·1 | 17·8 | 1·6 |
| 59 | 1 29·8 | 1 30·0 | 1 25·7 | 5·9 | 0·5 | 11·9 | 1·1 | 17·9 | 1·6 |
| 60 | 1 30·0 | 1 30·2 | 1 25·9 | 6·0 | 0·6 | 12·0 | 1·1 | 18·0 | 1·7 |

| $6^m$ | SUN PLANETS | ARIES | MOON | v or d | Corr$^n$ | v or d | Corr$^n$ | v or d | Corr$^n$ |
|---|---|---|---|---|---|---|---|---|---|
| s | ° ′ | ° ′ | ° ′ | ′ | ′ | ′ | ′ | ′ | ′ |
| 00 | 1 30·0 | 1 30·2 | 1 25·9 | 0·0 | 0·0 | 6·0 | 0·7 | 12·0 | 1·3 |
| 01 | 1 30·3 | 1 30·5 | 1 26·1 | 0·1 | 0·0 | 6·1 | 0·7 | 12·1 | 1·3 |
| 02 | 1 30·5 | 1 30·7 | 1 26·4 | 0·2 | 0·0 | 6·2 | 0·7 | 12·2 | 1·3 |
| 03 | 1 30·8 | 1 31·0 | 1 26·6 | 0·3 | 0·0 | 6·3 | 0·7 | 12·3 | 1·3 |
| 04 | 1 31·0 | 1 31·2 | 1 26·9 | 0·4 | 0·0 | 6·4 | 0·7 | 12·4 | 1·3 |
| 05 | 1 31·3 | 1 31·5 | 1 27·1 | 0·5 | 0·1 | 6·5 | 0·7 | 12·5 | 1·4 |
| 06 | 1 31·5 | 1 31·8 | 1 27·3 | 0·6 | 0·1 | 6·6 | 0·7 | 12·6 | 1·4 |
| 07 | 1 31·8 | 1 32·0 | 1 27·6 | 0·7 | 0·1 | 6·7 | 0·7 | 12·7 | 1·4 |
| 08 | 1 32·0 | 1 32·3 | 1 27·8 | 0·8 | 0·1 | 6·8 | 0·7 | 12·8 | 1·4 |
| 09 | 1 32·3 | 1 32·5 | 1 28·0 | 0·9 | 0·1 | 6·9 | 0·7 | 12·9 | 1·4 |
| 10 | 1 32·5 | 1 32·8 | 1 28·3 | 1·0 | 0·1 | 7·0 | 0·8 | 13·0 | 1·4 |
| 11 | 1 32·8 | 1 33·0 | 1 28·5 | 1·1 | 0·1 | 7·1 | 0·8 | 13·1 | 1·4 |
| 12 | 1 33·0 | 1 33·3 | 1 28·8 | 1·2 | 0·1 | 7·2 | 0·8 | 13·2 | 1·4 |
| 13 | 1 33·3 | 1 33·5 | 1 29·0 | 1·3 | 0·1 | 7·3 | 0·8 | 13·3 | 1·4 |
| 14 | 1 33·5 | 1 33·8 | 1 29·2 | 1·4 | 0·2 | 7·4 | 0·8 | 13·4 | 1·5 |
| 15 | 1 33·8 | 1 34·0 | 1 29·5 | 1·5 | 0·2 | 7·5 | 0·8 | 13·5 | 1·5 |
| 16 | 1 34·0 | 1 34·3 | 1 29·7 | 1·6 | 0·2 | 7·6 | 0·8 | 13·6 | 1·5 |
| 17 | 1 34·3 | 1 34·5 | 1 30·0 | 1·7 | 0·2 | 7·7 | 0·8 | 13·7 | 1·5 |
| 18 | 1 34·5 | 1 34·8 | 1 30·2 | 1·8 | 0·2 | 7·8 | 0·8 | 13·8 | 1·5 |
| 19 | 1 34·8 | 1 35·0 | 1 30·4 | 1·9 | 0·2 | 7·9 | 0·9 | 13·9 | 1·5 |
| 20 | 1 35·0 | 1 35·3 | 1 30·7 | 2·0 | 0·2 | 8·0 | 0·9 | 14·0 | 1·5 |
| 21 | 1 35·3 | 1 35·5 | 1 30·9 | 2·1 | 0·2 | 8·1 | 0·9 | 14·1 | 1·5 |
| 22 | 1 35·5 | 1 35·8 | 1 31·1 | 2·2 | 0·2 | 8·2 | 0·9 | 14·2 | 1·5 |
| 23 | 1 35·8 | 1 36·0 | 1 31·4 | 2·3 | 0·2 | 8·3 | 0·9 | 14·3 | 1·5 |
| 24 | 1 36·0 | 1 36·3 | 1 31·6 | 2·4 | 0·3 | 8·4 | 0·9 | 14·4 | 1·6 |
| 25 | 1 36·3 | 1 36·5 | 1 31·9 | 2·5 | 0·3 | 8·5 | 0·9 | 14·5 | 1·6 |
| 26 | 1 36·5 | 1 36·8 | 1 32·1 | 2·6 | 0·3 | 8·6 | 0·9 | 14·6 | 1·6 |
| 27 | 1 36·8 | 1 37·0 | 1 32·3 | 2·7 | 0·3 | 8·7 | 0·9 | 14·7 | 1·6 |
| 28 | 1 37·0 | 1 37·3 | 1 32·6 | 2·8 | 0·3 | 8·8 | 1·0 | 14·8 | 1·6 |
| 29 | 1 37·3 | 1 37·5 | 1 32·8 | 2·9 | 0·3 | 8·9 | 1·0 | 14·9 | 1·6 |
| 30 | 1 37·5 | 1 37·8 | 1 33·1 | 3·0 | 0·3 | 9·0 | 1·0 | 15·0 | 1·6 |
| 31 | 1 37·8 | 1 38·0 | 1 33·3 | 3·1 | 0·3 | 9·1 | 1·0 | 15·1 | 1·6 |
| 32 | 1 38·0 | 1 38·3 | 1 33·5 | 3·2 | 0·3 | 9·2 | 1·0 | 15·2 | 1·6 |
| 33 | 1 38·3 | 1 38·5 | 1 33·8 | 3·3 | 0·4 | 9·3 | 1·0 | 15·3 | 1·7 |
| 34 | 1 38·5 | 1 38·8 | 1 34·0 | 3·4 | 0·4 | 9·4 | 1·0 | 15·4 | 1·7 |
| 35 | 1 38·8 | 1 39·0 | 1 34·3 | 3·5 | 0·4 | 9·5 | 1·0 | 15·5 | 1·7 |
| 36 | 1 39·0 | 1 39·3 | 1 34·5 | 3·6 | 0·4 | 9·6 | 1·0 | 15·6 | 1·7 |
| 37 | 1 39·3 | 1 39·5 | 1 34·7 | 3·7 | 0·4 | 9·7 | 1·1 | 15·7 | 1·7 |
| 38 | 1 39·5 | 1 39·8 | 1 35·0 | 3·8 | 0·4 | 9·8 | 1·1 | 15·8 | 1·7 |
| 39 | 1 39·8 | 1 40·0 | 1 35·2 | 3·9 | 0·4 | 9·9 | 1·1 | 15·9 | 1·7 |
| 40 | 1 40·0 | 1 40·3 | 1 35·4 | 4·0 | 0·4 | 10·0 | 1·1 | 16·0 | 1·7 |
| 41 | 1 40·3 | 1 40·5 | 1 35·7 | 4·1 | 0·4 | 10·1 | 1·1 | 16·1 | 1·7 |
| 42 | 1 40·5 | 1 40·8 | 1 35·9 | 4·2 | 0·5 | 10·2 | 1·1 | 16·2 | 1·8 |
| 43 | 1 40·8 | 1 41·0 | 1 36·2 | 4·3 | 0·5 | 10·3 | 1·1 | 16·3 | 1·8 |
| 44 | 1 41·0 | 1 41·3 | 1 36·4 | 4·4 | 0·5 | 10·4 | 1·1 | 16·4 | 1·8 |
| 45 | 1 41·3 | 1 41·5 | 1 36·6 | 4·5 | 0·5 | 10·5 | 1·1 | 16·5 | 1·8 |
| 46 | 1 41·5 | 1 41·8 | 1 36·9 | 4·6 | 0·5 | 10·6 | 1·1 | 16·6 | 1·8 |
| 47 | 1 41·8 | 1 42·0 | 1 37·1 | 4·7 | 0·5 | 10·7 | 1·2 | 16·7 | 1·8 |
| 48 | 1 42·0 | 1 42·3 | 1 37·4 | 4·8 | 0·5 | 10·8 | 1·2 | 16·8 | 1·8 |
| 49 | 1 42·3 | 1 42·5 | 1 37·6 | 4·9 | 0·5 | 10·9 | 1·2 | 16·9 | 1·8 |
| 50 | 1 42·5 | 1 42·8 | 1 37·8 | 5·0 | 0·5 | 11·0 | 1·2 | 17·0 | 1·8 |
| 51 | 1 42·8 | 1 43·0 | 1 38·1 | 5·1 | 0·6 | 11·1 | 1·2 | 17·1 | 1·9 |
| 52 | 1 43·0 | 1 43·3 | 1 38·3 | 5·2 | 0·6 | 11·2 | 1·2 | 17·2 | 1·9 |
| 53 | 1 43·3 | 1 43·5 | 1 38·5 | 5·3 | 0·6 | 11·3 | 1·2 | 17·3 | 1·9 |
| 54 | 1 43·5 | 1 43·8 | 1 38·8 | 5·4 | 0·6 | 11·4 | 1·2 | 17·4 | 1·9 |
| 55 | 1 43·8 | 1 44·0 | 1 39·0 | 5·5 | 0·6 | 11·5 | 1·2 | 17·5 | 1·9 |
| 56 | 1 44·0 | 1 44·3 | 1 39·3 | 5·6 | 0·6 | 11·6 | 1·3 | 17·6 | 1·9 |
| 57 | 1 44·3 | 1 44·5 | 1 39·5 | 5·7 | 0·6 | 11·7 | 1·3 | 17·7 | 1·9 |
| 58 | 1 44·5 | 1 44·8 | 1 39·7 | 5·8 | 0·6 | 11·8 | 1·3 | 17·8 | 1·9 |
| 59 | 1 44·8 | 1 45·0 | 1 40·0 | 5·9 | 0·6 | 11·9 | 1·3 | 17·9 | 1·9 |
| 60 | 1 45·0 | 1 45·3 | 1 40·2 | 6·0 | 0·7 | 12·0 | 1·3 | 18·0 | 2·0 |

| $7^m$ | SUN PLANETS | ARIES | MOON | v or d | Corr$^n$ | v or d | Corr$^n$ | v or d | Corr$^n$ |
|---|---|---|---|---|---|---|---|---|---|
| s | ° ′ | ° ′ | ° ′ | ′ | ′ | ′ | ′ | ′ | ′ |
| 00 | 1 45·0 | 1 45·3 | 1 40·2 | 0·0 | 0·0 | 6·0 | 0·8 | 12·0 | 1·5 |
| 01 | 1 45·3 | 1 45·5 | 1 40·5 | 0·1 | 0·0 | 6·1 | 0·8 | 12·1 | 1·5 |
| 02 | 1 45·5 | 1 45·8 | 1 40·7 | 0·2 | 0·0 | 6·2 | 0·8 | 12·2 | 1·5 |
| 03 | 1 45·8 | 1 46·0 | 1 40·9 | 0·3 | 0·0 | 6·3 | 0·8 | 12·3 | 1·5 |
| 04 | 1 46·0 | 1 46·3 | 1 41·2 | 0·4 | 0·1 | 6·4 | 0·8 | 12·4 | 1·6 |
| 05 | 1 46·3 | 1 46·5 | 1 41·4 | 0·5 | 0·1 | 6·5 | 0·8 | 12·5 | 1·6 |
| 06 | 1 46·5 | 1 46·8 | 1 41·6 | 0·6 | 0·1 | 6·6 | 0·8 | 12·6 | 1·6 |
| 07 | 1 46·8 | 1 47·0 | 1 41·9 | 0·7 | 0·1 | 6·7 | 0·8 | 12·7 | 1·6 |
| 08 | 1 47·0 | 1 47·3 | 1 42·1 | 0·8 | 0·1 | 6·8 | 0·9 | 12·8 | 1·6 |
| 09 | 1 47·3 | 1 47·5 | 1 42·4 | 0·9 | 0·1 | 6·9 | 0·9 | 12·9 | 1·6 |
| 10 | 1 47·5 | 1 47·8 | 1 42·6 | 1·0 | 0·1 | 7·0 | 0·9 | 13·0 | 1·6 |
| 11 | 1 47·8 | 1 48·0 | 1 42·8 | 1·1 | 0·1 | 7·1 | 0·9 | 13·1 | 1·6 |
| 12 | 1 48·0 | 1 48·3 | 1 43·1 | 1·2 | 0·2 | 7·2 | 0·9 | 13·2 | 1·7 |
| 13 | 1 48·3 | 1 48·5 | 1 43·3 | 1·3 | 0·2 | 7·3 | 0·9 | 13·3 | 1·7 |
| 14 | 1 48·5 | 1 48·8 | 1 43·6 | 1·4 | 0·2 | 7·4 | 0·9 | 13·4 | 1·7 |
| 15 | 1 48·8 | 1 49·0 | 1 43·8 | 1·5 | 0·2 | 7·5 | 0·9 | 13·5 | 1·7 |
| 16 | 1 49·0 | 1 49·3 | 1 44·0 | 1·6 | 0·2 | 7·6 | 1·0 | 13·6 | 1·7 |
| 17 | 1 49·3 | 1 49·5 | 1 44·3 | 1·7 | 0·2 | 7·7 | 1·0 | 13·7 | 1·7 |
| 18 | 1 49·5 | 1 49·8 | 1 44·5 | 1·8 | 0·2 | 7·8 | 1·0 | 13·8 | 1·7 |
| 19 | 1 49·8 | 1 50·1 | 1 44·8 | 1·9 | 0·2 | 7·9 | 1·0 | 13·9 | 1·7 |
| 20 | 1 50·0 | 1 50·3 | 1 45·0 | 2·0 | 0·3 | 8·0 | 1·0 | 14·0 | 1·8 |
| 21 | 1 50·3 | 1 50·6 | 1 45·2 | 2·1 | 0·3 | 8·1 | 1·0 | 14·1 | 1·8 |
| 22 | 1 50·5 | 1 50·8 | 1 45·5 | 2·2 | 0·3 | 8·2 | 1·0 | 14·2 | 1·8 |
| 23 | 1 50·8 | 1 51·1 | 1 45·7 | 2·3 | 0·3 | 8·3 | 1·0 | 14·3 | 1·8 |
| 24 | 1 51·0 | 1 51·3 | 1 45·9 | 2·4 | 0·3 | 8·4 | 1·1 | 14·4 | 1·8 |
| 25 | 1 51·3 | 1 51·6 | 1 46·2 | 2·5 | 0·3 | 8·5 | 1·1 | 14·5 | 1·8 |
| 26 | 1 51·5 | 1 51·8 | 1 46·4 | 2·6 | 0·3 | 8·6 | 1·1 | 14·6 | 1·8 |
| 27 | 1 51·8 | 1 52·1 | 1 46·7 | 2·7 | 0·3 | 8·7 | 1·1 | 14·7 | 1·8 |
| 28 | 1 52·0 | 1 52·3 | 1 46·9 | 2·8 | 0·4 | 8·8 | 1·1 | 14·8 | 1·9 |
| 29 | 1 52·3 | 1 52·6 | 1 47·1 | 2·9 | 0·4 | 8·9 | 1·1 | 14·9 | 1·9 |
| 30 | 1 52·5 | 1 52·8 | 1 47·4 | 3·0 | 0·4 | 9·0 | 1·1 | 15·0 | 1·9 |
| 31 | 1 52·8 | 1 53·1 | 1 47·6 | 3·1 | 0·4 | 9·1 | 1·1 | 15·1 | 1·9 |
| 32 | 1 53·0 | 1 53·3 | 1 47·9 | 3·2 | 0·4 | 9·2 | 1·2 | 15·2 | 1·9 |
| 33 | 1 53·3 | 1 53·6 | 1 48·1 | 3·3 | 0·4 | 9·3 | 1·2 | 15·3 | 1·9 |
| 34 | 1 53·5 | 1 53·8 | 1 48·3 | 3·4 | 0·4 | 9·4 | 1·2 | 15·4 | 1·9 |
| 35 | 1 53·8 | 1 54·1 | 1 48·6 | 3·5 | 0·4 | 9·5 | 1·2 | 15·5 | 1·9 |
| 36 | 1 54·0 | 1 54·3 | 1 48·8 | 3·6 | 0·5 | 9·6 | 1·2 | 15·6 | 2·0 |
| 37 | 1 54·3 | 1 54·6 | 1 49·0 | 3·7 | 0·5 | 9·7 | 1·2 | 15·7 | 2·0 |
| 38 | 1 54·5 | 1 54·8 | 1 49·3 | 3·8 | 0·5 | 9·8 | 1·2 | 15·8 | 2·0 |
| 39 | 1 54·8 | 1 55·1 | 1 49·5 | 3·9 | 0·5 | 9·9 | 1·2 | 15·9 | 2·0 |
| 40 | 1 55·0 | 1 55·3 | 1 49·8 | 4·0 | 0·5 | 10·0 | 1·3 | 16·0 | 2·0 |
| 41 | 1 55·3 | 1 55·6 | 1 50·0 | 4·1 | 0·5 | 10·1 | 1·3 | 16·1 | 2·0 |
| 42 | 1 55·5 | 1 55·8 | 1 50·2 | 4·2 | 0·5 | 10·2 | 1·3 | 16·2 | 2·0 |
| 43 | 1 55·8 | 1 56·1 | 1 50·5 | 4·3 | 0·5 | 10·3 | 1·3 | 16·3 | 2·0 |
| 44 | 1 56·0 | 1 56·3 | 1 50·7 | 4·4 | 0·6 | 10·4 | 1·3 | 16·4 | 2·1 |
| 45 | 1 56·3 | 1 56·6 | 1 51·0 | 4·5 | 0·6 | 10·5 | 1·3 | 16·5 | 2·1 |
| 46 | 1 56·5 | 1 56·8 | 1 51·2 | 4·6 | 0·6 | 10·6 | 1·3 | 16·6 | 2·1 |
| 47 | 1 56·8 | 1 57·1 | 1 51·4 | 4·7 | 0·6 | 10·7 | 1·3 | 16·7 | 2·1 |
| 48 | 1 57·0 | 1 57·3 | 1 51·7 | 4·8 | 0·6 | 10·8 | 1·4 | 16·8 | 2·1 |
| 49 | 1 57·3 | 1 57·6 | 1 51·9 | 4·9 | 0·6 | 10·9 | 1·4 | 16·9 | 2·1 |
| 50 | 1 57·5 | 1 57·8 | 1 52·1 | 5·0 | 0·6 | 11·0 | 1·4 | 17·0 | 2·1 |
| 51 | 1 57·8 | 1 58·1 | 1 52·4 | 5·1 | 0·6 | 11·1 | 1·4 | 17·1 | 2·1 |
| 52 | 1 58·0 | 1 58·3 | 1 52·6 | 5·2 | 0·7 | 11·2 | 1·4 | 17·2 | 2·2 |
| 53 | 1 58·3 | 1 58·6 | 1 52·9 | 5·3 | 0·7 | 11·3 | 1·4 | 17·3 | 2·2 |
| 54 | 1 58·5 | 1 58·8 | 1 53·1 | 5·4 | 0·7 | 11·4 | 1·4 | 17·4 | 2·2 |
| 55 | 1 58·8 | 1 59·1 | 1 53·3 | 5·5 | 0·7 | 11·5 | 1·4 | 17·5 | 2·2 |
| 56 | 1 59·0 | 1 59·3 | 1 53·6 | 5·6 | 0·7 | 11·6 | 1·5 | 17·6 | 2·2 |
| 57 | 1 59·3 | 1 59·6 | 1 53·8 | 5·7 | 0·7 | 11·7 | 1·5 | 17·7 | 2·2 |
| 58 | 1 59·5 | 1 59·8 | 1 54·1 | 5·8 | 0·7 | 11·8 | 1·5 | 17·8 | 2·2 |
| 59 | 1 59·8 | 2 00·1 | 1 54·3 | 5·9 | 0·7 | 11·9 | 1·5 | 17·9 | 2·2 |
| 60 | 2 00·0 | 2 00·3 | 1 54·5 | 6·0 | 0·8 | 12·0 | 1·5 | 18·0 | 2·3 |

| $10^m$ | SUN PLANETS | ARIES | MOON | v or d | Corr$^n$ | v or d | Corr$^n$ | v or d | Corr$^n$ |
|---|---|---|---|---|---|---|---|---|---|
| s | ° ′ | ° ′ | ° ′ | ′ | ′ | ′ | ′ | ′ | ′ |
| 00 | 2 30·0 | 2 30·4 | 2 23·2 | 0·0 | 0·0 | 6·0 | 1·1 | 12·0 | 2·1 |
| 01 | 2 30·3 | 2 30·7 | 2 23·4 | 0·1 | 0·0 | 6·1 | 1·1 | 12·1 | 2·1 |
| 02 | 2 30·5 | 2 30·9 | 2 23·6 | 0·2 | 0·0 | 6·2 | 1·1 | 12·2 | 2·1 |
| 03 | 2 30·8 | 2 31·2 | 2 23·9 | 0·3 | 0·1 | 6·3 | 1·1 | 12·3 | 2·2 |
| 04 | 2 31·0 | 2 31·4 | 2 24·1 | 0·4 | 0·1 | 6·4 | 1·1 | 12·4 | 2·2 |
| 05 | 2 31·3 | 2 31·7 | 2 24·4 | 0·5 | 0·1 | 6·5 | 1·1 | 12·5 | 2·2 |
| 06 | 2 31·5 | 2 31·9 | 2 24·6 | 0·6 | 0·1 | 6·6 | 1·2 | 12·6 | 2·2 |
| 07 | 2 31·8 | 2 32·2 | 2 24·8 | 0·7 | 0·1 | 6·7 | 1·2 | 12·7 | 2·2 |
| 08 | 2 32·0 | 2 32·4 | 2 25·1 | 0·8 | 0·1 | 6·8 | 1·2 | 12·8 | 2·2 |
| 09 | 2 32·3 | 2 32·7 | 2 25·3 | 0·9 | 0·2 | 6·9 | 1·2 | 12·9 | 2·3 |
| 10 | 2 32·5 | 2 32·9 | 2 25·6 | 1·0 | 0·2 | 7·0 | 1·2 | 13·0 | 2·3 |
| 11 | 2 32·8 | 2 33·2 | 2 25·8 | 1·1 | 0·2 | 7·1 | 1·2 | 13·1 | 2·3 |
| 12 | 2 33·0 | 2 33·4 | 2 26·0 | 1·2 | 0·2 | 7·2 | 1·3 | 13·2 | 2·3 |
| 13 | 2 33·3 | 2 33·7 | 2 26·3 | 1·3 | 0·2 | 7·3 | 1·3 | 13·3 | 2·3 |
| 14 | 2 33·5 | 2 33·9 | 2 26·5 | 1·4 | 0·2 | 7·4 | 1·3 | 13·4 | 2·3 |
| 15 | 2 33·8 | 2 34·2 | 2 26·7 | 1·5 | 0·3 | 7·5 | 1·3 | 13·5 | 2·4 |
| 16 | 2 34·0 | 2 34·4 | 2 27·0 | 1·6 | 0·3 | 7·6 | 1·3 | 13·6 | 2·4 |
| 17 | 2 34·3 | 2 34·7 | 2 27·2 | 1·7 | 0·3 | 7·7 | 1·3 | 13·7 | 2·4 |
| 18 | 2 34·5 | 2 34·9 | 2 27·5 | 1·8 | 0·3 | 7·8 | 1·4 | 13·8 | 2·4 |
| 19 | 2 34·8 | 2 35·2 | 2 27·7 | 1·9 | 0·3 | 7·9 | 1·4 | 13·9 | 2·4 |
| 20 | 2 35·0 | 2 35·4 | 2 27·9 | 2·0 | 0·4 | 8·0 | 1·4 | 14·0 | 2·5 |
| 21 | 2 35·3 | 2 35·7 | 2 28·2 | 2·1 | 0·4 | 8·1 | 1·4 | 14·1 | 2·5 |
| 22 | 2 35·5 | 2 35·9 | 2 28·4 | 2·2 | 0·4 | 8·2 | 1·4 | 14·2 | 2·5 |
| 23 | 2 35·8 | 2 36·2 | 2 28·7 | 2·3 | 0·4 | 8·3 | 1·5 | 14·3 | 2·5 |
| 24 | 2 36·0 | 2 36·4 | 2 28·9 | 2·4 | 0·4 | 8·4 | 1·5 | 14·4 | 2·5 |
| 25 | 2 36·3 | 2 36·7 | 2 29·1 | 2·5 | 0·4 | 8·5 | 1·5 | 14·5 | 2·5 |
| 26 | 2 36·5 | 2 36·9 | 2 29·4 | 2·6 | 0·5 | 8·6 | 1·5 | 14·6 | 2·6 |
| 27 | 2 36·8 | 2 37·2 | 2 29·6 | 2·7 | 0·5 | 8·7 | 1·5 | 14·7 | 2·6 |
| 28 | 2 37·0 | 2 37·4 | 2 29·8 | 2·8 | 0·5 | 8·8 | 1·5 | 14·8 | 2·6 |
| 29 | 2 37·3 | 2 37·7 | 2 30·1 | 2·9 | 0·5 | 8·9 | 1·6 | 14·9 | 2·6 |
| 30 | 2 37·5 | 2 37·9 | 2 30·3 | 3·0 | 0·5 | 9·0 | 1·6 | 15·0 | 2·6 |
| 31 | 2 37·8 | 2 38·2 | 2 30·6 | 3·1 | 0·5 | 9·1 | 1·6 | 15·1 | 2·6 |
| 32 | 2 38·0 | 2 38·4 | 2 30·8 | 3·2 | 0·6 | 9·2 | 1·6 | 15·2 | 2·7 |
| 33 | 2 38·3 | 2 38·7 | 2 31·0 | 3·3 | 0·6 | 9·3 | 1·6 | 15·3 | 2·7 |
| 34 | 2 38·5 | 2 38·9 | 2 31·3 | 3·4 | 0·6 | 9·4 | 1·6 | 15·4 | 2·7 |
| 35 | 2 38·8 | 2 39·2 | 2 31·5 | 3·5 | 0·6 | 9·5 | 1·7 | 15·5 | 2·7 |
| 36 | 2 39·0 | 2 39·4 | 2 31·8 | 3·6 | 0·6 | 9·6 | 1·7 | 15·6 | 2·7 |
| 37 | 2 39·3 | 2 39·7 | 2 32·0 | 3·7 | 0·6 | 9·7 | 1·7 | 15·7 | 2·7 |
| 38 | 2 39·5 | 2 39·9 | 2 32·2 | 3·8 | 0·7 | 9·8 | 1·7 | 15·8 | 2·8 |
| 39 | 2 39·8 | 2 40·2 | 2 32·5 | 3·9 | 0·7 | 9·9 | 1·7 | 15·9 | 2·8 |
| 40 | 2 40·0 | 2 40·4 | 2 32·7 | 4·0 | 0·7 | 10·0 | 1·8 | 16·0 | 2·8 |
| 41 | 2 40·3 | 2 40·7 | 2 32·9 | 4·1 | 0·7 | 10·1 | 1·8 | 16·1 | 2·8 |
| 42 | 2 40·5 | 2 40·9 | 2 33·2 | 4·2 | 0·7 | 10·2 | 1·8 | 16·2 | 2·8 |
| 43 | 2 40·8 | 2 41·2 | 2 33·4 | 4·3 | 0·8 | 10·3 | 1·8 | 16·3 | 2·9 |
| 44 | 2 41·0 | 2 41·4 | 2 33·7 | 4·4 | 0·8 | 10·4 | 1·8 | 16·4 | 2·9 |
| 45 | 2 41·3 | 2 41·7 | 2 33·9 | 4·5 | 0·8 | 10·5 | 1·8 | 16·5 | 2·9 |
| 46 | 2 41·5 | 2 41·9 | 2 34·1 | 4·6 | 0·8 | 10·6 | 1·9 | 16·6 | 2·9 |
| 47 | 2 41·8 | 2 42·2 | 2 34·4 | 4·7 | 0·8 | 10·7 | 1·9 | 16·7 | 2·9 |
| 48 | 2 42·0 | 2 42·4 | 2 34·6 | 4·8 | 0·8 | 10·8 | 1·9 | 16·8 | 2·9 |
| 49 | 2 42·3 | 2 42·7 | 2 34·9 | 4·9 | 0·9 | 10·9 | 1·9 | 16·9 | 3·0 |
| 50 | 2 42·5 | 2 42·9 | 2 35·1 | 5·0 | 0·9 | 11·0 | 1·9 | 17·0 | 3·0 |
| 51 | 2 42·8 | 2 43·2 | 2 35·3 | 5·1 | 0·9 | 11·1 | 1·9 | 17·1 | 3·0 |
| 52 | 2 43·0 | 2 43·4 | 2 35·6 | 5·2 | 0·9 | 11·2 | 2·0 | 17·2 | 3·0 |
| 53 | 2 43·3 | 2 43·7 | 2 35·8 | 5·3 | 0·9 | 11·3 | 2·0 | 17·3 | 3·0 |
| 54 | 2 43·5 | 2 43·9 | 2 36·1 | 5·4 | 0·9 | 11·4 | 2·0 | 17·4 | 3·0 |
| 55 | 2 43·8 | 2 44·2 | 2 36·3 | 5·5 | 1·0 | 11·5 | 2·0 | 17·5 | 3·1 |
| 56 | 2 44·0 | 2 44·4 | 2 36·5 | 5·6 | 1·0 | 11·6 | 2·0 | 17·6 | 3·1 |
| 57 | 2 44·3 | 2 44·7 | 2 36·8 | 5·7 | 1·0 | 11·7 | 2·0 | 17·7 | 3·1 |
| 58 | 2 44·5 | 2 45·0 | 2 37·0 | 5·8 | 1·0 | 11·8 | 2·1 | 17·8 | 3·1 |
| 59 | 2 44·8 | 2 45·2 | 2 37·2 | 5·9 | 1·0 | 11·9 | 2·1 | 17·9 | 3·1 |
| 60 | 2 45·0 | 2 45·5 | 2 37·5 | 6·0 | 1·1 | 12·0 | 2·1 | 18·0 | 3·2 |

| $11^m$ | SUN PLANETS | ARIES | MOON | v or d | Corr$^n$ | v or d | Corr$^n$ | v or d | Corr$^n$ |
|---|---|---|---|---|---|---|---|---|---|
| s | ° ′ | ° ′ | ° ′ | ′ | ′ | ′ | ′ | ′ | ′ |
| 00 | 2 45·0 | 2 45·5 | 2 37·5 | 0·0 | 0·0 | 6·0 | 1·2 | 12·0 | 2·3 |
| 01 | 2 45·3 | 2 45·7 | 2 37·7 | 0·1 | 0·0 | 6·1 | 1·2 | 12·1 | 2·3 |
| 02 | 2 45·5 | 2 46·0 | 2 38·0 | 0·2 | 0·0 | 6·2 | 1·2 | 12·2 | 2·3 |
| 03 | 2 45·8 | 2 46·2 | 2 38·2 | 0·3 | 0·1 | 6·3 | 1·2 | 12·3 | 2·4 |
| 04 | 2 46·0 | 2 46·5 | 2 38·4 | 0·4 | 0·1 | 6·4 | 1·2 | 12·4 | 2·4 |
| 05 | 2 46·3 | 2 46·7 | 2 38·7 | 0·5 | 0·1 | 6·5 | 1·2 | 12·5 | 2·4 |
| 06 | 2 46·5 | 2 47·0 | 2 38·9 | 0·6 | 0·1 | 6·6 | 1·3 | 12·6 | 2·4 |
| 07 | 2 46·8 | 2 47·2 | 2 39·2 | 0·7 | 0·1 | 6·7 | 1·3 | 12·7 | 2·4 |
| 08 | 2 47·0 | 2 47·5 | 2 39·4 | 0·8 | 0·2 | 6·8 | 1·3 | 12·8 | 2·5 |
| 09 | 2 47·3 | 2 47·7 | 2 39·6 | 0·9 | 0·2 | 6·9 | 1·3 | 12·9 | 2·5 |
| 10 | 2 47·5 | 2 48·0 | 2 39·9 | 1·0 | 0·2 | 7·0 | 1·3 | 13·0 | 2·5 |
| 11 | 2 47·8 | 2 48·2 | 2 40·1 | 1·1 | 0·2 | 7·1 | 1·4 | 13·1 | 2·5 |
| 12 | 2 48·0 | 2 48·5 | 2 40·3 | 1·2 | 0·2 | 7·2 | 1·4 | 13·2 | 2·5 |
| 13 | 2 48·3 | 2 48·7 | 2 40·6 | 1·3 | 0·2 | 7·3 | 1·4 | 13·3 | 2·5 |
| 14 | 2 48·5 | 2 49·0 | 2 40·8 | 1·4 | 0·3 | 7·4 | 1·4 | 13·4 | 2·6 |
| 15 | 2 48·8 | 2 49·2 | 2 41·1 | 1·5 | 0·3 | 7·5 | 1·4 | 13·5 | 2·6 |
| 16 | 2 49·0 | 2 49·5 | 2 41·3 | 1·6 | 0·3 | 7·6 | 1·5 | 13·6 | 2·6 |
| 17 | 2 49·3 | 2 49·7 | 2 41·5 | 1·7 | 0·3 | 7·7 | 1·5 | 13·7 | 2·6 |
| 18 | 2 49·5 | 2 50·0 | 2 41·8 | 1·8 | 0·3 | 7·8 | 1·5 | 13·8 | 2·6 |
| 19 | 2 49·8 | 2 50·2 | 2 42·0 | 1·9 | 0·4 | 7·9 | 1·5 | 13·9 | 2·7 |
| 20 | 2 50·0 | 2 50·5 | 2 42·3 | 2·0 | 0·4 | 8·0 | 1·5 | 14·0 | 2·7 |
| 21 | 2 50·3 | 2 50·7 | 2 42·5 | 2·1 | 0·4 | 8·1 | 1·6 | 14·1 | 2·7 |
| 22 | 2 50·5 | 2 51·0 | 2 42·7 | 2·2 | 0·4 | 8·2 | 1·6 | 14·2 | 2·7 |
| 23 | 2 50·8 | 2 51·2 | 2 43·0 | 2·3 | 0·4 | 8·3 | 1·6 | 14·3 | 2·7 |
| 24 | 2 51·0 | 2 51·5 | 2 43·2 | 2·4 | 0·5 | 8·4 | 1·6 | 14·4 | 2·8 |
| 25 | 2 51·3 | 2 51·7 | 2 43·4 | 2·5 | 0·5 | 8·5 | 1·6 | 14·5 | 2·8 |
| 26 | 2 51·5 | 2 52·0 | 2 43·7 | 2·6 | 0·5 | 8·6 | 1·6 | 14·6 | 2·8 |
| 27 | 2 51·8 | 2 52·2 | 2 43·9 | 2·7 | 0·5 | 8·7 | 1·7 | 14·7 | 2·8 |
| 28 | 2 52·0 | 2 52·5 | 2 44·2 | 2·8 | 0·5 | 8·8 | 1·7 | 14·8 | 2·8 |
| 29 | 2 52·3 | 2 52·7 | 2 44·4 | 2·9 | 0·6 | 8·9 | 1·7 | 14·9 | 2·9 |
| 30 | 2 52·5 | 2 53·0 | 2 44·6 | 3·0 | 0·6 | 9·0 | 1·7 | 15·0 | 2·9 |
| 31 | 2 52·8 | 2 53·2 | 2 44·9 | 3·1 | 0·6 | 9·1 | 1·7 | 15·1 | 2·9 |
| 32 | 2 53·0 | 2 53·5 | 2 45·1 | 3·2 | 0·6 | 9·2 | 1·8 | 15·2 | 2·9 |
| 33 | 2 53·3 | 2 53·7 | 2 45·4 | 3·3 | 0·6 | 9·3 | 1·8 | 15·3 | 2·9 |
| 34 | 2 53·5 | 2 54·0 | 2 45·6 | 3·4 | 0·7 | 9·4 | 1·8 | 15·4 | 3·0 |
| 35 | 2 53·8 | 2 54·2 | 2 45·8 | 3·5 | 0·7 | 9·5 | 1·8 | 15·5 | 3·0 |
| 36 | 2 54·0 | 2 54·5 | 2 46·1 | 3·6 | 0·7 | 9·6 | 1·8 | 15·6 | 3·0 |
| 37 | 2 54·3 | 2 54·7 | 2 46·3 | 3·7 | 0·7 | 9·7 | 1·9 | 15·7 | 3·0 |
| 38 | 2 54·5 | 2 55·0 | 2 46·6 | 3·8 | 0·7 | 9·8 | 1·9 | 15·8 | 3·0 |
| 39 | 2 54·8 | 2 55·2 | 2 46·8 | 3·9 | 0·7 | 9·9 | 1·9 | 15·9 | 3·0 |
| 40 | 2 55·0 | 2 55·5 | 2 47·0 | 4·0 | 0·8 | 10·0 | 1·9 | 16·0 | 3·1 |
| 41 | 2 55·3 | 2 55·7 | 2 47·3 | 4·1 | 0·8 | 10·1 | 1·9 | 16·1 | 3·1 |
| 42 | 2 55·5 | 2 56·0 | 2 47·5 | 4·2 | 0·8 | 10·2 | 2·0 | 16·2 | 3·1 |
| 43 | 2 55·8 | 2 56·2 | 2 47·7 | 4·3 | 0·8 | 10·3 | 2·0 | 16·3 | 3·1 |
| 44 | 2 56·0 | 2 56·5 | 2 48·0 | 4·4 | 0·8 | 10·4 | 2·0 | 16·4 | 3·1 |
| 45 | 2 56·3 | 2 56·7 | 2 48·2 | 4·5 | 0·9 | 10·5 | 2·0 | 16·5 | 3·2 |
| 46 | 2 56·5 | 2 57·0 | 2 48·5 | 4·6 | 0·9 | 10·6 | 2·0 | 16·6 | 3·2 |
| 47 | 2 56·8 | 2 57·2 | 2 48·7 | 4·7 | 0·9 | 10·7 | 2·1 | 16·7 | 3·2 |
| 48 | 2 57·0 | 2 57·5 | 2 48·9 | 4·8 | 0·9 | 10·8 | 2·1 | 16·8 | 3·2 |
| 49 | 2 57·3 | 2 57·7 | 2 49·2 | 4·9 | 0·9 | 10·9 | 2·1 | 16·9 | 3·2 |
| 50 | 2 57·5 | 2 58·0 | 2 49·4 | 5·0 | 1·0 | 11·0 | 2·1 | 17·0 | 3·3 |
| 51 | 2 57·8 | 2 58·2 | 2 49·7 | 5·1 | 1·0 | 11·1 | 2·1 | 17·1 | 3·3 |
| 52 | 2 58·0 | 2 58·5 | 2 49·9 | 5·2 | 1·0 | 11·2 | 2·1 | 17·2 | 3·3 |
| 53 | 2 58·3 | 2 58·7 | 2 50·1 | 5·3 | 1·0 | 11·3 | 2·2 | 17·3 | 3·3 |
| 54 | 2 58·5 | 2 59·0 | 2 50·4 | 5·4 | 1·0 | 11·4 | 2·2 | 17·4 | 3·3 |
| 55 | 2 58·8 | 2 59·2 | 2 50·6 | 5·5 | 1·1 | 11·5 | 2·2 | 17·5 | 3·4 |
| 56 | 2 59·0 | 2 59·5 | 2 50·8 | 5·6 | 1·1 | 11·6 | 2·2 | 17·6 | 3·4 |
| 57 | 2 59·3 | 2 59·7 | 2 51·1 | 5·7 | 1·1 | 11·7 | 2·2 | 17·7 | 3·4 |
| 58 | 2 59·5 | 3 00·0 | 2 51·3 | 5·8 | 1·1 | 11·8 | 2·3 | 17·8 | 3·4 |
| 59 | 2 59·8 | 3 00·2 | 2 51·6 | 5·9 | 1·1 | 11·9 | 2·3 | 17·9 | 3·4 |
| 60 | 3 00·0 | 3 00·5 | 2 51·8 | 6·0 | 1·2 | 12·0 | 2·3 | 18·0 | 3·5 |

| $12^{m}$ | SUN PLANETS | ARIES | MOON | $v$ or $d$ | Corr$^{n}$ | $v$ or $d$ | Corr$^{n}$ | $v$ or $d$ | Corr$^{n}$ |
|---|---|---|---|---|---|---|---|---|---|
| s | ° ′ | ° ′ | ° ′ | ′ | ′ | ′ | ′ | ′ | ′ |
| 00 | 3 00·0 | 3 00·5 | 2 51·8 | 0·0 | 0·0 | 6·0 | 1·3 | 12·0 | 2·5 |
| 01 | 3 00·3 | 3 00·7 | 2 52·0 | 0·1 | 0·0 | 6·1 | 1·3 | 12·1 | 2·5 |
| 02 | 3 00·5 | 3 01·0 | 2 52·3 | 0·2 | 0·0 | 6·2 | 1·3 | 12·2 | 2·5 |
| 03 | 3 00·8 | 3 01·2 | 2 52·5 | 0·3 | 0·1 | 6·3 | 1·3 | 12·3 | 2·6 |
| 04 | 3 01·0 | 3 01·5 | 2 52·8 | 0·4 | 0·1 | 6·4 | 1·3 | 12·4 | 2·6 |
| 05 | 3 01·3 | 3 01·7 | 2 53·0 | 0·5 | 0·1 | 6·5 | 1·4 | 12·5 | 2·6 |
| 06 | 3 01·5 | 3 02·0 | 2 53·2 | 0·6 | 0·1 | 6·6 | 1·4 | 12·6 | 2·6 |
| 07 | 3 01·8 | 3 02·2 | 2 53·5 | 0·7 | 0·1 | 6·7 | 1·4 | 12·7 | 2·6 |
| 08 | 3 02·0 | 3 02·5 | 2 53·7 | 0·8 | 0·2 | 6·8 | 1·4 | 12·8 | 2·7 |
| 09 | 3 02·3 | 3 02·7 | 2 53·9 | 0·9 | 0·2 | 6·9 | 1·4 | 12·9 | 2·7 |
| 10 | 3 02·5 | 3 03·0 | 2 54·2 | 1·0 | 0·2 | 7·0 | 1·5 | 13·0 | 2·7 |
| 11 | 3 02·8 | 3 03·3 | 2 54·4 | 1·1 | 0·2 | 7·1 | 1·5 | 13·1 | 2·7 |
| 12 | 3 03·0 | 3 03·5 | 2 54·7 | 1·2 | 0·3 | 7·2 | 1·5 | 13·2 | 2·8 |
| 13 | 3 03·3 | 3 03·8 | 2 54·9 | 1·3 | 0·3 | 7·3 | 1·5 | 13·3 | 2·8 |
| 14 | 3 03·5 | 3 04·0 | 2 55·1 | 1·4 | 0·3 | 7·4 | 1·5 | 13·4 | 2·8 |
| 15 | 3 03·8 | 3 04·3 | 2 55·4 | 1·5 | 0·3 | 7·5 | 1·6 | 13·5 | 2·8 |
| 16 | 3 04·0 | 3 04·5 | 2 55·6 | 1·6 | 0·3 | 7·6 | 1·6 | 13·6 | 2·8 |
| 17 | 3 04·3 | 3 04·8 | 2 55·9 | 1·7 | 0·4 | 7·7 | 1·6 | 13·7 | 2·9 |
| 18 | 3 04·5 | 3 05·0 | 2 56·1 | 1·8 | 0·4 | 7·8 | 1·6 | 13·8 | 2·9 |
| 19 | 3 04·8 | 3 05·3 | 2 56·3 | 1·9 | 0·4 | 7·9 | 1·6 | 13·9 | 2·9 |
| 20 | 3 05·0 | 3 05·5 | 2 56·6 | 2·0 | 0·4 | 8·0 | 1·7 | 14·0 | 2·9 |
| 21 | 3 05·3 | 3 05·8 | 2 56·8 | 2·1 | 0·4 | 8·1 | 1·7 | 14·1 | 2·9 |
| 22 | 3 05·5 | 3 06·0 | 2 57·0 | 2·2 | 0·5 | 8·2 | 1·7 | 14·2 | 3·0 |
| 23 | 3 05·8 | 3 06·3 | 2 57·3 | 2·3 | 0·5 | 8·3 | 1·7 | 14·3 | 3·0 |
| 24 | 3 06·0 | 3 06·5 | 2 57·5 | 2·4 | 0·5 | 8·4 | 1·8 | 14·4 | 3·0 |
| 25 | 3 06·3 | 3 06·8 | 2 57·8 | 2·5 | 0·5 | 8·5 | 1·8 | 14·5 | 3·0 |
| 26 | 3 06·5 | 3 07·0 | 2 58·0 | 2·6 | 0·5 | 8·6 | 1·8 | 14·6 | 3·0 |
| 27 | 3 06·8 | 3 07·3 | 2 58·2 | 2·7 | 0·6 | 8·7 | 1·8 | 14·7 | 3·1 |
| 28 | 3 07·0 | 3 07·5 | 2 58·5 | 2·8 | 0·6 | 8·8 | 1·8 | 14·8 | 3·1 |
| 29 | 3 07·3 | 3 07·8 | 2 58·7 | 2·9 | 0·6 | 8·9 | 1·9 | 14·9 | 3·1 |
| 30 | 3 07·5 | 3 08·0 | 2 59·0 | 3·0 | 0·6 | 9·0 | 1·9 | 15·0 | 3·1 |
| 31 | 3 07·8 | 3 08·3 | 2 59·2 | 3·1 | 0·6 | 9·1 | 1·9 | 15·1 | 3·1 |
| 32 | 3 08·0 | 3 08·5 | 2 59·4 | 3·2 | 0·7 | 9·2 | 1·9 | 15·2 | 3·2 |
| 33 | 3 08·3 | 3 08·8 | 2 59·7 | 3·3 | 0·7 | 9·3 | 1·9 | 15·3 | 3·2 |
| 34 | 3 08·5 | 3 09·0 | 2 59·9 | 3·4 | 0·7 | 9·4 | 2·0 | 15·4 | 3·2 |
| 35 | 3 08·8 | 3 09·3 | 3 00·2 | 3·5 | 0·7 | 9·5 | 2·0 | 15·5 | 3·2 |
| 36 | 3 09·0 | 3 09·5 | 3 00·4 | 3·6 | 0·8 | 9·6 | 2·0 | 15·6 | 3·3 |
| 37 | 3 09·3 | 3 09·8 | 3 00·6 | 3·7 | 0·8 | 9·7 | 2·0 | 15·7 | 3·3 |
| 38 | 3 09·5 | 3 10·0 | 3 00·9 | 3·8 | 0·8 | 9·8 | 2·0 | 15·8 | 3·3 |
| 39 | 3 09·8 | 3 10·3 | 3 01·1 | 3·9 | 0·8 | 9·9 | 2·1 | 15·9 | 3·3 |
| 40 | 3 10·0 | 3 10·5 | 3 01·3 | 4·0 | 0·8 | 10·0 | 2·1 | 16·0 | 3·3 |
| 41 | 3 10·3 | 3 10·8 | 3 01·6 | 4·1 | 0·9 | 10·1 | 2·1 | 16·1 | 3·4 |
| 42 | 3 10·5 | 3 11·0 | 3 01·8 | 4·2 | 0·9 | 10·2 | 2·1 | 16·2 | 3·4 |
| 43 | 3 10·8 | 3 11·3 | 3 02·1 | 4·3 | 0·9 | 10·3 | 2·1 | 16·3 | 3·4 |
| 44 | 3 11·0 | 3 11·5 | 3 02·3 | 4·4 | 0·9 | 10·4 | 2·2 | 16·4 | 3·4 |
| 45 | 3 11·3 | 3 11·8 | 3 02·5 | 4·5 | 0·9 | 10·5 | 2·2 | 16·5 | 3·4 |
| 46 | 3 11·5 | 3 12·0 | 3 02·8 | 4·6 | 1·0 | 10·6 | 2·2 | 16·6 | 3·5 |
| 47 | 3 11·8 | 3 12·3 | 3 03·0 | 4·7 | 1·0 | 10·7 | 2·2 | 16·7 | 3·5 |
| 48 | 3 12·0 | 3 12·5 | 3 03·3 | 4·8 | 1·0 | 10·8 | 2·3 | 16·8 | 3·5 |
| 49 | 3 12·3 | 3 12·8 | 3 03·5 | 4·9 | 1·0 | 10·9 | 2·3 | 16·9 | 3·5 |
| 50 | 3 12·5 | 3 13·0 | 3 03·7 | 5·0 | 1·0 | 11·0 | 2·3 | 17·0 | 3·5 |
| 51 | 3 12·8 | 3 13·3 | 3 04·0 | 5·1 | 1·1 | 11·1 | 2·3 | 17·1 | 3·6 |
| 52 | 3 13·0 | 3 13·5 | 3 04·2 | 5·2 | 1·1 | 11·2 | 2·3 | 17·2 | 3·6 |
| 53 | 3 13·3 | 3 13·8 | 3 04·4 | 5·3 | 1·1 | 11·3 | 2·4 | 17·3 | 3·6 |
| 54 | 3 13·5 | 3 14·0 | 3 04·7 | 5·4 | 1·1 | 11·4 | 2·4 | 17·4 | 3·6 |
| 55 | 3 13·8 | 3 14·3 | 3 04·9 | 5·5 | 1·1 | 11·5 | 2·4 | 17·5 | 3·6 |
| 56 | 3 14·0 | 3 14·5 | 3 05·2 | 5·6 | 1·2 | 11·6 | 2·4 | 17·6 | 3·7 |
| 57 | 3 14·3 | 3 14·8 | 3 05·4 | 5·7 | 1·2 | 11·7 | 2·4 | 17·7 | 3·7 |
| 58 | 3 14·5 | 3 15·0 | 3 05·6 | 5·8 | 1·2 | 11·8 | 2·5 | 17·8 | 3·7 |
| 59 | 3 14·8 | 3 15·3 | 3 05·9 | 5·9 | 1·2 | 11·9 | 2·5 | 17·9 | 3·7 |
| 60 | 3 15·0 | 3 15·5 | 3 06·1 | 6·0 | 1·3 | 12·0 | 2·5 | 18·0 | 3·8 |

| $13^{m}$ | SUN PLANETS | ARIES | MOON | $v$ or $d$ | Corr$^{n}$ | $v$ or $d$ | Corr$^{n}$ | $v$ or $d$ | Corr$^{n}$ |
|---|---|---|---|---|---|---|---|---|---|
| s | ° ′ | ° ′ | ° ′ | ′ | ′ | ′ | ′ | ′ | ′ |
| 00 | 3 15·0 | 3 15·5 | 3 06·1 | 0·0 | 0·0 | 6·0 | 1·4 | 12·0 | 2·7 |
| 01 | 3 15·3 | 3 15·8 | 3 06·4 | 0·1 | 0·0 | 6·1 | 1·4 | 12·1 | 2·7 |
| 02 | 3 15·5 | 3 16·0 | 3 06·6 | 0·2 | 0·0 | 6·2 | 1·4 | 12·2 | 2·7 |
| 03 | 3 15·8 | 3 16·3 | 3 06·8 | 0·3 | 0·1 | 6·3 | 1·4 | 12·3 | 2·8 |
| 04 | 3 16·0 | 3 16·5 | 3 07·1 | 0·4 | 0·1 | 6·4 | 1·4 | 12·4 | 2·8 |
| 05 | 3 16·3 | 3 16·8 | 3 07·3 | 0·5 | 0·1 | 6·5 | 1·5 | 12·5 | 2·8 |
| 06 | 3 16·5 | 3 17·0 | 3 07·5 | 0·6 | 0·1 | 6·6 | 1·5 | 12·6 | 2·8 |
| 07 | 3 16·8 | 3 17·3 | 3 07·8 | 0·7 | 0·2 | 6·7 | 1·5 | 12·7 | 2·9 |
| 08 | 3 17·0 | 3 17·5 | 3 08·0 | 0·8 | 0·2 | 6·8 | 1·5 | 12·8 | 2·9 |
| 09 | 3 17·3 | 3 17·8 | 3 08·3 | 0·9 | 0·2 | 6·9 | 1·6 | 12·9 | 2·9 |
| 10 | 3 17·5 | 3 18·0 | 3 08·5 | 1·0 | 0·2 | 7·0 | 1·6 | 13·0 | 2·9 |
| 11 | 3 17·8 | 3 18·3 | 3 08·7 | 1·1 | 0·2 | 7·1 | 1·6 | 13·1 | 2·9 |
| 12 | 3 18·0 | 3 18·5 | 3 09·0 | 1·2 | 0·3 | 7·2 | 1·6 | 13·2 | 3·0 |
| 13 | 3 18·3 | 3 18·8 | 3 09·2 | 1·3 | 0·3 | 7·3 | 1·6 | 13·3 | 3·0 |
| 14 | 3 18·5 | 3 19·0 | 3 09·5 | 1·4 | 0·3 | 7·4 | 1·7 | 13·4 | 3·0 |
| 15 | 3 18·8 | 3 19·3 | 3 09·7 | 1·5 | 0·3 | 7·5 | 1·7 | 13·5 | 3·0 |
| 16 | 3 19·0 | 3 19·5 | 3 09·9 | 1·6 | 0·4 | 7·6 | 1·7 | 13·6 | 3·1 |
| 17 | 3 19·3 | 3 19·8 | 3 10·2 | 1·7 | 0·4 | 7·7 | 1·7 | 13·7 | 3·1 |
| 18 | 3 19·5 | 3 20·0 | 3 10·4 | 1·8 | 0·4 | 7·8 | 1·8 | 13·8 | 3·1 |
| 19 | 3 19·8 | 3 20·3 | 3 10·7 | 1·9 | 0·4 | 7·9 | 1·8 | 13·9 | 3·1 |
| 20 | 3 20·0 | 3 20·5 | 3 10·9 | 2·0 | 0·5 | 8·0 | 1·8 | 14·0 | 3·2 |
| 21 | 3 20·3 | 3 20·8 | 3 11·1 | 2·1 | 0·5 | 8·1 | 1·8 | 14·1 | 3·2 |
| 22 | 3 20·5 | 3 21·0 | 3 11·4 | 2·2 | 0·5 | 8·2 | 1·8 | 14·2 | 3·2 |
| 23 | 3 20·8 | 3 21·3 | 3 11·6 | 2·3 | 0·5 | 8·3 | 1·9 | 14·3 | 3·2 |
| 24 | 3 21·0 | 3 21·6 | 3 11·8 | 2·4 | 0·5 | 8·4 | 1·9 | 14·4 | 3·2 |
| 25 | 3 21·3 | 3 21·8 | 3 12·1 | 2·5 | 0·6 | 8·5 | 1·9 | 14·5 | 3·3 |
| 26 | 3 21·5 | 3 22·1 | 3 12·3 | 2·6 | 0·6 | 8·6 | 1·9 | 14·6 | 3·3 |
| 27 | 3 21·8 | 3 22·3 | 3 12·6 | 2·7 | 0·6 | 8·7 | 2·0 | 14·7 | 3·3 |
| 28 | 3 22·0 | 3 22·6 | 3 12·8 | 2·8 | 0·6 | 8·8 | 2·0 | 14·8 | 3·3 |
| 29 | 3 22·3 | 3 22·8 | 3 13·0 | 2·9 | 0·7 | 8·9 | 2·0 | 14·9 | 3·4 |
| 30 | 3 22·5 | 3 23·1 | 3 13·3 | 3·0 | 0·7 | 9·0 | 2·0 | 15·0 | 3·4 |
| 31 | 3 22·8 | 3 23·3 | 3 13·5 | 3·1 | 0·7 | 9·1 | 2·0 | 15·1 | 3·4 |
| 32 | 3 23·0 | 3 23·6 | 3 13·8 | 3·2 | 0·7 | 9·2 | 2·1 | 15·2 | 3·4 |
| 33 | 3 23·3 | 3 23·8 | 3 14·0 | 3·3 | 0·7 | 9·3 | 2·1 | 15·3 | 3·4 |
| 34 | 3 23·5 | 3 24·1 | 3 14·2 | 3·4 | 0·8 | 9·4 | 2·1 | 15·4 | 3·5 |
| 35 | 3 23·8 | 3 24·3 | 3 14·5 | 3·5 | 0·8 | 9·5 | 2·1 | 15·5 | 3·5 |
| 36 | 3 24·0 | 3 24·6 | 3 14·7 | 3·6 | 0·8 | 9·6 | 2·2 | 15·6 | 3·5 |
| 37 | 3 24·3 | 3 24·8 | 3 14·9 | 3·7 | 0·8 | 9·7 | 2·2 | 15·7 | 3·5 |
| 38 | 3 24·5 | 3 25·1 | 3 15·2 | 3·8 | 0·9 | 9·8 | 2·2 | 15·8 | 3·6 |
| 39 | 3 24·8 | 3 25·3 | 3 15·4 | 3·9 | 0·9 | 9·9 | 2·2 | 15·9 | 3·6 |
| 40 | 3 25·0 | 3 25·6 | 3 15·7 | 4·0 | 0·9 | 10·0 | 2·3 | 16·0 | 3·6 |
| 41 | 3 25·3 | 3 25·8 | 3 15·9 | 4·1 | 0·9 | 10·1 | 2·3 | 16·1 | 3·6 |
| 42 | 3 25·5 | 3 26·1 | 3 16·1 | 4·2 | 0·9 | 10·2 | 2·3 | 16·2 | 3·6 |
| 43 | 3 25·8 | 3 26·3 | 3 16·4 | 4·3 | 1·0 | 10·3 | 2·3 | 16·3 | 3·7 |
| 44 | 3 26·0 | 3 26·6 | 3 16·6 | 4·4 | 1·0 | 10·4 | 2·3 | 16·4 | 3·7 |
| 45 | 3 26·3 | 3 26·8 | 3 16·9 | 4·5 | 1·0 | 10·5 | 2·4 | 16·5 | 3·7 |
| 46 | 3 26·5 | 3 27·1 | 3 17·1 | 4·6 | 1·0 | 10·6 | 2·4 | 16·6 | 3·7 |
| 47 | 3 26·8 | 3 27·3 | 3 17·3 | 4·7 | 1·1 | 10·7 | 2·4 | 16·7 | 3·8 |
| 48 | 3 27·0 | 3 27·6 | 3 17·6 | 4·8 | 1·1 | 10·8 | 2·4 | 16·8 | 3·8 |
| 49 | 3 27·3 | 3 27·8 | 3 17·8 | 4·9 | 1·1 | 10·9 | 2·5 | 16·9 | 3·8 |
| 50 | 3 27·5 | 3 28·1 | 3 18·0 | 5·0 | 1·1 | 11·0 | 2·5 | 17·0 | 3·8 |
| 51 | 3 27·8 | 3 28·3 | 3 18·3 | 5·1 | 1·1 | 11·1 | 2·5 | 17·1 | 3·8 |
| 52 | 3 28·0 | 3 28·6 | 3 18·5 | 5·2 | 1·2 | 11·2 | 2·5 | 17·2 | 3·9 |
| 53 | 3 28·3 | 3 28·8 | 3 18·8 | 5·3 | 1·2 | 11·3 | 2·5 | 17·3 | 3·9 |
| 54 | 3 28·5 | 3 29·1 | 3 19·0 | 5·4 | 1·2 | 11·4 | 2·6 | 17·4 | 3·9 |
| 55 | 3 28·8 | 3 29·3 | 3 19·2 | 5·5 | 1·2 | 11·5 | 2·6 | 17·5 | 3·9 |
| 56 | 3 29·0 | 3 29·6 | 3 19·5 | 5·6 | 1·3 | 11·6 | 2·6 | 17·6 | 4·0 |
| 57 | 3 29·3 | 3 29·8 | 3 19·7 | 5·7 | 1·3 | 11·7 | 2·6 | 17·7 | 4·0 |
| 58 | 3 29·5 | 3 30·1 | 3 20·0 | 5·8 | 1·3 | 11·8 | 2·7 | 17·8 | 4·0 |
| 59 | 3 29·8 | 3 30·3 | 3 20·2 | 5·9 | 1·3 | 11·9 | 2·7 | 17·9 | 4·0 |
| 60 | 3 30·0 | 3 30·6 | 3 20·4 | 6·0 | 1·4 | 12·0 | 2·7 | 18·0 | 4·1 |

| 14m s | SUN PLANETS ° ′ | ARIES ° ′ | MOON ° ′ | v or d ′ | Corrn ′ | v or d ′ | Corrn ′ | v or d ′ | Corrn ′ |
|---|---|---|---|---|---|---|---|---|---|
| 00 | 3 30·0 | 3 30·6 | 3 20·4 | 0·0 | 0·0 | 6·0 | 1·5 | 12·0 | 2·9 |
| 01 | 3 30·3 | 3 30·8 | 3 20·7 | 0·1 | 0·0 | 6·1 | 1·5 | 12·1 | 2·9 |
| 02 | 3 30·5 | 3 31·1 | 3 20·9 | 0·2 | 0·0 | 6·2 | 1·5 | 12·2 | 2·9 |
| 03 | 3 30·8 | 3 31·3 | 3 21·1 | 0·3 | 0·1 | 6·3 | 1·5 | 12·3 | 3·0 |
| 04 | 3 31·0 | 3 31·6 | 3 21·4 | 0·4 | 0·1 | 6·4 | 1·5 | 12·4 | 3·0 |
| 05 | 3 31·3 | 3 31·8 | 3 21·6 | 0·5 | 0·1 | 6·5 | 1·6 | 12·5 | 3·0 |
| 06 | 3 31·5 | 3 32·1 | 3 21·9 | 0·6 | 0·1 | 6·6 | 1·6 | 12·6 | 3·0 |
| 07 | 3 31·8 | 3 32·3 | 3 22·1 | 0·7 | 0·2 | 6·7 | 1·6 | 12·7 | 3·1 |
| 08 | 3 32·0 | 3 32·6 | 3 22·3 | 0·8 | 0·2 | 6·8 | 1·6 | 12·8 | 3·1 |
| 09 | 3 32·3 | 3 32·8 | 3 22·6 | 0·9 | 0·2 | 6·9 | 1·7 | 12·9 | 3·1 |
| 10 | 3 32·5 | 3 33·1 | 3 22·8 | 1·0 | 0·2 | 7·0 | 1·7 | 13·0 | 3·1 |
| 11 | 3 32·8 | 3 33·3 | 3 23·1 | 1·1 | 0·3 | 7·1 | 1·7 | 13·1 | 3·2 |
| 12 | 3 33·0 | 3 33·6 | 3 23·3 | 1·2 | 0·3 | 7·2 | 1·7 | 13·2 | 3·2 |
| 13 | 3 33·3 | 3 33·8 | 3 23·5 | 1·3 | 0·3 | 7·3 | 1·8 | 13·3 | 3·2 |
| 14 | 3 33·5 | 3 34·1 | 3 23·8 | 1·4 | 0·3 | 7·4 | 1·8 | 13·4 | 3·2 |
| 15 | 3 33·8 | 3 34·3 | 3 24·0 | 1·5 | 0·4 | 7·5 | 1·8 | 13·5 | 3·3 |
| 16 | 3 34·0 | 3 34·6 | 3 24·3 | 1·6 | 0·4 | 7·6 | 1·8 | 13·6 | 3·3 |
| 17 | 3 34·3 | 3 34·8 | 3 24·5 | 1·7 | 0·4 | 7·7 | 1·9 | 13·7 | 3·3 |
| 18 | 3 34·5 | 3 35·1 | 3 24·7 | 1·8 | 0·4 | 7·8 | 1·9 | 13·8 | 3·3 |
| 19 | 3 34·8 | 3 35·3 | 3 25·0 | 1·9 | 0·5 | 7·9 | 1·9 | 13·9 | 3·4 |
| 20 | 3 35·0 | 3 35·6 | 3 25·2 | 2·0 | 0·5 | 8·0 | 1·9 | 14·0 | 3·4 |
| 21 | 3 35·3 | 3 35·8 | 3 25·4 | 2·1 | 0·5 | 8·1 | 2·0 | 14·1 | 3·4 |
| 22 | 3 35·5 | 3 36·1 | 3 25·7 | 2·2 | 0·5 | 8·2 | 2·0 | 14·2 | 3·4 |
| 23 | 3 35·8 | 3 36·3 | 3 25·9 | 2·3 | 0·6 | 8·3 | 2·0 | 14·3 | 3·5 |
| 24 | 3 36·0 | 3 36·6 | 3 26·2 | 2·4 | 0·6 | 8·4 | 2·0 | 14·4 | 3·5 |
| 25 | 3 36·3 | 3 36·8 | 3 26·4 | 2·5 | 0·6 | 8·5 | 2·1 | 14·5 | 3·5 |
| 26 | 3 36·5 | 3 37·1 | 3 26·6 | 2·6 | 0·6 | 8·6 | 2·1 | 14·6 | 3·5 |
| 27 | 3 36·8 | 3 37·3 | 3 26·9 | 2·7 | 0·7 | 8·7 | 2·1 | 14·7 | 3·6 |
| 28 | 3 37·0 | 3 37·6 | 3 27·1 | 2·8 | 0·7 | 8·8 | 2·1 | 14·8 | 3·6 |
| 29 | 3 37·3 | 3 37·8 | 3 27·4 | 2·9 | 0·7 | 8·9 | 2·2 | 14·9 | 3·6 |
| 30 | 3 37·5 | 3 38·1 | 3 27·6 | 3·0 | 0·7 | 9·0 | 2·2 | 15·0 | 3·6 |
| 31 | 3 37·8 | 3 38·3 | 3 27·8 | 3·1 | 0·7 | 9·1 | 2·2 | 15·1 | 3·6 |
| 32 | 3 38·0 | 3 38·6 | 3 28·1 | 3·2 | 0·8 | 9·2 | 2·2 | 15·2 | 3·7 |
| 33 | 3 38·3 | 3 38·8 | 3 28·3 | 3·3 | 0·8 | 9·3 | 2·2 | 15·3 | 3·7 |
| 34 | 3 38·5 | 3 39·1 | 3 28·5 | 3·4 | 0·8 | 9·4 | 2·3 | 15·4 | 3·7 |
| 35 | 3 38·8 | 3 39·3 | 3 28·8 | 3·5 | 0·8 | 9·5 | 2·3 | 15·5 | 3·7 |
| 36 | 3 39·0 | 3 39·6 | 3 29·0 | 3·6 | 0·9 | 9·6 | 2·3 | 15·6 | 3·8 |
| 37 | 3 39·3 | 3 39·9 | 3 29·3 | 3·7 | 0·9 | 9·7 | 2·3 | 15·7 | 3·8 |
| 38 | 3 39·5 | 3 40·1 | 3 29·5 | 3·8 | 0·9 | 9·8 | 2·4 | 15·8 | 3·8 |
| 39 | 3 39·8 | 3 40·4 | 3 29·7 | 3·9 | 0·9 | 9·9 | 2·4 | 15·9 | 3·8 |
| 40 | 3 40·0 | 3 40·6 | 3 30·0 | 4·0 | 1·0 | 10·0 | 2·4 | 16·0 | 3·9 |
| 41 | 3 40·3 | 3 40·9 | 3 30·2 | 4·1 | 1·0 | 10·1 | 2·4 | 16·1 | 3·9 |
| 42 | 3 40·5 | 3 41·1 | 3 30·5 | 4·2 | 1·0 | 10·2 | 2·5 | 16·2 | 3·9 |
| 43 | 3 40·8 | 3 41·4 | 3 30·7 | 4·3 | 1·0 | 10·3 | 2·5 | 16·3 | 3·9 |
| 44 | 3 41·0 | 3 41·6 | 3 30·9 | 4·4 | 1·1 | 10·4 | 2·5 | 16·4 | 4·0 |
| 45 | 3 41·3 | 3 41·9 | 3 31·2 | 4·5 | 1·1 | 10·5 | 2·5 | 16·5 | 4·0 |
| 46 | 3 41·5 | 3 42·1 | 3 31·4 | 4·6 | 1·1 | 10·6 | 2·6 | 16·6 | 4·0 |
| 47 | 3 41·8 | 3 42·4 | 3 31·6 | 4·7 | 1·1 | 10·7 | 2·6 | 16·7 | 4·0 |
| 48 | 3 42·0 | 3 42·6 | 3 31·9 | 4·8 | 1·2 | 10·8 | 2·6 | 16·8 | 4·1 |
| 49 | 3 42·3 | 3 42·9 | 3 32·1 | 4·9 | 1·2 | 10·9 | 2·6 | 16·9 | 4·1 |
| 50 | 3 42·5 | 3 43·1 | 3 32·4 | 5·0 | 1·2 | 11·0 | 2·7 | 17·0 | 4·1 |
| 51 | 3 42·8 | 3 43·4 | 3 32·6 | 5·1 | 1·2 | 11·1 | 2·7 | 17·1 | 4·1 |
| 52 | 3 43·0 | 3 43·6 | 3 32·8 | 5·2 | 1·3 | 11·2 | 2·7 | 17·2 | 4·2 |
| 53 | 3 43·3 | 3 43·9 | 3 33·1 | 5·3 | 1·3 | 11·3 | 2·7 | 17·3 | 4·2 |
| 54 | 3 43·5 | 3 44·1 | 3 33·3 | 5·4 | 1·3 | 11·4 | 2·8 | 17·4 | 4·2 |
| 55 | 3 43·8 | 3 44·4 | 3 33·6 | 5·5 | 1·3 | 11·5 | 2·8 | 17·5 | 4·2 |
| 56 | 3 44·0 | 3 44·6 | 3 33·8 | 5·6 | 1·4 | 11·6 | 2·8 | 17·6 | 4·3 |
| 57 | 3 44·3 | 3 44·9 | 3 34·0 | 5·7 | 1·4 | 11·7 | 2·8 | 17·7 | 4·3 |
| 58 | 3 44·5 | 3 45·1 | 3 34·3 | 5·8 | 1·4 | 11·8 | 2·9 | 17·8 | 4·3 |
| 59 | 3 44·8 | 3 45·4 | 3 34·5 | 5·9 | 1·4 | 11·9 | 2·9 | 17·9 | 4·3 |
| 60 | 3 45·0 | 3 45·6 | 3 34·8 | 6·0 | 1·5 | 12·0 | 2·9 | 18·0 | 4·4 |

| 15m s | SUN PLANETS ° ′ | ARIES ° ′ | MOON ° ′ | v or d ′ | Corrn ′ | v or d ′ | Corrn ′ | v or d ′ | Corrn ′ |
|---|---|---|---|---|---|---|---|---|---|
| 00 | 3 45·0 | 3 45·6 | 3 34·8 | 0·0 | 0·0 | 6·0 | 1·6 | 12·0 | 3·1 |
| 01 | 3 45·3 | 3 45·9 | 3 35·0 | 0·1 | 0·0 | 6·1 | 1·6 | 12·1 | 3·1 |
| 02 | 3 45·5 | 3 46·1 | 3 35·2 | 0·2 | 0·1 | 6·2 | 1·6 | 12·2 | 3·2 |
| 03 | 3 45·8 | 3 46·4 | 3 35·5 | 0·3 | 0·1 | 6·3 | 1·6 | 12·3 | 3·2 |
| 04 | 3 46·0 | 3 46·6 | 3 35·7 | 0·4 | 0·1 | 6·4 | 1·7 | 12·4 | 3·2 |
| 05 | 3 46·3 | 3 46·9 | 3 35·9 | 0·5 | 0·1 | 6·5 | 1·7 | 12·5 | 3·2 |
| 06 | 3 46·5 | 3 47·1 | 3 36·2 | 0·6 | 0·2 | 6·6 | 1·7 | 12·6 | 3·3 |
| 07 | 3 46·8 | 3 47·4 | 3 36·4 | 0·7 | 0·2 | 6·7 | 1·7 | 12·7 | 3·3 |
| 08 | 3 47·0 | 3 47·6 | 3 36·7 | 0·8 | 0·2 | 6·8 | 1·8 | 12·8 | 3·3 |
| 09 | 3 47·3 | 3 47·9 | 3 36·9 | 0·9 | 0·2 | 6·9 | 1·8 | 12·9 | 3·3 |
| 10 | 3 47·5 | 3 48·1 | 3 37·1 | 1·0 | 0·3 | 7·0 | 1·8 | 13·0 | 3·4 |
| 11 | 3 47·8 | 3 48·4 | 3 37·4 | 1·1 | 0·3 | 7·1 | 1·8 | 13·1 | 3·4 |
| 12 | 3 48·0 | 3 48·6 | 3 37·6 | 1·2 | 0·3 | 7·2 | 1·9 | 13·2 | 3·4 |
| 13 | 3 48·3 | 3 48·9 | 3 37·9 | 1·3 | 0·3 | 7·3 | 1·9 | 13·3 | 3·4 |
| 14 | 3 48·5 | 3 49·1 | 3 38·1 | 1·4 | 0·4 | 7·4 | 1·9 | 13·4 | 3·5 |
| 15 | 3 48·8 | 3 49·4 | 3 38·3 | 1·5 | 0·4 | 7·5 | 1·9 | 13·5 | 3·5 |
| 16 | 3 49·0 | 3 49·6 | 3 38·6 | 1·6 | 0·4 | 7·6 | 2·0 | 13·6 | 3·5 |
| 17 | 3 49·3 | 3 49·9 | 3 38·8 | 1·7 | 0·4 | 7·7 | 2·0 | 13·7 | 3·5 |
| 18 | 3 49·5 | 3 50·1 | 3 39·0 | 1·8 | 0·5 | 7·8 | 2·0 | 13·8 | 3·6 |
| 19 | 3 49·8 | 3 50·4 | 3 39·3 | 1·9 | 0·5 | 7·9 | 2·0 | 13·9 | 3·6 |
| 20 | 3 50·0 | 3 50·6 | 3 39·5 | 2·0 | 0·5 | 8·0 | 2·1 | 14·0 | 3·6 |
| 21 | 3 50·3 | 3 50·9 | 3 39·8 | 2·1 | 0·5 | 8·1 | 2·1 | 14·1 | 3·6 |
| 22 | 3 50·5 | 3 51·1 | 3 40·0 | 2·2 | 0·6 | 8·2 | 2·1 | 14·2 | 3·7 |
| 23 | 3 50·8 | 3 51·4 | 3 40·2 | 2·3 | 0·6 | 8·3 | 2·1 | 14·3 | 3·7 |
| 24 | 3 51·0 | 3 51·6 | 3 40·5 | 2·4 | 0·6 | 8·4 | 2·2 | 14·4 | 3·7 |
| 25 | 3 51·3 | 3 51·9 | 3 40·7 | 2·5 | 0·6 | 8·5 | 2·2 | 14·5 | 3·7 |
| 26 | 3 51·5 | 3 52·1 | 3 41·0 | 2·6 | 0·7 | 8·6 | 2·2 | 14·6 | 3·8 |
| 27 | 3 51·8 | 3 52·4 | 3 41·2 | 2·7 | 0·7 | 8·7 | 2·2 | 14·7 | 3·8 |
| 28 | 3 52·0 | 3 52·6 | 3 41·4 | 2·8 | 0·7 | 8·8 | 2·3 | 14·8 | 3·8 |
| 29 | 3 52·3 | 3 52·9 | 3 41·7 | 2·9 | 0·7 | 8·9 | 2·3 | 14·9 | 3·8 |
| 30 | 3 52·5 | 3 53·1 | 3 41·9 | 3·0 | 0·8 | 9·0 | 2·3 | 15·0 | 3·9 |
| 31 | 3 52·8 | 3 53·4 | 3 42·1 | 3·1 | 0·8 | 9·1 | 2·4 | 15·1 | 3·9 |
| 32 | 3 53·0 | 3 53·6 | 3 42·4 | 3·2 | 0·8 | 9·2 | 2·4 | 15·2 | 3·9 |
| 33 | 3 53·3 | 3 53·9 | 3 42·6 | 3·3 | 0·9 | 9·3 | 2·4 | 15·3 | 4·0 |
| 34 | 3 53·5 | 3 54·1 | 3 42·9 | 3·4 | 0·9 | 9·4 | 2·4 | 15·4 | 4·0 |
| 35 | 3 53·8 | 3 54·4 | 3 43·1 | 3·5 | 0·9 | 9·5 | 2·5 | 15·5 | 4·0 |
| 36 | 3 54·0 | 3 54·6 | 3 43·3 | 3·6 | 0·9 | 9·6 | 2·5 | 15·6 | 4·0 |
| 37 | 3 54·3 | 3 54·9 | 3 43·6 | 3·7 | 1·0 | 9·7 | 2·5 | 15·7 | 4·1 |
| 38 | 3 54·5 | 3 55·1 | 3 43·8 | 3·8 | 1·0 | 9·8 | 2·5 | 15·8 | 4·1 |
| 39 | 3 54·8 | 3 55·4 | 3 44·1 | 3·9 | 1·0 | 9·9 | 2·6 | 15·9 | 4·1 |
| 40 | 3 55·0 | 3 55·6 | 3 44·3 | 4·0 | 1·0 | 10·0 | 2·6 | 16·0 | 4·1 |
| 41 | 3 55·3 | 3 55·9 | 3 44·5 | 4·1 | 1·1 | 10·1 | 2·6 | 16·1 | 4·2 |
| 42 | 3 55·5 | 3 56·1 | 3 44·8 | 4·2 | 1·1 | 10·2 | 2·6 | 16·2 | 4·2 |
| 43 | 3 55·8 | 3 56·4 | 3 45·0 | 4·3 | 1·1 | 10·3 | 2·7 | 16·3 | 4·2 |
| 44 | 3 56·0 | 3 56·6 | 3 45·2 | 4·4 | 1·1 | 10·4 | 2·7 | 16·4 | 4·2 |
| 45 | 3 56·3 | 3 56·9 | 3 45·5 | 4·5 | 1·2 | 10·5 | 2·7 | 16·5 | 4·3 |
| 46 | 3 56·5 | 3 57·1 | 3 45·7 | 4·6 | 1·2 | 10·6 | 2·7 | 16·6 | 4·3 |
| 47 | 3 56·8 | 3 57·4 | 3 46·0 | 4·7 | 1·2 | 10·7 | 2·8 | 16·7 | 4·3 |
| 48 | 3 57·0 | 3 57·6 | 3 46·2 | 4·8 | 1·2 | 10·8 | 2·8 | 16·8 | 4·3 |
| 49 | 3 57·3 | 3 57·9 | 3 46·4 | 4·9 | 1·3 | 10·9 | 2·8 | 16·9 | 4·4 |
| 50 | 3 57·5 | 3 58·2 | 3 46·7 | 5·0 | 1·3 | 11·0 | 2·8 | 17·0 | 4·4 |
| 51 | 3 57·8 | 3 58·4 | 3 46·9 | 5·1 | 1·3 | 11·1 | 2·9 | 17·1 | 4·4 |
| 52 | 3 58·0 | 3 58·7 | 3 47·2 | 5·2 | 1·3 | 11·2 | 2·9 | 17·2 | 4·4 |
| 53 | 3 58·3 | 3 58·9 | 3 47·4 | 5·3 | 1·4 | 11·3 | 2·9 | 17·3 | 4·5 |
| 54 | 3 58·5 | 3 59·2 | 3 47·6 | 5·4 | 1·4 | 11·4 | 2·9 | 17·4 | 4·5 |
| 55 | 3 58·8 | 3 59·4 | 3 47·9 | 5·5 | 1·4 | 11·5 | 3·0 | 17·5 | 4·5 |
| 56 | 3 59·0 | 3 59·7 | 3 48·1 | 5·6 | 1·4 | 11·6 | 3·0 | 17·6 | 4·5 |
| 57 | 3 59·3 | 3 59·9 | 3 48·4 | 5·7 | 1·5 | 11·7 | 3·0 | 17·7 | 4·6 |
| 58 | 3 59·5 | 4 00·2 | 3 48·6 | 5·8 | 1·5 | 11·8 | 3·0 | 17·8 | 4·6 |
| 59 | 3 59·8 | 4 00·4 | 3 48·8 | 5·9 | 1·5 | 11·9 | 3·1 | 17·9 | 4·6 |
| 60 | 4 00·0 | 4 00·7 | 3 49·1 | 6·0 | 1·6 | 12·0 | 3·1 | 18·0 | 4·7 |

| $16^m$ | SUN PLANETS | ARIES | MOON | *v* or *d* | Corr$^n$ | *v* or *d* | Corr$^n$ | *v* or *d* | Corr$^n$ |
|---|---|---|---|---|---|---|---|---|---|
| s | ° ′ | ° ′ | ° ′ | ′ | ′ | ′ | ′ | ′ | ′ |
| 00 | 4 00·0 | 4 00·7 | 3 49·1 | 0·0 | 0·0 | 6·0 | 1·7 | 12·0 | 3·3 |
| 01 | 4 00·3 | 4 00·9 | 3 49·3 | 0·1 | 0·0 | 6·1 | 1·7 | 12·1 | 3·3 |
| 02 | 4 00·5 | 4 01·2 | 3 49·5 | 0·2 | 0·1 | 6·2 | 1·7 | 12·2 | 3·4 |
| 03 | 4 00·8 | 4 01·4 | 3 49·8 | 0·3 | 0·1 | 6·3 | 1·7 | 12·3 | 3·4 |
| 04 | 4 01·0 | 4 01·7 | 3 50·0 | 0·4 | 0·1 | 6·4 | 1·8 | 12·4 | 3·4 |
| 05 | 4 01·3 | 4 01·9 | 3 50·3 | 0·5 | 0·1 | 6·5 | 1·8 | 12·5 | 3·4 |
| 06 | 4 01·5 | 4 02·2 | 3 50·5 | 0·6 | 0·2 | 6·6 | 1·8 | 12·6 | 3·5 |
| 07 | 4 01·8 | 4 02·4 | 3 50·7 | 0·7 | 0·2 | 6·7 | 1·8 | 12·7 | 3·5 |
| 08 | 4 02·0 | 4 02·7 | 3 51·0 | 0·8 | 0·2 | 6·8 | 1·9 | 12·8 | 3·5 |
| 09 | 4 02·3 | 4 02·9 | 3 51·2 | 0·9 | 0·2 | 6·9 | 1·9 | 12·9 | 3·5 |
| 10 | 4 02·5 | 4 03·2 | 3 51·5 | 1·0 | 0·3 | 7·0 | 1·9 | 13·0 | 3·6 |
| 11 | 4 02·8 | 4 03·4 | 3 51·7 | 1·1 | 0·3 | 7·1 | 2·0 | 13·1 | 3·6 |
| 12 | 4 03·0 | 4 03·7 | 3 51·9 | 1·2 | 0·3 | 7·2 | 2·0 | 13·2 | 3·6 |
| 13 | 4 03·3 | 4 03·9 | 3 52·2 | 1·3 | 0·4 | 7·3 | 2·0 | 13·3 | 3·7 |
| 14 | 4 03·5 | 4 04·2 | 3 52·4 | 1·4 | 0·4 | 7·4 | 2·0 | 13·4 | 3·7 |
| 15 | 4 03·8 | 4 04·4 | 3 52·6 | 1·5 | 0·4 | 7·5 | 2·1 | 13·5 | 3·7 |
| 16 | 4 04·0 | 4 04·7 | 3 52·9 | 1·6 | 0·4 | 7·6 | 2·1 | 13·6 | 3·7 |
| 17 | 4 04·3 | 4 04·9 | 3 53·1 | 1·7 | 0·5 | 7·7 | 2·1 | 13·7 | 3·8 |
| 18 | 4 04·5 | 4 05·2 | 3 53·4 | 1·8 | 0·5 | 7·8 | 2·1 | 13·8 | 3·8 |
| 19 | 4 04·8 | 4 05·4 | 3 53·6 | 1·9 | 0·5 | 7·9 | 2·2 | 13·9 | 3·8 |
| 20 | 4 05·0 | 4 05·7 | 3 53·8 | 2·0 | 0·6 | 8·0 | 2·2 | 14·0 | 3·9 |
| 21 | 4 05·3 | 4 05·9 | 3 54·1 | 2·1 | 0·6 | 8·1 | 2·2 | 14·1 | 3·9 |
| 22 | 4 05·5 | 4 06·2 | 3 54·3 | 2·2 | 0·6 | 8·2 | 2·3 | 14·2 | 3·9 |
| 23 | 4 05·8 | 4 06·4 | 3 54·6 | 2·3 | 0·6 | 8·3 | 2·3 | 14·3 | 3·9 |
| 24 | 4 06·0 | 4 06·7 | 3 54·8 | 2·4 | 0·7 | 8·4 | 2·3 | 14·4 | 4·0 |
| 25 | 4 06·3 | 4 06·9 | 3 55·0 | 2·5 | 0·7 | 8·5 | 2·3 | 14·5 | 4·0 |
| 26 | 4 06·5 | 4 07·2 | 3 55·3 | 2·6 | 0·7 | 8·6 | 2·4 | 14·6 | 4·0 |
| 27 | 4 06·8 | 4 07·4 | 3 55·5 | 2·7 | 0·7 | 8·7 | 2·4 | 14·7 | 4·0 |
| 28 | 4 07·0 | 4 07·7 | 3 55·7 | 2·8 | 0·8 | 8·8 | 2·4 | 14·8 | 4·1 |
| 29 | 4 07·3 | 4 07·9 | 3 56·0 | 2·9 | 0·8 | 8·9 | 2·4 | 14·9 | 4·1 |
| 30 | 4 07·5 | 4 08·2 | 3 56·2 | 3·0 | 0·8 | 9·0 | 2·5 | 15·0 | 4·1 |
| 31 | 4 07·8 | 4 08·4 | 3 56·5 | 3·1 | 0·9 | 9·1 | 2·5 | 15·1 | 4·2 |
| 32 | 4 08·0 | 4 08·7 | 3 56·7 | 3·2 | 0·9 | 9·2 | 2·5 | 15·2 | 4·2 |
| 33 | 4 08·3 | 4 08·9 | 3 56·9 | 3·3 | 0·9 | 9·3 | 2·6 | 15·3 | 4·2 |
| 34 | 4 08·5 | 4 09·2 | 3 57·2 | 3·4 | 0·9 | 9·4 | 2·6 | 15·4 | 4·2 |
| 35 | 4 08·8 | 4 09·4 | 3 57·4 | 3·5 | 1·0 | 9·5 | 2·6 | 15·5 | 4·3 |
| 36 | 4 09·0 | 4 09·7 | 3 57·7 | 3·6 | 1·0 | 9·6 | 2·6 | 15·6 | 4·3 |
| 37 | 4 09·3 | 4 09·9 | 3 57·9 | 3·7 | 1·0 | 9·7 | 2·7 | 15·7 | 4·3 |
| 38 | 4 09·5 | 4 10·2 | 3 58·1 | 3·8 | 1·0 | 9·8 | 2·7 | 15·8 | 4·3 |
| 39 | 4 09·8 | 4 10·4 | 3 58·4 | 3·9 | 1·1 | 9·9 | 2·7 | 15·9 | 4·4 |
| 40 | 4 10·0 | 4 10·7 | 3 58·6 | 4·0 | 1·1 | 10·0 | 2·8 | 16·0 | 4·4 |
| 41 | 4 10·3 | 4 10·9 | 3 58·8 | 4·1 | 1·1 | 10·1 | 2·8 | 16·1 | 4·4 |
| 42 | 4 10·5 | 4 11·2 | 3 59·1 | 4·2 | 1·2 | 10·2 | 2·8 | 16·2 | 4·5 |
| 43 | 4 10·8 | 4 11·4 | 3 59·3 | 4·3 | 1·2 | 10·3 | 2·8 | 16·3 | 4·5 |
| 44 | 4 11·0 | 4 11·7 | 3 59·6 | 4·4 | 1·2 | 10·4 | 2·9 | 16·4 | 4·5 |
| 45 | 4 11·3 | 4 11·9 | 3 59·8 | 4·5 | 1·2 | 10·5 | 2·9 | 16·5 | 4·5 |
| 46 | 4 11·5 | 4 12·2 | 4 00·0 | 4·6 | 1·3 | 10·6 | 2·9 | 16·6 | 4·6 |
| 47 | 4 11·8 | 4 12·4 | 4 00·3 | 4·7 | 1·3 | 10·7 | 2·9 | 16·7 | 4·6 |
| 48 | 4 12·0 | 4 12·7 | 4 00·5 | 4·8 | 1·3 | 10·8 | 3·0 | 16·8 | 4·6 |
| 49 | 4 12·3 | 4 12·9 | 4 00·8 | 4·9 | 1·3 | 10·9 | 3·0 | 16·9 | 4·6 |
| 50 | 4 12·5 | 4 13·2 | 4 01·0 | 5·0 | 1·4 | 11·0 | 3·0 | 17·0 | 4·7 |
| 51 | 4 12·8 | 4 13·4 | 4 01·2 | 5·1 | 1·4 | 11·1 | 3·1 | 17·1 | 4·7 |
| 52 | 4 13·0 | 4 13·7 | 4 01·5 | 5·2 | 1·4 | 11·2 | 3·1 | 17·2 | 4·7 |
| 53 | 4 13·3 | 4 13·9 | 4 01·7 | 5·3 | 1·5 | 11·3 | 3·1 | 17·3 | 4·8 |
| 54 | 4 13·5 | 4 14·2 | 4 02·0 | 5·4 | 1·5 | 11·4 | 3·1 | 17·4 | 4·8 |
| 55 | 4 13·8 | 4 14·4 | 4 02·2 | 5·5 | 1·5 | 11·5 | 3·2 | 17·5 | 4·8 |
| 56 | 4 14·0 | 4 14·7 | 4 02·4 | 5·6 | 1·5 | 11·6 | 3·2 | 17·6 | 4·8 |
| 57 | 4 14·3 | 4 14·9 | 4 02·7 | 5·7 | 1·6 | 11·7 | 3·2 | 17·7 | 4·9 |
| 58 | 4 14·5 | 4 15·2 | 4 02·9 | 5·8 | 1·6 | 11·8 | 3·2 | 17·8 | 4·9 |
| 59 | 4 14·8 | 4 15·4 | 4 03·1 | 5·9 | 1·6 | 11·9 | 3·3 | 17·9 | 4·9 |
| 60 | 4 15·0 | 4 15·7 | 4 03·4 | 6·0 | 1·7 | 12·0 | 3·3 | 18·0 | 5·0 |

| $17^m$ | SUN PLANETS | ARIES | MOON | *v* or *d* | Corr$^n$ | *v* or *d* | Corr$^n$ | *v* or *d* | Corr$^n$ |
|---|---|---|---|---|---|---|---|---|---|
| s | ° ′ | ° ′ | ° ′ | ′ | ′ | ′ | ′ | ′ | ′ |
| 00 | 4 15·0 | 4 15·7 | 4 03·4 | 0·0 | 0·0 | 6·0 | 1·8 | 12·0 | 3·5 |
| 01 | 4 15·3 | 4 15·9 | 4 03·6 | 0·1 | 0·0 | 6·1 | 1·8 | 12·1 | 3·5 |
| 02 | 4 15·5 | 4 16·2 | 4 03·9 | 0·2 | 0·1 | 6·2 | 1·8 | 12·2 | 3·6 |
| 03 | 4 15·8 | 4 16·5 | 4 04·1 | 0·3 | 0·1 | 6·3 | 1·8 | 12·3 | 3·6 |
| 04 | 4 16·0 | 4 16·7 | 4 04·3 | 0·4 | 0·1 | 6·4 | 1·9 | 12·4 | 3·6 |
| 05 | 4 16·3 | 4 17·0 | 4 04·6 | 0·5 | 0·1 | 6·5 | 1·9 | 12·5 | 3·6 |
| 06 | 4 16·5 | 4 17·2 | 4 04·8 | 0·6 | 0·2 | 6·6 | 1·9 | 12·6 | 3·7 |
| 07 | 4 16·8 | 4 17·5 | 4 05·1 | 0·7 | 0·2 | 6·7 | 2·0 | 12·7 | 3·7 |
| 08 | 4 17·0 | 4 17·7 | 4 05·3 | 0·8 | 0·2 | 6·8 | 2·0 | 12·8 | 3·7 |
| 09 | 4 17·3 | 4 18·0 | 4 05·5 | 0·9 | 0·3 | 6·9 | 2·0 | 12·9 | 3·8 |
| 10 | 4 17·5 | 4 18·2 | 4 05·8 | 1·0 | 0·3 | 7·0 | 2·0 | 13·0 | 3·8 |
| 11 | 4 17·8 | 4 18·5 | 4 06·0 | 1·1 | 0·3 | 7·1 | 2·1 | 13·1 | 3·8 |
| 12 | 4 18·0 | 4 18·7 | 4 06·2 | 1·2 | 0·4 | 7·2 | 2·1 | 13·2 | 3·9 |
| 13 | 4 18·3 | 4 19·0 | 4 06·5 | 1·3 | 0·4 | 7·3 | 2·1 | 13·3 | 3·9 |
| 14 | 4 18·5 | 4 19·2 | 4 06·7 | 1·4 | 0·4 | 7·4 | 2·2 | 13·4 | 3·9 |
| 15 | 4 18·8 | 4 19·5 | 4 07·0 | 1·5 | 0·4 | 7·5 | 2·2 | 13·5 | 3·9 |
| 16 | 4 19·0 | 4 19·7 | 4 07·2 | 1·6 | 0·5 | 7·6 | 2·2 | 13·6 | 4·0 |
| 17 | 4 19·3 | 4 20·0 | 4 07·4 | 1·7 | 0·5 | 7·7 | 2·2 | 13·7 | 4·0 |
| 18 | 4 19·5 | 4 20·2 | 4 07·7 | 1·8 | 0·5 | 7·8 | 2·3 | 13·8 | 4·0 |
| 19 | 4 19·8 | 4 20·5 | 4 07·9 | 1·9 | 0·6 | 7·9 | 2·3 | 13·9 | 4·1 |
| 20 | 4 20·0 | 4 20·7 | 4 08·2 | 2·0 | 0·6 | 8·0 | 2·3 | 14·0 | 4·1 |
| 21 | 4 20·3 | 4 21·0 | 4 08·4 | 2·1 | 0·6 | 8·1 | 2·4 | 14·1 | 4·1 |
| 22 | 4 20·5 | 4 21·2 | 4 08·6 | 2·2 | 0·6 | 8·2 | 2·4 | 14·2 | 4·1 |
| 23 | 4 20·8 | 4 21·5 | 4 08·9 | 2·3 | 0·7 | 8·3 | 2·4 | 14·3 | 4·2 |
| 24 | 4 21·0 | 4 21·7 | 4 09·1 | 2·4 | 0·7 | 8·4 | 2·5 | 14·4 | 4·2 |
| 25 | 4 21·3 | 4 22·0 | 4 09·3 | 2·5 | 0·7 | 8·5 | 2·5 | 14·5 | 4·2 |
| 26 | 4 21·5 | 4 22·2 | 4 09·6 | 2·6 | 0·8 | 8·6 | 2·5 | 14·6 | 4·3 |
| 27 | 4 21·8 | 4 22·5 | 4 09·8 | 2·7 | 0·8 | 8·7 | 2·5 | 14·7 | 4·3 |
| 28 | 4 22·0 | 4 22·7 | 4 10·1 | 2·8 | 0·8 | 8·8 | 2·6 | 14·8 | 4·3 |
| 29 | 4 22·3 | 4 23·0 | 4 10·3 | 2·9 | 0·8 | 8·9 | 2·6 | 14·9 | 4·3 |
| 30 | 4 22·5 | 4 23·2 | 4 10·5 | 3·0 | 0·9 | 9·0 | 2·6 | 15·0 | 4·4 |
| 31 | 4 22·8 | 4 23·5 | 4 10·8 | 3·1 | 0·9 | 9·1 | 2·7 | 15·1 | 4·4 |
| 32 | 4 23·0 | 4 23·7 | 4 11·0 | 3·2 | 0·9 | 9·2 | 2·7 | 15·2 | 4·4 |
| 33 | 4 23·3 | 4 24·0 | 4 11·3 | 3·3 | 1·0 | 9·3 | 2·7 | 15·3 | 4·5 |
| 34 | 4 23·5 | 4 24·2 | 4 11·5 | 3·4 | 1·0 | 9·4 | 2·7 | 15·4 | 4·5 |
| 35 | 4 23·8 | 4 24·5 | 4 11·7 | 3·5 | 1·0 | 9·5 | 2·8 | 15·5 | 4·5 |
| 36 | 4 24·0 | 4 24·7 | 4 12·0 | 3·6 | 1·1 | 9·6 | 2·8 | 15·6 | 4·6 |
| 37 | 4 24·3 | 4 25·0 | 4 12·2 | 3·7 | 1·1 | 9·7 | 2·8 | 15·7 | 4·6 |
| 38 | 4 24·5 | 4 25·2 | 4 12·5 | 3·8 | 1·1 | 9·8 | 2·9 | 15·8 | 4·6 |
| 39 | 4 24·8 | 4 25·5 | 4 12·7 | 3·9 | 1·1 | 9·9 | 2·9 | 15·9 | 4·6 |
| 40 | 4 25·0 | 4 25·7 | 4 12·9 | 4·0 | 1·2 | 10·0 | 2·9 | 16·0 | 4·7 |
| 41 | 4 25·3 | 4 26·0 | 4 13·2 | 4·1 | 1·2 | 10·1 | 2·9 | 16·1 | 4·7 |
| 42 | 4 25·5 | 4 26·2 | 4 13·4 | 4·2 | 1·2 | 10·2 | 3·0 | 16·2 | 4·7 |
| 43 | 4 25·8 | 4 26·5 | 4 13·6 | 4·3 | 1·3 | 10·3 | 3·0 | 16·3 | 4·8 |
| 44 | 4 26·0 | 4 26·7 | 4 13·9 | 4·4 | 1·3 | 10·4 | 3·0 | 16·4 | 4·8 |
| 45 | 4 26·3 | 4 27·0 | 4 14·1 | 4·5 | 1·3 | 10·5 | 3·1 | 16·5 | 4·8 |
| 46 | 4 26·5 | 4 27·2 | 4 14·4 | 4·6 | 1·3 | 10·6 | 3·1 | 16·6 | 4·8 |
| 47 | 4 26·8 | 4 27·5 | 4 14·6 | 4·7 | 1·4 | 10·7 | 3·1 | 16·7 | 4·9 |
| 48 | 4 27·0 | 4 27·7 | 4 14·8 | 4·8 | 1·4 | 10·8 | 3·2 | 16·8 | 4·9 |
| 49 | 4 27·3 | 4 28·0 | 4 15·1 | 4·9 | 1·4 | 10·9 | 3·2 | 16·9 | 4·9 |
| 50 | 4 27·5 | 4 28·2 | 4 15·3 | 5·0 | 1·5 | 11·0 | 3·2 | 17·0 | 5·0 |
| 51 | 4 27·8 | 4 28·5 | 4 15·6 | 5·1 | 1·5 | 11·1 | 3·2 | 17·1 | 5·0 |
| 52 | 4 28·0 | 4 28·7 | 4 15·8 | 5·2 | 1·5 | 11·2 | 3·3 | 17·2 | 5·0 |
| 53 | 4 28·3 | 4 29·0 | 4 16·0 | 5·3 | 1·5 | 11·3 | 3·3 | 17·3 | 5·0 |
| 54 | 4 28·5 | 4 29·2 | 4 16·3 | 5·4 | 1·6 | 11·4 | 3·3 | 17·4 | 5·1 |
| 55 | 4 28·8 | 4 29·5 | 4 16·5 | 5·5 | 1·6 | 11·5 | 3·4 | 17·5 | 5·1 |
| 56 | 4 29·0 | 4 29·7 | 4 16·7 | 5·6 | 1·6 | 11·6 | 3·4 | 17·6 | 5·1 |
| 57 | 4 29·3 | 4 30·0 | 4 17·0 | 5·7 | 1·7 | 11·7 | 3·4 | 17·7 | 5·2 |
| 58 | 4 29·5 | 4 30·2 | 4 17·2 | 5·8 | 1·7 | 11·8 | 3·4 | 17·8 | 5·2 |
| 59 | 4 29·8 | 4 30·5 | 4 17·5 | 5·9 | 1·7 | 11·9 | 3·5 | 17·9 | 5·2 |
| 60 | 4 30·0 | 4 30·7 | 4 17·7 | 6·0 | 1·8 | 12·0 | 3·5 | 18·0 | 5·3 |

| 18ᵐ | SUN PLANETS | ARIES | MOON | v or d | Corrⁿ | v or d | Corrⁿ | v or d | Corrⁿ |
|---|---|---|---|---|---|---|---|---|---|
| s | ° ′ | ° ′ | ° ′ | ′ | ′ | ′ | ′ | ′ | ′ |
| 00 | 4 30·0 | 4 30·7 | 4 17·7 | 0·0 | 0·0 | 6·0 | 1·9 | 12·0 | 3·7 |
| 01 | 4 30·3 | 4 31·0 | 4 17·9 | 0·1 | 0·0 | 6·1 | 1·9 | 12·1 | 3·7 |
| 02 | 4 30·5 | 4 31·2 | 4 18·2 | 0·2 | 0·1 | 6·2 | 1·9 | 12·2 | 3·8 |
| 03 | 4 30·8 | 4 31·5 | 4 18·4 | 0·3 | 0·1 | 6·3 | 1·9 | 12·3 | 3·8 |
| 04 | 4 31·0 | 4 31·7 | 4 18·7 | 0·4 | 0·1 | 6·4 | 2·0 | 12·4 | 3·8 |
| 05 | 4 31·3 | 4 32·0 | 4 18·9 | 0·5 | 0·2 | 6·5 | 2·0 | 12·5 | 3·9 |
| 06 | 4 31·5 | 4 32·2 | 4 19·1 | 0·6 | 0·2 | 6·6 | 2·0 | 12·6 | 3·9 |
| 07 | 4 31·8 | 4 32·5 | 4 19·4 | 0·7 | 0·2 | 6·7 | 2·1 | 12·7 | 3·9 |
| 08 | 4 32·0 | 4 32·7 | 4 19·6 | 0·8 | 0·2 | 6·8 | 2·1 | 12·8 | 3·9 |
| 09 | 4 32·3 | 4 33·0 | 4 19·8 | 0·9 | 0·3 | 6·9 | 2·1 | 12·9 | 4·0 |
| 10 | 4 32·5 | 4 33·2 | 4 20·1 | 1·0 | 0·3 | 7·0 | 2·2 | 13·0 | 4·0 |
| 11 | 4 32·8 | 4 33·5 | 4 20·3 | 1·1 | 0·3 | 7·1 | 2·2 | 13·1 | 4·0 |
| 12 | 4 33·0 | 4 33·7 | 4 20·6 | 1·2 | 0·4 | 7·2 | 2·2 | 13·2 | 4·1 |
| 13 | 4 33·3 | 4 34·0 | 4 20·8 | 1·3 | 0·4 | 7·3 | 2·3 | 13·3 | 4·1 |
| 14 | 4 33·5 | 4 34·2 | 4 21·0 | 1·4 | 0·4 | 7·4 | 2·3 | 13·4 | 4·1 |
| 15 | 4 33·8 | 4 34·5 | 4 21·3 | 1·5 | 0·5 | 7·5 | 2·3 | 13·5 | 4·2 |
| 16 | 4 34·0 | 4 34·8 | 4 21·5 | 1·6 | 0·5 | 7·6 | 2·3 | 13·6 | 4·2 |
| 17 | 4 34·3 | 4 35·0 | 4 21·8 | 1·7 | 0·5 | 7·7 | 2·4 | 13·7 | 4·2 |
| 18 | 4 34·5 | 4 35·3 | 4 22·0 | 1·8 | 0·6 | 7·8 | 2·4 | 13·8 | 4·3 |
| 19 | 4 34·8 | 4 35·5 | 4 22·2 | 1·9 | 0·6 | 7·9 | 2·4 | 13·9 | 4·3 |
| 20 | 4 35·0 | 4 35·8 | 4 22·5 | 2·0 | 0·6 | 8·0 | 2·5 | 14·0 | 4·3 |
| 21 | 4 35·3 | 4 36·0 | 4 22·7 | 2·1 | 0·6 | 8·1 | 2·5 | 14·1 | 4·3 |
| 22 | 4 35·5 | 4 36·3 | 4 22·9 | 2·2 | 0·7 | 8·2 | 2·5 | 14·2 | 4·4 |
| 23 | 4 35·8 | 4 36·5 | 4 23·2 | 2·3 | 0·7 | 8·3 | 2·6 | 14·3 | 4·4 |
| 24 | 4 36·0 | 4 36·8 | 4 23·4 | 2·4 | 0·7 | 8·4 | 2·6 | 14·4 | 4·4 |
| 25 | 4 36·3 | 4 37·0 | 4 23·7 | 2·5 | 0·8 | 8·5 | 2·6 | 14·5 | 4·5 |
| 26 | 4 36·5 | 4 37·3 | 4 23·9 | 2·6 | 0·8 | 8·6 | 2·7 | 14·6 | 4·5 |
| 27 | 4 36·8 | 4 37·5 | 4 24·1 | 2·7 | 0·8 | 8·7 | 2·7 | 14·7 | 4·5 |
| 28 | 4 37·0 | 4 37·8 | 4 24·4 | 2·8 | 0·9 | 8·8 | 2·7 | 14·8 | 4·6 |
| 29 | 4 37·3 | 4 38·0 | 4 24·6 | 2·9 | 0·9 | 8·9 | 2·7 | 14·9 | 4·6 |
| 30 | 4 37·5 | 4 38·3 | 4 24·9 | 3·0 | 0·9 | 9·0 | 2·8 | 15·0 | 4·6 |
| 31 | 4 37·8 | 4 38·5 | 4 25·1 | 3·1 | 1·0 | 9·1 | 2·8 | 15·1 | 4·7 |
| 32 | 4 38·0 | 4 38·8 | 4 25·3 | 3·2 | 1·0 | 9·2 | 2·8 | 15·2 | 4·7 |
| 33 | 4 38·3 | 4 39·0 | 4 25·6 | 3·3 | 1·0 | 9·3 | 2·9 | 15·3 | 4·7 |
| 34 | 4 38·5 | 4 39·3 | 4 25·8 | 3·4 | 1·0 | 9·4 | 2·9 | 15·4 | 4·7 |
| 35 | 4 38·8 | 4 39·5 | 4 26·1 | 3·5 | 1·1 | 9·5 | 2·9 | 15·5 | 4·8 |
| 36 | 4 39·0 | 4 39·8 | 4 26·3 | 3·6 | 1·1 | 9·6 | 3·0 | 15·6 | 4·8 |
| 37 | 4 39·3 | 4 40·0 | 4 26·5 | 3·7 | 1·1 | 9·7 | 3·0 | 15·7 | 4·8 |
| 38 | 4 39·5 | 4 40·3 | 4 26·8 | 3·8 | 1·2 | 9·8 | 3·0 | 15·8 | 4·9 |
| 39 | 4 39·8 | 4 40·5 | 4 27·0 | 3·9 | 1·2 | 9·9 | 3·1 | 15·9 | 4·9 |
| 40 | 4 40·0 | 4 40·8 | 4 27·2 | 4·0 | 1·2 | 10·0 | 3·1 | 16·0 | 4·9 |
| 41 | 4 40·3 | 4 41·0 | 4 27·5 | 4·1 | 1·3 | 10·1 | 3·1 | 16·1 | 5·0 |
| 42 | 4 40·5 | 4 41·3 | 4 27·7 | 4·2 | 1·3 | 10·2 | 3·1 | 16·2 | 5·0 |
| 43 | 4 40·8 | 4 41·5 | 4 28·0 | 4·3 | 1·3 | 10·3 | 3·2 | 16·3 | 5·0 |
| 44 | 4 41·0 | 4 41·8 | 4 28·2 | 4·4 | 1·4 | 10·4 | 3·2 | 16·4 | 5·1 |
| 45 | 4 41·3 | 4 42·0 | 4 28·4 | 4·5 | 1·4 | 10·5 | 3·2 | 16·5 | 5·1 |
| 46 | 4 41·5 | 4 42·3 | 4 28·7 | 4·6 | 1·4 | 10·6 | 3·3 | 16·6 | 5·1 |
| 47 | 4 41·8 | 4 42·5 | 4 28·9 | 4·7 | 1·4 | 10·7 | 3·3 | 16·7 | 5·1 |
| 48 | 4 42·0 | 4 42·8 | 4 29·2 | 4·8 | 1·5 | 10·8 | 3·3 | 16·8 | 5·2 |
| 49 | 4 42·3 | 4 43·0 | 4 29·4 | 4·9 | 1·5 | 10·9 | 3·4 | 16·9 | 5·2 |
| 50 | 4 42·5 | 4 43·3 | 4 29·6 | 5·0 | 1·5 | 11·0 | 3·4 | 17·0 | 5·2 |
| 51 | 4 42·8 | 4 43·5 | 4 29·9 | 5·1 | 1·6 | 11·1 | 3·4 | 17·1 | 5·3 |
| 52 | 4 43·0 | 4 43·8 | 4 30·1 | 5·2 | 1·6 | 11·2 | 3·5 | 17·2 | 5·3 |
| 53 | 4 43·3 | 4 44·0 | 4 30·3 | 5·3 | 1·6 | 11·3 | 3·5 | 17·3 | 5·3 |
| 54 | 4 43·5 | 4 44·3 | 4 30·6 | 5·4 | 1·7 | 11·4 | 3·5 | 17·4 | 5·4 |
| 55 | 4 43·8 | 4 44·5 | 4 30·8 | 5·5 | 1·7 | 11·5 | 3·5 | 17·5 | 5·4 |
| 56 | 4 44·0 | 4 44·8 | 4 31·1 | 5·6 | 1·7 | 11·6 | 3·6 | 17·6 | 5·4 |
| 57 | 4 44·3 | 4 45·0 | 4 31·3 | 5·7 | 1·8 | 11·7 | 3·6 | 17·7 | 5·5 |
| 58 | 4 44·5 | 4 45·3 | 4 31·5 | 5·8 | 1·8 | 11·8 | 3·6 | 17·8 | 5·5 |
| 59 | 4 44·8 | 4 45·5 | 4 31·8 | 5·9 | 1·8 | 11·9 | 3·7 | 17·9 | 5·5 |
| 60 | 4 45·0 | 4 45·8 | 4 32·0 | 6·0 | 1·9 | 12·0 | 3·7 | 18·0 | 5·6 |

| 19ᵐ | SUN PLANETS | ARIES | MOON | v or d | Corrⁿ | v or d | Corrⁿ | v or d | Corrⁿ |
|---|---|---|---|---|---|---|---|---|---|
| s | ° ′ | ° ′ | ° ′ | ′ | ′ | ′ | ′ | ′ | ′ |
| 00 | 4 45·0 | 4 45·8 | 4 32·0 | 0·0 | 0·0 | 6·0 | 2·0 | 12·0 | 3·9 |
| 01 | 4 45·3 | 4 46·0 | 4 32·3 | 0·1 | 0·0 | 6·1 | 2·0 | 12·1 | 3·9 |
| 02 | 4 45·5 | 4 46·3 | 4 32·5 | 0·2 | 0·1 | 6·2 | 2·0 | 12·2 | 4·0 |
| 03 | 4 45·8 | 4 46·5 | 4 32·7 | 0·3 | 0·1 | 6·3 | 2·0 | 12·3 | 4·0 |
| 04 | 4 46·0 | 4 46·8 | 4 33·0 | 0·4 | 0·1 | 6·4 | 2·1 | 12·4 | 4·0 |
| 05 | 4 46·3 | 4 47·0 | 4 33·2 | 0·5 | 0·2 | 6·5 | 2·1 | 12·5 | 4·1 |
| 06 | 4 46·5 | 4 47·3 | 4 33·4 | 0·6 | 0·2 | 6·6 | 2·1 | 12·6 | 4·1 |
| 07 | 4 46·8 | 4 47·5 | 4 33·7 | 0·7 | 0·2 | 6·7 | 2·2 | 12·7 | 4·1 |
| 08 | 4 47·0 | 4 47·8 | 4 33·9 | 0·8 | 0·3 | 6·8 | 2·2 | 12·8 | 4·2 |
| 09 | 4 47·3 | 4 48·0 | 4 34·2 | 0·9 | 0·3 | 6·9 | 2·2 | 12·9 | 4·2 |
| 10 | 4 47·5 | 4 48·3 | 4 34·4 | 1·0 | 0·3 | 7·0 | 2·3 | 13·0 | 4·2 |
| 11 | 4 47·8 | 4 48·5 | 4 34·6 | 1·1 | 0·4 | 7·1 | 2·3 | 13·1 | 4·3 |
| 12 | 4 48·0 | 4 48·8 | 4 34·9 | 1·2 | 0·4 | 7·2 | 2·3 | 13·2 | 4·3 |
| 13 | 4 48·3 | 4 49·0 | 4 35·1 | 1·3 | 0·4 | 7·3 | 2·4 | 13·3 | 4·3 |
| 14 | 4 48·5 | 4 49·3 | 4 35·4 | 1·4 | 0·5 | 7·4 | 2·4 | 13·4 | 4·4 |
| 15 | 4 48·8 | 4 49·5 | 4 35·6 | 1·5 | 0·5 | 7·5 | 2·4 | 13·5 | 4·4 |
| 16 | 4 49·0 | 4 49·8 | 4 35·8 | 1·6 | 0·5 | 7·6 | 2·5 | 13·6 | 4·4 |
| 17 | 4 49·3 | 4 50·0 | 4 36·1 | 1·7 | 0·6 | 7·7 | 2·5 | 13·7 | 4·5 |
| 18 | 4 49·5 | 4 50·3 | 4 36·3 | 1·8 | 0·6 | 7·8 | 2·5 | 13·8 | 4·5 |
| 19 | 4 49·8 | 4 50·5 | 4 36·6 | 1·9 | 0·6 | 7·9 | 2·6 | 13·9 | 4·5 |
| 20 | 4 50·0 | 4 50·8 | 4 36·8 | 2·0 | 0·7 | 8·0 | 2·6 | 14·0 | 4·6 |
| 21 | 4 50·3 | 4 51·0 | 4 37·0 | 2·1 | 0·7 | 8·1 | 2·6 | 14·1 | 4·6 |
| 22 | 4 50·5 | 4 51·3 | 4 37·3 | 2·2 | 0·7 | 8·2 | 2·7 | 14·2 | 4·6 |
| 23 | 4 50·8 | 4 51·5 | 4 37·5 | 2·3 | 0·7 | 8·3 | 2·7 | 14·3 | 4·6 |
| 24 | 4 51·0 | 4 51·8 | 4 37·7 | 2·4 | 0·8 | 8·4 | 2·7 | 14·4 | 4·7 |
| 25 | 4 51·3 | 4 52·0 | 4 38·0 | 2·5 | 0·8 | 8·5 | 2·8 | 14·5 | 4·7 |
| 26 | 4 51·5 | 4 52·3 | 4 38·2 | 2·6 | 0·8 | 8·6 | 2·8 | 14·6 | 4·7 |
| 27 | 4 51·8 | 4 52·5 | 4 38·5 | 2·7 | 0·9 | 8·7 | 2·8 | 14·7 | 4·8 |
| 28 | 4 52·0 | 4 52·8 | 4 38·7 | 2·8 | 0·9 | 8·8 | 2·9 | 14·8 | 4·8 |
| 29 | 4 52·3 | 4 53·1 | 4 38·9 | 2·9 | 0·9 | 8·9 | 2·9 | 14·9 | 4·8 |
| 30 | 4 52·5 | 4 53·3 | 4 39·2 | 3·0 | 1·0 | 9·0 | 2·9 | 15·0 | 4·9 |
| 31 | 4 52·8 | 4 53·6 | 4 39·4 | 3·1 | 1·0 | 9·1 | 3·0 | 15·1 | 4·9 |
| 32 | 4 53·0 | 4 53·8 | 4 39·7 | 3·2 | 1·0 | 9·2 | 3·0 | 15·2 | 4·9 |
| 33 | 4 53·3 | 4 54·1 | 4 39·9 | 3·3 | 1·1 | 9·3 | 3·0 | 15·3 | 5·0 |
| 34 | 4 53·5 | 4 54·3 | 4 40·1 | 3·4 | 1·1 | 9·4 | 3·1 | 15·4 | 5·0 |
| 35 | 4 53·8 | 4 54·6 | 4 40·4 | 3·5 | 1·1 | 9·5 | 3·1 | 15·5 | 5·0 |
| 36 | 4 54·0 | 4 54·8 | 4 40·6 | 3·6 | 1·2 | 9·6 | 3·1 | 15·6 | 5·1 |
| 37 | 4 54·3 | 4 55·1 | 4 40·8 | 3·7 | 1·2 | 9·7 | 3·2 | 15·7 | 5·1 |
| 38 | 4 54·5 | 4 55·3 | 4 41·1 | 3·8 | 1·2 | 9·8 | 3·2 | 15·8 | 5·1 |
| 39 | 4 54·8 | 4 55·6 | 4 41·3 | 3·9 | 1·3 | 9·9 | 3·2 | 15·9 | 5·2 |
| 40 | 4 55·0 | 4 55·8 | 4 41·6 | 4·0 | 1·3 | 10·0 | 3·3 | 16·0 | 5·2 |
| 41 | 4 55·3 | 4 56·1 | 4 41·8 | 4·1 | 1·3 | 10·1 | 3·3 | 16·1 | 5·2 |
| 42 | 4 55·5 | 4 56·3 | 4 42·0 | 4·2 | 1·4 | 10·2 | 3·3 | 16·2 | 5·3 |
| 43 | 4 55·8 | 4 56·6 | 4 42·3 | 4·3 | 1·4 | 10·3 | 3·3 | 16·3 | 5·3 |
| 44 | 4 56·0 | 4 56·8 | 4 42·5 | 4·4 | 1·4 | 10·4 | 3·4 | 16·4 | 5·3 |
| 45 | 4 56·3 | 4 57·1 | 4 42·8 | 4·5 | 1·5 | 10·5 | 3·4 | 16·5 | 5·4 |
| 46 | 4 56·5 | 4 57·3 | 4 43·0 | 4·6 | 1·5 | 10·6 | 3·4 | 16·6 | 5·4 |
| 47 | 4 56·8 | 4 57·6 | 4 43·2 | 4·7 | 1·5 | 10·7 | 3·5 | 16·7 | 5·4 |
| 48 | 4 57·0 | 4 57·8 | 4 43·5 | 4·8 | 1·6 | 10·8 | 3·5 | 16·8 | 5·5 |
| 49 | 4 57·3 | 4 58·1 | 4 43·7 | 4·9 | 1·6 | 10·9 | 3·5 | 16·9 | 5·5 |
| 50 | 4 57·5 | 4 58·3 | 4 43·9 | 5·0 | 1·6 | 11·0 | 3·6 | 17·0 | 5·5 |
| 51 | 4 57·8 | 4 58·6 | 4 44·2 | 5·1 | 1·7 | 11·1 | 3·6 | 17·1 | 5·6 |
| 52 | 4 58·0 | 4 58·8 | 4 44·4 | 5·2 | 1·7 | 11·2 | 3·6 | 17·2 | 5·6 |
| 53 | 4 58·3 | 4 59·1 | 4 44·7 | 5·3 | 1·7 | 11·3 | 3·7 | 17·3 | 5·6 |
| 54 | 4 58·5 | 4 59·3 | 4 44·9 | 5·4 | 1·8 | 11·4 | 3·7 | 17·4 | 5·7 |
| 55 | 4 58·8 | 4 59·6 | 4 45·1 | 5·5 | 1·8 | 11·5 | 3·7 | 17·5 | 5·7 |
| 56 | 4 59·0 | 4 59·8 | 4 45·4 | 5·6 | 1·8 | 11·6 | 3·8 | 17·6 | 5·7 |
| 57 | 4 59·3 | 5 00·1 | 4 45·6 | 5·7 | 1·9 | 11·7 | 3·8 | 17·7 | 5·8 |
| 58 | 4 59·5 | 5 00·3 | 4 45·9 | 5·8 | 1·9 | 11·8 | 3·8 | 17·8 | 5·8 |
| 59 | 4 59·8 | 5 00·6 | 4 46·1 | 5·9 | 1·9 | 11·9 | 3·9 | 17·9 | 5·8 |
| 60 | 5 00·0 | 5 00·8 | 4 46·3 | 6·0 | 2·0 | 12·0 | 3·9 | 18·0 | 5·9 |

| 20ᵐ | SUN PLANETS | ARIES | MOON | v or d | Corrⁿ | v or d | Corrⁿ | v or d | Corrⁿ |
|---|---|---|---|---|---|---|---|---|---|
| s | ° ′ | ° ′ | ° ′ | ′ | ′ | ′ | ′ | ′ | ′ |
| 00 | 5 00·0 | 5 00·8 | 4 46·3 | 0·0 | 0·0 | 6·0 | 2·1 | 12·0 | 4·1 |
| 01 | 5 00·3 | 5 01·1 | 4 46·6 | 0·1 | 0·0 | 6·1 | 2·1 | 12·1 | 4·1 |
| 02 | 5 00·5 | 5 01·3 | 4 46·8 | 0·2 | 0·1 | 6·2 | 2·1 | 12·2 | 4·2 |
| 03 | 5 00·8 | 5 01·6 | 4 47·0 | 0·3 | 0·1 | 6·3 | 2·2 | 12·3 | 4·2 |
| 04 | 5 01·0 | 5 01·8 | 4 47·3 | 0·4 | 0·1 | 6·4 | 2·2 | 12·4 | 4·2 |
| 05 | 5 01·3 | 5 02·1 | 4 47·5 | 0·5 | 0·2 | 6·5 | 2·2 | 12·5 | 4·3 |
| 06 | 5 01·5 | 5 02·3 | 4 47·8 | 0·6 | 0·2 | 6·6 | 2·3 | 12·6 | 4·3 |
| 07 | 5 01·8 | 5 02·6 | 4 48·0 | 0·7 | 0·2 | 6·7 | 2·3 | 12·7 | 4·3 |
| 08 | 5 02·0 | 5 02·8 | 4 48·2 | 0·8 | 0·3 | 6·8 | 2·3 | 12·8 | 4·4 |
| 09 | 5 02·3 | 5 03·1 | 4 48·5 | 0·9 | 0·3 | 6·9 | 2·4 | 12·9 | 4·4 |
| 10 | 5 02·5 | 5 03·3 | 4 48·7 | 1·0 | 0·3 | 7·0 | 2·4 | 13·0 | 4·4 |
| 11 | 5 02·8 | 5 03·6 | 4 49·0 | 1·1 | 0·4 | 7·1 | 2·4 | 13·1 | 4·5 |
| 12 | 5 03·0 | 5 03·8 | 4 49·2 | 1·2 | 0·4 | 7·2 | 2·5 | 13·2 | 4·5 |
| 13 | 5 03·3 | 5 04·1 | 4 49·4 | 1·3 | 0·4 | 7·3 | 2·5 | 13·3 | 4·5 |
| 14 | 5 03·5 | 5 04·3 | 4 49·7 | 1·4 | 0·5 | 7·4 | 2·5 | 13·4 | 4·6 |
| 15 | 5 03·8 | 5 04·6 | 4 49·9 | 1·5 | 0·5 | 7·5 | 2·6 | 13·5 | 4·6 |
| 16 | 5 04·0 | 5 04·8 | 4 50·2 | 1·6 | 0·5 | 7·6 | 2·6 | 13·6 | 4·6 |
| 17 | 5 04·3 | 5 05·1 | 4 50·4 | 1·7 | 0·6 | 7·7 | 2·6 | 13·7 | 4·7 |
| 18 | 5 04·5 | 5 05·3 | 4 50·6 | 1·8 | 0·6 | 7·8 | 2·7 | 13·8 | 4·7 |
| 19 | 5 04·8 | 5 05·6 | 4 50·9 | 1·9 | 0·6 | 7·9 | 2·7 | 13·9 | 4·7 |
| 20 | 5 05·0 | 5 05·8 | 4 51·1 | 2·0 | 0·7 | 8·0 | 2·7 | 14·0 | 4·8 |
| 21 | 5 05·3 | 5 06·1 | 4 51·3 | 2·1 | 0·7 | 8·1 | 2·8 | 14·1 | 4·8 |
| 22 | 5 05·5 | 5 06·3 | 4 51·6 | 2·2 | 0·8 | 8·2 | 2·8 | 14·2 | 4·9 |
| 23 | 5 05·8 | 5 06·6 | 4 51·8 | 2·3 | 0·8 | 8·3 | 2·8 | 14·3 | 4·9 |
| 24 | 5 06·0 | 5 06·8 | 4 52·1 | 2·4 | 0·8 | 8·4 | 2·9 | 14·4 | 4·9 |
| 25 | 5 06·3 | 5 07·1 | 4 52·3 | 2·5 | 0·9 | 8·5 | 2·9 | 14·5 | 5·0 |
| 26 | 5 06·5 | 5 07·3 | 4 52·5 | 2·6 | 0·9 | 8·6 | 2·9 | 14·6 | 5·0 |
| 27 | 5 06·8 | 5 07·6 | 4 52·8 | 2·7 | 0·9 | 8·7 | 3·0 | 14·7 | 5·0 |
| 28 | 5 07·0 | 5 07·8 | 4 53·0 | 2·8 | 1·0 | 8·8 | 3·0 | 14·8 | 5·1 |
| 29 | 5 07·3 | 5 08·1 | 4 53·3 | 2·9 | 1·0 | 8·9 | 3·0 | 14·9 | 5·1 |
| 30 | 5 07·5 | 5 08·3 | 4 53·5 | 3·0 | 1·0 | 9·0 | 3·1 | 15·0 | 5·1 |
| 31 | 5 07·8 | 5 08·6 | 4 53·7 | 3·1 | 1·1 | 9·1 | 3·1 | 15·1 | 5·2 |
| 32 | 5 08·0 | 5 08·8 | 4 54·0 | 3·2 | 1·1 | 9·2 | 3·1 | 15·2 | 5·2 |
| 33 | 5 08·3 | 5 09·1 | 4 54·2 | 3·3 | 1·1 | 9·3 | 3·2 | 15·3 | 5·2 |
| 34 | 5 08·5 | 5 09·3 | 4 54·4 | 3·4 | 1·2 | 9·4 | 3·2 | 15·4 | 5·3 |
| 35 | 5 08·8 | 5 09·6 | 4 54·7 | 3·5 | 1·2 | 9·5 | 3·2 | 15·5 | 5·3 |
| 36 | 5 09·0 | 5 09·8 | 4 54·9 | 3·6 | 1·2 | 9·6 | 3·3 | 15·6 | 5·3 |
| 37 | 5 09·3 | 5 10·1 | 4 55·2 | 3·7 | 1·3 | 9·7 | 3·3 | 15·7 | 5·4 |
| 38 | 5 09·5 | 5 10·3 | 4 55·4 | 3·8 | 1·3 | 9·8 | 3·3 | 15·8 | 5·4 |
| 39 | 5 09·8 | 5 10·6 | 4 55·6 | 3·9 | 1·3 | 9·9 | 3·4 | 15·9 | 5·4 |
| 40 | 5 10·0 | 5 10·8 | 4 55·9 | 4·0 | 1·4 | 10·0 | 3·4 | 16·0 | 5·5 |
| 41 | 5 10·3 | 5 11·1 | 4 56·1 | 4·1 | 1·4 | 10·1 | 3·5 | 16·1 | 5·5 |
| 42 | 5 10·5 | 5 11·4 | 4 56·4 | 4·2 | 1·4 | 10·2 | 3·5 | 16·2 | 5·5 |
| 43 | 5 10·8 | 5 11·6 | 4 56·6 | 4·3 | 1·5 | 10·3 | 3·5 | 16·3 | 5·6 |
| 44 | 5 11·0 | 5 11·9 | 4 56·8 | 4·4 | 1·5 | 10·4 | 3·6 | 16·4 | 5·6 |
| 45 | 5 11·3 | 5 12·1 | 4 57·1 | 4·5 | 1·5 | 10·5 | 3·6 | 16·5 | 5·6 |
| 46 | 5 11·5 | 5 12·4 | 4 57·3 | 4·6 | 1·6 | 10·6 | 3·6 | 16·6 | 5·7 |
| 47 | 5 11·8 | 5 12·6 | 4 57·5 | 4·7 | 1·6 | 10·7 | 3·7 | 16·7 | 5·7 |
| 48 | 5 12·0 | 5 12·9 | 4 57·8 | 4·8 | 1·6 | 10·8 | 3·7 | 16·8 | 5·7 |
| 49 | 5 12·3 | 5 13·1 | 4 58·0 | 4·9 | 1·7 | 10·9 | 3·7 | 16·9 | 5·8 |
| 50 | 5 12·5 | 5 13·4 | 4 58·3 | 5·0 | 1·7 | 11·0 | 3·8 | 17·0 | 5·8 |
| 51 | 5 12·8 | 5 13·6 | 4 58·5 | 5·1 | 1·7 | 11·1 | 3·8 | 17·1 | 5·8 |
| 52 | 5 13·0 | 5 13·9 | 4 58·7 | 5·2 | 1·8 | 11·2 | 3·8 | 17·2 | 5·9 |
| 53 | 5 13·3 | 5 14·1 | 4 59·0 | 5·3 | 1·8 | 11·3 | 3·9 | 17·3 | 5·9 |
| 54 | 5 13·5 | 5 14·4 | 4 59·2 | 5·4 | 1·8 | 11·4 | 3·9 | 17·4 | 5·9 |
| 55 | 5 13·8 | 5 14·6 | 4 59·5 | 5·5 | 1·9 | 11·5 | 3·9 | 17·5 | 6·0 |
| 56 | 5 14·0 | 5 14·9 | 4 59·7 | 5·6 | 1·9 | 11·6 | 4·0 | 17·6 | 6·0 |
| 57 | 5 14·3 | 5 15·1 | 4 59·9 | 5·7 | 1·9 | 11·7 | 4·0 | 17·7 | 6·0 |
| 58 | 5 14·5 | 5 15·4 | 5 00·2 | 5·8 | 2·0 | 11·8 | 4·0 | 17·8 | 6·1 |
| 59 | 5 14·8 | 5 15·6 | 5 00·4 | 5·9 | 2·0 | 11·9 | 4·1 | 17·9 | 6·1 |
| 60 | 5 15·0 | 5 15·9 | 5 00·7 | 6·0 | 2·1 | 12·0 | 4·1 | 18·0 | 6·2 |

| 21ᵐ | SUN PLANETS | ARIES | MOON | v or d | Corrⁿ | v or d | Corrⁿ | v or d | Corrⁿ |
|---|---|---|---|---|---|---|---|---|---|
| s | ° ′ | ° ′ | ° ′ | ′ | ′ | ′ | ′ | ′ | ′ |
| 00 | 5 15·0 | 5 15·9 | 5 00·7 | 0·0 | 0·0 | 6·0 | 2·2 | 12·0 | 4·3 |
| 01 | 5 15·3 | 5 16·1 | 5 00·9 | 0·1 | 0·0 | 6·1 | 2·2 | 12·1 | 4·3 |
| 02 | 5 15·5 | 5 16·4 | 5 01·1 | 0·2 | 0·1 | 6·2 | 2·2 | 12·2 | 4·4 |
| 03 | 5 15·8 | 5 16·6 | 5 01·4 | 0·3 | 0·1 | 6·3 | 2·3 | 12·3 | 4·4 |
| 04 | 5 16·0 | 5 16·9 | 5 01·6 | 0·4 | 0·1 | 6·4 | 2·3 | 12·4 | 4·4 |
| 05 | 5 16·3 | 5 17·1 | 5 01·8 | 0·5 | 0·2 | 6·5 | 2·3 | 12·5 | 4·5 |
| 06 | 5 16·5 | 5 17·4 | 5 02·1 | 0·6 | 0·2 | 6·6 | 2·4 | 12·6 | 4·5 |
| 07 | 5 16·8 | 5 17·6 | 5 02·3 | 0·7 | 0·3 | 6·7 | 2·4 | 12·7 | 4·6 |
| 08 | 5 17·0 | 5 17·9 | 5 02·6 | 0·8 | 0·3 | 6·8 | 2·4 | 12·8 | 4·6 |
| 09 | 5 17·3 | 5 18·1 | 5 02·8 | 0·9 | 0·3 | 6·9 | 2·5 | 12·9 | 4·6 |
| 10 | 5 17·5 | 5 18·4 | 5 03·0 | 1·0 | 0·4 | 7·0 | 2·5 | 13·0 | 4·7 |
| 11 | 5 17·8 | 5 18·6 | 5 03·3 | 1·1 | 0·4 | 7·1 | 2·5 | 13·1 | 4·7 |
| 12 | 5 18·0 | 5 18·9 | 5 03·5 | 1·2 | 0·4 | 7·2 | 2·6 | 13·2 | 4·7 |
| 13 | 5 18·3 | 5 19·1 | 5 03·8 | 1·3 | 0·5 | 7·3 | 2·6 | 13·3 | 4·8 |
| 14 | 5 18·5 | 5 19·4 | 5 04·0 | 1·4 | 0·5 | 7·4 | 2·7 | 13·4 | 4·8 |
| 15 | 5 18·8 | 5 19·6 | 5 04·2 | 1·5 | 0·5 | 7·5 | 2·7 | 13·5 | 4·8 |
| 16 | 5 19·0 | 5 19·9 | 5 04·5 | 1·6 | 0·6 | 7·6 | 2·7 | 13·6 | 4·9 |
| 17 | 5 19·3 | 5 20·1 | 5 04·7 | 1·7 | 0·6 | 7·7 | 2·8 | 13·7 | 4·9 |
| 18 | 5 19·5 | 5 20·4 | 5 04·9 | 1·8 | 0·6 | 7·8 | 2·8 | 13·8 | 4·9 |
| 19 | 5 19·8 | 5 20·6 | 5 05·2 | 1·9 | 0·7 | 7·9 | 2·8 | 13·9 | 5·0 |
| 20 | 5 20·0 | 5 20·9 | 5 05·4 | 2·0 | 0·7 | 8·0 | 2·9 | 14·0 | 5·0 |
| 21 | 5 20·3 | 5 21·1 | 5 05·7 | 2·1 | 0·8 | 8·1 | 2·9 | 14·1 | 5·1 |
| 22 | 5 20·5 | 5 21·4 | 5 05·9 | 2·2 | 0·8 | 8·2 | 2·9 | 14·2 | 5·1 |
| 23 | 5 20·8 | 5 21·6 | 5 06·1 | 2·3 | 0·8 | 8·3 | 3·0 | 14·3 | 5·1 |
| 24 | 5 21·0 | 5 21·9 | 5 06·4 | 2·4 | 0·9 | 8·4 | 3·0 | 14·4 | 5·2 |
| 25 | 5 21·3 | 5 22·1 | 5 06·6 | 2·5 | 0·9 | 8·5 | 3·0 | 14·5 | 5·2 |
| 26 | 5 21·5 | 5 22·4 | 5 06·9 | 2·6 | 0·9 | 8·6 | 3·1 | 14·6 | 5·2 |
| 27 | 5 21·8 | 5 22·6 | 5 07·1 | 2·7 | 1·0 | 8·7 | 3·1 | 14·7 | 5·3 |
| 28 | 5 22·0 | 5 22·9 | 5 07·3 | 2·8 | 1·0 | 8·8 | 3·2 | 14·8 | 5·3 |
| 29 | 5 22·3 | 5 23·1 | 5 07·6 | 2·9 | 1·0 | 8·9 | 3·2 | 14·9 | 5·3 |
| 30 | 5 22·5 | 5 23·4 | 5 07·8 | 3·0 | 1·1 | 9·0 | 3·2 | 15·0 | 5·4 |
| 31 | 5 22·8 | 5 23·6 | 5 08·0 | 3·1 | 1·1 | 9·1 | 3·3 | 15·1 | 5·4 |
| 32 | 5 23·0 | 5 23·9 | 5 08·3 | 3·2 | 1·1 | 9·2 | 3·3 | 15·2 | 5·4 |
| 33 | 5 23·3 | 5 24·1 | 5 08·5 | 3·3 | 1·2 | 9·3 | 3·3 | 15·3 | 5·5 |
| 34 | 5 23·5 | 5 24·4 | 5 08·8 | 3·4 | 1·2 | 9·4 | 3·4 | 15·4 | 5·5 |
| 35 | 5 23·8 | 5 24·6 | 5 09·0 | 3·5 | 1·3 | 9·5 | 3·4 | 15·5 | 5·6 |
| 36 | 5 24·0 | 5 24·9 | 5 09·2 | 3·6 | 1·3 | 9·6 | 3·4 | 15·6 | 5·6 |
| 37 | 5 24·3 | 5 25·1 | 5 09·5 | 3·7 | 1·3 | 9·7 | 3·5 | 15·7 | 5·6 |
| 38 | 5 24·5 | 5 25·4 | 5 09·7 | 3·8 | 1·4 | 9·8 | 3·5 | 15·8 | 5·7 |
| 39 | 5 24·8 | 5 25·6 | 5 10·0 | 3·9 | 1·4 | 9·9 | 3·5 | 15·9 | 5·7 |
| 40 | 5 25·0 | 5 25·9 | 5 10·2 | 4·0 | 1·4 | 10·0 | 3·6 | 16·0 | 5·7 |
| 41 | 5 25·3 | 5 26·1 | 5 10·4 | 4·1 | 1·5 | 10·1 | 3·6 | 16·1 | 5·8 |
| 42 | 5 25·5 | 5 26·4 | 5 10·7 | 4·2 | 1·5 | 10·2 | 3·7 | 16·2 | 5·8 |
| 43 | 5 25·8 | 5 26·6 | 5 10·9 | 4·3 | 1·5 | 10·3 | 3·7 | 16·3 | 5·8 |
| 44 | 5 26·0 | 5 26·9 | 5 11·1 | 4·4 | 1·6 | 10·4 | 3·7 | 16·4 | 5·9 |
| 45 | 5 26·3 | 5 27·1 | 5 11·4 | 4·5 | 1·6 | 10·5 | 3·8 | 16·5 | 5·9 |
| 46 | 5 26·5 | 5 27·4 | 5 11·6 | 4·6 | 1·6 | 10·6 | 3·8 | 16·6 | 5·9 |
| 47 | 5 26·8 | 5 27·6 | 5 11·9 | 4·7 | 1·7 | 10·7 | 3·8 | 16·7 | 6·0 |
| 48 | 5 27·0 | 5 27·9 | 5 12·1 | 4·8 | 1·7 | 10·8 | 3·9 | 16·8 | 6·0 |
| 49 | 5 27·3 | 5 28·1 | 5 12·3 | 4·9 | 1·8 | 10·9 | 3·9 | 16·9 | 6·1 |
| 50 | 5 27·5 | 5 28·4 | 5 12·6 | 5·0 | 1·8 | 11·0 | 3·9 | 17·0 | 6·1 |
| 51 | 5 27·8 | 5 28·6 | 5 12·8 | 5·1 | 1·8 | 11·1 | 4·0 | 17·1 | 6·1 |
| 52 | 5 28·0 | 5 28·9 | 5 13·1 | 5·2 | 1·9 | 11·2 | 4·0 | 17·2 | 6·2 |
| 53 | 5 28·3 | 5 29·1 | 5 13·3 | 5·3 | 1·9 | 11·3 | 4·0 | 17·3 | 6·2 |
| 54 | 5 28·5 | 5 29·4 | 5 13·5 | 5·4 | 1·9 | 11·4 | 4·1 | 17·4 | 6·2 |
| 55 | 5 28·8 | 5 29·7 | 5 13·8 | 5·5 | 2·0 | 11·5 | 4·1 | 17·5 | 6·3 |
| 56 | 5 29·0 | 5 29·9 | 5 14·0 | 5·6 | 2·0 | 11·6 | 4·2 | 17·6 | 6·3 |
| 57 | 5 29·3 | 5 30·2 | 5 14·3 | 5·7 | 2·0 | 11·7 | 4·2 | 17·7 | 6·3 |
| 58 | 5 29·5 | 5 30·4 | 5 14·5 | 5·8 | 2·1 | 11·8 | 4·2 | 17·8 | 6·4 |
| 59 | 5 29·8 | 5 30·7 | 5 14·7 | 5·9 | 2·1 | 11·9 | 4·3 | 17·9 | 6·4 |
| 60 | 5 30·0 | 5 30·9 | 5 15·0 | 6·0 | 2·2 | 12·0 | 4·3 | 18·0 | 6·5 |

| 24ᵐ | SUN PLANETS | ARIES | MOON | v or d | Corrⁿ | v or d | Corrⁿ | v or d | Corrⁿ |
|---|---|---|---|---|---|---|---|---|---|
| s | ° ′ | ° ′ | ° ′ | ′ | ′ | ′ | ′ | ′ | ′ |
| 00 | 6 00·0 | 6 01·0 | 5 43·6 | 0·0 | 0·0 | 6·0 | 2·5 | 12·0 | 4·9 |
| 01 | 6 00·3 | 6 01·2 | 5 43·8 | 0·1 | 0·0 | 6·1 | 2·5 | 12·1 | 4·9 |
| 02 | 6 00·5 | 6 01·5 | 5 44·1 | 0·2 | 0·1 | 6·2 | 2·5 | 12·2 | 5·0 |
| 03 | 6 00·8 | 6 01·7 | 5 44·3 | 0·3 | 0·1 | 6·3 | 2·6 | 12·3 | 5·0 |
| 04 | 6 01·0 | 6 02·0 | 5 44·6 | 0·4 | 0·2 | 6·4 | 2·6 | 12·4 | 5·1 |
| 05 | 6 01·3 | 6 02·2 | 5 44·8 | 0·5 | 0·2 | 6·5 | 2·7 | 12·5 | 5·1 |
| 06 | 6 01·5 | 6 02·5 | 5 45·0 | 0·6 | 0·2 | 6·6 | 2·7 | 12·6 | 5·1 |
| 07 | 6 01·8 | 6 02·7 | 5 45·3 | 0·7 | 0·3 | 6·7 | 2·7 | 12·7 | 5·2 |
| 08 | 6 02·0 | 6 03·0 | 5 45·5 | 0·8 | 0·3 | 6·8 | 2·8 | 12·8 | 5·2 |
| 09 | 6 02·3 | 6 03·2 | 5 45·7 | 0·9 | 0·4 | 6·9 | 2·8 | 12·9 | 5·3 |
| 10 | 6 02·5 | 6 03·5 | 5 46·0 | 1·0 | 0·4 | 7·0 | 2·9 | 13·0 | 5·3 |
| 11 | 6 02·8 | 6 03·7 | 5 46·2 | 1·1 | 0·4 | 7·1 | 2·9 | 13·1 | 5·3 |
| 12 | 6 03·0 | 6 04·0 | 5 46·5 | 1·2 | 0·5 | 7·2 | 2·9 | 13·2 | 5·4 |
| 13 | 6 03·3 | 6 04·2 | 5 46·7 | 1·3 | 0·5 | 7·3 | 3·0 | 13·3 | 5·4 |
| 14 | 6 03·5 | 6 04·5 | 5 46·9 | 1·4 | 0·6 | 7·4 | 3·0 | 13·4 | 5·5 |
| 15 | 6 03·8 | 6 04·7 | 5 47·2 | 1·5 | 0·6 | 7·5 | 3·1 | 13·5 | 5·5 |
| 16 | 6 04·0 | 6 05·0 | 5 47·4 | 1·6 | 0·7 | 7·6 | 3·1 | 13·6 | 5·6 |
| 17 | 6 04·3 | 6 05·2 | 5 47·7 | 1·7 | 0·7 | 7·7 | 3·1 | 13·7 | 5·6 |
| 18 | 6 04·5 | 6 05·5 | 5 47·9 | 1·8 | 0·7 | 7·8 | 3·2 | 13·8 | 5·6 |
| 19 | 6 04·8 | 6 05·7 | 5 48·1 | 1·9 | 0·8 | 7·9 | 3·2 | 13·9 | 5·7 |
| 20 | 6 05·0 | 6 06·0 | 5 48·4 | 2·0 | 0·8 | 8·0 | 3·3 | 14·0 | 5·7 |
| 21 | 6 05·3 | 6 06·3 | 5 48·6 | 2·1 | 0·9 | 8·1 | 3·3 | 14·1 | 5·8 |
| 22 | 6 05·5 | 6 06·5 | 5 48·8 | 2·2 | 0·9 | 8·2 | 3·3 | 14·2 | 5·8 |
| 23 | 6 05·8 | 6 06·8 | 5 49·1 | 2·3 | 0·9 | 8·3 | 3·4 | 14·3 | 5·8 |
| 24 | 6 06·0 | 6 07·0 | 5 49·3 | 2·4 | 1·0 | 8·4 | 3·4 | 14·4 | 5·9 |
| 25 | 6 06·3 | 6 07·3 | 5 49·6 | 2·5 | 1·0 | 8·5 | 3·5 | 14·5 | 5·9 |
| 26 | 6 06·5 | 6 07·5 | 5 49·8 | 2·6 | 1·1 | 8·6 | 3·5 | 14·6 | 6·0 |
| 27 | 6 06·8 | 6 07·8 | 5 50·0 | 2·7 | 1·1 | 8·7 | 3·6 | 14·7 | 6·0 |
| 28 | 6 07·0 | 6 08·0 | 5 50·3 | 2·8 | 1·1 | 8·8 | 3·6 | 14·8 | 6·0 |
| 29 | 6 07·3 | 6 08·3 | 5 50·5 | 2·9 | 1·2 | 8·9 | 3·6 | 14·9 | 6·1 |
| 30 | 6 07·5 | 6 08·5 | 5 50·8 | 3·0 | 1·2 | 9·0 | 3·7 | 15·0 | 6·1 |
| 31 | 6 07·8 | 6 08·8 | 5 51·0 | 3·1 | 1·3 | 9·1 | 3·7 | 15·1 | 6·2 |
| 32 | 6 08·0 | 6 09·0 | 5 51·2 | 3·2 | 1·3 | 9·2 | 3·8 | 15·2 | 6·2 |
| 33 | 6 08·3 | 6 09·3 | 5 51·5 | 3·3 | 1·3 | 9·3 | 3·8 | 15·3 | 6·2 |
| 34 | 6 08·5 | 6 09·5 | 5 51·7 | 3·4 | 1·4 | 9·4 | 3·8 | 15·4 | 6·3 |
| 35 | 6 08·8 | 6 09·8 | 5 52·0 | 3·5 | 1·4 | 9·5 | 3·9 | 15·5 | 6·3 |
| 36 | 6 09·0 | 6 10·0 | 5 52·2 | 3·6 | 1·5 | 9·6 | 3·9 | 15·6 | 6·4 |
| 37 | 6 09·3 | 6 10·3 | 5 52·4 | 3·7 | 1·5 | 9·7 | 4·0 | 15·7 | 6·4 |
| 38 | 6 09·5 | 6 10·5 | 5 52·7 | 3·8 | 1·6 | 9·8 | 4·0 | 15·8 | 6·5 |
| 39 | 6 09·8 | 6 10·8 | 5 52·9 | 3·9 | 1·6 | 9·9 | 4·0 | 15·9 | 6·5 |
| 40 | 6 10·0 | 6 11·0 | 5 53·1 | 4·0 | 1·6 | 10·0 | 4·1 | 16·0 | 6·5 |
| 41 | 6 10·3 | 6 11·3 | 5 53·4 | 4·1 | 1·7 | 10·1 | 4·1 | 16·1 | 6·6 |
| 42 | 6 10·5 | 6 11·5 | 5 53·6 | 4·2 | 1·7 | 10·2 | 4·2 | 16·2 | 6·6 |
| 43 | 6 10·8 | 6 11·8 | 5 53·9 | 4·3 | 1·8 | 10·3 | 4·2 | 16·3 | 6·7 |
| 44 | 6 11·0 | 6 12·0 | 5 54·1 | 4·4 | 1·8 | 10·4 | 4·2 | 16·4 | 6·7 |
| 45 | 6 11·3 | 6 12·3 | 5 54·3 | 4·5 | 1·8 | 10·5 | 4·3 | 16·5 | 6·7 |
| 46 | 6 11·5 | 6 12·5 | 5 54·6 | 4·6 | 1·9 | 10·6 | 4·3 | 16·6 | 6·8 |
| 47 | 6 11·8 | 6 12·8 | 5 54·8 | 4·7 | 1·9 | 10·7 | 4·4 | 16·7 | 6·8 |
| 48 | 6 12·0 | 6 13·0 | 5 55·1 | 4·8 | 2·0 | 10·8 | 4·4 | 16·8 | 6·9 |
| 49 | 6 12·3 | 6 13·3 | 5 55·3 | 4·9 | 2·0 | 10·9 | 4·5 | 16·9 | 6·9 |
| 50 | 6 12·5 | 6 13·5 | 5 55·5 | 5·0 | 2·0 | 11·0 | 4·5 | 17·0 | 6·9 |
| 51 | 6 12·8 | 6 13·8 | 5 55·8 | 5·1 | 2·1 | 11·1 | 4·5 | 17·1 | 7·0 |
| 52 | 6 13·0 | 6 14·0 | 5 56·0 | 5·2 | 2·1 | 11·2 | 4·6 | 17·2 | 7·0 |
| 53 | 6 13·3 | 6 14·3 | 5 56·2 | 5·3 | 2·2 | 11·3 | 4·6 | 17·3 | 7·1 |
| 54 | 6 13·5 | 6 14·5 | 5 56·5 | 5·4 | 2·2 | 11·4 | 4·7 | 17·4 | 7·1 |
| 55 | 6 13·8 | 6 14·8 | 5 56·7 | 5·5 | 2·2 | 11·5 | 4·7 | 17·5 | 7·1 |
| 56 | 6 14·0 | 6 15·0 | 5 57·0 | 5·6 | 2·3 | 11·6 | 4·7 | 17·6 | 7·2 |
| 57 | 6 14·3 | 6 15·3 | 5 57·2 | 5·7 | 2·3 | 11·7 | 4·8 | 17·7 | 7·2 |
| 58 | 6 14·5 | 6 15·5 | 5 57·4 | 5·8 | 2·4 | 11·8 | 4·8 | 17·8 | 7·3 |
| 59 | 6 14·8 | 6 15·8 | 5 57·7 | 5·9 | 2·4 | 11·9 | 4·9 | 17·9 | 7·3 |
| 60 | 6 15·0 | 6 16·0 | 5 57·9 | 6·0 | 2·5 | 12·0 | 4·9 | 18·0 | 7·4 |

| 25ᵐ | SUN PLANETS | ARIES | MOON | v or d | Corrⁿ | v or d | Corrⁿ | v or d | Corrⁿ |
|---|---|---|---|---|---|---|---|---|---|
| s | ° ′ | ° ′ | ° ′ | ′ | ′ | ′ | ′ | ′ | ′ |
| 00 | 6 15·0 | 6 16·0 | 5 57·9 | 0·0 | 0·0 | 6·0 | 2·6 | 12·0 | 5·1 |
| 01 | 6 15·3 | 6 16·3 | 5 58·2 | 0·1 | 0·0 | 6·1 | 2·6 | 12·1 | 5·1 |
| 02 | 6 15·5 | 6 16·5 | 5 58·4 | 0·2 | 0·1 | 6·2 | 2·6 | 12·2 | 5·2 |
| 03 | 6 15·8 | 6 16·8 | 5 58·6 | 0·3 | 0·1 | 6·3 | 2·7 | 12·3 | 5·2 |
| 04 | 6 16·0 | 6 17·0 | 5 58·9 | 0·4 | 0·2 | 6·4 | 2·7 | 12·4 | 5·3 |
| 05 | 6 16·3 | 6 17·3 | 5 59·1 | 0·5 | 0·2 | 6·5 | 2·8 | 12·5 | 5·3 |
| 06 | 6 16·5 | 6 17·5 | 5 59·3 | 0·6 | 0·3 | 6·6 | 2·8 | 12·6 | 5·4 |
| 07 | 6 16·8 | 6 17·8 | 5 59·6 | 0·7 | 0·3 | 6·7 | 2·8 | 12·7 | 5·4 |
| 08 | 6 17·0 | 6 18·0 | 5 59·8 | 0·8 | 0·3 | 6·8 | 2·9 | 12·8 | 5·4 |
| 09 | 6 17·3 | 6 18·3 | 6 00·1 | 0·9 | 0·4 | 6·9 | 2·9 | 12·9 | 5·5 |
| 10 | 6 17·5 | 6 18·5 | 6 00·3 | 1·0 | 0·4 | 7·0 | 3·0 | 13·0 | 5·5 |
| 11 | 6 17·8 | 6 18·8 | 6 00·5 | 1·1 | 0·5 | 7·1 | 3·0 | 13·1 | 5·6 |
| 12 | 6 18·0 | 6 19·0 | 6 00·8 | 1·2 | 0·5 | 7·2 | 3·1 | 13·2 | 5·6 |
| 13 | 6 18·3 | 6 19·3 | 6 01·0 | 1·3 | 0·6 | 7·3 | 3·1 | 13·3 | 5·7 |
| 14 | 6 18·5 | 6 19·5 | 6 01·3 | 1·4 | 0·6 | 7·4 | 3·1 | 13·4 | 5·7 |
| 15 | 6 18·8 | 6 19·8 | 6 01·5 | 1·5 | 0·6 | 7·5 | 3·2 | 13·5 | 5·7 |
| 16 | 6 19·0 | 6 20·0 | 6 01·7 | 1·6 | 0·7 | 7·6 | 3·2 | 13·6 | 5·8 |
| 17 | 6 19·3 | 6 20·3 | 6 02·0 | 1·7 | 0·7 | 7·7 | 3·3 | 13·7 | 5·8 |
| 18 | 6 19·5 | 6 20·5 | 6 02·2 | 1·8 | 0·8 | 7·8 | 3·3 | 13·8 | 5·9 |
| 19 | 6 19·8 | 6 20·8 | 6 02·5 | 1·9 | 0·8 | 7·9 | 3·4 | 13·9 | 5·9 |
| 20 | 6 20·0 | 6 21·0 | 6 02·7 | 2·0 | 0·9 | 8·0 | 3·4 | 14·0 | 6·0 |
| 21 | 6 20·3 | 6 21·3 | 6 02·9 | 2·1 | 0·9 | 8·1 | 3·4 | 14·1 | 6·0 |
| 22 | 6 20·5 | 6 21·5 | 6 03·2 | 2·2 | 0·9 | 8·2 | 3·5 | 14·2 | 6·0 |
| 23 | 6 20·8 | 6 21·8 | 6 03·4 | 2·3 | 1·0 | 8·3 | 3·5 | 14·3 | 6·1 |
| 24 | 6 21·0 | 6 22·0 | 6 03·6 | 2·4 | 1·0 | 8·4 | 3·6 | 14·4 | 6·1 |
| 25 | 6 21·3 | 6 22·3 | 6 03·9 | 2·5 | 1·1 | 8·5 | 3·6 | 14·5 | 6·2 |
| 26 | 6 21·5 | 6 22·5 | 6 04·1 | 2·6 | 1·1 | 8·6 | 3·7 | 14·6 | 6·2 |
| 27 | 6 21·8 | 6 22·8 | 6 04·4 | 2·7 | 1·1 | 8·7 | 3·7 | 14·7 | 6·2 |
| 28 | 6 22·0 | 6 23·0 | 6 04·6 | 2·8 | 1·2 | 8·8 | 3·7 | 14·8 | 6·3 |
| 29 | 6 22·3 | 6 23·3 | 6 04·8 | 2·9 | 1·2 | 8·9 | 3·8 | 14·9 | 6·3 |
| 30 | 6 22·5 | 6 23·5 | 6 05·1 | 3·0 | 1·3 | 9·0 | 3·8 | 15·0 | 6·4 |
| 31 | 6 22·8 | 6 23·8 | 6 05·3 | 3·1 | 1·3 | 9·1 | 3·9 | 15·1 | 6·4 |
| 32 | 6 23·0 | 6 24·0 | 6 05·6 | 3·2 | 1·4 | 9·2 | 3·9 | 15·2 | 6·5 |
| 33 | 6 23·3 | 6 24·3 | 6 05·8 | 3·3 | 1·4 | 9·3 | 4·0 | 15·3 | 6·5 |
| 34 | 6 23·5 | 6 24·5 | 6 06·0 | 3·4 | 1·4 | 9·4 | 4·0 | 15·4 | 6·5 |
| 35 | 6 23·8 | 6 24·8 | 6 06·3 | 3·5 | 1·5 | 9·5 | 4·0 | 15·5 | 6·6 |
| 36 | 6 24·0 | 6 25·1 | 6 06·5 | 3·6 | 1·5 | 9·6 | 4·1 | 15·6 | 6·6 |
| 37 | 6 24·3 | 6 25·3 | 6 06·7 | 3·7 | 1·6 | 9·7 | 4·1 | 15·7 | 6·7 |
| 38 | 6 24·5 | 6 25·6 | 6 07·0 | 3·8 | 1·6 | 9·8 | 4·2 | 15·8 | 6·7 |
| 39 | 6 24·8 | 6 25·8 | 6 07·2 | 3·9 | 1·7 | 9·9 | 4·2 | 15·9 | 6·8 |
| 40 | 6 25·0 | 6 26·1 | 6 07·5 | 4·0 | 1·7 | 10·0 | 4·3 | 16·0 | 6·8 |
| 41 | 6 25·3 | 6 26·3 | 6 07·7 | 4·1 | 1·7 | 10·1 | 4·3 | 16·1 | 6·8 |
| 42 | 6 25·5 | 6 26·6 | 6 07·9 | 4·2 | 1·8 | 10·2 | 4·3 | 16·2 | 6·9 |
| 43 | 6 25·8 | 6 26·8 | 6 08·2 | 4·3 | 1·8 | 10·3 | 4·4 | 16·3 | 6·9 |
| 44 | 6 26·0 | 6 27·1 | 6 08·4 | 4·4 | 1·9 | 10·4 | 4·4 | 16·4 | 7·0 |
| 45 | 6 26·3 | 6 27·3 | 6 08·7 | 4·5 | 1·9 | 10·5 | 4·5 | 16·5 | 7·0 |
| 46 | 6 26·5 | 6 27·6 | 6 08·9 | 4·6 | 2·0 | 10·6 | 4·5 | 16·6 | 7·1 |
| 47 | 6 26·8 | 6 27·8 | 6 09·1 | 4·7 | 2·0 | 10·7 | 4·5 | 16·7 | 7·1 |
| 48 | 6 27·0 | 6 28·1 | 6 09·4 | 4·8 | 2·0 | 10·8 | 4·6 | 16·8 | 7·1 |
| 49 | 6 27·3 | 6 28·3 | 6 09·6 | 4·9 | 2·1 | 10·9 | 4·6 | 16·9 | 7·2 |
| 50 | 6 27·5 | 6 28·6 | 6 09·8 | 5·0 | 2·1 | 11·0 | 4·7 | 17·0 | 7·2 |
| 51 | 6 27·8 | 6 28·8 | 6 10·1 | 5·1 | 2·2 | 11·1 | 4·7 | 17·1 | 7·3 |
| 52 | 6 28·0 | 6 29·1 | 6 10·3 | 5·2 | 2·2 | 11·2 | 4·8 | 17·2 | 7·3 |
| 53 | 6 28·3 | 6 29·3 | 6 10·6 | 5·3 | 2·3 | 11·3 | 4·8 | 17·3 | 7·4 |
| 54 | 6 28·5 | 6 29·6 | 6 10·8 | 5·4 | 2·3 | 11·4 | 4·8 | 17·4 | 7·4 |
| 55 | 6 28·8 | 6 29·8 | 6 11·0 | 5·5 | 2·3 | 11·5 | 4·9 | 17·5 | 7·4 |
| 56 | 6 29·0 | 6 30·1 | 6 11·3 | 5·6 | 2·4 | 11·6 | 4·9 | 17·6 | 7·5 |
| 57 | 6 29·3 | 6 30·3 | 6 11·5 | 5·7 | 2·4 | 11·7 | 5·0 | 17·7 | 7·5 |
| 58 | 6 29·5 | 6 30·6 | 6 11·8 | 5·8 | 2·5 | 11·8 | 5·0 | 17·8 | 7·6 |
| 59 | 6 29·8 | 6 30·8 | 6 12·0 | 5·9 | 2·5 | 11·9 | 5·1 | 17·9 | 7·6 |
| 60 | 6 30·0 | 6 31·1 | 6 12·2 | 6·0 | 2·6 | 12·0 | 5·1 | 18·0 | 7·7 |

| 30ᵐ | SUN PLANETS | ARIES | MOON | v or d | Corrⁿ | v or d | Corrⁿ | v or d | Corrⁿ |
|---|---|---|---|---|---|---|---|---|---|
| s | ° ′ | ° ′ | ° ′ | ′ | ′ | ′ | ′ | ′ | ′ |
| 00 | 7 30·0 | 7 31·2 | 7 09·5 | 0·0 | 0·0 | 6·0 | 3·1 | 12·0 | 6·1 |
| 01 | 7 30·3 | 7 31·5 | 7 09·7 | 0·1 | 0·1 | 6·1 | 3·1 | 12·1 | 6·2 |
| 02 | 7 30·5 | 7 31·7 | 7 10·0 | 0·2 | 0·1 | 6·2 | 3·2 | 12·2 | 6·2 |
| 03 | 7 30·8 | 7 32·0 | 7 10·2 | 0·3 | 0·2 | 6·3 | 3·2 | 12·3 | 6·3 |
| 04 | 7 31·0 | 7 32·2 | 7 10·5 | 0·4 | 0·2 | 6·4 | 3·3 | 12·4 | 6·3 |
| 05 | 7 31·3 | 7 32·5 | 7 10·7 | 0·5 | 0·3 | 6·5 | 3·3 | 12·5 | 6·4 |
| 06 | 7 31·5 | 7 32·7 | 7 10·9 | 0·6 | 0·3 | 6·6 | 3·4 | 12·6 | 6·4 |
| 07 | 7 31·8 | 7 33·0 | 7 11·2 | 0·7 | 0·4 | 6·7 | 3·4 | 12·7 | 6·5 |
| 08 | 7 32·0 | 7 33·2 | 7 11·4 | 0·8 | 0·4 | 6·8 | 3·5 | 12·8 | 6·5 |
| 09 | 7 32·3 | 7 33·5 | 7 11·6 | 0·9 | 0·5 | 6·9 | 3·5 | 12·9 | 6·6 |
| 10 | 7 32·5 | 7 33·7 | 7 11·9 | 1·0 | 0·5 | 7·0 | 3·6 | 13·0 | 6·6 |
| 11 | 7 32·8 | 7 34·0 | 7 12·1 | 1·1 | 0·6 | 7·1 | 3·6 | 13·1 | 6·7 |
| 12 | 7 33·0 | 7 34·2 | 7 12·4 | 1·2 | 0·6 | 7·2 | 3·7 | 13·2 | 6·7 |
| 13 | 7 33·3 | 7 34·5 | 7 12·6 | 1·3 | 0·7 | 7·3 | 3·7 | 13·3 | 6·8 |
| 14 | 7 33·5 | 7 34·7 | 7 12·8 | 1·4 | 0·7 | 7·4 | 3·8 | 13·4 | 6·8 |
| 15 | 7 33·8 | 7 35·0 | 7 13·1 | 1·5 | 0·8 | 7·5 | 3·8 | 13·5 | 6·9 |
| 16 | 7 34·0 | 7 35·2 | 7 13·3 | 1·6 | 0·8 | 7·6 | 3·9 | 13·6 | 6·9 |
| 17 | 7 34·3 | 7 35·5 | 7 13·6 | 1·7 | 0·9 | 7·7 | 3·9 | 13·7 | 7·0 |
| 18 | 7 34·5 | 7 35·7 | 7 13·8 | 1·8 | 0·9 | 7·8 | 4·0 | 13·8 | 7·0 |
| 19 | 7 34·8 | 7 36·0 | 7 14·0 | 1·9 | 1·0 | 7·9 | 4·0 | 13·9 | 7·1 |
| 20 | 7 35·0 | 7 36·2 | 7 14·3 | 2·0 | 1·0 | 8·0 | 4·1 | 14·0 | 7·1 |
| 21 | 7 35·3 | 7 36·5 | 7 14·5 | 2·1 | 1·1 | 8·1 | 4·1 | 14·1 | 7·2 |
| 22 | 7 35·5 | 7 36·7 | 7 14·7 | 2·2 | 1·1 | 8·2 | 4·2 | 14·2 | 7·2 |
| 23 | 7 35·8 | 7 37·0 | 7 15·0 | 2·3 | 1·2 | 8·3 | 4·2 | 14·3 | 7·3 |
| 24 | 7 36·0 | 7 37·2 | 7 15·2 | 2·4 | 1·2 | 8·4 | 4·3 | 14·4 | 7·3 |
| 25 | 7 36·3 | 7 37·5 | 7 15·5 | 2·5 | 1·3 | 8·5 | 4·3 | 14·5 | 7·4 |
| 26 | 7 36·5 | 7 37·7 | 7 15·7 | 2·6 | 1·3 | 8·6 | 4·4 | 14·6 | 7·4 |
| 27 | 7 36·8 | 7 38·0 | 7 15·9 | 2·7 | 1·4 | 8·7 | 4·4 | 14·7 | 7·5 |
| 28 | 7 37·0 | 7 38·3 | 7 16·2 | 2·8 | 1·4 | 8·8 | 4·5 | 14·8 | 7·5 |
| 29 | 7 37·3 | 7 38·5 | 7 16·4 | 2·9 | 1·5 | 8·9 | 4·5 | 14·9 | 7·6 |
| 30 | 7 37·5 | 7 38·8 | 7 16·7 | 3·0 | 1·5 | 9·0 | 4·6 | 15·0 | 7·6 |
| 31 | 7 37·8 | 7 39·0 | 7 16·9 | 3·1 | 1·6 | 9·1 | 4·6 | 15·1 | 7·7 |
| 32 | 7 38·0 | 7 39·3 | 7 17·1 | 3·2 | 1·6 | 9·2 | 4·7 | 15·2 | 7·7 |
| 33 | 7 38·3 | 7 39·5 | 7 17·4 | 3·3 | 1·7 | 9·3 | 4·7 | 15·3 | 7·8 |
| 34 | 7 38·5 | 7 39·8 | 7 17·6 | 3·4 | 1·7 | 9·4 | 4·8 | 15·4 | 7·8 |
| 35 | 7 38·8 | 7 40·0 | 7 17·9 | 3·5 | 1·8 | 9·5 | 4·8 | 15·5 | 7·9 |
| 36 | 7 39·0 | 7 40·3 | 7 18·1 | 3·6 | 1·8 | 9·6 | 4·9 | 15·6 | 7·9 |
| 37 | 7 39·3 | 7 40·5 | 7 18·3 | 3·7 | 1·9 | 9·7 | 4·9 | 15·7 | 8·0 |
| 38 | 7 39·5 | 7 40·8 | 7 18·6 | 3·8 | 1·9 | 9·8 | 5·0 | 15·8 | 8·0 |
| 39 | 7 39·8 | 7 41·0 | 7 18·8 | 3·9 | 2·0 | 9·9 | 5·0 | 15·9 | 8·1 |
| 40 | 7 40·0 | 7 41·3 | 7 19·0 | 4·0 | 2·0 | 10·0 | 5·1 | 16·0 | 8·1 |
| 41 | 7 40·3 | 7 41·5 | 7 19·3 | 4·1 | 2·1 | 10·1 | 5·1 | 16·1 | 8·2 |
| 42 | 7 40·5 | 7 41·8 | 7 19·5 | 4·2 | 2·1 | 10·2 | 5·2 | 16·2 | 8·2 |
| 43 | 7 40·8 | 7 42·0 | 7 19·8 | 4·3 | 2·2 | 10·3 | 5·2 | 16·3 | 8·3 |
| 44 | 7 41·0 | 7 42·3 | 7 20·0 | 4·4 | 2·2 | 10·4 | 5·3 | 16·4 | 8·3 |
| 45 | 7 41·3 | 7 42·5 | 7 20·2 | 4·5 | 2·3 | 10·5 | 5·3 | 16·5 | 8·4 |
| 46 | 7 41·5 | 7 42·8 | 7 20·5 | 4·6 | 2·3 | 10·6 | 5·4 | 16·6 | 8·4 |
| 47 | 7 41·8 | 7 43·0 | 7 20·7 | 4·7 | 2·4 | 10·7 | 5·4 | 16·7 | 8·5 |
| 48 | 7 42·0 | 7 43·3 | 7 21·0 | 4·8 | 2·4 | 10·8 | 5·5 | 16·8 | 8·5 |
| 49 | 7 42·3 | 7 43·5 | 7 21·2 | 4·9 | 2·5 | 10·9 | 5·5 | 16·9 | 8·6 |
| 50 | 7 42·5 | 7 43·8 | 7 21·4 | 5·0 | 2·5 | 11·0 | 5·6 | 17·0 | 8·6 |
| 51 | 7 42·8 | 7 44·0 | 7 21·7 | 5·1 | 2·6 | 11·1 | 5·6 | 17·1 | 8·7 |
| 52 | 7 43·0 | 7 44·3 | 7 21·9 | 5·2 | 2·6 | 11·2 | 5·7 | 17·2 | 8·7 |
| 53 | 7 43·3 | 7 44·5 | 7 22·1 | 5·3 | 2·7 | 11·3 | 5·7 | 17·3 | 8·8 |
| 54 | 7 43·5 | 7 44·8 | 7 22·4 | 5·4 | 2·7 | 11·4 | 5·8 | 17·4 | 8·8 |
| 55 | 7 43·8 | 7 45·0 | 7 22·6 | 5·5 | 2·8 | 11·5 | 5·8 | 17·5 | 8·9 |
| 56 | 7 44·0 | 7 45·3 | 7 22·9 | 5·6 | 2·8 | 11·6 | 5·9 | 17·6 | 8·9 |
| 57 | 7 44·3 | 7 45·5 | 7 23·1 | 5·7 | 2·9 | 11·7 | 5·9 | 17·7 | 9·0 |
| 58 | 7 44·5 | 7 45·8 | 7 23·3 | 5·8 | 2·9 | 11·8 | 6·0 | 17·8 | 9·0 |
| 59 | 7 44·8 | 7 46·0 | 7 23·6 | 5·9 | 3·0 | 11·9 | 6·0 | 17·9 | 9·1 |
| 60 | 7 45·0 | 7 46·3 | 7 23·8 | 6·0 | 3·1 | 12·0 | 6·1 | 18·0 | 9·2 |

| 31ᵐ | SUN PLANETS | ARIES | MOON | v or d | Corrⁿ | v or d | Corrⁿ | v or d | Corrⁿ |
|---|---|---|---|---|---|---|---|---|---|
| s | ° ′ | ° ′ | ° ′ | ′ | ′ | ′ | ′ | ′ | ′ |
| 00 | 7 45·0 | 7 46·3 | 7 23·8 | 0·0 | 0·0 | 6·0 | 3·2 | 12·0 | 6·3 |
| 01 | 7 45·3 | 7 46·5 | 7 24·1 | 0·1 | 0·1 | 6·1 | 3·2 | 12·1 | 6·4 |
| 02 | 7 45·5 | 7 46·8 | 7 24·3 | 0·2 | 0·1 | 6·2 | 3·3 | 12·2 | 6·4 |
| 03 | 7 45·8 | 7 47·0 | 7 24·5 | 0·3 | 0·2 | 6·3 | 3·3 | 12·3 | 6·5 |
| 04 | 7 46·0 | 7 47·3 | 7 24·8 | 0·4 | 0·2 | 6·4 | 3·4 | 12·4 | 6·5 |
| 05 | 7 46·3 | 7 47·5 | 7 25·0 | 0·5 | 0·3 | 6·5 | 3·4 | 12·5 | 6·6 |
| 06 | 7 46·5 | 7 47·8 | 7 25·2 | 0·6 | 0·3 | 6·6 | 3·5 | 12·6 | 6·6 |
| 07 | 7 46·8 | 7 48·0 | 7 25·5 | 0·7 | 0·4 | 6·7 | 3·5 | 12·7 | 6·7 |
| 08 | 7 47·0 | 7 48·3 | 7 25·7 | 0·8 | 0·4 | 6·8 | 3·6 | 12·8 | 6·7 |
| 09 | 7 47·3 | 7 48·5 | 7 26·0 | 0·9 | 0·5 | 6·9 | 3·6 | 12·9 | 6·8 |
| 10 | 7 47·5 | 7 48·8 | 7 26·2 | 1·0 | 0·5 | 7·0 | 3·7 | 13·0 | 6·8 |
| 11 | 7 47·8 | 7 49·0 | 7 26·4 | 1·1 | 0·6 | 7·1 | 3·7 | 13·1 | 6·9 |
| 12 | 7 48·0 | 7 49·3 | 7 26·7 | 1·2 | 0·6 | 7·2 | 3·8 | 13·2 | 6·9 |
| 13 | 7 48·3 | 7 49·5 | 7 26·9 | 1·3 | 0·7 | 7·3 | 3·8 | 13·3 | 7·0 |
| 14 | 7 48·5 | 7 49·8 | 7 27·2 | 1·4 | 0·7 | 7·4 | 3·9 | 13·4 | 7·0 |
| 15 | 7 48·8 | 7 50·0 | 7 27·4 | 1·5 | 0·8 | 7·5 | 3·9 | 13·5 | 7·1 |
| 16 | 7 49·0 | 7 50·3 | 7 27·6 | 1·6 | 0·8 | 7·6 | 4·0 | 13·6 | 7·1 |
| 17 | 7 49·3 | 7 50·5 | 7 27·9 | 1·7 | 0·9 | 7·7 | 4·0 | 13·7 | 7·2 |
| 18 | 7 49·5 | 7 50·8 | 7 28·1 | 1·8 | 0·9 | 7·8 | 4·1 | 13·8 | 7·2 |
| 19 | 7 49·8 | 7 51·0 | 7 28·4 | 1·9 | 1·0 | 7·9 | 4·1 | 13·9 | 7·3 |
| 20 | 7 50·0 | 7 51·3 | 7 28·6 | 2·0 | 1·1 | 8·0 | 4·2 | 14·0 | 7·4 |
| 21 | 7 50·3 | 7 51·5 | 7 28·8 | 2·1 | 1·1 | 8·1 | 4·3 | 14·1 | 7·4 |
| 22 | 7 50·5 | 7 51·8 | 7 29·1 | 2·2 | 1·2 | 8·2 | 4·3 | 14·2 | 7·5 |
| 23 | 7 50·8 | 7 52·0 | 7 29·3 | 2·3 | 1·2 | 8·3 | 4·4 | 14·3 | 7·5 |
| 24 | 7 51·0 | 7 52·3 | 7 29·5 | 2·4 | 1·3 | 8·4 | 4·4 | 14·4 | 7·6 |
| 25 | 7 51·3 | 7 52·5 | 7 29·8 | 2·5 | 1·3 | 8·5 | 4·5 | 14·5 | 7·6 |
| 26 | 7 51·5 | 7 52·8 | 7 30·0 | 2·6 | 1·4 | 8·6 | 4·5 | 14·6 | 7·7 |
| 27 | 7 51·8 | 7 53·0 | 7 30·3 | 2·7 | 1·4 | 8·7 | 4·6 | 14·7 | 7·7 |
| 28 | 7 52·0 | 7 53·3 | 7 30·5 | 2·8 | 1·5 | 8·8 | 4·6 | 14·8 | 7·8 |
| 29 | 7 52·3 | 7 53·5 | 7 30·7 | 2·9 | 1·5 | 8·9 | 4·7 | 14·9 | 7·8 |
| 30 | 7 52·5 | 7 53·8 | 7 31·0 | 3·0 | 1·6 | 9·0 | 4·7 | 15·0 | 7·9 |
| 31 | 7 52·8 | 7 54·0 | 7 31·2 | 3·1 | 1·6 | 9·1 | 4·8 | 15·1 | 7·9 |
| 32 | 7 53·0 | 7 54·3 | 7 31·5 | 3·2 | 1·7 | 9·2 | 4·8 | 15·2 | 8·0 |
| 33 | 7 53·3 | 7 54·5 | 7 31·7 | 3·3 | 1·7 | 9·3 | 4·9 | 15·3 | 8·0 |
| 34 | 7 53·5 | 7 54·8 | 7 31·9 | 3·4 | 1·8 | 9·4 | 4·9 | 15·4 | 8·1 |
| 35 | 7 53·8 | 7 55·0 | 7 32·2 | 3·5 | 1·8 | 9·5 | 5·0 | 15·5 | 8·1 |
| 36 | 7 54·0 | 7 55·3 | 7 32·4 | 3·6 | 1·9 | 9·6 | 5·0 | 15·6 | 8·2 |
| 37 | 7 54·3 | 7 55·5 | 7 32·6 | 3·7 | 1·9 | 9·7 | 5·1 | 15·7 | 8·2 |
| 38 | 7 54·5 | 7 55·8 | 7 32·9 | 3·8 | 2·0 | 9·8 | 5·1 | 15·8 | 8·3 |
| 39 | 7 54·8 | 7 56·0 | 7 33·1 | 3·9 | 2·0 | 9·9 | 5·2 | 15·9 | 8·3 |
| 40 | 7 55·0 | 7 56·3 | 7 33·4 | 4·0 | 2·1 | 10·0 | 5·3 | 16·0 | 8·4 |
| 41 | 7 55·3 | 7 56·6 | 7 33·6 | 4·1 | 2·2 | 10·1 | 5·3 | 16·1 | 8·5 |
| 42 | 7 55·5 | 7 56·8 | 7 33·8 | 4·2 | 2·2 | 10·2 | 5·4 | 16·2 | 8·5 |
| 43 | 7 55·8 | 7 57·1 | 7 34·1 | 4·3 | 2·3 | 10·3 | 5·4 | 16·3 | 8·6 |
| 44 | 7 56·0 | 7 57·3 | 7 34·3 | 4·4 | 2·3 | 10·4 | 5·5 | 16·4 | 8·6 |
| 45 | 7 56·3 | 7 57·6 | 7 34·6 | 4·5 | 2·4 | 10·5 | 5·5 | 16·5 | 8·7 |
| 46 | 7 56·5 | 7 57·8 | 7 34·8 | 4·6 | 2·4 | 10·6 | 5·6 | 16·6 | 8·7 |
| 47 | 7 56·8 | 7 58·1 | 7 35·0 | 4·7 | 2·5 | 10·7 | 5·6 | 16·7 | 8·8 |
| 48 | 7 57·0 | 7 58·3 | 7 35·3 | 4·8 | 2·5 | 10·8 | 5·7 | 16·8 | 8·8 |
| 49 | 7 57·3 | 7 58·6 | 7 35·5 | 4·9 | 2·6 | 10·9 | 5·7 | 16·9 | 8·9 |
| 50 | 7 57·5 | 7 58·8 | 7 35·7 | 5·0 | 2·6 | 11·0 | 5·8 | 17·0 | 8·9 |
| 51 | 7 57·8 | 7 59·1 | 7 36·0 | 5·1 | 2·7 | 11·1 | 5·8 | 17·1 | 9·0 |
| 52 | 7 58·0 | 7 59·3 | 7 36·2 | 5·2 | 2·7 | 11·2 | 5·9 | 17·2 | 9·0 |
| 53 | 7 58·3 | 7 59·6 | 7 36·5 | 5·3 | 2·8 | 11·3 | 5·9 | 17·3 | 9·1 |
| 54 | 7 58·5 | 7 59·8 | 7 36·7 | 5·4 | 2·8 | 11·4 | 6·0 | 17·4 | 9·1 |
| 55 | 7 58·8 | 8 00·1 | 7 36·9 | 5·5 | 2·9 | 11·5 | 6·0 | 17·5 | 9·2 |
| 56 | 7 59·0 | 8 00·3 | 7 37·2 | 5·6 | 2·9 | 11·6 | 6·1 | 17·6 | 9·2 |
| 57 | 7 59·3 | 8 00·6 | 7 37·4 | 5·7 | 3·0 | 11·7 | 6·1 | 17·7 | 9·3 |
| 58 | 7 59·5 | 8 00·8 | 7 37·7 | 5·8 | 3·0 | 11·8 | 6·2 | 17·8 | 9·3 |
| 59 | 7 59·8 | 8 01·1 | 7 37·9 | 5·9 | 3·1 | 11·9 | 6·2 | 17·9 | 9·4 |
| 60 | 8 00·0 | 8 01·3 | 7 38·1 | 6·0 | 3·2 | 12·0 | 6·3 | 18·0 | 9·5 |

| 32m | SUN PLANETS | ARIES | MOON | v or d | Corrn | v or d | Corrn | v or d | Corrn |
|---|---|---|---|---|---|---|---|---|---|
| s | ° ′ | ° ′ | ° ′ | ′ | ′ | ′ | ′ | ′ | ′ |
| 00 | 8 00·0 | 8 01·3 | 7 38·1 | 0·0 | 0·0 | 6·0 | 3·3 | 12·0 | 6·5 |
| 01 | 8 00·3 | 8 01·6 | 7 38·4 | 0·1 | 0·1 | 6·1 | 3·3 | 12·1 | 6·6 |
| 02 | 8 00·5 | 8 01·8 | 7 38·6 | 0·2 | 0·1 | 6·2 | 3·4 | 12·2 | 6·6 |
| 03 | 8 00·8 | 8 02·1 | 7 38·8 | 0·3 | 0·2 | 6·3 | 3·4 | 12·3 | 6·7 |
| 04 | 8 01·0 | 8 02·3 | 7 39·1 | 0·4 | 0·2 | 6·4 | 3·5 | 12·4 | 6·7 |
| 05 | 8 01·3 | 8 02·6 | 7 39·3 | 0·5 | 0·3 | 6·5 | 3·5 | 12·5 | 6·8 |
| 06 | 8 01·5 | 8 02·8 | 7 39·6 | 0·6 | 0·3 | 6·6 | 3·6 | 12·6 | 6·8 |
| 07 | 8 01·8 | 8 03·1 | 7 39·8 | 0·7 | 0·4 | 6·7 | 3·6 | 12·7 | 6·9 |
| 08 | 8 02·0 | 8 03·3 | 7 40·0 | 0·8 | 0·4 | 6·8 | 3·7 | 12·8 | 6·9 |
| 09 | 8 02·3 | 8 03·6 | 7 40·3 | 0·9 | 0·5 | 6·9 | 3·7 | 12·9 | 7·0 |
| 10 | 8 02·5 | 8 03·8 | 7 40·5 | 1·0 | 0·5 | 7·0 | 3·8 | 13·0 | 7·0 |
| 11 | 8 02·8 | 8 04·1 | 7 40·8 | 1·1 | 0·6 | 7·1 | 3·8 | 13·1 | 7·1 |
| 12 | 8 03·0 | 8 04·3 | 7 41·0 | 1·2 | 0·7 | 7·2 | 3·9 | 13·2 | 7·2 |
| 13 | 8 03·3 | 8 04·6 | 7 41·2 | 1·3 | 0·7 | 7·3 | 4·0 | 13·3 | 7·2 |
| 14 | 8 03·5 | 8 04·8 | 7 41·5 | 1·4 | 0·8 | 7·4 | 4·0 | 13·4 | 7·3 |
| 15 | 8 03·8 | 8 05·1 | 7 41·7 | 1·5 | 0·8 | 7·5 | 4·1 | 13·5 | 7·3 |
| 16 | 8 04·0 | 8 05·3 | 7 42·0 | 1·6 | 0·9 | 7·6 | 4·1 | 13·6 | 7·4 |
| 17 | 8 04·3 | 8 05·6 | 7 42·2 | 1·7 | 0·9 | 7·7 | 4·2 | 13·7 | 7·4 |
| 18 | 8 04·5 | 8 05·8 | 7 42·4 | 1·8 | 1·0 | 7·8 | 4·2 | 13·8 | 7·5 |
| 19 | 8 04·8 | 8 06·1 | 7 42·7 | 1·9 | 1·0 | 7·9 | 4·3 | 13·9 | 7·5 |
| 20 | 8 05·0 | 8 06·3 | 7 42·9 | 2·0 | 1·1 | 8·0 | 4·3 | 14·0 | 7·6 |
| 21 | 8 05·3 | 8 06·6 | 7 43·1 | 2·1 | 1·1 | 8·1 | 4·4 | 14·1 | 7·6 |
| 22 | 8 05·5 | 8 06·8 | 7 43·4 | 2·2 | 1·2 | 8·2 | 4·4 | 14·2 | 7·7 |
| 23 | 8 05·8 | 8 07·1 | 7 43·6 | 2·3 | 1·2 | 8·3 | 4·5 | 14·3 | 7·7 |
| 24 | 8 06·0 | 8 07·3 | 7 43·9 | 2·4 | 1·3 | 8·4 | 4·6 | 14·4 | 7·8 |
| 25 | 8 06·3 | 8 07·6 | 7 44·1 | 2·5 | 1·4 | 8·5 | 4·6 | 14·5 | 7·9 |
| 26 | 8 06·5 | 8 07·8 | 7 44·3 | 2·6 | 1·4 | 8·6 | 4·7 | 14·6 | 7·9 |
| 27 | 8 06·8 | 8 08·1 | 7 44·6 | 2·7 | 1·5 | 8·7 | 4·7 | 14·7 | 8·0 |
| 28 | 8 07·0 | 8 08·3 | 7 44·8 | 2·8 | 1·5 | 8·8 | 4·8 | 14·8 | 8·0 |
| 29 | 8 07·3 | 8 08·6 | 7 45·1 | 2·9 | 1·6 | 8·9 | 4·8 | 14·9 | 8·1 |
| 30 | 8 07·5 | 8 08·8 | 7 45·3 | 3·0 | 1·6 | 9·0 | 4·9 | 15·0 | 8·1 |
| 31 | 8 07·8 | 8 09·1 | 7 45·5 | 3·1 | 1·7 | 9·1 | 4·9 | 15·1 | 8·2 |
| 32 | 8 08·0 | 8 09·3 | 7 45·8 | 3·2 | 1·7 | 9·2 | 5·0 | 15·2 | 8·2 |
| 33 | 8 08·3 | 8 09·6 | 7 46·0 | 3·3 | 1·8 | 9·3 | 5·0 | 15·3 | 8·3 |
| 34 | 8 08·5 | 8 09·8 | 7 46·2 | 3·4 | 1·8 | 9·4 | 5·1 | 15·4 | 8·3 |
| 35 | 8 08·8 | 8 10·1 | 7 46·5 | 3·5 | 1·9 | 9·5 | 5·1 | 15·5 | 8·4 |
| 36 | 8 09·0 | 8 10·3 | 7 46·7 | 3·6 | 2·0 | 9·6 | 5·2 | 15·6 | 8·5 |
| 37 | 8 09·3 | 8 10·6 | 7 47·0 | 3·7 | 2·0 | 9·7 | 5·3 | 15·7 | 8·5 |
| 38 | 8 09·5 | 8 10·8 | 7 47·2 | 3·8 | 2·1 | 9·8 | 5·3 | 15·8 | 8·6 |
| 39 | 8 09·8 | 8 11·1 | 7 47·4 | 3·9 | 2·1 | 9·9 | 5·4 | 15·9 | 8·6 |
| 40 | 8 10·0 | 8 11·3 | 7 47·7 | 4·0 | 2·2 | 10·0 | 5·4 | 16·0 | 8·7 |
| 41 | 8 10·3 | 8 11·6 | 7 47·9 | 4·1 | 2·2 | 10·1 | 5·5 | 16·1 | 8·7 |
| 42 | 8 10·5 | 8 11·8 | 7 48·2 | 4·2 | 2·3 | 10·2 | 5·5 | 16·2 | 8·8 |
| 43 | 8 10·8 | 8 12·1 | 7 48·4 | 4·3 | 2·3 | 10·3 | 5·6 | 16·3 | 8·8 |
| 44 | 8 11·0 | 8 12·3 | 7 48·6 | 4·4 | 2·4 | 10·4 | 5·6 | 16·4 | 8·9 |
| 45 | 8 11·3 | 8 12·6 | 7 48·9 | 4·5 | 2·4 | 10·5 | 5·7 | 16·5 | 8·9 |
| 46 | 8 11·5 | 8 12·8 | 7 49·1 | 4·6 | 2·5 | 10·6 | 5·7 | 16·6 | 9·0 |
| 47 | 8 11·8 | 8 13·1 | 7 49·3 | 4·7 | 2·5 | 10·7 | 5·8 | 16·7 | 9·0 |
| 48 | 8 12·0 | 8 13·3 | 7 49·6 | 4·8 | 2·6 | 10·8 | 5·9 | 16·8 | 9·1 |
| 49 | 8 12·3 | 8 13·6 | 7 49·8 | 4·9 | 2·7 | 10·9 | 5·9 | 16·9 | 9·2 |
| 50 | 8 12·5 | 8 13·8 | 7 50·1 | 5·0 | 2·7 | 11·0 | 6·0 | 17·0 | 9·2 |
| 51 | 8 12·8 | 8 14·1 | 7 50·3 | 5·1 | 2·8 | 11·1 | 6·0 | 17·1 | 9·3 |
| 52 | 8 13·0 | 8 14·3 | 7 50·5 | 5·2 | 2·8 | 11·2 | 6·1 | 17·2 | 9·3 |
| 53 | 8 13·3 | 8 14·6 | 7 50·8 | 5·3 | 2·9 | 11·3 | 6·1 | 17·3 | 9·4 |
| 54 | 8 13·5 | 8 14·9 | 7 51·0 | 5·4 | 2·9 | 11·4 | 6·2 | 17·4 | 9·4 |
| 55 | 8 13·8 | 8 15·1 | 7 51·3 | 5·5 | 3·0 | 11·5 | 6·2 | 17·5 | 9·5 |
| 56 | 8 14·0 | 8 15·4 | 7 51·5 | 5·6 | 3·0 | 11·6 | 6·3 | 17·6 | 9·5 |
| 57 | 8 14·3 | 8 15·6 | 7 51·7 | 5·7 | 3·1 | 11·7 | 6·3 | 17·7 | 9·6 |
| 58 | 8 14·5 | 8 15·9 | 7 52·0 | 5·8 | 3·1 | 11·8 | 6·4 | 17·8 | 9·6 |
| 59 | 8 14·8 | 8 16·1 | 7 52·2 | 5·9 | 3·2 | 11·9 | 6·4 | 17·9 | 9·7 |
| 60 | 8 15·0 | 8 16·4 | 7 52·5 | 6·0 | 3·3 | 12·0 | 6·5 | 18·0 | 9·8 |

| 33m | SUN PLANETS | ARIES | MOON | v or d | Corrn | v or d | Corrn | v or d | Corrn |
|---|---|---|---|---|---|---|---|---|---|
| s | ° ′ | ° ′ | ° ′ | ′ | ′ | ′ | ′ | ′ | ′ |
| 00 | 8 15·0 | 8 16·4 | 7 52·5 | 0·0 | 0·0 | 6·0 | 3·4 | 12·0 | 6·7 |
| 01 | 8 15·3 | 8 16·6 | 7 52·7 | 0·1 | 0·1 | 6·1 | 3·4 | 12·1 | 6·8 |
| 02 | 8 15·5 | 8 16·9 | 7 52·9 | 0·2 | 0·1 | 6·2 | 3·5 | 12·2 | 6·8 |
| 03 | 8 15·8 | 8 17·1 | 7 53·2 | 0·3 | 0·2 | 6·3 | 3·5 | 12·3 | 6·9 |
| 04 | 8 16·0 | 8 17·4 | 7 53·4 | 0·4 | 0·2 | 6·4 | 3·6 | 12·4 | 6·9 |
| 05 | 8 16·3 | 8 17·6 | 7 53·6 | 0·5 | 0·3 | 6·5 | 3·6 | 12·5 | 7·0 |
| 06 | 8 16·5 | 8 17·9 | 7 53·9 | 0·6 | 0·3 | 6·6 | 3·7 | 12·6 | 7·0 |
| 07 | 8 16·8 | 8 18·1 | 7 54·1 | 0·7 | 0·4 | 6·7 | 3·7 | 12·7 | 7·1 |
| 08 | 8 17·0 | 8 18·4 | 7 54·4 | 0·8 | 0·4 | 6·8 | 3·8 | 12·8 | 7·1 |
| 09 | 8 17·3 | 8 18·6 | 7 54·6 | 0·9 | 0·5 | 6·9 | 3·9 | 12·9 | 7·2 |
| 10 | 8 17·5 | 8 18·9 | 7 54·8 | 1·0 | 0·6 | 7·0 | 3·9 | 13·0 | 7·3 |
| 11 | 8 17·8 | 8 19·1 | 7 55·1 | 1·1 | 0·6 | 7·1 | 4·0 | 13·1 | 7·3 |
| 12 | 8 18·0 | 8 19·4 | 7 55·3 | 1·2 | 0·7 | 7·2 | 4·0 | 13·2 | 7·4 |
| 13 | 8 18·3 | 8 19·6 | 7 55·6 | 1·3 | 0·7 | 7·3 | 4·1 | 13·3 | 7·4 |
| 14 | 8 18·5 | 8 19·9 | 7 55·8 | 1·4 | 0·8 | 7·4 | 4·1 | 13·4 | 7·5 |
| 15 | 8 18·8 | 8 20·1 | 7 56·0 | 1·5 | 0·8 | 7·5 | 4·2 | 13·5 | 7·5 |
| 16 | 8 19·0 | 8 20·4 | 7 56·3 | 1·6 | 0·9 | 7·6 | 4·2 | 13·6 | 7·6 |
| 17 | 8 19·3 | 8 20·6 | 7 56·5 | 1·7 | 0·9 | 7·7 | 4·3 | 13·7 | 7·6 |
| 18 | 8 19·5 | 8 20·9 | 7 56·7 | 1·8 | 1·0 | 7·8 | 4·4 | 13·8 | 7·7 |
| 19 | 8 19·8 | 8 21·1 | 7 57·0 | 1·9 | 1·1 | 7·9 | 4·4 | 13·9 | 7·8 |
| 20 | 8 20·0 | 8 21·4 | 7 57·2 | 2·0 | 1·1 | 8·0 | 4·5 | 14·0 | 7·8 |
| 21 | 8 20·3 | 8 21·6 | 7 57·5 | 2·1 | 1·2 | 8·1 | 4·5 | 14·1 | 7·9 |
| 22 | 8 20·5 | 8 21·9 | 7 57·7 | 2·2 | 1·2 | 8·2 | 4·6 | 14·2 | 7·9 |
| 23 | 8 20·8 | 8 22·1 | 7 57·9 | 2·3 | 1·3 | 8·3 | 4·6 | 14·3 | 8·0 |
| 24 | 8 21·0 | 8 22·4 | 7 58·2 | 2·4 | 1·3 | 8·4 | 4·7 | 14·4 | 8·0 |
| 25 | 8 21·3 | 8 22·6 | 7 58·4 | 2·5 | 1·4 | 8·5 | 4·7 | 14·5 | 8·1 |
| 26 | 8 21·5 | 8 22·9 | 7 58·7 | 2·6 | 1·5 | 8·6 | 4·8 | 14·6 | 8·2 |
| 27 | 8 21·8 | 8 23·1 | 7 58·9 | 2·7 | 1·5 | 8·7 | 4·9 | 14·7 | 8·2 |
| 28 | 8 22·0 | 8 23·4 | 7 59·1 | 2·8 | 1·6 | 8·8 | 4·9 | 14·8 | 8·3 |
| 29 | 8 22·3 | 8 23·6 | 7 59·4 | 2·9 | 1·6 | 8·9 | 5·0 | 14·9 | 8·3 |
| 30 | 8 22·5 | 8 23·9 | 7 59·6 | 3·0 | 1·7 | 9·0 | 5·0 | 15·0 | 8·4 |
| 31 | 8 22·8 | 8 24·1 | 7 59·8 | 3·1 | 1·7 | 9·1 | 5·1 | 15·1 | 8·4 |
| 32 | 8 23·0 | 8 24·4 | 8 00·1 | 3·2 | 1·8 | 9·2 | 5·1 | 15·2 | 8·5 |
| 33 | 8 23·3 | 8 24·6 | 8 00·3 | 3·3 | 1·8 | 9·3 | 5·2 | 15·3 | 8·5 |
| 34 | 8 23·5 | 8 24·9 | 8 00·6 | 3·4 | 1·9 | 9·4 | 5·2 | 15·4 | 8·6 |
| 35 | 8 23·8 | 8 25·1 | 8 00·8 | 3·5 | 2·0 | 9·5 | 5·3 | 15·5 | 8·7 |
| 36 | 8 24·0 | 8 25·4 | 8 01·0 | 3·6 | 2·0 | 9·6 | 5·4 | 15·6 | 8·7 |
| 37 | 8 24·3 | 8 25·6 | 8 01·3 | 3·7 | 2·1 | 9·7 | 5·4 | 15·7 | 8·8 |
| 38 | 8 24·5 | 8 25·9 | 8 01·5 | 3·8 | 2·1 | 9·8 | 5·5 | 15·8 | 8·8 |
| 39 | 8 24·8 | 8 26·1 | 8 01·8 | 3·9 | 2·2 | 9·9 | 5·5 | 15·9 | 8·9 |
| 40 | 8 25·0 | 8 26·4 | 8 02·0 | 4·0 | 2·2 | 10·0 | 5·6 | 16·0 | 8·9 |
| 41 | 8 25·3 | 8 26·6 | 8 02·2 | 4·1 | 2·3 | 10·1 | 5·6 | 16·1 | 9·0 |
| 42 | 8 25·5 | 8 26·9 | 8 02·5 | 4·2 | 2·3 | 10·2 | 5·7 | 16·2 | 9·0 |
| 43 | 8 25·8 | 8 27·1 | 8 02·7 | 4·3 | 2·4 | 10·3 | 5·8 | 16·3 | 9·1 |
| 44 | 8 26·0 | 8 27·4 | 8 02·9 | 4·4 | 2·5 | 10·4 | 5·8 | 16·4 | 9·2 |
| 45 | 8 26·3 | 8 27·6 | 8 03·2 | 4·5 | 2·5 | 10·5 | 5·9 | 16·5 | 9·2 |
| 46 | 8 26·5 | 8 27·9 | 8 03·4 | 4·6 | 2·6 | 10·6 | 5·9 | 16·6 | 9·3 |
| 47 | 8 26·8 | 8 28·1 | 8 03·7 | 4·7 | 2·6 | 10·7 | 6·0 | 16·7 | 9·3 |
| 48 | 8 27·0 | 8 28·4 | 8 03·9 | 4·8 | 2·7 | 10·8 | 6·0 | 16·8 | 9·4 |
| 49 | 8 27·3 | 8 28·6 | 8 04·1 | 4·9 | 2·7 | 10·9 | 6·1 | 16·9 | 9·4 |
| 50 | 8 27·5 | 8 28·9 | 8 04·4 | 5·0 | 2·8 | 11·0 | 6·1 | 17·0 | 9·5 |
| 51 | 8 27·8 | 8 29·1 | 8 04·6 | 5·1 | 2·8 | 11·1 | 6·2 | 17·1 | 9·5 |
| 52 | 8 28·0 | 8 29·4 | 8 04·9 | 5·2 | 2·9 | 11·2 | 6·3 | 17·2 | 9·6 |
| 53 | 8 28·3 | 8 29·6 | 8 05·1 | 5·3 | 3·0 | 11·3 | 6·3 | 17·3 | 9·7 |
| 54 | 8 28·5 | 8 29·9 | 8 05·3 | 5·4 | 3·0 | 11·4 | 6·4 | 17·4 | 9·7 |
| 55 | 8 28·8 | 8 30·1 | 8 05·6 | 5·5 | 3·1 | 11·5 | 6·4 | 17·5 | 9·8 |
| 56 | 8 29·0 | 8 30·4 | 8 05·8 | 5·6 | 3·1 | 11·6 | 6·5 | 17·6 | 9·8 |
| 57 | 8 29·3 | 8 30·6 | 8 06·1 | 5·7 | 3·2 | 11·7 | 6·5 | 17·7 | 9·9 |
| 58 | 8 29·5 | 8 30·9 | 8 06·3 | 5·8 | 3·2 | 11·8 | 6·6 | 17·8 | 9·9 |
| 59 | 8 29·8 | 8 31·1 | 8 06·5 | 5·9 | 3·3 | 11·9 | 6·6 | 17·9 | 10·0 |
| 60 | 8 30·0 | 8 31·4 | 8 06·8 | 6·0 | 3·4 | 12·0 | 6·7 | 18·0 | 10·1 |

| 36m s | SUN PLANETS ° ′ | ARIES ° ′ | MOON ° ′ | v or d ′ | Corrn ′ | v or d ′ | Corrn ′ | v or d ′ | Corrn ′ |
|---|---|---|---|---|---|---|---|---|---|
| 00 | 9 00·0 | 9 01·5 | 8 35·4 | 0·0 | 0·0 | 6·0 | 3·7 | 12·0 | 7·3 |
| 01 | 9 00·3 | 9 01·7 | 8 35·6 | 0·1 | 0·1 | 6·1 | 3·7 | 12·1 | 7·4 |
| 02 | 9 00·5 | 9 02·0 | 8 35·9 | 0·2 | 0·1 | 6·2 | 3·8 | 12·2 | 7·4 |
| 03 | 9 00·8 | 9 02·2 | 8 36·1 | 0·3 | 0·2 | 6·3 | 3·8 | 12·3 | 7·5 |
| 04 | 9 01·0 | 9 02·5 | 8 36·4 | 0·4 | 0·2 | 6·4 | 3·9 | 12·4 | 7·5 |
| 05 | 9 01·3 | 9 02·7 | 8 36·6 | 0·5 | 0·3 | 6·5 | 4·0 | 12·5 | 7·6 |
| 06 | 9 01·5 | 9 03·0 | 8 36·8 | 0·6 | 0·4 | 6·6 | 4·0 | 12·6 | 7·7 |
| 07 | 9 01·8 | 9 03·2 | 8 37·1 | 0·7 | 0·4 | 6·7 | 4·1 | 12·7 | 7·7 |
| 08 | 9 02·0 | 9 03·5 | 8 37·3 | 0·8 | 0·5 | 6·8 | 4·1 | 12·8 | 7·8 |
| 09 | 9 02·3 | 9 03·7 | 8 37·5 | 0·9 | 0·5 | 6·9 | 4·2 | 12·9 | 7·8 |
| 10 | 9 02·5 | 9 04·0 | 8 37·8 | 1·0 | 0·6 | 7·0 | 4·3 | 13·0 | 7·9 |
| 11 | 9 02·8 | 9 04·2 | 8 38·0 | 1·1 | 0·7 | 7·1 | 4·3 | 13·1 | 8·0 |
| 12 | 9 03·0 | 9 04·5 | 8 38·3 | 1·2 | 0·7 | 7·2 | 4·4 | 13·2 | 8·0 |
| 13 | 9 03·3 | 9 04·7 | 8 38·5 | 1·3 | 0·8 | 7·3 | 4·4 | 13·3 | 8·1 |
| 14 | 9 03·5 | 9 05·0 | 8 38·7 | 1·4 | 0·9 | 7·4 | 4·5 | 13·4 | 8·2 |
| 15 | 9 03·8 | 9 05·2 | 8 39·0 | 1·5 | 0·9 | 7·5 | 4·6 | 13·5 | 8·2 |
| 16 | 9 04·0 | 9 05·5 | 8 39·2 | 1·6 | 1·0 | 7·6 | 4·6 | 13·6 | 8·3 |
| 17 | 9 04·3 | 9 05·7 | 8 39·5 | 1·7 | 1·0 | 7·7 | 4·7 | 13·7 | 8·3 |
| 18 | 9 04·5 | 9 06·0 | 8 39·7 | 1·8 | 1·1 | 7·8 | 4·7 | 13·8 | 8·4 |
| 19 | 9 04·8 | 9 06·2 | 8 39·9 | 1·9 | 1·2 | 7·9 | 4·8 | 13·9 | 8·5 |
| 20 | 9 05·0 | 9 06·5 | 8 40·2 | 2·0 | 1·2 | 8·0 | 4·9 | 14·0 | 8·5 |
| 21 | 9 05·3 | 9 06·7 | 8 40·4 | 2·1 | 1·3 | 8·1 | 4·9 | 14·1 | 8·6 |
| 22 | 9 05·5 | 9 07·0 | 8 40·6 | 2·2 | 1·3 | 8·2 | 5·0 | 14·2 | 8·6 |
| 23 | 9 05·8 | 9 07·2 | 8 40·9 | 2·3 | 1·4 | 8·3 | 5·0 | 14·3 | 8·7 |
| 24 | 9 06·0 | 9 07·5 | 8 41·1 | 2·4 | 1·5 | 8·4 | 5·1 | 14·4 | 8·8 |
| 25 | 9 06·3 | 9 07·7 | 8 41·4 | 2·5 | 1·5 | 8·5 | 5·2 | 14·5 | 8·8 |
| 26 | 9 06·5 | 9 08·0 | 8 41·6 | 2·6 | 1·6 | 8·6 | 5·2 | 14·6 | 8·9 |
| 27 | 9 06·8 | 9 08·2 | 8 41·8 | 2·7 | 1·6 | 8·7 | 5·3 | 14·7 | 8·9 |
| 28 | 9 07·0 | 9 08·5 | 8 42·1 | 2·8 | 1·7 | 8·8 | 5·4 | 14·8 | 9·0 |
| 29 | 9 07·3 | 9 08·7 | 8 42·3 | 2·9 | 1·8 | 8·9 | 5·4 | 14·9 | 9·1 |
| 30 | 9 07·5 | 9 09·0 | 8 42·6 | 3·0 | 1·8 | 9·0 | 5·5 | 15·0 | 9·1 |
| 31 | 9 07·8 | 9 09·2 | 8 42·8 | 3·1 | 1·9 | 9·1 | 5·5 | 15·1 | 9·2 |
| 32 | 9 08·0 | 9 09·5 | 8 43·0 | 3·2 | 1·9 | 9·2 | 5·6 | 15·2 | 9·2 |
| 33 | 9 08·3 | 9 09·8 | 8 43·3 | 3·3 | 2·0 | 9·3 | 5·7 | 15·3 | 9·3 |
| 34 | 9 08·5 | 9 10·0 | 8 43·5 | 3·4 | 2·1 | 9·4 | 5·7 | 15·4 | 9·4 |
| 35 | 9 08·8 | 9 10·3 | 8 43·8 | 3·5 | 2·1 | 9·5 | 5·8 | 15·5 | 9·4 |
| 36 | 9 09·0 | 9 10·5 | 8 44·0 | 3·6 | 2·2 | 9·6 | 5·8 | 15·6 | 9·5 |
| 37 | 9 09·3 | 9 10·8 | 8 44·2 | 3·7 | 2·3 | 9·7 | 5·9 | 15·7 | 9·6 |
| 38 | 9 09·5 | 9 11·0 | 8 44·5 | 3·8 | 2·3 | 9·8 | 6·0 | 15·8 | 9·6 |
| 39 | 9 09·8 | 9 11·3 | 8 44·7 | 3·9 | 2·4 | 9·9 | 6·0 | 15·9 | 9·7 |
| 40 | 9 10·0 | 9 11·5 | 8 44·9 | 4·0 | 2·4 | 10·0 | 6·1 | 16·0 | 9·7 |
| 41 | 9 10·3 | 9 11·8 | 8 45·2 | 4·1 | 2·5 | 10·1 | 6·1 | 16·1 | 9·8 |
| 42 | 9 10·5 | 9 12·0 | 8 45·4 | 4·2 | 2·6 | 10·2 | 6·2 | 16·2 | 9·9 |
| 43 | 9 10·8 | 9 12·3 | 8 45·7 | 4·3 | 2·6 | 10·3 | 6·3 | 16·3 | 9·9 |
| 44 | 9 11·0 | 9 12·5 | 8 45·9 | 4·4 | 2·7 | 10·4 | 6·3 | 16·4 | 10·0 |
| 45 | 9 11·3 | 9 12·8 | 8 46·1 | 4·5 | 2·7 | 10·5 | 6·4 | 16·5 | 10·0 |
| 46 | 9 11·5 | 9 13·0 | 8 46·4 | 4·6 | 2·8 | 10·6 | 6·4 | 16·6 | 10·1 |
| 47 | 9 11·8 | 9 13·3 | 8 46·6 | 4·7 | 2·9 | 10·7 | 6·5 | 16·7 | 10·2 |
| 48 | 9 12·0 | 9 13·5 | 8 46·9 | 4·8 | 2·9 | 10·8 | 6·6 | 16·8 | 10·2 |
| 49 | 9 12·3 | 9 13·8 | 8 47·1 | 4·9 | 3·0 | 10·9 | 6·6 | 16·9 | 10·3 |
| 50 | 9 12·5 | 9 14·0 | 8 47·3 | 5·0 | 3·0 | 11·0 | 6·7 | 17·0 | 10·3 |
| 51 | 9 12·8 | 9 14·3 | 8 47·6 | 5·1 | 3·1 | 11·1 | 6·8 | 17·1 | 10·4 |
| 52 | 9 13·0 | 9 14·5 | 8 47·8 | 5·2 | 3·2 | 11·2 | 6·8 | 17·2 | 10·5 |
| 53 | 9 13·3 | 9 14·8 | 8 48·0 | 5·3 | 3·2 | 11·3 | 6·9 | 17·3 | 10·5 |
| 54 | 9 13·5 | 9 15·0 | 8 48·3 | 5·4 | 3·3 | 11·4 | 6·9 | 17·4 | 10·6 |
| 55 | 9 13·8 | 9 15·3 | 8 48·5 | 5·5 | 3·3 | 11·5 | 7·0 | 17·5 | 10·6 |
| 56 | 9 14·0 | 9 15·5 | 8 48·8 | 5·6 | 3·4 | 11·6 | 7·1 | 17·6 | 10·7 |
| 57 | 9 14·3 | 9 15·8 | 8 49·0 | 5·7 | 3·5 | 11·7 | 7·1 | 17·7 | 10·8 |
| 58 | 9 14·5 | 9 16·0 | 8 49·2 | 5·8 | 3·5 | 11·8 | 7·2 | 17·8 | 10·8 |
| 59 | 9 14·8 | 9 16·3 | 8 49·5 | 5·9 | 3·6 | 11·9 | 7·2 | 17·9 | 10·9 |
| 60 | 9 15·0 | 9 16·5 | 8 49·7 | 6·0 | 3·7 | 12·0 | 7·3 | 18·0 | 11·0 |

| 37m s | SUN PLANETS ° ′ | ARIES ° ′ | MOON ° ′ | v or d ′ | Corrn ′ | v or d ′ | Corrn ′ | v or d ′ | Corrn ′ |
|---|---|---|---|---|---|---|---|---|---|
| 00 | 9 15·0 | 9 16·5 | 8 49·7 | 0·0 | 0·0 | 6·0 | 3·8 | 12·0 | 7·5 |
| 01 | 9 15·3 | 9 16·8 | 8 50·0 | 0·1 | 0·1 | 6·1 | 3·8 | 12·1 | 7·6 |
| 02 | 9 15·5 | 9 17·0 | 8 50·2 | 0·2 | 0·1 | 6·2 | 3·9 | 12·2 | 7·6 |
| 03 | 9 15·8 | 9 17·3 | 8 50·4 | 0·3 | 0·2 | 6·3 | 3·9 | 12·3 | 7·7 |
| 04 | 9 16·0 | 9 17·5 | 8 50·7 | 0·4 | 0·3 | 6·4 | 4·0 | 12·4 | 7·8 |
| 05 | 9 16·3 | 9 17·8 | 8 50·9 | 0·5 | 0·3 | 6·5 | 4·1 | 12·5 | 7·8 |
| 06 | 9 16·5 | 9 18·0 | 8 51·1 | 0·6 | 0·4 | 6·6 | 4·1 | 12·6 | 7·9 |
| 07 | 9 16·8 | 9 18·3 | 8 51·4 | 0·7 | 0·4 | 6·7 | 4·2 | 12·7 | 7·9 |
| 08 | 9 17·0 | 9 18·5 | 8 51·6 | 0·8 | 0·5 | 6·8 | 4·3 | 12·8 | 8·0 |
| 09 | 9 17·3 | 9 18·8 | 8 51·9 | 0·9 | 0·6 | 6·9 | 4·3 | 12·9 | 8·1 |
| 10 | 9 17·5 | 9 19·0 | 8 52·1 | 1·0 | 0·6 | 7·0 | 4·4 | 13·0 | 8·1 |
| 11 | 9 17·8 | 9 19·3 | 8 52·3 | 1·1 | 0·7 | 7·1 | 4·4 | 13·1 | 8·2 |
| 12 | 9 18·0 | 9 19·5 | 8 52·6 | 1·2 | 0·8 | 7·2 | 4·5 | 13·2 | 8·3 |
| 13 | 9 18·3 | 9 19·8 | 8 52·8 | 1·3 | 0·8 | 7·3 | 4·6 | 13·3 | 8·3 |
| 14 | 9 18·5 | 9 20·0 | 8 53·1 | 1·4 | 0·9 | 7·4 | 4·6 | 13·4 | 8·4 |
| 15 | 9 18·8 | 9 20·3 | 8 53·3 | 1·5 | 0·9 | 7·5 | 4·7 | 13·5 | 8·4 |
| 16 | 9 19·0 | 9 20·5 | 8 53·5 | 1·6 | 1·0 | 7·6 | 4·8 | 13·6 | 8·5 |
| 17 | 9 19·3 | 9 20·8 | 8 53·8 | 1·7 | 1·1 | 7·7 | 4·8 | 13·7 | 8·6 |
| 18 | 9 19·5 | 9 21·0 | 8 54·0 | 1·8 | 1·1 | 7·8 | 4·9 | 13·8 | 8·6 |
| 19 | 9 19·8 | 9 21·3 | 8 54·3 | 1·9 | 1·2 | 7·9 | 4·9 | 13·9 | 8·7 |
| 20 | 9 20·0 | 9 21·5 | 8 54·5 | 2·0 | 1·3 | 8·0 | 5·0 | 14·0 | 8·8 |
| 21 | 9 20·3 | 9 21·8 | 8 54·7 | 2·1 | 1·3 | 8·1 | 5·1 | 14·1 | 8·8 |
| 22 | 9 20·5 | 9 22·0 | 8 55·0 | 2·2 | 1·4 | 8·2 | 5·1 | 14·2 | 8·9 |
| 23 | 9 20·8 | 9 22·3 | 8 55·2 | 2·3 | 1·4 | 8·3 | 5·2 | 14·3 | 8·9 |
| 24 | 9 21·0 | 9 22·5 | 8 55·4 | 2·4 | 1·5 | 8·4 | 5·3 | 14·4 | 9·0 |
| 25 | 9 21·3 | 9 22·8 | 8 55·7 | 2·5 | 1·6 | 8·5 | 5·3 | 14·5 | 9·1 |
| 26 | 9 21·5 | 9 23·0 | 8 55·9 | 2·6 | 1·6 | 8·6 | 5·4 | 14·6 | 9·1 |
| 27 | 9 21·8 | 9 23·3 | 8 56·2 | 2·7 | 1·7 | 8·7 | 5·4 | 14·7 | 9·2 |
| 28 | 9 22·0 | 9 23·5 | 8 56·4 | 2·8 | 1·8 | 8·8 | 5·5 | 14·8 | 9·3 |
| 29 | 9 22·3 | 9 23·8 | 8 56·6 | 2·9 | 1·8 | 8·9 | 5·6 | 14·9 | 9·3 |
| 30 | 9 22·5 | 9 24·0 | 8 56·9 | 3·0 | 1·9 | 9·0 | 5·6 | 15·0 | 9·4 |
| 31 | 9 22·8 | 9 24·3 | 8 57·1 | 3·1 | 1·9 | 9·1 | 5·7 | 15·1 | 9·4 |
| 32 | 9 23·0 | 9 24·5 | 8 57·4 | 3·2 | 2·0 | 9·2 | 5·8 | 15·2 | 9·5 |
| 33 | 9 23·3 | 9 24·8 | 8 57·6 | 3·3 | 2·1 | 9·3 | 5·8 | 15·3 | 9·6 |
| 34 | 9 23·5 | 9 25·0 | 8 57·8 | 3·4 | 2·1 | 9·4 | 5·9 | 15·4 | 9·6 |
| 35 | 9 23·8 | 9 25·3 | 8 58·1 | 3·5 | 2·2 | 9·5 | 5·9 | 15·5 | 9·7 |
| 36 | 9 24·0 | 9 25·5 | 8 58·3 | 3·6 | 2·3 | 9·6 | 6·0 | 15·6 | 9·8 |
| 37 | 9 24·3 | 9 25·8 | 8 58·5 | 3·7 | 2·3 | 9·7 | 6·1 | 15·7 | 9·8 |
| 38 | 9 24·5 | 9 26·0 | 8 58·8 | 3·8 | 2·4 | 9·8 | 6·1 | 15·8 | 9·9 |
| 39 | 9 24·8 | 9 26·3 | 8 59·0 | 3·9 | 2·4 | 9·9 | 6·2 | 15·9 | 9·9 |
| 40 | 9 25·0 | 9 26·5 | 8 59·3 | 4·0 | 2·5 | 10·0 | 6·3 | 16·0 | 10·0 |
| 41 | 9 25·3 | 9 26·8 | 8 59·5 | 4·1 | 2·6 | 10·1 | 6·3 | 16·1 | 10·1 |
| 42 | 9 25·5 | 9 27·0 | 8 59·7 | 4·2 | 2·6 | 10·2 | 6·4 | 16·2 | 10·1 |
| 43 | 9 25·8 | 9 27·3 | 9 00·0 | 4·3 | 2·7 | 10·3 | 6·4 | 16·3 | 10·2 |
| 44 | 9 26·0 | 9 27·5 | 9 00·2 | 4·4 | 2·8 | 10·4 | 6·5 | 16·4 | 10·3 |
| 45 | 9 26·3 | 9 27·8 | 9 00·5 | 4·5 | 2·8 | 10·5 | 6·6 | 16·5 | 10·3 |
| 46 | 9 26·5 | 9 28·1 | 9 00·7 | 4·6 | 2·9 | 10·6 | 6·6 | 16·6 | 10·4 |
| 47 | 9 26·8 | 9 28·3 | 9 00·9 | 4·7 | 2·9 | 10·7 | 6·7 | 16·7 | 10·4 |
| 48 | 9 27·0 | 9 28·6 | 9 01·2 | 4·8 | 3·0 | 10·8 | 6·8 | 16·8 | 10·5 |
| 49 | 9 27·3 | 9 28·8 | 9 01·4 | 4·9 | 3·1 | 10·9 | 6·8 | 16·9 | 10·6 |
| 50 | 9 27·5 | 9 29·1 | 9 01·6 | 5·0 | 3·1 | 11·0 | 6·9 | 17·0 | 10·6 |
| 51 | 9 27·8 | 9 29·3 | 9 01·9 | 5·1 | 3·2 | 11·1 | 6·9 | 17·1 | 10·7 |
| 52 | 9 28·0 | 9 29·6 | 9 02·1 | 5·2 | 3·3 | 11·2 | 7·0 | 17·2 | 10·8 |
| 53 | 9 28·3 | 9 29·8 | 9 02·4 | 5·3 | 3·3 | 11·3 | 7·1 | 17·3 | 10·8 |
| 54 | 9 28·5 | 9 30·1 | 9 02·6 | 5·4 | 3·4 | 11·4 | 7·1 | 17·4 | 10·9 |
| 55 | 9 28·8 | 9 30·3 | 9 02·8 | 5·5 | 3·4 | 11·5 | 7·2 | 17·5 | 10·9 |
| 56 | 9 29·0 | 9 30·6 | 9 03·1 | 5·6 | 3·5 | 11·6 | 7·3 | 17·6 | 11·0 |
| 57 | 9 29·3 | 9 30·8 | 9 03·3 | 5·7 | 3·6 | 11·7 | 7·3 | 17·7 | 11·1 |
| 58 | 9 29·5 | 9 31·1 | 9 03·6 | 5·8 | 3·6 | 11·8 | 7·4 | 17·8 | 11·1 |
| 59 | 9 29·8 | 9 31·3 | 9 03·8 | 5·9 | 3·7 | 11·9 | 7·4 | 17·9 | 11·2 |
| 60 | 9 30·0 | 9 31·6 | 9 04·0 | 6·0 | 3·8 | 12·0 | 7·5 | 18·0 | 11·3 |

| 44m | SUN PLANETS | ARIES | MOON | v or d | Corrn | v or d | Corrn | v or d | Corrn |
|---|---|---|---|---|---|---|---|---|---|
| s | ° ′ | ° ′ | ° ′ | ′ | ′ | ′ | ′ | ′ | ′ |
| 00 | 11 00·0 | 11 01·8 | 10 29·9 | 0·0 | 0·0 | 6·0 | 4·5 | 12·0 | 8·9 |
| 01 | 11 00·3 | 11 02·1 | 10 30·2 | 0·1 | 0·1 | 6·1 | 4·5 | 12·1 | 9·0 |
| 02 | 11 00·5 | 11 02·3 | 10 30·4 | 0·2 | 0·1 | 6·2 | 4·6 | 12·2 | 9·0 |
| 03 | 11 00·8 | 11 02·6 | 10 30·6 | 0·3 | 0·2 | 6·3 | 4·7 | 12·3 | 9·1 |
| 04 | 11 01·0 | 11 02·8 | 10 30·9 | 0·4 | 0·3 | 6·4 | 4·7 | 12·4 | 9·2 |
| 05 | 11 01·3 | 11 03·1 | 10 31·1 | 0·5 | 0·4 | 6·5 | 4·8 | 12·5 | 9·3 |
| 06 | 11 01·5 | 11 03·3 | 10 31·4 | 0·6 | 0·4 | 6·6 | 4·9 | 12·6 | 9·3 |
| 07 | 11 01·8 | 11 03·6 | 10 31·6 | 0·7 | 0·5 | 6·7 | 5·0 | 12·7 | 9·4 |
| 08 | 11 02·0 | 11 03·8 | 10 31·8 | 0·8 | 0·6 | 6·8 | 5·0 | 12·8 | 9·5 |
| 09 | 11 02·3 | 11 04·1 | 10 32·1 | 0·9 | 0·7 | 6·9 | 5·1 | 12·9 | 9·6 |
| 10 | 11 02·5 | 11 04·3 | 10 32·3 | 1·0 | 0·7 | 7·0 | 5·2 | 13·0 | 9·6 |
| 11 | 11 02·8 | 11 04·6 | 10 32·6 | 1·1 | 0·8 | 7·1 | 5·3 | 13·1 | 9·7 |
| 12 | 11 03·0 | 11 04·8 | 10 32·8 | 1·2 | 0·9 | 7·2 | 5·3 | 13·2 | 9·8 |
| 13 | 11 03·3 | 11 05·1 | 10 33·0 | 1·3 | 1·0 | 7·3 | 5·4 | 13·3 | 9·9 |
| 14 | 11 03·5 | 11 05·3 | 10 33·3 | 1·4 | 1·0 | 7·4 | 5·5 | 13·4 | 9·9 |
| 15 | 11 03·8 | 11 05·6 | 10 33·5 | 1·5 | 1·1 | 7·5 | 5·6 | 13·5 | 10·0 |
| 16 | 11 04·0 | 11 05·8 | 10 33·8 | 1·6 | 1·2 | 7·6 | 5·6 | 13·6 | 10·1 |
| 17 | 11 04·3 | 11 06·1 | 10 34·0 | 1·7 | 1·3 | 7·7 | 5·7 | 13·7 | 10·2 |
| 18 | 11 04·5 | 11 06·3 | 10 34·2 | 1·8 | 1·3 | 7·8 | 5·8 | 13·8 | 10·2 |
| 19 | 11 04·8 | 11 06·6 | 10 34·5 | 1·9 | 1·4 | 7·9 | 5·9 | 13·9 | 10·3 |
| 20 | 11 05·0 | 11 06·8 | 10 34·7 | 2·0 | 1·5 | 8·0 | 5·9 | 14·0 | 10·4 |
| 21 | 11 05·3 | 11 07·1 | 10 34·9 | 2·1 | 1·6 | 8·1 | 6·0 | 14·1 | 10·5 |
| 22 | 11 05·5 | 11 07·3 | 10 35·2 | 2·2 | 1·6 | 8·2 | 6·1 | 14·2 | 10·5 |
| 23 | 11 05·8 | 11 07·6 | 10 35·4 | 2·3 | 1·7 | 8·3 | 6·2 | 14·3 | 10·6 |
| 24 | 11 06·0 | 11 07·8 | 10 35·7 | 2·4 | 1·8 | 8·4 | 6·2 | 14·4 | 10·7 |
| 25 | 11 06·3 | 11 08·1 | 10 35·9 | 2·5 | 1·9 | 8·5 | 6·3 | 14·5 | 10·8 |
| 26 | 11 06·5 | 11 08·3 | 10 36·1 | 2·6 | 1·9 | 8·6 | 6·4 | 14·6 | 10·8 |
| 27 | 11 06·8 | 11 08·6 | 10 36·4 | 2·7 | 2·0 | 8·7 | 6·5 | 14·7 | 10·9 |
| 28 | 11 07·0 | 11 08·8 | 10 36·6 | 2·8 | 2·1 | 8·8 | 6·5 | 14·8 | 11·0 |
| 29 | 11 07·3 | 11 09·1 | 10 36·9 | 2·9 | 2·2 | 8·9 | 6·6 | 14·9 | 11·1 |
| 30 | 11 07·5 | 11 09·3 | 10 37·1 | 3·0 | 2·2 | 9·0 | 6·7 | 15·0 | 11·1 |
| 31 | 11 07·8 | 11 09·6 | 10 37·3 | 3·1 | 2·3 | 9·1 | 6·7 | 15·1 | 11·2 |
| 32 | 11 08·0 | 11 09·8 | 10 37·6 | 3·2 | 2·4 | 9·2 | 6·8 | 15·2 | 11·3 |
| 33 | 11 08·3 | 11 10·1 | 10 37·8 | 3·3 | 2·4 | 9·3 | 6·9 | 15·3 | 11·3 |
| 34 | 11 08·5 | 11 10·3 | 10 38·0 | 3·4 | 2·5 | 9·4 | 7·0 | 15·4 | 11·4 |
| 35 | 11 08·8 | 11 10·6 | 10 38·3 | 3·5 | 2·6 | 9·5 | 7·0 | 15·5 | 11·5 |
| 36 | 11 09·0 | 11 10·8 | 10 38·5 | 3·6 | 2·7 | 9·6 | 7·1 | 15·6 | 11·6 |
| 37 | 11 09·3 | 11 11·1 | 10 38·8 | 3·7 | 2·7 | 9·7 | 7·2 | 15·7 | 11·6 |
| 38 | 11 09·5 | 11 11·3 | 10 39·0 | 3·8 | 2·8 | 9·8 | 7·3 | 15·8 | 11·7 |
| 39 | 11 09·8 | 11 11·6 | 10 39·2 | 3·9 | 2·9 | 9·9 | 7·3 | 15·9 | 11·8 |
| 40 | 11 10·0 | 11 11·8 | 10 39·5 | 4·0 | 3·0 | 10·0 | 7·4 | 16·0 | 11·9 |
| 41 | 11 10·3 | 11 12·1 | 10 39·7 | 4·1 | 3·0 | 10·1 | 7·5 | 16·1 | 11·9 |
| 42 | 11 10·5 | 11 12·3 | 10 40·0 | 4·2 | 3·1 | 10·2 | 7·6 | 16·2 | 12·0 |
| 43 | 11 10·8 | 11 12·6 | 10 40·2 | 4·3 | 3·2 | 10·3 | 7·6 | 16·3 | 12·1 |
| 44 | 11 11·0 | 11 12·8 | 10 40·4 | 4·4 | 3·3 | 10·4 | 7·7 | 16·4 | 12·2 |
| 45 | 11 11·3 | 11 13·1 | 10 40·7 | 4·5 | 3·3 | 10·5 | 7·8 | 16·5 | 12·2 |
| 46 | 11 11·5 | 11 13·3 | 10 40·9 | 4·6 | 3·4 | 10·6 | 7·9 | 16·6 | 12·3 |
| 47 | 11 11·8 | 11 13·6 | 10 41·1 | 4·7 | 3·5 | 10·7 | 7·9 | 16·7 | 12·4 |
| 48 | 11 12·0 | 11 13·8 | 10 41·4 | 4·8 | 3·6 | 10·8 | 8·0 | 16·8 | 12·5 |
| 49 | 11 12·3 | 11 14·1 | 10 41·6 | 4·9 | 3·6 | 10·9 | 8·1 | 16·9 | 12·5 |
| 50 | 11 12·5 | 11 14·3 | 10 41·9 | 5·0 | 3·7 | 11·0 | 8·2 | 17·0 | 12·6 |
| 51 | 11 12·8 | 11 14·6 | 10 42·1 | 5·1 | 3·8 | 11·1 | 8·2 | 17·1 | 12·7 |
| 52 | 11 13·0 | 11 14·8 | 10 42·3 | 5·2 | 3·9 | 11·2 | 8·3 | 17·2 | 12·8 |
| 53 | 11 13·3 | 11 15·1 | 10 42·6 | 5·3 | 3·9 | 11·3 | 8·4 | 17·3 | 12·8 |
| 54 | 11 13·5 | 11 15·3 | 10 42·8 | 5·4 | 4·0 | 11·4 | 8·5 | 17·4 | 12·9 |
| 55 | 11 13·8 | 11 15·6 | 10 43·1 | 5·5 | 4·1 | 11·5 | 8·5 | 17·5 | 13·0 |
| 56 | 11 14·0 | 11 15·8 | 10 43·3 | 5·6 | 4·2 | 11·6 | 8·6 | 17·6 | 13·1 |
| 57 | 11 14·3 | 11 16·1 | 10 43·5 | 5·7 | 4·2 | 11·7 | 8·7 | 17·7 | 13·1 |
| 58 | 11 14·5 | 11 16·3 | 10 43·8 | 5·8 | 4·3 | 11·8 | 8·8 | 17·8 | 13·2 |
| 59 | 11 14·8 | 11 16·6 | 10 44·0 | 5·9 | 4·4 | 11·9 | 8·8 | 17·9 | 13·3 |
| 60 | 11 15·0 | 11 16·8 | 10 44·3 | 6·0 | 4·5 | 12·0 | 8·9 | 18·0 | 13·4 |

| 45m | SUN PLANETS | ARIES | MOON | v or d | Corrn | v or d | Corrn | v or d | Corrn |
|---|---|---|---|---|---|---|---|---|---|
| s | ° ′ | ° ′ | ° ′ | ′ | ′ | ′ | ′ | ′ | ′ |
| 00 | 11 15·0 | 11 16·8 | 10 44·3 | 0·0 | 0·0 | 6·0 | 4·6 | 12·0 | 9·1 |
| 01 | 11 15·3 | 11 17·1 | 10 44·5 | 0·1 | 0·1 | 6·1 | 4·6 | 12·1 | 9·2 |
| 02 | 11 15·5 | 11 17·3 | 10 44·7 | 0·2 | 0·2 | 6·2 | 4·7 | 12·2 | 9·3 |
| 03 | 11 15·8 | 11 17·6 | 10 45·0 | 0·3 | 0·2 | 6·3 | 4·8 | 12·3 | 9·3 |
| 04 | 11 16·0 | 11 17·9 | 10 45·2 | 0·4 | 0·3 | 6·4 | 4·9 | 12·4 | 9·4 |
| 05 | 11 16·3 | 11 18·1 | 10 45·4 | 0·5 | 0·4 | 6·5 | 4·9 | 12·5 | 9·5 |
| 06 | 11 16·5 | 11 18·4 | 10 45·7 | 0·6 | 0·5 | 6·6 | 5·0 | 12·6 | 9·6 |
| 07 | 11 16·8 | 11 18·6 | 10 45·9 | 0·7 | 0·5 | 6·7 | 5·1 | 12·7 | 9·6 |
| 08 | 11 17·0 | 11 18·9 | 10 46·2 | 0·8 | 0·6 | 6·8 | 5·2 | 12·8 | 9·7 |
| 09 | 11 17·3 | 11 19·1 | 10 46·4 | 0·9 | 0·7 | 6·9 | 5·2 | 12·9 | 9·8 |
| 10 | 11 17·5 | 11 19·4 | 10 46·6 | 1·0 | 0·8 | 7·0 | 5·3 | 13·0 | 9·9 |
| 11 | 11 17·8 | 11 19·6 | 10 46·9 | 1·1 | 0·8 | 7·1 | 5·4 | 13·1 | 9·9 |
| 12 | 11 18·0 | 11 19·9 | 10 47·1 | 1·2 | 0·9 | 7·2 | 5·5 | 13·2 | 10·0 |
| 13 | 11 18·3 | 11 20·1 | 10 47·4 | 1·3 | 1·0 | 7·3 | 5·5 | 13·3 | 10·1 |
| 14 | 11 18·5 | 11 20·4 | 10 47·6 | 1·4 | 1·1 | 7·4 | 5·6 | 13·4 | 10·2 |
| 15 | 11 18·8 | 11 20·6 | 10 47·8 | 1·5 | 1·1 | 7·5 | 5·7 | 13·5 | 10·2 |
| 16 | 11 19·0 | 11 20·9 | 10 48·1 | 1·6 | 1·2 | 7·6 | 5·8 | 13·6 | 10·3 |
| 17 | 11 19·3 | 11 21·1 | 10 48·3 | 1·7 | 1·3 | 7·7 | 5·8 | 13·7 | 10·4 |
| 18 | 11 19·5 | 11 21·4 | 10 48·5 | 1·8 | 1·4 | 7·8 | 5·9 | 13·8 | 10·5 |
| 19 | 11 19·8 | 11 21·6 | 10 48·8 | 1·9 | 1·4 | 7·9 | 6·0 | 13·9 | 10·5 |
| 20 | 11 20·0 | 11 21·9 | 10 49·0 | 2·0 | 1·5 | 8·0 | 6·1 | 14·0 | 10·6 |
| 21 | 11 20·3 | 11 22·1 | 10 49·3 | 2·1 | 1·6 | 8·1 | 6·1 | 14·1 | 10·7 |
| 22 | 11 20·5 | 11 22·4 | 10 49·5 | 2·2 | 1·7 | 8·2 | 6·2 | 14·2 | 10·8 |
| 23 | 11 20·8 | 11 22·6 | 10 49·7 | 2·3 | 1·7 | 8·3 | 6·3 | 14·3 | 10·8 |
| 24 | 11 21·0 | 11 22·9 | 10 50·0 | 2·4 | 1·8 | 8·4 | 6·4 | 14·4 | 10·9 |
| 25 | 11 21·3 | 11 23·1 | 10 50·2 | 2·5 | 1·9 | 8·5 | 6·4 | 14·5 | 11·0 |
| 26 | 11 21·5 | 11 23·4 | 10 50·5 | 2·6 | 2·0 | 8·6 | 6·5 | 14·6 | 11·1 |
| 27 | 11 21·8 | 11 23·6 | 10 50·7 | 2·7 | 2·0 | 8·7 | 6·6 | 14·7 | 11·1 |
| 28 | 11 22·0 | 11 23·9 | 10 50·9 | 2·8 | 2·1 | 8·8 | 6·7 | 14·8 | 11·2 |
| 29 | 11 22·3 | 11 24·1 | 10 51·2 | 2·9 | 2·2 | 8·9 | 6·7 | 14·9 | 11·3 |
| 30 | 11 22·5 | 11 24·4 | 10 51·4 | 3·0 | 2·3 | 9·0 | 6·8 | 15·0 | 11·4 |
| 31 | 11 22·8 | 11 24·6 | 10 51·6 | 3·1 | 2·4 | 9·1 | 6·9 | 15·1 | 11·5 |
| 32 | 11 23·0 | 11 24·9 | 10 51·9 | 3·2 | 2·4 | 9·2 | 7·0 | 15·2 | 11·5 |
| 33 | 11 23·3 | 11 25·1 | 10 52·1 | 3·3 | 2·5 | 9·3 | 7·1 | 15·3 | 11·6 |
| 34 | 11 23·5 | 11 25·4 | 10 52·4 | 3·4 | 2·6 | 9·4 | 7·1 | 15·4 | 11·7 |
| 35 | 11 23·8 | 11 25·6 | 10 52·6 | 3·5 | 2·7 | 9·5 | 7·2 | 15·5 | 11·8 |
| 36 | 11 24·0 | 11 25·9 | 10 52·8 | 3·6 | 2·7 | 9·6 | 7·3 | 15·6 | 11·8 |
| 37 | 11 24·3 | 11 26·1 | 10 53·1 | 3·7 | 2·8 | 9·7 | 7·4 | 15·7 | 11·9 |
| 38 | 11 24·5 | 11 26·4 | 10 53·3 | 3·8 | 2·9 | 9·8 | 7·4 | 15·8 | 12·0 |
| 39 | 11 24·8 | 11 26·6 | 10 53·6 | 3·9 | 3·0 | 9·9 | 7·5 | 15·9 | 12·1 |
| 40 | 11 25·0 | 11 26·9 | 10 53·8 | 4·0 | 3·0 | 10·0 | 7·6 | 16·0 | 12·1 |
| 41 | 11 25·3 | 11 27·1 | 10 54·0 | 4·1 | 3·1 | 10·1 | 7·7 | 16·1 | 12·2 |
| 42 | 11 25·5 | 11 27·4 | 10 54·3 | 4·2 | 3·2 | 10·2 | 7·7 | 16·2 | 12·3 |
| 43 | 11 25·8 | 11 27·6 | 10 54·5 | 4·3 | 3·3 | 10·3 | 7·8 | 16·3 | 12·4 |
| 44 | 11 26·0 | 11 27·9 | 10 54·7 | 4·4 | 3·3 | 10·4 | 7·9 | 16·4 | 12·4 |
| 45 | 11 26·3 | 11 28·1 | 10 55·0 | 4·5 | 3·4 | 10·5 | 8·0 | 16·5 | 12·5 |
| 46 | 11 26·5 | 11 28·4 | 10 55·2 | 4·6 | 3·5 | 10·6 | 8·0 | 16·6 | 12·6 |
| 47 | 11 26·8 | 11 28·6 | 10 55·5 | 4·7 | 3·6 | 10·7 | 8·1 | 16·7 | 12·7 |
| 48 | 11 27·0 | 11 28·9 | 10 55·7 | 4·8 | 3·6 | 10·8 | 8·2 | 16·8 | 12·7 |
| 49 | 11 27·3 | 11 29·1 | 10 55·9 | 4·9 | 3·7 | 10·9 | 8·3 | 16·9 | 12·8 |
| 50 | 11 27·5 | 11 29·4 | 10 56·2 | 5·0 | 3·8 | 11·0 | 8·3 | 17·0 | 12·9 |
| 51 | 11 27·8 | 11 29·6 | 10 56·4 | 5·1 | 3·9 | 11·1 | 8·4 | 17·1 | 13·0 |
| 52 | 11 28·0 | 11 29·9 | 10 56·7 | 5·2 | 3·9 | 11·2 | 8·5 | 17·2 | 13·0 |
| 53 | 11 28·3 | 11 30·1 | 10 56·9 | 5·3 | 4·0 | 11·3 | 8·6 | 17·3 | 13·1 |
| 54 | 11 28·5 | 11 30·4 | 10 57·1 | 5·4 | 4·1 | 11·4 | 8·6 | 17·4 | 13·2 |
| 55 | 11 28·8 | 11 30·6 | 10 57·4 | 5·5 | 4·2 | 11·5 | 8·7 | 17·5 | 13·3 |
| 56 | 11 29·0 | 11 30·9 | 10 57·6 | 5·6 | 4·2 | 11·6 | 8·8 | 17·6 | 13·3 |
| 57 | 11 29·3 | 11 31·1 | 10 57·9 | 5·7 | 4·3 | 11·7 | 8·9 | 17·7 | 13·4 |
| 58 | 11 29·5 | 11 31·4 | 10 58·1 | 5·8 | 4·4 | 11·8 | 8·9 | 17·8 | 13·5 |
| 59 | 11 29·8 | 11 31·6 | 10 58·3 | 5·9 | 4·5 | 11·9 | 9·0 | 17·9 | 13·6 |
| 60 | 11 30·0 | 11 31·9 | 10 58·6 | 6·0 | 4·6 | 12·0 | 9·1 | 18·0 | 13·7 |

| 46m | SUN PLANETS | ARIES | MOON | v or d | Corr<sup>n</sup> | v or d | Corr<sup>n</sup> | v or d | Corr<sup>n</sup> |
|---|---|---|---|---|---|---|---|---|---|
| s | ° ′ | ° ′ | ° ′ | ′ | ′ | ′ | ′ | ′ | ′ |
| 00 | 11 30·0 | 11 31·9 | 10 58·6 | 0·0 | 0·0 | 6·0 | 4·7 | 12·0 | 9·3 |
| 01 | 11 30·3 | 11 32·1 | 10 58·8 | 0·1 | 0·1 | 6·1 | 4·7 | 12·1 | 9·4 |
| 02 | 11 30·5 | 11 32·4 | 10 59·0 | 0·2 | 0·2 | 6·2 | 4·8 | 12·2 | 9·5 |
| 03 | 11 30·8 | 11 32·6 | 10 59·3 | 0·3 | 0·2 | 6·3 | 4·9 | 12·3 | 9·5 |
| 04 | 11 31·0 | 11 32·9 | 10 59·5 | 0·4 | 0·3 | 6·4 | 5·0 | 12·4 | 9·6 |
| 05 | 11 31·3 | 11 33·1 | 10 59·8 | 0·5 | 0·4 | 6·5 | 5·0 | 12·5 | 9·7 |
| 06 | 11 31·5 | 11 33·4 | 11 00·0 | 0·6 | 0·5 | 6·6 | 5·1 | 12·6 | 9·8 |
| 07 | 11 31·8 | 11 33·6 | 11 00·2 | 0·7 | 0·5 | 6·7 | 5·2 | 12·7 | 9·8 |
| 08 | 11 32·0 | 11 33·9 | 11 00·5 | 0·8 | 0·6 | 6·8 | 5·3 | 12·8 | 9·9 |
| 09 | 11 32·3 | 11 34·1 | 11 00·7 | 0·9 | 0·7 | 6·9 | 5·3 | 12·9 | 10·0 |
| 10 | 11 32·5 | 11 34·4 | 11 01·0 | 1·0 | 0·8 | 7·0 | 5·4 | 13·0 | 10·1 |
| 11 | 11 32·8 | 11 34·6 | 11 01·2 | 1·1 | 0·9 | 7·1 | 5·5 | 13·1 | 10·2 |
| 12 | 11 33·0 | 11 34·9 | 11 01·4 | 1·2 | 0·9 | 7·2 | 5·6 | 13·2 | 10·2 |
| 13 | 11 33·3 | 11 35·1 | 11 01·7 | 1·3 | 1·0 | 7·3 | 5·7 | 13·3 | 10·3 |
| 14 | 11 33·5 | 11 35·4 | 11 01·9 | 1·4 | 1·1 | 7·4 | 5·7 | 13·4 | 10·4 |
| 15 | 11 33·8 | 11 35·6 | 11 02·1 | 1·5 | 1·2 | 7·5 | 5·8 | 13·5 | 10·5 |
| 16 | 11 34·0 | 11 35·9 | 11 02·4 | 1·6 | 1·2 | 7·6 | 5·9 | 13·6 | 10·5 |
| 17 | 11 34·3 | 11 36·2 | 11 02·6 | 1·7 | 1·3 | 7·7 | 6·0 | 13·7 | 10·6 |
| 18 | 11 34·5 | 11 36·4 | 11 02·9 | 1·8 | 1·4 | 7·8 | 6·0 | 13·8 | 10·7 |
| 19 | 11 34·8 | 11 36·7 | 11 03·1 | 1·9 | 1·5 | 7·9 | 6·1 | 13·9 | 10·8 |
| 20 | 11 35·0 | 11 36·9 | 11 03·3 | 2·0 | 1·6 | 8·0 | 6·2 | 14·0 | 10·9 |
| 21 | 11 35·3 | 11 37·2 | 11 03·6 | 2·1 | 1·6 | 8·1 | 6·3 | 14·1 | 10·9 |
| 22 | 11 35·5 | 11 37·4 | 11 03·8 | 2·2 | 1·7 | 8·2 | 6·4 | 14·2 | 11·0 |
| 23 | 11 35·8 | 11 37·7 | 11 04·1 | 2·3 | 1·8 | 8·3 | 6·4 | 14·3 | 11·1 |
| 24 | 11 36·0 | 11 37·9 | 11 04·3 | 2·4 | 1·9 | 8·4 | 6·5 | 14·4 | 11·2 |
| 25 | 11 36·3 | 11 38·2 | 11 04·5 | 2·5 | 1·9 | 8·5 | 6·6 | 14·5 | 11·2 |
| 26 | 11 36·5 | 11 38·4 | 11 04·8 | 2·6 | 2·0 | 8·6 | 6·7 | 14·6 | 11·3 |
| 27 | 11 36·8 | 11 38·7 | 11 05·0 | 2·7 | 2·1 | 8·7 | 6·7 | 14·7 | 11·4 |
| 28 | 11 37·0 | 11 38·9 | 11 05·2 | 2·8 | 2·2 | 8·8 | 6·8 | 14·8 | 11·5 |
| 29 | 11 37·3 | 11 39·2 | 11 05·5 | 2·9 | 2·2 | 8·9 | 6·9 | 14·9 | 11·5 |
| 30 | 11 37·5 | 11 39·4 | 11 05·7 | 3·0 | 2·3 | 9·0 | 7·0 | 15·0 | 11·6 |
| 31 | 11 37·8 | 11 39·7 | 11 06·0 | 3·1 | 2·4 | 9·1 | 7·1 | 15·1 | 11·7 |
| 32 | 11 38·0 | 11 39·9 | 11 06·2 | 3·2 | 2·5 | 9·2 | 7·1 | 15·2 | 11·8 |
| 33 | 11 38·3 | 11 40·2 | 11 06·4 | 3·3 | 2·6 | 9·3 | 7·2 | 15·3 | 11·9 |
| 34 | 11 38·5 | 11 40·4 | 11 06·7 | 3·4 | 2·6 | 9·4 | 7·3 | 15·4 | 11·9 |
| 35 | 11 38·8 | 11 40·7 | 11 06·9 | 3·5 | 2·7 | 9·5 | 7·4 | 15·5 | 12·0 |
| 36 | 11 39·0 | 11 40·9 | 11 07·2 | 3·6 | 2·8 | 9·6 | 7·4 | 15·6 | 12·1 |
| 37 | 11 39·3 | 11 41·2 | 11 07·4 | 3·7 | 2·9 | 9·7 | 7·5 | 15·7 | 12·2 |
| 38 | 11 39·5 | 11 41·4 | 11 07·6 | 3·8 | 2·9 | 9·8 | 7·6 | 15·8 | 12·2 |
| 39 | 11 39·8 | 11 41·7 | 11 07·9 | 3·9 | 3·0 | 9·9 | 7·7 | 15·9 | 12·3 |
| 40 | 11 40·0 | 11 41·9 | 11 08·1 | 4·0 | 3·1 | 10·0 | 7·8 | 16·0 | 12·4 |
| 41 | 11 40·3 | 11 42·2 | 11 08·3 | 4·1 | 3·2 | 10·1 | 7·8 | 16·1 | 12·5 |
| 42 | 11 40·5 | 11 42·4 | 11 08·6 | 4·2 | 3·3 | 10·2 | 7·9 | 16·2 | 12·6 |
| 43 | 11 40·8 | 11 42·7 | 11 08·8 | 4·3 | 3·3 | 10·3 | 8·0 | 16·3 | 12·6 |
| 44 | 11 41·0 | 11 42·9 | 11 09·1 | 4·4 | 3·4 | 10·4 | 8·1 | 16·4 | 12·7 |
| 45 | 11 41·3 | 11 43·2 | 11 09·3 | 4·5 | 3·5 | 10·5 | 8·1 | 16·5 | 12·8 |
| 46 | 11 41·5 | 11 43·4 | 11 09·5 | 4·6 | 3·6 | 10·6 | 8·2 | 16·6 | 12·9 |
| 47 | 11 41·8 | 11 43·7 | 11 09·8 | 4·7 | 3·6 | 10·7 | 8·3 | 16·7 | 12·9 |
| 48 | 11 42·0 | 11 43·9 | 11 10·0 | 4·8 | 3·7 | 10·8 | 8·4 | 16·8 | 13·0 |
| 49 | 11 42·3 | 11 44·2 | 11 10·3 | 4·9 | 3·8 | 10·9 | 8·4 | 16·9 | 13·1 |
| 50 | 11 42·5 | 11 44·4 | 11 10·5 | 5·0 | 3·9 | 11·0 | 8·5 | 17·0 | 13·2 |
| 51 | 11 42·8 | 11 44·7 | 11 10·7 | 5·1 | 4·0 | 11·1 | 8·6 | 17·1 | 13·3 |
| 52 | 11 43·0 | 11 44·9 | 11 11·0 | 5·2 | 4·0 | 11·2 | 8·7 | 17·2 | 13·3 |
| 53 | 11 43·3 | 11 45·2 | 11 11·2 | 5·3 | 4·1 | 11·3 | 8·8 | 17·3 | 13·4 |
| 54 | 11 43·5 | 11 45·4 | 11 11·5 | 5·4 | 4·2 | 11·4 | 8·8 | 17·4 | 13·5 |
| 55 | 11 43·8 | 11 45·7 | 11 11·7 | 5·5 | 4·3 | 11·5 | 8·9 | 17·5 | 13·6 |
| 56 | 11 44·0 | 11 45·9 | 11 11·9 | 5·6 | 4·3 | 11·6 | 9·0 | 17·6 | 13·6 |
| 57 | 11 44·3 | 11 46·2 | 11 12·2 | 5·7 | 4·4 | 11·7 | 9·1 | 17·7 | 13·7 |
| 58 | 11 44·5 | 11 46·4 | 11 12·4 | 5·8 | 4·5 | 11·8 | 9·1 | 17·8 | 13·8 |
| 59 | 11 44·8 | 11 46·7 | 11 12·6 | 5·9 | 4·6 | 11·9 | 9·2 | 17·9 | 13·9 |
| 60 | 11 45·0 | 11 46·9 | 11 12·9 | 6·0 | 4·7 | 12·0 | 9·3 | 18·0 | 14·0 |

| 47m | SUN PLANETS | ARIES | MOON | v or d | Corr<sup>n</sup> | v or d | Corr<sup>n</sup> | v or d | Corr<sup>n</sup> |
|---|---|---|---|---|---|---|---|---|---|
| s | ° ′ | ° ′ | ° ′ | ′ | ′ | ′ | ′ | ′ | ′ |
| 00 | 11 45·0 | 11 46·9 | 11 12·9 | 0·0 | 0·0 | 6·0 | 4·8 | 12·0 | 9·5 |
| 01 | 11 45·3 | 11 47·2 | 11 13·1 | 0·1 | 0·1 | 6·1 | 4·8 | 12·1 | 9·6 |
| 02 | 11 45·5 | 11 47·4 | 11 13·4 | 0·2 | 0·2 | 6·2 | 4·9 | 12·2 | 9·7 |
| 03 | 11 45·8 | 11 47·7 | 11 13·6 | 0·3 | 0·2 | 6·3 | 5·0 | 12·3 | 9·7 |
| 04 | 11 46·0 | 11 47·9 | 11 13·8 | 0·4 | 0·3 | 6·4 | 5·1 | 12·4 | 9·8 |
| 05 | 11 46·3 | 11 48·2 | 11 14·1 | 0·5 | 0·4 | 6·5 | 5·1 | 12·5 | 9·9 |
| 06 | 11 46·5 | 11 48·4 | 11 14·3 | 0·6 | 0·5 | 6·6 | 5·2 | 12·6 | 10·0 |
| 07 | 11 46·8 | 11 48·7 | 11 14·6 | 0·7 | 0·6 | 6·7 | 5·3 | 12·7 | 10·1 |
| 08 | 11 47·0 | 11 48·9 | 11 14·8 | 0·8 | 0·6 | 6·8 | 5·4 | 12·8 | 10·1 |
| 09 | 11 47·3 | 11 49·2 | 11 15·0 | 0·9 | 0·7 | 6·9 | 5·5 | 12·9 | 10·2 |
| 10 | 11 47·5 | 11 49·4 | 11 15·3 | 1·0 | 0·8 | 7·0 | 5·5 | 13·0 | 10·3 |
| 11 | 11 47·8 | 11 49·7 | 11 15·5 | 1·1 | 0·9 | 7·1 | 5·6 | 13·1 | 10·4 |
| 12 | 11 48·0 | 11 49·9 | 11 15·7 | 1·2 | 1·0 | 7·2 | 5·7 | 13·2 | 10·5 |
| 13 | 11 48·3 | 11 50·2 | 11 16·0 | 1·3 | 1·0 | 7·3 | 5·8 | 13·3 | 10·5 |
| 14 | 11 48·5 | 11 50·4 | 11 16·2 | 1·4 | 1·1 | 7·4 | 5·9 | 13·4 | 10·6 |
| 15 | 11 48·8 | 11 50·7 | 11 16·5 | 1·5 | 1·2 | 7·5 | 5·9 | 13·5 | 10·7 |
| 16 | 11 49·0 | 11 50·9 | 11 16·7 | 1·6 | 1·3 | 7·6 | 6·0 | 13·6 | 10·8 |
| 17 | 11 49·3 | 11 51·2 | 11 16·9 | 1·7 | 1·3 | 7·7 | 6·1 | 13·7 | 10·8 |
| 18 | 11 49·5 | 11 51·4 | 11 17·2 | 1·8 | 1·4 | 7·8 | 6·2 | 13·8 | 10·9 |
| 19 | 11 49·8 | 11 51·7 | 11 17·4 | 1·9 | 1·5 | 7·9 | 6·3 | 13·9 | 11·0 |
| 20 | 11 50·0 | 11 51·9 | 11 17·7 | 2·0 | 1·6 | 8·0 | 6·3 | 14·0 | 11·1 |
| 21 | 11 50·3 | 11 52·2 | 11 17·9 | 2·1 | 1·7 | 8·1 | 6·4 | 14·1 | 11·2 |
| 22 | 11 50·5 | 11 52·4 | 11 18·1 | 2·2 | 1·7 | 8·2 | 6·5 | 14·2 | 11·2 |
| 23 | 11 50·8 | 11 52·7 | 11 18·4 | 2·3 | 1·8 | 8·3 | 6·6 | 14·3 | 11·3 |
| 24 | 11 51·0 | 11 52·9 | 11 18·6 | 2·4 | 1·9 | 8·4 | 6·7 | 14·4 | 11·4 |
| 25 | 11 51·3 | 11 53·2 | 11 18·8 | 2·5 | 2·0 | 8·5 | 6·7 | 14·5 | 11·5 |
| 26 | 11 51·5 | 11 53·4 | 11 19·1 | 2·6 | 2·1 | 8·6 | 6·8 | 14·6 | 11·6 |
| 27 | 11 51·8 | 11 53·7 | 11 19·3 | 2·7 | 2·1 | 8·7 | 6·9 | 14·7 | 11·6 |
| 28 | 11 52·0 | 11 53·9 | 11 19·6 | 2·8 | 2·2 | 8·8 | 7·0 | 14·8 | 11·7 |
| 29 | 11 52·3 | 11 54·2 | 11 19·8 | 2·9 | 2·3 | 8·9 | 7·0 | 14·9 | 11·8 |
| 30 | 11 52·5 | 11 54·5 | 11 20·0 | 3·0 | 2·4 | 9·0 | 7·1 | 15·0 | 11·9 |
| 31 | 11 52·8 | 11 54·7 | 11 20·3 | 3·1 | 2·5 | 9·1 | 7·2 | 15·1 | 12·0 |
| 32 | 11 53·0 | 11 55·0 | 11 20·5 | 3·2 | 2·5 | 9·2 | 7·3 | 15·2 | 12·0 |
| 33 | 11 53·3 | 11 55·2 | 11 20·8 | 3·3 | 2·6 | 9·3 | 7·4 | 15·3 | 12·1 |
| 34 | 11 53·5 | 11 55·5 | 11 21·0 | 3·4 | 2·7 | 9·4 | 7·4 | 15·4 | 12·2 |
| 35 | 11 53·8 | 11 55·7 | 11 21·2 | 3·5 | 2·8 | 9·5 | 7·5 | 15·5 | 12·3 |
| 36 | 11 54·0 | 11 56·0 | 11 21·5 | 3·6 | 2·9 | 9·6 | 7·6 | 15·6 | 12·4 |
| 37 | 11 54·3 | 11 56·2 | 11 21·7 | 3·7 | 2·9 | 9·7 | 7·7 | 15·7 | 12·4 |
| 38 | 11 54·5 | 11 56·5 | 11 22·0 | 3·8 | 3·0 | 9·8 | 7·8 | 15·8 | 12·5 |
| 39 | 11 54·8 | 11 56·7 | 11 22·2 | 3·9 | 3·1 | 9·9 | 7·8 | 15·9 | 12·6 |
| 40 | 11 55·0 | 11 57·0 | 11 22·4 | 4·0 | 3·2 | 10·0 | 7·9 | 16·0 | 12·7 |
| 41 | 11 55·3 | 11 57·2 | 11 22·7 | 4·1 | 3·2 | 10·1 | 8·0 | 16·1 | 12·7 |
| 42 | 11 55·5 | 11 57·5 | 11 22·9 | 4·2 | 3·3 | 10·2 | 8·1 | 16·2 | 12·8 |
| 43 | 11 55·8 | 11 57·7 | 11 23·1 | 4·3 | 3·4 | 10·3 | 8·2 | 16·3 | 12·9 |
| 44 | 11 56·0 | 11 58·0 | 11 23·4 | 4·4 | 3·5 | 10·4 | 8·2 | 16·4 | 13·0 |
| 45 | 11 56·3 | 11 58·2 | 11 23·6 | 4·5 | 3·6 | 10·5 | 8·3 | 16·5 | 13·1 |
| 46 | 11 56·5 | 11 58·5 | 11 23·9 | 4·6 | 3·6 | 10·6 | 8·4 | 16·6 | 13·1 |
| 47 | 11 56·8 | 11 58·7 | 11 24·1 | 4·7 | 3·7 | 10·7 | 8·5 | 16·7 | 13·2 |
| 48 | 11 57·0 | 11 59·0 | 11 24·3 | 4·8 | 3·8 | 10·8 | 8·6 | 16·8 | 13·3 |
| 49 | 11 57·3 | 11 59·2 | 11 24·6 | 4·9 | 3·9 | 10·9 | 8·6 | 16·9 | 13·4 |
| 50 | 11 57·5 | 11 59·5 | 11 24·8 | 5·0 | 4·0 | 11·0 | 8·7 | 17·0 | 13·5 |
| 51 | 11 57·8 | 11 59·7 | 11 25·1 | 5·1 | 4·0 | 11·1 | 8·8 | 17·1 | 13·5 |
| 52 | 11 58·0 | 12 00·0 | 11 25·3 | 5·2 | 4·1 | 11·2 | 8·9 | 17·2 | 13·6 |
| 53 | 11 58·3 | 12 00·2 | 11 25·5 | 5·3 | 4·2 | 11·3 | 8·9 | 17·3 | 13·7 |
| 54 | 11 58·5 | 12 00·5 | 11 25·8 | 5·4 | 4·3 | 11·4 | 9·0 | 17·4 | 13·8 |
| 55 | 11 58·8 | 12 00·7 | 11 26·0 | 5·5 | 4·4 | 11·5 | 9·1 | 17·5 | 13·9 |
| 56 | 11 59·0 | 12 01·0 | 11 26·2 | 5·6 | 4·4 | 11·6 | 9·2 | 17·6 | 13·9 |
| 57 | 11 59·3 | 12 01·2 | 11 26·5 | 5·7 | 4·5 | 11·7 | 9·3 | 17·7 | 14·0 |
| 58 | 11 59·5 | 12 01·5 | 11 26·7 | 5·8 | 4·6 | 11·8 | 9·3 | 17·8 | 14·1 |
| 59 | 11 59·8 | 12 01·7 | 11 27·0 | 5·9 | 4·7 | 11·9 | 9·4 | 17·9 | 14·2 |
| 60 | 12 00·0 | 12 02·0 | 11 27·2 | 6·0 | 4·8 | 12·0 | 9·5 | 18·0 | 14·3 |

| 48ᵐ | SUN PLANETS | ARIES | MOON | v or d | Corrⁿ | v or d | Corrⁿ | v or d | Corrⁿ |
|---|---|---|---|---|---|---|---|---|---|
| s | ° ′ | ° ′ | ° ′ | ′ | ′ | ′ | ′ | ′ | ′ |
| 00 | 12 00·0 | 12 02·0 | 11 27·2 | 0·0 | 0·0 | 6·0 | 4·9 | 12·0 | 9·7 |
| 01 | 12 00·3 | 12 02·2 | 11 27·4 | 0·1 | 0·1 | 6·1 | 4·9 | 12·1 | 9·8 |
| 02 | 12 00·5 | 12 02·5 | 11 27·7 | 0·2 | 0·2 | 6·2 | 5·0 | 12·2 | 9·9 |
| 03 | 12 00·8 | 12 02·7 | 11 27·9 | 0·3 | 0·2 | 6·3 | 5·1 | 12·3 | 9·9 |
| 04 | 12 01·0 | 12 03·0 | 11 28·2 | 0·4 | 0·3 | 6·4 | 5·2 | 12·4 | 10·0 |
| 05 | 12 01·3 | 12 03·2 | 11 28·4 | 0·5 | 0·4 | 6·5 | 5·3 | 12·5 | 10·1 |
| 06 | 12 01·5 | 12 03·5 | 11 28·6 | 0·6 | 0·5 | 6·6 | 5·3 | 12·6 | 10·2 |
| 07 | 12 01·8 | 12 03·7 | 11 28·9 | 0·7 | 0·6 | 6·7 | 5·4 | 12·7 | 10·3 |
| 08 | 12 02·0 | 12 04·0 | 11 29·1 | 0·8 | 0·6 | 6·8 | 5·5 | 12·8 | 10·3 |
| 09 | 12 02·3 | 12 04·2 | 11 29·3 | 0·9 | 0·7 | 6·9 | 5·6 | 12·9 | 10·4 |
| 10 | 12 02·5 | 12 04·5 | 11 29·6 | 1·0 | 0·8 | 7·0 | 5·7 | 13·0 | 10·5 |
| 11 | 12 02·8 | 12 04·7 | 11 29·8 | 1·1 | 0·9 | 7·1 | 5·7 | 13·1 | 10·6 |
| 12 | 12 03·0 | 12 05·0 | 11 30·1 | 1·2 | 1·0 | 7·2 | 5·8 | 13·2 | 10·7 |
| 13 | 12 03·3 | 12 05·2 | 11 30·3 | 1·3 | 1·1 | 7·3 | 5·9 | 13·3 | 10·8 |
| 14 | 12 03·5 | 12 05·5 | 11 30·5 | 1·4 | 1·1 | 7·4 | 6·0 | 13·4 | 10·8 |
| 15 | 12 03·8 | 12 05·7 | 11 30·8 | 1·5 | 1·2 | 7·5 | 6·1 | 13·5 | 10·9 |
| 16 | 12 04·0 | 12 06·0 | 11 31·0 | 1·6 | 1·3 | 7·6 | 6·1 | 13·6 | 11·0 |
| 17 | 12 04·3 | 12 06·2 | 11 31·3 | 1·7 | 1·4 | 7·7 | 6·2 | 13·7 | 11·1 |
| 18 | 12 04·5 | 12 06·5 | 11 31·5 | 1·8 | 1·5 | 7·8 | 6·3 | 13·8 | 11·2 |
| 19 | 12 04·8 | 12 06·7 | 11 31·7 | 1·9 | 1·5 | 7·9 | 6·4 | 13·9 | 11·2 |
| 20 | 12 05·0 | 12 07·0 | 11 32·0 | 2·0 | 1·6 | 8·0 | 6·5 | 14·0 | 11·3 |
| 21 | 12 05·3 | 12 07·2 | 11 32·2 | 2·1 | 1·7 | 8·1 | 6·5 | 14·1 | 11·4 |
| 22 | 12 05·5 | 12 07·5 | 11 32·4 | 2·2 | 1·8 | 8·2 | 6·6 | 14·2 | 11·5 |
| 23 | 12 05·8 | 12 07·7 | 11 32·7 | 2·3 | 1·9 | 8·3 | 6·7 | 14·3 | 11·6 |
| 24 | 12 06·0 | 12 08·0 | 11 32·9 | 2·4 | 1·9 | 8·4 | 6·8 | 14·4 | 11·6 |
| 25 | 12 06·3 | 12 08·2 | 11 33·2 | 2·5 | 2·0 | 8·5 | 6·9 | 14·5 | 11·7 |
| 26 | 12 06·5 | 12 08·5 | 11 33·4 | 2·6 | 2·1 | 8·6 | 7·0 | 14·6 | 11·8 |
| 27 | 12 06·8 | 12 08·7 | 11 33·6 | 2·7 | 2·2 | 8·7 | 7·0 | 14·7 | 11·9 |
| 28 | 12 07·0 | 12 09·0 | 11 33·9 | 2·8 | 2·3 | 8·8 | 7·1 | 14·8 | 12·0 |
| 29 | 12 07·3 | 12 09·2 | 11 34·1 | 2·9 | 2·3 | 8·9 | 7·2 | 14·9 | 12·0 |
| 30 | 12 07·5 | 12 09·5 | 11 34·4 | 3·0 | 2·4 | 9·0 | 7·3 | 15·0 | 12·1 |
| 31 | 12 07·8 | 12 09·7 | 11 34·6 | 3·1 | 2·5 | 9·1 | 7·4 | 15·1 | 12·2 |
| 32 | 12 08·0 | 12 10·0 | 11 34·8 | 3·2 | 2·6 | 9·2 | 7·4 | 15·2 | 12·3 |
| 33 | 12 08·3 | 12 10·2 | 11 35·1 | 3·3 | 2·7 | 9·3 | 7·5 | 15·3 | 12·4 |
| 34 | 12 08·5 | 12 10·5 | 11 35·3 | 3·4 | 2·7 | 9·4 | 7·6 | 15·4 | 12·4 |
| 35 | 12 08·8 | 12 10·7 | 11 35·6 | 3·5 | 2·8 | 9·5 | 7·7 | 15·5 | 12·5 |
| 36 | 12 09·0 | 12 11·0 | 11 35·8 | 3·6 | 2·9 | 9·6 | 7·8 | 15·6 | 12·6 |
| 37 | 12 09·3 | 12 11·2 | 11 36·0 | 3·7 | 3·0 | 9·7 | 7·8 | 15·7 | 12·7 |
| 38 | 12 09·5 | 12 11·5 | 11 36·3 | 3·8 | 3·1 | 9·8 | 7·9 | 15·8 | 12·8 |
| 39 | 12 09·8 | 12 11·7 | 11 36·5 | 3·9 | 3·2 | 9·9 | 8·0 | 15·9 | 12·9 |
| 40 | 12 10·0 | 12 12·0 | 11 36·7 | 4·0 | 3·2 | 10·0 | 8·1 | 16·0 | 12·9 |
| 41 | 12 10·3 | 12 12·2 | 11 37·0 | 4·1 | 3·3 | 10·1 | 8·2 | 16·1 | 13·0 |
| 42 | 12 10·5 | 12 12·5 | 11 37·2 | 4·2 | 3·4 | 10·2 | 8·2 | 16·2 | 13·1 |
| 43 | 12 10·8 | 12 12·8 | 11 37·5 | 4·3 | 3·5 | 10·3 | 8·3 | 16·3 | 13·2 |
| 44 | 12 11·0 | 12 13·0 | 11 37·7 | 4·4 | 3·6 | 10·4 | 8·4 | 16·4 | 13·3 |
| 45 | 12 11·3 | 12 13·3 | 11 37·9 | 4·5 | 3·6 | 10·5 | 8·5 | 16·5 | 13·3 |
| 46 | 12 11·5 | 12 13·5 | 11 38·2 | 4·6 | 3·7 | 10·6 | 8·6 | 16·6 | 13·4 |
| 47 | 12 11·8 | 12 13·8 | 11 38·4 | 4·7 | 3·8 | 10·7 | 8·6 | 16·7 | 13·5 |
| 48 | 12 12·0 | 12 14·0 | 11 38·7 | 4·8 | 3·9 | 10·8 | 8·7 | 16·8 | 13·6 |
| 49 | 12 12·3 | 12 14·3 | 11 38·9 | 4·9 | 4·0 | 10·9 | 8·8 | 16·9 | 13·7 |
| 50 | 12 12·5 | 12 14·5 | 11 39·1 | 5·0 | 4·0 | 11·0 | 8·9 | 17·0 | 13·7 |
| 51 | 12 12·8 | 12 14·8 | 11 39·4 | 5·1 | 4·1 | 11·1 | 9·0 | 17·1 | 13·8 |
| 52 | 12 13·0 | 12 15·0 | 11 39·6 | 5·2 | 4·2 | 11·2 | 9·1 | 17·2 | 13·9 |
| 53 | 12 13·3 | 12 15·3 | 11 39·8 | 5·3 | 4·3 | 11·3 | 9·1 | 17·3 | 14·0 |
| 54 | 12 13·5 | 12 15·5 | 11 40·1 | 5·4 | 4·4 | 11·4 | 9·2 | 17·4 | 14·1 |
| 55 | 12 13·8 | 12 15·8 | 11 40·3 | 5·5 | 4·4 | 11·5 | 9·3 | 17·5 | 14·1 |
| 56 | 12 14·0 | 12 16·0 | 11 40·6 | 5·6 | 4·5 | 11·6 | 9·4 | 17·6 | 14·2 |
| 57 | 12 14·3 | 12 16·3 | 11 40·8 | 5·7 | 4·6 | 11·7 | 9·5 | 17·7 | 14·3 |
| 58 | 12 14·5 | 12 16·5 | 11 41·0 | 5·8 | 4·7 | 11·8 | 9·5 | 17·8 | 14·4 |
| 59 | 12 14·8 | 12 16·8 | 11 41·3 | 5·9 | 4·8 | 11·9 | 9·6 | 17·9 | 14·5 |
| 60 | 12 15·0 | 12 17·0 | 11 41·5 | 6·0 | 4·9 | 12·0 | 9·7 | 18·0 | 14·6 |

| 49ᵐ | SUN PLANETS | ARIES | MOON | v or d | Corrⁿ | v or d | Corrⁿ | v or d | Corrⁿ |
|---|---|---|---|---|---|---|---|---|---|
| s | ° ′ | ° ′ | ° ′ | ′ | ′ | ′ | ′ | ′ | ′ |
| 00 | 12 15·0 | 12 17·0 | 11 41·5 | 0·0 | 0·0 | 6·0 | 5·0 | 12·0 | 9·9 |
| 01 | 12 15·3 | 12 17·3 | 11 41·8 | 0·1 | 0·1 | 6·1 | 5·0 | 12·1 | 10·0 |
| 02 | 12 15·5 | 12 17·5 | 11 42·0 | 0·2 | 0·2 | 6·2 | 5·1 | 12·2 | 10·1 |
| 03 | 12 15·8 | 12 17·8 | 11 42·2 | 0·3 | 0·2 | 6·3 | 5·2 | 12·3 | 10·1 |
| 04 | 12 16·0 | 12 18·0 | 11 42·5 | 0·4 | 0·3 | 6·4 | 5·3 | 12·4 | 10·2 |
| 05 | 12 16·3 | 12 18·3 | 11 42·7 | 0·5 | 0·4 | 6·5 | 5·4 | 12·5 | 10·3 |
| 06 | 12 16·5 | 12 18·5 | 11 42·9 | 0·6 | 0·5 | 6·6 | 5·4 | 12·6 | 10·4 |
| 07 | 12 16·8 | 12 18·8 | 11 43·2 | 0·7 | 0·6 | 6·7 | 5·5 | 12·7 | 10·5 |
| 08 | 12 17·0 | 12 19·0 | 11 43·4 | 0·8 | 0·7 | 6·8 | 5·6 | 12·8 | 10·6 |
| 09 | 12 17·3 | 12 19·3 | 11 43·7 | 0·9 | 0·7 | 6·9 | 5·7 | 12·9 | 10·6 |
| 10 | 12 17·5 | 12 19·5 | 11 43·9 | 1·0 | 0·8 | 7·0 | 5·8 | 13·0 | 10·7 |
| 11 | 12 17·8 | 12 19·8 | 11 44·1 | 1·1 | 0·9 | 7·1 | 5·9 | 13·1 | 10·8 |
| 12 | 12 18·0 | 12 20·0 | 11 44·4 | 1·2 | 1·0 | 7·2 | 5·9 | 13·2 | 10·9 |
| 13 | 12 18·3 | 12 20·3 | 11 44·6 | 1·3 | 1·1 | 7·3 | 6·0 | 13·3 | 11·0 |
| 14 | 12 18·5 | 12 20·5 | 11 44·9 | 1·4 | 1·2 | 7·4 | 6·1 | 13·4 | 11·1 |
| 15 | 12 18·8 | 12 20·8 | 11 45·1 | 1·5 | 1·2 | 7·5 | 6·2 | 13·5 | 11·1 |
| 16 | 12 19·0 | 12 21·0 | 11 45·3 | 1·6 | 1·3 | 7·6 | 6·3 | 13·6 | 11·2 |
| 17 | 12 19·3 | 12 21·3 | 11 45·6 | 1·7 | 1·4 | 7·7 | 6·4 | 13·7 | 11·3 |
| 18 | 12 19·5 | 12 21·5 | 11 45·8 | 1·8 | 1·5 | 7·8 | 6·4 | 13·8 | 11·4 |
| 19 | 12 19·8 | 12 21·8 | 11 46·1 | 1·9 | 1·6 | 7·9 | 6·5 | 13·9 | 11·5 |
| 20 | 12 20·0 | 12 22·0 | 11 46·3 | 2·0 | 1·7 | 8·0 | 6·6 | 14·0 | 11·6 |
| 21 | 12 20·3 | 12 22·3 | 11 46·5 | 2·1 | 1·7 | 8·1 | 6·7 | 14·1 | 11·6 |
| 22 | 12 20·5 | 12 22·5 | 11 46·8 | 2·2 | 1·8 | 8·2 | 6·8 | 14·2 | 11·7 |
| 23 | 12 20·8 | 12 22·8 | 11 47·0 | 2·3 | 1·9 | 8·3 | 6·8 | 14·3 | 11·8 |
| 24 | 12 21·0 | 12 23·0 | 11 47·2 | 2·4 | 2·0 | 8·4 | 6·9 | 14·4 | 11·9 |
| 25 | 12 21·3 | 12 23·3 | 11 47·5 | 2·5 | 2·1 | 8·5 | 7·0 | 14·5 | 12·0 |
| 26 | 12 21·5 | 12 23·5 | 11 47·7 | 2·6 | 2·1 | 8·6 | 7·1 | 14·6 | 12·0 |
| 27 | 12 21·8 | 12 23·8 | 11 48·0 | 2·7 | 2·2 | 8·7 | 7·2 | 14·7 | 12·1 |
| 28 | 12 22·0 | 12 24·0 | 11 48·2 | 2·8 | 2·3 | 8·8 | 7·3 | 14·8 | 12·2 |
| 29 | 12 22·3 | 12 24·3 | 11 48·4 | 2·9 | 2·4 | 8·9 | 7·3 | 14·9 | 12·3 |
| 30 | 12 22·5 | 12 24·5 | 11 48·7 | 3·0 | 2·5 | 9·0 | 7·4 | 15·0 | 12·4 |
| 31 | 12 22·8 | 12 24·8 | 11 48·9 | 3·1 | 2·6 | 9·1 | 7·5 | 15·1 | 12·5 |
| 32 | 12 23·0 | 12 25·0 | 11 49·2 | 3·2 | 2·6 | 9·2 | 7·6 | 15·2 | 12·5 |
| 33 | 12 23·3 | 12 25·3 | 11 49·4 | 3·3 | 2·7 | 9·3 | 7·7 | 15·3 | 12·6 |
| 34 | 12 23·5 | 12 25·5 | 11 49·6 | 3·4 | 2·8 | 9·4 | 7·8 | 15·4 | 12·7 |
| 35 | 12 23·8 | 12 25·8 | 11 49·9 | 3·5 | 2·9 | 9·5 | 7·8 | 15·5 | 12·8 |
| 36 | 12 24·0 | 12 26·0 | 11 50·1 | 3·6 | 3·0 | 9·6 | 7·9 | 15·6 | 12·9 |
| 37 | 12 24·3 | 12 26·3 | 11 50·3 | 3·7 | 3·1 | 9·7 | 8·0 | 15·7 | 13·0 |
| 38 | 12 24·5 | 12 26·5 | 11 50·6 | 3·8 | 3·1 | 9·8 | 8·1 | 15·8 | 13·0 |
| 39 | 12 24·8 | 12 26·8 | 11 50·8 | 3·9 | 3·2 | 9·9 | 8·2 | 15·9 | 13·1 |
| 40 | 12 25·0 | 12 27·0 | 11 51·1 | 4·0 | 3·3 | 10·0 | 8·3 | 16·0 | 13·2 |
| 41 | 12 25·3 | 12 27·3 | 11 51·3 | 4·1 | 3·4 | 10·1 | 8·3 | 16·1 | 13·3 |
| 42 | 12 25·5 | 12 27·5 | 11 51·5 | 4·2 | 3·5 | 10·2 | 8·4 | 16·2 | 13·4 |
| 43 | 12 25·8 | 12 27·8 | 11 51·8 | 4·3 | 3·5 | 10·3 | 8·5 | 16·3 | 13·4 |
| 44 | 12 26·0 | 12 28·0 | 11 52·0 | 4·4 | 3·6 | 10·4 | 8·6 | 16·4 | 13·5 |
| 45 | 12 26·3 | 12 28·3 | 11 52·3 | 4·5 | 3·7 | 10·5 | 8·7 | 16·5 | 13·6 |
| 46 | 12 26·5 | 12 28·5 | 11 52·5 | 4·6 | 3·8 | 10·6 | 8·7 | 16·6 | 13·7 |
| 47 | 12 26·8 | 12 28·8 | 11 52·7 | 4·7 | 3·9 | 10·7 | 8·8 | 16·7 | 13·8 |
| 48 | 12 27·0 | 12 29·0 | 11 53·0 | 4·8 | 4·0 | 10·8 | 8·9 | 16·8 | 13·9 |
| 49 | 12 27·3 | 12 29·3 | 11 53·2 | 4·9 | 4·0 | 10·9 | 9·0 | 16·9 | 13·9 |
| 50 | 12 27·5 | 12 29·5 | 11 53·4 | 5·0 | 4·1 | 11·0 | 9·1 | 17·0 | 14·0 |
| 51 | 12 27·8 | 12 29·8 | 11 53·7 | 5·1 | 4·2 | 11·1 | 9·2 | 17·1 | 14·1 |
| 52 | 12 28·0 | 12 30·0 | 11 53·9 | 5·2 | 4·3 | 11·2 | 9·2 | 17·2 | 14·2 |
| 53 | 12 28·3 | 12 30·3 | 11 54·2 | 5·3 | 4·4 | 11·3 | 9·3 | 17·3 | 14·3 |
| 54 | 12 28·5 | 12 30·5 | 11 54·4 | 5·4 | 4·5 | 11·4 | 9·4 | 17·4 | 14·4 |
| 55 | 12 28·8 | 12 30·8 | 11 54·6 | 5·5 | 4·5 | 11·5 | 9·5 | 17·5 | 14·4 |
| 56 | 12 29·0 | 12 31·1 | 11 54·9 | 5·6 | 4·6 | 11·6 | 9·6 | 17·6 | 14·5 |
| 57 | 12 29·3 | 12 31·3 | 11 55·1 | 5·7 | 4·7 | 11·7 | 9·7 | 17·7 | 14·6 |
| 58 | 12 29·5 | 12 31·6 | 11 55·4 | 5·8 | 4·8 | 11·8 | 9·7 | 17·8 | 14·7 |
| 59 | 12 29·8 | 12 31·8 | 11 55·6 | 5·9 | 4·9 | 11·9 | 9·8 | 17·9 | 14·8 |
| 60 | 12 30·0 | 12 32·1 | 11 55·8 | 6·0 | 5·0 | 12·0 | 9·9 | 18·0 | 14·9 |

| $56^m$ | SUN PLANETS | ARIES | MOON | v or d | Corr^n | v or d | Corr^n | v or d | Corr^n |
|---|---|---|---|---|---|---|---|---|---|
| s | ° ′ | ° ′ | ° ′ | ′ | ′ | ′ | ′ | ′ | ′ |
| 00 | 14 00·0 | 14 02·3 | 13 21·7 | 0·0 | 0·0 | 6·0 | 5·7 | 12·0 | 11·3 |
| 01 | 14 00·3 | 14 02·6 | 13 22·0 | 0·1 | 0·1 | 6·1 | 5·7 | 12·1 | 11·4 |
| 02 | 14 00·5 | 14 02·8 | 13 22·2 | 0·2 | 0·2 | 6·2 | 5·8 | 12·2 | 11·5 |
| 03 | 14 00·8 | 14 03·1 | 13 22·4 | 0·3 | 0·3 | 6·3 | 5·9 | 12·3 | 11·6 |
| 04 | 14 01·0 | 14 03·3 | 13 22·7 | 0·4 | 0·4 | 6·4 | 6·0 | 12·4 | 11·7 |
| 05 | 14 01·3 | 14 03·6 | 13 22·9 | 0·5 | 0·5 | 6·5 | 6·1 | 12·5 | 11·8 |
| 06 | 14 01·5 | 14 03·8 | 13 23·2 | 0·6 | 0·6 | 6·6 | 6·2 | 12·6 | 11·9 |
| 07 | 14 01·8 | 14 04·1 | 13 23·4 | 0·7 | 0·7 | 6·7 | 6·3 | 12·7 | 12·0 |
| 08 | 14 02·0 | 14 04·3 | 13 23·6 | 0·8 | 0·8 | 6·8 | 6·4 | 12·8 | 12·1 |
| 09 | 14 02·3 | 14 04·6 | 13 23·9 | 0·9 | 0·8 | 6·9 | 6·5 | 12·9 | 12·1 |
| 10 | 14 02·5 | 14 04·8 | 13 24·1 | 1·0 | 0·9 | 7·0 | 6·6 | 13·0 | 12·2 |
| 11 | 14 02·8 | 14 05·1 | 13 24·4 | 1·1 | 1·0 | 7·1 | 6·7 | 13·1 | 12·3 |
| 12 | 14 03·0 | 14 05·3 | 13 24·6 | 1·2 | 1·1 | 7·2 | 6·8 | 13·2 | 12·4 |
| 13 | 14 03·3 | 14 05·6 | 13 24·8 | 1·3 | 1·2 | 7·3 | 6·9 | 13·3 | 12·5 |
| 14 | 14 03·5 | 14 05·8 | 13 25·1 | 1·4 | 1·3 | 7·4 | 7·0 | 13·4 | 12·6 |
| 15 | 14 03·8 | 14 06·1 | 13 25·3 | 1·5 | 1·4 | 7·5 | 7·1 | 13·5 | 12·7 |
| 16 | 14 04·0 | 14 06·3 | 13 25·6 | 1·6 | 1·5 | 7·6 | 7·2 | 13·6 | 12·8 |
| 17 | 14 04·3 | 14 06·6 | 13 25·8 | 1·7 | 1·6 | 7·7 | 7·3 | 13·7 | 12·9 |
| 18 | 14 04·5 | 14 06·8 | 13 26·0 | 1·8 | 1·7 | 7·8 | 7·3 | 13·8 | 13·0 |
| 19 | 14 04·8 | 14 07·1 | 13 26·3 | 1·9 | 1·8 | 7·9 | 7·4 | 13·9 | 13·1 |
| 20 | 14 05·0 | 14 07·3 | 13 26·5 | 2·0 | 1·9 | 8·0 | 7·5 | 14·0 | 13·2 |
| 21 | 14 05·3 | 14 07·6 | 13 26·7 | 2·1 | 2·0 | 8·1 | 7·6 | 14·1 | 13·3 |
| 22 | 14 05·5 | 14 07·8 | 13 27·0 | 2·2 | 2·1 | 8·2 | 7·7 | 14·2 | 13·4 |
| 23 | 14 05·8 | 14 08·1 | 13 27·2 | 2·3 | 2·2 | 8·3 | 7·8 | 14·3 | 13·5 |
| 24 | 14 06·0 | 14 08·3 | 13 27·5 | 2·4 | 2·3 | 8·4 | 7·9 | 14·4 | 13·6 |
| 25 | 14 06·3 | 14 08·6 | 13 27·7 | 2·5 | 2·4 | 8·5 | 8·0 | 14·5 | 13·7 |
| 26 | 14 06·5 | 14 08·8 | 13 27·9 | 2·6 | 2·4 | 8·6 | 8·1 | 14·6 | 13·7 |
| 27 | 14 06·8 | 14 09·1 | 13 28·2 | 2·7 | 2·5 | 8·7 | 8·2 | 14·7 | 13·8 |
| 28 | 14 07·0 | 14 09·3 | 13 28·4 | 2·8 | 2·6 | 8·8 | 8·3 | 14·8 | 13·9 |
| 29 | 14 07·3 | 14 09·6 | 13 28·7 | 2·9 | 2·7 | 8·9 | 8·4 | 14·9 | 14·0 |
| 30 | 14 07·5 | 14 09·8 | 13 28·9 | 3·0 | 2·8 | 9·0 | 8·5 | 15·0 | 14·1 |
| 31 | 14 07·8 | 14 10·1 | 13 29·1 | 3·1 | 2·9 | 9·1 | 8·6 | 15·1 | 14·2 |
| 32 | 14 08·0 | 14 10·3 | 13 29·4 | 3·2 | 3·0 | 9·2 | 8·7 | 15·2 | 14·3 |
| 33 | 14 08·3 | 14 10·6 | 13 29·6 | 3·3 | 3·1 | 9·3 | 8·8 | 15·3 | 14·4 |
| 34 | 14 08·5 | 14 10·8 | 13 29·8 | 3·4 | 3·2 | 9·4 | 8·9 | 15·4 | 14·5 |
| 35 | 14 08·8 | 14 11·1 | 13 30·1 | 3·5 | 3·3 | 9·5 | 8·9 | 15·5 | 14·6 |
| 36 | 14 09·0 | 14 11·3 | 13 30·3 | 3·6 | 3·4 | 9·6 | 9·0 | 15·6 | 14·7 |
| 37 | 14 09·3 | 14 11·6 | 13 30·6 | 3·7 | 3·5 | 9·7 | 9·1 | 15·7 | 14·8 |
| 38 | 14 09·5 | 14 11·8 | 13 30·8 | 3·8 | 3·6 | 9·8 | 9·2 | 15·8 | 14·9 |
| 39 | 14 09·8 | 14 12·1 | 13 31·0 | 3·9 | 3·7 | 9·9 | 9·3 | 15·9 | 15·0 |
| 40 | 14 10·0 | 14 12·3 | 13 31·3 | 4·0 | 3·8 | 10·0 | 9·4 | 16·0 | 15·1 |
| 41 | 14 10·3 | 14 12·6 | 13 31·5 | 4·1 | 3·9 | 10·1 | 9·5 | 16·1 | 15·2 |
| 42 | 14 10·5 | 14 12·8 | 13 31·8 | 4·2 | 4·0 | 10·2 | 9·6 | 16·2 | 15·3 |
| 43 | 14 10·8 | 14 13·1 | 13 32·0 | 4·3 | 4·0 | 10·3 | 9·7 | 16·3 | 15·3 |
| 44 | 14 11·0 | 14 13·3 | 13 32·2 | 4·4 | 4·1 | 10·4 | 9·8 | 16·4 | 15·4 |
| 45 | 14 11·3 | 14 13·6 | 13 32·5 | 4·5 | 4·2 | 10·5 | 9·9 | 16·5 | 15·5 |
| 46 | 14 11·5 | 14 13·8 | 13 32·7 | 4·6 | 4·3 | 10·6 | 10·0 | 16·6 | 15·6 |
| 47 | 14 11·8 | 14 14·1 | 13 32·9 | 4·7 | 4·4 | 10·7 | 10·1 | 16·7 | 15·7 |
| 48 | 14 12·0 | 14 14·3 | 13 33·2 | 4·8 | 4·5 | 10·8 | 10·2 | 16·8 | 15·8 |
| 49 | 14 12·3 | 14 14·6 | 13 33·4 | 4·9 | 4·6 | 10·9 | 10·3 | 16·9 | 15·9 |
| 50 | 14 12·5 | 14 14·8 | 13 33·7 | 5·0 | 4·7 | 11·0 | 10·4 | 17·0 | 16·0 |
| 51 | 14 12·8 | 14 15·1 | 13 33·9 | 5·1 | 4·8 | 11·1 | 10·5 | 17·1 | 16·1 |
| 52 | 14 13·0 | 14 15·3 | 13 34·1 | 5·2 | 4·9 | 11·2 | 10·5 | 17·2 | 16·2 |
| 53 | 14 13·3 | 14 15·6 | 13 34·4 | 5·3 | 5·0 | 11·3 | 10·6 | 17·3 | 16·3 |
| 54 | 14 13·5 | 14 15·8 | 13 34·6 | 5·4 | 5·1 | 11·4 | 10·7 | 17·4 | 16·4 |
| 55 | 14 13·8 | 14 16·1 | 13 34·9 | 5·5 | 5·2 | 11·5 | 10·8 | 17·5 | 16·5 |
| 56 | 14 14·0 | 14 16·3 | 13 35·1 | 5·6 | 5·3 | 11·6 | 10·9 | 17·6 | 16·6 |
| 57 | 14 14·3 | 14 16·6 | 13 35·3 | 5·7 | 5·4 | 11·7 | 11·0 | 17·7 | 16·7 |
| 58 | 14 14·5 | 14 16·8 | 13 35·6 | 5·8 | 5·5 | 11·8 | 11·1 | 17·8 | 16·8 |
| 59 | 14 14·8 | 14 17·1 | 13 35·8 | 5·9 | 5·6 | 11·9 | 11·2 | 17·9 | 16·9 |
| 60 | 14 15·0 | 14 17·3 | 13 36·1 | 6·0 | 5·7 | 12·0 | 11·3 | 18·0 | 17·0 |

| $57^m$ | SUN PLANETS | ARIES | MOON | v or d | Corr^n | v or d | Corr^n | v or d | Corr^n |
|---|---|---|---|---|---|---|---|---|---|
| s | ° ′ | ° ′ | ° ′ | ′ | ′ | ′ | ′ | ′ | ′ |
| 00 | 14 15·0 | 14 17·3 | 13 36·1 | 0·0 | 0·0 | 6·0 | 5·8 | 12·0 | 11·5 |
| 01 | 14 15·3 | 14 17·6 | 13 36·3 | 0·1 | 0·1 | 6·1 | 5·8 | 12·1 | 11·6 |
| 02 | 14 15·5 | 14 17·8 | 13 36·5 | 0·2 | 0·2 | 6·2 | 5·9 | 12·2 | 11·7 |
| 03 | 14 15·8 | 14 18·1 | 13 36·8 | 0·3 | 0·3 | 6·3 | 6·0 | 12·3 | 11·8 |
| 04 | 14 16·0 | 14 18·3 | 13 37·0 | 0·4 | 0·4 | 6·4 | 6·1 | 12·4 | 11·9 |
| 05 | 14 16·3 | 14 18·6 | 13 37·2 | 0·5 | 0·5 | 6·5 | 6·2 | 12·5 | 12·0 |
| 06 | 14 16·5 | 14 18·8 | 13 37·5 | 0·6 | 0·6 | 6·6 | 6·3 | 12·6 | 12·1 |
| 07 | 14 16·8 | 14 19·1 | 13 37·7 | 0·7 | 0·7 | 6·7 | 6·4 | 12·7 | 12·2 |
| 08 | 14 17·0 | 14 19·3 | 13 38·0 | 0·8 | 0·8 | 6·8 | 6·5 | 12·8 | 12·3 |
| 09 | 14 17·3 | 14 19·6 | 13 38·2 | 0·9 | 0·9 | 6·9 | 6·6 | 12·9 | 12·4 |
| 10 | 14 17·5 | 14 19·8 | 13 38·4 | 1·0 | 1·0 | 7·0 | 6·7 | 13·0 | 12·5 |
| 11 | 14 17·8 | 14 20·1 | 13 38·7 | 1·1 | 1·1 | 7·1 | 6·8 | 13·1 | 12·6 |
| 12 | 14 18·0 | 14 20·3 | 13 38·9 | 1·2 | 1·2 | 7·2 | 6·9 | 13·2 | 12·7 |
| 13 | 14 18·3 | 14 20·6 | 13 39·2 | 1·3 | 1·2 | 7·3 | 7·0 | 13·3 | 12·7 |
| 14 | 14 18·5 | 14 20·9 | 13 39·4 | 1·4 | 1·3 | 7·4 | 7·1 | 13·4 | 12·8 |
| 15 | 14 18·8 | 14 21·1 | 13 39·6 | 1·5 | 1·4 | 7·5 | 7·2 | 13·5 | 12·9 |
| 16 | 14 19·0 | 14 21·4 | 13 39·9 | 1·6 | 1·5 | 7·6 | 7·3 | 13·6 | 13·0 |
| 17 | 14 19·3 | 14 21·6 | 13 40·1 | 1·7 | 1·6 | 7·7 | 7·4 | 13·7 | 13·1 |
| 18 | 14 19·5 | 14 21·9 | 13 40·3 | 1·8 | 1·7 | 7·8 | 7·5 | 13·8 | 13·2 |
| 19 | 14 19·8 | 14 22·1 | 13 40·6 | 1·9 | 1·8 | 7·9 | 7·6 | 13·9 | 13·3 |
| 20 | 14 20·0 | 14 22·4 | 13 40·8 | 2·0 | 1·9 | 8·0 | 7·7 | 14·0 | 13·4 |
| 21 | 14 20·3 | 14 22·6 | 13 41·1 | 2·1 | 2·0 | 8·1 | 7·8 | 14·1 | 13·5 |
| 22 | 14 20·5 | 14 22·9 | 13 41·3 | 2·2 | 2·1 | 8·2 | 7·9 | 14·2 | 13·6 |
| 23 | 14 20·8 | 14 23·1 | 13 41·5 | 2·3 | 2·2 | 8·3 | 8·0 | 14·3 | 13·7 |
| 24 | 14 21·0 | 14 23·4 | 13 41·8 | 2·4 | 2·3 | 8·4 | 8·1 | 14·4 | 13·8 |
| 25 | 14 21·3 | 14 23·6 | 13 42·0 | 2·5 | 2·4 | 8·5 | 8·1 | 14·5 | 13·9 |
| 26 | 14 21·5 | 14 23·9 | 13 42·3 | 2·6 | 2·5 | 8·6 | 8·2 | 14·6 | 14·0 |
| 27 | 14 21·8 | 14 24·1 | 13 42·5 | 2·7 | 2·6 | 8·7 | 8·3 | 14·7 | 14·1 |
| 28 | 14 22·0 | 14 24·4 | 13 42·7 | 2·8 | 2·7 | 8·8 | 8·4 | 14·8 | 14·2 |
| 29 | 14 22·3 | 14 24·6 | 13 43·0 | 2·9 | 2·8 | 8·9 | 8·5 | 14·9 | 14·3 |
| 30 | 14 22·5 | 14 24·9 | 13 43·2 | 3·0 | 2·9 | 9·0 | 8·6 | 15·0 | 14·4 |
| 31 | 14 22·8 | 14 25·1 | 13 43·4 | 3·1 | 3·0 | 9·1 | 8·7 | 15·1 | 14·5 |
| 32 | 14 23·0 | 14 25·4 | 13 43·7 | 3·2 | 3·1 | 9·2 | 8·8 | 15·2 | 14·6 |
| 33 | 14 23·3 | 14 25·6 | 13 43·9 | 3·3 | 3·2 | 9·3 | 8·9 | 15·3 | 14·7 |
| 34 | 14 23·5 | 14 25·9 | 13 44·2 | 3·4 | 3·3 | 9·4 | 9·0 | 15·4 | 14·8 |
| 35 | 14 23·8 | 14 26·1 | 13 44·4 | 3·5 | 3·4 | 9·5 | 9·1 | 15·5 | 14·9 |
| 36 | 14 24·0 | 14 26·4 | 13 44·6 | 3·6 | 3·5 | 9·6 | 9·2 | 15·6 | 15·0 |
| 37 | 14 24·3 | 14 26·6 | 13 44·9 | 3·7 | 3·5 | 9·7 | 9·3 | 15·7 | 15·0 |
| 38 | 14 24·5 | 14 26·9 | 13 45·1 | 3·8 | 3·6 | 9·8 | 9·4 | 15·8 | 15·1 |
| 39 | 14 24·8 | 14 27·1 | 13 45·4 | 3·9 | 3·7 | 9·9 | 9·5 | 15·9 | 15·2 |
| 40 | 14 25·0 | 14 27·4 | 13 45·6 | 4·0 | 3·8 | 10·0 | 9·6 | 16·0 | 15·3 |
| 41 | 14 25·3 | 14 27·6 | 13 45·8 | 4·1 | 3·9 | 10·1 | 9·7 | 16·1 | 15·4 |
| 42 | 14 25·5 | 14 27·9 | 13 46·1 | 4·2 | 4·0 | 10·2 | 9·8 | 16·2 | 15·5 |
| 43 | 14 25·8 | 14 28·1 | 13 46·3 | 4·3 | 4·1 | 10·3 | 9·9 | 16·3 | 15·6 |
| 44 | 14 26·0 | 14 28·4 | 13 46·5 | 4·4 | 4·2 | 10·4 | 10·0 | 16·4 | 15·7 |
| 45 | 14 26·3 | 14 28·6 | 13 46·8 | 4·5 | 4·3 | 10·5 | 10·1 | 16·5 | 15·8 |
| 46 | 14 26·5 | 14 28·9 | 13 47·0 | 4·6 | 4·4 | 10·6 | 10·2 | 16·6 | 15·9 |
| 47 | 14 26·8 | 14 29·1 | 13 47·3 | 4·7 | 4·5 | 10·7 | 10·3 | 16·7 | 16·0 |
| 48 | 14 27·0 | 14 29·4 | 13 47·5 | 4·8 | 4·6 | 10·8 | 10·4 | 16·8 | 16·1 |
| 49 | 14 27·3 | 14 29·6 | 13 47·7 | 4·9 | 4·7 | 10·9 | 10·4 | 16·9 | 16·2 |
| 50 | 14 27·5 | 14 29·9 | 13 48·0 | 5·0 | 4·8 | 11·0 | 10·5 | 17·0 | 16·3 |
| 51 | 14 27·8 | 14 30·1 | 13 48·2 | 5·1 | 4·9 | 11·1 | 10·6 | 17·1 | 16·4 |
| 52 | 14 28·0 | 14 30·4 | 13 48·5 | 5·2 | 5·0 | 11·2 | 10·7 | 17·2 | 16·5 |
| 53 | 14 28·3 | 14 30·6 | 13 48·7 | 5·3 | 5·1 | 11·3 | 10·8 | 17·3 | 16·6 |
| 54 | 14 28·5 | 14 30·9 | 13 48·9 | 5·4 | 5·2 | 11·4 | 10·9 | 17·4 | 16·7 |
| 55 | 14 28·8 | 14 31·1 | 13 49·2 | 5·5 | 5·3 | 11·5 | 11·0 | 17·5 | 16·8 |
| 56 | 14 29·0 | 14 31·4 | 13 49·4 | 5·6 | 5·4 | 11·6 | 11·1 | 17·6 | 16·9 |
| 57 | 14 29·3 | 14 31·6 | 13 49·7 | 5·7 | 5·5 | 11·7 | 11·2 | 17·7 | 17·0 |
| 58 | 14 29·5 | 14 31·9 | 13 49·9 | 5·8 | 5·6 | 11·8 | 11·3 | 17·8 | 17·1 |
| 59 | 14 29·8 | 14 32·1 | 13 50·1 | 5·9 | 5·7 | 11·9 | 11·4 | 17·9 | 17·2 |
| 60 | 14 30·0 | 14 32·4 | 13 50·4 | 6·0 | 5·8 | 12·0 | 11·5 | 18·0 | 17·3 |

# EXPLANATION

## ALTITUDE CORRECTION TABLES

13. *General.* In general two corrections are given for application to altitudes observed with a marine sextant; additional corrections are required for Venus and Mars and also for very low altitudes.

Tables of the correction for dip of the horizon, due to height of eye above sea level, are given on pages A2 and xxxiv. Strictly this correction should be applied first and subtracted from the sextant altitude to give apparent altitude, which is the correct argument for the other tables.

Separate tables are given of the second correction for the Sun, for stars and planets (on pages A2 and A3), and for the Moon (on pages xxxiv and xxxv). For the Sun, values are given for both lower and upper limbs, for two periods of the year. The star tables are used for the planets, but additional corrections (page A2) are required for Venus and Mars. The Moon tables are in two parts: the main correction is a function of apparent altitude only and is tabulated for the lower limb (30′ must be subtracted to obtain the correction for the upper limb); the other, which is given for both lower and upper limbs, depends also on the horizontal parallax, which has to be taken from the daily pages.

An additional correction, given on page A4, is required for the change in the refraction, due to variations of pressure and temperature from the adopted standard conditions; it may generally be ignored for altitudes greater than 10°, except possibly in extreme conditions. The correction tables for the Sun, stars, and planets are in two parts; only those for altitudes greater than 10° are reprinted on the bookmark.

14. *Critical tables.* Some of the altitude correction tables are arranged as critical tables. In these an interval of apparent altitude (or height of eye) corresponds to a single value of the correction; no interpolation is required. At a "critical" entry the upper of the two possible values of the correction is to be taken. For example, in the table of dip, a correction of −4′·1 corresponds to all values of the height of eye from 5·3 to 5·5 metres (17·5 to 18·3 feet) inclusive.

15. *Examples.* The following examples illustrate the use of the altitude correction tables; the sextant altitudes given are assumed to be taken on 1983 January 22 with a marine sextant at height 5·4 metres (18 feet), temperature −3°C. and pressure 982 mb., the Moon sights being taken at about 10ʰ G.M.T.

| | SUN lower limb | SUN upper limb | MOON lower limb | MOON upper limb | VENUS | *Polaris* |
|---|---|---|---|---|---|---|
| | ° ′ | ° ′ | ° ′ | ° ′ | ° ′ | ° ′ |
| Sextant altitude | 21 19·7 | 3 20·2 | 33 27·6 | 26 06·7 | 4 32·6 | 49 36·5 |
| Dip, height 5·4 metres (18 feet) | −4·1 | −4·1 | −4·1 | −4·1 | −4·1 | −4·1 |
| Main correction | +13·8 | −29·6 | +57·4 | +60·5 | −10·8 | −0·8 |
| −30′ for upper limb (Moon) | — | — | — | −30·0 | — | — |
| L, U correction for Moon | — | — | +4·6 | +3·4 | — | — |
| Additional correction for Venus | — | — | — | — | +0·1 | — |
| Additional refraction correction | −0·1 | −0·3 | 0·0 | −0·1 | −0·3 | 0·0 |
| Corrected sextant altitude | 21 29·3 | 2 46·2 | 34 25·5 | 26 36·4 | 4 17·5 | 49 31·6 |

The main corrections have been taken out with apparent altitude (sextant altitude corrected for dip) as argument, interpolating where possible. These refinements are rarely necessary.

16. *Basis of the corrections.* The table for the dip of the sea horizon is based on the formula:

$$\text{Correction for dip} = -1'\cdot76\sqrt{(\text{height of eye in metres})} = -0'\cdot97\sqrt{(\text{height of eye in feet})}$$

The mean refraction, given explicitly in the correction table for the stars and planets and incorporated into those for the Sun and Moon, is based on Garfinkel's theory and is for a temperature of 10°C. (50°F.) and a pressure of 1010 mb. (29·83 inches). The additional corrections for variations of temperature and pressure from these adopted means are also based on Garfinkel's theory; there is no significant difference between the various theories to the accuracy given.

The correction table for the Sun includes the effects of semi-diameter and parallax, as well as the mean refraction; no correction for irradiation is included.

The additional corrections for Venus on page A2 and on the bookmark are designed for use only when the Sun is below the horizon; these corrections are mean values which allow for parallax and phase.

For daylight observations, the additional correction for Venus, which allows for parallax and phase, may be obtained from $p \cos H + k \cos \theta$, where $H$ is the altitude and $\theta$ is the angle at the planet between the vertical from the horizon to the planet and the direction from the planet to the Sun. An error of about 15° in $H$ or $\theta$ will produce a maximum error of about 0′·1 in the altitude correction. For 1983, $p$ and $k$ are:

| | Jan. 1 | | May 13 | | July 1 | | July 23 | | Aug. 8 | | Sept. 11 | | Sept. 28 | | Oct. 22 | | Dec. 13 | | Dec. 31 |
|---|---|---|---|---|---|---|---|---|---|---|---|---|---|---|---|---|---|---|---|
| | | ′ | | ′ | | ′ | | ′ | | ′ | | ′ | | ′ | | ′ | | ′ | |
| $p$ | | 0·1 | | 0·2 | | 0·3 | | 0·4 | | 0·5 | | 0·4 | | 0·3 | | 0·2 | | 0·1 | |
| $k$ | | 0·0 | | 0·1 | | 0·2 | | 0·3 | | 0·4 | | 0·3 | | 0·2 | | 0·1 | | 0·0 | |

The additional correction for Mars on page A2 and on the bookmark allows for parallax; no correction for phase is required.

In the case of the Moon the correction table includes the effects of semi-diameter, parallax and augmentation as well as the mean refraction; no correction for irradiation is included.

17. *Bubble sextant observations.* When observing with a bubble sextant no correction is necessary for dip, semi-diameter, or augmentation. The altitude corrections for the stars and planets on page A2 and on the bookmark should be used for the Sun as well as for the stars and planets; for the Moon it is easiest to take the mean of the corrections for lower and upper limbs and subtract 15′ from the altitude; the correction for dip must not be applied.

# ALTITUDE CORRECTION TABLES 0°–35°—MOON

| App. Alt. | 0°–4° | 5°–9° | 10°–14° | 15°–19° | 20°–24° | 25°–29° | 30°–34° | App. Alt. |
|---|---|---|---|---|---|---|---|---|
| | Corrⁿ | Corrⁿ | Corrⁿ | Corrⁿ | Corrⁿ | Corrⁿ | Corrⁿ | |
| ′ | ° ′ | ° ′ | ° ′ | ° ′ | ° ′ | ° ′ | ° ′ | ′ |
| 00 | 0 33·8 | 5 58·2 | 10 62·1 | 15 62·8 | 20 62·2 | 25 60·8 | 30 58·9 | 00 |
| 10 | 35·9 | 58·5 | 62·2 | 62·8 | 62·1 | 60·8 | 58·8 | 10 |
| 20 | 37·8 | 58·7 | 62·2 | 62·8 | 62·1 | 60·7 | 58·8 | 20 |
| 30 | 39·6 | 58·9 | 62·3 | 62·8 | 62·1 | 60·7 | 58·7 | 30 |
| 40 | 41·2 | 59·1 | 62·3 | 62·8 | 62·0 | 60·6 | 58·6 | 40 |
| 50 | 42·6 | 59·3 | 62·4 | 62·7 | 62·0 | 60·6 | 58·5 | 50 |
| 00 | 1 44·0 | 6 59·5 | 11 62·4 | 16 62·7 | 21 62·0 | 26 60·5 | 31 58·5 | 00 |
| 10 | 45·2 | 59·7 | 62·4 | 62·7 | 61·9 | 60·4 | 58·4 | 10 |
| 20 | 46·3 | 59·9 | 62·5 | 62·7 | 61·9 | 60·4 | 58·3 | 20 |
| 30 | 47·3 | 60·0 | 62·5 | 62·7 | 61·9 | 60·3 | 58·2 | 30 |
| 40 | 48·3 | 60·2 | 62·5 | 62·7 | 61·8 | 60·3 | 58·2 | 40 |
| 50 | 49·2 | 60·3 | 62·6 | 62·7 | 61·8 | 60·2 | 58·1 | 50 |
| 00 | 2 50·0 | 7 60·5 | 12 62·6 | 17 62·7 | 22 61·7 | 27 60·1 | 32 58·0 | 00 |
| 10 | 50·8 | 60·6 | 62·6 | 62·6 | 61·7 | 60·1 | 57·9 | 10 |
| 20 | 51·4 | 60·7 | 62·6 | 62·6 | 61·6 | 60·0 | 57·8 | 20 |
| 30 | 52·1 | 60·9 | 62·7 | 62·6 | 61·6 | 59·9 | 57·8 | 30 |
| 40 | 52·7 | 61·0 | 62·7 | 62·6 | 61·5 | 59·9 | 57·7 | 40 |
| 50 | 53·3 | 61·1 | 62·7 | 62·6 | 61·5 | 59·8 | 57·6 | 50 |
| 00 | 3 53·8 | 8 61·2 | 13 62·7 | 18 62·5 | 23 61·5 | 28 59·7 | 33 57·5 | 00 |
| 10 | 54·3 | 61·3 | 62·7 | 62·5 | 61·4 | 59·7 | 57·4 | 10 |
| 20 | 54·8 | 61·4 | 62·7 | 62·5 | 61·4 | 59·6 | 57·4 | 20 |
| 30 | 55·2 | 61·5 | 62·8 | 62·5 | 61·3 | 59·6 | 57·3 | 30 |
| 40 | 55·6 | 61·6 | 62·8 | 62·4 | 61·3 | 59·5 | 57·2 | 40 |
| 50 | 56·0 | 61·6 | 62·8 | 62·4 | 61·2 | 59·4 | 57·1 | 50 |
| 00 | 4 56·4 | 9 61·7 | 14 62·8 | 19 62·4 | 24 61·2 | 29 59·3 | 34 57·0 | 00 |
| 10 | 56·7 | 61·8 | 62·8 | 62·3 | 61·1 | 59·3 | 56·9 | 10 |
| 20 | 57·1 | 61·9 | 62·8 | 62·3 | 61·1 | 59·2 | 56·9 | 20 |
| 30 | 57·4 | 61·9 | 62·8 | 62·3 | 61·0 | 59·1 | 56·8 | 30 |
| 40 | 57·7 | 62·0 | 62·8 | 62·2 | 60·9 | 59·1 | 56·7 | 40 |
| 50 | 57·9 | 62·1 | 62·8 | 62·2 | 60·9 | 59·0 | 56·6 | 50 |

| H.P. | L | U | L | U | L | U | L | U | L | U | L | U | L | U | H.P. |
|---|---|---|---|---|---|---|---|---|---|---|---|---|---|---|---|
| ′ | ′ | ′ | ′ | ′ | ′ | ′ | ′ | ′ | ′ | ′ | ′ | ′ | ′ | ′ | ′ |
| 54·0 | 0·3 | 0·9 | 0·3 | 0·9 | 0·4 | 1·0 | 0·5 | 1·1 | 0·6 | 1·2 | 0·7 | 1·3 | 0·9 | 1·5 | 54·0 |
| 54·3 | 0·7 | 1·1 | 0·7 | 1·2 | 0·7 | 1·2 | 0·8 | 1·3 | 0·9 | 1·4 | 1·1 | 1·5 | 1·2 | 1·7 | 54·3 |
| 54·6 | 1·1 | 1·4 | 1·1 | 1·4 | 1·1 | 1·4 | 1·2 | 1·5 | 1·3 | 1·6 | 1·4 | 1·7 | 1·5 | 1·8 | 54·6 |
| 54·9 | 1·4 | 1·6 | 1·5 | 1·6 | 1·5 | 1·6 | 1·6 | 1·7 | 1·6 | 1·8 | 1·8 | 1·9 | 1·9 | 2·0 | 54·9 |
| 55·2 | 1·8 | 1·8 | 1·8 | 1·8 | 1·9 | 1·9 | 1·9 | 1·9 | 2·0 | 2·0 | 2·1 | 2·1 | 2·2 | 2·2 | 55·2 |
| 55·5 | 2·2 | 2·0 | 2·2 | 2·0 | 2·3 | 2·1 | 2·3 | 2·1 | 2·4 | 2·2 | 2·4 | 2·3 | 2·5 | 2·4 | 55·5 |
| 55·8 | 2·6 | 2·2 | 2·6 | 2·2 | 2·6 | 2·3 | 2·7 | 2·3 | 2·7 | 2·4 | 2·8 | 2·4 | 2·9 | 2·5 | 55·8 |
| 56·1 | 3·0 | 2·4 | 3·0 | 2·5 | 3·0 | 2·5 | 3·0 | 2·5 | 3·1 | 2·6 | 3·1 | 2·6 | 3·2 | 2·7 | 56·1 |
| 56·4 | 3·4 | 2·7 | 3·4 | 2·7 | 3·4 | 2·7 | 3·4 | 2·7 | 3·4 | 2·8 | 3·5 | 2·8 | 3·5 | 2·9 | 56·4 |
| 56·7 | 3·7 | 2·9 | 3·7 | 2·9 | 3·8 | 2·9 | 3·8 | 2·9 | 3·8 | 3·0 | 3·8 | 3·0 | 3·9 | 3·0 | 56·7 |
| 57·0 | 4·1 | 3·1 | 4·1 | 3·1 | 4·1 | 3·1 | 4·1 | 3·1 | 4·2 | 3·1 | 4·2 | 3·2 | 4·2 | 3·2 | 57·0 |
| 57·3 | 4·5 | 3·3 | 4·5 | 3·3 | 4·5 | 3·3 | 4·5 | 3·3 | 4·5 | 3·3 | 4·5 | 3·4 | 4·6 | 3·4 | 57·3 |
| 57·6 | 4·9 | 3·5 | 4·9 | 3·5 | 4·9 | 3·5 | 4·9 | 3·5 | 4·9 | 3·5 | 4·9 | 3·5 | 4·9 | 3·6 | 57·6 |
| 57·9 | 5·3 | 3·8 | 5·3 | 3·8 | 5·2 | 3·8 | 5·2 | 3·7 | 5·2 | 3·7 | 5·2 | 3·7 | 5·2 | 3·7 | 57·9 |
| 58·2 | 5·6 | 4·0 | 5·6 | 4·0 | 5·6 | 4·0 | 5·6 | 4·0 | 5·6 | 3·9 | 5·6 | 3·9 | 5·6 | 3·9 | 58·2 |
| 58·5 | 6·0 | 4·2 | 6·0 | 4·2 | 6·0 | 4·2 | 6·0 | 4·2 | 6·0 | 4·1 | 5·9 | 4·1 | 5·9 | 4·1 | 58·5 |
| 58·8 | 6·4 | 4·4 | 6·4 | 4·4 | 6·4 | 4·4 | 6·3 | 4·4 | 6·3 | 4·3 | 6·3 | 4·3 | 6·2 | 4·2 | 58·8 |
| 59·1 | 6·8 | 4·6 | 6·8 | 4·6 | 6·7 | 4·6 | 6·7 | 4·6 | 6·7 | 4·5 | 6·6 | 4·5 | 6·6 | 4·4 | 59·1 |
| 59·4 | 7·2 | 4·8 | 7·1 | 4·8 | 7·1 | 4·8 | 7·1 | 4·8 | 7·0 | 4·7 | 7·0 | 4·7 | 6·9 | 4·6 | 59·4 |
| 59·7 | 7·5 | 5·1 | 7·5 | 5·0 | 7·5 | 5·0 | 7·5 | 5·0 | 7·4 | 4·9 | 7·3 | 4·8 | 7·2 | 4·7 | 59·7 |
| 60·0 | 7·9 | 5·3 | 7·9 | 5·3 | 7·9 | 5·2 | 7·8 | 5·2 | 7·8 | 5·1 | 7·7 | 5·0 | 7·6 | 4·9 | 60·0 |
| 60·3 | 8·3 | 5·5 | 8·3 | 5·5 | 8·2 | 5·4 | 8·2 | 5·4 | 8·1 | 5·3 | 8·0 | 5·2 | 7·9 | 5·1 | 60·3 |
| 60·6 | 8·7 | 5·7 | 8·7 | 5·7 | 8·6 | 5·7 | 8·6 | 5·6 | 8·5 | 5·5 | 8·4 | 5·4 | 8·2 | 5·3 | 60·6 |
| 60·9 | 9·1 | 5·9 | 9·0 | 5·9 | 9·0 | 5·9 | 8·9 | 5·8 | 8·8 | 5·7 | 8·7 | 5·6 | 8·6 | 5·4 | 60·9 |
| 61·2 | 9·5 | 6·2 | 9·4 | 6·1 | 9·4 | 6·1 | 9·3 | 6·0 | 9·2 | 5·9 | 9·1 | 5·8 | 8·9 | 5·6 | 61·2 |
| 61·5 | 9·8 | 6·4 | 9·8 | 6·3 | 9·7 | 6·3 | 9·7 | 6·2 | 9·5 | 6·1 | 9·4 | 5·9 | 9·2 | 5·8 | 61·5 |

## DIP

| Ht. of Eye | Corrⁿ | Ht. of Eye | Corrⁿ | Ht. of Eye | Corrⁿ |
|---|---|---|---|---|---|
| ft. | ′ | ft. | ′ | ft. | ′ |
| 4·0 | −2·0 | 24 | −4·9 | 63 | − 7·8 |
| 4·4 | −2·1 | 26 | −5·0 | 65 | − 7·9 |
| 4·9 | −2·2 | 27 | −5·1 | 67 | − 8·0 |
| 5·3 | −2·3 | 28 | −5·2 | 68 | − 8·1 |
| 5·8 | −2·4 | 29 | −5·3 | 70 | − 8·2 |
| 6·3 | −2·5 | 30 | −5·4 | 72 | − 8·3 |
| 6·9 | −2·6 | 31 | −5·5 | 74 | − 8·4 |
| 7·4 | −2·7 | 32 | −5·6 | 75 | − 8·5 |
| 8·0 | −2·8 | 33 | −5·7 | 77 | − 8·6 |
| 8·6 | −2·9 | 35 | −5·8 | 79 | − 8·7 |
| 9·2 | −3·0 | 36 | −5·9 | 81 | − 8·8 |
| 9·8 | −3·1 | 37 | −6·0 | 83 | − 8·9 |
| 10·5 | −3·2 | 38 | −6·1 | 85 | − 9·0 |
| 11·2 | −3·3 | 40 | −6·2 | 87 | − 9·1 |
| 11·9 | −3·4 | 41 | −6·3 | 88 | − 9·2 |
| 12·6 | −3·5 | 42 | −6·4 | 90 | − 9·3 |
| 13·3 | −3·6 | 44 | −6·5 | 92 | − 9·4 |
| 14·1 | −3·7 | 45 | −6·6 | 94 | − 9·5 |
| 14·9 | −3·8 | 47 | −6·7 | 96 | − 9·6 |
| 15·7 | −3·9 | 48 | −6·8 | 98 | − 9·7 |
| 16·5 | −4·0 | 49 | −6·9 | 101 | − 9·8 |
| 17·4 | −4·1 | 51 | −7·0 | 103 | − 9·9 |
| 18·3 | −4·2 | 52 | −7·1 | 105 | −10·0 |
| 19·1 | −4·3 | 54 | −7·2 | 107 | −10·1 |
| 20·1 | −4·4 | 55 | −7·3 | 109 | −10·2 |
| 21·0 | −4·5 | 57 | −7·4 | 111 | −10·3 |
| 22·0 | −4·6 | 58 | −7·5 | 113 | −10·4 |
| 22·9 | −4·7 | 60 | −7·6 | 116 | −10·5 |
| 23·9 | −4·8 | 62 | −7·7 | 118 | −10·6 |
| 24·9 | | 63 | | 120 | |

## MOON CORRECTION TABLE

The correction is in two parts; the first correction is taken from the upper part of the table with argument apparent altitude, and the second from the lower part, with argument H.P., in the same column as that from which the first correction was taken. Separate corrections are given in the lower part for lower (L) and upper (U) limbs. All corrections are to be **added** to apparent altitude, *but* 30′ *is to be subtracted from the altitude of the upper limb.*

For corrections for pressure and temperature see page A4.

For bubble sextant observations ignore dip, take the mean of upper and lower limb corrections and subtract 15′ from the altitude.

App. Alt. = Apparent altitude = Sextant altitude corrected for index error and dip.

# C PAGE EXCERPTS FROM PUB. NO. 249, VOLUME 2

Material in this appendix is taken from Pub. No. 249 by permission of the U.S. Naval Oceanographic Office, now the Defense Mapping Agency Hydrographic/Topographic Center.

N. Lat. {LHA greater than 180°......Zn=Z; LHA less than 180°..........Zn=360−Z}

## DECLINATION (0°-14°) SAME NAME AS LATITUDE

| LHA | 0° Hc | 0° d | 0° Z | 1° Hc | 1° d | 1° Z | 2° Hc | 2° d | 2° Z | 3° Hc | 3° d | 3° Z | 4° Hc | 4° d | 4° Z |
|---|---|---|---|---|---|---|---|---|---|---|---|---|---|---|---|
| | ° ′ | ′ | ° | ° ′ | ′ | ° | ° ′ | ′ | ° | ° ′ | ′ | ° | ° ′ | ′ | ° |
| 0 | 73 00 | +60 | 180 | 74 00 | +60 | 180 | 75 00 | +60 | 180 | 76 00 | +60 | 180 | 77 00 | +60 | 180 |
| 1 | 72 58 | 60 | 177 | 73 58 | 60 | 176 | 74 58 | 60 | 176 | 75 58 | 60 | 176 | 76 58 | 60 | 176 |
| 2 | 72 53 | 60 | 173 | 73 53 | 59 | 173 | 74 52 | 60 | 172 | 75 52 | 59 | 172 | 76 51 | 60 | 171 |
| 3 | 72 45 | 59 | 170 | 73 44 | 59 | 169 | 74 43 | 59 | 169 | 75 42 | 58 | 168 | 76 40 | 59 | 167 |
| 4 | 72 33 | 58 | 167 | 73 31 | 59 | 166 | 74 30 | 58 | 165 | 75 28 | 57 | 164 | 76 25 | 58 | 163 |
| 5 | 72 18 | +58 | 163 | 73 16 | +57 | 162 | 74 13 | +57 | 161 | 75 10 | +56 | 160 | 76 06 | +56 | 159 |
| 6 | 72 00 | 57 | 160 | 72 57 | 56 | 159 | 73 53 | 56 | 158 | 74 49 | 55 | 157 | 75 44 | 54 | 155 |
| 7 | 71 39 | 56 | 157 | 72 35 | 55 | 156 | 73 30 | 54 | 155 | 74 24 | 54 | 153 | 75 18 | 53 | 151 |
| 8 | 71 16 | 54 | 154 | 72 10 | 54 | 153 | 73 04 | 53 | 152 | 73 57 | 52 | 150 | 74 49 | 51 | 148 |
| 9 | 70 50 | 53 | 152 | 71 43 | 52 | 150 | 72 35 | 52 | 149 | 73 27 | 50 | 147 | 74 17 | 50 | 145 |
| 10 | 70 21 | +52 | 149 | 71 13 | +51 | 147 | 72 04 | +50 | 146 | 72 54 | +49 | 144 | 73 43 | +48 | 142 |
| 11 | 69 50 | 51 | 146 | 70 41 | 50 | 145 | 71 31 | 48 | 143 | 72 19 | 47 | 141 | 73 06 | 46 | 139 |
| 12 | 69 18 | 49 | 144 | 70 07 | 48 | 142 | 70 55 | 47 | 141 | 71 42 | 46 | 139 | 72 28 | 44 | 137 |
| 13 | 68 43 | 48 | 142 | 69 31 | 47 | 140 | 70 18 | 45 | 138 | 71 03 | 44 | 136 | 71 47 | 43 | 134 |
| 14 | 68 07 | 46 | 140 | 68 53 | 46 | 138 | 69 39 | 44 | 136 | 70 23 | 42 | 134 | 71 05 | 42 | 132 |
| 15 | 67 29 | +45 | 138 | 68 14 | +44 | 136 | 68 58 | +43 | 134 | 69 41 | +41 | 132 | 70 22 | +40 | 130 |
| 16 | 66 49 | 44 | 136 | 67 33 | 43 | 134 | 68 16 | 41 | 132 | 68 57 | 40 | 130 | 69 37 | 39 | 128 |
| 17 | 66 08 | 43 | 134 | 66 51 | 42 | 132 | 67 33 | 40 | 130 | 68 13 | 38 | 128 | 68 51 | 38 | 126 |
| 18 | 65 26 | 42 | 132 | 66 08 | 40 | 130 | 66 48 | 39 | 128 | 67 27 | 38 | 126 | 68 05 | 35 | 124 |
| 19 | 64 43 | 40 | 130 | 65 23 | 40 | 129 | 66 03 | 37 | 127 | 66 40 | 37 | 125 | 67 17 | 34 | 123 |
| 20 | 63 59 | +39 | 129 | 64 38 | +38 | 127 | 65 16 | +37 | 125 | 65 53 | +35 | 123 | 66 28 | +34 | 121 |
| 21 | 63 14 | 38 | 127 | 63 52 | 37 | 126 | 64 29 | 36 | 124 | 65 05 | 34 | 122 | 65 39 | 32 | 120 |
| 22 | 62 28 | 37 | 126 | 63 05 | 36 | 124 | 63 41 | 35 | 122 | 64 16 | 33 | 121 | 64 49 | 31 | 119 |
| 23 | 61 41 | 36 | 125 | 62 17 | 35 | 123 | 62 52 | 34 | 121 | 63 26 | 32 | 119 | 63 58 | 31 | 117 |
| 24 | 60 53 | 35 | 123 | 61 28 | 35 | 122 | 62 03 | 32 | 120 | 62 35 | 32 | 118 | 63 07 | 30 | 116 |
| 25 | 60 05 | +34 | 122 | 60 39 | +34 | 120 | 61 13 | +32 | 119 | 61 45 | +30 | 117 | 62 15 | +29 | 115 |
| 26 | 59 16 | 34 | 121 | 59 50 | 32 | 119 | 60 22 | 31 | 118 | 60 53 | 30 | 116 | 61 23 | 28 | 114 |
| 27 | 58 26 | 33 | 120 | 58 59 | 32 | 118 | 59 31 | 30 | 117 | 60 01 | 29 | 115 | 60 30 | 28 | 113 |
| 28 | 57 36 | 32 | 119 | 58 08 | 31 | 117 | 58 39 | 30 | 116 | 59 09 | 28 | 114 | 59 37 | 27 | 112 |
| 29 | 56 46 | 31 | 118 | 57 17 | 30 | 116 | 57 47 | 29 | 115 | 58 16 | 28 | 113 | 58 44 | 26 | 111 |
| 30 | 55 55 | +31 | 117 | 56 26 | +29 | 115 | 56 55 | +28 | 114 | 57 23 | +27 | 112 | 57 50 | +26 | 110 |
| 31 | 55 03 | 30 | 116 | 55 33 | 29 | 114 | 56 02 | 28 | 113 | 56 30 | 27 | 111 | 56 57 | 25 | 110 |
| 32 | 54 12 | 29 | 115 | 54 41 | 28 | 114 | 55 09 | 27 | 112 | 55 36 | 26 | 111 | 56 02 | 25 | 109 |
| 33 | 53 19 | 29 | 114 | 53 48 | 28 | 113 | 54 16 | 27 | 111 | 54 43 | 25 | 110 | 55 08 | 24 | 108 |
| 34 | 52 27 | 28 | 113 | 52 55 | 27 | 112 | 53 22 | 26 | 111 | 53 48 | 25 | 109 | 54 13 | 24 | 107 |
| 35 | 51 34 | +28 | 113 | 52 02 | +27 | 111 | 52 29 | +25 | 110 | 52 54 | +24 | 108 | 53 18 | +24 | 107 |
| 36 | 50 41 | 27 | 112 | 51 08 | 26 | 111 | 51 34 | 25 | 109 | 51 59 | 24 | 108 | 52 23 | 23 | 106 |
| 37 | 49 48 | 26 | 111 | 50 14 | 26 | 110 | 50 40 | 25 | 108 | 51 05 | 23 | 107 | 51 28 | 23 | 106 |
| 38 | 48 54 | 26 | 111 | 49 20 | 26 | 109 | 49 46 | 24 | 108 | 50 10 | 23 | 106 | 50 33 | 22 | 105 |
| 39 | 48 00 | 26 | 110 | 48 26 | 25 | 109 | 48 51 | 24 | 107 | 49 15 | 22 | 106 | 49 37 | 22 | 104 |
| 40 | 47 06 | +26 | 109 | 47 32 | +24 | 108 | 47 56 | +23 | 107 | 48 19 | +23 | 105 | 48 42 | +21 | 104 |
| 41 | 46 12 | 25 | 109 | 46 37 | 24 | 107 | 47 01 | 23 | 106 | 47 24 | 22 | 105 | 47 46 | 21 | 103 |
| 42 | 45 17 | 25 | 108 | 45 42 | 24 | 107 | 46 06 | 22 | 105 | 46 28 | 22 | 104 | 46 50 | 21 | 103 |
| 43 | 44 23 | 24 | 107 | 44 47 | 23 | 106 | 45 10 | 22 | 105 | 45 32 | 22 | 104 | 45 54 | 20 | 102 |
| 44 | 43 28 | 24 | 107 | 43 52 | 23 | 106 | 44 15 | 22 | 104 | 44 37 | 21 | 103 | 44 58 | 20 | 102 |
| 45 | 42 33 | +23 | 106 | 42 56 | +23 | 105 | 43 19 | +22 | 104 | 43 41 | +20 | 103 | 44 01 | +20 | 101 |
| 46 | 41 38 | 23 | 106 | 42 01 | 22 | 105 | 42 23 | 22 | 103 | 42 45 | 20 | 102 | 43 05 | 20 | 101 |
| 47 | 40 43 | 22 | 105 | 41 05 | 22 | 104 | 41 27 | 21 | 103 | 41 48 | 21 | 102 | 42 09 | 19 | 100 |
| 48 | 39 47 | 23 | 105 | 40 10 | 21 | 104 | 40 31 | 21 | 102 | 40 52 | 20 | 101 | 41 12 | 20 | 100 |
| 49 | 38 52 | 22 | 104 | 39 14 | 21 | 103 | 39 35 | 21 | 102 | 39 56 | 20 | 101 | 40 16 | 19 | 99 |
| 50 | 37 56 | +22 | 104 | 38 18 | +21 | 103 | 38 39 | +20 | 101 | 38 59 | +20 | 100 | 39 19 | +19 | 99 |
| 51 | 37 00 | 22 | 103 | 37 22 | 21 | 102 | 37 43 | 20 | 101 | 38 03 | 19 | 100 | 38 22 | 19 | 99 |
| 52 | 36 04 | 22 | 103 | 36 26 | 20 | 102 | 36 46 | 20 | 101 | 37 06 | 20 | 99 | 37 26 | 18 | 98 |
| 53 | 35 08 | 21 | 102 | 35 29 | 21 | 101 | 35 50 | 20 | 100 | 36 10 | 19 | 99 | 36 29 | 18 | 98 |
| 54 | 34 12 | 21 | 102 | 34 33 | 20 | 101 | 34 53 | 20 | 100 | 35 13 | 19 | 99 | 35 32 | 18 | 97 |
| 55 | 33 16 | +21 | 102 | 33 37 | +20 | 100 | 33 57 | +19 | 99 | 34 16 | +19 | 98 | 34 35 | +18 | 97 |
| 56 | 32 20 | 20 | 101 | 32 40 | 20 | 100 | 33 00 | 19 | 99 | 33 19 | 19 | 98 | 33 38 | 18 | 97 |
| 57 | 31 23 | 21 | 101 | 31 44 | 19 | 100 | 32 03 | 19 | 98 | 32 22 | 19 | 97 | 32 41 | 18 | 96 |
| 58 | 30 27 | 20 | 100 | 30 47 | 20 | 99 | 31 07 | 19 | 98 | 31 26 | 18 | 97 | 31 44 | 18 | 96 |
| 59 | 29 30 | 20 | 100 | 29 50 | 20 | 99 | 30 10 | 19 | 98 | 30 29 | 18 | 97 | 30 47 | 17 | 96 |
| 60 | 28 34 | +20 | 100 | 28 54 | +19 | 99 | 29 13 | +19 | 97 | 29 32 | +18 | 96 | 29 50 | +17 | 95 |
| 61 | 27 37 | 20 | 99 | 27 57 | 19 | 98 | 28 16 | 18 | 97 | 28 34 | 19 | 96 | 28 53 | 17 | 95 |
| 62 | 26 41 | 19 | 99 | 27 00 | 19 | 98 | 27 19 | 18 | 97 | 27 37 | 18 | 96 | 27 55 | 18 | 95 |
| 63 | 25 44 | 19 | 99 | 26 03 | 19 | 97 | 26 22 | 18 | 96 | 26 40 | 18 | 95 | 26 58 | 17 | 94 |
| 64 | 24 47 | 19 | 98 | 25 06 | 19 | 97 | 25 25 | 18 | 96 | 25 43 | 18 | 95 | 26 01 | 17 | 94 |
| 65 | 23 50 | +19 | 98 | 24 09 | +19 | 97 | 24 28 | +18 | 96 | 24 46 | +18 | 95 | 25 04 | +17 | 94 |
| 66 | 22 53 | 19 | 97 | 23 12 | 19 | 96 | 23 31 | 18 | 95 | 23 49 | 17 | 94 | 24 06 | 18 | 93 |
| 67 | 21 57 | 18 | 97 | 22 15 | 19 | 96 | 22 34 | 18 | 95 | 22 52 | 17 | 94 | 23 09 | 17 | 93 |
| 68 | 21 00 | 18 | 97 | 21 18 | 18 | 96 | 21 36 | 18 | 95 | 21 54 | 18 | 94 | 22 12 | 17 | 93 |
| 69 | 20 03 | 18 | 96 | 20 21 | 18 | 95 | 20 39 | 18 | 94 | 20 57 | 17 | 93 | 21 14 | 17 | 92 |

| LHA | 5° Hc | 5° d | 5° Z | 6° Hc | 6° d | 6° Z | 7° Hc | 7° d | 7° Z | 8° Hc | 8° d | 8° Z | 9° Hc | 9° d | 9° Z |
|---|---|---|---|---|---|---|---|---|---|---|---|---|---|---|---|
| | ° ′ | ′ | ° | ° ′ | ′ | ° | ° ′ | ′ | ° | ° ′ | ′ | ° | ° ′ | ′ | ° |
| 0 | 78 00 | +60 | 180 | 79 00 | +60 | 180 | 80 00 | +60 | 180 | 81 00 | +60 | 180 | 82 00 | +60 | 180 |
| 1 | 77 58 | 59 | 175 | 78 57 | 60 | 175 | 79 57 | 60 | 174 | 80 57 | 60 | 174 | 81 57 | 59 | 173 |
| 2 | 77 51 | 59 | 171 | 78 50 | 59 | 170 | 79 49 | 59 | 169 | 80 48 | 58 | 168 | 81 46 | 58 | 166 |
| 3 | 77 39 | 58 | 166 | 78 37 | 59 | 165 | 79 35 | 57 | 163 | 80 32 | 57 | 162 | 81 29 | 56 | 160 |
| 4 | 77 23 | 57 | 162 | 78 20 | 56 | 160 | 79 16 | 56 | 158 | 80 12 | 54 | 156 | 81 06 | 54 | 154 |
| 5 | 77 02 | +56 | 157 | 77 58 | +54 | 155 | 78 52 | +54 | 153 | 79 46 | +52 | 151 | 80 38 | +51 | 148 |
| 6 | 76 38 | 54 | 153 | 77 32 | 53 | 151 | 78 25 | 51 | 149 | 79 16 | 50 | 146 | 80 06 | 47 | 143 |
| 7 | 76 11 | 52 | 150 | 77 03 | 50 | 147 | 77 53 | 49 | 145 | 78 42 | 48 | 142 | 79 30 | 44 | 139 |
| 8 | 75 40 | 50 | 146 | 76 30 | 49 | 144 | 77 19 | 46 | 141 | 78 05 | 45 | 138 | 78 50 | 42 | 135 |
| 9 | 75 07 | 48 | 143 | 75 55 | 46 | 140 | 76 41 | 45 | 138 | 77 26 | 42 | 135 | 78 08 | 40 | 131 |
| 10 | 74 31 | +46 | 140 | 75 17 | +44 | 137 | 76 01 | +43 | 135 | 76 44 | +40 | 132 | 77 24 | +38 | 128 |
| 11 | 73 52 | 45 | 137 | 74 37 | 42 | 134 | 75 19 | 41 | 132 | 76 00 | 38 | 129 | 76 38 | 36 | 125 |
| 12 | 73 12 | 43 | 134 | 73 55 | 41 | 132 | 74 36 | 38 | 129 | 75 14 | 37 | 126 | 75 51 | 33 | 123 |
| 13 | 72 30 | 41 | 132 | 73 11 | 39 | 129 | 73 50 | 37 | 127 | 74 27 | 35 | 124 | 75 02 | 32 | 121 |
| 14 | 71 47 | 39 | 130 | 72 26 | 38 | 127 | 73 04 | 35 | 125 | 73 39 | 33 | 122 | 74 12 | 30 | 119 |
| 15 | 71 02 | +38 | 128 | 71 40 | +36 | 125 | 72 16 | +34 | 123 | 72 50 | +31 | 120 | 73 21 | +29 | 117 |
| 16 | 70 16 | 36 | 126 | 70 52 | 35 | 123 | 71 27 | 32 | 121 | 71 59 | 31 | 118 | 72 30 | 27 | 115 |
| 17 | 69 29 | 35 | 124 | 70 04 | 33 | 122 | 70 37 | 31 | 119 | 71 08 | 29 | 116 | 71 37 | 27 | 114 |
| 18 | 68 40 | 35 | 122 | 69 15 | 32 | 120 | 69 47 | 30 | 118 | 70 17 | 28 | 115 | 70 45 | 25 | 112 |
| 19 | 67 51 | 33 | 121 | 68 24 | 31 | 118 | 68 55 | 29 | 116 | 69 24 | 27 | 114 | 69 51 | 25 | 111 |
| 20 | 67 02 | +32 | 119 | 67 34 | +30 | 117 | 68 04 | +28 | 115 | 68 32 | +25 | 112 | 68 57 | +24 | 110 |
| 21 | 66 11 | 31 | 118 | 66 42 | 29 | 116 | 67 11 | 27 | 114 | 67 38 | 25 | 111 | 68 03 | 23 | 109 |
| 22 | 65 20 | 30 | 117 | 65 50 | 28 | 115 | 66 18 | 27 | 112 | 66 45 | 24 | 110 | 67 09 | 22 | 108 |
| 23 | 64 29 | 29 | 115 | 64 58 | 27 | 113 | 65 25 | 25 | 111 | 65 50 | 24 | 109 | 66 14 | 21 | 107 |
| 24 | 63 37 | 28 | 114 | 64 05 | 26 | 112 | 64 31 | 25 | 110 | 64 56 | 23 | 108 | 65 19 | 21 | 106 |
| 25 | 62 44 | +28 | 113 | 63 12 | +25 | 111 | 63 37 | +24 | 109 | 64 01 | +23 | 107 | 64 24 | +20 | 105 |
| 26 | 61 51 | 27 | 112 | 62 18 | 25 | 110 | 62 43 | 23 | 108 | 63 06 | 22 | 106 | 63 28 | 20 | 104 |
| 27 | 60 58 | 26 | 111 | 61 24 | 24 | 109 | 61 48 | 23 | 108 | 62 11 | 21 | 106 | 62 32 | 20 | 104 |
| 28 | 60 04 | 26 | 110 | 60 30 | 24 | 109 | 60 54 | 22 | 107 | 61 16 | 20 | 105 | 61 36 | 19 | 103 |
| 29 | 59 10 | 25 | 110 | 59 35 | 23 | 108 | 59 58 | 22 | 106 | 60 20 | 20 | 104 | 60 40 | 19 | 102 |
| 30 | 58 16 | +24 | 109 | 58 40 | +23 | 107 | 59 03 | +22 | 105 | 59 25 | +19 | 103 | 59 44 | +18 | 102 |
| 31 | 57 22 | 23 | 108 | 57 45 | 23 | 106 | 58 08 | 21 | 105 | 58 29 | 19 | 103 | 58 48 | 18 | 101 |
| 32 | 56 27 | 23 | 107 | 56 50 | 22 | 106 | 57 12 | 21 | 104 | 57 33 | 19 | 102 | 57 52 | 17 | 100 |
| 33 | 55 32 | 23 | 107 | 55 55 | 21 | 105 | 56 16 | 20 | 103 | 56 36 | 19 | 102 | 56 55 | 17 | 100 |
| 34 | 54 37 | 22 | 106 | 54 59 | 21 | 104 | 55 20 | 20 | 103 | 55 40 | 19 | 101 | 55 59 | 16 | 99 |
| 35 | 53 42 | +22 | 105 | 54 04 | +20 | 104 | 54 24 | +20 | 102 | 54 44 | +18 | 100 | 55 02 | +17 | 99 |
| 36 | 52 46 | 22 | 105 | 53 08 | 20 | 103 | 53 28 | 19 | 102 | 53 47 | 18 | 100 | 54 05 | 17 | 98 |
| 37 | 51 51 | 21 | 104 | 52 12 | 20 | 103 | 52 32 | 19 | 101 | 52 51 | 17 | 99 | 53 08 | 17 | 98 |
| 38 | 50 55 | 21 | 103 | 51 16 | 19 | 102 | 51 35 | 19 | 100 | 51 54 | 17 | 99 | 52 11 | 16 | 97 |
| 39 | 49 59 | 21 | 103 | 50 20 | 19 | 101 | 50 39 | 18 | 100 | 50 57 | 17 | 98 | 51 14 | 16 | 97 |
| 40 | 49 03 | +20 | 102 | 49 23 | +19 | 101 | 49 42 | +19 | 99 | 50 01 | +16 | 98 | 50 17 | +16 | 96 |
| 41 | 48 07 | 20 | 102 | 48 27 | 19 | 100 | 48 46 | 18 | 99 | 49 04 | 16 | 98 | 49 20 | 16 | 96 |
| 42 | 47 11 | 19 | 101 | 47 30 | 19 | 100 | 47 49 | 18 | 99 | 48 07 | 16 | 97 | 48 23 | 16 | 96 |
| 43 | 46 14 | 20 | 101 | 46 34 | 18 | 99 | 46 52 | 18 | 98 | 47 10 | 16 | 97 | 47 26 | 16 | 95 |
| 44 | 45 18 | 19 | 100 | 45 37 | 18 | 99 | 45 55 | 18 | 98 | 46 13 | 16 | 96 | 46 29 | 15 | 95 |
| 45 | 44 21 | +19 | 100 | 44 40 | +19 | 99 | 44 59 | +17 | 97 | 45 16 | +16 | 96 | 45 32 | +15 | 95 |
| 46 | 43 25 | 19 | 99 | 43 44 | 18 | 98 | 44 02 | 17 | 97 | 44 19 | 16 | 96 | 44 35 | 15 | 94 |
| 47 | 42 28 | 19 | 99 | 42 47 | 18 | 98 | 43 05 | 17 | 96 | 43 21 | 16 | 95 | 43 37 | 15 | 94 |
| 48 | 41 32 | 18 | 99 | 41 50 | 18 | 98 | 42 08 | 16 | 96 | 42 24 | 16 | 95 | 42 40 | 15 | 93 |
| 49 | 40 35 | 18 | 98 | 40 53 | 17 | 97 | 41 10 | 17 | 96 | 41 27 | 16 | 94 | 41 43 | 15 | 93 |
| 50 | 39 38 | +18 | 98 | 39 56 | +17 | 97 | 40 13 | +17 | 95 | 40 30 | +16 | 94 | 40 46 | +14 | 93 |
| 51 | 38 41 | 18 | 97 | 38 59 | 17 | 96 | 39 16 | 17 | 95 | 39 33 | 15 | 94 | 39 48 | 15 | 92 |
| 52 | 37 44 | 18 | 97 | 38 02 | 17 | 96 | 38 19 | 16 | 95 | 38 35 | 16 | 93 | 38 51 | 15 | 92 |
| 53 | 36 47 | 18 | 97 | 37 05 | 17 | 95 | 37 22 | 16 | 94 | 37 38 | 16 | 93 | 37 54 | 14 | 92 |
| 54 | 35 50 | 18 | 96 | 36 08 | 17 | 95 | 36 25 | 16 | 94 | 36 41 | 15 | 93 | 36 56 | 15 | 91 |
| 55 | 34 53 | +18 | 96 | 35 11 | +16 | 95 | 35 27 | +16 | 94 | 35 43 | +16 | 92 | 35 59 | +15 | 91 |
| 56 | 33 56 | 17 | 95 | 34 13 | 17 | 94 | 34 30 | 16 | 93 | 34 46 | 15 | 92 | 35 01 | 15 | 91 |
| 57 | 32 59 | 17 | 95 | 33 16 | 17 | 94 | 33 33 | 16 | 93 | 33 49 | 15 | 92 | 34 04 | 15 | 91 |
| 58 | 32 02 | 17 | 95 | 32 19 | 16 | 94 | 32 35 | 16 | 93 | 32 51 | 16 | 91 | 33 07 | 14 | 90 |
| 59 | 31 04 | 18 | 95 | 31 22 | 16 | 93 | 31 38 | 16 | 92 | 31 54 | 15 | 91 | 32 09 | 15 | 90 |
| 60 | 30 07 | +17 | 94 | 30 24 | +17 | 93 | 30 41 | +16 | 92 | 30 57 | +15 | 91 | 31 12 | +15 | 90 |
| 61 | 29 10 | 17 | 94 | 29 27 | 16 | 93 | 29 43 | 16 | 92 | 29 59 | 16 | 91 | 30 15 | 14 | 89 |
| 62 | 28 13 | 17 | 94 | 28 30 | 16 | 93 | 28 46 | 16 | 91 | 29 02 | 15 | 90 | 29 17 | 15 | 89 |
| 63 | 27 15 | 17 | 93 | 27 32 | 17 | 92 | 27 49 | 16 | 91 | 28 05 | 15 | 90 | 28 20 | 15 | 89 |
| 64 | 26 18 | 17 | 93 | 26 35 | 16 | 92 | 26 51 | 16 | 91 | 27 07 | 15 | 90 | 27 22 | 15 | 89 |
| 65 | 25 21 | +17 | 93 | 25 38 | +16 | 92 | 25 54 | +16 | 90 | 26 10 | +15 | 89 | 26 25 | +15 | 88 |
| 66 | 24 24 | 16 | 92 | 24 40 | 17 | 91 | 24 57 | 15 | 90 | 25 12 | 16 | 89 | 25 28 | 15 | 88 |
| 67 | 23 26 | 17 | 92 | 23 43 | 16 | 91 | 23 59 | 16 | 90 | 24 15 | 15 | 89 | 24 30 | 15 | 88 |
| 68 | 22 29 | 17 | 92 | 22 46 | 16 | 91 | 23 02 | 16 | 90 | 23 18 | 15 | 89 | 23 33 | 15 | 87 |
| 69 | 21 31 | 17 | 91 | 21 48 | 16 | 90 | 22 04 | 16 | 89 | 22 20 | 16 | 88 | 22 36 | 15 | 87 |

| LHA | 10° Hc | 10° d | 10° Z | 11° Hc | 11° d | 11° Z | 12° Hc | 12° d | 12° Z | 13° Hc | 13° d | 13° Z | 14° Hc | 14° d | 14° Z | LHA |
|---|---|---|---|---|---|---|---|---|---|---|---|---|---|---|---|---|
| | ° ′ | ′ | ° | ° ′ | ′ | ° | ° ′ | ′ | ° | ° ′ | ′ | ° | ° ′ | ′ | ° | |
| 0 | 83 00 | +60 | 180 | 84 00 | +60 | 180 | 85 00 | +60 | 180 | 86 00 | +60 | 180 | 87 00 | +60 | 180 | 360 |
| 1 | 82 56 | 59 | 172 | 83 55 | 59 | 171 | 84 54 | 59 | 169 | 85 53 | 58 | 166 | 86 51 | 56 | 162 | 359 |
| 2 | 82 44 | 58 | 164 | 83 42 | 56 | 162 | 84 38 | 56 | 159 | 85 34 | 52 | 154 | 86 26 | 48 | 147 | 358 |
| 3 | 82 25 | 55 | 157 | 83 20 | 53 | 154 | 84 13 | 51 | 150 | 85 04 | 46 | 144 | 85 50 | 40 | 136 | 357 |
| 4 | 82 00 | 51 | 150 | 82 51 | 50 | 147 | 83 41 | 45 | 142 | 84 26 | 41 | 136 | 85 07 | 33 | 127 | 356 |
| 5 | 81 29 | +48 | 145 | 82 17 | +46 | 140 | 83 03 | +41 | 135 | 83 44 | +36 | 129 | 84 20 | +28 | 121 | 355 |
| 6 | 80 53 | 46 | 139 | 81 39 | 41 | 135 | 82 20 | 38 | 130 | 82 58 | 31 | 124 | 83 29 | 25 | 117 | 354 |
| 7 | 80 14 | 42 | 135 | 80 56 | 39 | 131 | 81 35 | 34 | 126 | 82 09 | 28 | 120 | 82 37 | 22 | 113 | 353 |
| 8 | 79 32 | 40 | 131 | 80 12 | 35 | 127 | 80 47 | 31 | 122 | 81 18 | 26 | 116 | 81 44 | 19 | 110 | 352 |
| 9 | 78 48 | 37 | 128 | 79 25 | 32 | 123 | 79 57 | 29 | 119 | 80 26 | 24 | 114 | 80 50 | 17 | 108 | 351 |
| 10 | 78 02 | +34 | 125 | 78 36 | +31 | 120 | 79 07 | +26 | 116 | 79 33 | +22 | 111 | 79 55 | +16 | 106 | 350 |
| 11 | 77 14 | 32 | 122 | 77 46 | 28 | 118 | 78 14 | 25 | 114 | 78 39 | 20 | 109 | 78 59 | 15 | 104 | 349 |
| 12 | 76 24 | 31 | 119 | 76 55 | 26 | 116 | 77 21 | 24 | 112 | 77 45 | 18 | 107 | 78 03 | 15 | 103 | 348 |
| 13 | 75 34 | 28 | 117 | 76 02 | 26 | 114 | 76 28 | 22 | 110 | 76 50 | 17 | 106 | 77 07 | 14 | 102 | 347 |
| 14 | 74 42 | 28 | 115 | 75 10 | 24 | 112 | 75 34 | 20 | 108 | 75 54 | 17 | 105 | 76 11 | 13 | 101 | 346 |
| 15 | 73 50 | +26 | 114 | 74 16 | +23 | 111 | 74 39 | +20 | 107 | 74 59 | +16 | 103 | 75 15 | +12 | 100 | 345 |
| 16 | 72 57 | 25 | 112 | 73 22 | 22 | 109 | 73 44 | 19 | 106 | 74 03 | 15 | 102 | 74 18 | 12 | 99 | 344 |
| 17 | 72 04 | 24 | 111 | 72 28 | 21 | 108 | 72 49 | 17 | 105 | 73 06 | 15 | 101 | 73 21 | 12 | 98 | 343 |
| 18 | 71 10 | 23 | 110 | 71 33 | 20 | 107 | 71 53 | 17 | 104 | 72 10 | 14 | 101 | 72 24 | 11 | 97 | 342 |
| 19 | 70 16 | 22 | 108 | 70 38 | 19 | 106 | 70 57 | 17 | 103 | 71 14 | 13 | 100 | 71 27 | 11 | 97 | 341 |
| 20 | 69 21 | +21 | 107 | 69 42 | +19 | 105 | 70 01 | +16 | 102 | 70 17 | +13 | 99 | 70 30 | +11 | 96 | 340 |
| 21 | 68 26 | 21 | 106 | 68 47 | 18 | 104 | 69 05 | 15 | 101 | 69 20 | 13 | 98 | 69 33 | 11 | 96 | 339 |
| 22 | 67 31 | 20 | 105 | 67 51 | 17 | 103 | 68 08 | 15 | 100 | 68 23 | 13 | 98 | 68 36 | 10 | 95 | 338 |
| 23 | 66 35 | 20 | 104 | 66 55 | 17 | 102 | 67 12 | 15 | 100 | 67 27 | 12 | 97 | 67 39 | 10 | 95 | 337 |
| 24 | 65 40 | 19 | 104 | 65 59 | 16 | 101 | 66 15 | 15 | 99 | 66 30 | 12 | 97 | 66 42 | 10 | 94 | 336 |
| 25 | 64 44 | +18 | 103 | 65 02 | +16 | 101 | 65 18 | +15 | 98 | 65 33 | +11 | 96 | 65 44 | +10 | 94 | 335 |
| 26 | 63 48 | 18 | 102 | 64 06 | 16 | 100 | 64 22 | 13 | 98 | 64 35 | 12 | 96 | 64 47 | 10 | 93 | 334 |
| 27 | 62 52 | 17 | 101 | 63 09 | 16 | 99 | 63 25 | 13 | 97 | 63 38 | 12 | 95 | 63 50 | 9 | 93 | 333 |
| 28 | 61 55 | 17 | 101 | 62 12 | 16 | 99 | 62 28 | 13 | 97 | 62 41 | 12 | 95 | 62 53 | 9 | 92 | 332 |
| 29 | 60 59 | 17 | 100 | 61 16 | 15 | 98 | 61 31 | 13 | 96 | 61 44 | 11 | 94 | 61 55 | 10 | 92 | 331 |
| 30 | 60 02 | +17 | 100 | 60 19 | +15 | 98 | 60 34 | +13 | 96 | 60 47 | +11 | 94 | 60 58 | +9 | 92 | 330 |
| 31 | 59 06 | 16 | 99 | 59 22 | 15 | 97 | 59 37 | 12 | 95 | 59 49 | 12 | 93 | 60 01 | 9 | 91 | 329 |
| 32 | 58 09 | 16 | 99 | 58 25 | 14 | 97 | 58 39 | 13 | 95 | 58 52 | 11 | 93 | 59 03 | 10 | 91 | 328 |
| 33 | 57 12 | 16 | 98 | 57 26 | 14 | 96 | 57 42 | 13 | 94 | 57 55 | 11 | 93 | 58 06 | 9 | 91 | 327 |
| 34 | 56 15 | 16 | 98 | 56 31 | 14 | 96 | 56 45 | 12 | 94 | 56 57 | 11 | 92 | 57 08 | 10 | 90 | 326 |
| 35 | 55 19 | +15 | 97 | 55 34 | +14 | 95 | 55 48 | +12 | 94 | 56 00 | +11 | 92 | 56 11 | +9 | 90 | 325 |
| 36 | 54 22 | 15 | 97 | 54 37 | 13 | 95 | 54 50 | 13 | 93 | 55 03 | 11 | 92 | 55 14 | 9 | 90 | 324 |
| 37 | 53 25 | 15 | 96 | 53 40 | 13 | 95 | 53 53 | 12 | 93 | 54 05 | 11 | 91 | 54 16 | 10 | 90 | 323 |
| 38 | 52 27 | 15 | 96 | 52 42 | 14 | 94 | 52 56 | 12 | 93 | 53 08 | 11 | 91 | 53 19 | 9 | 89 | 322 |
| 39 | 51 30 | 15 | 95 | 51 45 | 14 | 94 | 51 59 | 12 | 92 | 52 11 | 11 | 91 | 52 22 | 9 | 89 | 321 |
| 40 | 50 33 | +15 | 95 | 50 48 | +13 | 93 | 51 01 | +12 | 92 | 51 13 | +11 | 90 | 51 24 | +10 | 89 | 320 |
| 41 | 49 36 | 15 | 95 | 49 51 | 13 | 93 | 50 04 | 12 | 92 | 50 16 | 11 | 90 | 50 27 | 9 | 88 | 319 |
| 42 | 48 39 | 14 | 94 | 48 53 | 13 | 93 | 49 06 | 13 | 91 | 49 19 | 10 | 90 | 49 29 | 10 | 88 | 318 |
| 43 | 47 42 | 14 | 94 | 47 56 | 13 | 92 | 48 09 | 12 | 91 | 48 21 | 11 | 89 | 48 32 | 10 | 88 | 317 |
| 44 | 46 44 | 15 | 93 | 46 59 | 13 | 92 | 47 12 | 12 | 91 | 47 24 | 11 | 89 | 47 35 | 10 | 88 | 316 |
| 45 | 45 47 | +14 | 93 | 46 01 | +13 | 92 | 46 14 | +12 | 90 | 46 26 | +11 | 89 | 46 37 | +10 | 87 | 315 |
| 46 | 44 50 | 14 | 93 | 45 04 | 13 | 91 | 45 17 | 12 | 90 | 45 29 | 11 | 89 | 45 40 | 10 | 87 | 314 |
| 47 | 43 52 | 14 | 92 | 44 06 | 14 | 91 | 44 20 | 12 | 90 | 44 32 | 11 | 88 | 44 43 | 10 | 87 | 313 |
| 48 | 42 55 | 14 | 92 | 43 09 | 13 | 91 | 43 22 | 12 | 89 | 43 34 | 12 | 88 | 43 46 | 10 | 87 | 312 |
| 49 | 41 58 | 14 | 92 | 42 12 | 13 | 90 | 42 25 | 12 | 89 | 42 37 | 11 | 88 | 42 48 | 11 | 87 | 311 |
| 50 | 41 00 | +14 | 91 | 41 14 | +13 | 90 | 41 27 | +13 | 89 | 41 40 | +11 | 88 | 41 51 | +10 | 86 | 310 |
| 51 | 40 03 | 14 | 91 | 40 17 | 13 | 90 | 40 30 | 12 | 89 | 40 42 | 12 | 87 | 40 54 | 10 | 86 | 309 |
| 52 | 39 06 | 14 | 91 | 39 20 | 13 | 90 | 39 33 | 12 | 88 | 39 45 | 12 | 87 | 39 57 | 10 | 86 | 308 |
| 53 | 38 08 | 14 | 91 | 38 22 | 13 | 89 | 38 35 | 13 | 88 | 38 48 | 11 | 87 | 38 59 | 11 | 86 | 307 |
| 54 | 37 11 | 14 | 90 | 37 25 | 13 | 89 | 37 38 | 12 | 88 | 37 50 | 12 | 87 | 38 02 | 11 | 85 | 306 |
| 55 | 36 14 | +13 | 90 | 36 27 | +14 | 89 | 36 41 | +12 | 88 | 36 53 | +12 | 86 | 37 05 | +11 | 85 | 305 |
| 56 | 35 16 | 14 | 90 | 35 30 | 13 | 89 | 35 43 | 13 | 87 | 35 56 | 12 | 86 | 36 08 | 11 | 85 | 304 |
| 57 | 34 19 | 14 | 89 | 34 33 | 13 | 88 | 34 46 | 13 | 87 | 34 59 | 12 | 86 | 35 11 | 11 | 85 | 303 |
| 58 | 33 21 | 14 | 89 | 33 35 | 14 | 88 | 33 49 | 13 | 87 | 34 02 | 12 | 86 | 34 14 | 11 | 84 | 302 |
| 59 | 32 24 | 14 | 89 | 32 38 | 14 | 88 | 32 52 | 12 | 87 | 33 04 | 12 | 85 | 33 16 | 12 | 84 | 301 |
| 60 | 31 27 | +14 | 89 | 31 41 | +13 | 87 | 31 54 | +13 | 86 | 32 07 | +12 | 85 | 32 19 | +12 | 84 | 300 |
| 61 | 30 29 | 14 | 88 | 30 43 | 14 | 87 | 30 57 | 13 | 86 | 31 10 | 12 | 85 | 31 22 | 12 | 84 | 299 |
| 62 | 29 32 | 14 | 88 | 29 46 | 14 | 87 | 30 00 | 13 | 86 | 30 13 | 12 | 85 | 30 25 | 12 | 84 | 298 |
| 63 | 28 35 | 14 | 88 | 28 49 | 14 | 87 | 29 03 | 13 | 86 | 29 16 | 12 | 84 | 29 28 | 12 | 83 | 297 |
| 64 | 27 37 | 15 | 88 | 27 52 | 13 | 86 | 28 05 | 13 | 85 | 28 19 | 12 | 84 | 28 31 | 13 | 83 | 296 |
| 65 | 26 40 | +14 | 87 | 26 54 | +14 | 86 | 27 08 | +14 | 85 | 27 22 | +12 | 84 | 27 34 | +13 | 83 | 295 |
| 66 | 25 43 | 14 | 87 | 25 57 | 14 | 86 | 26 11 | 14 | 85 | 26 25 | 13 | 83 | 26 38 | 12 | 83 | 294 |
| 67 | 24 45 | 15 | 87 | 25 00 | 14 | 86 | 25 14 | 14 | 85 | 25 28 | 13 | 83 | 25 41 | 12 | 82 | 293 |
| 68 | 23 48 | 15 | 86 | 24 03 | 14 | 85 | 24 17 | 14 | 84 | 24 31 | 13 | 83 | 24 44 | 13 | 82 | 292 |
| 69 | 22 51 | 15 | 86 | 23 06 | 14 | 85 | 23 20 | 14 | 84 | 23 34 | 13 | 83 | 23 47 | 13 | 82 | 291 |

S. Lat. {LHA greater than 180°.......Zn=180−Z; LHA less than 180°..........Zn=180+Z}

DECLINATION (0°-14°) SAME NAME AS LATITUDE

# DECLINATION (0°-14°) CONTRARY NAME TO LATITUDE

**LAT 18°**

N. Lat. {LHA greater than 180° ....... Zn=Z; LHA less than 180° ........... Zn=360−Z}

| LHA | 0° Hc | d | Z | 1° Hc | d | Z | 2° Hc | d | Z | 3° Hc | d | Z | 4° Hc | d | Z | 5° Hc | d | Z | 6° Hc | d | Z | 7° Hc | d | Z |
|---|---|---|---|---|---|---|---|---|---|---|---|---|---|---|---|---|---|---|---|---|---|---|---|---|
| | ° ′ | ′ | ° | ° ′ | ′ | ° | ° ′ | ′ | ° | ° ′ | ′ | ° | ° ′ | ′ | ° | ° ′ | ′ | ° | ° ′ | ′ | ° | ° ′ | ′ | ° |
| 69 | 19 56 | 20 | 97 | 19 36 | 20 | 98 | 19 16 | 21 | 99 | 18 55 | 21 | 100 | 18 34 | 21 | 101 | 18 13 | 21 | 102 | 17 52 | 22 | 103 | 17 30 | 22 | 104 |
| 68 | 20 52 | 20 | 97 | 20 32 | 20 | 98 | 20 12 | 21 | 99 | 19 51 | 21 | 100 | 19 30 | 21 | 101 | 19 09 | 22 | 102 | 18 47 | 22 | 103 | 18 25 | 22 | 104 |
| 67 | 21 49 | 20 | 98 | 21 29 | 21 | 99 | 21 08 | 21 | 100 | 20 47 | 21 | 101 | 20 26 | 21 | 102 | 20 05 | 22 | 103 | 19 43 | 23 | 104 | 19 20 | 22 | 105 |
| 66 | 22 45 | 20 | 98 | 22 25 | 20 | 99 | 22 05 | 22 | 100 | 21 43 | 21 | 101 | 21 22 | 22 | 102 | 21 00 | 22 | 103 | 20 38 | 22 | 104 | 20 16 | 23 | 105 |
| 65 | 23 42 | −20 | 98 | 23 22 | −21 | 99 | 23 01 | −22 | 100 | 22 39 | −21 | 101 | 22 18 | −22 | 102 | 21 56 | −23 | 103 | 21 33 | −22 | 104 | 21 11 | −23 | 105 |
| 64 | 24 38 | 20 | 99 | 24 18 | 21 | 100 | 23 57 | 22 | 101 | 23 35 | 21 | 102 | 23 14 | 23 | 103 | 22 51 | 22 | 104 | 22 29 | 23 | 105 | 22 06 | 24 | 106 |
| 63 | 25 35 | 21 | 99 | 25 14 | 21 | 100 | 24 53 | 22 | 101 | 24 31 | 22 | 102 | 24 09 | 22 | 103 | 23 47 | 23 | 104 | 23 24 | 23 | 105 | 23 01 | 24 | 106 |
| 62 | 26 31 | 21 | 99 | 26 10 | 21 | 100 | 25 49 | 22 | 101 | 25 27 | 22 | 102 | 25 05 | 23 | 104 | 24 42 | 23 | 105 | 24 19 | 24 | 106 | 23 55 | 23 | 107 |
| 61 | 27 27 | 21 | 100 | 27 06 | 21 | 101 | 26 45 | 22 | 102 | 26 23 | 23 | 103 | 26 00 | 23 | 104 | 25 37 | 23 | 105 | 25 14 | 24 | 106 | 24 50 | 24 | 107 |
| 60 | 28 24 | −22 | 100 | 28 02 | −21 | 101 | 27 41 | −23 | 102 | 27 18 | −22 | 103 | 26 56 | −24 | 104 | 26 32 | −23 | 105 | 26 09 | −24 | 106 | 25 45 | −25 | 107 |
| 59 | 29 20 | 22 | 101 | 28 58 | 22 | 102 | 28 36 | 22 | 103 | 28 14 | 23 | 104 | 27 51 | 24 | 105 | 27 27 | 24 | 106 | 27 03 | 24 | 107 | 26 39 | 25 | 108 |
| 58 | 30 16 | 22 | 101 | 29 54 | 22 | 102 | 29 32 | 23 | 103 | 29 09 | 23 | 104 | 28 46 | 24 | 105 | 28 22 | 24 | 106 | 27 58 | 25 | 107 | 27 33 | 25 | 108 |
| 57 | 31 12 | 22 | 101 | 30 50 | 23 | 102 | 30 27 | 23 | 104 | 30 04 | 23 | 105 | 29 41 | 24 | 106 | 29 17 | 25 | 107 | 28 52 | 25 | 108 | 28 27 | 25 | 109 |
| 56 | 32 08 | 22 | 102 | 31 46 | 23 | 103 | 31 23 | 23 | 104 | 31 00 | 24 | 105 | 30 36 | 25 | 106 | 30 11 | 24 | 107 | 29 47 | 26 | 108 | 29 21 | 26 | 109 |
| 55 | 33 04 | −23 | 102 | 32 41 | −23 | 103 | 32 18 | −23 | 104 | 31 55 | −24 | 106 | 31 31 | −25 | 107 | 31 06 | −25 | 108 | 30 41 | −26 | 109 | 30 15 | −26 | 110 |
| 54 | 33 59 | 22 | 103 | 33 37 | 24 | 104 | 33 13 | 23 | 105 | 32 50 | 25 | 106 | 32 25 | 25 | 107 | 32 00 | 25 | 108 | 31 35 | 26 | 109 | 31 09 | 27 | 110 |
| 53 | 34 55 | 23 | 103 | 34 32 | 24 | 104 | 34 08 | 24 | 105 | 33 44 | 24 | 106 | 33 20 | 26 | 108 | 32 54 | 25 | 109 | 32 29 | 27 | 110 | 32 02 | 27 | 111 |
| 52 | 35 50 | 23 | 104 | 35 27 | 24 | 105 | 35 03 | 24 | 106 | 34 39 | 25 | 107 | 34 14 | 26 | 108 | 33 48 | 26 | 109 | 33 22 | 27 | 110 | 32 55 | 27 | 111 |
| 51 | 36 46 | 24 | 104 | 36 22 | 24 | 105 | 35 58 | 24 | 106 | 35 34 | 26 | 108 | 35 08 | 26 | 109 | 34 42 | 26 | 110 | 34 16 | 28 | 111 | 33 48 | 27 | 112 |
| 50 | 37 41 | −24 | 105 | 37 17 | −24 | 106 | 36 53 | −25 | 107 | 36 28 | −26 | 108 | 36 02 | −26 | 109 | 35 36 | −27 | 110 | 35 09 | −28 | 111 | 34 41 | −28 | 112 |
| 49 | 38 36 | 24 | 105 | 38 12 | 24 | 106 | 37 48 | 26 | 107 | 37 22 | 26 | 109 | 36 56 | 27 | 110 | 36 29 | 27 | 111 | 36 02 | 28 | 112 | 35 34 | 29 | 113 |
| 48 | 39 31 | 24 | 106 | 39 07 | 25 | 107 | 38 42 | 26 | 108 | 38 16 | 26 | 109 | 37 50 | 27 | 110 | 37 23 | 28 | 111 | 36 55 | 29 | 112 | 36 26 | 29 | 114 |
| 47 | 40 26 | 24 | 106 | 40 02 | 26 | 107 | 39 36 | 26 | 108 | 39 10 | 27 | 110 | 38 43 | 27 | 111 | 38 16 | 29 | 112 | 37 47 | 28 | 113 | 37 19 | 30 | 114 |
| 46 | 41 21 | 25 | 107 | 40 56 | 26 | 108 | 40 30 | 26 | 109 | 40 04 | 28 | 111 | 39 36 | 28 | 112 | 39 08 | 28 | 113 | 38 40 | 29 | 114 | 38 11 | 30 | 115 |
| 45 | 42 16 | −26 | 107 | 41 50 | −26 | 108 | 41 24 | −27 | 110 | 40 57 | −28 | 111 | 40 29 | −28 | 112 | 40 01 | −29 | 113 | 39 32 | −30 | 114 | 39 02 | −30 | 115 |
| 44 | 43 10 | 26 | 108 | 42 44 | 26 | 109 | 42 18 | 28 | 110 | 41 50 | 28 | 111 | 41 22 | 29 | 113 | 40 53 | 29 | 114 | 40 24 | 30 | 115 | 39 54 | 31 | 116 |
| 43 | 44 04 | 26 | 108 | 43 38 | 27 | 110 | 43 11 | 28 | 111 | 42 43 | 28 | 112 | 42 15 | 29 | 113 | 41 46 | 30 | 114 | 41 16 | 31 | 116 | 40 45 | 31 | 117 |
| 42 | 44 58 | 26 | 109 | 44 32 | 28 | 110 | 44 04 | 28 | 111 | 43 36 | 29 | 113 | 43 07 | 30 | 114 | 42 37 | 30 | 115 | 42 07 | 31 | 116 | 41 36 | 32 | 117 |
| 41 | 45 52 | 27 | 110 | 45 25 | 28 | 111 | 44 57 | 28 | 112 | 44 29 | 30 | 113 | 43 59 | 30 | 115 | 43 29 | 31 | 116 | 42 58 | 32 | 117 | 42 26 | 32 | 118 |
| 40 | 46 46 | −28 | 110 | 46 18 | −28 | 112 | 45 50 | −29 | 113 | 45 21 | −30 | 114 | 44 51 | −31 | 115 | 44 20 | −31 | 117 | 43 49 | −33 | 118 | 43 16 | −33 | 119 |
| 39 | 47 39 | 28 | 111 | 47 11 | 28 | 112 | 46 43 | 30 | 114 | 46 13 | 31 | 115 | 45 42 | 31 | 116 | 45 11 | 32 | 117 | 44 39 | 33 | 118 | 44 06 | 33 | 120 |
| 38 | 48 33 | 29 | 112 | 48 04 | 29 | 113 | 47 35 | 30 | 114 | 47 05 | 32 | 116 | 46 33 | 31 | 117 | 46 02 | 33 | 118 | 45 29 | 33 | 119 | 44 56 | 34 | 120 |
| 37 | 49 26 | 29 | 112 | 48 57 | 30 | 114 | 48 27 | 31 | 115 | 47 56 | 32 | 116 | 47 24 | 32 | 118 | 46 52 | 33 | 119 | 46 19 | 34 | 120 | 45 45 | 35 | 121 |
| 36 | 50 18 | 29 | 113 | 49 49 | 31 | 114 | 49 18 | 31 | 116 | 48 47 | 32 | 117 | 48 15 | 33 | 118 | 47 42 | 34 | 120 | 47 08 | 35 | 121 | 46 33 | 35 | 122 |
| 35 | 51 11 | −31 | 114 | 50 40 | −31 | 115 | 50 09 | −31 | 117 | 49 38 | −33 | 118 | 49 05 | −34 | 119 | 48 31 | −34 | 120 | 47 57 | −36 | 122 | 47 21 | −35 | 123 |
| 34 | 52 03 | 31 | 115 | 51 32 | 32 | 116 | 51 00 | 32 | 117 | 50 28 | 34 | 119 | 49 54 | 34 | 120 | 49 20 | 35 | 121 | 48 45 | 36 | 123 | 48 09 | 36 | 124 |
| 33 | 52 54 | 31 | 115 | 52 23 | 32 | 117 | 51 51 | 33 | 118 | 51 18 | 34 | 120 | 50 44 | 35 | 121 | 50 09 | 36 | 122 | 49 33 | 37 | 123 | 48 56 | 37 | 125 |
| 32 | 53 46 | 32 | 116 | 53 14 | 33 | 118 | 52 41 | 34 | 119 | 52 07 | 35 | 121 | 51 32 | 35 | 122 | 50 57 | 37 | 123 | 50 20 | 37 | 124 | 49 43 | 38 | 126 |
| 31 | 54 37 | 33 | 117 | 54 04 | 34 | 119 | 53 30 | 34 | 120 | 52 56 | 35 | 121 | 52 21 | 37 | 123 | 51 44 | 37 | 124 | 51 07 | 38 | 125 | 50 29 | 38 | 127 |
| 30 | 55 27 | −33 | 118 | 54 54 | −34 | 120 | 54 20 | −36 | 121 | 53 44 | −36 | 122 | 53 08 | −37 | 124 | 52 31 | −38 | 125 | 51 53 | −38 | 126 | 51 15 | −40 | 128 |
| 29 | 56 17 | 34 | 119 | 55 43 | 35 | 121 | 55 08 | 36 | 122 | 54 32 | 37 | 123 | 53 55 | 37 | 125 | 53 18 | 39 | 126 | 52 39 | 39 | 127 | 52 00 | 40 | 129 |
| 28 | 57 07 | 35 | 120 | 56 32 | 36 | 122 | 55 56 | 36 | 123 | 55 20 | 38 | 125 | 54 42 | 39 | 126 | 54 03 | 39 | 127 | 53 24 | 40 | 129 | 52 44 | 41 | 130 |
| 27 | 57 56 | 36 | 121 | 57 20 | 36 | 123 | 56 44 | 38 | 124 | 56 06 | 38 | 126 | 55 28 | 39 | 127 | 54 49 | 41 | 128 | 54 08 | 40 | 130 | 53 28 | 42 | 131 |
| 26 | 58 44 | 36 | 122 | 58 08 | 37 | 124 | 57 31 | 39 | 125 | 56 52 | 39 | 127 | 56 13 | 40 | 128 | 55 33 | 41 | 130 | 54 52 | 42 | 131 | 54 10 | 42 | 132 |
| 25 | 59 32 | −37 | 124 | 58 55 | −38 | 125 | 58 17 | −39 | 127 | 57 38 | −40 | 128 | 56 58 | −41 | 129 | 56 17 | −42 | 131 | 55 35 | −43 | 132 | 54 52 | −43 | 133 |
| 24 | 60 19 | 38 | 125 | 59 41 | 39 | 126 | 59 02 | 40 | 128 | 58 22 | 41 | 129 | 57 41 | 41 | 131 | 57 00 | 43 | 132 | 56 17 | 43 | 133 | 55 34 | 45 | 135 |
| 23 | 61 06 | 39 | 126 | 60 27 | 40 | 128 | 59 47 | 41 | 129 | 59 06 | 42 | 131 | 58 24 | 42 | 132 | 57 42 | 44 | 133 | 56 58 | 44 | 135 | 56 14 | 45 | 136 |
| 22 | 61 52 | 40 | 127 | 61 12 | 41 | 129 | 60 31 | 42 | 131 | 59 49 | 43 | 132 | 59 06 | 43 | 133 | 58 23 | 45 | 135 | 57 38 | 45 | 136 | 56 53 | 46 | 137 |
| 21 | 62 37 | 41 | 129 | 61 56 | 42 | 130 | 61 14 | 43 | 132 | 60 31 | 44 | 133 | 59 47 | 44 | 135 | 59 03 | 46 | 136 | 58 17 | 46 | 137 | 57 31 | 46 | 139 |
| 20 | 63 21 | −42 | 130 | 62 39 | −43 | 132 | 61 56 | −44 | 133 | 61 12 | −45 | 135 | 60 27 | −45 | 136 | 59 42 | −46 | 138 | 58 56 | −47 | 139 | 58 09 | −48 | 140 |
| 19 | 64 04 | 43 | 132 | 63 21 | 44 | 134 | 62 37 | 45 | 135 | 61 52 | 46 | 136 | 61 06 | 46 | 138 | 60 20 | 47 | 139 | 59 33 | 48 | 140 | 58 45 | 49 | 142 |
| 18 | 64 45 | 44 | 134 | 64 01 | 44 | 135 | 63 17 | 46 | 137 | 62 31 | 47 | 138 | 61 44 | 47 | 139 | 60 57 | 48 | 141 | 60 09 | 49 | 142 | 59 20 | 50 | 143 |
| 17 | 65 26 | 45 | 135 | 64 41 | 46 | 136 | 63 55 | 47 | 138 | 63 08 | 47 | 140 | 62 21 | 49 | 141 | 61 32 | 49 | 142 | 60 43 | 50 | 144 | 59 53 | 50 | 145 |
| 16 | 66 06 | 47 | 137 | 65 19 | 47 | 139 | 64 32 | 48 | 140 | 63 44 | 48 | 142 | 62 56 | 50 | 143 | 62 06 | 50 | 144 | 61 16 | 50 | 145 | 60 26 | 51 | 146 |
| 15 | 66 44 | −48 | 139 | 65 56 | −48 | 141 | 65 08 | −49 | 142 | 64 19 | −50 | 143 | 63 29 | −50 | 145 | 62 39 | −51 | 146 | 61 48 | −51 | 147 | 60 57 | −52 | 148 |
| 14 | 67 20 | 48 | 141 | 66 32 | 49 | 143 | 65 43 | 51 | 144 | 64 52 | 50 | 145 | 64 02 | 52 | 147 | 63 10 | 52 | 148 | 62 18 | 52 | 149 | 61 26 | 53 | 150 |
| 13 | 67 55 | 49 | 143 | 67 06 | 51 | 145 | 66 15 | 51 | 146 | 65 24 | 52 | 147 | 64 32 | 52 | 149 | 63 40 | 53 | 150 | 62 47 | 53 | 151 | 61 54 | 54 | 152 |
| 12 | 68 29 | 51 | 146 | 67 38 | 52 | 148 | 66 46 | 52 | 148 | 65 54 | 53 | 149 | 65 01 | 53 | 151 | 64 08 | 54 | 152 | 63 14 | 54 | 153 | 62 20 | 54 | 154 |
| 11 | 69 00 | 52 | 148 | 68 08 | 53 | 149 | 67 15 | 53 | 150 | 66 22 | 54 | 152 | 65 28 | 54 | 153 | 64 34 | 54 | 154 | 63 40 | 55 | 155 | 62 45 | 56 | 156 |
| 10 | 69 29 | −53 | 150 | 68 36 | −54 | 152 | 67 42 | −54 | 153 | 66 48 | −54 | 154 | 65 54 | −56 | 155 | 64 58 | −55 | 156 | 64 03 | −56 | 157 | 63 07 | −56 | 158 |
| 9 | 69 57 | 55 | 153 | 69 02 | 55 | 154 | 68 07 | 55 | 155 | 67 12 | 55 | 156 | 66 17 | 56 | 157 | 65 21 | 56 | 158 | 64 25 | 57 | 159 | 63 28 | 57 | 160 |
| 8 | 70 21 | 55 | 156 | 69 26 | 56 | 157 | 68 30 | 56 | 158 | 67 34 | 56 | 159 | 66 38 | 57 | 160 | 65 41 | 57 | 160 | 64 44 | 57 | 161 | 63 47 | 57 | 162 |
| 7 | 70 44 | 57 | 158 | 69 47 | 56 | 159 | 68 51 | 57 | 160 | 67 54 | 57 | 161 | 66 57 | 58 | 162 | 65 59 | 57 | 163 | 65 02 | 58 | 163 | 64 04 | 58 | 164 |
| 6 | 71 04 | 58 | 161 | 70 06 | 57 | 162 | 69 09 | 58 | 163 | 68 11 | 58 | 164 | 67 13 | 58 | 164 | 66 15 | 58 | 165 | 65 17 | 58 | 166 | 64 19 | 59 | 166 |
| 5 | 71 20 | −58 | 164 | 70 22 | −58 | 165 | 69 24 | −58 | 166 | 68 26 | −59 | 166 | 67 27 | −58 | 167 | 66 29 | −59 | 167 | 65 30 | −59 | 168 | 64 31 | −59 | 168 |
| 4 | 71 35 | 59 | 167 | 70 36 | 59 | 168 | 69 37 | 59 | 169 | 68 38 | 59 | 169 | 67 39 | 59 | 170 | 66 40 | 59 | 170 | 65 41 | 60 | 170 | 64 41 | 59 | 171 |
| 3 | 71 46 | 60 | 170 | 70 46 | 59 | 171 | 69 47 | 59 | 171 | 68 48 | 60 | 172 | 67 48 | 59 | 172 | 66 49 | 60 | 172 | 65 49 | 59 | 173 | 64 50 | 60 | 173 |
| 2 | 71 54 | 60 | 174 | 70 54 | 60 | 174 | 69 54 | 59 | 174 | 68 55 | 60 | 174 | 67 55 | 60 | 175 | 66 55 | 60 | 175 | 65 55 | 60 | 175 | 64 55 | 59 | 175 |
| 1 | 71 58 | 59 | 177 | 70 59 | 60 | 177 | 69 59 | 60 | 177 | 68 59 | 60 | 177 | 67 59 | 60 | 177 | 66 59 | 60 | 178 | 65 59 | 60 | 178 | 64 59 | 60 | 178 |
| 0 | 72 00 | −60 | 180 | 71 00 | −60 | 180 | 70 00 | −60 | 180 | 69 00 | −60 | 180 | 68 00 | −60 | 180 | 67 00 | −60 | 180 | 66 00 | −60 | 180 | 65 00 | −60 | 180 |

| 8° Hc | d | Z | 9° Hc | d | Z | 10° Hc | d | Z | 11° Hc | d | Z | 12° Hc | d | Z | 13° Hc | d | Z | 14° Hc | d | Z | LHA |
|---|---|---|---|---|---|---|---|---|---|---|---|---|---|---|---|---|---|---|---|---|---|
| ° ′ | ′ | ° | ° ′ | ′ | ° | ° ′ | ′ | ° | ° ′ | ′ | ° | ° ′ | ′ | ° | ° ′ | ′ | ° | ° ′ | ′ | ° | |
| 17 08 | 23 | 105 | 16 45 | 22 | 106 | 16 23 | 23 | 107 | 16 00 | 23 | 108 | 15 37 | 24 | 109 | 15 13 | 23 | 110 | 14 50 | 24 | 110 | 291 |
| 18 03 | 23 | 105 | 17 40 | 23 | 106 | 17 17 | 23 | 107 | 16 54 | 23 | 108 | 16 31 | 24 | 109 | 16 07 | 24 | 110 | 15 43 | 24 | 111 | 292 |
| 18 58 | 23 | 105 | 18 35 | 23 | 106 | 18 12 | 24 | 107 | 17 48 | 23 | 108 | 17 25 | 24 | 109 | 17 01 | 25 | 110 | 16 36 | 24 | 111 | 293 |
| 19 53 | 23 | 106 | 19 30 | 24 | 107 | 19 06 | 23 | 108 | 18 43 | 24 | 109 | 18 19 | 25 | 110 | 17 54 | 24 | 111 | 17 30 | 25 | 112 | 294 |
| 20 48 | −24 | 106 | 20 24 | −23 | 107 | 20 01 | −24 | 108 | 19 37 | −25 | 109 | 19 12 | −24 | 110 | 18 48 | −25 | 111 | 18 23 | −26 | 112 | 295 |
| 21 42 | 23 | 107 | 21 19 | 24 | 108 | 20 55 | 25 | 109 | 20 30 | 24 | 110 | 20 06 | 25 | 111 | 19 41 | 26 | 112 | 19 15 | 25 | 113 | 296 |
| 22 37 | 24 | 107 | 22 13 | 24 | 108 | 21 49 | 25 | 109 | 21 24 | 25 | 110 | 20 59 | 25 | 111 | 20 34 | 26 | 112 | 20 08 | 26 | 113 | 297 |
| 23 32 | 25 | 108 | 23 07 | 24 | 109 | 22 43 | 25 | 110 | 22 18 | 26 | 111 | 21 52 | 26 | 112 | 21 26 | 26 | 112 | 21 00 | 26 | 113 | 298 |
| 24 26 | 25 | 108 | 24 01 | 25 | 109 | 23 36 | 25 | 110 | 23 11 | 26 | 111 | 22 45 | 26 | 112 | 22 19 | 26 | 113 | 21 53 | 27 | 114 | 299 |
| 25 20 | −25 | 108 | 24 55 | −25 | 109 | 24 30 | −26 | 110 | 24 04 | −26 | 111 | 23 38 | −26 | 112 | 23 12 | −27 | 113 | 22 45 | −27 | 114 | 300 |
| 26 14 | 25 | 109 | 25 49 | 26 | 110 | 25 23 | 26 | 111 | 24 57 | 26 | 112 | 24 31 | 27 | 113 | 24 04 | 27 | 114 | 23 37 | 28 | 115 | 301 |
| 27 08 | 25 | 109 | 26 43 | 27 | 110 | 26 16 | 26 | 111 | 25 50 | 27 | 112 | 25 23 | 27 | 113 | 24 56 | 28 | 114 | 24 28 | 28 | 115 | 302 |
| 28 02 | 26 | 110 | 27 36 | 26 | 111 | 27 10 | 27 | 112 | 26 43 | 27 | 113 | 26 16 | 28 | 114 | 25 48 | 28 | 115 | 25 20 | 29 | 116 | 303 |
| 28 55 | 26 | 110 | 28 29 | 27 | 111 | 28 02 | 27 | 112 | 27 35 | 27 | 113 | 27 08 | 28 | 114 | 26 40 | 29 | 115 | 26 11 | 29 | 116 | 304 |
| 29 49 | −27 | 111 | 29 22 | −27 | 112 | 28 55 | −27 | 113 | 28 28 | −28 | 114 | 28 00 | −29 | 115 | 27 31 | −29 | 116 | 27 02 | −29 | 117 | 305 |
| 30 42 | 27 | 111 | 30 15 | 27 | 112 | 29 48 | 28 | 113 | 29 20 | 29 | 114 | 28 51 | 29 | 115 | 28 22 | 29 | 116 | 27 53 | 30 | 117 | 306 |
| 31 35 | 27 | 112 | 31 08 | 28 | 113 | 30 40 | 29 | 114 | 30 11 | 28 | 115 | 29 43 | 30 | 116 | 29 13 | 29 | 117 | 28 44 | 31 | 118 | 307 |
| 32 28 | 28 | 112 | 32 00 | 28 | 113 | 31 32 | 29 | 114 | 31 03 | 29 | 116 | 30 34 | 30 | 117 | 30 04 | 30 | 118 | 29 34 | 31 | 119 | 308 |
| 33 21 | 28 | 113 | 32 53 | 29 | 114 | 32 24 | 29 | 115 | 31 55 | 30 | 116 | 31 25 | 30 | 117 | 30 55 | 31 | 118 | 30 24 | 31 | 119 | 309 |
| 34 13 | −28 | 114 | 33 45 | −30 | 115 | 33 15 | −29 | 116 | 32 46 | −31 | 117 | 32 15 | −30 | 118 | 31 45 | −31 | 119 | 31 14 | −32 | 120 | 310 |
| 35 05 | 29 | 114 | 34 36 | 29 | 115 | 34 07 | 30 | 116 | 33 37 | 31 | 117 | 33 06 | 31 | 118 | 32 35 | 32 | 119 | 32 03 | 32 | 120 | 311 |
| 35 57 | 29 | 115 | 35 28 | 30 | 116 | 34 58 | 31 | 117 | 34 27 | 31 | 118 | 33 56 | 32 | 119 | 33 24 | 32 | 120 | 32 52 | 32 | 121 | 312 |
| 36 49 | 30 | 115 | 36 19 | 30 | 116 | 35 49 | 31 | 117 | 35 18 | 32 | 118 | 34 46 | 32 | 119 | 34 14 | 33 | 121 | 33 41 | 33 | 122 | 313 |
| 37 41 | 31 | 116 | 37 10 | 31 | 117 | 36 39 | 31 | 118 | 36 08 | 33 | 119 | 35 35 | 32 | 120 | 35 03 | 33 | 121 | 34 30 | 34 | 122 | 314 |
| 38 32 | −31 | 117 | 38 01 | −32 | 118 | 37 29 | −32 | 119 | 36 57 | −32 | 120 | 36 25 | −34 | 121 | 35 51 | −33 | 122 | 35 18 | −35 | 123 | 315 |
| 39 23 | 32 | 117 | 38 51 | 32 | 118 | 38 19 | 32 | 119 | 37 47 | 34 | 120 | 37 13 | 33 | 121 | 36 40 | 35 | 123 | 36 05 | 34 | 124 | 316 |
| 40 14 | 32 | 118 | 39 42 | 33 | 119 | 39 09 | 33 | 120 | 38 36 | 34 | 121 | 38 02 | 34 | 122 | 37 28 | 35 | 123 | 36 53 | 36 | 124 | 317 |
| 41 04 | 33 | 119 | 40 31 | 33 | 120 | 39 58 | 34 | 121 | 39 24 | 34 | 122 | 38 50 | 35 | 123 | 38 15 | 35 | 124 | 37 40 | 36 | 125 | 318 |
| 41 54 | 33 | 119 | 41 21 | 34 | 120 | 40 47 | 34 | 121 | 40 13 | 35 | 123 | 39 38 | 36 | 124 | 39 02 | 36 | 125 | 38 26 | 36 | 126 | 319 |
| 42 43 | −33 | 120 | 42 10 | −34 | 121 | 41 36 | −35 | 122 | 41 01 | −36 | 123 | 40 25 | −36 | 124 | 39 49 | −36 | 125 | 39 13 | −38 | 126 | 320 |
| 43 33 | 34 | 121 | 42 59 | 35 | 122 | 42 24 | 36 | 123 | 41 48 | 36 | 124 | 41 12 | 37 | 125 | 40 35 | 37 | 126 | 39 58 | 37 | 127 | 321 |
| 44 22 | 35 | 122 | 43 47 | 36 | 123 | 43 11 | 36 | 124 | 42 35 | 36 | 125 | 41 59 | 38 | 126 | 41 21 | 37 | 127 | 40 44 | 39 | 128 | 322 |
| 45 10 | 35 | 122 | 44 35 | 36 | 123 | 43 59 | 37 | 125 | 43 22 | 37 | 126 | 42 45 | 38 | 127 | 42 07 | 39 | 128 | 41 28 | 39 | 129 | 323 |
| 45 58 | 36 | 123 | 45 22 | 37 | 124 | 44 45 | 37 | 125 | 44 08 | 38 | 127 | 43 30 | 38 | 128 | 42 52 | 40 | 129 | 42 12 | 39 | 130 | 324 |
| 46 46 | −37 | 124 | 46 09 | −37 | 125 | 45 32 | −38 | 126 | 44 54 | −39 | 127 | 44 15 | −39 | 128 | 43 36 | −40 | 130 | 42 56 | −40 | 131 | 325 |
| 47 33 | 38 | 125 | 46 55 | 38 | 126 | 46 17 | 38 | 127 | 45 39 | 40 | 128 | 44 59 | 39 | 129 | 44 20 | 41 | 130 | 43 39 | 41 | 131 | 326 |
| 48 19 | 38 | 126 | 47 41 | 38 | 127 | 47 03 | 40 | 128 | 46 23 | 40 | 129 | 45 43 | 40 | 130 | 45 03 | 41 | 131 | 44 22 | 42 | 132 | 327 |
| 49 05 | 39 | 127 | 48 26 | 39 | 128 | 47 47 | 40 | 129 | 47 07 | 40 | 130 | 46 27 | 42 | 131 | 45 45 | 41 | 132 | 45 04 | 43 | 133 | 328 |
| 49 51 | 40 | 128 | 49 11 | 40 | 129 | 48 31 | 41 | 130 | 47 50 | 41 | 131 | 47 09 | 42 | 132 | 46 27 | 42 | 133 | 45 45 | 43 | 134 | 329 |
| 50 35 | −40 | 129 | 49 55 | −40 | 130 | 49 15 | −42 | 131 | 48 33 | −42 | 132 | 47 51 | −43 | 133 | 47 08 | −43 | 134 | 46 25 | −43 | 135 | 330 |
| 51 20 | 41 | 130 | 50 39 | 42 | 131 | 49 57 | 42 | 132 | 49 15 | 43 | 133 | 48 32 | 43 | 134 | 47 49 | 44 | 135 | 47 05 | 44 | 136 | 331 |
| 52 03 | 42 | 131 | 51 21 | 42 | 132 | 50 39 | 43 | 133 | 49 56 | 43 | 134 | 49 13 | 44 | 135 | 48 29 | 45 | 136 | 47 44 | 45 | 137 | 332 |
| 52 46 | 43 | 132 | 52 03 | 43 | 133 | 51 20 | 43 | 134 | 50 37 | 44 | 135 | 49 53 | 45 | 136 | 49 08 | 46 | 138 | 48 22 | 45 | 139 | 333 |
| 53 28 | 43 | 133 | 52 45 | 44 | 134 | 52 01 | 45 | 136 | 51 16 | 45 | 137 | 50 31 | 45 | 138 | 49 46 | 46 | 139 | 49 00 | 47 | 140 | 334 |
| 54 09 | −44 | 134 | 53 25 | −45 | 136 | 52 40 | −45 | 137 | 51 55 | −46 | 138 | 51 09 | −46 | 139 | 50 23 | −47 | 140 | 49 36 | −47 | 141 | 335 |
| 54 49 | 44 | 136 | 54 05 | 46 | 137 | 53 19 | 46 | 138 | 52 33 | 46 | 139 | 51 47 | 47 | 140 | 51 00 | 48 | 141 | 50 12 | 48 | 142 | 336 |
| 55 29 | 46 | 137 | 54 43 | 46 | 138 | 53 57 | 47 | 139 | 53 10 | 47 | 140 | 52 23 | 48 | 141 | 51 35 | 48 | 142 | 50 47 | 49 | 143 | 337 |
| 56 07 | 46 | 138 | 55 21 | 47 | 139 | 54 34 | 48 | 141 | 53 46 | 48 | 142 | 52 58 | 49 | 143 | 52 09 | 49 | 144 | 51 20 | 49 | 144 | 338 |
| 56 45 | 48 | 140 | 55 57 | 47 | 141 | 55 10 | 49 | 142 | 54 21 | 49 | 143 | 53 32 | 49 | 144 | 52 43 | 50 | 145 | 51 53 | 50 | 146 | 339 |
| 57 21 | −48 | 141 | 56 33 | −49 | 142 | 55 44 | −49 | 143 | 54 55 | −50 | 144 | 54 05 | −50 | 145 | 53 15 | −50 | 146 | 52 25 | −51 | 147 | 340 |
| 57 56 | 49 | 143 | 57 07 | 49 | 144 | 56 18 | 50 | 145 | 55 28 | 51 | 146 | 54 37 | 51 | 147 | 53 46 | 51 | 148 | 52 55 | 52 | 148 | 341 |
| 58 30 | 49 | 144 | 57 41 | 51 | 145 | 56 50 | 51 | 146 | 55 59 | 51 | 147 | 55 08 | 52 | 148 | 54 16 | 52 | 149 | 53 24 | 52 | 150 | 342 |
| 59 03 | 51 | 146 | 58 12 | 51 | 146 | 57 21 | 51 | 148 | 56 30 | 52 | 149 | 55 38 | 53 | 150 | 54 45 | 52 | 150 | 53 53 | 54 | 151 | 343 |
| 59 35 | 52 | 147 | 58 43 | 52 | 148 | 57 51 | 52 | 149 | 56 59 | 53 | 150 | 56 06 | 53 | 151 | 55 13 | 54 | 152 | 54 19 | 53 | 153 | 344 |
| 60 05 | −53 | 149 | 59 12 | −52 | 150 | 58 20 | −54 | 151 | 57 26 | −53 | 152 | 56 33 | −54 | 153 | 55 39 | −54 | 154 | 54 45 | −55 | 154 | 345 |
| 60 33 | 53 | 151 | 59 40 | 54 | 152 | 58 46 | 53 | 153 | 57 53 | 55 | 154 | 56 58 | 54 | 154 | 56 04 | 55 | 155 | 55 09 | 55 | 156 | 346 |
| 61 00 | 54 | 153 | 60 06 | 54 | 154 | 59 12 | 55 | 155 | 58 17 | 55 | 155 | 57 22 | 55 | 156 | 56 27 | 55 | 157 | 55 32 | 56 | 157 | 347 |
| 61 26 | 55 | 155 | 60 31 | 55 | 155 | 59 36 | 56 | 156 | 58 40 | 55 | 157 | 57 45 | 56 | 158 | 56 49 | 56 | 158 | 55 53 | 56 | 159 | 348 |
| 61 49 | 55 | 156 | 60 54 | 56 | 157 | 59 58 | 56 | 158 | 59 02 | 56 | 159 | 58 06 | 57 | 159 | 57 09 | 56 | 160 | 56 13 | 57 | 161 | 349 |
| 62 11 | −56 | 158 | 61 15 | −56 | 159 | 60 19 | −57 | 160 | 59 22 | −57 | 161 | 58 25 | −57 | 161 | 57 28 | −57 | 162 | 56 31 | −57 | 162 | 350 |
| 62 31 | 56 | 160 | 61 35 | 58 | 161 | 60 37 | 57 | 162 | 59 40 | 57 | 162 | 58 43 | 58 | 163 | 57 45 | 57 | 163 | 56 48 | 58 | 164 | 351 |
| 62 50 | 58 | 163 | 61 52 | 58 | 163 | 60 54 | 57 | 164 | 59 57 | 58 | 164 | 58 59 | 58 | 165 | 58 01 | 58 | 165 | 57 03 | 59 | 166 | 352 |
| 63 06 | 58 | 165 | 62 08 | 58 | 165 | 61 10 | 59 | 166 | 60 11 | 58 | 166 | 59 13 | 59 | 167 | 58 14 | 58 | 167 | 57 16 | 59 | 167 | 353 |
| 63 20 | 58 | 167 | 62 22 | 59 | 167 | 61 23 | 59 | 168 | 60 24 | 59 | 168 | 59 25 | 59 | 169 | 58 26 | 58 | 169 | 57 28 | 59 | 169 | 354 |
| 63 32 | −59 | 169 | 62 33 | −59 | 169 | 61 34 | −59 | 170 | 60 35 | −59 | 170 | 59 36 | −59 | 170 | 58 37 | −60 | 171 | 57 37 | −59 | 171 | 355 |
| 63 42 | 59 | 171 | 62 43 | 60 | 171 | 61 43 | 59 | 172 | 60 44 | 59 | 172 | 59 45 | 60 | 172 | 58 45 | 59 | 173 | 57 46 | 60 | 173 | 356 |
| 63 50 | 60 | 173 | 62 50 | 59 | 174 | 61 51 | 60 | 174 | 60 51 | 60 | 174 | 59 51 | 59 | 174 | 58 52 | 60 | 174 | 57 52 | 60 | 175 | 357 |
| 63 56 | 60 | 176 | 62 56 | 60 | 176 | 61 56 | 60 | 176 | 60 56 | 60 | 176 | 59 56 | 60 | 176 | 58 56 | 60 | 176 | 57 56 | 59 | 176 | 358 |
| 63 59 | 60 | 178 | 62 59 | 60 | 178 | 61 59 | 60 | 178 | 60 59 | 60 | 178 | 59 59 | 60 | 178 | 58 59 | 60 | 178 | 57 59 | 60 | 178 | 359 |
| 64 00 | −60 | 180 | 63 00 | −60 | 180 | 62 00 | −60 | 180 | 61 00 | −60 | 180 | 60 00 | −60 | 180 | 59 00 | −60 | 180 | 58 00 | −60 | 180 | 360 |

S. Lat. {LHA greater than 180° ....... Zn=180−Z; LHA less than 180° ........... Zn=180+Z}

DECLINATION (0°-14°) CONTRARY NAME TO LATITUDE

LAT 18°

# DECLINATION (0°-14°) CONTRARY NAME TO LATITUDE



| LHA | 0° Hc | 0° d | 0° Z | 1° Hc | 1° d | 1° Z | 2° Hc | 2° d | 2° Z | 3° Hc | 3° d | 3° Z | 4° Hc | 4° d | 4° Z |
|---|---|---|---|---|---|---|---|---|---|---|---|---|---|---|---|
| | ° ′ | ′ | ° | ° ′ | ′ | ° | ° ′ | ′ | ° | ° ′ | ′ | ° | ° ′ | ′ | ° |
| 69 | 19 48 | 20 | 97 | 19 28 | 22 | 98 | 19 06 | 21 | 99 | 18 45 | 22 | 100 | 18 23 | 22 | 101 |
| 68 | 20 45 | 21 | 98 | 20 24 | 22 | 99 | 20 02 | 21 | 100 | 19 41 | 23 | 101 | 19 18 | 22 | 102 |
| 67 | 21 41 | 21 | 98 | 21 20 | 22 | 99 | 20 58 | 22 | 100 | 20 36 | 22 | 101 | 20 14 | 23 | 102 |
| 66 | 22 37 | 21 | 98 | 22 16 | 22 | 99 | 21 54 | 22 | 100 | 21 32 | 23 | 101 | 21 09 | 22 | 102 |
| 65 | 23 33 | −21 | 99 | 23 12 | −22 | 100 | 22 50 | −22 | 101 | 22 28 | −23 | 102 | 22 05 | −23 | 103 |
| 64 | 24 29 | 21 | 99 | 24 08 | 22 | 100 | 23 46 | 23 | 101 | 23 23 | 23 | 102 | 23 00 | 23 | 103 |
| 63 | 25 25 | 22 | 99 | 25 03 | 22 | 100 | 24 41 | 23 | 102 | 24 18 | 23 | 103 | 23 55 | 23 | 104 |
| 62 | 26 21 | 22 | 100 | 25 59 | 22 | 101 | 25 37 | 23 | 102 | 25 14 | 23 | 103 | 24 51 | 24 | 104 |
| 61 | 27 17 | 22 | 100 | 26 55 | 23 | 101 | 26 32 | 23 | 102 | 26 09 | 23 | 103 | 25 46 | 24 | 104 |
| 60 | 28 13 | −23 | 101 | 27 50 | −22 | 102 | 27 28 | −24 | 103 | 27 04 | −24 | 104 | 26 40 | −24 | 105 |
| 59 | 29 09 | 23 | 101 | 28 46 | 23 | 102 | 28 23 | 24 | 103 | 27 59 | 24 | 104 | 27 35 | 24 | 105 |
| 58 | 30 04 | 23 | 102 | 29 41 | 23 | 103 | 29 18 | 24 | 104 | 28 54 | 24 | 105 | 28 30 | 25 | 106 |
| 57 | 31 00 | 23 | 102 | 30 37 | 24 | 103 | 30 13 | 24 | 104 | 29 49 | 25 | 105 | 29 24 | 25 | 106 |
| 56 | 31 55 | 23 | 102 | 31 32 | 24 | 104 | 31 08 | 24 | 105 | 30 44 | 25 | 106 | 30 19 | 26 | 107 |
| 55 | 32 51 | −24 | 103 | 32 27 | −24 | 104 | 32 03 | −25 | 105 | 31 38 | −25 | 106 | 31 13 | −26 | 107 |
| 54 | 33 46 | 24 | 103 | 33 22 | 24 | 104 | 32 58 | 25 | 106 | 32 33 | 26 | 107 | 32 07 | 26 | 108 |
| 53 | 34 41 | 24 | 104 | 34 17 | 25 | 105 | 33 52 | 25 | 106 | 33 27 | 26 | 107 | 33 01 | 26 | 108 |
| 52 | 35 36 | 24 | 104 | 35 12 | 25 | 105 | 34 47 | 26 | 107 | 34 21 | 26 | 108 | 33 55 | 27 | 109 |
| 51 | 36 31 | 25 | 105 | 36 06 | 25 | 106 | 35 41 | 26 | 107 | 35 15 | 26 | 108 | 34 49 | 27 | 109 |
| 50 | 37 26 | −25 | 105 | 37 01 | −26 | 106 | 36 35 | −26 | 108 | 36 09 | −27 | 109 | 35 42 | −27 | 110 |
| 49 | 38 20 | 25 | 106 | 37 55 | 26 | 107 | 37 29 | 26 | 108 | 37 03 | 28 | 109 | 36 35 | 27 | 110 |
| 48 | 39 15 | 26 | 106 | 38 49 | 26 | 108 | 38 23 | 27 | 109 | 37 56 | 27 | 110 | 37 29 | 29 | 111 |
| 47 | 40 09 | 26 | 107 | 39 43 | 26 | 108 | 39 17 | 28 | 109 | 38 49 | 28 | 110 | 38 21 | 28 | 112 |
| 46 | 41 03 | 26 | 108 | 40 37 | 27 | 109 | 40 10 | 27 | 110 | 39 43 | 29 | 111 | 39 14 | 29 | 112 |
| 45 | 41 58 | −27 | 108 | 41 31 | −28 | 109 | 41 03 | −28 | 110 | 40 35 | −28 | 112 | 40 07 | −30 | 113 |
| 44 | 42 51 | 27 | 109 | 42 24 | 27 | 110 | 41 57 | 29 | 111 | 41 28 | 29 | 112 | 40 59 | 30 | 113 |
| 43 | 43 45 | 27 | 109 | 43 18 | 29 | 111 | 42 49 | 29 | 112 | 42 20 | 29 | 113 | 41 51 | 31 | 114 |
| 42 | 44 38 | 27 | 110 | 44 11 | 29 | 111 | 43 42 | 29 | 112 | 43 13 | 31 | 114 | 42 42 | 30 | 115 |
| 41 | 45 32 | 29 | 111 | 45 03 | 29 | 112 | 44 34 | 30 | 113 | 44 04 | 30 | 114 | 43 34 | 32 | 115 |
| 40 | 46 25 | −29 | 111 | 45 56 | −30 | 113 | 45 26 | −30 | 114 | 44 56 | −31 | 115 | 44 25 | −32 | 116 |
| 39 | 47 17 | 29 | 112 | 46 48 | 30 | 113 | 46 18 | 31 | 114 | 45 47 | 31 | 116 | 45 16 | 33 | 117 |
| 38 | 48 10 | 30 | 113 | 47 40 | 30 | 114 | 47 10 | 32 | 115 | 46 38 | 32 | 116 | 46 06 | 33 | 118 |
| 37 | 49 02 | 30 | 113 | 48 32 | 31 | 115 | 48 01 | 32 | 116 | 47 29 | 33 | 117 | 46 56 | 33 | 119 |
| 36 | 49 54 | 31 | 114 | 49 23 | 31 | 116 | 48 52 | 33 | 117 | 48 19 | 33 | 118 | 47 46 | 34 | 119 |
| 35 | 50 46 | −32 | 115 | 50 14 | −32 | 116 | 49 42 | −33 | 118 | 49 09 | −34 | 119 | 48 35 | −35 | 120 |
| 34 | 51 37 | 32 | 116 | 51 05 | 33 | 117 | 50 32 | 33 | 118 | 49 59 | 35 | 120 | 49 24 | 35 | 121 |
| 33 | 52 28 | 33 | 117 | 51 55 | 33 | 118 | 51 22 | 34 | 119 | 50 48 | 36 | 121 | 50 12 | 36 | 122 |
| 32 | 53 18 | 33 | 118 | 52 45 | 34 | 119 | 52 11 | 35 | 120 | 51 36 | 36 | 122 | 51 00 | 36 | 123 |
| 31 | 54 09 | 34 | 119 | 53 35 | 35 | 120 | 53 00 | 36 | 121 | 52 24 | 36 | 123 | 51 48 | 38 | 124 |
| 30 | 54 58 | −34 | 119 | 54 24 | −36 | 121 | 53 48 | −36 | 122 | 53 12 | −38 | 124 | 52 34 | −38 | 125 |
| 29 | 55 47 | 35 | 120 | 55 12 | 36 | 122 | 54 36 | 37 | 123 | 53 59 | 38 | 125 | 53 21 | 39 | 126 |
| 28 | 56 36 | 36 | 122 | 56 00 | 37 | 123 | 55 23 | 38 | 124 | 54 45 | 39 | 126 | 54 06 | 39 | 127 |
| 27 | 57 24 | 37 | 123 | 56 47 | 37 | 124 | 56 10 | 39 | 125 | 55 31 | 40 | 127 | 54 51 | 40 | 128 |
| 26 | 58 12 | 38 | 124 | 57 34 | 39 | 125 | 56 55 | 39 | 127 | 56 16 | 40 | 128 | 55 36 | 42 | 129 |
| 25 | 58 58 | −38 | 125 | 58 20 | −39 | 126 | 57 41 | −41 | 128 | 57 00 | −41 | 129 | 56 19 | −42 | 131 |
| 24 | 59 45 | 40 | 126 | 59 05 | 40 | 128 | 58 25 | 41 | 129 | 57 44 | 42 | 131 | 57 02 | 43 | 132 |
| 23 | 60 30 | 40 | 128 | 59 50 | 41 | 129 | 59 09 | 42 | 130 | 58 27 | 43 | 132 | 57 44 | 44 | 133 |
| 22 | 61 15 | 41 | 129 | 60 34 | 43 | 130 | 59 51 | 43 | 132 | 59 08 | 43 | 133 | 58 25 | 45 | 135 |
| 21 | 61 58 | 42 | 130 | 61 16 | 43 | 132 | 60 33 | 44 | 133 | 59 49 | 44 | 135 | 59 05 | 46 | 136 |
| 20 | 62 41 | −43 | 132 | 61 58 | −44 | 133 | 61 14 | −45 | 135 | 60 29 | −45 | 136 | 59 44 | −47 | 137 |
| 19 | 63 23 | 44 | 133 | 62 39 | 45 | 135 | 61 54 | 46 | 136 | 61 08 | 47 | 138 | 60 21 | 47 | 139 |
| 18 | 64 04 | 46 | 135 | 63 18 | 46 | 137 | 62 32 | 46 | 138 | 61 46 | 48 | 139 | 60 58 | 48 | 141 |
| 17 | 64 43 | 46 | 137 | 63 57 | 47 | 138 | 63 10 | 48 | 140 | 62 22 | 49 | 141 | 61 33 | 49 | 142 |
| 16 | 65 21 | 47 | 139 | 64 34 | 48 | 140 | 63 46 | 49 | 142 | 62 57 | 49 | 143 | 62 08 | 51 | 144 |
| 15 | 65 58 | −48 | 141 | 65 10 | −50 | 142 | 64 20 | −49 | 143 | 63 31 | −51 | 145 | 62 40 | −51 | 146 |
| 14 | 66 33 | 49 | 143 | 65 44 | 50 | 144 | 64 54 | 51 | 145 | 64 03 | 52 | 147 | 63 11 | 52 | 148 |
| 13 | 67 07 | 51 | 145 | 66 16 | 51 | 146 | 65 25 | 52 | 147 | 64 33 | 52 | 149 | 63 41 | 53 | 150 |
| 12 | 67 39 | 52 | 147 | 66 47 | 52 | 148 | 65 55 | 53 | 149 | 65 02 | 53 | 151 | 64 09 | 54 | 152 |
| 11 | 68 09 | 53 | 149 | 67 16 | 53 | 150 | 66 23 | 54 | 152 | 65 29 | 54 | 153 | 64 35 | 55 | 154 |
| 10 | 68 37 | −54 | 152 | 67 43 | −54 | 153 | 66 49 | −55 | 154 | 65 54 | −55 | 155 | 64 59 | −55 | 156 |
| 9 | 69 03 | 55 | 154 | 68 08 | 55 | 155 | 67 13 | 56 | 156 | 66 17 | 56 | 157 | 65 21 | 56 | 158 |
| 8 | 69 27 | 56 | 157 | 68 31 | 56 | 158 | 67 35 | 57 | 159 | 66 38 | 57 | 160 | 65 41 | 57 | 160 |
| 7 | 69 48 | 57 | 159 | 68 51 | 57 | 160 | 67 54 | 57 | 161 | 66 57 | 58 | 162 | 65 59 | 57 | 163 |
| 6 | 70 07 | 58 | 162 | 69 09 | 58 | 163 | 68 11 | 58 | 164 | 67 13 | 58 | 164 | 66 15 | 58 | 165 |
| 5 | 70 23 | −59 | 165 | 69 24 | −58 | 166 | 68 26 | −59 | 166 | 67 27 | −58 | 167 | 66 29 | −59 | 167 |
| 4 | 70 36 | 59 | 168 | 69 37 | 59 | 168 | 68 38 | 59 | 169 | 67 39 | 59 | 169 | 66 40 | 59 | 170 |
| 3 | 70 46 | 59 | 171 | 69 47 | 59 | 171 | 68 48 | 60 | 172 | 67 48 | 59 | 172 | 66 49 | 60 | 172 |
| 2 | 70 54 | 60 | 174 | 69 54 | 59 | 174 | 68 55 | 60 | 174 | 67 55 | 60 | 175 | 66 55 | 60 | 175 |
| 1 | 70 59 | 60 | 177 | 69 59 | 60 | 177 | 68 59 | 60 | 177 | 67 59 | 60 | 177 | 66 59 | 60 | 177 |
| 0 | 71 00 | −60 | 180 | 70 00 | −60 | 180 | 69 00 | −60 | 180 | 68 00 | −60 | 180 | 67 00 | −60 | 180 |

| LHA | 5° Hc | 5° d | 5° Z | 6° Hc | 6° d | 6° Z | 7° Hc | 7° d | 7° Z | 8° Hc | 8° d | 8° Z | 9° Hc | 9° d | 9° Z |
|---|---|---|---|---|---|---|---|---|---|---|---|---|---|---|---|
| | ° ′ | ′ | ° | ° ′ | ′ | ° | ° ′ | ′ | ° | ° ′ | ′ | ° | ° ′ | ′ | ° |
| 69 | 18 01 | 23 | 102 | 17 38 | 23 | 103 | 17 15 | 23 | 104 | 16 52 | 23 | 105 | 16 29 | 23 | 106 |
| 68 | 18 56 | 23 | 102 | 18 33 | 23 | 103 | 18 10 | 23 | 104 | 17 47 | 23 | 105 | 17 24 | 24 | 106 |
| 67 | 19 51 | 22 | 103 | 19 29 | 24 | 104 | 19 05 | 23 | 105 | 18 42 | 24 | 106 | 18 18 | 24 | 107 |
| 66 | 20 47 | 23 | 103 | 20 24 | 24 | 104 | 20 00 | 24 | 105 | 19 36 | 24 | 106 | 19 12 | 24 | 107 |
| 65 | 21 42 | −24 | 104 | 21 18 | −23 | 105 | 20 55 | −24 | 106 | 20 31 | −25 | 107 | 20 06 | −24 | 108 |
| 64 | 22 37 | 24 | 104 | 22 13 | 24 | 105 | 21 49 | 24 | 106 | 21 25 | 25 | 107 | 21 00 | 25 | 108 |
| 63 | 23 32 | 24 | 105 | 23 08 | 24 | 106 | 22 44 | 25 | 107 | 22 19 | 25 | 108 | 21 54 | 25 | 109 |
| 62 | 24 27 | 24 | 105 | 24 03 | 25 | 106 | 23 38 | 25 | 107 | 23 13 | 25 | 108 | 22 48 | 26 | 109 |
| 61 | 25 22 | 25 | 105 | 24 57 | 25 | 106 | 24 32 | 25 | 107 | 24 07 | 25 | 108 | 23 42 | 26 | 109 |
| 60 | 26 16 | −24 | 106 | 25 52 | −26 | 107 | 25 26 | −25 | 108 | 25 01 | −26 | 109 | 24 35 | −26 | 110 |
| 59 | 27 11 | 25 | 106 | 26 46 | 26 | 107 | 26 20 | 25 | 108 | 25 55 | 27 | 109 | 25 28 | 26 | 110 |
| 58 | 28 05 | 25 | 107 | 27 40 | 26 | 108 | 27 14 | 26 | 109 | 26 48 | 27 | 110 | 26 21 | 27 | 111 |
| 57 | 28 59 | 25 | 107 | 28 34 | 26 | 108 | 28 08 | 27 | 109 | 27 41 | 27 | 110 | 27 14 | 27 | 111 |
| 56 | 29 53 | 25 | 108 | 29 28 | 27 | 109 | 29 01 | 27 | 110 | 28 34 | 27 | 111 | 28 07 | 28 | 112 |
| 55 | 30 47 | −26 | 108 | 30 21 | −26 | 109 | 29 55 | −28 | 110 | 29 27 | −27 | 111 | 29 00 | −28 | 112 |
| 54 | 31 41 | 26 | 109 | 31 15 | 27 | 110 | 30 48 | 28 | 111 | 30 20 | 28 | 112 | 29 52 | 28 | 113 |
| 53 | 32 35 | 27 | 109 | 32 08 | 27 | 110 | 31 41 | 28 | 111 | 31 13 | 29 | 112 | 30 44 | 29 | 113 |
| 52 | 33 28 | 27 | 110 | 33 01 | 28 | 111 | 32 33 | 28 | 112 | 32 05 | 29 | 113 | 31 36 | 29 | 114 |
| 51 | 34 22 | 28 | 110 | 33 54 | 28 | 111 | 33 26 | 29 | 112 | 32 57 | 29 | 114 | 32 28 | 30 | 115 |
| 50 | 35 15 | −28 | 111 | 34 47 | −29 | 112 | 34 18 | −29 | 113 | 33 49 | −30 | 114 | 33 19 | −30 | 115 |
| 49 | 36 08 | 29 | 111 | 35 39 | 29 | 113 | 35 10 | 29 | 114 | 34 41 | 30 | 115 | 34 11 | 31 | 116 |
| 48 | 37 00 | 28 | 112 | 36 32 | 30 | 113 | 36 02 | 30 | 114 | 35 32 | 30 | 115 | 35 02 | 31 | 116 |
| 47 | 37 53 | 29 | 113 | 37 24 | 30 | 114 | 36 54 | 31 | 115 | 36 23 | 31 | 116 | 35 52 | 31 | 117 |
| 46 | 38 45 | 30 | 113 | 38 15 | 30 | 114 | 37 45 | 31 | 115 | 37 14 | 31 | 117 | 36 43 | 32 | 118 |
| 45 | 39 37 | −30 | 114 | 39 07 | −31 | 115 | 38 36 | −31 | 116 | 38 05 | −32 | 117 | 37 33 | −33 | 118 |
| 44 | 40 29 | 31 | 115 | 39 58 | 31 | 116 | 39 27 | 32 | 117 | 38 55 | 32 | 118 | 38 23 | 33 | 119 |
| 43 | 41 20 | 31 | 115 | 40 49 | 31 | 116 | 40 18 | 33 | 117 | 39 45 | 33 | 119 | 39 12 | 33 | 120 |
| 42 | 42 12 | 32 | 116 | 41 40 | 32 | 117 | 41 08 | 33 | 118 | 40 35 | 34 | 119 | 40 01 | 34 | 120 |
| 41 | 43 02 | 32 | 117 | 42 30 | 32 | 118 | 41 58 | 34 | 119 | 41 24 | 34 | 120 | 40 50 | 35 | 121 |
| 40 | 43 53 | −33 | 117 | 43 20 | −33 | 119 | 42 47 | −34 | 120 | 42 13 | −34 | 121 | 41 39 | −36 | 122 |
| 39 | 44 43 | 33 | 118 | 44 10 | 34 | 119 | 43 36 | 34 | 120 | 43 02 | 35 | 122 | 42 27 | 36 | 123 |
| 38 | 45 33 | 34 | 119 | 44 59 | 34 | 120 | 44 25 | 35 | 121 | 43 50 | 36 | 122 | 43 14 | 36 | 123 |
| 37 | 46 23 | 35 | 120 | 45 48 | 35 | 121 | 45 13 | 35 | 122 | 44 38 | 37 | 123 | 44 01 | 37 | 124 |
| 36 | 47 12 | 35 | 121 | 46 37 | 36 | 122 | 46 01 | 36 | 123 | 45 25 | 37 | 124 | 44 48 | 38 | 125 |
| 35 | 48 00 | −35 | 121 | 47 25 | −36 | 123 | 46 49 | −37 | 124 | 46 12 | −38 | 125 | 45 34 | −38 | 126 |
| 34 | 48 49 | 37 | 122 | 48 12 | 36 | 123 | 47 36 | 38 | 125 | 46 58 | 38 | 126 | 46 20 | 39 | 127 |
| 33 | 49 36 | 37 | 123 | 48 59 | 37 | 124 | 48 22 | 38 | 126 | 47 44 | 39 | 127 | 47 05 | 40 | 128 |
| 32 | 50 24 | 38 | 124 | 49 46 | 38 | 125 | 49 08 | 39 | 127 | 48 29 | 40 | 128 | 47 49 | 40 | 129 |
| 31 | 51 10 | 38 | 125 | 50 32 | 39 | 126 | 49 53 | 40 | 128 | 49 13 | 40 | 129 | 48 33 | 41 | 130 |
| 30 | 51 56 | −39 | 126 | 51 17 | −39 | 127 | 50 38 | −41 | 129 | 49 57 | −41 | 130 | 49 16 | −41 | 131 |
| 29 | 52 42 | 40 | 127 | 52 02 | 40 | 128 | 51 22 | 41 | 130 | 50 41 | 42 | 131 | 49 59 | 42 | 132 |
| 28 | 53 27 | 41 | 128 | 52 46 | 41 | 130 | 52 05 | 42 | 131 | 51 23 | 42 | 132 | 50 41 | 43 | 133 |
| 27 | 54 11 | 41 | 129 | 53 30 | 42 | 131 | 52 48 | 43 | 132 | 52 05 | 43 | 133 | 51 22 | 44 | 134 |
| 26 | 54 54 | 42 | 131 | 54 12 | 42 | 132 | 53 30 | 44 | 133 | 52 46 | 44 | 134 | 52 02 | 44 | 135 |
| 25 | 55 37 | −43 | 132 | 54 54 | −43 | 133 | 54 11 | −44 | 134 | 53 27 | −45 | 135 | 52 42 | −45 | 137 |
| 24 | 56 19 | 44 | 133 | 55 35 | 44 | 134 | 54 51 | 45 | 136 | 54 06 | 45 | 137 | 53 21 | 47 | 138 |
| 23 | 57 00 | 44 | 134 | 56 16 | 46 | 136 | 55 30 | 45 | 137 | 54 45 | 47 | 138 | 53 58 | 47 | 139 |
| 22 | 57 40 | 45 | 136 | 56 55 | 46 | 137 | 56 09 | 47 | 138 | 55 22 | 47 | 139 | 54 35 | 48 | 140 |
| 21 | 58 19 | 46 | 137 | 57 33 | 47 | 138 | 56 46 | 47 | 140 | 55 59 | 48 | 141 | 55 11 | 49 | 142 |
| 20 | 58 57 | −47 | 139 | 58 10 | −48 | 140 | 57 22 | −48 | 141 | 56 34 | −49 | 142 | 55 45 | −49 | 143 |
| 19 | 59 34 | 48 | 140 | 58 46 | 48 | 141 | 57 58 | 50 | 143 | 57 08 | 49 | 144 | 56 19 | 50 | 145 |
| 18 | 60 10 | 49 | 142 | 59 21 | 49 | 143 | 58 32 | 50 | 144 | 57 42 | 51 | 145 | 56 51 | 51 | 146 |
| 17 | 60 44 | 49 | 143 | 59 55 | 51 | 145 | 59 04 | 51 | 146 | 58 13 | 51 | 147 | 57 22 | 52 | 148 |
| 16 | 61 17 | 50 | 145 | 60 27 | 51 | 146 | 59 36 | 52 | 147 | 58 44 | 52 | 148 | 57 52 | 53 | 149 |
| 15 | 61 49 | −51 | 147 | 60 58 | −53 | 148 | 60 05 | −52 | 149 | 59 13 | −53 | 150 | 58 20 | −53 | 151 |
| 14 | 62 19 | 52 | 149 | 61 27 | 53 | 150 | 60 34 | 53 | 151 | 59 41 | 54 | 152 | 58 47 | 54 | 153 |
| 13 | 62 48 | 53 | 151 | 61 55 | 54 | 152 | 61 01 | 54 | 153 | 60 07 | 55 | 153 | 59 12 | 54 | 154 |
| 12 | 63 15 | 54 | 153 | 62 21 | 55 | 154 | 61 26 | 55 | 154 | 60 31 | 55 | 155 | 59 36 | 55 | 156 |
| 11 | 63 40 | 55 | 155 | 62 45 | 55 | 156 | 61 50 | 56 | 156 | 60 54 | 56 | 157 | 59 58 | 56 | 158 |
| 10 | 64 04 | −56 | 157 | 63 08 | −56 | 158 | 62 12 | −57 | 158 | 61 15 | −56 | 159 | 60 19 | −57 | 160 |
| 9 | 64 25 | 57 | 159 | 63 28 | 56 | 160 | 62 32 | 57 | 160 | 61 35 | 57 | 161 | 60 38 | 58 | 162 |
| 8 | 64 44 | 57 | 161 | 63 47 | 57 | 162 | 62 50 | 58 | 162 | 61 52 | 57 | 163 | 60 55 | 58 | 164 |
| 7 | 65 02 | 58 | 163 | 64 04 | 58 | 164 | 63 06 | 58 | 165 | 62 08 | 58 | 165 | 61 10 | 59 | 166 |
| 6 | 65 17 | 58 | 166 | 64 19 | 59 | 166 | 63 20 | 58 | 167 | 62 22 | 59 | 167 | 61 23 | 59 | 168 |
| 5 | 65 30 | −59 | 168 | 64 31 | −59 | 168 | 63 32 | −59 | 169 | 62 33 | −59 | 169 | 61 34 | −59 | 170 |
| 4 | 65 41 | 59 | 170 | 64 42 | 60 | 171 | 63 42 | 59 | 171 | 62 43 | 60 | 171 | 61 43 | 59 | 172 |
| 3 | 65 49 | 59 | 173 | 64 50 | 60 | 173 | 63 50 | 60 | 173 | 62 50 | 59 | 174 | 61 51 | 60 | 174 |
| 2 | 65 55 | 60 | 175 | 64 55 | 59 | 175 | 63 56 | 60 | 176 | 62 56 | 60 | 176 | 61 56 | 60 | 176 |
| 1 | 65 59 | 60 | 178 | 64 59 | 60 | 178 | 63 59 | 60 | 178 | 62 59 | 60 | 178 | 61 59 | 60 | 178 |
| 0 | 66 00 | −60 | 180 | 65 00 | −60 | 180 | 64 00 | −60 | 180 | 63 00 | −60 | 180 | 62 00 | −60 | 180 |

| LHA | 10° Hc | 10° d | 10° Z | 11° Hc | 11° d | 11° Z | 12° Hc | 12° d | 12° Z | 13° Hc | 13° d | 13° Z | 14° Hc | 14° d | 14° Z | LHA |
|---|---|---|---|---|---|---|---|---|---|---|---|---|---|---|---|---|
| | ° ′ | ′ | ° | ° ′ | ′ | ° | ° ′ | ′ | ° | ° ′ | ′ | ° | ° ′ | ′ | ° | |
| 69 | 16 06 | 24 | 107 | 15 42 | 24 | 108 | 15 18 | 25 | 109 | 14 53 | 24 | 110 | 14 29 | 25 | 111 | 291 |
| 68 | 17 00 | 24 | 107 | 16 36 | 25 | 108 | 16 11 | 24 | 109 | 15 47 | 25 | 110 | 15 22 | 25 | 111 | 292 |
| 67 | 17 54 | 25 | 108 | 17 29 | 24 | 109 | 17 05 | 25 | 110 | 16 40 | 25 | 111 | 16 15 | 26 | 112 | 293 |
| 66 | 18 48 | 25 | 108 | 18 23 | 25 | 109 | 17 58 | 25 | 110 | 17 33 | 26 | 111 | 17 07 | 25 | 112 | 294 |
| 65 | 19 42 | −25 | 109 | 19 17 | −26 | 110 | 18 51 | −25 | 111 | 18 26 | −26 | 111 | 18 00 | −26 | 112 | 295 |
| 64 | 20 35 | 25 | 109 | 20 10 | 26 | 110 | 19 44 | 26 | 111 | 19 18 | 26 | 112 | 18 52 | 26 | 113 | 296 |
| 63 | 21 29 | 26 | 109 | 21 03 | 26 | 110 | 20 37 | 26 | 111 | 20 11 | 27 | 112 | 19 44 | 26 | 113 | 297 |
| 62 | 22 22 | 26 | 110 | 21 56 | 26 | 111 | 21 30 | 27 | 112 | 21 03 | 27 | 113 | 20 36 | 27 | 114 | 298 |
| 61 | 23 16 | 27 | 110 | 22 49 | 26 | 111 | 22 23 | 27 | 112 | 21 56 | 28 | 113 | 21 28 | 27 | 114 | 299 |
| 60 | 24 09 | −27 | 111 | 23 42 | −27 | 112 | 23 15 | −27 | 113 | 22 48 | −28 | 114 | 22 20 | −28 | 115 | 300 |
| 59 | 25 02 | 27 | 111 | 24 35 | 28 | 112 | 24 07 | 28 | 113 | 23 39 | 28 | 114 | 23 11 | 28 | 115 | 301 |
| 58 | 25 54 | 27 | 112 | 25 27 | 28 | 113 | 24 59 | 28 | 114 | 24 31 | 28 | 115 | 24 03 | 29 | 116 | 302 |
| 57 | 26 47 | 28 | 112 | 26 19 | 28 | 113 | 25 51 | 28 | 114 | 25 23 | 29 | 115 | 24 54 | 30 | 116 | 303 |
| 56 | 27 39 | 28 | 113 | 27 11 | 28 | 114 | 26 43 | 29 | 115 | 26 14 | 30 | 116 | 25 44 | 29 | 117 | 304 |
| 55 | 28 32 | −29 | 113 | 28 03 | −29 | 114 | 27 34 | −29 | 115 | 27 05 | −30 | 116 | 26 35 | −30 | 117 | 305 |
| 54 | 29 24 | 29 | 114 | 28 55 | 30 | 115 | 28 25 | 30 | 116 | 27 55 | 30 | 117 | 27 25 | 30 | 118 | 306 |
| 53 | 30 15 | 29 | 114 | 29 46 | 30 | 115 | 29 16 | 30 | 116 | 28 46 | 31 | 117 | 28 15 | 31 | 118 | 307 |
| 52 | 31 07 | 30 | 115 | 30 37 | 30 | 116 | 30 07 | 31 | 117 | 29 36 | 31 | 118 | 29 05 | 32 | 119 | 308 |
| 51 | 31 58 | 30 | 116 | 31 28 | 31 | 117 | 30 57 | 31 | 118 | 30 26 | 32 | 119 | 29 54 | 32 | 120 | 309 |
| 50 | 32 49 | −30 | 116 | 32 19 | −32 | 117 | 31 47 | −31 | 118 | 31 16 | −32 | 119 | 30 44 | −33 | 120 | 310 |
| 49 | 33 40 | 31 | 117 | 33 09 | 32 | 118 | 32 37 | 32 | 119 | 32 05 | 32 | 120 | 31 33 | 33 | 121 | 311 |
| 48 | 34 31 | 32 | 117 | 33 59 | 32 | 118 | 33 27 | 33 | 119 | 32 54 | 33 | 120 | 32 21 | 33 | 121 | 312 |
| 47 | 35 21 | 32 | 118 | 34 49 | 33 | 119 | 34 16 | 33 | 120 | 33 43 | 34 | 121 | 33 09 | 34 | 122 | 313 |
| 46 | 36 11 | 33 | 119 | 35 38 | 33 | 120 | 35 05 | 34 | 121 | 34 31 | 34 | 122 | 33 57 | 34 | 123 | 314 |
| 45 | 37 00 | −33 | 119 | 36 27 | −33 | 120 | 35 54 | −34 | 121 | 35 20 | −35 | 122 | 34 45 | −35 | 123 | 315 |
| 44 | 37 50 | 34 | 120 | 37 16 | 34 | 121 | 36 42 | 35 | 122 | 36 07 | 35 | 123 | 35 32 | 36 | 124 | 316 |
| 43 | 38 39 | 34 | 121 | 38 05 | 35 | 122 | 37 30 | 35 | 123 | 36 55 | 36 | 124 | 36 19 | 36 | 125 | 317 |
| 42 | 39 27 | 34 | 121 | 38 53 | 36 | 123 | 38 17 | 35 | 124 | 37 42 | 37 | 125 | 37 05 | 37 | 126 | 318 |
| 41 | 40 15 | 35 | 122 | 39 40 | 36 | 123 | 39 04 | 36 | 124 | 38 28 | 37 | 125 | 37 51 | 37 | 126 | 319 |
| 40 | 41 03 | −36 | 123 | 40 27 | −36 | 124 | 39 51 | −37 | 125 | 39 14 | −37 | 126 | 38 37 | −38 | 127 | 320 |
| 39 | 41 51 | 37 | 124 | 41 14 | 37 | 125 | 40 37 | 37 | 126 | 40 00 | 38 | 127 | 39 22 | 39 | 128 | 321 |
| 38 | 42 38 | 37 | 125 | 42 01 | 38 | 126 | 41 23 | 38 | 127 | 40 45 | 39 | 128 | 40 06 | 39 | 129 | 322 |
| 37 | 43 24 | 37 | 125 | 42 47 | 39 | 126 | 42 08 | 38 | 128 | 41 30 | 40 | 129 | 40 50 | 39 | 130 | 323 |
| 36 | 44 10 | 38 | 126 | 43 32 | 39 | 127 | 42 53 | 39 | 128 | 42 14 | 40 | 129 | 41 34 | 41 | 130 | 324 |
| 35 | 44 56 | −39 | 127 | 44 17 | −40 | 128 | 43 37 | −40 | 129 | 42 57 | −40 | 130 | 42 17 | −41 | 131 | 325 |
| 34 | 45 41 | 40 | 128 | 45 01 | 40 | 129 | 44 21 | 41 | 130 | 43 40 | 41 | 131 | 42 59 | 41 | 132 | 326 |
| 33 | 46 25 | 40 | 129 | 45 45 | 41 | 130 | 45 04 | 41 | 131 | 44 23 | 42 | 132 | 43 41 | 42 | 133 | 327 |
| 32 | 47 09 | 41 | 130 | 46 28 | 41 | 131 | 45 47 | 42 | 132 | 45 05 | 43 | 133 | 44 22 | 43 | 134 | 328 |
| 31 | 47 52 | 41 | 131 | 47 11 | 42 | 132 | 46 29 | 43 | 133 | 45 46 | 43 | 134 | 45 03 | 44 | 135 | 329 |
| 30 | 48 35 | −42 | 132 | 47 53 | −43 | 133 | 47 10 | −44 | 134 | 46 26 | −44 | 135 | 45 42 | −44 | 136 | 330 |
| 29 | 49 17 | 43 | 133 | 48 34 | 44 | 134 | 47 50 | 44 | 135 | 47 06 | 45 | 136 | 46 21 | 45 | 137 | 331 |
| 28 | 49 58 | 44 | 134 | 49 14 | 44 | 135 | 48 30 | 45 | 136 | 47 45 | 45 | 137 | 47 00 | 46 | 138 | 332 |
| 27 | 50 38 | 44 | 135 | 49 54 | 45 | 136 | 49 09 | 46 | 137 | 48 23 | 46 | 138 | 47 37 | 46 | 139 | 333 |
| 26 | 51 18 | 45 | 136 | 50 33 | 46 | 137 | 49 47 | 46 | 138 | 49 01 | 47 | 139 | 48 14 | 47 | 140 | 334 |
| 25 | 51 57 | −46 | 138 | 51 11 | −47 | 139 | 50 24 | −47 | 140 | 49 37 | −47 | 141 | 48 50 | −48 | 142 | 335 |
| 24 | 52 34 | 46 | 139 | 51 48 | 48 | 140 | 51 00 | 47 | 141 | 50 13 | 49 | 142 | 49 24 | 48 | 143 | 336 |
| 23 | 53 11 | 47 | 140 | 52 24 | 48 | 141 | 51 36 | 49 | 142 | 50 47 | 49 | 143 | 49 58 | 49 | 144 | 337 |
| 22 | 53 47 | 48 | 141 | 52 59 | 49 | 142 | 52 10 | 49 | 143 | 51 21 | 50 | 144 | 50 31 | 50 | 145 | 338 |
| 21 | 54 22 | 49 | 143 | 53 33 | 49 | 144 | 52 44 | 50 | 145 | 51 54 | 51 | 146 | 51 03 | 50 | 146 | 339 |
| 20 | 54 56 | −50 | 144 | 54 06 | −50 | 145 | 53 16 | −51 | 146 | 52 25 | −51 | 147 | 51 34 | −51 | 148 | 340 |
| 19 | 55 29 | 51 | 146 | 54 38 | 51 | 147 | 53 47 | 51 | 147 | 52 56 | 52 | 148 | 52 04 | 52 | 149 | 341 |
| 18 | 56 00 | 51 | 147 | 55 09 | 52 | 148 | 54 17 | 52 | 149 | 53 25 | 53 | 150 | 52 32 | 52 | 151 | 342 |
| 17 | 56 30 | 52 | 149 | 55 38 | 52 | 149 | 54 46 | 53 | 150 | 53 53 | 53 | 151 | 53 00 | 54 | 152 | 343 |
| 16 | 56 59 | 53 | 150 | 56 06 | 53 | 151 | 55 13 | 53 | 152 | 54 20 | 54 | 153 | 53 26 | 54 | 153 | 344 |
| 15 | 57 27 | −54 | 152 | 56 33 | −54 | 153 | 55 39 | −54 | 153 | 54 45 | −54 | 154 | 53 51 | −55 | 155 | 345 |
| 14 | 57 53 | 54 | 153 | 56 59 | 55 | 154 | 56 04 | 55 | 155 | 55 09 | 55 | 156 | 54 14 | 55 | 156 | 346 |
| 13 | 58 18 | 55 | 155 | 57 23 | 56 | 156 | 56 27 | 55 | 157 | 55 32 | 56 | 157 | 54 36 | 56 | 158 | 347 |
| 12 | 58 41 | 56 | 157 | 57 45 | 56 | 158 | 56 49 | 56 | 158 | 55 53 | 56 | 159 | 54 57 | 57 | 159 | 348 |
| 11 | 59 02 | 56 | 159 | 58 06 | 56 | 159 | 57 10 | 57 | 160 | 56 13 | 57 | 161 | 55 16 | 57 | 161 | 349 |
| 10 | 59 22 | −57 | 160 | 58 25 | −57 | 161 | 57 28 | −57 | 162 | 56 31 | −57 | 162 | 55 34 | −58 | 163 | 350 |
| 9 | 59 40 | 57 | 162 | 58 43 | 58 | 163 | 57 45 | 57 | 163 | 56 48 | 58 | 164 | 55 50 | 58 | 164 | 351 |
| 8 | 59 57 | 58 | 164 | 58 59 | 58 | 165 | 58 01 | 58 | 165 | 57 03 | 59 | 166 | 56 04 | 58 | 166 | 352 |
| 7 | 60 11 | 58 | 166 | 59 13 | 58 | 167 | 58 15 | 59 | 167 | 57 16 | 59 | 167 | 56 17 | 58 | 168 | 353 |
| 6 | 60 24 | 59 | 168 | 59 25 | 58 | 168 | 58 27 | 59 | 169 | 57 28 | 59 | 169 | 56 29 | 60 | 169 | 354 |
| 5 | 60 35 | −59 | 170 | 59 36 | −59 | 170 | 58 37 | −60 | 171 | 57 37 | −59 | 171 | 56 38 | −59 | 171 | 355 |
| 4 | 60 44 | 59 | 172 | 59 45 | 60 | 172 | 58 45 | 59 | 172 | 57 46 | 60 | 173 | 56 46 | 60 | 173 | 356 |
| 3 | 60 51 | 60 | 174 | 59 51 | 59 | 174 | 58 52 | 60 | 174 | 57 52 | 60 | 175 | 56 52 | 60 | 175 | 357 |
| 2 | 60 56 | 60 | 176 | 59 56 | 60 | 176 | 58 56 | 60 | 176 | 57 56 | 59 | 176 | 56 57 | 60 | 176 | 358 |
| 1 | 60 59 | 60 | 178 | 59 59 | 60 | 178 | 58 59 | 60 | 178 | 57 59 | 60 | 178 | 56 59 | 60 | 178 | 359 |
| 0 | 61 00 | −60 | 180 | 60 00 | −60 | 180 | 59 00 | −60 | 180 | 58 00 | −60 | 180 | 57 00 | −60 | 180 | 360 |

N. Lat. {LHA greater than 180°........ Zn=Z; LHA less than 180°........... Zn=360−Z}

# DECLINATION (0°-14°) SAME NAME AS LATITUDE



| LHA | 0° Hc | d | Z | 1° Hc | d | Z | 2° Hc | d | Z | 3° Hc | d | Z | 4° Hc | d | Z | 5° Hc | d | Z | 6° Hc | d | Z |
|---|---|---|---|---|---|---|---|---|---|---|---|---|---|---|---|---|---|---|---|---|---|
| | ° ′ | ′ | ° | ° ′ | ′ | ° | ° ′ | ′ | ° | ° ′ | ′ | ° | ° ′ | ′ | ° | ° ′ | ′ | ° | ° ′ | ′ | ° |
| 0 | 65 00 | +60 | 180 | 66 00 | +60 | 180 | 67 00 | +60 | 180 | 68 00 | +60 | 180 | 69 00 | +60 | 180 | 70 00 | +60 | 180 | 71 00 | +60 | 180 |
| 1 | 64 59 | 60 | 178 | 65 59 | 60 | 178 | 66 59 | 60 | 177 | 67 59 | 60 | 177 | 68 59 | 60 | 177 | 69 59 | 60 | 177 | 70 59 | 60 | 177 |
| 2 | 64 56 | 59 | 175 | 65 55 | 60 | 175 | 66 55 | 60 | 175 | 67 55 | 60 | 175 | 68 55 | 60 | 174 | 69 55 | 59 | 174 | 70 54 | 60 | 174 |
| 3 | 64 50 | 60 | 173 | 65 50 | 59 | 173 | 66 49 | 60 | 172 | 67 49 | 59 | 172 | 68 48 | 60 | 172 | 69 48 | 59 | 171 | 70 47 | 59 | 171 |
| 4 | 64 42 | 60 | 171 | 65 42 | 59 | 170 | 66 41 | 59 | 170 | 67 40 | 59 | 169 | 68 39 | 59 | 169 | 69 38 | 59 | 169 | 70 37 | 59 | 168 |
| 5 | 64 32 | +59 | 168 | 65 31 | +59 | 168 | 66 30 | +59 | 167 | 67 29 | +58 | 167 | 68 27 | +59 | 166 | 69 26 | +58 | 166 | 70 24 | +59 | 165 |
| 6 | 64 20 | 59 | 166 | 65 19 | 58 | 166 | 66 17 | 58 | 165 | 67 15 | 58 | 164 | 68 13 | 58 | 164 | 69 11 | 58 | 163 | 70 09 | 58 | 162 |
| 7 | 64 06 | 58 | 164 | 65 04 | 58 | 163 | 66 02 | 57 | 163 | 66 59 | 58 | 162 | 67 57 | 57 | 161 | 68 54 | 57 | 160 | 69 51 | 57 | 159 |
| 8 | 63 50 | 57 | 162 | 64 47 | 57 | 161 | 65 44 | 57 | 160 | 66 41 | 57 | 159 | 67 38 | 57 | 159 | 68 35 | 56 | 158 | 69 31 | 56 | 157 |
| 9 | 63 32 | 56 | 160 | 64 28 | 57 | 159 | 65 25 | 56 | 158 | 66 21 | 56 | 157 | 67 17 | 56 | 156 | 68 13 | 55 | 155 | 69 08 | 55 | 154 |
| 10 | 63 12 | +56 | 157 | 64 08 | +56 | 157 | 65 04 | +55 | 156 | 65 59 | +55 | 155 | 66 54 | +55 | 154 | 67 49 | +54 | 153 | 68 43 | +54 | 152 |
| 11 | 62 50 | 55 | 155 | 63 45 | 55 | 154 | 64 40 | 55 | 154 | 65 35 | 54 | 153 | 66 29 | 54 | 152 | 67 23 | 53 | 150 | 68 16 | 53 | 149 |
| 12 | 62 26 | 55 | 153 | 63 21 | 54 | 152 | 64 15 | 54 | 151 | 65 09 | 53 | 150 | 66 02 | 53 | 149 | 66 55 | 53 | 148 | 67 48 | 51 | 147 |
| 13 | 62 01 | 54 | 151 | 62 55 | 53 | 150 | 63 48 | 53 | 149 | 64 41 | 53 | 148 | 65 34 | 52 | 147 | 66 26 | 51 | 146 | 67 17 | 51 | 145 |
| 14 | 61 34 | 53 | 150 | 62 27 | 53 | 149 | 63 20 | 52 | 147 | 64 12 | 51 | 146 | 65 03 | 51 | 145 | 65 54 | 51 | 144 | 66 45 | 49 | 143 |
| 15 | 61 06 | +52 | 148 | 61 58 | +52 | 147 | 62 50 | +51 | 146 | 63 41 | +51 | 144 | 64 32 | +50 | 143 | 65 22 | +49 | 142 | 66 11 | +48 | 140 |
| 16 | 60 36 | 51 | 146 | 61 27 | 51 | 145 | 62 18 | 51 | 144 | 63 09 | 49 | 143 | 63 58 | 49 | 141 | 64 47 | 48 | 140 | 65 35 | 48 | 138 |
| 17 | 60 05 | 50 | 144 | 60 55 | 50 | 143 | 61 45 | 50 | 142 | 62 35 | 48 | 141 | 63 23 | 48 | 139 | 64 11 | 48 | 138 | 64 59 | 46 | 137 |
| 18 | 59 32 | 50 | 142 | 60 22 | 49 | 141 | 61 11 | 49 | 140 | 62 00 | 47 | 139 | 62 47 | 47 | 138 | 63 34 | 47 | 136 | 64 21 | 45 | 135 |
| 19 | 58 58 | 49 | 141 | 59 47 | 49 | 140 | 60 36 | 47 | 139 | 61 23 | 47 | 137 | 62 10 | 46 | •136 | 62 56 | 45 | 135 | 63 41 | 45 | 133 |
| 20 | 58 24 | +48 | 139 | 59 12 | +47 | 138 | 59 59 | +47 | 137 | 60 46 | +46 | 136 | 61 32 | +45 | 134 | 62 17 | +44 | 133 | 63 01 | +44 | 131 |
| 21 | 57 48 | 47 | 138 | 58 35 | 46 | 137 | 59 21 | 46 | 135 | 60 07 | 45 | 134 | 60 52 | 45 | 133 | 61 37 | 43 | 131 | 62 20 | 42 | 130 |
| 22 | 57 10 | 47 | 136 | 57 57 | 46 | 135 | 58 43 | 45 | 134 | 59 28 | 44 | 133 | 60 12 | 43 | 131 | 60 55 | 43 | 130 | 61 38 | 41 | 128 |
| 23 | 56 32 | 46 | 135 | 57 18 | 45 | 134 | 58 03 | 44 | 132 | 58 47 | 44 | 131 | 59 31 | 42 | 130 | 60 13 | 42 | 128 | 60 55 | 40 | 127 |
| 24 | 55 53 | 45 | 134 | 56 38 | 44 | 132 | 57 22 | 44 | 131 | 58 06 | 42 | 130 | 58 48 | 42 | 128 | 59 30 | 41 | 127 | 60 11 | 40 | 126 |
| 25 | 55 14 | +44 | 132 | 55 58 | +43 | 131 | 56 41 | +43 | 130 | 57 24 | +41 | 128 | 58 05 | +41 | 127 | 58 46 | +40 | 126 | 59 26 | +39 | 124 |
| 26 | 54 33 | 43 | 131 | 55 16 | 43 | 130 | 55 59 | 42 | 129 | 56 41 | 41 | 127 | 57 22 | 40 | 126 | 58 02 | 39 | 124 | 58 41 | 38 | 123 |
| 27 | 53 51 | 43 | 130 | 54 34 | 42 | 129 | 55 16 | 41 | 127 | 55 57 | 40 | 126 | 56 37 | 40 | 125 | 57 17 | 38 | 123 | 57 55 | 37 | 122 |
| 28 | 53 09 | 42 | 129 | 53 51 | 41 | 127 | 54 32 | 41 | 126 | 55 13 | 39 | 125 | 55 52 | 39 | 123 | 56 31 | 38 | 122 | 57 09 | 36 | 121 |
| 29 | 52 26 | 41 | 127 | 53 07 | 41 | 126 | 53 48 | 40 | 125 | 54 28 | 38 | 124 | 55 06 | 38 | 122 | 55 44 | 37 | 121 | 56 21 | 37 | 120 |
| 30 | 51 43 | +40 | 126 | 52 23 | +40 | 125 | 53 03 | +39 | 124 | 53 42 | +38 | 123 | 54 20 | +38 | 121 | 54 58 | +36 | 120 | 55 34 | +35 | 118 |
| 31 | 50 58 | 40 | 125 | 51 38 | 40 | 124 | 52 18 | 38 | 123 | 52 56 | 37 | 121 | 53 33 | 37 | 120 | 54 10 | 36 | 119 | 54 46 | 35 | 117 |
| 32 | 50 14 | 39 | 124 | 50 53 | 39 | 123 | 51 32 | 37 | 122 | 52 09 | 37 | 120 | 52 46 | 36 | 119 | 53 22 | 35 | 118 | 53 57 | 35 | 116 |
| 33 | 49 28 | 39 | 123 | 50 07 | 38 | 122 | 50 45 | 37 | 121 | 51 22 | 36 | 119 | 51 58 | 36 | 118 | 52 34 | 34 | 117 | 53 08 | 34 | 115 |
| 34 | 48 43 | 38 | 122 | 49 21 | 37 | 121 | 49 58 | 37 | 120 | 50 35 | 35 | 118 | 51 10 | 35 | 117 | 51 45 | 34 | 116 | 52 19 | 33 | 115 |
| 35 | 47 56 | +38 | 121 | 48 34 | +37 | 120 | 49 11 | +35 | 119 | 49 46 | +36 | 118 | 50 22 | +34 | 116 | 50 56 | +34 | 115 | 51 30 | +32 | 114 |
| 36 | 47 09 | 37 | 120 | 47 46 | 37 | 119 | 48 23 | 35 | 118 | 48 58 | 35 | 117 | 49 33 | 34 | 115 | 50 07 | 33 | 114 | 50 40 | 32 | 113 |
| 37 | 46 22 | 37 | 119 | 46 59 | 35 | 118 | 47 34 | 35 | 117 | 48 09 | 34 | 116 | 48 43 | 34 | 115 | 49 17 | 32 | 113 | 49 49 | 32 | 112 |
| 38 | 45 35 | 35 | 118 | 46 10 | 36 | 117 | 46 46 | 34 | 116 | 47 20 | 34 | 115 | 47 54 | 33 | 114 | 48 27 | 32 | 112 | 48 59 | 31 | 111 |
| 39 | 44 47 | 35 | 118 | 45 22 | 35 | 116 | 45 57 | 34 | 115 | 46 31 | 33 | 114 | 47 04 | 32 | 113 | 47 36 | 32 | 112 | 48 08 | 31 | 110 |
| 40 | 43 58 | +35 | 117 | 44 33 | +34 | 116 | 45 07 | +34 | 114 | 45 41 | +33 | 113 | 46 14 | +32 | 112 | 46 46 | +31 | 111 | 47 17 | +30 | 110 |
| 41 | 43 09 | 35 | 116 | 43 44 | 34 | 115 | 44 18 | 33 | 114 | 44 51 | 32 | 113 | 45 23 | 32 | 111 | 45 55 | 30 | 110 | 46 25 | 30 | 109 |
| 42 | 42 20 | 34 | 115 | 42 54 | 34 | 114 | 43 28 | 32 | 113 | 44 00 | 32 | 112 | 44 32 | 31 | 111 | 45 03 | 31 | 109 | 45 34 | 29 | 108 |
| 43 | 41 31 | 34 | 114 | 42 05 | 32 | 113 | 42 37 | 33 | 112 | 43 10 | 31 | 111 | 43 41 | 31 | 110 | 44 12 | 30 | 109 | 44 42 | 29 | 107 |
| 44 | 40 41 | 33 | 114 | 41 14 | 33 | 113 | 41 47 | 32 | 111 | 42 19 | 31 | 110 | 42 50 | 30 | 109 | 43 20 | 30 | 108 | 43 50 | 29 | 107 |
| 45 | 39 51 | +33 | 113 | 40 24 | +32 | 112 | 40 56 | +32 | 111 | 41 28 | +30 | 110 | 41 58 | +30 | 108 | 42 28 | +30 | 107 | 42 58 | +28 | 106 |
| 46 | 39 01 | 33 | 112 | 39 34 | 31 | 111 | 40 05 | 31 | 110 | 40 36 | 31 | 109 | 41 07 | 29 | 108 | 41 36 | 29 | 107 | 42 05 | 29 | 105 |
| 47 | 38 11 | 32 | 112 | 38 43 | 31 | 110 | 39 14 | 31 | 109 | 39 45 | 30 | 108 | 40 15 | 29 | 107 | 40 44 | 29 | 106 | 41 13 | 28 | 105 |
| 48 | 37 20 | 32 | 111 | 37 52 | 31 | 110 | 38 23 | 30 | 109 | 38 53 | 30 | 108 | 39 23 | 29 | 106 | 39 52 | 28 | 105 | 40 20 | 28 | 104 |
| 49 | 36 29 | 31 | 110 | 37 00 | 31 | 109 | 37 31 | 30 | 108 | 38 01 | 30 | 107 | 38 31 | 28 | 106 | 38 59 | 29 | 105 | 39 28 | 27 | 104 |
| 50 | 35 38 | +31 | 110 | 36 09 | +30 | 109 | 36 39 | +30 | 107 | 37 09 | +29 | 106 | 37 38 | +29 | 105 | 38 07 | +28 | 104 | 38 35 | +27 | 103 |
| 51 | 34 47 | 30 | 109 | 35 17 | 30 | 108 | 35 47 | 30 | 107 | 36 17 | 29 | 106 | 36 46 | 28 | 105 | 37 14 | 28 | 104 | 37 42 | 27 | 102 |
| 52 | 33 55 | 30 | 108 | 34 25 | 30 | 107 | 34 55 | 29 | 106 | 35 24 | 29 | 105 | 35 53 | 28 | 104 | 36 21 | 27 | 103 | 36 48 | 27 | 102 |
| 53 | 33 03 | 30 | 108 | 33 33 | 30 | 107 | 34 03 | 29 | 106 | 34 32 | 28 | 105 | 35 00 | 28 | 103 | 35 28 | 27 | 102 | 35 55 | 27 | 101 |
| 54 | 32 11 | 30 | 107 | 32 41 | 29 | 106 | 33 10 | 29 | 105 | 33 39 | 28 | 104 | 34 07 | 28 | 103 | 34 35 | 27 | 102 | 35 02 | 26 | 101 |
| 55 | 31 19 | +30 | 107 | 31 49 | +29 | 106 | 32 18 | +28 | 104 | 32 46 | +28 | 103 | 33 14 | +27 | 102 | 33 41 | +27 | 101 | 34 08 | +26 | 100 |
| 56 | 30 27 | 29 | 106 | 30 56 | 29 | 105 | 31 25 | 28 | 104 | 31 53 | 28 | 103 | 32 21 | 27 | 102 | 32 48 | 27 | 101 | 33 15 | 26 | 100 |
| 57 | 29 35 | 29 | 105 | 30 04 | 28 | 104 | 30 32 | 28 | 103 | 31 00 | 28 | 102 | 31 28 | 27 | 101 | 31 55 | 26 | 100 | 32 21 | 26 | 99 |
| 58 | 28 42 | 29 | 105 | 29 11 | 28 | 104 | 29 39 | 28 | 103 | 30 07 | 27 | 102 | 30 34 | 27 | 101 | 31 01 | 26 | 100 | 31 27 | 26 | 99 |
| 59 | 27 50 | 28 | 104 | 28 18 | 28 | 103 | 28 46 | 28 | 102 | 29 14 | 27 | 101 | 29 41 | 26 | 100 | 30 07 | 26 | 99 | 30 33 | 26 | 98 |
| 60 | 26 57 | +28 | 104 | 27 25 | +28 | 103 | 27 53 | +27 | 102 | 28 20 | +27 | 101 | 28 47 | +27 | 100 | 29 14 | +26 | 99 | 29 40 | +25 | 98 |
| 61 | 26 04 | 28 | 103 | 26 32 | 28 | 102 | 27 00 | 27 | 101 | 27 27 | 27 | 100 | 27 54 | 26 | 99 | 28 20 | 26 | 98 | 28 46 | 25 | 97 |
| 62 | 25 11 | 28 | 103 | 25 39 | 27 | 102 | 26 06 | 27 | 101 | 26 33 | 27 | 100 | 27 00 | 26 | 99 | 27 26 | 26 | 98 | 27 52 | 25 | 97 |
| 63 | 24 18 | 27 | 102 | 24 45 | 28 | 101 | 25 13 | 27 | 100 | 25 40 | 26 | 99 | 26 06 | 26 | 98 | 26 32 | 26 | 97 | 26 58 | 25 | 96 |
| 64 | 23 25 | 27 | 102 | 23 52 | 27 | 101 | 24 19 | 27 | 100 | 24 46 | 26 | 99 | 25 12 | 26 | 98 | 25 38 | 26 | 97 | 26 04 | 25 | 96 |
| 65 | 22 31 | +28 | 101 | 22 59 | +27 | 100 | 23 26 | +26 | 99 | 23 52 | +26 | 98 | 24 18 | +26 | 97 | 24 44 | +25 | 96 | 25 09 | +25 | 95 |
| 66 | 21 38 | 27 | 101 | 22 05 | 27 | 100 | 22 32 | 26 | 99 | 22 58 | 26 | 98 | 23 24 | 26 | 97 | 23 50 | 25 | 96 | 24 15 | 25 | 95 |
| 67 | 20 44 | 27 | 100 | 21 11 | 27 | 99 | 21 38 | 26 | 98 | 22 04 | 26 | 97 | 22 30 | 26 | 96 | 22 56 | 25 | 95 | 23 21 | 25 | 94 |
| 68 | 19 51 | 27 | 100 | 20 18 | 26 | 99 | 20 44 | 26 | 98 | 21 10 | 26 | 97 | 21 36 | 26 | 96 | 22 02 | 25 | 95 | 22 27 | 25 | 94 |
| 69 | 18 57 | 27 | 99 | 19 24 | 26 | 98 | 19 50 | 26 | 97 | 20 16 | 26 | 96 | 20 42 | 25 | 95 | 21 07 | 26 | 94 | 21 33 | 24 | 93 |

| 7° Hc | d | Z | 8° Hc | d | Z | 9° Hc | d | Z | 10° Hc | d | Z | 11° Hc | d | Z | 12° Hc | d | Z | 13° Hc | d | Z | 14° Hc | d | Z | LHA |
|---|---|---|---|---|---|---|---|---|---|---|---|---|---|---|---|---|---|---|---|---|---|---|---|---|
| ° ′ | ′ | ° | ° ′ | ′ | ° | ° ′ | ′ | ° | ° ′ | ′ | ° | ° ′ | ′ | ° | ° ′ | ′ | ° | ° ′ | ′ | ° | ° ′ | ′ | ° | |
| 72 00 | +60 | 180 | 73 00 | +60 | 180 | 74 00 | +60 | 180 | 75 00 | +60 | 180 | 76 00 | +60 | 180 | 77 00 | +60 | 180 | 78 00 | +60 | 180 | 79 00 | +60 | 180 | 360 |
| 71 59 | 59 | 177 | 72 58 | 60 | 177 | 73 58 | 60 | 176 | 74 58 | 60 | 176 | 75 58 | 60 | 176 | 76 58 | 60 | 176 | 77 58 | 60 | 175 | 78 58 | 59 | 175 | 359 |
| 71 54 | 60 | 174 | 72 54 | 59 | 173 | 73 53 | 60 | 173 | 74 53 | 59 | 172 | 75 52 | 60 | 172 | 76 52 | 59 | 171 | 77 51 | 59 | 171 | 78 50 | 60 | 170 | 358 |
| 71 46 | 60 | 170 | 72 46 | 59 | 170 | 73 45 | 59 | 169 | 74 44 | 59 | 169 | 75 43 | 59 | 168 | 76 42 | 58 | 167 | 77 40 | 59 | 166 | 78 39 | 58 | 165 | 357 |
| 71 36 | 59 | 167 | 72 35 | 58 | 167 | 73 33 | 59 | 166 | 74 32 | 58 | 165 | 75 30 | 58 | 164 | 76 28 | 57 | 163 | 77 25 | 58 | 162 | 78 23 | 56 | 160 | 356 |
| 71 23 | +58 | 164 | 72 21 | +57 | 164 | 73 18 | +58 | 163 | 74 16 | +57 | 162 | 75 13 | +57 | 160 | 76 10 | +56 | 159 | 77 06 | +56 | 158 | 78 02 | +55 | 156 | 355 |
| 71 07 | 57 | 161 | 72 04 | 57 | 160 | 73 01 | 56 | 159 | 73 57 | 56 | 158 | 74 53 | 56 | 157 | 75 49 | 55 | 155 | 76 44 | 54 | 154 | 77 38 | 54 | 152 | 354 |
| 70 48 | 56 | 158 | 71 44 | 56 | 157 | 72 40 | 56 | 156 | 73 36 | 54 | 155 | 74 30 | 55 | 153 | 75 25 | 53 | 152 | 76 18 | 53 | 150 | 77 11 | 51 | 148 | 353 |
| 70 27 | 55 | 156 | 71 22 | 55 | 154 | 72 17 | 54 | 153 | 73 11 | 54 | 152 | 74 05 | 53 | 150 | 74 58 | 52 | 148 | 75 50 | 50 | 146 | 76 40 | 50 | 144 | 352 |
| 70 03 | 54 | 153 | 70 57 | 54 | 152 | 71 51 | 53 | 150 | 72 44 | 52 | 149 | 73 36 | 52 | 147 | 74 28 | 50 | 145 | 75 18 | 49 | 143 | 76 07 | 48 | 141 | 351 |
| 69 37 | +53 | 150 | 70 30 | +53 | 149 | 71 23 | +52 | 148 | 72 15 | +51 | 146 | 73 06 | +50 | 144 | 73 56 | +48 | 142 | 74 44 | +48 | 140 | 75 32 | +45 | 138 | 350 |
| 69 09 | 52 | 148 | 70 01 | 52 | 146 | 70 53 | 50 | 145 | 71 43 | 50 | 143 | 72 33 | 48 | 141 | 73 21 | 47 | 139 | 74 08 | 46 | 137 | 74 54 | 44 | 135 | 349 |
| 68 39 | 51 | 146 | 69 30 | 50 | 144 | 70 20 | 50 | 142 | 71 10 | 48 | 141 | 71 58 | 47 | 139 | 72 45 | 45 | 137 | 73 30 | 44 | 135 | 74 14 | 43 | 132 | 348 |
| 68 08 | 49 | 143 | 68 57 | 49 | 142 | 69 46 | 48 | 140 | 70 34 | 47 | 138 | 71 21 | 46 | 136 | 72 07 | 44 | 134 | 72 51 | 42 | 132 | 73 33 | 41 | 130 | 347 |
| 67 34 | 49 | 141 | 68 23 | 48 | 139 | 69 11 | 46 | 138 | 69 57 | 46 | 136 | 70 43 | 44 | 134 | 71 27 | 43 | 132 | 72 10 | 41 | 130 | 72 51 | 39 | 127 | 346 |
| 66 59 | +48 | 139 | 67 47 | +46 | 137 | 68 33 | +46 | 136 | 69 19 | +44 | 134 | 70 03 | +43 | 132 | 70 46 | +41 | 130 | 71 27 | +40 | 128 | 72 07 | +38 | 125 | 345 |
| 66 23 | 46 | 137 | 67 09 | 46 | 135 | 67 55 | 44 | 134 | 68 39 | 43 | 132 | 69 22 | 41 | 130 | 70 03 | 40 | 128 | 70 43 | 39 | 126 | 71 22 | 36 | 123 | 344 |
| 65 45 | 45 | 135 | 66 30 | 45 | 133 | 67 15 | 43 | 132 | 67 58 | 42 | 130 | 68 40 | 40 | 128 | 69 20 | 39 | 126 | 69 59 | 37 | 124 | 70 36 | 35 | 121 | 343 |
| 65 06 | 44 | 133 | 65 50 | 43 | 132 | 66 33 | 42 | 130 | 67 15 | 41 | 128 | 67 56 | 39 | 126 | 68 35 | 38 | 124 | 69 13 | 36 | 122 | 69 49 | 34 | 120 | 342 |
| 64 26 | 43 | 132 | 65 09 | 42 | 130 | 65 51 | 41 | 128 | 66 32 | 40 | 126 | 67 12 | 38 | 125 | 67 50 | 36 | 122 | 68 26 | 35 | 120 | 69 01 | 33 | 118 | 341 |
| 63 45 | +42 | 130 | 64 27 | +41 | 128 | 65 08 | +40 | 127 | 65 48 | +38 | 125 | 66 26 | +38 | 123 | 67 04 | +35 | 121 | 67 39 | +34 | 119 | 68 13 | +32 | 117 | 340 |
| 63 02 | 42 | 128 | 63 44 | 40 | 127 | 64 24 | 39 | 125 | 65 03 | 37 | 123 | 65 40 | 37 | 121 | 66 17 | 34 | 119 | 66 51 | 33 | 117 | 67 24 | 31 | 115 | 339 |
| 62 19 | 41 | 127 | 63 00 | 39 | 125 | 63 39 | 38 | 124 | 64 17 | 37 | 122 | 64 54 | 35 | 120 | 65 29 | 34 | 118 | 66 03 | 32 | 116 | 66 35 | 30 | 114 | 338 |
| 61 35 | 40 | 125 | 62 15 | 38 | 124 | 62 53 | 37 | 122 | 63 30 | 36 | 120 | 64 06 | 35 | 119 | 64 41 | 32 | 117 | 65 13 | 32 | 115 | 65 45 | 29 | 113 | 337 |
| 60 51 | 38 | 124 | 61 29 | 38 | 123 | 62 07 | 36 | 121 | 62 43 | 35 | 119 | 63 18 | 34 | 117 | 63 52 | 32 | 115 | 64 24 | 30 | 114 | 64 54 | 29 | 112 | 336 |
| 60 05 | +38 | 123 | 60 43 | +37 | 121 | 61 20 | +35 | 120 | 61 55 | +35 | 118 | 62 30 | +32 | 116 | 63 02 | +32 | 114 | 63 34 | +30 | 112 | 64 04 | +28 | 110 | 335 |
| 59 19 | 37 | 122 | 59 56 | 36 | 120 | 60 32 | 35 | 118 | 61 07 | 33 | 117 | 61 40 | 33 | 115 | 62 13 | 30 | 113 | 62 43 | 29 | 111 | 63 12 | 28 | 109 | 334 |
| 58 32 | 37 | 120 | 59 09 | 35 | 119 | 59 44 | 34 | 117 | 60 18 | 33 | 116 | 60 51 | 31 | 114 | 61 22 | 30 | 112 | 61 52 | 29 | 110 | 62 21 | 27 | 108 | 333 |
| 57 45 | 36 | 119 | 58 21 | 34 | 118 | 58 55 | 34 | 116 | 59 29 | 32 | 114 | 60 01 | 31 | 113 | 60 32 | 29 | 111 | 61 01 | 28 | 109 | 61 29 | 27 | 107 | 332 |
| 56 58 | 35 | 118 | 57 33 | 33 | 117 | 58 06 | 33 | 115 | 58 39 | 32 | 113 | 59 11 | 30 | 112 | 59 41 | 29 | 110 | 60 10 | 27 | 108 | 60 37 | 26 | 107 | 331 |
| 56 09 | +35 | 117 | 56 44 | +33 | 116 | 57 17 | +32 | 114 | 57 49 | +31 | 112 | 58 20 | +30 | 111 | 58 50 | +28 | 109 | 59 18 | +27 | 107 | 59 45 | +25 | 106 | 330 |
| 55 21 | 33 | 116 | 55 54 | 33 | 115 | 56 27 | 32 | 113 | 56 59 | 30 | 112 | 57 29 | 29 | 110 | 57 58 | 28 | 108 | 58 26 | 26 | 107 | 58 52 | 25 | 105 | 329 |
| 54 32 | 33 | 115 | 55 05 | 32 | 114 | 55 37 | 31 | 112 | 56 08 | 30 | 111 | 56 38 | 28 | 109 | 57 06 | 28 | 107 | 57 34 | 26 | 106 | 58 00 | 24 | 104 | 328 |
| 53 42 | 33 | 114 | 54 15 | 31 | 113 | 54 46 | 31 | 111 | 55 17 | 29 | 110 | 55 46 | 28 | 108 | 56 14 | 27 | 107 | 56 41 | 26 | 105 | 57 07 | 24 | 103 | 327 |
| 52 52 | 32 | 113 | 53 24 | 31 | 112 | 53 55 | 30 | 110 | 54 25 | 29 | 109 | 54 54 | 28 | 107 | 55 22 | 27 | 106 | 55 49 | 25 | 104 | 56 14 | 24 | 103 | 326 |
| 52 02 | +32 | 112 | 52 34 | +30 | 111 | 53 04 | +30 | 109 | 53 34 | +28 | 108 | 54 02 | +28 | 107 | 54 30 | +26 | 105 | 54 56 | +25 | 103 | 55 21 | +23 | 102 | 325 |
| 51 12 | 31 | 111 | 51 43 | 30 | 110 | 52 13 | 29 | 109 | 52 42 | 28 | 107 | 53 10 | 27 | 106 | 53 37 | 26 | 104 | 54 03 | 24 | 103 | 54 27 | 24 | 101 | 324 |
| 50 21 | 31 | 111 | 50 52 | 29 | 109 | 51 21 | 29 | 108 | 51 50 | 28 | 107 | 52 18 | 26 | 105 | 52 44 | 26 | 104 | 53 10 | 24 | 102 | 53 34 | 23 | 101 | 323 |
| 49 30 | 30 | 110 | 50 00 | 29 | 109 | 50 29 | 29 | 107 | 50 58 | 27 | 106 | 51 25 | 26 | 104 | 51 51 | 25 | 103 | 52 16 | 25 | 101 | 52 41 | 22 | 100 | 322 |
| 48 39 | 29 | 109 | 49 08 | 29 | 108 | 49 37 | 28 | 106 | 50 05 | 27 | 105 | 50 32 | 26 | 104 | 50 58 | 25 | 102 | 51 23 | 24 | 101 | 51 47 | 23 | 99 | 321 |
| 47 47 | +29 | 108 | 48 16 | +29 | 107 | 48 45 | +28 | 106 | 49 13 | +26 | 104 | 49 39 | +26 | 103 | 50 05 | +25 | 102 | 50 30 | +23 | 100 | 50 53 | +23 | 99 | 320 |
| 46 55 | 29 | 108 | 47 24 | 29 | 106 | 47 53 | 27 | 105 | 48 20 | 26 | 104 | 48 46 | 26 | 102 | 49 12 | 24 | 101 | 49 36 | 23 | 100 | 49 59 | 23 | 98 | 319 |
| 46 03 | 29 | 107 | 46 32 | 28 | 106 | 47 00 | 27 | 104 | 47 27 | 26 | 103 | 47 53 | 25 | 102 | 48 18 | 24 | 100 | 48 42 | 24 | 99 | 49 06 | 22 | 98 | 318 |
| 45 11 | 29 | 106 | 45 40 | 27 | 105 | 46 07 | 27 | 104 | 46 34 | 26 | 102 | 47 00 | 25 | 101 | 47 25 | 24 | 100 | 47 49 | 23 | 98 | 48 12 | 22 | 97 | 317 |
| 44 19 | 28 | 106 | 44 47 | 27 | 104 | 45 14 | 27 | 103 | 45 41 | 25 | 102 | 46 06 | 25 | 100 | 46 31 | 24 | 99 | 46 55 | 23 | 98 | 47 18 | 21 | 96 | 316 |
| 43 26 | +28 | 105 | 43 54 | +27 | 104 | 44 21 | +26 | 102 | 44 47 | +26 | 101 | 45 13 | +24 | 100 | 45 37 | +24 | 99 | 46 01 | +22 | 97 | 46 23 | +22 | 96 | 315 |
| 42 34 | 27 | 104 | 43 01 | 27 | 103 | 43 28 | 26 | 102 | 43 54 | 25 | 101 | 44 19 | 24 | 99 | 44 43 | 24 | 98 | 45 07 | 22 | 97 | 45 29 | 22 | 95 | 314 |
| 41 41 | 28 | 104 | 42 08 | 27 | 102 | 42 35 | 26 | 101 | 43 01 | 24 | 100 | 43 25 | 25 | 99 | 43 50 | 23 | 97 | 44 13 | 22 | 96 | 44 35 | 22 | 95 | 313 |
| 40 48 | 27 | 103 | 41 15 | 26 | 102 | 41 41 | 26 | 101 | 42 07 | 25 | 99 | 42 32 | 24 | 98 | 42 56 | 23 | 97 | 43 19 | 22 | 96 | 43 41 | 21 | 94 | 312 |
| 39 55 | 27 | 102 | 40 22 | 26 | 101 | 40 48 | 25 | 100 | 41 13 | 25 | 99 | 41 38 | 24 | 98 | 42 02 | 23 | 96 | 42 25 | 22 | 95 | 42 47 | 21 | 94 | 311 |
| 39 02 | +26 | 102 | 39 28 | +26 | 101 | 39 54 | +25 | 100 | 40 19 | +25 | 98 | 40 44 | +24 | 97 | 41 08 | +22 | 96 | 41 30 | +23 | 95 | 41 53 | +21 | 93 | 310 |
| 38 09 | 26 | 101 | 38 35 | 26 | 100 | 39 01 | 25 | 99 | 39 26 | 24 | 98 | 39 50 | 23 | 97 | 40 13 | 23 | 95 | 40 36 | 22 | 94 | 40 58 | 21 | 93 | 309 |
| 37 15 | 26 | 100 | 37 41 | 26 | 100 | 38 07 | 25 | 98 | 38 32 | 24 | 97 | 38 56 | 23 | 96 | 39 19 | 23 | 95 | 39 42 | 22 | 94 | 40 04 | 21 | 93 | 308 |
| 36 22 | 26 | 100 | 36 48 | 25 | 99 | 37 13 | 25 | 98 | 37 38 | 24 | 97 | 38 02 | 23 | 96 | 38 25 | 23 | 94 | 38 48 | 22 | 93 | 39 10 | 21 | 92 | 307 |
| 35 28 | 26 | 100 | 35 54 | 25 | 99 | 36 19 | 25 | 97 | 36 44 | 24 | 96 | 37 08 | 23 | 95 | 37 31 | 22 | 94 | 37 53 | 22 | 93 | 38 15 | 21 | 92 | 306 |
| 34 34 | +26 | 99 | 35 00 | +25 | 98 | 35 25 | +25 | 97 | 35 50 | +23 | 96 | 36 13 | +24 | 95 | 36 37 | +22 | 94 | 36 59 | +22 | 92 | 37 21 | +21 | 91 | 305 |
| 33 41 | 25 | 98 | 34 06 | 25 | 98 | 34 31 | 24 | 96 | 34 55 | 24 | 95 | 35 19 | 23 | 94 | 35 42 | 23 | 93 | 36 05 | 21 | 92 | 36 26 | 22 | 91 | 304 |
| 32 47 | 25 | 98 | 33 12 | 25 | 97 | 33 37 | 24 | 96 | 34 01 | 24 | 95 | 34 25 | 23 | 94 | 34 48 | 22 | 93 | 35 10 | 22 | 91 | 35 32 | 21 | 90 | 303 |
| 31 53 | 25 | 97 | 32 18 | 25 | 97 | 32 43 | 24 | 95 | 33 07 | 24 | 94 | 33 31 | 23 | 93 | 33 54 | 22 | 92 | 34 16 | 22 | 91 | 34 38 | 21 | 90 | 302 |
| 30 59 | 25 | 97 | 31 24 | 25 | 96 | 31 49 | 24 | 95 | 32 13 | 23 | 94 | 32 36 | 23 | 93 | 32 59 | 23 | 92 | 33 22 | 21 | 91 | 33 43 | 21 | 89 | 301 |
| 30 05 | +25 | 97 | 30 30 | +25 | 96 | 30 55 | +24 | 95 | 31 19 | +23 | 93 | 31 42 | +23 | 92 | 32 05 | +22 | 91 | 32 27 | +22 | 90 | 32 49 | +21 | 89 | 300 |
| 29 11 | 25 | 96 | 29 36 | 24 | 95 | 30 00 | 24 | 94 | 30 24 | 24 | 93 | 30 48 | 23 | 92 | 31 11 | 22 | 91 | 31 33 | 22 | 90 | 31 55 | 21 | 89 | 299 |
| 28 17 | 25 | 95 | 28 42 | 24 | 95 | 29 06 | 24 | 94 | 29 30 | 23 | 93 | 29 53 | 23 | 91 | 30 16 | 22 | 90 | 30 38 | 22 | 89 | 31 00 | 21 | 88 | 298 |
| 27 23 | 25 | 95 | 27 48 | 24 | 94 | 28 12 | 24 | 93 | 28 36 | 23 | 92 | 28 59 | 23 | 91 | 29 22 | 22 | 90 | 29 44 | 22 | 89 | 30 06 | 21 | 88 | 297 |
| 26 29 | 24 | 95 | 26 53 | 25 | 94 | 27 18 | 23 | 93 | 27 41 | 24 | 92 | 28 05 | 22 | 91 | 28 27 | 23 | 90 | 28 50 | 22 | 89 | 29 12 | 21 | 87 | 296 |
| 25 34 | +25 | 94 | 25 59 | +24 | 93 | 26 23 | +24 | 92 | 26 47 | +23 | 91 | 27 10 | +23 | 90 | 27 33 | +22 | 89 | 27 55 | +22 | 88 | 28 17 | +22 | 87 | 295 |
| 24 40 | 25 | 94 | 25 05 | 24 | 93 | 25 29 | 24 | 92 | 25 53 | 23 | 91 | 26 16 | 23 | 90 | 26 39 | 22 | 89 | 27 01 | 22 | 88 | 27 23 | 21 | 87 | 294 |
| 23 46 | 24 | 93 | 24 10 | 25 | 92 | 24 35 | 23 | 91 | 24 58 | 23 | 90 | 25 21 | 23 | 89 | 25 44 | 23 | 88 | 26 07 | 22 | 87 | 26 29 | 21 | 86 | 293 |
| 22 52 | 24 | 93 | 23 16 | 24 | 92 | 23 40 | 24 | 91 | 24 04 | 23 | 90 | 24 27 | 23 | 89 | 24 50 | 22 | 88 | 25 12 | 22 | 87 | 25 34 | 22 | 86 | 292 |
| 21 57 | 25 | 93 | 22 22 | 24 | 92 | 22 46 | 23 | 91 | 23 09 | 24 | 90 | 23 33 | 23 | 89 | 23 56 | 22 | 88 | 24 18 | 22 | 87 | 24 40 | 22 | 85 | 291 |

S. Lat. {LHA greater than 180°........ Zn=180−Z; LHA less than 180°........... Zn=180+Z}

DECLINATION (0°-14°) SAME NAME AS LATITUDE

N. Lat. {LHA greater than 180°........ Zn=Z; LHA less than 180°........... Zn=360−Z}

# DECLINATION (0°-14°) CONTRARY NAME TO LATITUDE



| LHA | 0° | | | 1° | | | 2° | | | 3° | | | 4° | | |
|---|---|---|---|---|---|---|---|---|---|---|---|---|---|---|---|
| | Hc | d | Z | Hc | d | Z | Hc | d | Z | Hc | d | Z | Hc | d | Z |
| | ° ′ | ′ | ° | ° ′ | ′ | ° | ° ′ | ′ | ° | ° ′ | ′ | ° | ° ′ | ′ | ° |
| 69 | 18 57 | 27 | 99 | 18 30 | 27 | 100 | 18 03 | 27 | 101 | 17 36 | 28 | 102 | 17 08 | 28 | 103 |
| 68 | 19 51 | 27 | 100 | 19 24 | 28 | 101 | 18 56 | 27 | 102 | 18 29 | 28 | 103 | 18 01 | 29 | 103 |
| 67 | 20 44 | 27 | 100 | 20 17 | 27 | 101 | 19 50 | 28 | 102 | 19 22 | 28 | 103 | 18 54 | 29 | 104 |
| 66 | 21 38 | 28 | 101 | 21 10 | 27 | 102 | 20 43 | 28 | 103 | 20 15 | 29 | 104 | 19 46 | 28 | 104 |
| 65 | 22 31 | −27 | 101 | 22 04 | −28 | 102 | 21 36 | −29 | 103 | 21 07 | −28 | 104 | 20 39 | −29 | 105 |
| 64 | 23 25 | 28 | 102 | 22 57 | 28 | 103 | 22 29 | 29 | 104 | 22 00 | 29 | 105 | 21 31 | 29 | 106 |
| 63 | 24 18 | 28 | 102 | 23 50 | 29 | 103 | 23 21 | 28 | 104 | 22 53 | 29 | 105 | 22 24 | 30 | 106 |
| 62 | 25 11 | 28 | 103 | 24 43 | 29 | 104 | 24 14 | 29 | 105 | 23 45 | 29 | 106 | 23 16 | 30 | 107 |
| 61 | 26 04 | 28 | 103 | 25 36 | 29 | 104 | 25 07 | 29 | 105 | 24 38 | 30 | 106 | 24 08 | 30 | 107 |
| 60 | 26 57 | −29 | 104 | 26 28 | −29 | 105 | 25 59 | −29 | 106 | 25 30 | −30 | 107 | 25 00 | −30 | 108 |
| 59 | 27 50 | 29 | 104 | 27 21 | 30 | 105 | 26 51 | 29 | 106 | 26 22 | 30 | 107 | 25 52 | 31 | 108 |
| 58 | 28 42 | 29 | 105 | 28 13 | 29 | 106 | 27 44 | 30 | 107 | 27 14 | 31 | 108 | 26 43 | 31 | 109 |
| 57 | 29 35 | 30 | 105 | 29 05 | 29 | 106 | 28 36 | 31 | 107 | 28 05 | 30 | 108 | 27 35 | 31 | 109 |
| 56 | 30 27 | 30 | 106 | 29 57 | 30 | 107 | 29 27 | 30 | 108 | 28 57 | 31 | 109 | 28 26 | 32 | 110 |
| 55 | 31 19 | −30 | 107 | 30 49 | −30 | 108 | 30 19 | −31 | 109 | 29 48 | −31 | 110 | 29 17 | −32 | 111 |
| 54 | 32 11 | 30 | 107 | 31 41 | 31 | 108 | 31 10 | 31 | 109 | 30 39 | 31 | 110 | 30 08 | 32 | 111 |
| 53 | 33 03 | 30 | 108 | 32 33 | 31 | 109 | 32 02 | 32 | 110 | 31 30 | 32 | 111 | 30 58 | 32 | 112 |
| 52 | 33 55 | 31 | 108 | 33 24 | 31 | 109 | 32 53 | 32 | 110 | 32 21 | 32 | 111 | 31 49 | 33 | 112 |
| 51 | 34 47 | 32 | 109 | 34 15 | 31 | 110 | 33 44 | 32 | 111 | 33 12 | 33 | 112 | 32 39 | 33 | 113 |
| 50 | 35 38 | −32 | 110 | 35 06 | −32 | 111 | 34 34 | −32 | 112 | 34 02 | −33 | 113 | 33 29 | −34 | 114 |
| 49 | 36 29 | 32 | 110 | 35 57 | 32 | 111 | 35 25 | 33 | 112 | 34 52 | 33 | 113 | 34 19 | 34 | 114 |
| 48 | 37 20 | 32 | 111 | 36 48 | 33 | 112 | 36 15 | 33 | 113 | 35 42 | 34 | 114 | 35 08 | 34 | 115 |
| 47 | 38 11 | 33 | 112 | 37 38 | 33 | 113 | 37 05 | 34 | 114 | 36 31 | 34 | 115 | 35 57 | 34 | 116 |
| 46 | 39 01 | 33 | 112 | 38 28 | 33 | 113 | 37 55 | 34 | 114 | 37 21 | 35 | 115 | 36 46 | 35 | 116 |
| 45 | 39 51 | −33 | 113 | 39 18 | −34 | 114 | 38 44 | −34 | 115 | 38 10 | −35 | 116 | 37 35 | −36 | 117 |
| 44 | 40 41 | 33 | 114 | 40 08 | 35 | 115 | 39 33 | 35 | 116 | 38 58 | 35 | 117 | 38 23 | 36 | 118 |
| 43 | 41 31 | 34 | 114 | 40 57 | 35 | 116 | 40 22 | 35 | 117 | 39 47 | 36 | 118 | 39 11 | 37 | 119 |
| 42 | 42 20 | 34 | 115 | 41 46 | 35 | 116 | 41 11 | 36 | 117 | 40 35 | 37 | 118 | 39 58 | 37 | 119 |
| 41 | 43 09 | 35 | 116 | 42 34 | 35 | 117 | 41 59 | 37 | 118 | 41 22 | 37 | 119 | 40 45 | 37 | 120 |
| 40 | 43 58 | −35 | 117 | 43 23 | −37 | 118 | 42 46 | −36 | 119 | 42 10 | −38 | 120 | 41 32 | −38 | 121 |
| 39 | 44 47 | 36 | 118 | 44 11 | 37 | 119 | 43 34 | 37 | 120 | 42 57 | 38 | 121 | 42 19 | 39 | 122 |
| 38 | 45 35 | 37 | 118 | 44 58 | 37 | 120 | 44 21 | 38 | 121 | 43 43 | 38 | 122 | 43 05 | 39 | 123 |
| 37 | 46 22 | 37 | 119 | 45 45 | 38 | 120 | 45 07 | 38 | 122 | 44 29 | 39 | 123 | 43 50 | 39 | 124 |
| 36 | 47 09 | 37 | 120 | 46 32 | 38 | 121 | 45 54 | 39 | 122 | 45 15 | 40 | 124 | 44 35 | 40 | 125 |
| 35 | 47 56 | −38 | 121 | 47 18 | −39 | 122 | 46 39 | −39 | 123 | 46 00 | −40 | 125 | 45 20 | −41 | 126 |
| 34 | 48 43 | 39 | 122 | 48 04 | 40 | 123 | 47 24 | 40 | 124 | 46 44 | 40 | 125 | 46 04 | 42 | 127 |
| 33 | 49 28 | 39 | 123 | 48 49 | 40 | 124 | 48 09 | 41 | 125 | 47 28 | 41 | 126 | 46 47 | 42 | 128 |
| 32 | 50 14 | 40 | 124 | 49 34 | 41 | 125 | 48 53 | 41 | 126 | 48 12 | 42 | 128 | 47 30 | 43 | 129 |
| 31 | 50 58 | 40 | 125 | 50 18 | 41 | 126 | 49 37 | 42 | 127 | 48 55 | 43 | 129 | 48 12 | 43 | 130 |
| 30 | 51 43 | −42 | 126 | 51 01 | −42 | 127 | 50 19 | −42 | 129 | 49 37 | −43 | 130 | 48 54 | −44 | 131 |
| 29 | 52 26 | 42 | 127 | 51 44 | 42 | 129 | 51 02 | 44 | 130 | 50 18 | 43 | 131 | 49 35 | 45 | 132 |
| 28 | 53 09 | 42 | 129 | 52 27 | 44 | 130 | 51 43 | 44 | 131 | 50 59 | 44 | 132 | 50 15 | 45 | 133 |
| 27 | 53 51 | 43 | 130 | 53 08 | 44 | 131 | 52 24 | 45 | 132 | 51 39 | 45 | 133 | 50 54 | 46 | 134 |
| 26 | 54 33 | 44 | 131 | 53 49 | 45 | 132 | 53 04 | 45 | 133 | 52 19 | 46 | 134 | 51 33 | 47 | 135 |
| 25 | 55 14 | −45 | 132 | 54 29 | −46 | 133 | 53 43 | −46 | 135 | 52 57 | −46 | 136 | 52 11 | −47 | 137 |
| 24 | 55 53 | 45 | 134 | 55 08 | 46 | 135 | 54 22 | 47 | 136 | 53 35 | 47 | 137 | 52 48 | 48 | 138 |
| 23 | 56 32 | 46 | 135 | 55 46 | 47 | 136 | 54 59 | 47 | 137 | 54 12 | 48 | 138 | 53 24 | 49 | 139 |
| 22 | 57 10 | 47 | 136 | 56 23 | 47 | 137 | 55 36 | 49 | 139 | 54 47 | 48 | 140 | 53 59 | 50 | 141 |
| 21 | 57 48 | 48 | 138 | 57 00 | 49 | 139 | 56 11 | 49 | 140 | 55 22 | 49 | 141 | 54 33 | 50 | 142 |
| 20 | 58 24 | −49 | 139 | 57 35 | −49 | 140 | 56 46 | −50 | 141 | 55 56 | −50 | 142 | 55 06 | −51 | 143 |
| 19 | 58 58 | 49 | 141 | 58 09 | 50 | 142 | 57 19 | 50 | 143 | 56 29 | 51 | 144 | 55 38 | 52 | 145 |
| 18 | 59 32 | 50 | 142 | 58 42 | 51 | 144 | 57 51 | 51 | 145 | 57 00 | 52 | 146 | 56 08 | 52 | 146 |
| 17 | 60 05 | 51 | 144 | 59 14 | 52 | 145 | 58 22 | 52 | 146 | 57 30 | 52 | 147 | 56 38 | 53 | 148 |
| 16 | 60 36 | 52 | 146 | 59 44 | 52 | 147 | 58 52 | 53 | 148 | 57 59 | 53 | 149 | 57 06 | 54 | 150 |
| 15 | 61 06 | −53 | 148 | 60 13 | −53 | 149 | 59 20 | −53 | 150 | 58 27 | −54 | 150 | 57 33 | −54 | 151 |
| 14 | 61 34 | 53 | 150 | 60 41 | 54 | 150 | 59 47 | 54 | 151 | 58 53 | 55 | 152 | 57 58 | 55 | 153 |
| 13 | 62 01 | 54 | 151 | 61 07 | 55 | 152 | 60 12 | 55 | 153 | 59 17 | 55 | 154 | 58 22 | 56 | 155 |
| 12 | 62 26 | 55 | 153 | 61 31 | 55 | 154 | 60 36 | 55 | 155 | 59 41 | 56 | 156 | 58 45 | 56 | 156 |
| 11 | 62 50 | 56 | 155 | 61 54 | 56 | 156 | 60 58 | 56 | 157 | 60 02 | 56 | 158 | 59 06 | 57 | 158 |
| 10 | 63 12 | −57 | 157 | 62 15 | −56 | 158 | 61 19 | −57 | 159 | 60 22 | −57 | 160 | 59 25 | −57 | 160 |
| 9 | 63 32 | 57 | 160 | 62 35 | 57 | 160 | 61 38 | 58 | 161 | 60 40 | 57 | 161 | 59 43 | 58 | 162 |
| 8 | 63 50 | 58 | 162 | 62 52 | 57 | 162 | 61 55 | 58 | 163 | 60 57 | 58 | 163 | 59 59 | 58 | 164 |
| 7 | 64 06 | 58 | 164 | 63 08 | 58 | 164 | 62 10 | 59 | 165 | 61 11 | 58 | 165 | 60 13 | 59 | 166 |
| 6 | 64 20 | 58 | 166 | 63 22 | 59 | 167 | 62 23 | 59 | 167 | 61 24 | 59 | 167 | 60 25 | 59 | 168 |
| 5 | 64 32 | −59 | 168 | 63 33 | −59 | 169 | 62 34 | −59 | 169 | 61 35 | −59 | 170 | 60 36 | −59 | 170 |
| 4 | 64 42 | 59 | 171 | 63 43 | 60 | 171 | 62 43 | 59 | 171 | 61 44 | 60 | 172 | 60 44 | 59 | 172 |
| 3 | 64 50 | 60 | 173 | 63 50 | 59 | 173 | 62 51 | 60 | 173 | 61 51 | 60 | 174 | 60 51 | 59 | 174 |
| 2 | 64 56 | 60 | 175 | 63 56 | 60 | 175 | 62 56 | 60 | 176 | 61 56 | 60 | 176 | 60 56 | 60 | 176 |
| 1 | 64 59 | 60 | 178 | 63 59 | 60 | 178 | 62 59 | 60 | 178 | 61 59 | 60 | 178 | 60 59 | 60 | 178 |
| 0 | 65 00 | −60 | 180 | 64 00 | −60 | 180 | 63 00 | −60 | 180 | 62 00 | −60 | 180 | 61 00 | −60 | 180 |

| LHA | 5° | | | 6° | | | 7° | | | 8° | | | 9° | | |
|---|---|---|---|---|---|---|---|---|---|---|---|---|---|---|---|
| | Hc | d | Z | Hc | d | Z | Hc | d | Z | Hc | d | Z | Hc | d | Z |
| | ° ′ | ′ | ° | ° ′ | ′ | ° | ° ′ | ′ | ° | ° ′ | ′ | ° | ° ′ | ′ | ° |
| 69 | 16 40 | 29 | 104 | 16 11 | 28 | 105 | 15 43 | 29 | 106 | 15 14 | 29 | 107 | 14 45 | 29 | 108 |
| 68 | 17 32 | 28 | 104 | 17 04 | 29 | 105 | 16 35 | 29 | 106 | 16 06 | 29 | 107 | 15 37 | 29 | 108 |
| 67 | 18 25 | 29 | 105 | 17 56 | 29 | 106 | 17 27 | 29 | 107 | 16 58 | 29 | 108 | 16 29 | 30 | 109 |
| 66 | 19 18 | 29 | 105 | 18 49 | 30 | 106 | 18 19 | 29 | 107 | 17 50 | 30 | 108 | 17 20 | 30 | 109 |
| 65 | 20 10 | −29 | 106 | 19 41 | −30 | 107 | 19 11 | −29 | 108 | 18 42 | −30 | 109 | 18 12 | −31 | 110 |
| 64 | 21 02 | 29 | 106 | 20 33 | 30 | 107 | 20 03 | 30 | 108 | 19 33 | 30 | 109 | 19 03 | 31 | 110 |
| 63 | 21 54 | 29 | 107 | 21 25 | 30 | 108 | 20 55 | 31 | 109 | 20 24 | 30 | 110 | 19 54 | 31 | 111 |
| 62 | 22 46 | 30 | 108 | 22 16 | 30 | 108 | 21 46 | 31 | 109 | 21 15 | 31 | 110 | 20 44 | 31 | 111 |
| 61 | 23 38 | 30 | 108 | 23 08 | 30 | 109 | 22 37 | 31 | 110 | 22 06 | 31 | 111 | 21 35 | 31 | 112 |
| 60 | 24 30 | −31 | 109 | 23 59 | −31 | 110 | 23 28 | −31 | 110 | 22 57 | −31 | 111 | 22 26 | −32 | 112 |
| 59 | 25 21 | 31 | 109 | 24 50 | 31 | 110 | 24 19 | 31 | 111 | 23 48 | 32 | 112 | 23 16 | 32 | 113 |
| 58 | 26 12 | 31 | 110 | 25 41 | 31 | 111 | 25 10 | 32 | 112 | 24 38 | 32 | 113 | 24 06 | 33 | 113 |
| 57 | 27 04 | 32 | 110 | 26 32 | 32 | 111 | 26 00 | 32 | 112 | 25 28 | 32 | 113 | 24 56 | 33 | 114 |
| 56 | 27 54 | 31 | 111 | 27 23 | 32 | 112 | 26 51 | 33 | 113 | 26 18 | 33 | 114 | 25 45 | 33 | 115 |
| 55 | 28 45 | −32 | 111 | 28 13 | −32 | 112 | 27 41 | −33 | 113 | 27 08 | −34 | 114 | 26 34 | −33 | 115 |
| 54 | 29 36 | 33 | 112 | 29 03 | 33 | 113 | 28 30 | 33 | 114 | 27 57 | 34 | 115 | 27 23 | 34 | 116 |
| 53 | 30 26 | 33 | 113 | 29 53 | 33 | 114 | 29 20 | 34 | 115 | 28 46 | 34 | 116 | 28 12 | 34 | 117 |
| 52 | 31 16 | 33 | 113 | 30 43 | 34 | 114 | 30 09 | 34 | 115 | 29 35 | 34 | 116 | 29 01 | 35 | 117 |
| 51 | 32 06 | 34 | 114 | 31 32 | 34 | 115 | 30 58 | 34 | 116 | 30 24 | 35 | 117 | 29 49 | 35 | 118 |
| 50 | 32 55 | −34 | 115 | 32 21 | −34 | 116 | 31 47 | −35 | 117 | 31 12 | −35 | 118 | 30 37 | −36 | 119 |
| 49 | 33 45 | 35 | 115 | 33 10 | 34 | 116 | 32 36 | 36 | 117 | 32 00 | 35 | 118 | 31 25 | 36 | 119 |
| 48 | 34 34 | 35 | 116 | 33 59 | 35 | 117 | 33 24 | 36 | 118 | 32 48 | 36 | 119 | 32 12 | 37 | 120 |
| 47 | 35 23 | 36 | 117 | 34 47 | 35 | 118 | 34 12 | 36 | 119 | 33 36 | 37 | 120 | 32 59 | 37 | 121 |
| 46 | 36 11 | 36 | 117 | 35 35 | 36 | 118 | 34 59 | 36 | 119 | 34 23 | 37 | 120 | 33 46 | 38 | 121 |
| 45 | 36 59 | −36 | 118 | 36 23 | −37 | 119 | 35 46 | −37 | 120 | 35 09 | −37 | 121 | 34 32 | −38 | 122 |
| 44 | 37 47 | 37 | 119 | 37 10 | 37 | 120 | 36 33 | 37 | 121 | 35 56 | 38 | 122 | 35 18 | 39 | 123 |
| 43 | 38 34 | 37 | 120 | 37 57 | 37 | 121 | 37 20 | 38 | 122 | 36 42 | 39 | 123 | 36 03 | 39 | 124 |
| 42 | 39 21 | 37 | 120 | 38 44 | 38 | 122 | 38 06 | 39 | 122 | 37 27 | 39 | 123 | 36 48 | 39 | 124 |
| 41 | 40 08 | 38 | 121 | 39 30 | 38 | 122 | 38 52 | 39 | 123 | 38 13 | 40 | 124 | 37 33 | 40 | 125 |
| 40 | 40 54 | −38 | 122 | 40 16 | −39 | 123 | 39 37 | −40 | 124 | 38 57 | −40 | 125 | 38 17 | −40 | 126 |
| 39 | 41 40 | 39 | 123 | 41 01 | 39 | 124 | 40 22 | 40 | 125 | 39 42 | 41 | 126 | 39 01 | 41 | 127 |
| 38 | 42 26 | 40 | 124 | 41 46 | 40 | 125 | 41 06 | 41 | 126 | 40 25 | 41 | 127 | 39 44 | 41 | 128 |
| 37 | 43 11 | 41 | 125 | 42 30 | 40 | 126 | 41 50 | 41 | 127 | 41 09 | 42 | 128 | 40 27 | 42 | 129 |
| 36 | 43 55 | 41 | 126 | 43 14 | 41 | 127 | 42 33 | 42 | 128 | 41 51 | 42 | 129 | 41 09 | 42 | 130 |
| 35 | 44 39 | −41 | 127 | 43 58 | −42 | 128 | 43 16 | −42 | 129 | 42 34 | −43 | 130 | 41 51 | −43 | 131 |
| 34 | 45 22 | 42 | 128 | 44 40 | 42 | 129 | 43 58 | 43 | 130 | 43 15 | 43 | 131 | 42 32 | 44 | 132 |
| 33 | 46 05 | 42 | 129 | 45 23 | 43 | 130 | 44 40 | 44 | 131 | 43 56 | 44 | 132 | 43 12 | 44 | 132 |
| 32 | 46 47 | 43 | 130 | 46 04 | 43 | 131 | 45 21 | 44 | 132 | 44 37 | 45 | 133 | 43 52 | 45 | 133 |
| 31 | 47 29 | 44 | 131 | 46 45 | 44 | 132 | 46 01 | 45 | 133 | 45 16 | 45 | 134 | 44 31 | 45 | 135 |
| 30 | 48 10 | −44 | 132 | 47 26 | −45 | 133 | 46 41 | −45 | 134 | 45 56 | −46 | 135 | 45 10 | −46 | 136 |
| 29 | 48 50 | 45 | 133 | 48 05 | 45 | 134 | 47 20 | 46 | 135 | 46 34 | 46 | 136 | 45 48 | 47 | 137 |
| 28 | 49 30 | 46 | 134 | 48 44 | 46 | 135 | 47 58 | 47 | 136 | 47 11 | 47 | 137 | 46 24 | 47 | 138 |
| 27 | 50 08 | 46 | 135 | 49 22 | 47 | 136 | 48 35 | 47 | 137 | 47 48 | 47 | 138 | 47 01 | 48 | 139 |
| 26 | 50 46 | 46 | 136 | 50 00 | 48 | 137 | 49 12 | 48 | 138 | 48 24 | 48 | 139 | 47 36 | 49 | 140 |
| 25 | 51 24 | −48 | 138 | 50 36 | −48 | 139 | 49 48 | −49 | 140 | 48 59 | −49 | 140 | 48 10 | −49 | 141 |
| 24 | 52 00 | 48 | 139 | 51 12 | 49 | 140 | 50 23 | 49 | 141 | 49 34 | 50 | 142 | 48 44 | 50 | 143 |
| 23 | 52 35 | 49 | 140 | 51 46 | 49 | 141 | 50 57 | 50 | 142 | 50 07 | 50 | 143 | 49 17 | 51 | 144 |
| 22 | 53 09 | 49 | 142 | 52 20 | 50 | 142 | 51 30 | 51 | 143 | 50 39 | 51 | 144 | 49 48 | 51 | 145 |
| 21 | 53 43 | 51 | 143 | 52 52 | 50 | 144 | 52 02 | 51 | 145 | 51 11 | 52 | 146 | 50 19 | 52 | 146 |
| 20 | 54 15 | −51 | 144 | 53 24 | −51 | 145 | 52 33 | −52 | 146 | 51 41 | −52 | 147 | 50 49 | −53 | 148 |
| 19 | 54 46 | 52 | 146 | 53 54 | 52 | 147 | 53 02 | 52 | 148 | 52 10 | 53 | 148 | 51 17 | 53 | 149 |
| 18 | 55 16 | 52 | 147 | 54 24 | 53 | 148 | 53 31 | 53 | 149 | 52 38 | 53 | 150 | 51 45 | 54 | 151 |
| 17 | 55 45 | 53 | 149 | 54 52 | 54 | 150 | 53 58 | 53 | 150 | 53 05 | 54 | 151 | 52 11 | 54 | 152 |
| 16 | 56 12 | 53 | 150 | 55 19 | 54 | 151 | 54 25 | 55 | 152 | 53 30 | 54 | 153 | 52 36 | 55 | 153 |
| 15 | 56 39 | −55 | 152 | 55 44 | −54 | 153 | 54 50 | −55 | 154 | 53 55 | −56 | 154 | 52 59 | −55 | 155 |
| 14 | 57 03 | 55 | 154 | 56 08 | 55 | 154 | 55 13 | 55 | 155 | 54 18 | 56 | 156 | 53 22 | 56 | 156 |
| 13 | 57 27 | 56 | 155 | 56 31 | 56 | 156 | 55 35 | 56 | 157 | 54 39 | 56 | 157 | 53 43 | 56 | 158 |
| 12 | 57 49 | 57 | 157 | 56 52 | 56 | 158 | 55 56 | 57 | 158 | 54 59 | 56 | 159 | 54 03 | 57 | 160 |
| 11 | 58 09 | 57 | 159 | 57 12 | 57 | 160 | 56 15 | 57 | 160 | 55 18 | 57 | 161 | 54 21 | 57 | 161 |
| 10 | 58 28 | −57 | 161 | 57 31 | −58 | 161 | 56 33 | −57 | 162 | 55 36 | −58 | 163 | 54 38 | −58 | 163 |
| 9 | 58 45 | 58 | 163 | 57 47 | 58 | 163 | 56 49 | 58 | 164 | 55 51 | 58 | 164 | 54 53 | 58 | 164 |
| 8 | 59 01 | 59 | 164 | 58 02 | 58 | 165 | 57 04 | 58 | 165 | 56 06 | 59 | 166 | 55 07 | 58 | 166 |
| 7 | 59 14 | 58 | 166 | 58 16 | 59 | 167 | 57 17 | 59 | 167 | 56 18 | 59 | 167 | 55 19 | 58 | 168 |
| 6 | 59 26 | 59 | 168 | 58 27 | 59 | 169 | 57 28 | 59 | 169 | 56 29 | 59 | 169 | 55 30 | 59 | 170 |
| 5 | 59 37 | −60 | 170 | 58 37 | −59 | 170 | 57 38 | −59 | 171 | 56 39 | −60 | 171 | 55 39 | −59 | 171 |
| 4 | 59 45 | 60 | 172 | 58 45 | 59 | 172 | 57 46 | 60 | 173 | 56 46 | 59 | 173 | 55 47 | 60 | 173 |
| 3 | 59 52 | 60 | 174 | 58 52 | 60 | 174 | 57 52 | 60 | 174 | 56 52 | 59 | 175 | 55 53 | 60 | 175 |
| 2 | 59 56 | 60 | 176 | 58 56 | 60 | 176 | 57 56 | 59 | 176 | 56 57 | 60 | 176 | 55 57 | 60 | 177 |
| 1 | 59 59 | 60 | 178 | 58 59 | 60 | 178 | 57 59 | 60 | 178 | 56 59 | 60 | 178 | 55 59 | 60 | 178 |
| 0 | 60 00 | −60 | 180 | 59 00 | −60 | 180 | 58 00 | −60 | 180 | 57 00 | −60 | 180 | 56 00 | −60 | 180 |

| LHA | 10° | | | 11° | | | 12° | | | 13° | | | 14° | | | LHA |
|---|---|---|---|---|---|---|---|---|---|---|---|---|---|---|---|---|
| | Hc | d | Z | Hc | d | Z | Hc | d | Z | Hc | d | Z | Hc | d | Z | |
| | ° ′ | ′ | ° | ° ′ | ′ | ° | ° ′ | ′ | ° | ° ′ | ′ | ° | ° ′ | ′ | ° | |
| 69 | 14 16 | 29 | 108 | 13 47 | 30 | 109 | 13 17 | 29 | 110 | 12 48 | 30 | 111 | 12 18 | 30 | 112 | 291 |
| 68 | 15 08 | 30 | 109 | 14 38 | 30 | 110 | 14 08 | 30 | 111 | 13 38 | 30 | 112 | 13 08 | 30 | 113 | 292 |
| 67 | 15 59 | 30 | 109 | 15 29 | 30 | 110 | 14 59 | 30 | 111 | 14 29 | 31 | 112 | 13 58 | 31 | 113 | 293 |
| 66 | 16 50 | 30 | 110 | 16 20 | 30 | 111 | 15 50 | 31 | 112 | 15 19 | 31 | 113 | 14 48 | 31 | 114 | 294 |
| 65 | 17 41 | −30 | 111 | 17 11 | −31 | 111 | 16 40 | −31 | 112 | 16 09 | −31 | 113 | 15 38 | −32 | 114 | 295 |
| 64 | 18 32 | 31 | 111 | 18 01 | 31 | 112 | 17 30 | 31 | 113 | 16 59 | 32 | 114 | 16 27 | 31 | 115 | 296 |
| 63 | 19 23 | 31 | 112 | 18 52 | 32 | 112 | 18 20 | 31 | 113 | 17 49 | 32 | 114 | 17 17 | 32 | 115 | 297 |
| 62 | 20 13 | 31 | 112 | 19 42 | 32 | 113 | 19 10 | 32 | 114 | 18 38 | 32 | 115 | 18 06 | 33 | 116 | 298 |
| 61 | 21 04 | 32 | 113 | 20 32 | 32 | 114 | 20 00 | 33 | 114 | 19 27 | 32 | 115 | 18 55 | 33 | 116 | 299 |
| 60 | 21 54 | −32 | 113 | 21 22 | −33 | 114 | 20 49 | −33 | 115 | 20 16 | −33 | 116 | 19 43 | −33 | 117 | 300 |
| 59 | 22 44 | 33 | 114 | 22 11 | 33 | 115 | 21 38 | 33 | 116 | 21 05 | 33 | 117 | 20 32 | 34 | 117 | 301 |
| 58 | 23 33 | 33 | 114 | 23 00 | 33 | 115 | 22 27 | 33 | 116 | 21 54 | 34 | 117 | 21 20 | 34 | 118 | 302 |
| 57 | 24 23 | 34 | 115 | 23 49 | 33 | 116 | 23 16 | 34 | 117 | 22 42 | 34 | 118 | 22 08 | 35 | 119 | 303 |
| 56 | 25 12 | 34 | 116 | 24 38 | 34 | 117 | 24 04 | 34 | 117 | 23 30 | 35 | 118 | 22 55 | 34 | 119 | 304 |
| 55 | 26 01 | −34 | 116 | 25 27 | −35 | 117 | 24 52 | −34 | 118 | 24 18 | −35 | 119 | 23 43 | −35 | 120 | 305 |
| 54 | 26 49 | 34 | 117 | 26 15 | 35 | 118 | 25 40 | 35 | 119 | 25 05 | 35 | 120 | 24 30 | 36 | 120 | 306 |
| 53 | 27 38 | 35 | 117 | 27 03 | 35 | 118 | 26 28 | 36 | 119 | 25 52 | 35 | 120 | 25 17 | 36 | 121 | 307 |
| 52 | 28 26 | 35 | 118 | 27 51 | 36 | 119 | 27 15 | 36 | 120 | 26 39 | 36 | 121 | 26 03 | 36 | 122 | 308 |
| 51 | 29 14 | 36 | 119 | 28 38 | 36 | 120 | 28 02 | 36 | 121 | 27 26 | 37 | 121 | 26 49 | 37 | 122 | 309 |
| 50 | 30 01 | −36 | 119 | 29 25 | −36 | 120 | 28 49 | −37 | 121 | 28 12 | −37 | 122 | 27 35 | −37 | 123 | 310 |
| 49 | 30 49 | 37 | 120 | 30 12 | 37 | 121 | 29 35 | 37 | 122 | 28 58 | 38 | 123 | 28 20 | 38 | 124 | 311 |
| 48 | 31 35 | 36 | 121 | 30 59 | 38 | 122 | 30 21 | 37 | 123 | 29 44 | 39 | 124 | 29 05 | 38 | 124 | 312 |
| 47 | 32 22 | 37 | 122 | 31 45 | 38 | 122 | 31 07 | 38 | 123 | 30 29 | 39 | 124 | 29 50 | 39 | 125 | 313 |
| 46 | 33 08 | 38 | 122 | 32 30 | 38 | 123 | 31 52 | 39 | 124 | 31 13 | 39 | 125 | 30 34 | 39 | 126 | 314 |
| 45 | 33 54 | −38 | 123 | 33 16 | −39 | 124 | 32 37 | −39 | 125 | 31 58 | −40 | 126 | 31 18 | −40 | 127 | 315 |
| 44 | 34 39 | 38 | 124 | 34 01 | 40 | 125 | 33 21 | 39 | 126 | 32 42 | 40 | 127 | 32 02 | 41 | 127 | 316 |
| 43 | 35 24 | 39 | 125 | 34 45 | 40 | 125 | 34 05 | 40 | 126 | 33 25 | 40 | 127 | 32 45 | 41 | 128 | 317 |
| 42 | 36 09 | 40 | 125 | 35 29 | 40 | 126 | 34 49 | 41 | 127 | 34 08 | 41 | 128 | 33 27 | 41 | 129 | 318 |
| 41 | 36 53 | 40 | 126 | 36 13 | 41 | 127 | 35 32 | 41 | 128 | 34 51 | 42 | 129 | 34 09 | 42 | 130 | 319 |
| 40 | 37 37 | −41 | 127 | 36 56 | −41 | 128 | 36 15 | −42 | 129 | 35 33 | −42 | 130 | 34 51 | −42 | 131 | 320 |
| 39 | 38 20 | 41 | 128 | 37 39 | 42 | 129 | 36 57 | 42 | 130 | 36 15 | 43 | 131 | 35 32 | 43 | 131 | 321 |
| 38 | 39 03 | 42 | 129 | 38 21 | 43 | 130 | 37 38 | 42 | 131 | 36 56 | 43 | 131 | 36 13 | 44 | 132 | 322 |
| 37 | 39 45 | 43 | 130 | 39 02 | 42 | 131 | 38 20 | 44 | 131 | 37 36 | 43 | 132 | 36 53 | 45 | 133 | 323 |
| 36 | 40 27 | 43 | 131 | 39 44 | 44 | 131 | 39 00 | 44 | 132 | 38 16 | 44 | 133 | 37 32 | 45 | 134 | 324 |
| 35 | 41 08 | −44 | 131 | 40 24 | −44 | 132 | 39 40 | −44 | 133 | 38 56 | −45 | 134 | 38 11 | −45 | 135 | 325 |
| 34 | 41 48 | 44 | 132 | 41 04 | 45 | 133 | 40 19 | 45 | 134 | 39 34 | 45 | 135 | 38 49 | 46 | 136 | 326 |
| 33 | 42 28 | 45 | 133 | 41 43 | 45 | 134 | 40 58 | 46 | 135 | 40 12 | 46 | 136 | 39 26 | 46 | 137 | 327 |
| 32 | 43 07 | 45 | 134 | 42 22 | 46 | 135 | 41 36 | 46 | 136 | 40 50 | 47 | 137 | 40 03 | 47 | 138 | 328 |
| 31 | 43 46 | 46 | 135 | 43 00 | 47 | 136 | 42 13 | 46 | 137 | 41 27 | 47 | 138 | 40 40 | 48 | 139 | 329 |
| 30 | 44 24 | −47 | 137 | 43 37 | −47 | 137 | 42 50 | −47 | 138 | 42 03 | −48 | 139 | 41 15 | −48 | 140 | 330 |
| 29 | 45 01 | 47 | 138 | 44 14 | 48 | 138 | 43 26 | 48 | 139 | 42 38 | 48 | 140 | 41 50 | 49 | 141 | 331 |
| 28 | 45 37 | 48 | 139 | 44 49 | 48 | 140 | 44 01 | 48 | 140 | 43 13 | 49 | 141 | 42 24 | 50 | 142 | 332 |
| 27 | 46 13 | 49 | 140 | 45 24 | 49 | 141 | 44 35 | 49 | 141 | 43 46 | 49 | 142 | 42 57 | 50 | 143 | 333 |
| 26 | 46 47 | 49 | 141 | 45 58 | 49 | 142 | 45 09 | 50 | 143 | 44 19 | 50 | 143 | 43 29 | 50 | 144 | 334 |
| 25 | 47 21 | −50 | 142 | 46 31 | −50 | 143 | 45 41 | −50 | 144 | 44 51 | −50 | 145 | 44 01 | −51 | 145 | 335 |
| 24 | 47 54 | 50 | 143 | 47 04 | 51 | 144 | 46 13 | 51 | 145 | 45 22 | 51 | 146 | 44 31 | 51 | 146 | 336 |
| 23 | 48 26 | 51 | 145 | 47 35 | 51 | 145 | 46 44 | 52 | 146 | 45 52 | 51 | 147 | 45 01 | 52 | 148 | 337 |
| 22 | 48 57 | 51 | 146 | 48 06 | 52 | 147 | 47 14 | 52 | 147 | 46 22 | 53 | 148 | 45 29 | 52 | 149 | 338 |
| 21 | 49 27 | 52 | 147 | 48 35 | 52 | 148 | 47 43 | 53 | 149 | 46 50 | 53 | 149 | 45 57 | 53 | 150 | 339 |
| 20 | 49 56 | −53 | 148 | 49 03 | −53 | 149 | 48 10 | −53 | 150 | 47 17 | −53 | 151 | 46 24 | −54 | 151 | 340 |
| 19 | 50 24 | 53 | 150 | 49 31 | 54 | 151 | 48 37 | 54 | 151 | 47 43 | 54 | 152 | 46 49 | 54 | 153 | 341 |
| 18 | 50 51 | 54 | 151 | 49 57 | 54 | 152 | 49 03 | 54 | 153 | 48 09 | 55 | 153 | 47 14 | 54 | 154 | 342 |
| 17 | 51 17 | 55 | 153 | 50 22 | 55 | 153 | 49 27 | 55 | 154 | 48 32 | 55 | 155 | 47 37 | 55 | 155 | 343 |
| 16 | 51 41 | 55 | 154 | 50 46 | 55 | 155 | 49 51 | 56 | 155 | 48 55 | 55 | 156 | 48 00 | 56 | 156 | 344 |
| 15 | 52 04 | −55 | 156 | 51 09 | −56 | 156 | 50 13 | −56 | 157 | 49 17 | −56 | 157 | 48 21 | −56 | 158 | 345 |
| 14 | 52 26 | 56 | 157 | 51 30 | 56 | 158 | 50 34 | 57 | 158 | 49 37 | 56 | 159 | 48 41 | 57 | 159 | 346 |
| 13 | 52 47 | 57 | 159 | 51 50 | 57 | 159 | 50 53 | 57 | 160 | 49 56 | 56 | 160 | 49 00 | 58 | 161 | 347 |
| 12 | 53 06 | 57 | 160 | 52 09 | 57 | 161 | 51 12 | 58 | 161 | 50 14 | 57 | 162 | 49 17 | 58 | 162 | 348 |
| 11 | 53 24 | 58 | 162 | 52 26 | 57 | 162 | 51 29 | 58 | 163 | 50 31 | 58 | 163 | 49 33 | 58 | 163 | 349 |
| 10 | 53 40 | −58 | 164 | 52 42 | −58 | 164 | 51 44 | −58 | 164 | 50 46 | −58 | 165 | 49 48 | −58 | 165 | 350 |
| 9 | 53 55 | 58 | 165 | 52 57 | 59 | 165 | 51 58 | 58 | 166 | 51 00 | 58 | 166 | 50 02 | 59 | 166 | 351 |
| 8 | 54 09 | 59 | 167 | 53 10 | 59 | 167 | 52 11 | 59 | 167 | 51 12 | 58 | 168 | 50 14 | 59 | 168 | 352 |
| 7 | 54 21 | 59 | 168 | 53 22 | 59 | 168 | 52 23 | 59 | 169 | 51 24 | 60 | 169 | 50 24 | 59 | 169 | 353 |
| 6 | 54 31 | 59 | 170 | 53 32 | 60 | 170 | 52 32 | 59 | 170 | 51 33 | 59 | 171 | 50 34 | 59 | 171 | 354 |
| 5 | 54 40 | −60 | 172 | 53 40 | −59 | 172 | 52 41 | −60 | 172 | 51 41 | −59 | 172 | 50 42 | −60 | 172 | 355 |
| 4 | 54 47 | 60 | 173 | 53 47 | 59 | 173 | 52 48 | 60 | 174 | 51 48 | 60 | 174 | 50 48 | 59 | 174 | 356 |
| 3 | 54 53 | 60 | 175 | 53 53 | 60 | 175 | 52 53 | 60 | 175 | 51 53 | 60 | 175 | 50 53 | 59 | 175 | 357 |
| 2 | 54 57 | 60 | 177 | 53 57 | 60 | 177 | 52 57 | 60 | 177 | 51 57 | 60 | 177 | 50 57 | 60 | 177 | 358 |
| 1 | 54 59 | 60 | 178 | 53 59 | 60 | 178 | 52 59 | 60 | 178 | 51 59 | 60 | 178 | 50 59 | 60 | 179 | 359 |
| 0 | 55 00 | −60 | 180 | 54 00 | −60 | 180 | 53 00 | −60 | 180 | 52 00 | −60 | 180 | 51 00 | −60 | 180 | 360 |

S. Lat. {LHA greater than 180°........ Zn=180−Z; LHA less than 180°........... Zn=180+Z}

DECLINATION (0°-14°) CONTRARY NAME TO LATITUDE

N. Lat. {LHA greater than 180°........ Zn=Z; LHA less than 180°........... Zn=360−Z}

# DECLINATION (15°-29°) SAME NAME AS LATITUDE

| LHA | 15° Hc | 15° d | 15° Z | 16° Hc | 16° d | 16° Z | 17° Hc | 17° d | 17° Z | 18° Hc | 18° d | 18° Z | 19° Hc | 19° d | 19° Z |
|---|---|---|---|---|---|---|---|---|---|---|---|---|---|---|---|
| | ° ′ | ′ | ° | ° ′ | ′ | ° | ° ′ | ′ | ° | ° ′ | ′ | ° | ° ′ | ′ | ° |
| 0 | 80 00 | +60 | 180 | 81 00 | +60 | 180 | 82 00 | +60 | 180 | 83 00 | +60 | 180 | 84 00 | +60 | 180 |
| 1 | 79 57 | 60 | 175 | 80 57 | 60 | 174 | 81 57 | 59 | 173 | 82 56 | 60 | 172 | 83 56 | 59 | 171 |
| 2 | 79 50 | 59 | 169 | 80 49 | 58 | 168 | 81 47 | 58 | 167 | 82 45 | 58 | 165 | 83 43 | 57 | 162 |
| 3 | 79 37 | 57 | 164 | 80 34 | 58 | 162 | 81 32 | 56 | 160 | 82 28 | 55 | 158 | 83 23 | 54 | 155 |
| 4 | 79 19 | 56 | 159 | 80 15 | 55 | 157 | 81 10 | 54 | 154 | 82 04 | 53 | 151 | 82 57 | 50 | 148 |
| 5 | 78 57 | +55 | 154 | 79 52 | +52 | 152 | 80 44 | +52 | 149 | 81 36 | +49 | 145 | 82 25 | +47 | 141 |
| 6 | 78 32 | 52 | 150 | 79 24 | 50 | 147 | 80 14 | 49 | 144 | 81 03 | 46 | 140 | 81 49 | 43 | 136 |
| 7 | 78 02 | 50 | 145 | 78 52 | 49 | 143 | 79 41 | 46 | 139 | 80 27 | 43 | 136 | 81 10 | 40 | 131 |
| 8 | 77 30 | 48 | 142 | 78 18 | 46 | 139 | 79 04 | 43 | 136 | 79 47 | 41 | 132 | 80 28 | 37 | 127 |
| 9 | 76 55 | 46 | 138 | 77 41 | 43 | 135 | 78 24 | 42 | 132 | 79 06 | 38 | 128 | 79 44 | 34 | 124 |
| 10 | 76 17 | +44 | 135 | 77 01 | +42 | 132 | 77 43 | +39 | 129 | 78 22 | +36 | 125 | 78 58 | +32 | 121 |
| 11 | 75 38 | 42 | 132 | 76 20 | 40 | 129 | 77 00 | 37 | 126 | 77 37 | 34 | 122 | 78 11 | 30 | 118 |
| 12 | 74 57 | 40 | 129 | 75 37 | 38 | 126 | 76 15 | 35 | 123 | 76 50 | 32 | 120 | 77 22 | 29 | 116 |
| 13 | 74 14 | 39 | 127 | 74 53 | 36 | 124 | 75 29 | 33 | 121 | 76 02 | 31 | 118 | 76 33 | 27 | 114 |
| 14 | 73 30 | 37 | 125 | 74 07 | 35 | 122 | 74 42 | 32 | 119 | 75 14 | 29 | 116 | 75 43 | 26 | 112 |
| 15 | 72 45 | +35 | 123 | 73 20 | +34 | 120 | 73 54 | +30 | 117 | 74 24 | +28 | 114 | 74 52 | +25 | 110 |
| 16 | 71 58 | 35 | 121 | 72 33 | 32 | 118 | 73 05 | 29 | 115 | 73 34 | 27 | 112 | 74 01 | 24 | 109 |
| 17 | 71 11 | 33 | 119 | 71 44 | 31 | 116 | 72 15 | 28 | 114 | 72 43 | 26 | 111 | 73 09 | 23 | 108 |
| 18 | 70 23 | 32 | 117 | 70 55 | 30 | 115 | 71 25 | 27 | 112 | 71 52 | 25 | 109 | 72 17 | 22 | 106 |
| 19 | 69 34 | 31 | 116 | 70 05 | 29 | 113 | 70 34 | 27 | 111 | 71 01 | 24 | 108 | 71 25 | 21 | 105 |
| 20 | 68 45 | +30 | 114 | 69 15 | +28 | 112 | 69 43 | +26 | 109 | 70 09 | +23 | 107 | 70 32 | +21 | 104 |
| 21 | 67 55 | 29 | 113 | 68 24 | 28 | 111 | 68 52 | 25 | 108 | 69 17 | 22 | 106 | 69 39 | 20 | 103 |
| 22 | 67 05 | 28 | 112 | 67 33 | 27 | 109 | 68 00 | 24 | 107 | 68 24 | 22 | 105 | 68 46 | 20 | 102 |
| 23 | 66 14 | 28 | 111 | 66 42 | 26 | 108 | 67 08 | 23 | 106 | 67 31 | 22 | 104 | 67 53 | 19 | 101 |
| 24 | 65 23 | 27 | 109 | 65 50 | 25 | 107 | 66 15 | 23 | 105 | 66 38 | 21 | 103 | 66 59 | 19 | 100 |
| 25 | 64 32 | +26 | 108 | 64 58 | +25 | 106 | 65 23 | +22 | 104 | 65 45 | +21 | 102 | 66 06 | +18 | 100 |
| 26 | 63 40 | 26 | 107 | 64 06 | 24 | 105 | 64 30 | 22 | 103 | 64 52 | 20 | 101 | 65 12 | 18 | 99 |
| 27 | 62 48 | 25 | 106 | 63 13 | 24 | 104 | 63 37 | 22 | 102 | 63 59 | 19 | 100 | 64 18 | 18 | 98 |
| 28 | 61 56 | 24 | 106 | 62 20 | 24 | 104 | 62 44 | 21 | 102 | 63 05 | 20 | 100 | 63 25 | 17 | 97 |
| 29 | 61 03 | 24 | 105 | 61 27 | 23 | 103 | 61 50 | 21 | 101 | 62 11 | 20 | 99 | 62 31 | 17 | 97 |
| 30 | 60 10 | +24 | 104 | 60 34 | +23 | 102 | 60 57 | +20 | 100 | 61 17 | +20 | 98 | 61 37 | +17 | 96 |
| 31 | 59 17 | 24 | 103 | 59 41 | 22 | 101 | 60 03 | 21 | 99 | 60 24 | 18 | 98 | 60 42 | 18 | 96 |
| 32 | 58 24 | 24 | 102 | 58 48 | 21 | 101 | 59 09 | 21 | 99 | 59 30 | 18 | 97 | 59 48 | 17 | 95 |
| 33 | 57 31 | 23 | 102 | 57 54 | 22 | 100 | 58 16 | 20 | 98 | 58 36 | 18 | 96 | 58 54 | 17 | 94 |
| 34 | 56 38 | 23 | 101 | 57 01 | 21 | 99 | 57 22 | 20 | 98 | 57 42 | 18 | 96 | 58 00 | 17 | 94 |
| 35 | 55 44 | +23 | 100 | 56 07 | +21 | 99 | 56 28 | +19 | 97 | 56 47 | +19 | 95 | 57 06 | +16 | 93 |
| 36 | 54 51 | 22 | 100 | 55 13 | 21 | 98 | 55 34 | 19 | 96 | 55 53 | 18 | 95 | 56 11 | 17 | 93 |
| 37 | 53 57 | 22 | 99 | 54 19 | 21 | 97 | 54 40 | 19 | 96 | 54 59 | 18 | 94 | 55 17 | 17 | 92 |
| 38 | 53 03 | 22 | 98 | 53 25 | 21 | 97 | 53 46 | 19 | 95 | 54 05 | 18 | 94 | 54 23 | 16 | 92 |
| 39 | 52 10 | 21 | 98 | 52 31 | 20 | 96 | 52 51 | 19 | 95 | 53 10 | 18 | 93 | 53 28 | 17 | 92 |
| 40 | 51 16 | +21 | 97 | 51 37 | +20 | 96 | 51 57 | +19 | 94 | 52 16 | +18 | 93 | 52 34 | +16 | 91 |
| 41 | 50 22 | 21 | 97 | 50 43 | 20 | 95 | 51 03 | 19 | 94 | 51 22 | 18 | 92 | 51 40 | 16 | 91 |
| 42 | 49 28 | 21 | 96 | 49 49 | 20 | 95 | 50 09 | 18 | 93 | 50 27 | 18 | 92 | 50 45 | 17 | 90 |
| 43 | 48 34 | 20 | 96 | 48 54 | 20 | 94 | 49 14 | 19 | 93 | 49 33 | 18 | 91 | 49 51 | 16 | 90 |
| 44 | 47 39 | 21 | 95 | 48 00 | 20 | 94 | 48 20 | 19 | 92 | 48 39 | 17 | 91 | 48 56 | 17 | 89 |
| 45 | 46 45 | +21 | 95 | 47 06 | +20 | 93 | 47 26 | +18 | 92 | 47 44 | +18 | 90 | 48 02 | +17 | 89 |
| 46 | 45 51 | 21 | 94 | 46 12 | 19 | 93 | 46 31 | 19 | 91 | 46 50 | 18 | 90 | 47 08 | 16 | 89 |
| 47 | 44 57 | 20 | 94 | 45 17 | 20 | 92 | 45 37 | 19 | 91 | 45 56 | 17 | 90 | 46 13 | 17 | 88 |
| 48 | 44 02 | 21 | 93 | 44 23 | 20 | 92 | 44 43 | 18 | 90 | 45 01 | 18 | 89 | 45 19 | 17 | 88 |
| 49 | 43 08 | 21 | 93 | 43 29 | 19 | 91 | 43 48 | 19 | 90 | 44 07 | 18 | 89 | 44 25 | 16 | 87 |
| 50 | 42 14 | +20 | 92 | 42 34 | +20 | 91 | 42 54 | +19 | 90 | 43 13 | +17 | 88 | 43 30 | +17 | 87 |
| 51 | 41 19 | 21 | 92 | 41 40 | 19 | 90 | 41 59 | 19 | 89 | 42 18 | 18 | 88 | 42 36 | 17 | 87 |
| 52 | 40 25 | 20 | 91 | 40 45 | 20 | 90 | 41 05 | 19 | 89 | 41 24 | 18 | 88 | 41 42 | 17 | 86 |
| 53 | 39 31 | 20 | 91 | 39 51 | 20 | 90 | 40 11 | 19 | 88 | 40 30 | 18 | 87 | 40 48 | 17 | 86 |
| 54 | 38 36 | 21 | 90 | 38 57 | 19 | 89 | 39 16 | 19 | 88 | 39 35 | 18 | 87 | 39 53 | 18 | 86 |
| 55 | 37 42 | +20 | 90 | 38 02 | +20 | 89 | 38 22 | +19 | 88 | 38 41 | +18 | 86 | 38 59 | +17 | 85 |
| 56 | 36 48 | 20 | 90 | 37 08 | 20 | 88 | 37 28 | 19 | 87 | 37 47 | 18 | 86 | 38 05 | 17 | 85 |
| 57 | 35 53 | 21 | 89 | 36 14 | 19 | 88 | 36 33 | 19 | 87 | 36 52 | 19 | 86 | 37 11 | 17 | 84 |
| 58 | 34 59 | 20 | 89 | 35 19 | 20 | 88 | 35 39 | 19 | 86 | 35 58 | 19 | 85 | 36 17 | 17 | 84 |
| 59 | 34 04 | 21 | 88 | 34 25 | 20 | 87 | 34 45 | 19 | 86 | 35 04 | 19 | 85 | 35 23 | 18 | 84 |
| 60 | 33 10 | +21 | 88 | 33 31 | +20 | 87 | 33 51 | +19 | 86 | 34 10 | +19 | 85 | 34 29 | +18 | 83 |
| 61 | 32 16 | 20 | 88 | 32 36 | 20 | 86 | 32 56 | 20 | 85 | 33 16 | 19 | 84 | 33 35 | 18 | 83 |
| 62 | 31 21 | 21 | 87 | 31 42 | 20 | 86 | 32 02 | 20 | 85 | 32 22 | 19 | 84 | 32 41 | 18 | 83 |
| 63 | 30 27 | 21 | 87 | 30 48 | 20 | 86 | 31 08 | 20 | 85 | 31 28 | 19 | 83 | 31 47 | 18 | 82 |
| 64 | 29 33 | 21 | 86 | 29 54 | 20 | 85 | 30 14 | 20 | 84 | 30 34 | 19 | 83 | 30 53 | 18 | 82 |
| 65 | 28 39 | +21 | 86 | 29 00 | +20 | 85 | 29 20 | +20 | 84 | 29 40 | +19 | 83 | 29 59 | +19 | 82 |
| 66 | 27 44 | 21 | 86 | 28 05 | 21 | 85 | 28 26 | 20 | 83 | 28 46 | 19 | 82 | 29 05 | 19 | 81 |
| 67 | 26 50 | 21 | 85 | 27 11 | 21 | 84 | 27 32 | 20 | 83 | 27 52 | 20 | 82 | 28 12 | 19 | 81 |
| 68 | 25 56 | 21 | 85 | 26 17 | 21 | 84 | 26 38 | 20 | 83 | 26 58 | 20 | 82 | 27 18 | 19 | 81 |
| 69 | 25 02 | 21 | 84 | 25 23 | 21 | 83 | 25 44 | 20 | 82 | 26 04 | 20 | 81 | 26 24 | 20 | 80 |

| LHA | 20° Hc | 20° d | 20° Z | 21° Hc | 21° d | 21° Z | 22° Hc | 22° d | 22° Z | 23° Hc | 23° d | 23° Z | 24° Hc | 24° d | 24° Z |
|---|---|---|---|---|---|---|---|---|---|---|---|---|---|---|---|
| | ° ′ | ′ | ° | ° ′ | ′ | ° | ° ′ | ′ | ° | ° ′ | ′ | ° | ° ′ | ′ | ° |
| 0 | 85 00 | +60 | 180 | 86 00 | +60 | 180 | 87 00 | +60 | 180 | 88 00 | +60 | 180 | 89 00 | +60 | 180 |
| 1 | 84 55 | 59 | 169 | 85 54 | 58 | 167 | 86 52 | 56 | 163 | 87 48 | 51 | 155 | 88 39 | 27 | 138 |
| 2 | 84 40 | 56 | 159 | 85 36 | 53 | 155 | 86 29 | 49 | 148 | 87 18 | 37 | 137 | 87 55 | 16 | 118 |
| 3 | 84 17 | 51 | 150 | 85 08 | 48 | 145 | 85 56 | 40 | 137 | 86 36 | 30 | 126 | 87 06 | 11 | 110 |
| 4 | 83 47 | 47 | 143 | 84 34 | 42 | 137 | 85 16 | 34 | 129 | 85 50 | 24 | 118 | 86 14 | 9 | 105 |
| 5 | 83 12 | +42 | 136 | 83 54 | +37 | 130 | 84 31 | +30 | 122 | 85 01 | +20 | 113 | 85 21 | +7 | 101 |
| 6 | 82 32 | 39 | 131 | 83 11 | 33 | 125 | 83 44 | 26 | 117 | 84 10 | 17 | 109 | 84 27 | 7 | 99 |
| 7 | 81 50 | 35 | 126 | 82 25 | 30 | 120 | 82 55 | 23 | 114 | 83 18 | 15 | 106 | 83 33 | 6 | 98 |
| 8 | 81 05 | 32 | 123 | 81 37 | 28 | 117 | 82 05 | 21 | 111 | 82 26 | 13 | 104 | 82 39 | 6 | 96 |
| 9 | 80 18 | 30 | 119 | 80 48 | 25 | 114 | 81 13 | 19 | 108 | 81 32 | 13 | 102 | 81 45 | 6 | 95 |
| 10 | 79 30 | +28 | 116 | 79 58 | +23 | 112 | 80 21 | +18 | 106 | 80 39 | +12 | 100 | 80 51 | +5 | 94 |
| 11 | 78 41 | 26 | 114 | 79 07 | 22 | 109 | 79 29 | 16 | 104 | 79 45 | 12 | 99 | 79 57 | 5 | 93 |
| 12 | 77 51 | 25 | 112 | 78 16 | 20 | 108 | 78 36 | 16 | 103 | 78 52 | 10 | 98 | 79 02 | 6 | 93 |
| 13 | 77 00 | 23 | 110 | 77 23 | 20 | 106 | 77 43 | 15 | 102 | 77 58 | 10 | 97 | 78 08 | 5 | 92 |
| 14 | 76 09 | 22 | 108 | 76 31 | 18 | 104 | 76 49 | 15 | 100 | 77 04 | 10 | 96 | 77 14 | 5 | 92 |
| 15 | 75 17 | +21 | 107 | 75 38 | +18 | 103 | 75 56 | +14 | 99 | 76 10 | +9 | 95 | 76 19 | +6 | 91 |
| 16 | 74 25 | 20 | 106 | 74 45 | 17 | 102 | 75 02 | 13 | 98 | 75 15 | 10 | 95 | 75 25 | 6 | 91 |
| 17 | 73 32 | 20 | 104 | 73 52 | 16 | 101 | 74 08 | 13 | 97 | 74 21 | 10 | 94 | 74 31 | 5 | 90 |
| 18 | 72 39 | 19 | 103 | 72 58 | 16 | 100 | 73 14 | 13 | 97 | 73 27 | 9 | 93 | 73 36 | 6 | 90 |
| 19 | 71 46 | 19 | 102 | 72 05 | 15 | 99 | 72 20 | 13 | 96 | 72 33 | 9 | 93 | 72 42 | 6 | 89 |
| 20 | 70 53 | +18 | 101 | 71 11 | +15 | 98 | 71 26 | +12 | 95 | 71 38 | +9 | 92 | 71 47 | +6 | 89 |
| 21 | 69 59 | 18 | 100 | 70 17 | 15 | 97 | 70 32 | 12 | 95 | 70 44 | 9 | 92 | 70 53 | 6 | 89 |
| 22 | 69 06 | 17 | 99 | 69 23 | 15 | 97 | 69 38 | 12 | 94 | 69 50 | 9 | 91 | 69 59 | 6 | 88 |
| 23 | 68 12 | 17 | 99 | 68 29 | 14 | 96 | 68 43 | 12 | 93 | 68 55 | 9 | 91 | 69 04 | 7 | 88 |
| 24 | 67 18 | 17 | 98 | 67 35 | 14 | 95 | 67 49 | 12 | 93 | 68 01 | 9 | 90 | 68 10 | 7 | 88 |
| 25 | 66 24 | +17 | 97 | 66 41 | +14 | 95 | 66 55 | +11 | 92 | 67 06 | +10 | 90 | 67 16 | +7 | 87 |
| 26 | 65 30 | 16 | 97 | 65 46 | 14 | 94 | 66 00 | 12 | 92 | 66 12 | 9 | 89 | 66 21 | 7 | 87 |
| 27 | 64 36 | 16 | 96 | 64 52 | 14 | 94 | 65 06 | 12 | 91 | 65 18 | 9 | 89 | 65 27 | 7 | 87 |
| 28 | 63 42 | 16 | 95 | 63 58 | 14 | 93 | 64 12 | 11 | 91 | 64 23 | 10 | 89 | 64 33 | 7 | 86 |
| 29 | 62 48 | 16 | 95 | 63 04 | 13 | 93 | 63 17 | 12 | 90 | 63 29 | 10 | 88 | 63 39 | 7 | 86 |
| 30 | 61 54 | +15 | 94 | 62 09 | +14 | 92 | 62 23 | +12 | 90 | 62 35 | +9 | 88 | 62 44 | +8 | 86 |
| 31 | 61 00 | 15 | 94 | 61 15 | 14 | 92 | 61 29 | 11 | 90 | 61 40 | 10 | 88 | 61 50 | 8 | 85 |
| 32 | 60 05 | 16 | 93 | 60 21 | 13 | 91 | 60 34 | 12 | 89 | 60 46 | 10 | 87 | 60 56 | 8 | 85 |
| 33 | 59 11 | 15 | 93 | 59 26 | 14 | 91 | 59 40 | 12 | 89 | 59 52 | 10 | 87 | 60 02 | 8 | 85 |
| 34 | 58 17 | 15 | 92 | 58 32 | 13 | 90 | 58 45 | 12 | 88 | 58 57 | 11 | 87 | 59 08 | 8 | 85 |
| 35 | 57 22 | +15 | 92 | 57 37 | +14 | 90 | 57 51 | +12 | 88 | 58 03 | +10 | 86 | 58 13 | +9 | 84 |
| 36 | 56 28 | 15 | 91 | 56 43 | 14 | 89 | 56 57 | 12 | 88 | 57 09 | 10 | 86 | 57 19 | 9 | 84 |
| 37 | 55 34 | 15 | 91 | 55 49 | 13 | 89 | 56 02 | 13 | 87 | 56 15 | 10 | 86 | 56 25 | 10 | 84 |
| 38 | 54 39 | 15 | 90 | 54 54 | 14 | 89 | 55 08 | 12 | 87 | 55 20 | 11 | 85 | 55 31 | 10 | 84 |
| 39 | 53 45 | 15 | 90 | 54 00 | 14 | 88 | 54 14 | 12 | 87 | 54 26 | 11 | 85 | 54 37 | 10 | 83 |
| 40 | 52 50 | +16 | 90 | 53 06 | +14 | 88 | 53 20 | +12 | 86 | 53 32 | +11 | 85 | 53 43 | +10 | 83 |
| 41 | 51 56 | 15 | 89 | 52 11 | 14 | 88 | 52 25 | 13 | 86 | 52 38 | 11 | 84 | 52 49 | 10 | 83 |
| 42 | 51 02 | 15 | 89 | 51 17 | 14 | 87 | 51 31 | 13 | 86 | 51 44 | 11 | 84 | 51 55 | 11 | 82 |
| 43 | 50 07 | 16 | 88 | 50 23 | 14 | 87 | 50 37 | 13 | 85 | 50 50 | 12 | 84 | 51 02 | 10 | 82 |
| 44 | 49 13 | 15 | 88 | 49 28 | 15 | 86 | 49 43 | 13 | 85 | 49 56 | 12 | 83 | 50 08 | 10 | 82 |
| 45 | 48 19 | +15 | 88 | 48 34 | +15 | 86 | 48 49 | +13 | 85 | 49 02 | +12 | 83 | 49 14 | +11 | 82 |
| 46 | 47 24 | 16 | 87 | 47 40 | 14 | 86 | 47 54 | 14 | 84 | 48 08 | 12 | 83 | 48 20 | 11 | 81 |
| 47 | 46 30 | 16 | 87 | 46 46 | 14 | 85 | 47 00 | 14 | 84 | 47 14 | 12 | 83 | 47 26 | 12 | 81 |
| 48 | 45 36 | 16 | 86 | 45 52 | 14 | 85 | 46 06 | 14 | 84 | 46 20 | 13 | 82 | 46 33 | 11 | 81 |
| 49 | 44 41 | 16 | 86 | 44 57 | 15 | 85 | 45 12 | 14 | 83 | 45 26 | 13 | 82 | 45 39 | 12 | 81 |
| 50 | 43 47 | +16 | 86 | 44 03 | +15 | 84 | 44 18 | +14 | 83 | 44 32 | +13 | 82 | 44 45 | +13 | 80 |
| 51 | 42 53 | 16 | 85 | 43 09 | 15 | 84 | 43 24 | 15 | 83 | 43 39 | 13 | 81 | 43 52 | 12 | 80 |
| 52 | 41 59 | 16 | 85 | 42 15 | 15 | 84 | 42 30 | 15 | 82 | 42 45 | 13 | 81 | 42 58 | 13 | 80 |
| 53 | 41 05 | 16 | 85 | 41 21 | 16 | 83 | 41 37 | 14 | 82 | 41 51 | 14 | 81 | 42 05 | 13 | 79 |
| 54 | 40 11 | 16 | 84 | 40 27 | 16 | 83 | 40 43 | 14 | 82 | 40 57 | 14 | 80 | 41 11 | 13 | 79 |
| 55 | 39 16 | +17 | 84 | 39 33 | +16 | 83 | 39 49 | +15 | 81 | 40 04 | +14 | 80 | 40 18 | +13 | 79 |
| 56 | 38 22 | 17 | 84 | 38 39 | 16 | 82 | 38 55 | 15 | 81 | 39 10 | 15 | 80 | 39 25 | 13 | 79 |
| 57 | 37 28 | 17 | 83 | 37 45 | 16 | 82 | 38 01 | 16 | 81 | 38 17 | 14 | 80 | 38 31 | 14 | 78 |
| 58 | 36 34 | 17 | 83 | 36 51 | 17 | 82 | 37 08 | 15 | 81 | 37 23 | 15 | 79 | 37 38 | 14 | 78 |
| 59 | 35 41 | 17 | 83 | 35 58 | 16 | 81 | 36 14 | 16 | 80 | 36 30 | 15 | 79 | 36 45 | 14 | 78 |
| 60 | 34 47 | +17 | 82 | 35 04 | +17 | 81 | 35 21 | +16 | 80 | 35 37 | +15 | 79 | 35 52 | +15 | 78 |
| 61 | 33 53 | 17 | 82 | 34 10 | 17 | 81 | 34 27 | 16 | 80 | 34 43 | 16 | 78 | 34 59 | 15 | 77 |
| 62 | 32 59 | 18 | 82 | 33 17 | 17 | 80 | 33 34 | 16 | 79 | 33 50 | 16 | 78 | 34 06 | 15 | 77 |
| 63 | 32 05 | 18 | 81 | 32 23 | 17 | 80 | 32 40 | 17 | 79 | 32 57 | 16 | 78 | 33 13 | 15 | 77 |
| 64 | 31 11 | 19 | 81 | 31 30 | 17 | 80 | 31 47 | 17 | 79 | 32 04 | 16 | 78 | 32 20 | 16 | 76 |
| 65 | 30 18 | +18 | 81 | 30 36 | +18 | 79 | 30 54 | +17 | 78 | 31 11 | +16 | 77 | 31 27 | +16 | 76 |
| 66 | 29 24 | 19 | 80 | 29 43 | 17 | 79 | 30 00 | 18 | 78 | 30 18 | 16 | 77 | 30 34 | 17 | 76 |
| 67 | 28 31 | 18 | 80 | 28 49 | 18 | 79 | 29 07 | 18 | 78 | 29 25 | 17 | 77 | 29 42 | 16 | 76 |
| 68 | 27 37 | 19 | 80 | 27 56 | 18 | 78 | 28 14 | 18 | 77 | 28 32 | 17 | 76 | 28 49 | 17 | 75 |
| 69 | 26 44 | 19 | 79 | 27 03 | 18 | 78 | 27 21 | 18 | 77 | 27 39 | 18 | 76 | 27 57 | 17 | 75 |

| LHA | 25° Hc | 25° d | 25° Z | 26° Hc | 26° d | 26° Z | 27° Hc | 27° d | 27° Z | 28° Hc | 28° d | 28° Z | 29° Hc | 29° d | 29° Z | LHA |
|---|---|---|---|---|---|---|---|---|---|---|---|---|---|---|---|---|
| | ° ′ | ′ | ° | ° ′ | ′ | ° | ° ′ | ′ | ° | ° ′ | ′ | ° | ° ′ | ′ | ° | |
| 0 | 90 00 | −60 | .. | 89 00 | −60 | 0 | 88 00 | −60 | 0 | 87 00 | −60 | 0 | 86 00 | −60 | 0 | 360 |
| 1 | 89 06 | 27 | 90 | 88 39 | 51 | 42 | 87 48 | 56 | 24 | 86 52 | 58 | 16 | 85 54 | 59 | 12 | 359 |
| 2 | 88 11 | 15 | 90 | 87 56 | 37 | 61 | 87 19 | 49 | 42 | 86 30 | 53 | 30 | 85 37 | 55 | 24 | 358 |
| 3 | 87 17 | 10 | 89 | 87 07 | 28 | 69 | 86 39 | 40 | 53 | 85 59 | 48 | 41 | 85 11 | 51 | 33 | 357 |
| 4 | 86 23 | 8 | 89 | 86 15 | 22 | 74 | 85 53 | 33 | 60 | 85 20 | 41 | 49 | 84 39 | 47 | 41 | 356 |
| 5 | 85 28 | −5 | 89 | 85 23 | −18 | 76 | 85 05 | −28 | 65 | 84 37 | −36 | 55 | 84 01 | −42 | 47 | 355 |
| 6 | 84 34 | 4 | 89 | 84 30 | 15 | 78 | 84 15 | 24 | 68 | 83 51 | 31 | 60 | 83 20 | 38 | 52 | 354 |
| 7 | 83 39 | 3 | 89 | 83 36 | 12 | 80 | 83 24 | 21 | 71 | 83 03 | 27 | 63 | 82 36 | 34 | 56 | 353 |
| 8 | 82 45 | 2 | 88 | 82 43 | 11 | 80 | 82 32 | 18 | 73 | 82 14 | 24 | 66 | 81 50 | 31 | 59 | 352 |
| 9 | 81 51 | 2 | 88 | 81 49 | 9 | 81 | 81 40 | 15 | 74 | 81 25 | 22 | 68 | 81 03 | 28 | 62 | 351 |
| 10 | 80 56 | −1 | 88 | 80 55 | −7 | 82 | 80 48 | −14 | 75 | 80 34 | −20 | 69 | 80 14 | −24 | 64 | 350 |
| 11 | 80 02 | 0 | 88 | 80 02 | 7 | 82 | 79 55 | 12 | 76 | 79 43 | 18 | 71 | 79 25 | 22 | 65 | 349 |
| 12 | 79 08 | 0 | 88 | 79 08 | 6 | 82 | 79 02 | 11 | 77 | 78 51 | 15 | 72 | 78 36 | 21 | 67 | 348 |
| 13 | 78 13 | +1 | 87 | 78 14 | 5 | 82 | 78 09 | 9 | 78 | 78 00 | 15 | 73 | 77 45 | 18 | 68 | 347 |
| 14 | 77 19 | 1 | 87 | 77 20 | 4 | 83 | 77 16 | 8 | 78 | 77 08 | 13 | 74 | 76 55 | 17 | 69 | 346 |
| 15 | 76 25 | +1 | 87 | 76 26 | −3 | 83 | 76 23 | −8 | 78 | 76 15 | −11 | 74 | 76 04 | −16 | 70 | 345 |
| 16 | 75 31 | 1 | 87 | 75 32 | 2 | 83 | 75 30 | 7 | 79 | 75 23 | 10 | 75 | 75 13 | 15 | 71 | 344 |
| 17 | 74 36 | 2 | 86 | 74 38 | 2 | 83 | 74 36 | 6 | 79 | 74 30 | 9 | 75 | 74 21 | 13 | 71 | 343 |
| 18 | 73 42 | 2 | 86 | 73 44 | 1 | 83 | 73 43 | 5 | 79 | 73 38 | 9 | 76 | 73 29 | 11 | 72 | 342 |
| 19 | 72 48 | 2 | 86 | 72 50 | −1 | 83 | 72 49 | 4 | 79 | 72 45 | 7 | 76 | 72 38 | 11 | 73 | 341 |
| 20 | 71 53 | +3 | 86 | 71 56 | 0 | 83 | 71 56 | −4 | 79 | 71 52 | −6 | 76 | 71 46 | −10 | 73 | 340 |
| 21 | 70 59 | 3 | 86 | 71 02 | +1 | 83 | 71 03 | 3 | 79 | 71 00 | 6 | 76 | 70 54 | 9 | 73 | 339 |
| 22 | 70 05 | 4 | 85 | 70 09 | 0 | 82 | 70 09 | 2 | 79 | 70 07 | 5 | 77 | 70 02 | 8 | 74 | 338 |
| 23 | 69 11 | 4 | 85 | 69 15 | 1 | 82 | 69 16 | 2 | 80 | 69 14 | 5 | 77 | 69 09 | 7 | 74 | 337 |
| 24 | 68 17 | 4 | 85 | 68 21 | 1 | 82 | 68 22 | 1 | 80 | 68 21 | 4 | 77 | 68 17 | 6 | 74 | 336 |
| 25 | 67 23 | +4 | 85 | 67 27 | +2 | 82 | 67 29 | −1 | 79 | 67 28 | −3 | 77 | 67 25 | −6 | 74 | 335 |
| 26 | 66 28 | 5 | 84 | 66 33 | 2 | 82 | 66 35 | 0 | 79 | 66 35 | 3 | 77 | 66 32 | 5 | 74 | 334 |
| 27 | 65 34 | 5 | 84 | 65 39 | 3 | 82 | 65 42 | 0 | 79 | 65 42 | 2 | 77 | 65 40 | 4 | 75 | 333 |
| 28 | 64 40 | 5 | 84 | 64 45 | 3 | 82 | 64 48 | +1 | 79 | 64 49 | 1 | 77 | 64 48 | 4 | 75 | 332 |
| 29 | 63 46 | 6 | 84 | 63 52 | 3 | 82 | 63 55 | 1 | 79 | 63 56 | −1 | 77 | 63 55 | 3 | 75 | 331 |
| 30 | 62 52 | +6 | 84 | 62 58 | +4 | 81 | 63 02 | +1 | 79 | 63 03 | 0 | 77 | 63 03 | −3 | 75 | 330 |
| 31 | 61 58 | 6 | 83 | 62 04 | 4 | 81 | 62 08 | 2 | 79 | 62 10 | 0 | 77 | 62 10 | 2 | 75 | 329 |
| 32 | 61 04 | 6 | 83 | 61 10 | 5 | 81 | 61 15 | 2 | 79 | 61 17 | +1 | 77 | 61 18 | −1 | 75 | 328 |
| 33 | 60 10 | 7 | 83 | 60 17 | 4 | 81 | 60 21 | 3 | 79 | 60 24 | 1 | 77 | 60 25 | 0 | 75 | 327 |
| 34 | 59 16 | 7 | 83 | 59 23 | 5 | 81 | 59 28 | 3 | 79 | 59 31 | 2 | 77 | 59 33 | 0 | 75 | 326 |
| 35 | 58 22 | +7 | 82 | 58 29 | +6 | 81 | 58 35 | +3 | 79 | 58 38 | +2 | 77 | 58 40 | +1 | 75 | 325 |
| 36 | 57 28 | 8 | 82 | 57 36 | 5 | 80 | 57 41 | 5 | 79 | 57 46 | 2 | 77 | 57 48 | 1 | 75 | 324 |
| 37 | 56 35 | 7 | 82 | 56 42 | 6 | 80 | 56 48 | 5 | 78 | 56 53 | 2 | 77 | 56 55 | 2 | 75 | 323 |
| 38 | 55 41 | 8 | 82 | 55 49 | 6 | 80 | 55 55 | 5 | 78 | 56 00 | 3 | 76 | 56 03 | 2 | 75 | 322 |
| 39 | 54 47 | 8 | 82 | 54 55 | 7 | 80 | 55 02 | 5 | 78 | 55 07 | 4 | 76 | 55 11 | 2 | 75 | 321 |
| 40 | 53 53 | +9 | 81 | 54 02 | +7 | 80 | 54 09 | +5 | 78 | 54 14 | +4 | 76 | 54 18 | +3 | 75 | 320 |
| 41 | 52 59 | 9 | 81 | 53 08 | 7 | 79 | 53 15 | 6 | 78 | 53 21 | 5 | 76 | 53 26 | 3 | 74 | 319 |
| 42 | 52 06 | 9 | 81 | 52 15 | 7 | 79 | 52 22 | 7 | 78 | 52 29 | 4 | 76 | 52 33 | 4 | 74 | 318 |
| 43 | 51 12 | 9 | 81 | 51 21 | 8 | 79 | 51 29 | 7 | 77 | 51 36 | 5 | 76 | 51 41 | 4 | 74 | 317 |
| 44 | 50 18 | 10 | 80 | 50 28 | 8 | 79 | 50 36 | 7 | 77 | 50 43 | 6 | 76 | 50 49 | 4 | 74 | 316 |
| 45 | 49 25 | +10 | 80 | 49 35 | +8 | 79 | 49 43 | +7 | 77 | 49 50 | +7 | 76 | 49 57 | +4 | 74 | 315 |
| 46 | 48 31 | 10 | 80 | 48 41 | 9 | 78 | 48 50 | 8 | 77 | 48 58 | 6 | 75 | 49 04 | 6 | 74 | 314 |
| 47 | 47 38 | 10 | 80 | 47 48 | 9 | 78 | 47 57 | 8 | 77 | 48 05 | 7 | 75 | 48 12 | 6 | 74 | 313 |
| 48 | 46 44 | 11 | 79 | 46 55 | 9 | 78 | 47 04 | 9 | 77 | 47 13 | 7 | 75 | 47 20 | 6 | 74 | 312 |
| 49 | 45 51 | 11 | 79 | 46 02 | 9 | 78 | 46 11 | 9 | 76 | 46 20 | 8 | 75 | 46 28 | 6 | 73 | 311 |
| 50 | 44 58 | +11 | 79 | 45 09 | +10 | 78 | 45 19 | +9 | 76 | 45 28 | +8 | 75 | 45 36 | +7 | 73 | 310 |
| 51 | 44 04 | 12 | 79 | 44 16 | 10 | 77 | 44 26 | 9 | 76 | 44 35 | 9 | 75 | 44 44 | 7 | 73 | 309 |
| 52 | 43 11 | 12 | 78 | 43 23 | 10 | 77 | 43 33 | 10 | 76 | 43 43 | 9 | 74 | 43 52 | 7 | 73 | 308 |
| 53 | 42 18 | 12 | 78 | 42 30 | 11 | 77 | 42 41 | 10 | 75 | 42 51 | 9 | 74 | 43 00 | 8 | 73 | 307 |
| 54 | 41 24 | 13 | 78 | 41 37 | 11 | 77 | 41 48 | 10 | 75 | 41 58 | 10 | 74 | 42 08 | 8 | 73 | 306 |
| 55 | 40 31 | +13 | 78 | 40 44 | +11 | 76 | 40 55 | +11 | 75 | 41 06 | +10 | 74 | 41 16 | +9 | 72 | 305 |
| 56 | 39 38 | 13 | 77 | 39 51 | 12 | 76 | 40 03 | 11 | 75 | 40 14 | 10 | 74 | 40 24 | 10 | 72 | 304 |
| 57 | 38 45 | 13 | 77 | 38 58 | 12 | 76 | 39 10 | 12 | 75 | 39 22 | 10 | 73 | 39 32 | 10 | 72 | 303 |
| 58 | 37 52 | 14 | 77 | 38 06 | 12 | 76 | 38 18 | 12 | 74 | 38 30 | 11 | 73 | 38 41 | 10 | 72 | 302 |
| 59 | 36 59 | 14 | 77 | 37 13 | 13 | 75 | 37 26 | 12 | 74 | 37 38 | 11 | 73 | 37 49 | 11 | 72 | 301 |
| 60 | 36 07 | +13 | 76 | 36 20 | +13 | 75 | 36 33 | +13 | 74 | 36 46 | +12 | 73 | 36 58 | +10 | 71 | 300 |
| 61 | 35 14 | 14 | 76 | 35 28 | 13 | 75 | 35 41 | 13 | 74 | 35 54 | 12 | 72 | 36 06 | 11 | 71 | 299 |
| 62 | 34 21 | 14 | 76 | 34 35 | 14 | 75 | 34 49 | 13 | 73 | 35 02 | 13 | 72 | 35 15 | 11 | 71 | 298 |
| 63 | 33 28 | 15 | 76 | 33 43 | 14 | 74 | 33 57 | 13 | 73 | 34 10 | 13 | 72 | 34 23 | 12 | 71 | 297 |
| 64 | 32 36 | 15 | 75 | 32 51 | 14 | 74 | 33 05 | 14 | 73 | 33 19 | 13 | 72 | 33 32 | 12 | 71 | 296 |
| 65 | 31 43 | +15 | 75 | 31 58 | +15 | 74 | 32 13 | +14 | 73 | 32 27 | +14 | 72 | 32 41 | +12 | 70 | 295 |
| 66 | 30 51 | 15 | 75 | 31 06 | 15 | 74 | 31 21 | 15 | 72 | 31 36 | 13 | 71 | 31 49 | 14 | 70 | 294 |
| 67 | 29 58 | 16 | 74 | 30 14 | 15 | 73 | 30 29 | 15 | 72 | 30 44 | 14 | 71 | 30 58 | 14 | 70 | 293 |
| 68 | 29 06 | 16 | 74 | 29 22 | 16 | 73 | 29 38 | 15 | 72 | 29 53 | 14 | 71 | 30 07 | 14 | 70 | 292 |
| 69 | 28 14 | 16 | 74 | 28 30 | 16 | 73 | 28 46 | 16 | 72 | 29 02 | 14 | 71 | 29 16 | 15 | 69 | 291 |

S. Lat. {LHA greater than 180°........ Zn=180−Z; LHA less than 180°........... Zn=180+Z}

DECLINATION (15°-29°) SAME NAME AS LATITUDE



LAT 25°

N. Lat. {LHA greater than 180°....... Zn=Z; LHA less than 180°.......... Zn=360−Z}

# DECLINATION (15°-29°) CONTRARY NAME TO LATITUDE

| LHA | 15° Hc | d | Z | 16° Hc | d | Z | 17° Hc | d | Z | 18° Hc | d | Z | 19° Hc | d | Z |
|---|---|---|---|---|---|---|---|---|---|---|---|---|---|---|---|
| | ° ′ | ′ | ° | ° ′ | ′ | ° | ° ′ | ′ | ° | ° ′ | ′ | ° | ° ′ | ′ | ° |
| 69 | 11 48 | 31 | 113 | 11 17 | 30 | 114 | 10 47 | 31 | 115 | 10 16 | 30 | 116 | 09 46 | 31 | 116 |
| 68 | 12 38 | 31 | 113 | 12 07 | 31 | 114 | 11 36 | 31 | 115 | 11 05 | 31 | 116 | 10 34 | 31 | 117 |
| 67 | 13 27 | 31 | 114 | 12 56 | 31 | 115 | 12 25 | 31 | 116 | 11 54 | 31 | 117 | 11 23 | 32 | 117 |
| 66 | 14 17 | 31 | 114 | 13 46 | 32 | 115 | 13 14 | 31 | 116 | 12 43 | 32 | 117 | 12 11 | 32 | 118 |
| 65 | 15 06 | −31 | 115 | 14 35 | −32 | 116 | 14 03 | −32 | 117 | 13 31 | −32 | 118 | 12 59 | −33 | 118 |
| 64 | 15 56 | 32 | 116 | 15 24 | 33 | 116 | 14 51 | 32 | 117 | 14 19 | 33 | 118 | 13 46 | 32 | 119 |
| 63 | 16 45 | 33 | 116 | 16 12 | 32 | 117 | 15 40 | 33 | 118 | 15 07 | 33 | 119 | 14 34 | 33 | 120 |
| 62 | 17 33 | 32 | 117 | 17 01 | 33 | 117 | 16 28 | 34 | 118 | 15 54 | 33 | 119 | 15 21 | 34 | 120 |
| 61 | 18 22 | 33 | 117 | 17 49 | 34 | 118 | 17 15 | 33 | 119 | 16 42 | 34 | 120 | 16 08 | 34 | 121 |
| 60 | 19 10 | −33 | 118 | 18 37 | −34 | 119 | 18 03 | −34 | 119 | 17 29 | −34 | 120 | 16 55 | −35 | 121 |
| 59 | 19 58 | 34 | 118 | 19 24 | 34 | 119 | 18 50 | 34 | 120 | 18 16 | 35 | 121 | 17 41 | 35 | 122 |
| 58 | 20 46 | 34 | 119 | 20 12 | 35 | 120 | 19 37 | 35 | 121 | 19 02 | 35 | 121 | 18 27 | 35 | 122 |
| 57 | 21 33 | 34 | 119 | 20 59 | 35 | 120 | 20 24 | 36 | 121 | 19 48 | 35 | 122 | 19 13 | 36 | 123 |
| 56 | 22 21 | 36 | 120 | 21 45 | 35 | 121 | 21 10 | 36 | 122 | 20 34 | 35 | 123 | 19 59 | 37 | 124 |
| 55 | 23 08 | −36 | 121 | 22 32 | −36 | 122 | 21 56 | −36 | 122 | 21 20 | −36 | 123 | 20 44 | −37 | 124 |
| 54 | 23 54 | 36 | 121 | 23 18 | 36 | 122 | 22 42 | 37 | 123 | 22 05 | 36 | 124 | 21 29 | 38 | 125 |
| 53 | 24 41 | 37 | 122 | 24 04 | 37 | 123 | 23 27 | 37 | 124 | 22 50 | 37 | 125 | 22 13 | 37 | 125 |
| 52 | 25 27 | 37 | 123 | 24 50 | 38 | 123 | 24 12 | 37 | 124 | 23 35 | 38 | 125 | 22 57 | 38 | 126 |
| 51 | 26 12 | 37 | 123 | 25 35 | 38 | 124 | 24 57 | 38 | 125 | 24 19 | 38 | 126 | 23 41 | 38 | 127 |
| 50 | 26 58 | −38 | 124 | 26 20 | −38 | 125 | 25 42 | −39 | 126 | 25 03 | −38 | 127 | 24 25 | −39 | 127 |
| 49 | 27 42 | 38 | 125 | 27 04 | 38 | 125 | 26 26 | 39 | 126 | 25 47 | 39 | 127 | 25 08 | 40 | 128 |
| 48 | 28 27 | 39 | 125 | 27 48 | 39 | 126 | 27 09 | 39 | 127 | 26 30 | 40 | 128 | 25 50 | 40 | 129 |
| 47 | 29 11 | 39 | 126 | 28 32 | 40 | 127 | 27 52 | 39 | 128 | 27 13 | 41 | 129 | 26 32 | 40 | 129 |
| 46 | 29 55 | 40 | 127 | 29 15 | 40 | 128 | 28 35 | 40 | 128 | 27 55 | 41 | 129 | 27 14 | 41 | 130 |
| 45 | 30 38 | −40 | 128 | 29 58 | −40 | 128 | 29 18 | −41 | 129 | 28 37 | −41 | 130 | 27 56 | −42 | 131 |
| 44 | 31 21 | 40 | 128 | 30 41 | 41 | 129 | 30 00 | 42 | 130 | 29 18 | 41 | 131 | 28 37 | 42 | 132 |
| 43 | 32 04 | 41 | 129 | 31 23 | 42 | 130 | 30 41 | 42 | 131 | 29 59 | 42 | 132 | 29 17 | 42 | 132 |
| 42 | 32 46 | 42 | 130 | 32 04 | 42 | 131 | 31 22 | 42 | 132 | 30 40 | 43 | 132 | 29 57 | 43 | 133 |
| 41 | 33 27 | 42 | 131 | 32 45 | 42 | 131 | 32 03 | 43 | 132 | 31 20 | 44 | 133 | 30 36 | 43 | 134 |
| 40 | 34 09 | −43 | 131 | 33 26 | −43 | 132 | 32 43 | −44 | 133 | 31 59 | −44 | 134 | 31 15 | −44 | 135 |
| 39 | 34 49 | 43 | 132 | 34 06 | 44 | 133 | 33 22 | 44 | 134 | 32 38 | 44 | 135 | 31 54 | 45 | 136 |
| 38 | 35 29 | 44 | 133 | 34 45 | 44 | 134 | 34 01 | 45 | 135 | 33 16 | 44 | 136 | 32 32 | 46 | 136 |
| 37 | 36 08 | 44 | 134 | 35 24 | 45 | 135 | 34 39 | 45 | 136 | 33 54 | 45 | 136 | 33 09 | 46 | 137 |
| 36 | 36 47 | 45 | 135 | 36 02 | 45 | 136 | 35 17 | 46 | 137 | 34 31 | 45 | 137 | 33 46 | 47 | 138 |
| 35 | 37 26 | −46 | 136 | 36 40 | −46 | 137 | 35 54 | −46 | 137 | 35 08 | −46 | 138 | 34 22 | −47 | 139 |
| 34 | 38 03 | 46 | 137 | 37 17 | 46 | 138 | 36 31 | 47 | 138 | 35 44 | 47 | 139 | 34 57 | 47 | 140 |
| 33 | 38 40 | 46 | 138 | 37 54 | 47 | 138 | 37 07 | 48 | 139 | 36 19 | 47 | 140 | 35 32 | 48 | 141 |
| 32 | 39 16 | 47 | 139 | 38 29 | 47 | 139 | 37 42 | 48 | 140 | 36 54 | 48 | 141 | 36 06 | 49 | 142 |
| 31 | 39 52 | 48 | 140 | 39 04 | 48 | 140 | 38 16 | 48 | 141 | 37 28 | 49 | 142 | 36 39 | 49 | 143 |
| 30 | 40 27 | −48 | 141 | 39 39 | −49 | 141 | 38 50 | −49 | 142 | 38 01 | −49 | 143 | 37 12 | −50 | 144 |
| 29 | 41 01 | 49 | 142 | 40 12 | 49 | 142 | 39 23 | 50 | 143 | 38 33 | 49 | 144 | 37 44 | 50 | 145 |
| 28 | 41 34 | 49 | 143 | 40 45 | 50 | 143 | 39 55 | 50 | 144 | 39 05 | 50 | 145 | 38 15 | 51 | 146 |
| 27 | 42 07 | 50 | 144 | 41 17 | 50 | 145 | 40 27 | 51 | 145 | 39 36 | 51 | 146 | 38 45 | 51 | 147 |
| 26 | 42 39 | 51 | 145 | 41 48 | 51 | 146 | 40 57 | 51 | 146 | 40 06 | 51 | 147 | 39 15 | 52 | 148 |
| 25 | 43 10 | −52 | 146 | 42 18 | −51 | 147 | 41 27 | −52 | 147 | 40 35 | −52 | 148 | 39 43 | −52 | 149 |
| 24 | 43 40 | 52 | 147 | 42 48 | 52 | 148 | 41 56 | 52 | 149 | 41 04 | 53 | 149 | 40 11 | 53 | 150 |
| 23 | 44 09 | 53 | 148 | 43 16 | 52 | 149 | 42 24 | 53 | 150 | 41 31 | 53 | 150 | 40 38 | 53 | 151 |
| 22 | 44 37 | 53 | 150 | 43 44 | 53 | 150 | 42 51 | 53 | 151 | 41 58 | 54 | 151 | 41 04 | 54 | 152 |
| 21 | 45 04 | 53 | 151 | 44 11 | 54 | 151 | 43 17 | 54 | 152 | 42 23 | 54 | 153 | 41 29 | 54 | 153 |
| 20 | 45 30 | −54 | 152 | 44 36 | −54 | 153 | 43 42 | −54 | 153 | 42 48 | −55 | 154 | 41 53 | −54 | 154 |
| 19 | 45 55 | 54 | 153 | 45 01 | 55 | 154 | 44 06 | 55 | 154 | 43 11 | 55 | 155 | 42 16 | 55 | 155 |
| 18 | 46 19 | 55 | 154 | 45 24 | 55 | 155 | 44 29 | 55 | 156 | 43 34 | 56 | 156 | 42 38 | 55 | 157 |
| 17 | 46 42 | 55 | 156 | 45 47 | 56 | 156 | 44 51 | 56 | 157 | 43 55 | 55 | 157 | 43 00 | 56 | 158 |
| 16 | 47 04 | 56 | 157 | 46 08 | 56 | 158 | 45 12 | 56 | 158 | 44 16 | 56 | 159 | 43 20 | 57 | 159 |
| 15 | 47 25 | −57 | 158 | 46 28 | −56 | 159 | 45 32 | −57 | 159 | 44 35 | −56 | 160 | 43 39 | −57 | 160 |
| 14 | 47 44 | 57 | 160 | 46 47 | 57 | 160 | 45 50 | 57 | 161 | 44 53 | 57 | 161 | 43 56 | 57 | 162 |
| 13 | 48 02 | 57 | 161 | 47 05 | 57 | 162 | 46 08 | 57 | 162 | 45 11 | 58 | 162 | 44 13 | 57 | 163 |
| 12 | 48 19 | 57 | 162 | 47 22 | 58 | 163 | 46 24 | 58 | 163 | 45 26 | 57 | 164 | 44 29 | 58 | 164 |
| 11 | 48 35 | 58 | 164 | 47 37 | 58 | 164 | 46 39 | 58 | 165 | 45 41 | 58 | 165 | 44 43 | 58 | 165 |
| 10 | 48 50 | −59 | 165 | 47 51 | −58 | 166 | 46 53 | −58 | 166 | 45 55 | −59 | 166 | 44 56 | −58 | 167 |
| 9 | 49 03 | 59 | 167 | 48 04 | 58 | 167 | 47 06 | 59 | 167 | 46 07 | 59 | 168 | 45 08 | 59 | 168 |
| 8 | 49 15 | 59 | 168 | 48 16 | 59 | 168 | 47 17 | 59 | 169 | 46 18 | 59 | 169 | 45 19 | 59 | 169 |
| 7 | 49 25 | 59 | 170 | 48 26 | 59 | 170 | 47 27 | 59 | 170 | 46 28 | 59 | 170 | 45 29 | 60 | 171 |
| 6 | 49 35 | 60 | 171 | 48 35 | 59 | 171 | 47 36 | 60 | 172 | 46 36 | 59 | 172 | 45 37 | 60 | 172 |
| 5 | 49 42 | −59 | 173 | 48 43 | −60 | 173 | 47 43 | −59 | 173 | 46 44 | −60 | 173 | 45 44 | −60 | 173 |
| 4 | 49 49 | 60 | 174 | 48 49 | 60 | 174 | 47 49 | 60 | 174 | 46 49 | 59 | 174 | 45 50 | 60 | 175 |
| 3 | 49 54 | 60 | 176 | 48 54 | 60 | 176 | 47 54 | 60 | 176 | 46 54 | 60 | 176 | 45 54 | 60 | 176 |
| 2 | 49 57 | 60 | 177 | 48 57 | 60 | 177 | 47 57 | 60 | 177 | 46 57 | 60 | 177 | 45 57 | 59 | 177 |
| 1 | 49 59 | 60 | 179 | 48 59 | 60 | 179 | 47 59 | 60 | 179 | 46 59 | 60 | 179 | 45 59 | 60 | 179 |
| 0 | 50 00 | −60 | 180 | 49 00 | −60 | 180 | 48 00 | −60 | 180 | 47 00 | −60 | 180 | 46 00 | −60 | 180 |

| LHA | 20° Hc | d | Z | 21° Hc | d | Z | 22° Hc | d | Z | 23° Hc | d | Z | 24° Hc | d | Z |
|---|---|---|---|---|---|---|---|---|---|---|---|---|---|---|---|
| | ° ′ | ′ | ° | ° ′ | ′ | ° | ° ′ | ′ | ° | ° ′ | ′ | ° | ° ′ | ′ | ° |
| 69 | 09 15 | 31 | 117 | 08 44 | 31 | 118 | 08 13 | 31 | 119 | 07 42 | 32 | 120 | 07 10 | 31 | 121 |
| 68 | 10 03 | 31 | 118 | 09 32 | 32 | 119 | 09 00 | 31 | 120 | 08 29 | 32 | 120 | 07 57 | 32 | 121 |
| 67 | 10 51 | 32 | 118 | 10 19 | 32 | 119 | 09 47 | 32 | 120 | 09 15 | 32 | 121 | 08 43 | 32 | 122 |
| 66 | 11 39 | 32 | 119 | 11 07 | 33 | 120 | 10 34 | 32 | 121 | 10 02 | 33 | 121 | 09 29 | 32 | 122 |
| 65 | 12 26 | −32 | 119 | 11 54 | −33 | 120 | 11 21 | −33 | 121 | 10 48 | −33 | 122 | 10 15 | −33 | 123 |
| 64 | 13 14 | 33 | 120 | 12 41 | 33 | 121 | 12 08 | 34 | 122 | 11 34 | 33 | 122 | 11 01 | 34 | 123 |
| 63 | 14 01 | 34 | 120 | 13 27 | 33 | 121 | 12 54 | 34 | 122 | 12 20 | 34 | 123 | 11 46 | 34 | 124 |
| 62 | 14 47 | 33 | 121 | 14 14 | 34 | 122 | 13 40 | 34 | 123 | 13 06 | 35 | 123 | 12 31 | 34 | 124 |
| 61 | 15 34 | 34 | 121 | 15 00 | 35 | 122 | 14 25 | 34 | 123 | 13 51 | 35 | 124 | 13 16 | 35 | 125 |
| 60 | 16 20 | −34 | 122 | 15 46 | −35 | 123 | 15 11 | −35 | 124 | 14 36 | −35 | 125 | 14 01 | −36 | 125 |
| 59 | 17 06 | 35 | 123 | 16 31 | 35 | 123 | 15 56 | 36 | 124 | 15 20 | 35 | 125 | 14 45 | 36 | 126 |
| 58 | 17 52 | 36 | 123 | 17 16 | 35 | 124 | 16 41 | 36 | 125 | 16 05 | 36 | 126 | 15 29 | 37 | 127 |
| 57 | 18 37 | 36 | 124 | 18 01 | 36 | 125 | 17 25 | 36 | 125 | 16 49 | 37 | 126 | 16 12 | 37 | 127 |
| 56 | 19 22 | 36 | 124 | 18 46 | 37 | 125 | 18 09 | 37 | 126 | 17 32 | 37 | 127 | 16 55 | 37 | 128 |
| 55 | 20 07 | −37 | 125 | 19 30 | −37 | 126 | 18 53 | −37 | 127 | 18 16 | −38 | 127 | 17 38 | −37 | 128 |
| 54 | 20 51 | 37 | 126 | 20 14 | 37 | 126 | 19 37 | 38 | 127 | 18 59 | 38 | 128 | 18 21 | 38 | 129 |
| 53 | 21 36 | 38 | 126 | 20 58 | 38 | 127 | 20 20 | 39 | 128 | 19 41 | 38 | 129 | 19 03 | 39 | 130 |
| 52 | 22 19 | 38 | 127 | 21 41 | 38 | 128 | 21 03 | 39 | 129 | 20 24 | 39 | 129 | 19 45 | 39 | 130 |
| 51 | 23 03 | 39 | 128 | 22 24 | 39 | 128 | 21 45 | 39 | 129 | 21 06 | 40 | 130 | 20 26 | 39 | 131 |
| 50 | 23 46 | −40 | 128 | 23 06 | −39 | 129 | 22 27 | −40 | 130 | 21 47 | −40 | 131 | 21 07 | −40 | 131 |
| 49 | 24 28 | 40 | 129 | 23 48 | 40 | 130 | 23 08 | 40 | 130 | 22 28 | 40 | 131 | 21 48 | 41 | 132 |
| 48 | 25 10 | 40 | 130 | 24 30 | 40 | 130 | 23 50 | 41 | 131 | 23 09 | 41 | 132 | 22 28 | 41 | 133 |
| 47 | 25 52 | 41 | 130 | 25 11 | 41 | 131 | 24 30 | 41 | 132 | 23 49 | 41 | 133 | 23 08 | 42 | 133 |
| 46 | 26 33 | 41 | 131 | 25 52 | 41 | 132 | 25 11 | 42 | 133 | 24 29 | 42 | 133 | 23 47 | 42 | 134 |
| 45 | 27 14 | −41 | 132 | 26 33 | −42 | 132 | 25 51 | −43 | 133 | 25 08 | −42 | 134 | 24 26 | −43 | 135 |
| 44 | 27 55 | 43 | 132 | 27 12 | 42 | 133 | 26 30 | 43 | 134 | 25 47 | 43 | 135 | 25 04 | 43 | 136 |
| 43 | 28 35 | 43 | 133 | 27 52 | 43 | 134 | 27 09 | 44 | 135 | 26 25 | 43 | 136 | 25 42 | 44 | 136 |
| 42 | 29 14 | 43 | 134 | 28 31 | 44 | 135 | 27 47 | 44 | 136 | 27 03 | 44 | 136 | 26 19 | 44 | 137 |
| 41 | 29 53 | 44 | 135 | 29 09 | 44 | 136 | 28 25 | 44 | 136 | 27 41 | 45 | 137 | 26 56 | 45 | 138 |
| 40 | 30 31 | −44 | 136 | 29 47 | −45 | 136 | 29 02 | −44 | 137 | 28 18 | −46 | 138 | 27 32 | −45 | 139 |
| 39 | 31 09 | 45 | 136 | 30 24 | 45 | 137 | 29 39 | 45 | 138 | 28 54 | 46 | 139 | 28 08 | 46 | 139 |
| 38 | 31 46 | 45 | 137 | 31 01 | 46 | 138 | 30 15 | 46 | 139 | 29 29 | 46 | 139 | 28 43 | 46 | 140 |
| 37 | 32 23 | 46 | 138 | 31 37 | 46 | 139 | 30 51 | 46 | 140 | 30 05 | 47 | 140 | 29 18 | 47 | 141 |
| 36 | 32 59 | 46 | 139 | 32 13 | 47 | 140 | 31 26 | 47 | 140 | 30 39 | 47 | 141 | 29 52 | 48 | 142 |
| 35 | 33 35 | −47 | 140 | 32 48 | −48 | 140 | 32 00 | −47 | 141 | 31 13 | −48 | 142 | 30 25 | −48 | 143 |
| 34 | 34 10 | 48 | 141 | 33 22 | 48 | 141 | 32 34 | 48 | 142 | 31 46 | 48 | 143 | 30 58 | 49 | 143 |
| 33 | 34 44 | 48 | 142 | 33 56 | 49 | 142 | 33 07 | 48 | 143 | 32 19 | 49 | 144 | 31 30 | 49 | 144 |
| 32 | 35 17 | 48 | 142 | 34 29 | 49 | 143 | 33 40 | 49 | 144 | 32 51 | 50 | 145 | 32 01 | 49 | 145 |
| 31 | 35 50 | 49 | 143 | 35 01 | 49 | 144 | 34 12 | 50 | 145 | 33 22 | 50 | 145 | 32 32 | 50 | 146 |
| 30 | 36 22 | −49 | 144 | 35 33 | −50 | 145 | 34 43 | −51 | 146 | 33 52 | −50 | 146 | 33 02 | −50 | 147 |
| 29 | 36 54 | 51 | 145 | 36 03 | 50 | 146 | 35 13 | 51 | 147 | 34 22 | 51 | 147 | 33 31 | 51 | 148 |
| 28 | 37 24 | 51 | 146 | 36 33 | 51 | 147 | 35 42 | 51 | 148 | 34 51 | 51 | 148 | 34 00 | 52 | 149 |
| 27 | 37 54 | 51 | 147 | 37 03 | 52 | 148 | 36 11 | 52 | 149 | 35 19 | 51 | 149 | 34 28 | 52 | 150 |
| 26 | 38 23 | 52 | 148 | 37 31 | 52 | 149 | 36 39 | 52 | 150 | 35 47 | 52 | 150 | 34 55 | 53 | 151 |
| 25 | 38 51 | −52 | 149 | 37 59 | −53 | 150 | 37 06 | −52 | 151 | 36 14 | −53 | 151 | 35 21 | −53 | 152 |
| 24 | 39 18 | 52 | 150 | 38 26 | 53 | 151 | 37 33 | 54 | 152 | 36 39 | 53 | 152 | 35 46 | 54 | 153 |
| 23 | 39 45 | 53 | 152 | 38 52 | 54 | 152 | 37 58 | 54 | 153 | 37 04 | 54 | 153 | 36 10 | 53 | 154 |
| 22 | 40 10 | 53 | 153 | 39 17 | 54 | 153 | 38 23 | 55 | 154 | 37 28 | 54 | 154 | 36 34 | 54 | 155 |
| 21 | 40 35 | 54 | 154 | 39 41 | 55 | 154 | 38 46 | 54 | 155 | 37 52 | 55 | 155 | 36 57 | 55 | 156 |
| 20 | 40 59 | −55 | 155 | 40 04 | −55 | 155 | 39 09 | −55 | 156 | 38 14 | −55 | 156 | 37 19 | −56 | 157 |
| 19 | 41 21 | 55 | 156 | 40 26 | 55 | 157 | 39 31 | 56 | 157 | 38 35 | 55 | 158 | 37 40 | 56 | 158 |
| 18 | 41 43 | 56 | 157 | 40 47 | 56 | 158 | 39 51 | 55 | 158 | 38 56 | 56 | 159 | 38 00 | 57 | 159 |
| 17 | 42 04 | 57 | 158 | 41 07 | 56 | 159 | 40 11 | 56 | 159 | 39 15 | 57 | 160 | 38 18 | 56 | 160 |
| 16 | 42 23 | 56 | 160 | 41 27 | 57 | 160 | 40 30 | 57 | 160 | 39 33 | 56 | 161 | 38 37 | 57 | 161 |
| 15 | 42 42 | −57 | 161 | 41 45 | −57 | 161 | 40 48 | −57 | 162 | 39 51 | −57 | 162 | 38 54 | −58 | 162 |
| 14 | 42 59 | 57 | 162 | 42 02 | 57 | 162 | 41 05 | 58 | 163 | 40 07 | 57 | 163 | 39 10 | 58 | 163 |
| 13 | 43 16 | 58 | 163 | 42 18 | 58 | 164 | 41 20 | 58 | 164 | 40 22 | 57 | 164 | 39 25 | 58 | 165 |
| 12 | 43 31 | 58 | 164 | 42 33 | 58 | 165 | 41 35 | 58 | 165 | 40 37 | 59 | 165 | 39 38 | 58 | 166 |
| 11 | 43 45 | 58 | 166 | 42 47 | 59 | 166 | 41 48 | 58 | 166 | 40 50 | 59 | 167 | 39 51 | 58 | 167 |
| 10 | 43 58 | −59 | 167 | 42 59 | −58 | 167 | 42 01 | −59 | 168 | 41 02 | −59 | 168 | 40 03 | −59 | 168 |
| 9 | 44 09 | 58 | 168 | 43 11 | 59 | 168 | 42 12 | 59 | 169 | 41 13 | 59 | 169 | 40 14 | 59 | 169 |
| 8 | 44 20 | 59 | 170 | 43 21 | 59 | 170 | 42 22 | 59 | 170 | 41 23 | 59 | 170 | 40 24 | 60 | 170 |
| 7 | 44 29 | 59 | 171 | 43 30 | 59 | 171 | 42 31 | 60 | 171 | 41 31 | 59 | 171 | 40 32 | 59 | 172 |
| 6 | 44 37 | 59 | 172 | 43 38 | 60 | 172 | 42 38 | 59 | 172 | 41 39 | 60 | 173 | 40 39 | 59 | 173 |
| 5 | 44 44 | −59 | 173 | 43 45 | −60 | 174 | 42 45 | −60 | 174 | 41 45 | −59 | 174 | 40 46 | −60 | 174 |
| 4 | 44 50 | 60 | 175 | 43 50 | 60 | 175 | 42 50 | 59 | 175 | 41 51 | 60 | 175 | 40 51 | 60 | 175 |
| 3 | 44 54 | 59 | 176 | 43 55 | 60 | 176 | 42 55 | 60 | 176 | 41 55 | 60 | 176 | 40 55 | 60 | 176 |
| 2 | 44 58 | 60 | 177 | 43 58 | 60 | 177 | 42 58 | 60 | 178 | 41 58 | 60 | 178 | 40 58 | 60 | 178 |
| 1 | 44 59 | 60 | 179 | 43 59 | 60 | 179 | 42 59 | 60 | 179 | 41 59 | 60 | 179 | 40 59 | 60 | 179 |
| 0 | 45 00 | −60 | 180 | 44 00 | −60 | 180 | 43 00 | −60 | 180 | 42 00 | −60 | 180 | 41 00 | −60 | 180 |

| LHA | 25° Hc | d | Z | 26° Hc | d | Z | 27° Hc | d | Z | 28° Hc | d | Z | 29° Hc | d | Z | LHA |
|---|---|---|---|---|---|---|---|---|---|---|---|---|---|---|---|---|
| | ° ′ | ′ | ° | ° ′ | ′ | ° | ° ′ | ′ | ° | ° ′ | ′ | ° | ° ′ | ′ | ° | |
| 69 | 06 39 | 32 | 122 | 06 07 | 31 | 122 | 05 36 | 32 | 123 | 05 04 | 32 | 124 | 04 32 | 31 | 125 | 291 |
| 68 | 07 25 | 32 | 122 | 06 53 | 32 | 123 | 06 21 | 32 | 124 | 05 49 | 32 | 125 | 05 17 | 32 | 126 | 292 |
| 67 | 08 11 | 32 | 123 | 07 39 | 33 | 123 | 07 06 | 32 | 124 | 06 34 | 33 | 125 | 06 01 | 33 | 126 | 293 |
| 66 | 08 57 | 33 | 123 | 08 24 | 33 | 124 | 07 51 | 33 | 125 | 07 18 | 33 | 126 | 06 45 | 33 | 126 | 294 |
| 65 | 09 42 | −33 | 124 | 09 09 | −33 | 124 | 08 36 | −34 | 125 | 08 02 | −33 | 126 | 07 29 | −34 | 127 | 295 |
| 64 | 10 27 | 33 | 124 | 09 54 | 34 | 125 | 09 20 | 34 | 126 | 08 46 | 34 | 127 | 08 12 | 34 | 127 | 296 |
| 63 | 11 12 | 34 | 125 | 10 38 | 34 | 125 | 10 04 | 35 | 126 | 09 29 | 34 | 127 | 08 55 | 34 | 128 | 297 |
| 62 | 11 57 | 35 | 125 | 11 22 | 34 | 126 | 10 48 | 35 | 127 | 10 13 | 35 | 128 | 09 38 | 35 | 128 | 298 |
| 61 | 12 41 | 35 | 126 | 12 06 | 35 | 127 | 11 31 | 35 | 127 | 10 56 | 35 | 128 | 10 20 | 35 | 129 | 299 |
| 60 | 13 25 | −35 | 126 | 12 50 | −36 | 127 | 12 14 | −36 | 128 | 11 38 | −36 | 129 | 11 02 | −36 | 130 | 300 |
| 59 | 14 09 | 36 | 127 | 13 33 | 36 | 128 | 12 57 | 36 | 128 | 12 21 | 37 | 129 | 11 44 | 36 | 130 | 301 |
| 58 | 14 52 | 36 | 127 | 14 16 | 37 | 128 | 13 39 | 36 | 129 | 13 03 | 37 | 130 | 12 26 | 38 | 131 | 302 |
| 57 | 15 35 | 36 | 128 | 14 59 | 38 | 129 | 14 21 | 37 | 130 | 13 44 | 37 | 130 | 13 07 | 38 | 131 | 303 |
| 56 | 16 18 | 37 | 129 | 15 41 | 38 | 129 | 15 03 | 38 | 130 | 14 25 | 38 | 131 | 13 47 | 38 | 132 | 304 |
| 55 | 17 01 | −38 | 129 | 16 23 | −38 | 130 | 15 45 | −39 | 131 | 15 06 | −38 | 132 | 14 28 | −39 | 132 | 305 |
| 54 | 17 43 | 39 | 130 | 17 04 | 38 | 131 | 16 26 | 39 | 131 | 15 47 | 39 | 132 | 15 08 | 39 | 133 | 306 |
| 53 | 18 24 | 39 | 130 | 17 45 | 39 | 131 | 17 06 | 39 | 132 | 16 27 | 39 | 133 | 15 48 | 40 | 134 | 307 |
| 52 | 19 06 | 40 | 131 | 18 26 | 39 | 132 | 17 47 | 40 | 133 | 17 07 | 40 | 133 | 16 27 | 40 | 134 | 308 |
| 51 | 19 47 | 40 | 132 | 19 07 | 40 | 132 | 18 27 | 41 | 133 | 17 46 | 40 | 134 | 17 06 | 41 | 135 | 309 |
| 50 | 20 27 | −40 | 132 | 19 47 | −41 | 133 | 19 06 | −41 | 134 | 18 25 | −41 | 135 | 17 44 | −41 | 135 | 310 |
| 49 | 21 07 | 41 | 133 | 20 26 | 41 | 134 | 19 45 | 41 | 134 | 19 04 | 42 | 135 | 18 22 | 41 | 136 | 311 |
| 48 | 21 47 | 42 | 134 | 21 05 | 41 | 134 | 20 24 | 42 | 135 | 19 42 | 42 | 136 | 19 00 | 42 | 137 | 312 |
| 47 | 22 26 | 42 | 134 | 21 44 | 42 | 135 | 21 02 | 42 | 136 | 20 20 | 43 | 137 | 19 37 | 43 | 137 | 313 |
| 46 | 23 05 | 43 | 135 | 22 22 | 42 | 136 | 21 40 | 43 | 136 | 20 57 | 43 | 137 | 20 14 | 44 | 138 | 314 |
| 45 | 23 43 | −43 | 136 | 23 00 | −43 | 136 | 22 17 | −43 | 137 | 21 34 | −44 | 138 | 20 50 | −44 | 139 | 315 |
| 44 | 24 21 | 44 | 136 | 23 37 | 43 | 137 | 22 54 | 44 | 138 | 22 10 | 44 | 139 | 21 26 | 45 | 139 | 316 |
| 43 | 24 58 | 44 | 137 | 24 14 | 44 | 138 | 23 30 | 44 | 139 | 22 46 | 45 | 139 | 22 01 | 45 | 140 | 317 |
| 42 | 25 35 | 45 | 138 | 24 50 | 44 | 139 | 24 06 | 45 | 139 | 23 21 | 45 | 140 | 22 36 | 46 | 141 | 318 |
| 41 | 26 11 | 45 | 139 | 25 26 | 45 | 140 | 24 41 | 45 | 140 | 23 56 | 46 | 141 | 23 10 | 46 | 141 | 319 |
| 40 | 26 47 | −46 | 139 | 26 01 | −45 | 140 | 25 16 | −46 | 141 | 24 30 | −46 | 141 | 23 44 | −47 | 142 | 320 |
| 39 | 27 22 | 46 | 140 | 26 36 | 46 | 141 | 25 50 | 47 | 142 | 25 03 | 46 | 142 | 24 17 | 47 | 143 | 321 |
| 38 | 27 57 | 47 | 141 | 27 10 | 47 | 142 | 26 23 | 47 | 142 | 25 36 | 47 | 143 | 24 49 | 47 | 144 | 322 |
| 37 | 28 31 | 47 | 142 | 27 44 | 48 | 142 | 26 56 | 47 | 143 | 26 09 | 48 | 144 | 25 21 | 48 | 144 | 323 |
| 36 | 29 04 | 47 | 142 | 28 17 | 48 | 143 | 27 29 | 48 | 144 | 26 41 | 48 | 145 | 25 53 | 49 | 145 | 324 |
| 35 | 29 37 | −48 | 143 | 28 49 | −48 | 144 | 28 01 | −49 | 145 | 27 12 | −49 | 145 | 26 23 | −49 | 146 | 325 |
| 34 | 30 09 | 48 | 144 | 29 21 | 49 | 145 | 28 32 | 49 | 146 | 27 43 | 50 | 146 | 26 53 | 49 | 147 | 326 |
| 33 | 30 41 | 49 | 145 | 29 52 | 50 | 146 | 29 02 | 49 | 146 | 28 13 | 50 | 147 | 27 23 | 50 | 148 | 327 |
| 32 | 31 12 | 50 | 146 | 30 22 | 50 | 147 | 29 32 | 50 | 147 | 28 42 | 50 | 148 | 27 52 | 51 | 148 | 328 |
| 31 | 31 42 | 50 | 147 | 30 52 | 51 | 147 | 30 01 | 50 | 148 | 29 11 | 51 | 149 | 28 20 | 51 | 149 | 329 |
| 30 | 32 12 | −51 | 148 | 31 21 | −51 | 148 | 30 30 | −51 | 149 | 29 39 | −52 | 150 | 28 47 | −51 | 150 | 330 |
| 29 | 32 40 | 51 | 149 | 31 49 | 51 | 149 | 30 58 | 52 | 150 | 30 06 | 52 | 150 | 29 14 | 52 | 151 | 331 |
| 28 | 33 08 | 51 | 150 | 32 17 | 52 | 150 | 31 25 | 53 | 151 | 30 32 | 52 | 151 | 29 40 | 52 | 152 | 332 |
| 27 | 33 36 | 53 | 150 | 32 43 | 52 | 151 | 31 51 | 53 | 151 | 30 58 | 52 | 152 | 30 06 | 53 | 153 | 333 |
| 26 | 34 02 | 53 | 151 | 33 09 | 53 | 152 | 32 16 | 53 | 153 | 31 23 | 53 | 153 | 30 30 | 53 | 154 | 334 |
| 25 | 34 28 | −54 | 152 | 33 34 | −53 | 153 | 32 41 | −53 | 153 | 31 48 | −54 | 154 | 30 54 | −54 | 155 | 335 |
| 24 | 34 52 | 53 | 153 | 33 59 | 54 | 154 | 33 05 | 54 | 154 | 32 11 | 54 | 155 | 31 17 | 54 | 155 | 336 |
| 23 | 35 17 | 54 | 154 | 34 22 | 54 | 155 | 33 28 | 54 | 155 | 32 34 | 55 | 156 | 31 39 | 54 | 156 | 337 |
| 22 | 35 40 | 55 | 155 | 34 45 | 55 | 156 | 33 50 | 54 | 156 | 32 56 | 55 | 157 | 32 01 | 55 | 157 | 338 |
| 21 | 36 02 | 55 | 155 | 35 07 | 55 | 156 | 34 12 | 55 | 157 | 33 17 | 56 | 158 | 32 21 | 55 | 158 | 339 |
| 20 | 36 23 | −55 | 157 | 35 28 | −56 | 158 | 34 32 | −55 | 158 | 33 37 | −56 | 159 | 32 41 | −56 | 159 | 340 |
| 19 | 36 44 | 56 | 158 | 35 48 | 56 | 159 | 34 52 | 56 | 159 | 33 56 | 56 | 160 | 33 00 | 56 | 160 | 341 |
| 18 | 37 03 | 56 | 160 | 36 07 | 56 | 160 | 35 11 | 57 | 160 | 34 14 | 56 | 161 | 33 18 | 57 | 161 | 342 |
| 17 | 37 22 | 57 | 161 | 36 25 | 56 | 161 | 35 29 | 57 | 161 | 34 32 | 57 | 162 | 33 35 | 57 | 162 | 343 |
| 16 | 37 40 | 57 | 162 | 36 43 | 57 | 162 | 35 46 | 58 | 162 | 34 48 | 57 | 163 | 33 51 | 57 | 163 | 344 |
| 15 | 37 56 | −57 | 163 | 36 59 | −57 | 163 | 36 02 | −58 | 163 | 35 04 | −57 | 164 | 34 07 | −58 | 164 | 345 |
| 14 | 38 12 | 58 | 164 | 37 14 | 57 | 164 | 36 17 | 58 | 165 | 35 19 | 58 | 165 | 34 21 | 58 | 165 | 346 |
| 13 | 38 27 | 58 | 165 | 37 29 | 58 | 165 | 36 31 | 58 | 166 | 35 33 | 59 | 166 | 34 34 | 58 | 166 | 347 |
| 12 | 38 40 | 58 | 166 | 37 42 | 58 | 166 | 36 44 | 59 | 167 | 35 45 | 58 | 167 | 34 47 | 58 | 167 | 348 |
| 11 | 38 53 | 59 | 167 | 37 54 | 58 | 167 | 36 56 | 59 | 168 | 35 57 | 58 | 168 | 34 59 | 59 | 168 | 349 |
| 10 | 39 04 | −58 | 168 | 38 06 | −59 | 169 | 37 07 | −59 | 169 | 36 08 | −59 | 169 | 35 09 | −59 | 169 | 350 |
| 9 | 39 15 | 59 | 170 | 38 16 | 59 | 170 | 37 17 | 59 | 170 | 36 18 | 59 | 170 | 35 19 | 59 | 170 | 351 |
| 8 | 39 24 | 59 | 171 | 38 25 | 59 | 171 | 37 26 | 59 | 171 | 36 27 | 60 | 171 | 35 27 | 59 | 171 | 352 |
| 7 | 39 33 | 60 | 172 | 38 33 | 59 | 172 | 37 34 | 60 | 172 | 36 34 | 59 | 172 | 35 35 | 59 | 173 | 353 |
| 6 | 39 40 | 60 | 173 | 38 40 | 59 | 173 | 37 41 | 60 | 173 | 36 41 | 59 | 173 | 35 42 | 60 | 174 | 354 |
| 5 | 39 46 | −60 | 174 | 38 46 | −59 | 174 | 37 47 | −60 | 174 | 36 47 | −60 | 175 | 35 47 | −59 | 175 | 355 |
| 4 | 39 51 | 60 | 175 | 38 51 | 60 | 175 | 37 51 | 59 | 176 | 36 52 | 60 | 176 | 35 52 | 60 | 176 | 356 |
| 3 | 39 55 | 60 | 177 | 38 55 | 60 | 177 | 37 55 | 60 | 177 | 36 55 | 60 | 177 | 35 55 | 59 | 177 | 357 |
| 2 | 39 58 | 60 | 178 | 38 58 | 60 | 178 | 37 58 | 60 | 178 | 36 58 | 60 | 178 | 35 58 | 60 | 178 | 358 |
| 1 | 39 59 | 59 | 179 | 39 00 | 60 | 179 | 38 00 | 60 | 179 | 37 00 | 60 | 179 | 36 00 | 60 | 179 | 359 |
| 0 | 40 00 | −60 | 180 | 39 00 | −60 | 180 | 38 00 | −60 | 180 | 37 00 | −60 | 180 | 36 00 | −60 | 180 | 360 |

S. Lat. {LHA greater than 180°....... Zn=180−Z; LHA less than 180°.......... Zn=180+Z}

DECLINATION (15°-29°) CONTRARY NAME TO LATITUDE

LAT 25°

N. Lat. {LHA greater than 180°....... Zn=Z; LHA less than 180°........... Zn=360−Z}

# DECLINATION (0°-14°) CONTRARY NAME TO LATITUDE

| LHA | 0° Hc | 0° d | 0° Z | 1° Hc | 1° d | 1° Z | 2° Hc | 2° d | 2° Z | 3° Hc | 3° d | 3° Z | 4° Hc | 4° d | 4° Z |
|---|---|---|---|---|---|---|---|---|---|---|---|---|---|---|---|
| | ° ′ | ′ | ° | ° ′ | ′ | ° | ° ′ | ′ | ° | ° ′ | ′ | ° | ° ′ | ′ | ° |
| 69 | 18 47 | 27 | 100 | 18 20 | 29 | 101 | 17 51 | 28 | 101 | 17 23 | 29 | 102 | 16 54 | 29 | 103 |
| 68 | 19 41 | 28 | 100 | 19 13 | 29 | 101 | 18 44 | 28 | 102 | 18 16 | 29 | 103 | 17 47 | 30 | 104 |
| 67 | 20 34 | 29 | 101 | 20 05 | 28 | 102 | 19 37 | 29 | 102 | 19 08 | 29 | 103 | 18 39 | 29 | 104 |
| 66 | 21 27 | 29 | 101 | 20 58 | 29 | 102 | 20 29 | 29 | 103 | 20 00 | 29 | 104 | 19 31 | 29 | 105 |
| 65 | 22 20 | −29 | 102 | 21 51 | −29 | 103 | 21 22 | −29 | 103 | 20 53 | −30 | 104 | 20 23 | −30 | 105 |
| 64 | 23 12 | 28 | 102 | 22 44 | 30 | 103 | 22 14 | 29 | 104 | 21 45 | 30 | 105 | 21 15 | 30 | 106 |
| 63 | 24 05 | 29 | 103 | 23 36 | 29 | 104 | 23 07 | 30 | 105 | 22 37 | 30 | 105 | 22 07 | 30 | 106 |
| 62 | 24 58 | 30 | 103 | 24 28 | 29 | 104 | 23 59 | 30 | 105 | 23 29 | 30 | 106 | 22 59 | 31 | 107 |
| 61 | 25 50 | 29 | 104 | 25 21 | 30 | 105 | 24 51 | 30 | 106 | 24 21 | 31 | 107 | 23 50 | 31 | 108 |
| 60 | 26 42 | −29 | 104 | 26 13 | −30 | 105 | 25 43 | −31 | 106 | 25 12 | −30 | 107 | 24 42 | −32 | 108 |
| 59 | 27 35 | 30 | 105 | 27 05 | 31 | 106 | 26 34 | 30 | 107 | 26 04 | 31 | 108 | 25 33 | 32 | 109 |
| 58 | 28 27 | 30 | 105 | 27 57 | 31 | 106 | 27 26 | 31 | 107 | 26 55 | 31 | 108 | 26 24 | 32 | 109 |
| 57 | 29 19 | 31 | 106 | 28 48 | 31 | 107 | 28 17 | 31 | 108 | 27 46 | 31 | 109 | 27 15 | 32 | 110 |
| 56 | 30 10 | 30 | 107 | 29 40 | 31 | 108 | 29 09 | 32 | 108 | 28 37 | 32 | 109 | 28 05 | 32 | 110 |
| 55 | 31 02 | −31 | 107 | 30 31 | −31 | 108 | 30 00 | −32 | 109 | 29 28 | −32 | 110 | 28 56 | −33 | 111 |
| 54 | 31 53 | 31 | 108 | 31 22 | 31 | 109 | 30 51 | 33 | 110 | 30 18 | 32 | 111 | 29 46 | 33 | 112 |
| 53 | 32 45 | 32 | 108 | 32 13 | 32 | 109 | 31 41 | 32 | 110 | 31 09 | 33 | 111 | 30 36 | 33 | 112 |
| 52 | 33 36 | 32 | 109 | 33 04 | 32 | 110 | 32 32 | 33 | 111 | 31 59 | 33 | 112 | 31 26 | 34 | 113 |
| 51 | 34 27 | 32 | 110 | 33 55 | 33 | 111 | 33 22 | 33 | 112 | 32 49 | 34 | 113 | 32 15 | 34 | 114 |
| 50 | 35 18 | −33 | 110 | 34 45 | −33 | 111 | 34 12 | −33 | 112 | 33 39 | −34 | 113 | 33 05 | −35 | 114 |
| 49 | 36 08 | 33 | 111 | 35 35 | 33 | 112 | 35 02 | 34 | 113 | 34 28 | 34 | 114 | 33 54 | 35 | 115 |
| 48 | 36 58 | 33 | 112 | 36 25 | 34 | 113 | 35 51 | 34 | 114 | 35 17 | 35 | 115 | 34 42 | 35 | 116 |
| 47 | 37 48 | 33 | 112 | 37 15 | 34 | 113 | 36 41 | 35 | 114 | 36 06 | 35 | 115 | 35 31 | 36 | 116 |
| 46 | 38 38 | 34 | 113 | 38 04 | 34 | 114 | 37 30 | 35 | 115 | 36 55 | 36 | 116 | 36 19 | 36 | 117 |
| 45 | 39 28 | −35 | 114 | 38 53 | −35 | 115 | 38 18 | −35 | 116 | 37 43 | −36 | 117 | 37 07 | −37 | 118 |
| 44 | 40 17 | 35 | 114 | 39 42 | 35 | 116 | 39 07 | 36 | 117 | 38 31 | 37 | 118 | 37 54 | 36 | 119 |
| 43 | 41 06 | 35 | 115 | 40 31 | 36 | 116 | 39 55 | 36 | 117 | 39 19 | 37 | 118 | 38 42 | 38 | 119 |
| 42 | 41 55 | 36 | 116 | 41 19 | 36 | 117 | 40 43 | 37 | 118 | 40 06 | 38 | 119 | 39 28 | 37 | 120 |
| 41 | 42 43 | 36 | 117 | 42 07 | 37 | 118 | 41 30 | 37 | 119 | 40 53 | 38 | 120 | 40 15 | 38 | 121 |
| 40 | 43 31 | −37 | 118 | 42 54 | −37 | 119 | 42 17 | −38 | 120 | 41 39 | −38 | 121 | 41 01 | −39 | 122 |
| 39 | 44 18 | 37 | 118 | 43 41 | 37 | 120 | 43 04 | 39 | 121 | 42 25 | 38 | 122 | 41 47 | 40 | 123 |
| 38 | 45 06 | 38 | 119 | 44 28 | 38 | 120 | 43 50 | 39 | 122 | 43 11 | 39 | 123 | 42 32 | 40 | 124 |
| 37 | 45 52 | 38 | 120 | 45 14 | 38 | 121 | 44 36 | 40 | 122 | 43 56 | 40 | 123 | 43 16 | 40 | 125 |
| 36 | 46 39 | 39 | 121 | 46 00 | 39 | 122 | 45 21 | 40 | 123 | 44 41 | 40 | 124 | 44 01 | 41 | 125 |
| 35 | 47 25 | −39 | 122 | 46 46 | −40 | 123 | 46 06 | −41 | 124 | 45 25 | −41 | 125 | 44 44 | −41 | 126 |
| 34 | 48 10 | 39 | 123 | 47 31 | 41 | 124 | 46 50 | 41 | 125 | 46 09 | 41 | 126 | 45 28 | 43 | 127 |
| 33 | 48 55 | 40 | 124 | 48 15 | 41 | 125 | 47 34 | 42 | 126 | 46 52 | 42 | 127 | 46 10 | 43 | 128 |
| 32 | 49 40 | 41 | 125 | 48 59 | 42 | 126 | 48 17 | 42 | 127 | 47 35 | 43 | 128 | 46 52 | 43 | 129 |
| 31 | 50 24 | 42 | 126 | 49 42 | 42 | 127 | 49 00 | 43 | 128 | 48 17 | 43 | 129 | 47 34 | 44 | 130 |
| 30 | 51 07 | −42 | 127 | 50 25 | −43 | 128 | 49 42 | −44 | 129 | 48 58 | −44 | 131 | 48 14 | −44 | 132 |
| 29 | 51 49 | 42 | 128 | 51 07 | 44 | 130 | 50 23 | 44 | 131 | 49 39 | 45 | 132 | 48 54 | 45 | 133 |
| 28 | 52 31 | 43 | 130 | 51 48 | 44 | 131 | 51 04 | 45 | 132 | 50 19 | 45 | 133 | 49 34 | 46 | 134 |
| 27 | 53 13 | 45 | 131 | 52 28 | 44 | 132 | 51 44 | 46 | 133 | 50 58 | 46 | 134 | 50 12 | 46 | 135 |
| 26 | 53 53 | 45 | 132 | 53 08 | 45 | 133 | 52 23 | 46 | 134 | 51 37 | 47 | 135 | 50 50 | 47 | 136 |
| 25 | 54 33 | −46 | 133 | 53 47 | −46 | 134 | 53 01 | −47 | 135 | 52 14 | −47 | 136 | 51 27 | −48 | 137 |
| 24 | 55 12 | 47 | 135 | 54 25 | 47 | 136 | 53 38 | 47 | 137 | 52 51 | 48 | 138 | 52 03 | 49 | 139 |
| 23 | 55 50 | 47 | 136 | 55 03 | 48 | 137 | 54 15 | 48 | 138 | 53 27 | 49 | 139 | 52 38 | 49 | 140 |
| 22 | 56 27 | 48 | 137 | 55 39 | 49 | 138 | 54 50 | 49 | 140 | 54 01 | 49 | 140 | 53 12 | 50 | 141 |
| 21 | 57 03 | 49 | 139 | 56 14 | 49 | 140 | 55 25 | 50 | 141 | 54 35 | 50 | 142 | 53 45 | 50 | 143 |
| 20 | 57 38 | −50 | 140 | 56 48 | −50 | 141 | 55 58 | −50 | 142 | 55 08 | −51 | 143 | 54 17 | −51 | 144 |
| 19 | 58 12 | 51 | 142 | 57 21 | 50 | 143 | 56 31 | 51 | 144 | 55 40 | 52 | 145 | 54 48 | 52 | 146 |
| 18 | 58 44 | 51 | 144 | 57 53 | 51 | 145 | 57 02 | 52 | 145 | 56 10 | 52 | 146 | 55 18 | 53 | 147 |
| 17 | 59 16 | 52 | 145 | 58 24 | 52 | 146 | 57 32 | 52 | 147 | 56 40 | 53 | 148 | 55 47 | 54 | 149 |
| 16 | 59 46 | 52 | 147 | 58 54 | 53 | 148 | 58 01 | 53 | 149 | 57 08 | 54 | 150 | 56 14 | 54 | 150 |
| 15 | 60 15 | −53 | 149 | 59 22 | −54 | 150 | 58 28 | −54 | 150 | 57 34 | −54 | 151 | 56 40 | −55 | 152 |
| 14 | 60 42 | 54 | 150 | 59 48 | 54 | 151 | 58 54 | 55 | 152 | 57 59 | 54 | 153 | 57 05 | 56 | 154 |
| 13 | 61 08 | 55 | 152 | 60 13 | 54 | 153 | 59 19 | 56 | 154 | 58 23 | 55 | 155 | 57 28 | 56 | 155 |
| 12 | 61 32 | 55 | 154 | 60 37 | 55 | 155 | 59 42 | 56 | 156 | 58 46 | 56 | 156 | 57 50 | 57 | 157 |
| 11 | 61 55 | 56 | 156 | 60 59 | 56 | 157 | 60 03 | 57 | 158 | 59 06 | 56 | 158 | 58 10 | 57 | 159 |
| 10 | 62 16 | −56 | 158 | 61 20 | −57 | 159 | 60 23 | −57 | 159 | 59 26 | −58 | 160 | 58 28 | −57 | 161 |
| 9 | 62 35 | 57 | 160 | 61 38 | 57 | 161 | 60 41 | 58 | 161 | 59 43 | 57 | 162 | 58 46 | 58 | 163 |
| 8 | 62 53 | 58 | 162 | 61 55 | 58 | 163 | 60 57 | 58 | 163 | 59 59 | 58 | 164 | 59 01 | 58 | 164 |
| 7 | 63 08 | 58 | 164 | 62 10 | 58 | 165 | 61 12 | 59 | 165 | 60 13 | 58 | 166 | 59 15 | 59 | 166 |
| 6 | 63 22 | 59 | 167 | 62 23 | 59 | 167 | 61 24 | 59 | 167 | 60 25 | 58 | 168 | 59 27 | 59 | 168 |
| 5 | 63 33 | −59 | 169 | 62 34 | −59 | 169 | 61 35 | −59 | 170 | 60 36 | −59 | 170 | 59 37 | −60 | 170 |
| 4 | 63 43 | 59 | 171 | 62 44 | 60 | 171 | 61 44 | 59 | 172 | 60 45 | 60 | 172 | 59 45 | 59 | 172 |
| 3 | 63 50 | 59 | 173 | 62 51 | 60 | 173 | 61 51 | 60 | 174 | 60 51 | 59 | 174 | 59 52 | 60 | 174 |
| 2 | 63 56 | 60 | 175 | 62 56 | 60 | 176 | 61 56 | 60 | 176 | 60 56 | 60 | 176 | 59 56 | 60 | 176 |
| 1 | 63 59 | 60 | 178 | 62 59 | 60 | 178 | 61 59 | 60 | 178 | 60 59 | 60 | 178 | 59 59 | 60 | 178 |
| 0 | 64 00 | −60 | 180 | 63 00 | −60 | 180 | 62 00 | −60 | 180 | 61 00 | −60 | 180 | 60 00 | −60 | 180 |

| LHA | 5° Hc | 5° d | 5° Z | 6° Hc | 6° d | 6° Z | 7° Hc | 7° d | 7° Z | 8° Hc | 8° d | 8° Z | 9° Hc | 9° d | 9° Z |
|---|---|---|---|---|---|---|---|---|---|---|---|---|---|---|---|
| | ° ′ | ′ | ° | ° ′ | ′ | ° | ° ′ | ′ | ° | ° ′ | ′ | ° | ° ′ | ′ | ° |
| 69 | 16 25 | 29 | 104 | 15 56 | 29 | 105 | 15 27 | 30 | 106 | 14 57 | 30 | 107 | 14 27 | 30 | 108 |
| 68 | 17 17 | 29 | 105 | 16 48 | 30 | 106 | 16 18 | 29 | 107 | 15 49 | 31 | 107 | 15 18 | 30 | 108 |
| 67 | 18 10 | 30 | 105 | 17 40 | 30 | 106 | 17 10 | 30 | 107 | 16 40 | 30 | 108 | 16 10 | 31 | 109 |
| 66 | 19 02 | 30 | 106 | 18 32 | 30 | 107 | 18 02 | 31 | 108 | 17 31 | 31 | 108 | 17 00 | 30 | 109 |
| 65 | 19 53 | −30 | 106 | 19 23 | −30 | 107 | 18 53 | −31 | 108 | 18 22 | −31 | 109 | 17 51 | −31 | 110 |
| 64 | 20 45 | 30 | 107 | 20 15 | 31 | 108 | 19 44 | 31 | 109 | 19 13 | 31 | 110 | 18 42 | 32 | 110 |
| 63 | 21 37 | 31 | 107 | 21 06 | 31 | 108 | 20 35 | 31 | 109 | 20 04 | 32 | 110 | 19 32 | 31 | 111 |
| 62 | 22 28 | 31 | 108 | 21 57 | 31 | 109 | 21 26 | 32 | 110 | 20 54 | 31 | 111 | 20 23 | 32 | 112 |
| 61 | 23 19 | 31 | 108 | 22 48 | 31 | 109 | 22 17 | 32 | 110 | 21 45 | 32 | 111 | 21 13 | 33 | 112 |
| 60 | 24 10 | −31 | 109 | 23 39 | −32 | 110 | 23 07 | −32 | 111 | 22 35 | −32 | 112 | 22 03 | −33 | 113 |
| 59 | 25 01 | 31 | 110 | 24 30 | 33 | 111 | 23 57 | 32 | 111 | 23 25 | 33 | 112 | 22 52 | 33 | 113 |
| 58 | 25 52 | 32 | 110 | 25 20 | 33 | 111 | 24 48 | 33 | 112 | 24 15 | 33 | 113 | 23 42 | 34 | 114 |
| 57 | 26 43 | 33 | 111 | 26 10 | 33 | 112 | 25 37 | 33 | 113 | 25 04 | 33 | 114 | 24 31 | 34 | 114 |
| 56 | 27 33 | 33 | 111 | 27 00 | 33 | 112 | 26 27 | 33 | 113 | 25 54 | 34 | 114 | 25 20 | 34 | 115 |
| 55 | 28 23 | −33 | 112 | 27 50 | −33 | 113 | 27 17 | −34 | 114 | 26 43 | −34 | 115 | 26 09 | −35 | 116 |
| 54 | 29 13 | 33 | 113 | 28 40 | 34 | 114 | 28 06 | 34 | 115 | 27 32 | 35 | 115 | 26 57 | 35 | 116 |
| 53 | 30 03 | 34 | 113 | 29 29 | 34 | 114 | 28 55 | 35 | 115 | 28 20 | 35 | 116 | 27 45 | 35 | 117 |
| 52 | 30 52 | 34 | 114 | 30 18 | 35 | 115 | 29 43 | 34 | 116 | 29 09 | 36 | 117 | 28 33 | 35 | 118 |
| 51 | 31 41 | 34 | 115 | 31 07 | 35 | 116 | 30 32 | 35 | 116 | 29 57 | 36 | 117 | 29 21 | 36 | 118 |
| 50 | 32 30 | −35 | 115 | 31 55 | −35 | 116 | 31 20 | −36 | 117 | 30 44 | −36 | 118 | 30 08 | −36 | 119 |
| 49 | 33 19 | 35 | 116 | 32 44 | 36 | 117 | 32 08 | 36 | 118 | 31 32 | 37 | 119 | 30 55 | 37 | 120 |
| 48 | 34 07 | 35 | 117 | 33 32 | 37 | 118 | 32 55 | 36 | 119 | 32 19 | 37 | 120 | 31 42 | 37 | 120 |
| 47 | 34 55 | 36 | 117 | 34 19 | 36 | 118 | 33 43 | 37 | 119 | 33 06 | 38 | 120 | 32 28 | 38 | 121 |
| 46 | 35 43 | 37 | 118 | 35 06 | 36 | 119 | 34 30 | 38 | 120 | 33 52 | 38 | 121 | 33 14 | 38 | 122 |
| 45 | 36 30 | −37 | 119 | 35 53 | −37 | 120 | 35 16 | −38 | 121 | 34 38 | −38 | 122 | 34 00 | −39 | 123 |
| 44 | 37 18 | 38 | 120 | 36 40 | 38 | 121 | 36 02 | 38 | 122 | 35 24 | 39 | 122 | 34 45 | 39 | 123 |
| 43 | 38 04 | 38 | 120 | 37 26 | 38 | 121 | 36 48 | 39 | 122 | 36 09 | 39 | 123 | 35 30 | 40 | 124 |
| 42 | 38 51 | 39 | 121 | 38 12 | 39 | 122 | 37 33 | 39 | 123 | 36 54 | 40 | 124 | 36 14 | 40 | 125 |
| 41 | 39 37 | 39 | 122 | 38 58 | 40 | 123 | 38 18 | 39 | 124 | 37 39 | 41 | 125 | 36 58 | 40 | 126 |
| 40 | 40 22 | −39 | 123 | 39 43 | −40 | 124 | 39 03 | −40 | 125 | 38 23 | −41 | 126 | 37 42 | −41 | 127 |
| 39 | 41 07 | 40 | 124 | 40 27 | 40 | 125 | 39 47 | 41 | 126 | 39 06 | 41 | 127 | 38 25 | 42 | 128 |
| 38 | 41 52 | 41 | 125 | 41 11 | 40 | 126 | 40 31 | 42 | 127 | 39 49 | 42 | 128 | 39 07 | 42 | 128 |
| 37 | 42 36 | 41 | 126 | 41 55 | 41 | 127 | 41 14 | 42 | 127 | 40 32 | 43 | 128 | 39 49 | 42 | 129 |
| 36 | 43 20 | 42 | 126 | 42 38 | 42 | 127 | 41 56 | 42 | 128 | 41 14 | 43 | 129 | 40 31 | 44 | 130 |
| 35 | 44 03 | −42 | 127 | 43 21 | −43 | 128 | 42 38 | −43 | 129 | 41 55 | −43 | 130 | 41 12 | −44 | 131 |
| 34 | 44 45 | 42 | 128 | 44 03 | 43 | 129 | 43 20 | 44 | 130 | 42 36 | 44 | 131 | 41 52 | 45 | 132 |
| 33 | 45 27 | 43 | 129 | 44 44 | 44 | 130 | 44 00 | 44 | 131 | 43 16 | 44 | 132 | 42 32 | 45 | 133 |
| 32 | 46 09 | 44 | 130 | 45 25 | 44 | 131 | 44 41 | 45 | 132 | 43 56 | 45 | 133 | 43 11 | 46 | 134 |
| 31 | 46 50 | 45 | 131 | 46 05 | 45 | 132 | 45 20 | 45 | 133 | 44 35 | 46 | 134 | 43 49 | 46 | 135 |
| 30 | 47 30 | −45 | 133 | 46 45 | −46 | 134 | 45 59 | −46 | 134 | 45 13 | −46 | 135 | 44 27 | −47 | 136 |
| 29 | 48 09 | 46 | 134 | 47 23 | 46 | 135 | 46 37 | 46 | 136 | 45 51 | 47 | 136 | 45 04 | 48 | 137 |
| 28 | 48 48 | 47 | 135 | 48 01 | 46 | 136 | 47 15 | 48 | 137 | 46 27 | 47 | 138 | 45 40 | 48 | 138 |
| 27 | 49 26 | 47 | 136 | 48 39 | 48 | 137 | 47 51 | 48 | 138 | 47 03 | 48 | 139 | 46 15 | 48 | 140 |
| 26 | 50 03 | 48 | 137 | 49 15 | 48 | 138 | 48 27 | 48 | 139 | 47 39 | 49 | 140 | 46 50 | 49 | 141 |
| 25 | 50 39 | −48 | 138 | 49 51 | −49 | 139 | 49 02 | −49 | 140 | 48 13 | −50 | 141 | 47 23 | −49 | 142 |
| 24 | 51 14 | 49 | 140 | 50 25 | 49 | 141 | 49 36 | 50 | 142 | 48 46 | 50 | 142 | 47 56 | 50 | 143 |
| 23 | 51 49 | 50 | 141 | 50 59 | 50 | 142 | 50 09 | 50 | 143 | 49 19 | 51 | 144 | 48 28 | 51 | 144 |
| 22 | 52 22 | 50 | 142 | 51 32 | 51 | 143 | 50 41 | 51 | 144 | 49 50 | 51 | 145 | 48 59 | 52 | 146 |
| 21 | 52 55 | 51 | 144 | 52 04 | 52 | 145 | 51 12 | 51 | 145 | 50 21 | 52 | 146 | 49 29 | 52 | 147 |
| 20 | 53 26 | −51 | 145 | 52 35 | −52 | 146 | 51 43 | −53 | 147 | 50 50 | −52 | 148 | 49 58 | −53 | 148 |
| 19 | 53 56 | 52 | 147 | 53 04 | 52 | 147 | 52 12 | 53 | 148 | 51 19 | 53 | 149 | 50 26 | 54 | 150 |
| 18 | 54 25 | 52 | 148 | 53 33 | 54 | 149 | 52 39 | 53 | 150 | 51 46 | 54 | 150 | 50 52 | 54 | 151 |
| 17 | 54 53 | 53 | 150 | 54 00 | 54 | 150 | 53 06 | 54 | 151 | 52 12 | 54 | 152 | 51 18 | 55 | 153 |
| 16 | 55 20 | 54 | 151 | 54 26 | 54 | 152 | 53 32 | 55 | 153 | 52 37 | 55 | 153 | 51 42 | 55 | 154 |
| 15 | 55 45 | −54 | 153 | 54 51 | −55 | 153 | 53 56 | −56 | 154 | 53 00 | −55 | 155 | 52 05 | −56 | 155 |
| 14 | 56 09 | 55 | 154 | 55 14 | 55 | 155 | 54 19 | 56 | 156 | 53 23 | 56 | 156 | 52 27 | 56 | 157 |
| 13 | 56 32 | 56 | 156 | 55 36 | 56 | 157 | 54 40 | 56 | 157 | 53 44 | 57 | 158 | 52 47 | 56 | 158 |
| 12 | 56 53 | 56 | 158 | 55 57 | 57 | 158 | 55 00 | 57 | 159 | 54 03 | 57 | 160 | 53 06 | 57 | 160 |
| 11 | 57 13 | 57 | 159 | 56 16 | 57 | 160 | 55 19 | 57 | 161 | 54 22 | 58 | 161 | 53 24 | 57 | 162 |
| 10 | 57 31 | −57 | 161 | 56 34 | −58 | 162 | 55 36 | −58 | 162 | 54 38 | −57 | 163 | 53 41 | −58 | 163 |
| 9 | 57 48 | 58 | 163 | 56 50 | 58 | 164 | 55 52 | 58 | 164 | 54 54 | 59 | 164 | 53 55 | 58 | 165 |
| 8 | 58 03 | 59 | 165 | 57 04 | 58 | 165 | 56 06 | 59 | 166 | 55 07 | 59 | 166 | 54 09 | 59 | 166 |
| 7 | 58 16 | 59 | 167 | 57 17 | 59 | 167 | 56 18 | 58 | 167 | 55 20 | 59 | 168 | 54 21 | 59 | 168 |
| 6 | 58 28 | 59 | 169 | 57 29 | 60 | 169 | 56 29 | 59 | 169 | 55 30 | 59 | 170 | 54 31 | 59 | 170 |
| 5 | 58 37 | −59 | 170 | 57 38 | −59 | 171 | 56 39 | −60 | 171 | 55 39 | −59 | 171 | 54 40 | −60 | 171 |
| 4 | 58 46 | 60 | 172 | 57 46 | 60 | 173 | 56 46 | 59 | 173 | 55 47 | 60 | 173 | 54 47 | 60 | 173 |
| 3 | 58 52 | 60 | 174 | 57 52 | 60 | 174 | 56 52 | 59 | 175 | 55 53 | 60 | 175 | 54 53 | 60 | 175 |
| 2 | 58 56 | 59 | 176 | 57 57 | 60 | 176 | 56 57 | 60 | 176 | 55 57 | 60 | 177 | 54 57 | 60 | 177 |
| 1 | 58 59 | 60 | 178 | 57 59 | 60 | 178 | 56 59 | 60 | 178 | 55 59 | 60 | 178 | 54 59 | 60 | 178 |
| 0 | 59 00 | −60 | 180 | 58 00 | −60 | 180 | 57 00 | −60 | 180 | 56 00 | −60 | 180 | 55 00 | −60 | 180 |

| LHA | 10° Hc | 10° d | 10° Z | 11° Hc | 11° d | 11° Z | 12° Hc | 12° d | 12° Z | 13° Hc | 13° d | 13° Z | 14° Hc | 14° d | 14° Z | LHA |
|---|---|---|---|---|---|---|---|---|---|---|---|---|---|---|---|---|
| | ° ′ | ′ | ° | ° ′ | ′ | ° | ° ′ | ′ | ° | ° ′ | ′ | ° | ° ′ | ′ | ° | |
| 69 | 13 57 | 30 | 109 | 13 27 | 31 | 110 | 12 56 | 30 | 111 | 12 26 | 31 | 111 | 11 55 | 31 | 112 | 291 |
| 68 | 14 48 | 30 | 109 | 14 18 | 31 | 110 | 13 47 | 31 | 111 | 13 16 | 31 | 112 | 12 45 | 31 | 113 | 292 |
| 67 | 15 39 | 31 | 110 | 15 08 | 31 | 111 | 14 37 | 31 | 112 | 14 06 | 32 | 112 | 13 34 | 31 | 113 | 293 |
| 66 | 16 30 | 31 | 110 | 15 59 | 32 | 111 | 15 27 | 31 | 112 | 14 56 | 32 | 113 | 14 24 | 32 | 114 | 294 |
| 65 | 17 20 | −31 | 111 | 16 49 | −32 | 112 | 16 17 | −32 | 113 | 15 45 | −32 | 113 | 15 13 | −32 | 114 | 295 |
| 64 | 18 10 | 31 | 111 | 17 39 | 32 | 112 | 17 07 | 32 | 113 | 16 35 | 33 | 114 | 16 02 | 32 | 115 | 296 |
| 63 | 19 01 | 32 | 112 | 18 29 | 33 | 113 | 17 56 | 32 | 114 | 17 24 | 33 | 115 | 16 51 | 33 | 115 | 297 |
| 62 | 19 51 | 33 | 112 | 19 18 | 32 | 113 | 18 46 | 33 | 114 | 18 13 | 33 | 115 | 17 40 | 34 | 116 | 298 |
| 61 | 20 40 | 32 | 113 | 20 08 | 33 | 114 | 19 35 | 34 | 115 | 19 01 | 33 | 116 | 18 28 | 34 | 117 | 299 |
| 60 | 21 30 | −33 | 114 | 20 57 | −33 | 115 | 20 24 | −34 | 115 | 19 50 | −34 | 116 | 19 16 | −34 | 117 | 300 |
| 59 | 22 19 | 33 | 114 | 21 46 | 34 | 115 | 21 12 | 34 | 116 | 20 38 | 34 | 117 | 20 04 | 34 | 118 | 301 |
| 58 | 23 08 | 33 | 115 | 22 35 | 34 | 116 | 22 01 | 35 | 117 | 21 26 | 34 | 117 | 20 52 | 35 | 118 | 302 |
| 57 | 23 57 | 34 | 115 | 23 23 | 34 | 116 | 22 49 | 35 | 117 | 22 14 | 35 | 118 | 21 39 | 35 | 119 | 303 |
| 56 | 24 46 | 35 | 116 | 24 11 | 34 | 117 | 23 37 | 36 | 118 | 23 01 | 35 | 119 | 22 26 | 36 | 120 | 304 |
| 55 | 25 34 | −35 | 117 | 24 59 | −35 | 118 | 24 24 | −35 | 118 | 23 49 | −36 | 119 | 23 13 | −36 | 120 | 305 |
| 54 | 26 22 | 35 | 117 | 25 47 | 36 | 118 | 25 11 | 35 | 119 | 24 36 | 37 | 120 | 23 59 | 36 | 121 | 306 |
| 53 | 27 10 | 36 | 118 | 26 34 | 36 | 119 | 25 58 | 36 | 120 | 25 22 | 36 | 121 | 24 46 | 37 | 121 | 307 |
| 52 | 27 58 | 36 | 119 | 27 22 | 37 | 119 | 26 45 | 37 | 120 | 26 08 | 37 | 121 | 25 31 | 37 | 122 | 308 |
| 51 | 28 45 | 37 | 119 | 28 08 | 36 | 120 | 27 32 | 38 | 121 | 26 54 | 37 | 122 | 26 17 | 38 | 123 | 309 |
| 50 | 29 32 | −37 | 120 | 28 55 | −37 | 121 | 28 18 | −38 | 122 | 27 40 | −38 | 123 | 27 02 | −38 | 123 | 310 |
| 49 | 30 18 | 37 | 121 | 29 41 | 38 | 122 | 29 03 | 38 | 122 | 28 25 | 38 | 123 | 27 47 | 39 | 124 | 311 |
| 48 | 31 05 | 38 | 121 | 30 27 | 38 | 122 | 29 49 | 39 | 123 | 29 10 | 39 | 124 | 28 31 | 39 | 125 | 312 |
| 47 | 31 50 | 38 | 122 | 31 12 | 38 | 123 | 30 34 | 39 | 124 | 29 55 | 40 | 125 | 29 15 | 39 | 126 | 313 |
| 46 | 32 36 | 39 | 123 | 31 57 | 39 | 124 | 31 18 | 39 | 125 | 30 39 | 40 | 125 | 29 59 | 40 | 126 | 314 |
| 45 | 33 21 | −39 | 124 | 32 42 | −40 | 124 | 32 02 | −39 | 125 | 31 23 | −41 | 126 | 30 42 | −40 | 127 | 315 |
| 44 | 34 06 | 40 | 124 | 33 26 | 40 | 125 | 32 46 | 40 | 126 | 32 06 | 41 | 127 | 31 25 | 41 | 128 | 316 |
| 43 | 34 50 | 40 | 125 | 34 10 | 40 | 126 | 33 30 | 41 | 127 | 32 49 | 41 | 128 | 32 08 | 42 | 129 | 317 |
| 42 | 35 34 | 40 | 126 | 34 54 | 41 | 127 | 34 13 | 42 | 128 | 33 31 | 42 | 129 | 32 49 | 42 | 129 | 318 |
| 41 | 36 18 | 42 | 127 | 35 36 | 41 | 128 | 34 55 | 42 | 129 | 34 13 | 42 | 129 | 33 31 | 43 | 130 | 319 |
| 40 | 37 01 | −42 | 128 | 36 19 | −42 | 129 | 35 37 | −42 | 129 | 34 55 | −43 | 130 | 34 12 | −43 | 131 | 320 |
| 39 | 37 43 | 42 | 128 | 37 01 | 43 | 129 | 36 18 | 43 | 130 | 35 35 | 43 | 131 | 34 52 | 43 | 132 | 321 |
| 38 | 38 25 | 43 | 129 | 37 42 | 43 | 130 | 36 59 | 43 | 131 | 36 16 | 44 | 132 | 35 32 | 44 | 133 | 322 |
| 37 | 39 07 | 44 | 130 | 38 23 | 43 | 131 | 37 40 | 44 | 132 | 36 56 | 45 | 133 | 36 11 | 44 | 134 | 323 |
| 36 | 39 47 | 43 | 131 | 39 04 | 45 | 132 | 38 19 | 44 | 133 | 37 35 | 45 | 134 | 36 50 | 45 | 135 | 324 |
| 35 | 40 28 | −45 | 132 | 39 43 | −44 | 133 | 38 59 | −45 | 134 | 38 14 | −46 | 135 | 37 28 | −46 | 136 | 325 |
| 34 | 41 07 | 44 | 133 | 40 23 | 46 | 134 | 39 37 | 45 | 135 | 38 52 | 46 | 136 | 38 06 | 47 | 136 | 326 |
| 33 | 41 47 | 46 | 134 | 41 01 | 46 | 135 | 40 15 | 46 | 136 | 39 29 | 46 | 137 | 38 43 | 47 | 137 | 327 |
| 32 | 42 25 | 46 | 135 | 41 39 | 46 | 136 | 40 53 | 47 | 137 | 40 06 | 47 | 138 | 39 19 | 48 | 138 | 328 |
| 31 | 43 03 | 47 | 136 | 42 16 | 47 | 137 | 41 29 | 47 | 138 | 40 42 | 48 | 139 | 39 54 | 48 | 139 | 329 |
| 30 | 43 40 | −47 | 137 | 42 53 | −48 | 138 | 42 05 | −48 | 139 | 41 17 | −48 | 140 | 40 29 | −49 | 140 | 330 |
| 29 | 44 16 | 48 | 138 | 43 28 | 48 | 139 | 42 40 | 48 | 140 | 41 52 | 49 | 141 | 41 03 | 49 | 141 | 331 |
| 28 | 44 52 | 49 | 139 | 44 03 | 48 | 140 | 43 15 | 49 | 141 | 42 26 | 50 | 142 | 41 36 | 49 | 143 | 332 |
| 27 | 45 27 | 49 | 140 | 44 38 | 50 | 141 | 43 48 | 49 | 142 | 42 59 | 50 | 143 | 42 09 | 51 | 144 | 333 |
| 26 | 46 01 | 50 | 142 | 45 11 | 50 | 142 | 44 21 | 50 | 143 | 43 31 | 51 | 144 | 42 40 | 51 | 145 | 334 |
| 25 | 46 34 | −51 | 143 | 45 43 | −50 | 144 | 44 53 | −51 | 144 | 44 02 | −51 | 145 | 43 11 | −51 | 146 | 335 |
| 24 | 47 06 | 51 | 144 | 46 15 | 51 | 145 | 45 24 | 51 | 146 | 44 33 | 52 | 146 | 43 41 | 52 | 147 | 336 |
| 23 | 47 37 | 51 | 145 | 46 46 | 52 | 146 | 45 54 | 52 | 147 | 45 02 | 52 | 147 | 44 10 | 53 | 148 | 337 |
| 22 | 48 07 | 52 | 147 | 47 15 | 52 | 147 | 46 23 | 52 | 148 | 45 31 | 53 | 149 | 44 38 | 53 | 149 | 338 |
| 21 | 48 37 | 53 | 148 | 47 44 | 53 | 149 | 46 51 | 53 | 149 | 45 58 | 53 | 150 | 45 05 | 54 | 151 | 339 |
| 20 | 49 05 | −53 | 149 | 48 12 | −54 | 150 | 47 18 | −53 | 150 | 46 25 | −54 | 151 | 45 31 | −54 | 152 | 340 |
| 19 | 49 32 | 54 | 150 | 48 38 | 53 | 151 | 47 45 | 55 | 152 | 46 50 | 54 | 152 | 45 56 | 54 | 153 | 341 |
| 18 | 49 58 | 54 | 152 | 49 04 | 54 | 152 | 48 10 | 55 | 153 | 47 15 | 55 | 154 | 46 20 | 55 | 154 | 342 |
| 17 | 50 23 | 55 | 153 | 49 28 | 55 | 154 | 48 33 | 55 | 154 | 47 38 | 55 | 155 | 46 43 | 56 | 156 | 343 |
| 16 | 50 47 | 55 | 155 | 49 52 | 56 | 155 | 48 56 | 56 | 156 | 48 00 | 55 | 156 | 47 05 | 56 | 157 | 344 |
| 15 | 51 09 | −55 | 156 | 50 14 | −56 | 157 | 49 18 | −57 | 157 | 48 21 | −56 | 158 | 47 25 | −56 | 158 | 345 |
| 14 | 51 31 | 57 | 158 | 50 34 | 56 | 158 | 49 38 | 57 | 159 | 48 41 | 56 | 159 | 47 45 | 57 | 160 | 346 |
| 13 | 51 51 | 57 | 159 | 50 54 | 57 | 160 | 49 57 | 57 | 160 | 49 00 | 57 | 161 | 48 03 | 57 | 161 | 347 |
| 12 | 52 09 | 57 | 161 | 51 12 | 57 | 161 | 50 15 | 58 | 162 | 49 17 | 57 | 162 | 48 20 | 58 | 162 | 348 |
| 11 | 52 27 | 58 | 162 | 51 29 | 58 | 163 | 50 31 | 58 | 163 | 49 33 | 57 | 163 | 48 36 | 58 | 164 | 349 |
| 10 | 52 43 | −58 | 164 | 51 45 | −59 | 164 | 50 46 | −58 | 164 | 49 48 | −58 | 165 | 48 50 | −58 | 165 | 350 |
| 9 | 52 57 | 58 | 165 | 51 59 | 59 | 166 | 51 00 | 58 | 166 | 50 02 | 59 | 166 | 49 03 | 58 | 167 | 351 |
| 8 | 53 10 | 59 | 167 | 52 11 | 58 | 167 | 51 13 | 59 | 167 | 50 14 | 59 | 168 | 49 15 | 59 | 168 | 352 |
| 7 | 53 22 | 59 | 168 | 52 23 | 59 | 169 | 51 24 | 59 | 169 | 50 25 | 60 | 169 | 49 25 | 59 | 170 | 353 |
| 6 | 53 32 | 59 | 170 | 52 33 | 60 | 170 | 51 33 | 59 | 171 | 50 34 | 59 | 171 | 49 35 | 60 | 171 | 354 |
| 5 | 53 40 | −59 | 172 | 52 41 | −60 | 172 | 51 41 | −59 | 172 | 50 42 | −60 | 172 | 49 42 | −59 | 173 | 355 |
| 4 | 53 47 | 59 | 173 | 52 48 | 60 | 174 | 51 48 | 60 | 174 | 50 48 | 59 | 174 | 49 49 | 60 | 174 | 356 |
| 3 | 53 53 | 60 | 175 | 52 53 | 60 | 175 | 51 53 | 59 | 175 | 50 54 | 60 | 175 | 49 54 | 60 | 176 | 357 |
| 2 | 53 57 | 60 | 177 | 52 57 | 60 | 177 | 51 57 | 60 | 177 | 50 57 | 60 | 177 | 49 57 | 60 | 177 | 358 |
| 1 | 53 59 | 60 | 178 | 52 59 | 60 | 178 | 51 59 | 60 | 178 | 50 59 | 60 | 179 | 49 59 | 60 | 179 | 359 |
| 0 | 54 00 | −60 | 180 | 53 00 | −60 | 180 | 52 00 | −60 | 180 | 51 00 | −60 | 180 | 50 00 | −60 | 180 | 360 |

S. Lat. {LHA greater than 180°........ Zn=180−Z; LHA less than 180°........... Zn=180+Z}

DECLINATION (0°-14°) CONTRARY NAME TO LATITUDE

N. Lat. {LHA greater than 180°.......Zn=Z
{LHA less than 180°...........Zn=360−Z

# DECLINATION (15°-29°) CONTRARY NAME TO LATITUDE

| LHA | 15° Hc | d | Z | 16° Hc | d | Z | 17° Hc | d | Z | 18° Hc | d | Z | 19° Hc | d | Z |
|---|---|---|---|---|---|---|---|---|---|---|---|---|---|---|---|
| | ° ′ | ′ | ° | ° ′ | ′ | ° | ° ′ | ′ | ° | ° ′ | ′ | ° | ° ′ | ′ | ° |
| 69 | 11 24 | 31 | 113 | 10 53 | 31 | 114 | 10 22 | 32 | 115 | 09 50 | 31 | 116 | 09 19 | 32 | 117 |
| 68 | 12 14 | 32 | 114 | 11 42 | 31 | 115 | 11 11 | 32 | 115 | 10 39 | 32 | 116 | 10 07 | 32 | 117 |
| 67 | 13 03 | 32 | 114 | 12 31 | 32 | 115 | 11 59 | 32 | 116 | 11 27 | 32 | 117 | 10 55 | 32 | 118 |
| 66 | 13 52 | 32 | 115 | 13 20 | 32 | 116 | 12 48 | 33 | 116 | 12 15 | 32 | 117 | 11 43 | 33 | 118 |
| 65 | 14 41 | −33 | 115 | 14 08 | −32 | 116 | 13 36 | −33 | 117 | 13 03 | −33 | 118 | 12 30 | −33 | 119 |
| 64 | 15 30 | 33 | 116 | 14 57 | 33 | 117 | 14 24 | 33 | 118 | 13 51 | 34 | 118 | 13 17 | 33 | 119 |
| 63 | 16 18 | 33 | 116 | 15 45 | 33 | 117 | 15 12 | 34 | 118 | 14 38 | 34 | 119 | 14 04 | 34 | 120 |
| 62 | 17 06 | 33 | 117 | 16 33 | 34 | 118 | 15 59 | 34 | 119 | 15 25 | 34 | 119 | 14 51 | 34 | 120 |
| 61 | 17 54 | 34 | 117 | 17 20 | 34 | 118 | 16 46 | 34 | 119 | 16 12 | 35 | 120 | 15 37 | 34 | 121 |
| 60 | 18 42 | −34 | 118 | 18 08 | −35 | 119 | 17 33 | −35 | 120 | 16 58 | −34 | 121 | 16 24 | −36 | 121 |
| 59 | 19 30 | 35 | 119 | 18 55 | 35 | 119 | 18 20 | 35 | 120 | 17 45 | 36 | 121 | 17 09 | 35 | 122 |
| 58 | 20 17 | 35 | 119 | 19 42 | 36 | 120 | 19 06 | 35 | 121 | 18 31 | 36 | 122 | 17 55 | 36 | 123 |
| 57 | 21 04 | 36 | 120 | 20 28 | 35 | 121 | 19 53 | 36 | 122 | 19 17 | 37 | 122 | 18 40 | 36 | 123 |
| 56 | 21 50 | 35 | 120 | 21 15 | 37 | 121 | 20 38 | 36 | 122 | 20 02 | 37 | 123 | 19 25 | 37 | 124 |
| 55 | 22 37 | −37 | 121 | 22 00 | −36 | 122 | 21 24 | −37 | 123 | 20 47 | −37 | 124 | 20 10 | −37 | 124 |
| 54 | 23 23 | 37 | 122 | 22 46 | 37 | 123 | 22 09 | 37 | 123 | 21 32 | 38 | 124 | 20 54 | 38 | 125 |
| 53 | 24 09 | 38 | 122 | 23 31 | 37 | 123 | 22 54 | 38 | 124 | 22 16 | 38 | 125 | 21 38 | 38 | 126 |
| 52 | 24 54 | 38 | 123 | 24 16 | 37 | 124 | 23 39 | 39 | 125 | 23 00 | 38 | 126 | 22 22 | 39 | 126 |
| 51 | 25 39 | 38 | 124 | 25 01 | 38 | 125 | 24 23 | 39 | 125 | 23 44 | 39 | 126 | 23 05 | 39 | 127 |
| 50 | 26 24 | −39 | 124 | 25 45 | −38 | 125 | 25 07 | −40 | 126 | 24 27 | −39 | 127 | 23 48 | −40 | 128 |
| 49 | 27 08 | 39 | 125 | 26 29 | 39 | 126 | 25 50 | 40 | 127 | 25 10 | 39 | 128 | 24 31 | 41 | 128 |
| 48 | 27 52 | 39 | 126 | 27 13 | 40 | 127 | 26 33 | 40 | 127 | 25 53 | 40 | 128 | 25 13 | 41 | 129 |
| 47 | 28 36 | 40 | 126 | 27 56 | 40 | 127 | 27 16 | 41 | 128 | 26 35 | 41 | 129 | 25 54 | 41 | 130 |
| 46 | 29 19 | 40 | 127 | 28 39 | 41 | 128 | 27 58 | 41 | 129 | 27 17 | 41 | 130 | 26 36 | 42 | 131 |
| 45 | 30 02 | −41 | 128 | 29 21 | −41 | 129 | 28 40 | −42 | 130 | 27 58 | −42 | 130 | 27 16 | −42 | 131 |
| 44 | 30 44 | 41 | 129 | 30 03 | 42 | 130 | 29 21 | 42 | 130 | 28 39 | 42 | 131 | 27 57 | 43 | 132 |
| 43 | 31 26 | 42 | 130 | 30 44 | 42 | 130 | 30 02 | 43 | 131 | 29 19 | 43 | 132 | 28 36 | 43 | 133 |
| 42 | 32 07 | 42 | 130 | 31 25 | 43 | 131 | 30 42 | 43 | 132 | 29 59 | 43 | 133 | 29 16 | 44 | 134 |
| 41 | 32 48 | 43 | 131 | 32 05 | 43 | 132 | 31 22 | 43 | 133 | 30 39 | 44 | 134 | 29 55 | 44 | 134 |
| 40 | 33 29 | −44 | 132 | 32 45 | −44 | 133 | 32 01 | −44 | 134 | 31 17 | −44 | 134 | 30 33 | −45 | 135 |
| 39 | 34 09 | 44 | 133 | 33 25 | 44 | 134 | 32 40 | 44 | 134 | 31 56 | 45 | 135 | 31 11 | 45 | 136 |
| 38 | 34 48 | 45 | 134 | 34 03 | 44 | 134 | 33 19 | 46 | 135 | 32 33 | 45 | 136 | 31 48 | 46 | 137 |
| 37 | 35 27 | 45 | 135 | 34 42 | 46 | 135 | 33 56 | 45 | 136 | 33 11 | 46 | 137 | 32 25 | 47 | 138 |
| 36 | 36 05 | 46 | 135 | 35 19 | 46 | 136 | 34 33 | 46 | 137 | 33 47 | 46 | 138 | 33 01 | 47 | 139 |
| 35 | 36 42 | −46 | 136 | 35 56 | −46 | 137 | 35 10 | −47 | 138 | 34 23 | −47 | 139 | 33 36 | −47 | 139 |
| 34 | 37 19 | 46 | 137 | 36 33 | 47 | 138 | 35 46 | 48 | 139 | 34 58 | 47 | 140 | 34 11 | 48 | 140 |
| 33 | 37 56 | 48 | 138 | 37 08 | 47 | 139 | 36 21 | 48 | 140 | 35 33 | 48 | 141 | 34 45 | 48 | 141 |
| 32 | 38 31 | 47 | 139 | 37 44 | 49 | 140 | 36 55 | 48 | 141 | 36 07 | 48 | 141 | 35 19 | 49 | 142 |
| 31 | 39 06 | 48 | 140 | 38 18 | 49 | 141 | 37 29 | 49 | 142 | 36 40 | 49 | 142 | 35 51 | 49 | 143 |
| 30 | 39 40 | −48 | 141 | 38 52 | −50 | 142 | 38 02 | −49 | 143 | 37 13 | −50 | 143 | 36 23 | −50 | 144 |
| 29 | 40 14 | 50 | 142 | 39 24 | 49 | 143 | 38 35 | 50 | 144 | 37 45 | 50 | 144 | 36 55 | 51 | 145 |
| 28 | 40 47 | 50 | 143 | 39 57 | 51 | 144 | 39 06 | 50 | 145 | 38 16 | 51 | 145 | 37 25 | 51 | 146 |
| 27 | 41 18 | 50 | 144 | 40 28 | 51 | 145 | 39 37 | 51 | 146 | 38 46 | 51 | 146 | 37 55 | 52 | 147 |
| 26 | 41 49 | 51 | 145 | 40 58 | 51 | 146 | 40 07 | 51 | 147 | 39 16 | 52 | 147 | 38 24 | 52 | 148 |
| 25 | 42 20 | −52 | 147 | 41 28 | −52 | 147 | 40 36 | −52 | 148 | 39 44 | −52 | 149 | 38 52 | −53 | 149 |
| 24 | 42 49 | 52 | 148 | 41 57 | 52 | 148 | 41 05 | 53 | 149 | 40 12 | 53 | 150 | 39 19 | 53 | 150 |
| 23 | 43 17 | 52 | 149 | 42 25 | 53 | 149 | 41 32 | 53 | 150 | 40 39 | 53 | 151 | 39 46 | 54 | 151 |
| 22 | 43 45 | 53 | 150 | 42 52 | 54 | 151 | 41 58 | 53 | 151 | 41 05 | 54 | 152 | 40 11 | 54 | 152 |
| 21 | 44 11 | 53 | 151 | 43 18 | 54 | 152 | 42 24 | 54 | 152 | 41 30 | 54 | 153 | 40 36 | 55 | 154 |
| 20 | 44 37 | −54 | 152 | 43 43 | −55 | 153 | 42 48 | −54 | 154 | 41 54 | −55 | 154 | 40 59 | −55 | 155 |
| 19 | 45 02 | 55 | 154 | 44 07 | 55 | 154 | 43 12 | 55 | 155 | 42 17 | 55 | 155 | 41 22 | 56 | 156 |
| 18 | 45 25 | 55 | 155 | 44 30 | 56 | 155 | 43 34 | 55 | 156 | 42 39 | 56 | 156 | 41 43 | 55 | 157 |
| 17 | 45 47 | 55 | 156 | 44 52 | 56 | 157 | 43 56 | 56 | 157 | 43 00 | 56 | 158 | 42 04 | 56 | 158 |
| 16 | 46 09 | 56 | 157 | 45 13 | 57 | 158 | 44 16 | 56 | 158 | 43 20 | 56 | 159 | 42 24 | 57 | 159 |
| 15 | 46 29 | −57 | 159 | 45 32 | −56 | 159 | 44 36 | −57 | 160 | 43 39 | −57 | 160 | 42 42 | −57 | 161 |
| 14 | 46 48 | 57 | 160 | 45 51 | 57 | 161 | 44 54 | 57 | 161 | 43 57 | 58 | 161 | 42 59 | 57 | 162 |
| 13 | 47 06 | 58 | 161 | 46 08 | 57 | 162 | 45 11 | 58 | 162 | 44 13 | 57 | 163 | 43 16 | 58 | 163 |
| 12 | 47 22 | 57 | 163 | 46 25 | 58 | 163 | 45 27 | 58 | 164 | 44 29 | 58 | 164 | 43 31 | 58 | 164 |
| 11 | 47 38 | 58 | 164 | 46 40 | 59 | 165 | 45 41 | 58 | 165 | 44 43 | 58 | 165 | 43 45 | 58 | 166 |
| 10 | 47 52 | −59 | 166 | 46 53 | −58 | 166 | 45 55 | −59 | 166 | 44 56 | −58 | 167 | 43 58 | −59 | 167 |
| 9 | 48 05 | 59 | 167 | 47 06 | 59 | 167 | 46 07 | 59 | 168 | 45 08 | 58 | 168 | 44 10 | 59 | 168 |
| 8 | 48 16 | 59 | 168 | 47 17 | 59 | 169 | 46 18 | 59 | 169 | 45 19 | 59 | 169 | 44 20 | 59 | 169 |
| 7 | 48 26 | 59 | 170 | 47 27 | 59 | 170 | 46 28 | 59 | 170 | 45 29 | 60 | 171 | 44 29 | 59 | 171 |
| 6 | 48 35 | 59 | 171 | 47 36 | 60 | 171 | 46 36 | 59 | 172 | 45 37 | 60 | 172 | 44 37 | 59 | 172 |
| 5 | 48 43 | −60 | 173 | 47 43 | −59 | 173 | 46 44 | −60 | 173 | 45 44 | −60 | 173 | 44 44 | −59 | 173 |
| 4 | 48 49 | 60 | 174 | 47 49 | 59 | 174 | 46 50 | 60 | 174 | 45 50 | 60 | 175 | 44 50 | 60 | 175 |
| 3 | 48 54 | 60 | 176 | 47 54 | 60 | 176 | 46 54 | 60 | 176 | 45 54 | 60 | 176 | 44 54 | 59 | 176 |
| 2 | 48 57 | 60 | 177 | 47 57 | 60 | 177 | 46 57 | 60 | 177 | 45 57 | 59 | 177 | 44 58 | 60 | 177 |
| 1 | 48 59 | 60 | 179 | 47 59 | 60 | 179 | 46 59 | 60 | 179 | 45 59 | 60 | 179 | 44 59 | 60 | 179 |
| 0 | 49 00 | −60 | 180 | 48 00 | −60 | 180 | 47 00 | −60 | 180 | 46 00 | −60 | 180 | 45 00 | −60 | 180 |

| LHA | 20° Hc | d | Z | 21° Hc | d | Z | 22° Hc | d | Z | 23° Hc | d | Z | 24° Hc | d | Z |
|---|---|---|---|---|---|---|---|---|---|---|---|---|---|---|---|
| | ° ′ | ′ | ° | ° ′ | ′ | ° | ° ′ | ′ | ° | ° ′ | ′ | ° | ° ′ | ′ | ° |
| 69 | 08 47 | 32 | 117 | 08 15 | 31 | 118 | 07 44 | 32 | 119 | 07 12 | 32 | 120 | 06 40 | 33 | 121 |
| 68 | 09 35 | 32 | 118 | 09 03 | 32 | 119 | 08 31 | 33 | 120 | 07 58 | 32 | 121 | 07 26 | 33 | 121 |
| 67 | 10 23 | 33 | 118 | 09 50 | 33 | 119 | 09 17 | 32 | 120 | 08 45 | 33 | 121 | 08 12 | 33 | 122 |
| 66 | 11 10 | 33 | 119 | 10 37 | 33 | 120 | 10 04 | 33 | 121 | 09 31 | 34 | 122 | 08 57 | 33 | 122 |
| 65 | 11 57 | −33 | 120 | 11 24 | −34 | 120 | 10 50 | −33 | 121 | 10 17 | −34 | 122 | 09 43 | −34 | 123 |
| 64 | 12 44 | 34 | 120 | 12 10 | 34 | 121 | 11 36 | 34 | 122 | 11 02 | 34 | 123 | 10 28 | 34 | 123 |
| 63 | 13 30 | 34 | 121 | 12 56 | 34 | 121 | 12 22 | 35 | 122 | 11 47 | 34 | 123 | 11 13 | 35 | 124 |
| 62 | 14 17 | 35 | 121 | 13 42 | 35 | 122 | 13 07 | 35 | 123 | 12 32 | 35 | 124 | 11 57 | 35 | 125 |
| 61 | 15 03 | 35 | 122 | 14 28 | 35 | 123 | 13 53 | 36 | 123 | 13 17 | 35 | 124 | 12 42 | 36 | 125 |
| 60 | 15 48 | −35 | 122 | 15 13 | −36 | 123 | 14 37 | −35 | 124 | 14 02 | −36 | 125 | 13 26 | −36 | 126 |
| 59 | 16 34 | 36 | 123 | 15 58 | 36 | 124 | 15 22 | 36 | 125 | 14 46 | 36 | 125 | 14 10 | 37 | 126 |
| 58 | 17 19 | 36 | 123 | 16 43 | 37 | 124 | 16 06 | 36 | 125 | 15 30 | 37 | 126 | 14 53 | 37 | 127 |
| 57 | 18 04 | 37 | 124 | 17 27 | 37 | 125 | 16 50 | 37 | 126 | 16 13 | 37 | 127 | 15 36 | 37 | 127 |
| 56 | 18 48 | 37 | 125 | 18 11 | 37 | 125 | 17 34 | 38 | 126 | 16 56 | 37 | 127 | 16 19 | 38 | 128 |
| 55 | 19 33 | −38 | 125 | 18 55 | −38 | 126 | 18 17 | −38 | 127 | 17 39 | −38 | 128 | 17 01 | −38 | 129 |
| 54 | 20 16 | 38 | 126 | 19 38 | 38 | 127 | 19 00 | 38 | 128 | 18 22 | 39 | 128 | 17 43 | 39 | 129 |
| 53 | 21 00 | 38 | 127 | 20 22 | 39 | 127 | 19 43 | 39 | 128 | 19 04 | 39 | 129 | 18 25 | 40 | 130 |
| 52 | 21 43 | 39 | 127 | 21 04 | 39 | 128 | 20 25 | 39 | 129 | 19 46 | 40 | 130 | 19 06 | 40 | 130 |
| 51 | 22 26 | 39 | 128 | 21 47 | 40 | 129 | 21 07 | 40 | 129 | 20 27 | 40 | 130 | 19 47 | 40 | 131 |
| 50 | 23 08 | −40 | 129 | 22 28 | −40 | 129 | 21 48 | −40 | 130 | 21 08 | −41 | 131 | 20 27 | −40 | 132 |
| 49 | 23 50 | 40 | 129 | 23 10 | 41 | 130 | 22 29 | 40 | 131 | 21 49 | 42 | 132 | 21 07 | 41 | 132 |
| 48 | 24 32 | 41 | 130 | 23 51 | 41 | 131 | 23 10 | 41 | 132 | 22 29 | 42 | 132 | 21 47 | 42 | 133 |
| 47 | 25 13 | 41 | 131 | 24 32 | 41 | 131 | 23 50 | 42 | 132 | 23 08 | 42 | 133 | 22 26 | 42 | 134 |
| 46 | 25 54 | 42 | 131 | 25 12 | 42 | 132 | 24 30 | 42 | 133 | 23 48 | 43 | 134 | 23 05 | 43 | 134 |
| 45 | 26 34 | −42 | 132 | 25 52 | −43 | 133 | 25 09 | −43 | 134 | 24 26 | −43 | 134 | 23 43 | −43 | 135 |
| 44 | 27 14 | 43 | 133 | 26 31 | 43 | 134 | 25 48 | 43 | 134 | 25 05 | 44 | 135 | 24 21 | 44 | 136 |
| 43 | 27 53 | 43 | 134 | 27 10 | 44 | 134 | 26 26 | 43 | 135 | 25 43 | 45 | 136 | 24 58 | 44 | 137 |
| 42 | 28 32 | 44 | 134 | 27 48 | 44 | 135 | 27 04 | 44 | 136 | 26 20 | 45 | 137 | 25 35 | 45 | 137 |
| 41 | 29 11 | 44 | 135 | 28 27 | 45 | 136 | 27 42 | 45 | 137 | 26 57 | 45 | 137 | 26 12 | 46 | 138 |
| 40 | 29 48 | −44 | 136 | 29 04 | −46 | 137 | 28 18 | −45 | 137 | 27 33 | −46 | 138 | 26 47 | −46 | 139 |
| 39 | 30 26 | 46 | 137 | 29 40 | 45 | 138 | 28 55 | 46 | 138 | 28 09 | 46 | 139 | 27 23 | 47 | 140 |
| 38 | 31 02 | 46 | 138 | 30 16 | 46 | 138 | 29 30 | 46 | 139 | 28 44 | 47 | 140 | 27 57 | 47 | 141 |
| 37 | 31 38 | 46 | 138 | 30 52 | 47 | 139 | 30 05 | 47 | 140 | 29 18 | 47 | 141 | 28 31 | 47 | 141 |
| 36 | 32 14 | 47 | 139 | 31 27 | 47 | 140 | 30 40 | 48 | 141 | 29 52 | 47 | 141 | 29 05 | 48 | 142 |
| 35 | 32 49 | −48 | 140 | 32 01 | −47 | 141 | 31 14 | −48 | 142 | 30 26 | −49 | 142 | 29 37 | −48 | 143 |
| 34 | 33 23 | 48 | 141 | 32 35 | 48 | 142 | 31 47 | 49 | 142 | 30 58 | 48 | 143 | 30 10 | 49 | 144 |
| 33 | 33 57 | 49 | 142 | 33 08 | 49 | 143 | 32 19 | 49 | 143 | 31 30 | 49 | 144 | 30 41 | 49 | 145 |
| 32 | 34 30 | 49 | 143 | 33 41 | 50 | 144 | 32 51 | 49 | 144 | 32 02 | 50 | 145 | 31 12 | 50 | 146 |
| 31 | 35 02 | 50 | 144 | 34 12 | 50 | 145 | 33 22 | 50 | 145 | 32 32 | 50 | 146 | 31 42 | 50 | 146 |
| 30 | 35 33 | −50 | 145 | 34 43 | −50 | 145 | 33 53 | −51 | 146 | 33 02 | −50 | 147 | 32 12 | −51 | 147 |
| 29 | 36 04 | 50 | 146 | 35 14 | 51 | 146 | 34 23 | 51 | 147 | 33 32 | 52 | 148 | 32 40 | 51 | 148 |
| 28 | 36 34 | 51 | 147 | 35 43 | 51 | 147 | 34 52 | 52 | 148 | 34 00 | 52 | 149 | 33 08 | 51 | 149 |
| 27 | 37 03 | 51 | 148 | 36 12 | 52 | 148 | 35 20 | 52 | 149 | 34 28 | 52 | 150 | 33 36 | 53 | 150 |
| 26 | 37 32 | 52 | 149 | 36 40 | 53 | 149 | 35 47 | 52 | 150 | 34 55 | 53 | 151 | 34 02 | 53 | 151 |
| 25 | 37 59 | −52 | 150 | 37 07 | −53 | 150 | 36 14 | −53 | 151 | 35 21 | −53 | 152 | 34 28 | −54 | 152 |
| 24 | 38 26 | 53 | 151 | 37 33 | 53 | 151 | 36 40 | 54 | 152 | 35 46 | 53 | 153 | 34 53 | 54 | 153 |
| 23 | 38 52 | 54 | 152 | 37 58 | 53 | 152 | 37 05 | 54 | 153 | 36 11 | 54 | 154 | 35 17 | 55 | 154 |
| 22 | 39 17 | 54 | 153 | 38 23 | 54 | 154 | 37 29 | 55 | 154 | 36 34 | 54 | 155 | 35 40 | 55 | 155 |
| 21 | 39 41 | 54 | 154 | 38 47 | 55 | 155 | 37 52 | 55 | 155 | 36 57 | 55 | 156 | 36 02 | 55 | 156 |
| 20 | 40 04 | −55 | 155 | 39 09 | −55 | 156 | 38 14 | −55 | 156 | 37 19 | −56 | 157 | 36 23 | −55 | 157 |
| 19 | 40 26 | 55 | 156 | 39 31 | 56 | 157 | 38 35 | 55 | 157 | 37 40 | 56 | 158 | 36 44 | 56 | 158 |
| 18 | 40 48 | 56 | 157 | 39 52 | 56 | 158 | 38 56 | 56 | 158 | 38 00 | 57 | 159 | 37 03 | 56 | 159 |
| 17 | 41 08 | 57 | 159 | 40 11 | 56 | 159 | 39 15 | 56 | 160 | 38 19 | 57 | 160 | 37 22 | 57 | 160 |
| 16 | 41 27 | 57 | 160 | 40 30 | 57 | 160 | 39 33 | 56 | 161 | 38 37 | 57 | 161 | 37 40 | 57 | 162 |
| 15 | 41 45 | −57 | 161 | 40 48 | −57 | 161 | 39 51 | −57 | 162 | 38 54 | −58 | 162 | 37 56 | −57 | 163 |
| 14 | 42 02 | 57 | 162 | 41 05 | 58 | 163 | 40 07 | 57 | 163 | 39 10 | 58 | 163 | 38 12 | 58 | 164 |
| 13 | 42 18 | 58 | 163 | 41 20 | 58 | 164 | 40 22 | 57 | 164 | 39 25 | 58 | 165 | 38 27 | 58 | 165 |
| 12 | 42 33 | 58 | 165 | 41 35 | 58 | 165 | 40 37 | 58 | 165 | 39 39 | 59 | 166 | 38 40 | 58 | 166 |
| 11 | 42 47 | 59 | 166 | 41 48 | 58 | 166 | 40 50 | 59 | 167 | 39 51 | 58 | 167 | 38 53 | 59 | 167 |
| 10 | 42 59 | −58 | 167 | 42 01 | −59 | 167 | 41 02 | −59 | 168 | 40 03 | −59 | 168 | 39 04 | −58 | 168 |
| 9 | 43 11 | 59 | 168 | 42 12 | 59 | 169 | 41 13 | 59 | 169 | 40 14 | 59 | 169 | 39 15 | 59 | 169 |
| 8 | 43 21 | 59 | 170 | 42 22 | 59 | 170 | 41 23 | 59 | 170 | 40 24 | 60 | 170 | 39 24 | 59 | 171 |
| 7 | 43 30 | 59 | 171 | 42 31 | 60 | 171 | 41 31 | 59 | 171 | 40 32 | 59 | 172 | 39 33 | 60 | 172 |
| 6 | 43 38 | 59 | 172 | 42 39 | 60 | 172 | 41 39 | 60 | 173 | 40 39 | 59 | 173 | 39 40 | 60 | 173 |
| 5 | 43 45 | −60 | 174 | 42 45 | −60 | 174 | 41 45 | −59 | 174 | 40 46 | −60 | 174 | 39 46 | −60 | 174 |
| 4 | 43 50 | 60 | 175 | 42 50 | 59 | 175 | 41 51 | 60 | 175 | 40 51 | 60 | 175 | 39 51 | 60 | 175 |
| 3 | 43 55 | 60 | 176 | 42 55 | 60 | 176 | 41 55 | 60 | 176 | 40 55 | 60 | 176 | 39 55 | 60 | 176 |
| 2 | 43 58 | 60 | 177 | 42 58 | 60 | 177 | 41 58 | 60 | 178 | 40 58 | 60 | 178 | 39 58 | 60 | 178 |
| 1 | 43 59 | 60 | 179 | 42 59 | 60 | 179 | 41 59 | 60 | 179 | 40 59 | 60 | 179 | 39 59 | 59 | 179 |
| 0 | 44 00 | −60 | 180 | 43 00 | −60 | 180 | 42 00 | −60 | 180 | 41 00 | −60 | 180 | 40 00 | −60 | 180 |

| LHA | 25° Hc | d | Z | 26° Hc | d | Z | 27° Hc | d | Z | 28° Hc | d | Z | 29° Hc | d | Z | LHA |
|---|---|---|---|---|---|---|---|---|---|---|---|---|---|---|---|---|
| | ° ′ | ′ | ° | ° ′ | ′ | ° | ° ′ | ′ | ° | ° ′ | ′ | ° | ° ′ | ′ | ° | |
| 69 | 06 07 | 32 | 122 | 05 35 | 32 | 123 | 05 03 | 33 | 123 | 04 30 | 32 | 124 | 03 58 | 32 | 125 | 291 |
| 68 | 06 53 | 32 | 122 | 06 21 | 33 | 123 | 05 48 | 33 | 124 | 05 15 | 33 | 125 | 04 42 | 33 | 126 | 292 |
| 67 | 07 39 | 33 | 123 | 07 06 | 34 | 124 | 06 32 | 33 | 124 | 05 59 | 33 | 125 | 05 26 | 34 | 126 | 293 |
| 66 | 08 24 | 34 | 123 | 07 50 | 33 | 124 | 07 17 | 34 | 125 | 06 43 | 34 | 126 | 06 09 | 34 | 127 | 294 |
| 65 | 09 09 | −34 | 124 | 08 35 | −34 | 125 | 08 01 | −34 | 125 | 07 27 | −34 | 126 | 06 53 | −35 | 127 | 295 |
| 64 | 09 54 | 35 | 124 | 09 19 | 34 | 125 | 08 45 | 35 | 126 | 08 10 | 35 | 127 | 07 35 | 34 | 128 | 296 |
| 63 | 10 38 | 35 | 125 | 10 03 | 35 | 126 | 09 28 | 35 | 126 | 08 53 | 35 | 127 | 08 18 | 35 | 128 | 297 |
| 62 | 11 22 | 35 | 125 | 10 47 | 35 | 126 | 10 12 | 36 | 127 | 09 36 | 36 | 128 | 09 00 | 35 | 129 | 298 |
| 61 | 12 06 | 36 | 126 | 11 30 | 35 | 127 | 10 55 | 36 | 128 | 10 19 | 37 | 128 | 09 42 | 36 | 129 | 299 |
| 60 | 12 50 | −36 | 126 | 12 14 | −37 | 127 | 11 37 | −36 | 128 | 11 01 | −37 | 129 | 10 24 | −37 | 130 | 300 |
| 59 | 13 33 | 37 | 127 | 12 56 | 37 | 128 | 12 19 | 36 | 129 | 11 43 | 38 | 129 | 11 05 | 37 | 130 | 301 |
| 58 | 14 16 | 37 | 128 | 13 39 | 38 | 128 | 13 01 | 37 | 129 | 12 24 | 38 | 130 | 11 46 | 37 | 131 | 302 |
| 57 | 14 59 | 38 | 128 | 14 21 | 38 | 129 | 13 43 | 38 | 130 | 13 05 | 38 | 131 | 12 27 | 38 | 131 | 303 |
| 56 | 15 41 | 38 | 129 | 15 03 | 39 | 130 | 14 24 | 38 | 130 | 13 46 | 39 | 131 | 13 07 | 38 | 132 | 304 |
| 55 | 16 23 | −39 | 129 | 15 44 | −39 | 130 | 15 05 | −38 | 131 | 14 27 | −40 | 132 | 13 47 | −39 | 133 | 305 |
| 54 | 17 04 | 39 | 130 | 16 25 | 39 | 131 | 15 46 | 39 | 132 | 15 07 | 40 | 132 | 14 27 | 40 | 133 | 306 |
| 53 | 17 45 | 39 | 131 | 17 06 | 40 | 131 | 16 26 | 40 | 132 | 15 46 | 40 | 133 | 15 06 | 40 | 134 | 307 |
| 52 | 18 26 | 40 | 131 | 17 46 | 40 | 132 | 17 06 | 40 | 133 | 16 26 | 41 | 134 | 15 45 | 41 | 134 | 308 |
| 51 | 19 07 | 41 | 132 | 18 26 | 41 | 133 | 17 45 | 40 | 133 | 17 05 | 42 | 134 | 16 23 | 41 | 135 | 309 |
| 50 | 19 47 | −41 | 133 | 19 06 | −42 | 133 | 18 24 | −41 | 134 | 17 43 | −42 | 135 | 17 01 | −41 | 136 | 310 |
| 49 | 20 26 | 41 | 133 | 19 45 | 42 | 134 | 19 03 | 42 | 135 | 18 21 | 42 | 135 | 17 39 | 42 | 136 | 311 |
| 48 | 21 05 | 42 | 134 | 20 23 | 42 | 135 | 19 41 | 42 | 135 | 18 59 | 43 | 136 | 18 16 | 43 | 137 | 312 |
| 47 | 21 44 | 42 | 135 | 21 02 | 43 | 135 | 20 19 | 43 | 136 | 19 36 | 43 | 137 | 18 53 | 43 | 138 | 313 |
| 46 | 22 22 | 43 | 135 | 21 39 | 43 | 136 | 20 56 | 43 | 137 | 20 13 | 44 | 137 | 19 29 | 44 | 138 | 314 |
| 45 | 23 00 | −43 | 136 | 22 17 | −44 | 137 | 21 33 | −44 | 137 | 20 49 | −44 | 138 | 20 05 | −44 | 139 | 315 |
| 44 | 23 37 | 44 | 137 | 22 53 | 44 | 137 | 22 09 | 44 | 138 | 21 25 | 45 | 139 | 20 40 | 45 | 140 | 316 |
| 43 | 24 14 | 44 | 137 | 23 30 | 45 | 138 | 22 45 | 45 | 139 | 22 00 | 45 | 140 | 21 15 | 45 | 140 | 317 |
| 42 | 24 50 | 45 | 138 | 24 05 | 45 | 139 | 23 20 | 45 | 140 | 22 35 | 46 | 140 | 21 49 | 46 | 141 | 318 |
| 41 | 25 26 | 45 | 139 | 24 41 | 46 | 140 | 23 55 | 46 | 140 | 23 09 | 46 | 141 | 22 23 | 46 | 142 | 319 |
| 40 | 26 01 | −46 | 140 | 25 15 | −46 | 140 | 24 29 | −46 | 141 | 23 43 | −47 | 142 | 22 56 | −47 | 142 | 320 |
| 39 | 26 36 | 46 | 140 | 25 50 | 47 | 141 | 25 03 | 47 | 142 | 24 16 | 47 | 142 | 23 29 | 48 | 143 | 321 |
| 38 | 27 10 | 47 | 141 | 26 23 | 47 | 142 | 25 36 | 48 | 143 | 24 48 | 47 | 143 | 24 01 | 48 | 144 | 322 |
| 37 | 27 44 | 48 | 142 | 26 56 | 48 | 143 | 26 08 | 48 | 143 | 25 20 | 48 | 144 | 24 32 | 48 | 145 | 323 |
| 36 | 28 17 | 48 | 143 | 27 29 | 49 | 144 | 26 40 | 48 | 144 | 25 52 | 49 | 145 | 25 03 | 49 | 145 | 324 |
| 35 | 28 49 | −49 | 144 | 28 00 | −48 | 144 | 27 12 | −49 | 145 | 26 23 | −50 | 146 | 25 33 | −49 | 146 | 325 |
| 34 | 29 21 | 49 | 145 | 28 32 | 50 | 145 | 27 42 | 49 | 146 | 26 53 | 50 | 146 | 26 03 | 50 | 147 | 326 |
| 33 | 29 52 | 50 | 145 | 29 02 | 50 | 146 | 28 12 | 50 | 147 | 27 22 | 50 | 147 | 26 32 | 50 | 148 | 327 |
| 32 | 30 22 | 50 | 146 | 29 32 | 50 | 147 | 28 42 | 51 | 147 | 27 51 | 50 | 148 | 27 01 | 51 | 149 | 328 |
| 31 | 30 52 | 51 | 147 | 30 01 | 51 | 148 | 29 10 | 51 | 148 | 28 19 | 51 | 149 | 27 28 | 51 | 150 | 329 |
| 30 | 31 21 | −51 | 148 | 30 30 | −52 | 149 | 29 38 | −51 | 149 | 28 47 | −52 | 150 | 27 55 | −51 | 150 | 330 |
| 29 | 31 49 | 52 | 149 | 30 57 | 51 | 150 | 30 06 | 52 | 150 | 29 14 | 52 | 151 | 28 22 | 53 | 151 | 331 |
| 28 | 32 17 | 53 | 150 | 31 24 | 52 | 150 | 30 32 | 52 | 151 | 29 40 | 53 | 152 | 28 47 | 52 | 152 | 332 |
| 27 | 32 43 | 52 | 151 | 31 51 | 53 | 151 | 30 58 | 53 | 152 | 30 05 | 53 | 152 | 29 12 | 53 | 153 | 333 |
| 26 | 33 09 | 53 | 152 | 32 16 | 53 | 152 | 31 23 | 53 | 153 | 30 30 | 54 | 153 | 29 36 | 53 | 154 | 334 |
| 25 | 33 34 | −53 | 153 | 32 41 | −54 | 153 | 31 47 | −53 | 154 | 30 54 | −54 | 154 | 30 00 | −54 | 155 | 335 |
| 24 | 33 59 | 54 | 154 | 33 05 | 54 | 154 | 32 11 | 54 | 155 | 31 17 | 55 | 155 | 30 22 | 54 | 156 | 336 |
| 23 | 34 22 | 54 | 155 | 33 28 | 54 | 155 | 32 34 | 55 | 156 | 31 39 | 55 | 156 | 30 44 | 55 | 157 | 337 |
| 22 | 34 45 | 55 | 156 | 33 50 | 55 | 156 | 32 55 | 55 | 157 | 32 00 | 55 | 157 | 31 05 | 55 | 158 | 338 |
| 21 | 35 07 | 55 | 157 | 34 12 | 56 | 157 | 33 16 | 55 | 158 | 32 21 | 56 | 158 | 31 25 | 55 | 159 | 339 |
| 20 | 35 28 | −56 | 158 | 34 32 | −55 | 158 | 33 37 | −56 | 159 | 32 41 | −56 | 159 | 31 45 | −56 | 159 | 340 |
| 19 | 35 48 | 56 | 159 | 34 52 | 56 | 159 | 33 56 | 56 | 160 | 33 00 | 57 | 160 | 32 03 | 56 | 160 | 341 |
| 18 | 36 07 | 56 | 160 | 35 11 | 57 | 160 | 34 14 | 56 | 161 | 33 18 | 57 | 161 | 32 21 | 57 | 161 | 342 |
| 17 | 36 25 | 56 | 161 | 35 29 | 57 | 161 | 34 32 | 57 | 162 | 33 35 | 57 | 162 | 32 38 | 57 | 162 | 343 |
| 16 | 36 43 | 57 | 162 | 35 46 | 58 | 162 | 34 48 | 57 | 163 | 33 51 | 57 | 163 | 32 54 | 58 | 163 | 344 |
| 15 | 36 59 | −57 | 163 | 36 02 | −58 | 163 | 35 04 | −58 | 164 | 34 06 | −57 | 164 | 33 09 | −58 | 164 | 345 |
| 14 | 37 14 | 57 | 164 | 36 17 | 58 | 164 | 35 19 | 58 | 165 | 34 21 | 58 | 165 | 33 23 | 58 | 165 | 346 |
| 13 | 37 29 | 58 | 165 | 36 31 | 58 | 165 | 35 33 | 59 | 166 | 34 34 | 58 | 166 | 33 36 | 58 | 166 | 347 |
| 12 | 37 42 | 58 | 166 | 36 44 | 59 | 167 | 35 45 | 58 | 167 | 34 47 | 59 | 167 | 33 48 | 58 | 167 | 348 |
| 11 | 37 54 | 58 | 167 | 36 56 | 59 | 168 | 35 57 | 59 | 168 | 34 58 | 58 | 168 | 34 00 | 59 | 168 | 349 |
| 10 | 38 06 | −59 | 169 | 37 07 | −59 | 169 | 36 08 | −59 | 169 | 35 09 | −59 | 169 | 34 10 | −59 | 169 | 350 |
| 9 | 38 16 | 59 | 170 | 37 17 | 59 | 170 | 36 18 | 59 | 170 | 35 19 | 59 | 170 | 34 20 | 60 | 171 | 351 |
| 8 | 38 25 | 59 | 171 | 37 26 | 59 | 171 | 36 27 | 60 | 171 | 35 27 | 59 | 171 | 34 28 | 59 | 172 | 352 |
| 7 | 38 33 | 59 | 172 | 37 34 | 60 | 172 | 36 34 | 59 | 172 | 35 35 | 59 | 172 | 34 36 | 60 | 173 | 353 |
| 6 | 38 40 | 59 | 173 | 37 41 | 60 | 173 | 36 41 | 59 | 173 | 35 42 | 60 | 174 | 34 42 | 60 | 174 | 354 |
| 5 | 38 46 | −59 | 174 | 37 47 | −60 | 174 | 36 47 | −60 | 174 | 35 47 | −59 | 175 | 34 48 | −60 | 175 | 355 |
| 4 | 38 51 | 60 | 175 | 37 51 | 59 | 175 | 36 52 | 60 | 176 | 35 52 | 60 | 176 | 34 52 | 60 | 176 | 356 |
| 3 | 38 55 | 60 | 177 | 37 55 | 60 | 177 | 36 55 | 60 | 177 | 35 55 | 59 | 177 | 34 56 | 60 | 177 | 357 |
| 2 | 38 58 | 60 | 178 | 37 58 | 60 | 178 | 36 58 | 60 | 178 | 35 58 | 60 | 178 | 34 58 | 60 | 178 | 358 |
| 1 | 39 00 | 60 | 179 | 38 00 | 60 | 179 | 37 00 | 60 | 179 | 36 00 | 60 | 179 | 35 00 | 60 | 179 | 359 |
| 0 | 39 00 | −60 | 180 | 38 00 | −60 | 180 | 37 00 | −60 | 180 | 36 00 | −60 | 180 | 35 00 | −60 | 180 | 360 |

S. Lat. {LHA greater than 180°.......Zn=180−Z
{LHA less than 180°...........Zn=180+Z

DECLINATION (15°-29°) CONTRARY NAME TO LATITUDE

LAT 26°

N. Lat. {LHA greater than 180°........ Zn=Z / LHA less than 180°........... Zn=360−Z}

# DECLINATION (0°-14°) SAME NAME AS LATITUDE

## LAT 27°

| LHA | 0° Hc | d | Z | 1° Hc | d | Z | 2° Hc | d | Z | 3° Hc | d | Z | 4° Hc | d | Z | 5° Hc | d | Z | 6° Hc | d | Z | 7° Hc | d | Z |
|---|---|---|---|---|---|---|---|---|---|---|---|---|---|---|---|---|---|---|---|---|---|---|---|---|
| | ° ′ | ′ | ° | ° ′ | ′ | ° | ° ′ | ′ | ° | ° ′ | ′ | ° | ° ′ | ′ | ° | ° ′ | ′ | ° | ° ′ | ′ | ° | ° ′ | ′ | ° |
| 0 | 63 00 | +60 | 180 | 64 00 | +60 | 180 | 65 00 | +60 | 180 | 66 00 | +60 | 180 | 67 00 | +60 | 180 | 68 00 | +60 | 180 | 69 00 | +60 | 180 | 70 00 | +60 | 180 |
| 1 | 62 59 | 60 | 178 | 63 59 | 60 | 178 | 64 59 | 60 | 178 | 65 59 | 60 | 178 | 66 59 | 60 | 177 | 67 59 | 60 | 177 | 68 59 | 60 | 177 | 69 59 | 60 | 177 |
| 2 | 62 56 | 60 | 176 | 63 56 | 60 | 175 | 64 56 | 59 | 175 | 65 55 | 60 | 175 | 66 55 | 60 | 175 | 67 55 | 60 | 175 | 68 55 | 60 | 175 | 69 55 | 59 | 174 |
| 3 | 62 51 | 60 | 173 | 63 51 | 59 | 173 | 64 50 | 60 | 173 | 65 50 | 59 | 173 | 66 49 | 60 | 172 | 67 49 | 59 | 172 | 68 48 | 60 | 172 | 69 48 | 59 | 171 |
| 4 | 62 44 | 59 | 171 | 63 43 | 60 | 171 | 64 43 | 59 | 171 | 65 42 | 59 | 170 | 66 41 | 59 | 170 | 67 40 | 60 | 170 | 68 40 | 59 | 169 | 69 39 | 59 | 169 |
| 5 | 62 35 | +59 | 169 | 63 34 | +59 | 169 | 64 33 | +59 | 168 | 65 32 | +59 | 168 | 66 31 | +58 | 167 | 67 29 | +59 | 167 | 68 28 | +59 | 166 | 69 27 | +58 | 166 |
| 6 | 62 23 | 59 | 167 | 63 22 | 59 | 167 | 64 21 | 58 | 166 | 65 19 | 59 | 166 | 66 18 | 58 | 165 | 67 16 | 58 | 164 | 68 14 | 58 | 164 | 69 12 | 58 | 163 |
| 7 | 62 10 | 59 | 165 | 63 09 | 58 | 164 | 64 07 | 58 | 164 | 65 05 | 58 | 163 | 66 03 | 58 | 163 | 67 01 | 57 | 162 | 67 58 | 57 | 161 | 68 55 | 58 | 160 |
| 8 | 61 56 | 57 | 163 | 62 53 | 58 | 162 | 63 51 | 58 | 162 | 64 49 | 57 | 161 | 65 46 | 57 | 160 | 66 43 | 57 | 160 | 67 40 | 56 | 159 | 68 36 | 57 | 158 |
| 9 | 61 39 | 57 | 161 | 62 36 | 57 | 160 | 63 33 | 57 | 159 | 64 30 | 57 | 159 | 65 27 | 56 | 158 | 66 23 | 56 | 157 | 67 19 | 56 | 156 | 68 15 | 55 | 155 |
| 10 | 61 20 | +57 | 159 | 62 17 | +57 | 158 | 63 14 | +56 | 157 | 64 10 | +56 | 157 | 65 06 | +55 | 156 | 66 01 | +56 | 155 | 66 57 | +55 | 154 | 67 52 | +54 | 153 |
| 11 | 61 00 | 56 | 157 | 61 56 | 56 | 156 | 62 52 | 56 | 155 | 63 48 | 55 | 154 | 64 43 | 55 | 154 | 65 38 | 54 | 153 | 66 32 | 54 | 152 | 67 26 | 54 | 150 |
| 12 | 60 38 | 56 | 155 | 61 34 | 55 | 154 | 62 29 | 55 | 153 | 63 24 | 54 | 152 | 64 18 | 54 | 151 | 65 12 | 54 | 150 | 66 06 | 53 | 149 | 66 59 | 52 | 148 |
| 13 | 60 15 | 55 | 153 | 61 10 | 54 | 152 | 62 04 | 54 | 151 | 62 58 | 54 | 150 | 63 52 | 53 | 149 | 64 45 | 53 | 148 | 65 38 | 52 | 147 | 66 30 | 51 | 146 |
| 14 | 59 50 | 54 | 151 | 60 44 | 54 | 150 | 61 38 | 53 | 149 | 62 31 | 53 | 148 | 63 24 | 52 | 147 | 64 16 | 52 | 146 | 65 08 | 51 | 145 | 65 59 | 51 | 144 |
| 15 | 59 23 | +54 | 150 | 60 17 | +53 | 149 | 61 10 | +52 | 148 | 62 02 | +52 | 147 | 62 54 | +52 | 146 | 63 46 | +50 | 144 | 64 36 | +51 | 143 | 65 27 | +49 | 142 |
| 16 | 58 56 | 52 | 148 | 59 48 | 52 | 147 | 60 40 | 52 | 146 | 61 32 | 51 | 145 | 62 23 | 51 | 144 | 63 14 | 50 | 142 | 64 04 | 49 | 141 | 64 53 | 49 | 140 |
| 17 | 58 26 | 52 | 146 | 59 18 | 52 | 145 | 60 10 | 50 | 144 | 61 00 | 51 | 143 | 61 51 | 49 | 142 | 62 40 | 49 | 141 | 63 29 | 49 | 139 | 64 18 | 47 | 138 |
| 18 | 57 56 | 51 | 144 | 58 47 | 50 | 143 | 59 37 | 51 | 142 | 60 28 | 49 | 141 | 61 17 | 49 | 140 | 62 06 | 48 | 139 | 62 54 | 47 | 138 | 63 41 | 47 | 136 |
| 19 | 57 24 | 50 | 143 | 58 14 | 50 | 142 | 59 04 | 49 | 141 | 59 53 | 49 | 140 | 60 42 | 48 | 138 | 61 30 | 47 | 137 | 62 17 | 47 | 136 | 63 04 | 46 | 135 |
| 20 | 56 51 | +50 | 141 | 57 41 | +49 | 140 | 58 30 | +48 | 139 | 59 18 | +48 | 138 | 60 06 | +47 | 137 | 60 53 | +47 | 136 | 61 40 | +45 | 134 | 62 25 | +45 | 133 |
| 21 | 56 17 | 49 | 140 | 57 06 | 48 | 139 | 57 54 | 48 | 138 | 58 42 | 47 | 137 | 59 29 | 46 | 135 | 60 15 | 46 | 134 | 61 01 | 44 | 133 | 61 45 | 44 | 131 |
| 22 | 55 42 | 48 | 138 | 56 30 | 48 | 137 | 57 18 | 47 | 136 | 58 05 | 46 | 135 | 58 51 | 45 | 134 | 59 36 | 45 | 133 | 60 21 | 44 | 131 | 61 05 | 43 | 130 |
| 23 | 55 06 | 48 | 137 | 55 54 | 46 | 136 | 56 40 | 46 | 135 | 57 26 | 46 | 134 | 58 12 | 44 | 132 | 58 56 | 44 | 131 | 59 40 | 43 | 130 | 60 23 | 42 | 128 |
| 24 | 54 29 | 47 | 136 | 55 16 | 46 | 135 | 56 02 | 45 | 133 | 56 47 | 45 | 132 | 57 32 | 44 | 131 | 58 16 | 43 | 130 | 58 59 | 42 | 128 | 59 41 | 41 | 127 |
| 25 | 53 51 | +46 | 134 | 54 37 | +45 | 133 | 55 22 | +45 | 132 | 56 07 | +44 | 131 | 56 51 | +43 | 130 | 57 34 | +42 | 128 | 58 16 | +42 | 127 | 58 58 | +40 | 126 |
| 26 | 53 13 | 45 | 133 | 53 58 | 44 | 132 | 54 42 | 44 | 131 | 55 26 | 43 | 130 | 56 09 | 43 | 128 | 56 52 | 41 | 127 | 57 33 | 41 | 126 | 58 14 | 40 | 124 |
| 27 | 52 33 | 45 | 132 | 53 18 | 43 | 131 | 54 01 | 44 | 129 | 54 45 | 42 | 128 | 55 27 | 42 | 127 | 56 09 | 41 | 126 | 56 50 | 39 | 124 | 57 29 | 40 | 123 |
| 28 | 51 53 | 44 | 131 | 52 37 | 43 | 129 | 53 20 | 42 | 128 | 54 02 | 42 | 127 | 54 44 | 41 | 126 | 55 25 | 40 | 125 | 56 05 | 39 | 123 | 56 44 | 39 | 122 |
| 29 | 51 12 | 43 | 130 | 51 55 | 42 | 128 | 52 37 | 42 | 127 | 53 19 | 41 | 126 | 54 00 | 41 | 125 | 54 41 | 39 | 123 | 55 20 | 39 | 122 | 55 59 | 37 | 121 |
| 30 | 50 30 | +43 | 128 | 51 13 | +41 | 127 | 51 54 | +42 | 126 | 52 36 | +40 | 125 | 53 16 | +40 | 124 | 53 56 | +38 | 122 | 54 34 | +38 | 121 | 55 12 | +37 | 120 |
| 31 | 49 48 | 42 | 127 | 50 30 | 41 | 126 | 51 11 | 40 | 125 | 51 51 | 40 | 124 | 52 31 | 39 | 122 | 53 10 | 38 | 121 | 53 48 | 38 | 120 | 54 26 | 36 | 119 |
| 32 | 49 05 | 41 | 126 | 49 46 | 41 | 125 | 50 27 | 40 | 124 | 51 07 | 39 | 123 | 51 46 | 38 | 121 | 52 24 | 38 | 120 | 53 02 | 36 | 119 | 53 38 | 36 | 118 |
| 33 | 48 21 | 41 | 125 | 49 02 | 40 | 124 | 49 42 | 39 | 123 | 50 21 | 39 | 122 | 51 00 | 38 | 120 | 51 38 | 37 | 119 | 52 15 | 36 | 118 | 52 51 | 35 | 117 |
| 34 | 47 37 | 40 | 124 | 48 17 | 40 | 123 | 48 57 | 38 | 122 | 49 35 | 38 | 121 | 50 13 | 38 | 119 | 50 51 | 36 | 118 | 51 27 | 36 | 117 | 52 03 | 35 | 116 |
| 35 | 46 53 | +39 | 123 | 47 32 | +39 | 122 | 48 11 | +38 | 121 | 48 49 | +38 | 120 | 49 27 | +36 | 118 | 50 03 | +36 | 117 | 50 39 | +35 | 116 | 51 14 | +35 | 115 |
| 36 | 46 07 | 39 | 122 | 46 46 | 39 | 121 | 47 25 | 37 | 120 | 48 02 | 37 | 119 | 48 39 | 37 | 117 | 49 16 | 35 | 116 | 49 51 | 35 | 115 | 50 26 | 33 | 114 |
| 37 | 45 22 | 38 | 121 | 46 00 | 38 | 120 | 46 38 | 37 | 119 | 47 15 | 37 | 118 | 47 52 | 35 | 117 | 48 27 | 35 | 115 | 49 02 | 34 | 114 | 49 36 | 34 | 113 |
| 38 | 44 36 | 38 | 120 | 45 14 | 37 | 119 | 45 51 | 37 | 118 | 46 28 | 36 | 117 | 47 04 | 35 | 116 | 47 39 | 34 | 114 | 48 13 | 34 | 113 | 48 47 | 33 | 112 |
| 39 | 43 49 | 38 | 119 | 44 27 | 37 | 118 | 45 04 | 36 | 117 | 45 40 | 35 | 116 | 46 15 | 35 | 115 | 46 50 | 34 | 114 | 47 24 | 33 | 112 | 47 57 | 33 | 111 |
| 40 | 43 03 | +37 | 118 | 43 40 | +36 | 117 | 44 16 | +36 | 116 | 44 52 | +35 | 115 | 45 27 | +34 | 114 | 46 01 | +34 | 113 | 46 35 | +32 | 112 | 47 07 | +32 | 110 |
| 41 | 42 15 | 37 | 118 | 42 52 | 36 | 117 | 43 28 | 35 | 115 | 44 03 | 35 | 114 | 44 38 | 34 | 113 | 45 12 | 33 | 112 | 45 45 | 32 | 111 | 46 17 | 32 | 110 |
| 42 | 41 28 | 36 | 117 | 42 04 | 35 | 116 | 42 39 | 35 | 115 | 43 14 | 34 | 114 | 43 48 | 34 | 112 | 44 22 | 33 | 111 | 44 55 | 32 | 110 | 45 27 | 31 | 109 |
| 43 | 40 40 | 36 | 116 | 41 16 | 35 | 115 | 41 51 | 34 | 114 | 42 25 | 34 | 113 | 42 59 | 33 | 112 | 43 32 | 32 | 110 | 44 04 | 32 | 109 | 44 36 | 31 | 108 |
| 44 | 39 52 | 35 | 115 | 40 27 | 35 | 114 | 41 02 | 34 | 113 | 41 36 | 33 | 112 | 42 09 | 33 | 111 | 42 42 | 32 | 110 | 43 14 | 31 | 109 | 43 45 | 30 | 107 |
| 45 | 39 03 | +35 | 114 | 39 38 | +34 | 113 | 40 12 | +34 | 112 | 40 46 | +33 | 111 | 41 19 | +32 | 110 | 41 51 | +32 | 109 | 42 23 | +31 | 108 | 42 54 | +30 | 107 |
| 46 | 38 14 | 35 | 114 | 38 49 | 34 | 113 | 39 23 | 33 | 112 | 39 56 | 33 | 111 | 40 29 | 31 | 109 | 41 00 | 32 | 108 | 41 32 | 31 | 107 | 42 03 | 29 | 106 |
| 47 | 37 25 | 34 | 113 | 37 59 | 34 | 112 | 38 33 | 33 | 111 | 39 06 | 32 | 110 | 39 38 | 32 | 109 | 40 10 | 31 | 108 | 40 41 | 30 | 107 | 41 11 | 30 | 105 |
| 48 | 36 36 | 34 | 112 | 37 10 | 33 | 111 | 37 43 | 32 | 110 | 38 15 | 32 | 109 | 38 47 | 32 | 108 | 39 19 | 30 | 107 | 39 49 | 30 | 106 | 40 19 | 30 | 105 |
| 49 | 35 46 | 34 | 112 | 36 20 | 32 | 111 | 36 52 | 33 | 110 | 37 25 | 31 | 108 | 37 56 | 31 | 107 | 38 27 | 31 | 106 | 38 58 | 30 | 105 | 39 28 | 29 | 104 |
| 50 | 34 56 | +33 | 111 | 35 29 | +33 | 110 | 36 02 | +32 | 109 | 36 34 | +31 | 108 | 37 05 | +31 | 107 | 37 36 | +30 | 106 | 38 06 | +30 | 105 | 38 36 | +29 | 103 |
| 51 | 34 06 | 33 | 110 | 34 39 | 32 | 109 | 35 11 | 32 | 108 | 35 43 | 31 | 107 | 36 14 | 30 | 106 | 36 44 | 30 | 105 | 37 14 | 30 | 104 | 37 44 | 28 | 103 |
| 52 | 33 16 | 32 | 110 | 33 48 | 32 | 109 | 34 20 | 32 | 108 | 34 52 | 30 | 107 | 35 22 | 31 | 105 | 35 53 | 29 | 104 | 36 22 | 29 | 103 | 36 51 | 29 | 102 |
| 53 | 32 26 | 32 | 109 | 32 58 | 31 | 108 | 33 29 | 31 | 107 | 34 00 | 31 | 106 | 34 31 | 30 | 105 | 35 01 | 29 | 104 | 35 30 | 29 | 103 | 35 59 | 28 | 102 |
| 54 | 31 35 | 32 | 108 | 32 07 | 31 | 107 | 32 38 | 31 | 106 | 33 09 | 30 | 105 | 33 39 | 30 | 104 | 34 09 | 29 | 103 | 34 38 | 29 | 102 | 35 07 | 28 | 101 |
| 55 | 30 44 | +32 | 108 | 31 16 | +31 | 107 | 31 47 | +30 | 106 | 32 17 | +30 | 105 | 32 47 | +30 | 104 | 33 17 | +29 | 103 | 33 46 | +28 | 102 | 34 14 | +28 | 100 |
| 56 | 29 53 | 31 | 107 | 30 24 | 31 | 106 | 30 55 | 30 | 105 | 31 25 | 30 | 104 | 31 55 | 29 | 103 | 32 24 | 29 | 102 | 32 53 | 28 | 101 | 33 21 | 28 | 100 |
| 57 | 29 02 | 31 | 106 | 29 33 | 30 | 105 | 30 03 | 30 | 105 | 30 33 | 30 | 103 | 31 03 | 29 | 102 | 31 32 | 29 | 101 | 32 01 | 28 | 100 | 32 29 | 27 | 99 |
| 58 | 28 11 | 30 | 106 | 28 41 | 30 | 105 | 29 11 | 30 | 104 | 29 41 | 30 | 103 | 30 11 | 29 | 102 | 30 40 | 28 | 101 | 31 08 | 28 | 100 | 31 36 | 27 | 99 |
| 59 | 27 19 | 30 | 105 | 27 49 | 31 | 104 | 28 20 | 29 | 103 | 28 49 | 29 | 102 | 29 18 | 29 | 101 | 29 47 | 28 | 100 | 30 15 | 28 | 99 | 30 43 | 27 | 98 |
| 60 | 26 27 | +31 | 105 | 26 58 | +29 | 104 | 27 27 | +30 | 103 | 27 57 | +29 | 102 | 28 26 | +28 | 101 | 28 54 | +29 | 100 | 29 23 | +27 | 99 | 29 50 | +27 | 98 |
| 61 | 25 36 | 30 | 104 | 26 06 | 29 | 103 | 26 35 | 29 | 102 | 27 04 | 29 | 101 | 27 33 | 29 | 100 | 28 02 | 28 | 99 | 28 30 | 27 | 98 | 28 57 | 27 | 97 |
| 62 | 24 44 | 30 | 104 | 25 14 | 29 | 103 | 25 43 | 29 | 102 | 26 12 | 29 | 101 | 26 41 | 28 | 100 | 27 09 | 28 | 99 | 27 37 | 27 | 98 | 28 04 | 27 | 97 |
| 63 | 23 52 | 29 | 103 | 24 21 | 30 | 102 | 24 51 | 28 | 101 | 25 19 | 29 | 100 | 25 48 | 28 | 99 | 26 16 | 28 | 98 | 26 44 | 27 | 97 | 27 11 | 27 | 96 |
| 64 | 23 00 | 29 | 103 | 23 29 | 29 | 102 | 23 58 | 29 | 101 | 24 27 | 28 | 100 | 24 55 | 28 | 99 | 25 23 | 28 | 98 | 25 51 | 27 | 97 | 26 18 | 27 | 96 |
| 65 | 22 07 | +30 | 102 | 22 37 | +28 | 101 | 23 05 | +29 | 100 | 23 34 | +28 | 99 | 24 02 | +28 | 98 | 24 30 | +28 | 97 | 24 58 | +27 | 96 | 25 25 | +26 | 95 |
| 66 | 21 15 | 29 | 101 | 21 44 | 29 | 101 | 22 13 | 28 | 100 | 22 41 | 28 | 99 | 23 09 | 28 | 98 | 23 37 | 27 | 97 | 24 04 | 27 | 96 | 24 31 | 27 | 95 |
| 67 | 20 22 | 29 | 101 | 20 51 | 29 | 100 | 21 20 | 28 | 99 | 21 48 | 28 | 98 | 22 16 | 28 | 97 | 22 44 | 27 | 96 | 23 11 | 27 | 95 | 23 38 | 27 | 94 |
| 68 | 19 30 | 29 | 100 | 19 59 | 28 | 100 | 20 27 | 28 | 99 | 20 55 | 28 | 98 | 21 23 | 28 | 97 | 21 51 | 27 | 96 | 22 18 | 27 | 95 | 22 45 | 26 | 94 |
| 69 | 18 37 | 29 | 100 | 19 06 | 28 | 99 | 19 34 | 28 | 98 | 20 02 | 28 | 97 | 20 30 | 27 | 96 | 20 57 | 28 | 95 | 21 25 | 26 | 94 | 21 51 | 27 | 93 |

| LHA | 8° Hc | d | Z | 9° Hc | d | Z | 10° Hc | d | Z | 11° Hc | d | Z | 12° Hc | d | Z | 13° Hc | d | Z | 14° Hc | d | Z | LHA |
|---|---|---|---|---|---|---|---|---|---|---|---|---|---|---|---|---|---|---|---|---|---|---|
| | ° ′ | ′ | ° | ° ′ | ′ | ° | ° ′ | ′ | ° | ° ′ | ′ | ° | ° ′ | ′ | ° | ° ′ | ′ | ° | ° ′ | ′ | ° | |
| 0 | 71 00 | +60 | 180 | 72 00 | +60 | 180 | 73 00 | +60 | 180 | 74 00 | +60 | 180 | 75 00 | +60 | 180 | 76 00 | +60 | 180 | 77 00 | +60 | 180 | 360 |
| 1 | 70 59 | 60 | 177 | 71 59 | 59 | 177 | 72 58 | 60 | 177 | 73 58 | 60 | 176 | 74 58 | 60 | 176 | 75 58 | 60 | 176 | 76 58 | 60 | 176 | 359 |
| 2 | 70 54 | 60 | 174 | 71 54 | 60 | 174 | 72 54 | 59 | 173 | 73 53 | 60 | 173 | 74 53 | 60 | 173 | 75 53 | 59 | 172 | 76 52 | 59 | 171 | 358 |
| 3 | 70 47 | 60 | 171 | 71 47 | 59 | 171 | 72 46 | 59 | 170 | 73 45 | 59 | 169 | 74 44 | 59 | 169 | 75 43 | 59 | 168 | 76 42 | 59 | 167 | 357 |
| 4 | 70 38 | 58 | 168 | 71 36 | 59 | 167 | 72 35 | 59 | 167 | 73 34 | 58 | 166 | 74 32 | 59 | 165 | 75 31 | 57 | 164 | 76 28 | 58 | 163 | 356 |
| 5 | 70 25 | +58 | 165 | 71 23 | +58 | 164 | 72 21 | +58 | 164 | 73 19 | +58 | 163 | 74 17 | +57 | 162 | 75 14 | +57 | 161 | 76 11 | +57 | 159 | 355 |
| 6 | 70 10 | 58 | 162 | 71 08 | 57 | 161 | 72 05 | 57 | 161 | 73 02 | 57 | 159 | 73 59 | 56 | 158 | 74 55 | 56 | 157 | 75 51 | 55 | 156 | 354 |
| 7 | 69 53 | 56 | 160 | 70 49 | 57 | 159 | 71 46 | 56 | 157 | 72 42 | 55 | 156 | 73 37 | 56 | 155 | 74 33 | 54 | 154 | 75 27 | 54 | 152 | 353 |
| 8 | 69 33 | 56 | 157 | 70 29 | 55 | 156 | 71 24 | 55 | 155 | 72 19 | 55 | 153 | 73 14 | 53 | 152 | 74 07 | 53 | 150 | 75 00 | 53 | 149 | 352 |
| 9 | 69 10 | 55 | 154 | 70 05 | 55 | 153 | 71 00 | 54 | 152 | 71 54 | 53 | 150 | 72 47 | 53 | 149 | 73 40 | 51 | 147 | 74 31 | 51 | 145 | 351 |
| 10 | 68 46 | +54 | 152 | 69 40 | +53 | 150 | 70 33 | +53 | 149 | 71 26 | +52 | 148 | 72 18 | +52 | 146 | 73 10 | +50 | 144 | 74 00 | +49 | 142 | 350 |
| 11 | 68 20 | 53 | 149 | 69 13 | 52 | 148 | 70 05 | 52 | 147 | 70 57 | 50 | 145 | 71 47 | 50 | 143 | 72 37 | 49 | 142 | 73 26 | 48 | 140 | 349 |
| 12 | 67 51 | 52 | 147 | 68 43 | 52 | 146 | 69 35 | 50 | 144 | 70 25 | 50 | 143 | 71 15 | 48 | 141 | 72 03 | 47 | 139 | 72 50 | 46 | 137 | 348 |
| 13 | 67 21 | 51 | 145 | 68 12 | 50 | 143 | 69 02 | 50 | 142 | 69 52 | 48 | 140 | 70 40 | 47 | 138 | 71 27 | 46 | 137 | 72 13 | 45 | 134 | 347 |
| 14 | 66 50 | 49 | 143 | 67 39 | 49 | 141 | 68 28 | 48 | 140 | 69 16 | 48 | 138 | 70 04 | 45 | 136 | 70 49 | 45 | 134 | 71 34 | 43 | 132 | 346 |
| 15 | 66 16 | +49 | 140 | 67 05 | +48 | 139 | 67 53 | +47 | 137 | 68 40 | +46 | 136 | 69 26 | +44 | 134 | 70 10 | +44 | 132 | 70 54 | +41 | 130 | 345 |
| 16 | 65 42 | 47 | 139 | 66 29 | 47 | 137 | 67 16 | 46 | 135 | 68 02 | 45 | 134 | 68 47 | 43 | 132 | 69 30 | 42 | 130 | 70 12 | 41 | 128 | 344 |
| 17 | 65 05 | 47 | 137 | 65 52 | 46 | 135 | 66 38 | 45 | 134 | 67 23 | 43 | 132 | 68 06 | 42 | 130 | 68 48 | 41 | 128 | 69 29 | 40 | 126 | 343 |
| 18 | 64 28 | 46 | 135 | 65 14 | 44 | 133 | 65 58 | 44 | 132 | 66 42 | 43 | 130 | 67 25 | 41 | 128 | 68 06 | 40 | 126 | 68 46 | 38 | 124 | 342 |
| 19 | 63 50 | 44 | 133 | 64 34 | 44 | 132 | 65 18 | 43 | 130 | 66 01 | 41 | 128 | 66 42 | 40 | 126 | 67 22 | 39 | 125 | 68 01 | 37 | 123 | 341 |
| 20 | 63 10 | +44 | 131 | 63 54 | +42 | 130 | 64 36 | +42 | 128 | 65 18 | +41 | 127 | 65 59 | +39 | 125 | 66 38 | +37 | 123 | 67 15 | +36 | 121 | 340 |
| 21 | 62 29 | 43 | 130 | 63 12 | 42 | 128 | 63 54 | 41 | 127 | 64 35 | 39 | 125 | 65 14 | 38 | 123 | 65 52 | 37 | 121 | 66 29 | 35 | 119 | 339 |
| 22 | 61 48 | 42 | 128 | 62 30 | 41 | 127 | 63 11 | 40 | 125 | 63 51 | 38 | 124 | 64 29 | 37 | 122 | 65 06 | 36 | 120 | 65 42 | 34 | 118 | 338 |
| 23 | 61 05 | 42 | 127 | 61 47 | 40 | 126 | 62 27 | 39 | 124 | 63 06 | 37 | 122 | 63 43 | 37 | 120 | 64 20 | 35 | 119 | 64 55 | 33 | 117 | 337 |
| 24 | 60 22 | 41 | 125 | 61 03 | 39 | 124 | 61 42 | 38 | 122 | 62 20 | 37 | 121 | 62 57 | 35 | 119 | 63 32 | 35 | 117 | 64 07 | 32 | 115 | 336 |
| 25 | 59 38 | +40 | 124 | 60 18 | +38 | 123 | 60 56 | +38 | 121 | 61 34 | +36 | 119 | 62 10 | +35 | 118 | 62 45 | +33 | 116 | 63 18 | +32 | 114 | 335 |
| 26 | 58 54 | 39 | 123 | 59 33 | 37 | 121 | 60 10 | 37 | 120 | 60 47 | 35 | 118 | 61 22 | 34 | 117 | 61 56 | 33 | 115 | 62 29 | 31 | 113 | 334 |
| 27 | 58 09 | 38 | 122 | 58 47 | 37 | 120 | 59 24 | 35 | 119 | 59 59 | 35 | 117 | 60 34 | 34 | 115 | 61 08 | 32 | 114 | 61 40 | 30 | 112 | 333 |
| 28 | 57 23 | 37 | 120 | 58 00 | 36 | 119 | 58 36 | 36 | 117 | 59 12 | 34 | 116 | 59 46 | 32 | 114 | 60 18 | 32 | 113 | 60 50 | 30 | 111 | 332 |
| 29 | 56 36 | 37 | 119 | 57 13 | 36 | 118 | 57 49 | 34 | 116 | 58 23 | 34 | 115 | 58 57 | 32 | 113 | 59 29 | 31 | 112 | 60 00 | 29 | 110 | 331 |
| 30 | 55 49 | +36 | 118 | 56 25 | +36 | 117 | 57 01 | +33 | 115 | 57 34 | +33 | 114 | 58 07 | +32 | 112 | 58 39 | +30 | 111 | 59 09 | +29 | 109 | 330 |
| 31 | 55 02 | 36 | 117 | 55 38 | 34 | 116 | 56 12 | 33 | 114 | 56 45 | 33 | 113 | 57 18 | 31 | 111 | 57 49 | 30 | 110 | 58 19 | 28 | 108 | 329 |
| 32 | 54 14 | 35 | 116 | 54 49 | 34 | 115 | 55 23 | 33 | 113 | 55 56 | 32 | 112 | 56 28 | 30 | 110 | 56 58 | 30 | 109 | 57 28 | 28 | 107 | 328 |
| 33 | 53 26 | 34 | 115 | 54 00 | 34 | 114 | 54 34 | 32 | 112 | 55 06 | 31 | 111 | 55 37 | 30 | 109 | 56 07 | 29 | 108 | 56 36 | 28 | 106 | 327 |
| 34 | 52 38 | 33 | 114 | 53 11 | 33 | 113 | 53 44 | 32 | 111 | 54 16 | 31 | 110 | 54 47 | 29 | 109 | 55 16 | 29 | 107 | 55 45 | 27 | 105 | 326 |
| 35 | 51 49 | +33 | 113 | 52 22 | +32 | 112 | 52 54 | +32 | 111 | 53 26 | +30 | 109 | 53 56 | +29 | 108 | 54 25 | +28 | 106 | 54 53 | +27 | 105 | 325 |
| 36 | 50 59 | 33 | 112 | 51 32 | 32 | 111 | 52 04 | 31 | 110 | 52 35 | 30 | 108 | 53 05 | 29 | 107 | 53 34 | 27 | 105 | 54 01 | 27 | 104 | 324 |
| 37 | 50 10 | 32 | 112 | 50 42 | 32 | 110 | 51 14 | 30 | 109 | 51 44 | 30 | 108 | 52 14 | 28 | 106 | 52 42 | 27 | 105 | 53 09 | 27 | 103 | 323 |
| 38 | 49 20 | 32 | 111 | 49 52 | 31 | 109 | 50 23 | 30 | 108 | 50 53 | 29 | 107 | 51 22 | 28 | 105 | 51 50 | 27 | 104 | 52 17 | 26 | 102 | 322 |
| 39 | 48 30 | 31 | 110 | 49 01 | 31 | 109 | 49 32 | 30 | 107 | 50 02 | 28 | 106 | 50 30 | 28 | 105 | 50 58 | 27 | 103 | 51 25 | 26 | 102 | 321 |
| 40 | 47 39 | +31 | 109 | 48 10 | +31 | 108 | 48 41 | +29 | 107 | 49 10 | +29 | 105 | 49 39 | +27 | 104 | 50 06 | +27 | 103 | 50 33 | +25 | 101 | 320 |
| 41 | 46 49 | 30 | 108 | 47 19 | 30 | 107 | 47 49 | 29 | 106 | 48 18 | 29 | 105 | 48 47 | 27 | 103 | 49 14 | 26 | 102 | 49 40 | 25 | 100 | 319 |
| 42 | 45 58 | 30 | 108 | 46 28 | 30 | 106 | 46 58 | 29 | 105 | 47 27 | 28 | 104 | 47 55 | 26 | 103 | 48 21 | 26 | 101 | 48 47 | 26 | 100 | 318 |
| 43 | 45 07 | 30 | 107 | 45 37 | 29 | 106 | 46 06 | 29 | 104 | 46 35 | 27 | 103 | 47 02 | 27 | 102 | 47 29 | 26 | 101 | 47 55 | 25 | 99 | 317 |
| 44 | 44 15 | 30 | 106 | 44 45 | 29 | 105 | 45 14 | 29 | 104 | 45 43 | 27 | 103 | 46 10 | 26 | 101 | 46 36 | 26 | 100 | 47 02 | 25 | 99 | 316 |
| 45 | 43 24 | +30 | 106 | 43 54 | +28 | 104 | 44 22 | +28 | 103 | 44 50 | +27 | 102 | 45 17 | +27 | 101 | 45 44 | +25 | 99 | 46 09 | +25 | 98 | 315 |
| 46 | 42 32 | 30 | 105 | 43 02 | 28 | 104 | 43 30 | 28 | 102 | 43 58 | 27 | 101 | 44 25 | 26 | 100 | 44 51 | 25 | 99 | 45 16 | 24 | 97 | 314 |
| 47 | 41 41 | 29 | 104 | 42 10 | 28 | 103 | 42 38 | 27 | 102 | 43 05 | 27 | 101 | 43 32 | 26 | 99 | 43 58 | 25 | 98 | 44 23 | 24 | 97 | 313 |
| 48 | 40 49 | 28 | 104 | 41 17 | 28 | 102 | 41 45 | 28 | 101 | 42 13 | 26 | 100 | 42 39 | 26 | 99 | 43 05 | 25 | 98 | 43 30 | 24 | 96 | 312 |
| 49 | 39 57 | 28 | 103 | 40 25 | 28 | 102 | 40 53 | 27 | 101 | 41 20 | 26 | 99 | 41 46 | 26 | 98 | 42 12 | 25 | 97 | 42 37 | 24 | 96 | 311 |
| 50 | 39 05 | +28 | 102 | 39 33 | +27 | 101 | 40 00 | +27 | 100 | 40 27 | +26 | 99 | 40 53 | +26 | 98 | 41 19 | +25 | 96 | 41 44 | +23 | 95 | 310 |
| 51 | 38 12 | 28 | 102 | 38 40 | 28 | 101 | 39 08 | 26 | 99 | 39 34 | 26 | 98 | 40 00 | 26 | 97 | 40 26 | 24 | 96 | 40 50 | 24 | 95 | 309 |
| 52 | 37 20 | 28 | 101 | 37 48 | 27 | 100 | 38 15 | 26 | 99 | 38 41 | 26 | 98 | 39 07 | 25 | 97 | 39 32 | 25 | 95 | 39 57 | 24 | 94 | 308 |
| 53 | 36 27 | 28 | 101 | 36 55 | 27 | 99 | 37 22 | 26 | 98 | 37 48 | 26 | 97 | 38 14 | 25 | 96 | 38 39 | 25 | 95 | 39 04 | 23 | 94 | 307 |
| 54 | 35 35 | 27 | 100 | 36 02 | 27 | 99 | 36 29 | 26 | 98 | 36 55 | 26 | 97 | 37 21 | 25 | 96 | 37 46 | 24 | 94 | 38 10 | 24 | 93 | 306 |
| 55 | 34 42 | +27 | 99 | 35 09 | +27 | 98 | 35 36 | +26 | 97 | 36 02 | +26 | 96 | 36 28 | +25 | 95 | 36 53 | +24 | 94 | 37 17 | +23 | 93 | 305 |
| 56 | 33 49 | 27 | 99 | 34 16 | 27 | 98 | 34 43 | 26 | 97 | 35 09 | 25 | 96 | 35 34 | 25 | 94 | 35 59 | 24 | 93 | 36 23 | 24 | 92 | 304 |
| 57 | 32 56 | 27 | 98 | 33 23 | 27 | 97 | 33 50 | 26 | 96 | 34 16 | 25 | 95 | 34 41 | 25 | 94 | 35 06 | 24 | 93 | 35 30 | 23 | 92 | 303 |
| 58 | 32 03 | 27 | 98 | 32 30 | 27 | 97 | 32 57 | 26 | 96 | 33 23 | 25 | 95 | 33 48 | 24 | 94 | 34 12 | 25 | 92 | 34 37 | 23 | 91 | 302 |
| 59 | 31 10 | 27 | 97 | 31 37 | 26 | 96 | 32 03 | 26 | 95 | 32 29 | 25 | 94 | 32 54 | 25 | 93 | 33 19 | 24 | 92 | 33 43 | 24 | 91 | 301 |
| 60 | 30 17 | +27 | 97 | 30 44 | +26 | 96 | 31 10 | +26 | 95 | 31 36 | +25 | 94 | 32 01 | +25 | 93 | 32 26 | +24 | 91 | 32 50 | +23 | 90 | 300 |
| 61 | 29 24 | 27 | 96 | 29 51 | 26 | 95 | 30 17 | 26 | 94 | 30 43 | 25 | 93 | 31 08 | 24 | 92 | 31 32 | 24 | 91 | 31 56 | 24 | 90 | 299 |
| 62 | 28 31 | 27 | 96 | 28 58 | 26 | 95 | 29 24 | 25 | 94 | 29 49 | 25 | 93 | 30 14 | 25 | 92 | 30 39 | 24 | 91 | 31 03 | 23 | 89 | 298 |
| 63 | 27 38 | 26 | 95 | 28 04 | 26 | 94 | 28 30 | 26 | 93 | 28 56 | 25 | 92 | 29 21 | 24 | 91 | 29 45 | 24 | 90 | 30 09 | 24 | 89 | 297 |
| 64 | 26 45 | 26 | 95 | 27 11 | 26 | 94 | 27 37 | 25 | 93 | 28 02 | 25 | 92 | 28 27 | 25 | 91 | 28 52 | 24 | 90 | 29 16 | 23 | 89 | 296 |
| 65 | 25 51 | +27 | 94 | 26 18 | +25 | 93 | 26 43 | +26 | 92 | 27 09 | +25 | 91 | 27 34 | +24 | 90 | 27 58 | +24 | 89 | 28 22 | +24 | 88 | 295 |
| 66 | 24 58 | 26 | 94 | 25 24 | 26 | 93 | 25 50 | 25 | 92 | 26 15 | 25 | 91 | 26 40 | 25 | 90 | 27 05 | 24 | 89 | 27 29 | 24 | 88 | 294 |
| 67 | 24 05 | 26 | 93 | 24 31 | 26 | 92 | 24 57 | 25 | 91 | 25 22 | 25 | 90 | 25 47 | 24 | 89 | 26 11 | 25 | 88 | 26 36 | 23 | 87 | 293 |
| 68 | 23 11 | 26 | 93 | 23 37 | 26 | 92 | 24 03 | 25 | 91 | 24 28 | 25 | 90 | 24 53 | 25 | 89 | 25 18 | 24 | 88 | 25 42 | 24 | 87 | 292 |
| 69 | 22 18 | 26 | 92 | 22 44 | 26 | 91 | 23 10 | 25 | 90 | 23 35 | 25 | 89 | 24 00 | 25 | 88 | 24 25 | 24 | 87 | 24 49 | 24 | 86 | 291 |

S. Lat. {LHA greater than 180°........ Zn=180−Z / LHA less than 180°........... Zn=180+Z}

DECLINATION (0°-14°) SAME NAME AS LATITUDE

LAT 27°

N. Lat. {LHA greater than 180°....... Zn=Z; LHA less than 180°........... Zn=360−Z}

# DECLINATION (0°-14°) CONTRARY NAME TO LATITUDE

| LHA | 0° Hc | 0° d | 0° Z | 1° Hc | 1° d | 1° Z | 2° Hc | 2° d | 2° Z | 3° Hc | 3° d | 3° Z | 4° Hc | 4° d | 4° Z |
|---|---|---|---|---|---|---|---|---|---|---|---|---|---|---|---|
| | ° ′ | ′ | ° | ° ′ | ′ | ° | ° ′ | ′ | ° | ° ′ | ′ | ° | ° ′ | ′ | ° |
| 69 | 18 37 | 29 | 100 | 18 08 | 29 | 101 | 17 39 | 29 | 102 | 17 10 | 30 | 103 | 16 40 | 30 | 104 |
| 68 | 19 30 | 29 | 100 | 19 01 | 29 | 101 | 18 32 | 30 | 102 | 18 02 | 30 | 103 | 17 32 | 30 | 104 |
| 67 | 20 22 | 29 | 101 | 19 53 | 29 | 102 | 19 24 | 30 | 103 | 18 54 | 30 | 104 | 18 24 | 30 | 105 |
| 66 | 21 15 | 29 | 101 | 20 46 | 30 | 102 | 20 16 | 30 | 103 | 19 46 | 30 | 104 | 19 16 | 31 | 105 |
| 65 | 22 07 | −29 | 102 | 21 38 | −30 | 103 | 21 08 | −30 | 104 | 20 38 | −31 | 105 | 20 07 | −31 | 106 |
| 64 | 23 00 | 30 | 103 | 22 30 | 30 | 103 | 22 00 | 31 | 104 | 21 29 | 30 | 105 | 20 59 | 31 | 106 |
| 63 | 23 52 | 30 | 103 | 23 22 | 31 | 104 | 22 51 | 30 | 105 | 22 21 | 31 | 106 | 21 50 | 31 | 107 |
| 62 | 24 44 | 30 | 104 | 24 14 | 31 | 105 | 23 43 | 31 | 106 | 23 12 | 31 | 106 | 22 41 | 32 | 107 |
| 61 | 25 36 | 31 | 104 | 25 05 | 31 | 105 | 24 34 | 31 | 106 | 24 03 | 31 | 107 | 23 32 | 32 | 108 |
| 60 | 26 27 | −30 | 105 | 25 57 | −31 | 106 | 25 26 | −32 | 107 | 24 54 | −31 | 108 | 24 23 | −32 | 109 |
| 59 | 27 19 | 31 | 105 | 26 48 | 31 | 106 | 26 17 | 32 | 107 | 25 45 | 32 | 108 | 25 13 | 32 | 109 |
| 58 | 28 11 | 32 | 106 | 27 39 | 31 | 107 | 27 08 | 32 | 108 | 26 36 | 32 | 109 | 26 04 | 33 | 110 |
| 57 | 29 02 | 31 | 106 | 28 31 | 32 | 107 | 27 59 | 32 | 108 | 27 27 | 33 | 109 | 26 54 | 33 | 110 |
| 56 | 29 53 | 32 | 107 | 29 21 | 32 | 108 | 28 49 | 32 | 109 | 28 17 | 33 | 110 | 27 44 | 33 | 111 |
| 55 | 30 44 | −32 | 108 | 30 12 | −32 | 109 | 29 40 | −33 | 110 | 29 07 | −33 | 111 | 28 34 | −34 | 112 |
| 54 | 31 35 | 32 | 108 | 31 03 | 33 | 109 | 30 30 | 33 | 110 | 29 57 | 33 | 111 | 29 24 | 34 | 112 |
| 53 | 32 26 | 33 | 109 | 31 53 | 33 | 110 | 31 20 | 33 | 111 | 30 47 | 34 | 112 | 30 13 | 34 | 113 |
| 52 | 33 16 | 33 | 110 | 32 43 | 33 | 111 | 32 10 | 34 | 112 | 31 36 | 34 | 113 | 31 02 | 34 | 113 |
| 51 | 34 06 | 33 | 110 | 33 33 | 33 | 111 | 33 00 | 34 | 112 | 32 26 | 35 | 113 | 31 51 | 35 | 114 |
| 50 | 34 56 | −33 | 111 | 34 23 | −34 | 112 | 33 49 | −34 | 113 | 33 15 | −35 | 114 | 32 40 | −36 | 115 |
| 49 | 35 46 | 33 | 112 | 35 13 | 35 | 113 | 34 38 | 35 | 114 | 34 03 | 35 | 115 | 33 28 | 36 | 116 |
| 48 | 36 36 | 34 | 112 | 36 02 | 35 | 113 | 35 27 | 35 | 114 | 34 52 | 36 | 115 | 34 16 | 36 | 116 |
| 47 | 37 25 | 34 | 113 | 36 51 | 35 | 114 | 36 16 | 36 | 115 | 35 40 | 36 | 116 | 35 04 | 37 | 117 |
| 46 | 38 14 | 35 | 114 | 37 39 | 35 | 115 | 37 04 | 36 | 116 | 36 28 | 37 | 117 | 35 51 | 36 | 118 |
| 45 | 39 03 | −35 | 114 | 38 28 | −36 | 116 | 37 52 | −36 | 117 | 37 16 | −37 | 118 | 36 39 | −38 | 119 |
| 44 | 39 52 | 36 | 115 | 39 16 | 36 | 116 | 38 40 | 37 | 117 | 38 03 | 38 | 118 | 37 25 | 37 | 119 |
| 43 | 40 40 | 36 | 116 | 40 04 | 37 | 117 | 39 27 | 37 | 118 | 38 50 | 38 | 119 | 38 12 | 38 | 120 |
| 42 | 41 28 | 37 | 117 | 40 51 | 37 | 118 | 40 14 | 38 | 119 | 39 36 | 38 | 120 | 38 58 | 39 | 121 |
| 41 | 42 15 | 37 | 118 | 41 38 | 37 | 119 | 41 01 | 39 | 120 | 40 22 | 38 | 121 | 39 44 | 40 | 122 |
| 40 | 43 03 | −38 | 118 | 42 25 | −38 | 120 | 41 47 | −39 | 121 | 41 08 | −39 | 122 | 40 29 | −40 | 123 |
| 39 | 43 49 | 38 | 119 | 43 11 | 38 | 120 | 42 33 | 39 | 121 | 41 54 | 40 | 122 | 41 14 | 40 | 123 |
| 38 | 44 36 | 39 | 120 | 43 57 | 39 | 121 | 43 18 | 39 | 122 | 42 39 | 41 | 123 | 41 58 | 40 | 124 |
| 37 | 45 22 | 39 | 121 | 44 43 | 40 | 122 | 44 03 | 40 | 123 | 43 23 | 41 | 124 | 42 42 | 41 | 125 |
| 36 | 46 07 | 39 | 122 | 45 28 | 40 | 123 | 44 48 | 41 | 124 | 44 07 | 41 | 125 | 43 26 | 42 | 126 |
| 35 | 46 53 | −41 | 123 | 46 12 | −40 | 124 | 45 32 | −42 | 125 | 44 50 | −42 | 126 | 44 08 | −42 | 127 |
| 34 | 47 37 | 41 | 124 | 46 56 | 41 | 125 | 46 15 | 42 | 126 | 45 33 | 42 | 127 | 44 51 | 43 | 128 |
| 33 | 48 21 | 41 | 125 | 47 40 | 42 | 126 | 46 58 | 42 | 127 | 46 16 | 43 | 128 | 45 33 | 44 | 129 |
| 32 | 49 05 | 42 | 126 | 48 23 | 43 | 127 | 47 40 | 43 | 128 | 46 57 | 43 | 129 | 46 14 | 44 | 130 |
| 31 | 49 48 | 43 | 127 | 49 05 | 43 | 128 | 48 22 | 44 | 129 | 47 38 | 44 | 130 | 46 54 | 44 | 131 |
| 30 | 50 30 | −43 | 128 | 49 47 | −44 | 129 | 49 03 | −44 | 130 | 48 19 | −45 | 131 | 47 34 | −45 | 132 |
| 29 | 51 12 | 44 | 129 | 50 28 | 44 | 130 | 49 44 | 45 | 132 | 48 59 | 46 | 133 | 48 13 | 46 | 134 |
| 28 | 51 53 | 45 | 131 | 51 08 | 45 | 132 | 50 23 | 45 | 133 | 49 38 | 46 | 134 | 48 52 | 47 | 135 |
| 27 | 52 33 | 45 | 132 | 51 48 | 46 | 133 | 51 02 | 46 | 134 | 50 16 | 47 | 135 | 49 29 | 47 | 136 |
| 26 | 53 13 | 46 | 133 | 52 27 | 46 | 134 | 51 41 | 47 | 135 | 50 54 | 48 | 136 | 50 06 | 48 | 137 |
| 25 | 53 51 | −46 | 134 | 53 05 | −47 | 135 | 52 18 | −48 | 136 | 51 30 | −48 | 137 | 50 42 | −48 | 138 |
| 24 | 54 29 | 47 | 136 | 53 42 | 48 | 137 | 52 54 | 48 | 138 | 52 06 | 49 | 139 | 51 17 | 49 | 140 |
| 23 | 55 06 | 48 | 137 | 54 18 | 48 | 138 | 53 30 | 49 | 139 | 52 41 | 49 | 140 | 51 52 | 50 | 141 |
| 22 | 55 42 | 48 | 138 | 54 54 | 49 | 139 | 54 05 | 50 | 140 | 53 15 | 50 | 141 | 52 25 | 50 | 142 |
| 21 | 56 17 | 49 | 140 | 55 28 | 50 | 141 | 54 38 | 50 | 142 | 53 48 | 51 | 143 | 52 57 | 51 | 144 |
| 20 | 56 51 | −50 | 141 | 56 01 | −50 | 142 | 55 11 | −51 | 143 | 54 20 | −52 | 144 | 53 28 | −51 | 145 |
| 19 | 57 24 | 51 | 143 | 56 33 | 51 | 144 | 55 42 | 52 | 145 | 54 50 | 52 | 146 | 53 58 | 52 | 147 |
| 18 | 57 56 | 52 | 144 | 57 04 | 52 | 145 | 56 12 | 52 | 146 | 55 20 | 53 | 147 | 54 27 | 53 | 148 |
| 17 | 58 26 | 52 | 146 | 57 34 | 53 | 147 | 56 41 | 53 | 148 | 55 48 | 53 | 149 | 54 55 | 54 | 150 |
| 16 | 58 56 | 53 | 148 | 58 03 | 54 | 149 | 57 09 | 53 | 150 | 56 16 | 54 | 150 | 55 22 | 55 | 151 |
| 15 | 59 23 | −53 | 150 | 58 30 | −54 | 150 | 57 36 | −55 | 151 | 56 41 | −54 | 152 | 55 47 | −55 | 153 |
| 14 | 59 50 | 55 | 151 | 58 55 | 54 | 152 | 58 01 | 55 | 153 | 57 06 | 55 | 154 | 56 11 | 56 | 154 |
| 13 | 60 15 | 55 | 153 | 59 20 | 55 | 154 | 58 25 | 56 | 155 | 57 29 | 56 | 155 | 56 33 | 56 | 156 |
| 12 | 60 38 | 55 | 155 | 59 43 | 56 | 156 | 58 47 | 56 | 156 | 57 51 | 57 | 157 | 56 54 | 56 | 158 |
| 11 | 61 00 | 56 | 157 | 60 04 | 57 | 158 | 59 07 | 56 | 158 | 58 11 | 57 | 159 | 57 14 | 57 | 159 |
| 10 | 61 20 | −57 | 159 | 60 23 | −57 | 159 | 59 26 | −57 | 160 | 58 29 | −57 | 161 | 57 32 | −58 | 161 |
| 9 | 61 39 | 58 | 161 | 60 41 | 57 | 161 | 59 44 | 58 | 162 | 58 46 | 58 | 163 | 57 48 | 58 | 163 |
| 8 | 61 56 | 58 | 163 | 60 58 | 58 | 163 | 60 00 | 59 | 164 | 59 01 | 58 | 164 | 58 03 | 58 | 165 |
| 7 | 62 10 | 58 | 165 | 61 12 | 58 | 165 | 60 14 | 59 | 166 | 59 15 | 59 | 166 | 58 16 | 58 | 167 |
| 6 | 62 23 | 58 | 167 | 61 25 | 59 | 167 | 60 26 | 59 | 168 | 59 27 | 59 | 168 | 58 28 | 59 | 169 |
| 5 | 62 35 | −60 | 169 | 61 35 | −59 | 169 | 60 36 | −59 | 170 | 59 37 | −59 | 170 | 58 38 | −60 | 170 |
| 4 | 62 44 | 60 | 171 | 61 44 | 59 | 172 | 60 45 | 60 | 172 | 59 45 | 59 | 172 | 58 46 | 60 | 172 |
| 3 | 62 51 | 60 | 173 | 61 51 | 60 | 174 | 60 51 | 59 | 174 | 59 52 | 60 | 174 | 58 52 | 60 | 174 |
| 2 | 62 56 | 60 | 176 | 61 56 | 60 | 176 | 60 56 | 60 | 176 | 59 56 | 60 | 176 | 58 56 | 59 | 176 |
| 1 | 62 59 | 60 | 178 | 61 59 | 60 | 178 | 60 59 | 60 | 178 | 59 59 | 60 | 178 | 58 59 | 60 | 178 |
| 0 | 63 00 | −60 | 180 | 62 00 | −60 | 180 | 61 00 | −60 | 180 | 60 00 | −60 | 180 | 59 00 | −60 | 180 |

| LHA | 5° Hc | 5° d | 5° Z | 6° Hc | 6° d | 6° Z | 7° Hc | 7° d | 7° Z | 8° Hc | 8° d | 8° Z | 9° Hc | 9° d | 9° Z |
|---|---|---|---|---|---|---|---|---|---|---|---|---|---|---|---|
| | ° ′ | ′ | ° | ° ′ | ′ | ° | ° ′ | ′ | ° | ° ′ | ′ | ° | ° ′ | ′ | ° |
| 69 | 16 10 | 30 | 105 | 15 40 | 30 | 105 | 15 10 | 31 | 106 | 14 39 | 30 | 107 | 14 09 | 31 | 108 |
| 68 | 17 02 | 30 | 105 | 16 32 | 31 | 106 | 16 01 | 31 | 107 | 15 30 | 31 | 108 | 14 59 | 31 | 109 |
| 67 | 17 54 | 31 | 106 | 17 23 | 31 | 106 | 16 52 | 31 | 107 | 16 21 | 31 | 108 | 15 50 | 31 | 109 |
| 66 | 18 45 | 31 | 106 | 18 14 | 31 | 107 | 17 43 | 31 | 108 | 17 12 | 32 | 109 | 16 40 | 31 | 110 |
| 65 | 19 36 | −31 | 107 | 19 05 | −31 | 108 | 18 34 | −31 | 108 | 18 03 | −32 | 109 | 17 31 | −32 | 110 |
| 64 | 20 28 | 32 | 107 | 19 56 | 31 | 108 | 19 25 | 32 | 109 | 18 53 | 32 | 110 | 18 21 | 32 | 111 |
| 63 | 21 19 | 32 | 108 | 20 47 | 32 | 109 | 20 15 | 32 | 110 | 19 43 | 32 | 110 | 19 11 | 33 | 111 |
| 62 | 22 09 | 31 | 108 | 21 38 | 32 | 109 | 21 06 | 33 | 110 | 20 33 | 33 | 111 | 20 00 | 32 | 112 |
| 61 | 23 00 | 32 | 109 | 22 28 | 32 | 110 | 21 56 | 33 | 111 | 21 23 | 33 | 112 | 20 50 | 33 | 112 |
| 60 | 23 51 | −33 | 109 | 23 18 | −32 | 110 | 22 46 | −33 | 111 | 22 13 | −34 | 112 | 21 39 | −33 | 113 |
| 59 | 24 41 | 33 | 110 | 24 08 | 33 | 111 | 23 35 | 33 | 112 | 23 02 | 34 | 113 | 22 28 | 34 | 114 |
| 58 | 25 31 | 33 | 111 | 24 58 | 33 | 112 | 24 25 | 34 | 112 | 23 51 | 34 | 113 | 23 17 | 34 | 114 |
| 57 | 26 21 | 33 | 111 | 25 48 | 34 | 112 | 25 14 | 34 | 113 | 24 40 | 34 | 114 | 24 06 | 35 | 115 |
| 56 | 27 11 | 34 | 112 | 26 37 | 34 | 113 | 26 03 | 34 | 114 | 25 29 | 35 | 115 | 24 54 | 35 | 116 |
| 55 | 28 00 | −34 | 112 | 27 26 | −34 | 113 | 26 52 | −35 | 114 | 26 17 | −35 | 115 | 25 42 | −35 | 116 |
| 54 | 28 50 | 35 | 113 | 28 15 | 34 | 114 | 27 41 | 35 | 115 | 27 06 | 36 | 116 | 26 30 | 35 | 117 |
| 53 | 29 39 | 35 | 114 | 29 04 | 35 | 115 | 28 29 | 35 | 116 | 27 54 | 36 | 117 | 27 18 | 36 | 117 |
| 52 | 30 28 | 35 | 114 | 29 53 | 36 | 115 | 29 17 | 36 | 116 | 28 41 | 36 | 117 | 28 05 | 36 | 118 |
| 51 | 31 16 | 35 | 115 | 30 41 | 36 | 116 | 30 05 | 36 | 117 | 29 29 | 37 | 118 | 28 52 | 37 | 119 |
| 50 | 32 04 | −35 | 116 | 31 29 | −37 | 117 | 30 52 | −36 | 118 | 30 16 | −37 | 119 | 29 39 | −37 | 120 |
| 49 | 32 52 | 36 | 117 | 32 16 | 36 | 117 | 31 40 | 37 | 118 | 31 03 | 38 | 119 | 30 25 | 37 | 120 |
| 48 | 33 40 | 37 | 117 | 33 03 | 37 | 118 | 32 26 | 37 | 119 | 31 49 | 38 | 120 | 31 11 | 38 | 121 |
| 47 | 34 27 | 37 | 118 | 33 50 | 37 | 119 | 33 13 | 38 | 120 | 32 35 | 38 | 121 | 31 57 | 39 | 122 |
| 46 | 35 15 | 38 | 119 | 34 37 | 38 | 120 | 33 59 | 38 | 121 | 33 21 | 39 | 122 | 32 42 | 39 | 122 |
| 45 | 36 01 | −38 | 119 | 35 23 | −38 | 120 | 34 45 | −39 | 121 | 34 06 | −39 | 122 | 33 27 | −39 | 123 |
| 44 | 36 48 | 39 | 120 | 36 09 | 38 | 121 | 35 31 | 40 | 122 | 34 51 | 39 | 123 | 34 12 | 40 | 124 |
| 43 | 37 34 | 39 | 121 | 36 55 | 39 | 122 | 36 16 | 40 | 123 | 35 36 | 40 | 124 | 34 56 | 41 | 125 |
| 42 | 38 19 | 39 | 122 | 37 40 | 40 | 123 | 37 00 | 40 | 124 | 36 20 | 40 | 125 | 35 40 | 41 | 126 |
| 41 | 39 04 | 39 | 123 | 38 25 | 40 | 124 | 37 45 | 41 | 125 | 37 04 | 41 | 126 | 36 23 | 42 | 126 |
| 40 | 39 49 | −40 | 124 | 39 09 | −41 | 125 | 38 28 | −41 | 125 | 37 47 | −41 | 126 | 37 06 | −42 | 127 |
| 39 | 40 34 | 41 | 124 | 39 53 | 41 | 125 | 39 12 | 42 | 126 | 38 30 | 42 | 127 | 37 48 | 42 | 128 |
| 38 | 41 18 | 42 | 125 | 40 36 | 41 | 126 | 39 55 | 43 | 127 | 39 12 | 42 | 128 | 38 30 | 43 | 129 |
| 37 | 42 01 | 42 | 126 | 41 19 | 42 | 127 | 40 37 | 43 | 128 | 39 54 | 43 | 129 | 39 11 | 43 | 130 |
| 36 | 42 44 | 43 | 127 | 42 01 | 42 | 128 | 41 19 | 44 | 129 | 40 35 | 43 | 130 | 39 52 | 44 | 131 |
| 35 | 43 26 | −43 | 128 | 42 43 | −43 | 129 | 42 00 | −44 | 130 | 41 16 | −44 | 131 | 40 32 | −45 | 132 |
| 34 | 44 08 | 44 | 129 | 43 24 | 43 | 130 | 42 41 | 45 | 131 | 41 56 | 45 | 132 | 41 11 | 45 | 133 |
| 33 | 44 49 | 44 | 130 | 44 05 | 44 | 131 | 43 21 | 45 | 132 | 42 36 | 46 | 133 | 41 50 | 45 | 134 |
| 32 | 45 30 | 45 | 131 | 44 45 | 45 | 132 | 44 00 | 45 | 133 | 43 15 | 46 | 134 | 42 29 | 47 | 135 |
| 31 | 46 10 | 46 | 132 | 45 24 | 45 | 133 | 44 39 | 46 | 134 | 43 53 | 47 | 135 | 43 06 | 47 | 136 |
| 30 | 46 49 | −46 | 133 | 46 03 | −46 | 134 | 45 17 | −47 | 135 | 44 30 | −47 | 136 | 43 43 | −47 | 137 |
| 29 | 47 27 | 46 | 134 | 46 41 | 47 | 135 | 45 54 | 47 | 136 | 45 07 | 48 | 137 | 44 19 | 48 | 138 |
| 28 | 48 05 | 47 | 136 | 47 18 | 47 | 137 | 46 31 | 48 | 137 | 45 43 | 48 | 138 | 44 55 | 49 | 139 |
| 27 | 48 42 | 47 | 137 | 47 55 | 48 | 138 | 47 07 | 49 | 139 | 46 18 | 49 | 139 | 45 29 | 49 | 140 |
| 26 | 49 18 | 48 | 138 | 48 30 | 48 | 139 | 47 42 | 49 | 140 | 46 53 | 50 | 141 | 46 03 | 50 | 141 |
| 25 | 49 54 | −49 | 139 | 49 05 | −49 | 140 | 48 16 | −50 | 141 | 47 26 | −50 | 142 | 46 36 | −50 | 143 |
| 24 | 50 28 | 49 | 141 | 49 39 | 50 | 141 | 48 49 | 50 | 142 | 47 59 | 51 | 143 | 47 08 | 51 | 144 |
| 23 | 51 02 | 50 | 142 | 50 12 | 51 | 143 | 49 21 | 51 | 144 | 48 30 | 51 | 144 | 47 39 | 51 | 145 |
| 22 | 51 35 | 51 | 143 | 50 44 | 51 | 144 | 49 53 | 52 | 145 | 49 01 | 52 | 146 | 48 09 | 52 | 146 |
| 21 | 52 06 | 51 | 145 | 51 15 | 52 | 145 | 50 23 | 52 | 146 | 49 31 | 53 | 147 | 48 38 | 52 | 148 |
| 20 | 52 37 | −52 | 146 | 51 45 | −53 | 147 | 50 52 | −53 | 148 | 49 59 | −52 | 148 | 49 07 | −54 | 149 |
| 19 | 53 06 | 53 | 147 | 52 13 | 53 | 148 | 51 20 | 53 | 149 | 50 27 | 53 | 150 | 49 34 | 54 | 150 |
| 18 | 53 34 | 53 | 149 | 52 41 | 54 | 150 | 51 47 | 53 | 150 | 50 54 | 54 | 151 | 50 00 | 55 | 152 |
| 17 | 54 01 | 53 | 150 | 53 08 | 55 | 151 | 52 13 | 54 | 152 | 51 19 | 55 | 152 | 50 24 | 55 | 153 |
| 16 | 54 27 | 54 | 152 | 53 33 | 55 | 153 | 52 38 | 55 | 153 | 51 43 | 55 | 154 | 50 48 | 55 | 155 |
| 15 | 54 52 | −55 | 153 | 53 57 | −55 | 154 | 53 02 | −56 | 155 | 52 06 | −56 | 155 | 51 10 | −56 | 156 |
| 14 | 55 15 | 55 | 155 | 54 20 | 56 | 156 | 53 24 | 56 | 156 | 52 28 | 56 | 157 | 51 32 | 57 | 157 |
| 13 | 55 37 | 56 | 157 | 54 41 | 56 | 157 | 53 45 | 57 | 158 | 52 48 | 57 | 158 | 51 51 | 56 | 159 |
| 12 | 55 58 | 57 | 158 | 55 01 | 57 | 159 | 54 04 | 57 | 159 | 53 07 | 57 | 160 | 52 10 | 57 | 160 |
| 11 | 56 17 | 57 | 160 | 55 20 | 58 | 161 | 54 22 | 57 | 161 | 53 25 | 58 | 162 | 52 27 | 57 | 162 |
| 10 | 56 34 | −57 | 162 | 55 37 | −58 | 162 | 54 39 | −58 | 163 | 53 41 | −58 | 163 | 52 43 | −58 | 164 |
| 9 | 56 50 | 58 | 163 | 55 52 | 58 | 164 | 54 54 | 58 | 164 | 53 56 | 59 | 165 | 52 57 | 58 | 165 |
| 8 | 57 05 | 59 | 165 | 56 06 | 58 | 166 | 55 08 | 59 | 166 | 54 09 | 59 | 166 | 53 10 | 58 | 167 |
| 7 | 57 18 | 59 | 167 | 56 19 | 59 | 167 | 55 20 | 59 | 168 | 54 21 | 59 | 168 | 53 22 | 59 | 168 |
| 6 | 57 29 | 59 | 169 | 56 30 | 60 | 169 | 55 30 | 59 | 169 | 54 31 | 59 | 170 | 53 32 | 59 | 170 |
| 5 | 57 38 | −59 | 171 | 56 39 | −60 | 171 | 55 39 | −59 | 171 | 54 40 | −59 | 171 | 53 41 | −60 | 172 |
| 4 | 57 46 | 60 | 173 | 56 46 | 59 | 173 | 55 47 | 60 | 173 | 54 47 | 59 | 173 | 53 48 | 60 | 173 |
| 3 | 57 52 | 60 | 174 | 56 52 | 59 | 175 | 55 53 | 60 | 175 | 54 53 | 60 | 175 | 53 53 | 60 | 175 |
| 2 | 57 57 | 60 | 176 | 56 57 | 60 | 176 | 55 57 | 60 | 177 | 54 57 | 60 | 177 | 53 57 | 60 | 177 |
| 1 | 57 59 | 60 | 178 | 56 59 | 60 | 178 | 55 59 | 60 | 178 | 54 59 | 60 | 178 | 53 59 | 60 | 178 |
| 0 | 58 00 | −60 | 180 | 57 00 | −60 | 180 | 56 00 | −60 | 180 | 55 00 | −60 | 180 | 54 00 | −60 | 180 |

| LHA | 10° Hc | 10° d | 10° Z | 11° Hc | 11° d | 11° Z | 12° Hc | 12° d | 12° Z | 13° Hc | 13° d | 13° Z | 14° Hc | 14° d | 14° Z | LHA |
|---|---|---|---|---|---|---|---|---|---|---|---|---|---|---|---|---|
| | ° ′ | ′ | ° | ° ′ | ′ | ° | ° ′ | ′ | ° | ° ′ | ′ | ° | ° ′ | ′ | ° | |
| 69 | 13 38 | 31 | 109 | 13 07 | 32 | 110 | 12 35 | 31 | 111 | 12 04 | 32 | 112 | 11 32 | 32 | 112 | 291 |
| 68 | 14 28 | 31 | 109 | 13 57 | 32 | 110 | 13 25 | 31 | 111 | 12 54 | 32 | 113 | 12 22 | 33 | 113 | 292 |
| 67 | 15 19 | 32 | 110 | 14 47 | 32 | 111 | 14 15 | 32 | 112 | 13 43 | 32 | 113 | 13 11 | 33 | 114 | 293 |
| 66 | 16 09 | 32 | 111 | 15 37 | 32 | 111 | 15 05 | 33 | 112 | 14 32 | 33 | 113 | 14 00 | 33 | 114 | 294 |
| 65 | 16 59 | −33 | 111 | 16 26 | −32 | 112 | 15 54 | −33 | 113 | 15 21 | −33 | 114 | 14 48 | −33 | 115 | 295 |
| 64 | 17 49 | 33 | 112 | 17 16 | 33 | 113 | 16 43 | 33 | 113 | 16 10 | 33 | 114 | 15 37 | 34 | 115 | 296 |
| 63 | 18 38 | 33 | 112 | 18 05 | 33 | 113 | 17 32 | 33 | 114 | 16 59 | 34 | 115 | 16 25 | 34 | 116 | 297 |
| 62 | 19 28 | 34 | 113 | 18 54 | 33 | 114 | 18 21 | 34 | 115 | 17 47 | 34 | 115 | 17 13 | 34 | 116 | 298 |
| 61 | 20 17 | 34 | 113 | 19 43 | 34 | 114 | 19 09 | 34 | 115 | 18 35 | 34 | 116 | 18 01 | 34 | 117 | 299 |
| 60 | 21 06 | −34 | 114 | 20 32 | −34 | 115 | 19 58 | −35 | 116 | 19 23 | −34 | 117 | 18 49 | −35 | 117 | 300 |
| 59 | 21 54 | 34 | 115 | 21 20 | 34 | 115 | 20 46 | 35 | 116 | 20 11 | 35 | 117 | 19 36 | 35 | 118 | 301 |
| 58 | 22 43 | 35 | 115 | 22 08 | 34 | 116 | 21 34 | 36 | 117 | 20 58 | 35 | 118 | 20 23 | 36 | 119 | 302 |
| 57 | 23 31 | 35 | 116 | 22 56 | 35 | 117 | 22 21 | 35 | 118 | 21 46 | 36 | 118 | 21 10 | 36 | 119 | 303 |
| 56 | 24 19 | 35 | 116 | 23 44 | 35 | 117 | 23 08 | 35 | 118 | 22 33 | 37 | 119 | 21 56 | 36 | 120 | 304 |
| 55 | 25 07 | −36 | 117 | 24 31 | −36 | 118 | 23 55 | −36 | 119 | 23 19 | −36 | 120 | 22 43 | −37 | 121 | 305 |
| 54 | 25 55 | 37 | 118 | 25 18 | 36 | 119 | 24 42 | 37 | 119 | 24 05 | 37 | 120 | 23 28 | 37 | 121 | 306 |
| 53 | 26 42 | 37 | 118 | 26 05 | 36 | 119 | 25 29 | 38 | 120 | 24 51 | 37 | 121 | 24 14 | 38 | 122 | 307 |
| 52 | 27 29 | 37 | 119 | 26 52 | 37 | 120 | 26 15 | 38 | 121 | 25 37 | 38 | 122 | 24 59 | 38 | 123 | 308 |
| 51 | 28 15 | 37 | 120 | 27 38 | 38 | 121 | 27 00 | 37 | 121 | 26 23 | 39 | 122 | 25 44 | 38 | 123 | 309 |
| 50 | 29 02 | −38 | 120 | 28 24 | −38 | 121 | 27 46 | −38 | 122 | 27 08 | −39 | 123 | 26 29 | −39 | 124 | 310 |
| 49 | 29 48 | 39 | 121 | 29 09 | 38 | 122 | 28 31 | 39 | 123 | 27 52 | 39 | 124 | 27 13 | 39 | 125 | 311 |
| 48 | 30 33 | 38 | 122 | 29 55 | 39 | 123 | 29 16 | 40 | 124 | 28 36 | 39 | 124 | 27 57 | 40 | 125 | 312 |
| 47 | 31 18 | 39 | 123 | 30 39 | 39 | 123 | 30 00 | 40 | 124 | 29 20 | 40 | 125 | 28 40 | 40 | 126 | 313 |
| 46 | 32 03 | 39 | 123 | 31 24 | 40 | 124 | 30 44 | 40 | 125 | 30 04 | 41 | 126 | 29 23 | 40 | 127 | 314 |
| 45 | 32 48 | −40 | 124 | 32 08 | −40 | 125 | 31 28 | −41 | 126 | 30 47 | −41 | 127 | 30 06 | −41 | 128 | 315 |
| 44 | 33 32 | 41 | 125 | 32 51 | 40 | 126 | 32 11 | 41 | 127 | 31 30 | 42 | 128 | 30 48 | 42 | 128 | 316 |
| 43 | 34 15 | 40 | 126 | 33 35 | 42 | 127 | 32 53 | 41 | 127 | 32 12 | 42 | 128 | 31 30 | 42 | 129 | 317 |
| 42 | 34 59 | 42 | 127 | 34 17 | 41 | 127 | 33 36 | 42 | 128 | 32 54 | 43 | 129 | 32 11 | 43 | 130 | 318 |
| 41 | 35 41 | 41 | 127 | 35 00 | 43 | 128 | 34 17 | 42 | 129 | 33 35 | 43 | 130 | 32 52 | 43 | 131 | 319 |
| 40 | 36 24 | −43 | 128 | 35 41 | −42 | 129 | 34 59 | −43 | 130 | 34 16 | −44 | 131 | 33 32 | −44 | 132 | 320 |
| 39 | 37 06 | 43 | 129 | 36 23 | 44 | 130 | 35 39 | 43 | 131 | 34 56 | 44 | 132 | 34 12 | 44 | 132 | 321 |
| 38 | 37 47 | 44 | 130 | 37 03 | 43 | 131 | 36 20 | 44 | 132 | 35 36 | 45 | 133 | 34 51 | 45 | 133 | 322 |
| 37 | 38 28 | 44 | 131 | 37 44 | 45 | 132 | 36 59 | 44 | 133 | 36 15 | 45 | 133 | 35 30 | 46 | 134 | 323 |
| 36 | 39 08 | 45 | 132 | 38 23 | 45 | 133 | 37 38 | 45 | 133 | 36 53 | 45 | 134 | 36 08 | 46 | 135 | 324 |
| 35 | 39 47 | −45 | 133 | 39 02 | −45 | 134 | 38 17 | −46 | 134 | 37 31 | −46 | 135 | 36 45 | −46 | 136 | 325 |
| 34 | 40 26 | 45 | 134 | 39 41 | 46 | 135 | 38 55 | 46 | 135 | 38 09 | 47 | 136 | 37 22 | 47 | 137 | 326 |
| 33 | 41 05 | 46 | 135 | 40 19 | 47 | 136 | 39 32 | 47 | 136 | 38 45 | 47 | 137 | 37 58 | 47 | 138 | 327 |
| 32 | 41 42 | 46 | 136 | 40 56 | 47 | 137 | 40 09 | 48 | 137 | 39 21 | 47 | 138 | 38 34 | 48 | 139 | 328 |
| 31 | 42 19 | 47 | 137 | 41 32 | 47 | 138 | 40 45 | 48 | 138 | 39 57 | 49 | 139 | 39 08 | 48 | 140 | 329 |
| 30 | 42 56 | −48 | 138 | 42 08 | −48 | 139 | 41 20 | −49 | 139 | 40 31 | −48 | 140 | 39 43 | −49 | 141 | 330 |
| 29 | 43 31 | 48 | 139 | 42 43 | 49 | 140 | 41 54 | 49 | 140 | 41 05 | 49 | 141 | 40 16 | 50 | 142 | 331 |
| 28 | 44 06 | 49 | 140 | 43 17 | 49 | 141 | 42 28 | 50 | 142 | 41 38 | 50 | 142 | 40 48 | 50 | 143 | 332 |
| 27 | 44 40 | 49 | 141 | 43 51 | 50 | 142 | 43 01 | 50 | 143 | 42 11 | 51 | 143 | 41 20 | 50 | 144 | 333 |
| 26 | 45 13 | 50 | 142 | 44 23 | 50 | 143 | 43 33 | 51 | 144 | 42 42 | 51 | 145 | 41 51 | 51 | 145 | 334 |
| 25 | 45 46 | −51 | 143 | 44 55 | −51 | 144 | 44 04 | −51 | 145 | 43 13 | −52 | 146 | 42 21 | −51 | 146 | 335 |
| 24 | 46 17 | 51 | 145 | 45 26 | 52 | 145 | 44 34 | 51 | 146 | 43 43 | 52 | 147 | 42 51 | 53 | 147 | 336 |
| 23 | 46 48 | 52 | 146 | 45 56 | 52 | 147 | 45 04 | 53 | 147 | 44 11 | 52 | 148 | 43 19 | 53 | 149 | 337 |
| 22 | 47 17 | 52 | 147 | 46 25 | 53 | 148 | 45 32 | 53 | 149 | 44 39 | 53 | 150 | 43 46 | 53 | 150 | 338 |
| 21 | 47 46 | 53 | 148 | 46 53 | 53 | 149 | 46 00 | 54 | 150 | 45 06 | 53 | 150 | 44 13 | 54 | 151 | 339 |
| 20 | 48 13 | −53 | 150 | 47 20 | −54 | 150 | 46 26 | −54 | 151 | 45 32 | −54 | 152 | 44 38 | −54 | 152 | 340 |
| 19 | 48 40 | 54 | 151 | 47 46 | 54 | 152 | 46 52 | 55 | 152 | 45 57 | 55 | 153 | 45 02 | 54 | 153 | 341 |
| 18 | 49 05 | 54 | 152 | 48 11 | 55 | 153 | 47 16 | 55 | 154 | 46 21 | 55 | 154 | 45 26 | 55 | 155 | 342 |
| 17 | 49 29 | 55 | 154 | 48 34 | 55 | 154 | 47 39 | 55 | 155 | 46 44 | 56 | 155 | 45 48 | 56 | 156 | 343 |
| 16 | 49 53 | 56 | 155 | 48 57 | 56 | 156 | 48 01 | 56 | 156 | 47 05 | 56 | 157 | 46 09 | 56 | 157 | 344 |
| 15 | 50 14 | −56 | 157 | 49 18 | −56 | 157 | 48 22 | −56 | 158 | 47 26 | −57 | 158 | 46 29 | −56 | 159 | 345 |
| 14 | 50 35 | 56 | 158 | 49 39 | 57 | 159 | 48 42 | 57 | 159 | 47 45 | 57 | 160 | 46 48 | 57 | 160 | 346 |
| 13 | 50 55 | 57 | 159 | 49 58 | 57 | 160 | 49 01 | 58 | 160 | 48 03 | 57 | 161 | 47 06 | 57 | 161 | 347 |
| 12 | 51 13 | 58 | 161 | 50 15 | 57 | 161 | 49 18 | 58 | 162 | 48 20 | 57 | 162 | 47 23 | 58 | 163 | 348 |
| 11 | 51 30 | 58 | 162 | 50 32 | 58 | 163 | 49 34 | 58 | 163 | 48 36 | 58 | 164 | 47 38 | 58 | 164 | 349 |
| 10 | 51 45 | −58 | 164 | 50 47 | −58 | 164 | 49 49 | −59 | 165 | 48 50 | −58 | 165 | 47 52 | −58 | 166 | 350 |
| 9 | 51 59 | 58 | 166 | 51 01 | 59 | 166 | 50 02 | 59 | 166 | 49 03 | 58 | 167 | 48 05 | 59 | 167 | 351 |
| 8 | 52 12 | 59 | 167 | 51 13 | 59 | 167 | 50 14 | 59 | 168 | 49 15 | 59 | 168 | 48 16 | 59 | 168 | 352 |
| 7 | 52 23 | 59 | 169 | 51 24 | 59 | 169 | 50 25 | 59 | 169 | 49 26 | 60 | 170 | 48 26 | 59 | 170 | 353 |
| 6 | 52 33 | 60 | 170 | 51 33 | 59 | 171 | 50 34 | 59 | 171 | 49 35 | 60 | 171 | 48 35 | 59 | 171 | 354 |
| 5 | 52 41 | −59 | 172 | 51 42 | −60 | 172 | 50 42 | −60 | 172 | 49 42 | −59 | 173 | 48 43 | −60 | 173 | 355 |
| 4 | 52 48 | 60 | 174 | 51 48 | 60 | 174 | 50 48 | 59 | 174 | 49 49 | 60 | 174 | 48 49 | 60 | 174 | 356 |
| 3 | 52 53 | 60 | 175 | 51 53 | 59 | 175 | 50 54 | 60 | 175 | 49 54 | 60 | 176 | 48 54 | 60 | 176 | 357 |
| 2 | 52 57 | 60 | 177 | 51 57 | 60 | 177 | 50 57 | 60 | 177 | 49 57 | 60 | 177 | 48 57 | 60 | 177 | 358 |
| 1 | 52 59 | 60 | 178 | 51 59 | 60 | 178 | 50 59 | 60 | 178 | 49 59 | 60 | 179 | 48 59 | 60 | 179 | 359 |
| 0 | 53 00 | −60 | 180 | 52 00 | −60 | 180 | 51 00 | −60 | 180 | 50 00 | −60 | 180 | 49 00 | −60 | 180 | 360 |

S. Lat. {LHA greater than 180°....... Zn=180−Z; LHA less than 180°........... Zn=180+Z}

DECLINATION (0°-14°) CONTRARY NAME TO LATITUDE

N. Lat. {LHA greater than 180°........ Zn=Z
LHA less than 180°............ Zn=360−Z}

# DECLINATION (15°-29°) CONTRARY NAME TO LATITUDE

| LHA | 15° Hc | 15° d | 15° Z | 16° Hc | 16° d | 16° Z | 17° Hc | 17° d | 17° Z | 18° Hc | 18° d | 18° Z | 19° Hc | 19° d | 19° Z | 20° Hc | 20° d | 20° Z | 21° Hc | 21° d | 21° Z |
|---|---|---|---|---|---|---|---|---|---|---|---|---|---|---|---|---|---|---|---|---|---|
| | ° ′ | ′ | ° | ° ′ | ′ | ° | ° ′ | ′ | ° | ° ′ | ′ | ° | ° ′ | ′ | ° | ° ′ | ′ | ° | ° ′ | ′ | ° |
| 69 | 10 13 | 34 | 114 | 09 39 | 34 | 115 | 09 05 | 33 | 115 | 08 32 | 34 | 116 | 07 58 | 34 | 117 | 07 24 | 34 | 118 | 06 50 | 34 | 119 |
| 68 | 11 01 | 34 | 114 | 10 27 | 34 | 115 | 09 53 | 34 | 116 | 09 19 | 35 | 117 | 08 44 | 34 | 118 | 08 10 | 34 | 118 | 07 36 | 35 | 119 |
| 67 | 11 48 | 34 | 115 | 11 14 | 34 | 116 | 10 40 | 35 | 116 | 10 05 | 34 | 117 | 09 31 | 35 | 118 | 08 56 | 35 | 119 | 08 21 | 35 | 120 |
| 66 | 12 36 | 34 | 115 | 12 02 | 35 | 116 | 11 27 | 35 | 117 | 10 52 | 35 | 118 | 10 17 | 35 | 119 | 09 42 | 35 | 119 | 09 07 | 36 | 120 |
| 65 | 13 23 | −34 | 116 | 12 49 | −36 | 117 | 12 13 | −35 | 118 | 11 38 | −35 | 118 | 11 03 | −35 | 119 | 10 28 | −36 | 120 | 09 52 | −36 | 121 |
| 64 | 14 10 | 35 | 116 | 13 35 | 35 | 117 | 13 00 | 36 | 118 | 12 24 | 35 | 119 | 11 49 | 36 | 120 | 11 13 | 36 | 121 | 10 37 | 36 | 121 |
| 63 | 14 57 | 35 | 117 | 14 22 | 36 | 118 | 13 46 | 36 | 119 | 13 10 | 36 | 120 | 12 34 | 36 | 120 | 11 58 | 36 | 121 | 11 22 | 37 | 122 |
| 62 | 15 44 | 36 | 118 | 15 08 | 36 | 119 | 14 32 | 36 | 119 | 13 56 | 37 | 120 | 13 19 | 36 | 121 | 12 43 | 37 | 122 | 12 06 | 37 | 123 |
| 61 | 16 30 | 36 | 118 | 15 54 | 36 | 119 | 15 18 | 37 | 120 | 14 41 | 37 | 121 | 14 04 | 37 | 122 | 13 27 | 37 | 122 | 12 50 | 37 | 123 |
| 60 | 17 16 | −36 | 119 | 16 40 | −37 | 120 | 16 03 | −37 | 121 | 15 26 | −37 | 121 | 14 49 | −38 | 122 | 14 11 | −37 | 123 | 13 34 | −38 | 124 |
| 59 | 18 02 | 37 | 120 | 17 25 | 37 | 120 | 16 48 | 37 | 121 | 16 11 | 38 | 122 | 15 33 | 38 | 123 | 14 55 | 38 | 124 | 14 17 | 38 | 124 |
| 58 | 18 48 | 38 | 120 | 18 10 | 37 | 121 | 17 33 | 38 | 122 | 16 55 | 38 | 123 | 16 17 | 38 | 123 | 15 39 | 38 | 124 | 15 01 | 39 | 125 |
| 57 | 19 33 | 38 | 121 | 18 55 | 38 | 122 | 18 17 | 38 | 122 | 17 39 | 38 | 123 | 17 01 | 39 | 124 | 16 22 | 39 | 125 | 15 43 | 39 | 126 |
| 56 | 20 18 | 38 | 121 | 19 40 | 39 | 122 | 19 01 | 38 | 123 | 18 23 | 39 | 124 | 17 44 | 39 | 125 | 17 05 | 39 | 125 | 16 26 | 39 | 126 |
| 55 | 21 03 | −39 | 122 | 20 24 | −39 | 123 | 19 45 | −39 | 124 | 19 06 | −39 | 125 | 18 27 | −39 | 125 | 17 48 | −40 | 126 | 17 08 | −40 | 127 |
| 54 | 21 47 | 39 | 123 | 21 08 | 39 | 124 | 20 29 | 40 | 124 | 19 49 | 39 | 125 | 19 10 | 40 | 126 | 18 30 | 40 | 127 | 17 50 | 40 | 128 |
| 53 | 22 31 | 39 | 123 | 21 52 | 40 | 124 | 21 12 | 40 | 125 | 20 32 | 40 | 126 | 19 52 | 40 | 127 | 19 12 | 41 | 127 | 18 31 | 40 | 128 |
| 52 | 23 15 | 40 | 124 | 22 35 | 40 | 125 | 21 55 | 40 | 126 | 21 15 | 41 | 127 | 20 34 | 41 | 127 | 19 53 | 41 | 128 | 19 12 | 41 | 129 |
| 51 | 23 58 | 40 | 125 | 23 18 | 41 | 126 | 22 37 | 40 | 126 | 21 57 | 41 | 127 | 21 16 | 42 | 128 | 20 34 | 41 | 129 | 19 53 | 41 | 130 |
| 50 | 24 41 | −41 | 126 | 24 00 | −41 | 126 | 23 19 | −41 | 127 | 22 38 | −41 | 128 | 21 57 | −42 | 129 | 21 15 | −42 | 129 | 20 33 | −42 | 130 |
| 49 | 25 23 | 41 | 126 | 24 42 | 41 | 127 | 24 01 | 42 | 128 | 23 19 | 41 | 129 | 22 38 | 42 | 129 | 21 56 | 43 | 130 | 21 13 | 42 | 131 |
| 48 | 26 06 | 42 | 127 | 25 24 | 42 | 128 | 24 42 | 42 | 129 | 24 00 | 42 | 129 | 23 18 | 43 | 130 | 22 35 | 42 | 131 | 21 53 | 43 | 132 |
| 47 | 26 47 | 42 | 128 | 26 05 | 42 | 129 | 25 23 | 42 | 129 | 24 41 | 43 | 130 | 23 58 | 43 | 131 | 23 15 | 43 | 132 | 22 32 | 44 | 132 |
| 46 | 27 29 | 43 | 128 | 26 46 | 43 | 129 | 26 03 | 42 | 130 | 25 21 | 44 | 131 | 24 37 | 43 | 132 | 23 54 | 44 | 132 | 23 10 | 44 | 133 |
| 45 | 28 10 | −43 | 129 | 27 27 | −44 | 130 | 26 43 | −43 | 131 | 26 00 | −44 | 132 | 25 16 | −44 | 132 | 24 32 | −44 | 133 | 23 48 | −44 | 134 |
| 44 | 28 50 | 43 | 130 | 28 07 | 44 | 131 | 27 23 | 44 | 132 | 26 39 | 44 | 132 | 25 55 | 44 | 133 | 25 11 | 45 | 134 | 24 26 | 45 | 135 |
| 43 | 29 30 | 44 | 131 | 28 46 | 44 | 132 | 28 02 | 44 | 132 | 27 18 | 45 | 133 | 26 33 | 45 | 134 | 25 48 | 45 | 135 | 25 03 | 45 | 135 |
| 42 | 30 09 | 44 | 132 | 29 25 | 45 | 132 | 28 40 | 44 | 133 | 27 56 | 45 | 134 | 27 11 | 46 | 135 | 26 25 | 45 | 135 | 25 40 | 46 | 136 |
| 41 | 30 48 | 44 | 133 | 30 04 | 46 | 133 | 29 18 | 45 | 134 | 28 33 | 45 | 135 | 27 48 | 46 | 136 | 27 02 | 46 | 136 | 26 16 | 46 | 137 |
| 40 | 31 27 | −45 | 133 | 30 42 | −46 | 134 | 29 56 | −46 | 135 | 29 10 | −46 | 136 | 28 24 | −46 | 136 | 27 38 | −47 | 137 | 26 51 | −46 | 138 |
| 39 | 32 05 | 46 | 134 | 31 19 | 46 | 135 | 30 33 | 46 | 136 | 29 47 | 47 | 136 | 29 00 | 47 | 137 | 28 13 | 47 | 138 | 27 26 | 47 | 139 |
| 38 | 32 42 | 46 | 135 | 31 56 | 47 | 136 | 31 09 | 46 | 137 | 30 23 | 47 | 137 | 29 36 | 48 | 138 | 28 48 | 47 | 139 | 28 01 | 48 | 139 |
| 37 | 33 19 | 47 | 136 | 32 32 | 47 | 137 | 31 45 | 47 | 137 | 30 58 | 48 | 138 | 30 10 | 47 | 139 | 29 23 | 48 | 140 | 28 35 | 48 | 140 |
| 36 | 33 55 | 47 | 137 | 33 08 | 48 | 138 | 32 20 | 47 | 138 | 31 33 | 48 | 139 | 30 45 | 49 | 140 | 29 56 | 48 | 140 | 29 08 | 49 | 141 |
| 35 | 34 31 | −48 | 138 | 33 43 | −48 | 139 | 32 55 | −48 | 139 | 32 07 | −49 | 140 | 31 18 | −48 | 141 | 30 30 | −49 | 141 | 29 41 | −49 | 142 |
| 34 | 35 06 | 49 | 139 | 34 17 | 48 | 139 | 33 29 | 49 | 140 | 32 40 | 49 | 141 | 31 51 | 49 | 142 | 31 02 | 49 | 142 | 30 13 | 50 | 143 |
| 33 | 35 40 | 49 | 140 | 34 51 | 49 | 140 | 34 02 | 49 | 141 | 33 13 | 49 | 142 | 32 24 | 50 | 142 | 31 34 | 50 | 143 | 30 44 | 50 | 144 |
| 32 | 36 14 | 50 | 141 | 35 24 | 49 | 141 | 34 35 | 50 | 142 | 33 45 | 50 | 143 | 32 55 | 50 | 143 | 32 05 | 50 | 144 | 31 15 | 51 | 145 |
| 31 | 36 47 | 50 | 142 | 35 57 | 50 | 142 | 35 07 | 50 | 143 | 34 17 | 51 | 144 | 33 26 | 50 | 144 | 32 36 | 51 | 145 | 31 45 | 51 | 146 |
| 30 | 37 19 | −50 | 143 | 36 29 | −51 | 143 | 35 38 | −51 | 144 | 34 47 | −50 | 145 | 33 57 | −52 | 145 | 33 05 | −51 | 146 | 32 14 | −51 | 147 |
| 29 | 37 50 | 50 | 144 | 37 00 | 51 | 144 | 36 09 | 52 | 145 | 35 17 | 51 | 146 | 34 26 | 52 | 146 | 33 34 | 51 | 147 | 32 43 | 52 | 148 |
| 28 | 38 21 | 51 | 145 | 37 30 | 52 | 145 | 36 38 | 51 | 146 | 35 47 | 52 | 147 | 34 55 | 52 | 147 | 34 03 | 52 | 148 | 33 11 | 53 | 148 |
| 27 | 38 51 | 52 | 146 | 37 59 | 52 | 146 | 37 07 | 52 | 147 | 36 15 | 52 | 148 | 35 23 | 53 | 148 | 34 30 | 52 | 149 | 33 38 | 53 | 149 |
| 26 | 39 20 | 52 | 147 | 38 28 | 53 | 147 | 37 35 | 52 | 148 | 36 43 | 53 | 149 | 35 50 | 53 | 149 | 34 57 | 53 | 150 | 34 04 | 53 | 150 |
| 25 | 39 48 | −52 | 148 | 38 56 | −53 | 149 | 38 03 | −53 | 149 | 37 10 | −53 | 150 | 36 17 | −54 | 150 | 35 23 | −53 | 151 | 34 30 | −54 | 151 |
| 24 | 40 16 | 53 | 149 | 39 23 | 54 | 150 | 38 29 | 53 | 150 | 37 36 | 54 | 151 | 36 42 | 54 | 151 | 35 48 | 54 | 152 | 34 54 | 54 | 152 |
| 23 | 40 42 | 53 | 150 | 39 49 | 54 | 151 | 38 55 | 54 | 151 | 38 01 | 54 | 152 | 37 07 | 54 | 152 | 36 13 | 55 | 153 | 35 18 | 54 | 154 |
| 22 | 41 08 | 54 | 151 | 40 14 | 54 | 152 | 39 20 | 55 | 152 | 38 25 | 54 | 153 | 37 31 | 55 | 154 | 36 36 | 55 | 154 | 35 41 | 55 | 155 |
| 21 | 41 33 | 55 | 153 | 40 38 | 54 | 153 | 39 44 | 55 | 154 | 38 49 | 55 | 154 | 37 54 | 55 | 155 | 36 59 | 56 | 155 | 36 03 | 55 | 156 |
| 20 | 41 57 | −55 | 154 | 41 02 | −56 | 154 | 40 06 | −55 | 155 | 39 11 | −55 | 155 | 38 16 | −56 | 156 | 37 20 | −55 | 156 | 36 25 | −56 | 157 |
| 19 | 42 19 | 55 | 155 | 41 24 | 56 | 155 | 40 28 | 55 | 156 | 39 33 | 56 | 156 | 38 37 | 56 | 157 | 37 41 | 56 | 157 | 36 45 | 56 | 158 |
| 18 | 42 41 | 56 | 156 | 41 45 | 56 | 157 | 40 49 | 56 | 157 | 39 53 | 56 | 158 | 38 57 | 56 | 158 | 38 01 | 57 | 158 | 37 04 | 56 | 159 |
| 17 | 43 02 | 56 | 157 | 42 06 | 57 | 158 | 41 09 | 56 | 158 | 40 13 | 57 | 159 | 39 16 | 56 | 159 | 38 20 | 57 | 160 | 37 23 | 57 | 160 |
| 16 | 43 22 | 57 | 159 | 42 25 | 57 | 159 | 41 28 | 56 | 159 | 40 32 | 57 | 160 | 39 35 | 57 | 160 | 38 38 | 58 | 161 | 37 40 | 57 | 161 |
| 15 | 43 40 | −57 | 160 | 42 43 | −57 | 160 | 41 46 | −57 | 161 | 40 49 | −57 | 161 | 39 52 | −58 | 161 | 38 54 | −57 | 162 | 37 57 | −57 | 162 |
| 14 | 43 58 | 57 | 161 | 43 01 | 58 | 162 | 42 03 | 57 | 162 | 41 06 | 58 | 162 | 40 08 | 58 | 163 | 39 10 | 57 | 163 | 38 13 | 58 | 163 |
| 13 | 44 15 | 58 | 162 | 43 17 | 58 | 163 | 42 19 | 58 | 163 | 41 21 | 58 | 163 | 40 23 | 58 | 164 | 39 25 | 58 | 164 | 38 27 | 58 | 164 |
| 12 | 44 30 | 58 | 164 | 43 32 | 58 | 164 | 42 34 | 58 | 164 | 41 36 | 59 | 165 | 40 37 | 58 | 165 | 39 39 | 58 | 165 | 38 41 | 59 | 166 |
| 11 | 44 44 | 58 | 165 | 43 46 | 59 | 165 | 42 47 | 58 | 166 | 41 49 | 59 | 166 | 40 50 | 58 | 166 | 39 52 | 59 | 167 | 38 53 | 58 | 167 |
| 10 | 44 57 | −58 | 166 | 43 59 | −59 | 167 | 43 00 | −59 | 167 | 42 01 | −59 | 167 | 41 02 | −58 | 167 | 40 04 | −59 | 168 | 39 05 | −59 | 168 |
| 9 | 45 09 | 59 | 168 | 44 10 | 59 | 168 | 43 11 | 59 | 168 | 42 12 | 59 | 168 | 41 13 | 59 | 169 | 40 14 | 59 | 169 | 39 15 | 59 | 169 |
| 8 | 45 20 | 60 | 169 | 44 20 | 59 | 169 | 43 21 | 59 | 170 | 42 22 | 59 | 170 | 41 23 | 59 | 170 | 40 24 | 59 | 170 | 39 25 | 60 | 170 |
| 7 | 45 29 | 59 | 170 | 44 30 | 60 | 171 | 43 30 | 59 | 171 | 42 31 | 59 | 171 | 41 32 | 60 | 171 | 40 32 | 59 | 171 | 39 33 | 60 | 172 |
| 6 | 45 37 | 59 | 172 | 44 38 | 60 | 172 | 43 38 | 59 | 172 | 42 39 | 60 | 172 | 41 39 | 59 | 172 | 40 40 | 60 | 173 | 39 40 | 60 | 173 |
| 5 | 45 44 | −59 | 173 | 44 45 | −60 | 173 | 43 45 | −60 | 173 | 42 45 | −59 | 174 | 41 46 | −60 | 174 | 40 46 | −60 | 174 | 39 46 | −60 | 174 |
| 4 | 45 50 | 60 | 175 | 44 50 | 60 | 175 | 43 50 | 59 | 175 | 42 51 | 60 | 175 | 41 51 | 60 | 175 | 40 51 | 60 | 175 | 39 51 | 60 | 175 |
| 3 | 45 54 | 60 | 176 | 44 54 | 59 | 176 | 43 55 | 60 | 176 | 42 55 | 60 | 176 | 41 55 | 60 | 176 | 40 55 | 60 | 176 | 39 55 | 60 | 176 |
| 2 | 45 58 | 60 | 177 | 44 58 | 60 | 177 | 43 58 | 60 | 177 | 42 58 | 60 | 177 | 41 58 | 60 | 178 | 40 58 | 60 | 178 | 39 58 | 60 | 178 |
| 1 | 45 59 | 60 | 179 | 44 59 | 60 | 179 | 43 59 | 60 | 179 | 42 59 | 60 | 179 | 41 59 | 60 | 179 | 40 59 | 60 | 179 | 39 59 | 59 | 179 |
| 0 | 46 00 | −60 | 180 | 45 00 | −60 | 180 | 44 00 | −60 | 180 | 43 00 | −60 | 180 | 42 00 | −60 | 180 | 41 00 | −60 | 180 | 40 00 | −60 | 180 |

| LHA | 22° Hc | 22° d | 22° Z | 23° Hc | 23° d | 23° Z | 24° Hc | 24° d | 24° Z | 25° Hc | 25° d | 25° Z | 26° Hc | 26° d | 26° Z | 27° Hc | 27° d | 27° Z | 28° Hc | 28° d | 28° Z | 29° Hc | 29° d | 29° Z | LHA |
|---|---|---|---|---|---|---|---|---|---|---|---|---|---|---|---|---|---|---|---|---|---|---|---|---|---|
| | ° ′ | ′ | ° | ° ′ | ′ | ° | ° ′ | ′ | ° | ° ′ | ′ | ° | ° ′ | ′ | ° | ° ′ | ′ | ° | ° ′ | ′ | ° | ° ′ | ′ | ° | |
| 69 | 06 16 | 35 | 119 | 05 41 | 34 | 120 | 05 07 | 35 | 121 | 04 32 | 34 | 122 | 03 58 | 34 | 123 | 03 24 | 35 | 124 | 02 49 | 35 | 124 | 02 14 | 34 | 125 | 291 |
| 68 | 07 01 | 35 | 120 | 06 26 | 34 | 121 | 05 52 | 35 | 122 | 05 17 | 35 | 122 | 04 42 | 35 | 123 | 04 07 | 35 | 124 | 03 32 | 35 | 125 | 02 57 | 35 | 126 | 292 |
| 67 | 07 46 | 35 | 121 | 07 11 | 35 | 121 | 06 36 | 35 | 122 | 06 01 | 35 | 123 | 05 26 | 36 | 124 | 04 50 | 35 | 125 | 04 15 | 35 | 125 | 03 40 | 36 | 126 | 293 |
| 66 | 08 31 | 35 | 121 | 07 56 | 35 | 122 | 07 21 | 36 | 123 | 06 45 | 36 | 124 | 06 09 | 35 | 124 | 05 34 | 36 | 125 | 04 58 | 36 | 126 | 04 22 | 36 | 127 | 294 |
| 65 | 09 16 | −35 | 122 | 08 41 | −36 | 122 | 08 05 | −36 | 123 | 07 29 | −36 | 124 | 06 53 | −37 | 125 | 06 16 | −36 | 126 | 05 40 | −36 | 127 | 05 04 | −37 | 127 | 295 |
| 64 | 10 01 | 36 | 122 | 09 25 | 37 | 123 | 08 48 | 36 | 124 | 08 12 | 37 | 125 | 07 35 | 36 | 125 | 06 59 | 37 | 126 | 06 22 | 37 | 127 | 05 45 | 36 | 128 | 296 |
| 63 | 10 45 | 36 | 123 | 10 09 | 37 | 124 | 09 32 | 37 | 124 | 08 55 | 37 | 125 | 08 18 | 37 | 126 | 07 41 | 37 | 127 | 07 04 | 37 | 128 | 06 27 | 38 | 128 | 297 |
| 62 | 11 29 | 37 | 123 | 10 52 | 37 | 124 | 10 15 | 37 | 125 | 09 38 | 38 | 126 | 09 00 | 37 | 127 | 08 23 | 38 | 127 | 07 45 | 37 | 128 | 07 08 | 38 | 129 | 298 |
| 61 | 12 13 | 38 | 124 | 11 35 | 37 | 125 | 10 58 | 38 | 126 | 10 20 | 38 | 126 | 09 42 | 38 | 127 | 09 04 | 38 | 128 | 08 26 | 38 | 129 | 07 48 | 38 | 130 | 299 |
| 60 | 12 56 | −38 | 125 | 12 18 | −38 | 125 | 11 40 | −38 | 126 | 11 02 | −38 | 127 | 10 24 | −38 | 128 | 09 46 | −39 | 129 | 09 07 | −38 | 129 | 08 29 | −39 | 130 | 300 |
| 59 | 13 39 | 38 | 125 | 13 01 | 38 | 126 | 12 23 | 39 | 127 | 11 44 | 39 | 128 | 11 05 | 38 | 128 | 10 27 | 39 | 129 | 09 48 | 39 | 130 | 09 09 | 39 | 131 | 301 |
| 58 | 14 22 | 39 | 126 | 13 43 | 38 | 127 | 13 05 | 39 | 127 | 12 26 | 40 | 128 | 11 46 | 39 | 129 | 11 07 | 39 | 130 | 10 28 | 40 | 130 | 09 48 | 39 | 131 | 302 |
| 57 | 15 04 | 39 | 126 | 14 25 | 39 | 127 | 13 46 | 39 | 128 | 13 07 | 40 | 129 | 12 27 | 40 | 130 | 11 47 | 39 | 130 | 11 08 | 40 | 131 | 10 28 | 40 | 132 | 303 |
| 56 | 15 47 | 40 | 127 | 15 07 | 40 | 128 | 14 27 | 40 | 129 | 13 47 | 40 | 129 | 13 07 | 40 | 130 | 12 27 | 40 | 131 | 11 47 | 40 | 132 | 11 07 | 41 | 132 | 304 |
| 55 | 16 28 | −40 | 128 | 15 48 | −40 | 128 | 15 08 | −40 | 129 | 14 28 | −41 | 130 | 13 47 | −40 | 131 | 13 07 | −41 | 132 | 12 26 | −41 | 132 | 11 45 | −41 | 133 | 305 |
| 54 | 17 10 | 41 | 128 | 16 29 | 40 | 129 | 15 49 | 41 | 130 | 15 08 | 41 | 131 | 14 27 | 41 | 131 | 13 46 | 41 | 132 | 13 05 | 41 | 133 | 12 24 | 42 | 134 | 306 |
| 53 | 17 51 | 41 | 129 | 17 10 | 41 | 130 | 16 29 | 41 | 131 | 15 48 | 42 | 131 | 15 06 | 41 | 132 | 14 25 | 42 | 133 | 13 43 | 42 | 134 | 13 01 | 42 | 134 | 307 |
| 52 | 18 31 | 41 | 130 | 17 50 | 41 | 130 | 17 09 | 42 | 131 | 16 27 | 42 | 132 | 15 45 | 42 | 133 | 15 03 | 42 | 133 | 14 21 | 42 | 134 | 13 39 | 43 | 135 | 308 |
| 51 | 19 12 | 42 | 130 | 18 30 | 42 | 131 | 17 48 | 42 | 132 | 17 06 | 43 | 133 | 16 23 | 42 | 133 | 15 41 | 42 | 134 | 14 59 | 43 | 135 | 14 16 | 43 | 136 | 309 |
| 50 | 19 51 | −42 | 131 | 19 09 | −42 | 132 | 18 27 | −43 | 133 | 17 44 | −43 | 133 | 17 01 | −42 | 134 | 16 19 | −43 | 135 | 15 36 | −44 | 135 | 14 52 | −43 | 136 | 310 |
| 49 | 20 31 | 43 | 132 | 19 48 | 43 | 132 | 19 05 | 43 | 133 | 18 22 | 43 | 134 | 17 39 | 43 | 135 | 16 56 | 44 | 135 | 16 12 | 43 | 136 | 15 29 | 44 | 137 | 311 |
| 48 | 21 10 | 43 | 132 | 20 27 | 44 | 133 | 19 43 | 43 | 134 | 19 00 | 44 | 135 | 18 16 | 44 | 135 | 17 32 | 44 | 136 | 16 48 | 44 | 137 | 16 04 | 44 | 137 | 312 |
| 47 | 21 48 | 43 | 133 | 21 05 | 44 | 134 | 20 21 | 44 | 135 | 19 37 | 44 | 135 | 18 53 | 44 | 136 | 18 09 | 45 | 137 | 17 24 | 44 | 137 | 16 40 | 45 | 138 | 313 |
| 46 | 22 26 | 44 | 134 | 21 42 | 44 | 135 | 20 58 | 44 | 135 | 20 14 | 45 | 136 | 19 29 | 45 | 137 | 18 44 | 45 | 137 | 17 59 | 45 | 138 | 17 14 | 45 | 139 | 314 |
| 45 | 23 04 | −44 | 135 | 22 20 | −45 | 135 | 21 35 | −45 | 136 | 20 50 | −45 | 137 | 20 05 | −45 | 137 | 19 20 | −46 | 138 | 18 34 | −45 | 139 | 17 49 | −46 | 140 | 315 |
| 44 | 23 41 | 45 | 135 | 22 56 | 45 | 136 | 22 11 | 45 | 137 | 21 26 | 46 | 137 | 20 40 | 46 | 138 | 19 54 | 45 | 139 | 19 09 | 46 | 140 | 18 23 | 47 | 140 | 316 |
| 43 | 24 18 | 46 | 136 | 23 32 | 45 | 137 | 22 47 | 46 | 138 | 22 01 | 46 | 138 | 21 15 | 46 | 139 | 20 29 | 47 | 140 | 19 42 | 46 | 140 | 18 56 | 47 | 141 | 317 |
| 42 | 24 54 | 46 | 137 | 24 08 | 46 | 138 | 23 22 | 46 | 138 | 22 36 | 47 | 139 | 21 49 | 47 | 140 | 21 02 | 46 | 140 | 20 16 | 47 | 141 | 19 29 | 47 | 142 | 318 |
| 41 | 25 30 | 47 | 138 | 24 43 | 46 | 138 | 23 57 | 47 | 139 | 23 10 | 47 | 140 | 22 23 | 47 | 140 | 21 36 | 48 | 141 | 20 48 | 47 | 142 | 20 01 | 48 | 142 | 319 |
| 40 | 26 05 | −47 | 138 | 25 18 | −47 | 139 | 24 31 | −47 | 140 | 23 44 | −48 | 141 | 22 56 | −48 | 141 | 22 08 | −47 | 142 | 21 21 | −48 | 143 | 20 33 | −48 | 143 | 320 |
| 39 | 26 39 | 47 | 139 | 25 52 | 48 | 140 | 25 04 | 47 | 141 | 24 17 | 48 | 141 | 23 29 | 48 | 142 | 22 41 | 49 | 143 | 21 52 | 48 | 143 | 21 04 | 49 | 144 | 321 |
| 38 | 27 13 | 48 | 140 | 26 25 | 48 | 141 | 25 37 | 48 | 141 | 24 49 | 48 | 142 | 24 01 | 49 | 143 | 23 12 | 48 | 143 | 22 24 | 49 | 144 | 21 35 | 49 | 145 | 322 |
| 37 | 27 47 | 49 | 141 | 26 58 | 48 | 142 | 26 10 | 49 | 142 | 25 21 | 49 | 143 | 24 32 | 49 | 144 | 23 43 | 49 | 144 | 22 54 | 49 | 145 | 22 05 | 50 | 145 | 323 |
| 36 | 28 19 | 48 | 142 | 27 31 | 49 | 142 | 26 42 | 49 | 143 | 25 53 | 50 | 144 | 25 03 | 49 | 144 | 24 14 | 50 | 145 | 23 24 | 50 | 146 | 22 34 | 50 | 146 | 324 |
| 35 | 28 52 | −50 | 143 | 28 02 | −49 | 143 | 27 13 | −50 | 144 | 26 23 | −50 | 145 | 25 33 | −49 | 145 | 24 44 | −51 | 146 | 23 53 | −50 | 146 | 23 03 | −50 | 147 | 325 |
| 34 | 29 23 | 50 | 144 | 28 33 | 50 | 144 | 27 43 | 50 | 145 | 26 53 | 50 | 145 | 26 03 | 50 | 146 | 25 13 | 51 | 147 | 24 22 | 50 | 147 | 23 32 | 51 | 148 | 326 |
| 33 | 29 54 | 50 | 144 | 29 04 | 51 | 145 | 28 13 | 50 | 146 | 27 23 | 51 | 146 | 26 32 | 51 | 147 | 25 41 | 51 | 147 | 24 50 | 51 | 148 | 23 59 | 51 | 149 | 327 |
| 32 | 30 24 | 50 | 145 | 29 34 | 51 | 146 | 28 43 | 51 | 147 | 27 52 | 51 | 147 | 27 01 | 52 | 148 | 26 09 | 51 | 148 | 25 18 | 52 | 149 | 24 26 | 51 | 149 | 328 |
| 31 | 30 54 | 51 | 146 | 30 03 | 52 | 147 | 29 11 | 51 | 147 | 28 20 | 52 | 148 | 27 28 | 51 | 149 | 26 37 | 52 | 149 | 25 45 | 52 | 150 | 24 53 | 53 | 150 | 329 |
| 30 | 31 23 | −52 | 147 | 30 31 | −52 | 148 | 29 39 | −52 | 148 | 28 47 | −52 | 149 | 27 55 | −52 | 149 | 27 03 | −52 | 150 | 26 11 | −53 | 151 | 25 18 | −52 | 151 | 330 |
| 29 | 31 51 | 52 | 148 | 30 59 | 52 | 149 | 30 07 | 53 | 149 | 29 14 | 52 | 150 | 28 22 | 53 | 150 | 27 29 | 53 | 151 | 26 36 | 53 | 151 | 25 43 | 53 | 152 | 331 |
| 28 | 32 18 | 52 | 149 | 31 26 | 53 | 150 | 30 33 | 53 | 150 | 29 40 | 53 | 151 | 28 47 | 53 | 151 | 27 54 | 53 | 152 | 27 01 | 53 | 152 | 26 08 | 54 | 153 | 332 |
| 27 | 32 45 | 53 | 150 | 31 52 | 53 | 151 | 30 59 | 53 | 151 | 30 06 | 54 | 152 | 29 12 | 53 | 152 | 28 19 | 54 | 153 | 27 25 | 54 | 153 | 26 31 | 53 | 154 | 333 |
| 26 | 33 11 | 54 | 151 | 32 17 | 53 | 152 | 31 24 | 54 | 152 | 30 30 | 54 | 153 | 29 36 | 54 | 153 | 28 42 | 54 | 154 | 27 48 | 54 | 154 | 26 54 | 54 | 155 | 334 |
| 25 | 33 36 | −54 | 152 | 32 42 | −54 | 153 | 31 48 | −54 | 153 | 30 54 | −54 | 154 | 30 00 | −55 | 154 | 29 05 | −54 | 155 | 28 11 | −55 | 155 | 27 16 | −54 | 155 | 335 |
| 24 | 34 00 | 54 | 153 | 33 06 | 55 | 154 | 32 11 | 54 | 154 | 31 17 | 55 | 154 | 30 22 | 54 | 155 | 29 28 | 55 | 155 | 28 33 | 55 | 156 | 27 38 | 55 | 156 | 336 |
| 23 | 34 24 | 55 | 154 | 33 29 | 55 | 155 | 32 34 | 55 | 155 | 31 39 | 55 | 155 | 30 44 | 55 | 156 | 29 49 | 55 | 156 | 28 54 | 55 | 157 | 27 59 | 56 | 157 | 337 |
| 22 | 34 46 | 55 | 155 | 33 51 | 55 | 156 | 32 56 | 55 | 156 | 32 01 | 56 | 156 | 31 05 | 55 | 157 | 30 10 | 56 | 157 | 29 14 | 55 | 158 | 28 19 | 56 | 158 | 338 |
| 21 | 35 08 | 56 | 156 | 34 12 | 55 | 157 | 33 17 | 56 | 157 | 32 21 | 56 | 157 | 31 25 | 55 | 158 | 30 30 | 56 | 158 | 29 34 | 56 | 159 | 28 38 | 56 | 159 | 339 |
| 20 | 35 29 | −56 | 157 | 34 33 | −56 | 158 | 33 37 | −56 | 158 | 32 41 | −56 | 158 | 31 45 | −56 | 159 | 30 49 | −57 | 159 | 29 52 | −56 | 160 | 28 56 | −56 | 160 | 340 |
| 19 | 35 49 | 56 | 158 | 34 53 | 57 | 159 | 33 56 | 56 | 159 | 33 00 | 57 | 159 | 32 03 | 56 | 160 | 31 07 | 57 | 160 | 30 10 | 56 | 161 | 29 14 | 57 | 161 | 341 |
| 18 | 36 08 | 57 | 159 | 35 11 | 56 | 160 | 34 15 | 57 | 160 | 33 18 | 57 | 160 | 32 21 | 57 | 161 | 31 24 | 57 | 161 | 30 27 | 57 | 162 | 29 30 | 57 | 162 | 342 |
| 17 | 36 26 | 57 | 160 | 35 29 | 57 | 161 | 34 32 | 57 | 161 | 33 35 | 57 | 162 | 32 38 | 57 | 162 | 31 41 | 58 | 162 | 30 43 | 57 | 163 | 29 46 | 57 | 163 | 343 |
| 16 | 36 43 | 57 | 161 | 35 46 | 57 | 162 | 34 49 | 58 | 162 | 33 51 | 57 | 163 | 32 54 | 58 | 163 | 31 56 | 57 | 163 | 30 59 | 58 | 164 | 30 01 | 57 | 164 | 344 |
| 15 | 37 00 | −58 | 163 | 36 02 | −58 | 163 | 35 04 | −57 | 163 | 34 07 | −58 | 164 | 33 09 | −58 | 164 | 32 11 | −58 | 164 | 31 13 | −58 | 165 | 30 15 | −58 | 165 | 345 |
| 14 | 37 15 | 58 | 164 | 36 17 | 58 | 164 | 35 19 | 58 | 164 | 34 21 | 58 | 165 | 33 23 | 58 | 165 | 32 25 | 58 | 165 | 31 27 | 58 | 166 | 30 29 | 58 | 166 | 346 |
| 13 | 37 29 | 58 | 165 | 36 31 | 58 | 165 | 35 33 | 59 | 165 | 34 34 | 58 | 166 | 33 36 | 58 | 166 | 32 38 | 58 | 166 | 31 40 | 59 | 167 | 30 41 | 58 | 167 | 347 |
| 12 | 37 42 | 58 | 166 | 36 44 | 59 | 166 | 35 45 | 58 | 167 | 34 47 | 59 | 167 | 33 48 | 58 | 167 | 32 50 | 59 | 167 | 31 51 | 58 | 168 | 30 53 | 59 | 168 | 348 |
| 11 | 37 55 | 59 | 167 | 36 56 | 59 | 167 | 35 57 | 58 | 168 | 34 59 | 59 | 168 | 34 00 | 59 | 168 | 33 01 | 59 | 168 | 32 02 | 59 | 169 | 31 03 | 58 | 169 | 349 |
| 10 | 38 06 | −59 | 168 | 37 07 | −59 | 168 | 36 08 | −59 | 169 | 35 09 | −59 | 169 | 34 10 | −59 | 169 | 33 11 | −59 | 169 | 32 12 | −59 | 170 | 31 13 | −59 | 170 | 350 |
| 9 | 38 16 | 59 | 169 | 37 17 | 59 | 170 | 36 18 | 59 | 170 | 35 19 | 59 | 170 | 34 20 | 60 | 170 | 33 20 | 59 | 170 | 32 21 | 59 | 171 | 31 22 | 59 | 171 | 351 |
| 8 | 38 25 | 59 | 171 | 37 26 | 59 | 171 | 36 27 | 60 | 171 | 35 27 | 59 | 171 | 34 28 | 59 | 171 | 33 29 | 60 | 172 | 32 29 | 59 | 172 | 31 30 | 59 | 172 | 352 |
| 7 | 38 33 | 59 | 172 | 37 34 | 60 | 172 | 36 34 | 59 | 172 | 35 35 | 59 | 172 | 34 36 | 60 | 172 | 33 36 | 59 | 173 | 32 37 | 60 | 173 | 31 37 | 60 | 173 | 353 |
| 6 | 38 40 | 59 | 173 | 37 41 | 60 | 173 | 36 41 | 59 | 173 | 35 42 | 60 | 173 | 34 42 | 60 | 173 | 33 42 | 59 | 174 | 32 43 | 60 | 174 | 31 43 | 60 | 174 | 354 |
| 5 | 38 46 | −59 | 174 | 37 47 | −60 | 174 | 36 47 | −60 | 174 | 35 47 | −59 | 174 | 34 48 | −60 | 175 | 33 48 | −60 | 175 | 32 48 | −60 | 175 | 31 48 | −59 | 175 | 355 |
| 4 | 38 51 | 59 | 175 | 37 52 | 60 | 175 | 36 52 | 60 | 175 | 35 52 | 60 | 176 | 34 52 | 60 | 176 | 33 52 | 60 | 176 | 32 52 | 59 | 176 | 31 53 | 60 | 176 | 356 |
| 3 | 38 55 | 60 | 176 | 37 55 | 60 | 177 | 36 55 | 60 | 177 | 35 55 | 59 | 177 | 34 56 | 60 | 177 | 33 56 | 60 | 177 | 32 56 | 60 | 177 | 31 56 | 60 | 177 | 357 |
| 2 | 38 58 | 60 | 178 | 37 58 | 60 | 178 | 36 58 | 60 | 178 | 35 58 | 60 | 178 | 34 58 | 60 | 178 | 33 58 | 60 | 178 | 32 58 | 60 | 178 | 31 58 | 60 | 178 | 358 |
| 1 | 39 00 | 60 | 179 | 38 00 | 60 | 179 | 37 00 | 60 | 179 | 36 00 | 60 | 179 | 35 00 | 60 | 179 | 34 00 | 60 | 179 | 33 00 | 60 | 179 | 32 00 | 60 | 179 | 359 |
| 0 | 39 00 | −60 | 180 | 38 00 | −60 | 180 | 37 00 | −60 | 180 | 36 00 | −60 | 180 | 35 00 | −60 | 180 | 34 00 | −60 | 180 | 33 00 | −60 | 180 | 32 00 | −60 | 180 | 360 |

S. Lat. {LHA greater than 180°........ Zn=180−Z
LHA less than 180°............ Zn=180+Z}

DECLINATION (15°-29°) CONTRARY NAME TO LATITUDE

LAT 29°

## TABLE 5.—Correction to Tabulated Altitude for Minutes of Declination

| d / ′ | 1 | 2 | 3 | 4 | 5 | 6 | 7 | 8 | 9 | 10 | 11 | 12 | 13 | 14 | 15 | 16 | 17 | 18 | 19 | 20 | 21 | 22 | 23 | 24 | 25 | 26 | 27 | 28 | 29 | 30 |
|---|---|---|---|---|---|---|---|---|---|---|---|---|---|---|---|---|---|---|---|---|---|---|---|---|---|---|---|---|---|---|
| 0 | 0 | 0 | 0 | 0 | 0 | 0 | 0 | 0 | 0 | 0 | 0 | 0 | 0 | 0 | 0 | 0 | 0 | 0 | 0 | 0 | 0 | 0 | 0 | 0 | 0 | 0 | 0 | 0 | 0 | 0 |
| 1 | 0 | 0 | 0 | 0 | 0 | 0 | 0 | 0 | 0 | 0 | 0 | 0 | 0 | 0 | 0 | 0 | 0 | 0 | 0 | 0 | 0 | 0 | 0 | 0 | 0 | 0 | 0 | 0 | 0 | 0 |
| 2 | 0 | 0 | 0 | 0 | 0 | 0 | 0 | 0 | 0 | 0 | 0 | 0 | 0 | 0 | 0 | 1 | 1 | 1 | 1 | 1 | 1 | 1 | 1 | 1 | 1 | 1 | 1 | 1 | 1 | 1 |
| 3 | 0 | 0 | 0 | 0 | 0 | 0 | 0 | 0 | 0 | 0 | 1 | 1 | 1 | 1 | 1 | 1 | 1 | 1 | 1 | 1 | 1 | 1 | 1 | 1 | 1 | 1 | 1 | 1 | 1 | 2 |
| 4 | 0 | 0 | 0 | 0 | 0 | 0 | 0 | 1 | 1 | 1 | 1 | 1 | 1 | 1 | 1 | 1 | 1 | 1 | 1 | 1 | 1 | 1 | 2 | 2 | 2 | 2 | 2 | 2 | 2 | 2 |
| 5 | 0 | 0 | 0 | 0 | 0 | 0 | 1 | 1 | 1 | 1 | 1 | 1 | 1 | 1 | 1 | 1 | 1 | 2 | 2 | 2 | 2 | 2 | 2 | 2 | 2 | 2 | 2 | 2 | 2 | 2 |
| 6 | 0 | 0 | 0 | 0 | 0 | 1 | 1 | 1 | 1 | 1 | 1 | 1 | 1 | 1 | 2 | 2 | 2 | 2 | 2 | 2 | 2 | 2 | 2 | 2 | 2 | 3 | 3 | 3 | 3 | 3 |
| 7 | 0 | 0 | 0 | 0 | 1 | 1 | 1 | 1 | 1 | 1 | 1 | 1 | 2 | 2 | 2 | 2 | 2 | 2 | 2 | 2 | 2 | 3 | 3 | 3 | 3 | 3 | 3 | 3 | 3 | 4 |
| 8 | 0 | 0 | 0 | 1 | 1 | 1 | 1 | 1 | 1 | 1 | 1 | 2 | 2 | 2 | 2 | 2 | 2 | 2 | 3 | 3 | 3 | 3 | 3 | 3 | 3 | 3 | 4 | 4 | 4 | 4 |
| 9 | 0 | 0 | 0 | 1 | 1 | 1 | 1 | 1 | 1 | 2 | 2 | 2 | 2 | 2 | 2 | 2 | 3 | 3 | 3 | 3 | 3 | 3 | 3 | 4 | 4 | 4 | 4 | 4 | 4 | 4 |
| 10 | 0 | 0 | 0 | 1 | 1 | 1 | 1 | 1 | 2 | 2 | 2 | 2 | 2 | 2 | 2 | 3 | 3 | 3 | 3 | 3 | 4 | 4 | 4 | 4 | 4 | 4 | 4 | 5 | 5 | 5 |
| 11 | 0 | 0 | 1 | 1 | 1 | 1 | 1 | 1 | 2 | 2 | 2 | 2 | 2 | 3 | 3 | 3 | 3 | 3 | 3 | 4 | 4 | 4 | 4 | 4 | 5 | 5 | 5 | 5 | 5 | 6 |
| 12 | 0 | 0 | 1 | 1 | 1 | 1 | 1 | 2 | 2 | 2 | 2 | 2 | 3 | 3 | 3 | 3 | 3 | 4 | 4 | 4 | 4 | 4 | 5 | 5 | 5 | 5 | 5 | 6 | 6 | 6 |
| 13 | 0 | 0 | 1 | 1 | 1 | 1 | 2 | 2 | 2 | 2 | 2 | 3 | 3 | 3 | 3 | 3 | 4 | 4 | 4 | 4 | 5 | 5 | 5 | 5 | 5 | 6 | 6 | 6 | 6 | 6 |
| 14 | 0 | 0 | 1 | 1 | 1 | 1 | 2 | 2 | 2 | 2 | 3 | 3 | 3 | 3 | 4 | 4 | 4 | 4 | 4 | 5 | 5 | 5 | 5 | 6 | 6 | 6 | 6 | 7 | 7 | 7 |
| 15 | 0 | 0 | 1 | 1 | 1 | 2 | 2 | 2 | 2 | 2 | 3 | 3 | 3 | 4 | 4 | 4 | 4 | 4 | 5 | 5 | 5 | 6 | 6 | 6 | 6 | 6 | 7 | 7 | 7 | 8 |
| 16 | 0 | 1 | 1 | 1 | 1 | 2 | 2 | 2 | 2 | 3 | 3 | 3 | 3 | 4 | 4 | 4 | 5 | 5 | 5 | 5 | 6 | 6 | 6 | 6 | 7 | 7 | 7 | 7 | 8 | 8 |
| 17 | 0 | 1 | 1 | 1 | 1 | 2 | 2 | 2 | 3 | 3 | 3 | 3 | 4 | 4 | 4 | 5 | 5 | 5 | 5 | 6 | 6 | 6 | 7 | 7 | 7 | 7 | 8 | 8 | 8 | 8 |
| 18 | 0 | 1 | 1 | 1 | 2 | 2 | 2 | 2 | 3 | 3 | 3 | 4 | 4 | 4 | 4 | 5 | 5 | 5 | 6 | 6 | 6 | 7 | 7 | 7 | 8 | 8 | 8 | 8 | 9 | 9 |
| 19 | 0 | 1 | 1 | 1 | 2 | 2 | 2 | 3 | 3 | 3 | 3 | 4 | 4 | 4 | 5 | 5 | 5 | 6 | 6 | 6 | 7 | 7 | 7 | 8 | 8 | 8 | 9 | 9 | 9 | 10 |
| 20 | 0 | 1 | 1 | 1 | 2 | 2 | 2 | 3 | 3 | 3 | 4 | 4 | 4 | 5 | 5 | 5 | 6 | 6 | 6 | 7 | 7 | 7 | 8 | 8 | 8 | 9 | 9 | 9 | 10 | 10 |
| 21 | 0 | 1 | 1 | 1 | 2 | 2 | 2 | 3 | 3 | 4 | 4 | 4 | 5 | 5 | 5 | 6 | 6 | 6 | 7 | 7 | 7 | 8 | 8 | 8 | 9 | 9 | 9 | 10 | 10 | 10 |
| 22 | 0 | 1 | 1 | 1 | 2 | 2 | 3 | 3 | 3 | 4 | 4 | 4 | 5 | 5 | 6 | 6 | 6 | 7 | 7 | 7 | 8 | 8 | 8 | 9 | 9 | 10 | 10 | 10 | 11 | 11 |
| 23 | 0 | 1 | 1 | 2 | 2 | 2 | 3 | 3 | 3 | 4 | 4 | 5 | 5 | 5 | 6 | 6 | 7 | 7 | 7 | 8 | 8 | 8 | 9 | 9 | 10 | 10 | 10 | 11 | 11 | 12 |
| 24 | 0 | 1 | 1 | 2 | 2 | 2 | 3 | 3 | 4 | 4 | 4 | 5 | 5 | 6 | 6 | 6 | 7 | 7 | 8 | 8 | 8 | 9 | 9 | 10 | 10 | 10 | 11 | 11 | 12 | 12 |
| 25 | 0 | 1 | 1 | 2 | 2 | 2 | 3 | 3 | 4 | 4 | 5 | 5 | 5 | 6 | 6 | 7 | 7 | 8 | 8 | 8 | 9 | 9 | 10 | 10 | 10 | 11 | 11 | 12 | 12 | 12 |
| 26 | 0 | 1 | 1 | 2 | 2 | 3 | 3 | 3 | 4 | 4 | 5 | 5 | 6 | 6 | 6 | 7 | 7 | 8 | 8 | 9 | 9 | 10 | 10 | 10 | 11 | 11 | 12 | 12 | 13 | 13 |
| 27 | 0 | 1 | 1 | 2 | 2 | 3 | 3 | 4 | 4 | 4 | 5 | 5 | 6 | 6 | 7 | 7 | 8 | 8 | 9 | 9 | 9 | 10 | 10 | 11 | 11 | 12 | 12 | 13 | 13 | 14 |
| 28 | 0 | 1 | 1 | 2 | 2 | 3 | 3 | 4 | 4 | 5 | 5 | 6 | 6 | 7 | 7 | 7 | 8 | 8 | 9 | 9 | 10 | 10 | 11 | 11 | 12 | 12 | 13 | 13 | 14 | 14 |
| 29 | 0 | 1 | 1 | 2 | 2 | 3 | 3 | 4 | 4 | 5 | 5 | 6 | 6 | 7 | 7 | 8 | 8 | 9 | 9 | 10 | 10 | 11 | 11 | 12 | 12 | 13 | 13 | 14 | 14 | 14 |
| 30 | 0 | 1 | 2 | 2 | 2 | 3 | 4 | 4 | 4 | 5 | 6 | 6 | 6 | 7 | 8 | 8 | 8 | 9 | 10 | 10 | 10 | 11 | 12 | 12 | 12 | 13 | 14 | 14 | 14 | 15 |
| 31 | 1 | 1 | 2 | 2 | 3 | 3 | 4 | 4 | 5 | 5 | 6 | 6 | 7 | 7 | 8 | 8 | 9 | 9 | 10 | 10 | 11 | 11 | 12 | 12 | 13 | 13 | 14 | 14 | 15 | 16 |
| 32 | 1 | 1 | 2 | 2 | 3 | 3 | 4 | 4 | 5 | 5 | 6 | 6 | 7 | 7 | 8 | 9 | 9 | 10 | 10 | 11 | 11 | 12 | 12 | 13 | 13 | 14 | 14 | 15 | 15 | 16 |
| 33 | 1 | 1 | 2 | 2 | 3 | 3 | 4 | 4 | 5 | 6 | 6 | 7 | 7 | 8 | 8 | 9 | 9 | 10 | 10 | 11 | 12 | 12 | 13 | 13 | 14 | 14 | 15 | 15 | 16 | 16 |
| 34 | 1 | 1 | 2 | 2 | 3 | 3 | 4 | 5 | 5 | 6 | 6 | 7 | 7 | 8 | 8 | 9 | 10 | 10 | 11 | 11 | 12 | 12 | 13 | 13 | 14 | 15 | 15 | 16 | 16 | 17 |
| 35 | 1 | 1 | 2 | 2 | 3 | 4 | 4 | 5 | 5 | 6 | 6 | 7 | 8 | 8 | 9 | 9 | 10 | 10 | 11 | 12 | 12 | 13 | 13 | 14 | 15 | 15 | 16 | 16 | 17 | 18 |
| 36 | 1 | 1 | 2 | 2 | 3 | 4 | 4 | 5 | 5 | 6 | 7 | 7 | 8 | 8 | 9 | 10 | 10 | 11 | 11 | 12 | 13 | 13 | 14 | 14 | 15 | 16 | 16 | 17 | 17 | 18 |
| 37 | 1 | 1 | 2 | 2 | 3 | 4 | 4 | 5 | 6 | 6 | 7 | 7 | 8 | 9 | 9 | 10 | 10 | 11 | 12 | 12 | 13 | 14 | 14 | 15 | 15 | 16 | 17 | 17 | 18 | 18 |
| 38 | 1 | 1 | 2 | 3 | 3 | 4 | 4 | 5 | 6 | 6 | 7 | 8 | 8 | 9 | 10 | 10 | 11 | 11 | 12 | 13 | 13 | 14 | 15 | 15 | 16 | 16 | 17 | 18 | 18 | 19 |
| 39 | 1 | 1 | 2 | 3 | 3 | 4 | 5 | 5 | 6 | 6 | 7 | 8 | 8 | 9 | 10 | 10 | 11 | 12 | 12 | 13 | 14 | 14 | 15 | 16 | 16 | 17 | 18 | 18 | 19 | 20 |
| 40 | 1 | 1 | 2 | 3 | 3 | 4 | 5 | 5 | 6 | 7 | 7 | 8 | 9 | 9 | 10 | 11 | 11 | 12 | 13 | 13 | 14 | 15 | 15 | 16 | 17 | 17 | 18 | 19 | 19 | 20 |
| 41 | 1 | 1 | 2 | 3 | 3 | 4 | 5 | 5 | 6 | 7 | 8 | 8 | 9 | 10 | 10 | 11 | 12 | 12 | 13 | 14 | 14 | 15 | 16 | 16 | 17 | 18 | 18 | 19 | 20 | 20 |
| 42 | 1 | 1 | 2 | 3 | 4 | 4 | 5 | 6 | 6 | 7 | 8 | 8 | 9 | 10 | 10 | 11 | 12 | 13 | 13 | 14 | 15 | 15 | 16 | 17 | 18 | 18 | 19 | 20 | 20 | 21 |
| 43 | 1 | 1 | 2 | 3 | 4 | 4 | 5 | 6 | 6 | 7 | 8 | 9 | 9 | 10 | 11 | 11 | 12 | 13 | 14 | 14 | 15 | 16 | 16 | 17 | 18 | 19 | 19 | 20 | 21 | 22 |
| 44 | 1 | 1 | 2 | 3 | 4 | 4 | 5 | 6 | 7 | 7 | 8 | 9 | 10 | 10 | 11 | 12 | 12 | 13 | 14 | 15 | 15 | 16 | 17 | 18 | 18 | 19 | 20 | 21 | 21 | 22 |
| 45 | 1 | 2 | 2 | 3 | 4 | 4 | 5 | 6 | 7 | 8 | 8 | 9 | 10 | 10 | 11 | 12 | 13 | 14 | 14 | 15 | 16 | 16 | 17 | 18 | 19 | 20 | 20 | 21 | 22 | 22 |
| 46 | 1 | 2 | 2 | 3 | 4 | 5 | 5 | 6 | 7 | 8 | 8 | 9 | 10 | 11 | 12 | 12 | 13 | 14 | 15 | 15 | 16 | 17 | 18 | 18 | 19 | 20 | 21 | 21 | 22 | 23 |
| 47 | 1 | 2 | 2 | 3 | 4 | 5 | 5 | 6 | 7 | 8 | 9 | 9 | 10 | 11 | 12 | 13 | 13 | 14 | 15 | 16 | 16 | 17 | 18 | 19 | 20 | 20 | 21 | 22 | 23 | 24 |
| 48 | 1 | 2 | 2 | 3 | 4 | 5 | 6 | 6 | 7 | 8 | 9 | 10 | 10 | 11 | 12 | 13 | 14 | 14 | 15 | 16 | 17 | 18 | 18 | 19 | 20 | 21 | 22 | 22 | 23 | 24 |
| 49 | 1 | 2 | 2 | 3 | 4 | 5 | 6 | 7 | 7 | 8 | 9 | 10 | 11 | 11 | 12 | 13 | 14 | 15 | 16 | 16 | 17 | 18 | 19 | 20 | 20 | 21 | 22 | 23 | 24 | 24 |
| 50 | 1 | 2 | 2 | 3 | 4 | 5 | 6 | 7 | 8 | 8 | 9 | 10 | 11 | 12 | 12 | 13 | 14 | 15 | 16 | 17 | 18 | 18 | 19 | 20 | 21 | 22 | 22 | 23 | 24 | 25 |
| 51 | 1 | 2 | 3 | 3 | 4 | 5 | 6 | 7 | 8 | 8 | 9 | 10 | 11 | 12 | 13 | 14 | 14 | 15 | 16 | 17 | 18 | 19 | 20 | 20 | 21 | 22 | 23 | 24 | 25 | 26 |
| 52 | 1 | 2 | 3 | 3 | 4 | 5 | 6 | 7 | 8 | 9 | 10 | 10 | 11 | 12 | 13 | 14 | 15 | 16 | 16 | 17 | 18 | 19 | 20 | 21 | 22 | 23 | 23 | 24 | 25 | 26 |
| 53 | 1 | 2 | 3 | 4 | 4 | 5 | 6 | 7 | 8 | 9 | 10 | 11 | 11 | 12 | 13 | 14 | 15 | 16 | 17 | 18 | 19 | 19 | 20 | 21 | 22 | 23 | 24 | 25 | 26 | 26 |
| 54 | 1 | 2 | 3 | 4 | 4 | 5 | 6 | 7 | 8 | 9 | 10 | 11 | 12 | 13 | 14 | 14 | 15 | 16 | 17 | 18 | 19 | 20 | 21 | 22 | 22 | 23 | 24 | 25 | 26 | 27 |
| 55 | 1 | 2 | 3 | 4 | 5 | 6 | 6 | 7 | 8 | 9 | 10 | 11 | 12 | 13 | 14 | 15 | 16 | 16 | 17 | 18 | 19 | 20 | 21 | 22 | 23 | 24 | 25 | 26 | 27 | 28 |
| 56 | 1 | 2 | 3 | 4 | 5 | 6 | 7 | 7 | 8 | 9 | 10 | 11 | 12 | 13 | 14 | 15 | 16 | 17 | 18 | 19 | 20 | 21 | 21 | 22 | 23 | 24 | 25 | 26 | 27 | 28 |
| 57 | 1 | 2 | 3 | 4 | 5 | 6 | 7 | 8 | 9 | 10 | 10 | 11 | 12 | 13 | 14 | 15 | 16 | 17 | 18 | 19 | 20 | 21 | 22 | 23 | 24 | 25 | 26 | 27 | 28 | 28 |
| 58 | 1 | 2 | 3 | 4 | 5 | 6 | 7 | 8 | 9 | 10 | 11 | 12 | 13 | 14 | 14 | 15 | 16 | 17 | 18 | 19 | 20 | 21 | 22 | 23 | 24 | 25 | 26 | 27 | 28 | 29 |
| 59 | 1 | 2 | 3 | 4 | 5 | 6 | 7 | 8 | 9 | 10 | 11 | 12 | 13 | 14 | 15 | 16 | 17 | 18 | 19 | 20 | 21 | 22 | 23 | 24 | 25 | 26 | 27 | 28 | 29 | 30 |

| d / ′ | 31 | 32 | 33 | 34 | 35 | 36 | 37 | 38 | 39 | 40 | 41 | 42 | 43 | 44 | 45 | 46 | 47 | 48 | 49 | 50 | 51 | 52 | 53 | 54 | 55 | 56 | 57 | 58 | 59 | 60 | d / ′ |
|---|---|---|---|---|---|---|---|---|---|---|---|---|---|---|---|---|---|---|---|---|---|---|---|---|---|---|---|---|---|---|---|
| 0 | 0 | 0 | 0 | 0 | 0 | 0 | 0 | 0 | 0 | 0 | 0 | 0 | 0 | 0 | 0 | 0 | 0 | 0 | 0 | 0 | 0 | 0 | 0 | 0 | 0 | 0 | 0 | 0 | 0 | 0 | 0 |
| 1 | 1 | 1 | 1 | 1 | 1 | 1 | 1 | 1 | 1 | 1 | 1 | 1 | 1 | 1 | 1 | 1 | 1 | 1 | 1 | 1 | 1 | 1 | 1 | 1 | 1 | 1 | 1 | 1 | 1 | 1 | 1 |
| 2 | 1 | 1 | 1 | 1 | 1 | 1 | 1 | 1 | 1 | 1 | 1 | 1 | 1 | 1 | 2 | 2 | 2 | 2 | 2 | 2 | 2 | 2 | 2 | 2 | 2 | 2 | 2 | 2 | 2 | 2 | 2 |
| 3 | 2 | 2 | 2 | 2 | 2 | 2 | 2 | 2 | 2 | 2 | 2 | 2 | 2 | 2 | 2 | 2 | 2 | 2 | 2 | 2 | 3 | 3 | 3 | 3 | 3 | 3 | 3 | 3 | 3 | 3 | 3 |
| 4 | 2 | 2 | 2 | 2 | 2 | 2 | 2 | 3 | 3 | 3 | 3 | 3 | 3 | 3 | 3 | 3 | 3 | 3 | 3 | 3 | 3 | 3 | 4 | 4 | 4 | 4 | 4 | 4 | 4 | 4 | 4 |
| 5 | 3 | 3 | 3 | 3 | 3 | 3 | 3 | 3 | 3 | 3 | 3 | 4 | 4 | 4 | 4 | 4 | 4 | 4 | 4 | 4 | 4 | 4 | 4 | 4 | 5 | 5 | 5 | 5 | 5 | 5 | 5 |
| 6 | 3 | 3 | 3 | 3 | 4 | 4 | 4 | 4 | 4 | 4 | 4 | 4 | 4 | 4 | 4 | 5 | 5 | 5 | 5 | 5 | 5 | 5 | 5 | 5 | 6 | 6 | 6 | 6 | 6 | 6 | 6 |
| 7 | 4 | 4 | 4 | 4 | 4 | 4 | 4 | 4 | 5 | 5 | 5 | 5 | 5 | 5 | 5 | 5 | 5 | 6 | 6 | 6 | 6 | 6 | 6 | 6 | 6 | 7 | 7 | 7 | 7 | 7 | 7 |
| 8 | 4 | 4 | 4 | 5 | 5 | 5 | 5 | 5 | 5 | 5 | 5 | 6 | 6 | 6 | 6 | 6 | 6 | 6 | 7 | 7 | 7 | 7 | 7 | 7 | 7 | 7 | 8 | 8 | 8 | 8 | 8 |
| 9 | 5 | 5 | 5 | 5 | 5 | 5 | 6 | 6 | 6 | 6 | 6 | 6 | 6 | 7 | 7 | 7 | 7 | 7 | 7 | 8 | 8 | 8 | 8 | 8 | 8 | 8 | 9 | 9 | 9 | 9 | 9 |
| 10 | 5 | 5 | 6 | 6 | 6 | 6 | 6 | 6 | 6 | 7 | 7 | 7 | 7 | 7 | 8 | 8 | 8 | 8 | 8 | 8 | 8 | 9 | 9 | 9 | 9 | 9 | 10 | 10 | 10 | 10 | 10 |
| 11 | 6 | 6 | 6 | 6 | 6 | 7 | 7 | 7 | 7 | 7 | 8 | 8 | 8 | 8 | 8 | 8 | 9 | 9 | 9 | 9 | 9 | 10 | 10 | 10 | 10 | 10 | 10 | 11 | 11 | 11 | 11 |
| 12 | 6 | 6 | 7 | 7 | 7 | 7 | 7 | 8 | 8 | 8 | 8 | 8 | 9 | 9 | 9 | 9 | 9 | 10 | 10 | 10 | 10 | 10 | 11 | 11 | 11 | 11 | 11 | 12 | 12 | 12 | 12 |
| 13 | 7 | 7 | 7 | 7 | 8 | 8 | 8 | 8 | 8 | 9 | 9 | 9 | 9 | 10 | 10 | 10 | 10 | 10 | 11 | 11 | 11 | 11 | 11 | 12 | 12 | 12 | 12 | 13 | 13 | 13 | 13 |
| 14 | 7 | 7 | 8 | 8 | 8 | 8 | 9 | 9 | 9 | 9 | 10 | 10 | 10 | 10 | 10 | 11 | 11 | 11 | 11 | 12 | 12 | 12 | 12 | 13 | 13 | 13 | 13 | 14 | 14 | 14 | 14 |
| 15 | 8 | 8 | 8 | 8 | 9 | 9 | 9 | 10 | 10 | 10 | 10 | 10 | 11 | 11 | 11 | 12 | 12 | 12 | 12 | 12 | 13 | 13 | 13 | 14 | 14 | 14 | 14 | 14 | 15 | 15 | 15 |
| 16 | 8 | 9 | 9 | 9 | 9 | 10 | 10 | 10 | 10 | 11 | 11 | 11 | 11 | 12 | 12 | 12 | 13 | 13 | 13 | 13 | 14 | 14 | 14 | 14 | 15 | 15 | 15 | 15 | 16 | 16 | 16 |
| 17 | 9 | 9 | 9 | 10 | 10 | 10 | 10 | 11 | 11 | 11 | 12 | 12 | 12 | 12 | 13 | 13 | 13 | 14 | 14 | 14 | 14 | 15 | 15 | 15 | 16 | 16 | 16 | 16 | 17 | 17 | 17 |
| 18 | 9 | 10 | 10 | 10 | 10 | 11 | 11 | 11 | 12 | 12 | 12 | 13 | 13 | 13 | 14 | 14 | 14 | 14 | 15 | 15 | 15 | 16 | 16 | 16 | 16 | 17 | 17 | 17 | 18 | 18 | 18 |
| 19 | 10 | 10 | 10 | 11 | 11 | 11 | 12 | 12 | 12 | 13 | 13 | 13 | 14 | 14 | 14 | 15 | 15 | 15 | 16 | 16 | 16 | 16 | 17 | 17 | 17 | 18 | 18 | 18 | 19 | 19 | 19 |
| 20 | 10 | 11 | 11 | 11 | 12 | 12 | 12 | 13 | 13 | 13 | 14 | 14 | 14 | 15 | 15 | 15 | 16 | 16 | 16 | 17 | 17 | 17 | 18 | 18 | 18 | 19 | 19 | 19 | 20 | 20 | 20 |
| 21 | 11 | 11 | 12 | 12 | 12 | 13 | 13 | 13 | 14 | 14 | 14 | 15 | 15 | 15 | 16 | 16 | 16 | 17 | 17 | 18 | 18 | 18 | 19 | 19 | 19 | 20 | 20 | 20 | 21 | 21 | 21 |
| 22 | 11 | 12 | 12 | 12 | 13 | 13 | 14 | 14 | 14 | 15 | 15 | 15 | 16 | 16 | 16 | 17 | 17 | 18 | 18 | 18 | 19 | 19 | 19 | 20 | 20 | 21 | 21 | 21 | 22 | 22 | 22 |
| 23 | 12 | 12 | 13 | 13 | 13 | 14 | 14 | 15 | 15 | 15 | 16 | 16 | 16 | 17 | 17 | 18 | 18 | 18 | 19 | 19 | 20 | 20 | 20 | 21 | 21 | 21 | 22 | 22 | 23 | 23 | 23 |
| 24 | 12 | 13 | 13 | 14 | 14 | 14 | 15 | 15 | 16 | 16 | 16 | 17 | 17 | 18 | 18 | 18 | 19 | 19 | 20 | 20 | 20 | 21 | 21 | 22 | 22 | 22 | 23 | 23 | 24 | 24 | 24 |
| 25 | 13 | 13 | 14 | 14 | 15 | 15 | 15 | 16 | 16 | 17 | 17 | 18 | 18 | 18 | 19 | 19 | 20 | 20 | 20 | 21 | 21 | 22 | 22 | 22 | 23 | 23 | 24 | 24 | 25 | 25 | 25 |
| 26 | 13 | 14 | 14 | 15 | 15 | 16 | 16 | 16 | 17 | 17 | 18 | 18 | 19 | 19 | 20 | 20 | 20 | 21 | 21 | 22 | 22 | 23 | 23 | 23 | 24 | 24 | 25 | 25 | 26 | 26 | 26 |
| 27 | 14 | 14 | 15 | 15 | 16 | 16 | 17 | 17 | 18 | 18 | 18 | 19 | 19 | 20 | 20 | 21 | 21 | 22 | 22 | 22 | 23 | 23 | 24 | 24 | 25 | 25 | 26 | 26 | 27 | 27 | 27 |
| 28 | 14 | 15 | 15 | 16 | 16 | 17 | 17 | 18 | 18 | 19 | 19 | 20 | 20 | 21 | 21 | 21 | 22 | 22 | 23 | 23 | 24 | 24 | 25 | 25 | 26 | 26 | 27 | 27 | 28 | 28 | 28 |
| 29 | 15 | 15 | 16 | 16 | 17 | 17 | 18 | 18 | 19 | 19 | 20 | 20 | 21 | 21 | 22 | 22 | 23 | 23 | 24 | 24 | 25 | 25 | 26 | 26 | 27 | 27 | 28 | 28 | 29 | 29 | 29 |
| 30 | 16 | 16 | 16 | 17 | 18 | 18 | 18 | 19 | 20 | 20 | 20 | 21 | 22 | 22 | 22 | 23 | 24 | 24 | 24 | 25 | 26 | 26 | 26 | 27 | 28 | 28 | 28 | 29 | 30 | 30 | 30 |
| 31 | 16 | 17 | 17 | 18 | 18 | 19 | 19 | 20 | 20 | 21 | 21 | 22 | 22 | 23 | 23 | 24 | 24 | 25 | 25 | 26 | 26 | 27 | 27 | 28 | 28 | 29 | 29 | 30 | 30 | 31 | 31 |
| 32 | 17 | 17 | 18 | 18 | 19 | 19 | 20 | 20 | 21 | 21 | 22 | 22 | 23 | 23 | 24 | 25 | 25 | 26 | 26 | 27 | 27 | 28 | 28 | 29 | 29 | 30 | 30 | 31 | 31 | 32 | 32 |
| 33 | 17 | 18 | 18 | 19 | 19 | 20 | 20 | 21 | 21 | 22 | 23 | 23 | 24 | 24 | 25 | 25 | 26 | 26 | 27 | 28 | 28 | 29 | 29 | 30 | 30 | 31 | 31 | 32 | 32 | 33 | 33 |
| 34 | 18 | 18 | 19 | 19 | 20 | 20 | 21 | 22 | 22 | 23 | 23 | 24 | 24 | 25 | 26 | 26 | 27 | 27 | 28 | 28 | 29 | 29 | 30 | 31 | 31 | 32 | 32 | 33 | 33 | 34 | 34 |
| 35 | 18 | 19 | 19 | 20 | 20 | 21 | 22 | 22 | 23 | 23 | 24 | 24 | 25 | 26 | 26 | 27 | 27 | 28 | 29 | 29 | 30 | 30 | 31 | 32 | 32 | 33 | 33 | 34 | 34 | 35 | 35 |
| 36 | 19 | 19 | 20 | 20 | 21 | 22 | 22 | 23 | 23 | 24 | 25 | 25 | 26 | 26 | 27 | 28 | 28 | 29 | 29 | 30 | 31 | 31 | 32 | 32 | 33 | 34 | 34 | 35 | 35 | 36 | 36 |
| 37 | 19 | 20 | 20 | 21 | 22 | 22 | 23 | 23 | 24 | 25 | 25 | 26 | 27 | 27 | 28 | 28 | 29 | 30 | 30 | 31 | 31 | 32 | 33 | 33 | 34 | 35 | 35 | 36 | 36 | 37 | 37 |
| 38 | 20 | 20 | 21 | 22 | 22 | 23 | 23 | 24 | 25 | 25 | 26 | 27 | 27 | 28 | 28 | 29 | 30 | 30 | 31 | 32 | 32 | 33 | 34 | 34 | 35 | 35 | 36 | 37 | 37 | 38 | 38 |
| 39 | 20 | 21 | 21 | 22 | 23 | 23 | 24 | 25 | 25 | 26 | 27 | 27 | 28 | 29 | 29 | 30 | 31 | 31 | 32 | 32 | 33 | 34 | 34 | 35 | 36 | 36 | 37 | 38 | 38 | 39 | 39 |
| 40 | 21 | 21 | 22 | 23 | 23 | 24 | 25 | 25 | 26 | 27 | 27 | 28 | 29 | 29 | 30 | 31 | 31 | 32 | 33 | 33 | 34 | 35 | 35 | 36 | 37 | 37 | 38 | 39 | 39 | 40 | 40 |
| 41 | 21 | 22 | 23 | 23 | 24 | 25 | 25 | 26 | 27 | 27 | 28 | 29 | 29 | 30 | 31 | 31 | 32 | 33 | 33 | 34 | 35 | 36 | 36 | 37 | 38 | 38 | 39 | 40 | 40 | 41 | 41 |
| 42 | 22 | 22 | 23 | 24 | 24 | 25 | 26 | 27 | 27 | 28 | 29 | 29 | 30 | 31 | 32 | 32 | 33 | 34 | 34 | 35 | 36 | 36 | 37 | 38 | 38 | 39 | 40 | 41 | 41 | 42 | 42 |
| 43 | 22 | 23 | 24 | 24 | 25 | 26 | 27 | 27 | 28 | 29 | 29 | 30 | 31 | 32 | 32 | 33 | 34 | 34 | 35 | 36 | 37 | 37 | 38 | 39 | 39 | 40 | 41 | 42 | 42 | 43 | 43 |
| 44 | 23 | 23 | 24 | 25 | 26 | 26 | 27 | 28 | 29 | 29 | 30 | 31 | 32 | 32 | 33 | 34 | 34 | 35 | 36 | 37 | 37 | 38 | 39 | 40 | 40 | 41 | 42 | 43 | 43 | 44 | 44 |
| 45 | 23 | 24 | 25 | 26 | 26 | 27 | 28 | 28 | 29 | 30 | 31 | 32 | 32 | 33 | 34 | 34 | 35 | 36 | 37 | 38 | 38 | 39 | 40 | 40 | 41 | 42 | 43 | 44 | 44 | 45 | 45 |
| 46 | 24 | 25 | 25 | 26 | 27 | 28 | 28 | 29 | 30 | 31 | 31 | 32 | 33 | 34 | 34 | 35 | 36 | 37 | 38 | 38 | 39 | 40 | 41 | 41 | 42 | 43 | 44 | 44 | 45 | 46 | 46 |
| 47 | 24 | 25 | 26 | 27 | 27 | 28 | 29 | 30 | 31 | 31 | 32 | 33 | 34 | 34 | 35 | 36 | 37 | 38 | 38 | 39 | 40 | 41 | 42 | 42 | 43 | 44 | 45 | 45 | 46 | 47 | 47 |
| 48 | 25 | 26 | 26 | 27 | 28 | 29 | 30 | 30 | 31 | 32 | 33 | 34 | 34 | 35 | 36 | 37 | 38 | 38 | 39 | 40 | 41 | 42 | 42 | 43 | 44 | 45 | 46 | 46 | 47 | 48 | 48 |
| 49 | 25 | 26 | 27 | 28 | 29 | 29 | 30 | 31 | 32 | 33 | 33 | 34 | 35 | 36 | 37 | 38 | 38 | 39 | 40 | 41 | 42 | 42 | 43 | 44 | 45 | 46 | 47 | 47 | 48 | 49 | 49 |
| 50 | 26 | 27 | 28 | 28 | 29 | 30 | 31 | 32 | 32 | 33 | 34 | 35 | 36 | 37 | 38 | 38 | 39 | 40 | 41 | 42 | 42 | 43 | 44 | 45 | 46 | 47 | 48 | 48 | 49 | 50 | 50 |
| 51 | 26 | 27 | 28 | 29 | 30 | 31 | 31 | 32 | 33 | 34 | 35 | 36 | 37 | 37 | 38 | 39 | 40 | 41 | 42 | 42 | 43 | 44 | 45 | 46 | 47 | 48 | 48 | 49 | 50 | 51 | 51 |
| 52 | 27 | 28 | 29 | 29 | 30 | 31 | 32 | 33 | 34 | 35 | 36 | 36 | 37 | 38 | 39 | 40 | 41 | 42 | 42 | 43 | 44 | 45 | 46 | 47 | 48 | 49 | 49 | 50 | 51 | 52 | 52 |
| 53 | 27 | 28 | 29 | 30 | 31 | 32 | 33 | 34 | 34 | 35 | 36 | 37 | 38 | 39 | 40 | 41 | 42 | 42 | 43 | 44 | 45 | 46 | 47 | 48 | 49 | 49 | 50 | 51 | 52 | 53 | 53 |
| 54 | 28 | 29 | 30 | 31 | 32 | 32 | 33 | 34 | 35 | 36 | 37 | 38 | 39 | 40 | 40 | 41 | 42 | 43 | 44 | 45 | 46 | 47 | 48 | 49 | 50 | 50 | 51 | 52 | 53 | 54 | 54 |
| 55 | 28 | 29 | 30 | 31 | 32 | 33 | 34 | 35 | 36 | 37 | 38 | 38 | 39 | 40 | 41 | 42 | 43 | 44 | 45 | 46 | 47 | 48 | 49 | 50 | 50 | 51 | 52 | 53 | 54 | 55 | 55 |
| 56 | 29 | 30 | 31 | 32 | 33 | 34 | 35 | 35 | 36 | 37 | 38 | 39 | 40 | 41 | 42 | 43 | 44 | 45 | 46 | 47 | 48 | 49 | 49 | 50 | 51 | 52 | 53 | 54 | 55 | 56 | 56 |
| 57 | 29 | 30 | 31 | 32 | 33 | 34 | 35 | 36 | 37 | 38 | 39 | 40 | 41 | 42 | 43 | 44 | 45 | 46 | 47 | 48 | 48 | 49 | 50 | 51 | 52 | 53 | 54 | 55 | 56 | 57 | 57 |
| 58 | 30 | 31 | 32 | 33 | 34 | 35 | 36 | 37 | 38 | 39 | 40 | 41 | 42 | 43 | 44 | 44 | 45 | 46 | 47 | 48 | 49 | 50 | 51 | 52 | 53 | 54 | 55 | 56 | 57 | 58 | 58 |
| 59 | 30 | 31 | 32 | 33 | 34 | 35 | 36 | 37 | 38 | 39 | 40 | 41 | 42 | 43 | 44 | 45 | 46 | 47 | 48 | 49 | 50 | 51 | 52 | 53 | 54 | 55 | 56 | 57 | 58 | 59 | 59 |

# D PAGE EXCERPTS FROM THE *AIR ALMANAC*

Material in this appendix is taken from the *Air Almanac* by permission of the Controller of Her Majesty's Stationery Office, Royal Greenwich Observatory, Sussex, England, and by permission of the Almanac Office, U.S. Naval Observatory, Washington, D.C.

## STARS, JAN.—JUNE, 1983

| No. | Name | | Mag. | S.H.A. | Dec. |
|---|---|---|---|---|---|
| | | | | ° ′ | ° ′ |
| 7* | *Acamar* | | 3·1 | 315 36 | S. 40 23 |
| 5* | *Achernar* | | 0·6 | 335 44 | S. 57 19 |
| 30* | *Acrux* | | 1·1 | 173 35 | S. 63 00 |
| 19 | *Adhara* | † | 1·6 | 255 31 | S. 28 57 |
| 10* | *Aldebaran* | † | 1·1 | 291 16 | N. 16 29 |
| 32* | *Alioth* | | 1·7 | 166 40 | N. 56 03 |
| 34* | *Alkaid* | | 1·9 | 153 17 | N. 49 24 |
| 55 | *Al Na'ir* | | 2·2 | 28 13 | S. 47 03 |
| 15 | *Alnilam* | † | 1·8 | 276 10 | S. 1 13 |
| 25* | *Alphard* | † | 2·2 | 218 19 | S. 8 35 |
| 41* | *Alphecca* | † | 2·3 | 126 30 | N. 26 46 |
| 1* | *Alpheratz* | † | 2·2 | 358 08 | N. 29 00 |
| 51* | *Altair* | † | 0·9 | 62 31 | N. 8 49 |
| 2 | *Ankaa* | | 2·4 | 353 39 | S. 42 24 |
| 42* | *Antares* | † | 1·2 | 112 55 | S. 26 24 |
| 37* | *Arcturus* | † | 0·2 | 146 17 | N. 19 16 |
| 43 | *Atria* | | 1·9 | 108 17 | S. 69 00 |
| 22 | *Avior* | | 1·7 | 234 27 | S. 59 27 |
| 13 | *Bellatrix* | † | 1·7 | 278 57 | N. 6 20 |
| 16* | *Betelgeuse* | † | 0·1–1·2 | 271 26 | N. 7 24 |
| 17* | *Canopus* | | −0·9 | 264 06 | S. 52 41 |
| 12* | *Capella* | | 0·2 | 281 09 | N. 45 59 |
| 53* | *Deneb* | | 1·3 | 49 47 | N. 45 13 |
| 28* | *Denebola* | † | 2·2 | 182 57 | N. 14 40 |
| 4* | *Diphda* | † | 2·2 | 349 19 | S. 18 05 |
| 27* | *Dubhe* | | 2·0 | 194 19 | N. 61 51 |
| 14 | *Elnath* | † | 1·8 | 278 42 | N. 28 36 |
| 47 | *Eltanin* | | 2·4 | 90 57 | N. 51 29 |
| 54* | *Enif* | † | 2·5 | 34 10 | N. 9 48 |
| 56* | *Fomalhaut* | † | 1·3 | 15 50 | S. 29 43 |
| 31 | *Gacrux* | | 1·6 | 172 26 | S. 57 01 |
| 29* | *Gienah* | † | 2·8 | 176 16 | S. 17 27 |
| 35 | *Hadar* | | 0·9 | 149 21 | S. 60 17 |
| 6* | *Hamal* | † | 2·2 | 328 27 | N. 23 23 |
| 48 | *Kaus Aust.* | | 2·0 | 84 14 | S. 34 24 |
| 40* | *Kochab* | | 2·2 | 137 18 | N. 74 13 |
| 57 | *Markab* | † | 2·6 | 14 02 | N. 15 07 |
| 8* | *Menkar* | † | 2·8 | 314 39 | N. 4 01 |
| 36 | *Menkent* | | 2·3 | 148 35 | S. 36 17 |
| 24* | *Miaplacidus* | | 1·8 | 221 44 | S. 69 39 |
| 9* | *Mirfak* | | 1·9 | 309 14 | N. 49 48 |
| 50* | *Nunki* | † | 2·1 | 76 27 | S. 26 19 |
| 52* | *Peacock* | | 2·1 | 53 56 | S. 56 47 |
| 21* | *Pollux* | † | 1·2 | 243 56 | N. 28 04 |
| 20* | *Procyon* | † | 0·5 | 245 24 | N. 5 16 |
| 46* | *Rasalhague* | † | 2·1 | 96 28 | N. 12 34 |
| 26* | *Regulus* | † | 1·3 | 208 08 | N. 12 03 |
| 11* | *Rigel* | † | 0·3 | 281 34 | S. 8 13 |
| 38* | *Rigil Kent.* | | 0·1 | 140 23 | S. 60 46 |
| 44 | *Sabik* | † | 2·6 | 102 39 | S. 15 42 |
| 3* | *Schedar* | | 2·5 | 350 07 | N. 56 27 |
| 45* | *Shaula* | | 1·7 | 96 53 | S. 37 06 |
| 18* | *Sirius* | † | −1·6 | 258 54 | S. 16 42 |
| 33* | *Spica* | † | 1·2 | 158 55 | S. 11 04 |
| 23* | *Suhail* | | 2·2 | 223 09 | S. 43 22 |
| 49* | *Vega* | | 0·1 | 80 55 | N. 38 46 |
| 39 | *Zuben'ubi* | † | 2·9 | 137 31 | S. 15 58 |

## INTERPOLATION OF G.H.A.

Increment to be added for intervals of G.M.T. to G.H.A. of: Sun, Aries (♈) and planets; Moon

| SUN, etc. | | MOON | SUN, etc. | | MOON | SUN, etc. | | MOON |
|---|---|---|---|---|---|---|---|---|
| m s | ° ′ | m s | m s | ° ′ | m s | m s | ° ′ | m s |
| 00 00 | 0 00 | 00 00 | 03 17 | 0 50 | 03 25 | 06 37 | 1 40 | 06 52 |
| 01 | 0 01 | 00 02 | 21 | 0 51 | 03 29 | 41 | 1 41 | 06 56 |
| 05 | 0 02 | 00 06 | 25 | 0 52 | 03 33 | 45 | 1 42 | 07 00 |
| 09 | 0 03 | 00 10 | 29 | 0 53 | 03 37 | 49 | 1 43 | 07 04 |
| 13 | 0 04 | 00 14 | 33 | 0 54 | 03 41 | 53 | 1 44 | 07 08 |
| 17 | 0 05 | 00 18 | 37 | 0 55 | 03 45 | 06 57 | 1 45 | 07 13 |
| 21 | 0 06 | 00 22 | 41 | 0 56 | 03 49 | 07 01 | 1 46 | 07 17 |
| 25 | 0 07 | 00 26 | 45 | 0 57 | 03 54 | 05 | 1 47 | 07 21 |
| 29 | 0 08 | 00 31 | 49 | 0 58 | 03 58 | 09 | 1 48 | 07 25 |
| 33 | 0 09 | 00 35 | 53 | 0 59 | 04 02 | 13 | 1 49 | 07 29 |
| 37 | 0 10 | 00 39 | 03 57 | 1 00 | 04 06 | 17 | 1 50 | 07 33 |
| 41 | 0 11 | 00 43 | 04 01 | 1 01 | 04 10 | 21 | 1 51 | 07 37 |
| 45 | 0 12 | 00 47 | 05 | 1 02 | 04 14 | 25 | 1 52 | 07 42 |
| 49 | 0 13 | 00 51 | 09 | 1 03 | 04 19 | 29 | 1 53 | 07 46 |
| 53 | 0 14 | 00 55 | 13 | 1 04 | 04 23 | 33 | 1 54 | 07 50 |
| 00 57 | 0 15 | 01 00 | 17 | 1 05 | 04 27 | 37 | 1 55 | 07 54 |
| 01 01 | 0 16 | 01 04 | 21 | 1 06 | 04 31 | 41 | 1 56 | 07 58 |
| 05 | 0 17 | 01 08 | 25 | 1 07 | 04 35 | 45 | 1 57 | 08 02 |
| 09 | 0 18 | 01 12 | 29 | 1 08 | 04 39 | 49 | 1 58 | 08 06 |
| 13 | 0 19 | 01 16 | 33 | 1 09 | 04 43 | 53 | 1 59 | 08 11 |
| 17 | 0 20 | 01 20 | 37 | 1 10 | 04 48 | 07 57 | 2 00 | 08 15 |
| 21 | 0 21 | 01 24 | 41 | 1 11 | 04 52 | 08 01 | 2 01 | 08 19 |
| 25 | 0 22 | 01 29 | 45 | 1 12 | 04 56 | 05 | 2 02 | 08 23 |
| 29 | 0 23 | 01 33 | 49 | 1 13 | 05 00 | 09 | 2 03 | 08 27 |
| 33 | 0 24 | 01 37 | 53 | 1 14 | 05 04 | 13 | 2 04 | 08 31 |
| 37 | 0 25 | 01 41 | 04 57 | 1 15 | 05 08 | 17 | 2 05 | 08 35 |
| 41 | 0 26 | 01 45 | 05 01 | 1 16 | 05 12 | 21 | 2 06 | 08 40 |
| 45 | 0 27 | 01 49 | 05 | 1 17 | 05 17 | 25 | 2 07 | 08 44 |
| 49 | 0 28 | 01 53 | 09 | 1 18 | 05 21 | 29 | 2 08 | 08 48 |
| 53 | 0 29 | 01 58 | 13 | 1 19 | 05 25 | 33 | 2 09 | 08 52 |
| 01 57 | 0 30 | 02 02 | 17 | 1 20 | 05 29 | 37 | 2 10 | 08 56 |
| 02 01 | 0 31 | 02 06 | 21 | 1 21 | 05 33 | 41 | 2 11 | 09 00 |
| 05 | 0 32 | 02 10 | 25 | 1 22 | 05 37 | 45 | 2 12 | 09 04 |
| 09 | 0 33 | 02 14 | 29 | 1 23 | 05 41 | 49 | 2 13 | 09 09 |
| 13 | 0 34 | 02 18 | 33 | 1 24 | 05 46 | 53 | 2 14 | 09 13 |
| 17 | 0 35 | 02 22 | 37 | 1 25 | 05 50 | 08 57 | 2 15 | 09 17 |
| 21 | 0 36 | 02 27 | 41 | 1 26 | 05 54 | 09 01 | 2 16 | 09 21 |
| 25 | 0 37 | 02 31 | 45 | 1 27 | 05 58 | 05 | 2 17 | 09 25 |
| 29 | 0 38 | 02 35 | 49 | 1 28 | 06 02 | 09 | 2 18 | 09 29 |
| 33 | 0 39 | 02 39 | 53 | 1 29 | 06 06 | 13 | 2 19 | 09 33 |
| 37 | 0 40 | 02 43 | 05 57 | 1 30 | 06 10 | 17 | 2 20 | 09 38 |
| 41 | 0 41 | 02 47 | 06 01 | 1 31 | 06 15 | 21 | 2 21 | 09 42 |
| 45 | 0 42 | 02 51 | 05 | 1 32 | 06 19 | 25 | 2 22 | 09 46 |
| 49 | 0 43 | 02 56 | 09 | 1 33 | 06 23 | 29 | 2 23 | 09 50 |
| 53 | 0 44 | 03 00 | 13 | 1 34 | 06 27 | 33 | 2 24 | 09 54 |
| 02 57 | 0 45 | 03 04 | 17 | 1 35 | 06 31 | 37 | 2 25 | 09 58 |
| 03 01 | 0 46 | 03 08 | 21 | 1 36 | 06 35 | 41 | 2 26 | 10 00 |
| 05 | 0 47 | 03 12 | 25 | 1 37 | 06 39 | 45 | 2 27 | |
| 09 | 0 48 | 03 16 | 29 | 1 38 | 06 44 | 49 | 2 28 | |
| 13 | 0 49 | 03 20 | 33 | 1 39 | 06 48 | 53 | 2 29 | |
| 17 | 0 50 | 03 25 | 37 | 1 40 | 06 52 | 09 57 | 2 30 | |
| 03 21 | | 03 29 | 06 41 | | 06 56 | 10 00 | | |

* Stars used in H.O. 249 (A.P. 3270) Vol. 1.
† Stars that may be used with Vols. 2 and 3.

| GMT | ☉ SUN GHA | SUN Dec. | ARIES GHA ♈ | VENUS −3.9 GHA | VENUS Dec. | JUPITER −2.1 GHA | JUPITER Dec. | SATURN 0.7 GHA | SATURN Dec. | ◐ MOON GHA | MOON Dec. |
|---|---|---|---|---|---|---|---|---|---|---|---|
| h m | ° ′ | ° ′ | ° ′ | ° ′ | ° ′ | ° ′ | ° ′ | ° ′ | ° ′ | ° ′ | ° ′ |
| 12 00 | 359 58.6 | N23 15.0 | 82 13.1 | 311 00 | N20 16 | 200 09 | S20 05 | 235 19 | S 8 18 | 310 24 | N21 40 |
| 10 | 2 28.6 | 15.0 | 84 43.5 | 313 30 | | 202 39 | | 237 49 | | 312 48 | 39 |
| 20 | 4 58.6 | 15.0 | 87 13.9 | 316 00 | | 205 10 | | 240 19 | | 315 12 | 37 |
| 30 | 7 28.6 | · 15.1 | 89 44.3 | 318 30 | · · | 207 40 | · · | 242 50 | · · | 317 36 | · 36 |
| 40 | 9 58.5 | 15.1 | 92 14.7 | 321 00 | | 210 11 | | 245 20 | | 320 00 | 35 |
| 50 | 12 28.5 | 15.1 | 94 45.1 | 323 30 | | 212 41 | | 247 51 | | 322 24 | 34 |
| 13 00 | 14 58.5 | N23 15.1 | 97 15.5 | 326 00 | N20 15 | 215 12 | S20 05 | 250 21 | S 8 18 | 324 48 | N21 32 |
| 10 | 17 28.5 | 15.1 | 99 45.9 | 328 30 | | 217 42 | | 252 51 | | 327 12 | 31 |
| 20 | 19 58.5 | 15.2 | 102 16.3 | 331 00 | | 220 13 | | 255 22 | | 329 36 | 30 |
| 30 | 22 28.4 | · 15.2 | 104 46.8 | 333 30 | · · | 222 43 | · · | 257 52 | · · | 332 00 | · 29 |
| 40 | 24 58.4 | 15.2 | 107 17.2 | 336 00 | | 225 14 | | 260 23 | | 334 24 | 27 |
| 50 | 27 28.4 | 15.2 | 109 47.6 | 338 30 | | 227 44 | | 262 53 | | 336 48 | 26 |
| 14 00 | 29 58.4 | N23 15.3 | 112 18.0 | 341 00 | N20 15 | 230 15 | S20 05 | 265 24 | S 8 18 | 339 12 | N21 25 |
| 10 | 32 28.3 | 15.3 | 114 48.4 | 343 30 | | 232 45 | | 267 54 | | 341 36 | 24 |
| 20 | 34 58.3 | 15.3 | 117 18.8 | 346 00 | | 235 15 | | 270 24 | | 344 00 | 22 |
| 30 | 37 28.3 | · 15.3 | 119 49.2 | 348 30 | · · | 237 46 | · · | 272 55 | · · | 346 24 | · 21 |
| 40 | 39 58.3 | 15.3 | 122 19.6 | 351 00 | | 240 16 | | 275 25 | | 348 48 | 20 |
| 50 | 42 28.3 | 15.4 | 124 50.0 | 353 30 | | 242 47 | | 277 56 | | 351 12 | 18 |
| 15 00 | 44 58.2 | N23 15.4 | 127 20.5 | 356 00 | N20 14 | 245 17 | S20 05 | 280 26 | S 8 18 | 353 36 | N21 17 |
| 10 | 47 28.2 | 15.4 | 129 50.9 | 358 30 | | 247 48 | | 282 57 | | 356 00 | 16 |
| 20 | 49 58.2 | 15.4 | 132 21.3 | 1 00 | | 250 18 | | 285 27 | | 358 24 | 15 |
| 30 | 52 28.2 | · 15.4 | 134 51.7 | 3 30 | · · | 252 49 | · · | 287 57 | · · | 0 48 | · 13 |
| 40 | 54 58.2 | 15.5 | 137 22.1 | 6 00 | | 255 19 | | 290 28 | | 3 12 | 12 |
| 50 | 57 28.1 | 15.5 | 139 52.5 | 8 30 | | 257 50 | | 292 58 | | 5 36 | 11 |
| 16 00 | 59 58.1 | N23 15.5 | 142 22.9 | 11 00 | N20 13 | 260 20 | S20 05 | 295 29 | S 8 18 | 8 00 | N21 09 |
| 10 | 62 28.1 | 15.5 | 144 53.3 | 13 30 | | 262 50 | | 297 59 | | 10 24 | 08 |
| 20 | 64 58.1 | 15.5 | 147 23.7 | 16 00 | | 265 21 | | 300 29 | | 12 48 | 07 |
| 30 | 67 28.0 | · 15.6 | 149 54.1 | 18 30 | · · | 267 51 | · · | 303 00 | · · | 15 12 | · 05 |
| 40 | 69 58.0 | 15.6 | 152 24.6 | 21 00 | | 270 22 | | 305 30 | | 17 36 | 04 |
| 50 | 72 28.0 | 15.6 | 154 55.0 | 23 30 | | 272 52 | | 308 01 | | 20 00 | 03 |
| 17 00 | 74 58.0 | N23 15.6 | 157 25.4 | 26 00 | N20 12 | 275 23 | S20 05 | 310 31 | S 8 18 | 22 24 | N21 01 |
| 10 | 77 28.0 | 15.6 | 159 55.8 | 28 30 | | 277 53 | | 313 02 | | 24 48 | 21 00 |
| 20 | 79 57.9 | 15.7 | 162 26.2 | 31 00 | | 280 24 | | 315 32 | | 27 12 | 20 59 |
| 30 | 82 27.9 | · 15.7 | 164 56.6 | 33 30 | · · | 282 54 | · · | 318 02 | · · | 29 36 | · 57 |
| 40 | 84 57.9 | 15.7 | 167 27.0 | 36 00 | | 285 25 | | 320 33 | | 32 00 | 56 |
| 50 | 87 27.9 | 15.7 | 169 57.4 | 38 30 | | 287 55 | | 323 03 | | 34 24 | 54 |
| 18 00 | 89 57.8 | N23 15.8 | 172 27.8 | 41 00 | N20 12 | 290 26 | S20 05 | 325 34 | S 8 18 | 36 48 | N20 53 |
| 10 | 92 27.8 | 15.8 | 174 58.3 | 43 30 | | 292 56 | | 328 04 | | 39 12 | 52 |
| 20 | 94 57.8 | 15.8 | 177 28.7 | 46 00 | | 295 27 | | 330 34 | | 41 36 | 50 |
| 30 | 97 27.8 | · 15.8 | 179 59.1 | 48 30 | · · | 297 57 | · · | 333 05 | · · | 44 00 | · 49 |
| 40 | 99 57.8 | 15.8 | 182 29.5 | 51 00 | | 300 27 | | 335 35 | | 46 24 | 48 |
| 50 | 102 27.7 | 15.9 | 184 59.9 | 53 30 | | 302 58 | | 338 06 | | 48 48 | 46 |
| 19 00 | 104 57.7 | N23 15.9 | 187 30.3 | 56 00 | N20 11 | 305 28 | S20 04 | 340 36 | S 8 18 | 51 12 | N20 45 |
| 10 | 107 27.7 | 15.9 | 190 00.7 | 58 30 | | 307 59 | | 343 07 | | 53 36 | 43 |
| 20 | 109 57.7 | 15.9 | 192 31.1 | 61 00 | | 310 29 | | 345 37 | | 56 00 | 42 |
| 30 | 112 27.6 | · 15.9 | 195 01.5 | 63 30 | · · | 313 00 | · · | 348 07 | · · | 58 24 | · 41 |
| 40 | 114 57.6 | 16.0 | 197 32.0 | 66 00 | | 315 30 | | 350 38 | | 60 48 | 39 |
| 50 | 117 27.6 | 16.0 | 200 02.4 | 68 30 | | 318 01 | | 353 08 | | 63 12 | 38 |
| 20 00 | 119 57.6 | N23 16.0 | 202 32.8 | 71 00 | N20 10 | 320 31 | S20 04 | 355 39 | S 8 18 | 65 37 | N20 36 |
| 10 | 122 27.6 | 16.0 | 205 03.2 | 73 30 | | 323 02 | | 358 09 | | 68 01 | 35 |
| 20 | 124 57.5 | 16.0 | 207 33.6 | 76 00 | | 325 32 | | 0 40 | | 70 25 | 34 |
| 30 | 127 27.5 | · 16.1 | 210 04.0 | 78 30 | · · | 328 02 | · · | 3 10 | · · | 72 49 | · 32 |
| 40 | 129 57.5 | 16.1 | 212 34.4 | 81 00 | | 330 33 | | 5 40 | | 75 13 | 31 |
| 50 | 132 27.5 | 16.1 | 215 04.8 | 83 30 | | 333 03 | | 8 11 | | 77 37 | 29 |
| 21 00 | 134 57.4 | N23 16.1 | 217 35.2 | 86 00 | N20 09 | 335 34 | S20 04 | 10 41 | S 8 18 | 80 01 | N20 28 |
| 10 | 137 27.4 | 16.1 | 220 05.6 | 88 30 | | 338 04 | | 13 12 | | 82 25 | 26 |
| 20 | 139 57.4 | 16.2 | 222 36.1 | 91 00 | | 340 35 | | 15 42 | | 84 49 | 25 |
| 30 | 142 27.4 | · 16.2 | 225 06.5 | 93 30 | · · | 343 05 | · · | 18 13 | · · | 87 13 | · 23 |
| 40 | 144 57.4 | 16.2 | 227 36.9 | 96 00 | | 345 36 | | 20 43 | | 89 37 | 22 |
| 50 | 147 27.3 | 16.2 | 230 07.3 | 98 30 | | 348 06 | | 23 13 | | 92 02 | 21 |
| 22 00 | 149 57.3 | N23 16.2 | 232 37.7 | 101 00 | N20 09 | 350 37 | S20 04 | 25 44 | S 8 18 | 94 26 | N20 19 |
| 10 | 152 27.3 | 16.3 | 235 08.1 | 103 30 | | 353 07 | | 28 14 | | 96 50 | 18 |
| 20 | 154 57.3 | 16.3 | 237 38.5 | 106 00 | | 355 38 | | 30 45 | | 99 14 | 16 |
| 30 | 157 27.2 | · 16.3 | 240 08.9 | 108 30 | · · | 358 08 | · · | 33 15 | · · | 101 38 | · 15 |
| 40 | 159 57.2 | 16.3 | 242 39.3 | 111 00 | | 0 38 | | 35 45 | | 104 02 | 13 |
| 50 | 162 27.2 | 16.3 | 245 09.8 | 113 30 | | 3 09 | | 38 16 | | 106 26 | 12 |
| 23 00 | 164 57.2 | N23 16.4 | 247 40.2 | 116 00 | N20 08 | 5 39 | S20 04 | 40 46 | S 8 18 | 108 50 | N20 10 |
| 10 | 167 27.2 | 16.4 | 250 10.6 | 118 30 | | 8 10 | | 43 17 | | 111 14 | 09 |
| 20 | 169 57.1 | 16.4 | 252 41.0 | 121 00 | | 10 40 | | 45 47 | | 113 39 | 07 |
| 30 | 172 27.1 | · 16.4 | 255 11.4 | 123 30 | · · | 13 11 | · · | 48 18 | · · | 116 03 | · 06 |
| 40 | 174 57.1 | 16.4 | 257 41.8 | 126 00 | | 15 41 | | 50 48 | | 118 27 | 04 |
| 50 | 177 27.1 | 16.5 | 260 12.2 | 128 30 | | 18 12 | | 53 18 | | 120 51 | 03 |
| Rate | 14 59.9 | N0 00.1 | | 15 00.0 | S0 00.8 | 15 02.8 | N0 00.1 | 15 02.5 | 0 00.0 | 14 24.3 | S0 08.2* |

| Lat. | Moon-set | Diff. |
|---|---|---|
| N | | |
| ° | h m | m |
| 72 | □ | * |
| 70 | □ | * |
| 68 | □ | * |
| 66 | 01 31 | * |
| 64 | 00 48 | 10 |
| 62 | 00 18 | 15 |
| 60 | 24 22 | 11 |
| 58 | 24 08 | 13 |
| 56 | 23 56 | 15 |
| 54 | 23 46 | 16 |
| 52 | 23 36 | 17 |
| 50 | 23 28 | 18 |
| 45 | 23 10 | 20 |
| 40 | 22 55 | 22 |
| 35 | 22 43 | 24 |
| 30 | 22 32 | 24 |
| 20 | 22 13 | 27 |
| 10 | 21 56 | 28 |
| 0 | 21 41 | 30 |
| 10 | 21 25 | 31 |
| 20 | 21 08 | 33 |
| 30 | 20 49 | 34 |
| 35 | 20 37 | 36 |
| 40 | 20 24 | 37 |
| 45 | 20 09 | 39 |
| 50 | 19 49 | 41 |
| 52 | 19 40 | 42 |
| 54 | 19 30 | 43 |
| 56 | 19 18 | 44 |
| 58 | 19 04 | 46 |
| 60 | 18 48 | 49 |
| S | | |

Moon's P. in A.

| Alt. | Corr. | Alt. | Corr. |
|---|---|---|---|
| ° | + ′ | ° | + ′ |
| 0 | | 53 | |
| | 60 | | 35 |
| 8 | | 55 | |
| | 59 | | 34 |
| 13 | | 56 | |
| | 58 | | 33 |
| 17 | | 57 | |
| | 57 | | 32 |
| 20 | | 58 | |
| | 56 | | 31 |
| 22 | | 59 | |
| | 55 | | 30 |
| 25 | | 60 | |
| | 54 | | 29 |
| 27 | | 61 | |
| | 53 | | 28 |
| 29 | | 62 | |
| | 52 | | 27 |
| 31 | | 63 | |
| | 51 | | 26 |
| 33 | | 64 | |
| | 50 | | 25 |
| 34 | | 65 | |
| | 49 | | 24 |
| 36 | | 67 | |
| | 48 | | 23 |
| 37 | | 68 | |
| | 47 | | 22 |
| 39 | | 69 | |
| | 46 | | 21 |
| 40 | | 70 | |
| | 45 | | 20 |
| 42 | | 71 | |
| | 44 | | 19 |
| 43 | | 72 | |
| | 43 | | 18 |
| 45 | | 73 | |
| | 42 | | 17 |
| 46 | | 74 | |
| | 41 | | 16 |
| 47 | | 75 | |
| | 40 | | 15 |
| 49 | | 76 | |
| | 39 | | 14 |
| 50 | | 77 | |
| | 38 | | 13 |
| 51 | | 78 | |
| | 37 | | 12 |
| 52 | | 79 | |
| | 36 | | 11 |
| 53 | | 80 | |
| | 35 | | 10 |
| 55 | | 81 | |

Sun SD 15′.8

Moon SD 16′

Age 4d

| GMT | ☉ SUN GHA | SUN Dec. | ARIES GHA ♈ | VENUS −3.9 GHA | VENUS Dec. | JUPITER −2.1 GHA | JUPITER Dec. | SATURN 0.7 GHA | SATURN Dec. | ◐ MOON GHA | MOON Dec. |
|---|---|---|---|---|---|---|---|---|---|---|---|
| h m | ° ′ | ° ′ | ° ′ | ° ′ | ° ′ | ° ′ | ° ′ | ° ′ | ° ′ | ° ′ | ° ′ |
| 00 00 | 179 57.0 | N23 16.5 | 262 42.6 | 131 00 | N20 07 | 20 42 | S20 04 | 55 49 | S 8 18 | 123 15 | N20 01 |
| 10 | 182 27.0 | 16.5 | 265 13.0 | 133 30 | | 23 13 | | 58 19 | | 125 39 | 20 00 |
| 20 | 184 57.0 | 16.5 | 267 43.5 | 136 00 | | 25 43 | | 60 50 | | 128 03 | 19 58 |
| 30 | 187 27.0 | · 16.5 | 270 13.9 | 138 30 | · · | 28 14 | · · | 63 20 | · · | 130 27 | · 57 |
| 40 | 189 57.0 | 16.6 | 272 44.3 | 141 00 | | 30 44 | | 65 51 | | 132 52 | 55 |
| 50 | 192 26.9 | 16.6 | 275 14.7 | 143 30 | | 33 14 | | 68 21 | | 135 16 | 54 |
| 01 00 | 194 56.9 | N23 16.6 | 277 45.1 | 146 00 | N20 06 | 35 45 | S20 04 | 70 51 | S 8 18 | 137 40 | N19 52 |
| 10 | 197 26.9 | 16.6 | 280 15.5 | 148 30 | | 38 15 | | 73 22 | | 140 04 | 51 |
| 20 | 199 56.9 | 16.6 | 282 45.9 | 151 00 | | 40 46 | | 75 52 | | 142 28 | 49 |
| 30 | 202 26.8 | · 16.7 | 285 16.3 | 153 30 | · · | 43 16 | · · | 78 23 | · · | 144 52 | · 48 |
| 40 | 204 56.8 | 16.7 | 287 46.7 | 156 00 | | 45 47 | | 80 53 | | 147 17 | 46 |
| 50 | 207 26.8 | 16.7 | 290 17.1 | 158 30 | | 48 17 | | 83 23 | | 149 41 | 45 |
| 02 00 | 209 56.8 | N23 16.7 | 292 47.6 | 161 00 | N20 06 | 50 48 | S20 04 | 85 54 | S 8 18 | 152 05 | N19 43 |
| 10 | 212 26.8 | 16.7 | 295 18.0 | 163 30 | | 53 18 | | 88 24 | | 154 29 | 41 |
| 20 | 214 56.7 | 16.8 | 297 48.4 | 166 00 | | 55 49 | | 90 55 | | 156 53 | 40 |
| 30 | 217 26.7 | · 16.8 | 300 18.8 | 168 30 | · · | 58 19 | · · | 93 25 | · · | 159 17 | · 38 |
| 40 | 219 56.7 | 16.8 | 302 49.2 | 171 00 | | 60 50 | | 95 56 | | 161 42 | 37 |
| 50 | 222 26.7 | 16.8 | 305 19.6 | 173 30 | | 63 20 | | 98 26 | | 164 06 | 35 |
| 03 00 | 224 56.6 | N23 16.8 | 307 50.0 | 176 00 | N20 05 | 65 50 | S20 04 | 100 56 | S 8 18 | 166 30 | N19 34 |
| 10 | 227 26.6 | 16.9 | 310 20.4 | 178 30 | | 68 21 | | 103 27 | | 168 54 | 32 |
| 20 | 229 56.6 | 16.9 | 312 50.8 | 181 00 | | 70 51 | | 105 57 | | 171 18 | 30 |
| 30 | 232 26.6 | · 16.9 | 315 21.3 | 183 30 | · · | 73 22 | · · | 108 28 | · · | 173 42 | · 29 |
| 40 | 234 56.6 | 16.9 | 317 51.7 | 186 00 | | 75 52 | | 110 58 | | 176 07 | 27 |
| 50 | 237 26.5 | 16.9 | 320 22.1 | 188 30 | | 78 23 | | 113 29 | | 178 31 | 26 |
| 04 00 | 239 56.5 | N23 17.0 | 322 52.5 | 191 00 | N20 04 | 80 53 | S20 04 | 115 59 | S 8 18 | 180 55 | N19 24 |
| 10 | 242 26.5 | 17.0 | 325 22.9 | 193 30 | | 83 24 | | 118 29 | | 183 19 | 23 |
| 20 | 244 56.5 | 17.0 | 327 53.3 | 196 00 | | 85 54 | | 121 00 | | 185 43 | 21 |
| 30 | 247 26.4 | · 17.0 | 330 23.7 | 198 30 | · · | 88 25 | · · | 123 30 | · · | 188 08 | · 19 |
| 40 | 249 56.4 | 17.0 | 332 54.1 | 201 00 | | 90 55 | | 126 01 | | 190 32 | 18 |
| 50 | 252 26.4 | 17.1 | 335 24.5 | 203 30 | | 93 25 | | 128 31 | | 192 56 | 16 |
| 05 00 | 254 56.4 | N23 17.1 | 337 55.0 | 206 00 | N20 03 | 95 56 | S20 04 | 131 02 | S 8 18 | 195 20 | N19 15 |
| 10 | 257 26.4 | 17.1 | 340 25.4 | 208 30 | | 98 26 | | 133 32 | | 197 44 | 13 |
| 20 | 259 56.3 | 17.1 | 342 55.8 | 211 00 | | 100 57 | | 136 02 | | 200 09 | 11 |
| 30 | 262 26.3 | · 17.1 | 345 26.2 | 213 30 | · · | 103 27 | · · | 138 33 | · · | 202 33 | · 10 |
| 40 | 264 56.3 | 17.2 | 347 56.6 | 216 00 | | 105 58 | | 141 03 | | 204 57 | 08 |
| 50 | 267 26.3 | 17.2 | 350 27.0 | 218 30 | | 108 28 | | 143 34 | | 207 21 | 07 |
| 06 00 | 269 56.2 | N23 17.2 | 352 57.4 | 221 00 | N20 02 | 110 59 | S20 04 | 146 04 | S 8 18 | 209 46 | N19 05 |
| 10 | 272 26.2 | 17.2 | 355 27.8 | 223 30 | | 113 29 | | 148 34 | | 212 10 | 03 |
| 20 | 274 56.2 | 17.2 | 357 58.2 | 226 00 | | 116 00 | | 151 05 | | 214 34 | 02 |
| 30 | 277 26.2 | · 17.2 | 0 28.6 | 228 30 | · · | 118 30 | · · | 153 35 | · · | 216 58 | 19 00 |
| 40 | 279 56.2 | 17.3 | 2 59.1 | 231 00 | | 121 01 | | 156 06 | | 219 23 | 18 58 |
| 50 | 282 26.1 | 17.3 | 5 29.5 | 233 30 | | 123 31 | | 158 36 | | 221 47 | 57 |
| 07 00 | 284 56.1 | N23 17.3 | 7 59.9 | 236 00 | N20 02 | 126 02 | S20 04 | 161 07 | S 8 18 | 224 11 | N18 55 |
| 10 | 287 26.1 | 17.3 | 10 30.3 | 238 30 | | 128 32 | | 163 37 | | 226 35 | 53 |
| 20 | 289 56.1 | 17.3 | 13 00.7 | 241 00 | | 131 02 | | 166 07 | | 229 00 | 52 |
| 30 | 292 26.0 | · 17.4 | 15 31.1 | 243 30 | · · | 133 33 | · · | 168 38 | · · | 231 24 | · 50 |
| 40 | 294 56.0 | 17.4 | 18 01.5 | 246 00 | | 136 03 | | 171 08 | | 233 48 | 48 |
| 50 | 297 26.0 | 17.4 | 20 31.9 | 248 30 | | 138 34 | | 173 39 | | 236 12 | 47 |
| 08 00 | 299 56.0 | N23 17.4 | 23 02.3 | 251 00 | N20 01 | 141 04 | S20 04 | 176 09 | S 8 18 | 238 37 | N18 45 |
| 10 | 302 26.0 | 17.4 | 25 32.8 | 253 30 | | 143 35 | | 178 40 | | 241 01 | 44 |
| 20 | 304 55.9 | 17.5 | 28 03.2 | 256 00 | | 146 05 | | 181 10 | | 243 25 | 42 |
| 30 | 307 25.9 | · 17.5 | 30 33.6 | 258 30 | · · | 148 36 | · · | 183 40 | · · | 245 49 | · 40 |
| 40 | 309 55.9 | 17.5 | 33 04.0 | 261 00 | | 151 06 | | 186 11 | | 248 14 | 39 |
| 50 | 312 25.9 | 17.5 | 35 34.4 | 263 30 | | 153 37 | | 188 41 | | 250 38 | 37 |
| 09 00 | 314 55.8 | N23 17.5 | 38 04.8 | 266 00 | N20 00 | 156 07 | S20 04 | 191 12 | S 8 18 | 253 02 | N18 35 |
| 10 | 317 25.8 | 17.6 | 40 35.2 | 268 30 | | 158 37 | | 193 42 | | 255 26 | 33 |
| 20 | 319 55.8 | 17.6 | 43 05.6 | 271 00 | | 161 08 | | 196 12 | | 257 51 | 32 |
| 30 | 322 25.8 | · 17.6 | 45 36.0 | 273 30 | · · | 163 38 | · · | 198 43 | · · | 260 15 | · 30 |
| 40 | 324 55.8 | 17.6 | 48 06.4 | 276 00 | | 166 09 | | 201 13 | | 262 39 | 28 |
| 50 | 327 25.7 | 17.6 | 50 36.9 | 278 30 | | 168 39 | | 203 44 | | 265 04 | 27 |
| 10 00 | 329 55.7 | N23 17.6 | 53 07.3 | 281 00 | N19 59 | 171 10 | S20 04 | 206 14 | S 8 18 | 267 28 | N18 25 |
| 10 | 332 25.7 | 17.7 | 55 37.7 | 283 30 | | 173 40 | | 208 45 | | 269 52 | 23 |
| 20 | 334 55.7 | 17.7 | 58 08.1 | 286 00 | | 176 11 | | 211 15 | | 272 16 | 22 |
| 30 | 337 25.6 | · 17.7 | 60 38.5 | 288 30 | · · | 178 41 | · · | 213 45 | · · | 274 41 | · 20 |
| 40 | 339 55.6 | 17.7 | 63 08.9 | 291 00 | | 181 12 | | 216 16 | | 277 05 | 18 |
| 50 | 342 25.6 | 17.7 | 65 39.3 | 293 30 | | 183 42 | | 218 46 | | 279 29 | 16 |
| 11 00 | 344 55.6 | N23 17.8 | 68 09.7 | 296 00 | N19 59 | 186 13 | S20 04 | 221 17 | S 8 18 | 281 54 | N18 15 |
| 10 | 347 25.6 | 17.8 | 70 40.1 | 298 30 | | 188 43 | | 223 47 | | 284 18 | 13 |
| 20 | 349 55.5 | 17.8 | 73 10.6 | 301 00 | | 191 13 | | 226 18 | | 286 42 | 11 |
| 30 | 352 25.5 | · 17.8 | 75 41.0 | 303 30 | · · | 193 44 | · · | 228 48 | · · | 289 07 | · 10 |
| 40 | 354 55.5 | 17.8 | 78 11.4 | 306 00 | | 196 14 | | 231 18 | | 291 31 | 08 |
| 50 | 357 25.5 | 17.9 | 80 41.8 | 308 30 | | 198 45 | | 233 49 | | 293 55 | 06 |
| Rate | 14 59.9 | N0 00.1 | | 15 00.0 | S0 00.8 | 15 02.8 | 0 00.0 | 15 02.5 | 0 00.0 | 14 25.4 | S0 09.7* |

| Lat. | Moonrise | Diff. |
|---|---|---|
| N | | |
| ° | h m | m |
| 72 | ▭ | * |
| 70 | 05 12 | * |
| 68 | 06 09 | * |
| 66 | 06 43 | 64 |
| 64 | 07 07 | 57 |
| 62 | 07 26 | 52 |
| 60 | 07 42 | 49 |
| 58 | 07 55 | 47 |
| 56 | 08 06 | 45 |
| 54 | 08 16 | 44 |
| 52 | 08 25 | 42 |
| 50 | 08 33 | 41 |
| 45 | 08 50 | 39 |
| 40 | 09 03 | 37 |
| 35 | 09 15 | 36 |
| 30 | 09 25 | 34 |
| 20 | 09 42 | 32 |
| 10 | 09 57 | 30 |
| 0 | 10 11 | 29 |
| 10 | 10 25 | 27 |
| 20 | 10 40 | 25 |
| 30 | 10 57 | 23 |
| 35 | 11 07 | 21 |
| 40 | 11 18 | 20 |
| 45 | 11 31 | 18 |
| 50 | 11 47 | 16 |
| 52 | 11 55 | 15 |
| 54 | 12 03 | 14 |
| 56 | 12 12 | 12 |
| 58 | 12 23 | 10 |
| 60 | 12 34 | 08 |
| S | | |

Moon's P. in A.

| Alt. | Corr. | Alt. | Corr. |
|---|---|---|---|
| ° | + ′ | ° | + ′ |
| 0 | | 53 | |
| | 60 | | 35 |
| 7 | | 54 | |
| | 59 | | 34 |
| 13 | | 56 | |
| | 58 | | 33 |
| 16 | | 57 | |
| | 57 | | 32 |
| 19 | | 58 | |
| | 56 | | 31 |
| 22 | | 59 | |
| | 55 | | 30 |
| 24 | | 60 | |
| | 54 | | 29 |
| 26 | | 61 | |
| | 53 | | 28 |
| 29 | | 62 | |
| | 52 | | 27 |
| 30 | | 63 | |
| | 51 | | 26 |
| 32 | | 64 | |
| | 50 | | 25 |
| 34 | | 65 | |
| | 49 | | 24 |
| 36 | | 66 | |
| | 48 | | 23 |
| 37 | | 67 | |
| | 47 | | 22 |
| 39 | | 69 | |
| | 46 | | 21 |
| 40 | | 70 | |
| | 45 | | 20 |
| 42 | | 71 | |
| | 44 | | 19 |
| 43 | | 72 | |
| | 43 | | 18 |
| 44 | | 73 | |
| | 42 | | 17 |
| 46 | | 74 | |
| | 41 | | 16 |
| 47 | | 75 | |
| | 40 | | 15 |
| 48 | | 76 | |
| | 39 | | 14 |
| 50 | | 77 | |
| | 38 | | 13 |
| 51 | | 78 | |
| | 37 | | 12 |
| 52 | | 79 | |
| | 36 | | 11 |
| 53 | | 80 | |
| | 35 | | 10 |
| 54 | | 81 | |

Sun SD 15′.8

Moon SD 16′

Age 4d

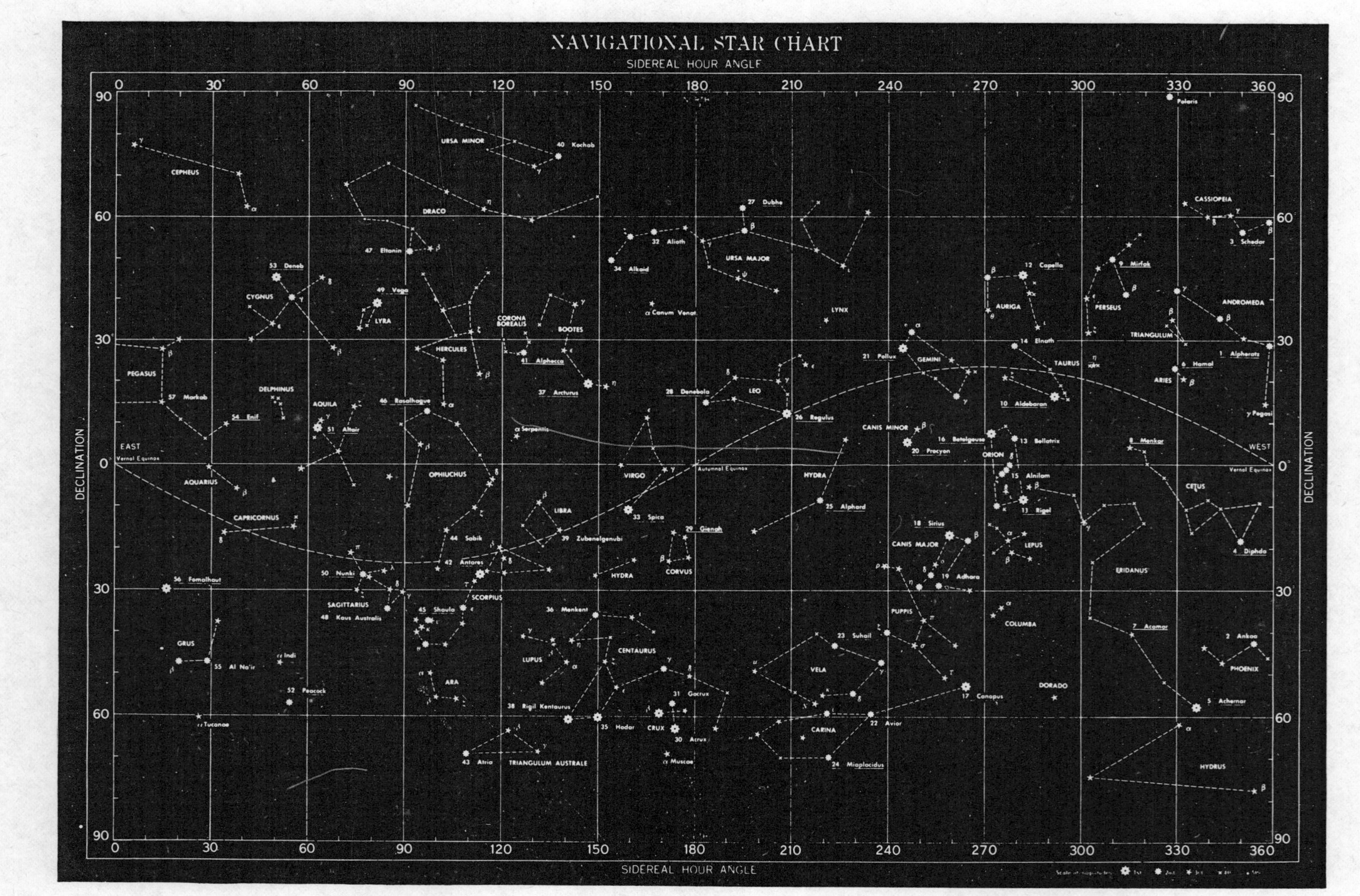
NAVIGATIONAL STAR CHART
SIDEREAL HOUR ANGLE
DECLINATION
EAST
Vernal Equinox
WEST
Vernal Equinox
Autumnal Equinox
40 Kochab
URSA MINOR
CEPHEUS
DRACO
47 Eltanin
53 Deneb
CYGNUS
49 Vega
LYRA
HERCULES
CORONA BOREALIS
41 Alphecca
BOOTES
37 Arcturus
32 Alioth
34 Alkaid
27 Dubhe
URSA MAJOR
α Canum Venat.
LYNX
PEGASUS
57 Markab
54 Enif
DELPHINUS
AQUILA
51 Altair
46 Rasalhague
OPHIUCHUS
α Serpentis
AQUARIUS
CAPRICORNUS
56 Fomalhaut
GRUS
55 Al Na'ir
α Indi
52 Peacock
α Tucanae
50 Nunki
SAGITTARIUS
48 Kaus Australis
45 Shaula
44 Sabik
42 Antares
SCORPIUS
ARA
LIBRA
39 Zubenelgenubi
36 Menkent
LUPUS
CENTAURUS
38 Rigil Kentaurus
35 Hadar
43 Atria
TRIANGULUM AUSTRALE
VIRGO
33 Spica
29 Gienah
CORVUS
HYDRA
28 Denebola
LEO
26 Regulus
31 Gacrux
CRUX
30 Acrux
α Muscae
HYDRA
25 Alphard
23 Suhail
VELA
CARINA
24 Miaplacidus
22 Avior
21 Pollux
GEMINI
CANIS MINOR
20 Procyon
16 Betelgeuse
18 Sirius
CANIS MAJOR
19 Adhara
PUPPIS
17 Canopus
12 Capella
AURIGA
14 Elnath
TAURUS
10 Aldebaran
13 Bellatrix
ORION
15 Alnilam
11 Rigel
LEPUS
COLUMBA
DORADO
9 Mirfak
PERSEUS
8 Menkar
CETUS
ERIDANUS
7 Acamar
Polaris
CASSIOPEIA
3 Schedar
ANDROMEDA
TRIANGULUM
1 Alpheratz
6 Hamal
ARIES
γ Pegasi
4 Diphda
2 Ankaa
PHOENIX
5 Achernar
HYDRUS

## How to Use the Navigational Star Chart

The navigational star chart reproduced here is a Mercator projection of the entire heavens. It shows the navigational stars with the constellations in which they are found. Mercator distortion must be taken into account when visualizing the heavens as displayed on the chart for comparison with the stars visible at your position at star time.

To determine what stars should be visible and in good position for observation at star time, it is first necessary to locate the position of your zenith on the star chart. Most navigators find eyeball accuracy good enough for this. There is no need to mark up your chart. Your latitude equals the declination of your zenith. The chart is marked west of Aries, the point at which the plane of the ecliptic intersects the vernal equinox, in sidereal hour angle through 360°. The SHA of Aries is 0°. To find your position west of Aries, simply find the LHA of Aries and subtract this from 360°. This gives you the SHA of your zenith. Actually you could locate your zenith with the LHA of Aries without this extra step by simply noting the point on the chart where this LHA value appears. For example, assume the LHA of Aries at your position is 137°. This means Aries is 137° west of your meridian. Note that Aries is shown on both the right and left margins of the star chart, in much the same way that the Greenwich meridian may be shown on both the right and left margins of a Mercator chart of the world. Since you are east of Aries, count off four of the charted sidereal meridians to the left of the right margin. The charted meridians are 30° apart. Four times 30 is 120. The longitudinal distance between charted meridians is divided into three parts of 10° each. Keep on going left through the first of these intermediate marks until you get to a point about three quarters of the way between the first and second intermediate marks. You have added another 10°, plus, approximately, another 7° to the charted distance east of Aries. You have located your position 137° east of Aries. Remember, eyeball accuracy is all you normally need. This is the meridian of your zenith, its SHA. And look what the chart reads—223°. Subtract 137° from 360° and the remainder is 223°, and 223° is the SHA of your zenith. Simple as that. Keep your eye on this meridian, and run up to a declination that approximates the DR latitude of your vessel, which is the declination of your zenith. You have your zenith located.

Now take the chart and position it with the north side up, as you would any other chart. The position of your zenith shows what stars should be directly overhead. Assuming you can see all the way down to the horizon, you should be able to see any star in the heavens within a radius of 90° on the chart from the center of the circle of your zenith. Actually, when Mercator distortion is accounted for, any star in the direction of the elevated pole whose codeclination is less than your latitude—the declination of your zenith—theoretically would be visible. Of course, the heavens are often obscured by haze to a height 10 or 15 degrees above the horizon.

If you can't quite visualize the lay of the heavens with the star chart lying on the chart table, take it out on deck, hold it over your head properly oriented north and south, and face the elevated pole. With your zenith as the point overhead, what you see in the heavens will be what you see on the star chart after Mercator distortion is taken into account.

# CORRECTIONS TO BE APPLIED TO SEXTANT ALTITUDE

## REFRACTION

To be subtracted from sextant altitude (referred to as observed altitude in A.P. 3270)

| $R_o$ | 0 | 5 | 10 | 15 | 20 | 25 | 30 | 35 | 40 | 45 | 50 | 55 | $R_o$ | 0·9 | 1·0 | 1·1 | 1·2 |
|---|---|---|---|---|---|---|---|---|---|---|---|---|---|---|---|---|---|
| | Height above sea level in units of 1 000 ft. — Sextant Altitude | | | | | | | | | | | | | $R = R_o \times f$ | | | |
| ′ | ° ′ | ° ′ | ° ′ | ° ′ | ° ′ | ° ′ | ° ′ | ° ′ | ° ′ | ° ′ | ° ′ | ° ′ | ′ | ′ | ′ | ′ | ′ |
| | 90 | 90 | 90 | 90 | 90 | 90 | 90 | 90 | 90 | 90 | 90 | 90 | | | | | |
| 0 | | | | | | | | | | | | | 0 | 0 | 0 | 0 | 0 |
| | 63 | 59 | 55 | 51 | 46 | 41 | 36 | 31 | 26 | 20 | 17 | 13 | | | | | |
| 1 | | | | | | | | | | | | | 1 | 1 | 1 | 1 | 1 |
| | 33 | 29 | 26 | 22 | 19 | 16 | 14 | 11 | 9 | 7 | 6 | 4 | | | | | |
| 2 | | | | | | | | | | | | | 2 | 2 | 2 | 2 | 2 |
| | 21 | 19 | 16 | 14 | 12 | 10 | 8 | 7 | 5 | 4 | 2 40 | 1 40 | | | | | |
| 3 | | | | | | | | | | | | | 3 | 3 | 3 | 3 | 4 |
| | 16 | 14 | 12 | 10 | 8 | 7 | 6 | 5 | 3 10 | 2 20 | 1 30 | 0 40 | | | | | |
| 4 | | | | | | | | | | | | | 4 | 4 | 4 | 4 | 5 |
| | 12 | 11 | 9 | 8 | 7 | 5 | 4 00 | 3 10 | 2 10 | 1 30 | 0 39 | +0 05 | | | | | |
| 5 | | | | | | | | | | | | | 5 | 5 | 5 | 5 | 6 |
| | 10 | 9 | 7 | 5 50 | 4 50 | 3 50 | 3 10 | 2 20 | 1 30 | 0 49 | +0 11 | −0 19 | | | | | |
| 6 | | | | | | | | | | | | | 6 | 5 | 6 | 7 | 7 |
| | 8 10 | 6 50 | 5 50 | 4 50 | 4 00 | 3 00 | 2 20 | 1 50 | 1 10 | 0 24 | −0 11 | −0 38 | | | | | |
| 7 | | | | | | | | | | | | | 7 | 6 | 7 | 8 | 8 |
| | 6 50 | 5 50 | 5 00 | 4 00 | 3 10 | 2 30 | 1 50 | 1 20 | 0 38 | +0 04 | −0 28 | −0 54 | | | | | |
| 8 | | | | | | | | | | | | | 8 | 7 | 8 | 9 | 10 |
| | 6 00 | 5 10 | 4 10 | 3 20 | 2 40 | 2 00 | 1 30 | 1 00 | 0 19 | −0 13 | −0 42 | −1 08 | | | | | |
| 9 | | | | | | | | | | | | | 9 | 8 | 9 | 10 | 11 |
| | 5 20 | 4 30 | 3 40 | 2 50 | 2 10 | 1 40 | 1 10 | 0 35 | +0 03 | −0 27 | −0 53 | −1 18 | | | | | |
| 10 | | | | | | | | | | | | | 10 | 9 | 10 | 11 | 12 |
| | 4 30 | 3 40 | 2 50 | 2 20 | 1 40 | 1 10 | 0 37 | +0 11 | −0 16 | −0 43 | −1 08 | −1 31 | | | | | |
| 12 | | | | | | | | | | | | | 12 | 11 | 12 | 13 | 14 |
| | 3 30 | 2 50 | 2 10 | 1 40 | 1 10 | 0 34 | +0 09 | −0 14 | −0 37 | −1 00 | −1 23 | −1 44 | | | | | |
| 14 | | | | | | | | | | | | | 14 | 13 | 14 | 15 | 17 |
| | 2 50 | 2 10 | 1 40 | 1 10 | 0 37 | +0 10 | −0 13 | −0 34 | −0 53 | −1 14 | −1 35 | −1 56 | | | | | |
| 16 | | | | | | | | | | | | | 16 | 14 | 16 | 18 | 19 |
| | 2 20 | 1 40 | 1 20 | 0 43 | +0 15 | −0 08 | −0 31 | −0 52 | −1 08 | −1 27 | −1 46 | −2 05 | | | | | |
| 18 | | | | | | | | | | | | | 18 | 16 | 18 | 20 | 22 |
| | 1 50 | 1 20 | 0 49 | +0 23 | −0 02 | −0 26 | −0 46 | −1 06 | −1 22 | −1 39 | −1 57 | −2 14 | | | | | |
| 20 | | | | | | | | | | | | | 20 | 18 | 20 | 22 | 24 |
| | 1 12 | 0 44 | +0 19 | −0 06 | −0 28 | −0 48 | −1 09 | −1 27 | −1 42 | −1 58 | −2 14 | −2 30 | | | | | |
| 25 | | | | | | | | | | | | | 25 | 22 | 25 | 28 | 30 |
| | 0 34 | +0 10 | −0 13 | −0 36 | −0 55 | −1 14 | −1 32 | −1 51 | −2 06 | −2 21 | −2 34 | −2 49 | | | | | |
| 30 | | | | | | | | | | | | | 30 | 27 | 30 | 33 | 36 |
| | +0 06 | −0 16 | −0 37 | −0 59 | −1 17 | −1 33 | −1 51 | −2 07 | −2 23 | −2 37 | −2 51 | −3 04 | | | | | |
| 35 | | | | | | | | | | | | | 35 | 31 | 35 | 38 | 42 |
| | −0 18 | −0 37 | −0 58 | −1 16 | −1 34 | −1 49 | −2 06 | −2 22 | −2 35 | −2 49 | −3 03 | −3 16 | | | | | |
| 40 | | | | | | | | | | | | | 40 | 36 | 40 | 44 | 48 |
| | | −0 53 | −1 14 | −1 31 | −1 47 | −2 03 | −2 18 | −2 33 | −2 47 | −2 59 | −3 13 | −3 25 | | | | | |
| 45 | | | | | | | | | | | | | 45 | 40 | 45 | 50 | 54 |
| | | −1 10 | −1 28 | −1 44 | −1 59 | −2 15 | −2 28 | −2 43 | −2 56 | −3 08 | −3 22 | −3 33 | | | | | |
| 50 | | | | | | | | | | | | | 50 | 45 | 50 | 55 | 60 |
| | | | −1 40 | −1 53 | −2 09 | −2 24 | −2 38 | −2 52 | −3 04 | −3 17 | −3 29 | −3 41 | | | | | |
| 55 | | | | | | | | | | | | | 55 | 49 | 55 | 60 | 66 |
| | | | | −2 03 | −2 18 | −2 33 | −2 46 | −3 01 | −3 12 | −3 25 | −3 37 | −3 48 | | | | | |
| 60 | | | | | | | | | | | | | 60 | 54 | 60 | 66 | 72 |
| | | | | | | | −2 53 | −3 07 | −3 19 | −3 31 | −3 42 | −3 53 | | | | | |

| $f$ | 0 | 5 | 10 | 15 | 20 | 25 | 30 | 35 | 40 | 45 | 50 | 55 | $f$ | 0·9 | 1·0 | 1·1 | 1·2 |
|---|---|---|---|---|---|---|---|---|---|---|---|---|---|---|---|---|---|
| | Temperature in °C. | | | | | | | | | | | | | $f$ | | | |
| | +47 | +36 | +27 | +18 | +10 | + 3 | − 5 | −13 | For these heights no temperature correction is necessary, so use $R = R_o$ | | | | | Where $R_o$ is less than 10′ or the height greater than 35 000 ft. use $R = R_o$ | | | |
| 0·9 | | | | | | | | | | | | | 0·9 | | | | |
| | +26 | +16 | + 6 | − 4 | −13 | −22 | −31 | −40 | | | | | | | | | |
| 1·0 | | | | | | | | | | | | | 1·0 | | | | |
| | + 5 | − 5 | −15 | −25 | −36 | −46 | −57 | −68 | | | | | | | | | |
| 1·1 | | | | | | | | | | | | | 1·1 | | | | |
| | −16 | −25 | −36 | −46 | −58 | −71 | −83 | −95 | | | | | | | | | |
| 1·2 | | | | | | | | | | | | | 1·2 | | | | |
| | −37 | −45 | −56 | −67 | −81 | −95 | | | | | | | | | | | |

Choose the column appropriate to height, in units of 1 000 ft., and find the range of altitude in which the sextant altitude lies; the corresponding value of $R_o$ is the refraction, to be subtracted from sextant altitude, unless conditions are extreme. In that case find $f$ from the lower table, with critical argument temperature. Use the table on the right to form the refraction, $R = R_o \times f$.

## CORIOLIS ($Z$) CORRECTION

To be applied by moving the position line a distance $Z$ to starboard (right) of the track in northern latitudes and to port (left) in southern latitudes.

| G/S KNOTS | 0° | 10° | 20° | 30° | 40° | 50° | 60° | 70° | 80° | 90° | G/S KNOTS | 0° | 10° | 20° | 30° | 40° | 50° | 60° | 70° | 80° | 90° |
|---|---|---|---|---|---|---|---|---|---|---|---|---|---|---|---|---|---|---|---|---|---|
| | ′ | ′ | ′ | ′ | ′ | ′ | ′ | ′ | ′ | ′ | | ′ | ′ | ′ | ′ | ′ | ′ | ′ | ′ | ′ | ′ |
| 150 | 0 | 1 | 1 | 2 | 3 | 3 | 3 | 4 | 4 | 4 | 550 | 0 | 3 | 5 | 7 | 9 | 11 | 12 | 14 | 14 | 14 |
| 200 | 0 | 1 | 2 | 3 | 3 | 4 | 5 | 5 | 5 | 5 | 600 | 0 | 3 | 5 | 8 | 10 | 12 | 14 | 15 | 16 | 16 |
| 250 | 0 | 1 | 2 | 3 | 4 | 5 | 6 | 6 | 6 | 7 | 650 | 0 | 3 | 6 | 9 | 11 | 13 | 15 | 16 | 17 | 17 |
| 300 | 0 | 1 | 3 | 4 | 5 | 6 | 7 | 7 | 8 | 8 | 700 | 0 | 3 | 6 | 9 | 12 | 14 | 16 | 17 | 18 | 18 |
| 350 | 0 | 2 | 3 | 5 | 6 | 7 | 8 | 9 | 9 | 9 | 750 | 0 | 3 | 7 | 10 | 13 | 15 | 17 | 18 | 19 | 20 |
| 400 | 0 | 2 | 4 | 5 | 7 | 8 | 9 | 10 | 10 | 10 | 800 | 0 | 4 | 7 | 10 | 13 | 16 | 18 | 20 | 21 | 21 |
| 450 | 0 | 2 | 4 | 6 | 8 | 9 | 10 | 11 | 12 | 12 | 850 | 0 | 4 | 8 | 11 | 14 | 17 | 19 | 21 | 22 | 22 |
| 500 | 0 | 2 | 4 | 7 | 8 | 10 | 11 | 12 | 13 | 13 | 900 | 0 | 4 | 8 | 12 | 15 | 18 | 20 | 22 | 23 | 24 |

## STANDARD DOME REFRACTION

To be *subtracted* from sextant altitude when using sextant suspension in a perspex dome

| Alt. | Refn. | Alt. | Refn. |
|---|---|---|---|
| ° | ′ | ° | ′ |
| 10 | 8 | 50 | 4 |
| 20 | 7 | 60 | 4 |
| 30 | 6 | 70 | 3 |
| 40 | 5 | 80 | 3 |

This table must not be used if a calibration table is fitted to the dome, or if a flat glass plate is provided, or for non-standard domes.

## BUBBLE SEXTANT ERROR

| Sextant Number | Alt. | Corr. |
|---|---|---|
| | ° | ′ |

# CORRECTIONS TO BE APPLIED TO MARINE SEXTANT ALTITUDES

## CORRECTION FOR DIP OF THE HORIZON

To be subtracted from sextant altitude

| Ht. | Dip | Ht. | Dip | Ht. | Dip | Ht. | Dip | Ht. | Dip |
|---|---|---|---|---|---|---|---|---|---|
| Ft. | ′ | Ft. | ′ | Ft. | ′ | Ft. | ′ | Ft. | ′ |
| 0 | | 114 | | 437 | | 968 | | 1 707 | |
| | 1 | | 11 | | 21 | | 31 | | 41 |
| 2 | | 137 | | 481 | | 1 033 | | 1 792 | |
| | 2 | | 12 | | 22 | | 32 | | 42 |
| 6 | | 162 | | 527 | | 1 099 | | 1 880 | |
| | 3 | | 13 | | 23 | | 33 | | 43 |
| 12 | | 189 | | 575 | | 1 168 | | 1 970 | |
| | 4 | | 14 | | 24 | | 34 | | 44 |
| 21 | | 218 | | 625 | | 1 239 | | 2 061 | |
| | 5 | | 15 | | 25 | | 35 | | 45 |
| 31 | | 250 | | 677 | | 1 311 | | 2 155 | |
| | 6 | | 16 | | 26 | | 36 | | 46 |
| 43 | | 283 | | 731 | | 1 386 | | 2 251 | |
| | 7 | | 17 | | 27 | | 37 | | 47 |
| 58 | | 318 | | 787 | | 1 463 | | 2 349 | |
| | 8 | | 18 | | 28 | | 38 | | 48 |
| 75 | | 356 | | 845 | | 1 543 | | 2 449 | |
| | 9 | | 19 | | 29 | | 39 | | 49 |
| 93 | | 395 | | 906 | | 1 624 | | 2 551 | |
| | 10 | | 20 | | 30 | | 40 | | 50 |
| 114 | | 437 | | 968 | | 1 707 | | 2 655 | |

### CORRECTIONS

In addition to sextant error and dip, corrections are to be applied for:
- Refraction
- Semi-diameter (for the Sun and Moon)
- Parallax (for the Moon)
- Dome refraction (if applicable)

### MARINE SEXTANT ERROR

Sextant Number

Index Error

# LIST OF CONTENTS

# E PAGE EXCERPTS FROM PUB. NO. 249, VOLUME I

Material in this appendix is taken from Pub. No. 249 by permission of the U.S. Naval Oceanographic Office, now the Defense Mapping Agency Hydrographic/Topographic Center.

 LAT 25°N

| LHA ♈ | Hc Zn | Hc Zn | Hc Zn | Hc Zn | Hc Zn | Hc Zn | Hc Zn |
|---|---|---|---|---|---|---|---|
| | •Alkaid | Alphecca | •SPICA | Gienah | REGULUS | •POLLUX | Dubhe |
| 180 | 57 58 034 | 42 11 076 | 48 28 148 | 47 23 175 | 60 28 249 | 33 25 288 | 51 55 349 |
| 181 | 58 27 033 | 43 04 076 | 48 56 149 | 47 27 176 | 59 37 250 | 32 33 289 | 51 45 348 |
| 182 | 58 56 032 | 43 56 076 | 49 24 150 | 47 30 178 | 58 45 251 | 31 42 289 | 51 33 348 |
| 183 | 59 25 031 | 44 49 076 | 49 50 152 | 47 32 179 | 57 54 252 | 30 50 289 | 51 21 347 |
| 184 | 59 52 030 | 45 42 076 | 50 15 153 | 47 32 180 | 57 02 253 | 29 59 289 | 51 09 346 |
| 185 | 60 19 029 | 46 35 077 | 50 39 155 | 47 31 182 | 56 10 254 | 29 08 290 | 50 56 346 |
| 186 | 60 45 028 | 47 28 077 | 51 02 156 | 47 29 183 | 55 18 254 | 28 16 290 | 50 42 345 |
| 187 | 61 11 027 | 48 21 077 | 51 23 157 | 47 25 185 | 54 25 255 | 27 25 290 | 50 28 344 |
| 188 | 61 35 026 | 49 14 077 | 51 43 159 | 47 20 186 | 53 32 256 | 26 34 290 | 50 13 344 |
| 189 | 61 58 025 | 50 07 077 | 52 02 160 | 47 14 187 | 52 40 257 | 25 43 291 | 49 57 343 |
| 190 | 62 21 024 | 51 00 078 | 52 20 162 | 47 06 189 | 51 47 257 | 24 52 291 | 49 41 343 |
| 191 | 62 42 023 | 51 53 078 | 52 36 164 | 46 58 190 | 50 54 258 | 24 02 291 | 49 25 342 |
| 192 | 63 03 021 | 52 46 078 | 52 50 165 | 46 47 192 | 50 00 259 | 23 11 291 | 49 07 341 |
| 193 | 63 22 020 | 53 40 078 | 53 04 167 | 46 36 193 | 49 07 259 | 22 20 292 | 48 50 341 |
| 194 | 63 40 019 | 54 33 078 | 53 16 168 | 46 23 194 | 48 13 260 | 21 30 292 | 48 32 340 |
| | •Kochab | VEGA | Rasalhague | •ANTARES | SPICA | •REGULUS | Dubhe |
| 195 | 38 41 009 | 19 42 055 | 24 32 087 | 18 05 132 | 53 26 170 | 47 20 260 | 48 13 340 |
| 196 | 38 50 009 | 20 27 056 | 25 27 087 | 18 46 133 | 53 35 172 | 46 26 261 | 47 54 339 |
| 197 | 38 58 009 | 21 12 056 | 26 21 088 | 19 26 133 | 53 42 173 | 45 32 262 | 47 34 339 |
| 198 | 39 06 008 | 21 57 056 | 27 15 088 | 20 05 134 | 53 47 175 | 44 39 262 | 47 14 338 |
| 199 | 39 14 008 | 22 42 056 | 28 10 089 | 20 44 135 | 53 52 177 | 43 45 263 | 46 54 338 |
| 200 | 39 21 008 | 23 28 057 | 29 04 089 | 21 23 135 | 53 54 178 | 42 51 263 | 46 33 337 |
| 201 | 39 29 007 | 24 13 057 | 29 59 090 | 22 01 136 | 53 55 180 | 41 57 264 | 46 12 337 |
| 202 | 39 35 007 | 24 59 057 | 30 53 090 | 22 38 137 | 53 54 181 | 41 03 264 | 45 50 336 |
| 203 | 39 42 007 | 25 44 057 | 31 47 090 | 23 15 137 | 53 52 183 | 40 09 265 | 45 28 336 |
| 204 | 39 48 007 | 26 30 057 | 32 42 091 | 23 52 138 | 53 48 185 | 39 14 265 | 45 06 336 |
| 205 | 39 54 006 | 27 16 058 | 33 36 091 | 24 28 139 | 53 43 186 | 38 20 266 | 44 43 335 |
| 206 | 40 00 006 | 28 02 058 | 34 30 092 | 25 04 139 | 53 36 188 | 37 26 266 | 44 20 335 |
| 207 | 40 06 006 | 28 48 058 | 35 25 092 | 25 39 140 | 53 28 190 | 36 32 267 | 43 57 334 |
| 208 | 40 11 005 | 29 34 058 | 36 19 093 | 26 13 141 | 53 18 191 | 35 37 267 | 43 33 334 |
| 209 | 40 15 005 | 30 20 058 | 37 13 093 | 26 47 142 | 53 06 193 | 34 43 268 | 43 09 334 |
| | •Kochab | VEGA | Rasalhague | •ANTARES | SPICA | •REGULUS | Dubhe |
| 210 | 40 20 004 | 31 06 058 | 38 08 093 | 27 21 143 | 52 53 195 | 33 49 268 | 42 45 333 |
| 211 | 40 24 004 | 31 53 058 | 39 02 094 | 27 54 143 | 52 39 196 | 32 54 268 | 42 20 333 |
| 212 | 40 28 004 | 32 39 059 | 39 56 094 | 28 26 144 | 52 23 198 | 32 00 269 | 41 56 333 |
| 213 | 40 31 003 | 33 25 059 | 40 50 095 | 28 57 145 | 52 06 199 | 31 06 269 | 41 30 332 |
| 214 | 40 34 003 | 34 12 059 | 41 45 095 | 29 28 146 | 51 47 201 | 30 11 270 | 41 05 332 |
| 215 | 40 37 003 | 34 58 059 | 42 39 096 | 29 58 147 | 51 27 202 | 29 17 270 | 40 40 332 |
| 216 | 40 39 002 | 35 45 059 | 43 33 096 | 30 28 148 | 51 06 204 | 28 22 271 | 40 14 332 |
| 217 | 40 41 002 | 36 32 059 | 44 27 097 | 30 57 148 | 50 44 205 | 27 28 271 | 39 48 331 |
| 218 | 40 43 002 | 37 18 059 | 45 21 097 | 31 25 149 | 50 20 207 | 26 34 271 | 39 22 331 |
| 219 | 40 45 001 | 38 05 059 | 46 15 098 | 31 52 150 | 49 55 208 | 25 39 272 | 38 55 331 |
| 220 | 40 46 001 | 38 52 059 | 47 08 099 | 32 19 151 | 49 29 209 | 24 45 272 | 38 29 331 |
| 221 | 40 47 001 | 39 38 059 | 48 02 099 | 32 45 152 | 49 02 211 | 23 51 273 | 38 02 330 |
| 222 | 40 47 000 | 40 25 059 | 48 56 100 | 33 10 153 | 48 34 212 | 22 56 273 | 37 35 330 |
| 223 | 40 47 000 | 41 12 059 | 49 49 100 | 33 34 154 | 48 04 213 | 22 02 273 | 37 08 330 |
| 224 | 40 47 000 | 41 59 059 | 50 43 101 | 33 58 155 | 47 34 214 | 21 08 274 | 36 41 330 |
| | VEGA | •ALTAIR | Nunki | ANTARES | •SPICA | Denebola | •Alkaid |
| 225 | 42 46 059 | 19 31 089 | 13 39 128 | 34 21 156 | 47 03 216 | 43 59 266 | 61 46 334 |
| 226 | 43 32 059 | 20 25 090 | 14 22 129 | 34 42 157 | 46 31 217 | 43 05 267 | 61 22 333 |
| 227 | 44 19 059 | 21 19 090 | 15 04 129 | 35 03 158 | 45 57 218 | 42 11 267 | 60 57 332 |
| 228 | 45 06 059 | 22 14 090 | 15 46 130 | 35 23 159 | 45 24 219 | 41 16 268 | 60 32 331 |
| 229 | 45 53 059 | 23 08 091 | 16 28 130 | 35 42 160 | 44 49 220 | 40 22 268 | 60 05 330 |
| 230 | 46 39 059 | 24 02 091 | 17 09 131 | 36 01 161 | 44 13 221 | 39 27 269 | 59 38 329 |
| 231 | 47 26 059 | 24 57 092 | 17 50 132 | 36 18 162 | 43 37 223 | 38 33 269 | 59 10 329 |
| 232 | 48 13 059 | 25 51 092 | 18 30 132 | 36 34 163 | 43 00 224 | 37 39 270 | 58 41 328 |
| 233 | 48 59 059 | 26 46 093 | 19 11 133 | 36 49 164 | 42 22 225 | 36 44 270 | 58 12 327 |
| 234 | 49 46 059 | 27 40 093 | 19 50 133 | 37 04 165 | 41 43 226 | 35 50 270 | 57 42 326 |
| 235 | 50 32 059 | 28 34 094 | 20 29 134 | 37 17 166 | 41 04 227 | 34 56 271 | 57 11 325 |
| 236 | 51 19 058 | 29 28 094 | 21 08 135 | 37 29 167 | 40 24 228 | 34 01 271 | 56 40 325 |
| 237 | 52 05 058 | 30 23 094 | 21 47 135 | 37 41 169 | 39 44 228 | 33 07 272 | 56 08 324 |
| 238 | 52 51 058 | 31 17 095 | 22 24 136 | 37 51 170 | 39 03 229 | 32 13 272 | 55 36 323 |
| 239 | 53 37 058 | 32 11 095 | 23 02 137 | 38 00 171 | 38 21 230 | 31 18 272 | 55 03 323 |
| | •VEGA | ALTAIR | Nunki | •ANTARES | SPICA | ARCTURUS | •Alkaid |
| 240 | 54 23 058 | 33 05 096 | 23 39 138 | 38 09 172 | 37 39 231 | 65 03 262 | 54 30 322 |
| 241 | 55 09 057 | 33 59 096 | 24 15 138 | 38 16 173 | 36 57 232 | 64 09 263 | 53 56 321 |
| 242 | 55 55 057 | 34 53 097 | 24 51 139 | 38 22 174 | 36 14 233 | 63 15 263 | 53 22 321 |
| 243 | 56 40 057 | 35 47 097 | 25 27 140 | 38 27 175 | 35 30 234 | 62 21 264 | 52 48 320 |
| 244 | 57 26 056 | 36 41 098 | 26 01 141 | 38 31 176 | 34 46 234 | 61 27 265 | 52 13 320 |
| 245 | 58 11 056 | 37 35 098 | 26 36 141 | 38 34 178 | 34 02 235 | 60 33 265 | 51 38 320 |
| 246 | 58 56 056 | 38 29 099 | 27 09 142 | 38 35 179 | 33 17 236 | 59 39 266 | 51 02 319 |
| 247 | 59 41 055 | 39 22 099 | 27 43 143 | 38 36 180 | 32 32 237 | 58 45 266 | 50 27 319 |
| 248 | 60 25 055 | 40 16 100 | 28 15 144 | 38 36 181 | 31 46 237 | 57 50 267 | 49 51 318 |
| 249 | 61 09 054 | 41 09 101 | 28 47 145 | 38 34 182 | 31 00 238 | 56 56 267 | 49 14 318 |
| 250 | 61 53 054 | 42 03 101 | 29 18 145 | 38 32 183 | 30 14 239 | 56 02 268 | 48 38 318 |
| 251 | 62 37 053 | 42 56 102 | 29 49 146 | 38 28 184 | 29 27 240 | 55 07 268 | 48 01 317 |
| 252 | 63 20 052 | 43 49 102 | 30 19 147 | 38 23 186 | 28 40 240 | 54 13 269 | 47 24 317 |
| 253 | 64 03 052 | 44 42 103 | 30 48 148 | 38 17 187 | 27 53 241 | 53 19 269 | 46 47 317 |
| 254 | 64 45 051 | 45 35 104 | 31 16 149 | 38 10 188 | 27 05 242 | 52 24 269 | 46 09 316 |
| | •DENEB | ALTAIR | •Nunki | ANTARES | SPICA | •ARCTURUS | Alkaid |
| 255 | 41 37 051 | 46 28 104 | 31 44 150 | 38 02 189 | 26 17 242 | 51 30 270 | 45 32 316 |
| 256 | 42 19 051 | 47 21 105 | 32 11 151 | 37 53 190 | 25 29 243 | 50 36 270 | 44 54 316 |
| 257 | 43 01 050 | 48 13 106 | 32 37 152 | 37 43 191 | 24 40 243 | 49 41 271 | 44 16 316 |
| 258 | 43 42 050 | 49 06 106 | 33 03 152 | 37 32 192 | 23 51 244 | 48 47 271 | 43 38 315 |
| 259 | 44 24 050 | 49 58 107 | 33 28 153 | 37 20 193 | 23 02 245 | 47 52 272 | 43 00 315 |
| 260 | 45 06 050 | 50 50 108 | 33 52 154 | 37 07 194 | 22 13 245 | 46 58 272 | 42 21 315 |
| 261 | 45 48 050 | 51 41 108 | 34 15 155 | 36 53 196 | 21 24 246 | 46 04 272 | 41 43 315 |
| 262 | 46 29 050 | 52 33 109 | 34 37 156 | 36 38 197 | 20 34 246 | 45 09 273 | 41 04 315 |
| 263 | 47 11 050 | 53 24 110 | 34 58 157 | 36 22 198 | 19 44 247 | 44 15 273 | 40 26 315 |
| 264 | 47 52 049 | 54 15 111 | 35 19 158 | 36 05 199 | 18 54 247 | 43 21 273 | 39 47 315 |
| 265 | 48 33 049 | 55 06 112 | 35 38 159 | 35 47 200 | 18 04 248 | 42 27 274 | 39 08 314 |
| 266 | 49 14 049 | 55 56 113 | 35 57 160 | 35 28 201 | 17 13 248 | 41 32 274 | 38 29 314 |
| 267 | 49 55 049 | 56 46 114 | 36 15 162 | 35 08 202 | 16 23 249 | 40 38 275 | 37 50 314 |
| 268 | 50 36 048 | 57 35 115 | 36 31 163 | 34 47 203 | 15 32 250 | 39 44 275 | 37 11 314 |
| 269 | 51 16 048 | 58 25 116 | 36 47 164 | 34 26 204 | 14 41 250 | 38 50 275 | 36 32 314 |

| LHA ♈ | Hc Zn | Hc Zn | Hc Zn | Hc Zn | Hc Zn | Hc Zn | Hc Zn |
|---|---|---|---|---|---|---|---|
| | •DENEB | Enif | •Nunki | ANTARES | •ARCTURUS | Alkaid | Kochab |
| 270 | 51 56 048 | 34 58 096 | 37 02 165 | 34 03 205 | 37 56 276 | 35 53 314 | 35 01 346 |
| 271 | 52 36 047 | 35 52 096 | 37 16 166 | 33 40 206 | 37 02 276 | 35 14 314 | 34 48 346 |
| 272 | 53 16 047 | 36 46 097 | 37 29 167 | 33 16 207 | 36 07 276 | 34 35 314 | 34 34 345 |
| 273 | 53 56 046 | 37 40 097 | 37 41 168 | 32 51 208 | 35 13 277 | 33 56 314 | 34 20 345 |
| 274 | 54 35 046 | 38 34 098 | 37 51 169 | 32 25 209 | 34 19 277 | 33 17 314 | 34 07 345 |
| 275 | 55 14 045 | 39 28 098 | 38 01 170 | 31 59 210 | 33 26 277 | 32 38 314 | 33 52 345 |
| 276 | 55 52 045 | 40 22 099 | 38 10 171 | 31 32 210 | 32 32 278 | 31 59 314 | 33 38 345 |
| 277 | 56 30 044 | 41 16 099 | 38 17 172 | 31 04 211 | 31 38 278 | 31 20 314 | 33 24 345 |
| 278 | 57 08 044 | 42 09 100 | 38 24 174 | 30 35 212 | 30 44 278 | 30 41 314 | 33 10 345 |
| 279 | 57 46 043 | 43 03 100 | 38 30 175 | 30 06 213 | 29 50 279 | 30 02 314 | 32 55 344 |
| 280 | 58 23 043 | 43 56 101 | 38 34 176 | 29 36 214 | 28 56 279 | 29 23 314 | 32 40 344 |
| 281 | 58 59 042 | 44 50 101 | 38 37 177 | 29 05 215 | 28 03 279 | 28 44 314 | 32 25 344 |
| 282 | 59 35 041 | 45 43 102 | 38 40 178 | 28 34 216 | 27 09 280 | 28 06 314 | 32 10 344 |
| 283 | 60 11 040 | 46 36 103 | 38 41 179 | 28 01 216 | 26 16 280 | 27 27 315 | 31 55 344 |
| 284 | 60 46 040 | 47 29 103 | 38 41 180 | 27 29 217 | 25 22 280 | 26 48 315 | 31 40 344 |
| | DENEB | •Alpheratz | FOMALHAUT | •Nunki | ANTARES | •ARCTURUS | Kochab |
| 285 | 61 20 039 | 22 37 067 | 11 10 130 | 38 40 182 | 26 56 218 | 24 29 281 | 31 25 344 |
| 286 | 61 54 038 | 23 27 068 | 11 51 131 | 38 38 183 | 26 22 219 | 23 35 281 | 31 09 344 |
| 287 | 62 27 037 | 24 18 068 | 12 32 132 | 38 35 184 | 25 47 220 | 22 42 282 | 30 54 343 |
| 288 | 62 59 036 | 25 08 068 | 13 12 132 | 38 30 185 | 25 12 220 | 21 49 282 | 30 38 343 |
| 289 | 63 31 035 | 25 58 068 | 13 53 133 | 38 25 186 | 24 37 221 | 20 56 282 | 30 23 343 |
| 290 | 64 01 034 | 26 49 069 | 14 32 133 | 38 19 187 | 24 01 222 | 20 02 283 | 30 07 343 |
| 291 | 64 31 033 | 27 40 069 | 15 12 134 | 38 11 188 | 23 24 223 | 19 09 283 | 29 51 343 |
| 292 | 65 00 031 | 28 31 069 | 15 51 134 | 38 03 190 | 22 47 223 | 18 16 283 | 29 35 343 |
| 293 | 65 28 030 | 29 21 069 | 16 29 135 | 37 53 191 | 22 10 224 | 17 23 284 | 29 19 343 |
| 294 | 65 55 029 | 30 12 070 | 17 07 136 | 37 42 192 | 21 32 225 | 16 31 284 | 29 03 343 |
| 295 | 66 20 027 | 31 03 070 | 17 45 136 | 37 31 193 | 20 53 225 | 15 38 284 | 28 47 343 |
| 296 | 66 45 026 | 31 54 070 | 18 23 137 | 37 18 194 | 20 14 226 | 14 45 285 | 28 31 343 |
| 297 | 67 08 025 | 32 46 070 | 18 59 138 | 37 04 195 | 19 35 227 | 13 53 285 | 28 15 343 |
| 298 | 67 30 023 | 33 37 071 | 19 36 138 | 36 50 196 | 18 55 227 | 13 00 285 | 27 59 343 |
| 299 | 67 50 021 | 34 28 071 | 20 12 139 | 36 34 197 | 18 15 228 | 12 08 286 | 27 43 343 |
| | •DENEB | Schedar | Alpheratz | •FOMALHAUT | Nunki | •Rasalhague | VEGA |
| 300 | 68 09 020 | 31 37 038 | 35 20 071 | 20 47 140 | 36 18 198 | 53 29 257 | 67 38 313 |
| 301 | 68 27 018 | 32 10 038 | 36 11 071 | 21 22 140 | 36 00 199 | 52 35 258 | 66 58 312 |
| 302 | 68 43 016 | 32 43 037 | 37 03 071 | 21 57 141 | 35 41 200 | 51 42 258 | 66 18 311 |
| 303 | 68 57 014 | 33 16 037 | 37 54 072 | 22 31 142 | 35 22 201 | 50 49 259 | 65 36 310 |
| 304 | 69 09 012 | 33 49 037 | 38 46 072 | 23 04 142 | 35 02 202 | 49 56 260 | 64 54 309 |
| 305 | 69 20 010 | 34 22 037 | 39 38 072 | 23 37 143 | 34 41 203 | 49 02 260 | 64 12 309 |
| 306 | 69 29 009 | 34 55 037 | 40 29 072 | 24 09 144 | 34 18 204 | 48 08 261 | 63 29 308 |
| 307 | 69 36 007 | 35 28 037 | 41 21 072 | 24 41 145 | 33 56 205 | 47 15 261 | 62 46 307 |
| 308 | 69 42 005 | 36 01 037 | 42 13 073 | 25 12 145 | 33 32 206 | 46 21 262 | 62 03 307 |
| 309 | 69 45 003 | 36 34 037 | 43 05 073 | 25 43 146 | 33 07 207 | 45 27 262 | 61 19 306 |
| 310 | 69 46 000 | 37 06 037 | 43 57 073 | 26 13 147 | 32 42 208 | 44 33 263 | 60 35 305 |
| 311 | 69 46 358 | 37 39 037 | 44 49 073 | 26 42 148 | 32 16 209 | 43 39 264 | 59 50 305 |
| 312 | 69 43 356 | 38 11 037 | 45 41 073 | 27 10 149 | 31 49 210 | 42 45 264 | 59 05 304 |
| 313 | 69 39 354 | 38 44 036 | 46 33 073 | 27 38 149 | 31 21 211 | 41 51 265 | 58 21 304 |
| 314 | 69 33 352 | 39 16 036 | 47 25 074 | 28 06 150 | 30 53 212 | 40 57 265 | 57 35 304 |
| | •Alpheratz | Diphda | •FOMALHAUT | ALTAIR | Rasalhague | •VEGA | DENEB |
| 315 | 48 17 074 | 20 45 123 | 28 32 151 | 66 47 229 | 40 03 266 | 56 50 303 | 69 25 350 |
| 316 | 49 09 074 | 21 31 124 | 28 58 152 | 66 06 231 | 39 08 266 | 56 04 303 | 69 15 348 |
| 317 | 50 02 074 | 22 16 124 | 29 23 153 | 65 23 232 | 38 14 266 | 55 19 303 | 69 03 347 |
| 318 | 50 54 074 | 23 01 125 | 29 48 154 | 64 40 234 | 37 20 267 | 54 33 302 | 68 49 345 |
| 319 | 51 46 074 | 23 45 125 | 30 11 155 | 63 55 235 | 36 25 267 | 53 47 302 | 68 34 343 |
| 320 | 52 39 074 | 24 29 126 | 30 34 156 | 63 10 237 | 35 31 268 | 53 01 302 | 68 17 341 |
| 321 | 53 31 074 | 25 13 127 | 30 56 156 | 62 24 238 | 34 37 268 | 52 15 302 | 67 59 339 |
| 322 | 54 23 074 | 25 57 127 | 31 18 157 | 61 38 240 | 33 42 269 | 51 28 302 | 67 39 338 |
| 323 | 55 16 075 | 26 40 128 | 31 38 158 | 60 50 241 | 32 48 269 | 50 42 301 | 67 18 336 |
| 324 | 56 08 075 | 27 22 129 | 31 58 159 | 60 03 242 | 31 54 270 | 49 56 301 | 66 55 335 |
| 325 | 57 01 075 | 28 04 130 | 32 17 160 | 59 14 243 | 30 59 270 | 49 09 301 | 66 32 333 |
| 326 | 57 53 075 | 28 46 130 | 32 35 161 | 58 26 244 | 30 05 270 | 48 23 301 | 66 06 332 |
| 327 | 58 45 075 | 29 27 131 | 32 52 162 | 57 37 245 | 29 11 271 | 47 36 301 | 65 40 330 |
| 328 | 59 38 075 | 30 08 132 | 33 08 163 | 56 47 246 | 28 16 271 | 46 49 301 | 65 13 329 |
| 329 | 60 30 075 | 30 48 133 | 33 23 164 | 55 57 247 | 27 22 272 | 46 03 301 | 64 44 328 |
| | •Mirfak | Hamal | Diphda | •FOMALHAUT | ALTAIR | •VEGA | DENEB |
| 330 | 24 36 044 | 34 19 078 | 31 28 133 | 33 38 165 | 55 07 248 | 45 16 301 | 64 15 327 |
| 331 | 25 14 045 | 35 12 078 | 32 07 134 | 33 51 166 | 54 16 249 | 44 29 301 | 63 45 326 |
| 332 | 25 52 045 | 36 05 078 | 32 46 135 | 34 03 167 | 53 25 250 | 43 42 301 | 63 14 325 |
| 333 | 26 30 045 | 36 58 079 | 33 24 136 | 34 15 168 | 52 34 251 | 42 56 301 | 62 42 324 |
| 334 | 27 09 045 | 37 52 079 | 34 02 137 | 34 26 169 | 51 43 251 | 42 09 301 | 62 09 323 |
| 335 | 27 47 045 | 38 45 079 | 34 39 138 | 34 35 170 | 50 51 252 | 41 22 301 | 61 36 322 |
| 336 | 28 26 045 | 39 38 079 | 35 15 138 | 34 44 171 | 49 59 253 | 40 35 301 | 61 02 321 |
| 337 | 29 04 045 | 40 32 080 | 35 51 139 | 34 52 172 | 49 07 254 | 39 48 301 | 60 27 320 |
| 338 | 29 43 045 | 41 25 080 | 36 26 140 | 34 58 173 | 48 15 254 | 39 02 301 | 59 52 319 |
| 339 | 30 21 045 | 42 19 080 | 37 00 141 | 35 04 174 | 47 22 255 | 38 15 301 | 59 16 318 |
| 340 | 31 00 045 | 43 13 081 | 37 34 142 | 35 09 176 | 46 30 256 | 37 28 301 | 58 40 318 |
| 341 | 31 39 045 | 44 06 081 | 38 07 143 | 35 13 177 | 45 37 256 | 36 42 301 | 58 03 317 |
| 342 | 32 18 045 | 45 00 081 | 38 39 144 | 35 15 178 | 44 44 257 | 35 55 301 | 57 26 316 |
| 343 | 32 56 045 | 45 54 081 | 39 11 145 | 35 17 179 | 43 51 258 | 35 08 301 | 56 48 316 |
| 344 | 33 35 045 | 46 48 082 | 39 41 146 | 35 18 180 | 42 58 258 | 34 22 301 | 56 10 315 |
| | Schedar | CAPELLA | •ALDEBARAN | Diphda | •FOMALHAUT | ALTAIR | •DENEB |
| 345 | 53 45 023 | 15 08 046 | 12 22 077 | 40 11 147 | 35 18 181 | 42 04 259 | 55 31 315 |
| 346 | 54 06 022 | 15 47 046 | 13 16 078 | 40 40 148 | 35 16 182 | 41 11 259 | 54 53 314 |
| 347 | 54 27 022 | 16 27 046 | 14 09 078 | 41 08 150 | 35 14 183 | 40 17 260 | 54 14 314 |
| 348 | 54 46 021 | 17 06 047 | 15 02 079 | 41 35 151 | 35 11 184 | 39 24 261 | 53 34 313 |
| 349 | 55 05 020 | 17 46 047 | 15 55 079 | 42 01 152 | 35 06 185 | 38 30 261 | 52 55 313 |
| 350 | 55 24 019 | 18 26 047 | 16 49 079 | 42 27 153 | 35 01 186 | 37 36 262 | 52 15 313 |
| 351 | 55 41 019 | 19 05 047 | 17 42 080 | 42 51 154 | 34 55 187 | 36 42 262 | 51 35 312 |
| 352 | 55 58 018 | 19 45 047 | 18 36 080 | 43 14 155 | 34 47 188 | 35 49 263 | 50 54 312 |
| 353 | 56 14 017 | 20 26 048 | 19 29 080 | 43 36 156 | 34 39 189 | 34 55 263 | 50 14 312 |
| 354 | 56 30 016 | 21 06 048 | 20 23 081 | 43 58 158 | 34 30 190 | 34 01 264 | 49 33 311 |
| 355 | 56 44 015 | 21 46 048 | 21 17 081 | 44 18 159 | 34 20 191 | 33 06 264 | 48 52 311 |
| 356 | 56 58 014 | 22 27 048 | 22 10 081 | 44 37 160 | 34 08 192 | 32 12 265 | 48 11 311 |
| 357 | 57 11 013 | 23 07 048 | 23 04 082 | 44 55 161 | 33 56 193 | 31 18 265 | 47 30 311 |
| 358 | 57 23 012 | 23 48 049 | 23 58 082 | 45 11 163 | 33 43 194 | 30 24 266 | 46 48 310 |
| 359 | 57 34 011 | 24 29 049 | 24 52 083 | 45 27 164 | 33 29 195 | 29 30 266 | 46 07 310 |

# F PAGE EXCERPTS FROM PUB. NO. 229, VOLUME 2

Material in this appendix is taken from Pub. No. 229 by permission of the U.S. Naval Oceanographic Office, now the Defense Mapping Agency Hydrographic/Topographic Center.

# INTERPOLATION TABLE

| Dec. Inc. | Altitude Difference (d) Tens 10′ | 20′ | 30′ | 40′ | 50′ | Decimals | Units 0′ | 1′ | 2′ | 3′ | 4′ | 5′ | 6′ | 7′ | 8′ | 9′ | Double Second Diff. and Corr. |
|---|---|---|---|---|---|---|---|---|---|---|---|---|---|---|---|---|---|
| ′ | ′ | ′ | ′ | ′ | ′ | | ′ | ′ | ′ | ′ | ′ | ′ | ′ | ′ | ′ | ′ | ′ ′ |
| 0.0 | 0.0 | 0.0 | 0.0 | 0.0 | 0.0 | .0 | 0.0 | 0.0 | 0.0 | 0.0 | 0.0 | 0.0 | 0.0 | 0.1 | 0.1 | 0.1 | 0.0<br>0.0<br>48.2 |
| 0.1 | 0.0 | 0.0 | 0.0 | 0.0 | 0.1 | .1 | 0.0 | 0.0 | 0.0 | 0.0 | 0.0 | 0.0 | 0.1 | 0.1 | 0.1 | 0.1 | |
| 0.2 | 0.0 | 0.0 | 0.1 | 0.1 | 0.1 | .2 | 0.0 | 0.0 | 0.0 | 0.0 | 0.0 | 0.0 | 0.1 | 0.1 | 0.1 | 0.1 | |
| 0.3 | 0.0 | 0.1 | 0.1 | 0.2 | 0.2 | .3 | 0.0 | 0.0 | 0.0 | 0.0 | 0.0 | 0.0 | 0.1 | 0.1 | 0.1 | 0.1 | |
| 0.4 | 0.1 | 0.1 | 0.2 | 0.3 | 0.3 | .4 | 0.0 | 0.0 | 0.0 | 0.0 | 0.0 | 0.0 | 0.1 | 0.1 | 0.1 | 0.1 | |
| 0.5 | 0.1 | 0.2 | 0.3 | 0.3 | 0.4 | .5 | 0.0 | 0.0 | 0.0 | 0.0 | 0.0 | 0.0 | 0.1 | 0.1 | 0.1 | 0.1 | 16.2<br>0.1<br>48.6 |
| 0.6 | 0.1 | 0.2 | 0.3 | 0.4 | 0.5 | .6 | 0.0 | 0.0 | 0.0 | 0.0 | 0.0 | 0.0 | 0.1 | 0.1 | 0.1 | 0.1 | |
| 0.7 | 0.1 | 0.3 | 0.4 | 0.5 | 0.6 | .7 | 0.0 | 0.0 | 0.0 | 0.0 | 0.0 | 0.0 | 0.1 | 0.1 | 0.1 | 0.1 | |
| 0.8 | 0.2 | 0.3 | 0.4 | 0.6 | 0.7 | .8 | 0.0 | 0.0 | 0.0 | 0.0 | 0.0 | 0.0 | 0.1 | 0.1 | 0.1 | 0.1 | |
| 0.9 | 0.2 | 0.3 | 0.5 | 0.6 | 0.8 | .9 | 0.0 | 0.0 | 0.0 | 0.0 | 0.0 | 0.0 | 0.1 | 0.1 | 0.1 | 0.1 | |
| 1.0 | 0.1 | 0.3 | 0.5 | 0.6 | 0.8 | .0 | 0.0 | 0.0 | 0.0 | 0.1 | 0.1 | 0.1 | 0.1 | 0.2 | 0.2 | 0.2 | 8.2<br>0.1<br>24.6<br>0.2<br>41.0 |
| 1.1 | 0.2 | 0.3 | 0.5 | 0.7 | 0.9 | .1 | 0.0 | 0.0 | 0.1 | 0.1 | 0.1 | 0.1 | 0.2 | 0.2 | 0.2 | 0.2 | |
| 1.2 | 0.2 | 0.4 | 0.6 | 0.8 | 1.0 | .2 | 0.0 | 0.0 | 0.1 | 0.1 | 0.1 | 0.1 | 0.2 | 0.2 | 0.2 | 0.2 | |
| 1.3 | 0.2 | 0.4 | 0.6 | 0.9 | 1.1 | .3 | 0.0 | 0.0 | 0.1 | 0.1 | 0.1 | 0.1 | 0.2 | 0.2 | 0.2 | 0.2 | |
| 1.4 | 0.2 | 0.5 | 0.7 | 0.9 | 1.2 | .4 | 0.0 | 0.0 | 0.1 | 0.1 | 0.1 | 0.1 | 0.2 | 0.2 | 0.2 | 0.2 | |
| 1.5 | 0.3 | 0.5 | 0.8 | 1.0 | 1.3 | .5 | 0.0 | 0.0 | 0.1 | 0.1 | 0.1 | 0.1 | 0.2 | 0.2 | 0.2 | 0.2 | |
| 1.6 | 0.3 | 0.5 | 0.8 | 1.1 | 1.3 | .6 | 0.0 | 0.0 | 0.1 | 0.1 | 0.1 | 0.1 | 0.2 | 0.2 | 0.2 | 0.2 | |
| 1.7 | 0.3 | 0.6 | 0.9 | 1.2 | 1.4 | .7 | 0.0 | 0.0 | 0.1 | 0.1 | 0.1 | 0.1 | 0.2 | 0.2 | 0.2 | 0.2 | |
| 1.8 | 0.3 | 0.6 | 0.9 | 1.2 | 1.5 | .8 | 0.0 | 0.0 | 0.1 | 0.1 | 0.1 | 0.1 | 0.2 | 0.2 | 0.2 | 0.2 | |
| 1.9 | 0.4 | 0.7 | 1.0 | 1.3 | 1.6 | .9 | 0.0 | 0.0 | 0.1 | 0.1 | 0.1 | 0.1 | 0.2 | 0.2 | 0.2 | 0.2 | |
| 2.0 | 0.3 | 0.6 | 1.0 | 1.3 | 1.6 | .0 | 0.0 | 0.0 | 0.1 | 0.1 | 0.2 | 0.2 | 0.2 | 0.3 | 0.3 | 0.4 | 5.0<br>0.1<br>15.0<br>0.2<br>25.0<br>0.3<br>35.1 |
| 2.1 | 0.3 | 0.7 | 1.0 | 1.4 | 1.7 | .1 | 0.0 | 0.0 | 0.1 | 0.1 | 0.2 | 0.2 | 0.3 | 0.3 | 0.3 | 0.4 | |
| 2.2 | 0.3 | 0.7 | 1.1 | 1.4 | 1.8 | .2 | 0.0 | 0.0 | 0.1 | 0.1 | 0.2 | 0.2 | 0.3 | 0.3 | 0.3 | 0.4 | |
| 2.3 | 0.4 | 0.8 | 1.1 | 1.5 | 1.9 | .3 | 0.0 | 0.1 | 0.1 | 0.1 | 0.2 | 0.2 | 0.3 | 0.3 | 0.3 | 0.4 | |
| 2.4 | 0.4 | 0.8 | 1.2 | 1.6 | 2.0 | .4 | 0.0 | 0.1 | 0.1 | 0.1 | 0.2 | 0.2 | 0.3 | 0.3 | 0.3 | 0.4 | |
| 2.5 | 0.4 | 0.8 | 1.3 | 1.7 | 2.1 | .5 | 0.0 | 0.1 | 0.1 | 0.1 | 0.2 | 0.2 | 0.3 | 0.3 | 0.4 | 0.4 | |
| 2.6 | 0.4 | 0.9 | 1.3 | 1.7 | 2.2 | .6 | 0.0 | 0.1 | 0.1 | 0.1 | 0.2 | 0.2 | 0.3 | 0.3 | 0.4 | 0.4 | |
| 2.7 | 0.5 | 0.9 | 1.4 | 1.8 | 2.3 | .7 | 0.0 | 0.1 | 0.1 | 0.2 | 0.2 | 0.2 | 0.3 | 0.3 | 0.4 | 0.4 | |
| 2.8 | 0.5 | 1.0 | 1.4 | 1.9 | 2.4 | .8 | 0.0 | 0.1 | 0.1 | 0.2 | 0.2 | 0.2 | 0.3 | 0.3 | 0.4 | 0.4 | |
| 2.9 | 0.5 | 1.0 | 1.5 | 2.0 | 2.5 | .9 | 0.0 | 0.1 | 0.1 | 0.2 | 0.2 | 0.2 | 0.3 | 0.3 | 0.4 | 0.4 | |
| 3.0 | 0.5 | 1.0 | 1.5 | 2.0 | 2.5 | .0 | 0.0 | 0.1 | 0.1 | 0.2 | 0.2 | 0.3 | 0.3 | 0.4 | 0.5 | 0.5 | 3.6<br>0.1<br>10.9<br>0.2<br>18.2<br>0.3<br>25.5<br>0.4<br>32.8<br>0.5<br>40.1 |
| 3.1 | 0.5 | 1.0 | 1.5 | 2.0 | 2.6 | .1 | 0.0 | 0.1 | 0.1 | 0.2 | 0.2 | 0.3 | 0.4 | 0.4 | 0.5 | 0.5 | |
| 3.2 | 0.5 | 1.0 | 1.6 | 2.1 | 2.6 | .2 | 0.0 | 0.1 | 0.1 | 0.2 | 0.2 | 0.3 | 0.4 | 0.4 | 0.5 | 0.5 | |
| 3.3 | 0.5 | 1.1 | 1.6 | 2.2 | 2.7 | .3 | 0.0 | 0.1 | 0.1 | 0.2 | 0.3 | 0.3 | 0.4 | 0.4 | 0.5 | 0.5 | |
| 3.4 | 0.6 | 1.1 | 1.7 | 2.3 | 2.8 | .4 | 0.0 | 0.1 | 0.1 | 0.2 | 0.3 | 0.3 | 0.4 | 0.4 | 0.5 | 0.5 | |
| 3.5 | 0.6 | 1.2 | 1.8 | 2.3 | 2.9 | .5 | 0.0 | 0.1 | 0.1 | 0.2 | 0.3 | 0.3 | 0.4 | 0.4 | 0.5 | 0.6 | |
| 3.6 | 0.6 | 1.2 | 1.8 | 2.4 | 3.0 | .6 | 0.0 | 0.1 | 0.2 | 0.2 | 0.3 | 0.3 | 0.4 | 0.4 | 0.5 | 0.6 | |
| 3.7 | 0.6 | 1.3 | 1.9 | 2.5 | 3.1 | .7 | 0.0 | 0.1 | 0.2 | 0.2 | 0.3 | 0.3 | 0.4 | 0.4 | 0.5 | 0.6 | |
| 3.8 | 0.7 | 1.3 | 1.9 | 2.6 | 3.2 | .8 | 0.0 | 0.1 | 0.2 | 0.2 | 0.3 | 0.3 | 0.4 | 0.5 | 0.5 | 0.6 | |
| 3.9 | 0.7 | 1.3 | 2.0 | 2.6 | 3.3 | .9 | 0.1 | 0.1 | 0.2 | 0.2 | 0.3 | 0.3 | 0.4 | 0.5 | 0.5 | 0.6 | |
| 4.0 | 0.6 | 1.3 | 2.0 | 2.6 | 3.3 | .0 | 0.0 | 0.1 | 0.1 | 0.2 | 0.3 | 0.4 | 0.4 | 0.5 | 0.6 | 0.7 | 2.9<br>0.1<br>8.6<br>0.2<br>14.4<br>0.3<br>20.2<br>0.4<br>25.9<br>0.5<br>31.7<br>0.6<br>37.5 |
| 4.1 | 0.7 | 1.3 | 2.0 | 2.7 | 3.4 | .1 | 0.0 | 0.1 | 0.2 | 0.2 | 0.3 | 0.4 | 0.5 | 0.5 | 0.6 | 0.7 | |
| 4.2 | 0.7 | 1.4 | 2.1 | 2.8 | 3.5 | .2 | 0.0 | 0.1 | 0.2 | 0.2 | 0.3 | 0.4 | 0.5 | 0.5 | 0.6 | 0.7 | |
| 4.3 | 0.7 | 1.4 | 2.1 | 2.9 | 3.6 | .3 | 0.0 | 0.1 | 0.2 | 0.2 | 0.3 | 0.4 | 0.5 | 0.5 | 0.6 | 0.7 | |
| 4.4 | 0.7 | 1.5 | 2.2 | 2.9 | 3.7 | .4 | 0.0 | 0.1 | 0.2 | 0.3 | 0.3 | 0.4 | 0.5 | 0.6 | 0.6 | 0.7 | |
| 4.5 | 0.8 | 1.5 | 2.3 | 3.0 | 3.8 | .5 | 0.0 | 0.1 | 0.2 | 0.3 | 0.3 | 0.4 | 0.5 | 0.6 | 0.6 | 0.7 | |
| 4.6 | 0.8 | 1.5 | 2.3 | 3.1 | 3.8 | .6 | 0.0 | 0.1 | 0.2 | 0.3 | 0.3 | 0.4 | 0.5 | 0.6 | 0.6 | 0.7 | |
| 4.7 | 0.8 | 1.6 | 2.4 | 3.2 | 3.9 | .7 | 0.1 | 0.1 | 0.2 | 0.3 | 0.4 | 0.4 | 0.5 | 0.6 | 0.7 | 0.7 | |
| 4.8 | 0.8 | 1.6 | 2.4 | 3.2 | 4.0 | .8 | 0.1 | 0.1 | 0.2 | 0.3 | 0.4 | 0.4 | 0.5 | 0.6 | 0.7 | 0.7 | |
| 4.9 | 0.9 | 1.7 | 2.5 | 3.3 | 4.1 | .9 | 0.1 | 0.1 | 0.2 | 0.3 | 0.4 | 0.4 | 0.5 | 0.6 | 0.7 | 0.7 | |
| 5.0 | 0.8 | 1.6 | 2.5 | 3.3 | 4.1 | .0 | 0.0 | 0.1 | 0.2 | 0.3 | 0.4 | 0.5 | 0.5 | 0.6 | 0.7 | 0.8 | 2.4<br>0.1<br>7.2<br>0.2<br>12.0<br>0.3<br>16.8<br>0.4<br>21.6<br>0.5<br>26.4<br>0.6<br>31.2<br>0.7<br>36.0 |
| 5.1 | 0.8 | 1.7 | 2.5 | 3.4 | 4.2 | .1 | 0.0 | 0.1 | 0.2 | 0.3 | 0.4 | 0.5 | 0.6 | 0.7 | 0.7 | 0.8 | |
| 5.2 | 0.8 | 1.7 | 2.6 | 3.4 | 4.3 | .2 | 0.0 | 0.1 | 0.2 | 0.3 | 0.4 | 0.5 | 0.6 | 0.7 | 0.8 | 0.8 | |
| 5.3 | 0.9 | 1.8 | 2.6 | 3.5 | 4.4 | .3 | 0.0 | 0.1 | 0.2 | 0.3 | 0.4 | 0.5 | 0.6 | 0.7 | 0.8 | 0.9 | |
| 5.4 | 0.9 | 1.8 | 2.7 | 3.6 | 4.5 | .4 | 0.0 | 0.1 | 0.2 | 0.3 | 0.4 | 0.5 | 0.6 | 0.7 | 0.8 | 0.9 | |
| 5.5 | 0.9 | 1.8 | 2.8 | 3.7 | 4.6 | .5 | 0.0 | 0.1 | 0.2 | 0.3 | 0.4 | 0.5 | 0.6 | 0.7 | 0.8 | 0.9 | |
| 5.6 | 0.9 | 1.9 | 2.8 | 3.7 | 4.7 | .6 | 0.1 | 0.1 | 0.2 | 0.3 | 0.4 | 0.5 | 0.6 | 0.7 | 0.8 | 0.9 | |
| 5.7 | 1.0 | 1.9 | 2.9 | 3.8 | 4.8 | .7 | 0.1 | 0.2 | 0.2 | 0.3 | 0.4 | 0.5 | 0.6 | 0.7 | 0.8 | 0.9 | |
| 5.8 | 1.0 | 2.0 | 2.9 | 3.9 | 4.9 | .8 | 0.1 | 0.2 | 0.3 | 0.3 | 0.4 | 0.5 | 0.6 | 0.7 | 0.8 | 0.9 | |
| 5.9 | 1.0 | 2.0 | 3.0 | 4.0 | 5.0 | .9 | 0.1 | 0.2 | 0.3 | 0.4 | 0.4 | 0.5 | 0.6 | 0.7 | 0.8 | 0.9 | |
| 6.0 | 1.0 | 2.0 | 3.0 | 4.0 | 5.0 | .0 | 0.0 | 0.1 | 0.2 | 0.3 | 0.4 | 0.5 | 0.6 | 0.8 | 0.9 | 1.0 | 2.1<br>0.1<br>6.2<br>0.2<br>10.4<br>0.3<br>14.5<br>0.4<br>18.6<br>0.5<br>22.8<br>0.6<br>26.9<br>0.7<br>31.1<br>0.8<br>35.2 |
| 6.1 | 1.0 | 2.0 | 3.0 | 4.0 | 5.1 | .1 | 0.0 | 0.1 | 0.2 | 0.3 | 0.4 | 0.6 | 0.7 | 0.8 | 0.9 | 1.0 | |
| 6.2 | 1.0 | 2.0 | 3.1 | 4.1 | 5.1 | .2 | 0.0 | 0.1 | 0.2 | 0.3 | 0.5 | 0.6 | 0.7 | 0.8 | 0.9 | 1.0 | |
| 6.3 | 1.0 | 2.1 | 3.1 | 4.2 | 5.2 | .3 | 0.0 | 0.1 | 0.2 | 0.4 | 0.5 | 0.6 | 0.7 | 0.8 | 0.9 | 1.0 | |
| 6.4 | 1.1 | 2.1 | 3.2 | 4.3 | 5.3 | .4 | 0.0 | 0.2 | 0.3 | 0.4 | 0.5 | 0.6 | 0.7 | 0.8 | 0.9 | 1.0 | |
| 6.5 | 1.1 | 2.2 | 3.3 | 4.3 | 5.4 | .5 | 0.1 | 0.2 | 0.3 | 0.4 | 0.5 | 0.6 | 0.7 | 0.8 | 0.9 | 1.0 | |
| 6.6 | 1.1 | 2.2 | 3.3 | 4.4 | 5.5 | .6 | 0.1 | 0.2 | 0.3 | 0.4 | 0.5 | 0.6 | 0.7 | 0.8 | 0.9 | 1.0 | |
| 6.7 | 1.1 | 2.3 | 3.4 | 4.5 | 5.6 | .7 | 0.1 | 0.2 | 0.3 | 0.4 | 0.5 | 0.6 | 0.7 | 0.8 | 0.9 | 1.1 | |
| 6.8 | 1.2 | 2.3 | 3.4 | 4.6 | 5.7 | .8 | 0.1 | 0.2 | 0.3 | 0.4 | 0.5 | 0.6 | 0.7 | 0.8 | 1.0 | 1.1 | |
| 6.9 | 1.2 | 2.3 | 3.5 | 4.6 | 5.8 | .9 | 0.1 | 0.2 | 0.3 | 0.4 | 0.5 | 0.6 | 0.7 | 0.9 | 1.0 | 1.1 | |
| 7.0 | 1.1 | 2.3 | 3.5 | 4.6 | 5.8 | .0 | 0.0 | 0.1 | 0.2 | 0.4 | 0.5 | 0.6 | 0.7 | 0.9 | 1.0 | 1.1 | 1.8<br>0.1<br>5.5<br>0.2<br>9.1<br>0.3<br>12.8<br>0.4<br>16.5<br>0.5<br>20.1<br>0.6<br>23.8<br>0.7<br>27.4<br>0.8<br>31.1<br>0.9<br>34.7 |
| 7.1 | 1.2 | 2.3 | 3.5 | 4.7 | 5.9 | .1 | 0.0 | 0.1 | 0.3 | 0.4 | 0.5 | 0.6 | 0.8 | 0.9 | 1.0 | 1.1 | |
| 7.2 | 1.2 | 2.4 | 3.6 | 4.8 | 6.0 | .2 | 0.0 | 0.1 | 0.3 | 0.4 | 0.5 | 0.6 | 0.8 | 0.9 | 1.0 | 1.1 | |
| 7.3 | 1.2 | 2.4 | 3.6 | 4.9 | 6.1 | .3 | 0.0 | 0.2 | 0.3 | 0.4 | 0.5 | 0.7 | 0.8 | 0.9 | 1.0 | 1.2 | |
| 7.4 | 1.2 | 2.5 | 3.7 | 4.9 | 6.2 | .4 | 0.0 | 0.2 | 0.3 | 0.4 | 0.5 | 0.7 | 0.8 | 0.9 | 1.0 | 1.2 | |
| 7.5 | 1.3 | 2.5 | 3.8 | 5.0 | 6.3 | .5 | 0.1 | 0.2 | 0.3 | 0.4 | 0.6 | 0.7 | 0.8 | 0.9 | 1.1 | 1.2 | |
| 7.6 | 1.3 | 2.5 | 3.8 | 5.1 | 6.3 | .6 | 0.1 | 0.2 | 0.3 | 0.4 | 0.6 | 0.7 | 0.8 | 0.9 | 1.1 | 1.2 | |
| 7.7 | 1.3 | 2.6 | 3.9 | 5.2 | 6.4 | .7 | 0.1 | 0.2 | 0.3 | 0.5 | 0.6 | 0.7 | 0.8 | 1.0 | 1.1 | 1.2 | |
| 7.8 | 1.3 | 2.6 | 3.9 | 5.2 | 6.5 | .8 | 0.1 | 0.2 | 0.3 | 0.5 | 0.6 | 0.7 | 0.8 | 1.0 | 1.1 | 1.2 | |
| 7.9 | 1.4 | 2.7 | 4.0 | 5.3 | 6.6 | .9 | 0.1 | 0.2 | 0.4 | 0.5 | 0.6 | 0.7 | 0.9 | 1.0 | 1.1 | 1.2 | |
| | 10′ | 20′ | 30′ | 40′ | 50′ | | 0′ | 1′ | 2′ | 3′ | 4′ | 5′ | 6′ | 7′ | 8′ | 9′ | |

| Dec. Inc. | Altitude Difference (d) Tens 10′ | 20′ | 30′ | 40′ | 50′ | Decimals | Units 0′ | 1′ | 2′ | 3′ | 4′ | 5′ | 6′ | 7′ | 8′ | 9′ | Double Second Diff. and Corr. |
|---|---|---|---|---|---|---|---|---|---|---|---|---|---|---|---|---|---|
| ′ | ′ | ′ | ′ | ′ | ′ | | ′ | ′ | ′ | ′ | ′ | ′ | ′ | ′ | ′ | ′ | ′ ′ |
| 8.0 | 1.3 | 2.6 | 4.0 | 5.3 | 6.6 | .0 | 0.0 | 0.1 | 0.3 | 0.4 | 0.6 | 0.7 | 0.8 | 1.0 | 1.1 | 1.3 | 1.6<br>0.1<br>4.8<br>0.2<br>8.0<br>0.3<br>11.2<br>0.4<br>14.5<br>0.5<br>17.7<br>0.6<br>20.9<br>0.7<br>24.1<br>0.8<br>27.3<br>0.9<br>30.5<br>1.0<br>33.7<br>1.1<br>36.9 |
| 8.1 | 1.3 | 2.7 | 4.0 | 5.4 | 6.7 | .1 | 0.0 | 0.2 | 0.3 | 0.4 | 0.6 | 0.7 | 0.9 | 1.0 | 1.1 | 1.3 | |
| 8.2 | 1.3 | 2.7 | 4.1 | 5.4 | 6.8 | .2 | 0.0 | 0.2 | 0.3 | 0.5 | 0.6 | 0.7 | 0.9 | 1.0 | 1.2 | 1.3 | |
| 8.3 | 1.4 | 2.8 | 4.1 | 5.5 | 6.9 | .3 | 0.0 | 0.2 | 0.3 | 0.5 | 0.6 | 0.8 | 0.9 | 1.0 | 1.2 | 1.3 | |
| 8.4 | 1.4 | 2.8 | 4.2 | 5.6 | 7.0 | .4 | 0.1 | 0.2 | 0.3 | 0.5 | 0.6 | 0.8 | 0.9 | 1.0 | 1.2 | 1.3 | |
| 8.5 | 1.4 | 2.8 | 4.3 | 5.7 | 7.1 | .5 | 0.1 | 0.2 | 0.4 | 0.5 | 0.6 | 0.8 | 0.9 | 1.1 | 1.2 | 1.3 | |
| 8.6 | 1.4 | 2.9 | 4.3 | 5.7 | 7.2 | .6 | 0.1 | 0.2 | 0.4 | 0.5 | 0.7 | 0.8 | 0.9 | 1.1 | 1.2 | 1.4 | |
| 8.7 | 1.5 | 2.9 | 4.4 | 5.8 | 7.3 | .7 | 0.1 | 0.2 | 0.4 | 0.5 | 0.7 | 0.8 | 0.9 | 1.1 | 1.2 | 1.4 | |
| 8.8 | 1.5 | 3.0 | 4.4 | 5.9 | 7.4 | .8 | 0.1 | 0.3 | 0.4 | 0.5 | 0.7 | 0.8 | 1.0 | 1.1 | 1.2 | 1.4 | |
| 8.9 | 1.5 | 3.0 | 4.5 | 6.0 | 7.5 | .9 | 0.1 | 0.3 | 0.4 | 0.6 | 0.7 | 0.8 | 1.0 | 1.1 | 1.3 | 1.4 | |
| 9.0 | 1.5 | 3.0 | 4.5 | 6.0 | 7.5 | .0 | 0.0 | 0.2 | 0.3 | 0.5 | 0.6 | 0.8 | 0.9 | 1.1 | 1.3 | 1.4 | |
| 9.1 | 1.5 | 3.0 | 4.5 | 6.0 | 7.6 | .1 | 0.0 | 0.2 | 0.3 | 0.5 | 0.6 | 0.8 | 1.0 | 1.1 | 1.3 | 1.4 | |
| 9.2 | 1.5 | 3.0 | 4.6 | 6.1 | 7.6 | .2 | 0.0 | 0.2 | 0.3 | 0.5 | 0.7 | 0.8 | 1.0 | 1.1 | 1.3 | 1.5 | |
| 9.3 | 1.5 | 3.1 | 4.6 | 6.2 | 7.7 | .3 | 0.0 | 0.2 | 0.4 | 0.5 | 0.7 | 0.8 | 1.0 | 1.2 | 1.3 | 1.5 | |
| 9.4 | 1.6 | 3.1 | 4.7 | 6.3 | 7.8 | .4 | 0.1 | 0.2 | 0.4 | 0.5 | 0.7 | 0.9 | 1.0 | 1.2 | 1.3 | 1.5 | |
| 9.5 | 1.6 | 3.2 | 4.8 | 6.3 | 7.9 | .5 | 0.1 | 0.2 | 0.4 | 0.6 | 0.7 | 0.9 | 1.0 | 1.2 | 1.3 | 1.5 | 1.4<br>0.1<br>4.2<br>0.2<br>7.1<br>0.3<br>9.9<br>0.4<br>12.7<br>0.5<br>15.5<br>0.6<br>18.4<br>0.7<br>21.2<br>0.8<br>24.0<br>0.9<br>26.8<br>1.0<br>29.7<br>1.1<br>32.5<br>1.2<br>35.3 |
| 9.6 | 1.6 | 3.2 | 4.8 | 6.4 | 8.0 | .6 | 0.1 | 0.3 | 0.4 | 0.6 | 0.7 | 0.9 | 1.0 | 1.2 | 1.4 | 1.5 | |
| 9.7 | 1.6 | 3.3 | 4.9 | 6.5 | 8.1 | .7 | 0.1 | 0.3 | 0.4 | 0.6 | 0.7 | 0.9 | 1.1 | 1.2 | 1.4 | 1.5 | |
| 9.8 | 1.7 | 3.3 | 4.9 | 6.6 | 8.2 | .8 | 0.1 | 0.3 | 0.4 | 0.6 | 0.8 | 0.9 | 1.1 | 1.2 | 1.4 | 1.6 | |
| 9.9 | 1.7 | 3.3 | 5.0 | 6.6 | 8.3 | .9 | 0.1 | 0.3 | 0.5 | 0.6 | 0.8 | 0.9 | 1.1 | 1.3 | 1.4 | 1.6 | |
| 10.0 | 1.6 | 3.3 | 5.0 | 6.6 | 8.3 | .0 | 0.0 | 0.2 | 0.3 | 0.5 | 0.7 | 0.9 | 1.0 | 1.2 | 1.4 | 1.6 | |
| 10.1 | 1.7 | 3.3 | 5.0 | 6.7 | 8.4 | .1 | 0.0 | 0.2 | 0.4 | 0.5 | 0.7 | 0.9 | 1.1 | 1.2 | 1.4 | 1.6 | |
| 10.2 | 1.7 | 3.4 | 5.1 | 6.8 | 8.5 | .2 | 0.0 | 0.2 | 0.4 | 0.6 | 0.7 | 0.9 | 1.1 | 1.3 | 1.4 | 1.6 | |
| 10.3 | 1.7 | 3.4 | 5.1 | 6.9 | 8.6 | .3 | 0.1 | 0.2 | 0.4 | 0.6 | 0.8 | 0.9 | 1.1 | 1.3 | 1.5 | 1.6 | |
| 10.4 | 1.7 | 3.5 | 5.2 | 6.9 | 8.7 | .4 | 0.1 | 0.2 | 0.4 | 0.6 | 0.8 | 0.9 | 1.1 | 1.3 | 1.5 | 1.6 | |
| 10.5 | 1.8 | 3.5 | 5.3 | 7.0 | 8.8 | .5 | 0.1 | 0.3 | 0.4 | 0.6 | 0.8 | 1.0 | 1.1 | 1.3 | 1.5 | 1.7 | |
| 10.6 | 1.8 | 3.5 | 5.3 | 7.1 | 8.8 | .6 | 0.1 | 0.3 | 0.5 | 0.6 | 0.8 | 1.0 | 1.2 | 1.3 | 1.5 | 1.7 | |
| 10.7 | 1.8 | 3.6 | 5.4 | 7.2 | 8.9 | .7 | 0.1 | 0.3 | 0.5 | 0.6 | 0.8 | 1.0 | 1.2 | 1.3 | 1.5 | 1.7 | |
| 10.8 | 1.8 | 3.6 | 5.4 | 7.2 | 9.0 | .8 | 0.1 | 0.3 | 0.5 | 0.7 | 0.8 | 1.0 | 1.2 | 1.4 | 1.5 | 1.7 | |
| 10.9 | 1.9 | 3.7 | 5.5 | 7.3 | 9.1 | .9 | 0.2 | 0.3 | 0.5 | 0.7 | 0.9 | 1.0 | 1.2 | 1.4 | 1.6 | 1.7 | |
| 11.0 | 1.8 | 3.6 | 5.5 | 7.3 | 9.1 | .0 | 0.0 | 0.2 | 0.4 | 0.6 | 0.8 | 1.0 | 1.1 | 1.3 | 1.5 | 1.7 | 1.3<br>0.1<br>3.8<br>0.2<br>6.3<br>0.3<br>8.9<br>0.4<br>11.4<br>0.5<br>14.0<br>0.6<br>16.5<br>0.7<br>19.0<br>0.8<br>21.6<br>0.9<br>24.1<br>1.0<br>26.7<br>1.1<br>29.2<br>1.2<br>31.7<br>1.3<br>34.3 |
| 11.1 | 1.8 | 3.7 | 5.5 | 7.4 | 9.2 | .1 | 0.0 | 0.2 | 0.4 | 0.6 | 0.8 | 1.0 | 1.2 | 1.4 | 1.6 | 1.7 | |
| 11.2 | 1.8 | 3.7 | 5.6 | 7.4 | 9.3 | .2 | 0.0 | 0.2 | 0.4 | 0.6 | 0.8 | 1.0 | 1.2 | 1.4 | 1.6 | 1.8 | |
| 11.3 | 1.9 | 3.8 | 5.6 | 7.5 | 9.4 | .3 | 0.1 | 0.2 | 0.4 | 0.6 | 0.8 | 1.0 | 1.2 | 1.4 | 1.6 | 1.8 | |
| 11.4 | 1.9 | 3.8 | 5.7 | 7.6 | 9.5 | .4 | 0.1 | 0.3 | 0.5 | 0.7 | 0.8 | 1.0 | 1.2 | 1.4 | 1.6 | 1.8 | |
| 11.5 | 1.9 | 3.8 | 5.8 | 7.7 | 9.6 | .5 | 0.1 | 0.3 | 0.5 | 0.7 | 0.9 | 1.1 | 1.2 | 1.4 | 1.6 | 1.8 | |
| 11.6 | 1.9 | 3.9 | 5.8 | 7.7 | 9.7 | .6 | 0.1 | 0.3 | 0.5 | 0.7 | 0.9 | 1.1 | 1.3 | 1.5 | 1.6 | 1.8 | |
| 11.7 | 2.0 | 3.9 | 5.9 | 7.8 | 9.8 | .7 | 0.1 | 0.3 | 0.5 | 0.7 | 0.9 | 1.1 | 1.3 | 1.5 | 1.7 | 1.9 | |
| 11.8 | 2.0 | 4.0 | 5.9 | 7.9 | 9.9 | .8 | 0.2 | 0.3 | 0.5 | 0.7 | 0.9 | 1.1 | 1.3 | 1.5 | 1.7 | 1.9 | |
| 11.9 | 2.0 | 4.0 | 6.0 | 8.0 | 10.0 | .9 | 0.2 | 0.4 | 0.6 | 0.7 | 0.9 | 1.1 | 1.3 | 1.5 | 1.7 | 1.9 | |
| 12.0 | 2.0 | 4.0 | 6.0 | 8.0 | 10.0 | .0 | 0.0 | 0.2 | 0.4 | 0.6 | 0.8 | 1.0 | 1.2 | 1.5 | 1.7 | 1.9 | |
| 12.1 | 2.0 | 4.0 | 6.0 | 8.0 | 10.1 | .1 | 0.0 | 0.2 | 0.4 | 0.6 | 0.9 | 1.1 | 1.3 | 1.5 | 1.7 | 1.9 | |
| 12.2 | 2.0 | 4.0 | 6.1 | 8.1 | 10.1 | .2 | 0.0 | 0.2 | 0.5 | 0.7 | 0.9 | 1.1 | 1.3 | 1.5 | 1.7 | 1.9 | |
| 12.3 | 2.0 | 4.1 | 6.1 | 8.2 | 10.2 | .3 | 0.1 | 0.3 | 0.5 | 0.7 | 0.9 | 1.1 | 1.3 | 1.5 | 1.7 | 1.9 | |
| 12.4 | 2.1 | 4.1 | 6.2 | 8.3 | 10.3 | .4 | 0.1 | 0.3 | 0.5 | 0.7 | 0.9 | 1.1 | 1.3 | 1.5 | 1.7 | 2.0 | |
| 12.5 | 2.1 | 4.2 | 6.3 | 8.3 | 10.4 | .5 | 0.1 | 0.3 | 0.5 | 0.7 | 0.9 | 1.1 | 1.4 | 1.6 | 1.8 | 2.0 | 1.2<br>0.1<br>3.5<br>0.2<br>5.8<br>0.3<br>8.1<br>0.4<br>10.5<br>0.5<br>12.8<br>0.6<br>15.1<br>0.7<br>17.4<br>0.8<br>19.8<br>0.9<br>22.1<br>1.0<br>24.4<br>1.1<br>26.7<br>1.2<br>29.1<br>1.3<br>31.4<br>1.4<br>33.7<br>1.5<br>36.0 |
| 12.6 | 2.1 | 4.2 | 6.3 | 8.4 | 10.5 | .6 | 0.1 | 0.3 | 0.5 | 0.7 | 1.0 | 1.2 | 1.4 | 1.6 | 1.8 | 2.0 | |
| 12.7 | 2.1 | 4.3 | 6.4 | 8.5 | 10.6 | .7 | 0.1 | 0.4 | 0.6 | 0.8 | 1.0 | 1.2 | 1.4 | 1.6 | 1.8 | 2.0 | |
| 12.8 | 2.2 | 4.3 | 6.4 | 8.6 | 10.7 | .8 | 0.2 | 0.4 | 0.6 | 0.8 | 1.0 | 1.2 | 1.4 | 1.6 | 1.8 | 2.0 | |
| 12.9 | 2.2 | 4.3 | 6.5 | 8.6 | 10.8 | .9 | 0.2 | 0.4 | 0.6 | 0.8 | 1.0 | 1.2 | 1.4 | 1.6 | 1.9 | 2.1 | |
| 13.0 | 2.1 | 4.3 | 6.5 | 8.6 | 10.8 | .0 | 0.0 | 0.2 | 0.4 | 0.7 | 0.9 | 1.1 | 1.3 | 1.6 | 1.8 | 2.0 | |
| 13.1 | 2.2 | 4.3 | 6.5 | 8.7 | 10.9 | .1 | 0.0 | 0.2 | 0.5 | 0.7 | 0.9 | 1.1 | 1.4 | 1.6 | 1.8 | 2.0 | |
| 13.2 | 2.2 | 4.4 | 6.6 | 8.8 | 11.0 | .2 | 0.0 | 0.3 | 0.5 | 0.7 | 0.9 | 1.2 | 1.4 | 1.6 | 1.8 | 2.1 | |
| 13.3 | 2.2 | 4.4 | 6.6 | 8.9 | 11.1 | .3 | 0.1 | 0.3 | 0.5 | 0.7 | 1.0 | 1.2 | 1.4 | 1.6 | 1.9 | 2.1 | |
| 13.4 | 2.2 | 4.5 | 6.7 | 8.9 | 11.2 | .4 | 0.1 | 0.3 | 0.5 | 0.8 | 1.0 | 1.2 | 1.4 | 1.7 | 1.9 | 2.1 | |
| 13.5 | 2.3 | 4.5 | 6.8 | 9.0 | 11.3 | .5 | 0.1 | 0.3 | 0.6 | 0.8 | 1.0 | 1.2 | 1.5 | 1.7 | 1.9 | 2.1 | |
| 13.6 | 2.3 | 4.5 | 6.8 | 9.1 | 11.3 | .6 | 0.1 | 0.4 | 0.6 | 0.8 | 1.0 | 1.3 | 1.5 | 1.7 | 1.9 | 2.2 | |
| 13.7 | 2.3 | 4.6 | 6.9 | 9.2 | 11.4 | .7 | 0.2 | 0.4 | 0.6 | 0.8 | 1.1 | 1.3 | 1.5 | 1.7 | 2.0 | 2.2 | |
| 13.8 | 2.3 | 4.6 | 6.9 | 9.2 | 11.5 | .8 | 0.2 | 0.4 | 0.6 | 0.9 | 1.1 | 1.3 | 1.5 | 1.8 | 2.0 | 2.2 | |
| 13.9 | 2.4 | 4.7 | 7.0 | 9.3 | 11.6 | .9 | 0.2 | 0.4 | 0.7 | 0.9 | 1.1 | 1.3 | 1.6 | 1.8 | 2.0 | 2.2 | |
| 14.0 | 2.3 | 4.6 | 7.0 | 9.3 | 11.6 | .0 | 0.0 | 0.2 | 0.5 | 0.7 | 1.0 | 1.2 | 1.4 | 1.7 | 1.9 | 2.2 | 1.1<br>0.1<br>3.2<br>0.2<br>5.3<br>0.3<br>7.5<br>0.4<br>9.6<br>0.5<br>11.7<br>0.6<br>13.9<br>0.7<br>16.0<br>0.8<br>18.1<br>0.9<br>20.3<br>1.0<br>22.4<br>1.1<br>24.5<br>1.2<br>26.7<br>1.3<br>28.8<br>1.4<br>30.9<br>1.5<br>33.1<br>1.6<br>35.2 |
| 14.1 | 2.3 | 4.7 | 7.0 | 9.4 | 11.7 | .1 | 0.0 | 0.3 | 0.5 | 0.7 | 1.0 | 1.2 | 1.5 | 1.7 | 2.0 | 2.2 | |
| 14.2 | 2.3 | 4.7 | 7.1 | 9.4 | 11.8 | .2 | 0.0 | 0.3 | 0.5 | 0.8 | 1.0 | 1.3 | 1.5 | 1.7 | 2.0 | 2.2 | |
| 14.3 | 2.4 | 4.8 | 7.1 | 9.5 | 11.9 | .3 | 0.1 | 0.3 | 0.6 | 0.8 | 1.0 | 1.3 | 1.5 | 1.8 | 2.0 | 2.2 | |
| 14.4 | 2.4 | 4.8 | 7.2 | 9.6 | 12.0 | .4 | 0.1 | 0.3 | 0.6 | 0.8 | 1.1 | 1.3 | 1.5 | 1.8 | 2.0 | 2.3 | |
| 14.5 | 2.4 | 4.8 | 7.3 | 9.7 | 12.1 | .5 | 0.1 | 0.4 | 0.6 | 0.8 | 1.1 | 1.3 | 1.6 | 1.8 | 2.1 | 2.3 | |
| 14.6 | 2.4 | 4.9 | 7.3 | 9.7 | 12.2 | .6 | 0.1 | 0.4 | 0.6 | 0.9 | 1.1 | 1.4 | 1.6 | 1.8 | 2.1 | 2.3 | |
| 14.7 | 2.5 | 4.9 | 7.4 | 9.8 | 12.3 | .7 | 0.2 | 0.4 | 0.7 | 0.9 | 1.1 | 1.4 | 1.6 | 1.9 | 2.1 | 2.3 | |
| 14.8 | 2.5 | 5.0 | 7.4 | 9.9 | 12.4 | .8 | 0.2 | 0.4 | 0.7 | 0.9 | 1.2 | 1.4 | 1.6 | 1.9 | 2.1 | 2.4 | |
| 14.9 | 2.5 | 5.0 | 7.5 | 10.0 | 12.5 | .9 | 0.2 | 0.5 | 0.7 | 0.9 | 1.2 | 1.4 | 1.7 | 1.9 | 2.2 | 2.4 | |
| 15.0 | 2.5 | 5.0 | 7.5 | 10.0 | 12.5 | .0 | 0.0 | 0.3 | 0.5 | 0.8 | 1.0 | 1.3 | 1.5 | 1.8 | 2.1 | 2.3 | |
| 15.1 | 2.5 | 5.0 | 7.5 | 10.0 | 12.6 | .1 | 0.0 | 0.3 | 0.5 | 0.8 | 1.1 | 1.3 | 1.6 | 1.8 | 2.1 | 2.4 | |
| 15.2 | 2.5 | 5.0 | 7.6 | 10.1 | 12.6 | .2 | 0.1 | 0.3 | 0.6 | 0.8 | 1.1 | 1.3 | 1.6 | 1.9 | 2.1 | 2.4 | |
| 15.3 | 2.5 | 5.1 | 7.6 | 10.2 | 12.7 | .3 | 0.1 | 0.3 | 0.6 | 0.9 | 1.1 | 1.4 | 1.6 | 1.9 | 2.1 | 2.4 | |
| 15.4 | 2.6 | 5.1 | 7.7 | 10.3 | 12.8 | .4 | 0.1 | 0.4 | 0.6 | 0.9 | 1.1 | 1.4 | 1.7 | 1.9 | 2.2 | 2.4 | |
| 15.5 | 2.6 | 5.2 | 7.8 | 10.3 | 12.9 | .5 | 0.1 | 0.4 | 0.6 | 0.9 | 1.2 | 1.4 | 1.7 | 1.9 | 2.2 | 2.5 | |
| 15.6 | 2.6 | 5.2 | 7.8 | 10.4 | 13.0 | .6 | 0.2 | 0.4 | 0.7 | 0.9 | 1.2 | 1.4 | 1.7 | 2.0 | 2.2 | 2.5 | |
| 15.7 | 2.6 | 5.3 | 7.9 | 10.5 | 13.1 | .7 | 0.2 | 0.4 | 0.7 | 1.0 | 1.2 | 1.5 | 1.7 | 2.0 | 2.2 | 2.5 | |
| 15.8 | 2.7 | 5.3 | 7.9 | 10.6 | 13.2 | .8 | 0.2 | 0.5 | 0.7 | 1.0 | 1.2 | 1.5 | 1.8 | 2.0 | 2.3 | 2.5 | |
| 15.9 | 2.7 | 5.3 | 8.0 | 10.6 | 13.3 | .9 | 0.2 | 0.5 | 0.7 | 1.0 | 1.3 | 1.5 | 1.8 | 2.0 | 2.3 | 2.6 | |
| | 10′ | 20′ | 30′ | 40′ | 50′ | | 0′ | 1′ | 2′ | 3′ | 4′ | 5′ | 6′ | 7′ | 8′ | 9′ | |

The Double-Second-Difference correction (Corr.) is always to be added to the tabulated altitude.

# INTERPOLATION TABLE

| Dec. Inc. | Tens 10' | 20' | 30' | 40' | 50' | Decimals | Units 0' | 1' | 2' | 3' | 4' | 5' | 6' | 7' | 8' | 9' |
|---|---|---|---|---|---|---|---|---|---|---|---|---|---|---|---|---|
| ' | ' | ' | ' | ' | ' | ' | ' | ' | ' | ' | ' | ' | ' | ' | ' | ' |
| **16.0** | 2.6 | 5.3 | 8.0 | 10.6 | 13.3 | .0 | 0.0 | 0.3 | 0.5 | 0.8 | 1.1 | 1.4 | 1.6 | 1.9 | 2.2 | 2.5 |
| **16.1** | 2.7 | 5.3 | 8.0 | 10.7 | 13.4 | .1 | 0.0 | 0.3 | 0.6 | 0.9 | 1.1 | 1.4 | 1.7 | 2.0 | 2.2 | 2.5 |
| **16.2** | 2.7 | 5.4 | 8.1 | 10.8 | 13.5 | .2 | 0.1 | 0.3 | 0.6 | 0.9 | 1.2 | 1.4 | 1.7 | 2.0 | 2.3 | 2.5 |
| **16.3** | 2.7 | 5.4 | 8.1 | 10.9 | 13.6 | .3 | 0.1 | 0.4 | 0.6 | 0.9 | 1.2 | 1.5 | 1.7 | 2.0 | 2.3 | 2.6 |
| **16.4** | 2.7 | 5.5 | 8.2 | 10.9 | 13.7 | .4 | 0.1 | 0.4 | 0.7 | 0.9 | 1.2 | 1.5 | 1.8 | 2.0 | 2.3 | 2.6 |
| **16.5** | 2.8 | 5.5 | 8.3 | 11.0 | 13.8 | .5 | 0.1 | 0.4 | 0.7 | 1.0 | 1.2 | 1.5 | 1.8 | 2.1 | 2.3 | 2.6 |
| **16.6** | 2.8 | 5.5 | 8.3 | 11.1 | 13.8 | .6 | 0.2 | 0.4 | 0.7 | 1.0 | 1.3 | 1.5 | 1.8 | 2.1 | 2.4 | 2.6 |
| **16.7** | 2.8 | 5.6 | 8.4 | 11.2 | 13.9 | .7 | 0.2 | 0.5 | 0.7 | 1.0 | 1.3 | 1.6 | 1.8 | 2.1 | 2.4 | 2.7 |
| **16.8** | 2.8 | 5.6 | 8.4 | 11.2 | 14.0 | .8 | 0.2 | 0.5 | 0.8 | 1.0 | 1.3 | 1.6 | 1.9 | 2.1 | 2.4 | 2.7 |
| **16.9** | 2.9 | 5.7 | 8.5 | 11.3 | 14.1 | .9 | 0.2 | 0.5 | 0.8 | 1.1 | 1.3 | 1.6 | 1.9 | 2.2 | 2.4 | 2.7 |
| **17.0** | 2.8 | 5.6 | 8.5 | 11.3 | 14.1 | .0 | 0.0 | 0.3 | 0.6 | 0.9 | 1.2 | 1.5 | 1.7 | 2.0 | 2.3 | 2.6 |
| **17.1** | 2.8 | 5.7 | 8.5 | 11.4 | 14.2 | .1 | 0.0 | 0.3 | 0.6 | 0.9 | 1.2 | 1.5 | 1.8 | 2.1 | 2.4 | 2.7 |
| **17.2** | 2.8 | 5.7 | 8.6 | 11.4 | 14.3 | .2 | 0.1 | 0.3 | 0.6 | 0.9 | 1.2 | 1.5 | 1.8 | 2.1 | 2.4 | 2.7 |
| **17.3** | 2.9 | 5.8 | 8.6 | 11.5 | 14.4 | .3 | 0.1 | 0.4 | 0.7 | 1.0 | 1.3 | 1.5 | 1.8 | 2.1 | 2.4 | 2.7 |
| **17.4** | 2.9 | 5.8 | 8.7 | 11.6 | 14.5 | .4 | 0.1 | 0.4 | 0.7 | 1.0 | 1.3 | 1.6 | 1.9 | 2.2 | 2.4 | 2.7 |
| **17.5** | 2.9 | 5.8 | 8.8 | 11.7 | 14.6 | .5 | 0.1 | 0.4 | 0.7 | 1.0 | 1.3 | 1.6 | 1.9 | 2.2 | 2.5 | 2.8 |
| **17.6** | 2.9 | 5.9 | 8.8 | 11.7 | 14.7 | .6 | 0.2 | 0.5 | 0.8 | 1.0 | 1.3 | 1.6 | 1.9 | 2.2 | 2.5 | 2.8 |
| **17.7** | 3.0 | 5.9 | 8.9 | 11.8 | 14.8 | .7 | 0.2 | 0.5 | 0.8 | 1.1 | 1.4 | 1.7 | 2.0 | 2.2 | 2.5 | 2.8 |
| **17.8** | 3.0 | 6.0 | 8.9 | 11.9 | 14.9 | .8 | 0.2 | 0.5 | 0.8 | 1.1 | 1.4 | 1.7 | 2.0 | 2.3 | 2.6 | 2.9 |
| **17.9** | 3.0 | 6.0 | 9.0 | 12.0 | 15.0 | .9 | 0.3 | 0.6 | 0.8 | 1.1 | 1.4 | 1.7 | 2.0 | 2.3 | 2.6 | 2.9 |
| **18.0** | 3.0 | 6.0 | 9.0 | 12.0 | 15.0 | .0 | 0.0 | 0.3 | 0.6 | 0.9 | 1.2 | 1.5 | 1.8 | 2.2 | 2.5 | 2.8 |
| **18.1** | 3.0 | 6.0 | 9.0 | 12.0 | 15.1 | .1 | 0.0 | 0.3 | 0.6 | 1.0 | 1.3 | 1.6 | 1.9 | 2.2 | 2.5 | 2.8 |
| **18.2** | 3.0 | 6.0 | 9.1 | 12.1 | 15.1 | .2 | 0.1 | 0.4 | 0.7 | 1.0 | 1.3 | 1.6 | 1.9 | 2.2 | 2.5 | 2.8 |
| **18.3** | 3.0 | 6.1 | 9.1 | 12.2 | 15.2 | .3 | 0.1 | 0.4 | 0.7 | 1.0 | 1.3 | 1.6 | 1.9 | 2.3 | 2.6 | 2.9 |
| **18.4** | 3.1 | 6.1 | 9.2 | 12.3 | 15.3 | .4 | 0.1 | 0.4 | 0.7 | 1.0 | 1.4 | 1.7 | 2.0 | 2.3 | 2.6 | 2.9 |
| **18.5** | 3.1 | 6.2 | 9.3 | 12.3 | 15.4 | .5 | 0.2 | 0.5 | 0.8 | 1.1 | 1.4 | 1.7 | 2.0 | 2.3 | 2.6 | 2.9 |
| **18.6** | 3.1 | 6.2 | 9.3 | 12.4 | 15.5 | .6 | 0.2 | 0.5 | 0.8 | 1.1 | 1.4 | 1.7 | 2.0 | 2.3 | 2.7 | 3.0 |
| **18.7** | 3.1 | 6.3 | 9.4 | 12.5 | 15.6 | .7 | 0.2 | 0.5 | 0.8 | 1.1 | 1.4 | 1.8 | 2.1 | 2.4 | 2.7 | 3.0 |
| **18.8** | 3.2 | 6.3 | 9.4 | 12.6 | 15.7 | .8 | 0.2 | 0.6 | 0.9 | 1.2 | 1.5 | 1.8 | 2.1 | 2.4 | 2.7 | 3.0 |
| **18.9** | 3.2 | 6.3 | 9.5 | 12.6 | 15.8 | .9 | 0.3 | 0.6 | 0.9 | 1.2 | 1.5 | 1.8 | 2.1 | 2.4 | 2.7 | 3.1 |
| **19.0** | 3.1 | 6.3 | 9.5 | 12.6 | 15.8 | .0 | 0.0 | 0.3 | 0.6 | 1.0 | 1.3 | 1.6 | 1.9 | 2.3 | 2.6 | 2.9 |
| **19.1** | 3.2 | 6.3 | 9.5 | 12.7 | 15.9 | .1 | 0.0 | 0.4 | 0.7 | 1.0 | 1.3 | 1.7 | 2.0 | 2.3 | 2.6 | 3.0 |
| **19.2** | 3.2 | 6.4 | 9.6 | 12.8 | 16.0 | .2 | 0.1 | 0.4 | 0.7 | 1.0 | 1.4 | 1.7 | 2.0 | 2.3 | 2.7 | 3.0 |
| **19.3** | 3.2 | 6.4 | 9.6 | 12.9 | 16.1 | .3 | 0.1 | 0.4 | 0.7 | 1.1 | 1.4 | 1.7 | 2.0 | 2.4 | 2.7 | 3.0 |
| **19.4** | 3.2 | 6.5 | 9.7 | 12.9 | 16.2 | .4 | 0.1 | 0.5 | 0.8 | 1.1 | 1.4 | 1.8 | 2.1 | 2.4 | 2.7 | 3.1 |
| **19.5** | 3.3 | 6.5 | 9.8 | 13.0 | 16.3 | .5 | 0.2 | 0.5 | 0.8 | 1.1 | 1.5 | 1.8 | 2.1 | 2.4 | 2.8 | 3.1 |
| **19.6** | 3.3 | 6.5 | 9.8 | 13.1 | 16.3 | .6 | 0.2 | 0.5 | 0.8 | 1.2 | 1.5 | 1.8 | 2.1 | 2.5 | 2.8 | 3.1 |
| **19.7** | 3.3 | 6.6 | 9.9 | 13.2 | 16.4 | .7 | 0.2 | 0.6 | 0.9 | 1.2 | 1.5 | 1.9 | 2.2 | 2.5 | 2.8 | 3.2 |
| **19.8** | 3.3 | 6.6 | 9.9 | 13.2 | 16.5 | .8 | 0.3 | 0.6 | 0.9 | 1.2 | 1.6 | 1.9 | 2.2 | 2.5 | 2.9 | 3.2 |
| **19.9** | 3.4 | 6.7 | 10.0 | 13.3 | 16.6 | .9 | 0.3 | 0.6 | 0.9 | 1.3 | 1.6 | 1.9 | 2.2 | 2.6 | 2.9 | 3.2 |
| **20.0** | 3.3 | 6.6 | 10.0 | 13.3 | 16.6 | .0 | 0.0 | 0.3 | 0.7 | 1.0 | 1.4 | 1.7 | 2.0 | 2.4 | 2.7 | 3.1 |
| **20.1** | 3.3 | 6.7 | 10.0 | 13.4 | 16.7 | .1 | 0.0 | 0.4 | 0.7 | 1.1 | 1.4 | 1.7 | 2.1 | 2.4 | 2.8 | 3.1 |
| **20.2** | 3.3 | 6.7 | 10.1 | 13.4 | 16.8 | .2 | 0.1 | 0.4 | 0.8 | 1.1 | 1.4 | 1.8 | 2.1 | 2.5 | 2.8 | 3.1 |
| **20.3** | 3.4 | 6.8 | 10.1 | 13.5 | 16.9 | .3 | 0.1 | 0.4 | 0.8 | 1.1 | 1.5 | 1.8 | 2.2 | 2.5 | 2.8 | 3.2 |
| **20.4** | 3.4 | 6.8 | 10.2 | 13.6 | 17.0 | .4 | 0.1 | 0.5 | 0.8 | 1.2 | 1.5 | 1.8 | 2.2 | 2.5 | 2.9 | 3.2 |
| **20.5** | 3.4 | 6.8 | 10.3 | 13.7 | 17.1 | .5 | 0.2 | 0.5 | 0.9 | 1.2 | 1.5 | 1.9 | 2.2 | 2.6 | 2.9 | 3.2 |
| **20.6** | 3.4 | 6.9 | 10.3 | 13.7 | 17.2 | .6 | 0.2 | 0.5 | 0.9 | 1.2 | 1.6 | 1.9 | 2.3 | 2.6 | 2.9 | 3.3 |
| **20.7** | 3.5 | 6.9 | 10.4 | 13.8 | 17.3 | .7 | 0.2 | 0.6 | 0.9 | 1.3 | 1.6 | 1.9 | 2.3 | 2.6 | 3.0 | 3.3 |
| **20.8** | 3.5 | 7.0 | 10.4 | 13.9 | 17.4 | .8 | 0.3 | 0.6 | 1.0 | 1.3 | 1.6 | 2.0 | 2.3 | 2.7 | 3.0 | 3.3 |
| **20.9** | 3.5 | 7.0 | 10.5 | 14.0 | 17.5 | .9 | 0.3 | 0.6 | 1.0 | 1.3 | 1.7 | 2.0 | 2.4 | 2.7 | 3.0 | 3.4 |
| **21.0** | 3.5 | 7.0 | 10.5 | 14.0 | 17.5 | .0 | 0.0 | 0.4 | 0.7 | 1.1 | 1.4 | 1.8 | 2.1 | 2.5 | 2.9 | 3.2 |
| **21.1** | 3.5 | 7.0 | 10.5 | 14.0 | 17.6 | .1 | 0.0 | 0.4 | 0.8 | 1.1 | 1.5 | 1.8 | 2.2 | 2.5 | 2.9 | 3.3 |
| **21.2** | 3.5 | 7.0 | 10.6 | 14.1 | 17.6 | .2 | 0.1 | 0.4 | 0.8 | 1.1 | 1.5 | 1.9 | 2.2 | 2.6 | 2.9 | 3.3 |
| **21.3** | 3.5 | 7.1 | 10.6 | 14.2 | 17.7 | .3 | 0.1 | 0.5 | 0.8 | 1.2 | 1.5 | 1.9 | 2.3 | 2.6 | 3.0 | 3.3 |
| **21.4** | 3.6 | 7.1 | 10.7 | 14.3 | 17.8 | .4 | 0.1 | 0.5 | 0.9 | 1.2 | 1.6 | 1.9 | 2.3 | 2.7 | 3.0 | 3.4 |
| **21.5** | 3.6 | 7.2 | 10.8 | 14.3 | 17.9 | .5 | 0.2 | 0.5 | 0.9 | 1.3 | 1.6 | 2.0 | 2.3 | 2.7 | 3.0 | 3.4 |
| **21.6** | 3.6 | 7.2 | 10.8 | 14.4 | 18.0 | .6 | 0.2 | 0.6 | 0.9 | 1.3 | 1.6 | 2.0 | 2.4 | 2.7 | 3.1 | 3.4 |
| **21.7** | 3.6 | 7.3 | 10.9 | 14.5 | 18.1 | .7 | 0.3 | 0.6 | 1.0 | 1.3 | 1.7 | 2.0 | 2.4 | 2.8 | 3.1 | 3.5 |
| **21.8** | 3.7 | 7.3 | 10.9 | 14.6 | 18.2 | .8 | 0.3 | 0.6 | 1.0 | 1.4 | 1.7 | 2.1 | 2.4 | 2.8 | 3.2 | 3.5 |
| **21.9** | 3.7 | 7.3 | 11.0 | 14.6 | 18.3 | .9 | 0.3 | 0.7 | 1.0 | 1.4 | 1.8 | 2.1 | 2.5 | 2.8 | 3.2 | 3.5 |
| **22.0** | 3.6 | 7.3 | 11.0 | 14.6 | 18.3 | .0 | 0.0 | 0.4 | 0.7 | 1.1 | 1.5 | 1.9 | 2.2 | 2.6 | 3.0 | 3.4 |
| **22.1** | 3.7 | 7.3 | 11.0 | 14.7 | 18.4 | .1 | 0.0 | 0.4 | 0.8 | 1.2 | 1.5 | 1.9 | 2.3 | 2.7 | 3.0 | 3.4 |
| **22.2** | 3.7 | 7.4 | 11.1 | 14.8 | 18.5 | .2 | 0.1 | 0.4 | 0.8 | 1.2 | 1.6 | 1.9 | 2.3 | 2.7 | 3.1 | 3.4 |
| **22.3** | 3.7 | 7.4 | 11.1 | 14.9 | 18.6 | .3 | 0.1 | 0.5 | 0.9 | 1.2 | 1.6 | 2.0 | 2.4 | 2.7 | 3.1 | 3.5 |
| **22.4** | 3.7 | 7.5 | 11.2 | 14.9 | 18.7 | .4 | 0.1 | 0.5 | 0.9 | 1.3 | 1.6 | 2.0 | 2.4 | 2.8 | 3.1 | 3.5 |
| **22.5** | 3.8 | 7.5 | 11.3 | 15.0 | 18.8 | .5 | 0.2 | 0.6 | 0.9 | 1.3 | 1.7 | 2.1 | 2.4 | 2.8 | 3.2 | 3.6 |
| **22.6** | 3.8 | 7.5 | 11.3 | 15.1 | 18.8 | .6 | 0.2 | 0.6 | 1.0 | 1.3 | 1.7 | 2.1 | 2.5 | 2.8 | 3.2 | 3.6 |
| **22.7** | 3.8 | 7.6 | 11.4 | 15.2 | 18.9 | .7 | 0.3 | 0.6 | 1.0 | 1.4 | 1.8 | 2.1 | 2.5 | 2.9 | 3.3 | 3.6 |
| **22.8** | 3.8 | 7.6 | 11.4 | 15.2 | 19.0 | .8 | 0.3 | 0.7 | 1.0 | 1.4 | 1.8 | 2.2 | 2.5 | 2.9 | 3.3 | 3.7 |
| **22.9** | 3.9 | 7.7 | 11.5 | 15.3 | 19.1 | .9 | 0.3 | 0.7 | 1.1 | 1.5 | 1.8 | 2.2 | 2.6 | 3.0 | 3.3 | 3.7 |
| **23.0** | 3.8 | 7.6 | 11.5 | 15.3 | 19.1 | .0 | 0.0 | 0.4 | 0.8 | 1.2 | 1.6 | 2.0 | 2.3 | 2.7 | 3.1 | 3.5 |
| **23.1** | 3.8 | 7.7 | 11.5 | 15.4 | 19.2 | .1 | 0.0 | 0.4 | 0.8 | 1.2 | 1.6 | 2.0 | 2.4 | 2.8 | 3.2 | 3.6 |
| **23.2** | 3.8 | 7.7 | 11.6 | 15.4 | 19.3 | .2 | 0.1 | 0.5 | 0.9 | 1.3 | 1.6 | 2.0 | 2.4 | 2.8 | 3.2 | 3.6 |
| **23.3** | 3.9 | 7.8 | 11.6 | 15.5 | 19.4 | .3 | 0.1 | 0.5 | 0.9 | 1.3 | 1.7 | 2.1 | 2.5 | 2.9 | 3.3 | 3.6 |
| **23.4** | 3.9 | 7.8 | 11.7 | 15.6 | 19.5 | .4 | 0.2 | 0.5 | 0.9 | 1.3 | 1.7 | 2.1 | 2.5 | 2.9 | 3.3 | 3.7 |
| **23.5** | 3.9 | 7.8 | 11.8 | 15.7 | 19.6 | .5 | 0.2 | 0.6 | 1.0 | 1.4 | 1.8 | 2.2 | 2.5 | 2.9 | 3.3 | 3.7 |
| **23.6** | 3.9 | 7.9 | 11.8 | 15.7 | 19.7 | .6 | 0.2 | 0.6 | 1.0 | 1.4 | 1.8 | 2.2 | 2.6 | 3.0 | 3.4 | 3.8 |
| **23.7** | 4.0 | 7.9 | 11.9 | 15.8 | 19.8 | .7 | 0.3 | 0.7 | 1.1 | 1.4 | 1.8 | 2.2 | 2.6 | 3.0 | 3.4 | 3.8 |
| **23.8** | 4.0 | 8.0 | 11.9 | 15.9 | 19.9 | .8 | 0.3 | 0.7 | 1.1 | 1.5 | 1.9 | 2.3 | 2.7 | 3.1 | 3.4 | 3.8 |
| **23.9** | 4.0 | 8.0 | 12.0 | 16.0 | 20.0 | .9 | 0.4 | 0.7 | 1.1 | 1.5 | 1.9 | 2.3 | 2.7 | 3.1 | 3.5 | 3.9 |
| | 10' | 20' | 30' | 40' | 50' | | 0' | 1' | 2' | 3' | 4' | 5' | 6' | 7' | 8' | 9' |

Double Second Diff. and Corr. (16.0–17.9)

| Diff. | Corr. |
|---|---|
| ' | ' |
| 1.0 | |
| | 0.1 |
| 3.0 | |
| | 0.2 |
| 4.9 | |
| | 0.3 |
| 6.9 | |
| | 0.4 |
| 8.9 | |
| | 0.5 |
| 10.8 | |
| | 0.6 |
| 12.8 | |
| | 0.7 |
| 14.8 | |
| | 0.8 |
| 16.7 | |
| | 0.9 |
| 18.7 | |
| | 1.0 |
| 20.7 | |
| | 1.1 |
| 22.7 | |
| | 1.2 |
| 24.6 | |
| | 1.3 |
| 26.6 | |
| | 1.4 |
| 28.6 | |
| | 1.5 |
| 30.5 | |
| | 1.6 |
| 32.5 | |
| | 1.7 |
| 34.5 | |

Double Second Diff. and Corr. (18.0–19.9)

| Diff. | Corr. |
|---|---|
| 0.9 | |
| | 0.1 |
| 2.8 | |
| | 0.2 |
| 4.6 | |
| | 0.3 |
| 6.5 | |
| | 0.4 |
| 8.3 | |
| | 0.5 |
| 10.2 | |
| | 0.6 |
| 12.0 | |
| | 0.7 |
| 13.9 | |
| | 0.8 |
| 15.7 | |
| | 0.9 |
| 17.6 | |
| | 1.0 |
| 19.4 | |
| | 1.1 |
| 21.3 | |
| | 1.2 |
| 23.1 | |
| | 1.3 |
| 25.0 | |
| | 1.4 |
| 26.8 | |
| | 1.5 |
| 28.7 | |
| | 1.6 |
| 30.5 | |
| | 1.7 |
| 32.3 | |
| | 1.8 |
| 34.2 | |

Double Second Diff. and Corr. (20.0–21.9)

| Diff. | Corr. |
|---|---|
| 0.9 | |
| | 0.1 |
| 2.6 | |
| | 0.2 |
| 4.4 | |
| | 0.3 |
| 6.2 | |
| | 0.4 |
| 7.9 | |
| | 0.5 |
| 9.7 | |
| | 0.6 |
| 11.4 | |
| | 0.7 |
| 13.2 | |
| | 0.8 |
| 14.9 | |
| | 0.9 |
| 16.7 | |
| | 1.0 |
| 18.5 | |
| | 1.1 |
| 20.2 | |
| | 1.2 |
| 22.0 | |
| | 1.3 |
| 23.7 | |
| | 1.4 |
| 25.5 | |
| | 1.5 |
| 27.3 | |
| | 1.6 |
| 29.0 | |
| | 1.7 |
| 30.8 | |
| | 1.8 |
| 32.5 | |
| | 1.9 |
| 34.3 | |

Double Second Diff. and Corr. (22.0–23.9)

| Diff. | Corr. |
|---|---|
| 0.8 | |
| | 0.1 |
| 2.5 | |
| | 0.2 |
| 4.2 | |
| | 0.3 |
| 5.9 | |
| | 0.4 |
| 7.6 | |
| | 0.5 |
| 9.3 | |
| | 0.6 |
| 11.0 | |
| | 0.7 |
| 12.7 | |
| | 0.8 |
| 14.4 | |
| | 0.9 |
| 16.1 | |
| | 1.0 |
| 17.8 | |
| | 1.1 |
| 19.5 | |
| | 1.2 |
| 21.2 | |
| | 1.3 |
| 22.8 | |
| | 1.4 |
| 24.5 | |
| | 1.5 |
| 26.2 | |
| | 1.6 |
| 27.9 | |
| | 1.7 |
| 29.6 | |
| | 1.8 |
| 31.3 | |
| | 1.9 |
| 33.0 | |
| | 2.0 |
| 34.7 | |

| Dec. Inc. | Tens 10' | 20' | 30' | 40' | 50' | Decimals | Units 0' | 1' | 2' | 3' | 4' | 5' | 6' | 7' | 8' | 9' |
|---|---|---|---|---|---|---|---|---|---|---|---|---|---|---|---|---|
| **24.0** | 4.0 | 8.0 | 12.0 | 16.0 | 20.0 | .0 | 0.0 | 0.4 | 0.8 | 1.2 | 1.6 | 2.0 | 2.4 | 2.9 | 3.3 | 3.7 |
| **24.1** | 4.0 | 8.0 | 12.0 | 16.0 | 20.1 | .1 | 0.0 | 0.4 | 0.9 | 1.3 | 1.7 | 2.1 | 2.5 | 2.9 | 3.3 | 3.7 |
| **24.2** | 4.0 | 8.0 | 12.1 | 16.1 | 20.1 | .2 | 0.1 | 0.5 | 0.9 | 1.3 | 1.7 | 2.1 | 2.5 | 2.9 | 3.3 | 3.8 |
| **24.3** | 4.0 | 8.1 | 12.1 | 16.2 | 20.2 | .3 | 0.1 | 0.5 | 0.9 | 1.3 | 1.8 | 2.2 | 2.6 | 3.0 | 3.4 | 3.8 |
| **24.4** | 4.1 | 8.1 | 12.2 | 16.3 | 20.3 | .4 | 0.2 | 0.6 | 1.0 | 1.4 | 1.8 | 2.2 | 2.6 | 3.0 | 3.4 | 3.8 |
| **24.5** | 4.1 | 8.2 | 12.3 | 16.3 | 20.4 | .5 | 0.2 | 0.6 | 1.0 | 1.4 | 1.8 | 2.2 | 2.7 | 3.1 | 3.5 | 3.9 |
| **24.6** | 4.1 | 8.2 | 12.3 | 16.4 | 20.5 | .6 | 0.2 | 0.7 | 1.1 | 1.5 | 1.9 | 2.3 | 2.7 | 3.1 | 3.5 | 3.9 |
| **24.7** | 4.1 | 8.3 | 12.4 | 16.5 | 20.6 | .7 | 0.3 | 0.7 | 1.1 | 1.5 | 1.9 | 2.3 | 2.7 | 3.1 | 3.6 | 4.0 |
| **24.8** | 4.2 | 8.3 | 12.4 | 16.6 | 20.7 | .8 | 0.3 | 0.7 | 1.1 | 1.6 | 2.0 | 2.4 | 2.8 | 3.2 | 3.6 | 4.0 |
| **24.9** | 4.2 | 8.3 | 12.5 | 16.6 | 20.8 | .9 | 0.4 | 0.8 | 1.2 | 1.6 | 2.0 | 2.4 | 2.8 | 3.2 | 3.6 | 4.0 |
| **25.0** | 4.1 | 8.3 | 12.5 | 16.6 | 20.8 | .0 | 0.0 | 0.4 | 0.8 | 1.3 | 1.7 | 2.1 | 2.5 | 3.0 | 3.4 | 3.8 |
| **25.1** | 4.2 | 8.3 | 12.5 | 16.7 | 20.9 | .1 | 0.0 | 0.5 | 0.9 | 1.3 | 1.7 | 2.2 | 2.6 | 3.0 | 3.4 | 3.9 |
| **25.2** | 4.2 | 8.4 | 12.6 | 16.8 | 21.0 | .2 | 0.1 | 0.5 | 0.9 | 1.4 | 1.8 | 2.2 | 2.6 | 3.1 | 3.5 | 3.9 |
| **25.3** | 4.2 | 8.4 | 12.6 | 16.9 | 21.1 | .3 | 0.1 | 0.6 | 1.0 | 1.4 | 1.8 | 2.3 | 2.7 | 3.1 | 3.5 | 4.0 |
| **25.4** | 4.2 | 8.5 | 12.7 | 16.9 | 21.2 | .4 | 0.2 | 0.6 | 1.0 | 1.4 | 1.9 | 2.3 | 2.7 | 3.1 | 3.6 | 4.0 |
| **25.5** | 4.3 | 8.5 | 12.8 | 17.0 | 21.3 | .5 | 0.2 | 0.6 | 1.1 | 1.5 | 1.9 | 2.3 | 2.8 | 3.2 | 3.6 | 4.0 |
| **25.6** | 4.3 | 8.5 | 12.8 | 17.1 | 21.3 | .6 | 0.3 | 0.7 | 1.1 | 1.5 | 2.0 | 2.4 | 2.8 | 3.2 | 3.7 | 4.1 |
| **25.7** | 4.3 | 8.6 | 12.9 | 17.2 | 21.4 | .7 | 0.3 | 0.7 | 1.1 | 1.6 | 2.0 | 2.4 | 2.8 | 3.3 | 3.7 | 4.1 |
| **25.8** | 4.3 | 8.6 | 12.9 | 17.2 | 21.5 | .8 | 0.3 | 0.8 | 1.2 | 1.6 | 2.0 | 2.5 | 2.9 | 3.3 | 3.7 | 4.2 |
| **25.9** | 4.4 | 8.7 | 13.0 | 17.3 | 21.6 | .9 | 0.4 | 0.8 | 1.2 | 1.7 | 2.1 | 2.5 | 2.9 | 3.4 | 3.8 | 4.2 |
| **26.0** | 4.3 | 8.6 | 13.0 | 17.3 | 21.6 | .0 | 0.0 | 0.4 | 0.9 | 1.3 | 1.8 | 2.2 | 2.6 | 3.1 | 3.5 | 4.0 |
| **26.1** | 4.3 | 8.7 | 13.0 | 17.4 | 21.7 | .1 | 0.0 | 0.5 | 0.9 | 1.4 | 1.8 | 2.3 | 2.7 | 3.1 | 3.6 | 4.0 |
| **26.2** | 4.3 | 8.7 | 13.1 | 17.4 | 21.8 | .2 | 0.1 | 0.5 | 1.0 | 1.4 | 1.9 | 2.3 | 2.7 | 3.2 | 3.6 | 4.1 |
| **26.3** | 4.4 | 8.8 | 13.1 | 17.5 | 21.9 | .3 | 0.1 | 0.6 | 1.0 | 1.5 | 1.9 | 2.3 | 2.8 | 3.2 | 3.7 | 4.1 |
| **26.4** | 4.4 | 8.8 | 13.2 | 17.6 | 22.0 | .4 | 0.2 | 0.6 | 1.1 | 1.5 | 1.9 | 2.4 | 2.8 | 3.3 | 3.7 | 4.2 |
| **26.5** | 4.4 | 8.8 | 13.3 | 17.7 | 22.1 | .5 | 0.2 | 0.7 | 1.1 | 1.5 | 2.0 | 2.4 | 2.9 | 3.3 | 3.8 | 4.2 |
| **26.6** | 4.4 | 8.9 | 13.3 | 17.7 | 22.2 | .6 | 0.3 | 0.7 | 1.1 | 1.6 | 2.0 | 2.5 | 2.9 | 3.4 | 3.8 | 4.2 |
| **26.7** | 4.5 | 8.9 | 13.4 | 17.8 | 22.3 | .7 | 0.3 | 0.8 | 1.2 | 1.6 | 2.1 | 2.5 | 3.0 | 3.4 | 3.8 | 4.3 |
| **26.8** | 4.5 | 9.0 | 13.4 | 17.9 | 22.4 | .8 | 0.4 | 0.8 | 1.2 | 1.7 | 2.1 | 2.6 | 3.0 | 3.4 | 3.9 | 4.3 |
| **26.9** | 4.5 | 9.0 | 13.5 | 18.0 | 22.5 | .9 | 0.4 | 0.8 | 1.3 | 1.7 | 2.2 | 2.6 | 3.0 | 3.5 | 3.9 | 4.4 |
| **27.0** | 4.5 | 9.0 | 13.5 | 18.0 | 22.5 | .0 | 0.0 | 0.5 | 0.9 | 1.4 | 1.8 | 2.3 | 2.7 | 3.2 | 3.7 | 4.1 |
| **27.1** | 4.5 | 9.0 | 13.5 | 18.0 | 22.6 | .1 | 0.0 | 0.5 | 1.0 | 1.4 | 1.9 | 2.3 | 2.8 | 3.3 | 3.7 | 4.2 |
| **27.2** | 4.5 | 9.0 | 13.6 | 18.1 | 22.6 | .2 | 0.1 | 0.5 | 1.0 | 1.5 | 1.9 | 2.4 | 2.8 | 3.3 | 3.8 | 4.2 |
| **27.3** | 4.5 | 9.1 | 13.6 | 18.2 | 22.7 | .3 | 0.1 | 0.6 | 1.1 | 1.5 | 2.0 | 2.4 | 2.9 | 3.3 | 3.8 | 4.3 |
| **27.4** | 4.6 | 9.1 | 13.7 | 18.3 | 22.8 | .4 | 0.2 | 0.6 | 1.1 | 1.6 | 2.0 | 2.5 | 2.9 | 3.4 | 3.8 | 4.3 |
| **27.5** | 4.6 | 9.2 | 13.8 | 18.3 | 22.9 | .5 | 0.2 | 0.7 | 1.1 | 1.6 | 2.1 | 2.5 | 3.0 | 3.4 | 3.9 | 4.4 |
| **27.6** | 4.6 | 9.2 | 13.8 | 18.4 | 23.0 | .6 | 0.3 | 0.7 | 1.2 | 1.6 | 2.1 | 2.6 | 3.0 | 3.5 | 3.9 | 4.4 |
| **27.7** | 4.6 | 9.3 | 13.9 | 18.5 | 23.1 | .7 | 0.3 | 0.8 | 1.2 | 1.7 | 2.2 | 2.6 | 3.1 | 3.5 | 4.0 | 4.4 |
| **27.8** | 4.7 | 9.3 | 13.9 | 18.6 | 23.2 | .8 | 0.4 | 0.8 | 1.3 | 1.7 | 2.2 | 2.7 | 3.1 | 3.6 | 4.0 | 4.5 |
| **27.9** | 4.7 | 9.3 | 14.0 | 18.6 | 23.3 | .9 | 0.4 | 0.9 | 1.3 | 1.8 | 2.2 | 2.7 | 3.2 | 3.6 | 4.1 | 4.5 |
| **28.0** | 4.6 | 9.3 | 14.0 | 18.6 | 23.3 | .0 | 0.0 | 0.5 | 0.9 | 1.4 | 1.9 | 2.4 | 2.8 | 3.3 | 3.8 | 4.3 |
| **28.1** | 4.7 | 9.3 | 14.0 | 18.7 | 23.4 | .1 | 0.0 | 0.5 | 1.0 | 1.5 | 1.9 | 2.4 | 2.9 | 3.4 | 3.8 | 4.3 |
| **28.2** | 4.7 | 9.4 | 14.1 | 18.8 | 23.5 | .2 | 0.1 | 0.6 | 1.0 | 1.5 | 2.0 | 2.5 | 2.9 | 3.4 | 3.9 | 4.4 |
| **28.3** | 4.7 | 9.4 | 14.1 | 18.9 | 23.6 | .3 | 0.1 | 0.6 | 1.1 | 1.6 | 2.0 | 2.5 | 3.0 | 3.5 | 3.9 | 4.4 |
| **28.4** | 4.7 | 9.5 | 14.2 | 18.9 | 23.7 | .4 | 0.2 | 0.7 | 1.1 | 1.6 | 2.1 | 2.6 | 3.0 | 3.5 | 4.0 | 4.5 |
| **28.5** | 4.8 | 9.5 | 14.3 | 19.0 | 23.8 | .5 | 0.2 | 0.7 | 1.2 | 1.7 | 2.1 | 2.6 | 3.1 | 3.6 | 4.0 | 4.5 |
| **28.6** | 4.8 | 9.5 | 14.3 | 19.1 | 23.8 | .6 | 0.3 | 0.8 | 1.2 | 1.7 | 2.2 | 2.7 | 3.1 | 3.6 | 4.1 | 4.6 |
| **28.7** | 4.8 | 9.6 | 14.4 | 19.2 | 23.9 | .7 | 0.3 | 0.8 | 1.3 | 1.8 | 2.2 | 2.7 | 3.2 | 3.7 | 4.1 | 4.6 |
| **28.8** | 4.8 | 9.6 | 14.4 | 19.2 | 24.0 | .8 | 0.4 | 0.9 | 1.3 | 1.8 | 2.3 | 2.8 | 3.2 | 3.7 | 4.2 | 4.7 |
| **28.9** | 4.9 | 9.7 | 14.5 | 19.3 | 24.1 | .9 | 0.4 | 0.9 | 1.4 | 1.9 | 2.3 | 2.8 | 3.3 | 3.8 | 4.2 | 4.7 |
| **29.0** | 4.8 | 9.6 | 14.5 | 19.3 | 24.1 | .0 | 0.0 | 0.5 | 1.0 | 1.5 | 2.0 | 2.5 | 2.9 | 3.4 | 3.9 | 4.4 |
| **29.1** | 4.8 | 9.7 | 14.5 | 19.4 | 24.2 | .1 | 0.0 | 0.5 | 1.0 | 1.5 | 2.0 | 2.5 | 3.0 | 3.5 | 4.0 | 4.5 |
| **29.2** | 4.8 | 9.7 | 14.6 | 19.4 | 24.3 | .2 | 0.1 | 0.6 | 1.1 | 1.6 | 2.1 | 2.6 | 3.0 | 3.5 | 4.0 | 4.5 |
| **29.3** | 4.9 | 9.8 | 14.6 | 19.5 | 24.4 | .3 | 0.1 | 0.6 | 1.1 | 1.6 | 2.1 | 2.6 | 3.1 | 3.6 | 4.1 | 4.6 |
| **29.4** | 4.9 | 9.8 | 14.7 | 19.6 | 24.5 | .4 | 0.2 | 0.7 | 1.2 | 1.7 | 2.2 | 2.7 | 3.1 | 3.6 | 4.1 | 4.6 |
| **29.5** | 4.9 | 9.8 | 14.8 | 19.7 | 24.6 | .5 | 0.2 | 0.7 | 1.2 | 1.7 | 2.2 | 2.7 | 3.2 | 3.7 | 4.2 | 4.7 |
| **29.6** | 4.9 | 9.9 | 14.8 | 19.7 | 24.7 | .6 | 0.3 | 0.8 | 1.3 | 1.8 | 2.3 | 2.8 | 3.2 | 3.7 | 4.2 | 4.7 |
| **29.7** | 5.0 | 9.9 | 14.9 | 19.8 | 24.8 | .7 | 0.3 | 0.8 | 1.3 | 1.8 | 2.3 | 2.8 | 3.3 | 3.8 | 4.3 | 4.8 |
| **29.8** | 5.0 | 10.0 | 14.9 | 19.9 | 24.9 | .8 | 0.4 | 0.9 | 1.4 | 1.9 | 2.4 | 2.9 | 3.3 | 3.8 | 4.3 | 4.8 |
| **29.9** | 5.0 | 10.0 | 15.0 | 20.0 | 25.0 | .9 | 0.4 | 0.9 | 1.4 | 1.9 | 2.4 | 2.9 | 3.4 | 3.9 | 4.4 | 4.9 |
| **30.0** | 5.0 | 10.0 | 15.0 | 20.0 | 25.0 | .0 | 0.0 | 0.5 | 1.0 | 1.5 | 2.0 | 2.5 | 3.0 | 3.6 | 4.1 | 4.6 |
| **30.1** | 5.0 | 10.0 | 15.0 | 20.0 | 25.1 | .1 | 0.1 | 0.6 | 1.1 | 1.6 | 2.1 | 2.6 | 3.1 | 3.6 | 4.1 | 4.6 |
| **30.2** | 5.0 | 10.0 | 15.1 | 20.1 | 25.1 | .2 | 0.1 | 0.6 | 1.1 | 1.6 | 2.1 | 2.6 | 3.2 | 3.7 | 4.2 | 4.7 |
| **30.3** | 5.0 | 10.1 | 15.1 | 20.2 | 25.2 | .3 | 0.2 | 0.7 | 1.2 | 1.7 | 2.2 | 2.7 | 3.2 | 3.7 | 4.2 | 4.7 |
| **30.4** | 5.1 | 10.1 | 15.2 | 20.3 | 25.3 | .4 | 0.2 | 0.7 | 1.2 | 1.7 | 2.2 | 2.7 | 3.3 | 3.8 | 4.3 | 4.8 |
| **30.5** | 5.1 | 10.2 | 15.3 | 20.3 | 25.4 | .5 | 0.3 | 0.8 | 1.3 | 1.8 | 2.3 | 2.8 | 3.3 | 3.8 | 4.3 | 4.8 |
| **30.6** | 5.1 | 10.2 | 15.3 | 20.4 | 25.5 | .6 | 0.3 | 0.8 | 1.3 | 1.8 | 2.3 | 2.8 | 3.4 | 3.9 | 4.4 | 4.9 |
| **30.7** | 5.1 | 10.3 | 15.4 | 20.5 | 25.6 | .7 | 0.4 | 0.9 | 1.4 | 1.9 | 2.4 | 2.9 | 3.4 | 3.9 | 4.4 | 4.9 |
| **30.8** | 5.2 | 10.3 | 15.4 | 20.6 | 25.7 | .8 | 0.4 | 0.9 | 1.4 | 1.9 | 2.4 | 2.9 | 3.5 | 4.0 | 4.5 | 5.0 |
| **30.9** | 5.2 | 10.3 | 15.5 | 20.6 | 25.8 | .9 | 0.5 | 1.0 | 1.5 | 2.0 | 2.5 | 3.0 | 3.5 | 4.0 | 4.5 | 5.0 |
| **31.0** | 5.1 | 10.3 | 15.5 | 20.6 | 25.8 | .0 | 0.0 | 0.5 | 1.0 | 1.6 | 2.1 | 2.6 | 3.1 | 3.7 | 4.2 | 4.7 |
| **31.1** | 5.2 | 10.3 | 15.5 | 20.7 | 25.9 | .1 | 0.1 | 0.6 | 1.1 | 1.6 | 2.2 | 2.7 | 3.2 | 3.7 | 4.3 | 4.8 |
| **31.2** | 5.2 | 10.4 | 15.6 | 20.8 | 26.0 | .2 | 0.1 | 0.6 | 1.2 | 1.7 | 2.2 | 2.7 | 3.3 | 3.8 | 4.3 | 4.8 |
| **31.3** | 5.2 | 10.4 | 15.6 | 20.9 | 26.1 | .3 | 0.2 | 0.7 | 1.2 | 1.7 | 2.3 | 2.8 | 3.3 | 3.8 | 4.4 | 4.9 |
| **31.4** | 5.2 | 10.5 | 15.7 | 20.9 | 26.2 | .4 | 0.2 | 0.7 | 1.3 | 1.8 | 2.3 | 2.8 | 3.4 | 3.9 | 4.4 | 4.9 |
| **31.5** | 5.3 | 10.5 | 15.8 | 21.0 | 26.3 | .5 | 0.3 | 0.8 | 1.3 | 1.8 | 2.4 | 2.9 | 3.4 | 3.9 | 4.5 | 5.0 |
| **31.6** | 5.3 | 10.5 | 15.8 | 21.1 | 26.3 | .6 | 0.3 | 0.8 | 1.4 | 1.9 | 2.4 | 2.9 | 3.5 | 4.0 | 4.5 | 5.0 |
| **31.7** | 5.3 | 10.6 | 15.9 | 21.2 | 26.4 | .7 | 0.4 | 0.9 | 1.4 | 1.9 | 2.5 | 3.0 | 3.5 | 4.0 | 4.6 | 5.1 |
| **31.8** | 5.3 | 10.6 | 15.9 | 21.2 | 26.5 | .8 | 0.4 | 0.9 | 1.5 | 2.0 | 2.5 | 3.0 | 3.6 | 4.1 | 4.6 | 5.1 |
| **31.9** | 5.4 | 10.7 | 16.0 | 21.3 | 26.6 | .9 | 0.5 | 1.0 | 1.5 | 2.0 | 2.6 | 3.1 | 3.6 | 4.1 | 4.7 | 5.2 |
| | 10' | 20' | 30' | 40' | 50' | | 0' | 1' | 2' | 3' | 4' | 5' | 6' | 7' | 8' | 9' |

Double Second Diff. and Corr. (24.0–25.9)

| Diff. | Corr. |
|---|---|
| ' | ' |
| 0.8 | |
| | 0.1 |
| 2.5 | |
| | 0.2 |
| 4.1 | |
| | 0.3 |
| 5.8 | |
| | 0.4 |
| 7.4 | |
| | 0.5 |
| 9.1 | |
| | 0.6 |
| 10.7 | |
| | 0.7 |
| 12.3 | |
| | 0.8 |
| 14.0 | |
| | 0.9 |
| 15.6 | |
| | 1.0 |
| 17.3 | |
| | 1.1 |
| 18.9 | |
| | 1.2 |
| 20.6 | |
| | 1.3 |
| 22.2 | |
| | 1.4 |
| 23.9 | |
| | 1.5 |
| 25.5 | |
| | 1.6 |
| 27.2 | |
| | 1.7 |
| 28.8 | |
| | 1.8 |
| 30.4 | |
| | 1.9 |
| 32.1 | |
| | 2.0 |
| 33.7 | |
| | 2.1 |
| 35.4 | |

Double Second Diff. and Corr. (26.0–27.9)

| Diff. | Corr. |
|---|---|
| 0.8 | |
| | 0.1 |
| 2.4 | |
| | 0.2 |
| 4.0 | |
| | 0.3 |
| 5.7 | |
| | 0.4 |
| 7.3 | |
| | 0.5 |
| 8.9 | |
| | 0.6 |
| 10.5 | |
| | 0.7 |
| 12.1 | |
| | 0.8 |
| 13.7 | |
| | 0.9 |
| 15.4 | |
| | 1.0 |
| 17.0 | |
| | 1.1 |
| 18.6 | |
| | 1.2 |
| 20.2 | |
| | 1.3 |
| 21.8 | |
| | 1.4 |
| 23.4 | |
| | 1.5 |
| 25.1 | |
| | 1.6 |
| 26.7 | |
| | 1.7 |
| 28.3 | |
| | 1.8 |
| 29.9 | |
| | 1.9 |
| 31.5 | |
| | 2.0 |
| 33.1 | |
| | 2.1 |
| 34.7 | |

Double Second Diff. and Corr. (28.0–29.9)

| Diff. | Corr. |
|---|---|
| 0.8 | |
| | 0.1 |
| 2.4 | |
| | 0.2 |
| 4.0 | |
| | 0.3 |
| 5.6 | |
| | 0.4 |
| 7.2 | |
| | 0.5 |
| 8.8 | |
| | 0.6 |
| 10.4 | |
| | 0.7 |
| 12.0 | |
| | 0.8 |
| 13.6 | |
| | 0.9 |
| 15.2 | |
| | 1.0 |
| 16.8 | |
| | 1.1 |
| 18.4 | |
| | 1.2 |
| 20.0 | |
| | 1.3 |
| 21.6 | |
| | 1.4 |
| 23.2 | |
| | 1.5 |
| 24.8 | |
| | 1.6 |
| 26.4 | |
| | 1.7 |
| 28.0 | |
| | 1.8 |
| 29.6 | |
| | 1.9 |
| 31.2 | |
| | 2.0 |
| 32.8 | |
| | 2.1 |
| 34.4 | |

Double Second Diff. and Corr. (30.0–31.9)

| Diff. | Corr. |
|---|---|
| 0.8 | |
| | 0.1 |
| 2.4 | |
| | 0.2 |
| 4.0 | |
| | 0.3 |
| 5.6 | |
| | 0.4 |
| 7.2 | |
| | 0.5 |
| 8.8 | |
| | 0.6 |
| 10.4 | |
| | 0.7 |
| 12.0 | |
| | 0.8 |
| 13.6 | |
| | 0.9 |
| 15.2 | |
| | 1.0 |
| 16.8 | |
| | 1.1 |
| 18.4 | |
| | 1.2 |
| 20.0 | |
| | 1.3 |
| 21.6 | |
| | 1.4 |
| 23.2 | |
| | 1.5 |
| 24.8 | |
| | 1.6 |
| 26.4 | |
| | 1.7 |
| 28.0 | |
| | 1.8 |
| 29.6 | |
| | 1.9 |
| 31.2 | |
| | 2.0 |
| 32.8 | |
| | 2.1 |
| 34.4 | |

The Double-Second-Difference correction (Corr.) is always to be added to the tabulated altitude.

## LATITUDE CONTRARY NAME TO DECLINATION L.H.A. 13°, 347°

| Dec. | 15° Hc | d | Z | 16° Hc | d | Z | 17° Hc | d | Z | 18° Hc | d | Z | Dec. |
|---|---|---|---|---|---|---|---|---|---|---|---|---|---|
| ° | ° ′ | ′ | ° | ° ′ | ′ | ° | ° ′ | ′ | ° | ° ′ | ′ | ° | ° |
| 0 | 70 14.9 | -46.5 | 138.3 | 69 29.5 | -47.7 | 140.1 | 68 43.0 | -48.8 | 141.7 | 67 55.4 | -49.7 | 143.2 | 0 |
| 1 | 69 28.4 | 47.6 | 140.1 | 68 41.8 | 48.7 | 141.8 | 67 54.2 | 49.6 | 143.3 | 67 05.7 | 50.5 | 144.7 | 1 |
| 2 | 68 40.8 | 48.6 | 141.8 | 67 53.1 | 49.5 | 143.3 | 67 04.6 | 50.4 | 144.7 | 66 15.2 | 51.2 | 146.1 | 2 |
| 3 | 67 52.2 | 49.4 | 143.4 | 67 03.6 | 50.3 | 144.8 | 66 14.2 | 51.1 | 146.1 | 65 24.0 | 51.7 | 147.3 | 3 |
| 4 | 67 02.8 | 50.2 | 144.9 | 66 13.3 | 50.9 | 146.2 | 65 23.1 | 51.6 | 147.4 | 64 32.3 | 52.3 | 148.5 | 4 |
| 5 | 66 12.6 | -50.9 | 146.3 | 65 22.4 | -51.6 | 147.5 | 64 31.5 | -52.3 | 148.6 | 63 40.0 | -52.8 | 149.7 | 5 |
| 6 | 65 21.7 | 51.5 | 147.5 | 64 30.8 | 52.2 | 148.7 | 63 39.2 | 52.7 | 149.7 | 62 47.2 | 53.3 | 150.7 | 6 |
| 7 | 64 30.2 | 52.1 | 148.8 | 63 38.6 | 52.6 | 149.8 | 62 46.5 | 53.2 | 150.8 | 61 53.9 | 53.6 | 151.7 | 7 |
| 8 | 63 38.1 | 52.5 | 149.9 | 62 46.0 | 53.1 | 150.9 | 61 53.3 | 53.5 | 151.8 | 61 00.3 | 54.1 | 152.6 | 8 |
| 9 | 62 45.6 | 53.1 | 151.0 | 61 52.9 | 53.5 | 151.9 | 60 59.8 | 54.0 | 152.7 | 60 06.2 | 54.3 | 153.5 | 9 |
| 10 | 61 52.5 | -53.4 | 152.0 | 60 59.4 | -53.9 | 152.8 | 60 05.8 | -54.3 | 153.6 | 59 11.9 | -54.7 | 154.4 | 10 |
| 11 | 60 59.1 | 53.8 | 152.9 | 60 05.5 | 54.2 | 153.7 | 59 11.5 | 54.6 | 154.5 | 58 17.2 | 54.9 | 155.2 | 11 |
| 12 | 60 05.3 | 54.2 | 153.8 | 59 11.3 | 54.6 | 154.6 | 58 16.9 | 54.9 | 155.3 | 57 22.3 | 55.2 | 155.9 | 12 |
| 13 | 59 11.1 | 54.5 | 154.7 | 58 16.7 | 54.8 | 155.4 | 57 22.0 | 55.1 | 156.0 | 56 27.1 | 55.5 | 156.6 | 13 |
| 14 | 58 16.6 | 54.8 | 155.5 | 57 21.9 | 55.1 | 156.1 | 56 26.9 | 55.4 | 156.7 | 55 31.6 | 55.6 | 157.3 | 14 |
| 15 | 57 21.8 | -55.0 | 156.2 | 56 26.8 | -55.3 | 156.9 | 55 31.5 | -55.6 | 157.4 | 54 36.0 | -55.9 | 158.0 | 15 |
| 16 | 56 26.8 | 55.3 | 157.0 | 55 31.5 | 55.6 | 157.5 | 54 35.9 | 55.8 | 158.1 | 53 40.1 | 56.0 | 158.6 | 16 |
| 17 | 55 31.5 | 55.5 | 157.7 | 54 35.9 | 55.8 | 158.2 | 53 40.1 | 56.0 | 158.7 | 52 44.1 | 56.2 | 159.2 | 17 |
| 18 | 54 36.0 | 55.7 | 158.3 | 53 40.1 | 55.9 | 158.8 | 52 44.1 | 56.2 | 159.3 | 51 47.9 | 56.4 | 159.8 | 18 |
| 19 | 53 40.3 | 55.9 | 159.0 | 52 44.2 | 56.2 | 159.4 | 51 47.9 | 56.3 | 159.9 | 50 51.5 | 56.5 | 160.3 | 19 |
| 20 | 52 44.4 | -56.1 | 159.6 | 51 48.0 | -56.2 | 160.0 | 50 51.6 | -56.5 | 160.4 | 49 55.0 | -56.7 | 160.8 | 20 |
| 21 | 51 48.3 | 56.3 | 160.1 | 50 51.8 | 56.5 | 160.6 | 49 55.1 | 56.6 | 161.0 | 48 58.3 | 56.8 | 161.3 | 21 |
| 22 | 50 52.0 | 56.4 | 160.7 | 49 55.3 | 56.6 | 161.1 | 48 58.5 | 56.8 | 161.5 | 48 01.5 | 56.9 | 161.8 | 22 |
| 23 | 49 55.6 | 56.5 | 161.2 | 48 58.7 | 56.7 | 161.6 | 48 01.7 | 56.8 | 162.0 | 47 04.6 | 57.0 | 162.3 | 23 |
| 24 | 48 59.1 | 56.7 | 161.8 | 48 02.0 | 56.8 | 162.1 | 47 04.9 | 57.0 | 162.4 | 46 07.6 | 57.1 | 162.8 | 24 |
| 25 | 48 02.4 | -56.8 | 162.2 | 47 05.2 | -57.0 | 162.6 | 46 07.9 | -57.1 | 162.9 | 45 10.5 | -57.2 | 163.2 | 25 |
| 26 | 47 05.6 | 56.9 | 162.7 | 46 08.2 | 57.0 | 163.0 | 45 10.8 | 57.2 | 163.3 | 44 13.3 | 57.3 | 163.6 | 26 |
| 27 | 46 08.7 | 57.1 | 163.2 | 45 11.2 | 57.2 | 163.5 | 44 13.6 | 57.3 | 163.8 | 43 16.0 | 57.4 | 164.0 | 27 |
| 28 | 45 11.6 | 57.1 | 163.6 | 44 14.0 | 57.2 | 163.9 | 43 16.3 | 57.3 | 164.2 | 42 18.6 | 57.5 | 164.4 | 28 |
| 29 | 44 14.5 | 57.2 | 164.1 | 43 16.8 | 57.4 | 164.3 | 42 19.0 | 57.5 | 164.6 | 41 21.1 | 57.5 | 164.8 | 29 |
| 30 | 43 17.3 | -57.3 | 164.5 | 42 19.4 | -57.4 | 164.7 | 41 21.5 | -57.5 | 165.0 | 40 23.6 | -57.7 | 165.2 | 30 |
| 31 | 42 20.0 | 57.4 | 164.9 | 41 22.0 | 57.5 | 165.1 | 40 24.0 | 57.6 | 165.3 | 39 25.9 | 57.7 | 165.5 | 31 |
| 32 | 41 22.6 | 57.5 | 165.3 | 40 24.5 | 57.5 | 165.5 | 39 26.4 | 57.6 | 165.7 | 38 28.2 | 57.7 | 165.9 | 32 |
| 33 | 40 25.1 | 57.5 | 165.7 | 39 27.0 | 57.7 | 165.9 | 38 28.8 | 57.8 | 166.1 | 37 30.5 | 57.8 | 166.2 | 33 |
| 34 | 39 27.6 | 57.6 | 166.0 | 38 29.3 | 57.7 | 166.2 | 37 31.0 | 57.7 | 166.4 | 36 32.7 | 57.9 | 166.6 | 34 |
| 35 | 38 30.0 | -57.7 | 166.4 | 37 31.6 | -57.7 | 166.6 | 36 33.3 | -57.9 | 166.7 | 35 34.8 | -57.9 | 166.9 | 35 |
| 36 | 37 32.3 | 57.7 | 166.7 | 36 33.9 | 57.8 | 166.9 | 35 35.4 | 57.9 | 167.1 | 34 36.9 | 57.9 | 167.2 | 36 |
| 37 | 36 34.6 | 57.8 | 167.1 | 35 36.1 | 57.9 | 167.2 | 34 37.5 | 57.9 | 167.4 | 33 39.0 | 58.0 | 167.5 | 37 |
| 38 | 35 36.8 | 57.9 | 167.4 | 34 38.2 | 57.9 | 167.6 | 33 39.6 | 58.0 | 167.7 | 32 41.0 | 58.1 | 167.8 | 38 |
| 39 | 34 38.9 | 57.9 | 167.7 | 33 40.3 | 58.0 | 167.9 | 32 41.6 | 58.0 | 168.0 | 31 42.9 | 58.1 | 168.1 | 39 |
| 40 | 33 41.0 | -57.9 | 168.0 | 32 42.3 | -58.0 | 168.2 | 31 43.6 | -58.1 | 168.3 | 30 44.8 | -58.1 | 168.4 | 40 |
| 41 | 32 43.1 | 58.0 | 168.4 | 31 44.3 | 58.0 | 168.5 | 30 45.5 | 58.1 | 168.6 | 29 46.7 | 58.2 | 168.7 | 41 |
| 42 | 31 45.1 | 58.0 | 168.7 | 30 46.3 | 58.1 | 168.8 | 29 47.4 | 58.1 | 168.9 | 28 48.5 | 58.2 | 169.0 | 42 |
| 43 | 30 47.1 | 58.1 | 169.0 | 29 48.2 | 58.1 | 169.1 | 28 49.3 | 58.2 | 169.2 | 27 50.3 | 58.2 | 169.3 | 43 |
| 44 | 29 49.0 | 58.1 | 169.3 | 28 50.1 | 58.2 | 169.4 | 27 51.1 | 58.2 | 169.5 | 26 52.1 | 58.3 | 169.5 | 44 |
| 45 | 28 50.9 | -58.1 | 169.5 | 27 51.9 | -58.2 | 169.6 | 26 52.9 | -58.3 | 169.7 | 25 53.8 | -58.3 | 169.8 | 45 |
| 46 | 27 52.8 | 58.2 | 169.8 | 26 53.7 | 58.2 | 169.9 | 25 54.6 | 58.3 | 170.0 | 24 55.5 | 58.3 | 170.1 | 46 |
| 47 | 26 54.6 | 58.2 | 170.1 | 25 55.5 | 58.3 | 170.2 | 24 56.3 | 58.3 | 170.3 | 23 57.2 | 58.3 | 170.3 | 47 |
| 48 | 25 56.4 | 58.3 | 170.4 | 24 57.2 | 58.3 | 170.4 | 23 58.0 | 58.3 | 170.5 | 22 58.9 | 58.4 | 170.6 | 48 |
| 49 | 24 58.1 | 58.2 | 170.6 | 23 58.9 | 58.3 | 170.7 | 22 59.7 | 58.3 | 170.8 | 22 00.5 | 58.4 | 170.8 | 49 |
| 50 | 23 59.9 | -58.3 | 170.9 | 23 00.6 | -58.3 | 171.0 | 22 01.4 | -58.4 | 171.0 | 21 02.1 | -58.4 | 171.1 | 50 |
| 51 | 23 01.6 | 58.3 | 171.2 | 22 02.3 | 58.4 | 171.2 | 21 03.0 | 58.4 | 171.3 | 20 03.7 | 58.4 | 171.3 | 51 |
| 52 | 22 03.3 | 58.4 | 171.4 | 21 03.9 | 58.3 | 171.5 | 20 04.6 | 58.4 | 171.5 | 19 05.3 | 58.5 | 171.6 | 52 |
| 53 | 21 04.9 | 58.3 | 171.7 | 20 05.6 | 58.4 | 171.7 | 19 06.2 | 58.4 | 171.8 | 18 06.8 | 58.5 | 171.8 | 53 |
| 54 | 20 06.6 | 58.4 | 171.9 | 19 07.2 | 58.4 | 172.0 | 18 07.8 | 58.5 | 172.0 | 17 08.3 | 58.5 | 172.0 | 54 |
| 55 | 19 08.2 | -58.4 | 172.2 | 18 08.8 | -58.5 | 172.2 | 17 09.3 | -58.5 | 172.2 | 16 09.8 | -58.4 | 172.3 | 55 |
| 56 | 18 09.8 | 58.4 | 172.4 | 17 10.3 | 58.4 | 172.4 | 16 10.8 | 58.4 | 172.5 | 15 11.4 | 58.6 | 172.5 | 56 |
| 57 | 17 11.4 | 58.5 | 172.6 | 16 11.9 | 58.5 | 172.7 | 15 12.4 | 58.5 | 172.7 | 14 12.8 | 58.5 | 172.7 | 57 |
| 58 | 16 12.9 | 58.4 | 172.9 | 15 13.4 | 58.5 | 172.9 | 14 13.9 | 58.6 | 172.9 | 13 14.3 | 58.5 | 173.0 | 58 |
| 59 | 15 14.5 | 58.5 | 173.1 | 14 14.9 | 58.5 | 173.1 | 13 15.3 | 58.5 | 173.2 | 12 15.8 | 58.6 | 173.2 | 59 |
| 60 | 14 16.0 | -58.5 | 173.3 | 13 16.4 | -58.5 | 173.4 | 12 16.8 | -58.5 | 173.4 | 11 17.2 | -58.5 | 173.4 | 60 |
| 61 | 13 17.5 | 58.5 | 173.6 | 12 17.9 | 58.5 | 173.6 | 11 18.3 | 58.6 | 173.6 | 10 18.7 | 58.6 | 173.6 | 61 |
| 62 | 12 19.0 | 58.5 | 173.8 | 11 19.4 | 58.5 | 173.8 | 10 19.7 | 58.5 | 173.8 | 9 20.1 | 58.6 | 173.9 | 62 |
| 63 | 11 20.5 | 58.5 | 174.0 | 10 20.9 | 58.6 | 174.0 | 9 21.2 | 58.6 | 174.1 | 8 21.5 | 58.6 | 174.1 | 63 |
| 64 | 10 22.0 | 58.5 | 174.2 | 9 22.3 | 58.5 | 174.3 | 8 22.6 | 58.5 | 174.3 | 7 22.9 | 58.6 | 174.3 | 64 |
| 65 | 9 23.5 | -58.5 | 174.5 | 8 23.8 | -58.6 | 174.5 | 7 24.1 | -58.6 | 174.5 | 6 24.3 | -58.6 | 174.5 | 65 |
| 66 | 8 25.0 | 58.6 | 174.7 | 7 25.2 | 58.5 | 174.7 | 6 25.5 | 58.6 | 174.7 | 5 25.7 | 58.6 | 174.7 | 66 |
| 67 | 7 26.4 | 58.5 | 174.9 | 6 26.7 | 58.6 | 174.9 | 5 26.9 | 58.6 | 174.9 | 4 27.1 | 58.6 | 174.9 | 67 |
| 68 | 6 27.9 | 58.6 | 175.1 | 5 28.1 | 58.6 | 175.1 | 4 28.3 | 58.6 | 175.2 | 3 28.5 | 58.6 | 175.2 | 68 |
| 69 | 5 29.3 | 58.5 | 175.4 | 4 29.5 | 58.5 | 175.4 | 3 29.7 | 58.6 | 175.4 | 2 29.9 | 58.6 | 175.4 | 69 |
| 70 | 4 30.8 | -58.6 | 175.6 | 3 31.0 | -58.6 | 175.6 | 2 31.1 | -58.5 | 175.6 | 1 31.3 | -58.6 | 175.6 | 70 |
| 71 | 3 32.2 | 58.5 | 175.8 | 2 32.4 | 58.6 | 175.8 | 1 32.6 | 58.6 | 175.8 | 0 32.7 | -58.6 | 175.8 | 71 |
| 72 | 2 33.7 | 58.6 | 176.0 | 1 33.8 | 58.6 | 176.0 | 0 34.0 | -58.6 | 176.0 | 0 25.9 | +58.6 | 4.0 | 72 |
| 73 | 1 35.1 | 58.6 | 176.2 | 0 35.2 | -58.5 | 176.2 | 0 24.6 | +58.6 | 3.8 | 1 24.5 | 58.6 | 3.8 | 73 |
| 74 | 0 36.5 | -58.5 | 176.4 | 0 23.3 | +58.6 | 3.6 | 1 23.2 | 58.6 | 3.6 | 2 23.1 | 58.6 | 3.6 | 74 |
| 75 | 0 22.0 | +58.6 | 3.3 | 1 21.9 | +58.6 | 3.3 | 2 21.8 | +58.6 | 3.3 | 3 21.7 | +58.6 | 3.3 | 75 |
| 76 | 1 20.6 | 58.6 | 3.1 | 2 20.5 | 58.6 | 3.1 | 3 20.4 | 58.6 | 3.1 | 4 20.3 | 58.6 | 3.1 | 76 |
| 77 | 2 19.2 | 58.5 | 2.9 | 3 19.1 | 58.6 | 2.9 | 4 19.0 | 58.6 | 2.9 | 5 18.9 | 58.6 | 2.9 | 77 |
| 78 | 3 17.7 | 58.6 | 2.7 | 4 17.7 | 58.5 | 2.7 | 5 17.6 | 58.6 | 2.7 | 6 17.5 | 58.6 | 2.7 | 78 |
| 79 | 4 16.3 | 58.5 | 2.5 | 5 16.2 | 58.6 | 2.5 | 6 16.2 | 58.5 | 2.5 | 7 16.1 | 58.6 | 2.5 | 79 |
| 80 | 5 14.8 | +58.6 | 2.2 | 6 14.8 | +58.6 | 2.3 | 7 14.7 | +58.6 | 2.3 | 8 14.7 | +58.6 | 2.3 | 80 |
| 81 | 6 13.4 | 58.5 | 2.0 | 7 13.4 | 58.5 | 2.0 | 8 13.3 | 58.6 | 2.0 | 9 13.3 | 58.5 | 2.0 | 81 |
| 82 | 7 11.9 | 58.6 | 1.8 | 8 11.9 | 58.6 | 1.8 | 9 11.9 | 58.5 | 1.8 | 10 11.8 | 58.6 | 1.8 | 82 |
| 83 | 8 10.5 | 58.5 | 1.6 | 9 10.5 | 58.5 | 1.6 | 10 10.4 | 58.6 | 1.6 | 11 10.4 | 58.6 | 1.6 | 83 |
| 84 | 9 09.0 | 58.5 | 1.4 | 10 09.0 | 58.5 | 1.4 | 11 09.0 | 58.5 | 1.4 | 12 09.0 | 58.5 | 1.4 | 84 |
| 85 | 10 07.5 | +58.5 | 1.1 | 11 07.5 | +58.5 | 1.1 | 12 07.5 | +58.5 | 1.1 | 13 07.5 | +58.5 | 1.2 | 85 |
| 86 | 11 06.0 | 58.6 | 0.9 | 12 06.0 | 58.5 | 0.9 | 13 06.0 | 58.5 | 0.9 | 14 06.0 | 58.5 | 0.9 | 86 |
| 87 | 12 04.6 | 58.4 | 0.7 | 13 04.5 | 58.5 | 0.7 | 14 04.5 | 58.5 | 0.7 | 15 04.5 | 58.5 | 0.7 | 87 |
| 88 | 13 03.0 | 58.5 | 0.5 | 14 03.0 | 58.5 | 0.5 | 15 03.0 | 58.5 | 0.5 | 16 03.0 | 58.5 | 0.5 | 88 |
| 89 | 14 01.5 | 58.5 | 0.2 | 15 01.5 | 58.5 | 0.2 | 16 01.5 | 58.5 | 0.2 | 17 01.5 | 58.5 | 0.2 | 89 |
| 90 | 15 00.0 | +58.5 | 0.0 | 16 00.0 | +58.5 | 0.0 | 17 00.0 | +58.5 | 0.0 | 18 00.0 | +58.5 | 0.0 | 90 |
| | 15° | | | 16° | | | 17° | | | 18° | | | |

| Dec. | 19° Hc | d | Z | 20° Hc | d | Z | 21° Hc | d | Z | 22° Hc | d | Z | Dec. |
|---|---|---|---|---|---|---|---|---|---|---|---|---|---|
| ° | ° ′ | ′ | ° | ° ′ | ′ | ° | ° ′ | ′ | ° | ° ′ | ′ | ° | ° |
| 0 | 67 06.9 | -50.6 | 144.7 | 66 17.5 | -51.3 | 146.0 | 65 27.4 | -52.0 | 147.2 | 64 36.7 | -52.7 | 148.4 | 0 |
| 1 | 66 16.3 | 51.3 | 146.0 | 65 26.2 | 52.0 | 147.2 | 64 35.4 | 52.6 | 148.4 | 63 44.0 | 53.1 | 149.5 | 1 |
| 2 | 65 25.0 | 51.8 | 147.3 | 64 34.2 | 52.5 | 148.4 | 63 42.8 | 53.0 | 149.5 | 62 50.9 | 53.6 | 150.5 | 2 |
| 3 | 64 33.2 | 52.4 | 148.5 | 63 41.7 | 52.9 | 149 5 | 62 49.8 | 53.5 | 150.5 | 61 57.3 | 54.0 | 151.5 | 3 |
| 4 | 63 40.8 | 52.9 | 149.6 | 62 48.8 | 53.4 | 150.6 | 61 56.3 | 53.9 | 151.5 | 61 03.3 | 54.3 | 152.4 | 4 |
| 5 | 62 47.9 | -53.3 | 150.6 | 61 55.4 | -53.8 | 151.6 | 61 02.4 | -54.2 | 152.4 | 60 09.0 | -54.6 | 153.2 | 5 |
| 6 | 61 54.6 | 53.7 | 151.6 | 61 01.6 | 54.2 | 152.5 | 60 08.2 | 54.6 | 153.3 | 59 14.4 | 54.9 | 154.1 | 6 |
| 7 | 61 00.9 | 54.1 | 152.6 | 60 07.4 | 54.5 | 153.4 | 59 13.6 | 54.9 | 154.1 | 58 19.5 | 55.2 | 154.8 | 7 |
| 8 | 60 06.8 | 54.4 | 153.4 | 59 12.9 | 54.8 | 154.2 | 58 18.7 | 55.1 | 154.9 | 57 24.3 | 55.5 | 155.6 | 8 |
| 9 | 59 12.4 | 54.8 | 154.3 | 58 18.1 | 55.0 | 155.0 | 57 23.6 | 55.4 | 155.6 | 56 28.8 | 55.6 | 156.3 | 9 |
| 10 | 58 17.6 | -55.0 | 155.1 | 57 23.1 | -55.3 | 155.7 | 56 28.2 | -55.6 | 156.4 | 55 33.2 | -55.9 | 156.9 | 10 |
| 11 | 57 22.6 | 55.2 | 155.8 | 56 27.8 | 55.6 | 156.4 | 55 32.6 | 55.8 | 157.0 | 54 37.3 | 56.1 | 157.6 | 11 |
| 12 | 56 27.4 | 55.5 | 156.5 | 55 32.2 | 55.8 | 157.1 | 54 36.8 | 56.0 | 157.7 | 53 41.2 | 56.2 | 158.2 | 12 |
| 13 | 55 31.9 | 55.7 | 157.2 | 54 36.4 | 55.9 | 157.8 | 53 40.8 | 56.2 | 158.3 | 52 45.0 | 56.4 | 158.8 | 13 |
| 14 | 54 36.2 | 55.9 | 157.9 | 53 40.5 | 56.1 | 158.4 | 52 44.6 | 56.3 | 158.9 | 51 48.6 | 56.6 | 159.3 | 14 |
| 15 | 53 40.3 | -56.1 | 158.5 | 52 44.4 | -56.4 | 159.0 | 51 48.3 | -56.5 | 159.4 | 50 52.0 | -56.7 | 159.9 | 15 |
| 16 | 52 44.2 | 56.3 | 159.1 | 51 48.0 | 56.4 | 159.5 | 50 51.8 | 56.7 | 160.0 | 49 55.3 | 56.8 | 160.4 | 16 |
| 17 | 51 47.9 | 56.4 | 159.6 | 50 51.6 | 56.6 | 160.1 | 49 55.1 | 56.8 | 160.5 | 48 58.5 | 57.0 | 160.9 | 17 |
| 18 | 50 51.5 | 56.6 | 160.2 | 49 55.0 | 56.8 | 160.6 | 48 58.3 | 56.9 | 161.0 | 48 01.5 | 57.0 | 161.3 | 18 |
| 19 | 49 54.9 | 56.7 | 160.7 | 48 58.2 | 56.8 | 161.1 | 48 01.4 | 57.0 | 161.5 | 47 04.5 | 57.2 | 161.8 | 19 |
| 20 | 48 58.2 | -56.8 | 161.2 | 48 01.4 | -57.0 | 161.6 | 47 04.4 | -57.1 | 161.9 | 46 07.3 | -57.3 | 162.2 | 20 |
| 21 | 48 01.4 | 56.9 | 161.7 | 47 04.4 | 57.1 | 162.0 | 46 07.3 | 57.3 | 162.4 | 45 10.0 | 57.3 | 162.7 | 21 |
| 22 | 47 04.5 | 57.1 | 162.2 | 46 07.3 | 57.2 | 162.5 | 45 10.0 | 57.3 | 162.8 | 44 12.7 | 57.5 | 163.1 | 22 |
| 23 | 46 07.4 | 57.1 | 162.6 | 45 10.1 | 57.3 | 162.9 | 44 12.7 | 57.4 | 163.2 | 43 15.2 | 57.5 | 163.5 | 23 |
| 24 | 45 10.3 | 57.3 | 163.1 | 44 12.8 | 57.3 | 163.3 | 43 15.3 | 57.5 | 163.6 | 42 17.7 | 57.6 | 163.9 | 24 |
| 25 | 44 13.0 | -57.3 | 163.5 | 43 15.5 | -57.5 | 163.7 | 42 17.8 | -57.6 | 164.0 | 41 20.1 | -57.7 | 164.2 | 25 |
| 26 | 43 15.7 | 57.5 | 163.9 | 42 18.0 | 57.5 | 164.1 | 41 20.2 | 57.6 | 164.4 | 40 22.4 | 57.7 | 164.6 | 26 |
| 27 | 42 18.2 | 57.5 | 164.3 | 41 20.5 | 57.6 | 164.5 | 40 22.6 | 57.7 | 164.7 | 39 24.7 | 57.8 | 165.0 | 27 |
| 28 | 41 20.7 | 57.5 | 164.7 | 40 22.9 | 57.7 | 164.9 | 39 24.9 | 57.8 | 165.1 | 38 26.9 | 57.9 | 165.3 | 28 |
| 29 | 40 23.2 | 57.7 | 165.0 | 39 25.2 | 57.8 | 165.2 | 38 27.1 | 57.8 | 165.4 | 37 29.0 | 57.9 | 165.6 | 29 |
| 30 | 39 25.5 | -57.7 | 165.4 | 38 27.4 | -57.8 | 165.6 | 37 29.3 | -57.9 | 165.8 | 36 31.1 | -58.0 | 166.0 | 30 |
| 31 | 38 27.8 | 57.8 | 165.7 | 37 29.6 | 57.8 | 165.9 | 36 31.4 | 57.9 | 166.1 | 35 33.1 | 58.0 | 166.3 | 31 |
| 32 | 37 30.0 | 57.8 | 166.1 | 36 31.8 | 57.9 | 166.3 | 35 33.5 | 58.0 | 166.4 | 34 35.1 | 58.0 | 166.6 | 32 |
| 33 | 36 32.2 | 57.9 | 166.4 | 35 33.9 | 58.0 | 166.6 | 34 35.5 | 58.1 | 166.8 | 33 37.1 | 58.2 | 166.9 | 33 |
| 34 | 35 34.3 | 57.9 | 166.7 | 34 35.9 | 58.0 | 166.9 | 33 37.4 | 58.0 | 167.1 | 32 38.9 | 58.1 | 167.2 | 34 |
| 35 | 34 36.4 | -58.0 | 167.1 | 33 37.9 | -58.1 | 167.2 | 32 39.4 | -58.2 | 167.4 | 31 40.8 | -58.2 | 167.5 | 35 |
| 36 | 33 38.4 | 58.0 | 167.4 | 32 39.8 | 58.1 | 167.5 | 31 41.2 | 58.1 | 167.7 | 30 42.6 | 58.2 | 167.8 | 36 |
| 37 | 32 40.4 | 58.1 | 167.7 | 31 41.7 | 58.1 | 167.8 | 30 43.1 | 58.2 | 167.9 | 29 44.4 | 58.3 | 168.1 | 37 |
| 38 | 31 42.3 | 58.1 | 168.0 | 30 43.6 | 58.2 | 168.1 | 29 44.9 | 58.3 | 168.2 | 28 46.1 | 58.3 | 168.3 | 38 |
| 39 | 30 44.2 | 58.2 | 168.3 | 29 45.4 | 58.2 | 168.4 | 28 46.6 | 58.2 | 168.5 | 27 47.8 | 58.3 | 168.6 | 39 |
| 40 | 29 46.0 | -58.2 | 168.5 | 28 47.2 | -58.2 | 168.7 | 27 48.4 | -58.3 | 168.8 | 26 49.5 | -58.4 | 168.9 | 40 |
| 41 | 28 47.8 | 58.2 | 168.8 | 27 49.0 | 58.3 | 168.9 | 26 50.1 | 58.4 | 169.0 | 25 51.1 | 58.3 | 169.1 | 41 |
| 42 | 27 49.6 | 58.2 | 169.1 | 26 50.7 | 58.3 | 169.2 | 25 51.7 | 58.3 | 169.3 | 24 52.8 | 58.4 | 169.4 | 42 |
| 43 | 26 51.4 | 58.3 | 169.4 | 25 52.4 | 58.4 | 169.5 | 24 53.4 | 58.4 | 169.6 | 23 54.4 | 58.5 | 169.6 | 43 |
| 44 | 25 53.1 | 58.3 | 169.6 | 24 54.0 | 58.3 | 169.7 | 23 55.0 | 58.4 | 169.8 | 22 55.9 | 58.4 | 169.9 | 44 |
| 45 | 24 54.8 | -58.4 | 169.9 | 23 55.7 | -58.4 | 170.0 | 22 56.6 | -58.4 | 170.1 | 21 57.5 | -58.5 | 170.1 | 45 |
| 46 | 23 56.4 | 58.3 | 170.2 | 22 57.3 | 58.4 | 170.2 | 21 58.2 | 58.5 | 170.3 | 20 59.0 | 58.5 | 170.4 | 46 |
| 47 | 22 58.1 | 58.4 | 170.4 | 21 58.9 | 58.4 | 170.5 | 20 59.7 | 58.5 | 170.5 | 20 00.5 | 58.5 | 170.6 | 47 |
| 48 | 21 59.7 | 58.4 | 170.7 | 21 00.5 | 58.5 | 170.7 | 20 01.2 | 58.5 | 170.8 | 19 02.0 | 58.5 | 170.8 | 48 |
| 49 | 21 01.3 | 58.5 | 170.9 | 20 02.0 | 58.5 | 171.0 | 19 02.7 | 58.5 | 171.0 | 18 03.5 | 58.6 | 171.1 | 49 |
| 50 | 20 02.8 | -58.4 | 171.1 | 19 03.5 | -58.5 | 171.2 | 18 04.2 | -58.5 | 171.3 | 17 04.9 | -58.5 | 171.3 | 50 |
| 51 | 19 04.4 | 58.5 | 171.4 | 18 05.0 | 58.5 | 171.4 | 17 05.7 | 58.5 | 171.5 | 16 06.4 | 58.6 | 171.5 | 51 |
| 52 | 18 05.9 | 58.5 | 171.6 | 17 06.5 | 58.5 | 171.7 | 16 07.2 | 58.6 | 171.7 | 15 07.8 | 58.6 | 171.8 | 52 |
| 53 | 17 07.4 | 58.5 | 171.9 | 16 08.0 | 58.5 | 171.9 | 15 08.6 | 58.6 | 171.9 | 14 09.2 | 58.6 | 172.0 | 53 |
| 54 | 16 08.9 | 58.5 | 172.1 | 15 09.5 | 58.6 | 172.1 | 14 10.0 | 58.5 | 172.2 | 13 10.6 | 58.6 | 172.2 | 54 |
| 55 | 15 10.4 | -58.5 | 172.3 | 14 10.9 | -58.5 | 172.4 | 13 11.5 | -58.6 | 172.4 | 12 12.0 | -58.6 | 172.4 | 55 |
| 56 | 14 11.9 | 58.6 | 172.5 | 13 12.4 | 58.6 | 172.6 | 12 12.9 | 58.6 | 172.6 | 11 13.4 | 58.7 | 172.6 | 56 |
| 57 | 13 13.3 | 58.5 | 172.8 | 12 13.8 | 58.6 | 172.8 | 11 14.3 | 58.7 | 172.8 | 10 14.7 | 58.6 | 172.8 | 57 |
| 58 | 12 14.8 | 58.6 | 173.0 | 11 15.2 | 58.6 | 173.0 | 10 15.6 | 58.6 | 173.0 | 9 16.1 | 58.7 | 173.1 | 58 |
| 59 | 11 16.2 | 58.6 | 173.2 | 10 16.6 | 58.6 | 173.2 | 9 17.0 | 58.6 | 173.3 | 8 17.4 | 58.6 | 173.3 | 59 |
| 60 | 10 17.6 | -58.6 | 173.4 | 9 18.0 | -58.6 | 173.5 | 8 18.4 | -58.6 | 173.5 | 7 18.8 | -58.7 | 173.5 | 60 |
| 61 | 9 19.0 | 58.6 | 173.7 | 8 19.4 | 58.6 | 173.7 | 7 19.8 | 58.7 | 173.7 | 6 20.1 | 58.6 | 173.7 | 61 |
| 62 | 8 20.4 | 58.6 | 173.9 | 7 20.8 | 58.6 | 173.9 | 6 21.1 | 58.6 | 173.9 | 5 21.5 | 58.7 | 173.9 | 62 |
| 63 | 7 21.8 | 58.6 | 174.1 | 6 22.2 | 58.7 | 174.1 | 5 22.5 | 58.7 | 174.1 | 4 22.8 | 58.7 | 174.1 | 63 |
| 64 | 6 23.2 | 58.6 | 174.3 | 5 23.5 | 58.6 | 174.3 | 4 23.8 | 58.6 | 174.3 | 3 24.1 | 58.7 | 174.3 | 64 |
| 65 | 5 24.6 | -58.6 | 174.5 | 4 24.9 | -58.6 | 174.5 | 3 25.2 | -58.7 | 174.5 | 2 25.4 | -58.6 | 174.5 | 65 |
| 66 | 4 26.0 | 58.6 | 174.7 | 3 26.3 | 58.7 | 174.7 | 2 26.5 | 58.7 | 174.7 | 1 26.8 | 58.7 | 174.7 | 66 |
| 67 | 3 27.4 | 58.6 | 174.9 | 2 27.6 | 58.6 | 175.0 | 1 27.8 | 58.6 | 175.0 | 0 28.1 | -58.7 | 175.0 | 67 |
| 68 | 2 28.8 | 58.7 | 175.2 | 1 29.0 | 58.7 | 175.2 | 0 29.2 | -58.7 | 175.2 | 0 30.6 | +58.7 | 4.8 | 68 |
| 69 | 1 30.1 | 58.6 | 175.4 | 0 30.3 | -58.6 | 175.4 | 0 29.5 | +58.6 | 4.6 | 1 29.3 | 58.7 | 4.6 | 69 |
| 70 | 0 31.5 | -58.6 | 175.6 | 0 28.3 | +58.7 | 4.4 | 1 28.1 | +58.7 | 4.4 | 2 28.0 | +58.6 | 4.4 | 70 |
| 71 | 0 27.1 | +58.6 | 4.2 | 1 27.0 | 58.6 | 4.2 | 2 26.8 | 58.7 | 4.2 | 3 26.6 | 58.7 | 4.2 | 71 |
| 72 | 1 25.7 | 58.7 | 4.0 | 2 25.6 | 58.6 | 4.0 | 3 25.5 | 58.6 | 4.0 | 4 25.3 | 58.7 | 4.0 | 72 |
| 73 | 2 24.4 | 58.6 | 3.8 | 3 24.2 | 58.7 | 3.8 | 4 24.1 | 58.7 | 3.8 | 5 24.0 | 58.6 | 3.8 | 73 |
| 74 | 3 23.0 | 58.6 | 3.6 | 4 22.9 | 58.6 | 3.6 | 5 22.8 | 58.6 | 3.6 | 6 22.6 | 58.7 | 3.6 | 74 |
| 75 | 4 21.6 | +58.6 | 3.3 | 5 21.5 | +58.6 | 3.4 | 6 21.4 | +58.7 | 3.4 | 7 21.3 | +58.7 | 3.4 | 75 |
| 76 | 5 20.2 | 58.6 | 3.1 | 6 20.1 | 58.7 | 3.1 | 7 20.1 | 58.6 | 3.1 | 8 20.0 | 58.6 | 3.2 | 76 |
| 77 | 6 18.8 | 58.7 | 2.9 | 7 18.8 | 58.6 | 2.9 | 8 18.7 | 58.6 | 2.9 | 9 18.6 | 58.7 | 2.9 | 77 |
| 78 | 7 17.5 | 58.6 | 2.7 | 8 17.4 | 58.6 | 2.7 | 9 17.3 | 58.6 | 2.7 | 10 17.3 | 58.6 | 2.7 | 78 |
| 79 | 8 16.1 | 58.6 | 2.5 | 9 16.0 | 58.6 | 2.5 | 10 15.9 | 58.7 | 2.5 | 11 15.9 | 58.6 | 2.5 | 79 |
| 80 | 9 14.7 | +58.5 | 2.3 | 10 14.6 | +58.6 | 2.3 | 11 14.6 | +58.6 | 2.3 | 12 14.5 | +58.6 | 2.3 | 80 |
| 81 | 10 13.2 | 58.6 | 2.0 | 11 13.2 | 58.6 | 2.1 | 12 13.2 | 58.6 | 2.1 | 13 13.1 | 58.6 | 2.1 | 81 |
| 82 | 11 11.8 | 58.6 | 1.8 | 12 11.8 | 58.6 | 1.8 | 13 11.8 | 58.5 | 1.8 | 14 11.7 | 58.6 | 1.9 | 82 |
| 83 | 12 10.4 | 58.5 | 1.6 | 13 10.4 | 58.5 | 1.6 | 14 10.3 | 58.6 | 1.6 | 15 10.3 | 58.6 | 1.6 | 83 |
| 84 | 13 08.9 | 58.6 | 1.4 | 14 08.9 | 58.6 | 1.4 | 15 08.9 | 58.6 | 1.4 | 16 08.9 | 58.5 | 1.4 | 84 |
| 85 | 14 07.5 | +58.5 | 1.2 | 15 07.5 | +58.5 | 1.2 | 16 07.5 | +58.5 | 1.2 | 17 07.4 | +58.6 | 1.2 | 85 |
| 86 | 15 06.0 | 58.5 | 0.9 | 16 06.0 | 58.5 | 0.9 | 17 06.0 | 58.5 | 0.9 | 18 06.0 | 58.5 | 0.9 | 86 |
| 87 | 16 04.5 | 58.5 | 0.7 | 17 04.5 | 58.5 | 0.7 | 18 04.5 | 58.5 | 0.7 | 19 04.5 | 58.5 | 0.7 | 87 |
| 88 | 17 03.0 | 58.5 | 0.5 | 18 03.0 | 58.5 | 0.5 | 19 03.0 | 58.5 | 0.5 | 20 03.0 | 58.5 | 0.5 | 88 |
| 89 | 18 01.5 | 58.5 | 0.2 | 19 01.5 | 58.5 | 0.2 | 20 01.5 | 58.5 | 0.2 | 21 01.5 | 58.5 | 0.2 | 89 |
| 90 | 19 00.0 | +58.5 | 0.0 | 20 00.0 | +58.5 | 0.0 | 21 00.0 | +58.5 | 0.0 | 22 00.0 | +58.5 | 0.0 | 90 |
| | 19° | | | 20° | | | 21° | | | 22° | | | |

S. Lat. { L.H.A. greater than 180°......Zn=180°−Z; L.H.A. less than 180°............Zn=180°+Z }

## LATITUDE SAME NAME AS DECLINATION L.H.A. 167°, 193°

N. Lat. {L.H.A. greater than 180°......Zn=Z; L.H.A. less than 180°............Zn=360°−Z}

| Dec. | 23° Hc | d | Z | 24° Hc | d | Z | 25° Hc | d | Z | 26° Hc | d | Z | Dec. |
|---|---|---|---|---|---|---|---|---|---|---|---|---|---|
| ° | ° ′ | ′ | ° | ° ′ | ′ | ° | ° ′ | ′ | ° | ° ′ | ′ | ° | ° |
| 0 | 38 53.2 | +29.8 | 110.0 | 38 32.3 | +30.9 | 110.8 | 38 10.7 | +31.9 | 111.5 | 37 48.3 | +33.0 | 112.2 | 0 |
| 1 | 39 23.0 | 29.1 | 108.9 | 39 03.2 | 30.2 | 109.7 | 38 42.6 | 31.4 | 110.4 | 38 21.3 | 32.4 | 111.2 | 1 |
| 2 | 39 52.1 | 28.5 | 107.8 | 39 33.4 | 29.7 | 108.6 | 39 14.0 | 30.7 | 109.3 | 38 53.7 | 31.8 | 110.1 | 2 |
| 3 | 40 20.6 | 27.8 | 106.6 | 40 03.1 | 28.9 | 107.4 | 39 44.7 | 30.1 | 108.2 | 39 25.5 | 31.2 | 109.0 | 3 |
| 4 | 40 48.4 | 27.1 | 105.4 | 40 32.0 | 28.3 | 106.3 | 40 14.8 | 29.4 | 107.1 | 39 56.7 | 30.6 | 107.9 | 4 |
| 5 | 41 15.5 | +26.3 | 104.3 | 41 00.3 | +27.5 | 105.1 | 40 44.2 | +28.7 | 105.9 | 40 27.3 | +29.9 | 106.8 | 5 |
| 6 | 41 41.8 | 25.6 | 103.1 | 41 27.8 | 26.8 | 103.9 | 41 12.9 | 28.0 | 104.8 | 40 57.2 | 29.2 | 105.6 | 6 |
| 7 | 42 07.4 | 24.8 | 101.9 | 41 54.6 | 26.0 | 102.7 | 41 40.9 | 27.3 | 103.6 | 41 26.4 | 28.5 | 104.5 | 7 |
| 8 | 42 32.2 | 23.9 | 100.6 | 42 20.6 | 25.3 | 101.5 | 42 08.2 | 26.5 | 102.4 | 41 54.9 | 27.7 | 103.3 | 8 |
| 9 | 42 56.1 | 23.2 | 99.4 | 42 45.9 | 24.5 | 100.3 | 42 34.7 | 25.8 | 101.2 | 42 22.6 | 27.0 | 102.1 | 9 |
| 10 | 43 19.3 | +22.3 | 98.1 | 43 10.4 | +23.6 | 99.0 | 43 00.5 | +24.9 | 100.0 | 42 49.6 | +26.2 | 100.9 | 10 |
| 11 | 43 41.6 | 21.5 | 96.8 | 43 34.0 | 22.8 | 97.8 | 43 25.4 | 24.1 | 98.7 | 43 15.8 | 25.4 | 99.6 | 11 |
| 12 | 44 03.1 | 20.5 | 95.5 | 43 56.8 | 21.9 | 96.5 | 43 49.5 | 23.3 | 97.4 | 43 41.2 | 24.6 | 98.4 | 12 |
| 13 | 44 23.6 | 19.7 | 94.2 | 44 18.7 | 21.0 | 95.2 | 44 12.8 | 22.4 | 96.2 | 44 05.8 | 23.8 | 97.1 | 13 |
| 14 | 44 43.3 | 18.7 | 92.9 | 44 39.7 | 20.2 | 93.9 | 44 35.2 | 21.5 | 94.9 | 44 29.6 | 22.8 | 95.8 | 14 |
| 15 | 45 02.0 | +17.8 | 91.6 | 44 59.9 | +19.1 | 92.6 | 44 56.7 | +20.5 | 93.6 | 44 52.4 | +22.0 | 94.5 | 15 |
| 16 | 45 19.8 | 16.8 | 90.2 | 45 19.0 | 18.3 | 91.2 | 45 17.2 | 19.7 | 92.2 | 45 14.4 | 21.0 | 93.2 | 16 |
| 17 | 45 36.6 | 15.8 | 88.8 | 45 37.3 | 17.2 | 89.9 | 45 36.9 | 18.7 | 90.9 | 45 35.4 | 20.1 | 91.9 | 17 |
| 18 | 45 52.4 | 14.8 | 87.4 | 45 54.5 | 16.3 | 88.5 | 45 55.6 | 17.7 | 89.5 | 45 55.5 | 19.2 | 90.5 | 18 |
| 19 | 46 07.2 | 13.8 | 86.1 | 46 10.8 | 15.2 | 87.1 | 46 13.3 | 16.7 | 88.1 | 46 14.7 | 18.1 | 89.2 | 19 |
| 20 | 46 21.0 | +12.7 | 84.6 | 46 26.0 | +14.2 | 85.7 | 46 30.0 | +15.6 | 86.7 | 46 32.8 | +17.2 | 87.8 | 20 |
| 21 | 46 33.7 | 11.6 | 83.2 | 46 40.2 | 13.2 | 84.3 | 46 45.6 | 14.7 | 85.3 | 46 50.0 | 16.1 | 86.4 | 21 |
| 22 | 46 45.3 | 10.6 | 81.8 | 46 53.4 | 12.0 | 82.9 | 47 00.3 | 13.5 | 83.9 | 47 06.1 | 15.0 | 85.0 | 22 |
| 23 | 46 55.9 | 9.5 | 80.4 | 47 05.4 | 11.0 | 81.4 | 47 13.8 | 12.5 | 82.5 | 47 21.1 | 14.0 | 83.6 | 23 |
| 24 | 47 05.4 | 8.4 | 78.9 | 47 16.4 | 9.9 | 80.0 | 47 26.3 | 11.4 | 81.0 | 47 35.1 | 12.9 | 82.1 | 24 |
| 25 | 47 13.8 | +7.3 | 77.5 | 47 26.3 | +8.8 | 78.5 | 47 37.7 | +10.3 | 79.6 | 47 48.0 | +11.8 | 80.7 | 25 |
| 26 | 47 21.1 | 6.2 | 76.0 | 47 35.1 | 7.7 | 77.1 | 47 48.0 | 9.2 | 78.1 | 47 59.8 | 10.7 | 79.2 | 26 |
| 27 | 47 27.3 | 5.0 | 74.5 | 47 42.8 | 6.5 | 75.6 | 47 57.2 | 8.0 | 76.6 | 48 10.5 | 9.5 | 77.7 | 27 |
| 28 | 47 32.3 | 3.9 | 73.0 | 47 49.3 | 5.4 | 74.1 | 48 05.2 | 6.9 | 75.2 | 48 20.0 | 8.4 | 76.3 | 28 |
| 29 | 47 36.2 | 2.8 | 71.6 | 47 54.7 | 4.2 | 72.6 | 48 12.1 | 5.7 | 73.7 | 48 28.4 | 7.2 | 74.8 | 29 |
| 30 | 47 39.0 | +1.6 | 70.1 | 47 58.9 | +3.1 | 71.1 | 48 17.8 | +4.6 | 72.2 | 48 35.6 | +6.1 | 73.3 | 30 |
| 31 | 47 40.6 | +0.5 | 68.6 | 48 02.0 | 1.9 | 69.6 | 48 22.4 | 3.3 | 70.7 | 48 41.7 | 4.8 | 71.8 | 31 |
| 32 | 47 41.1 | −0.7 | 67.1 | 48 03.9 | +0.7 | 68.1 | 48 25.7 | 2.2 | 69.2 | 48 46.5 | 3.7 | 70.2 | 32 |
| 33 | 47 40.4 | 1.9 | 65.6 | 48 04.6 | −0.4 | 66.6 | 48 27.9 | +1.1 | 67.7 | 48 50.2 | 2.5 | 68.7 | 33 |
| 34 | 47 38.5 | 3.0 | 64.1 | 48 04.2 | 1.6 | 65.1 | 48 29.0 | −0.2 | 66.2 | 48 52.7 | 1.3 | 67.2 | 34 |
| 35 | 47 35.5 | −4.1 | 62.7 | 48 02.6 | −2.7 | 63.6 | 48 28.8 | −1.4 | 64.7 | 48 54.0 | +0.1 | 65.7 | 35 |
| 36 | 47 31.4 | 5.3 | 61.2 | 47 59.9 | 3.9 | 62.2 | 48 27.4 | 2.5 | 63.1 | 48 54.1 | −1.1 | 64.2 | 36 |
| 37 | 47 26.1 | 6.4 | 59.7 | 47 56.0 | 5.1 | 60.7 | 48 24.9 | 3.7 | 61.6 | 48 53.0 | 2.4 | 62.6 | 37 |
| 38 | 47 19.7 | 7.5 | 58.2 | 47 50.9 | 6.2 | 59.2 | 48 21.2 | 4.9 | 60.1 | 48 50.6 | 3.5 | 61.1 | 38 |
| 39 | 47 12.2 | 8.6 | 56.8 | 47 44.7 | 7.4 | 57.7 | 48 16.3 | 6.0 | 58.6 | 48 47.1 | 4.7 | 59.6 | 39 |
| 40 | 47 03.6 | −9.7 | 55.3 | 47 37.3 | −8.4 | 56.2 | 48 10.3 | −7.2 | 57.1 | 48 42.4 | −5.9 | 58.1 | 40 |
| 41 | 46 53.9 | 10.8 | 53.9 | 47 28.9 | 9.6 | 54.8 | 48 03.1 | 8.3 | 55.7 | 48 36.5 | 7.0 | 56.6 | 41 |
| 42 | 46 43.1 | 11.9 | 52.4 | 47 19.3 | 10.7 | 53.3 | 47 54.8 | 9.5 | 54.2 | 48 29.5 | 8.2 | 55.1 | 42 |
| 43 | 46 31.2 | 13.0 | 51.0 | 47 08.6 | 11.8 | 51.8 | 47 45.3 | 10.6 | 52.7 | 48 21.3 | 9.4 | 53.6 | 43 |
| 44 | 46 18.2 | 13.9 | 49.6 | 46 56.8 | 12.9 | 50.4 | 47 34.7 | 11.7 | 51.2 | 48 11.9 | 10.5 | 52.1 | 44 |
| 45 | 46 04.3 | −15.1 | 48.2 | 46 43.9 | −13.9 | 49.0 | 47 23.0 | −12.8 | 49.8 | 48 01.4 | −11.6 | 50.6 | 45 |
| 46 | 45 49.2 | 16.0 | 46.8 | 46 30.0 | 14.9 | 47.6 | 47 10.2 | 13.9 | 48.4 | 47 49.8 | 12.8 | 49.2 | 46 |
| 47 | 45 33.2 | 17.0 | 45.4 | 46 15.1 | 16.0 | 46.2 | 46 56.3 | 14.9 | 46.9 | 47 37.0 | 13.8 | 47.7 | 47 |
| 48 | 45 16.2 | 17.9 | 44.1 | 45 59.1 | 17.0 | 44.8 | 46 41.4 | 15.9 | 45.5 | 47 23.2 | 14.9 | 46.3 | 48 |
| 49 | 44 58.3 | 18.9 | 42.7 | 45 42.1 | 17.9 | 43.4 | 46 25.5 | 17.0 | 44.1 | 47 08.3 | 16.0 | 44.9 | 49 |
| 50 | 44 39.4 | −19.9 | 41.4 | 45 24.2 | −19.0 | 42.0 | 46 08.5 | −18.0 | 42.7 | 46 52.3 | −17.0 | 43.4 | 50 |
| 51 | 44 19.5 | 20.7 | 40.0 | 45 05.2 | 19.8 | 40.7 | 45 50.5 | 18.9 | 41.4 | 46 35.3 | 18.0 | 42.0 | 51 |
| 52 | 43 58.8 | 21.7 | 38.7 | 44 45.4 | 20.8 | 39.4 | 45 31.6 | 19.9 | 40.0 | 46 17.3 | 19.0 | 40.7 | 52 |
| 53 | 43 37.1 | 22.4 | 37.4 | 44 24.6 | 21.7 | 38.0 | 45 11.7 | 20.9 | 38.7 | 45 58.3 | 20.0 | 39.3 | 53 |
| 54 | 43 14.7 | 23.4 | 36.2 | 44 02.9 | 22.5 | 36.7 | 44 50.8 | 21.7 | 37.3 | 45 38.3 | 20.9 | 37.9 | 54 |
| 55 | 42 51.3 | −24.1 | 34.9 | 43 40.4 | −23.4 | 35.4 | 44 29.1 | −22.7 | 36.0 | 45 17.4 | −21.8 | 36.6 | 55 |
| 56 | 42 27.2 | 25.0 | 33.7 | 43 17.0 | 24.3 | 34.2 | 44 06.4 | 23.5 | 34.7 | 44 55.6 | 22.8 | 35.3 | 56 |
| 57 | 42 02.2 | 25.7 | 32.4 | 42 52.7 | 25.0 | 32.9 | 43 42.9 | 24.3 | 33.4 | 44 32.8 | 23.6 | 34.0 | 57 |
| 58 | 41 36.5 | 26.5 | 31.2 | 42 27.7 | 25.9 | 31.7 | 43 18.6 | 25.2 | 32.2 | 44 09.2 | 24.4 | 32.7 | 58 |
| 59 | 41 10.0 | 27.2 | 30.0 | 42 01.8 | 26.6 | 30.5 | 42 53.4 | 25.9 | 30.9 | 43 44.8 | 25.3 | 31.4 | 59 |
| 60 | 40 42.8 | −27.9 | 28.8 | 41 35.2 | −27.3 | 29.3 | 42 27.5 | −26.8 | 29.7 | 43 19.5 | −26.1 | 30.2 | 60 |
| 61 | 40 14.9 | 28.6 | 27.7 | 41 07.9 | 28.0 | 28.1 | 42 00.7 | 27.4 | 28.5 | 42 53.4 | 26.9 | 28.9 | 61 |
| 62 | 39 46.3 | 29.3 | 26.5 | 40 39.9 | 28.8 | 26.9 | 41 33.3 | 28.2 | 27.3 | 42 26.5 | 27.7 | 27.7 | 62 |
| 63 | 39 17.0 | 29.9 | 25.4 | 40 11.1 | 29.4 | 25.8 | 41 05.1 | 29.0 | 26.1 | 41 58.8 | 28.4 | 26.5 | 63 |
| 64 | 38 47.1 | 30.6 | 24.3 | 39 41.7 | 30.1 | 24.6 | 40 36.1 | 29.6 | 25.0 | 41 30.4 | 29.1 | 25.3 | 64 |
| 65 | 38 16.5 | −31.2 | 23.2 | 39 11.6 | −30.7 | 23.5 | 40 06.5 | −30.2 | 23.8 | 41 01.3 | −29.7 | 24.2 | 65 |
| 66 | 37 45.3 | 31.7 | 22.1 | 38 40.9 | 31.4 | 22.4 | 39 36.3 | 30.9 | 22.7 | 40 31.6 | 30.5 | 23.0 | 66 |
| 67 | 37 13.6 | 32.3 | 21.0 | 38 09.5 | 31.9 | 21.3 | 39 05.4 | 31.6 | 21.6 | 40 01.1 | 31.1 | 21.9 | 67 |
| 68 | 36 41.3 | 32.9 | 20.0 | 37 37.6 | 32.5 | 20.2 | 38 33.8 | 32.1 | 20.5 | 39 30.0 | 31.7 | 20.8 | 68 |
| 69 | 36 08.4 | 33.4 | 18.9 | 37 05.1 | 33.1 | 19.2 | 38 01.7 | 32.7 | 19.4 | 38 58.3 | 32.4 | 19.7 | 69 |
| 70 | 35 35.0 | −33.9 | 17.9 | 36 32.0 | −33.6 | 18.1 | 37 29.0 | −33.3 | 18.4 | 38 25.9 | −32.9 | 18.6 | 70 |
| 71 | 35 01.1 | 34.4 | 16.9 | 35 58.4 | 34.1 | 17.1 | 36 55.7 | 33.8 | 17.3 | 37 53.0 | 33.5 | 17.6 | 71 |
| 72 | 34 26.7 | 34.9 | 15.9 | 35 24.3 | 34.6 | 16.1 | 36 21.9 | 34.3 | 16.3 | 37 19.5 | 34.0 | 16.5 | 72 |
| 73 | 33 51.8 | 35.4 | 14.9 | 34 49.7 | 35.1 | 15.1 | 35 47.6 | 34.8 | 15.3 | 36 45.5 | 34.6 | 15.5 | 73 |
| 74 | 33 16.4 | 35.8 | 14.0 | 34 14.6 | 35.5 | 14.1 | 35 12.8 | 35.3 | 14.3 | 36 10.9 | 35.1 | 14.5 | 74 |
| 75 | 32 40.6 | −36.2 | 13.0 | 33 39.1 | −36.1 | 13.1 | 34 37.5 | −35.8 | 13.3 | 35 35.8 | −35.5 | 13.5 | 75 |
| 76 | 32 04.4 | 36.7 | 12.1 | 33 03.0 | 36.4 | 12.2 | 34 01.7 | 36.3 | 12.3 | 35 00.3 | 36.1 | 12.5 | 76 |
| 77 | 31 27.7 | 37.0 | 11.1 | 32 26.6 | 36.9 | 11.2 | 33 25.4 | 36.6 | 11.4 | 34 24.2 | 36.4 | 11.5 | 77 |
| 78 | 30 50.7 | 37.4 | 10.2 | 31 49.7 | 37.2 | 10.3 | 32 48.8 | 37.1 | 10.4 | 33 47.8 | 37.0 | 10.5 | 78 |
| 79 | 30 13.3 | 37.8 | 9.3 | 31 12.5 | 37.7 | 9.4 | 32 11.7 | 37.5 | 9.5 | 33 10.8 | 37.3 | 9.6 | 79 |
| 80 | 29 35.5 | −38.2 | 8.4 | 30 34.8 | −38.0 | 8.5 | 31 34.2 | −37.9 | 8.6 | 32 33.5 | −37.8 | 8.7 | 80 |
| 81 | 28 57.3 | 38.5 | 7.5 | 29 56.8 | 38.4 | 7.6 | 30 56.3 | 38.3 | 7.7 | 31 55.7 | 38.1 | 7.7 | 81 |
| 82 | 28 18.8 | 38.8 | 6.6 | 29 18.4 | 38.7 | 6.7 | 30 18.0 | 38.6 | 6.8 | 31 17.6 | 38.5 | 6.8 | 82 |
| 83 | 27 40.0 | 39.1 | 5.8 | 28 39.7 | 39.1 | 5.8 | 29 39.4 | 39.0 | 5.9 | 30 39.1 | 38.9 | 5.9 | 83 |
| 84 | 27 00.9 | 39.5 | 4.9 | 28 00.6 | 39.3 | 5.0 | 29 00.4 | 39.3 | 5.0 | 30 00.2 | 39.3 | 5.1 | 84 |
| 85 | 26 21.4 | −39.7 | 4.1 | 27 21.3 | −39.7 | 4.1 | 28 21.1 | −39.6 | 4.2 | 29 20.9 | −39.5 | 4.2 | 85 |
| 86 | 25 41.7 | 40.1 | 3.2 | 26 41.6 | 40.0 | 3.3 | 27 41.5 | 40.0 | 3.3 | 28 41.4 | 39.9 | 3.3 | 86 |
| 87 | 25 01.6 | 40.2 | 2.4 | 26 01.6 | 40.3 | 2.4 | 27 01.5 | 40.2 | 2.5 | 28 01.5 | 40.2 | 2.5 | 87 |
| 88 | 24 21.4 | 40.6 | 1.6 | 25 21.3 | 40.5 | 1.6 | 26 21.3 | 40.5 | 1.6 | 27 21.3 | 40.5 | 1.6 | 88 |
| 89 | 23 40.8 | 40.8 | 0.8 | 24 40.8 | 40.8 | 0.8 | 25 40.8 | 40.8 | 0.8 | 26 40.8 | 40.8 | 0.8 | 89 |
| 90 | 23 00.0 | −41.0 | 0.0 | 24 00.0 | −41.0 | 0.0 | 25 00.0 | −41.0 | 0.0 | 26 00.0 | −41.1 | 0.0 | 90 |
| | 23° | | | 24° | | | 25° | | | 26° | | | |

| Dec. | 27° Hc | d | Z | 28° Hc | d | Z | 29° Hc | d | Z | 30° Hc | d | Z | Dec. |
|---|---|---|---|---|---|---|---|---|---|---|---|---|---|
| ° | ° ′ | ′ | ° | ° ′ | ′ | ° | ° ′ | ′ | ° | ° ′ | ′ | ° | ° |
| 0 | 37 25.3 | +34.0 | 112.9 | 37 01.5 | +35.0 | 113.6 | 36 37.1 | +36.0 | 114.3 | 36 12.1 | +36.9 | 115.0 | 0 |
| 1 | 37 59.3 | 33.4 | 111.9 | 37 36.5 | 34.5 | 112.6 | 37 13.1 | 35.5 | 113.3 | 36 49.0 | 36.5 | 114.0 | 1 |
| 2 | 38 32.7 | 32.9 | 110.8 | 38 11.0 | 34.0 | 111.6 | 37 48.6 | 34.9 | 112.3 | 37 25.5 | 35.9 | 113.0 | 2 |
| 3 | 39 05.6 | 32.3 | 109.8 | 38 45.0 | 33.3 | 110.5 | 38 23.5 | 34.4 | 111.3 | 38 01.4 | 35.4 | 112.0 | 3 |
| 4 | 39 37.9 | 31.7 | 108.7 | 39 18.3 | 32.8 | 109.5 | 38 57.9 | 33.9 | 110.2 | 38 36.8 | 34.9 | 111.0 | 4 |
| 5 | 40 09.6 | +31.0 | 107.6 | 39 51.1 | +32.1 | 108.4 | 39 31.8 | +33.2 | 109.2 | 39 11.7 | +34.3 | 109.9 | 5 |
| 6 | 40 40.6 | 30.4 | 106.5 | 40 23.2 | 31.5 | 107.3 | 40 05.0 | 32.6 | 108.1 | 39 46.0 | 33.7 | 108.9 | 6 |
| 7 | 41 11.0 | 29.6 | 105.3 | 40 54.7 | 30.8 | 106.1 | 40 37.6 | 32.0 | 107.0 | 40 19.7 | 33.1 | 107.8 | 7 |
| 8 | 41 40.6 | 29.0 | 104.2 | 41 25.5 | 30.2 | 105.0 | 41 09.6 | 31.3 | 105.9 | 40 52.8 | 32.4 | 106.7 | 8 |
| 9 | 42 09.6 | 28.2 | 103.0 | 41 55.7 | 29.4 | 103.9 | 41 40.9 | 30.6 | 104.7 | 41 25.2 | 31.8 | 105.6 | 9 |
| 10 | 42 37.8 | +27.5 | 101.8 | 42 25.1 | +28.7 | 102.7 | 42 11.5 | +29.9 | 103.6 | 41 57.0 | +31.1 | 104.4 | 10 |
| 11 | 43 05.3 | 26.7 | 100.6 | 42 53.8 | 28.0 | 101.5 | 42 41.4 | 29.2 | 102.4 | 42 28.1 | 30.4 | 103.3 | 11 |
| 12 | 43 32.0 | 25.9 | 99.3 | 43 21.8 | 27.2 | 100.3 | 43 10.6 | 28.4 | 101.2 | 42 58.5 | 29.7 | 102.1 | 12 |
| 13 | 43 57.9 | 25.0 | 98.1 | 43 49.0 | 26.3 | 99.0 | 43 39.0 | 27.7 | 100.0 | 43 28.2 | 28.9 | 100.9 | 13 |
| 14 | 44 22.9 | 24.3 | 96.8 | 44 15.3 | 25.6 | 97.8 | 44 06.7 | 26.8 | 98.8 | 43 57.1 | 28.1 | 99.7 | 14 |
| 15 | 44 47.2 | +23.3 | 95.5 | 44 40.9 | +24.6 | 96.5 | 44 33.5 | +26.0 | 97.5 | 44 25.2 | +27.3 | 98.5 | 15 |
| 16 | 45 10.5 | 22.4 | 94.2 | 45 05.5 | 23.8 | 95.2 | 44 59.5 | 25.2 | 96.2 | 44 52.5 | 26.5 | 97.2 | 16 |
| 17 | 45 32.9 | 21.5 | 92.9 | 45 29.3 | 22.9 | 93.9 | 45 24.7 | 24.3 | 94.9 | 45 19.0 | 25.6 | 96.0 | 17 |
| 18 | 45 54.4 | 20.6 | 91.6 | 45 52.2 | 22.0 | 92.6 | 45 49.0 | 23.3 | 93.6 | 45 44.6 | 24.8 | 94.7 | 18 |
| 19 | 46 15.0 | 19.6 | 90.2 | 46 14.2 | 21.0 | 91.3 | 46 12.3 | 22.5 | 92.3 | 46 09.4 | 23.8 | 93.3 | 19 |
| 20 | 46 34.6 | +18.6 | 88.9 | 46 35.2 | +20.1 | 89.9 | 46 34.8 | +21.5 | 91.0 | 46 33.2 | +22.9 | 92.0 | 20 |
| 21 | 46 53.2 | 17.6 | 87.5 | 46 55.3 | 19.0 | 88.5 | 46 56.3 | 20.5 | 89.6 | 46 56.1 | 21.9 | 90.7 | 21 |
| 22 | 47 10.8 | 16.5 | 86.1 | 47 14.3 | 18.0 | 87.1 | 47 16.8 | 19.4 | 88.2 | 47 18.0 | 21.0 | 89.3 | 22 |
| 23 | 47 27.3 | 15.5 | 84.6 | 47 32.3 | 17.0 | 85.7 | 47 36.2 | 18.5 | 86.8 | 47 39.0 | 19.9 | 87.9 | 23 |
| 24 | 47 42.8 | 14.4 | 83.2 | 47 49.3 | 15.9 | 84.3 | 47 54.7 | 17.4 | 85.4 | 47 58.9 | 18.9 | 86.5 | 24 |
| 25 | 47 57.2 | +13.3 | 81.8 | 48 05.2 | +14.8 | 82.9 | 48 12.1 | +16.3 | 84.0 | 48 17.8 | +17.8 | 85.1 | 25 |
| 26 | 48 10.5 | 12.2 | 80.3 | 48 20.0 | 13.7 | 81.4 | 48 28.4 | 15.2 | 82.5 | 48 35.6 | 16.8 | 83.6 | 26 |
| 27 | 48 22.7 | 11.0 | 78.8 | 48 33.7 | 12.6 | 79.9 | 48 43.6 | 14.1 | 81.1 | 48 52.4 | 15.6 | 82.2 | 27 |
| 28 | 48 33.7 | 9.9 | 77.4 | 48 46.3 | 11.4 | 78.5 | 48 57.7 | 13.0 | 79.6 | 49 08.0 | 14.5 | 80.7 | 28 |
| 29 | 48 43.6 | 8.8 | 75.9 | 48 57.7 | 10.3 | 77.0 | 49 10.7 | 11.8 | 78.1 | 49 22.5 | 13.3 | 79.2 | 29 |
| 30 | 48 52.4 | +7.5 | 74.4 | 49 08.0 | +9.1 | 75.5 | 49 22.5 | +10.6 | 76.6 | 49 35.8 | +12.1 | 77.7 | 30 |
| 31 | 48 59.9 | 6.4 | 72.8 | 49 17.1 | 7.8 | 74.0 | 49 33.1 | 9.4 | 75.1 | 49 47.9 | 11.0 | 76.2 | 31 |
| 32 | 49 06.3 | 5.2 | 71.3 | 49 24.9 | 6.7 | 72.4 | 49 42.5 | 8.2 | 73.6 | 49 58.9 | 9.8 | 74.7 | 32 |
| 33 | 49 11.5 | 3.9 | 69.8 | 49 31.6 | 5.5 | 70.9 | 49 50.7 | 7.0 | 72.0 | 50 08.7 | 8.5 | 73.2 | 33 |
| 34 | 49 15.4 | 2.8 | 68.3 | 49 37.1 | 4.2 | 69.4 | 49 57.7 | 5.8 | 70.5 | 50 17.2 | 7.3 | 71.6 | 34 |
| 35 | 49 18.2 | +1.5 | 66.7 | 49 41.3 | +3.1 | 67.8 | 50 03.5 | +4.5 | 68.9 | 50 24.5 | +6.0 | 70.1 | 35 |
| 36 | 49 19.7 | +0.3 | 65.2 | 49 44.4 | 1.7 | 66.3 | 50 08.0 | 3.2 | 67.4 | 50 30.5 | 4.8 | 68.5 | 36 |
| 37 | 49 20.0 | −0.9 | 63.7 | 49 46.1 | +0.6 | 64.7 | 50 11.2 | 2.1 | 65.8 | 50 35.3 | 3.5 | 66.9 | 37 |
| 38 | 49 19.1 | 2.1 | 62.1 | 49 46.7 | −0.7 | 63.2 | 50 13.3 | +0.7 | 64.3 | 50 38.8 | 2.2 | 65.3 | 38 |
| 39 | 49 17.0 | 3.3 | 60.6 | 49 46.0 | 1.9 | 61.6 | 50 14.0 | −0.5 | 62.7 | 50 41.0 | +1.0 | 63.8 | 39 |
| 40 | 49 13.7 | −4.5 | 59.1 | 49 44.1 | −3.2 | 60.1 | 50 13.5 | −1.7 | 61.1 | 50 42.0 | −0.3 | 62.2 | 40 |
| 41 | 49 09.2 | 5.8 | 57.6 | 49 40.9 | 4.4 | 58.5 | 50 11.8 | 3.0 | 59.6 | 50 41.7 | 1.6 | 60.6 | 41 |
| 42 | 49 03.4 | 6.9 | 56.0 | 49 36.5 | 5.6 | 57.0 | 50 08.8 | 4.3 | 58.0 | 50 40.1 | 2.9 | 59.0 | 42 |
| 43 | 48 56.5 | 8.1 | 54.5 | 49 30.9 | 6.8 | 55.5 | 50 04.5 | 5.5 | 56.5 | 50 37.2 | 4.1 | 57.5 | 43 |
| 44 | 48 48.4 | 9.3 | 53.0 | 49 24.1 | 8.1 | 53.9 | 49 59.0 | 6.7 | 54.9 | 50 33.1 | 5.4 | 55.9 | 44 |
| 45 | 48 39.1 | −10.5 | 51.5 | 49 16.0 | −9.2 | 52.4 | 49 52.3 | −8.0 | 53.4 | 50 27.7 | −6.7 | 54.3 | 45 |
| 46 | 48 28.6 | 11.6 | 50.0 | 49 06.8 | 10.4 | 50.9 | 49 44.3 | 9.2 | 51.8 | 50 21.0 | 7.9 | 52.8 | 46 |
| 47 | 48 17.0 | 12.7 | 48.6 | 48 56.4 | 11.5 | 49.4 | 49 35.1 | 10.4 | 50.3 | 50 13.1 | 9.2 | 51.2 | 47 |
| 48 | 48 04.3 | 13.8 | 47.1 | 48 44.9 | 12.7 | 47.9 | 49 24.7 | 11.5 | 48.8 | 50 03.9 | 10.3 | 49.7 | 48 |
| 49 | 47 50.5 | 14.9 | 45.6 | 48 32.2 | 13.9 | 46.4 | 49 13.2 | 12.7 | 47.3 | 49 53.6 | 11.6 | 48.1 | 49 |
| 50 | 47 35.6 | −16.0 | 44.2 | 48 18.3 | −14.9 | 45.0 | 49 00.5 | −13.9 | 45.8 | 49 42.0 | −12.7 | 46.6 | 50 |
| 51 | 47 19.6 | 17.0 | 42.8 | 48 03.4 | 16.0 | 43.5 | 48 46.6 | 15.0 | 44.3 | 49 29.3 | 13.9 | 45.1 | 51 |
| 52 | 47 02.6 | 18.1 | 41.4 | 47 47.4 | 17.1 | 42.1 | 48 31.6 | 16.1 | 42.8 | 49 15.4 | 15.1 | 43.6 | 52 |
| 53 | 46 44.5 | 19.1 | 40.0 | 47 30.3 | 18.2 | 40.7 | 48 15.5 | 17.1 | 41.4 | 49 00.3 | 16.2 | 42.1 | 53 |
| 54 | 46 25.4 | 20.0 | 38.6 | 47 12.1 | 19.1 | 39.3 | 47 58.4 | 18.3 | 39.9 | 48 44.1 | 17.3 | 40.7 | 54 |
| 55 | 46 05.4 | −21.0 | 37.2 | 46 53.0 | −20.2 | 37.9 | 47 40.1 | −19.2 | 38.5 | 48 26.8 | −18.3 | 39.2 | 55 |
| 56 | 45 44.4 | 22.0 | 35.9 | 46 32.8 | 21.1 | 36.5 | 47 20.9 | 20.3 | 37.1 | 48 08.5 | 19.4 | 37.8 | 56 |
| 57 | 45 22.4 | 22.8 | 34.5 | 46 11.7 | 22.1 | 35.1 | 47 00.6 | 21.3 | 35.7 | 47 49.1 | 20.5 | 36.4 | 57 |
| 58 | 44 59.6 | 23.8 | 33.2 | 45 49.6 | 23.0 | 33.8 | 46 39.3 | 22.2 | 34.4 | 47 28.6 | 21.4 | 35.0 | 58 |
| 59 | 44 35.8 | 24.6 | 31.9 | 45 26.6 | 23.9 | 32.5 | 46 17.1 | 23.2 | 33.0 | 47 07.2 | 22.4 | 33.6 | 59 |
| 60 | 44 11.2 | −25.4 | 30.7 | 45 02.7 | −24.8 | 31.2 | 45 53.9 | −24.1 | 31.7 | 46 44.8 | −23.4 | 32.3 | 60 |
| 61 | 43 45.8 | 26.3 | 29.4 | 44 37.9 | 25.6 | 29.9 | 45 29.8 | 25.0 | 30.4 | 46 21.4 | 24.3 | 30.9 | 61 |
| 62 | 43 19.5 | 27.1 | 28.2 | 44 12.3 | 26.5 | 28.6 | 45 04.8 | 25.8 | 29.1 | 45 57.1 | 25.2 | 29.6 | 62 |
| 63 | 42 52.4 | 27.8 | 26.9 | 43 45.8 | 27.3 | 27.4 | 44 39.0 | 26.7 | 27.8 | 45 31.9 | 26.0 | 28.3 | 63 |
| 64 | 42 24.6 | 28.6 | 25.7 | 43 18.5 | 28.0 | 26.1 | 44 12.3 | 27.5 | 26.6 | 45 05.9 | 26.9 | 27.0 | 64 |
| 65 | 41 56.0 | −29.3 | 24.5 | 42 50.5 | −28.8 | 24.9 | 43 44.8 | −28.3 | 25.3 | 44 39.0 | −27.8 | 25.8 | 65 |
| 66 | 41 26.7 | 30.0 | 23.4 | 42 21.7 | 29.5 | 23.7 | 43 16.5 | 29.0 | 24.1 | 44 11.2 | 28.5 | 24.5 | 66 |
| 67 | 40 56.7 | 30.7 | 22.2 | 41 52.2 | 30.3 | 22.6 | 42 47.5 | 29.8 | 22.9 | 43 42.7 | 29.3 | 23.3 | 67 |
| 68 | 40 26.0 | 31.3 | 21.1 | 41 21.9 | 30.9 | 21.4 | 42 17.7 | 30.4 | 21.7 | 43 13.4 | 30.0 | 22.1 | 68 |
| 69 | 39 54.7 | 32.0 | 20.0 | 40 51.0 | 31.5 | 20.3 | 41 47.3 | 31.2 | 20.6 | 42 43.4 | 30.8 | 20.9 | 69 |
| 70 | 39 22.7 | −32.5 | 18.9 | 40 19.5 | −32.3 | 19.2 | 41 16.1 | −31.9 | 19.4 | 42 12.6 | −31.4 | 19.7 | 70 |
| 71 | 38 50.2 | 33.2 | 17.8 | 39 47.2 | 32.8 | 18.1 | 40 44.2 | 32.4 | 18.3 | 41 41.2 | 32.2 | 18.6 | 71 |
| 72 | 38 17.0 | 33.7 | 16.7 | 39 14.4 | 33.4 | 17.0 | 40 11.8 | 33.1 | 17.2 | 41 09.0 | 32.7 | 17.5 | 72 |
| 73 | 37 43.3 | 34.3 | 15.7 | 38 41.0 | 34.0 | 15.9 | 39 38.7 | 33.7 | 16.1 | 40 36.3 | 33.4 | 16.4 | 73 |
| 74 | 37 09.0 | 34.8 | 14.7 | 38 07.0 | 34.5 | 14.8 | 39 05.0 | 34.3 | 15.1 | 40 02.9 | 34.0 | 15.3 | 74 |
| 75 | 36 34.2 | −35.4 | 13.6 | 37 32.5 | −35.1 | 13.8 | 38 30.7 | −34.8 | 14.0 | 39 28.9 | −34.6 | 14.2 | 75 |
| 76 | 35 58.8 | 35.8 | 12.6 | 36 57.4 | 35.6 | 12.8 | 37 55.9 | 35.4 | 13.0 | 38 54.3 | 35.1 | 13.1 | 76 |
| 77 | 35 23.0 | 36.3 | 11.6 | 36 21.8 | 36.1 | 11.8 | 37 20.5 | 35.9 | 11.9 | 38 19.2 | 35.7 | 12.1 | 77 |
| 78 | 34 46.7 | 36.7 | 10.7 | 35 45.7 | 36.6 | 10.8 | 36 44.6 | 36.4 | 11.0 | 37 43.5 | 36.2 | 11.1 | 78 |
| 79 | 34 10.0 | 37.2 | 9.7 | 35 09.1 | 37.0 | 9.8 | 36 08.2 | 36.8 | 10.0 | 37 07.3 | 36.7 | 10.1 | 79 |
| 80 | 33 32.8 | −37.6 | 8.8 | 34 32.1 | −37.5 | 8.9 | 35 31.4 | −37.4 | 9.0 | 36 30.6 | −37.2 | 9.1 | 80 |
| 81 | 32 55.2 | 38.1 | 7.8 | 33 54.6 | 37.9 | 7.9 | 34 54.0 | 37.7 | 8.0 | 35 53.4 | 37.6 | 8.1 | 81 |
| 82 | 32 17.1 | 38.4 | 6.9 | 33 16.7 | 38.3 | 7.0 | 34 16.3 | 38.2 | 7.1 | 35 15.8 | 38.1 | 7.2 | 82 |
| 83 | 31 38.7 | 38.8 | 6.0 | 32 38.4 | 38.7 | 6.1 | 33 38.1 | 38.7 | 6.1 | 34 37.7 | 38.5 | 6.2 | 83 |
| 84 | 30 59.9 | 39.1 | 5.1 | 31 59.7 | 39.1 | 5.2 | 32 59.4 | 39.0 | 5.2 | 33 59.2 | 38.9 | 5.3 | 84 |
| 85 | 30 20.8 | −39.5 | 4.2 | 31 20.6 | −39.4 | 4.3 | 32 20.4 | −39.3 | 4.3 | 33 20.3 | −39.4 | 4.4 | 85 |
| 86 | 29 41.3 | 39.9 | 3.4 | 30 41.2 | 39.8 | 3.4 | 31 41.1 | 39.8 | 3.4 | 32 40.9 | 39.7 | 3.5 | 86 |
| 87 | 29 01.4 | 40.1 | 2.5 | 30 01.4 | 40.2 | 2.5 | 31 01.3 | 40.1 | 2.6 | 32 01.2 | 40.0 | 2.6 | 87 |
| 88 | 28 21.3 | 40.5 | 1.7 | 29 21.2 | 40.4 | 1.7 | 30 21.2 | 40.4 | 1.7 | 31 21.2 | 40.4 | 1.7 | 88 |
| 89 | 27 40.8 | 40.8 | 0.8 | 28 40.8 | 40.8 | 0.8 | 29 40.8 | 40.8 | 0.8 | 30 40.8 | 40.8 | 0.9 | 89 |
| 90 | 27 00.0 | −41.1 | 0.0 | 28 00.0 | −41.1 | 0.0 | 29 00.0 | −41.1 | 0.0 | 30 00.0 | −41.1 | 0.0 | 90 |
| | 27° | | | 28° | | | 29° | | | 30° | | | |

# G THE AGETON METHOD

Material in this appendix is taken from Table 35 in the *American Practical Navigator,* Pub. No. 9, Vol. II, by permission of the Defense Mapping Agency Hydrographic/Topographic Center.

## TABLE 35

### The Ageton Method

**When Meridian Angle is Greater Than 90° Take "K" From Bottom of Table**

| ′ | 0° 00′ | | 0° 30′ | | 1° 00′ | | 1° 30′ | | 2° 00′ | | ′ |
|---|---|---|---|---|---|---|---|---|---|---|---|
| | A | B | A | B | A | B | A | B | A | B | |
| 0 | -------- | 0.0 | 205916 | 1.7 | 175814 | 6.6 | 158208 | 14.9 | 145718 | 26.5 | 30 |
| | 383730 | 0.0 | 205198 | 1.7 | 175454 | 6.7 | 157967 | 15.1 | 145538 | 26.7 | |
| 1 | 353627 | 0.0 | 204492 | 1.8 | 175097 | 6.8 | 157728 | 15.2 | 145358 | 26.9 | 29 |
| | 336018 | 0.0 | 203797 | 1.8 | 174742 | 7.0 | 157490 | 15.4 | 145179 | 27.1 | |
| 2 | 323524 | 0.0 | 203113 | 1.9 | 174391 | 7.1 | 157254 | 15.6 | 145000 | 27.3 | 28 |
| | 313833 | 0.0 | 202440 | 1.9 | 174042 | 7.2 | 157019 | 15.7 | 144823 | 27.6 | |
| 3 | 305915 | 0.0 | 201777 | 2.0 | 173696 | 7.3 | 156784 | 15.9 | 144646 | 27.8 | 27 |
| | 299221 | 0.0 | 201124 | 2.1 | 173352 | 7.4 | 156552 | 16.1 | 144470 | 28.0 | |
| 4 | 293421 | 0.0 | 200480 | 2.1 | 173012 | 7.5 | 156320 | 16.2 | 144295 | 28.3 | 26 |
| | 288306 | 0.0 | 199846 | 2.2 | 172674 | 7.6 | 156090 | 16.4 | 144120 | 28.5 | |
| 5 | 283730 | 0.0 | 199221 | 2.3 | 172339 | 7.8 | 155861 | 16.6 | 143946 | 28.7 | 25 |
| | 279591 | 0.1 | 198605 | 2.3 | 172006 | 7.9 | 155633 | 16.8 | 143773 | 28.9 | |
| 6 | 275812 | 0.1 | 197998 | 2.4 | 171676 | 8.0 | 155406 | 16.9 | 143600 | 29.2 | 24 |
| | 272336 | 0.1 | 197399 | 2.4 | 171348 | 8.1 | 155180 | 17.1 | 143428 | 29.4 | |
| 7 | 269118 | 0.1 | 196808 | 2.5 | 171023 | 8.2 | 154956 | 17.3 | 143257 | 29.6 | 23 |
| | 266121 | 0.1 | 196225 | 2.6 | 170700 | 8.4 | 154733 | 17.5 | 143086 | 29.9 | |
| 8 | 263318 | 0.1 | 195650 | 2.7 | 170379 | 8.5 | 154511 | 17.6 | 142916 | 30.1 | 22 |
| | 260685 | 0.1 | 195082 | 2.7 | 170061 | 8.6 | 154290 | 17.8 | 142747 | 30.4 | |
| 9 | 258203 | 0.1 | 194522 | 2.8 | 169745 | 8.7 | 154070 | 18.0 | 142579 | 30.6 | 21 |
| | 255855 | 0.2 | 193969 | 2.9 | 169432 | 8.9 | 153851 | 18.2 | 142411 | 30.8 | |
| 10 | 253627 | 0.2 | 193422 | 2.9 | 169121 | 9.0 | 153633 | 18.4 | 142243 | 31.1 | 20 |
| | 251508 | 0.2 | 192883 | 3.0 | 168811 | 9.1 | 153417 | 18.6 | 142077 | 31.3 | |
| 11 | 249488 | 0.2 | 192350 | 3.1 | 168505 | 9.3 | 153201 | 18.7 | 141911 | 31.5 | 19 |
| | 247558 | 0.2 | 191824 | 3.2 | 168200 | 9.4 | 152987 | 18.9 | 141745 | 31.8 | |
| 12 | 245709 | 0.3 | 191303 | 3.2 | 167897 | 9.5 | 152774 | 19.1 | 141581 | 32.0 | 18 |
| | 243936 | 0.3 | 190790 | 3.3 | 167597 | 9.7 | 152561 | 19.3 | 141417 | 32.3 | |
| 13 | 242233 | 0.3 | 190282 | 3.4 | 167298 | 9.8 | 152350 | 19.5 | 141253 | 32.5 | 17 |
| | 240594 | 0.3 | 189780 | 3.5 | 167002 | 9.9 | 152140 | 19.7 | 141090 | 32.8 | |
| 14 | 239015 | 0.4 | 189283 | 3.6 | 166708 | 10.1 | 151931 | 19.9 | 140928 | 33.0 | 16 |
| | 237491 | 0.4 | 188793 | 3.6 | 166415 | 10.2 | 151722 | 20.1 | 140766 | 33.3 | |
| 15 | 236018 | 0.4 | 188307 | 3.7 | 166125 | 10.3 | 151515 | 20.3 | 140605 | 33.5 | 15 |
| | 234594 | 0.4 | 187827 | 3.8 | 165836 | 10.5 | 151309 | 20.5 | 140445 | 33.7 | |
| 16 | 233215 | 0.5 | 187353 | 3.9 | 165550 | 10.6 | 151104 | 20.6 | 140285 | 34.0 | 14 |
| | 231879 | 0.5 | 186883 | 4.0 | 165265 | 10.8 | 150899 | 20.8 | 140125 | 34.2 | |
| 17 | 230583 | 0.5 | 186419 | 4.1 | 164982 | 10.9 | 150696 | 21.0 | 139967 | 34.5 | 13 |
| | 229324 | 0.6 | 185959 | 4.1 | 164701 | 11.0 | 150494 | 21.2 | 139809 | 34.7 | |
| 18 | 228100 | 0.6 | 185505 | 4.2 | 164422 | 11.2 | 150292 | 21.4 | 139651 | 35.0 | 12 |
| | 226910 | 0.6 | 185055 | 4.3 | 164144 | 11.3 | 150092 | 21.6 | 139494 | 35.3 | |
| 19 | 225752 | 0.7 | 184609 | 4.4 | 163868 | 11.5 | 149892 | 21.8 | 139338 | 35.5 | 11 |
| | 224624 | 0.7 | 184168 | 4.5 | 163594 | 11.6 | 149693 | 22.0 | 139182 | 35.8 | |
| 20 | 223525 | 0.7 | 183732 | 4.6 | 163322 | 11.8 | 149495 | 22.2 | 139027 | 36.0 | 10 |
| | 222452 | 0.8 | 183300 | 4.7 | 163052 | 11.9 | 149299 | 22.4 | 138872 | 36.3 | |
| 21 | 221406 | 0.8 | 182872 | 4.8 | 162783 | 12.1 | 149103 | 22.6 | 138718 | 36.5 | 9 |
| | 220384 | 0.9 | 182448 | 4.9 | 162516 | 12.2 | 148907 | 22.9 | 138564 | 36.8 | |
| 22 | 219385 | 0.9 | 182029 | 5.0 | 162250 | 12.4 | 148713 | 23.1 | 138411 | 37.1 | 8 |
| | 218409 | 0.9 | 181613 | 5.1 | 161986 | 12.5 | 148520 | 23.3 | 138258 | 37.3 | |
| 23 | 217455 | 1.0 | 181201 | 5.2 | 161724 | 12.7 | 148327 | 23.5 | 138106 | 37.6 | 7 |
| | 216521 | 1.0 | 180794 | 5.3 | 161463 | 12.8 | 148135 | 23.7 | 137955 | 37.9 | |
| 24 | 215607 | 1.1 | 180390 | 5.4 | 161204 | 13.0 | 147945 | 23.9 | 137804 | 38.1 | 6 |
| | 214711 | 1.1 | 179990 | 5.5 | 160946 | 13.1 | 147755 | 24.1 | 137653 | 38.4 | |
| 25 | 213834 | 1.1 | 179593 | 5.6 | 160690 | 13.3 | 147566 | 24.3 | 137504 | 38.6 | 5 |
| | 212974 | 1.2 | 179200 | 5.7 | 160435 | 13.4 | 147377 | 24.5 | 137354 | 38.9 | |
| 26 | 212130 | 1.2 | 178810 | 5.8 | 160182 | 13.6 | 147190 | 24.7 | 137205 | 39.2 | 4 |
| | 211303 | 1.3 | 178424 | 5.9 | 159930 | 13.8 | 147003 | 24.9 | 137057 | 39.4 | |
| 27 | 210491 | 1.3 | 178042 | 6.0 | 159680 | 13.9 | 146817 | 25.2 | 136909 | 39.7 | 3 |
| | 209695 | 1.4 | 177663 | 6.1 | 159431 | 14.1 | 146632 | 25.4 | 136761 | 40.0 | |
| 28 | 208912 | 1.4 | 177287 | 6.2 | 159184 | 14.2 | 146448 | 25.6 | 136615 | 40.3 | 2 |
| | 208143 | 1.5 | 176914 | 6.3 | 158938 | 14.4 | 146264 | 25.8 | 136468 | 40.5 | |
| 29 | 207388 | 1.5 | 176544 | 6.4 | 158693 | 14.6 | 146081 | 26.0 | 136322 | 40.8 | 1 |
| | 206646 | 1.6 | 176178 | 6.5 | 158450 | 14.7 | 145899 | 26.2 | 136177 | 41.1 | |
| 30 | 205916 | 1.7 | 175814 | 6.6 | 158208 | 14.9 | 145718 | 26.5 | 136032 | 41.4 | 0 |
| | A | B | A | B | A | B | A | B | A | B | |
| ′ | 179° 30′ | | 179° 00′ | | 178° 30′ | | 178° 00′ | | 177° 30′ | | ′ |

# TABLE 35

## The Ageton Method

**Always Take "Z" From Bottom of Table, Except When "K" is Same Name and Greater Than Latitude, in Which Case Take "Z" From Top of Table**

| ′ | 2° 30′ | | 3° 00′ | | 3° 30′ | | 4° 00′ | | 4° 30′ | | ′ |
|---|---|---|---|---|---|---|---|---|---|---|---|
| | A | B | A | B | A | B | A | B | A | B | |
| 0 | 136032 | 41.4 | 128120 | 59.6 | 121432 | 81.1 | 115641 | 105.9 | 110536 | 134.1 | 30 |
| | 135888 | 41.6 | 128000 | 59.9 | 121329 | 81.5 | 115551 | 106.4 | 110455 | 134.6 | |
| 1 | 135744 | 41.9 | 127880 | 60.2 | 121226 | 81.9 | 115461 | 106.8 | 110375 | 135.1 | 29 |
| | 135600 | 42.2 | 127760 | 60.6 | 121124 | 82.2 | 115371 | 107.3 | 110296 | 135.6 | |
| 2 | 135457 | 42.5 | 127640 | 60.9 | 121021 | 82.6 | 115282 | 107.7 | 110216 | 136.1 | 28 |
| | 135315 | 42.7 | 127521 | 61.2 | 120919 | 83.0 | 115192 | 108.1 | 110136 | 136.6 | |
| 3 | 135173 | 43.0 | 127403 | 61.6 | 120817 | 83.4 | 115103 | 108.6 | 110057 | 137.1 | 27 |
| | 135031 | 43.3 | 127284 | 61.9 | 120715 | 83.8 | 115014 | 109.0 | 109977 | 137.6 | |
| 4 | 134890 | 43.6 | 127166 | 62.2 | 120614 | 84.2 | 114925 | 109.5 | 109898 | 138.1 | 26 |
| | 134749 | 43.9 | 127049 | 62.6 | 120513 | 84.6 | 114836 | 109.9 | 109819 | 138.6 | |
| 5 | 134609 | 44.2 | 126931 | 62.9 | 120412 | 85.0 | 114747 | 110.4 | 109740 | 139.1 | 25 |
| | 134469 | 44.4 | 126814 | 63.3 | 120311 | 85.4 | 114659 | 110.8 | 109662 | 139.6 | |
| 6 | 134330 | 44.7 | 126697 | 63.6 | 120211 | 85.8 | 114571 | 111.3 | 109583 | 140.1 | 24 |
| | 134191 | 45.0 | 126581 | 63.9 | 120110 | 86.2 | 114483 | 111.7 | 109505 | 140.6 | |
| 7 | 134052 | 45.3 | 126465 | 64.3 | 120010 | 86.6 | 114395 | 112.2 | 109426 | 141.1 | 23 |
| | 133914 | 45.6 | 126349 | 64.6 | 119910 | 87.0 | 114307 | 112.7 | 109348 | 141.7 | |
| 8 | 133777 | 45.9 | 126233 | 65.0 | 119811 | 87.4 | 114220 | 113.1 | 109270 | 142.2 | 22 |
| | 133640 | 46.2 | 126118 | 65.3 | 119711 | 87.8 | 114133 | 113.6 | 109192 | 142.7 | |
| 9 | 133503 | 46.5 | 126003 | 65.7 | 119612 | 88.2 | 114045 | 114.0 | 109115 | 143.2 | 21 |
| | 133367 | 46.8 | 125888 | 66.0 | 119513 | 88.6 | 113958 | 114.5 | 109037 | 143.7 | |
| 10 | 133231 | 47.1 | 125774 | 66.4 | 119415 | 89.0 | 113872 | 114.9 | 108960 | 144.2 | 20 |
| | 133096 | 47.4 | 125660 | 66.7 | 119316 | 89.4 | 113785 | 115.4 | 108882 | 144.7 | |
| 11 | 132961 | 47.6 | 125546 | 67.1 | 119218 | 89.8 | 113699 | 115.9 | 108805 | 145.2 | 19 |
| | 132826 | 47.9 | 125433 | 67.4 | 119120 | 90.2 | 113612 | 116.3 | 108728 | 145.8 | |
| 12 | 132692 | 48.2 | 125320 | 67.8 | 119022 | 90.6 | 113526 | 116.8 | 108651 | 146.3 | 18 |
| | 132558 | 48.5 | 125207 | 68.1 | 118925 | 91.0 | 113440 | 117.3 | 108574 | 146.8 | |
| 13 | 132425 | 48.8 | 125094 | 68.5 | 118827 | 91.4 | 113354 | 117.7 | 108498 | 147.3 | 17 |
| | 132292 | 49.1 | 124982 | 68.8 | 118730 | 91.8 | 113269 | 118.2 | 108421 | 147.8 | |
| 14 | 132159 | 49.4 | 124870 | 69.2 | 118633 | 92.3 | 113183 | 118.7 | 108345 | 148.4 | 16 |
| | 132027 | 49.7 | 124759 | 69.6 | 118537 | 92.7 | 113098 | 119.1 | 108269 | 148.9 | |
| 15 | 131896 | 50.0 | 124647 | 69.9 | 118440 | 93.1 | 113013 | 119.6 | 108193 | 149.4 | 15 |
| | 131764 | 50.3 | 124536 | 70.3 | 118344 | 93.5 | 112928 | 120.1 | 108117 | 149.9 | |
| 16 | 131633 | 50.7 | 124425 | 70.6 | 118248 | 93.9 | 112843 | 120.5 | 108041 | 150.5 | 14 |
| | 131503 | 51.0 | 124315 | 71.0 | 118152 | 94.3 | 112759 | 121.0 | 107965 | 151.0 | |
| 17 | 131373 | 51.3 | 124204 | 71.3 | 118056 | 94.7 | 112674 | 121.5 | 107890 | 151.5 | 13 |
| | 131243 | 51.6 | 124095 | 71.7 | 117961 | 95.2 | 112590 | 121.9 | 107814 | 152.1 | |
| 18 | 131114 | 51.9 | 123985 | 72.1 | 117866 | 95.6 | 112506 | 122.4 | 107739 | 152.6 | 12 |
| | 130985 | 52.2 | 123875 | 72.4 | 117771 | 96.0 | 112422 | 122.9 | 107664 | 153.1 | |
| 19 | 130856 | 52.5 | 123766 | 72.8 | 117676 | 96.4 | 112338 | 123.4 | 107589 | 153.6 | 11 |
| | 130728 | 52.8 | 123657 | 73.2 | 117581 | 96.9 | 112255 | 123.9 | 107514 | 154.2 | |
| 20 | 130600 | 53.1 | 123549 | 73.5 | 117487 | 97.3 | 112171 | 124.3 | 107439 | 154.7 | 10 |
| | 130473 | 53.4 | 123441 | 73.9 | 117393 | 97.7 | 112088 | 124.8 | 107364 | 155.2 | |
| 21 | 130346 | 53.7 | 123332 | 74.3 | 117299 | 98.1 | 112005 | 125.3 | 107290 | 155.8 | 9 |
| | 130219 | 54.1 | 123225 | 74.6 | 117205 | 98.5 | 111922 | 125.8 | 107216 | 156.3 | |
| 22 | 130093 | 54.4 | 123117 | 75.0 | 117112 | 99.0 | 111839 | 126.2 | 107141 | 156.9 | 8 |
| | 129967 | 54.7 | 123010 | 75.4 | 117018 | 99.4 | 111757 | 126.7 | 107067 | 157.4 | |
| 23 | 129841 | 55.0 | 122903 | 75.8 | 116925 | 99.8 | 111674 | 127.2 | 106993 | 157.9 | 7 |
| | 129716 | 55.3 | 122796 | 76.1 | 116832 | 100.3 | 111592 | 127.7 | 106919 | 158.5 | |
| 24 | 129591 | 55.7 | 122690 | 76.5 | 116739 | 100.7 | 111510 | 128.2 | 106846 | 159.0 | 6 |
| | 129466 | 56.0 | 122584 | 76.9 | 116647 | 101.1 | 111428 | 128.7 | 106772 | 159.6 | |
| 25 | 129342 | 56.3 | 122478 | 77.3 | 116554 | 101.6 | 111346 | 129.2 | 106698 | 160.1 | 5 |
| | 129218 | 56.6 | 122372 | 77.6 | 116462 | 102.0 | 111264 | 129.7 | 106625 | 160.6 | |
| 26 | 129095 | 56.9 | 122267 | 78.0 | 116370 | 102.4 | 111183 | 130.1 | 106552 | 161.2 | 4 |
| | 128972 | 57.3 | 122161 | 78.4 | 116278 | 102.9 | 111101 | 130.6 | 106479 | 161.7 | |
| 27 | 128849 | 57.6 | 122057 | 78.8 | 116187 | 103.3 | 111020 | 131.1 | 106406 | 162.3 | 3 |
| | 128727 | 57.9 | 121952 | 79.2 | 116096 | 103.7 | 110939 | 131.6 | 106333 | 162.8 | |
| 28 | 128605 | 58.2 | 121848 | 79.5 | 116004 | 104.2 | 110858 | 132.1 | 106260 | 163.4 | 2 |
| | 128483 | 58.6 | 121743 | 79.9 | 115913 | 104.6 | 110777 | 132.6 | 106187 | 163.9 | |
| 29 | 128362 | 58.9 | 121639 | 80.3 | 115823 | 105.0 | 110696 | 133.1 | 106115 | 164.5 | 1 |
| | 128240 | 59.2 | 121536 | 80.7 | 115732 | 105.5 | 110616 | 133.6 | 106043 | 165.0 | |
| 30 | 128120 | 59.6 | 121432 | 81.1 | 115641 | 105.9 | 110536 | 134.1 | 105970 | 165.6 | 0 |
| | A | B | A | B | A | B | A | B | A | B | |
| ′ | 177° 00′ | | 176° 30′ | | 176° 00′ | | 175° 30′ | | 175° 00′ | | ′ |

# TABLE 35

## The Ageton Method

When Meridian Angle is Greater Than 90° Take "K" From Bottom of Table

| ′ | 5° 00′ A | 5° 00′ B | 5° 30′ A | 5° 30′ B | 6° 00′ A | 6° 00′ B | 6° 30′ A | 6° 30′ B | 7° 00′ A | 7° 00′ B | ′ |
|---|---|---|---|---|---|---|---|---|---|---|---|
| 0 | 105970 | **165. 6** | 101843 | **200. 4** | 98076 | **239** | 94614 | **280** | 91411 | **325** | 30 |
| | 105898 | **166. 1** | 101777 | **201. 0** | 98017 | **239** | 94559 | **281** | 91359 | **326** | |
| 1 | 105826 | **166. 7** | 101712 | **201. 6** | 97957 | **240** | 94503 | **281** | 91308 | **326** | 29 |
| | 105754 | **167. 2** | 101646 | **202. 2** | 97897 | **241** | 94448 | **282** | 91257 | **327** | |
| 2 | 105683 | **167. 8** | 101581 | **202. 8** | 97837 | **241** | 94393 | **283** | 91205 | **328** | 28 |
| | 105611 | **168. 4** | 101516 | **203. 5** | 97777 | **242** | 94338 | **284** | 91154 | **329** | |
| 3 | 105539 | **168. 9** | 101451 | **204. 1** | 97717 | **243** | 94283 | **284** | 91103 | **330** | 27 |
| | 105468 | **169. 4** | 101386 | **204. 7** | 97658 | **243** | 94228 | **285** | 91052 | **330** | |
| 4 | 105397 | **170. 0** | 101321 | **205. 3** | 97598 | **244** | 94173 | **286** | 91001 | **331** | 26 |
| | 105325 | **170. 6** | 101256 | **205. 9** | 97539 | **245** | 94118 | **287** | 90950 | **332** | |
| 5 | 105254 | **171. 1** | 101192 | **206. 5** | 97480 | **245** | 94063 | **287** | 90899 | **333** | 25 |
| | 105183 | **171. 7** | 101127 | **207. 1** | 97420 | **246** | 94009 | **288** | 90848 | **333** | |
| 6 | 105113 | **172. 3** | 101063 | **207. 8** | 97361 | **247** | 93954 | **289** | 90798 | **334** | 24 |
| | 105042 | **172. 8** | 100998 | **208. 4** | 97302 | **247** | 93899 | **289** | 90747 | **335** | |
| 7 | 104971 | **173. 4** | 100934 | **209. 0** | 97243 | **248** | 93845 | **290** | 90696 | **336** | 23 |
| | 104901 | **174. 0** | 100870 | **209. 6** | 97184 | **249** | 93790 | **291** | 90646 | **337** | |
| 8 | 104830 | **174. 5** | 100806 | **210. 3** | 97126 | **249** | 93736 | **292** | 90595 | **337** | 22 |
| | 104760 | **175. 1** | 100742 | **210. 9** | 97067 | **250** | 93682 | **292** | 90545 | **338** | |
| 9 | 104690 | **175. 7** | 100678 | **211. 5** | 97008 | **251** | 93628 | **293** | 90494 | **339** | 21 |
| | 104620 | **176. 2** | 100614 | **212. 1** | 96950 | **251** | 93573 | **294** | 90444 | **340** | |
| 10 | 104550 | **176. 8** | 100550 | **212. 8** | 96891 | **252** | 93519 | **295** | 90394 | **341** | 20 |
| | 104480 | **177. 4** | 100487 | **213. 4** | 96833 | **253** | 93465 | **295** | 90344 | **341** | |
| 11 | 104411 | **178. 0** | 100423 | **214. 0** | 96774 | **253** | 93411 | **296** | 90293 | **342** | 19 |
| | 104341 | **178. 5** | 100360 | **214. 6** | 96716 | **254** | 93358 | **297** | 90243 | **343** | |
| 12 | 104272 | **179. 1** | 100296 | **215. 3** | 96658 | **255** | 93304 | **298** | 90193 | **344** | 18 |
| | 104202 | **179. 7** | 100233 | **215. 9** | 96600 | **255** | 93250 | **298** | 90143 | **345** | |
| 13 | 104133 | **180. 3** | 100170 | **216. 5** | 96542 | **256** | 93196 | **299** | 90093 | **345** | 17 |
| | 104064 | **180. 8** | 100107 | **217. 2** | 96484 | **257** | 93143 | **300** | 90044 | **346** | |
| 14 | 103995 | **181. 4** | 100044 | **217. 8** | 96426 | **257** | 93089 | **301** | 89994 | **347** | 16 |
| | 103926 | **182. 0** | 99981 | **218. 4** | 96368 | **258** | 93036 | **301** | 89944 | **348** | |
| 15 | 103857 | **182. 6** | 99918 | **219. 1** | 96310 | **259** | 92982 | **302** | 89894 | **349** | 15 |
| | 103788 | **183. 2** | 99856 | **219. 7** | 96253 | **260** | 92929 | **303** | 89845 | **349** | |
| 16 | 103720 | **183. 7** | 99793 | **220. 3** | 96195 | **260** | 92876 | **304** | 89795 | **350** | 14 |
| | 103651 | **184. 3** | 99731 | **221. 0** | 96138 | **261** | 92823 | **304** | 89746 | **351** | |
| 17 | 103583 | **184. 9** | 99668 | **221. 6** | 96080 | **262** | 92769 | **305** | 89696 | **352** | 13 |
| | 103515 | **185. 5** | 99606 | **222. 3** | 96023 | **262** | 92716 | **306** | 89647 | **353** | |
| 18 | 103447 | **186. 1** | 99544 | **222. 9** | 95966 | **263** | 92663 | **307** | 89597 | **353** | 12 |
| | 103379 | **186. 7** | 99481 | **223. 5** | 95909 | **264** | 92610 | **307** | 89548 | **354** | |
| 19 | 103311 | **187. 2** | 99420 | **224. 2** | 95851 | **264** | 92558 | **308** | 89499 | **355** | 11 |
| | 103243 | **187. 8** | 99357 | **224. 8** | 95795 | **265** | 92505 | **309** | 89450 | **356** | |
| 20 | 103175 | **188. 4** | 99296 | **225. 5** | 95737 | **266** | 92452 | **310** | 89401 | **357** | 10 |
| | 103107 | **189. 0** | 99234 | **226. 1** | 95681 | **267** | 92399 | **310** | 89352 | **357** | |
| 21 | 103040 | **189. 6** | 99172 | **226. 8** | 95624 | **267** | 92347 | **311** | 89303 | **358** | 9 |
| | 102973 | **190. 2** | 99110 | **227. 4** | 95567 | **268** | 92294 | **312** | 89254 | **359** | |
| 22 | 102905 | **190. 8** | 99049 | **228. 1** | 95510 | **269** | 92242 | **313** | 89205 | **360** | 8 |
| | 102838 | **191. 4** | 98988 | **228. 7** | 95454 | **269** | 92189 | **313** | 89156 | **361** | |
| 23 | 102771 | **192. 0** | 98926 | **229. 4** | 95397 | **270** | 92137 | **314** | 89107 | **362** | 7 |
| | 102704 | **192. 6** | 98865 | **230. 0** | 95341 | **271** | 92085 | **315** | 89059 | **362** | |
| 24 | 102637 | **193. 2** | 98804 | **230. 7** | 95285 | **271** | 92032 | **316** | 89010 | **363** | 6 |
| | 102570 | **193. 8** | 98743 | **231. 3** | 95228 | **272** | 91980 | **316** | 88961 | **364** | |
| 25 | 102504 | **194. 4** | 98682 | **232. 0** | 95172 | **273** | 91928 | **317** | 88913 | **365** | 5 |
| | 102437 | **195. 0** | 98621 | **232. 6** | 95116 | **274** | 91876 | **318** | 88864 | **366** | |
| 26 | 102371 | **195. 6** | 98560 | **233. 3** | 95060 | **274** | 91824 | **319** | 88816 | **366** | 4 |
| | 102304 | **196. 2** | 98499 | **233. 9** | 95004 | **275** | 91772 | **319** | 88767 | **367** | |
| 27 | 102238 | **196. 8** | 98439 | **234. 6** | 94948 | **276** | 91720 | **320** | 88719 | **368** | 3 |
| | 102172 | **197. 4** | 98378 | **235. 3** | 94892 | **276** | 91668 | **321** | 88671 | **369** | |
| 28 | 102106 | **198. 0** | 98318 | **235. 9** | 94836 | **277** | 91617 | **322** | 88623 | **370** | 2 |
| | 102040 | **198. 6** | 98257 | **236. 6** | 94781 | **278** | 91565 | **323** | 88574 | **371** | |
| 29 | 101974 | **199. 2** | 98197 | **237. 2** | 94725 | **279** | 91514 | **323** | 88526 | **371** | 1 |
| | 101908 | **199. 8** | 98137 | **237. 9** | 94670 | **279** | 91462 | **324** | 88478 | **372** | |
| 30 | 101843 | **200. 4** | 98076 | **238. 6** | 94614 | **280** | 91411 | **325** | 88430 | **373** | 0 |
| ′ | A | B | A | B | A | B | A | B | A | B | ′ |
| | 174° 30′ | | 174° 00′ | | 173° 30′ | | 173° 00′ | | 172° 30′ | | |

# TABLE 35

## The Ageton Method

**Always Take "Z" From Bottom of Table, Except When "K" is Same Name and Greater Than Latitude, in Which Case Take "Z" From Top of Table**

| ′ | 7° 30′ | | 8° 00′ | | 8° 30′ | | 9° 00′ | | 9° 30′ | | ′ |
|---|---|---|---|---|---|---|---|---|---|---|---|
| | A | B | A | B | A | B | A | B | A | B | |
| 0 | 88430 | 373 | 85644 | 425 | 83030 | 480 | 80567 | 538 | 78239 | 600 | 30 |
| | 88382 | 374 | 85599 | 426 | 82987 | 481 | 80527 | 539 | 78201 | 601 | |
| 1 | 88334 | 375 | 85555 | 426 | 82945 | 482 | 80487 | 540 | 78164 | 602 | 29 |
| | 88286 | 376 | 85510 | 427 | 82903 | 482 | 80447 | 541 | 78126 | 603 | |
| 2 | 88239 | 376 | 85465 | 428 | 82861 | 483 | 80407 | 542 | 78088 | 604 | 28 |
| | 88191 | 377 | 85420 | 429 | 82819 | 484 | 80368 | 543 | 78051 | 605 | |
| 3 | 88143 | 378 | 85376 | 430 | 82777 | 485 | 80328 | 544 | 78013 | 606 | 27 |
| | 88096 | 379 | 85331 | 431 | 82735 | 486 | 80288 | 545 | 77976 | 607 | |
| 4 | 88048 | 380 | 85286 | 432 | 82693 | 487 | 80249 | 546 | 77938 | 608 | 26 |
| | 88001 | 381 | 85242 | 433 | 82651 | 488 | 80209 | 547 | 77901 | 609 | |
| 5 | 87953 | 381 | 85197 | 434 | 82609 | 489 | 80170 | 548 | 77863 | 610 | 25 |
| | 87906 | 382 | 85153 | 434 | 82567 | 490 | 80130 | 549 | 77826 | 611 | |
| 6 | 87858 | 383 | 85108 | 435 | 82526 | 491 | 80091 | 550 | 77788 | 612 | 24 |
| | 87811 | 384 | 85064 | 436 | 82484 | 492 | 80051 | 551 | 77751 | 614 | |
| 7 | 87764 | 385 | 85020 | 437 | 82442 | 493 | 80012 | 552 | 77714 | 615 | 23 |
| | 87716 | 386 | 84976 | 438 | 82400 | 494 | 79973 | 553 | 77677 | 616 | |
| 8 | 87669 | 387 | 84931 | 439 | 82359 | 495 | 79933 | 554 | 77639 | 617 | 22 |
| | 87622 | 387 | 84887 | 440 | 82317 | 496 | 79894 | 555 | 77602 | 618 | |
| 9 | 87575 | 388 | 84843 | 441 | 82276 | 497 | 79855 | 556 | 77565 | 619 | 21 |
| | 87528 | 389 | 84799 | 442 | 82234 | 498 | 79816 | 557 | 77528 | 620 | |
| 10 | 87481 | 390 | 84755 | 443 | 82193 | 499 | 79777 | 558 | 77491 | 621 | 20 |
| | 87434 | 391 | 84711 | 444 | 82151 | 500 | 79737 | 559 | 77454 | 622 | |
| 11 | 87387 | 392 | 84667 | 444 | 82110 | 501 | 79698 | 560 | 77417 | 623 | 19 |
| | 87341 | 392 | 84623 | 445 | 82069 | 502 | 79659 | 561 | 77380 | 624 | |
| 12 | 87294 | 393 | 84579 | 446 | 82027 | 503 | 79620 | 562 | 77343 | 625 | 18 |
| | 87247 | 394 | 84535 | 447 | 81986 | 504 | 79581 | 563 | 77306 | 626 | |
| 13 | 87201 | 395 | 84492 | 448 | 81945 | 504 | 79542 | 564 | 77269 | 627 | 17 |
| | 87154 | 396 | 84448 | 449 | 81904 | 505 | 79503 | 565 | 77232 | 629 | |
| 14 | 87107 | 397 | 84404 | 450 | 81863 | 506 | 79465 | 566 | 77195 | 630 | 16 |
| | 87061 | 398 | 84361 | 451 | 81821 | 507 | 79426 | 567 | 77158 | 631 | |
| 15 | 87015 | 399 | 84317 | 452 | 81780 | 508 | 79387 | 568 | 77122 | 632 | 15 |
| | 86968 | 399 | 84273 | 453 | 81739 | 509 | 79348 | 569 | 77085 | 633 | |
| 16 | 86922 | 400 | 84230 | 454 | 81698 | 510 | 79309 | 570 | 77048 | 634 | 14 |
| | 86876 | 401 | 84186 | 454 | 81657 | 511 | 79271 | 571 | 77011 | 635 | |
| 17 | 86829 | 402 | 84143 | 455 | 81617 | 512 | 79232 | 573 | 76975 | 636 | 13 |
| | 86783 | 403 | 84100 | 456 | 81576 | 513 | 79193 | 574 | 76938 | 637 | |
| 18 | 86737 | 404 | 84056 | 457 | 81535 | 514 | 79155 | 575 | 76902 | 638 | 12 |
| | 86691 | 405 | 84013 | 458 | 81494 | 515 | 79116 | 576 | 76865 | 639 | |
| 19 | 86645 | 405 | 83970 | 459 | 81453 | 516 | 79078 | 577 | 76828 | 641 | 11 |
| | 86599 | 406 | 83927 | 460 | 81413 | 517 | 79039 | 578 | 76792 | 642 | |
| 20 | 86553 | 407 | 83884 | 461 | 81372 | 518 | 79001 | 579 | 76756 | 643 | 10 |
| | 86507 | 408 | 83840 | 462 | 81331 | 519 | 78962 | 580 | 76719 | 644 | |
| 21 | 86461 | 409 | 83797 | 463 | 81291 | 520 | 78924 | 581 | 76683 | 645 | 9 |
| | 86415 | 410 | 83754 | 464 | 81250 | 521 | 78886 | 582 | 76646 | 646 | |
| 22 | 86370 | 411 | 83711 | 465 | 81210 | 522 | 78847 | 583 | 76610 | 647 | 8 |
| | 86324 | 411 | 83668 | 466 | 81169 | 523 | 78809 | 584 | 76574 | 648 | |
| 23 | 86278 | 412 | 83626 | 467 | 81129 | 524 | 78771 | 585 | 76537 | 649 | 7 |
| | 86233 | 413 | 83583 | 467 | 81088 | 525 | 78733 | 586 | 76501 | 650 | |
| 24 | 86187 | 414 | 83540 | 468 | 81048 | 526 | 78694 | 587 | 76465 | 652 | 6 |
| | 86142 | 415 | 83497 | 469 | 81008 | 527 | 78656 | 588 | 76429 | 653 | |
| 25 | 86096 | 416 | 83455 | 470 | 80967 | 528 | 78618 | 589 | 76393 | 654 | 5 |
| | 86051 | 417 | 83412 | 471 | 80927 | 529 | 78580 | 590 | 76357 | 655 | |
| 26 | 86006 | 418 | 83369 | 472 | 80887 | 530 | 78542 | 591 | 76320 | 656 | 4 |
| | 85960 | 418 | 83327 | 473 | 80847 | 531 | 78504 | 592 | 76284 | 657 | |
| 27 | 85915 | 419 | 83284 | 474 | 80807 | 532 | 78466 | 593 | 76248 | 658 | 3 |
| | 85870 | 420 | 83242 | 475 | 80767 | 533 | 78428 | 594 | 76212 | 659 | |
| 28 | 85825 | 421 | 83199 | 476 | 80727 | 534 | 78390 | 595 | 76176 | 660 | 2 |
| | 85779 | 422 | 83157 | 477 | 80687 | 535 | 78352 | 597 | 76141 | 661 | |
| 29 | 85734 | 423 | 83114 | 478 | 80647 | 536 | 78315 | 598 | 76105 | 663 | 1 |
| | 85689 | 424 | 83072 | 479 | 80607 | 537 | 78277 | 599 | 76069 | 664 | |
| 30 | 85644 | 425 | 83030 | 480 | 80567 | 538 | 78239 | 600 | 76033 | 665 | 0 |
| | A | B | A | B | A | B | A | B | A | B | |
| ′ | 172° 00′ | | 171° 30′ | | 171° 00′ | | 170° 30′ | | 170° 00′ | | ′ |

# TABLE 35

## The Ageton Method

When Meridian Angle is Greater Than 90° Take "K" From Bottom of Table

| ′ | 10° 00′ | | 10° 30′ | | 11° 00′ | | 11° 30′ | | 12° 00′ | | ′ |
|---|---|---|---|---|---|---|---|---|---|---|---|
| | A | B | A | B | A | B | A | B | A | B | |
| 0 | 76033 | 665 | 73937 | 733 | 71940 | 805 | 70034 | 881 | 68212 | 960 | 30 |
| | 75997 | 666 | 73903 | 735 | 71908 | 807 | 70003 | 882 | 68182 | 961 | |
| 1 | 75961 | 667 | 73869 | 736 | 71875 | 808 | 69972 | 883 | 68153 | 962 | 29 |
| | 75926 | 668 | 73835 | 737 | 71843 | 809 | 69941 | 885 | 68123 | 964 | |
| 2 | 75890 | 669 | 73801 | 738 | 71810 | 810 | 69910 | 886 | 68093 | 965 | 28 |
| | 75854 | 670 | 73767 | 739 | 71778 | 811 | 69879 | 887 | 68064 | 966 | |
| 3 | 75819 | 672 | 73733 | 740 | 71746 | 813 | 69849 | 888 | 68034 | 968 | 27 |
| | 75783 | 673 | 73699 | 742 | 71713 | 814 | 69818 | 890 | 68005 | 969 | |
| 4 | 75747 | 674 | 73665 | 743 | 71681 | 815 | 69787 | 891 | 67975 | 970 | 26 |
| | 75712 | 675 | 73631 | 744 | 71649 | 816 | 69756 | 892 | 67945 | 972 | |
| 5 | 75676 | 676 | 73597 | 745 | 71616 | 818 | 69725 | 894 | 67916 | 973 | 25 |
| | 75641 | 677 | 73563 | 746 | 71584 | 819 | 69694 | 895 | 67886 | 974 | |
| 6 | 75605 | 678 | 73530 | 747 | 71552 | 820 | 69664 | 896 | 67857 | 976 | 24 |
| | 75570 | 679 | 73496 | 749 | 71520 | 821 | 69633 | 897 | 67828 | 977 | |
| 7 | 75534 | 680 | 73462 | 750 | 71488 | 823 | 69602 | 899 | 67798 | 978 | 23 |
| | 75499 | 682 | 73429 | 751 | 71455 | 824 | 69571 | 900 | 67769 | 980 | |
| 8 | 75464 | 683 | 73395 | 752 | 71423 | 825 | 69541 | 901 | 67739 | 981 | 22 |
| | 75428 | 684 | 73361 | 753 | 71391 | 826 | 69510 | 903 | 67710 | 982 | |
| 9 | 75393 | 685 | 73328 | 755 | 71359 | 828 | 69479 | 904 | 67681 | 984 | 21 |
| | 75358 | 686 | 73294 | 756 | 71327 | 829 | 69449 | 905 | 67651 | 985 | |
| 10 | 75322 | 687 | 73260 | 757 | 71295 | 830 | 69418 | 907 | 67622 | 987 | 20 |
| | 75287 | 688 | 73227 | 758 | 71263 | 831 | 69387 | 908 | 67593 | 988 | |
| 11 | 75252 | 690 | 73193 | 759 | 71231 | 833 | 69357 | 909 | 67563 | 989 | 19 |
| | 75217 | 691 | 73160 | 761 | 71199 | 834 | 69326 | 910 | 67534 | 991 | |
| 12 | 75182 | 692 | 73127 | 762 | 71167 | 835 | 69296 | 912 | 67505 | 992 | 18 |
| | 75147 | 693 | 73093 | 763 | 71135 | 836 | 69265 | 913 | 67476 | 993 | |
| 13 | 75112 | 694 | 73060 | 764 | 71104 | 838 | 69235 | 914 | 67447 | 995 | 17 |
| | 75077 | 695 | 73026 | 765 | 71072 | 839 | 69204 | 916 | 67417 | 996 | |
| 14 | 75042 | 696 | 72993 | 766 | 71040 | 840 | 69174 | 917 | 67388 | 997 | 16 |
| | 75007 | 698 | 72960 | 768 | 71008 | 841 | 69144 | 918 | 67359 | 999 | |
| 15 | 74972 | 699 | 72926 | 769 | 70976 | 843 | 69113 | 920 | 67330 | 1000 | 15 |
| | 74937 | 700 | 72893 | 770 | 70945 | 844 | 69083 | 921 | 67301 | 1002 | |
| 16 | 74902 | 701 | 72860 | 771 | 70913 | 845 | 69053 | 922 | 67272 | 1003 | 14 |
| | 74867 | 702 | 72827 | 772 | 70881 | 846 | 69022 | 924 | 67243 | 1004 | |
| 17 | 74832 | 703 | 72794 | 774 | 70850 | 848 | 68992 | 925 | 67214 | 1006 | 13 |
| | 74797 | 704 | 72760 | 775 | 70818 | 849 | 68962 | 926 | 67185 | 1007 | |
| 18 | 74763 | 706 | 72727 | 776 | 70786 | 850 | 68931 | 928 | 67156 | 1008 | 12 |
| | 74728 | 707 | 72694 | 777 | 70755 | 851 | 68901 | 929 | 67127 | 1010 | |
| 19 | 74693 | 708 | 72661 | 779 | 70723 | 853 | 68871 | 930 | 67098 | 1011 | 11 |
| | 74659 | 709 | 72628 | 780 | 70692 | 854 | 68841 | 932 | 67069 | 1013 | |
| 20 | 74624 | 710 | 72595 | 781 | 70660 | 855 | 68811 | 933 | 67040 | 1014 | 10 |
| | 74589 | 711 | 72562 | 782 | 70629 | 856 | 68781 | 934 | 67011 | 1015 | |
| 21 | 74555 | 712 | 72529 | 783 | 70597 | 858 | 68750 | 935 | 66982 | 1017 | 9 |
| | 74520 | 714 | 72496 | 785 | 70566 | 859 | 68720 | 937 | 66953 | 1018 | |
| 22 | 74486 | 715 | 72463 | 786 | 70534 | 860 | 68690 | 938 | 66925 | 1020 | 8 |
| | 74451 | 716 | 72430 | 787 | 70503 | 862 | 68660 | 939 | 66896 | 1021 | |
| 23 | 74417 | 717 | 72397 | 788 | 70471 | 863 | 68630 | 941 | 66867 | 1022 | 7 |
| | 74382 | 718 | 72365 | 790 | 70440 | 864 | 68600 | 942 | 66838 | 1024 | |
| 24 | 74348 | 719 | 72332 | 791 | 70409 | 865 | 68570 | 943 | 66810 | 1025 | 6 |
| | 74313 | 721 | 72299 | 792 | 70377 | 867 | 68540 | 945 | 66781 | 1026 | |
| 25 | 74279 | 722 | 72266 | 793 | 70346 | 868 | 68510 | 946 | 66752 | 1028 | 5 |
| | 74245 | 723 | 72234 | 794 | 70315 | 869 | 68480 | 947 | 66724 | 1029 | |
| 26 | 74210 | 724 | 72201 | 796 | 70284 | 870 | 68450 | 949 | 66695 | 1031 | 4 |
| | 74176 | 725 | 72168 | 797 | 70252 | 872 | 68421 | 950 | 66666 | 1032 | |
| 27 | 74142 | 726 | 72135 | 798 | 70221 | 873 | 68391 | 951 | 66638 | 1033 | 3 |
| | 74107 | 728 | 72103 | 799 | 70190 | 874 | 68361 | 953 | 66609 | 1035 | |
| 28 | 74073 | 729 | 72070 | 800 | 70159 | 876 | 68331 | 954 | 66580 | 1036 | 2 |
| | 74039 | 730 | 72038 | 802 | 70128 | 877 | 68301 | 955 | 66552 | 1038 | |
| 29 | 74005 | 731 | 72005 | 803 | 70097 | 878 | 68272 | 957 | 66523 | 1039 | 1 |
| | 73971 | 732 | 71973 | 804 | 70065 | 879 | 68242 | 958 | 66495 | 1040 | |
| 30 | 73937 | 733 | 71940 | 805 | 70034 | 881 | 68212 | 960 | 66466 | 1042 | 0 |
| | A | B | A | B | A | B | A | B | A | B | |
| ′ | 169° 30′ | | 169° 00′ | | 168° 30′ | | 168° 00′ | | 167° 30′ | | ′ |

## TABLE 35

### The Ageton Method

**Always Take "Z" From Bottom of Table, Except When "K" is Same Name and Greater Than Latitude, in Which Case Take "Z" From Top of Table**

| ′ | 12° 30′ | | 13° 00′ | | 13° 30′ | | 14° 00′ | | 14° 30′ | | ′ |
|---|---|---|---|---|---|---|---|---|---|---|---|
| | A | B | A | B | A | B | A | B | A | B | |
| 0 | 66466 | 1042 | 64791 | 1128 | 63181 | 1217 | 61632 | 1310 | 60140 | 1406 | 30 |
| | 66438 | 1043 | 64764 | 1129 | 63155 | 1218 | 61607 | 1311 | 60116 | 1407 | |
| 1 | 66409 | 1045 | 64736 | 1130 | 63129 | 1220 | 61582 | 1313 | 60091 | 1409 | 29 |
| | 66381 | 1046 | 64709 | 1132 | 63103 | 1221 | 61556 | 1314 | 60067 | 1411 | |
| 2 | 66352 | 1047 | 64682 | 1133 | 63076 | 1223 | 61531 | 1316 | 60042 | 1412 | 28 |
| | 66324 | 1049 | 64655 | 1135 | 63050 | 1224 | 61506 | 1317 | 60018 | 1414 | |
| 3 | 66296 | 1050 | 64627 | 1136 | 63024 | 1226 | 61481 | 1319 | 59994 | 1416 | 27 |
| | 66267 | 1052 | 64600 | 1138 | 62998 | 1227 | 61455 | 1321 | 59969 | 1417 | |
| 4 | 66239 | 1053 | 64573 | 1139 | 62971 | 1229 | 61430 | 1322 | 59945 | 1419 | 26 |
| | 66211 | 1054 | 64546 | 1141 | 62945 | 1230 | 61405 | 1324 | 59921 | 1421 | |
| 5 | 66182 | 1056 | 64518 | 1142 | 62919 | 1232 | 61380 | 1325 | 59896 | 1422 | 25 |
| | 66154 | 1057 | 64491 | 1144 | 62893 | 1234 | 61355 | 1327 | 59872 | 1424 | |
| 6 | 66126 | 1059 | 64464 | 1145 | 62867 | 1235 | 61330 | 1329 | 59848 | 1425 | 24 |
| | 66098 | 1060 | 64437 | 1147 | 62841 | 1237 | 61304 | 1330 | 59824 | 1427 | |
| 7 | 66069 | 1061 | 64410 | 1148 | 62815 | 1238 | 61279 | 1332 | 59800 | 1429 | 23 |
| | 66041 | 1063 | 64383 | 1150 | 62789 | 1240 | 61254 | 1333 | 59775 | 1430 | |
| 8 | 66013 | 1064 | 64356 | 1151 | 62763 | 1241 | 61229 | 1335 | 59751 | 1432 | 22 |
| | 65985 | 1066 | 64329 | 1152 | 62737 | 1243 | 61204 | 1336 | 59727 | 1434 | |
| 9 | 65957 | 1067 | 64302 | 1154 | 62711 | 1244 | 61179 | 1338 | 59703 | 1435 | 21 |
| | 65928 | 1069 | 64275 | 1155 | 62685 | 1246 | 61154 | 1340 | 59679 | 1437 | |
| 10 | 65900 | 1070 | 64248 | 1157 | 62659 | 1247 | 61129 | 1341 | 59654 | 1439 | 20 |
| | 65872 | 1071 | 64221 | 1158 | 62633 | 1249 | 61104 | 1343 | 59630 | 1440 | |
| 11 | 65844 | 1073 | 64194 | 1160 | 62607 | 1250 | 61079 | 1344 | 59606 | 1442 | 19 |
| | 65816 | 1074 | 64167 | 1161 | 62581 | 1252 | 61054 | 1346 | 59582 | 1444 | |
| 12 | 65788 | 1076 | 64140 | 1163 | 62555 | 1253 | 61029 | 1348 | 59558 | 1445 | 18 |
| | 65760 | 1077 | 64113 | 1164 | 62529 | 1255 | 61004 | 1349 | 59534 | 1447 | |
| 13 | 65732 | 1079 | 64086 | 1166 | 62503 | 1257 | 60979 | 1351 | 59510 | 1449 | 17 |
| | 65704 | 1080 | 64059 | 1167 | 62477 | 1258 | 60954 | 1352 | 59486 | 1450 | |
| 14 | 65676 | 1081 | 64032 | 1169 | 62451 | 1260 | 60929 | 1354 | 59462 | 1452 | 16 |
| | 65648 | 1083 | 64005 | 1170 | 62425 | 1261 | 60904 | 1356 | 59438 | 1454 | |
| 15 | 65620 | 1084 | 63978 | 1172 | 62400 | 1263 | 60879 | 1357 | 59414 | 1455 | 15 |
| | 65592 | 1086 | 63952 | 1173 | 62374 | 1264 | 60855 | 1359 | 59390 | 1457 | |
| 16 | 65564 | 1087 | 63925 | 1175 | 62348 | 1266 | 60830 | 1360 | 59366 | 1459 | 14 |
| | 65537 | 1089 | 63898 | 1176 | 62322 | 1267 | 60805 | 1362 | 59342 | 1460 | |
| 17 | 65509 | 1090 | 63871 | 1178 | 62296 | 1269 | 60780 | 1364 | 59318 | 1462 | 13 |
| | 65481 | 1091 | 63845 | 1179 | 62271 | 1270 | 60755 | 1365 | 59294 | 1464 | |
| 18 | 65453 | 1093 | 63818 | 1181 | 62245 | 1272 | 60730 | 1367 | 59270 | 1465 | 12 |
| | 65425 | 1094 | 63791 | 1182 | 62219 | 1274 | 60706 | 1368 | 59246 | 1467 | |
| 19 | 65398 | 1096 | 63764 | 1184 | 62194 | 1275 | 60681 | 1370 | 59222 | 1469 | 11 |
| | 65370 | 1097 | 63738 | 1185 | 62168 | 1277 | 60656 | 1372 | 59198 | 1470 | |
| 20 | 65342 | 1099 | 63711 | 1187 | 62142 | 1278 | 60631 | 1373 | 59175 | 1472 | 10 |
| | 65314 | 1100 | 63684 | 1188 | 62117 | 1280 | 60607 | 1375 | 59151 | 1474 | |
| 21 | 65287 | 1101 | 63658 | 1190 | 62091 | 1281 | 60582 | 1377 | 59127 | 1475 | 9 |
| | 65259 | 1103 | 63631 | 1191 | 62065 | 1283 | 60557 | 1378 | 59103 | 1477 | |
| 22 | 65231 | 1104 | 63605 | 1193 | 62040 | 1284 | 60533 | 1380 | 59079 | 1479 | 8 |
| | 65204 | 1106 | 63578 | 1194 | 62014 | 1286 | 60508 | 1381 | 59055 | 1480 | |
| 23 | 65176 | 1107 | 63551 | 1196 | 61989 | 1288 | 60483 | 1383 | 59032 | 1482 | 7 |
| | 65148 | 1109 | 63525 | 1197 | 61963 | 1289 | 60459 | 1385 | 59008 | 1484 | |
| 24 | 65121 | 1110 | 63498 | 1199 | 61938 | 1291 | 60434 | 1386 | 58984 | 1485 | 6 |
| | 65093 | 1112 | 63472 | 1200 | 61912 | 1292 | 60410 | 1388 | 58960 | 1487 | |
| 25 | 65066 | 1113 | 63445 | 1202 | 61887 | 1294 | 60385 | 1390 | 58937 | 1489 | 5 |
| | 65038 | 1114 | 63419 | 1203 | 61861 | 1295 | 60360 | 1391 | 58913 | 1490 | |
| 26 | 65011 | 1116 | 63392 | 1205 | 61836 | 1297 | 60336 | 1393 | 58889 | 1492 | 4 |
| | 64983 | 1117 | 63366 | 1206 | 61810 | 1299 | 60311 | 1394 | 58866 | 1494 | |
| 27 | 64956 | 1119 | 63340 | 1208 | 61785 | 1300 | 60287 | 1396 | 58842 | 1495 | 3 |
| | 64928 | 1120 | 63313 | 1209 | 61759 | 1301 | 60262 | 1398 | 58818 | 1497 | |
| 28 | 64901 | 1122 | 63287 | 1211 | 61734 | 1303 | 60238 | 1399 | 58795 | 1499 | 2 |
| | 64873 | 1123 | 63260 | 1212 | 61709 | 1305 | 60213 | 1401 | 58771 | 1500 | |
| 29 | 64846 | 1125 | 63234 | 1214 | 61683 | 1306 | 60189 | 1403 | 58748 | 1502 | 1 |
| | 64819 | 1126 | 63208 | 1215 | 61658 | 1308 | 60164 | 1404 | 58724 | 1504 | |
| 30 | 64791 | 1128 | 63181 | 1217 | 61632 | 1310 | 60140 | 1406 | 58700 | 1506 | 0 |
| | A | B | A | B | A | B | A | B | A | B | |
| ′ | 167° 00′ | | 166° 30′ | | 166° 00′ | | 165° 30′ | | 165° 00′ | | ′ |

# TABLE 35

## The Ageton Method

When Meridian Angle is Greater Than 90° Take "K" From Bottom of Table

| ′ | 15° 00′ | | 15° 30′ | | 16° 00′ | | 16° 30′ | | 17° 00′ | | ′ |
|---|---|---|---|---|---|---|---|---|---|---|---|
| ′ | A | B | A | B | A | B | A | B | A | B | ′ |
| 0 | 58700 | **1506** | 57310 | **1609** | 55966 | **1716** | 54666 | **1826** | 53406 | **1940** | 30 |
| | 58677 | **1507** | 57287 | **1611** | 55944 | **1718** | 54644 | **1828** | 53386 | **1942** | |
| 1 | 58653 | **1509** | 57265 | **1612** | 55922 | **1719** | 54623 | **1830** | 53365 | **1944** | 29 |
| | 58630 | **1511** | 57242 | **1614** | 55900 | **1721** | 54602 | **1832** | 53344 | **1946** | |
| 2 | 58606 | **1512** | 57219 | **1616** | 55878 | **1723** | 54581 | **1834** | 53324 | **1948** | 28 |
| | 58583 | **1514** | 57196 | **1618** | 55856 | **1725** | 54559 | **1836** | 53303 | **1950** | |
| 3 | 58559 | **1516** | 57174 | **1619** | 55834 | **1727** | 54538 | **1837** | 53283 | **1952** | 27 |
| | 58536 | **1517** | 57151 | **1621** | 55812 | **1728** | 54517 | **1839** | 53262 | **1954** | |
| 4 | 58512 | **1519** | 57128 | **1623** | 55790 | **1730** | 54496 | **1841** | 53241 | **1956** | 26 |
| | 58489 | **1521** | 57106 | **1625** | 55768 | **1732** | 54474 | **1843** | 53221 | **1958** | |
| 5 | 58465 | **1523** | 57083 | **1627** | 55746 | **1734** | 54453 | **1845** | 53200 | **1960** | 25 |
| | 58442 | **1524** | 57060 | **1628** | 55725 | **1736** | 54432 | **1847** | 53180 | **1962** | |
| 6 | 58418 | **1526** | 57038 | **1630** | 55703 | **1738** | 54411 | **1849** | 53159 | **1964** | 24 |
| | 58395 | **1528** | 57015 | **1632** | 55681 | **1739** | 54390 | **1851** | 53139 | **1966** | |
| 7 | 58372 | **1529** | 56992 | **1634** | 55659 | **1741** | 54368 | **1853** | 53118 | **1967** | 23 |
| | 58348 | **1531** | 56970 | **1635** | 55637 | **1743** | 54347 | **1854** | 53098 | **1969** | |
| 8 | 58325 | **1533** | 56947 | **1637** | 55615 | **1745** | 54326 | **1856** | 53077 | **1971** | 22 |
| | 58302 | **1534** | 56925 | **1639** | 55593 | **1747** | 54305 | **1858** | 53057 | **1973** | |
| 9 | 58278 | **1536** | 56902 | **1641** | 55572 | **1749** | 54284 | **1860** | 53036 | **1975** | 21 |
| | 58255 | **1538** | 56880 | **1642** | 55550 | **1750** | 54263 | **1862** | 53016 | **1977** | |
| 10 | 58232 | **1540** | 56857 | **1644** | 55528 | **1752** | 54242 | **1864** | 52995 | **1979** | 20 |
| | 58208 | **1541** | 56835 | **1646** | 55506 | **1754** | 54220 | **1866** | 52975 | **1981** | |
| 11 | 58185 | **1543** | 56812 | **1648** | 55484 | **1756** | 54199 | **1868** | 52954 | **1983** | 19 |
| | 58162 | **1545** | 56790 | **1649** | 55463 | **1758** | 54178 | **1870** | 52934 | **1985** | |
| 12 | 58138 | **1546** | 56767 | **1651** | 55441 | **1760** | 54157 | **1871** | 52914 | **1987** | 18 |
| | 58115 | **1548** | 56745 | **1653** | 55419 | **1761** | 54136 | **1873** | 52893 | **1989** | |
| 13 | 58092 | **1550** | 56722 | **1655** | 55397 | **1763** | 54115 | **1875** | 52873 | **1991** | 17 |
| | 58069 | **1552** | 56700 | **1657** | 55376 | **1765** | 54094 | **1877** | 52852 | **1993** | |
| 14 | 58046 | **1553** | 56677 | **1658** | 55354 | **1767** | 54073 | **1879** | 52832 | **1995** | 16 |
| | 58022 | **1555** | 56655 | **1660** | 55332 | **1769** | 54052 | **1881** | 52812 | **1997** | |
| 15 | 57999 | **1557** | 56632 | **1662** | 55311 | **1771** | 54031 | **1883** | 52791 | **1999** | 15 |
| | 57976 | **1559** | 56610 | **1664** | 55289 | **1772** | 54010 | **1885** | 52771 | **2001** | |
| 16 | 57953 | **1560** | 56588 | **1665** | 55267 | **1774** | 53989 | **1887** | 52751 | **2003** | 14 |
| | 57930 | **1562** | 56565 | **1667** | 55246 | **1776** | 53968 | **1889** | 52730 | **2005** | |
| 17 | 57907 | **1564** | 56543 | **1669** | 55224 | **1778** | 53947 | **1890** | 52710 | **2007** | 13 |
| | 57884 | **1565** | 56521 | **1671** | 55202 | **1780** | 53926 | **1892** | 52690 | **2009** | |
| 18 | 57860 | **1567** | 56498 | **1673** | 55181 | **1782** | 53905 | **1894** | 52670 | **2010** | 12 |
| | 57837 | **1569** | 56476 | **1674** | 55159 | **1783** | 53884 | **1896** | 52649 | **2012** | |
| 19 | 57814 | **1571** | 56454 | **1676** | 55138 | **1785** | 53864 | **1898** | 52629 | **2014** | 11 |
| | 57791 | **1572** | 56431 | **1678** | 55116 | **1787** | 53843 | **1900** | 52609 | **2016** | |
| 20 | 57768 | **1574** | 56409 | **1680** | 55095 | **1789** | 53822 | **1902** | 52588 | **2018** | 10 |
| | 57745 | **1576** | 56387 | **1682** | 55073 | **1791** | 53801 | **1904** | 52568 | **2020** | |
| 21 | 57722 | **1578** | 56365 | **1683** | 55051 | **1793** | 53780 | **1906** | 52548 | **2022** | 9 |
| | 57699 | **1579** | 56342 | **1685** | 55030 | **1795** | 53759 | **1908** | 52528 | **2024** | |
| 22 | 57676 | **1581** | 56320 | **1687** | 55008 | **1796** | 53738 | **1910** | 52508 | **2026** | 8 |
| | 57653 | **1583** | 56298 | **1689** | 54987 | **1798** | 53718 | **1911** | 52487 | **2028** | |
| 23 | 57630 | **1584** | 56276 | **1691** | 54965 | **1800** | 53697 | **1913** | 52467 | **2030** | 7 |
| | 57607 | **1586** | 56254 | **1692** | 54944 | **1802** | 53676 | **1915** | 52447 | **2032** | |
| 24 | 57584 | **1588** | 56231 | **1694** | 54922 | **1804** | 53655 | **1917** | 52427 | **2034** | 6 |
| | 57561 | **1590** | 56209 | **1696** | 54901 | **1806** | 53634 | **1919** | 52407 | **2036** | |
| 25 | 57538 | **1591** | 56187 | **1698** | 54880 | **1808** | 53614 | **1921** | 52387 | **2038** | 5 |
| | 57516 | **1593** | 56165 | **1700** | 54858 | **1809** | 53593 | **1923** | 52366 | **2040** | |
| 26 | 57493 | **1595** | 56143 | **1701** | 54837 | **1811** | 53572 | **1925** | 52346 | **2042** | 4 |
| | 57470 | **1597** | 56121 | **1703** | 54815 | **1813** | 53551 | **1927** | 52326 | **2044** | |
| 27 | 57447 | **1598** | 56099 | **1705** | 54794 | **1815** | 53531 | **1929** | 52306 | **2046** | 3 |
| | 57424 | **1600** | 56076 | **1707** | 54773 | **1817** | 53510 | **1931** | 52286 | **2048** | |
| 28 | 57401 | **1602** | 56054 | **1709** | 54751 | **1819** | 53489 | **1933** | 52266 | **2050** | 2 |
| | 57378 | **1604** | 56032 | **1710** | 54730 | **1821** | 53468 | **1935** | 52246 | **2052** | |
| 29 | 57356 | **1605** | 56010 | **1712** | 54708 | **1823** | 53448 | **1936** | 52226 | **2054** | 1 |
| | 57333 | **1607** | 55988 | **1714** | 54687 | **1824** | 53427 | **1938** | 52206 | **2056** | |
| 30 | 57310 | **1609** | 55966 | **1716** | 54666 | **1826** | 53406 | **1940** | 52186 | **2058** | 0 |
| | A | B | A | B | A | B | A | B | A | B | |
| ′ | 164° 30′ | | 164° 00′ | | 163° 30′ | | 163° 00′ | | 162° 30′ | | ′ |

# TABLE 35

## The Ageton Method

**Always Take "Z" From Bottom of Table, Except When "K" is Same Name and Greater Than Latitude, in Which Case Take "Z" From Top of Table**

| | 17° 30′ | | 18° 00′ | | 18° 30′ | | 19° 00′ | | 19° 30′ | | |
|---|---|---|---|---|---|---|---|---|---|---|---|
| ′ | A | B | A | B | A | B | A | B | A | B | ′ |
| 0 | 52186 | 2058 | 51002 | 2179 | 49852 | 2304 | 48736 | 2433 | 47650 | 2565 | 30 |
| | 52166 | 2060 | 50982 | 2181 | 49833 | 2306 | 48717 | 2435 | 47633 | 2568 | |
| 1 | 52146 | 2062 | 50963 | 2183 | 49815 | 2309 | 48699 | 2437 | 47615 | 2570 | 29 |
| | 52126 | 2064 | 50943 | 2185 | 49796 | 2311 | 48681 | 2439 | 47597 | 2572 | |
| 2 | 52106 | 2066 | 50924 | 2188 | 49777 | 2313 | 48662 | 2442 | 47579 | 2574 | 28 |
| | 52086 | 2068 | 50905 | 2190 | 49758 | 2315 | 48644 | 2444 | 47561 | 2576 | |
| 3 | 52066 | 2070 | 50885 | 2192 | 49739 | 2317 | 48626 | 2446 | 47544 | 2579 | 27 |
| | 52046 | 2072 | 50866 | 2194 | 49720 | 2319 | 48608 | 2448 | 47526 | 2581 | |
| 4 | 52026 | 2074 | 50846 | 2196 | 49702 | 2321 | 48589 | 2450 | 47508 | 2583 | 26 |
| | 52006 | 2076 | 50827 | 2198 | 49683 | 2323 | 48571 | 2453 | 47490 | 2585 | |
| 5 | 51986 | 2078 | 50808 | 2200 | 49664 | 2325 | 48553 | 2455 | 47472 | 2588 | 25 |
| | 51966 | 2080 | 50788 | 2202 | 49645 | 2328 | 48534 | 2457 | 47455 | 2590 | |
| 6 | 51946 | 2082 | 50769 | 2204 | 49626 | 2330 | 48516 | 2459 | 47437 | 2592 | 24 |
| | 51926 | 2084 | 50750 | 2206 | 49608 | 2332 | 48498 | 2461 | 47419 | 2594 | |
| 7 | 51906 | 2086 | 50730 | 2208 | 49589 | 2334 | 48480 | 2463 | 47402 | 2597 | 23 |
| | 51886 | 2088 | 50711 | 2210 | 49570 | 2336 | 48462 | 2466 | 47384 | 2599 | |
| 8 | 51867 | 2090 | 50692 | 2212 | 49551 | 2338 | 48443 | 2468 | 47366 | 2601 | 22 |
| | 51847 | 2092 | 50673 | 2214 | 49533 | 2340 | 48425 | 2470 | 47348 | 2603 | |
| 9 | 51827 | 2094 | 50653 | 2216 | 49514 | 2343 | 48407 | 2472 | 47331 | 2606 | 21 |
| | 51807 | 2096 | 50634 | 2218 | 49495 | 2345 | 48389 | 2474 | 47313 | 2608 | |
| 10 | 51787 | 2098 | 50615 | 2221 | 49477 | 2347 | 48371 | 2477 | 47295 | 2610 | 20 |
| | 51767 | 2100 | 50596 | 2223 | 49458 | 2349 | 48352 | 2479 | 47278 | 2613 | |
| 11 | 51747 | 2102 | 50576 | 2225 | 49439 | 2351 | 48334 | 2481 | 47260 | 2615 | 19 |
| | 51728 | 2104 | 50557 | 2227 | 49421 | 2353 | 48316 | 2483 | 47242 | 2617 | |
| 12 | 51708 | 2106 | 50538 | 2229 | 49402 | 2355 | 48298 | 2485 | 47225 | 2619 | 18 |
| | 51688 | 2108 | 50519 | 2231 | 49383 | 2357 | 48280 | 2488 | 47207 | 2622 | |
| 13 | 51668 | 2110 | 50499 | 2233 | 49365 | 2360 | 48262 | 2490 | 47189 | 2624 | 17 |
| | 51649 | 2112 | 50480 | 2235 | 49346 | 2362 | 48244 | 2492 | 47172 | 2626 | |
| 14 | 51629 | 2114 | 50461 | 2237 | 49327 | 2364 | 48225 | 2494 | 47154 | 2628 | 16 |
| | 51609 | 2116 | 50442 | 2239 | 49309 | 2366 | 48207 | 2496 | 47137 | 2631 | |
| 15 | 51589 | 2118 | 50423 | 2241 | 49290 | 2368 | 48189 | 2499 | 47119 | 2633 | 15 |
| | 51570 | 2120 | 50404 | 2243 | 49271 | 2370 | 48171 | 2501 | 47101 | 2635 | |
| 16 | 51550 | 2122 | 50385 | 2246 | 49253 | 2372 | 48153 | 2503 | 47084 | 2637 | 14 |
| | 51530 | 2124 | 50365 | 2248 | 49234 | 2375 | 48135 | 2505 | 47066 | 2640 | |
| 17 | 51510 | 2126 | 50346 | 2250 | 49216 | 2377 | 48117 | 2507 | 47049 | 2642 | 13 |
| | 51491 | 2128 | 50327 | 2252 | 49197 | 2379 | 48099 | 2510 | 47031 | 2644 | |
| 18 | 51471 | 2130 | 50308 | 2254 | 49179 | 2381 | 48081 | 2512 | 47014 | 2646 | 12 |
| | 51451 | 2132 | 50289 | 2256 | 49160 | 2383 | 48063 | 2514 | 46996 | 2649 | |
| 19 | 51432 | 2134 | 50270 | 2258 | 49141 | 2385 | 48045 | 2516 | 46978 | 2651 | 11 |
| | 51412 | 2136 | 50251 | 2260 | 49123 | 2387 | 48027 | 2519 | 46961 | 2653 | |
| 20 | 51392 | 2138 | 50232 | 2262 | 49104 | 2390 | 48009 | 2521 | 46943 | 2656 | 10 |
| | 51373 | 2141 | 50213 | 2264 | 49086 | 2392 | 47991 | 2523 | 46926 | 2658 | |
| 21 | 51353 | 2143 | 50194 | 2266 | 49067 | 2394 | 47973 | 2525 | 46908 | 2660 | 9 |
| | 51334 | 2145 | 50175 | 2269 | 49049 | 2396 | 47955 | 2527 | 46891 | 2662 | |
| 22 | 51314 | 2147 | 50156 | 2271 | 49030 | 2398 | 47937 | 2530 | 46873 | 2665 | 8 |
| | 51294 | 2149 | 50137 | 2273 | 49012 | 2400 | 47919 | 2532 | 46856 | 2667 | |
| 23 | 51275 | 2151 | 50117 | 2275 | 48993 | 2403 | 47901 | 2534 | 46839 | 2669 | 7 |
| | 51255 | 2153 | 50098 | 2277 | 48975 | 2405 | 47883 | 2536 | 46821 | 2672 | |
| 24 | 51236 | 2155 | 50080 | 2279 | 48957 | 2407 | 47865 | 2539 | 46804 | 2674 | 6 |
| | 51216 | 2157 | 50061 | 2281 | 48938 | 2409 | 47847 | 2541 | 46786 | 2676 | |
| 25 | 51197 | 2159 | 50042 | 2283 | 48920 | 2411 | 47829 | 2543 | 46769 | 2678 | 5 |
| | 51177 | 2161 | 50023 | 2285 | 48901 | 2413 | 47811 | 2545 | 46751 | 2681 | |
| 26 | 51158 | 2163 | 50004 | 2287 | 48883 | 2416 | 47793 | 2547 | 46734 | 2683 | 4 |
| | 51138 | 2165 | 49985 | 2290 | 48864 | 2418 | 47775 | 2550 | 46716 | 2685 | |
| 27 | 51119 | 2167 | 49966 | 2292 | 48846 | 2420 | 47758 | 2552 | 46699 | 2688 | 3 |
| | 51099 | 2169 | 49947 | 2294 | 48828 | 2422 | 47740 | 2554 | 46682 | 2690 | |
| 28 | 51080 | 2171 | 49928 | 2296 | 48809 | 2424 | 47722 | 2556 | 46664 | 2692 | 2 |
| | 51060 | 2173 | 49909 | 2298 | 48791 | 2426 | 47704 | 2559 | 46647 | 2694 | |
| 29 | 51041 | 2175 | 49890 | 2300 | 48772 | 2429 | 47686 | 2561 | 46630 | 2697 | 1 |
| | 51021 | 2177 | 49871 | 2302 | 48754 | 2431 | 47668 | 2563 | 46612 | 2699 | |
| 30 | 51002 | 2179 | 49852 | 2304 | 48736 | 2433 | 47650 | 2565 | 46595 | 2701 | 0 |
| | A | B | A | B | A | B | A | B | A | B | |
| ′ | 162° 00′ | | 161° 30′ | | 161° 00′ | | 160° 30′ | | 160° 00′ | | ′ |

# TABLE 35

## The Ageton Method

**When Meridian Angle is Greater Than 90° Take "K" From Bottom of Table**

| ′ | 20° 00′ A | 20° 00′ B | 20° 30′ A | 20° 30′ B | 21° 00′ A | 21° 00′ B | 21° 30′ A | 21° 30′ B | 22° 00′ A | 22° 00′ B | ′ |
|---|---|---|---|---|---|---|---|---|---|---|---|
| 0 | 46595 | 2701 | 45567 | 2841 | 44567 | 2985 | 43592 | 3132 | 42642 | 3283 | 30 |
| | 46577 | 2704 | 45551 | 2844 | 44551 | 2988 | 43576 | 3135 | 42627 | 3286 | |
| 1 | 46560 | 2706 | 45534 | 2846 | 44534 | 2990 | 43560 | 3137 | 42611 | 3288 | 29 |
| | 46543 | 2708 | 45517 | 2848 | 44518 | 2992 | 43544 | 3140 | 42596 | 3291 | |
| 2 | 46525 | 2711 | 45500 | 2851 | 44501 | 2994 | 43528 | 3142 | 42580 | 3294 | 28 |
| | 46508 | 2713 | 45483 | 2853 | 44485 | 2997 | 43512 | 3145 | 42564 | 3296 | |
| 3 | 46491 | 2715 | 45466 | 2855 | 44468 | 2999 | 43496 | 3147 | 42549 | 3299 | 27 |
| | 46473 | 2717 | 45449 | 2858 | 44452 | 3002 | 43480 | 3150 | 42533 | 3301 | |
| 4 | 46456 | 2720 | 45433 | 2860 | 44436 | 3004 | 43464 | 3152 | 42518 | 3304 | 26 |
| | 46439 | 2722 | 45416 | 2862 | 44419 | 3007 | 43448 | 3155 | 42502 | 3306 | |
| 5 | 46422 | 2724 | 45399 | 2865 | 44403 | 3009 | 43432 | 3157 | 42486 | 3309 | 25 |
| | 46404 | 2727 | 45382 | 2867 | 44386 | 3012 | 43416 | 3160 | 42471 | 3312 | |
| 6 | 46387 | 2729 | 45365 | 2870 | 44370 | 3014 | 43400 | 3162 | 42455 | 3314 | 24 |
| | 46370 | 2731 | 45348 | 2872 | 44354 | 3016 | 43385 | 3165 | 42440 | 3317 | |
| 7 | 46353 | 2734 | 45332 | 2874 | 44337 | 3019 | 43369 | 3167 | 42424 | 3319 | 23 |
| | 46335 | 2736 | 45315 | 2877 | 44321 | 3021 | 43353 | 3170 | 42409 | 3322 | |
| 8 | 46318 | 2738 | 45298 | 2879 | 44305 | 3024 | 43337 | 3172 | 42393 | 3324 | 22 |
| | 46301 | 2741 | 45281 | 2881 | 44288 | 3026 | 43321 | 3175 | 42378 | 3327 | |
| 9 | 46284 | 2743 | 45265 | 2884 | 44272 | 3029 | 43305 | 3177 | 42362 | 3329 | 21 |
| | 46266 | 2745 | 45248 | 2886 | 44256 | 3031 | 43289 | 3180 | 42347 | 3332 | |
| 10 | 46249 | 2748 | 45231 | 2889 | 44239 | 3033 | 43273 | 3182 | 42331 | 3335 | 20 |
| | 46232 | 2750 | 45214 | 2891 | 44223 | 3036 | 43257 | 3185 | 42316 | 3337 | |
| 11 | 46215 | 2752 | 45198 | 2893 | 44207 | 3038 | 43241 | 3187 | 42300 | 3340 | 19 |
| | 46198 | 2755 | 45181 | 2896 | 44190 | 3041 | 43225 | 3190 | 42285 | 3342 | |
| 12 | 46181 | 2757 | 45164 | 2898 | 44174 | 3043 | 43210 | 3192 | 42269 | 3345 | 18 |
| | 46163 | 2759 | 45147 | 2901 | 44158 | 3046 | 43194 | 3195 | 42254 | 3347 | |
| 13 | 46146 | 2761 | 45131 | 2903 | 44142 | 3048 | 43178 | 3197 | 42238 | 3350 | 17 |
| | 46129 | 2764 | 45114 | 2905 | 44125 | 3051 | 43162 | 3200 | 42223 | 3353 | |
| 14 | 46112 | 2766 | 45097 | 2908 | 44109 | 3053 | 43146 | 3202 | 42207 | 3355 | 16 |
| | 46095 | 2768 | 45081 | 2910 | 44093 | 3056 | 43130 | 3205 | 42192 | 3358 | |
| 15 | 46078 | 2771 | 45064 | 2913 | 44077 | 3058 | 43114 | 3207 | 42176 | 3360 | 15 |
| | 46061 | 2773 | 45047 | 2915 | 44060 | 3060 | 43099 | 3210 | 42161 | 3363 | |
| 16 | 46043 | 2775 | 45031 | 2917 | 44044 | 3063 | 43083 | 3212 | 42145 | 3366 | 14 |
| | 46026 | 2778 | 45014 | 2920 | 44028 | 3065 | 43067 | 3215 | 42130 | 3368 | |
| 17 | 46009 | 2780 | 44997 | 2922 | 44012 | 3068 | 43051 | 3217 | 42115 | 3371 | 13 |
| | 45992 | 2782 | 44981 | 2924 | 43995 | 3070 | 43035 | 3220 | 42099 | 3373 | |
| 18 | 45975 | 2785 | 44964 | 2927 | 43979 | 3073 | 43020 | 3222 | 42084 | 3376 | 12 |
| | 45958 | 2787 | 44947 | 2929 | 43963 | 3075 | 43004 | 3225 | 42068 | 3379 | |
| 19 | 45941 | 2789 | 44931 | 2932 | 43947 | 3078 | 42988 | 3227 | 42053 | 3381 | 11 |
| | 45924 | 2792 | 44914 | 2934 | 43931 | 3080 | 42972 | 3230 | 42038 | 3384 | |
| 20 | 45907 | 2794 | 44898 | 2936 | 43914 | 3083 | 42956 | 3233 | 42022 | 3386 | 10 |
| | 45890 | 2797 | 44881 | 2939 | 43898 | 3085 | 42941 | 3235 | 42007 | 3389 | |
| 21 | 45873 | 2799 | 44864 | 2941 | 43882 | 3088 | 42925 | 3238 | 41991 | 3391 | 9 |
| | 45856 | 2801 | 44848 | 2944 | 43866 | 3090 | 42909 | 3240 | 41976 | 3394 | |
| 22 | 45839 | 2804 | 44831 | 2946 | 43850 | 3092 | 42893 | 3243 | 41961 | 3397 | 8 |
| | 45822 | 2806 | 44815 | 2949 | 43834 | 3095 | 42878 | 3245 | 41945 | 3399 | |
| 23 | 45805 | 2808 | 44798 | 2951 | 43818 | 3097 | 42862 | 3248 | 41930 | 3402 | 7 |
| | 45788 | 2811 | 44782 | 2953 | 43801 | 3100 | 42846 | 3250 | 41915 | 3404 | |
| 24 | 45771 | 2813 | 44765 | 2956 | 43785 | 3102 | 42830 | 3253 | 41899 | 3407 | 6 |
| | 45754 | 2815 | 44748 | 2958 | 43769 | 3105 | 42815 | 3255 | 41884 | 3410 | |
| 25 | 45737 | 2818 | 44732 | 2961 | 43753 | 3107 | 42799 | 3258 | 41869 | 3412 | 5 |
| | 45720 | 2820 | 44715 | 2963 | 43737 | 3110 | 42783 | 3260 | 41853 | 3415 | |
| 26 | 45703 | 2822 | 44699 | 2965 | 43721 | 3112 | 42768 | 3263 | 41838 | 3418 | 4 |
| | 45686 | 2825 | 44682 | 2968 | 43705 | 3115 | 42752 | 3266 | 41823 | 3420 | |
| 27 | 45669 | 2827 | 44666 | 2970 | 43689 | 3117 | 42736 | 3268 | 41808 | 3423 | 3 |
| | 45652 | 2829 | 44649 | 2973 | 43673 | 3120 | 42721 | 3271 | 41792 | 3425 | |
| 28 | 45635 | 2832 | 44633 | 2975 | 43657 | 3122 | 42705 | 3273 | 41777 | 3428 | 2 |
| | 45618 | 2834 | 44616 | 2978 | 43641 | 3125 | 42689 | 3276 | 41762 | 3431 | |
| 29 | 45601 | 2836 | 44600 | 2980 | 43624 | 3127 | 42674 | 3278 | 41746 | 3433 | 1 |
| | 45584 | 2839 | 44583 | 2982 | 43608 | 3130 | 42658 | 3281 | 41731 | 3436 | |
| 30 | 45567 | 2841 | 44567 | 2985 | 43592 | 3132 | 42642 | 3283 | 41716 | 3438 | 0 |
| | A | B | A | B | A | B | A | B | A | B | |
| ′ | 159° 30′ | | 159° 00′ | | 158° 30′ | | 158° 00′ | | 157° 30′ | | ′ |

## TABLE 35

### The Ageton Method

**Always Take "Z" From Bottom of Table, Except When "K" is Same Name and Greater Than Latitude, in Which Case Take "Z" From Top of Table**

| ′ | 22° 30′ | | 23° 00′ | | 23° 30′ | | 24° 00′ | | 24° 30′ | | ′ |
|---|---|---|---|---|---|---|---|---|---|---|---|
| | A | B | A | B | A | B | A | B | A | B | |
| 0 | 41716 | 3438 | 40812 | 3597 | 39930 | 3760 | 39069 | 3927 | 38227 | 4098 | 30 |
| | 41701 | 3441 | 40797 | 3600 | 39915 | 3763 | 39054 | 3930 | 38213 | 4101 | |
| 1 | 41685 | 3444 | 40782 | 3603 | 39901 | 3766 | 39040 | 3932 | 38200 | 4103 | 29 |
| | 41670 | 3446 | 40768 | 3605 | 39886 | 3768 | 39026 | 3935 | 38186 | 4106 | |
| 2 | 41655 | 3449 | 40753 | 3608 | 39872 | 3771 | 39012 | 3938 | 38172 | 4109 | 28 |
| | 41640 | 3452 | 40738 | 3611 | 39857 | 3774 | 38998 | 3941 | 38158 | 4112 | |
| 3 | 41625 | 3454 | 40723 | 3613 | 39843 | 3777 | 38984 | 3944 | 38144 | 4115 | 27 |
| | 41609 | 3457 | 40708 | 3616 | 39828 | 3779 | 38969 | 3947 | 38130 | 4118 | |
| 4 | 41594 | 3459 | 40693 | 3619 | 39814 | 3782 | 38955 | 3949 | 38117 | 4121 | 26 |
| | 41579 | 3462 | 40678 | 3622 | 39799 | 3785 | 38941 | 3952 | 38103 | 4124 | |
| 5 | 41564 | 3465 | 40664 | 3624 | 39785 | 3788 | 38927 | 3955 | 38089 | 4127 | 25 |
| | 41549 | 3467 | 40649 | 3627 | 39771 | 3790 | 38913 | 3958 | 38075 | 4129 | |
| 6 | 41533 | 3470 | 40634 | 3630 | 39756 | 3793 | 38899 | 3961 | 38061 | 4132 | 24 |
| | 41518 | 3473 | 40619 | 3632 | 39742 | 3796 | 38885 | 3964 | 38048 | 4135 | |
| 7 | 41503 | 3475 | 40604 | 3635 | 39727 | 3799 | 38871 | 3966 | 38034 | 4138 | 23 |
| | 41488 | 3478 | 40590 | 3638 | 39713 | 3801 | 38856 | 3969 | 38020 | 4141 | |
| 8 | 41473 | 3480 | 40575 | 3640 | 39698 | 3804 | 38842 | 3972 | 38006 | 4144 | 22 |
| | 41458 | 3483 | 40560 | 3643 | 39684 | 3807 | 38828 | 3975 | 37992 | 4147 | |
| 9 | 41443 | 3486 | 40545 | 3646 | 39669 | 3810 | 38814 | 3978 | 37979 | 4150 | 21 |
| | 41427 | 3488 | 40530 | 3648 | 39655 | 3813 | 38800 | 3981 | 37965 | 4153 | |
| 10 | 41412 | 3491 | 40516 | 3651 | 39641 | 3815 | 38786 | 3983 | 37951 | 4155 | 20 |
| | 41397 | 3494 | 40501 | 3654 | 39626 | 3818 | 38772 | 3986 | 37937 | 4158 | |
| 11 | 41382 | 3496 | 40486 | 3657 | 39612 | 3821 | 38758 | 3989 | 37924 | 4161 | 19 |
| | 41367 | 3499 | 40471 | 3659 | 39597 | 3824 | 38744 | 3992 | 37910 | 4164 | |
| 12 | 41352 | 3502 | 40457 | 3662 | 39583 | 3826 | 38730 | 3995 | 37896 | 4167 | 18 |
| | 41337 | 3504 | 40442 | 3665 | 39569 | 3829 | 38716 | 3998 | 37882 | 4170 | |
| 13 | 41322 | 3507 | 40427 | 3667 | 39554 | 3832 | 38702 | 4000 | 37869 | 4173 | 17 |
| | 41307 | 3509 | 40413 | 3670 | 39540 | 3835 | 38688 | 4003 | 37855 | 4176 | |
| 14 | 41291 | 3512 | 40398 | 3673 | 39525 | 3838 | 38674 | 4006 | 37841 | 4179 | 16 |
| | 41276 | 3515 | 40383 | 3676 | 39511 | 3840 | 38660 | 4009 | 37828 | 4182 | |
| 15 | 41261 | 3517 | 40368 | 3678 | 39497 | 3843 | 38645 | 4012 | 37814 | 4185 | 15 |
| | 41246 | 3520 | 40354 | 3681 | 39482 | 3846 | 38631 | 4015 | 37800 | 4187 | |
| 16 | 41231 | 3523 | 40339 | 3684 | 39468 | 3849 | 38617 | 4017 | 37786 | 4190 | 14 |
| | 41216 | 3525 | 40324 | 3686 | 39454 | 3851 | 38603 | 4020 | 37773 | 4193 | |
| 17 | 41201 | 3528 | 40310 | 3689 | 39439 | 3854 | 38589 | 4023 | 37759 | 4196 | 13 |
| | 41186 | 3531 | 40295 | 3692 | 39425 | 3857 | 38575 | 4026 | 37745 | 4199 | |
| 18 | 41171 | 3533 | 40280 | 3695 | 39411 | 3860 | 38561 | 4029 | 37732 | 4202 | 12 |
| | 41156 | 3536 | 40266 | 3697 | 39396 | 3863 | 38547 | 4032 | 37718 | 4205 | |
| 19 | 41141 | 3539 | 40251 | 3700 | 39382 | 3865 | 38533 | 4035 | 37704 | 4208 | 11 |
| | 41126 | 3541 | 40236 | 3703 | 39368 | 3868 | 38520 | 4037 | 37691 | 4211 | |
| 20 | 41111 | 3544 | 40222 | 3705 | 39353 | 3871 | 38506 | 4040 | 37677 | 4214 | 10 |
| | 41096 | 3547 | 40207 | 3708 | 39339 | 3874 | 38492 | 4043 | 37663 | 4217 | |
| 21 | 41081 | 3549 | 40192 | 3711 | 39325 | 3876 | 38478 | 4046 | 37650 | 4220 | 9 |
| | 41066 | 3552 | 40178 | 3714 | 39311 | 3879 | 38464 | 4049 | 37636 | 4222 | |
| 22 | 41051 | 3555 | 40163 | 3716 | 39296 | 3882 | 38450 | 4052 | 37623 | 4225 | 8 |
| | 41036 | 3557 | 40149 | 3719 | 39282 | 3885 | 38436 | 4055 | 37609 | 4228 | |
| 23 | 41021 | 3560 | 40134 | 3722 | 39268 | 3888 | 38422 | 4057 | 37595 | 4231 | 7 |
| | 41006 | 3563 | 40119 | 3725 | 39254 | 3890 | 38408 | 4060 | 37582 | 4234 | |
| 24 | 40991 | 3565 | 40105 | 3727 | 39239 | 3893 | 38394 | 4063 | 37568 | 4237 | 6 |
| | 40976 | 3568 | 40090 | 3730 | 39225 | 3896 | 38380 | 4066 | 37554 | 4240 | |
| 25 | 40961 | 3571 | 40076 | 3733 | 39211 | 3899 | 38366 | 4069 | 37541 | 4243 | 5 |
| | 40946 | 3573 | 40061 | 3735 | 39197 | 3902 | 38352 | 4072 | 37527 | 4246 | |
| 26 | 40931 | 3576 | 40046 | 3738 | 39182 | 3904 | 38338 | 4075 | 37514 | 4249 | 4 |
| | 40916 | 3579 | 40032 | 3741 | 39168 | 3907 | 38324 | 4078 | 37500 | 4252 | |
| 27 | 40902 | 3581 | 40017 | 3744 | 39154 | 3910 | 38311 | 4080 | 37486 | 4255 | 3 |
| | 40887 | 3584 | 40003 | 3746 | 39140 | 3913 | 38297 | 4083 | 37473 | 4258 | |
| 28 | 40872 | 3587 | 39988 | 3749 | 39125 | 3916 | 38283 | 4086 | 37459 | 4261 | 2 |
| | 40857 | 3589 | 39974 | 3752 | 39111 | 3918 | 38269 | 4089 | 37446 | 4264 | |
| 29 | 40842 | 3592 | 39959 | 3755 | 39097 | 3921 | 38255 | 4092 | 37432 | 4266 | 1 |
| | 40827 | 3595 | 39945 | 3757 | 39083 | 3924 | 38241 | 4095 | 37419 | 4269 | |
| 30 | 40812 | 3597 | 39930 | 3760 | 39069 | 3927 | 38227 | 4098 | 37405 | 4272 | 0 |
| | A | B | A | B | A | B | A | B | A | B | |
| ′ | 157° 00′ | | 156° 30′ | | 156° 00′ | | 155° 30′ | | 155° 00′ | | ′ |

## TABLE 35

### The Ageton Method

**When Meridian Angle is Greater Than 90° Take "K" From Bottom of Table**

| ′ | 25° 00′ | | 25° 30′ | | 26° 00′ | | 26° 30′ | | 27° 00′ | | ′ |
|---|---|---|---|---|---|---|---|---|---|---|---|
| ′ | A | B | A | B | A | B | A | B | A | B | ′ |
| 0 | 37405 | **4272** | 36602 | **4451** | 35816 | **4634** | 35047 | **4821** | 34295 | **5012** | 30 |
| | 37392 | **4275** | 36588 | **4454** | 35803 | **4637** | 35035 | **4824** | 34283 | **5015** | |
| 1 | 37378 | **4278** | 36575 | **4457** | 35790 | **4640** | 35022 | **4827** | 34270 | **5018** | 29 |
| | 37365 | **4281** | 36562 | **4460** | 35777 | **4643** | 35009 | **4830** | 34258 | **5022** | |
| 2 | 37351 | **4284** | 36549 | **4463** | 35764 | **4646** | 34997 | **4833** | 34246 | **5025** | 28 |
| | 37337 | **4287** | 36535 | **4466** | 35751 | **4649** | 34984 | **4837** | 34233 | **5028** | |
| 3 | 37324 | **4290** | 36522 | **4469** | 35738 | **4651** | 34971 | **4840** | 34221 | **5031** | 27 |
| | 37310 | **4293** | 36509 | **4472** | 35725 | **4656** | 34959 | **4843** | 34209 | **5034** | |
| 4 | 37297 | **4296** | 36496 | **4475** | 35712 | **4659** | 34946 | **4846** | 34196 | **5038** | 26 |
| | 37283 | **4299** | 36483 | **4478** | 35699 | **4662** | 34933 | **4849** | 34184 | **5041** | |
| 5 | 37270 | **4302** | 36469 | **4481** | 35686 | **4665** | 34921 | **4852** | 34172 | **5044** | 25 |
| | 37256 | **4305** | 36456 | **4484** | 35674 | **4668** | 34908 | **4856** | 34159 | **5047** | |
| 6 | 37243 | **4308** | 36443 | **4487** | 35661 | **4671** | 34896 | **4859** | 34147 | **5051** | 24 |
| | 37229 | **4311** | 36430 | **4490** | 35648 | **4674** | 34883 | **4862** | 34134 | **5054** | |
| 7 | 37216 | **4314** | 36417 | **4493** | 35635 | **4677** | 34870 | **4865** | 34122 | **5057** | 23 |
| | 37203 | **4317** | 36403 | **4496** | 35622 | **4680** | 34858 | **4868** | 34110 | **5060** | |
| 8 | 37189 | **4320** | 36390 | **4499** | 35609 | **4683** | 34845 | **4871** | 34097 | **5064** | 22 |
| | 37176 | **4323** | 36377 | **4503** | 35596 | **4686** | 34832 | **4875** | 34085 | **5067** | |
| 9 | 37162 | **4326** | 36364 | **4506** | 35583 | **4690** | 34820 | **4878** | 34073 | **5070** | 21 |
| | 37149 | **4329** | 36351 | **4509** | 35571 | **4693** | 34807 | **4881** | 34061 | **5073** | |
| 10 | 37135 | **4332** | 36338 | **4512** | 35558 | **4696** | 34795 | **4884** | 34048 | **5076** | 20 |
| | 37122 | **4334** | 36325 | **4515** | 35545 | **4699** | 34782 | **4887** | 34036 | **5080** | |
| 11 | 37108 | **4337** | 36311 | **4518** | 35532 | **4702** | 34770 | **4890** | 34024 | **5083** | 19 |
| | 37095 | **4340** | 36298 | **4521** | 35519 | **4705** | 34757 | **4894** | 34011 | **5086** | |
| 12 | 37081 | **4343** | 36285 | **4524** | 35506 | **4708** | 34744 | **4897** | 33999 | **5089** | 18 |
| | 37068 | **4346** | 36272 | **4527** | 35493 | **4711** | 34732 | **4900** | 33987 | **5093** | |
| 13 | 37055 | **4349** | 36259 | **4530** | 35481 | **4714** | 34719 | **4903** | 33974 | **5096** | 17 |
| | 37041 | **4352** | 36246 | **4533** | 35468 | **4718** | 34707 | **4906** | 33962 | **5099** | |
| 14 | 37028 | **4355** | 36233 | **4536** | 35455 | **4721** | 34694 | **4910** | 33950 | **5102** | 16 |
| | 37014 | **4358** | 36220 | **4539** | 35442 | **4724** | 34682 | **4913** | 33938 | **5106** | |
| 15 | 37001 | **4361** | 36206 | **4542** | 35429 | **4727** | 34669 | **4916** | 33925 | **5109** | 15 |
| | 36988 | **4364** | 36193 | **4545** | 35417 | **4730** | 34657 | **4919** | 33913 | **5112** | |
| 16 | 36974 | **4367** | 36180 | **4548** | 35404 | **4733** | 34644 | **4922** | 33901 | **5115** | 14 |
| | 36961 | **4370** | 36167 | **4551** | 35391 | **4736** | 34632 | **4925** | 33889 | **5119** | |
| 17 | 36948 | **4373** | 36154 | **4554** | 35378 | **4739** | 34619 | **4929** | 33876 | **5122** | 13 |
| | 36934 | **4376** | 36141 | **4557** | 35365 | **4742** | 34607 | **4932** | 33864 | **5125** | |
| 18 | 36921 | **4379** | 36128 | **4560** | 35353 | **4746** | 34594 | **4935** | 33852 | **5128** | 12 |
| | 36907 | **4382** | 36115 | **4563** | 35340 | **4749** | 34582 | **4938** | 33840 | **5132** | |
| 19 | 36894 | **4385** | 36102 | **4566** | 35327 | **4752** | 34569 | **4941** | 33827 | **5135** | 11 |
| | 36881 | **4388** | 36089 | **4569** | 35314 | **4755** | 34557 | **4945** | 33815 | **5138** | |
| 20 | 36867 | **4391** | 36076 | **4573** | 35302 | **4758** | 34544 | **4948** | 33803 | **5142** | 10 |
| | 36854 | **4394** | 36063 | **4576** | 35289 | **4761** | 34532 | **4951** | 33791 | **5145** | |
| 21 | 36841 | **4397** | 36050 | **4579** | 35276 | **4764** | 34519 | **4954** | 33779 | **5148** | 9 |
| | 36827 | **4400** | 36037 | **4582** | 35263 | **4769** | 34507 | **4957** | 33766 | **5151** | |
| 22 | 36814 | **4403** | 36024 | **4585** | 35251 | **4771** | 34494 | **4961** | 33754 | **5155** | 8 |
| | 36801 | **4406** | 36011 | **4588** | 35238 | **4774** | 34482 | **4964** | 33742 | **5158** | |
| 23 | 36787 | **4409** | 35998 | **4591** | 35225 | **4777** | 34469 | **4967** | 33730 | **5161** | 7 |
| | 36774 | **4412** | 35985 | **4594** | 35212 | **4780** | 34457 | **4970** | 33717 | **5164** | |
| 24 | 36761 | **4415** | 35972 | **4597** | 35200 | **4783** | 34445 | **4973** | 33705 | **5168** | 6 |
| | 36747 | **4418** | 35959 | **4600** | 35187 | **4786** | 34432 | **4977** | 33693 | **5171** | |
| 25 | 36734 | **4421** | 35946 | **4603** | 35174 | **4789** | 34420 | **4980** | 33681 | **5174** | 5 |
| | 36721 | **4424** | 35933 | **4606** | 35161 | **4793** | 34407 | **4983** | 33669 | **5178** | |
| 26 | 36708 | **4427** | 35920 | **4609** | 35149 | **4796** | 34395 | **4986** | 33657 | **5181** | 4 |
| | 36694 | **4430** | 35907 | **4612** | 35136 | **4799** | 34382 | **4989** | 33644 | **5184** | |
| 27 | 36681 | **4433** | 35894 | **4615** | 35123 | **4802** | 34370 | **4993** | 33632 | **5187** | 3 |
| | 36668 | **4436** | 35881 | **4619** | 35111 | **4805** | 34357 | **4996** | 33620 | **5191** | |
| 28 | 36655 | **4439** | 35868 | **4622** | 35098 | **4808** | 34345 | **4999** | 33608 | **5194** | 2 |
| | 36641 | **4442** | 35855 | **4625** | 35085 | **4811** | 34332 | **5002** | 33596 | **5197** | |
| 29 | 36628 | **4445** | 35842 | **4628** | 35073 | **4815** | 34320 | **5005** | 33584 | **5200** | 1 |
| | 36615 | **4448** | 35829 | **4631** | 35060 | **4818** | 34308 | **5009** | 33572 | **5204** | |
| 30 | 36602 | **4451** | 35816 | **4634** | 35047 | **4821** | 34295 | **5012** | 33559 | **5207** | 0 |
| | A | B | A | B | A | B | A | B | A | B | |
| ′ | 154° 30′ | | 154° 00′ | | 153° 30′ | | 153° 00′ | | 152° 30′ | | ′ |

## TABLE 35

### The Ageton Method

**Always Take "Z" From Bottom of Table, Except When "K" is Same Name and Greater Than Latitude, in Which Case Take "Z" From Top of Table**

| ′ | 27° 30′ | | 28° 00′ | | 28° 30′ | | 29° 00′ | | 29° 30′ | | ′ |
|---|---|---|---|---|---|---|---|---|---|---|---|
| | A | B | A | B | A | B | A | B | A | B | |
| 0 | 33559 | 5207 | 32839 | 5406 | 32134 | 5610 | 31443 | 5818 | 30766 | 6030 | 30 |
| | 33547 | 5210 | 32827 | 5410 | 32122 | 5614 | 31431 | 5822 | 30755 | 6034 | |
| 1 | 33535 | 5214 | 32815 | 5413 | 32110 | 5617 | 31420 | 5825 | 30744 | 6038 | 29 |
| | 33523 | 5217 | 32803 | 5417 | 32099 | 5620 | 31409 | 5829 | 30733 | 6041 | |
| 2 | 33511 | 5220 | 32792 | 5420 | 32087 | 5624 | 31397 | 5832 | 30721 | 6045 | 28 |
| | 33499 | 5224 | 32780 | 5423 | 32076 | 5627 | 31386 | 5836 | 30710 | 6048 | |
| 3 | 33487 | 5227 | 32768 | 5426 | 32064 | 5631 | 31375 | 5839 | 30699 | 6052 | 27 |
| | 33475 | 5230 | 32756 | 5430 | 32052 | 5634 | 31363 | 5843 | 30688 | 6055 | |
| 4 | 33462 | 5233 | 32744 | 5433 | 32041 | 5638 | 31352 | 5846 | 30677 | 6059 | 26 |
| | 33450 | 5237 | 32732 | 5437 | 32029 | 5641 | 31340 | 5850 | 30666 | 6062 | |
| 5 | 33438 | 5240 | 32720 | 5440 | 32018 | 5645 | 31329 | 5853 | 30655 | 6066 | 25 |
| | 33426 | 5243 | 32709 | 5443 | 32006 | 5648 | 31318 | 5857 | 30643 | 6070 | |
| 6 | 33414 | 5247 | 32697 | 5447 | 31994 | 5651 | 31306 | 5860 | 30632 | 6073 | 24 |
| | 33402 | 5250 | 32685 | 5450 | 31983 | 5655 | 31295 | 5864 | 30621 | 6077 | |
| 7 | 33390 | 5253 | 32673 | 5454 | 31971 | 5658 | 31284 | 5867 | 30610 | 6080 | 23 |
| | 33378 | 5257 | 32661 | 5457 | 31960 | 5662 | 31272 | 5871 | 30599 | 6084 | |
| 8 | 33366 | 5260 | 32649 | 5460 | 31948 | 5665 | 31261 | 5874 | 30588 | 6088 | 22 |
| | 33354 | 5263 | 32638 | 5464 | 31936 | 5669 | 31250 | 5878 | 30577 | 6091 | |
| 9 | 33342 | 5266 | 32625 | 5467 | 31925 | 5672 | 31238 | 5881 | 30566 | 6095 | 21 |
| | 33330 | 5270 | 32614 | 5470 | 31913 | 5675 | 31227 | 5885 | 30555 | 6098 | |
| 10 | 33318 | 5273 | 32602 | 5474 | 31902 | 5679 | 31216 | 5888 | 30544 | 6102 | 20 |
| | 33306 | 5277 | 32590 | 5477 | 31890 | 5682 | 31204 | 5892 | 30532 | 6106 | |
| 11 | 33293 | 5280 | 32579 | 5481 | 31879 | 5686 | 31193 | 5895 | 30521 | 6109 | 19 |
| | 33281 | 5283 | 32567 | 5484 | 31867 | 5689 | 31182 | 5899 | 30510 | 6113 | |
| 12 | 33269 | 5287 | 32555 | 5487 | 31856 | 5693 | 31170 | 5902 | 30499 | 6116 | 18 |
| | 33257 | 5290 | 32543 | 5491 | 31844 | 5696 | 31159 | 5906 | 30488 | 6120 | |
| 13 | 33245 | 5293 | 32532 | 5494 | 31833 | 5700 | 31148 | 5909 | 30477 | 6124 | 17 |
| | 33233 | 5296 | 32520 | 5498 | 31821 | 5703 | 31137 | 5913 | 30466 | 6127 | |
| 14 | 33221 | 5300 | 32508 | 5501 | 31809 | 5707 | 31125 | 5917 | 30455 | 6131 | 16 |
| | 33209 | 5303 | 32496 | 5504 | 31798 | 5710 | 31114 | 5920 | 30444 | 6134 | |
| 15 | 33197 | 5306 | 32484 | 5508 | 31786 | 5714 | 31103 | 5924 | 30433 | 6138 | 15 |
| | 33185 | 5310 | 32473 | 5511 | 31775 | 5717 | 31091 | 5927 | 30422 | 6142 | |
| 16 | 33173 | 5313 | 32461 | 5515 | 31763 | 5720 | 31080 | 5931 | 30411 | 6145 | 14 |
| | 33161 | 5316 | 32449 | 5518 | 31752 | 5724 | 31069 | 5934 | 30400 | 6149 | |
| 17 | 33149 | 5320 | 32438 | 5521 | 31740 | 5727 | 31058 | 5938 | 30389 | 6153 | 13 |
| | 33137 | 5323 | 32426 | 5525 | 31729 | 5731 | 31046 | 5941 | 30378 | 6156 | |
| 18 | 33125 | 5326 | 32414 | 5528 | 31717 | 5734 | 31035 | 5945 | 30367 | 6160 | 12 |
| | 33113 | 5330 | 32402 | 5532 | 31706 | 5738 | 31024 | 5948 | 30356 | 6163 | |
| 19 | 33101 | 5333 | 32391 | 5535 | 31694 | 5741 | 31013 | 5952 | 30345 | 6167 | 11 |
| | 33089 | 5336 | 32379 | 5538 | 31683 | 5745 | 31001 | 5955 | 30334 | 6171 | |
| 20 | 33077 | 5340 | 32367 | 5542 | 31672 | 5748 | 30990 | 5959 | 30322 | 6174 | 10 |
| | 33065 | 5343 | 32355 | 5545 | 31660 | 5752 | 30979 | 5963 | 30311 | 6178 | |
| 21 | 33054 | 5346 | 32344 | 5549 | 31648 | 5755 | 30968 | 5966 | 30300 | 6181 | 9 |
| | 33042 | 5350 | 32332 | 5552 | 31637 | 5759 | 30956 | 5970 | 30289 | 6185 | |
| 22 | 33030 | 5353 | 32320 | 5555 | 31626 | 5762 | 30945 | 5973 | 30278 | 6189 | 8 |
| | 33018 | 5356 | 32309 | 5559 | 31614 | 5766 | 30934 | 5977 | 30267 | 6192 | |
| 23 | 33006 | 5360 | 32297 | 5562 | 31603 | 5769 | 30923 | 5980 | 30256 | 6196 | 7 |
| | 32994 | 5363 | 32285 | 5566 | 31591 | 5773 | 30912 | 5984 | 30245 | 6200 | |
| 24 | 32982 | 5366 | 32274 | 5569 | 31580 | 5776 | 30900 | 5988 | 30235 | 6203 | 6 |
| | 32970 | 5370 | 32262 | 5572 | 31569 | 5780 | 30889 | 5991 | 30224 | 6207 | |
| 25 | 32958 | 5373 | 32250 | 5576 | 31557 | 5783 | 30878 | 5995 | 30213 | 6210 | 5 |
| | 32946 | 5376 | 32239 | 5579 | 31546 | 5787 | 30867 | 5998 | 30202 | 6214 | |
| 26 | 32934 | 5380 | 32227 | 5583 | 31534 | 5790 | 30856 | 6002 | 30191 | 6218 | 4 |
| | 32922 | 5383 | 32215 | 5586 | 31523 | 5794 | 30844 | 6005 | 30180 | 6221 | |
| 27 | 32910 | 5386 | 32204 | 5590 | 31511 | 5797 | 30833 | 6009 | 30169 | 6225 | 3 |
| | 32898 | 5390 | 32192 | 5593 | 31500 | 5801 | 30822 | 6012 | 30158 | 6229 | |
| 28 | 32887 | 5393 | 32180 | 5596 | 31488 | 5804 | 30811 | 6016 | 30147 | 6232 | 2 |
| | 32875 | 5396 | 32169 | 5600 | 31477 | 5808 | 30800 | 6020 | 30136 | 6236 | |
| 29 | 32863 | 5400 | 32157 | 5603 | 31466 | 5811 | 30788 | 6023 | 30125 | 6240 | 1 |
| | 32851 | 5403 | 32145 | 5607 | 31454 | 5815 | 30777 | 6027 | 30114 | 6243 | |
| 30 | 32839 | 5406 | 32134 | 5610 | 31443 | 5818 | 30766 | 6030 | 30103 | 6247 | 0 |
| | A | B | A | B | A | B | A | B | A | B | |
| ′ | 152° 00′ | | 151° 30′ | | 151° 00′ | | 150° 30′ | | 150° 00′ | | ′ |

# TABLE 35

## The Ageton Method

When Meridian Angle is Greater Than 90° Take "K" From Bottom of Table

| ′ | 30° 00′ A | 30° 00′ B | 30° 30′ A | 30° 30′ B | 31° 00′ A | 31° 00′ B | 31° 30′ A | 31° 30′ B | 32° 00′ A | 32° 00′ B | ′ |
|---|---|---|---|---|---|---|---|---|---|---|---|
| 0 | 30103 | **6247** | 29453 | **6468** | 28816 | **6693** | 28191 | **6923** | 27579 | **7158** | 30 |
| | 30092 | **6251** | 29442 | **6472** | 28806 | **6697** | 28181 | **6927** | 27569 | **7162** | |
| 1 | 30081 | **6254** | 29432 | **6475** | 28795 | **6701** | 28171 | **6931** | 27559 | **7166** | 29 |
| | 30070 | **6258** | 29421 | **6479** | 28785 | **6705** | 28161 | **6935** | 27549 | **7170** | |
| 2 | 30059 | **6262** | 29410 | **6483** | 28774 | **6709** | 28150 | **6939** | 27539 | **7174** | 28 |
| | 30048 | **6265** | 29399 | **6487** | 28763 | **6712** | 28140 | **6943** | 27528 | **7178** | |
| 3 | 30037 | **6269** | 29389 | **6490** | 28753 | **6716** | 28130 | **6947** | 27518 | **7182** | 27 |
| | 30026 | **6273** | 29378 | **6494** | 28743 | **6720** | 28119 | **6951** | 27508 | **7186** | |
| 4 | 30016 | **6276** | 29367 | **6498** | 28732 | **6724** | 28109 | **6954** | 27498 | **7190** | 26 |
| | 30005 | **6280** | 29357 | **6501** | 28722 | **6728** | 28099 | **6958** | 27488 | **7193** | |
| 5 | 29994 | **6284** | 29346 | **6505** | 28711 | **6731** | 28089 | **6962** | 27478 | **7197** | 25 |
| | 29983 | **6287** | 29335 | **6509** | 28701 | **6735** | 28078 | **6966** | 27468 | **7201** | |
| 6 | 29972 | **6291** | 29325 | **6513** | 28690 | **6739** | 28068 | **6970** | 27458 | **7205** | 24 |
| | 29961 | **6294** | 29314 | **6516** | 28680 | **6743** | 28058 | **6974** | 27448 | **7209** | |
| 7 | 29950 | **6298** | 29303 | **6520** | 28669 | **6747** | 28047 | **6978** | 27438 | **7213** | 23 |
| | 29939 | **6302** | 29293 | **6524** | 28659 | **6750** | 28037 | **6982** | 27428 | **7217** | |
| 8 | 29928 | **6305** | 29282 | **6528** | 28648 | **6754** | 28027 | **6985** | 27418 | **7221** | 22 |
| | 29917 | **6309** | 29271 | **6531** | 28638 | **6758** | 28017 | **6989** | 27408 | **7225** | |
| 9 | 29907 | **6313** | 29261 | **6535** | 28627 | **6762** | 28006 | **6993** | 27398 | **7229** | 21 |
| | 29896 | **6316** | 29250 | **6539** | 28617 | **6766** | 27996 | **6997** | 27387 | **7233** | |
| 10 | 29885 | **6320** | 29239 | **6543** | 28606 | **6770** | 27986 | **7001** | 27377 | **7237** | 20 |
| | 29874 | **6324** | 29229 | **6546** | 28596 | **6773** | 27976 | **7005** | 27367 | **7241** | |
| 11 | 29863 | **6328** | 29218 | **6550** | 28586 | **6777** | 27965 | **7009** | 27357 | **7245** | 19 |
| | 29852 | **6331** | 29207 | **6554** | 28575 | **6781** | 27955 | **7013** | 27347 | **7249** | |
| 12 | 29841 | **6335** | 29197 | **6558** | 28565 | **6785** | 27945 | **7017** | 27337 | **7253** | 18 |
| | 29831 | **6339** | 29186 | **6561** | 28554 | **6789** | 27935 | **7021** | 27327 | **7257** | |
| 13 | 29820 | **6342** | 29175 | **6565** | 28544 | **6793** | 27925 | **7024** | 27317 | **7261** | 17 |
| | 29809 | **6346** | 29165 | **6569** | 28533 | **6796** | 27914 | **7028** | 27307 | **7265** | |
| 14 | 29798 | **6350** | 29154 | **6573** | 28523 | **6800** | 27904 | **7032** | 27297 | **7269** | 16 |
| | 29787 | **6353** | 29144 | **6576** | 28513 | **6804** | 27894 | **7036** | 27287 | **7273** | |
| 15 | 29776 | **6357** | 29133 | **6580** | 28502 | **6808** | 27884 | **7040** | 27277 | **7277** | 15 |
| | 29766 | **6361** | 29122 | **6584** | 28492 | **6812** | 27874 | **7044** | 27267 | **7281** | |
| 16 | 29755 | **6364** | 29112 | **6588** | 28481 | **6815** | 27863 | **7048** | 27257 | **7285** | 14 |
| | 29744 | **6368** | 29101 | **6591** | 28471 | **6819** | 27853 | **7052** | 27247 | **7289** | |
| 17 | 29733 | **6372** | 29091 | **6595** | 28461 | **6823** | 27843 | **7056** | 27237 | **7293** | 13 |
| | 29722 | **6375** | 29080 | **6599** | 28450 | **6827** | 27833 | **7060** | 27227 | **7297** | |
| 18 | 29711 | **6379** | 29069 | **6603** | 28440 | **6831** | 27823 | **7064** | 27217 | **7301** | 12 |
| | 29701 | **6383** | 29059 | **6606** | 28429 | **6835** | 27812 | **7067** | 27207 | **7305** | |
| 19 | 29690 | **6386** | 29048 | **6610** | 28419 | **6839** | 27802 | **7071** | 27197 | **7309** | 11 |
| | 29679 | **6390** | 29038 | **6614** | 28409 | **6842** | 27792 | **7075** | 27187 | **7313** | |
| 20 | 29668 | **6394** | 29027 | **6618** | 28398 | **6846** | 27782 | **7079** | 27177 | **7317** | 10 |
| | 29657 | **6398** | 29016 | **6622** | 28388 | **6850** | 27772 | **7083** | 27167 | **7321** | |
| 21 | 29647 | **6401** | 29006 | **6625** | 28378 | **6854** | 27761 | **7087** | 27157 | **7325** | 9 |
| | 29636 | **6405** | 28995 | **6629** | 28367 | **6858** | 27751 | **7091** | 27147 | **7329** | |
| 22 | 29625 | **6409** | 28985 | **6633** | 28357 | **6862** | 27741 | **7095** | 27137 | **7333** | 8 |
| | 29614 | **6412** | 28974 | **6637** | 28346 | **6865** | 27731 | **7099** | 27127 | **7337** | |
| 23 | 29604 | **6416** | 28964 | **6640** | 28336 | **6869** | 27721 | **7103** | 27117 | **7341** | 7 |
| | 29593 | **6420** | 28953 | **6644** | 28326 | **6873** | 27711 | **7107** | 27107 | **7345** | |
| 24 | 29582 | **6423** | 28942 | **6648** | 28315 | **6877** | 27701 | **7111** | 27098 | **7349** | 6 |
| | 29571 | **6427** | 28932 | **6652** | 28305 | **6881** | 27690 | **7115** | 27088 | **7353** | |
| 25 | 29560 | **6431** | 28921 | **6655** | 28295 | **6885** | 27680 | **7118** | 27078 | **7357** | 5 |
| | 29550 | **6435** | 28911 | **6659** | 28284 | **6889** | 27670 | **7122** | 27068 | **7361** | |
| 26 | 29539 | **6438** | 28900 | **6663** | 28274 | **6893** | 27660 | **7126** | 27058 | **7365** | 4 |
| | 29528 | **6442** | 28890 | **6667** | 28264 | **6896** | 27650 | **7130** | 27048 | **7369** | |
| 27 | 29517 | **6446** | 28879 | **6671** | 28253 | **6900** | 27640 | **7134** | 27038 | **7373** | 3 |
| | 29507 | **6449** | 28869 | **6674** | 28243 | **6904** | 27630 | **7138** | 27028 | **7377** | |
| 28 | 29496 | **6453** | 28858 | **6678** | 28233 | **6908** | 27619 | **7142** | 27018 | **7381** | 2 |
| | 29485 | **6457** | 28848 | **6682** | 28222 | **6912** | 27609 | **7146** | 27008 | **7385** | |
| 29 | 29475 | **6461** | 28837 | **6686** | 28212 | **6916** | 27599 | **7150** | 26998 | **7389** | 1 |
| | 29464 | **6464** | 28827 | **6690** | 28202 | **6920** | 27589 | **7154** | 26988 | **7393** | |
| 30 | 29453 | **6468** | 28816 | **6693** | 28191 | **6923** | 27579 | **7158** | 26978 | **7397** | 0 |
| | A | B | A | B | A | B | A | B | A | B | |
| ′ | 149° 30′ | | 149° 00′ | | 148° 30′ | | 148° 00′ | | 147° 30′ | | ′ |

# TABLE 35

## The Ageton Method

**Always Take "Z" From Bottom of Table, Except When "K" is Same Name and Greater Than Latitude, in Which Case Take "Z" From Top of Table**

| ′ | 32° 30′ | | 33° 00′ | | 33° 30′ | | 34° 00′ | | 34° 30′ | | ′ |
|---|---|---|---|---|---|---|---|---|---|---|---|
| | A | B | A | B | A | B | A | B | A | B | |
| 0 | 26978 | 7397 | 26389 | 7641 | 25811 | 7889 | 25244 | 8143 | 24687 | 8401 | 30 |
| | 26968 | 7401 | 26379 | 7645 | 25801 | 7893 | 25235 | 8147 | 24678 | 8405 | |
| 1 | 26958 | 7405 | 26370 | 7649 | 25792 | 7898 | 25225 | 8151 | 24669 | 8409 | 29 |
| | 26949 | 7409 | 26360 | 7653 | 25782 | 7902 | 25216 | 8155 | 24660 | 8414 | |
| 2 | 26939 | 7413 | 26350 | 7657 | 25773 | 7906 | 25206 | 8160 | 24650 | 8418 | 28 |
| | 26929 | 7417 | 26340 | 7661 | 25763 | 7910 | 25197 | 8164 | 24641 | 8422 | |
| 3 | 26919 | 7421 | 26331 | 7665 | 25754 | 7914 | 25188 | 8168 | 24632 | 8427 | 27 |
| | 26909 | 7425 | 26321 | 7670 | 25744 | 7919 | 25178 | 8172 | 24623 | 8431 | |
| 4 | 26899 | 7429 | 26311 | 7674 | 25735 | 7923 | 25169 | 8177 | 24614 | 8435 | 26 |
| | 26889 | 7433 | 26302 | 7678 | 25725 | 7927 | 25160 | 8181 | 24605 | 8440 | |
| 5 | 26879 | 7437 | 26292 | 7682 | 25716 | 7931 | 25150 | 8185 | 24595 | 8444 | 25 |
| | 26869 | 7441 | 26282 | 7686 | 25706 | 7935 | 25141 | 8189 | 24586 | 8448 | |
| 6 | 26860 | 7445 | 26273 | 7690 | 25697 | 7940 | 25132 | 8194 | 24577 | 8453 | 24 |
| | 26850 | 7449 | 26263 | 7694 | 25687 | 7944 | 25122 | 8198 | 24568 | 8457 | |
| 7 | 26840 | 7453 | 26253 | 7698 | 25678 | 7948 | 25113 | 8202 | 24559 | 8461 | 23 |
| | 26830 | 7458 | 26244 | 7702 | 25668 | 7952 | 25104 | 8207 | 24550 | 8466 | |
| 8 | 26820 | 7462 | 26234 | 7707 | 25659 | 7956 | 25094 | 8211 | 24540 | 8470 | 22 |
| | 26810 | 7466 | 26224 | 7711 | 25649 | 7961 | 25085 | 8215 | 24531 | 8475 | |
| 9 | 26800 | 7470 | 26214 | 7715 | 25640 | 7965 | 25076 | 8219 | 24522 | 8479 | 21 |
| | 26790 | 7474 | 26205 | 7719 | 25630 | 7969 | 25066 | 8224 | 24513 | 8483 | |
| 10 | 26781 | 7478 | 26195 | 7723 | 25621 | 7973 | 25057 | 8228 | 24504 | 8488 | 20 |
| | 26771 | 7482 | 26185 | 7727 | 25611 | 7977 | 25048 | 8232 | 24495 | 8492 | |
| 11 | 26761 | 7486 | 26176 | 7731 | 25602 | 7982 | 25038 | 8237 | 24486 | 8496 | 19 |
| | 26751 | 7490 | 26166 | 7736 | 25592 | 7986 | 25029 | 8241 | 24477 | 8501 | |
| 12 | 26741 | 7494 | 26157 | 7740 | 25583 | 7990 | 25020 | 8245 | 24467 | 8505 | 18 |
| | 26731 | 7498 | 26147 | 7744 | 25573 | 7994 | 25011 | 8249 | 24458 | 8510 | |
| 13 | 26722 | 7502 | 26137 | 7748 | 25564 | 7998 | 25001 | 8254 | 24449 | 8514 | 17 |
| | 26712 | 7506 | 26128 | 7752 | 25554 | 8003 | 24992 | 8258 | 24440 | 8518 | |
| 14 | 26702 | 7510 | 26118 | 7756 | 25545 | 8007 | 24983 | 8262 | 24431 | 8523 | 16 |
| | 26692 | 7514 | 26108 | 7760 | 25536 | 8011 | 24973 | 8267 | 24422 | 8527 | |
| 15 | 26682 | 7518 | 26099 | 7764 | 25526 | 8015 | 24964 | 8271 | 24413 | 8531 | 15 |
| | 26672 | 7522 | 26089 | 7769 | 25517 | 8020 | 24955 | 8275 | 24404 | 8536 | |
| 16 | 26663 | 7526 | 26079 | 7773 | 25507 | 8024 | 24946 | 8280 | 24395 | 8540 | 14 |
| | 26653 | 7531 | 26070 | 7777 | 25498 | 8028 | 24936 | 8284 | 24385 | 8545 | |
| 17 | 26643 | 7535 | 26060 | 7781 | 25488 | 8032 | 24927 | 8288 | 24376 | 8549 | 13 |
| | 26633 | 7539 | 26051 | 7785 | 25479 | 8037 | 24918 | 8292 | 24367 | 8553 | |
| 18 | 26623 | 7543 | 26041 | 7789 | 25469 | 8041 | 24909 | 8297 | 24358 | 8558 | 12 |
| | 26614 | 7547 | 26031 | 7793 | 25460 | 8045 | 24899 | 8301 | 24349 | 8562 | |
| 19 | 26604 | 7551 | 26022 | 7798 | 25451 | 8049 | 24890 | 8305 | 24340 | 8567 | 11 |
| | 26594 | 7555 | 26012 | 7802 | 25441 | 8053 | 24881 | 8310 | 24331 | 8571 | |
| 20 | 26584 | 7559 | 26002 | 7806 | 25432 | 8058 | 24872 | 8314 | 24322 | 8575 | 10 |
| | 26574 | 7563 | 25993 | 7810 | 25422 | 8062 | 24862 | 8318 | 24313 | 8580 | |
| 21 | 26565 | 7567 | 25983 | 7814 | 25413 | 8066 | 24853 | 8323 | 24304 | 8584 | 9 |
| | 26555 | 7571 | 25974 | 7818 | 25403 | 8070 | 24844 | 8327 | 24295 | 8589 | |
| 22 | 26545 | 7575 | 25964 | 7823 | 25394 | 8075 | 24835 | 8331 | 24286 | 8593 | 8 |
| | 26535 | 7579 | 25954 | 7827 | 25385 | 8079 | 24825 | 8336 | 24276 | 8597 | |
| 23 | 26526 | 7584 | 25945 | 7831 | 25375 | 8083 | 24816 | 8340 | 24267 | 8602 | 7 |
| | 26516 | 7588 | 25935 | 7835 | 25366 | 8087 | 24807 | 8344 | 24258 | 8606 | |
| 24 | 26506 | 7592 | 25926 | 7839 | 25356 | 8091 | 24798 | 8349 | 24249 | 8611 | 6 |
| | 26496 | 7596 | 25916 | 7843 | 25347 | 8096 | 24788 | 8353 | 24240 | 8615 | |
| 25 | 26486 | 7600 | 25907 | 7848 | 25338 | 8100 | 24779 | 8357 | 24231 | 8619 | 5 |
| | 26477 | 7604 | 25897 | 7852 | 25328 | 8104 | 24770 | 8362 | 24222 | 8624 | |
| 26 | 26467 | 7608 | 25887 | 7856 | 25319 | 8108 | 24761 | 8366 | 24213 | 8628 | 4 |
| | 26457 | 7612 | 25878 | 7860 | 25309 | 8113 | 24752 | 8370 | 24204 | 8633 | |
| 27 | 26447 | 7616 | 25868 | 7864 | 25300 | 8117 | 24742 | 8375 | 24195 | 8637 | 3 |
| | 26438 | 7620 | 25859 | 7868 | 25291 | 8121 | 24733 | 8379 | 24186 | 8641 | |
| 28 | 26428 | 7625 | 25849 | 7873 | 25281 | 8125 | 24724 | 8383 | 24177 | 8646 | 2 |
| | 26418 | 7629 | 25840 | 7877 | 25272 | 8130 | 24715 | 8388 | 24168 | 8650 | |
| 29 | 26409 | 7633 | 25830 | 7881 | 25263 | 8134 | 24706 | 8392 | 24159 | 8655 | 1 |
| | 26399 | 7637 | 25821 | 7885 | 25253 | 8138 | 24696 | 8396 | 24150 | 8659 | |
| 30 | 26389 | 7641 | 25811 | 7889 | 25244 | 8143 | 24687 | 8401 | 24141 | 8663 | 0 |
| | A | B | A | B | A | B | A | B | A | B | |
| ′ | 147° 00′ | | 146° 30′ | | 146° 00′ | | 145° 30′ | | 145° 00′ | | ′ |

# TABLE 35

## The Ageton Method

**When Meridian Angle is Greater Than 90° Take "K" From Bottom of Table**

| ′ | 35° 00′ | | 35° 30′ | | 36° 00′ | | 36° 30′ | | 37° 00′ | | ′ |
|---|---|---|---|---|---|---|---|---|---|---|---|
| | A | B | A | B | A | B | A | B | A | B | |
| 0 | 24141 | 8663 | 23605 | 8931 | 23078 | 9204 | 22561 | 9482 | 22054 | 9765 | 30 |
| | 24132 | 8668 | 23596 | 8936 | 23069 | 9209 | 22553 | 9487 | 22045 | 9770 | |
| 1 | 24123 | 8672 | 23587 | 8940 | 23061 | 9213 | 22544 | 9492 | 22037 | 9775 | 29 |
| | 24114 | 8677 | 23578 | 8945 | 23052 | 9218 | 22536 | 9496 | 22029 | 9779 | |
| 2 | 24105 | 8681 | 23569 | 8949 | 23043 | 9223 | 22527 | 9501 | 22020 | 9784 | 28 |
| | 24096 | 8686 | 23560 | 8954 | 23035 | 9227 | 22519 | 9505 | 22012 | 9789 | |
| 3 | 24087 | 8690 | 23551 | 8958 | 23026 | 9232 | 22510 | 9510 | 22003 | 9794 | 27 |
| | 24078 | 8694 | 23543 | 8963 | 23017 | 9236 | 22501 | 9515 | 21995 | 9798 | |
| 4 | 24069 | 8699 | 23534 | 8967 | 23009 | 9241 | 22493 | 9520 | 21987 | 9803 | 26 |
| | 24060 | 8703 | 23525 | 8972 | 23000 | 9246 | 22484 | 9524 | 21978 | 9808 | |
| 5 | 24051 | 8708 | 23516 | 8976 | 22991 | 9250 | 22476 | 9529 | 21970 | 9813 | 25 |
| | 24042 | 8712 | 23507 | 8981 | 22983 | 9255 | 22467 | 9534 | 21962 | 9818 | |
| 6 | 24033 | 8717 | 23498 | 8986 | 22974 | 9259 | 22459 | 9538 | 21953 | 9822 | 24 |
| | 24024 | 8721 | 23490 | 8990 | 22965 | 9264 | 22450 | 9543 | 21945 | 9827 | |
| 7 | 24015 | 8726 | 23481 | 8995 | 22957 | 9269 | 22442 | 9548 | 21937 | 9832 | 23 |
| | 24006 | 8730 | 23472 | 8999 | 22948 | 9273 | 22433 | 9552 | 21928 | 9837 | |
| 8 | 23997 | 8734 | 23463 | 9004 | 22939 | 9278 | 22425 | 9557 | 21920 | 9841 | 22 |
| | 23988 | 8739 | 23454 | 9008 | 22931 | 9282 | 22416 | 9562 | 21912 | 9846 | |
| 9 | 23979 | 8743 | 23446 | 9013 | 22922 | 9287 | 22408 | 9566 | 21903 | 9851 | 21 |
| | 23970 | 8748 | 23437 | 9017 | 22913 | 9292 | 22399 | 9571 | 21895 | 9856 | |
| 10 | 23961 | 8752 | 23428 | 9022 | 22905 | 9296 | 22391 | 9576 | 21887 | 9861 | 20 |
| | 23952 | 8757 | 23419 | 9026 | 22896 | 9301 | 22382 | 9581 | 21878 | 9865 | |
| 11 | 23943 | 8761 | 23410 | 9031 | 22887 | 9305 | 22374 | 9585 | 21870 | 9870 | 19 |
| | 23934 | 8766 | 23402 | 9035 | 22879 | 9310 | 22366 | 9590 | 21862 | 9875 | |
| 12 | 23925 | 8770 | 23393 | 9040 | 22870 | 9315 | 22357 | 9595 | 21853 | 9880 | 18 |
| | 23916 | 8775 | 23384 | 9044 | 22862 | 9319 | 22349 | 9599 | 21845 | 9885 | |
| 13 | 23907 | 8779 | 23375 | 9049 | 22853 | 9324 | 22340 | 9604 | 21837 | 9889 | 17 |
| | 23898 | 8783 | 23366 | 9054 | 22844 | 9329 | 22332 | 9609 | 21828 | 9894 | |
| 14 | 23889 | 8788 | 23358 | 9058 | 22836 | 9333 | 22323 | 9614 | 21820 | 9899 | 16 |
| | 23880 | 8792 | 23349 | 9063 | 22827 | 9338 | 22315 | 9618 | 21812 | 9904 | |
| 15 | 23871 | 8797 | 23340 | 9067 | 22818 | 9342 | 22306 | 9623 | 21803 | 9909 | 15 |
| | 23863 | 8801 | 23331 | 9072 | 22810 | 9347 | 22298 | 9628 | 21795 | 9913 | |
| 16 | 23854 | 8806 | 23323 | 9076 | 22801 | 9352 | 22289 | 9632 | 21787 | 9918 | 14 |
| | 23845 | 8810 | 23314 | 9081 | 22793 | 9356 | 22281 | 9637 | 21778 | 9923 | |
| 17 | 23836 | 8815 | 23305 | 9085 | 22784 | 9361 | 22272 | 9642 | 21770 | 9928 | 13 |
| | 23827 | 8819 | 23296 | 9090 | 22775 | 9366 | 22264 | 9647 | 21762 | 9933 | |
| 18 | 23818 | 8824 | 23288 | 9094 | 22767 | 9370 | 22256 | 9651 | 21754 | 9937 | 12 |
| | 23809 | 8828 | 23279 | 9099 | 22758 | 9375 | 22247 | 9656 | 21745 | 9942 | |
| 19 | 23800 | 8833 | 23270 | 9104 | 22750 | 9380 | 22239 | 9661 | 21737 | 9947 | 11 |
| | 23791 | 8837 | 23261 | 9108 | 22741 | 9384 | 22230 | 9665 | 21729 | 9952 | |
| 20 | 23782 | 8842 | 23252 | 9113 | 22732 | 9389 | 22222 | 9670 | 21720 | 9957 | 10 |
| | 23773 | 8846 | 23244 | 9117 | 22724 | 9394 | 22213 | 9675 | 21712 | 9962 | |
| 21 | 23764 | 8850 | 23235 | 9122 | 22715 | 9398 | 22205 | 9680 | 21704 | 9966 | 9 |
| | 23755 | 8855 | 23226 | 9126 | 22707 | 9403 | 22197 | 9684 | 21696 | 9971 | |
| 22 | 23747 | 8859 | 23218 | 9131 | 22698 | 9407 | 22188 | 9689 | 21687 | 9976 | 8 |
| | 23738 | 8864 | 23209 | 9136 | 22690 | 9412 | 22180 | 9694 | 21679 | 9981 | |
| 23 | 23729 | 8868 | 23200 | 9140 | 22681 | 9417 | 22171 | 9699 | 21671 | 9986 | 7 |
| | 23720 | 8873 | 23191 | 9145 | 22672 | 9421 | 22163 | 9703 | 21662 | 9990 | |
| 24 | 23711 | 8877 | 23183 | 9149 | 22664 | 9426 | 22154 | 9708 | 21654 | 9995 | 6 |
| | 23702 | 8882 | 23174 | 9154 | 22655 | 9431 | 22146 | 9713 | 21646 | 10000 | |
| 25 | 23693 | 8886 | 23165 | 9158 | 22647 | 9435 | 22138 | 9718 | 21638 | 10005 | 5 |
| | 23684 | 8891 | 23156 | 9163 | 22638 | 9440 | 22129 | 9722 | 21629 | 10010 | |
| 26 | 23675 | 8895 | 23148 | 9168 | 22630 | 9445 | 22121 | 9727 | 21621 | 10015 | 4 |
| | 23667 | 8900 | 23139 | 9172 | 22621 | 9449 | 22112 | 9732 | 21613 | 10019 | |
| 27 | 23658 | 8904 | 23130 | 9177 | 22612 | 9454 | 22104 | 9737 | 21605 | 10024 | 3 |
| | 23649 | 8909 | 23122 | 9181 | 22604 | 9459 | 22096 | 9741 | 21596 | 10029 | |
| 28 | 23640 | 8913 | 23113 | 9186 | 22595 | 9463 | 22087 | 9746 | 21588 | 10034 | 2 |
| | 23631 | 8918 | 23104 | 9190 | 22587 | 9468 | 22079 | 9751 | 21580 | 10039 | |
| 29 | 23622 | 8922 | 23095 | 9195 | 22578 | 9473 | 22070 | 9756 | 21572 | 10044 | 1 |
| | 23613 | 8927 | 23087 | 9200 | 22570 | 9477 | 22062 | 9760 | 21563 | 10049 | |
| 30 | 23605 | 8931 | 23078 | 9204 | 22561 | 9482 | 22054 | 9765 | 21555 | 10053 | 0 |
| | A | B | A | B | A | B | A | B | A | B | |
| ′ | 144° 30′ | | 144° 00′ | | 143° 30′ | | 143° 00′ | | 142° 30′ | | ′ |

## TABLE 35

### The Ageton Method

**Always Take "Z" From Bottom of Table, Except When "K" is Same Name and Greater Than Latitude, in Which Case Take "Z" From Top of Table**

| | 37° 30′ | | 38° 00′ | | 38° 30′ | | 39° 00′ | | 39° 30′ | | |
|---|---|---|---|---|---|---|---|---|---|---|---|
| ′ | A | B | A | B | A | B | A | B | A | B | ′ |
| 0 | 21555 | 10053 | 21066 | 10347 | 20585 | 10646 | 20113 | 10950 | 19649 | 11259 | 30 |
| | 21547 | 10058 | 21058 | 10352 | 20577 | 10651 | 20105 | 10955 | 19641 | 11265 | |
| 1 | 21539 | 10063 | 21050 | 10357 | 20569 | 10656 | 20097 | 10960 | 19634 | 11270 | 29 |
| | 21531 | 10068 | 21042 | 10362 | 20561 | 10661 | 20089 | 10965 | 19626 | 11275 | |
| 2 | 21522 | 10073 | 21033 | 10367 | 20553 | 10666 | 20082 | 10970 | 19618 | 11280 | 28 |
| | 21514 | 10078 | 21025 | 10372 | 20545 | 10671 | 20074 | 10975 | 19611 | 11285 | |
| 3 | 21506 | 10082 | 21017 | 10376 | 20537 | 10676 | 20066 | 10980 | 19603 | 11291 | 27 |
| | 21498 | 10087 | 21009 | 10381 | 20529 | 10681 | 20058 | 10986 | 19595 | 11296 | |
| 4 | 21489 | 10092 | 21001 | 10386 | 20522 | 10686 | 20050 | 10991 | 19588 | 11301 | 26 |
| | 21481 | 10097 | 20993 | 10391 | 20514 | 10691 | 20043 | 10996 | 19580 | 11306 | |
| 5 | 21473 | 10102 | 20985 | 10396 | 20506 | 10696 | 20035 | 11001 | 19572 | 11311 | 25 |
| | 21465 | 10107 | 20977 | 10401 | 20498 | 10701 | 20027 | 11006 | 19565 | 11317 | |
| 6 | 21457 | 10112 | 20969 | 10406 | 20490 | 10706 | 20019 | 11011 | 19557 | 11322 | 24 |
| | 21448 | 10116 | 20961 | 10411 | 20482 | 10711 | 20012 | 11016 | 19549 | 11327 | |
| 7 | 21440 | 10121 | 20953 | 10416 | 20474 | 10716 | 20004 | 11021 | 19541 | 11332 | 23 |
| | 21432 | 10126 | 20945 | 10421 | 20466 | 10721 | 19996 | 11027 | 19534 | 11338 | |
| 8 | 21424 | 10131 | 20937 | 10426 | 20458 | 10726 | 19988 | 11032 | 19527 | 11343 | 22 |
| | 21416 | 10136 | 20929 | 10431 | 20450 | 10731 | 19980 | 11037 | 19519 | 11348 | |
| 9 | 21407 | 10141 | 20921 | 10436 | 20442 | 10736 | 19973 | 11042 | 19511 | 11353 | 21 |
| | 21399 | 10146 | 20913 | 10441 | 20435 | 10741 | 19965 | 11047 | 19504 | 11359 | |
| 10 | 21391 | 10151 | 20905 | 10446 | 20427 | 10746 | 19957 | 11052 | 19496 | 11364 | 20 |
| | 21383 | 10155 | 20897 | 10451 | 20419 | 10751 | 19949 | 11057 | 19488 | 11369 | |
| 11 | 21375 | 10160 | 20888 | 10456 | 20411 | 10756 | 19942 | 11063 | 19481 | 11374 | 19 |
| | 21367 | 10165 | 20880 | 10461 | 20403 | 10761 | 19934 | 11068 | 19473 | 11380 | |
| 12 | 21358 | 10170 | 20872 | 10466 | 20395 | 10767 | 19926 | 11073 | 19466 | 11385 | 18 |
| | 21350 | 10175 | 20864 | 10471 | 20387 | 10772 | 19919 | 11078 | 19458 | 11390 | |
| 13 | 21342 | 10180 | 20856 | 10476 | 20379 | 10777 | 19911 | 11083 | 19450 | 11395 | 17 |
| | 21334 | 10185 | 20848 | 10481 | 20371 | 10782 | 19903 | 11088 | 19443 | 11400 | |
| 14 | 21326 | 10190 | 20840 | 10486 | 20364 | 10787 | 19895 | 11094 | 19435 | 11406 | 16 |
| | 21318 | 10195 | 20832 | 10491 | 20356 | 10792 | 19888 | 11099 | 19428 | 11411 | |
| 15 | 21309 | 10199 | 20824 | 10496 | 20348 | 10797 | 19880 | 11104 | 19420 | 11416 | 15 |
| | 21301 | 10204 | 20816 | 10500 | 20340 | 10802 | 19872 | 11109 | 19412 | 11422 | |
| 16 | 21293 | 10209 | 20808 | 10505 | 20332 | 10807 | 19864 | 11114 | 19405 | 11427 | 14 |
| | 21285 | 10214 | 20800 | 10510 | 20324 | 10812 | 19857 | 11119 | 19397 | 11432 | |
| 17 | 21277 | 10219 | 20792 | 10515 | 20316 | 10817 | 19849 | 11124 | 19390 | 11437 | 13 |
| | 21269 | 10224 | 20784 | 10520 | 20309 | 10822 | 19841 | 11130 | 19382 | 11443 | |
| 18 | 21260 | 10229 | 20776 | 10525 | 20301 | 10827 | 19834 | 11135 | 19375 | 11448 | 12 |
| | 21252 | 10234 | 20768 | 10530 | 20293 | 10832 | 19826 | 11140 | 19367 | 11453 | |
| 19 | 21244 | 10239 | 20760 | 10535 | 20285 | 10838 | 19818 | 11145 | 19359 | 11458 | 11 |
| | 21236 | 10243 | 20752 | 10540 | 20277 | 10843 | 19810 | 11150 | 19352 | 11464 | |
| 20 | 21228 | 10248 | 20744 | 10545 | 20269 | 10848 | 19803 | 11156 | 19344 | 11469 | 10 |
| | 21220 | 10253 | 20736 | 10550 | 20261 | 10853 | 19795 | 11161 | 19337 | 11474 | |
| 21 | 21212 | 10258 | 20728 | 10555 | 20254 | 10858 | 19787 | 11166 | 19329 | 11479 | 9 |
| | 21204 | 10263 | 20720 | 10560 | 20246 | 10863 | 19779 | 11171 | 19321 | 11485 | |
| 22 | 21195 | 10268 | 20712 | 10565 | 20238 | 10868 | 19772 | 11176 | 19314 | 11490 | 8 |
| | 21187 | 10273 | 20704 | 10570 | 20230 | 10873 | 19764 | 11181 | 19306 | 11495 | |
| 23 | 21179 | 10278 | 20696 | 10575 | 20222 | 10878 | 19756 | 11187 | 19299 | 11501 | 7 |
| | 21171 | 10283 | 20688 | 10580 | 20214 | 10883 | 19749 | 11192 | 19291 | 11506 | |
| 24 | 21163 | 10288 | 20680 | 10585 | 20207 | 10888 | 19741 | 11197 | 19284 | 11511 | 6 |
| | 21155 | 10293 | 20672 | 10590 | 20199 | 10894 | 19733 | 11202 | 19276 | 11516 | |
| 25 | 21147 | 10298 | 20665 | 10595 | 20191 | 10899 | 19726 | 11207 | 19269 | 11522 | 5 |
| | 21139 | 10302 | 20657 | 10600 | 20183 | 10904 | 19718 | 11213 | 19261 | 11527 | |
| 26 | 21131 | 10307 | 20649 | 10605 | 20175 | 10909 | 19710 | 11218 | 19253 | 11532 | 4 |
| | 21122 | 10312 | 20641 | 10610 | 20167 | 10914 | 19703 | 11223 | 19246 | 11537 | |
| 27 | 21114 | 10317 | 20633 | 10615 | 20160 | 10919 | 19695 | 11228 | 19238 | 11543 | 3 |
| | 21106 | 10322 | 20625 | 10620 | 20152 | 10924 | 19687 | 11233 | 19231 | 11548 | |
| 28 | 21098 | 10327 | 20617 | 10625 | 20144 | 10929 | 19680 | 11239 | 19223 | 11553 | 2 |
| | 21090 | 10332 | 20609 | 10630 | 20136 | 10934 | 19672 | 11244 | 19216 | 11559 | |
| 29 | 21082 | 10337 | 20601 | 10635 | 20128 | 10939 | 19664 | 11249 | 19208 | 11564 | 1 |
| | 21074 | 10342 | 20593 | 10640 | 20121 | 10945 | 19657 | 11254 | 19201 | 11569 | |
| 30 | 21066 | 10347 | 20585 | 10646 | 20113 | 10950 | 19649 | 11259 | 19193 | 11575 | 0 |
| | A | B | A | B | A | B | A | B | A | B | |
| ′ | 142° 00′ | | 141° 30′ | | 141° 00′ | | 140° 30′ | | 140° 00′ | | ′ |

## TABLE 35

### The Ageton Method

When Meridian Angle is Greater Than 90° Take "K" From Bottom of Table

| ′ | 40° 00′ | | 40° 30′ | | 41° 00′ | | 41° 30′ | | 42° 00′ | | ′ |
|---|---|---|---|---|---|---|---|---|---|---|---|
| | A | B | A | B | A | B | A | B | A | B | |
| 0 | 19193 | **11575** | 18746 | **11895** | 18306 | **12222** | 17873 | **12554** | 17449 | **12893** | 30 |
| | 19186 | **11580** | 18738 | **11901** | 18298 | **12228** | 17866 | **12560** | 17442 | **12898** | |
| 1 | 19178 | **11585** | 18731 | **11906** | 18291 | **12233** | 17859 | **12566** | 17435 | **12904** | 29 |
| | 19171 | **11590** | 18723 | **11912** | 18284 | **12238** | 17852 | **12571** | 17428 | **12910** | |
| 2 | 19163 | **11596** | 18716 | **11917** | 18277 | **12244** | 17845 | **12577** | 17421 | **12915** | 28 |
| | 19156 | **11601** | 18709 | **11922** | 18269 | **12249** | 17838 | **12582** | 17414 | **12921** | |
| 3 | 19148 | **11606** | 18701 | **11928** | 18262 | **12255** | 17831 | **12588** | 17407 | **12927** | 27 |
| | 19141 | **11612** | 18694 | **11933** | 18255 | **12260** | 17824 | **12593** | 17400 | **12932** | |
| 4 | 19133 | **11617** | 18686 | **11939** | 18248 | **12266** | 17816 | **12599** | 17393 | **12938** | 26 |
| | 19126 | **11622** | 18679 | **11944** | 18240 | **12271** | 17809 | **12605** | 17386 | **12944** | |
| 5 | 19118 | **11628** | 18672 | **11949** | 18233 | **12277** | 17802 | **12610** | 17379 | **12950** | 25 |
| | 19111 | **11633** | 18664 | **11955** | 18226 | **12282** | 17795 | **12616** | 17372 | **12955** | |
| 6 | 19103 | **11638** | 18657 | **11960** | 18219 | **12288** | 17788 | **12622** | 17365 | **12961** | 24 |
| | 19096 | **11644** | 18650 | **11966** | 18211 | **12293** | 17781 | **12627** | 17358 | **12967** | |
| 7 | 19088 | **11649** | 18642 | **11971** | 18204 | **12299** | 17774 | **12633** | 17351 | **12972** | 23 |
| | 19081 | **11654** | 18635 | **11977** | 18197 | **12305** | 17767 | **12638** | 17344 | **12978** | |
| 8 | 19073 | **11660** | 18627 | **11982** | 18190 | **12310** | 17760 | **12644** | 17337 | **12984** | 22 |
| | 19066 | **11665** | 18620 | **11987** | 18182 | **12316** | 17752 | **12650** | 17330 | **12990** | |
| 9 | 19058 | **11670** | 18613 | **11993** | 18175 | **12321** | 17745 | **12655** | 17323 | **12995** | 21 |
| | 19051 | **11676** | 18605 | **11998** | 18168 | **12327** | 17738 | **12661** | 17316 | **13001** | |
| 10 | 19043 | **11681** | 18598 | **12004** | 18161 | **12332** | 17731 | **12667** | 17309 | **13007** | 20 |
| | 19036 | **11686** | 18591 | **12009** | 18154 | **12338** | 17724 | **12672** | 17302 | **13012** | |
| 11 | 19028 | **11692** | 18583 | **12014** | 18146 | **12343** | 17717 | **12678** | 17295 | **13018** | 19 |
| | 19021 | **11697** | 18576 | **12020** | 18139 | **12349** | 17710 | **12683** | 17288 | **13024** | |
| 12 | 19013 | **11702** | 18569 | **12025** | 18132 | **12354** | 17703 | **12689** | 17281 | **13030** | 18 |
| | 19006 | **11708** | 18561 | **12031** | 18125 | **12360** | 17696 | **12695** | 17274 | **13035** | |
| 13 | 18998 | **11713** | 18554 | **12036** | 18117 | **12365** | 17689 | **12700** | 17267 | **13041** | 17 |
| | 18991 | **11718** | 18547 | **12042** | 18110 | **12371** | 17681 | **12706** | 17260 | **13047** | |
| 14 | 18983 | **11724** | 18539 | **12047** | 18103 | **12376** | 17674 | **12711** | 17253 | **13053** | 16 |
| | 18976 | **11729** | 18532 | **12053** | 18096 | **12382** | 17667 | **12717** | 17246 | **13058** | |
| 15 | 18968 | **11734** | 18525 | **12058** | 18089 | **12387** | 17660 | **12723** | 17239 | **13064** | 15 |
| | 18961 | **11740** | 18517 | **12063** | 18081 | **12393** | 17653 | **12728** | 17232 | **13070** | |
| 16 | 18953 | **11745** | 18510 | **12069** | 18074 | **12398** | 17646 | **12734** | 17225 | **13075** | 14 |
| | 18946 | **11750** | 18503 | **12074** | 18067 | **12404** | 17639 | **12740** | 17218 | **13081** | |
| 17 | 18939 | **11756** | 18495 | **12080** | 18060 | **12410** | 17632 | **12745** | 17212 | **13087** | 13 |
| | 18931 | **11761** | 18488 | **12085** | 18053 | **12415** | 17625 | **12751** | 17205 | **13093** | |
| 18 | 18924 | **11766** | 18481 | **12091** | 18045 | **12421** | 17618 | **12757** | 17198 | **13098** | 12 |
| | 18916 | **11772** | 18473 | **12096** | 18038 | **12426** | 17611 | **12762** | 17191 | **13104** | |
| 19 | 18909 | **11777** | 18466 | **12102** | 18031 | **12432** | 17604 | **12768** | 17184 | **13110** | 11 |
| | 18901 | **11782** | 18459 | **12107** | 18024 | **12437** | 17597 | **12774** | 17177 | **13116** | |
| 20 | 18894 | **11788** | 18451 | **12112** | 18017 | **12443** | 17590 | **12779** | 17170 | **13121** | 10 |
| | 18886 | **11793** | 18444 | **12118** | 18010 | **12448** | 17583 | **12785** | 17163 | **13127** | |
| 21 | 18879 | **11799** | 18437 | **12123** | 18002 | **12454** | 17575 | **12790** | 17156 | **13133** | 9 |
| | 18872 | **11804** | 18429 | **12129** | 17995 | **12460** | 17568 | **12796** | 17149 | **13139** | |
| 22 | 18864 | **11809** | 18422 | **12134** | 17988 | **12465** | 17561 | **12802** | 17142 | **13144** | 8 |
| | 18857 | **11815** | 18415 | **12140** | 17981 | **12471** | 17554 | **12807** | 17135 | **13150** | |
| 23 | 18849 | **11820** | 18408 | **12145** | 17974 | **12476** | 17547 | **12813** | 17128 | **13156** | 7 |
| | 18842 | **11825** | 18400 | **12151** | 17966 | **12482** | 17540 | **12819** | 17121 | **13162** | |
| 24 | 18834 | **11831** | 18393 | **12156** | 17959 | **12487** | 17533 | **12824** | 17114 | **13168** | 6 |
| | 18827 | **11836** | 18386 | **12162** | 17952 | **12493** | 17526 | **12830** | 17108 | **13173** | |
| 25 | 18820 | **11842** | 18378 | **12167** | 17945 | **12499** | 17519 | **12836** | 17101 | **13179** | 5 |
| | 18812 | **11847** | 18371 | **12173** | 17938 | **12504** | 17512 | **12841** | 17094 | **13185** | |
| 26 | 18805 | **11852** | 18364 | **12178** | 17931 | **12510** | 17505 | **12847** | 17087 | **13191** | 4 |
| | 18797 | **11858** | 18357 | **12184** | 17924 | **12515** | 17498 | **12853** | 17080 | **13196** | |
| 27 | 18790 | **11863** | 18349 | **12189** | 17916 | **12521** | 17491 | **12859** | 17073 | **13202** | 3 |
| | 18783 | **11868** | 18342 | **12195** | 17909 | **12526** | 17484 | **12864** | 17066 | **13208** | |
| 28 | 18775 | **11874** | 18335 | **12200** | 17902 | **12532** | 17477 | **12870** | 17059 | **13214** | 2 |
| | 18768 | **11879** | 18327 | **12205** | 17895 | **12538** | 17470 | **12876** | 17052 | **13220** | |
| 29 | 18760 | **11885** | 18320 | **12211** | 17888 | **12543** | 17463 | **12881** | 17046 | **13225** | 1 |
| | 18753 | **11890** | 18313 | **12216** | 17881 | **12549** | 17456 | **12887** | 17039 | **13231** | |
| 30 | 18746 | **11895** | 18306 | **12222** | 17873 | **12554** | 17449 | **12893** | 17032 | **13237** | 0 |
| | A | B | A | B | A | B | A | B | A | B | |
| ′ | 139° 30′ | | 139° 00′ | | 138° 30′ | | 138° 00′ | | 137° 30′ | | ′ |

## TABLE 35

### The Ageton Method

**Always Take "Z" From Bottom of Table, Except When "K" is Same Name and Greater Than Latitude, in Which Case Take "Z" From Top of Table**

| ′ | 42° 30′ | | 43° 00′ | | 43° 30′ | | 44° 00′ | | 44° 30′ | | ′ |
|---|---|---|---|---|---|---|---|---|---|---|---|
| ′ | A | B | A | B | A | B | A | B | A | B | ′ |
| 0 | 17032 | **13237** | 16622 | **13587** | 16219 | **13944** | 15823 | **14307** | 15434 | **14676** | 30 |
| | 17025 | **13243** | 16615 | **13593** | 16212 | **13950** | 15816 | **14313** | 15427 | **14682** | |
| 1 | 17018 | **13248** | 16608 | **13599** | 16205 | **13956** | 15810 | **14319** | 15421 | **14688** | 29 |
| | 17011 | **13254** | 16601 | **13605** | 16199 | **13962** | 15803 | **14325** | 15414 | **14694** | |
| 2 | 17004 | **13260** | 16595 | **13611** | 16192 | **13968** | 15797 | **14331** | 15408 | **14701** | 28 |
| | 16997 | **13266** | 16588 | **13617** | 16186 | **13974** | 15790 | **14337** | 15402 | **14707** | |
| 3 | 16990 | **13272** | 16581 | **13623** | 16179 | **13980** | 15784 | **14343** | 15395 | **14713** | 27 |
| | 16983 | **13277** | 16574 | **13628** | 16172 | **13986** | 15777 | **14349** | 15389 | **14719** | |
| 4 | 16977 | **13283** | 16567 | **13634** | 16166 | **13992** | 15771 | **14355** | 15382 | **14726** | 26 |
| | 16970 | **13289** | 16561 | **13640** | 16159 | **13998** | 15764 | **14362** | 15376 | **14732** | |
| 5 | 16963 | **13295** | 16554 | **13646** | 16152 | **14004** | 15758 | **14368** | 15370 | **14738** | 25 |
| | 16956 | **13301** | 16547 | **13652** | 16146 | **14010** | 15751 | **14374** | 15363 | **14744** | |
| 6 | 16949 | **13306** | 16540 | **13658** | 16139 | **14016** | 15744 | **14380** | 15357 | **14750** | 24 |
| | 16942 | **13312** | 16534 | **13664** | 16132 | **14022** | 15738 | **14386** | 15350 | **14757** | |
| 7 | 16935 | **13318** | 16527 | **13670** | 16126 | **14028** | 15731 | **14392** | 15344 | **14763** | 23 |
| | 16928 | **13324** | 16520 | **13676** | 16119 | **14034** | 15725 | **14398** | 15338 | **14769** | |
| 8 | 16922 | **13330** | 16513 | **13682** | 16112 | **14040** | 15718 | **14404** | 15331 | **14775** | 22 |
| | 16915 | **13336** | 16507 | **13688** | 16106 | **14046** | 15712 | **14411** | 15325 | **14782** | |
| 9 | 16908 | **13341** | 16500 | **13694** | 16099 | **14052** | 15705 | **14417** | 15318 | **14788** | 21 |
| | 16901 | **13347** | 16493 | **13700** | 16093 | **14058** | 15699 | **14423** | 15312 | **14794** | |
| 10 | 16894 | **13353** | 16487 | **13705** | 16086 | **14064** | 15692 | **14429** | 15306 | **14800** | 20 |
| | 16887 | **13359** | 16480 | **13711** | 16079 | **14070** | 15686 | **14435** | 15299 | **14807** | |
| 11 | 16880 | **13365** | 16473 | **13717** | 16073 | **14076** | 15679 | **14441** | 15293 | **14813** | 19 |
| | 16874 | **13370** | 16466 | **13723** | 16066 | **14082** | 15673 | **14447** | 15286 | **14819** | |
| 12 | 16867 | **13376** | 16460 | **13729** | 16060 | **14088** | 15666 | **14453** | 15280 | **14825** | 18 |
| | 16860 | **13382** | 16453 | **13735** | 16053 | **14094** | 15660 | **14460** | 15274 | **14831** | |
| 13 | 16853 | **13388** | 16446 | **13741** | 16046 | **14100** | 15653 | **14466** | 15267 | **14838** | 17 |
| | 16846 | **13394** | 16439 | **13747** | 16040 | **14106** | 15647 | **14472** | 15261 | **14844** | |
| 14 | 16839 | **13400** | 16433 | **13753** | 16033 | **14112** | 15640 | **14478** | 15255 | **14850** | 16 |
| | 16833 | **13405** | 16426 | **13759** | 16027 | **14118** | 15634 | **14484** | 15248 | **14857** | |
| 15 | 16826 | **13411** | 16419 | **13765** | 16020 | **14124** | 15627 | **14490** | 15242 | **14863** | 15 |
| | 16819 | **13417** | 16413 | **13771** | 16013 | **14130** | 15621 | **14496** | 15235 | **14869** | |
| 16 | 16812 | **13423** | 16406 | **13777** | 16007 | **14136** | 15614 | **14503** | 15229 | **14875** | 14 |
| | 16805 | **13429** | 16399 | **13783** | 16000 | **14142** | 15608 | **14509** | 15223 | **14882** | |
| 17 | 16798 | **13435** | 16392 | **13789** | 15994 | **14149** | 15602 | **14515** | 15216 | **14888** | 13 |
| | 16792 | **13440** | 16386 | **13794** | 15987 | **14155** | 15595 | **14521** | 15210 | **14894** | |
| 18 | 16785 | **13446** | 16379 | **13800** | 15980 | **14161** | 15589 | **14527** | 15204 | **14900** | 12 |
| | 16778 | **13452** | 16372 | **13806** | 15974 | **14167** | 15582 | **14533** | 15197 | **14907** | |
| 19 | 16771 | **13458** | 16366 | **13812** | 15967 | **14173** | 15576 | **14540** | 15191 | **14913** | 11 |
| | 16764 | **13464** | 16359 | **13818** | 15961 | **14179** | 15569 | **14546** | 15184 | **14919** | |
| 20 | 16757 | **13470** | 16352 | **13824** | 15954 | **14185** | 15563 | **14552** | 15178 | **14925** | 10 |
| | 16751 | **13476** | 16346 | **13830** | 15947 | **14191** | 15556 | **14558** | 15172 | **14932** | |
| 21 | 16744 | **13481** | 16339 | **13836** | 15941 | **14197** | 15550 | **14564** | 15165 | **14938** | 9 |
| | 16737 | **13487** | 16332 | **13842** | 15934 | **14203** | 15543 | **14570** | 15159 | **14944** | |
| 22 | 16730 | **13493** | 16325 | **13848** | 15928 | **14209** | 15537 | **14577** | 15153 | **14951** | 8 |
| | 16723 | **13499** | 16319 | **13854** | 15921 | **14215** | 15530 | **14583** | 15146 | **14957** | |
| 23 | 16717 | **13505** | 16312 | **13860** | 15915 | **14221** | 15524 | **14589** | 15140 | **14963** | 7 |
| | 16710 | **13511** | 16305 | **13866** | 15908 | **14227** | 15517 | **14595** | 15134 | **14969** | |
| 24 | 16703 | **13517** | 16299 | **13872** | 15901 | **14233** | 15511 | **14601** | 15127 | **14976** | 6 |
| | 16696 | **13523** | 16292 | **13878** | 15895 | **14240** | 15505 | **14608** | 15121 | **14982** | |
| 25 | 16689 | **13528** | 16285 | **13884** | 15888 | **14246** | 15498 | **14614** | 15115 | **14988** | 5 |
| | 16683 | **13534** | 16279 | **13890** | 15882 | **14252** | 15492 | **14620** | 15108 | **14995** | |
| 26 | 16676 | **13540** | 16272 | **13896** | 15875 | **14258** | 15485 | **14626** | 15102 | **15001** | 4 |
| | 16669 | **13546** | 16265 | **13902** | 15869 | **14264** | 15479 | **14632** | 15096 | **15007** | |
| 27 | 16662 | **13552** | 16259 | **13908** | 15862 | **14270** | 15472 | **14639** | 15089 | **15014** | 3 |
| | 16656 | **13558** | 16252 | **13914** | 15856 | **14276** | 15466 | **14645** | 15083 | **15020** | |
| 28 | 16649 | **13564** | 16245 | **13920** | 15849 | **14282** | 15459 | **14651** | 15077 | **15026** | 2 |
| | 16642 | **13570** | 16239 | **13926** | 15842 | **14288** | 15453 | **14657** | 15070 | **15033** | |
| 29 | 16635 | **13575** | 16232 | **13932** | 15836 | **14294** | 15447 | **14663** | 15064 | **15039** | 1 |
| | 16628 | **13581** | 16225 | **13938** | 15829 | **14300** | 15440 | **14670** | 15058 | **15045** | |
| 30 | 16622 | **13587** | 16219 | **13944** | 15823 | **14307** | 15434 | **14676** | 15051 | **15051** | 0 |
| | A | B | A | B | A | B | A | B | A | B | |
| ′ | 137° 00′ | | 136° 30′ | | 136° 00′ | | 135° 30′ | | 135° 00′ | | ′ |

# TABLE 35

## The Ageton Method

When Meridian Angle is Greater Than 90° Take "K" From Bottom of Table

| ′ | 45° 00′ | | 45° 30′ | | 46° 00′ | | 46° 30′ | | 47° 00′ | | ′ |
|---|---|---|---|---|---|---|---|---|---|---|---|
| | A | B | A | B | A | B | A | B | A | B | |
| 0 | 15051 | **15051** | 14676 | **15434** | 14307 | **15823** | 13944 | **16219** | 13587 | **16622** | 30 |
| | 15045 | **15058** | 14670 | **15440** | 14300 | **15829** | 13938 | **16225** | 13581 | **16628** | |
| 1 | 15039 | **15064** | 14663 | **15447** | 14294 | **15836** | 13932 | **16232** | 13575 | **16635** | 29 |
| | 15033 | **15070** | 14657 | **15453** | 14288 | **15842** | 13926 | **16239** | 13570 | **16642** | |
| 2 | 15026 | **15077** | 14651 | **15459** | 14282 | **15849** | 13920 | **16245** | 13564 | **16649** | 28 |
| | 15020 | **15083** | 14645 | **15466** | 14276 | **15856** | 13914 | **16252** | 13558 | **16656** | |
| 3 | 15014 | **15089** | 14639 | **15472** | 14270 | **15862** | 13908 | **16259** | 13552 | **16662** | 27 |
| | 15007 | **15096** | 14632 | **15479** | 14264 | **15869** | 13902 | **16265** | 13546 | **16669** | |
| 4 | 15001 | **15102** | 14626 | **15485** | 14258 | **15875** | 13896 | **16272** | 13540 | **16676** | 26 |
| | 14995 | **15108** | 14620 | **15492** | 14252 | **15882** | 13890 | **16279** | 13534 | **16683** | |
| 5 | 14988 | **15115** | 14614 | **15498** | 14246 | **15888** | 13884 | **16285** | 13528 | **16689** | 25 |
| | 14982 | **15121** | 14608 | **15505** | 14240 | **15895** | 13878 | **16292** | 13523 | **16696** | |
| 6 | 14976 | **15127** | 14601 | **15511** | 14233 | **15901** | 13872 | **16299** | 13517 | **16703** | 24 |
| | 14969 | **15134** | 14595 | **15517** | 14227 | **15908** | 13866 | **16305** | 13511 | **16710** | |
| 7 | 14963 | **15140** | 14589 | **15524** | 14221 | **15915** | 13860 | **16312** | 13505 | **16717** | 23 |
| | 14957 | **15146** | 14583 | **15530** | 14215 | **15921** | 13854 | **16319** | 13499 | **16723** | |
| 8 | 14951 | **15153** | 14577 | **15537** | 14209 | **15928** | 13848 | **16325** | 13493 | **16730** | 22 |
| | 14944 | **15159** | 14570 | **15543** | 14203 | **15934** | 13842 | **16332** | 13487 | **16737** | |
| 9 | 14938 | **15165** | 14564 | **15550** | 14197 | **15941** | 13836 | **16339** | 13481 | **16744** | 21 |
| | 14932 | **15172** | 14558 | **15556** | 14191 | **15947** | 13830 | **16346** | 13476 | **16751** | |
| 10 | 14925 | **15178** | 14552 | **15563** | 14185 | **15954** | 13824 | **16352** | 13470 | **16757** | 20 |
| | 14919 | **15184** | 14546 | **15569** | 14179 | **15961** | 13818 | **16359** | 13464 | **16764** | |
| 11 | 14913 | **15191** | 14540 | **15576** | 14173 | **15967** | 13812 | **16366** | 13458 | **16771** | 19 |
| | 14907 | **15197** | 14533 | **15582** | 14167 | **15974** | 13806 | **16372** | 13452 | **16778** | |
| 12 | 14900 | **15204** | 14527 | **15589** | 14161 | **15980** | 13800 | **16379** | 13446 | **16785** | 18 |
| | 14894 | **15210** | 14521 | **15595** | 14155 | **15987** | 13794 | **16386** | 13440 | **16792** | |
| 13 | 14888 | **15216** | 14515 | **15602** | 14149 | **15994** | 13788 | **16392** | 13435 | **16798** | 17 |
| | 14882 | **15223** | 14509 | **15608** | 14142 | **16000** | 13783 | **16399** | 13429 | **16805** | |
| 14 | 14875 | **15229** | 14503 | **15614** | 14136 | **16007** | 13777 | **16406** | 13423 | **16812** | 16 |
| | 14869 | **15235** | 14496 | **15621** | 14130 | **16013** | 13771 | **16413** | 13417 | **16819** | |
| 15 | 14863 | **15242** | 14490 | **15627** | 14124 | **16020** | 13765 | **16419** | 13411 | **16826** | 15 |
| | 14857 | **15248** | 14484 | **15634** | 14118 | **16027** | 13759 | **16426** | 13405 | **16833** | |
| 16 | 14850 | **15255** | 14478 | **15640** | 14112 | **16033** | 13753 | **16433** | 13400 | **16839** | 14 |
| | 14844 | **15261** | 14472 | **15647** | 14106 | **16040** | 13747 | **16439** | 13394 | **16846** | |
| 17 | 14838 | **15267** | 14466 | **15653** | 14100 | **16046** | 13741 | **16446** | 13388 | **16853** | 13 |
| | 14831 | **15274** | 14460 | **15660** | 14094 | **16053** | 13735 | **16453** | 13382 | **16860** | |
| 18 | 14825 | **15280** | 14453 | **15666** | 14088 | **16060** | 13729 | **16460** | 13376 | **16867** | 12 |
| | 14819 | **15286** | 14447 | **15673** | 14082 | **16066** | 13723 | **16466** | 13370 | **16874** | |
| 19 | 14813 | **15293** | 14441 | **15679** | 14076 | **16073** | 13717 | **16473** | 13365 | **16880** | 11 |
| | 14807 | **15299** | 14435 | **15686** | 14070 | **16079** | 13711 | **16480** | 13359 | **16887** | |
| 20 | 14800 | **15306** | 14429 | **15692** | 14064 | **16086** | 13705 | **16487** | 13353 | **16894** | 10 |
| | 14794 | **15312** | 14423 | **15699** | 14058 | **16093** | 13699 | **16493** | 13347 | **16901** | |
| 21 | 14788 | **15318** | 14417 | **15705** | 14052 | **16099** | 13694 | **16500** | 13341 | **16908** | 9 |
| | 14782 | **15325** | 14411 | **15712** | 14046 | **16105** | 13688 | **16507** | 13336 | **16915** | |
| 22 | 14775 | **15331** | 14404 | **15718** | 14040 | **16112** | 13682 | **16513** | 13330 | **16922** | 8 |
| | 14769 | **15338** | 14398 | **15725** | 14034 | **16119** | 13676 | **16520** | 13324 | **16928** | |
| 23 | 14763 | **15344** | 14392 | **15731** | 14028 | **16126** | 13670 | **16527** | 13318 | **16935** | 7 |
| | 14757 | **15350** | 14386 | **15738** | 14022 | **16132** | 13664 | **16534** | 13312 | **16942** | |
| 24 | 14750 | **15357** | 14380 | **15744** | 14016 | **16139** | 13658 | **16540** | 13306 | **16949** | 6 |
| | 14744 | **15363** | 14374 | **15751** | 14010 | **16146** | 13652 | **16547** | 13301 | **16956** | |
| 25 | 14738 | **15370** | 14368 | **15758** | 14004 | **16152** | 13646 | **16554** | 13295 | **16963** | 5 |
| | 14732 | **15376** | 14362 | **15764** | 13998 | **16159** | 13640 | **16561** | 13289 | **16970** | |
| 26 | 14725 | **15382** | 14355 | **15771** | 13992 | **16166** | 13634 | **16567** | 13283 | **16977** | 4 |
| | 14719 | **15389** | 14349 | **15777** | 13986 | **16172** | 13628 | **16574** | 13277 | **16983** | |
| 27 | 14713 | **15395** | 14343 | **15784** | 13980 | **16179** | 13623 | **16581** | 13272 | **16990** | 3 |
| | 14707 | **15402** | 14337 | **15790** | 13974 | **16185** | 13617 | **16588** | 13266 | **16997** | |
| 28 | 14701 | **15408** | 14331 | **15797** | 13968 | **16192** | 13611 | **16595** | 13260 | **17004** | 2 |
| | 14694 | **15414** | 14325 | **15803** | 13962 | **16199** | 13605 | **16601** | 13254 | **17011** | |
| 29 | 14688 | **15421** | 14319 | **15810** | 13956 | **16205** | 13599 | **16608** | 13248 | **17018** | 1 |
| | 14682 | **15427** | 14313 | **15816** | 13950 | **16212** | 13593 | **16615** | 13243 | **17025** | |
| 30 | 14676 | **15434** | 14307 | **15823** | 13944 | **16219** | 13587 | **16622** | 13237 | **17032** | 0 |
| | A | B | A | B | A | B | A | B | A | B | |
| ′ | 134° 30′ | | 134° 00′ | | 133° 30′ | | 133° 00′ | | 132° 30′ | | ′ |

## TABLE 35

### The Ageton Method

**Always Take "Z" From Bottom of Table, Except When "K" is Same Name and Greater Than Latitude, in Which Case Take "Z" From Top of Table**

| | 47° 30′ | | 48° 00′ | | 48° 30′ | | 49° 00′ | | 49° 30′ | | |
|---|---|---|---|---|---|---|---|---|---|---|---|
| ′ | A | B | A | B | A | B | A | B | A | B | ′ |
| 0 | 13237 | **17032** | 12893 | **17449** | 12554 | **17873** | 12222 | **18306** | 11895 | **18746** | 30 |
| | 13231 | **17039** | 12887 | **17456** | 12549 | **17881** | 12216 | **18313** | 11890 | **18753** | |
| 1 | 13225 | **17045** | 12881 | **17463** | 12543 | **17888** | 12211 | **18320** | 11885 | **18760** | 29 |
| | 13220 | **17052** | 12876 | **17470** | 12538 | **17895** | 12205 | **18327** | 11879 | **18768** | |
| 2 | 13214 | **17059** | 12870 | **17477** | 12532 | **17902** | 12200 | **18335** | 11874 | **18775** | 28 |
| | 13208 | **17066** | 12864 | **17484** | 12526 | **17909** | 12195 | **18342** | 11868 | **18783** | |
| 3 | 13202 | **17073** | 12859 | **17491** | 12521 | **17916** | 12189 | **18349** | 11863 | **18790** | 27 |
| | 13196 | **17080** | 12853 | **17498** | 12515 | **17924** | 12184 | **18357** | 11858 | **18797** | |
| 4 | 13191 | **17087** | 12847 | **17505** | 12510 | **17931** | 12178 | **18364** | 11852 | **18805** | 26 |
| | 13185 | **17094** | 12841 | **17512** | 12504 | **17938** | 12173 | **18371** | 11847 | **18812** | |
| 5 | 13179 | **17101** | 12836 | **17519** | 12499 | **17945** | 12167 | **18378** | 11842 | **18820** | 25 |
| | 13173 | **17108** | 12830 | **17526** | 12493 | **17952** | 12162 | **18386** | 11836 | **18827** | |
| 6 | 13168 | **17114** | 12824 | **17533** | 12487 | **17959** | 12156 | **18393** | 11831 | **18834** | 24 |
| | 13162 | **17121** | 12819 | **17540** | 12482 | **17966** | 12151 | **18400** | 11825 | **18842** | |
| 7 | 13156 | **17128** | 12813 | **17547** | 12476 | **17974** | 12145 | **18408** | 11820 | **18849** | 23 |
| | 13150 | **17135** | 12807 | **17554** | 12471 | **17981** | 12140 | **18415** | 11815 | **18857** | |
| 8 | 13144 | **17142** | 12802 | **17561** | 12465 | **17988** | 12134 | **18422** | 11809 | **18864** | 22 |
| | 13139 | **17149** | 12796 | **17568** | 12460 | **17995** | 12129 | **18429** | 11804 | **18872** | |
| 9 | 13133 | **17156** | 12790 | **17576** | 12454 | **18002** | 12123 | **18437** | 11799 | **18879** | 21 |
| | 13127 | **17163** | 12785 | **17583** | 12448 | **18010** | 12118 | **18444** | 11793 | **18886** | |
| 10 | 13121 | **17170** | 12779 | **17590** | 12443 | **18017** | 12112 | **18451** | 11788 | **18894** | 20 |
| | 13116 | **17177** | 12774 | **17597** | 12437 | **18024** | 12107 | **18459** | 11782 | **18901** | |
| 11 | 13110 | **17184** | 12768 | **17604** | 12432 | **18031** | 12102 | **18466** | 11777 | **18909** | 19 |
| | 13104 | **17191** | 12762 | **17611** | 12426 | **18038** | 12096 | **18473** | 11772 | **18916** | |
| 12 | 13098 | **17198** | 12757 | **17618** | 12421 | **18045** | 12091 | **18481** | 11766 | **18924** | 18 |
| | 13093 | **17205** | 12751 | **17625** | 12415 | **18053** | 12085 | **18488** | 11761 | **18931** | |
| 13 | 13087 | **17212** | 12745 | **17632** | 12410 | **18060** | 12080 | **18495** | 11756 | **18939** | 17 |
| | 13081 | **17218** | 12740 | **17639** | 12404 | **18067** | 12074 | **18503** | 11750 | **18946** | |
| 14 | 13075 | **17225** | 12734 | **17646** | 12398 | **18074** | 12069 | **18510** | 11745 | **18953** | 16 |
| | 13070 | **17232** | 12728 | **17653** | 12393 | **18081** | 12063 | **18517** | 11740 | **18961** | |
| 15 | 13064 | **17239** | 12723 | **17660** | 12387 | **18089** | 12058 | **18525** | 11734 | **18968** | 15 |
| | 13058 | **17246** | 12717 | **17667** | 12382 | **18096** | 12053 | **18532** | 11729 | **18976** | |
| 16 | 13053 | **17253** | 12711 | **17674** | 12376 | **18103** | 12047 | **18539** | 11724 | **18983** | 14 |
| | 13047 | **17260** | 12706 | **17681** | 12371 | **18110** | 12042 | **18547** | 11718 | **18991** | |
| 17 | 13041 | **17267** | 12700 | **17689** | 12365 | **18117** | 12036 | **18554** | 11713 | **18998** | 13 |
| | 13035 | **17274** | 12695 | **17696** | 12360 | **18125** | 12031 | **18561** | 11708 | **19006** | |
| 18 | 13030 | **17281** | 12689 | **17703** | 12354 | **18132** | 12025 | **18569** | 11702 | **19013** | 12 |
| | 13024 | **17288** | 12683 | **17710** | 12349 | **18139** | 12020 | **18576** | 11697 | **19021** | |
| 19 | 13018 | **17295** | 12678 | **17717** | 12343 | **18146** | 12014 | **18583** | 11692 | **19028** | 11 |
| | 13012 | **17302** | 12672 | **17724** | 12338 | **18154** | 12009 | **18591** | 11686 | **19036** | |
| 20 | 13007 | **17309** | 12666 | **17731** | 12332 | **18161** | 12004 | **18598** | 11681 | **19043** | 10 |
| | 13001 | **17316** | 12661 | **17738** | 12327 | **18168** | 11998 | **18605** | 11676 | **19051** | |
| 21 | 12995 | **17323** | 12655 | **17745** | 12321 | **18175** | 11993 | **18613** | 11670 | **19058** | 9 |
| | 12990 | **17330** | 12650 | **17752** | 12316 | **18182** | 11987 | **18620** | 11665 | **19066** | |
| 22 | 12984 | **17337** | 12644 | **17760** | 12310 | **18190** | 11982 | **18627** | 11660 | **19073** | 8 |
| | 12978 | **17344** | 12638 | **17767** | 12305 | **18197** | 11976 | **18635** | 11654 | **19081** | |
| 23 | 12972 | **17351** | 12633 | **17774** | 12299 | **18204** | 11971 | **18642** | 11649 | **19088** | 7 |
| | 12967 | **17358** | 12627 | **17781** | 12293 | **18211** | 11966 | **18650** | 11644 | **19096** | |
| 24 | 12961 | **17365** | 12622 | **17788** | 12288 | **18219** | 11960 | **18657** | 11638 | **19103** | 6 |
| | 12955 | **17372** | 12616 | **17795** | 12282 | **18226** | 11955 | **18664** | 11633 | **19111** | |
| 25 | 12950 | **17379** | 12610 | **17802** | 12277 | **18233** | 11949 | **18672** | 11628 | **19118** | 5 |
| | 12944 | **17386** | 12605 | **17809** | 12271 | **18240** | 11944 | **18679** | 11622 | **19126** | |
| 26 | 12938 | **17393** | 12599 | **17816** | 12266 | **18248** | 11939 | **18686** | 11617 | **19133** | 4 |
| | 12932 | **17400** | 12593 | **17824** | 12260 | **18255** | 11933 | **18694** | 11612 | **19141** | |
| 27 | 12927 | **17407** | 12588 | **17831** | 12255 | **18262** | 11928 | **18701** | 11606 | **19148** | 3 |
| | 12921 | **17414** | 12582 | **17838** | 12249 | **18269** | 11922 | **18709** | 11601 | **19156** | |
| 28 | 12915 | **17421** | 12577 | **17845** | 12244 | **18277** | 11917 | **18716** | 11596 | **19163** | 2 |
| | 12910 | **17428** | 12571 | **17852** | 12238 | **18284** | 11912 | **18723** | 11590 | **19171** | |
| 29 | 12904 | **17435** | 12566 | **17859** | 12233 | **18291** | 11906 | **18731** | 11585 | **19178** | 1 |
| | 12898 | **17442** | 12560 | **17866** | 12227 | **18298** | 11901 | **18738** | 11580 | **19186** | |
| 30 | 12893 | **17449** | 12554 | **17873** | 12222 | **18306** | 11895 | **18746** | 11575 | **19193** | 0 |
| | A | B | A | B | A | B | A | B | A | B | |
| ′ | 132° 00′ | | 131° 30′ | | 131° 00′ | | 130° 30′ | | 130° 00′ | | ′ |

# TABLE 35

## The Ageton Method

When Meridian Angle is Greater Than 90° Take "K" From Bottom of Table

| ′ | 50° 00′ | | 50° 30′ | | 51° 00′ | | 51° 30′ | | 52° 00′ | | ′ |
|---|---|---|---|---|---|---|---|---|---|---|---|
| | A | B | A | B | A | B | A | B | A | B | |
| 0 | 11575 | **19193** | 11259 | **19649** | 10950 | **20113** | 10646 | **20585** | 10347 | **21066** | 30 |
| | 11569 | **19201** | 11254 | **19657** | 10945 | **20121** | 10640 | **20593** | 10342 | **21074** | |
| 1 | 11564 | **19208** | 11249 | **19664** | 10939 | **20128** | 10635 | **20601** | 10337 | **21082** | 29 |
| | 11559 | **19216** | 11244 | **19672** | 10934 | **20136** | 10630 | **20609** | 10332 | **21090** | |
| 2 | 11553 | **19223** | 11239 | **19680** | 10929 | **20144** | 10625 | **20617** | 10327 | **21098** | 28 |
| | 11548 | **19231** | 11233 | **19687** | 10924 | **20152** | 10620 | **20625** | 10322 | **21106** | |
| 3 | 11543 | **19238** | 11228 | **19695** | 10919 | **20160** | 10615 | **20633** | 10317 | **21114** | 27 |
| | 11537 | **19246** | 11223 | **19703** | 10914 | **20167** | 10610 | **20641** | 10312 | **21122** | |
| 4 | 11532 | **19253** | 11218 | **19710** | 10909 | **20175** | 10605 | **20649** | 10307 | **21131** | 26 |
| | 11527 | **19261** | 11213 | **19718** | 10904 | **20183** | 10600 | **20657** | 10302 | **21139** | |
| 5 | 11522 | **19269** | 11207 | **19726** | 10899 | **20191** | 10595 | **20665** | 10298 | **21147** | 25 |
| | 11516 | **19276** | 11202 | **19733** | 10894 | **20199** | 10590 | **20672** | 10293 | **21155** | |
| 6 | 11511 | **19284** | 11197 | **19741** | 10888 | **20207** | 10585 | **20680** | 10288 | **21163** | 24 |
| | 11506 | **19291** | 11192 | **19749** | 10883 | **20214** | 10580 | **20688** | 10283 | **21171** | |
| 7 | 11501 | **19299** | 11187 | **19756** | 10878 | **20222** | 10575 | **20696** | 10278 | **21179** | 23 |
| | 11495 | **19306** | 11181 | **19764** | 10873 | **20230** | 10570 | **20704** | 10273 | **21187** | |
| 8 | 11490 | **19314** | 11176 | **19772** | 10868 | **20238** | 10565 | **20712** | 10268 | **21195** | 22 |
| | 11485 | **19321** | 11171 | **19779** | 10863 | **20246** | 10560 | **20720** | 10263 | **21204** | |
| 9 | 11479 | **19329** | 11166 | **19787** | 10858 | **20254** | 10555 | **20728** | 10258 | **21212** | 21 |
| | 11474 | **19337** | 11161 | **19795** | 10853 | **20261** | 10550 | **20736** | 10253 | **21220** | |
| 10 | 11469 | **19344** | 11156 | **19803** | 10848 | **20269** | 10545 | **20744** | 10248 | **21228** | 20 |
| | 11464 | **19352** | 11150 | **19810** | 10843 | **20277** | 10540 | **20752** | 10243 | **21236** | |
| 11 | 11458 | **19359** | 11145 | **19818** | 10838 | **20285** | 10535 | **20760** | 10239 | **21244** | 19 |
| | 11453 | **19367** | 11140 | **19826** | 10832 | **20293** | 10530 | **20768** | 10234 | **21252** | |
| 12 | 11448 | **19375** | 11135 | **19834** | 10827 | **20301** | 10525 | **20776** | 10229 | **21260** | 18 |
| | 11443 | **19382** | 11130 | **19841** | 10822 | **20308** | 10520 | **20784** | 10224 | **21269** | |
| 13 | 11437 | **19390** | 11124 | **19849** | 10817 | **20316** | 10515 | **20792** | 10219 | **21277** | 17 |
| | 11432 | **19397** | 11119 | **19857** | 10812 | **20324** | 10510 | **20800** | 10214 | **21285** | |
| 14 | 11427 | **19405** | 11114 | **19864** | 10807 | **20332** | 10505 | **20808** | 10209 | **21293** | 16 |
| | 11421 | **19412** | 11109 | **19872** | 10802 | **20340** | 10500 | **20816** | 10204 | **21301** | |
| 15 | 11416 | **19420** | 11104 | **19880** | 10797 | **20348** | 10496 | **20824** | 10199 | **21309** | 15 |
| | 11411 | **19428** | 11099 | **19888** | 10792 | **20356** | 10491 | **20832** | 10195 | **21318** | |
| 16 | 11406 | **19435** | 11094 | **19895** | 10787 | **20364** | 10486 | **20840** | 10190 | **21326** | 14 |
| | 11400 | **19443** | 11088 | **19903** | 10782 | **20371** | 10481 | **20848** | 10185 | **21334** | |
| 17 | 11395 | **19450** | 11083 | **19911** | 10777 | **20379** | 10476 | **20856** | 10180 | **21342** | 13 |
| | 11390 | **19458** | 11078 | **19918** | 10772 | **20387** | 10471 | **20864** | 10175 | **21350** | |
| 18 | 11385 | **19466** | 11073 | **19926** | 10767 | **20395** | 10466 | **20872** | 10170 | **21358** | 12 |
| | 11380 | **19473** | 11068 | **19934** | 10761 | **20403** | 10461 | **20880** | 10165 | **21367** | |
| 19 | 11374 | **19481** | 11063 | **19942** | 10756 | **20411** | 10456 | **20888** | 10160 | **21375** | 11 |
| | 11369 | **19488** | 11057 | **19949** | 10751 | **20419** | 10451 | **20897** | 10155 | **21383** | |
| 20 | 11364 | **19496** | 11052 | **19957** | 10746 | **20427** | 10446 | **20905** | 10151 | **21391** | 10 |
| | 11359 | **19504** | 11047 | **19965** | 10741 | **20435** | 10441 | **20913** | 10146 | **21399** | |
| 21 | 11353 | **19511** | 11042 | **19973** | 10736 | **20442** | 10436 | **20921** | 10141 | **21407** | 9 |
| | 11348 | **19519** | 11037 | **19980** | 10731 | **20450** | 10431 | **20929** | 10136 | **21416** | |
| 22 | 11343 | **19527** | 11032 | **19988** | 10726 | **20458** | 10426 | **20937** | 10131 | **21424** | 8 |
| | 11338 | **19534** | 11027 | **19996** | 10721 | **20466** | 10421 | **20945** | 10126 | **21432** | |
| 23 | 11332 | **19542** | 11021 | **20004** | 10716 | **20474** | 10416 | **20953** | 10121 | **21440** | 7 |
| | 11327 | **19549** | 11016 | **20012** | 10711 | **20482** | 10411 | **20961** | 10116 | **21448** | |
| 24 | 11322 | **19557** | 11011 | **20019** | 10706 | **20490** | 10406 | **20969** | 10112 | **21457** | 6 |
| | 11317 | **19565** | 11006 | **20027** | 10701 | **20498** | 10401 | **20977** | 10107 | **21465** | |
| 25 | 11311 | **19572** | 11001 | **20035** | 10696 | **20506** | 10396 | **20985** | 10102 | **21473** | 5 |
| | 11306 | **19580** | 10996 | **20043** | 10691 | **20514** | 10391 | **20993** | 10097 | **21481** | |
| 26 | 11301 | **19588** | 10991 | **20050** | 10686 | **20522** | 10386 | **21001** | 10092 | **21489** | 4 |
| | 11296 | **19595** | 10986 | **20058** | 10681 | **20529** | 10381 | **21009** | 10087 | **21498** | |
| 27 | 11291 | **19603** | 10980 | **20066** | 10676 | **20537** | 10376 | **21017** | 10082 | **21506** | 3 |
| | 11285 | **19611** | 10975 | **20074** | 10671 | **20545** | 10372 | **21025** | 10078 | **21514** | |
| 28 | 11280 | **19618** | 10970 | **20082** | 10666 | **20553** | 10367 | **21033** | 10073 | **21522** | 2 |
| | 11275 | **19626** | 10965 | **20089** | 10661 | **20561** | 10362 | **21042** | 10068 | **21531** | |
| 29 | 11270 | **19634** | 10960 | **20097** | 10656 | **20569** | 10357 | **21050** | 10063 | **21539** | 1 |
| | 11265 | **19641** | 10955 | **20105** | 10651 | **20577** | 10352 | **21058** | 10058 | **21547** | |
| 30 | 11259 | **19649** | 10950 | **20113** | 10646 | **20585** | 10347 | **21066** | 10053 | **21555** | 0 |
| | A | B | A | B | A | B | A | B | A | B | |
| ′ | 129° 30′ | | 129° 00′ | | 128° 30′ | | 128° 00′ | | 127° 30′ | | ′ |

# TABLE 35

## The Ageton Method

**Always Take "Z" From Bottom of Table, Except When "K" is Same Name and Greater Than Latitude, in Which Case Take "Z" From Top of Table**

| ′ | 52° 30′ | | 53° 00′ | | 53° 30′ | | 54° 00′ | | 54° 30′ | | ′ |
|---|---|---|---|---|---|---|---|---|---|---|---|
| | A | B | A | B | A | B | A | B | A | B | |
| 0 | 10053 | 21555 | 9765 | 22054 | 9482 | 22561 | 9204 | 23078 | 8931 | 23605 | 30 |
| | 10049 | 21563 | 9760 | 22062 | 9477 | 22570 | 9200 | 23087 | 8927 | 23613 | |
| 1 | 10044 | 21572 | 9756 | 22070 | 9473 | 22578 | 9195 | 23095 | 8922 | 23622 | 29 |
| | 10039 | 21580 | 9751 | 22079 | 9468 | 22587 | 9190 | 23104 | 8918 | 23631 | |
| 2 | 10034 | 21588 | 9746 | 22087 | 9463 | 22595 | 9186 | 23113 | 8913 | 23640 | 28 |
| | 10029 | 21596 | 9741 | 22096 | 9459 | 22604 | 9181 | 23122 | 8909 | 23649 | |
| 3 | 10024 | 21605 | 9737 | 22104 | 9454 | 22612 | 9177 | 23130 | 8904 | 23658 | 27 |
| | 10019 | 21613 | 9732 | 22112 | 9449 | 22621 | 9172 | 23139 | 8900 | 23667 | |
| 4 | 10015 | 21621 | 9727 | 22121 | 9445 | 22630 | 9168 | 23148 | 8895 | 23675 | 26 |
| | 10010 | 21629 | 9722 | 22129 | 9440 | 22638 | 9163 | 23156 | 8891 | 23684 | |
| 5 | 10005 | 21638 | 9718 | 22138 | 9435 | 22647 | 9158 | 23165 | 8886 | 23693 | 25 |
| | 10000 | 21646 | 9713 | 22146 | 9431 | 22655 | 9154 | 23174 | 8882 | 23702 | |
| 6 | 9995 | 21654 | 9708 | 22154 | 9426 | 22664 | 9149 | 23183 | 8877 | 23711 | 24 |
| | 9990 | 21662 | 9703 | 22163 | 9421 | 22672 | 9145 | 23191 | 8873 | 23720 | |
| 7 | 9986 | 21671 | 9699 | 22171 | 9417 | 22681 | 9140 | 23200 | 8868 | 23729 | 23 |
| | 9981 | 21679 | 9694 | 22180 | 9412 | 22690 | 9136 | 23209 | 8864 | 23738 | |
| 8 | 9976 | 21687 | 9689 | 22188 | 9407 | 22698 | 9131 | 23218 | 8859 | 23747 | 22 |
| | 9971 | 21696 | 9684 | 22197 | 9403 | 22707 | 9126 | 23226 | 8855 | 23755 | |
| 9 | 9966 | 21704 | 9680 | 22205 | 9398 | 22715 | 9122 | 23235 | 8850 | 23764 | 21 |
| | 9962 | 21712 | 9675 | 22213 | 9394 | 22724 | 9117 | 23244 | 8846 | 23773 | |
| 10 | 9957 | 21720 | 9670 | 22222 | 9389 | 22732 | 9113 | 23252 | 8842 | 23782 | 20 |
| | 9952 | 21729 | 9665 | 22230 | 9384 | 22741 | 9108 | 23261 | 8837 | 23791 | |
| 11 | 9947 | 21737 | 9661 | 22239 | 9380 | 22750 | 9104 | 23270 | 8833 | 23800 | 19 |
| | 9942 | 21745 | 9656 | 22247 | 9375 | 22758 | 9099 | 23279 | 8828 | 23809 | |
| 12 | 9937 | 21754 | 9651 | 22256 | 9370 | 22767 | 9094 | 23288 | 8824 | 23818 | 18 |
| | 9933 | 21762 | 9647 | 22264 | 9366 | 22775 | 9090 | 23296 | 8819 | 23827 | |
| 13 | 9928 | 21770 | 9642 | 22272 | 9361 | 22784 | 9085 | 23305 | 8815 | 23836 | 17 |
| | 9923 | 21778 | 9637 | 22281 | 9356 | 22793 | 9081 | 23314 | 8810 | 23845 | |
| 14 | 9918 | 21787 | 9632 | 22289 | 9352 | 22801 | 9076 | 23323 | 8806 | 23854 | 16 |
| | 9913 | 21795 | 9628 | 22298 | 9347 | 22810 | 9072 | 23331 | 8801 | 23863 | |
| 15 | 9909 | 21803 | 9623 | 22306 | 9342 | 22818 | 9067 | 23340 | 8797 | 23871 | 15 |
| | 9904 | 21812 | 9618 | 22315 | 9338 | 22827 | 9063 | 23349 | 8792 | 23880 | |
| 16 | 9899 | 21820 | 9614 | 22323 | 9333 | 22836 | 9058 | 23358 | 8788 | 23889 | 14 |
| | 9894 | 21828 | 9609 | 22332 | 9329 | 22844 | 9054 | 23366 | 8783 | 23898 | |
| 17 | 9889 | 21837 | 9604 | 22340 | 9324 | 22853 | 9049 | 23375 | 8779 | 23907 | 13 |
| | 9885 | 21845 | 9599 | 22349 | 9319 | 22862 | 9044 | 23384 | 8775 | 23916 | |
| 18 | 9880 | 21853 | 9595 | 22357 | 9315 | 22870 | 9040 | 23393 | 8770 | 23925 | 12 |
| | 9875 | 21862 | 9590 | 22366 | 9310 | 22879 | 9035 | 23402 | 8766 | 23934 | |
| 19 | 9870 | 21870 | 9585 | 22374 | 9305 | 22887 | 9031 | 23410 | 8761 | 23943 | 11 |
| | 9865 | 21878 | 9581 | 22382 | 9301 | 22896 | 9026 | 23419 | 8757 | 23952 | |
| 20 | 9861 | 21887 | 9576 | 22391 | 9296 | 22905 | 9022 | 23428 | 8752 | 23961 | 10 |
| | 9856 | 21895 | 9571 | 22399 | 9292 | 22913 | 9017 | 23437 | 8748 | 23970 | |
| 21 | 9851 | 21903 | 9566 | 22408 | 9287 | 22922 | 9013 | 23446 | 8743 | 23979 | 9 |
| | 9846 | 21912 | 9562 | 22416 | 9282 | 22931 | 9008 | 23454 | 8739 | 23988 | |
| 22 | 9841 | 21920 | 9557 | 22425 | 9278 | 22939 | 9004 | 23463 | 8734 | 23997 | 8 |
| | 9837 | 21928 | 9552 | 22433 | 9273 | 22948 | 8999 | 23472 | 8730 | 24006 | |
| 23 | 9832 | 21937 | 9548 | 22442 | 9269 | 22957 | 8995 | 23481 | 8726 | 24015 | 7 |
| | 9827 | 21945 | 9543 | 22450 | 9264 | 22965 | 8990 | 23490 | 8721 | 24024 | |
| 24 | 9822 | 21953 | 9538 | 22459 | 9259 | 22974 | 8985 | 23498 | 8717 | 24033 | 6 |
| | 9818 | 21962 | 9534 | 22467 | 9255 | 22983 | 8981 | 23507 | 8712 | 24042 | |
| 25 | 9813 | 21970 | 9529 | 22476 | 9250 | 22991 | 8976 | 23516 | 8708 | 24051 | 5 |
| | 9808 | 21978 | 9524 | 22484 | 9246 | 23000 | 8972 | 23525 | 8703 | 24060 | |
| 26 | 9803 | 21987 | 9520 | 22493 | 9241 | 23009 | 8967 | 23534 | 8699 | 24069 | 4 |
| | 9798 | 21995 | 9515 | 22501 | 9236 | 23017 | 8963 | 23543 | 8694 | 24078 | |
| 27 | 9794 | 22003 | 9510 | 22510 | 9232 | 23026 | 8958 | 23551 | 8690 | 24037 | 3 |
| | 9789 | 22012 | 9505 | 22519 | 9227 | 23035 | 8954 | 23560 | 8686 | 24096 | |
| 28 | 9784 | 22020 | 9501 | 22527 | 9223 | 23043 | 8949 | 23569 | 8681 | 24105 | 2 |
| | 9779 | 22029 | 9496 | 22536 | 9218 | 23052 | 8945 | 23578 | 8677 | 24114 | |
| 29 | 9775 | 22037 | 9491 | 22544 | 9213 | 23061 | 8940 | 23587 | 8672 | 24123 | 1 |
| | 9770 | 22045 | 9487 | 22553 | 9209 | 23069 | 8936 | 23596 | 8668 | 24132 | |
| 30 | 9765 | 22054 | 9482 | 22561 | 9204 | 23078 | 8931 | 23605 | 8663 | 24141 | 0 |
| | A | B | A | B | A | B | A | B | A | B | |
| ′ | 127° 00′ | | 126° 30′ | | 126° 00′ | | 125° 30′ | | 125° 00′ | | ′ |

## TABLE 35

### The Ageton Method

**When Meridian Angle is Greater Than 90° Take "K" From Bottom of Table**

| | 55° 00′ | | 55° 30′ | | 56° 00′ | | 56° 30′ | | 57° 00′ | | |
|---|---|---|---|---|---|---|---|---|---|---|---|
| ′ | A | B | A | B | A | B | A | B | A | B | ′ |
| 0 | 8663 | 24141 | 8401 | 24687 | 8143 | 25244 | 7889 | 25811 | 7641 | 26389 | 30 |
| | 8659 | 24150 | 8396 | 24696 | 8138 | 25253 | 7885 | 25821 | 7637 | 26399 | |
| 1 | 8655 | 24159 | 8392 | 24706 | 8134 | 25263 | 7881 | 25830 | 7633 | 26409 | 29 |
| | 8650 | 24168 | 8388 | 24715 | 8130 | 25272 | 7877 | 25840 | 7629 | 26418 | |
| 2 | 8646 | 24177 | 8383 | 24724 | 8125 | 25281 | 7873 | 25849 | 7624 | 26428 | 28 |
| | 8641 | 24186 | 8379 | 24733 | 8121 | 25291 | 7868 | 25859 | 7620 | 26438 | |
| 3 | 8637 | 24195 | 8375 | 24742 | 8117 | 25300 | 7864 | 25868 | 7616 | 26447 | 27 |
| | 8633 | 24204 | 8370 | 24752 | 8113 | 25309 | 7860 | 25878 | 7612 | 26457 | |
| 4 | 8628 | 24213 | 8366 | 24761 | 8108 | 25319 | 7856 | 25887 | 7608 | 26467 | 26 |
| | 8624 | 24222 | 8362 | 24770 | 8104 | 25328 | 7852 | 25897 | 7604 | 26477 | |
| 5 | 8619 | 24231 | 8357 | 24779 | 8100 | 25338 | 7848 | 25907 | 7600 | 26486 | 25 |
| | 8615 | 24240 | 8353 | 24788 | 8096 | 25347 | 7843 | 25916 | 7596 | 26496 | |
| 6 | 8611 | 24249 | 8349 | 24798 | 8092 | 25356 | 7839 | 25926 | 7592 | 26506 | 24 |
| | 8606 | 24258 | 8344 | 24807 | 8087 | 25366 | 7835 | 25935 | 7588 | 26516 | |
| 7 | 8602 | 24267 | 8340 | 24816 | 8083 | 25375 | 7831 | 25945 | 7584 | 26526 | 23 |
| | 8597 | 24276 | 8336 | 24825 | 8079 | 25385 | 7827 | 25954 | 7579 | 26535 | |
| 8 | 8593 | 24286 | 8331 | 24835 | 8075 | 25394 | 7823 | 25964 | 7575 | 26545 | 22 |
| | 8589 | 24295 | 8327 | 24844 | 8070 | 25403 | 7818 | 25974 | 7571 | 26555 | |
| 9 | 8584 | 24304 | 8323 | 24853 | 8066 | 25413 | 7814 | 25983 | 7567 | 26565 | 21 |
| | 8580 | 24313 | 8318 | 24862 | 8062 | 25422 | 7810 | 25993 | 7563 | 26574 | |
| 10 | 8575 | 24322 | 8314 | 24872 | 8058 | 25432 | 7806 | 26002 | 7559 | 26584 | 20 |
| | 8571 | 24331 | 8310 | 24881 | 8053 | 25441 | 7802 | 26012 | 7555 | 26594 | |
| 11 | 8567 | 24340 | 8305 | 24890 | 8049 | 25451 | 7798 | 26022 | 7551 | 26604 | 19 |
| | 8562 | 24349 | 8301 | 24899 | 8045 | 25460 | 7793 | 26031 | 7547 | 26614 | |
| 12 | 8558 | 24358 | 8297 | 24909 | 8041 | 25469 | 7789 | 26041 | 7543 | 26623 | 18 |
| | 8553 | 24367 | 8292 | 24918 | 8036 | 25479 | 7785 | 26051 | 7539 | 26633 | |
| 13 | 8549 | 24376 | 8288 | 24927 | 8032 | 25488 | 7781 | 26060 | 7535 | 26643 | 17 |
| | 8545 | 24385 | 8284 | 24936 | 8028 | 25498 | 7777 | 26070 | 7531 | 26653 | |
| 14 | 8540 | 24395 | 8280 | 24946 | 8024 | 25507 | 7773 | 26079 | 7526 | 26663 | 16 |
| | 8536 | 24404 | 8275 | 24955 | 8020 | 25517 | 7769 | 26089 | 7522 | 26672 | |
| 15 | 8531 | 24413 | 8271 | 24964 | 8015 | 25526 | 7764 | 26099 | 7518 | 26682 | 15 |
| | 8527 | 24422 | 8267 | 24973 | 8011 | 25536 | 7760 | 26108 | 7514 | 26692 | |
| 16 | 8523 | 24431 | 8262 | 24983 | 8007 | 25545 | 7756 | 26118 | 7510 | 26702 | 14 |
| | 8518 | 24440 | 8258 | 24992 | 8003 | 25554 | 7752 | 26128 | 7506 | 26712 | |
| 17 | 8514 | 24449 | 8254 | 25001 | 7998 | 25564 | 7748 | 26137 | 7502 | 26722 | 13 |
| | 8510 | 24458 | 8249 | 25011 | 7994 | 25573 | 7744 | 26147 | 7498 | 26731 | |
| 18 | 8505 | 24467 | 8245 | 25020 | 7990 | 25583 | 7740 | 26157 | 7494 | 26741 | 12 |
| | 8501 | 24477 | 8241 | 25029 | 7986 | 25592 | 7736 | 26166 | 7490 | 26751 | |
| 19 | 8496 | 24486 | 8237 | 25038 | 7982 | 25602 | 7731 | 26176 | 7486 | 26761 | 11 |
| | 8492 | 24495 | 8232 | 25048 | 7977 | 25611 | 7727 | 26185 | 7482 | 26771 | |
| 20 | 8488 | 24504 | 8228 | 25057 | 7973 | 25621 | 7723 | 26195 | 7478 | 26781 | 10 |
| | 8483 | 24513 | 8224 | 25066 | 7969 | 25630 | 7719 | 26205 | 7474 | 26790 | |
| 21 | 8479 | 24522 | 8219 | 25076 | 7965 | 25640 | 7715 | 26214 | 7470 | 26800 | 9 |
| | 8475 | 24531 | 8215 | 25085 | 7961 | 25649 | 7711 | 26224 | 7466 | 26810 | |
| 22 | 8470 | 24540 | 8211 | 25094 | 7956 | 25659 | 7707 | 26234 | 7462 | 26820 | 8 |
| | 8466 | 24550 | 8207 | 25104 | 7952 | 25668 | 7702 | 26244 | 7458 | 26830 | |
| 23 | 8461 | 24559 | 8202 | 25113 | 7948 | 25678 | 7698 | 26253 | 7453 | 26840 | 7 |
| | 8457 | 24568 | 8198 | 25122 | 7944 | 25687 | 7694 | 26263 | 7449 | 26850 | |
| 24 | 8453 | 24577 | 8194 | 25132 | 7940 | 25697 | 7690 | 26273 | 7445 | 26860 | 6 |
| | 8448 | 24586 | 8189 | 25141 | 7935 | 25706 | 7686 | 26282 | 7441 | 26869 | |
| 25 | 8444 | 24595 | 8185 | 25150 | 7931 | 25716 | 7682 | 26292 | 7437 | 26879 | 5 |
| | 8440 | 24605 | 8181 | 25160 | 7927 | 25725 | 7678 | 26302 | 7433 | 26889 | |
| 26 | 8435 | 24614 | 8177 | 25169 | 7923 | 25735 | 7674 | 26311 | 7429 | 26899 | 4 |
| | 8431 | 24623 | 8172 | 25178 | 7919 | 25744 | 7670 | 26321 | 7425 | 26909 | |
| 27 | 8427 | 24632 | 8168 | 25188 | 7914 | 25754 | 7665 | 26331 | 7421 | 26919 | 3 |
| | 8422 | 24641 | 8164 | 25197 | 7910 | 25763 | 7661 | 26340 | 7417 | 26929 | |
| 28 | 8418 | 24650 | 8160 | 25206 | 7906 | 25773 | 7657 | 26350 | 7413 | 26939 | 2 |
| | 8414 | 24660 | 8155 | 25216 | 7902 | 25782 | 7653 | 26360 | 7409 | 26949 | |
| 29 | 8409 | 24669 | 8151 | 25225 | 7898 | 25792 | 7649 | 26370 | 7405 | 26958 | 1 |
| | 8405 | 24678 | 8147 | 25234 | 7893 | 25801 | 7645 | 26379 | 7401 | 26968 | |
| 30 | 8401 | 24687 | 8143 | 25244 | 7889 | 25811 | 7641 | 26389 | 7397 | 26978 | 0 |
| | A | B | A | B | A | B | A | B | A | B | |
| ′ | 124° 30′ | | 124° 00′ | | 123° 30′ | | 123° 00′ | | 122° 30′ | | ′ |

# TABLE 35

## The Ageton Method

**Always Take "Z" From Bottom of Table, Except When "K" is Same Name and Greater Than Latitude, in Which Case Take "Z" From Top of Table**

| ′ | 57° 30′ | | 58° 00′ | | 58° 30′ | | 59° 00′ | | 59° 30′ | | ′ |
|---|---|---|---|---|---|---|---|---|---|---|---|
| ′ | A | B | A | B | A | B | A | B | A | B | ′ |
| 0 | 7397 | 26978 | 7158 | 27579 | 6923 | 28191 | 6693 | 28816 | 6468 | 29453 | 30 |
| | 7393 | 26988 | 7154 | 27589 | 6920 | 28202 | 6690 | 28827 | 6464 | 29464 | |
| 1 | 7389 | 26998 | 7150 | 27599 | 6916 | 28212 | 6686 | 28837 | 6460 | 29475 | 29 |
| | 7385 | 27008 | 7146 | 27609 | 6912 | 28222 | 6682 | 28848 | 6457 | 29485 | |
| 2 | 7381 | 27018 | 7142 | 27619 | 6908 | 28233 | 6678 | 28858 | 6453 | 29496 | 28 |
| | 7377 | 27028 | 7138 | 27630 | 6904 | 28243 | 6674 | 28869 | 6449 | 29507 | |
| 3 | 7373 | 27038 | 7134 | 27640 | 6900 | 28253 | 6671 | 28879 | 6446 | 29517 | 27 |
| | 7369 | 27048 | 7130 | 27650 | 6896 | 28264 | 6667 | 28890 | 6442 | 29528 | |
| 4 | 7365 | 27058 | 7126 | 27660 | 6892 | 28274 | 6663 | 28900 | 6438 | 29539 | 26 |
| | 7361 | 27068 | 7122 | 27670 | 6889 | 28284 | 6659 | 28911 | 6434 | 29550 | |
| 5 | 7357 | 27078 | 7118 | 27680 | 6885 | 28295 | 6655 | 28921 | 6431 | 29560 | 25 |
| | 7353 | 27088 | 7115 | 27690 | 6881 | 28305 | 6652 | 28932 | 6427 | 29571 | |
| 6 | 7349 | 27098 | 7111 | 27701 | 6877 | 28315 | 6648 | 28942 | 6423 | 29582 | 24 |
| | 7345 | 27107 | 7107 | 27711 | 6873 | 28326 | 6644 | 28953 | 6420 | 29593 | |
| 7 | 7341 | 27117 | 7103 | 27721 | 6869 | 28336 | 6640 | 28964 | 6416 | 29604 | 23 |
| | 7337 | 27127 | 7099 | 27731 | 6865 | 28346 | 6637 | 28974 | 6412 | 29614 | |
| 8 | 7333 | 27137 | 7095 | 27741 | 6862 | 28357 | 6633 | 28985 | 6409 | 29625 | 22 |
| | 7329 | 27147 | 7091 | 27751 | 6858 | 28367 | 6629 | 28995 | 6405 | 29636 | |
| 9 | 7325 | 27157 | 7087 | 27761 | 6854 | 28378 | 6625 | 29006 | 6401 | 29647 | 21 |
| | 7321 | 27167 | 7083 | 27772 | 6850 | 28388 | 6622 | 29016 | 6397 | 29657 | |
| 10 | 7317 | 27177 | 7079 | 27782 | 6846 | 28398 | 6618 | 29027 | 6394 | 29668 | 20 |
| | 7313 | 27187 | 7075 | 27792 | 6842 | 28409 | 6614 | 29038 | 6390 | 29679 | |
| 11 | 7309 | 27197 | 7071 | 27802 | 6839 | 28419 | 6610 | 29048 | 6386 | 29690 | 19 |
| | 7305 | 27207 | 7068 | 27812 | 6835 | 28429 | 6607 | 29059 | 6383 | 29701 | |
| 12 | 7301 | 27217 | 7064 | 27823 | 6831 | 28440 | 6603 | 29069 | 6379 | 29711 | 18 |
| | 7297 | 27227 | 7060 | 27833 | 6827 | 28450 | 6599 | 29080 | 6375 | 29722 | |
| 13 | 7293 | 27237 | 7056 | 27843 | 6823 | 28461 | 6595 | 29091 | 6372 | 29733 | 17 |
| | 7289 | 27247 | 7052 | 27853 | 6819 | 28471 | 6591 | 29101 | 6368 | 29744 | |
| 14 | 7285 | 27257 | 7048 | 27863 | 6815 | 28481 | 6588 | 29112 | 6364 | 29755 | 16 |
| | 7281 | 27267 | 7044 | 27874 | 6812 | 28492 | 6584 | 29122 | 6361 | 29766 | |
| 15 | 7277 | 27277 | 7040 | 27884 | 6808 | 28502 | 6580 | 29133 | 6357 | 29776 | 15 |
| | 7273 | 27287 | 7036 | 27894 | 6804 | 28513 | 6576 | 29144 | 6353 | 29787 | |
| 16 | 7269 | 27297 | 7032 | 27904 | 6800 | 28523 | 6573 | 29154 | 6349 | 29798 | 14 |
| | 7265 | 27307 | 7028 | 27914 | 6796 | 28533 | 6569 | 29165 | 6346 | 29809 | |
| 17 | 7261 | 27317 | 7024 | 27925 | 6792 | 28544 | 6565 | 29175 | 6342 | 29820 | 13 |
| | 7257 | 27327 | 7021 | 27935 | 6789 | 28554 | 6561 | 29186 | 6338 | 29831 | |
| 18 | 7253 | 27337 | 7017 | 27945 | 6785 | 28565 | 6558 | 29197 | 6335 | 29841 | 12 |
| | 7249 | 27347 | 7013 | 27955 | 6781 | 28575 | 6554 | 29207 | 6331 | 29852 | |
| 19 | 7245 | 27357 | 7009 | 27965 | 6777 | 28586 | 6550 | 29218 | 6327 | 29863 | 11 |
| | 7241 | 27367 | 7005 | 27976 | 6773 | 28596 | 6546 | 29229 | 6324 | 29874 | |
| 20 | 7237 | 27377 | 7001 | 27986 | 6770 | 28607 | 6543 | 29239 | 6320 | 29885 | 10 |
| | 7233 | 27387 | 6997 | 27996 | 6766 | 28617 | 6539 | 29250 | 6316 | 29896 | |
| 21 | 7229 | 27398 | 6993 | 28006 | 6762 | 28627 | 6535 | 29261 | 6313 | 29907 | 9 |
| | 7225 | 27408 | 6989 | 28017 | 6758 | 28638 | 6531 | 29271 | 6309 | 29917 | |
| 22 | 7221 | 27418 | 6985 | 28027 | 6754 | 28648 | 6528 | 29282 | 6305 | 29929 | 8 |
| | 7217 | 27428 | 6982 | 28037 | 6750 | 28659 | 6524 | 29293 | 6302 | 29939 | |
| 23 | 7213 | 27438 | 6978 | 28047 | 6747 | 28669 | 6520 | 29303 | 6298 | 29950 | 7 |
| | 7209 | 27448 | 6974 | 28058 | 6743 | 28680 | 6516 | 29314 | 6294 | 29961 | |
| 24 | 7205 | 27458 | 6970 | 28068 | 6739 | 28690 | 6513 | 29325 | 6291 | 29972 | 6 |
| | 7201 | 27468 | 6966 | 28078 | 6735 | 28701 | 6509 | 29335 | 6287 | 29983 | |
| 25 | 7197 | 27478 | 6962 | 28089 | 6731 | 28711 | 6505 | 29346 | 6283 | 29994 | 5 |
| | 7193 | 27488 | 6958 | 28099 | 6728 | 28722 | 6502 | 29357 | 6280 | 30005 | |
| 26 | 7190 | 27498 | 6954 | 28109 | 6724 | 28732 | 6498 | 29367 | 6276 | 30015 | 4 |
| | 7186 | 27508 | 6951 | 28119 | 6720 | 28743 | 6494 | 29378 | 6272 | 30026 | |
| 27 | 7182 | 27518 | 6947 | 28130 | 6716 | 28753 | 6490 | 29389 | 6269 | 30037 | 3 |
| | 7178 | 27528 | 6943 | 28140 | 6712 | 28763 | 6487 | 29399 | 6265 | 30048 | |
| 28 | 7174 | 27539 | 6939 | 28150 | 6709 | 28774 | 6483 | 29410 | 6261 | 30059 | 2 |
| | 7170 | 27549 | 6935 | 28161 | 6705 | 28784 | 6479 | 29421 | 6258 | 30070 | |
| 29 | 7166 | 27559 | 6931 | 28171 | 6701 | 28795 | 6475 | 29432 | 6254 | 30081 | 1 |
| | 7162 | 27569 | 6927 | 28181 | 6697 | 28806 | 6472 | 29442 | 6251 | 30092 | |
| 30 | 7158 | 27579 | 6923 | 28191 | 6693 | 28816 | 6468 | 29453 | 6247 | 30103 | 0 |
| ′ | A | B | A | B | A | B | A | B | A | B | ′ |
| ′ | 122° 00′ | | 121° 30′ | | 121° 00′ | | 120° 30′ | | 120° 00′ | | ′ |

## TABLE 35

### The Ageton Method

**When Meridian Angle is Greater Than 90° Take "K" From Bottom of Table**

| ′ | 60° 00′ | | 60° 30′ | | 61° 00′ | | 61° 30′ | | 62° 00′ | | ′ |
|---|---|---|---|---|---|---|---|---|---|---|---|
| | A | B | A | B | A | B | A | B | A | B | |
| 0 | 6247 | 30103 | 6030 | 30766 | 5818 | 31443 | 5610 | 32134 | 5406 | 32839 | 30 |
| | 6243 | 30114 | 6027 | 30777 | 5815 | 31454 | 5607 | 32145 | 5403 | 32851 | |
| 1 | 6240 | 30125 | 6023 | 30788 | 5811 | 31466 | 5603 | 32157 | 5400 | 32863 | 29 |
| | 6236 | 30136 | 6020 | 30800 | 5808 | 31477 | 5600 | 32169 | 5396 | 32875 | |
| 2 | 6232 | 30147 | 6016 | 30811 | 5804 | 31488 | 5596 | 32180 | 5393 | 32887 | 28 |
| | 6229 | 30158 | 6012 | 30822 | 5801 | 31500 | 5593 | 32192 | 5390 | 32898 | |
| 3 | 6225 | 30169 | 6009 | 30833 | 5797 | 31511 | 5590 | 32204 | 5386 | 32910 | 27 |
| | 6221 | 30180 | 6005 | 30844 | 5794 | 31523 | 5586 | 32215 | 5383 | 32922 | |
| 4 | 6218 | 30191 | 6002 | 30856 | 5790 | 31534 | 5583 | 32227 | 5380 | 32934 | 26 |
| | 6214 | 30202 | 5998 | 30867 | 5787 | 31546 | 5579 | 32239 | 5376 | 32946 | |
| 5 | 6210 | 30213 | 5995 | 30878 | 5783 | 31557 | 5575 | 32250 | 5373 | 32958 | 25 |
| | 6207 | 30224 | 5991 | 30889 | 5780 | 31569 | 5572 | 32262 | 5370 | 32970 | |
| 6 | 6203 | 30235 | 5987 | 30900 | 5776 | 31580 | 5569 | 32274 | 5366 | 32982 | 24 |
| | 6200 | 30245 | 5984 | 30912 | 5773 | 31591 | 5566 | 32285 | 5363 | 32994 | |
| 7 | 6196 | 30256 | 5980 | 30923 | 5769 | 31603 | 5562 | 32297 | 5360 | 33006 | 23 |
| | 6192 | 30267 | 5977 | 30934 | 5766 | 31614 | 5559 | 32309 | 5356 | 33018 | |
| 8 | 6189 | 30278 | 5973 | 30945 | 5762 | 31626 | 5555 | 32320 | 5353 | 33030 | 22 |
| | 6185 | 30289 | 5970 | 30956 | 5759 | 31637 | 5552 | 32332 | 5350 | 33042 | |
| 9 | 6181 | 30300 | 5966 | 30968 | 5755 | 31649 | 5549 | 32344 | 5346 | 33054 | 21 |
| | 6178 | 30311 | 5963 | 30979 | 5752 | 31660 | 5545 | 32355 | 5343 | 33065 | |
| 10 | 6174 | 30322 | 5959 | 30990 | 5748 | 31672 | 5542 | 32367 | 5340 | 33077 | 20 |
| | 6171 | 30334 | 5955 | 31001 | 5745 | 31683 | 5538 | 32379 | 5336 | 33089 | |
| 11 | 6167 | 30345 | 5952 | 31013 | 5741 | 31694 | 5535 | 32391 | 5333 | 33101 | 19 |
| | 6163 | 30355 | 5948 | 31024 | 5738 | 31706 | 5532 | 32402 | 5330 | 33113 | |
| 12 | 6160 | 30367 | 5945 | 31035 | 5734 | 31717 | 5528 | 32414 | 5326 | 33125 | 18 |
| | 6156 | 30378 | 5941 | 31046 | 5731 | 31729 | 5525 | 32426 | 5323 | 33137 | |
| 13 | 6152 | 30389 | 5938 | 31058 | 5727 | 31740 | 5521 | 32438 | 5320 | 33149 | 17 |
| | 6149 | 30400 | 5934 | 31069 | 5724 | 31752 | 5518 | 32449 | 5316 | 33161 | |
| 14 | 6145 | 30411 | 5931 | 31080 | 5720 | 31763 | 5515 | 32461 | 5313 | 33173 | 16 |
| | 6142 | 30422 | 5927 | 31091 | 5717 | 31775 | 5511 | 32473 | 5310 | 33185 | |
| 15 | 6138 | 30433 | 5924 | 31103 | 5714 | 31786 | 5508 | 32484 | 5306 | 33197 | 15 |
| | 6134 | 30444 | 5920 | 31114 | 5710 | 31798 | 5504 | 32496 | 5303 | 33209 | |
| 16 | 6131 | 30455 | 5917 | 31125 | 5707 | 31809 | 5501 | 32508 | 5300 | 33221 | 14 |
| | 6127 | 30466 | 5913 | 31137 | 5703 | 31821 | 5498 | 32520 | 5296 | 33233 | |
| 17 | 6124 | 30477 | 5909 | 31148 | 5700 | 31833 | 5494 | 32532 | 5293 | 33245 | 13 |
| | 6120 | 30488 | 5906 | 31159 | 5696 | 31844 | 5491 | 32543 | 5290 | 33257 | |
| 18 | 6116 | 30499 | 5902 | 31170 | 5693 | 31856 | 5487 | 32555 | 5286 | 33269 | 12 |
| | 6113 | 30510 | 5899 | 31182 | 5689 | 31867 | 5484 | 32567 | 5283 | 33281 | |
| 19 | 6109 | 30521 | 5895 | 31193 | 5686 | 31879 | 5481 | 32579 | 5280 | 33293 | 11 |
| | 6106 | 30532 | 5892 | 31204 | 5682 | 31890 | 5477 | 32590 | 5276 | 33306 | |
| 20 | 6102 | 30544 | 5888 | 31216 | 5679 | 31902 | 5474 | 32602 | 5273 | 33318 | 10 |
| | 6098 | 30555 | 5885 | 31227 | 5675 | 31913 | 5470 | 32614 | 5270 | 33330 | |
| 21 | 6095 | 30566 | 5881 | 31238 | 5672 | 31925 | 5467 | 32625 | 5266 | 33342 | 9 |
| | 6091 | 30577 | 5878 | 31250 | 5669 | 31936 | 5464 | 32638 | 5263 | 33354 | |
| 22 | 6088 | 30588 | 5874 | 31261 | 5665 | 31948 | 5460 | 32649 | 5260 | 33366 | 8 |
| | 6084 | 30599 | 5871 | 31272 | 5662 | 31960 | 5457 | 32661 | 5257 | 33378 | |
| 23 | 6080 | 30610 | 5867 | 31284 | 5658 | 31971 | 5454 | 32673 | 5253 | 33390 | 7 |
| | 6077 | 30621 | 5864 | 31295 | 5655 | 31983 | 5450 | 32685 | 5250 | 33402 | |
| 24 | 6073 | 30632 | 5860 | 31306 | 5651 | 31994 | 5447 | 32697 | 5247 | 33414 | 6 |
| | 6070 | 30643 | 5857 | 31318 | 5648 | 32006 | 5443 | 32709 | 5243 | 33426 | |
| 25 | 6066 | 30655 | 5853 | 31329 | 5644 | 32018 | 5440 | 32720 | 5240 | 33438 | 5 |
| | 6062 | 30666 | 5850 | 31340 | 5641 | 32029 | 5437 | 32732 | 5237 | 33450 | |
| 26 | 6059 | 30677 | 5846 | 31352 | 5638 | 32041 | 5433 | 32744 | 5233 | 33462 | 4 |
| | 6055 | 30688 | 5843 | 31363 | 5634 | 32052 | 5430 | 32756 | 5230 | 33475 | |
| 27 | 6052 | 30699 | 5839 | 31375 | 5631 | 32064 | 5427 | 32768 | 5227 | 33487 | 3 |
| | 6048 | 30710 | 5836 | 31386 | 5627 | 32076 | 5423 | 32780 | 5224 | 33499 | |
| 28 | 6045 | 30721 | 5832 | 31397 | 5624 | 32087 | 5420 | 32792 | 5220 | 33511 | 2 |
| | 6041 | 30733 | 5829 | 31409 | 5620 | 32099 | 5417 | 32803 | 5217 | 33523 | |
| 29 | 6037 | 30744 | 5825 | 31420 | 5617 | 32110 | 5413 | 32815 | 5214 | 33535 | 1 |
| | 6034 | 30755 | 5822 | 31431 | 5614 | 32122 | 5410 | 32827 | 5210 | 33547 | |
| 30 | 6030 | 30766 | 5818 | 31443 | 5610 | 32134 | 5406 | 32839 | 5207 | 33559 | 0 |
| | A | B | A | B | A | B | A | B | A | B | |
| ′ | 119° 30′ | | 119° 00′ | | 118° 30′ | | 118° 00′ | | 117° 30′ | | ′ |

## TABLE 35

### The Ageton Method

**Always Take "Z" From Bottom of Table, Except When "K" is Same Name and Greater Than Latitude, in Which Case Take "Z" From Top of Table**

| ′ | 62° 30′ | | 63° 00′ | | 63° 30′ | | 64° 00′ | | 64° 30′ | | ′ |
|---|---|---|---|---|---|---|---|---|---|---|---|
| ′ | A | B | A | B | A | B | A | B | A | B | ′ |
| 0 | 5207 | 33559 | 5012 | 34295 | 4821 | 35047 | 4634 | 35816 | 4451 | 36602 | 30 |
| | 5204 | 33572 | 5009 | 34308 | 4818 | 35060 | 4631 | 35829 | 4448 | 36615 | |
| 1 | 5200 | 33584 | 5005 | 34320 | 4815 | 35073 | 4628 | 35842 | 4445 | 36628 | 29 |
| | 5197 | 33596 | 5002 | 34332 | 4811 | 35085 | 4625 | 35855 | 4442 | 36641 | |
| 2 | 5194 | 33608 | 4999 | 34345 | 4808 | 35098 | 4622 | 35868 | 4439 | 36655 | 28 |
| | 5191 | 33620 | 4996 | 34357 | 4805 | 35111 | 4619 | 35881 | 4436 | 36668 | |
| 3 | 5187 | 33632 | 4993 | 34370 | 4802 | 35123 | 4615 | 35894 | 4433 | 36681 | 27 |
| | 5184 | 33644 | 4989 | 34382 | 4799 | 35136 | 4612 | 35907 | 4430 | 36694 | |
| 4 | 5181 | 33657 | 4986 | 34395 | 4796 | 35149 | 4609 | 35920 | 4427 | 36708 | 26 |
| | 5178 | 33669 | 4983 | 34407 | 4793 | 35161 | 4606 | 35933 | 4424 | 36721 | |
| 5 | 5174 | 33681 | 4980 | 34420 | 4789 | 35174 | 4603 | 35946 | 4421 | 36734 | 25 |
| | 5171 | 33693 | 4977 | 34432 | 4786 | 35187 | 4600 | 35959 | 4418 | 36747 | |
| 6 | 5168 | 33705 | 4973 | 34444 | 4783 | 35200 | 4597 | 35972 | 4415 | 36761 | 24 |
| | 5164 | 33717 | 4970 | 34457 | 4780 | 35212 | 4594 | 35985 | 4412 | 36774 | |
| 7 | 5161 | 33730 | 4967 | 34469 | 4777 | 35225 | 4591 | 35998 | 4409 | 36787 | 23 |
| | 5158 | 33742 | 4964 | 34482 | 4774 | 35238 | 4588 | 36011 | 4406 | 36801 | |
| 8 | 5155 | 33754 | 4961 | 34494 | 4771 | 35251 | 4585 | 36024 | 4403 | 36814 | 22 |
| | 5151 | 33766 | 4957 | 34507 | 4767 | 35263 | 4582 | 36037 | 4400 | 36827 | |
| 9 | 5148 | 33779 | 4954 | 34519 | 4764 | 35276 | 4579 | 36050 | 4397 | 36841 | 21 |
| | 5145 | 33791 | 4951 | 34532 | 4761 | 35289 | 4576 | 36063 | 4394 | 36854 | |
| 10 | 5142 | 33803 | 4948 | 34544 | 4758 | 35302 | 4573 | 36076 | 4391 | 36867 | 20 |
| | 5138 | 33815 | 4945 | 34557 | 4755 | 35314 | 4569 | 36089 | 4388 | 36881 | |
| 11 | 5135 | 33827 | 4941 | 34569 | 4752 | 35327 | 4566 | 36102 | 4385 | 36894 | 19 |
| | 5132 | 33840 | 4938 | 34582 | 4749 | 35340 | 4563 | 36115 | 4382 | 36907 | |
| 12 | 5128 | 33852 | 4935 | 34594 | 4746 | 35353 | 4560 | 36128 | 4379 | 36921 | 18 |
| | 5125 | 33864 | 4932 | 34607 | 4742 | 35365 | 4557 | 36141 | 4376 | 36934 | |
| 13 | 5122 | 33876 | 4929 | 34619 | 4739 | 35378 | 4554 | 36154 | 4373 | 36948 | 17 |
| | 5119 | 33889 | 4925 | 34632 | 4736 | 35391 | 4551 | 36167 | 4370 | 36961 | |
| 14 | 5115 | 33901 | 4922 | 34644 | 4733 | 35404 | 4548 | 36180 | 4367 | 36974 | 16 |
| | 5112 | 33913 | 4919 | 34657 | 4730 | 35417 | 4545 | 36193 | 4364 | 36988 | |
| 15 | 5109 | 33925 | 4916 | 34669 | 4727 | 35429 | 4542 | 36206 | 4361 | 37001 | 15 |
| | 5106 | 33938 | 4913 | 34682 | 4724 | 35442 | 4539 | 36220 | 4358 | 37014 | |
| 16 | 5102 | 33950 | 4910 | 34694 | 4721 | 35455 | 4536 | 36233 | 4355 | 37028 | 14 |
| | 5099 | 33962 | 4906 | 34707 | 4718 | 35468 | 4533 | 36246 | 4352 | 37041 | |
| 17 | 5096 | 33974 | 4903 | 34719 | 4714 | 35481 | 4530 | 36259 | 4349 | 37055 | 13 |
| | 5093 | 33987 | 4900 | 34732 | 4711 | 35493 | 4527 | 36272 | 4346 | 37068 | |
| 18 | 5089 | 33999 | 4897 | 34744 | 4708 | 35506 | 4524 | 36285 | 4343 | 37081 | 12 |
| | 5086 | 34011 | 4894 | 34757 | 4705 | 35519 | 4521 | 36298 | 4340 | 37095 | |
| 19 | 5083 | 34024 | 4890 | 34770 | 4702 | 35532 | 4518 | 36311 | 4337 | 37108 | 11 |
| | 5080 | 34036 | 4887 | 34782 | 4699 | 35545 | 4515 | 36325 | 4334 | 37122 | |
| 20 | 5076 | 34048 | 4884 | 34795 | 4696 | 35558 | 4512 | 36338 | 4332 | 37135 | 10 |
| | 5073 | 34061 | 4881 | 34807 | 4693 | 35571 | 4509 | 36351 | 4329 | 37149 | |
| 21 | 5070 | 34073 | 4878 | 34820 | 4690 | 35583 | 4506 | 36364 | 4326 | 37162 | 9 |
| | 5067 | 34085 | 4875 | 34832 | 4686 | 35596 | 4503 | 36377 | 4323 | 37176 | |
| 22 | 5064 | 34097 | 4871 | 34845 | 4683 | 35609 | 4500 | 36390 | 4320 | 37189 | 8 |
| | 5060 | 34110 | 4868 | 34858 | 4680 | 35622 | 4497 | 36403 | 4317 | 37203 | |
| 23 | 5057 | 34122 | 4865 | 34870 | 4677 | 35635 | 4493 | 36417 | 4314 | 37216 | 7 |
| | 5054 | 34134 | 4862 | 34883 | 4674 | 35648 | 4490 | 36430 | 4311 | 37229 | |
| 24 | 5051 | 34147 | 4859 | 34896 | 4671 | 35661 | 4487 | 36443 | 4308 | 37243 | 6 |
| | 5047 | 34159 | 4856 | 34908 | 4668 | 35674 | 4484 | 36456 | 4305 | 37256 | |
| 25 | 5044 | 34172 | 4852 | 34921 | 4665 | 35686 | 4481 | 36469 | 4302 | 37270 | 5 |
| | 5041 | 34184 | 4849 | 34933 | 4662 | 35699 | 4478 | 36483 | 4299 | 37283 | |
| 26 | 5038 | 34196 | 4846 | 34946 | 4659 | 35712 | 4475 | 36496 | 4296 | 37297 | 4 |
| | 5034 | 34209 | 4843 | 34959 | 4656 | 35725 | 4472 | 36509 | 4293 | 37310 | |
| 27 | 5031 | 34221 | 4840 | 34971 | 4652 | 35738 | 4469 | 36522 | 4290 | 37324 | 3 |
| | 5028 | 34233 | 4837 | 34984 | 4649 | 35751 | 4466 | 36535 | 4287 | 37337 | |
| 28 | 5025 | 34246 | 4833 | 34997 | 4646 | 35764 | 4463 | 36549 | 4284 | 37351 | 2 |
| | 5022 | 34258 | 4830 | 35009 | 4643 | 35777 | 4460 | 36562 | 4281 | 37365 | |
| 29 | 5018 | 34270 | 4827 | 35022 | 4640 | 35790 | 4457 | 36575 | 4278 | 37378 | 1 |
| | 5015 | 34283 | 4824 | 35035 | 4637 | 35803 | 4454 | 36588 | 4275 | 37392 | |
| 30 | 5012 | 34295 | 4821 | 35047 | 4634 | 35816 | 4451 | 36602 | 4272 | 37405 | 0 |
| | A | B | A | B | A | B | A | B | A | B | |
| ′ | 117° 00′ | | 116° 30′ | | 116° 00′ | | 115° 30′ | | 115° 00′ | | ′ |

## TABLE 35

### The Ageton Method

**When Meridian Angle is Greater Than 90° Take "K" From Bottom of Table**

| ′ | 65° 00′ | | 65° 30′ | | 66° 00′ | | 66° 30′ | | 67° 00′ | | ′ |
|---|---|---|---|---|---|---|---|---|---|---|---|
| | A | B | A | B | A | B | A | B | A | B | |
| 0 | 4272 | 37405 | 4098 | 38227 | 3927 | 39069 | 3760 | 39930 | 3597 | 40812 | 30 |
| | 4269 | 37419 | 4095 | 38241 | 3924 | 39083 | 3757 | 39945 | 3595 | 40827 | |
| 1 | 4266 | 37432 | 4092 | 38255 | 3921 | 39097 | 3755 | 39959 | 3592 | 40842 | 29 |
| | 4264 | 37446 | 4089 | 38269 | 3918 | 39111 | 3752 | 39974 | 3589 | 40857 | |
| 2 | 4261 | 37459 | 4086 | 38283 | 3916 | 39125 | 3749 | 39988 | 3587 | 40872 | 28 |
| | 4258 | 37473 | 4083 | 38297 | 3913 | 39140 | 3746 | 40003 | 3584 | 40887 | |
| 3 | 4255 | 37487 | 4080 | 38311 | 3910 | 39154 | 3744 | 40017 | 3581 | 40902 | 27 |
| | 4252 | 37500 | 4078 | 38324 | 3907 | 39168 | 3741 | 40032 | 3579 | 40916 | |
| 4 | 4249 | 37514 | 4075 | 38338 | 3904 | 39182 | 3738 | 40046 | 3576 | 40931 | 26 |
| | 4246 | 37527 | 4072 | 38352 | 3902 | 39197 | 3735 | 40061 | 3573 | 40946 | |
| 5 | 4243 | 37541 | 4069 | 38366 | 3899 | 39211 | 3733 | 40076 | 3571 | 40961 | 25 |
| | 4240 | 37554 | 4066 | 38380 | 3896 | 39225 | 3730 | 40090 | 3568 | 40976 | |
| 6 | 4237 | 37568 | 4063 | 38394 | 3893 | 39239 | 3727 | 40105 | 3565 | 40991 | 24 |
| | 4234 | 37582 | 4060 | 38408 | 3890 | 39254 | 3725 | 40119 | 3563 | 41006 | |
| 7 | 4231 | 37595 | 4057 | 38422 | 3888 | 39268 | 3722 | 40134 | 3560 | 41021 | 23 |
| | 4228 | 37609 | 4055 | 38436 | 3885 | 39282 | 3719 | 40149 | 3557 | 41036 | |
| 8 | 4225 | 37623 | 4052 | 38450 | 3882 | 39296 | 3716 | 40163 | 3555 | 41051 | 22 |
| | 4222 | 37636 | 4049 | 38464 | 3879 | 39311 | 3714 | 40178 | 3552 | 41066 | |
| 9 | 4220 | 37650 | 4046 | 38478 | 3876 | 39325 | 3711 | 40192 | 3549 | 41081 | 21 |
| | 4217 | 37663 | 4043 | 38492 | 3874 | 39339 | 3708 | 40207 | 3547 | 41096 | |
| 10 | 4214 | 37677 | 4040 | 38506 | 3871 | 39353 | 3705 | 40222 | 3544 | 41111 | 20 |
| | 4211 | 37691 | 4037 | 38520 | 3868 | 39368 | 3703 | 40236 | 3541 | 41126 | |
| 11 | 4208 | 37704 | 4035 | 38533 | 3865 | 39382 | 3700 | 40251 | 3539 | 41141 | 19 |
| | 4205 | 37718 | 4032 | 38547 | 3863 | 39396 | 3697 | 40266 | 3536 | 41156 | |
| 12 | 4202 | 37732 | 4029 | 38561 | 3860 | 39411 | 3695 | 40280 | 3533 | 41171 | 18 |
| | 4199 | 37745 | 4026 | 38575 | 3857 | 39425 | 3692 | 40295 | 3531 | 41186 | |
| 13 | 4196 | 37759 | 4023 | 38589 | 3854 | 39439 | 3689 | 40310 | 3528 | 41201 | 17 |
| | 4193 | 37773 | 4020 | 38603 | 3851 | 39454 | 3686 | 40324 | 3525 | 41216 | |
| 14 | 4190 | 37786 | 4017 | 38617 | 3849 | 39468 | 3684 | 40339 | 3523 | 41231 | 16 |
| | 4187 | 37800 | 4015 | 38631 | 3846 | 39482 | 3681 | 40354 | 3520 | 41246 | |
| 15 | 4185 | 37814 | 4012 | 38645 | 3843 | 39497 | 3678 | 40368 | 3517 | 41261 | 15 |
| | 4182 | 37828 | 4009 | 38660 | 3840 | 39511 | 3676 | 40383 | 3515 | 41276 | |
| 16 | 4179 | 37841 | 4006 | 38674 | 3838 | 39525 | 3673 | 40398 | 3512 | 41291 | 14 |
| | 4176 | 37855 | 4003 | 38688 | 3835 | 39540 | 3670 | 40413 | 3509 | 41307 | |
| 17 | 4173 | 37869 | 4000 | 38702 | 3832 | 39554 | 3667 | 40427 | 3507 | 41322 | 13 |
| | 4170 | 37882 | 3998 | 38716 | 3829 | 39569 | 3665 | 40442 | 3504 | 41337 | |
| 18 | 4167 | 37896 | 3995 | 38730 | 3826 | 39583 | 3662 | 40457 | 3502 | 41352 | 12 |
| | 4164 | 37910 | 3992 | 38744 | 3824 | 39597 | 3659 | 40471 | 3499 | 41367 | |
| 19 | 4161 | 37924 | 3989 | 38758 | 3821 | 39612 | 3657 | 40486 | 3496 | 41382 | 11 |
| | 4158 | 37937 | 3986 | 38772 | 3818 | 39626 | 3654 | 40501 | 3494 | 41397 | |
| 20 | 4155 | 37951 | 3983 | 38786 | 3815 | 39641 | 3651 | 40516 | 3491 | 41412 | 10 |
| | 4153 | 37965 | 3981 | 38800 | 3813 | 39655 | 3648 | 40530 | 3488 | 41427 | |
| 21 | 4150 | 37979 | 3978 | 38814 | 3810 | 39669 | 3646 | 40545 | 3486 | 41443 | 9 |
| | 4147 | 37992 | 3975 | 38828 | 3807 | 39684 | 3643 | 40560 | 3483 | 41458 | |
| 22 | 4144 | 38006 | 3972 | 38842 | 3804 | 39698 | 3640 | 40575 | 3480 | 41473 | 8 |
| | 4141 | 38020 | 3969 | 38856 | 3801 | 39713 | 3638 | 40590 | 3478 | 41488 | |
| 23 | 4138 | 38034 | 3966 | 38871 | 3799 | 39727 | 3635 | 40604 | 3475 | 41503 | 7 |
| | 4135 | 38048 | 3964 | 38885 | 3796 | 39742 | 3632 | 40619 | 3473 | 41518 | |
| 24 | 4132 | 38061 | 3961 | 38899 | 3793 | 39756 | 3630 | 40634 | 3470 | 41533 | 6 |
| | 4129 | 38075 | 3958 | 38913 | 3790 | 39771 | 3627 | 40649 | 3467 | 41549 | |
| 25 | 4127 | 38089 | 3955 | 38927 | 3788 | 39785 | 3624 | 40664 | 3465 | 41564 | 5 |
| | 4124 | 38103 | 3952 | 38941 | 3785 | 39799 | 3622 | 40678 | 3462 | 41579 | |
| 26 | 4121 | 38117 | 3949 | 38955 | 3782 | 39814 | 3619 | 40693 | 3459 | 41594 | 4 |
| | 4118 | 38130 | 3947 | 38969 | 3779 | 39828 | 3616 | 40708 | 3457 | 41609 | |
| 27 | 4115 | 38144 | 3944 | 38984 | 3777 | 39843 | 3613 | 40723 | 3454 | 41625 | 3 |
| | 4112 | 38158 | 3941 | 38998 | 3774 | 39857 | 3611 | 40738 | 3452 | 41640 | |
| 28 | 4109 | 38172 | 3938 | 39012 | 3771 | 39872 | 3608 | 40753 | 3449 | 41655 | 2 |
| | 4106 | 38186 | 3935 | 39026 | 3768 | 39886 | 3605 | 40768 | 3446 | 41670 | |
| 29 | 4103 | 38200 | 3933 | 39040 | 3766 | 39901 | 3603 | 40782 | 3444 | 41685 | 1 |
| | 4101 | 38213 | 3930 | 39054 | 3763 | 39915 | 3600 | 40797 | 3441 | 41701 | |
| 30 | 4098 | 38227 | 3927 | 39069 | 3760 | 39930 | 3597 | 40812 | 3438 | 41716 | 0 |
| | A | B | A | B | A | B | A | B | A | B | |
| ′ | 114° 30′ | | 114° 00′ | | 113° 30′ | | 113° 00′ | | 112° 30′ | | ′ |

## TABLE 35

### The Ageton Method

**Always Take "Z" From Bottom of Table, Except When "K" is Same Name and Greater Than Latitude, in Which Case Take "Z" From Top of Table**

| ′ | 67° 30′ | | 68° 00′ | | 68° 30′ | | 69° 00′ | | 69° 30′ | | ′ |
|---|---|---|---|---|---|---|---|---|---|---|---|
| | A | B | A | B | A | B | A | B | A | B | |
| 0 | 3438 | 41716 | 3283 | 42642 | 3132 | 43592 | 2985 | 44567 | 2841 | 45567 | 30 |
| | 3436 | 41731 | 3281 | 42658 | 3130 | 43608 | 2982 | 44583 | 2839 | 45584 | |
| 1 | 3433 | 41746 | 3278 | 42674 | 3127 | 43624 | 2980 | 44600 | 2836 | 45601 | 29 |
| | 3431 | 41762 | 3276 | 42689 | 3125 | 43641 | 2978 | 44616 | 2834 | 45618 | |
| 2 | 3428 | 41777 | 3273 | 42705 | 3122 | 43657 | 2975 | 44633 | 2832 | 45635 | 28 |
| | 3425 | 41792 | 3271 | 42721 | 3120 | 43673 | 2973 | 44649 | 2829 | 45652 | |
| 3 | 3423 | 41808 | 3268 | 42736 | 3117 | 43689 | 2970 | 44666 | 2827 | 45669 | 27 |
| | 3420 | 41823 | 3266 | 42752 | 3115 | 43705 | 2968 | 44682 | 2825 | 45686 | |
| 4 | 3418 | 41838 | 3263 | 42768 | 3112 | 43721 | 2965 | 44699 | 2822 | 45703 | 26 |
| | 3415 | 41853 | 3260 | 42783 | 3110 | 43737 | 2963 | 44715 | 2820 | 45720 | |
| 5 | 3412 | 41869 | 3258 | 42799 | 3107 | 43753 | 2961 | 44732 | 2818 | 45737 | 25 |
| | 3410 | 41884 | 3255 | 42815 | 3105 | 43769 | 2958 | 44748 | 2815 | 45754 | |
| 6 | 3407 | 41899 | 3253 | 42830 | 3102 | 43785 | 2956 | 44765 | 2813 | 45771 | 24 |
| | 3404 | 41915 | 3250 | 42846 | 3100 | 43801 | 2953 | 44782 | 2811 | 45788 | |
| 7 | 3402 | 41930 | 3248 | 42862 | 3097 | 43818 | 2951 | 44798 | 2808 | 45805 | 23 |
| | 3399 | 41945 | 3245 | 42878 | 3095 | 43834 | 2949 | 44815 | 2806 | 45822 | |
| 8 | 3397 | 41961 | 3243 | 42893 | 3092 | 43850 | 2946 | 44831 | 2804 | 45839 | 22 |
| | 3394 | 41976 | 3240 | 42909 | 3090 | 43866 | 2944 | 44848 | 2801 | 45856 | |
| 9 | 3391 | 41991 | 3237 | 42925 | 3088 | 43882 | 2941 | 44864 | 2799 | 45873 | 21 |
| | 3389 | 42007 | 3235 | 42941 | 3085 | 43898 | 2939 | 44881 | 2797 | 45890 | |
| 10 | 3386 | 42022 | 3233 | 42956 | 3083 | 43914 | 2936 | 44898 | 2794 | 45907 | 20 |
| | 3384 | 42038 | 3230 | 42972 | 3080 | 43931 | 2934 | 44914 | 2792 | 45924 | |
| 11 | 3381 | 42053 | 3227 | 42988 | 3078 | 43947 | 2932 | 44931 | 2789 | 45941 | 19 |
| | 3379 | 42068 | 3225 | 43004 | 3075 | 43963 | 2929 | 44947 | 2787 | 45958 | |
| 12 | 3376 | 42084 | 3222 | 43020 | 3073 | 43979 | 2927 | 44964 | 2785 | 45975 | 18 |
| | 3373 | 42099 | 3220 | 43035 | 3070 | 43995 | 2924 | 44981 | 2782 | 45992 | |
| 13 | 3371 | 42115 | 3217 | 43051 | 3068 | 44012 | 2922 | 44997 | 2780 | 46009 | 17 |
| | 3368 | 42130 | 3215 | 43067 | 3065 | 44028 | 2920 | 45014 | 2778 | 46026 | |
| 14 | 3366 | 42145 | 3212 | 43083 | 3063 | 44044 | 2917 | 45031 | 2775 | 46043 | 16 |
| | 3363 | 42161 | 3210 | 43099 | 3060 | 44060 | 2915 | 45047 | 2773 | 46061 | |
| 15 | 3360 | 42176 | 3207 | 43114 | 3058 | 44077 | 2913 | 45064 | 2771 | 46078 | 15 |
| | 3358 | 42192 | 3205 | 43130 | 3056 | 44093 | 2910 | 45081 | 2768 | 46095 | |
| 16 | 3355 | 42207 | 3202 | 43146 | 3053 | 44109 | 2908 | 45097 | 2766 | 46112 | 14 |
| | 3353 | 42223 | 3200 | 43162 | 3051 | 44125 | 2905 | 45114 | 2764 | 46129 | |
| 17 | 3350 | 42238 | 3197 | 43178 | 3048 | 44142 | 2903 | 45131 | 2761 | 46146 | 13 |
| | 3348 | 42254 | 3195 | 43194 | 3046 | 44158 | 2901 | 45147 | 2759 | 46163 | |
| 18 | 3345 | 42269 | 3192 | 43210 | 3043 | 44174 | 2898 | 45164 | 2757 | 46181 | 12 |
| | 3342 | 42285 | 3190 | 43225 | 3041 | 44190 | 2896 | 45181 | 2755 | 46198 | |
| 19 | 3340 | 42300 | 3187 | 43241 | 3038 | 44207 | 2893 | 45198 | 2752 | 46215 | 11 |
| | 3337 | 42316 | 3185 | 43257 | 3036 | 44223 | 2891 | 45214 | 2750 | 46232 | |
| 20 | 3335 | 42331 | 3182 | 43273 | 3033 | 44239 | 2889 | 45231 | 2748 | 46249 | 10 |
| | 3332 | 42347 | 3180 | 43289 | 3031 | 44256 | 2886 | 45248 | 2745 | 46266 | |
| 21 | 3329 | 42362 | 3177 | 43305 | 3029 | 44272 | 2884 | 45265 | 2743 | 46284 | 9 |
| | 3327 | 42378 | 3175 | 43321 | 3026 | 44288 | 2881 | 45281 | 2741 | 46301 | |
| 22 | 3324 | 42393 | 3172 | 43337 | 3024 | 44305 | 2879 | 45298 | 2738 | 46318 | 8 |
| | 3322 | 42409 | 3170 | 43353 | 3021 | 44321 | 2877 | 45315 | 2736 | 46335 | |
| 23 | 3319 | 42424 | 3167 | 43369 | 3019 | 44337 | 2874 | 45332 | 2734 | 46353 | 7 |
| | 3317 | 42440 | 3165 | 43385 | 3016 | 44354 | 2872 | 45348 | 2731 | 46370 | |
| 24 | 3314 | 42455 | 3162 | 43400 | 3014 | 44370 | 2870 | 45365 | 2729 | 46387 | 6 |
| | 3312 | 42471 | 3160 | 43416 | 3012 | 44386 | 2867 | 45382 | 2727 | 46404 | |
| 25 | 3309 | 42486 | 3157 | 43432 | 3009 | 44403 | 2865 | 45399 | 2724 | 46422 | 5 |
| | 3306 | 42502 | 3155 | 43448 | 3007 | 44419 | 2862 | 45416 | 2722 | 46439 | |
| 26 | 3304 | 42518 | 3152 | 43464 | 3004 | 44436 | 2860 | 45433 | 2720 | 46456 | 4 |
| | 3301 | 42533 | 3150 | 43480 | 3002 | 44452 | 2858 | 45449 | 2717 | 46473 | |
| 27 | 3299 | 42549 | 3147 | 43496 | 2999 | 44468 | 2855 | 45466 | 2715 | 46491 | 3 |
| | 3296 | 42564 | 3145 | 43512 | 2997 | 44485 | 2853 | 45483 | 2713 | 46508 | |
| 28 | 3294 | 42580 | 3142 | 43528 | 2994 | 44501 | 2851 | 45500 | 2711 | 46525 | 2 |
| | 3291 | 42596 | 3140 | 43544 | 2992 | 44518 | 2848 | 45517 | 2708 | 46543 | |
| 29 | 3289 | 42611 | 3137 | 43560 | 2990 | 44534 | 2846 | 45534 | 2706 | 46560 | 1 |
| | 3286 | 42627 | 3135 | 43576 | 2987 | 44551 | 2844 | 45551 | 2704 | 46577 | |
| 30 | 3283 | 42642 | 3132 | 43592 | 2985 | 44567 | 2841 | 45567 | 2701 | 46595 | 0 |
| | A | B | A | B | A | B | A | B | A | B | |
| ′ | 112° 00′ | | 111° 30′ | | 111° 00′ | | 110° 30′ | | 110° 00′ | | ′ |

# TABLE 35

## The Ageton Method

When Meridian Angle is Greater Than 90° Take "K" From Bottom of Table

| ′ | 70° 00′ | | 70° 30′ | | 71° 00′ | | 71° 30′ | | 72° 00′ | | ′ |
|---|---|---|---|---|---|---|---|---|---|---|---|
| | A | B | A | B | A | B | A | B | A | B | |
| 0 | 2701 | 46595 | 2565 | 47650 | 2433 | 48736 | 2304 | 49852 | 2179 | 51002 | 30 |
| | 2699 | 46612 | 2563 | 47668 | 2431 | 48754 | 2302 | 49871 | 2177 | 51021 | |
| 1 | 2697 | 46630 | 2561 | 47686 | 2429 | 48772 | 2300 | 49890 | 2175 | 51041 | 29 |
| | 2694 | 46647 | 2559 | 47704 | 2427 | 48791 | 2298 | 49909 | 2173 | 51060 | |
| 2 | 2692 | 46664 | 2556 | 47722 | 2424 | 48809 | 2296 | 49928 | 2171 | 51080 | 28 |
| | 2690 | 46682 | 2554 | 47740 | 2422 | 48828 | 2294 | 49947 | 2169 | 51099 | |
| 3 | 2688 | 46699 | 2552 | 47758 | 2420 | 48846 | 2292 | 49966 | 2167 | 51119 | 27 |
| | 2685 | 46716 | 2550 | 47775 | 2418 | 48864 | 2290 | 49985 | 2165 | 51138 | |
| 4 | 2683 | 46734 | 2547 | 47793 | 2416 | 48883 | 2287 | 50004 | 2163 | 51158 | 26 |
| | 2681 | 46751 | 2545 | 47811 | 2413 | 48901 | 2285 | 50023 | 2161 | 51177 | |
| 5 | 2678 | 46769 | 2543 | 47829 | 2411 | 48920 | 2283 | 50042 | 2159 | 51197 | 25 |
| | 2676 | 46786 | 2541 | 47847 | 2409 | 48938 | 2281 | 50061 | 2157 | 51216 | |
| 6 | 2674 | 46804 | 2539 | 47865 | 2407 | 48957 | 2279 | 50080 | 2155 | 51236 | 24 |
| | 2672 | 46821 | 2536 | 47883 | 2405 | 48975 | 2277 | 50098 | 2153 | 51255 | |
| 7 | 2669 | 46839 | 2534 | 47901 | 2403 | 48993 | 2275 | 50117 | 2151 | 51275 | 23 |
| | 2667 | 46856 | 2532 | 47919 | 2400 | 49012 | 2273 | 50137 | 2149 | 51294 | |
| 8 | 2665 | 46873 | 2530 | 47937 | 2398 | 49030 | 2271 | 50156 | 2147 | 51314 | 22 |
| | 2662 | 46891 | 2528 | 47955 | 2396 | 49049 | 2269 | 50175 | 2145 | 51334 | |
| 9 | 2660 | 46908 | 2525 | 47973 | 2394 | 49067 | 2266 | 50194 | 2143 | 51353 | 21 |
| | 2658 | 46926 | 2523 | 47991 | 2392 | 49086 | 2264 | 50213 | 2141 | 51373 | |
| 10 | 2656 | 46943 | 2521 | 48009 | 2390 | 49104 | 2262 | 50232 | 2138 | 51392 | 20 |
| | 2653 | 46961 | 2519 | 48027 | 2387 | 49123 | 2260 | 50251 | 2136 | 51412 | |
| 11 | 2651 | 46978 | 2516 | 48045 | 2385 | 49141 | 2258 | 50270 | 2134 | 51432 | 19 |
| | 2649 | 46996 | 2514 | 48063 | 2383 | 49160 | 2256 | 50289 | 2132 | 51451 | |
| 12 | 2646 | 47014 | 2512 | 48081 | 2381 | 49179 | 2254 | 50308 | 2130 | 51471 | 18 |
| | 2644 | 47031 | 2510 | 48099 | 2379 | 49197 | 2252 | 50327 | 2128 | 51491 | |
| 13 | 2642 | 47049 | 2507 | 48117 | 2377 | 49216 | 2250 | 50346 | 2126 | 51510 | 17 |
| | 2640 | 47066 | 2505 | 48135 | 2375 | 49234 | 2248 | 50365 | 2124 | 51530 | |
| 14 | 2637 | 47084 | 2503 | 48153 | 2372 | 49253 | 2246 | 50385 | 2122 | 51550 | 16 |
| | 2635 | 47101 | 2501 | 48171 | 2370 | 49271 | 2243 | 50404 | 2120 | 51570 | |
| 15 | 2633 | 47119 | 2499 | 48189 | 2368 | 49290 | 2241 | 50423 | 2118 | 51589 | 15 |
| | 2631 | 47137 | 2496 | 48207 | 2366 | 49309 | 2239 | 50442 | 2116 | 51609 | |
| 16 | 2628 | 47154 | 2494 | 48226 | 2364 | 49327 | 2237 | 50461 | 2114 | 51629 | 14 |
| | 2626 | 47172 | 2492 | 48244 | 2362 | 49346 | 2235 | 50480 | 2112 | 51649 | |
| 17 | 2624 | 47189 | 2490 | 48262 | 2360 | 49365 | 2233 | 50499 | 2110 | 51668 | 13 |
| | 2622 | 47207 | 2488 | 48280 | 2358 | 49383 | 2231 | 50519 | 2108 | 51688 | |
| 18 | 2619 | 47225 | 2485 | 48298 | 2355 | 49402 | 2229 | 50538 | 2106 | 51708 | 12 |
| | 2617 | 47242 | 2483 | 48316 | 2353 | 49421 | 2227 | 50557 | 2104 | 51728 | |
| 19 | 2615 | 47260 | 2481 | 48334 | 2351 | 49439 | 2225 | 50576 | 2102 | 51747 | 11 |
| | 2613 | 47278 | 2479 | 48352 | 2349 | 49458 | 2223 | 50596 | 2100 | 51767 | |
| 20 | 2610 | 47295 | 2477 | 48371 | 2347 | 49477 | 2221 | 50615 | 2098 | 51787 | 10 |
| | 2608 | 47313 | 2474 | 48389 | 2345 | 49495 | 2218 | 50634 | 2096 | 51807 | |
| 21 | 2606 | 47331 | 2472 | 48407 | 2343 | 49514 | 2216 | 50653 | 2094 | 51827 | 9 |
| | 2604 | 47348 | 2470 | 48425 | 2340 | 49533 | 2214 | 50673 | 2092 | 51847 | |
| 22 | 2601 | 47366 | 2468 | 48443 | 2338 | 49551 | 2212 | 50692 | 2090 | 51867 | 8 |
| | 2599 | 47384 | 2466 | 48462 | 2336 | 49570 | 2210 | 50711 | 2088 | 51886 | |
| 23 | 2597 | 47402 | 2463 | 48480 | 2334 | 49589 | 2208 | 50730 | 2086 | 51906 | 7 |
| | 2594 | 47419 | 2461 | 48498 | 2332 | 49608 | 2206 | 50750 | 2084 | 51926 | |
| 24 | 2592 | 47437 | 2459 | 48516 | 2330 | 49626 | 2204 | 50769 | 2082 | 51946 | 6 |
| | 2590 | 47455 | 2457 | 48534 | 2328 | 49645 | 2202 | 50788 | 2080 | 51966 | |
| 25 | 2588 | 47472 | 2455 | 48553 | 2325 | 49664 | 2200 | 50808 | 2078 | 51986 | 5 |
| | 2585 | 47490 | 2453 | 48571 | 2323 | 49683 | 2198 | 50827 | 2076 | 52006 | |
| 26 | 2583 | 47508 | 2450 | 48589 | 2321 | 49702 | 2196 | 50846 | 2074 | 52026 | 4 |
| | 2581 | 47526 | 2448 | 48608 | 2319 | 49720 | 2194 | 50866 | 2072 | 52046 | |
| 27 | 2579 | 47544 | 2446 | 48626 | 2317 | 49739 | 2192 | 50885 | 2070 | 52066 | 3 |
| | 2576 | 47561 | 2444 | 48644 | 2315 | 49758 | 2190 | 50905 | 2068 | 52086 | |
| 28 | 2574 | 47579 | 2442 | 48662 | 2313 | 49777 | 2188 | 50924 | 2066 | 52106 | 2 |
| | 2572 | 47597 | 2439 | 48681 | 2311 | 49796 | 2185 | 50943 | 2064 | 52126 | |
| 29 | 2570 | 47615 | 2437 | 48699 | 2309 | 49815 | 2183 | 50963 | 2062 | 52146 | 1 |
| | 2568 | 47633 | 2435 | 48717 | 2306 | 49833 | 2181 | 50982 | 2060 | 52166 | |
| 30 | 2565 | 47650 | 2433 | 48736 | 2304 | 49852 | 2179 | 51002 | 2058 | 52186 | 0 |
| | A | B | A | B | A | B | A | B | A | B | |
| ′ | 109° 30′ | | 109° 00′ | | 108° 30′ | | 108° 00′ | | 107° 30′ | | ′ |

# TABLE 35

## The Ageton Method

**Always Take "Z" From Bottom of Table, Except When "K" is Same Name and Greater Than Latitude, in Which Case Take "Z" From Top of Table**

| ′ | 72° 30′ | | 73° 00′ | | 73° 30′ | | 74° 00′ | | 74° 30′ | | ′ |
|---|---|---|---|---|---|---|---|---|---|---|---|
| ′ | A | B | A | B | A | B | A | B | A | B | ′ |
| 0 | 2058 | 52186 | 1940 | 53406 | 1826 | 54666 | 1716 | 55966 | 1609 | 57310 | 30 |
| | 2056 | 52206 | 1938 | 53427 | 1824 | 54687 | 1714 | 55988 | 1607 | 57333 | |
| 1 | 2054 | 52226 | 1936 | 53448 | 1823 | 54708 | 1712 | 56010 | 1605 | 57356 | 29 |
| | 2052 | 52246 | 1935 | 53468 | 1821 | 54730 | 1710 | 56032 | 1604 | 57378 | |
| 2 | 2050 | 52266 | 1933 | 53489 | 1819 | 54751 | 1709 | 56054 | 1602 | 57401 | 28 |
| | 2048 | 52286 | 1931 | 53510 | 1817 | 54773 | 1707 | 56076 | 1600 | 57424 | |
| 3 | 2046 | 52306 | 1929 | 53531 | 1815 | 54794 | 1705 | 56099 | 1598 | 57447 | 27 |
| | 2044 | 52326 | 1927 | 53551 | 1813 | 54815 | 1703 | 56121 | 1597 | 57470 | |
| 4 | 2042 | 52346 | 1925 | 53572 | 1811 | 54837 | 1701 | 56143 | 1595 | 57493 | 26 |
| | 2040 | 52366 | 1923 | 53593 | 1809 | 54858 | 1700 | 56165 | 1593 | 57516 | |
| 5 | 2038 | 52387 | 1921 | 53614 | 1808 | 54880 | 1698 | 56187 | 1591 | 57538 | 25 |
| | 2036 | 52407 | 1919 | 53634 | 1806 | 54901 | 1696 | 56209 | 1590 | 57561 | |
| 6 | 2034 | 52427 | 1917 | 53655 | 1804 | 54922 | 1694 | 56231 | 1588 | 57584 | 24 |
| | 2032 | 52447 | 1915 | 53676 | 1802 | 54944 | 1692 | 56254 | 1586 | 57607 | |
| 7 | 2030 | 52467 | 1913 | 53697 | 1800 | 54965 | 1691 | 56276 | 1584 | 57630 | 23 |
| | 2028 | 52487 | 1911 | 53718 | 1798 | 54987 | 1689 | 56298 | 1583 | 57653 | |
| 8 | 2026 | 52508 | 1910 | 53738 | 1796 | 55008 | 1687 | 56320 | 1581 | 57676 | 22 |
| | 2024 | 52528 | 1908 | 53759 | 1795 | 55030 | 1685 | 56342 | 1579 | 57699 | |
| 9 | 2022 | 52548 | 1906 | 53780 | 1793 | 55051 | 1683 | 56365 | 1578 | 57722 | 21 |
| | 2020 | 52568 | 1904 | 53801 | 1791 | 55073 | 1682 | 56387 | 1576 | 57745 | |
| 10 | 2018 | 52588 | 1902 | 53822 | 1789 | 55095 | 1680 | 56409 | 1574 | 57768 | 20 |
| | 2016 | 52609 | 1900 | 53843 | 1787 | 55116 | 1678 | 56431 | 1572 | 57791 | |
| 11 | 2014 | 52629 | 1898 | 53864 | 1785 | 55138 | 1676 | 56454 | 1571 | 57814 | 19 |
| | 2012 | 52649 | 1896 | 53884 | 1783 | 55159 | 1674 | 56476 | 1569 | 57837 | |
| 12 | 2010 | 52670 | 1894 | 53905 | 1782 | 55181 | 1673 | 56498 | 1567 | 57860 | 18 |
| | 2009 | 52690 | 1892 | 53926 | 1780 | 55202 | 1671 | 56521 | 1565 | 57884 | |
| 13 | 2007 | 52710 | 1890 | 53947 | 1778 | 55224 | 1669 | 56543 | 1564 | 57907 | 17 |
| | 2005 | 52730 | 1889 | 53968 | 1776 | 55246 | 1667 | 56565 | 1562 | 57930 | |
| 14 | 2003 | 52751 | 1887 | 53989 | 1774 | 55267 | 1665 | 56588 | 1560 | 57953 | 16 |
| | 2001 | 52771 | 1885 | 54010 | 1772 | 55289 | 1664 | 56610 | 1559 | 57976 | |
| 15 | 1999 | 52791 | 1883 | 54031 | 1771 | 55311 | 1662 | 56632 | 1557 | 57999 | 15 |
| | 1997 | 52812 | 1881 | 54052 | 1769 | 55332 | 1660 | 56655 | 1555 | 58022 | |
| 16 | 1995 | 52832 | 1879 | 54073 | 1767 | 55354 | 1658 | 56677 | 1553 | 58046 | 14 |
| | 1993 | 52852 | 1877 | 54094 | 1765 | 55376 | 1657 | 56700 | 1552 | 58069 | |
| 17 | 1991 | 52873 | 1875 | 54115 | 1763 | 55397 | 1655 | 56722 | 1550 | 58092 | 13 |
| | 1989 | 52893 | 1873 | 54136 | 1761 | 55419 | 1653 | 56745 | 1548 | 58115 | |
| 18 | 1987 | 52914 | 1871 | 54157 | 1760 | 55441 | 1651 | 56767 | 1546 | 58138 | 12 |
| | 1985 | 52934 | 1870 | 54178 | 1758 | 55463 | 1650 | 56790 | 1545 | 58162 | |
| 19 | 1983 | 52954 | 1868 | 54199 | 1756 | 55484 | 1648 | 56812 | 1543 | 58185 | 11 |
| | 1981 | 52975 | 1866 | 54220 | 1754 | 55506 | 1646 | 56835 | 1541 | 58208 | |
| 20 | 1979 | 52995 | 1864 | 54242 | 1752 | 55528 | 1644 | 56857 | 1540 | 58232 | 10 |
| | 1977 | 53016 | 1862 | 54263 | 1750 | 55550 | 1642 | 56880 | 1538 | 58255 | |
| 21 | 1975 | 53036 | 1860 | 54284 | 1749 | 55572 | 1641 | 56902 | 1536 | 58278 | 9 |
| | 1973 | 53057 | 1858 | 54305 | 1747 | 55593 | 1639 | 56925 | 1534 | 58302 | |
| 22 | 1971 | 53077 | 1856 | 54326 | 1745 | 55615 | 1637 | 56947 | 1533 | 58325 | 8 |
| | 1969 | 53098 | 1854 | 54347 | 1743 | 55637 | 1635 | 56970 | 1531 | 58348 | |
| 23 | 1967 | 53118 | 1853 | 54368 | 1741 | 55659 | 1634 | 56992 | 1529 | 58372 | 7 |
| | 1966 | 53139 | 1851 | 54390 | 1739 | 55681 | 1632 | 57015 | 1528 | 58395 | |
| 24 | 1964 | 53159 | 1849 | 54411 | 1738 | 55703 | 1630 | 57038 | 1526 | 58418 | 6 |
| | 1962 | 53180 | 1847 | 54432 | 1736 | 55725 | 1628 | 57060 | 1524 | 58442 | |
| 25 | 1960 | 53200 | 1845 | 54453 | 1734 | 55746 | 1627 | 57083 | 1523 | 58465 | 5 |
| | 1958 | 53221 | 1843 | 54474 | 1732 | 55768 | 1625 | 57106 | 1521 | 58489 | |
| 26 | 1956 | 53241 | 1841 | 54496 | 1730 | 55790 | 1623 | 57128 | 1519 | 58512 | 4 |
| | 1954 | 53262 | 1839 | 54517 | 1728 | 55812 | 1621 | 57151 | 1517 | 58536 | |
| 27 | 1952 | 53283 | 1837 | 54538 | 1727 | 55834 | 1619 | 57174 | 1516 | 58559 | 3 |
| | 1950 | 53303 | 1836 | 54559 | 1725 | 55856 | 1618 | 57196 | 1514 | 58583 | |
| 28 | 1948 | 53324 | 1834 | 54581 | 1723 | 55878 | 1616 | 57219 | 1512 | 58606 | 2 |
| | 1946 | 53344 | 1832 | 54602 | 1721 | 55900 | 1614 | 57242 | 1511 | 58630 | |
| 29 | 1944 | 53365 | 1830 | 54623 | 1719 | 55922 | 1612 | 57265 | 1509 | 58653 | 1 |
| | 1942 | 53386 | 1828 | 54644 | 1718 | 55944 | 1611 | 57287 | 1507 | 58677 | |
| 30 | 1940 | 53406 | 1826 | 54666 | 1716 | 55966 | 1609 | 57310 | 1506 | 58700 | 0 |
| ′ | A | B | A | B | A | B | A | B | A | B | ′ |
| ′ | 107° 00′ | | 106° 30′ | | 106° 00′ | | 105° 30′ | | 105° 00′ | | ′ |

# TABLE 35

## The Ageton Method

**When Meridian Angle is Greater Than 90° Take "K" From Bottom of Table**

| ′ | 75° 00′ | | 75° 30′ | | 76° 00′ | | 76° 30′ | | 77° 00′ | | ′ |
|---|---|---|---|---|---|---|---|---|---|---|---|
| | A | B | A | B | A | B | A | B | A | B | |
| 0 | 1506 | 58700 | 1406 | 60140 | 1310 | 61632 | 1217 | 63181 | 1128 | 64791 | 30 |
| | 1504 | 58724 | 1404 | 60164 | 1308 | 61658 | 1215 | 63208 | 1126 | 64819 | |
| 1 | 1502 | 58748 | 1403 | 60189 | 1306 | 61683 | 1214 | 63234 | 1125 | 64846 | 29 |
| | 1500 | 58771 | 1401 | 60213 | 1305 | 61709 | 1212 | 63260 | 1123 | 64873 | |
| 2 | 1499 | 58795 | 1399 | 60238 | 1303 | 61734 | 1211 | 63287 | 1122 | 64901 | 28 |
| | 1497 | 58818 | 1398 | 60262 | 1301 | 61759 | 1209 | 63313 | 1120 | 64928 | |
| 3 | 1495 | 58842 | 1396 | 60287 | 1300 | 61785 | 1208 | 63340 | 1119 | 64956 | 27 |
| | 1494 | 58866 | 1394 | 60311 | 1299 | 61810 | 1206 | 63366 | 1117 | 64983 | |
| 4 | 1492 | 58889 | 1393 | 60336 | 1297 | 61836 | 1205 | 63392 | 1116 | 65011 | 26 |
| | 1490 | 58913 | 1391 | 60360 | 1295 | 61861 | 1203 | 63419 | 1114 | 65038 | |
| 5 | 1489 | 58937 | 1390 | 60385 | 1294 | 61887 | 1202 | 63445 | 1113 | 65066 | 25 |
| | 1487 | 58960 | 1388 | 60410 | 1292 | 61912 | 1200 | 63472 | 1112 | 65093 | |
| 6 | 1485 | 58984 | 1386 | 60434 | 1291 | 61938 | 1199 | 63498 | 1110 | 65121 | 24 |
| | 1484 | 59008 | 1385 | 60459 | 1289 | 61963 | 1197 | 63525 | 1109 | 65148 | |
| 7 | 1482 | 59032 | 1383 | 60483 | 1288 | 61989 | 1196 | 63551 | 1107 | 65176 | 23 |
| | 1480 | 59055 | 1381 | 60508 | 1286 | 62014 | 1194 | 63578 | 1106 | 65204 | |
| 8 | 1479 | 59079 | 1380 | 60533 | 1284 | 62040 | 1193 | 63605 | 1104 | 65231 | 22 |
| | 1477 | 59103 | 1378 | 60557 | 1283 | 62065 | 1191 | 63631 | 1103 | 65259 | |
| 9 | 1475 | 59127 | 1377 | 60582 | 1281 | 62091 | 1190 | 63658 | 1101 | 65287 | 21 |
| | 1474 | 59151 | 1375 | 60607 | 1280 | 62117 | 1188 | 63684 | 1100 | 65314 | |
| 10 | 1472 | 59175 | 1373 | 60631 | 1278 | 62142 | 1187 | 63711 | 1099 | 65342 | 20 |
| | 1470 | 59198 | 1372 | 60656 | 1277 | 62168 | 1185 | 63738 | 1097 | 65370 | |
| 11 | 1469 | 59222 | 1370 | 60681 | 1275 | 62194 | 1184 | 63764 | 1096 | 65398 | 19 |
| | 1467 | 59246 | 1368 | 60706 | 1274 | 62219 | 1182 | 63791 | 1094 | 65425 | |
| 12 | 1465 | 59270 | 1367 | 60730 | 1272 | 62245 | 1181 | 63818 | 1093 | 65453 | 18 |
| | 1464 | 59294 | 1365 | 60755 | 1270 | 62271 | 1179 | 63845 | 1091 | 65481 | |
| 13 | 1462 | 59318 | 1364 | 60780 | 1269 | 62296 | 1178 | 63871 | 1090 | 65509 | 17 |
| | 1460 | 59342 | 1362 | 60805 | 1267 | 62322 | 1176 | 63898 | 1089 | 65537 | |
| 14 | 1459 | 59366 | 1360 | 60830 | 1266 | 62348 | 1175 | 63925 | 1087 | 65564 | 16 |
| | 1457 | 59390 | 1359 | 60855 | 1264 | 62374 | 1173 | 63952 | 1086 | 65592 | |
| 15 | 1455 | 59414 | 1357 | 60879 | 1263 | 62400 | 1172 | 63978 | 1084 | 65620 | 15 |
| | 1454 | 59438 | 1356 | 60904 | 1261 | 62425 | 1170 | 64005 | 1083 | 65648 | |
| 16 | 1452 | 59462 | 1354 | 60929 | 1260 | 62451 | 1169 | 64032 | 1081 | 65676 | 14 |
| | 1450 | 59486 | 1352 | 60954 | 1258 | 62477 | 1167 | 64059 | 1080 | 65704 | |
| 17 | 1449 | 59510 | 1351 | 60979 | 1257 | 62503 | 1166 | 64086 | 1079 | 65732 | 13 |
| | 1447 | 59534 | 1349 | 61004 | 1255 | 62529 | 1164 | 64113 | 1077 | 65760 | |
| 18 | 1445 | 59558 | 1348 | 61029 | 1253 | 62555 | 1163 | 64140 | 1076 | 65788 | 12 |
| | 1444 | 59582 | 1346 | 61054 | 1252 | 62581 | 1161 | 64167 | 1074 | 65816 | |
| 19 | 1442 | 59606 | 1344 | 61079 | 1250 | 62607 | 1160 | 64194 | 1073 | 65844 | 11 |
| | 1440 | 59630 | 1343 | 61104 | 1249 | 62633 | 1158 | 64221 | 1071 | 65872 | |
| 20 | 1439 | 59654 | 1341 | 61129 | 1247 | 62659 | 1157 | 64248 | 1070 | 65900 | 10 |
| | 1437 | 59679 | 1340 | 61154 | 1246 | 62685 | 1155 | 64275 | 1069 | 65928 | |
| 21 | 1435 | 59703 | 1338 | 61179 | 1244 | 62711 | 1154 | 64302 | 1067 | 65957 | 9 |
| | 1434 | 59727 | 1336 | 61204 | 1243 | 62737 | 1152 | 64329 | 1066 | 65985 | |
| 22 | 1432 | 59751 | 1335 | 61229 | 1241 | 62763 | 1151 | 64356 | 1064 | 66013 | 8 |
| | 1430 | 59775 | 1333 | 61254 | 1240 | 62789 | 1150 | 64383 | 1063 | 66041 | |
| 23 | 1429 | 59800 | 1332 | 61279 | 1238 | 62815 | 1148 | 64410 | 1061 | 66069 | 7 |
| | 1427 | 59824 | 1330 | 61304 | 1237 | 62841 | 1147 | 64437 | 1060 | 66098 | |
| 24 | 1425 | 59848 | 1329 | 61330 | 1235 | 62867 | 1145 | 64464 | 1059 | 66126 | 6 |
| | 1424 | 59872 | 1327 | 61355 | 1234 | 62893 | 1144 | 64491 | 1057 | 66154 | |
| 25 | 1422 | 59896 | 1325 | 61380 | 1232 | 62919 | 1142 | 64518 | 1056 | 66182 | 5 |
| | 1421 | 59921 | 1324 | 61405 | 1230 | 62945 | 1141 | 64546 | 1054 | 66211 | |
| 26 | 1419 | 59945 | 1322 | 61430 | 1229 | 62971 | 1139 | 64573 | 1053 | 66239 | 4 |
| | 1417 | 59969 | 1321 | 61456 | 1227 | 62998 | 1138 | 64600 | 1052 | 66267 | |
| 27 | 1416 | 59994 | 1319 | 61481 | 1226 | 63024 | 1136 | 64627 | 1050 | 66296 | 3 |
| | 1414 | 60018 | 1317 | 61506 | 1224 | 63050 | 1135 | 64655 | 1049 | 66324 | |
| 28 | 1412 | 60042 | 1316 | 61531 | 1223 | 63076 | 1133 | 64682 | 1047 | 66352 | 2 |
| | 1411 | 60067 | 1314 | 61556 | 1221 | 63103 | 1132 | 64709 | 1046 | 66381 | |
| 29 | 1409 | 60091 | 1313 | 61582 | 1220 | 63129 | 1130 | 64736 | 1045 | 66409 | 1 |
| | 1407 | 60116 | 1311 | 61607 | 1218 | 63155 | 1129 | 64764 | 1043 | 66438 | |
| 30 | 1406 | 60140 | 1310 | 61632 | 1217 | 63181 | 1128 | 64791 | 1042 | 66466 | 0 |
| | A | B | A | B | A | B | A | B | A | B | |
| ′ | 104° 30′ | | 104° 00′ | | 103° 30′ | | 103° 00′ | | 102° 30′ | | ′ |

# TABLE 35

The Ageton Method

Always Take "Z" From Bottom of Table, Except When "K" is Same Name and Greater Than Latitude, in Which Case Take "Z" From Top of Table

| ′ | 77° 30′ | | 78° 00′ | | 78° 30′ | | 79° 00′ | | 79° 30′ | | ′ |
|---|---|---|---|---|---|---|---|---|---|---|---|
| | A | B | A | B | A | B | A | B | A | B | |
| 0 | 1042 | 66466 | 960 | 68212 | 881 | 70034 | 805 | 71940 | 733 | 73937 | 30 |
| | 1040 | 66495 | 958 | 68242 | 879 | 70065 | 804 | 71973 | 732 | 73971 | |
| 1 | 1039 | 66523 | 957 | 68272 | 878 | 70097 | 803 | 72005 | 731 | 74005 | 29 |
| | 1038 | 66552 | 955 | 68301 | 877 | 70128 | 802 | 72038 | 730 | 74039 | |
| 2 | 1036 | 66580 | 954 | 68331 | 876 | 70159 | 800 | 72070 | 729 | 74073 | 28 |
| | 1035 | 66609 | 953 | 68361 | 874 | 70190 | 799 | 72103 | 728 | 74107 | |
| 3 | 1033 | 66638 | 951 | 68391 | 873 | 70221 | 798 | 72136 | 726 | 74142 | 27 |
| | 1032 | 66666 | 950 | 68421 | 872 | 70252 | 797 | 72168 | 725 | 74176 | |
| 4 | 1031 | 66695 | 949 | 68450 | 870 | 70284 | 796 | 72201 | 724 | 74210 | 26 |
| | 1029 | 66724 | 947 | 68480 | 869 | 70315 | 794 | 72234 | 723 | 74245 | |
| 5 | 1028 | 66752 | 946 | 68510 | 868 | 70346 | 793 | 72266 | 722 | 74279 | 25 |
| | 1026 | 66781 | 945 | 68540 | 867 | 70377 | 792 | 72299 | 721 | 74313 | |
| 6 | 1025 | 66810 | 943 | 68570 | 865 | 70409 | 791 | 72332 | 719 | 74348 | 24 |
| | 1024 | 66838 | 942 | 68600 | 864 | 70440 | 790 | 72365 | 718 | 74382 | |
| 7 | 1022 | 66867 | 941 | 68630 | 863 | 70471 | 788 | 72397 | 717 | 74417 | 23 |
| | 1021 | 66896 | 939 | 68660 | 862 | 70503 | 787 | 72430 | 716 | 74451 | |
| 8 | 1020 | 66925 | 938 | 68690 | 860 | 70534 | 786 | 72463 | 715 | 74486 | 22 |
| | 1018 | 66953 | 937 | 68720 | 859 | 70566 | 785 | 72496 | 714 | 74520 | |
| 9 | 1017 | 66982 | 935 | 68750 | 858 | 70597 | 783 | 72529 | 712 | 74555 | 21 |
| | 1015 | 67011 | 934 | 68781 | 856 | 70629 | 782 | 72562 | 711 | 74589 | |
| 10 | 1014 | 67040 | 933 | 68811 | 855 | 70660 | 781 | 72595 | 710 | 74624 | 20 |
| | 1013 | 67069 | 932 | 68841 | 854 | 70692 | 780 | 72628 | 709 | 74659 | |
| 11 | 1011 | 67098 | 930 | 68871 | 853 | 70723 | 779 | 72661 | 708 | 74693 | 19 |
| | 1010 | 67127 | 929 | 68901 | 851 | 70755 | 777 | 72694 | 707 | 74728 | |
| 12 | 1008 | 67156 | 928 | 68931 | 850 | 70786 | 776 | 72727 | 706 | 74763 | 18 |
| | 1007 | 67185 | 926 | 68962 | 849 | 70818 | 775 | 72760 | 704 | 74797 | |
| 13 | 1006 | 67214 | 925 | 68992 | 848 | 70850 | 774 | 72794 | 703 | 74832 | 17 |
| | 1004 | 67243 | 924 | 69022 | 846 | 70881 | 772 | 72827 | 702 | 74867 | |
| 14 | 1003 | 67272 | 922 | 69053 | 845 | 70913 | 771 | 72860 | 701 | 74902 | 16 |
| | 1002 | 67301 | 921 | 69083 | 844 | 70945 | 770 | 72893 | 700 | 74937 | |
| 15 | 1000 | 67330 | 920 | 69113 | 843 | 70976 | 769 | 72926 | 699 | 74972 | 15 |
| | 999 | 67359 | 918 | 69144 | 841 | 71008 | 768 | 72960 | 698 | 75007 | |
| 16 | 997 | 67388 | 917 | 69174 | 840 | 71040 | 767 | 72993 | 696 | 75042 | 14 |
| | 996 | 67417 | 916 | 69204 | 839 | 71072 | 765 | 73026 | 695 | 75077 | |
| 17 | 995 | 67447 | 914 | 69235 | 838 | 71104 | 764 | 73060 | 694 | 75112 | 13 |
| | 993 | 67476 | 913 | 69265 | 836 | 71135 | 763 | 73093 | 693 | 75147 | |
| 18 | 992 | 67505 | 912 | 69296 | 835 | 71167 | 762 | 73127 | 692 | 75182 | 12 |
| | 991 | 67534 | 910 | 69326 | 834 | 71199 | 761 | 73160 | 691 | 75217 | |
| 19 | 989 | 67563 | 909 | 69357 | 833 | 71231 | 759 | 73193 | 690 | 75252 | 11 |
| | 988 | 67593 | 908 | 69387 | 831 | 71263 | 758 | 73227 | 688 | 75287 | |
| 20 | 987 | 67622 | 907 | 69418 | 830 | 71295 | 757 | 73260 | 687 | 75322 | 10 |
| | 985 | 67651 | 905 | 69449 | 829 | 71327 | 756 | 73294 | 686 | 75358 | |
| 21 | 984 | 67681 | 904 | 69479 | 828 | 71359 | 755 | 73328 | 685 | 75393 | 9 |
| | 982 | 67710 | 903 | 69510 | 826 | 71391 | 753 | 73361 | 684 | 75428 | |
| 22 | 981 | 67739 | 901 | 69541 | 825 | 71423 | 752 | 73395 | 683 | 75464 | 8 |
| | 980 | 67769 | 900 | 69571 | 824 | 71455 | 751 | 73429 | 682 | 75499 | |
| 23 | 978 | 67798 | 899 | 69602 | 823 | 71488 | 750 | 73462 | 680 | 75534 | 7 |
| | 977 | 67828 | 897 | 69633 | 821 | 71520 | 749 | 73496 | 679 | 75570 | |
| 24 | 976 | 67857 | 896 | 69664 | 820 | 71552 | 747 | 73530 | 678 | 75605 | 6 |
| | 974 | 67886 | 895 | 69694 | 819 | 71584 | 746 | 73563 | 677 | 75641 | |
| 25 | 973 | 67916 | 894 | 69725 | 818 | 71616 | 745 | 73597 | 676 | 75676 | 5 |
| | 972 | 67945 | 892 | 69756 | 816 | 71649 | 744 | 73631 | 675 | 75712 | |
| 26 | 970 | 67975 | 891 | 69787 | 815 | 71681 | 743 | 73665 | 674 | 75747 | 4 |
| | 969 | 68005 | 890 | 69818 | 814 | 71713 | 742 | 73699 | 673 | 75783 | |
| 27 | 968 | 68034 | 888 | 69849 | 813 | 71746 | 740 | 73733 | 672 | 75819 | 3 |
| | 966 | 68064 | 887 | 69879 | 811 | 71778 | 739 | 73767 | 670 | 75854 | |
| 28 | 965 | 68093 | 886 | 69910 | 810 | 71810 | 738 | 73801 | 669 | 75890 | 2 |
| | 964 | 68123 | 885 | 69941 | 809 | 71843 | 737 | 73835 | 668 | 75926 | |
| 29 | 962 | 68153 | 883 | 69972 | 808 | 71875 | 736 | 73869 | 667 | 75961 | 1 |
| | 961 | 68182 | 882 | 70003 | 807 | 71908 | 735 | 73903 | 666 | 75997 | |
| 30 | 960 | 68212 | 881 | 70034 | 805 | 71940 | 733 | 73937 | 665 | 76033 | 0 |
| | A | B | A | B | A | B | A | B | A | B | |
| ′ | 102° 00′ | | 101° 30′ | | 101° 00′ | | 100° 30′ | | 100° 00′ | | ′ |

## TABLE 35

### The Ageton Method

**When Meridian Angle is Greater Than 90° Take "K" From Bottom of Table**

| ′ | 80° 00′ | | 80° 30′ | | 81° 00′ | | 81° 30′ | | 82° 00′ | | ′ |
|---|---|---|---|---|---|---|---|---|---|---|---|
| ′ | A | B | A | B | A | B | A | B | A | B | ′ |
| 0 | 665 | 76033 | 600 | 78239 | 538 | 80567 | 480 | 83030 | 425 | 85644 | 30 |
| | 664 | 76069 | 599 | 78277 | 537 | 80607 | 479 | 83072 | 424 | 85689 | |
| 1 | 663 | 76105 | 598 | 78315 | 536 | 80647 | 478 | 83114 | 423 | 85734 | 29 |
| | 661 | 76141 | 597 | 78352 | 535 | 80687 | 477 | 83157 | 422 | 85779 | |
| 2 | 660 | 76176 | 595 | 78390 | 534 | 80727 | 476 | 83199 | 421 | 85825 | 28 |
| | 659 | 76212 | 594 | 78428 | 533 | 80767 | 475 | 83242 | 420 | 85870 | |
| 3 | 658 | 76248 | 593 | 78466 | 532 | 80807 | 474 | 83284 | 419 | 85915 | 27 |
| | 657 | 76284 | 592 | 78504 | 531 | 80847 | 473 | 83327 | 418 | 85960 | |
| 4 | 656 | 76320 | 591 | 78542 | 530 | 80887 | 472 | 83369 | 418 | 86006 | 26 |
| | 655 | 76357 | 590 | 78580 | 529 | 80927 | 471 | 83412 | 417 | 86051 | |
| 5 | 654 | 76393 | 589 | 78618 | 528 | 80967 | 470 | 83455 | 416 | 86096 | 25 |
| | 653 | 76429 | 588 | 78656 | 527 | 81008 | 469 | 83497 | 415 | 86142 | |
| 6 | 652 | 76465 | 587 | 78694 | 526 | 81048 | 468 | 83540 | 414 | 86187 | 24 |
| | 650 | 76501 | 586 | 78733 | 525 | 81088 | 467 | 83583 | 413 | 86233 | |
| 7 | 649 | 76537 | 585 | 78771 | 524 | 81129 | 467 | 83626 | 412 | 86278 | 23 |
| | 648 | 76574 | 584 | 78809 | 523 | 81169 | 466 | 83668 | 411 | 86324 | |
| 8 | 647 | 76610 | 583 | 78847 | 522 | 81210 | 465 | 83711 | 411 | 86370 | 22 |
| | 646 | 76646 | 582 | 78886 | 521 | 81250 | 464 | 83754 | 410 | 86415 | |
| 9 | 645 | 76683 | 581 | 78924 | 520 | 81291 | 463 | 83797 | 409 | 86461 | 21 |
| | 644 | 76719 | 580 | 78962 | 519 | 81331 | 462 | 83840 | 408 | 86507 | |
| 10 | 643 | 76756 | 579 | 79001 | 518 | 81372 | 461 | 83884 | 407 | 86553 | 20 |
| | 642 | 76792 | 578 | 79039 | 517 | 81413 | 460 | 83927 | 406 | 86599 | |
| 11 | 641 | 76828 | 577 | 79078 | 516 | 81453 | 459 | 83970 | 405 | 86645 | 19 |
| | 639 | 76865 | 576 | 79116 | 515 | 81494 | 458 | 84013 | 405 | 86691 | |
| 12 | 638 | 76902 | 575 | 79155 | 514 | 81535 | 457 | 84056 | 404 | 86737 | 18 |
| | 637 | 76938 | 574 | 79193 | 513 | 81576 | 456 | 84100 | 403 | 86783 | |
| 13 | 636 | 76975 | 573 | 79232 | 512 | 81617 | 455 | 84143 | 402 | 86829 | 17 |
| | 635 | 77011 | 571 | 79271 | 511 | 81657 | 454 | 84186 | 401 | 86876 | |
| 14 | 634 | 77048 | 570 | 79309 | 510 | 81698 | 454 | 84230 | 400 | 86922 | 16 |
| | 633 | 77085 | 569 | 79348 | 509 | 81739 | 453 | 84273 | 399 | 86968 | |
| 15 | 632 | 77122 | 568 | 79387 | 508 | 81780 | 452 | 84317 | 399 | 87015 | 15 |
| | 631 | 77158 | 567 | 79426 | 507 | 81821 | 451 | 84361 | 398 | 87061 | |
| 16 | 630 | 77195 | 566 | 79465 | 506 | 81863 | 450 | 84404 | 397 | 87107 | 14 |
| | 629 | 77232 | 565 | 79503 | 505 | 81904 | 449 | 84448 | 396 | 87154 | |
| 17 | 627 | 77269 | 564 | 79542 | 504 | 81945 | 448 | 84492 | 395 | 87201 | 13 |
| | 626 | 77306 | 563 | 79581 | 504 | 81986 | 447 | 84535 | 394 | 87247 | |
| 18 | 625 | 77343 | 562 | 79620 | 503 | 82027 | 446 | 84579 | 393 | 87294 | 12 |
| | 624 | 77380 | 561 | 79659 | 502 | 82069 | 445 | 84623 | 392 | 87341 | |
| 19 | 623 | 77417 | 560 | 79698 | 501 | 82110 | 444 | 84667 | 392 | 87387 | 11 |
| | 622 | 77454 | 559 | 79737 | 500 | 82151 | 444 | 84711 | 391 | 87434 | |
| 20 | 621 | 77491 | 558 | 79777 | 499 | 82193 | 443 | 84755 | 390 | 87481 | 10 |
| | 620 | 77528 | 557 | 79816 | 498 | 82234 | 442 | 84799 | 389 | 87528 | |
| 21 | 619 | 77565 | 556 | 79855 | 497 | 82276 | 441 | 84843 | 388 | 87575 | 9 |
| | 618 | 77602 | 555 | 79894 | 496 | 82317 | 440 | 84887 | 387 | 87622 | |
| 22 | 617 | 77639 | 554 | 79933 | 495 | 82359 | 439 | 84931 | 387 | 87669 | 8 |
| | 616 | 77677 | 553 | 79973 | 494 | 82400 | 438 | 84976 | 386 | 87716 | |
| 23 | 615 | 77714 | 552 | 80012 | 493 | 82442 | 437 | 85020 | 385 | 87764 | 7 |
| | 614 | 77751 | 551 | 80051 | 492 | 82484 | 436 | 85064 | 384 | 87811 | |
| 24 | 612 | 77788 | 550 | 80091 | 491 | 82526 | 435 | 85109 | 383 | 87858 | 6 |
| | 611 | 77826 | 549 | 80130 | 490 | 82567 | 434 | 85153 | 382 | 87906 | |
| 25 | 610 | 77863 | 548 | 80170 | 489 | 82609 | 434 | 85197 | 381 | 87953 | 5 |
| | 609 | 77901 | 547 | 80209 | 488 | 82651 | 433 | 85242 | 381 | 88001 | |
| 26 | 608 | 77938 | 546 | 80249 | 487 | 82693 | 432 | 85286 | 380 | 88048 | 4 |
| | 607 | 77976 | 545 | 80288 | 486 | 82735 | 431 | 85331 | 379 | 88096 | |
| 27 | 606 | 78013 | 544 | 80328 | 485 | 82777 | 430 | 85376 | 378 | 88143 | 3 |
| | 605 | 78051 | 543 | 80368 | 484 | 82819 | 429 | 85420 | 377 | 88191 | |
| 28 | 604 | 78088 | 542 | 80407 | 483 | 82861 | 428 | 85465 | 376 | 88239 | 2 |
| | 603 | 78126 | 541 | 80447 | 482 | 82903 | 427 | 85510 | 376 | 88286 | |
| 29 | 602 | 78164 | 540 | 80487 | 482 | 82945 | 426 | 85555 | 375 | 88334 | 1 |
| | 601 | 78201 | 539 | 80527 | 481 | 82987 | 426 | 85599 | 374 | 88382 | |
| 30 | 600 | 78239 | 538 | 80567 | 480 | 83030 | 425 | 85644 | 373 | 88430 | 0 |
| | A | B | A | B | A | B | A | B | A | B | |
| ′ | 99° 30′ | | 99° 00′ | | 98° 30′ | | 98° 00′ | | 97° 30′ | | ′ |

# TABLE 35

The Ageton Method

Always Take "Z" From Bottom of Table, Except When "K" is Same Name and Greater Than Latitude, in Which Case Take "Z" From Top of Table

| ′ | 82° 30′ A | 82° 30′ B | 83° 00′ A | 83° 00′ B | 83° 30′ A | 83° 30′ B | 84° 00′ A | 84° 00′ B | 84° 30′ A | 84° 30′ B | ′ |
|---|---|---|---|---|---|---|---|---|---|---|---|
| 0 | 373 | 88430 | 325 | 91411 | 280 | 94614 | 238.6 | 98076 | 200.4 | 101843 | 30 |
| | 372 | 88478 | 324 | 91462 | 279 | 94670 | 237.9 | 98137 | 199.8 | 101908 | |
| 1 | 371 | 88526 | 323 | 91514 | 279 | 94725 | 237.2 | 98197 | 199.2 | 101974 | 29 |
| | 371 | 88574 | 323 | 91565 | 278 | 94781 | 236.6 | 98257 | 198.6 | 102040 | |
| 2 | 370 | 88623 | 322 | 91617 | 277 | 94836 | 235.9 | 98318 | 198.0 | 102106 | 28 |
| | 369 | 88671 | 321 | 91668 | 276 | 94892 | 235.3 | 98378 | 197.4 | 102172 | |
| 3 | 368 | 88719 | 320 | 91720 | 276 | 94948 | 234.6 | 98439 | 196.8 | 102238 | 27 |
| | 367 | 88767 | 319 | 91772 | 275 | 95004 | 233.9 | 98499 | 196.2 | 102304 | |
| 4 | 366 | 88816 | 319 | 91824 | 274 | 95060 | 233.3 | 98560 | 195.6 | 102371 | 26 |
| | 366 | 88864 | 318 | 91876 | 274 | 95116 | 232.6 | 98621 | 195.0 | 102437 | |
| 5 | 365 | 88913 | 317 | 91928 | 273 | 95172 | 232.0 | 98682 | 194.4 | 102504 | 25 |
| | 364 | 88961 | 316 | 91980 | 272 | 95228 | 231.3 | 98743 | 193.8 | 102570 | |
| 6 | 363 | 89010 | 316 | 92032 | 271 | 95285 | 230.7 | 98804 | 193.2 | 102637 | 24 |
| | 362 | 89059 | 315 | 92085 | 271 | 95341 | 230.0 | 98865 | 192.6 | 102704 | |
| 7 | 362 | 89107 | 314 | 92137 | 270 | 95397 | 229.4 | 98926 | 192.0 | 102771 | 23 |
| | 361 | 89156 | 313 | 92189 | 269 | 95454 | 228.7 | 98988 | 191.4 | 102838 | |
| 8 | 360 | 89205 | 313 | 92242 | 269 | 95510 | 228.1 | 99049 | 190.8 | 102905 | 22 |
| | 359 | 89254 | 312 | 92294 | 268 | 95567 | 227.4 | 99111 | 190.2 | 102973 | |
| 9 | 358 | 89303 | 311 | 92347 | 267 | 95624 | 226.8 | 99172 | 189.6 | 103040 | 21 |
| | 357 | 89352 | 310 | 92399 | 267 | 95681 | 226.1 | 99234 | 189.0 | 103107 | |
| 10 | 357 | 89401 | 310 | 92452 | 266 | 95737 | 225.5 | 99296 | 188.4 | 103175 | 20 |
| | 356 | 89450 | 309 | 92505 | 265 | 95795 | 224.8 | 99357 | 187.8 | 103243 | |
| 11 | 355 | 89499 | 308 | 92558 | 264 | 95851 | 224.2 | 99419 | 187.2 | 103311 | 19 |
| | 354 | 89548 | 307 | 92610 | 264 | 95909 | 223.5 | 99482 | 186.7 | 103379 | |
| 12 | 353 | 89597 | 307 | 92663 | 263 | 95966 | 222.9 | 99544 | 186.1 | 103447 | 18 |
| | 353 | 89647 | 306 | 92716 | 262 | 96023 | 222.3 | 99606 | 185.5 | 103515 | |
| 13 | 352 | 89696 | 305 | 92769 | 262 | 96080 | 221.6 | 99668 | 184.9 | 103583 | 17 |
| | 351 | 89746 | 304 | 92823 | 261 | 96138 | 221.0 | 99731 | 184.3 | 103651 | |
| 14 | 350 | 89795 | 304 | 92876 | 260 | 96195 | 220.3 | 99793 | 183.7 | 103720 | 16 |
| | 349 | 89845 | 303 | 92929 | 260 | 96253 | 219.7 | 99856 | 183.2 | 103788 | |
| 15 | 349 | 89894 | 302 | 92982 | 259 | 96310 | 219.1 | 99918 | 182.6 | 103857 | 15 |
| | 348 | 89944 | 301 | 93036 | 258 | 96368 | 218.4 | 99981 | 182.0 | 103926 | |
| 16 | 347 | 89994 | 301 | 93089 | 257 | 96426 | 217.8 | 100044 | 181.4 | 103995 | 14 |
| | 346 | 90044 | 300 | 93143 | 257 | 96484 | 217.2 | 100107 | 180.8 | 104064 | |
| 17 | 345 | 90093 | 299 | 93196 | 256 | 96542 | 216.5 | 100170 | 130.3 | 104133 | 13 |
| | 345 | 90143 | 298 | 93250 | 255 | 96600 | 215.9 | 100233 | 179.7 | 104202 | |
| 18 | 344 | 90193 | 298 | 93304 | 255 | 96658 | 215.3 | 100296 | 179.1 | 104272 | 12 |
| | 343 | 90243 | 297 | 93358 | 254 | 96716 | 214.6 | 100360 | 178.5 | 104341 | |
| 19 | 342 | 90293 | 296 | 93411 | 253 | 96774 | 214.0 | 100423 | 178.0 | 104411 | 11 |
| | 341 | 90344 | 295 | 93465 | 253 | 96833 | 213.4 | 100487 | 177.4 | 104480 | |
| 20 | 341 | 90394 | 295 | 93519 | 252 | 96891 | 212.8 | 100550 | 176.8 | 104550 | 10 |
| | 340 | 90444 | 294 | 93573 | 251 | 96950 | 212.1 | 100614 | 176.2 | 104620 | |
| 21 | 339 | 90494 | 293 | 93628 | 251 | 97008 | 211.5 | 100678 | 175.7 | 104690 | 9 |
| | 338 | 90545 | 292 | 93682 | 250 | 97067 | 210.9 | 100742 | 175.1 | 104760 | |
| 22 | 337 | 90595 | 292 | 93736 | 249 | 97126 | 210.3 | 100806 | 174.5 | 104830 | 8 |
| | 337 | 90646 | 291 | 93790 | 249 | 97184 | 209.6 | 100870 | 174.0 | 104901 | |
| 23 | 336 | 90696 | 290 | 93845 | 248 | 97243 | 209.0 | 100934 | 173.4 | 104971 | 7 |
| | 335 | 90747 | 289 | 93899 | 247 | 97302 | 208.4 | 100998 | 172.8 | 105042 | |
| 24 | 334 | 90798 | 289 | 93954 | 247 | 97361 | 207.8 | 101063 | 172.3 | 105113 | 6 |
| | 333 | 90848 | 288 | 94009 | 246 | 97420 | 207.1 | 101127 | 171.7 | 105183 | |
| 25 | 333 | 90899 | 287 | 94063 | 245 | 97480 | 206.5 | 101192 | 171.1 | 105254 | 5 |
| | 332 | 90950 | 287 | 94118 | 245 | 97539 | 205.9 | 101256 | 170.6 | 105325 | |
| 26 | 331 | 91001 | 286 | 94173 | 244 | 97598 | 205.3 | 101321 | 170.0 | 105397 | 4 |
| | 330 | 91052 | 285 | 94228 | 243 | 97658 | 204.7 | 101386 | 169.5 | 105468 | |
| 27 | 330 | 91103 | 284 | 94283 | 243 | 97717 | 204.1 | 101451 | 168.9 | 105539 | 3 |
| | 329 | 91154 | 284 | 94338 | 242 | 97777 | 203.5 | 101516 | 168.4 | 105611 | |
| 28 | 328 | 91205 | 283 | 94393 | 241 | 97837 | 202.8 | 101581 | 167.8 | 105683 | 2 |
| | 327 | 91257 | 282 | 94448 | 241 | 97897 | 202.2 | 101646 | 167.2 | 105754 | |
| 29 | 326 | 91308 | 281 | 94503 | 240 | 97957 | 201.6 | 101712 | 166.7 | 105826 | 1 |
| | 326 | 91359 | 281 | 94559 | 239 | 98017 | 201.0 | 101777 | 166.0 | 105898 | |
| 30 | 325 | 91411 | 280 | 94614 | 239 | 98076 | 200.4 | 101843 | 165.6 | 105970 | 0 |
| | A | B | A | B | A | B | A | B | A | B | |
| ′ | 97° 00′ | | 96° 30′ | | 96° 00′ | | 95° 30′ | | 95° 00′ | | ′ |

# TABLE 35

## The Ageton Method

When Meridian Angle is Greater Than 90° Take "K" From Bottom of Table

| | 85° 00′ | | 85° 30′ | | 86° 00′ | | 86° 30′ | | 87° 00′ | | |
|---|---|---|---|---|---|---|---|---|---|---|---|
| ′ | A | B | A | B | A | B | A | B | A | B | ′ |
| 0 | 165.6 | 105970 | 134.1 | 110536 | 105.9 | 115641 | 81.1 | 121432 | 59.6 | 128120 | 30 |
| | 165.0 | 106043 | 133.6 | 110616 | 105.5 | 115732 | 80.7 | 121536 | 59.2 | 128241 | |
| 1 | 164.5 | 106115 | 133.1 | 110696 | 105.0 | 115823 | 80.3 | 121639 | 58.9 | 128362 | 29 |
| | 163.9 | 106187 | 132.6 | 110777 | 104.6 | 115913 | 79.9 | 121743 | 58.6 | 128483 | |
| 2 | 163.4 | 106260 | 132.1 | 110858 | 104.2 | 116004 | 79.5 | 121848 | 58.2 | 128605 | 28 |
| | 162.8 | 106333 | 131.6 | 110939 | 103.7 | 116096 | 79.2 | 121952 | 57.9 | 128727 | |
| 3 | 162.3 | 106406 | 131.1 | 111020 | 103.3 | 116187 | 78.8 | 122057 | 57.6 | 128849 | 27 |
| | 161.7 | 106479 | 130.6 | 111101 | 102.9 | 116278 | 78.4 | 122161 | 57.3 | 128972 | |
| 4 | 161.2 | 106552 | 130.1 | 111183 | 102.4 | 116370 | 78.0 | 122267 | 56.9 | 129095 | 26 |
| | 160.6 | 106625 | 129.6 | 111264 | 102.0 | 116462 | 77.6 | 122372 | 56.6 | 129218 | |
| 5 | 160.1 | 106698 | 129.2 | 111346 | 101.6 | 116554 | 77.3 | 122478 | 56.3 | 129342 | 25 |
| | 159.6 | 106772 | 128.7 | 111428 | 101.1 | 116647 | 76.9 | 122584 | 56.0 | 129466 | |
| 6 | 159.0 | 106846 | 128.2 | 111510 | 100.7 | 116739 | 76.5 | 122690 | 55.7 | 129591 | 24 |
| | 158.5 | 106919 | 127.7 | 111592 | 100.3 | 116832 | 76.1 | 122796 | 55.3 | 129716 | |
| 7 | 157.9 | 106993 | 127.2 | 111674 | 99.8 | 116925 | 75.8 | 122903 | 55.0 | 129841 | 23 |
| | 157.4 | 107067 | 126.7 | 111757 | 99.4 | 117018 | 75.4 | 123010 | 54.7 | 129967 | |
| 8 | 156.9 | 107141 | 126.2 | 111839 | 99.0 | 117112 | 75.0 | 123117 | 54.4 | 130093 | 22 |
| | 156.3 | 107216 | 125.8 | 111922 | 98.5 | 117205 | 74.6 | 123225 | 54.1 | 130219 | |
| 9 | 155.8 | 107290 | 125.3 | 112005 | 98.1 | 117299 | 74.3 | 123332 | 53.7 | 130346 | 21 |
| | 155.2 | 107364 | 124.8 | 112088 | 97.7 | 117393 | 73.9 | 123441 | 53.4 | 130473 | |
| 10 | 154.7 | 107439 | 124.3 | 112171 | 97.3 | 117487 | 73.5 | 123549 | 53.1 | 130600 | 20 |
| | 154.2 | 107514 | 123.8 | 112255 | 96.8 | 117581 | 73.2 | 123657 | 52.8 | 130728 | |
| 11 | 153.6 | 107589 | 123.4 | 112338 | 96.4 | 117676 | 72.8 | 123766 | 52.5 | 130856 | 19 |
| | 153.1 | 107664 | 122.9 | 112422 | 96.0 | 117771 | 72.4 | 123875 | 52.2 | 130985 | |
| 12 | 152.6 | 107739 | 122.4 | 112506 | 95.6 | 117866 | 72.1 | 123985 | 51.9 | 131114 | 18 |
| | 152.1 | 107814 | 121.9 | 112590 | 95.2 | 117961 | 71.7 | 124095 | 51.6 | 131243 | |
| 13 | 151.5 | 107890 | 121.5 | 112674 | 94.7 | 118056 | 71.3 | 124204 | 51.3 | 131373 | 17 |
| | 151.0 | 107965 | 121.0 | 112759 | 94.3 | 118152 | 71.0 | 124315 | 51.0 | 131503 | |
| 14 | 150.5 | 108041 | 120.5 | 112843 | 93.9 | 118248 | 70.6 | 124425 | 50.7 | 131633 | 16 |
| | 149.9 | 108117 | 120.1 | 112928 | 93.5 | 118344 | 70.3 | 124536 | 50.3 | 131764 | |
| 15 | 149.4 | 108193 | 119.6 | 113013 | 93.1 | 118440 | 69.9 | 124647 | 50.0 | 131896 | 15 |
| | 148.9 | 108269 | 119.1 | 113098 | 92.7 | 118537 | 69.5 | 124759 | 49.7 | 132027 | |
| 16 | 148.4 | 108345 | 118.7 | 113183 | 92.3 | 118633 | 69.2 | 124870 | 49.4 | 132159 | 14 |
| | 147.8 | 108421 | 118.2 | 113269 | 91.8 | 118730 | 68.8 | 124982 | 49.1 | 132292 | |
| 17 | 147.3 | 108498 | 117.7 | 113354 | 91.4 | 118827 | 68.5 | 125094 | 48.8 | 132425 | 13 |
| | 146.8 | 108574 | 117.3 | 113440 | 91.0 | 118925 | 68.1 | 125207 | 48.5 | 132558 | |
| 18 | 146.3 | 108651 | 116.8 | 113526 | 90.6 | 119022 | 67.8 | 125320 | 48.2 | 132692 | 12 |
| | 145.8 | 108728 | 116.3 | 113612 | 90.2 | 119120 | 67.4 | 125433 | 47.9 | 132826 | |
| 19 | 145.2 | 108805 | 115.9 | 113699 | 89.8 | 119218 | 67.1 | 125546 | 47.6 | 132961 | 11 |
| | 144.7 | 108882 | 115.4 | 113785 | 89.4 | 119316 | 66.7 | 125660 | 47.3 | 133096 | |
| 20 | 144.2 | 108960 | 114.9 | 113872 | 89.0 | 119415 | 66.4 | 125774 | 47.1 | 133231 | 10 |
| | 143.7 | 109037 | 114.5 | 113958 | 88.6 | 119513 | 66.0 | 125888 | 46.8 | 133367 | |
| 21 | 143.2 | 109115 | 114.0 | 114045 | 88.2 | 119612 | 65.7 | 126003 | 46.5 | 133503 | 9 |
| | 142.7 | 109192 | 113.6 | 114133 | 87.8 | 119711 | 65.3 | 126118 | 46.2 | 133640 | |
| 22 | 142.2 | 109270 | 113.1 | 114220 | 87.4 | 119811 | 65.0 | 126233 | 45.9 | 133777 | 8 |
| | 141.6 | 109348 | 112.7 | 114307 | 87.0 | 119910 | 64.6 | 126349 | 45.6 | 133914 | |
| 23 | 141.1 | 109426 | 112.2 | 114395 | 86.6 | 120010 | 64.3 | 126465 | 45.3 | 134052 | 7 |
| | 140.6 | 109505 | 111.7 | 114483 | 86.2 | 120110 | 63.9 | 126581 | 45.0 | 134191 | |
| 24 | 140.1 | 109583 | 111.3 | 114571 | 85.8 | 120211 | 63.6 | 126697 | 44.7 | 134330 | 6 |
| | 139.6 | 109662 | 110.8 | 114659 | 85.4 | 120311 | 63.3 | 126814 | 44.4 | 134469 | |
| 25 | 139.1 | 109740 | 110.4 | 114747 | 85.0 | 120412 | 62.9 | 126931 | 44.2 | 134609 | 5 |
| | 138.6 | 109819 | 109.9 | 114836 | 84.6 | 120513 | 62.6 | 127049 | 43.9 | 134749 | |
| 26 | 138.1 | 109898 | 109.5 | 114925 | 84.2 | 120614 | 62.2 | 127166 | 43.6 | 134890 | 4 |
| | 137.6 | 109978 | 109.0 | 115014 | 83.8 | 120715 | 61.9 | 127284 | 43.3 | 135031 | |
| 27 | 137.1 | 110057 | 108.6 | 115103 | 83.4 | 120817 | 61.6 | 127403 | 43.0 | 135173 | 3 |
| | 136.6 | 110136 | 108.1 | 115192 | 83.0 | 120919 | 61.2 | 127521 | 42.7 | 135315 | |
| 28 | 136.1 | 110216 | 107.7 | 115282 | 82.6 | 121021 | 60.9 | 127640 | 42.5 | 135457 | 2 |
| | 135.6 | 110296 | 107.3 | 115371 | 82.2 | 121124 | 60.6 | 127760 | 42.2 | 135600 | |
| 29 | 135.1 | 110375 | 106.8 | 115461 | 81.9 | 121226 | 60.2 | 127880 | 41.9 | 135744 | 1 |
| | 134.6 | 110455 | 106.4 | 115551 | 81.5 | 121329 | 59.9 | 128000 | 41.6 | 135888 | |
| 30 | 134.1 | 110536 | 105.9 | 115641 | 81.1 | 121432 | 59.6 | 128120 | 41.4 | 136032 | 0 |
| | A | B | A | B | A | B | A | B | A | B | |
| ′ | 94° 30′ | | 94° 00′ | | 93° 30′ | | 93° 00′ | | 92° 30′ | | ′ |

# TABLE 35

## The Ageton Method

**Always Take "Z" From Bottom of Table, Except When "K" is Same Name and Greater Than Latitude, in Which Case Take "Z" From Top of Table**

| ′ | 87° 30′ | | 88° 00′ | | 88° 30′ | | 89° 00′ | | 89° 30′ | | ′ |
|---|---|---|---|---|---|---|---|---|---|---|---|
| | A | B | A | B | A | B | A | B | A | B | |
| 0 | 41.4 | 136032 | 26.5 | 145718 | 14.9 | 158208 | 6.6 | 175814 | 1.7 | 205916 | 30 |
| | 41.1 | 136177 | 26.2 | 145899 | 14.7 | 158450 | 6.5 | 176178 | 1.6 | 206646 | |
| 1 | 40.8 | 136322 | 26.0 | 146081 | 14.6 | 158693 | 6.4 | 176544 | 1.5 | 207388 | 29 |
| | 40.5 | 136468 | 25.8 | 146264 | 14.4 | 158938 | 6.3 | 176914 | 1.5 | 208143 | |
| 2 | 40.3 | 136615 | 25.6 | 146448 | 14.2 | 159184 | 6.2 | 177287 | 1.4 | 208912 | 28 |
| | 40.0 | 136761 | 25.4 | 146632 | 14.1 | 159431 | 6.1 | 177663 | 1.4 | 209695 | |
| 3 | 39.7 | 136909 | 25.2 | 146817 | 13.9 | 159680 | 6.0 | 178042 | 1.3 | 210491 | 27 |
| | 39.4 | 137057 | 24.9 | 147003 | 13.7 | 159930 | 5.9 | 178424 | 1.3 | 211303 | |
| 4 | 39.2 | 137205 | 24.7 | 147190 | 13.6 | 160182 | 5.8 | 178810 | 1.2 | 212130 | 26 |
| | 38.9 | 137354 | 24.5 | 147377 | 13.4 | 160435 | 5.7 | 179200 | 1.2 | 212974 | |
| 5 | 38.6 | 137503 | 24.3 | 147566 | 13.3 | 160690 | 5.6 | 179593 | 1.1 | 213834 | 25 |
| | 38.4 | 137653 | 24.1 | 147755 | 13.1 | 160946 | 5.5 | 179990 | 1.1 | 214711 | |
| 6 | 38.1 | 137804 | 23.9 | 147945 | 13.0 | 161204 | 5.4 | 180390 | 1.1 | 215607 | 24 |
| | 37.8 | 137955 | 23.7 | 148135 | 12.8 | 161463 | 5.3 | 180794 | 1.0 | 216521 | |
| 7 | 37.6 | 138106 | 23.5 | 148327 | 12.7 | 161724 | 5.2 | 181201 | 1.0 | 217455 | 23 |
| | 37.3 | 138258 | 23.3 | 148520 | 12.5 | 161986 | 5.1 | 181613 | 0.9 | 218409 | |
| 8 | 37.1 | 138411 | 23.1 | 148713 | 12.4 | 162250 | 5.0 | 182029 | 0.9 | 219385 | 22 |
| | 36.8 | 138564 | 22.8 | 148907 | 12.2 | 162516 | 4.9 | 182448 | 0.9 | 220384 | |
| 9 | 36.5 | 138718 | 22.6 | 149103 | 12.1 | 162783 | 4.8 | 182872 | 0.8 | 221406 | 21 |
| | 36.3 | 138872 | 22.4 | 149299 | 11.9 | 163052 | 4.7 | 183300 | 0.8 | 222452 | |
| 10 | 36.0 | 139027 | 22.2 | 149495 | 11.8 | 163322 | 4.6 | 183732 | 0.7 | 223525 | 20 |
| | 35.8 | 139182 | 22.0 | 149693 | 11.6 | 163594 | 4.5 | 184168 | 0.7 | 224624 | |
| 11 | 35.5 | 139338 | 21.8 | 149892 | 11.5 | 163868 | 4.4 | 184609 | 0.7 | 225752 | 19 |
| | 35.3 | 139494 | 21.6 | 150092 | 11.3 | 164144 | 4.3 | 185055 | 0.6 | 226910 | |
| 12 | 35.0 | 139651 | 21.4 | 150292 | 11.2 | 164422 | 4.2 | 185505 | 0.6 | 228100 | 18 |
| | 34.7 | 139809 | 21.2 | 150494 | 11.0 | 164701 | 4.1 | 185959 | 0.6 | 229324 | |
| 13 | 34.5 | 139967 | 21.0 | 150696 | 10.9 | 164982 | 4.1 | 186419 | 0.5 | 230583 | 17 |
| | 34.2 | 140125 | 20.8 | 150899 | 10.8 | 165265 | 4.0 | 186883 | 0.5 | 231879 | |
| 14 | 34.0 | 140285 | 20.6 | 151104 | 10.6 | 165550 | 3.9 | 187353 | 0.5 | 233215 | 16 |
| | 33.7 | 140445 | 20.5 | 151309 | 10.5 | 165836 | 3.8 | 187827 | 0.4 | 234594 | |
| 15 | 33.5 | 140605 | 20.3 | 151515 | 10.3 | 166125 | 3.7 | 188307 | 0.4 | 236018 | 15 |
| | 33.2 | 140766 | 20.1 | 151722 | 10.2 | 166415 | 3.6 | 188793 | 0.4 | 237491 | |
| 16 | 33.0 | 140928 | 19.9 | 151931 | 10.1 | 166708 | 3.6 | 189283 | 0.4 | 239015 | 14 |
| | 32.8 | 141090 | 19.7 | 152140 | 9.9 | 167002 | 3.5 | 189780 | 0.3 | 240594 | |
| 17 | 32.5 | 141253 | 19.5 | 152350 | 9.8 | 167298 | 3.4 | 190282 | 0.3 | 242233 | 13 |
| | 32.3 | 141417 | 19.3 | 152561 | 9.7 | 167597 | 3.3 | 190790 | 0.3 | 243936 | |
| 18 | 32.0 | 141581 | 19.1 | 152774 | 9.5 | 167897 | 3.2 | 191303 | 0.3 | 245709 | 12 |
| | 31.8 | 141745 | 18.9 | 152987 | 9.4 | 168200 | 3.2 | 191824 | 0.2 | 247558 | |
| 19 | 31.5 | 141911 | 18.7 | 153201 | 9.3 | 168505 | 3.1 | 192350 | 0.2 | 249488 | 11 |
| | 31.3 | 142077 | 18.6 | 153417 | 9.1 | 168811 | 3.0 | 192883 | 0.2 | 251508 | |
| 20 | 31.1 | 142243 | 18.4 | 153633 | 9.0 | 169121 | 2.9 | 193422 | 0.2 | 253627 | 10 |
| | 30.8 | 142411 | 18.2 | 153851 | 8.9 | 169432 | 2.9 | 193969 | 0.2 | 255855 | |
| 21 | 30.6 | 142579 | 18.0 | 154070 | 8.7 | 169745 | 2.8 | 194522 | 0.1 | 258203 | 9 |
| | 30.4 | 142747 | 17.8 | 154290 | 8.6 | 170061 | 2.7 | 195082 | 0.1 | 260685 | |
| 22 | 30.1 | 142916 | 17.6 | 154511 | 8.5 | 170379 | 2.7 | 195650 | 0.1 | 263318 | 8 |
| | 29.9 | 143086 | 17.5 | 154733 | 8.4 | 170700 | 2.6 | 196225 | 0.1 | 266121 | |
| 23 | 29.6 | 143257 | 17.3 | 154956 | 8.2 | 171023 | 2.5 | 196808 | 0.1 | 269118 | 7 |
| | 29.4 | 143428 | 17.1 | 155180 | 8.1 | 171348 | 2.4 | 197399 | 0.1 | 272336 | |
| 24 | 29.2 | 143600 | 16.9 | 155406 | 8.0 | 171676 | 2.4 | 197998 | 0.1 | 275812 | 6 |
| | 28.9 | 143773 | 16.8 | 155633 | 7.9 | 172006 | 2.3 | 198605 | 0.1 | 279591 | |
| 25 | 28.7 | 143946 | 16.6 | 155861 | 7.8 | 172339 | 2.3 | 199221 | 0.0 | 283730 | 5 |
| | 28.5 | 144120 | 16.4 | 156090 | 7.6 | 172674 | 2.2 | 199846 | 0.0 | 288306 | |
| 26 | 28.3 | 144295 | 16.2 | 156320 | 7.5 | 173012 | 2.1 | 200480 | 0.0 | 293421 | 4 |
| | 28.0 | 144470 | 16.1 | 156552 | 7.4 | 173352 | 2.1 | 201124 | 0.0 | 299221 | |
| 27 | 27.8 | 144646 | 15.9 | 156784 | 7.3 | 173696 | 2.0 | 201777 | 0.0 | 305915 | 3 |
| | 27.6 | 144823 | 15.7 | 157019 | 7.2 | 174042 | 1.9 | 202440 | 0.0 | 313833 | |
| 28 | 27.4 | 145000 | 15.6 | 157254 | 7.1 | 174391 | 1.9 | 203113 | 0.0 | 323524 | 2 |
| | 27.1 | 145179 | 15.4 | 157490 | 6.9 | 174742 | 1.8 | 203797 | 0.0 | 336018 | |
| 29 | 26.9 | 145358 | 15.2 | 157728 | 6.8 | 175097 | 1.8 | 204492 | 0.0 | 353627 | 1 |
| | 26.7 | 145538 | 15.1 | 157967 | 6.7 | 175454 | 1.7 | 205198 | 0.0 | 383730 | |
| 30 | 26.5 | 145718 | 14.9 | 158208 | 6.6 | 175814 | 1.7 | 205916 | 0.0 | -------- | 0 |
| | A | B | A | B | A | B | A | B | A | B | |
| | 92° 00′ | | 91° 30′ | | 91° 00′ | | 90° 30′ | | 90° 00′ | | ′ |

# H NATURAL TRIGONOMETRIC FUNCTIONS

Material in this appendix is taken from Table 31 in the *American Practical Navigator,* Pub. No. 9, Vol. II, by permission of the Defense Mapping Agency Hydrographic/Topographic Center.

# TABLE 31

Natural Trigonometric Functions

| 30° → ↓ ′ | sin | Diff. 1′ | csc | Diff. 1′ | tan | Diff. 1′ | cot | Diff. 1′ | sec | Diff. 1′ | cos | Diff. 1′ | ← 149° ↓ ′ |
|---|---|---|---|---|---|---|---|---|---|---|---|---|---|
| 0 | 0.50000 | 25 | 2.00000 | 101 | 0.57735 | 39 | 1.73205 | 116 | 1.15470 | 19 | 0.86603 | 15 | 60 |
| 1 | .50025 | 25 | 1.99899 | 100 | .57774 | 39 | .73089 | 116 | .15489 | 20 | .86588 | 15 | 59 |
| 2 | .50050 | 26 | .99799 | 101 | .57813 | 38 | .72973 | 116 | .15509 | 19 | .86573 | 14 | 58 |
| 3 | .50076 | 25 | .99698 | 100 | .57851 | 39 | .72857 | 116 | .15528 | 20 | .86559 | 15 | 57 |
| 4 | .50101 | 25 | .99598 | 100 | .57890 | 39 | .72741 | 116 | .15548 | 19 | .86544 | 14 | 56 |
| 5 | 0.50126 | 25 | 1.99498 | 100 | 0.57929 | 39 | 1.72625 | 116 | 1.15567 | 20 | 0.86530 | 15 | 55 |
| 6 | .50151 | 25 | .99398 | 100 | .57968 | 39 | .72509 | 116 | .15587 | 19 | .86515 | 14 | 54 |
| 7 | .50176 | 25 | .99298 | 100 | .58007 | 39 | .72393 | 115 | .15606 | 20 | .86501 | 15 | 53 |
| 8 | .50201 | 26 | .99198 | 100 | .58046 | 39 | .72278 | 115 | .15626 | 19 | .86486 | 15 | 52 |
| 9 | .50227 | 25 | .99098 | 100 | .58085 | 39 | .72163 | 116 | .15645 | 20 | .86471 | 14 | 51 |
| 10 | 0.50252 | 25 | 1.98998 | 99 | 0.58124 | 38 | 1.72047 | 115 | 1.15665 | 19 | 0.86457 | 15 | 50 |
| 11 | .50277 | 25 | .98899 | 100 | .58162 | 39 | .71932 | 115 | .15684 | 20 | .86442 | 15 | 49 |
| 12 | .50302 | 25 | .98799 | 99 | .58201 | 39 | .71817 | 115 | .15704 | 20 | .86427 | 14 | 48 |
| 13 | .50327 | 25 | .98700 | 99 | .58240 | 39 | .71702 | 114 | .15724 | 19 | .86413 | 15 | 47 |
| 14 | .50352 | 25 | .98601 | 99 | .58279 | 39 | .71588 | 115 | .15743 | 20 | .86398 | 14 | 46 |
| 15 | 0.50377 | 26 | 1.98502 | 99 | 0.58318 | 39 | 1.71473 | 115 | 1.15763 | 19 | 0.86384 | 15 | 45 |
| 16 | .50403 | 25 | .98403 | 99 | .58357 | 39 | .71358 | 114 | .15782 | 20 | .86369 | 15 | 44 |
| 17 | .50428 | 25 | .98304 | 99 | .58396 | 39 | .71244 | 115 | .15802 | 20 | .86354 | 14 | 43 |
| 18 | .50453 | 25 | .98205 | 98 | .58435 | 39 | .71129 | 114 | .15822 | 19 | .86340 | 15 | 42 |
| 19 | .50478 | 25 | .98107 | 99 | .58474 | 39 | .71015 | 114 | .15841 | 20 | .86325 | 15 | 41 |
| 20 | 0.50503 | 25 | 1.98008 | 98 | 0.58513 | 39 | 1.70901 | 114 | 1.15861 | 20 | 0.86310 | 15 | 40 |
| 21 | .50528 | 25 | .97910 | 99 | .58552 | 39 | .70787 | 114 | .15881 | 20 | .86295 | 14 | 39 |
| 22 | .50553 | 25 | .97811 | 98 | .58591 | 40 | .70673 | 113 | .15901 | 19 | .86281 | 15 | 38 |
| 23 | .50578 | 25 | .97713 | 98 | .58631 | 39 | .70560 | 114 | .15920 | 20 | .86266 | 15 | 37 |
| 24 | .50603 | 25 | .97615 | 98 | .58670 | 39 | .70446 | 114 | .15940 | 20 | .86251 | 14 | 36 |
| 25 | 0.50628 | 26 | 1.97517 | 97 | 0.58709 | 39 | 1.70332 | 113 | 1.15960 | 20 | 0.86237 | 15 | 35 |
| 26 | .50654 | 25 | .97420 | 98 | .58748 | 39 | .70219 | 113 | .15980 | 20 | .86222 | 15 | 34 |
| 27 | .50679 | 25 | .97322 | 98 | .58787 | 39 | .70106 | 114 | .16000 | 19 | .86207 | 15 | 33 |
| 28 | .50704 | 25 | .97224 | 97 | .58826 | 39 | .69992 | 113 | .16019 | 20 | .86192 | 14 | 32 |
| 29 | .50729 | 25 | .97127 | 98 | .58865 | 40 | .69879 | 113 | .16039 | 20 | .86178 | 15 | 31 |
| 30 | 0.50754 | 25 | 1.97029 | 97 | 0.58905 | 39 | 1.69766 | 113 | 1.16059 | 20 | 0.86163 | 15 | 30 |
| 31 | .50779 | 25 | .96932 | 97 | .58944 | 39 | .69653 | 112 | .16079 | 20 | .86148 | 15 | 29 |
| 32 | .50804 | 25 | .96835 | 97 | .58983 | 39 | .69541 | 113 | .16099 | 20 | .86133 | 14 | 28 |
| 33 | .50829 | 25 | .96738 | 97 | .59022 | 39 | .69428 | 112 | .16119 | 20 | .86119 | 15 | 27 |
| 34 | .50854 | 25 | .96641 | 97 | .59061 | 40 | .69316 | 113 | .16139 | 20 | .86104 | 15 | 26 |
| 35 | 0.50879 | 25 | 1.96544 | 96 | 0.59101 | 39 | 1.69203 | 112 | 1.16159 | 20 | 0.86089 | 15 | 25 |
| 36 | .50904 | 25 | .96448 | 97 | .59140 | 39 | .69091 | 112 | .16179 | 20 | .86074 | 15 | 24 |
| 37 | .50929 | 25 | .96351 | 96 | .59179 | 39 | .68979 | 113 | .16199 | 20 | .86059 | 14 | 23 |
| 38 | .50954 | 25 | .96255 | 97 | .59218 | 40 | .68866 | 112 | .16219 | 20 | .86045 | 15 | 22 |
| 39 | .50979 | 25 | .96158 | 96 | .59258 | 39 | .68754 | 111 | .16239 | 20 | .86030 | 15 | 21 |
| 40 | 0.51004 | 25 | 1.96062 | 96 | 0.59297 | 39 | 1.68643 | 112 | 1.16259 | 20 | 0.86015 | 15 | 20 |
| 41 | .51029 | 25 | .95966 | 96 | .59336 | 40 | .68531 | 112 | .16279 | 20 | .86000 | 15 | 19 |
| 42 | .51054 | 25 | .95870 | 96 | .59376 | 39 | .68419 | 111 | .16299 | 20 | .85985 | 15 | 18 |
| 43 | .51079 | 25 | .95774 | 96 | .59415 | 39 | .68308 | 112 | .16319 | 20 | .85970 | 14 | 17 |
| 44 | .51104 | 25 | .95678 | 95 | .59454 | 40 | .68196 | 111 | .16339 | 20 | .85956 | 15 | 16 |
| 45 | 0.51129 | 25 | 1.95583 | 96 | 0.59494 | 39 | 1.68085 | 111 | 1.16359 | 21 | 0.85941 | 15 | 15 |
| 46 | .51154 | 25 | .95487 | 95 | .59533 | 40 | .67974 | 111 | .16380 | 20 | .85926 | 15 | 14 |
| 47 | .51179 | 25 | .95392 | 96 | .59573 | 39 | .67863 | 111 | .16400 | 20 | .85911 | 15 | 13 |
| 48 | .51204 | 25 | .95296 | 95 | .59612 | 39 | .67752 | 111 | .16420 | 20 | .85896 | 15 | 12 |
| 49 | .51229 | 25 | .95201 | 95 | .59651 | 40 | .67641 | 111 | .16440 | 20 | .85881 | 15 | 11 |
| 50 | 0.51254 | 25 | 1.95106 | 95 | 0.59691 | 39 | 1.67530 | 111 | 1.16460 | 21 | 0.85866 | 15 | 10 |
| 51 | .51279 | 25 | .95011 | 95 | .59730 | 40 | .67419 | 110 | .16481 | 20 | .85851 | 15 | 9 |
| 52 | .51304 | 25 | .94916 | 95 | .59770 | 39 | .67309 | 111 | .16501 | 20 | .85836 | 15 | 8 |
| 53 | .51329 | 25 | .94821 | 95 | .59809 | 40 | .67198 | 110 | .16521 | 20 | .85821 | 15 | 7 |
| 54 | .51354 | 25 | .94726 | 94 | .59849 | 39 | .67088 | 110 | .16541 | 21 | .85806 | 14 | 6 |
| 55 | 0.51379 | 25 | 1.94632 | 95 | 0.59888 | 40 | 1.66978 | 111 | 1.16562 | 20 | 0.85792 | 15 | 5 |
| 56 | .51404 | 25 | .94537 | 94 | .59928 | 39 | .66867 | 110 | .16582 | 20 | .85777 | 15 | 4 |
| 57 | .51429 | 25 | .94443 | 94 | .59967 | 40 | .66757 | 110 | .16602 | 21 | .85762 | 15 | 3 |
| 58 | .51454 | 25 | .94349 | 95 | .60007 | 39 | .66647 | 109 | .16623 | 20 | .85747 | 15 | 2 |
| 59 | .51479 | 25 | .94254 | 94 | .60046 | 40 | .66538 | 110 | .16643 | 20 | .85732 | 15 | 1 |
| 60 | 0.51504 | | 1.94160 | | 0.60086 | | 1.66428 | | 1.16663 | | 0.85717 | | 0 |
| ↑ 120° → | cos | Diff. 1′ | sec | Diff. 1′ | cot | Diff. 1′ | tan | Diff. 1′ | csc | Diff. 1′ | sin | Diff. 1′ | ← 59° ↑ |

# TABLE 31

Natural Trigonometric Functions

| 34°→ ↓ ′ | sin | Diff. 1′ | csc | Diff. 1′ | tan | Diff. 1′ | cot | Diff. 1′ | sec | Diff. 1′ | cos | Diff. 1′ | ←145° ↓ ′ |
|---|---|---|---|---|---|---|---|---|---|---|---|---|---|
| 0 | 0.55919 | 24 | 1.78829 | 77 | 0.67451 | 42 | 1.48256 | 93 | 1.20622 | 23 | 0.82904 | 17 | 60 |
| 1 | .55943 | 25 | .78752 | 77 | .67493 | 43 | .48163 | 93 | .20645 | 24 | .82887 | 16 | 59 |
| 2 | .55968 | 24 | .78675 | 77 | .67536 | 42 | .48070 | 93 | .20669 | 24 | .82871 | 16 | 58 |
| 3 | .55992 | 24 | .78598 | 77 | .67578 | 42 | .47977 | 92 | .20693 | 24 | .82855 | 16 | 57 |
| 4 | .56016 | 24 | .78521 | 76 | .67620 | 43 | .47885 | 93 | .20717 | 23 | .82839 | 17 | 56 |
| 5 | 0.56040 | 24 | 1.78445 | 77 | 0.67663 | 42 | 1.47792 | 93 | 1.20740 | 24 | 0.82822 | 16 | 55 |
| 6 | .56064 | 24 | .78368 | 77 | .67705 | 43 | .47699 | 92 | .20764 | 24 | .82806 | 16 | 54 |
| 7 | .56088 | 24 | .78291 | 76 | .67748 | 42 | .47607 | 93 | .20788 | 24 | .82790 | 17 | 53 |
| 8 | .56112 | 24 | .78215 | 77 | .67790 | 42 | .47514 | 92 | .20812 | 24 | .82773 | 16 | 52 |
| 9 | .56136 | 24 | .78138 | 76 | .67832 | 43 | .47422 | 92 | .20836 | 23 | .82757 | 16 | 51 |
| 10 | 0.56160 | 24 | 1.78062 | 76 | 0.67875 | 42 | 1.47330 | 92 | 1.20859 | 24 | 0.82741 | 17 | 50 |
| 11 | .56184 | 24 | .77986 | 76 | .67917 | 43 | .47238 | 92 | .20883 | 24 | .82724 | 16 | 49 |
| 12 | .56208 | 24 | .77910 | 77 | .67960 | 42 | .47146 | 93 | .20907 | 24 | .82708 | 16 | 48 |
| 13 | .56232 | 24 | .77833 | 76 | .68002 | 43 | .47053 | 91 | .20931 | 24 | .82692 | 17 | 47 |
| 14 | .56256 | 24 | .77757 | 76 | .68045 | 43 | .46962 | 92 | .20955 | 24 | .82675 | 16 | 46 |
| 15 | 0.56280 | 25 | 1.77681 | 75 | 0.68088 | 42 | 1.46870 | 92 | 1.20979 | 24 | 0.82659 | 16 | 45 |
| 16 | .56305 | 24 | .77606 | 76 | .68130 | 43 | .46778 | 92 | .21003 | 24 | .82643 | 17 | 44 |
| 17 | .56329 | 24 | .77530 | 76 | .68173 | 42 | .46686 | 91 | .21027 | 24 | .82626 | 16 | 43 |
| 18 | .56353 | 24 | .77454 | 76 | .68215 | 43 | .46595 | 92 | .21051 | 24 | .82610 | 17 | 42 |
| 19 | .56377 | 24 | .77378 | 75 | .68258 | 43 | .46503 | 92 | .21075 | 24 | .82593 | 16 | 41 |
| 20 | 0.56401 | 24 | 1.77303 | 76 | 0.68301 | 42 | 1.46411 | 91 | 1.21099 | 24 | 0.82577 | 16 | 40 |
| 21 | .56425 | 24 | .77227 | 75 | .68343 | 43 | .46320 | 91 | .21123 | 24 | .82561 | 17 | 39 |
| 22 | .56449 | 24 | .77152 | 75 | .68386 | 43 | .46229 | 92 | .21147 | 24 | .82544 | 16 | 38 |
| 23 | .56473 | 24 | .77077 | 76 | .68429 | 42 | .46137 | 91 | .21171 | 24 | .82528 | 17 | 37 |
| 24 | .56497 | 24 | .77001 | 75 | .68471 | 43 | .46046 | 91 | .21195 | 25 | .82511 | 16 | 36 |
| 25 | 0.56521 | 24 | 1.76926 | 75 | 0.68514 | 43 | 1.45955 | 91 | 1.21220 | 24 | 0.82495 | 17 | 35 |
| 26 | .56545 | 24 | .76851 | 75 | .68557 | 43 | .45864 | 91 | .21244 | 24 | .82478 | 16 | 34 |
| 27 | .56569 | 24 | .76776 | 75 | .68600 | 42 | .45773 | 91 | .21268 | 24 | .82462 | 16 | 33 |
| 28 | .56593 | 24 | .76701 | 75 | .68642 | 43 | .45682 | 90 | .21292 | 24 | .82446 | 17 | 32 |
| 29 | .56617 | 24 | .76626 | 74 | .68685 | 43 | .45592 | 91 | .21316 | 25 | .82429 | 16 | 31 |
| 30 | 0.56641 | 24 | 1.76552 | 75 | 0.68728 | 43 | 1.45501 | 91 | 1.21341 | 24 | 0.82413 | 17 | 30 |
| 31 | .56665 | 24 | .76477 | 75 | .68771 | 43 | .45410 | 90 | .21365 | 24 | .82396 | 16 | 29 |
| 32 | .56689 | 24 | .76402 | 74 | .68814 | 43 | .45320 | 91 | .21389 | 25 | .82380 | 17 | 28 |
| 33 | .56713 | 23 | .76328 | 75 | .68857 | 43 | .45229 | 90 | .21414 | 24 | .82363 | 16 | 27 |
| 34 | .56736 | 24 | .76253 | 74 | .68900 | 42 | .45139 | 90 | .21438 | 24 | .82347 | 17 | 26 |
| 35 | 0.56760 | 24 | 1.76179 | 74 | 0.68942 | 43 | 1.45049 | 91 | 1.21462 | 25 | 0.82330 | 16 | 25 |
| 36 | .56784 | 24 | .76105 | 74 | .68985 | 43 | .44958 | 90 | .21487 | 24 | .82314 | 17 | 24 |
| 37 | .56808 | 24 | .76031 | 75 | .69028 | 43 | .44868 | 90 | .21511 | 24 | .82297 | 16 | 23 |
| 38 | .56832 | 24 | .75956 | 74 | .69071 | 43 | .44778 | 90 | .21535 | 25 | .82281 | 17 | 22 |
| 39 | .56856 | 24 | .75882 | 74 | .69114 | 43 | .44688 | 90 | .21560 | 24 | .82264 | 16 | 21 |
| 40 | 0.56880 | 24 | 1.75808 | 74 | 0.69157 | 43 | 1.44598 | 90 | 1.21584 | 25 | 0.82248 | 17 | 20 |
| 41 | .56904 | 24 | .75734 | 73 | .69200 | 43 | .44508 | 90 | .21609 | 24 | .82231 | 17 | 19 |
| 42 | .56928 | 24 | .75661 | 74 | .69243 | 43 | .44418 | 89 | .21633 | 25 | .82214 | 16 | 18 |
| 43 | .56952 | 24 | .75587 | 74 | .69286 | 43 | .44329 | 90 | .21658 | 24 | .82198 | 17 | 17 |
| 44 | .56976 | 24 | .75513 | 73 | .69329 | 43 | .44239 | 90 | .21682 | 25 | .82181 | 16 | 16 |
| 45 | 0.57000 | 24 | 1.75440 | 74 | 0.69372 | 44 | 1.44149 | 89 | 1.21707 | 24 | 0.82165 | 17 | 15 |
| 46 | .57024 | 23 | .75366 | 73 | .69416 | 43 | .44060 | 90 | .21731 | 25 | .82148 | 16 | 14 |
| 47 | .57047 | 24 | .75293 | 74 | .69459 | 43 | .43970 | 89 | .21756 | 25 | .82132 | 17 | 13 |
| 48 | .57071 | 24 | .75219 | 73 | .69502 | 43 | .43881 | 89 | .21781 | 24 | .82115 | 17 | 12 |
| 49 | .57095 | 24 | .75146 | 73 | .69545 | 43 | .43792 | 89 | .21805 | 25 | .82098 | 16 | 11 |
| 50 | 0.57119 | 24 | 1.75073 | 73 | 0.69588 | 43 | 1.43703 | 89 | 1.21830 | 25 | 0.82082 | 17 | 10 |
| 51 | .57143 | 24 | .75000 | 73 | .69631 | 44 | .43614 | 89 | .21855 | 24 | .82065 | 17 | 9 |
| 52 | .57167 | 24 | .74927 | 73 | .69675 | 43 | .43525 | 89 | .21879 | 25 | .82048 | 16 | 8 |
| 53 | .57191 | 24 | .74854 | 73 | .69718 | 43 | .43436 | 89 | .21904 | 25 | .82032 | 17 | 7 |
| 54 | .57215 | 23 | .74781 | 73 | .69761 | 43 | .43347 | 89 | .21929 | 24 | .82015 | 16 | 6 |
| 55 | 0.57238 | 24 | 1.74708 | 73 | 0.69804 | 43 | 1.43258 | 89 | 1.21953 | 25 | 0.81999 | 17 | 5 |
| 56 | .57262 | 24 | .74635 | 73 | .69847 | 44 | .43169 | 89 | .21978 | 25 | .81982 | 17 | 4 |
| 57 | .57286 | 24 | .74562 | 72 | .69891 | 43 | .43080 | 88 | .22003 | 25 | .81965 | 16 | 3 |
| 58 | .57310 | 24 | .74490 | 73 | .69934 | 43 | .42992 | 89 | .22028 | 25 | .81949 | 17 | 2 |
| 59 | .57334 | 24 | .74417 | 72 | .69977 | 44 | .42903 | 88 | .22053 | 24 | .81932 | 17 | 1 |
| 60 | 0.57358 | | 1.74345 | | 0.70021 | | 1.42815 | | 1.22077 | | 0.81915 | | 0 |
| ↑ 124°→ | cos | Diff. 1′ | sec | Diff. 1′ | cot | Diff. 1′ | tan | Diff. 1′ | csc | Diff. 1′ | sin | Diff. 1′ | ←55° ↑ |

# I LOGARITHMS OF TRIGONOMETRIC FUNCTIONS

Material in this appendix is taken from Table 33 in the *American Practical Navigator,* Pub. No. 9, Vol. II, by permission of the Defense Mapping Agency Hydrographic/Topographic Center.

## TABLE 33

Logarithms of Trigonometric Functions

| 6° → ↓ ′ | sin | Diff. 1′ | csc | tan | Diff. 1′ | cot | sec | Diff. 1′ | cos | ← 173° ↓ ′ |
|---|---|---|---|---|---|---|---|---|---|---|
| 0 | 9.01923 | 120 | 10.98077 | 9.02162 | 121 | 10.97838 | 10.00239 | 1 | 9.99761 | 60 |
| 1 | .02043 | 120 | .97957 | .02283 | 121 | .97717 | .00240 | 1 | .99760 | 59 |
| 2 | .02163 | 120 | .97837 | .02404 | 121 | .97596 | .00241 | 2 | .99759 | 58 |
| 3 | .02283 | 119 | .97717 | .02525 | 120 | .97475 | .00243 | 1 | .99757 | 57 |
| 4 | .02402 | 118 | .97598 | .02645 | 121 | .97355 | .00244 | 1 | .99756 | 56 |
| 5 | 9.02520 | 119 | 10.97480 | 9.02766 | 119 | 10.97234 | 10.00245 | 2 | 9.99755 | 55 |
| 6 | .02639 | 118 | .97361 | .02885 | 120 | .97115 | .00247 | 1 | .99753 | 54 |
| 7 | .02757 | 117 | .97243 | .03005 | 119 | .96995 | .00248 | 1 | .99752 | 53 |
| 8 | .02874 | 118 | .97126 | .03124 | 118 | .96876 | .00249 | 2 | .99751 | 52 |
| 9 | .02992 | 117 | .97008 | .03242 | 119 | .96758 | .00251 | 1 | .99749 | 51 |
| 10 | 9.03109 | 117 | 10.96891 | 9.03361 | 118 | 10.96639 | 10.00252 | 1 | 9.99748 | 50 |
| 11 | .03226 | 116 | .96774 | .03479 | 118 | .96521 | .00253 | 2 | .99747 | 49 |
| 12 | .03342 | 116 | .96658 | .03597 | 117 | .96403 | .00255 | 1 | .99745 | 48 |
| 13 | .03458 | 116 | .96542 | .03714 | 118 | .96286 | .00256 | 2 | .99744 | 47 |
| 14 | .03574 | 116 | .96426 | .03832 | 116 | .96168 | .00258 | 1 | .99742 | 46 |
| 15 | 9.03690 | 115 | 10.96310 | 9.03948 | 117 | 10.96052 | 10.00259 | 1 | 9.99741 | 45 |
| 16 | .03805 | 115 | .96195 | .04065 | 116 | .95935 | .00260 | 2 | .99740 | 44 |
| 17 | .03920 | 114 | .96080 | .04181 | 116 | .95819 | .00262 | 1 | .99738 | 43 |
| 18 | .04034 | 115 | .95966 | .04297 | 116 | .95703 | .00263 | 1 | .99737 | 42 |
| 19 | .04149 | 113 | .95851 | .04413 | 115 | .95587 | .00264 | 2 | .99736 | 41 |
| 20 | 9.04262 | 114 | 10.95738 | 9.04528 | 115 | 10.95472 | 10.00266 | 1 | 9.99734 | 40 |
| 21 | .04376 | 114 | .95624 | .04643 | 115 | .95357 | .00267 | 2 | .99733 | 39 |
| 22 | .04490 | 113 | .95510 | .04758 | 115 | .95242 | .00269 | 1 | .99731 | 38 |
| 23 | .04603 | 112 | .95397 | .04873 | 114 | .95127 | .00270 | 2 | .99730 | 37 |
| 24 | .04715 | 113 | .95285 | .04987 | 114 | .95013 | .00272 | 1 | .99728 | 36 |
| 25 | 9.04828 | 112 | 10.95172 | 9.05101 | 113 | 10.94899 | 10.00273 | 1 | 9.99727 | 35 |
| 26 | .04940 | 112 | .95060 | .05214 | 114 | .94786 | .00274 | 2 | .99726 | 34 |
| 27 | .05052 | 112 | .94948 | .05328 | 113 | .94672 | .00276 | 1 | .99724 | 33 |
| 28 | .05164 | 111 | .94836 | .05441 | 112 | .94559 | .00277 | 2 | .99723 | 32 |
| 29 | .05275 | 111 | .94725 | .05553 | 113 | .94447 | .00279 | 1 | .99721 | 31 |
| 30 | 9.05386 | 111 | 10.94614 | 9.05666 | 112 | 10.94334 | 10.00280 | 2 | 9.99720 | 30 |
| 31 | .05497 | 110 | .94503 | .05778 | 112 | .94222 | .00282 | 1 | .99718 | 29 |
| 32 | .05607 | 110 | .94393 | .05890 | 112 | .94110 | .00283 | 1 | .99717 | 28 |
| 33 | .05717 | 110 | .94283 | .06002 | 111 | .93998 | .00284 | 2 | .99716 | 27 |
| 34 | .05827 | 110 | .94173 | .06113 | 111 | .93887 | .00286 | 1 | .99714 | 26 |
| 35 | 9.05937 | 109 | 10.94063 | 9.06224 | 111 | 10.93776 | 10.00287 | 2 | 9.99713 | 25 |
| 36 | .06046 | 109 | .93954 | .06335 | 110 | .93665 | .00289 | 1 | .99711 | 24 |
| 37 | .06155 | 109 | .93845 | .06445 | 111 | .93555 | .00290 | 2 | .99710 | 23 |
| 38 | .06264 | 108 | .93736 | .06556 | 110 | .93444 | .00292 | 1 | .99708 | 22 |
| 39 | .06372 | 109 | .93628 | .06666 | 109 | .93334 | .00293 | 2 | .99707 | 21 |
| 40 | 9.06481 | 108 | 10.93519 | 9.06775 | 110 | 10.93225 | 10.00295 | 1 | 9.99705 | 20 |
| 41 | .06589 | 107 | .93411 | .06885 | 109 | .93115 | .00296 | 2 | .99704 | 19 |
| 42 | .06696 | 108 | .93304 | .06994 | 109 | .93006 | .00298 | 1 | .99702 | 18 |
| 43 | .06804 | 107 | .93196 | .07103 | 108 | .92897 | .00299 | 2 | .99701 | 17 |
| 44 | .06911 | 107 | .93089 | .07211 | 109 | .92789 | .00301 | 1 | .99699 | 16 |
| 45 | 9.07018 | 106 | 10.92982 | 9.07320 | 108 | 10.92680 | 10.00302 | 2 | 9.99698 | 15 |
| 46 | .07124 | 107 | .92876 | .07428 | 108 | .92572 | .00304 | 1 | .99696 | 14 |
| 47 | .07231 | 106 | .92769 | .07536 | 107 | .92464 | .00305 | 2 | .99695 | 13 |
| 48 | .07337 | 105 | .92663 | .07643 | 108 | .92357 | .00307 | 1 | .99693 | 12 |
| 49 | .07442 | 106 | .92558 | .07751 | 107 | .92249 | .00308 | 2 | .99692 | 11 |
| 50 | 9.07548 | 105 | 10.92452 | 9.07858 | 106 | 10.92142 | 10.00310 | 1 | 9.99690 | 10 |
| 51 | .07653 | 105 | .92347 | .07964 | 107 | .92036 | .00311 | 2 | .99689 | 9 |
| 52 | .07758 | 105 | .92242 | .08071 | 106 | .91929 | .00313 | 1 | .99687 | 8 |
| 53 | .07863 | 105 | .92137 | .08177 | 106 | .91823 | .00314 | 2 | .99686 | 7 |
| 54 | .07968 | 104 | .92032 | .08283 | 106 | .91717 | .00316 | 1 | .99684 | 6 |
| 55 | 9.08072 | 104 | 10.91928 | 9.08389 | 106 | 10.91611 | 10.00317 | 2 | 9.99683 | 5 |
| 56 | .08176 | 104 | .91824 | .08495 | 105 | .91505 | .00319 | 1 | .99681 | 4 |
| 57 | .08280 | 103 | .91720 | .08600 | 105 | .91400 | .00320 | 2 | .99680 | 3 |
| 58 | .08383 | 103 | .91617 | .08705 | 105 | .91295 | .00322 | 1 | .99678 | 2 |
| 59 | .08486 | 103 | .91514 | .08810 | 104 | .91190 | .00323 | 2 | .99677 | 1 |
| 60 | 9.08589 | | 10.91411 | 9.08914 | | 10.91086 | 10.00325 | | 9.99675 | 0 |
| ↑ 96° → | cos | Diff. 1′ | sec | cot | Diff. 1′ | tan | csc | Diff. 1′ | sin | ← 83° ↑ |

# TABLE 33

Logarithms of Trigonometric Functions

| 13°→ ′ | sin | Diff. 1′ | csc | tan | Diff. 1′ | cot | sec | Diff. 1′ | cos | ←166° ′ |
|---|---|---|---|---|---|---|---|---|---|---|
| 0 | 9.35209 | 54 | 10.64791 | 9.36336 | 58 | 10.63664 | 10.01128 | 3 | 9.98872 | 60 |
| 1 | .35263 | 55 | .64737 | .36394 | 58 | .63606 | .01131 | 2 | .98869 | 59 |
| 2 | .35318 | 55 | .64682 | .36452 | 57 | .63548 | .01133 | 3 | .98867 | 58 |
| 3 | .35373 | 54 | .64627 | .36509 | 57 | .63491 | .01136 | 3 | .98864 | 57 |
| 4 | .35427 | 54 | .64573 | .36566 | 58 | .63434 | .01139 | 3 | .98861 | 56 |
| 5 | 9.35481 | 55 | 10.64519 | 9.36624 | 57 | 10.63376 | 10.01142 | 3 | 9.98858 | 55 |
| 6 | .35536 | 54 | .64464 | .36681 | 57 | .63319 | .01145 | 3 | .98855 | 54 |
| 7 | .35590 | 54 | .64410 | .36738 | 57 | .63262 | .01148 | 3 | .98852 | 53 |
| 8 | .35644 | 54 | .64356 | .36795 | 57 | .63205 | .01151 | 3 | .98849 | 52 |
| 9 | .35698 | 54 | .64302 | .36852 | 57 | .63148 | .01154 | 3 | .98846 | 51 |
| 10 | 9.35752 | 54 | 10.64248 | 9.36909 | 57 | 10.63091 | 10.01157 | 3 | 9.98843 | 50 |
| 11 | .35806 | 54 | .64194 | .36966 | 57 | .63034 | .01160 | 3 | .98840 | 49 |
| 12 | .35860 | 54 | .64140 | .37023 | 57 | .62977 | .01163 | 3 | .98837 | 48 |
| 13 | .35914 | 54 | .64086 | .37080 | 57 | .62920 | .01166 | 3 | .98834 | 47 |
| 14 | .35968 | 54 | .64032 | .37137 | 56 | .62863 | .01169 | 3 | .98831 | 46 |
| 15 | 9.36022 | 53 | 10.63978 | 9.37193 | 57 | 10.62807 | 10.01172 | 3 | 9.98828 | 45 |
| 16 | .36075 | 54 | .63925 | .37250 | 56 | .62750 | .01175 | 3 | .98825 | 44 |
| 17 | .36129 | 53 | .63871 | .37306 | 57 | .62694 | .01178 | 3 | .98822 | 43 |
| 18 | .36182 | 54 | .63818 | .37363 | 56 | .62637 | .01181 | 3 | .98819 | 42 |
| 19 | .36236 | 53 | .63764 | .37419 | 57 | .62581 | .01184 | 3 | .98816 | 41 |
| 20 | 9.36289 | 53 | 10.63711 | 9.37476 | 56 | 10.62524 | 10.01187 | 3 | 9.98813 | 40 |
| 21 | .36342 | 53 | .63658 | .37532 | 56 | .62468 | .01190 | 3 | .98810 | 39 |
| 22 | .36395 | 54 | .63605 | .37588 | 56 | .62412 | .01193 | 3 | .98807 | 38 |
| 23 | .36449 | 53 | .63551 | .37644 | 56 | .62356 | .01196 | 3 | .98804 | 37 |
| 24 | .36502 | 53 | .63498 | .37700 | 56 | .62300 | .01199 | 3 | .98801 | 36 |
| 25 | 9.36555 | 53 | 10.63445 | 9.37756 | 56 | 10.62244 | 10.01202 | 3 | 9.98798 | 35 |
| 26 | .36608 | 52 | .63392 | .37812 | 56 | .62188 | .01205 | 3 | .98795 | 34 |
| 27 | .36660 | 53 | .63340 | .37868 | 56 | .62132 | .01208 | 3 | .98792 | 33 |
| 28 | .36713 | 53 | .63287 | .37924 | 56 | .62076 | .01211 | 3 | .98789 | 32 |
| 29 | .36766 | 53 | .63234 | .37980 | 55 | .62020 | .01214 | 3 | .98786 | 31 |
| 30 | 9.36819 | 52 | 10.63181 | 9.38035 | 56 | 10.61965 | 10.01217 | 3 | 9.98783 | 30 |
| 31 | .36871 | 53 | .63129 | .38091 | 56 | .61909 | .01220 | 3 | .98780 | 29 |
| 32 | .36924 | 52 | .63076 | .38147 | 55 | .61853 | .01223 | 3 | .98777 | 28 |
| 33 | .36976 | 52 | .63024 | .38202 | 55 | .61798 | .01226 | 3 | .98774 | 27 |
| 34 | .37028 | 53 | .62972 | .38257 | 56 | .61743 | .01229 | 3 | .98771 | 26 |
| 35 | 9.37081 | 52 | 10.62919 | 9.38313 | 55 | 10.61687 | 10.01232 | 3 | 9.98768 | 25 |
| 36 | .37133 | 52 | .62867 | .38368 | 55 | .61632 | .01235 | 3 | .98765 | 24 |
| 37 | .37185 | 52 | .62815 | .38423 | 56 | .61577 | .01238 | 3 | .98762 | 23 |
| 38 | .37237 | 52 | .62763 | .38479 | 55 | .61521 | .01241 | 3 | .98759 | 22 |
| 39 | .37289 | 52 | .62711 | .38534 | 55 | .61466 | .01244 | 3 | .98756 | 21 |
| 40 | 9.37341 | 52 | 10.62659 | 9.38589 | 55 | 10.61411 | 10.01247 | 3 | 9.98753 | 20 |
| 41 | .37393 | 52 | .62607 | .38644 | 55 | .61356 | .01250 | 4 | .98750 | 19 |
| 42 | .37445 | 52 | .62555 | .38699 | 55 | .61301 | .01254 | 3 | .98746 | 18 |
| 43 | .37497 | 52 | .62503 | .38754 | 54 | .61246 | .01257 | 3 | .98743 | 17 |
| 44 | .37549 | 51 | .62451 | .38808 | 55 | .61192 | .01260 | 3 | .98740 | 16 |
| 45 | 9.37600 | 52 | 10.62400 | 9.38863 | 55 | 10.61137 | 10.01263 | 3 | 9.98737 | 15 |
| 46 | .37652 | 51 | .62348 | .38918 | 54 | .61082 | .01266 | 3 | .98734 | 14 |
| 47 | .37703 | 52 | .62297 | .38972 | 55 | .61028 | .01269 | 3 | .98731 | 13 |
| 48 | .37755 | 51 | .62245 | .39027 | 55 | .60973 | .01272 | 3 | .98728 | 12 |
| 49 | .37806 | 52 | .62194 | .39082 | 54 | .60918 | .01275 | 3 | .98725 | 11 |
| 50 | 9.37858 | 51 | 10.62142 | 9.39136 | 54 | 10.60864 | 10.01278 | 3 | 9.98722 | 10 |
| 51 | .37909 | 51 | .62091 | .39190 | 55 | .60810 | .01281 | 4 | .98719 | 9 |
| 52 | .37960 | 51 | .62040 | .39245 | 54 | .60755 | .01285 | 3 | .98715 | 8 |
| 53 | .38011 | 51 | .61989 | .39299 | 54 | .60701 | .01288 | 3 | .98712 | 7 |
| 54 | .38062 | 51 | .61938 | .39353 | 54 | .60647 | .01291 | 3 | .98709 | 6 |
| 55 | 9.38113 | 51 | 10.61887 | 9.39407 | 54 | 10.60593 | 10.01294 | 3 | 9.98706 | 5 |
| 56 | .38164 | 51 | .61836 | .39461 | 54 | .60539 | .01297 | 3 | .98703 | 4 |
| 57 | .38215 | 51 | .61785 | .39515 | 54 | .60485 | .01300 | 3 | .98700 | 3 |
| 58 | .38266 | 51 | .61734 | .39569 | 54 | .60431 | .01303 | 3 | .98697 | 2 |
| 59 | .38317 | 51 | .61683 | .39623 | 54 | .60377 | .01306 | 4 | .98694 | 1 |
| 60 | 9.38368 | | 10.61632 | 9.39677 | | 10.60323 | 10.01310 | | 9.98690 | 0 |
| 103°→ | cos | Diff. 1′ | sec | cot | Diff. 1′ | tan | csc | Diff. 1′ | sin | ←76° |

## TABLE 33
Logarithms of Trigonometric Functions

| 24° → ′ | sin | Diff. 1′ | csc | tan | Diff. 1′ | cot | sec | Diff. 1′ | cos | ← 155° ′ |
|---|---|---|---|---|---|---|---|---|---|---|
| 0 | 9. 60931 | 29 | 10. 39069 | 9. 64858 | 34 | 10. 35142 | 10. 03927 | 6 | 9. 96073 | 60 |
| 1 | . 60960 | 28 | . 39040 | . 64892 | 34 | . 35108 | . 03933 | 5 | . 96067 | 59 |
| 2 | . 60988 | 28 | . 39012 | . 64926 | 34 | . 35074 | . 03938 | 6 | . 96062 | 58 |
| 3 | . 61016 | 29 | . 38984 | . 64960 | 34 | . 35040 | . 03944 | 6 | . 96056 | 57 |
| 4 | . 61045 | 28 | . 38955 | . 64994 | 34 | . 35006 | . 03950 | 5 | . 96050 | 56 |
| 5 | 9. 61073 | 28 | 10. 38927 | 9. 65028 | 34 | 10. 34972 | 10. 03955 | 6 | 9. 96045 | 55 |
| 6 | . 61101 | 28 | . 38899 | . 65062 | 34 | . 34938 | . 03961 | 5 | . 96039 | 54 |
| 7 | . 61129 | 29 | . 38871 | . 65096 | 34 | . 34904 | . 03966 | 6 | . 96034 | 53 |
| 8 | . 61158 | 28 | . 38842 | . 65130 | 34 | . 34870 | . 03972 | 6 | . 96028 | 52 |
| 9 | . 61186 | 28 | . 38814 | . 65164 | 33 | . 34836 | . 03978 | 5 | . 96022 | 51 |
| 10 | 9. 61214 | 28 | 10. 38786 | 9. 65197 | 34 | 10. 34803 | 10. 03983 | 6 | 9. 96017 | 50 |
| 11 | . 61242 | 28 | . 38758 | . 65231 | 34 | . 34769 | . 03989 | 6 | . 96011 | 49 |
| 12 | . 61270 | 28 | . 38730 | . 65265 | 34 | . 34735 | . 03995 | 5 | . 96005 | 48 |
| 13 | . 61298 | 28 | . 38702 | . 65299 | 34 | . 34701 | . 04000 | 6 | . 96000 | 47 |
| 14 | . 61326 | 28 | . 38674 | . 65333 | 33 | . 34667 | . 04006 | 6 | . 95994 | 46 |
| 15 | 9. 61354 | 28 | 10. 38646 | 9. 65366 | 34 | 10. 34634 | 10. 04012 | 6 | 9. 95988 | 45 |
| 16 | . 61382 | 29 | . 38618 | . 65400 | 34 | . 34600 | . 04018 | 5 | . 95982 | 44 |
| 17 | . 61411 | 27 | . 38589 | . 65434 | 33 | . 34566 | . 04023 | 6 | . 95977 | 43 |
| 18 | . 61438 | 28 | . 38562 | . 65467 | 34 | . 34533 | . 04029 | 6 | . 95971 | 42 |
| 19 | . 61466 | 28 | . 38534 | . 65501 | 34 | . 34499 | . 04035 | 5 | . 95965 | 41 |
| 20 | 9. 61494 | 28 | 10. 38506 | 9. 65535 | 33 | 10. 34465 | 10. 04040 | 6 | 9. 95960 | 40 |
| 21 | . 61522 | 28 | . 38478 | . 65568 | 34 | . 34432 | . 04046 | 6 | . 95954 | 39 |
| 22 | . 61550 | 28 | . 38450 | . 65602 | 34 | . 34398 | . 04052 | 6 | . 95948 | 38 |
| 23 | . 61578 | 28 | . 38422 | . 65636 | 33 | . 34364 | . 04058 | 5 | . 95942 | 37 |
| 24 | . 61606 | 28 | . 38394 | . 65669 | 34 | . 34331 | . 04063 | 6 | . 95937 | 36 |
| 25 | 9. 61634 | 28 | 10. 38366 | 9. 65703 | 33 | 10. 34297 | 10. 04069 | 6 | 9. 95931 | 35 |
| 26 | . 61662 | 27 | . 38338 | . 65736 | 34 | . 34264 | . 04075 | 5 | . 95925 | 34 |
| 27 | . 61689 | 28 | . 38311 | . 65770 | 33 | . 34230 | . 04080 | 6 | . 95920 | 33 |
| 28 | . 61717 | 28 | . 38283 | . 65803 | 34 | . 34197 | . 04086 | 6 | . 95914 | 32 |
| 29 | . 61745 | 28 | . 38255 | . 65837 | 33 | . 34163 | . 04092 | 6 | . 95908 | 31 |
| 30 | 9. 61773 | 27 | 10. 38227 | 9. 65870 | 34 | 10. 34130 | 10. 04098 | 5 | 9. 95902 | 30 |
| 31 | . 61800 | 28 | . 38200 | . 65904 | 33 | . 34096 | . 04103 | 6 | . 95897 | 29 |
| 32 | . 61828 | 28 | . 38172 | . 65937 | 34 | . 34063 | . 04109 | 6 | . 95891 | 28 |
| 33 | . 61856 | 27 | . 38144 | . 65971 | 33 | . 34029 | . 04115 | 6 | . 95885 | 27 |
| 34 | . 61883 | 28 | . 38117 | . 66004 | 34 | . 33996 | . 04121 | 6 | . 95879 | 26 |
| 35 | 9. 61911 | 28 | 10. 38089 | 9. 66038 | 33 | 10. 33962 | 10. 04127 | 5 | 9. 95873 | 25 |
| 36 | . 61939 | 27 | . 38061 | . 66071 | 33 | . 33929 | . 04132 | 6 | . 95868 | 24 |
| 37 | . 61966 | 28 | . 38034 | . 66104 | 34 | . 33896 | . 04138 | 6 | . 95862 | 23 |
| 38 | . 61994 | 27 | . 38006 | . 66138 | 33 | . 33862 | . 04144 | 6 | . 95856 | 22 |
| 39 | . 62021 | 28 | . 37979 | . 66171 | 33 | . 33829 | . 04150 | 6 | . 95850 | 21 |
| 40 | 9. 62049 | 27 | 10. 37951 | 9. 66204 | 34 | 10. 33796 | 10. 04156 | 5 | 9. 95844 | 20 |
| 41 | . 62076 | 28 | . 37924 | . 66238 | 33 | . 33762 | . 04161 | 6 | . 95839 | 19 |
| 42 | . 62104 | 27 | . 37896 | . 66271 | 33 | . 33729 | . 04167 | 6 | . 95833 | 18 |
| 43 | . 62131 | 28 | . 37869 | . 66304 | 33 | . 33696 | . 04173 | 6 | . 95827 | 17 |
| 44 | . 62159 | 27 | . 37841 | . 66337 | 34 | . 33663 | . 04179 | 6 | . 95821 | 16 |
| 45 | 9. 62186 | 28 | 10. 37814 | 9. 66371 | 33 | 10. 33629 | 10. 04185 | 5 | 9. 95815 | 15 |
| 46 | . 62214 | 27 | . 37786 | . 66404 | 33 | . 33596 | . 04190 | 6 | . 95810 | 14 |
| 47 | . 62241 | 27 | . 37759 | . 66437 | 33 | . 33563 | . 04196 | 6 | . 95804 | 13 |
| 48 | . 62268 | 28 | . 37732 | . 66470 | 33 | . 33530 | . 04202 | 6 | . 95798 | 12 |
| 49 | . 62296 | 27 | . 37704 | . 66503 | 34 | . 33497 | . 04208 | 6 | . 95792 | 11 |
| 50 | 9. 62323 | 27 | 10. 37677 | 9. 66537 | 33 | 10. 33463 | 10. 04214 | 6 | 9. 95786 | 10 |
| 51 | . 62350 | 27 | . 37650 | . 66570 | 33 | . 33430 | . 04220 | 5 | . 95780 | 9 |
| 52 | . 62377 | 28 | . 37623 | . 66603 | 33 | . 33397 | . 04225 | 6 | . 95775 | 8 |
| 53 | . 62405 | 27 | . 37595 | . 66636 | 33 | . 33364 | . 04231 | 6 | . 95769 | 7 |
| 54 | . 62432 | 27 | . 37568 | . 66669 | 33 | . 33331 | . 04237 | 6 | . 95763 | 6 |
| 55 | 9. 62459 | 27 | 10. 37541 | 9. 66702 | 33 | 10. 33298 | 10. 04243 | 6 | 9. 95757 | 5 |
| 56 | . 62486 | 27 | . 37514 | . 66735 | 33 | . 33265 | . 04249 | 6 | . 95751 | 4 |
| 57 | . 62513 | 28 | . 37487 | . 66768 | 33 | . 33232 | . 04255 | 6 | . 95745 | 3 |
| 58 | . 62541 | 27 | . 37459 | . 66801 | 33 | . 33199 | . 04261 | 6 | . 95739 | 2 |
| 59 | . 62568 | 27 | . 37432 | . 66834 | 33 | . 33166 | . 04267 | 5 | . 95733 | 1 |
| 60 | 9. 62595 | | 10. 37405 | 9. 66867 | | 10. 33133 | 10. 04272 | | 9. 95728 | 0 |
| 114° → | cos | Diff. 1′ | sec | cot | Diff. 1′ | tan | csc | Diff. 1′ | sin | ← 65° |

# TABLE 33

Logarithms of Trigonometric Functions

| 25°→ ↓ | sin | Diff. 1′ | csc | tan | Diff. 1′ | cot | sec | Diff. 1′ | cos | ←154° ↓ |
|---|---|---|---|---|---|---|---|---|---|---|
| ′ | | | | | | | | | | ′ |
| 0 | 9. 62595 | 27 | 10. 37405 | 9. 66867 | 33 | 10. 33133 | 10. 04272 | 6 | 9. 95728 | 60 |
| 1 | . 62622 | 27 | . 37378 | . 66900 | 33 | . 33100 | . 04278 | 6 | . 95722 | 59 |
| 2 | . 62649 | 27 | . 37351 | . 66933 | 33 | . 33067 | . 04284 | 6 | . 95716 | 58 |
| 3 | . 62676 | 27 | . 37324 | . 66966 | 33 | . 33034 | . 04290 | 6 | . 95710 | 57 |
| 4 | . 62703 | 27 | . 37297 | . 66999 | 33 | . 33001 | . 04296 | 6 | . 95704 | 56 |
| 5 | 9. 62730 | 27 | 10. 37270 | 9. 67032 | 33 | 10. 32968 | 10. 04302 | 6 | 9. 95698 | 55 |
| 6 | . 62757 | 27 | . 37243 | . 67065 | 33 | . 32935 | . 04308 | 6 | . 95692 | 54 |
| 7 | . 62784 | 27 | . 37216 | . 67098 | 33 | . 32902 | . 04314 | 6 | . 95686 | 53 |
| 8 | . 62811 | 27 | . 37189 | . 67131 | 32 | . 32869 | . 04320 | 6 | . 95680 | 52 |
| 9 | . 62838 | 27 | . 37162 | . 67163 | 33 | . 32837 | . 04326 | 6 | . 95674 | 51 |
| 10 | 9. 62865 | 27 | 10. 37135 | 9. 67196 | 33 | 10. 32804 | 10. 04332 | 5 | 9. 95668 | 50 |
| 11 | . 62892 | 26 | . 37108 | . 67229 | 33 | . 32771 | . 04337 | 6 | . 95663 | 49 |
| 12 | . 62918 | 27 | . 37082 | . 67262 | 33 | . 32738 | . 04343 | 6 | . 95657 | 48 |
| 13 | . 62945 | 27 | . 37055 | . 67295 | 32 | . 32705 | . 04349 | 6 | . 95651 | 47 |
| 14 | . 62972 | 27 | . 37028 | . 67327 | 33 | . 32673 | . 04355 | 6 | . 95645 | 46 |
| 15 | 9. 62999 | 27 | 10. 37001 | 9. 67360 | 33 | 10. 32640 | 10. 04361 | 6 | 9. 95639 | 45 |
| 16 | . 63026 | 26 | . 36974 | . 67393 | 33 | . 32607 | . 04367 | 6 | . 95633 | 44 |
| 17 | . 63052 | 27 | . 36948 | . 67426 | 32 | . 32574 | . 04373 | 6 | . 95627 | 43 |
| 18 | . 63079 | 27 | . 36921 | . 67458 | 33 | . 32542 | . 04379 | 6 | . 95621 | 42 |
| 19 | . 63106 | 27 | . 36894 | . 67491 | 33 | . 32509 | . 04385 | 6 | . 95615 | 41 |
| 20 | 9. 63133 | 26 | 10. 36867 | 9. 67524 | 32 | 10. 32476 | 10. 04391 | 6 | 9. 95609 | 40 |
| 21 | . 63159 | 27 | . 36841 | . 67556 | 33 | . 32444 | . 04397 | 6 | . 95603 | 39 |
| 22 | . 63186 | 27 | . 36814 | . 67589 | 33 | . 32411 | . 04403 | 6 | . 95597 | 38 |
| 23 | . 63213 | 26 | . 36787 | . 67622 | 32 | . 32378 | . 04409 | 6 | . 95591 | 37 |
| 24 | . 63239 | 27 | . 36761 | . 67654 | 33 | . 32346 | . 04415 | 6 | . 95585 | 36 |
| 25 | 9. 63266 | 26 | 10. 36734 | 9. 67687 | 32 | 10. 32313 | 10. 04421 | 6 | 9. 95579 | 35 |
| 26 | . 63292 | 27 | . 36708 | . 67719 | 33 | . 32281 | . 04427 | 6 | . 95573 | 34 |
| 27 | . 63319 | 26 | . 36681 | . 67752 | 33 | . 32248 | . 04433 | 6 | . 95567 | 33 |
| 28 | . 63345 | 27 | . 36655 | . 67785 | 32 | . 32215 | . 04439 | 6 | . 95561 | 32 |
| 29 | . 63372 | 26 | . 36628 | . 67817 | 33 | . 32183 | . 04445 | 6 | . 95555 | 31 |
| 30 | 9. 63398 | 27 | 10. 36602 | 9. 67850 | 32 | 10. 32150 | 10. 04451 | 6 | 9. 95549 | 30 |
| 31 | . 63425 | 26 | . 36575 | . 67882 | 33 | . 32118 | . 04457 | 6 | . 95543 | 29 |
| 32 | . 63451 | 27 | . 36549 | . 67915 | 32 | . 32085 | . 04463 | 6 | . 95537 | 28 |
| 33 | . 63478 | 26 | . 36522 | . 67947 | 33 | . 32053 | . 04469 | 6 | . 95531 | 27 |
| 34 | . 63504 | 27 | . 36496 | . 67980 | 32 | . 32020 | . 04475 | 6 | . 95525 | 26 |
| 35 | 9. 63531 | 26 | 10. 36469 | 9. 68012 | 32 | 10. 31988 | 10. 04481 | 6 | 9. 95519 | 25 |
| 36 | . 63557 | 26 | . 36443 | . 68044 | 33 | . 31956 | . 04487 | 6 | . 95513 | 24 |
| 37 | . 63583 | 27 | . 36417 | . 68077 | 32 | . 31923 | . 04493 | 7 | . 95507 | 23 |
| 38 | . 63610 | 26 | . 36390 | . 68109 | 33 | . 31891 | . 04500 | 6 | . 95500 | 22 |
| 39 | . 63636 | 26 | . 36364 | . 68142 | 32 | . 31858 | . 04506 | 6 | . 95494 | 21 |
| 40 | 9. 63662 | 27 | 10. 36338 | 9. 68174 | 32 | 10. 31826 | 10. 04512 | 6 | 9. 95488 | 20 |
| 41 | . 63689 | 26 | . 36311 | . 68206 | 33 | . 31794 | . 04518 | 6 | . 95482 | 19 |
| 42 | . 63715 | 26 | . 36285 | . 68239 | 32 | . 31761 | . 04524 | 6 | . 95476 | 18 |
| 43 | . 63741 | 26 | . 36259 | . 68271 | 32 | . 31729 | . 04530 | 6 | . 95470 | 17 |
| 44 | . 63767 | 27 | . 36233 | . 68303 | 33 | . 31697 | . 04536 | 6 | . 95464 | 16 |
| 45 | 9. 63794 | 26 | 10. 36206 | 9. 68336 | 32 | 10. 31664 | 10. 04542 | 6 | 9. 95458 | 15 |
| 46 | . 63820 | 26 | . 36180 | . 68368 | 32 | . 31632 | . 04548 | 6 | . 95452 | 14 |
| 47 | . 63846 | 26 | . 36154 | . 68400 | 32 | . 31600 | . 04554 | 6 | . 95446 | 13 |
| 48 | . 63872 | 26 | . 36128 | . 68432 | 33 | . 31568 | . 04560 | 6 | . 95440 | 12 |
| 49 | . 63898 | 26 | . 36102 | . 68465 | 32 | . 31535 | . 04566 | 7 | . 95434 | 11 |
| 50 | 9. 63924 | 26 | 10. 36076 | 9. 68497 | 32 | 10. 31503 | 10. 04573 | 6 | 9. 95427 | 10 |
| 51 | . 63950 | 26 | . 36050 | . 68529 | 32 | . 31471 | . 04579 | 6 | . 95421 | 9 |
| 52 | . 63976 | 26 | . 36024 | . 68561 | 32 | . 31439 | . 04585 | 6 | . 95415 | 8 |
| 53 | . 64002 | 26 | . 35998 | . 68593 | 33 | . 31407 | . 04591 | 6 | . 95409 | 7 |
| 54 | . 64028 | 26 | . 35972 | . 68626 | 32 | . 31374 | . 04597 | 6 | . 95403 | 6 |
| 55 | 9. 64054 | 26 | 10. 35946 | 9. 68658 | 32 | 10. 31342 | 10. 04603 | 6 | 9. 95397 | 5 |
| 56 | . 64080 | 26 | . 35920 | . 68690 | 32 | . 31310 | . 04609 | 7 | . 95391 | 4 |
| 57 | . 64106 | 26 | . 35894 | . 68722 | 32 | . 31278 | . 04616 | 6 | . 95384 | 3 |
| 58 | . 64132 | 26 | . 35868 | . 68754 | 32 | . 31246 | . 04622 | 6 | . 95378 | 2 |
| 59 | . 64158 | 26 | . 35842 | . 68786 | 32 | . 31214 | . 04628 | 6 | . 95372 | 1 |
| 60 | 9. 64184 | | 10. 35816 | 9. 68818 | | 10. 31182 | 10. 04634 | | 9. 95366 | 0 |
| ↑ 115°→ | cos | Diff. 1′ | sec | cot | Diff. 1′ | tan | csc | Diff. 1′ | sin | ↑ ←64° |

## TABLE 33

Logarithms of Trigonometric Functions

| 30° → ↓ ′ | sin | Diff. 1′ | csc | tan | Diff. 1′ | cot | sec | Diff. 1′ | cos | ← 149° ↓ ′ |
|---|---|---|---|---|---|---|---|---|---|---|
| 0 | 9. 69897 | 22 | 10. 30103 | 9. 76144 | 29 | 10. 23856 | 10. 06247 | 7 | 9. 93753 | 60 |
| 1 | . 69919 | 22 | . 30081 | . 76173 | 29 | . 23827 | . 06254 | 8 | . 93746 | 59 |
| 2 | . 69941 | 22 | . 30059 | . 76202 | 29 | . 23798 | . 06262 | 7 | . 93738 | 58 |
| 3 | . 69963 | 21 | . 30037 | . 76231 | 30 | . 23769 | . 06269 | 7 | . 93731 | 57 |
| 4 | . 69984 | 22 | . 30016 | . 76261 | 29 | . 23739 | . 06276 | 7 | . 93724 | 56 |
| 5 | 9. 70006 | 22 | 10. 29994 | 9. 76290 | 29 | 10. 23710 | 10. 06283 | 8 | 9. 93717 | 55 |
| 6 | . 70028 | 22 | . 29972 | . 76319 | 29 | . 23681 | . 06291 | 7 | . 93709 | 54 |
| 7 | . 70050 | 22 | . 29950 | . 76348 | 29 | . 23652 | . 06298 | 7 | . 93702 | 53 |
| 8 | . 70072 | 21 | . 29928 | . 76377 | 29 | . 23623 | . 06305 | 8 | . 93695 | 52 |
| 9 | . 70093 | 22 | . 29907 | . 76406 | 29 | . 23594 | . 06313 | 7 | . 93687 | 51 |
| 10 | 9. 70115 | 22 | 10. 29885 | 9. 76435 | 29 | 10. 23565 | 10. 06320 | 7 | 9. 93680 | 50 |
| 11 | . 70137 | 22 | . 29863 | . 76464 | 29 | . 23536 | . 06327 | 8 | . 93673 | 49 |
| 12 | . 70159 | 21 | . 29841 | . 76493 | 29 | . 23507 | . 06335 | 7 | . 93665 | 48 |
| 13 | . 70180 | 22 | . 29820 | . 76522 | 29 | . 23478 | . 06342 | 8 | . 93658 | 47 |
| 14 | . 70202 | 22 | . 29798 | . 76551 | 29 | . 23449 | . 06350 | 7 | . 93650 | 46 |
| 15 | 9. 70224 | 21 | 10. 29776 | 9. 76580 | 29 | 10. 23420 | 10. 06357 | 7 | 9. 93643 | 45 |
| 16 | . 70245 | 22 | . 29755 | . 76609 | 30 | . 23391 | . 06364 | 8 | . 93636 | 44 |
| 17 | . 70267 | 21 | . 29733 | . 76639 | 29 | . 23361 | . 06372 | 7 | . 93628 | 43 |
| 18 | . 70288 | 22 | . 29712 | . 76668 | 29 | . 23332 | . 06379 | 7 | . 93621 | 42 |
| 19 | . 70310 | 22 | . 29690 | . 76697 | 28 | . 23303 | . 06386 | 8 | . 93614 | 41 |
| 20 | 9. 70332 | 21 | 10. 29668 | 9. 76725 | 29 | 10. 23275 | 10. 06394 | 7 | 9. 93606 | 40 |
| 21 | . 70353 | 22 | . 29647 | . 76754 | 29 | . 23246 | . 06401 | 8 | . 93599 | 39 |
| 22 | . 70375 | 21 | . 29625 | . 76783 | 29 | . 23217 | . 06409 | 7 | . 93591 | 38 |
| 23 | . 70396 | 22 | . 29604 | . 76812 | 29 | . 23188 | . 06416 | 7 | . 93584 | 37 |
| 24 | . 70418 | 21 | . 29582 | . 76841 | 29 | . 23159 | . 06423 | 8 | . 93577 | 36 |
| 25 | 9. 70439 | 22 | 10. 29561 | 9. 76870 | 29 | 10. 23130 | 10. 06431 | 7 | 9. 93569 | 35 |
| 26 | . 70461 | 21 | . 29539 | . 76899 | 29 | . 23101 | . 06438 | 8 | . 93562 | 34 |
| 27 | . 70482 | 22 | . 29518 | . 76928 | 29 | . 23072 | . 06446 | 7 | . 93554 | 33 |
| 28 | . 70504 | 21 | . 29496 | . 76957 | 29 | . 23043 | . 06453 | 8 | . 93547 | 32 |
| 29 | . 70525 | 22 | . 29475 | . 76986 | 29 | . 23014 | . 06461 | 7 | . 93539 | 31 |
| 30 | 9. 70547 | 21 | 10. 29453 | 9. 77015 | 29 | 10. 22985 | 10. 06468 | 7 | 9. 93532 | 30 |
| 31 | . 70568 | 22 | . 29432 | . 77044 | 29 | . 22956 | . 06475 | 8 | . 93525 | 29 |
| 32 | . 70590 | 21 | . 29410 | . 77073 | 28 | . 22927 | . 06483 | 7 | . 93517 | 28 |
| 33 | . 70611 | 22 | . 29389 | . 77101 | 29 | . 22899 | . 06490 | 8 | . 93510 | 27 |
| 34 | . 70633 | 21 | . 29367 | . 77130 | 29 | . 22870 | . 06498 | 7 | . 93502 | 26 |
| 35 | 9. 70654 | 21 | 10. 29346 | 9. 77159 | 29 | 10. 22841 | 10. 06505 | 8 | 9. 93495 | 25 |
| 36 | . 70675 | 22 | . 29325 | . 77188 | 29 | . 22812 | . 06513 | 7 | . 93487 | 24 |
| 37 | . 70697 | 21 | . 29303 | . 77217 | 29 | . 22783 | . 06520 | 8 | . 93480 | 23 |
| 38 | . 70718 | 21 | . 29282 | . 77246 | 28 | . 22754 | . 06528 | 7 | . 93472 | 22 |
| 39 | . 70739 | 22 | . 29261 | . 77274 | 29 | . 22726 | . 06535 | 8 | . 93465 | 21 |
| 40 | 9. 70761 | 21 | 10. 29239 | 9. 77303 | 29 | 10. 22697 | 10. 06543 | 7 | 9. 93457 | 20 |
| 41 | . 70782 | 21 | . 29218 | . 77332 | 29 | . 22668 | . 06550 | 8 | . 93450 | 19 |
| 42 | . 70803 | 21 | . 29197 | . 77361 | 29 | . 22639 | . 06558 | 7 | . 93442 | 18 |
| 43 | . 70824 | 22 | . 29176 | . 77390 | 28 | . 22610 | . 06565 | 8 | . 93435 | 17 |
| 44 | . 70846 | 21 | . 29154 | . 77418 | 29 | . 22582 | . 06573 | 7 | . 93427 | 16 |
| 45 | 9. 70867 | 21 | 10. 29133 | 9. 77447 | 29 | 10. 22553 | 10. 06580 | 8 | 9. 93420 | 15 |
| 46 | . 70888 | 21 | . 29112 | . 77476 | 29 | . 22524 | . 06588 | 7 | . 93412 | 14 |
| 47 | . 70909 | 22 | . 29091 | . 77505 | 28 | . 22495 | . 06595 | 8 | . 93405 | 13 |
| 48 | . 70931 | 21 | . 29069 | . 77533 | 29 | . 22467 | . 06603 | 7 | . 93397 | 12 |
| 49 | . 70952 | 21 | . 29048 | . 77562 | 29 | . 22438 | . 06610 | 8 | . 93390 | 11 |
| 50 | 9. 70973 | 21 | 10. 29027 | 9. 77591 | 28 | 10. 22409 | 10. 06618 | 7 | 9. 93382 | 10 |
| 51 | . 70994 | 21 | . 29006 | . 77619 | 29 | . 22381 | . 06625 | 8 | . 93375 | 9 |
| 52 | . 71015 | 21 | . 28985 | . 77648 | 29 | . 22352 | . 06633 | 7 | . 93367 | 8 |
| 53 | . 71036 | 22 | . 28964 | . 77677 | 29 | . 22323 | . 06640 | 8 | . 93360 | 7 |
| 54 | . 71058 | 21 | . 28942 | . 77706 | 28 | . 22294 | . 06648 | 8 | . 93352 | 6 |
| 55 | 9. 71079 | 21 | 10. 28921 | 9. 77734 | 29 | 10. 22266 | 10. 06656 | 7 | 9. 93344 | 5 |
| 56 | . 71100 | 21 | . 28900 | . 77763 | 28 | . 22237 | . 06663 | 8 | . 93337 | 4 |
| 57 | . 71121 | 21 | . 28879 | . 77791 | 29 | . 22209 | . 06671 | 7 | . 93329 | 3 |
| 58 | . 71142 | 21 | . 28858 | . 77820 | 29 | . 22180 | . 06678 | 8 | . 93322 | 2 |
| 59 | . 71163 | 21 | . 28837 | . 77849 | 28 | . 22151 | . 06686 | 7 | . 93314 | 1 |
| 60 | 9. 71184 | | 10. 28816 | 9. 77877 | | 10. 22123 | 10. 06693 | | 9. 93307 | 0 |
| ↑ 120° → | cos | Diff. 1′ | sec | cot | Diff. 1′ | tan | csc | Diff. 1′ | sin | ← 59° ↑ |

## TABLE 33

Logarithms of Trigonometric Functions

| 34° → ↓ ′ | sin | Diff. 1′ | csc | tan | Diff. 1′ | cot | sec | Diff. 1′ | cos | ← 145° ↓ ′ |
|---|---|---|---|---|---|---|---|---|---|---|
| 0 | 9. 74756 | 19 | 10. 25244 | 9. 82899 | 27 | 10. 17101 | 10. 08143 | 8 | 9. 91857 | 60 |
| 1 | . 74775 | 19 | . 25225 | . 82926 | 27 | . 17074 | . 08151 | 9 | . 91849 | 59 |
| 2 | . 74794 | 18 | . 25206 | . 82953 | 27 | . 17047 | . 08160 | 8 | . 91840 | 58 |
| 3 | . 74812 | 19 | . 25188 | . 82980 | 28 | . 17020 | . 08168 | 9 | . 91832 | 57 |
| 4 | . 74831 | 19 | . 25169 | . 83008 | 27 | . 16992 | . 08177 | 8 | . 91823 | 56 |
| 5 | 9. 74850 | 18 | 10. 25150 | 9. 83035 | 27 | 10. 16965 | 10. 08185 | 9 | 9. 91815 | 55 |
| 6 | . 74868 | 19 | . 25132 | . 83062 | 27 | . 16938 | . 08194 | 8 | . 91806 | 54 |
| 7 | . 74887 | 19 | . 25113 | . 83089 | 28 | . 16911 | . 08202 | 9 | . 91798 | 53 |
| 8 | . 74906 | 18 | . 25094 | . 83117 | 27 | . 16883 | . 08211 | 8 | . 91789 | 52 |
| 9 | . 74924 | 19 | . 25076 | . 83144 | 27 | . 16856 | . 08219 | 9 | . 91781 | 51 |
| 10 | 9. 74943 | 18 | 10. 25057 | 9. 83171 | 27 | 10. 16829 | 10. 08228 | 9 | 9. 91772 | 50 |
| 11 | . 74961 | 19 | . 25039 | . 83198 | 27 | . 16802 | . 08237 | 8 | . 91763 | 49 |
| 12 | . 74980 | 19 | . 25020 | . 83225 | 27 | . 16775 | . 08245 | 9 | . 91755 | 48 |
| 13 | . 74999 | 18 | . 25001 | . 83252 | 28 | . 16748 | . 08254 | 8 | . 91746 | 47 |
| 14 | . 75017 | 19 | . 24983 | . 83280 | 27 | . 16720 | . 08262 | 9 | . 91738 | 46 |
| 15 | 9. 75036 | 18 | 10. 24964 | 9. 83307 | 27 | 10. 16693 | 10. 08271 | 9 | 9. 91729 | 45 |
| 16 | . 75054 | 19 | . 24946 | . 83334 | 27 | . 16666 | . 08280 | 8 | . 91720 | 44 |
| 17 | . 75073 | 18 | . 24927 | . 83361 | 27 | . 16639 | . 08288 | 9 | . 91712 | 43 |
| 18 | . 75091 | 19 | . 24909 | . 83388 | 27 | . 16612 | . 08297 | 8 | . 91703 | 42 |
| 19 | . 75110 | 18 | . 24890 | . 83415 | 27 | . 16585 | . 08305 | 9 | . 91695 | 41 |
| 20 | 9. 75128 | 19 | 10. 24872 | 9. 83442 | 28 | 10. 16558 | 10. 08314 | 9 | 9. 91686 | 40 |
| 21 | . 75147 | 18 | . 24853 | . 83470 | 27 | . 16530 | . 08323 | 8 | . 91677 | 39 |
| 22 | . 75165 | 19 | . 24835 | . 83497 | 27 | . 16503 | . 08331 | 9 | . 91669 | 38 |
| 23 | . 75184 | 18 | . 24816 | . 83524 | 27 | . 16476 | . 08340 | 9 | . 91660 | 37 |
| 24 | . 75202 | 19 | . 24798 | . 83551 | 27 | . 16449 | . 08349 | 8 | . 91651 | 36 |
| 25 | 9. 75221 | 18 | 10. 24779 | 9. 83578 | 27 | 10. 16422 | 10. 08357 | 9 | 9. 91643 | 35 |
| 26 | . 75239 | 19 | . 24761 | . 83605 | 27 | . 16395 | . 08366 | 9 | . 91634 | 34 |
| 27 | . 75258 | 18 | . 24742 | . 83632 | 27 | . 16368 | . 08375 | 8 | . 91625 | 33 |
| 28 | . 75276 | 18 | . 24724 | . 83659 | 27 | . 16341 | . 08383 | 9 | . 91617 | 32 |
| 29 | . 75294 | 19 | . 24706 | . 83686 | 27 | . 16314 | . 08392 | 9 | . 91608 | 31 |
| 30 | 9. 75313 | 18 | 10. 24687 | 9. 83713 | 27 | 10. 16287 | 10. 08401 | 8 | 9. 91599 | 30 |
| 31 | . 75331 | 19 | . 24669 | . 83740 | 28 | . 16260 | . 08409 | 9 | . 91591 | 29 |
| 32 | . 75350 | 18 | . 24650 | . 83768 | 27 | . 16232 | . 08418 | 9 | . 91582 | 28 |
| 33 | . 75368 | 18 | . 24632 | . 83795 | 27 | . 16205 | . 08427 | 8 | . 91573 | 27 |
| 34 | . 75386 | 19 | . 24614 | . 83822 | 27 | . 16178 | . 08435 | 9 | . 91565 | 26 |
| 35 | 9. 75405 | 18 | 10. 24595 | 9. 83849 | 27 | 10. 16151 | 10. 08444 | 9 | 9. 91556 | 25 |
| 36 | . 75423 | 18 | . 24577 | . 83876 | 27 | . 16124 | . 08453 | 9 | . 91547 | 24 |
| 37 | . 75441 | 18 | . 24559 | . 83903 | 27 | . 16097 | . 08462 | 8 | . 91538 | 23 |
| 38 | . 75459 | 19 | . 24541 | . 83930 | 27 | . 16070 | . 08470 | 9 | . 91530 | 22 |
| 39 | . 75478 | 18 | . 24522 | . 83957 | 27 | . 16043 | . 08479 | 9 | . 91521 | 21 |
| 40 | 9. 75496 | 18 | 10. 24504 | 9. 83984 | 27 | 10. 16016 | 10. 08488 | 8 | 9. 91512 | 20 |
| 41 | . 75514 | 19 | . 24486 | . 84011 | 27 | . 15989 | . 08496 | 9 | . 91504 | 19 |
| 42 | . 75533 | 18 | . 24467 | . 84038 | 27 | . 15962 | . 08505 | 9 | . 91495 | 18 |
| 43 | . 75551 | 18 | . 24449 | . 84065 | 27 | . 15935 | . 08514 | 9 | . 91486 | 17 |
| 44 | . 75569 | 18 | . 24431 | . 84092 | 27 | . 15908 | . 08523 | 8 | . 91477 | 16 |
| 45 | 9. 75587 | 18 | 10. 24413 | 9. 84119 | 27 | 10. 15881 | 10. 08531 | 9 | 9. 91469 | 15 |
| 46 | . 75605 | 19 | . 24395 | . 84146 | 27 | . 15854 | . 08540 | 9 | . 91460 | 14 |
| 47 | . 75624 | 18 | . 24376 | . 84173 | 27 | . 15827 | . 08549 | 9 | . 91451 | 13 |
| 48 | . 75642 | 18 | . 24358 | . 84200 | 27 | . 15800 | . 08558 | 9 | . 91442 | 12 |
| 49 | . 75660 | 18 | . 24340 | . 84227 | 27 | . 15773 | . 08567 | 8 | . 91433 | 11 |
| 50 | 9. 75678 | 18 | 10. 24322 | 9. 84254 | 26 | 10. 15746 | 10. 08575 | 9 | 9. 91425 | 10 |
| 51 | . 75696 | 18 | . 24304 | . 84280 | 27 | . 15720 | . 08584 | 9 | . 91416 | 9 |
| 52 | . 75714 | 19 | . 24286 | . 84307 | 27 | . 15693 | . 08593 | 9 | . 91407 | 8 |
| 53 | . 75733 | 18 | . 24267 | . 84334 | 27 | . 15666 | . 08602 | 9 | . 91398 | 7 |
| 54 | . 75751 | 18 | . 24249 | . 84361 | 27 | . 15639 | . 08611 | 8 | . 91389 | 6 |
| 55 | 9. 75769 | 18 | 10. 24231 | 9. 84388 | 27 | 10. 15612 | 10. 08619 | 9 | 9. 91381 | 5 |
| 56 | . 75787 | 18 | . 24213 | . 84415 | 27 | . 15585 | . 08628 | 9 | . 91372 | 4 |
| 57 | . 75805 | 18 | . 24195 | . 84442 | 27 | . 15558 | . 08637 | 9 | . 91363 | 3 |
| 58 | . 75823 | 18 | . 24177 | . 84469 | 27 | . 15531 | . 08646 | 9 | . 91354 | 2 |
| 59 | . 75841 | 18 | . 24159 | . 84496 | 27 | . 15504 | . 08655 | 9 | . 91345 | 1 |
| 60 | 9. 75859 | | 10. 24141 | 9. 84523 | | 10. 15477 | 10. 08664 | | 9. 91336 | 0 |
| ↑ 124° → | cos | Diff. 1′ | sec | cot | Diff. 1′ | tan | csc | Diff. 1′ | sin | ← 55° ↑ |

## TABLE 33
Logarithms of Trigonometric Functions

| 38° ↓ ′ | sin | Diff. 1′ | csc | tan | Diff. 1′ | cot | sec | Diff. 1′ | cos | 141° ↓ ′ |
|---|---|---|---|---|---|---|---|---|---|---|
| 0 | 9. 78934 | 16 | 10. 21066 | 9. 89281 | 26 | 10. 10719 | 10. 10347 | 10 | 9. 89653 | 60 |
| 1 | . 78950 | 17 | . 21050 | . 89307 | 26 | . 10693 | . 10357 | 10 | . 89643 | 59 |
| 2 | . 78967 | 16 | . 21033 | . 89333 | 26 | . 10667 | . 10367 | 9 | . 89633 | 58 |
| 3 | . 78983 | 16 | . 21017 | . 89359 | 26 | . 10641 | . 10376 | 10 | . 89624 | 57 |
| 4 | . 78999 | 16 | . 21001 | . 89385 | 26 | . 10615 | . 10386 | 10 | . 89614 | 56 |
| 5 | 9. 79015 | 16 | 10. 20985 | 9. 89411 | 26 | 10. 10589 | 10. 10396 | 10 | 9. 89604 | 55 |
| 6 | . 79031 | 16 | . 20969 | . 89437 | 26 | . 10563 | . 10406 | 10 | . 89594 | 54 |
| 7 | . 79047 | 16 | . 20953 | . 89463 | 26 | . 10537 | . 10416 | 10 | . 89584 | 53 |
| 8 | . 79063 | 16 | . 20937 | . 89489 | 26 | . 10511 | . 10426 | 10 | . 89574 | 52 |
| 9 | . 79079 | 16 | . 20921 | . 89515 | 26 | . 10485 | . 10436 | 10 | . 89564 | 51 |
| 10 | 9. 79095 | 16 | 10. 20905 | 9. 89541 | 26 | 10. 10459 | 10. 10446 | 10 | 9. 89554 | 50 |
| 11 | . 79111 | 17 | . 20889 | . 89567 | 26 | . 10433 | . 10456 | 10 | . 89544 | 49 |
| 12 | . 79128 | 16 | . 20872 | . 89593 | 26 | . 10407 | . 10466 | 10 | . 89534 | 48 |
| 13 | . 79144 | 16 | . 20856 | . 89619 | 26 | . 10381 | . 10476 | 10 | . 89524 | 47 |
| 14 | . 79160 | 16 | . 20840 | . 89645 | 26 | . 10355 | . 10486 | 10 | . 89514 | 46 |
| 15 | 9. 79176 | 16 | 10. 20824 | 9. 89671 | 26 | 10. 10329 | 10. 10496 | 9 | 9. 89504 | 45 |
| 16 | . 79192 | 16 | . 20808 | . 89697 | 26 | . 10303 | . 10505 | 10 | . 89495 | 44 |
| 17 | . 79208 | 16 | . 20792 | . 89723 | 26 | . 10277 | . 10515 | 10 | . 89485 | 43 |
| 18 | . 79224 | 16 | . 20776 | . 89749 | 26 | . 10251 | . 10525 | 10 | . 89475 | 42 |
| 19 | . 79240 | 16 | . 20760 | . 89775 | 26 | . 10225 | . 10535 | 10 | . 89465 | 41 |
| 20 | 9. 79256 | 16 | 10. 20744 | 9. 89801 | 26 | 10. 10199 | 10. 10545 | 10 | 9. 89455 | 40 |
| 21 | . 79272 | 16 | . 20728 | . 89827 | 26 | . 10173 | . 10555 | 10 | . 89445 | 39 |
| 22 | . 79288 | 16 | . 20712 | . 89853 | 26 | . 10147 | . 10565 | 10 | . 89435 | 38 |
| 23 | . 79304 | 15 | . 20696 | . 89879 | 26 | . 10121 | . 10575 | 10 | . 89425 | 37 |
| 24 | . 79319 | 16 | . 20681 | . 89905 | 26 | . 10095 | . 10585 | 10 | . 89415 | 36 |
| 25 | 9. 79335 | 16 | 10. 20665 | 9. 89931 | 26 | 10. 10069 | 10. 10595 | 10 | 9. 89405 | 35 |
| 26 | . 79351 | 16 | . 20649 | . 89957 | 26 | . 10043 | . 10605 | 10 | . 89395 | 34 |
| 27 | . 79367 | 16 | . 20633 | . 89983 | 26 | . 10017 | . 10615 | 10 | . 89385 | 33 |
| 28 | . 79383 | 16 | . 20617 | . 90009 | 26 | . 09991 | . 10625 | 11 | . 89375 | 32 |
| 29 | . 79399 | 16 | . 20601 | . 90035 | 26 | . 09965 | . 10636 | 10 | . 89364 | 31 |
| 30 | 9. 79415 | 16 | 10. 20585 | 9. 90061 | 25 | 10. 09939 | 10. 10646 | 10 | 9. 89354 | 30 |
| 31 | . 79431 | 16 | . 20569 | . 90086 | 26 | . 09914 | . 10656 | 10 | . 89344 | 29 |
| 32 | . 79447 | 16 | . 20553 | . 90112 | 26 | . 09888 | . 10666 | 10 | . 89334 | 28 |
| 33 | . 79463 | 15 | . 20537 | . 90138 | 26 | . 09862 | . 10676 | 10 | . 89324 | 27 |
| 34 | . 79478 | 16 | . 20522 | . 90164 | 26 | . 09836 | . 10686 | 10 | . 89314 | 26 |
| 35 | 9. 79494 | 16 | 10. 20506 | 9. 90190 | 26 | 10. 09810 | 10. 10696 | 10 | 9. 89304 | 25 |
| 36 | . 79510 | 16 | . 20490 | . 90216 | 26 | . 09784 | . 10706 | 10 | . 89294 | 24 |
| 37 | . 79526 | 16 | . 20474 | . 90242 | 26 | . 09758 | . 10716 | 10 | . 89284 | 23 |
| 38 | . 79542 | 16 | . 20458 | . 90268 | 26 | . 09732 | . 10726 | 10 | . 89274 | 22 |
| 39 | . 79558 | 15 | . 20442 | . 90294 | 26 | . 09706 | . 10736 | 10 | . 89264 | 21 |
| 40 | 9. 79573 | 16 | 10. 20427 | 9. 90320 | 26 | 10. 09680 | 10. 10746 | 10 | 9. 89254 | 20 |
| 41 | . 79589 | 16 | . 20411 | . 90346 | 25 | . 09654 | . 10756 | 11 | . 89244 | 19 |
| 42 | . 79605 | 16 | . 20395 | . 90371 | 26 | . 09629 | . 10767 | 10 | . 89233 | 18 |
| 43 | . 79621 | 15 | . 20379 | . 90397 | 26 | . 09603 | . 10777 | 10 | . 89223 | 17 |
| 44 | . 79636 | 16 | . 20364 | . 90423 | 26 | . 09577 | . 10787 | 10 | . 89213 | 16 |
| 45 | 9. 79652 | 16 | 10. 20348 | 9. 90449 | 26 | 10. 09551 | 10. 10797 | 10 | 9. 89203 | 15 |
| 46 | . 79668 | 16 | . 20332 | . 90475 | 26 | . 09525 | . 10807 | 10 | . 89193 | 14 |
| 47 | . 79684 | 15 | . 20316 | . 90501 | 26 | . 09499 | . 10817 | 10 | . 89183 | 13 |
| 48 | . 79699 | 16 | . 20301 | . 90527 | 26 | . 09473 | . 10827 | 11 | . 89173 | 12 |
| 49 | . 79715 | 16 | . 20285 | . 90553 | 25 | . 09447 | . 10838 | 10 | . 89162 | 11 |
| 50 | 9. 79731 | 15 | 10. 20269 | 9. 90578 | 26 | 10. 09422 | 10. 10848 | 10 | 9. 89152 | 10 |
| 51 | . 79746 | 16 | . 20254 | . 90604 | 26 | . 09396 | . 10858 | 10 | . 89142 | 9 |
| 52 | . 79762 | 16 | . 20238 | . 90630 | 26 | . 09370 | . 10868 | 10 | . 89132 | 8 |
| 53 | . 79778 | 15 | . 20222 | . 90656 | 26 | . 09344 | . 10878 | 10 | . 89122 | 7 |
| 54 | . 79793 | 16 | . 20207 | . 90682 | 26 | . 09318 | . 10888 | 11 | . 89112 | 6 |
| 55 | 9. 79809 | 16 | 10. 20191 | 9. 90708 | 26 | 10. 09292 | 10. 10899 | 10 | 9. 89101 | 5 |
| 56 | . 79825 | 15 | . 20175 | . 90734 | 25 | . 09266 | . 10909 | 10 | . 89091 | 4 |
| 57 | . 79840 | 16 | . 20160 | . 90759 | 26 | . 09241 | . 10919 | 10 | . 89081 | 3 |
| 58 | . 79856 | 16 | . 20144 | . 90785 | 26 | . 09215 | . 10929 | 11 | . 89071 | 2 |
| 59 | . 79872 | 15 | . 20128 | . 90811 | 26 | . 09189 | . 10940 | 10 | . 89060 | 1 |
| 60 | 9. 79887 | | 10. 20113 | 9. 90837 | | 10. 09163 | 10. 10950 | | 9. 89050 | 0 |
| ↑ 128° → | cos | Diff. 1′ | sec | cot | Diff. 1′ | tan | csc | Diff. 1′ | sin | ↑ ← 51° |

# Index

Sailings, 109
St.-Hilaire, Commander Marcq, 109
Semidiameter correction, 15, 16
Sextant: described, 14; in piloting, 19-21
Sextant altitude: correcting of, with *Air Almanac,* 83, 85; correcting of, with *Nautical Almanac,* 14-17, 30-31, 33-34, 39, 44; defined, 14
Sextant error. *See* Index error
Sidereal hour angle (SHA), 42
Sight reduction: defined, 31; process of, outlined, 17, 80
Sight reduction forms: limitations of, 91
Sight reduction formulas. *See* Ageton's formulas; Trigonometric formulas